Instructor's Solutions Manual

Michael Butros
Victor Valley College

Calculus for the Life Sciences

Marvin L. Bittinger
Indiana University Purdue University Indianapolis

Neal Brand
University of North Texas

John Quintanilla
University of North Texas

Boston San Francisco New York
London Toronto Sydney Tokyo Singapore Madrid
Mexico City Munich Paris Cape Town Hong Kong Montreal

Reproduced by Pearson Addison-Wesley from electronic files supplied by the author.

Publishing as Pearson Addison-Wesley, 75 Arlington Street, Boston, MA 02116.

ISBN 0-321-28604-9

1 2 3 4 5 6 BB 09 08 07 06

Contents

Chapter 1

Functions and Graphs

Exercise Set 1.1

1. Graph $y = -4$.

Note that y is constant and therefore any value of x we choose will yield the same value for y, which is -4. Thus, we will have a horizontal line at $y = -4$.

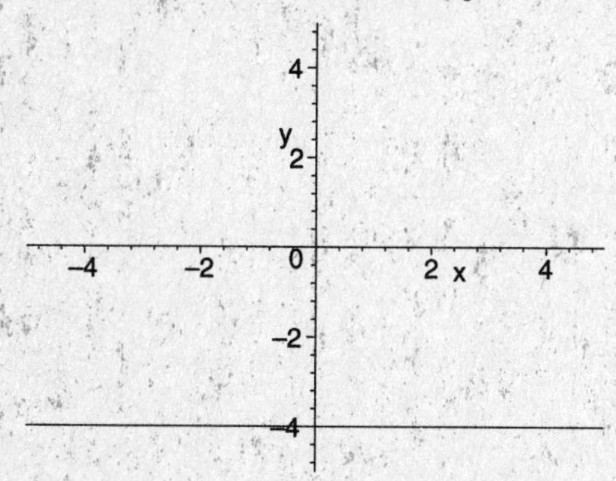

2. Horizontal line at $y = -3.5$

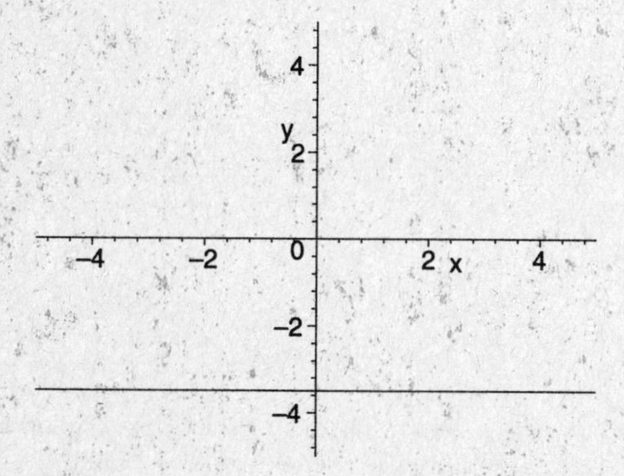

3. Graph $x = -4.5$.

Note that x is constant and therefore any value of y we choose will yield the same value for x, which is 4.5. Thus, we will have a vertical line at $x = -4.5$.

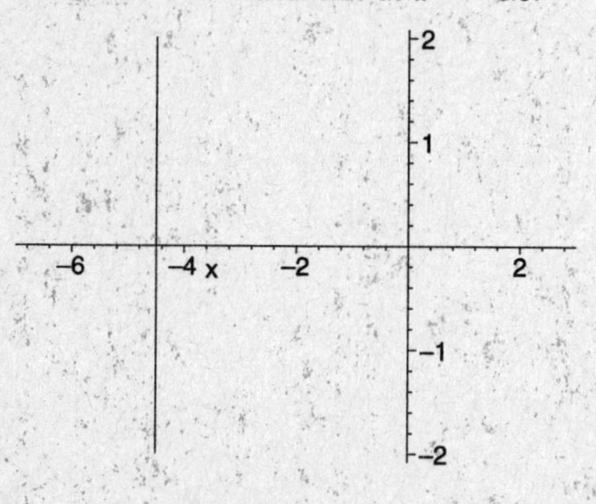

4. A vertical line at $x = 10$

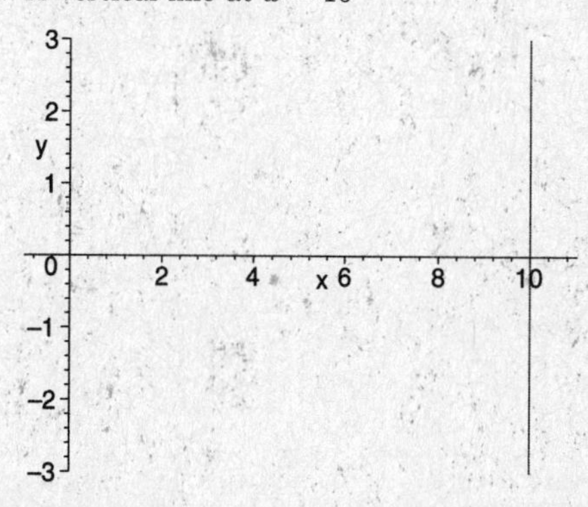

5. Graph. Find the slope and the y-intercept of $y = -3x$.

First, we find some points that satisfy the equation, then we plot the ordered pairs and connect the plotted points to get the graph.

When $x = 0$, $y = -3(0) = 0$, ordered pair $(0, 0)$

When $x = 1$, $y = -3(1) = -3$, ordered pair $(1, -3)$

When $x = -1$, $y = -3(-1) = 3$, ordered pair $(-1, 3)$

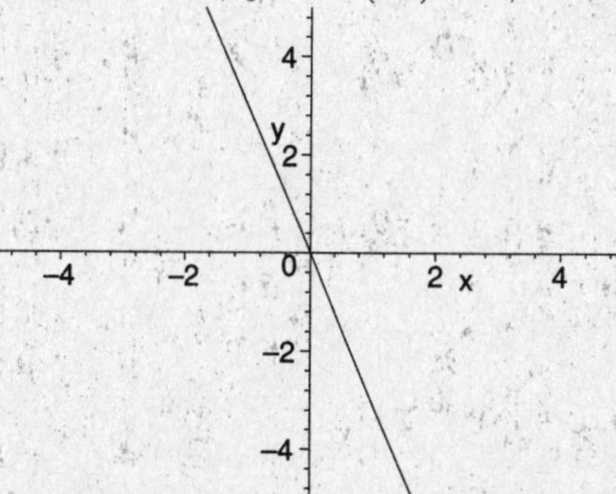

Compare the equation $y = -3x$ to the general linear equation form of $y = mx + b$ to conclude the equation has a slope of $m = -3$ and a y-intercept of $(0, 0)$.

6. Slope of $m = -0.5$ and y-intercept of $(0, 0)$

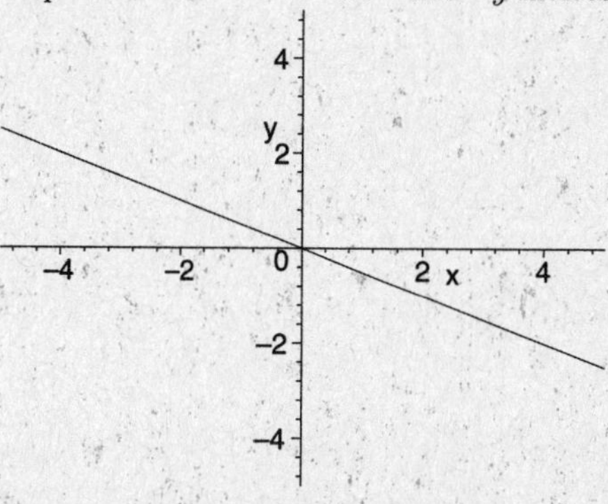

7. Graph. Find the slope and the y-intercept of $y = 0.5x$.

First, we find some points that satisfy the equation, then we plot the ordered pairs and connect the plotted points to get the graph.

When $x = 0$, $y = 0.5(0) = 0$, ordered pair $(0, 0)$

When $x = 6$, $y = 0.5(6) = 3$, ordered pair $(6, 3)$

When $x = -2$, $y = 0.5(-2) = -1$, ordered pair $(-2, -1)$

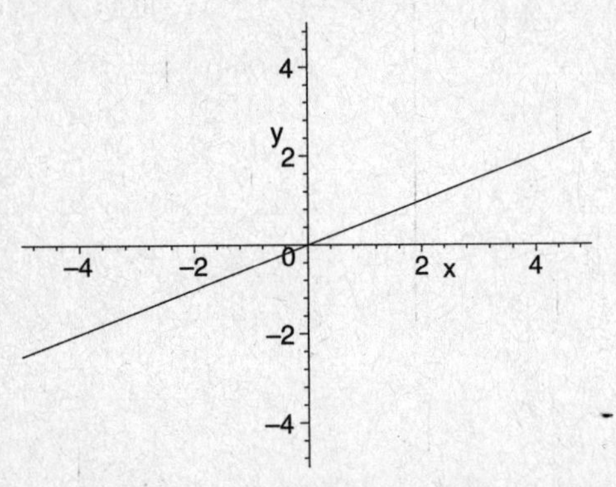

Compare the equation $y = 0.5x$ to the general linear equation form of $y = mx + b$ to conclude the equation has a slope of $m = 0.5$ and a y-intercept of $(0, 0)$.

8. Slope of $m = 3$ and y-intercept of $(0, 0)$

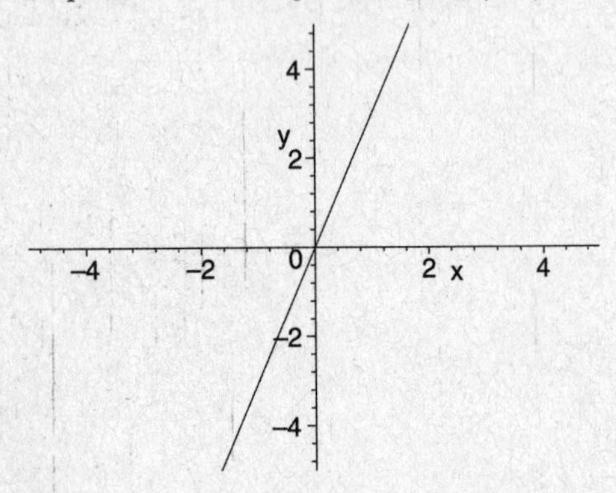

9. Graph. Find the slope and the y-intercept of $y = -2x + 3$.

First, we find some points that satisfy the equation, then we plot the ordered pairs and connect the plotted points to get the graph.

When $x = 0$, $y = -2(0) + 3 = 3$, ordered pair $(0, 3)$

When $x = 2$, $y = -2(2) + 3 = -1$, ordered pair $(2, -1)$

When $x = -2$, $y = -2(-2) + 3 = 7$, ordered pair $(-2, 7)$

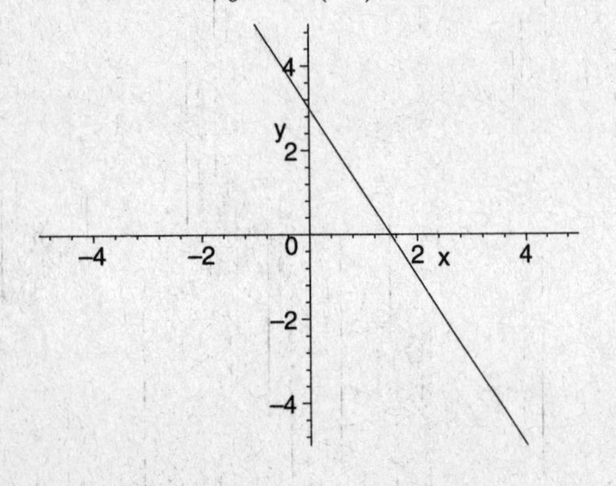

Compare the equation $y = -2x + 3$ to the general linear equation form of $y = mx + b$ to conclude the equation has a slope of $m = -2$ and a y-intercept of $(0, 3)$.

10. Slope of $m = -1$ and y-intercept of $(0, 4)$

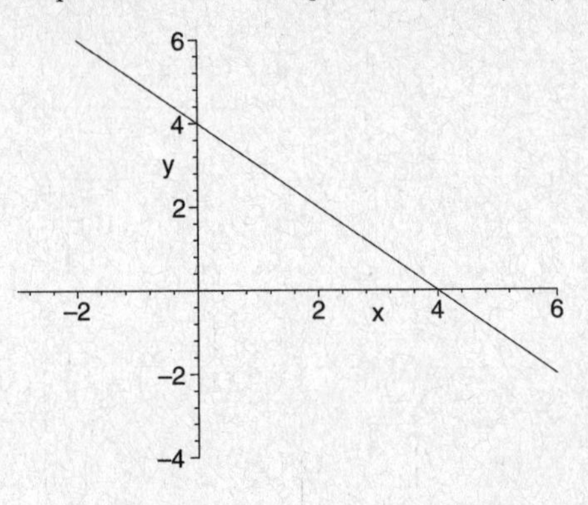

11. Graph. Find the slope and the y-intercept of $y = -x - 2$.

First, we find some points that satisfy the equation, then we plot the ordered pairs and connect the plotted points to get the graph.

When $x = 0$, $y = -(0) - 2 = -2$, ordered pair $(0, -2)$

When $x = 3$, $y = -(3) - 2 = -5$, ordered pair $(3, -5)$

When $x = -2$, $y = -(-2) - 2 = 0$, ordered pair $(-2, 0)$

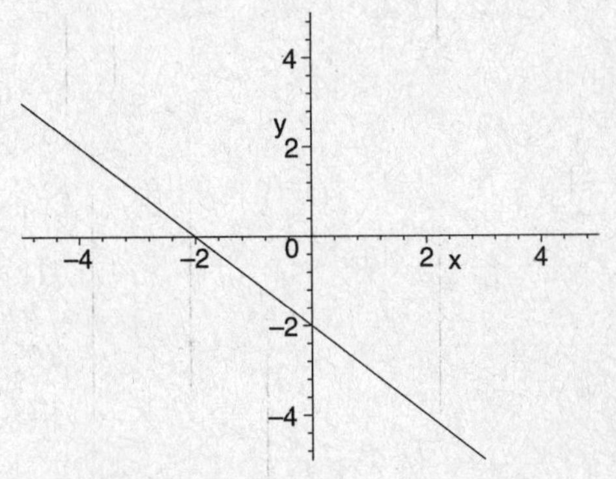

Compare the equation $y = -x - 2$ to the general linear equation form of $y = mx + b$ to conclude the equation has a slope of $m = -1$ and a y-intercept of $(0, -2)$.

12. Slope of $m = -3$ and y-intercept of $(0, 2)$

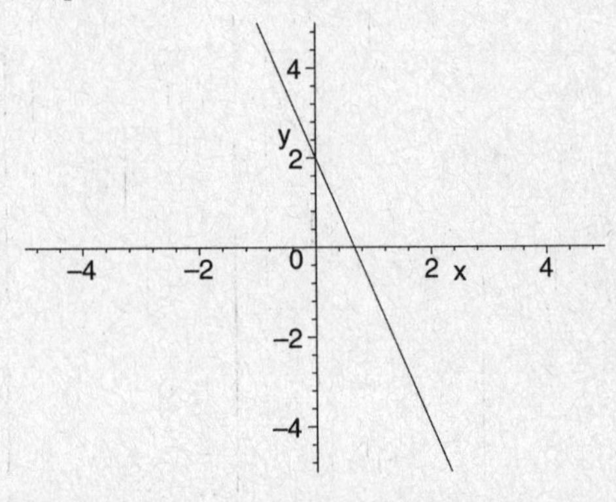

13. Find the slope and y-intercept of $2x + y - 2 = 0$.

Solve the equation for y.

$$\begin{aligned} 2x + y - 2 &= 0 \\ y &= -2x + 2 \end{aligned}$$

Compare to $y = mx + b$ to conclude the equation has a slope of $m = -2$ and a y-intercept of $(0, 2)$.

14. $y = 2x + 3$, slope of $m = 2$ and y-intercept of $(0, 3)$

15. Find the slope and y-intercept of $2x + 2y + 5 = 0$.

Solve the equation for y.

$$\begin{aligned} 2x + 2y + 5 &= 0 \\ 2y &= -2x - 5 \\ y &= -x - \frac{5}{2} \end{aligned}$$

Compare to $y = mx + b$ to conclude the equation has a slope of $m = -1$ and a y-intercept of $(0, -\frac{5}{2})$.

16. $y = x + 2$, slope of $m = 1$ and y-intercept of $(0, 2)$.

17. Find the slope and y-intercept of $x = 2y + 8$.

Solve the equation for y.

$$\begin{aligned} x &= 2y + 8 \\ x - 8 &= 2y \\ \frac{1}{2}x - 4 &= y \end{aligned}$$

Compare to $y = mx + b$ to conclude the equation has a slope of $m = \frac{1}{2}$ and a y-intercept of $(0, -4)$.

18. $y = -\frac{1}{4}x + \frac{3}{4}$, slope of $m = -\frac{1}{4}$ and y-intercept of $(0, \frac{3}{4})$

19. Find the equation of the line: with $m = -5$, containing $(1, -5)$

Plug the given information into equation $y - y_1 = m(x - x_1)$ and solve for y

$$\begin{aligned} y - y_1 &= m(x - x_1) \\ y - (-5) &= -5(x - 1) \\ y + 5 &= -5x + 5 \\ y &= -5x + 5 - 5 \\ y &= -5x \end{aligned}$$

20.

$$\begin{aligned} y - 7 &= 7(x - 1) \\ y - 7 &= 7x - 7 \\ y &= 7x \end{aligned}$$

21. Find the equation of line: with $m = -2$, containing $(2, 3)$

Plug the given information into the equation $y - y_1 = m(x - x_1)$ and solve for y

$$\begin{aligned} y - 3 &= -2(x - 2) \\ y - 3 &= -2x + 4 \\ y &= -2x + 4 + 3 \\ y &= -2x + 7 \end{aligned}$$

22.

$$\begin{aligned} y - (-2) &= -3(x - 5) \\ y + 2 &= -3x + 15 \end{aligned}$$
$$y = -3x + 13$$

23. Find the equation of line: with $m = 2$, containing $(3, 0)$

Plug the given information into the equation $y - y_1 = m(x - x_1)$ and solve for y

$$\begin{aligned} y - 0 &= 2(x - 3) \\ y &= 2x - 6 \end{aligned}$$

24.

$$\begin{aligned} y - 0 &= -5(x - 5) \\ y &= -5x + 25 \end{aligned}$$

25. Find the equation of line: with y-intercept $(0, -6)$ and $m = \frac{1}{2}$

Plug the given information into the equation $y = mx + b$

$$\begin{aligned} y &= mx + b \\ y &= \frac{1}{2}x + (-6) \\ y &= \frac{1}{2}x - 6 \end{aligned}$$

26. $y = \frac{4}{3}x + 7$

27. Find the equation of line: with $m = 0$, containing $(2, 3)$

Plug the given information into the equation $y - y_1 = m(x - x_1)$ and solve for y

$$\begin{aligned} y - 3 &= 0(x - 2) \\ y - 3 &= 0 \\ y &= 3 \end{aligned}$$

28.

$$\begin{aligned} y - 8 &= 0(x - 4) \\ y - 8 &= 0 \\ y &= 8 \end{aligned}$$

29. Find the slope given $(-4, -2)$ and $(-2, 1)$

Use the slope equation $m = \frac{y_2 - y_1}{x_2 - x_1}$. **NOTE:** It does not matter which point is chosen as (x_1, y_1) and which is chosen as (x_2, y_2) as long as the order the point coordinates are subtracted in the same order as illustrated below

$$\begin{aligned} m &= \frac{1 - (-2)}{-2 - (-4)} \\ &= \frac{1 + 2}{-2 + 4} \\ &= \frac{3}{2} \end{aligned}$$

$$\begin{aligned} m &= \frac{-2-1)}{-4-(-2)} \\ &= \frac{-3}{-2} \\ &= \frac{3}{2} \end{aligned}$$

30. $m = \frac{3-1}{6-(-2)} = \frac{2}{8} = \frac{1}{4}$

31. Find the slope given $(\frac{2}{5}, \frac{1}{2})$ and $(-3, \frac{4}{5})$

$$\begin{aligned} m &= \frac{\frac{4}{5} - \frac{1}{2}}{-3 - \frac{2}{5}} \\ &= \frac{\frac{8}{10} - \frac{5}{10}}{\frac{-15}{5} - \frac{10}{5}} \\ &= \frac{\frac{3}{10}}{\frac{17}{5}} \\ &= \frac{3}{10} \cdot \frac{5}{17} \\ &= \frac{15}{170} \\ &= \frac{3}{34} \end{aligned}$$

32. $m = \frac{-\frac{3}{16} - \frac{5}{6}}{-\frac{1}{2} - (-\frac{3}{4})} = \frac{-\frac{3}{16}}{\frac{1}{4}} = -\frac{3}{16} \cdot \frac{4}{1} = -\frac{3}{4}$

33. Find the slope given $(3, -7)$ and $(3, -9)$

$$\begin{aligned} m &= \frac{-9-(-7)}{3-3} \\ &= \frac{-2}{0} \quad \text{undefined quantity} \end{aligned}$$

This line has no slope

34. $m = \frac{10-2}{-4-(-4)} = \frac{8}{0}$ This line has no slope

35. Find the slope given $(2, 3)$ and $(-1, 3)$

$$\begin{aligned} m &= \frac{3-3}{-1-2} \\ &= \frac{0}{-3} \\ &= 0 \end{aligned}$$

36. $m = \frac{\frac{1}{2} - \frac{1}{2}}{-7-(-6)} = \frac{0}{-1} = 0$

37. Find the slope given $(x, 3x)$ and $(x+h, 3(x+h))$

$$\begin{aligned} m &= \frac{3(x+h) - 3x}{x+h-x} \\ &= \frac{3x + 3h - 3x}{h} \\ &= \frac{3h}{h} \\ &= 3 \end{aligned}$$

38. $m = \frac{4(x+h)-4x}{x+h-x} = \frac{4x+4h-4x}{h} = \frac{4h}{h} = 4$

39. Find the slope given $(x, 2x+3)$ and $(x+h, 2(x+h)+3)$

$$\begin{aligned} m &= \frac{[2(x+h)+3] - (2x+3)}{x+h-x} \\ &= \frac{2x + 2h + 3 - 2x - 3}{h} \\ &= \frac{2h}{h} \\ &= 2 \end{aligned}$$

40. $m = \frac{[3(x+h)-1]-(3x-1)}{x+h-x} = \frac{3x+3h-1-3x+1}{h} = \frac{3h}{h} = 3$

41. Find equation of line containing $(-4, -2)$ and $(-2, 1)$

From Exercise 29, we know that the slope of the line is $\frac{3}{2}$. Using the point$(-2, 1)$ and the value of the slope in the point-slope formula $y - y_1 = m(x - x_1)$ and solving for y we get:

$$\begin{aligned} y - 1 &= \frac{3}{2}(x - (-2)) \\ y - 1 &= \frac{3}{2}(x + 2) \\ y - 1 &= \frac{3}{2}x + 3 \\ y &= \frac{3}{2}x + 3 + 1 \\ y &= \frac{3}{2}x + 4 \end{aligned}$$

NOTE: You could use either of the given points and you would reach the final equation.

42. Using $m = \frac{1}{4}$ and the point $(6, 3)$

$$\begin{aligned} y - 3 &= \frac{1}{4}(x - 6) \\ y - 3 &= \frac{1}{4}x - \frac{6}{4} \\ y &= \frac{1}{4}x - \frac{3}{2} + 3 \\ y &= \frac{1}{4}x + \frac{3}{2} \end{aligned}$$

43. Find equation of line containing $(\frac{2}{5}, \frac{1}{2})$ and $(-3, \frac{4}{5})$

From Exercise 31, we know that the slope of the line is $-\frac{3}{34}$ and using the point $(-3, \frac{4}{5})$

$$\begin{aligned} y - \frac{4}{5} &= -\frac{3}{34}(x - (-3)) \\ y - \frac{4}{5} &= -\frac{3}{34}(x + 3) \\ y - \frac{4}{5} &= -\frac{3}{34}x - \frac{9}{34} \\ y &= -\frac{3}{34}x - \frac{9}{34} + \frac{4}{5} \\ y &= -\frac{3}{34}x - \frac{45}{170} + \frac{136}{170} \\ y &= -\frac{3}{34}x + \frac{91}{170} \end{aligned}$$

44. Using $m = -\frac{13}{4}$ and the point $\left(-\frac{3}{4}, \frac{5}{8}\right)$

$$\begin{aligned} y - \frac{5}{8} &= -\frac{13}{4}\left(x - \left(-\frac{3}{4}\right)\right) \\ y - \frac{5}{8} &= -\frac{13}{4}x - \frac{39}{16} \\ y &= -\frac{13}{4}x - \frac{39}{16} + \frac{5}{8} \\ y &= -\frac{13}{4}x - \frac{39}{16} + \frac{10}{16} \\ y &= -\frac{13}{4}x - \frac{29}{16} \end{aligned}$$

45. Find equation of line containing $(3, -7)$ and $(3, -9)$

From Exercise 33, we found that the line containing $(3, -7)$ and $(3, -9)$ has no slope. We notice that the x-coordinate does not change regardless of the y-value. Therefore, the line in vertical and has the equation $x = 3$.

46. Since the line has no slope, it is vertical. The equation of the line is $x = -4$.

47. Find equation of line containing $(2, 3)$ and $(-1, 3)$

From Exercise 35, we found that the line containing $(2, 3)$ and $(-1, 3)$ has a slope of $m = 0$. We notice that the y-coordinate does not change regardless of the x-value. Therefore, the line in horizontal and has the equation $y = 3$.

48. Since the line has a slope of $m = 0$, it is horizontal. The equation of the line is $y = \frac{1}{2}$

49. Find equation of line containing $(x, 3x)$ and $(x+h, 3(x+h))$

From Exercise 37, we found that the line containing $(x, 3x)$ and $(x+h, 3(x+h))$ had a slope of $m = 3$. Using the point $(x, 3x)$ and the value of the slope in the point-slope formula

$$\begin{aligned} y - 3x &= 3(x - x) \\ y - 3x &= 3(0) \\ y - 3x &= = 0 \\ y &= 3x \end{aligned}$$

50. Using $m = 4$ and the point $(x, 4x)$

$$\begin{aligned} y - 4x &= 4(x - x) \\ y - 4x &= 0 \\ y &= 4x \end{aligned}$$

51. Find equation of line containing $(x, 2x+3)$ and $(x+h, 2(x+h)+3)$

From Exercise 37, we found that the line containing $(x, 2x+3)$ and $(x+h, 2(x+h)+3)$ had a slope of $m = 2$. Using the point $(x, 3x)$ and the value of the slope in the point-slope formula

$$\begin{aligned} y - (2x+3) &= 2(x - x) \\ y - (2x+3) &= 2(0) \\ y - (2x+3) &= 0 \\ y &= 2x + 3 \end{aligned}$$

52. Using $m = 3$ and the point $(x, 3x - 1)$

$$\begin{aligned} y - (3x - 1) &= 3(x - x) \\ y - (3x - 1) &= 0 \\ y &= 3x - 1 \end{aligned}$$

53. Slope $= \frac{0.4}{5} = 0.08$. This means the treadmill has a grade of 8%.

54. The roof has a slope of $\frac{2.6}{6.2} \approx 0.3171$, or 31.71%

55. The slope (or head) of the river is $\frac{43.33}{1238} = 0.035 = 3.5\%$

56. The stairs have a maximum grade of $\frac{8.25}{9} = 0.91\overline{6} \approx 0.9167 = 91.67\%$

57. The average rate of change of life expectancy at birth is computed by finding the slope of the line containing the two points $(1990, 73.7)$ and $(2000, 76.9)$, which is given by

$$\begin{aligned} \text{Rate} &= \frac{\text{Change in Life expectancy}}{\text{Change in Time}} \\ &= \frac{76.9 - 73.7}{2000 - 1990} \\ &= \frac{3.2}{10} \\ &= 0.32 \text{ per year} \end{aligned}$$

58. **a)** $F(-10) = \frac{9}{5} \cdot (-10) + 32 = -18 + 32 = 14^oF$
$F(0) = \frac{9}{5} \cdot (0) + 32 = 0 + 32 = 32^oF$
$F(10) = \frac{9}{5} \cdot (10) + 32 = 18 + 32 = 50^oF$
$F(40) = \frac{9}{5} \cdot (40) + 32 = 72 + 32 = 104^oF$

b) $F(30) = \frac{9}{5} \cdot (30) + 32 = 54 + 32 = 86^oF$

c) Same temperature in both means $F(x) = x$. So

$$\begin{aligned} F(x) &= x \\ \frac{9}{5}x + 32 &= x \\ \frac{9}{5}x - x &= -32 \\ \frac{4}{5}x &= -32 \\ x &= -32 \cdot \frac{5}{4} \\ x &= -40^o \end{aligned}$$

59. **a)** Since R and T are directly proportional we can write that $R = kT$, where k is a constant of proportionality. Using $R = 12.51$ when $T = 3$ we can find k.

$$\begin{aligned} R &= kT \\ 12.51 &= k(3) \\ \frac{12.51}{3} &= k \\ 4.17 &= k \end{aligned}$$

Thus, we can write the equation of variation as $R = 4.17T$

b) This is the same as asking: find R when $T = 6$. So, we use the variation equation

$$\begin{aligned} R &= 4.17T \\ &= 4.17(6) \\ &= 25.02 \end{aligned}$$

60. We need to find t when $D = 6$.

$$\begin{aligned} D &= 293t \\ 6 &= 293t \\ \frac{6}{293} &= t \\ 0.0205 \text{ seconds} &\approx t \end{aligned}$$

61. **a)** Since B s directly proportional to W we can write $B = kW$.

b) When $W = 200$ $B = 5$ means that

$$\begin{aligned} B &= kW \\ 5 &= k(200) \\ \frac{5}{200} &= k \\ 0.025 &= k \\ 2.5\% &= k \end{aligned}$$

This means that the weight of the brain is 2.5% the weight of the person.

c) Find B when $W = 120$

$$\begin{aligned} B &= 0.025W \\ &= 0.025(120\ lbs) \\ &= 3\ lbs \end{aligned}$$

62. **a)**

$$\begin{aligned} M &= kW \\ 80 &= k(200) \\ 0.4 &= k \end{aligned}$$

Thus, the equation of variation is $M = 0.4W$

b) $k = 0.4 = 40\%$ means that 40% of the body weight is the weight of muscles.

c)

$$\begin{aligned} M &= 0.4(120) \\ &= 48 \text{ lb} \end{aligned}$$

63. **a)**

$$\begin{aligned} D(0) &= 2(0) + 115 = 0 + 115\ ft \\ D(-20) &= 2(-20) + 115 = -40 + 115 = 75\ ft \\ D(10) &= 2(10) + 115 = 20 + 115 = 135\ ft \\ D(32) &= 2(32) + 115 = 64 + 115 = 179\ ft \end{aligned}$$

b) The stopping distance has to be a non-negative value. Therefore we need to solve the inequality

$$\begin{aligned} 0 &\le 2F + 115 \\ -115 &\le 2F \\ -57.5 &\le F \end{aligned}$$

The 32° limit comes from the fact that for any temperature above that there would be no ice. Thus, the domain of the function is restricted in the interval $[-57.5, 32]$.

64. **a)**

$$\begin{aligned} D(5) &= \frac{11\cdot 0+5}{10} = \frac{5}{10} = 0.5\ ft \\ D(10) &= \frac{11\cdot 10+5}{10} = \frac{115}{10} = 11.5\ ft \\ D(20) &= \frac{11\cdot 20+5}{10} = \frac{225}{10} = 22.5\ ft \\ D(50) &= \frac{11\cdot 50+5}{10} = \frac{555}{10} = 55.5\ ft \\ D(65) &= \frac{11\cdot 65+5}{10} = \frac{720}{10} = 72\ ft \end{aligned}$$

b)

c) Since cars cannot have negative speed, and since the car will not need to stop if it has speed of 0 then the domain is any positive real number. **NOTE:** The domain will have an upper bound since cars have a top speed limit, depending on the make and model of the car.

65. **a)**

$$\begin{aligned} M(x) &= 2.89x + 70.64 \\ M(26) &= 2.89(26) + 70.64 \\ &= 75.14 + 70.64 \\ &= 145.78 \end{aligned}$$

The male was 145.78 cm tall.

b)

$$\begin{aligned} F(x) &= 2.75x + 71.48 \\ F(26) &= 2.75(26) + 71.48 \\ &= 71.5 + 71.48 \\ &= 142.98 \end{aligned}$$

The female was 142.98 cm tall.

66. **a)** The equation of variation is given by $N = P + 0.02P = 1.02P$.

b) $N = 1.02(200000) = 204000$

c)

$$\begin{aligned} 367200 &= 1.02P \\ \frac{367200}{1.02} &= P \\ 360000 &= P \end{aligned}$$

67. **a)**

$$\begin{aligned} A(0) &= 0.08(0) + 19.7 = 0 + 19.7 = 19.7 \\ A(1) &= 0.08(1) + 19.7 = 0.08 + 19.7 = 19.78 \\ A(10) &= 0.08(10) + 19.7 = 0.8 + 19.7 = 20.5 \\ A(30) &= 0.08(30) + 19.7 = 2.4 + 19.7 = 22.1 \\ A(50) &= 0.08(50) + 19.7 = 4 + 19.7 = 23.7 \end{aligned}$$

b) First we find the value of $t4$, which is $2003 - 1950 = 53$. So, we have to find $A(53)$.

$$A(53) = 0.08(53) + 19.7 = 4.24 + 19.8 = 23.94$$

The median age of women at first marriage in the year 2003 is 23.94 years.

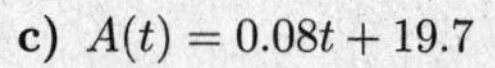

c) $A(t) = 0.08t + 19.7$

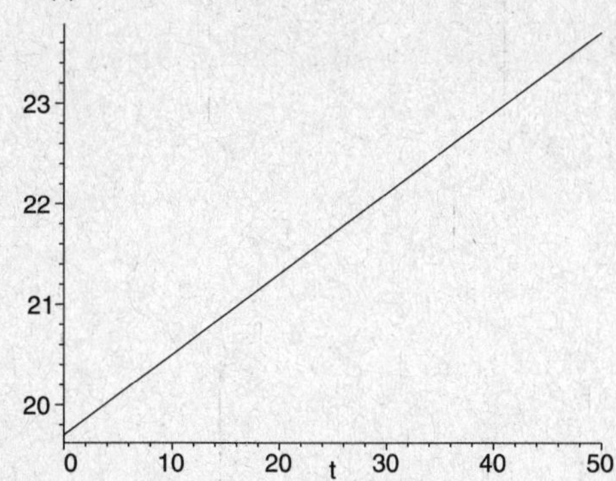

68. The use of the slope-intercept equation or the point-slope equation depends on the problem. If the problem gives the slope and the y-intercept then one should use the slope-intercept equation. If the problem gives the slope and a point that falls on the line, or two points that fall on the line then the point-slope equation should be used.

Exercise Set 1.2

1. $y = \frac{1}{2}x^2$ and $y = -\frac{1}{2}x^2$

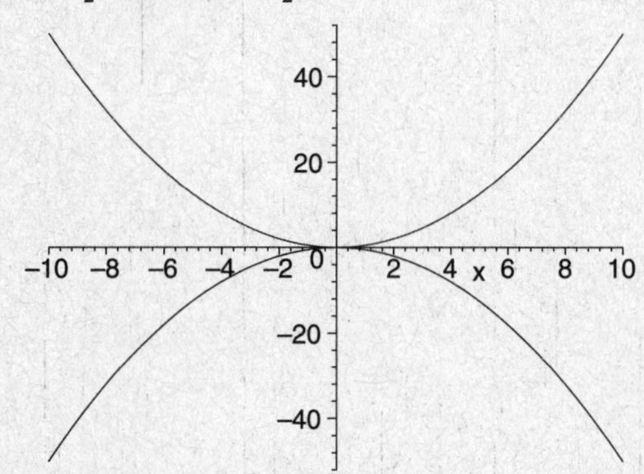

2. $y = \frac{1}{4}x^2$ and $y = -\frac{1}{4}x^2$

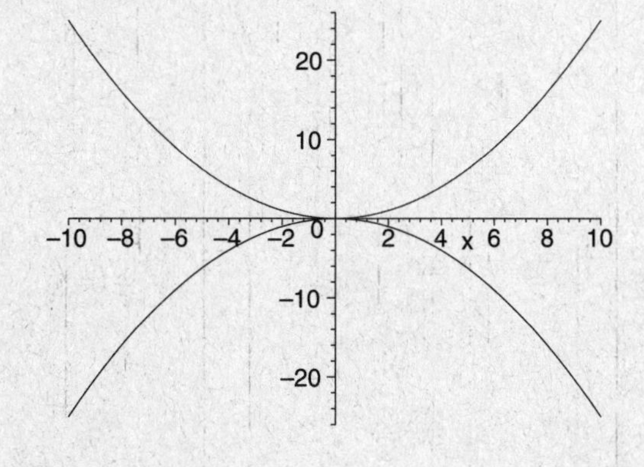

3. $y = x^2$ and $y = (x-1)^2$

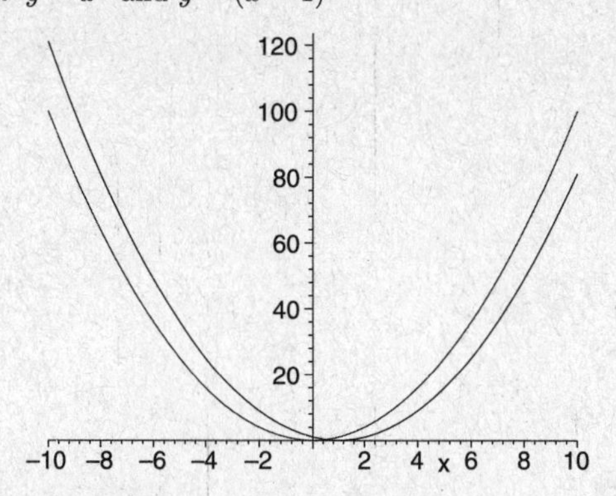

4. $y = x^2$ and $y = (x-3)^2$

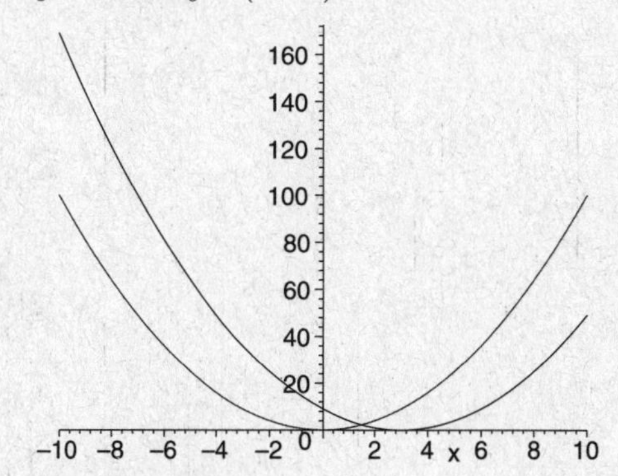

5. $y = x^2$ and $y = (x+1)^2$

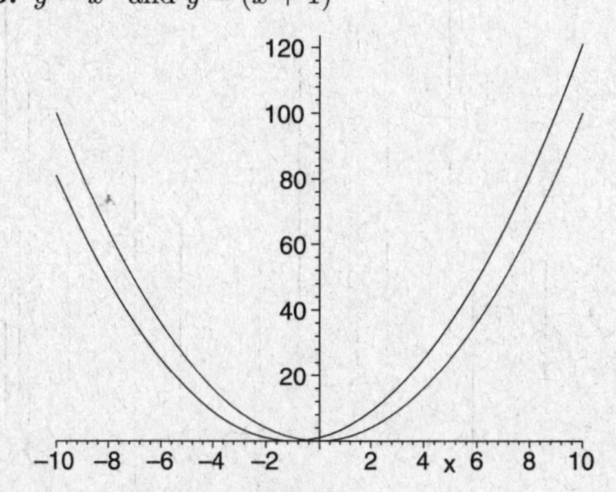

6. $y = x^2$ and $y = (x+3)^2$

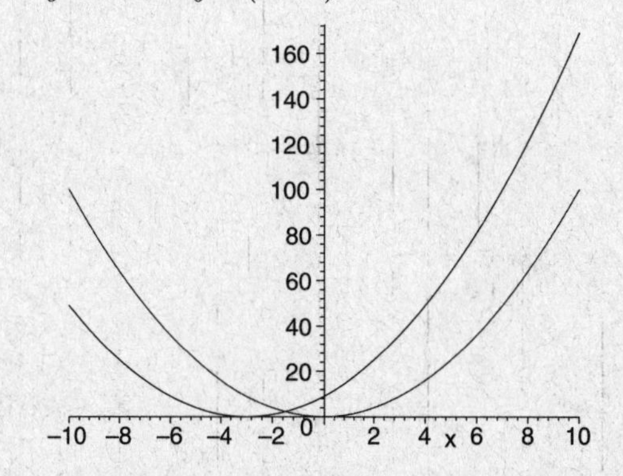

7. $y = x^3$ and $y = x^3 + 1$

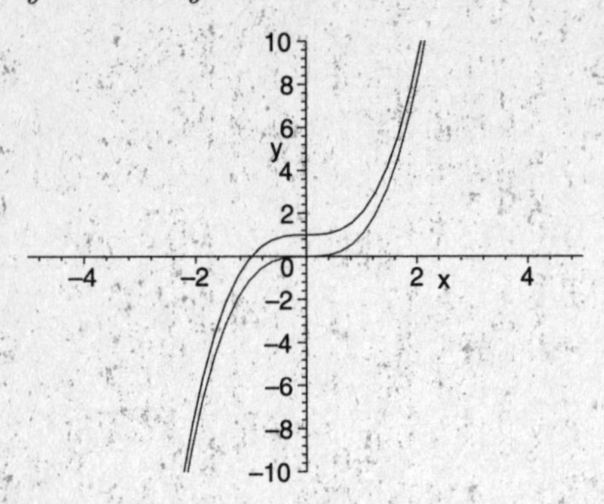

8. $y = x^3$ and $y = x^3 - 1$

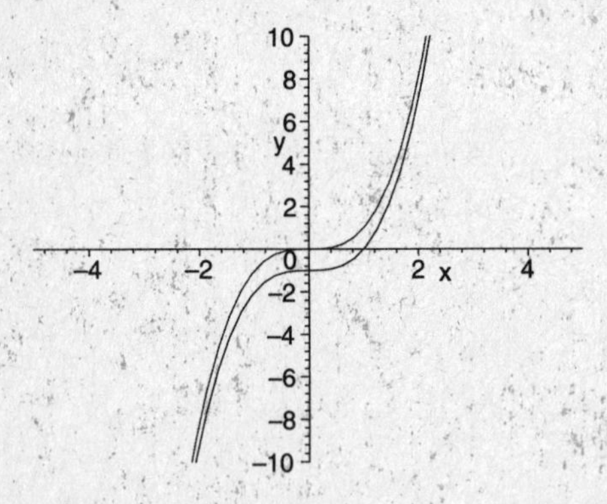

9. Since the equation has the form $ax^2 + bx + c$, with $a \neq 0$, the graph of the function is a parabola. The x-value of the vertex is given by

$$x = -\frac{b}{2a} = -\frac{4}{2(1)} = -2$$

The y-value of the vertex is given by

$$\begin{aligned} y &= (-2)^2 + 4(-2) - 7 \\ &= 4 - 8 - 7 \\ &= -11 \end{aligned}$$

Therefore, the vertex is $(-2, 11)$.

10. Since the equation is not in the form of $ax^2 + bx + c$, the graph of the function is not a parabola.

11. Since the equation is not in the form of $ax^2 + bx + c$, the graph of the function is not a parabola.

12. Since the equation has the form $ax^2 + bx + c$, with $a \neq 0$, the graph of the function is a parabola. The x-value of the vertex is given by

$$x = -\frac{b}{2a} = -\frac{-6}{2(3)} = 1$$

The y-value of the vertex is given by

$$\begin{aligned} y &= 3(1)^2 - 6(1) \\ &= 3 - 6 \\ &= -3 \end{aligned}$$

Therefore, the vertex is $(1, -3)$.

13. $y = x^2 - 4x + 3$

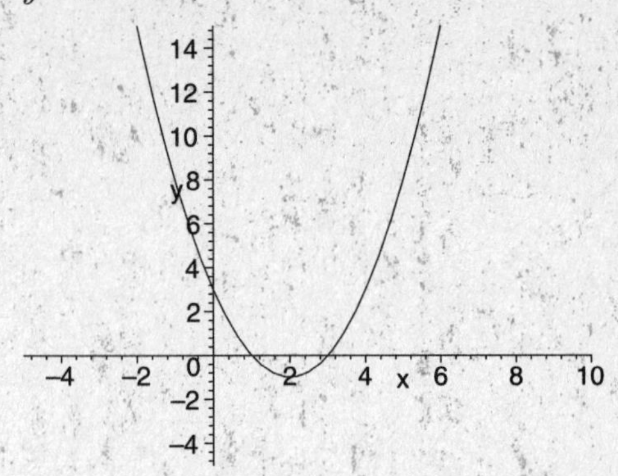

14. $y = x^2 - 6x + 5$

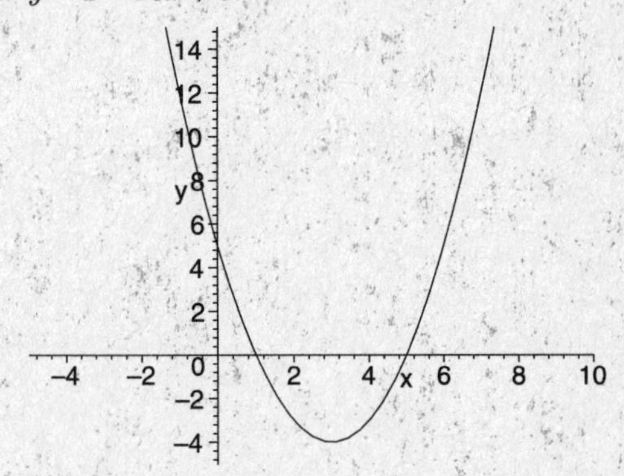

15. $y = -x^2 + 2x - 1$

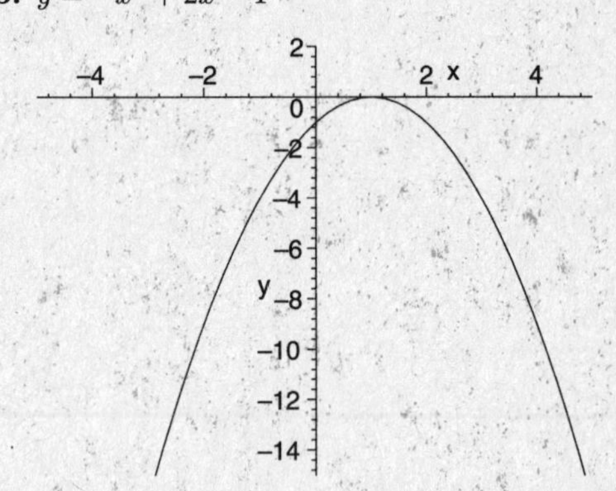

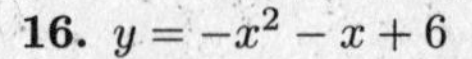

16. $y = -x^2 - x + 6$

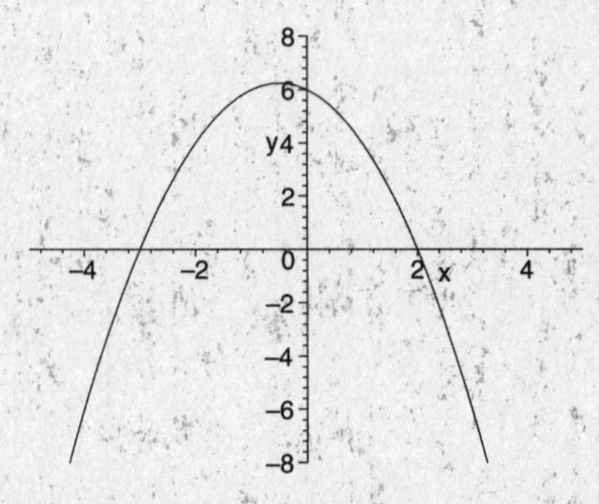

17. $y = 2x^2 + 4x - 7$

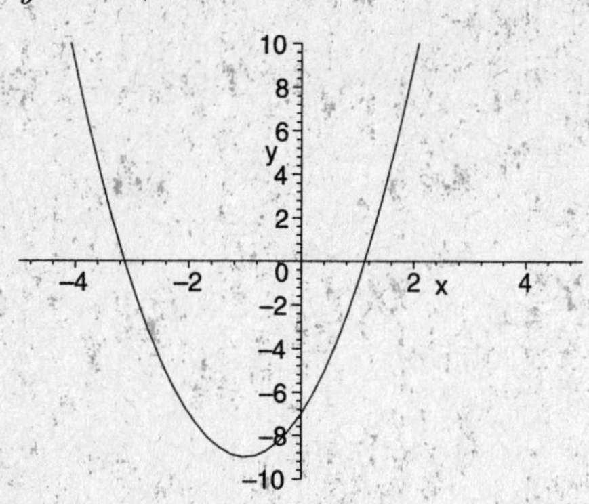

18. $y = 3x^2 - 9x + 2$

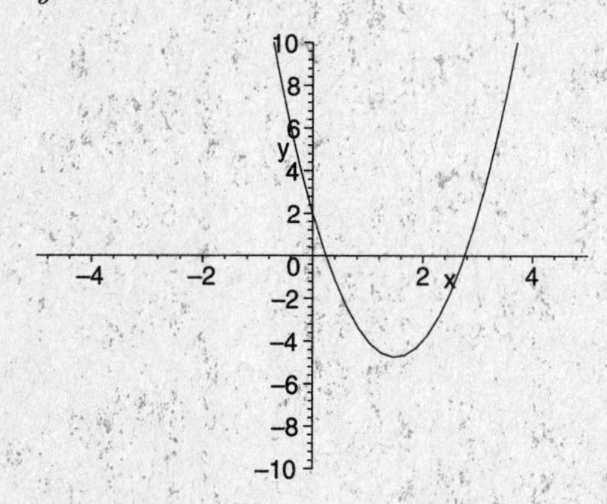

19. $y = \frac{1}{2}x^2 + 3x - 5$

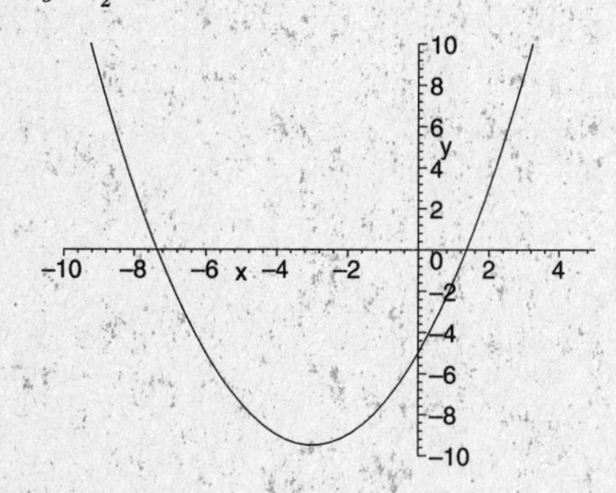

20. $y = \frac{1}{3}x^2 + 4x - 2$

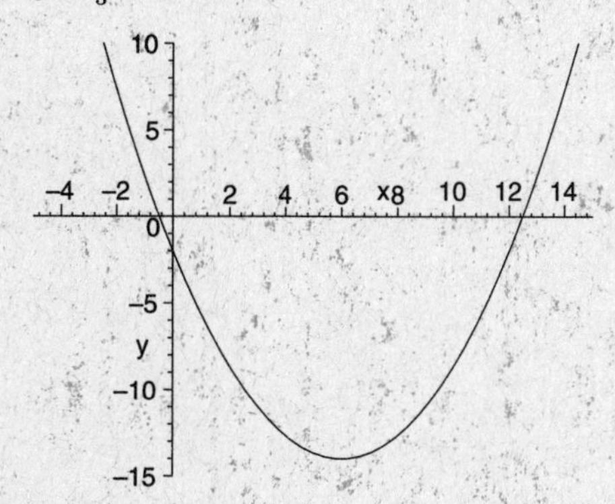

21. Solve $x^2 - 2x = 2$
Write the equation so that one side equals zero, that is $x^2 - 2x - 2 = 0$, then use the quadratic formula, with $a = 1$, $b = -2$, and $c = -2$, to solve for x.

$$\begin{aligned} x &= \frac{-b \pm \sqrt{b^2 - 4ac}}{2a} \\ x &= \frac{-(-2) \pm \sqrt{(-2)^2 - 4(1)(-2)}}{2(1)} \\ &= \frac{2 \pm \sqrt{4+8}}{2} \\ &= \frac{2 \pm \sqrt{12}}{2} \\ &= \frac{2 \pm 2\sqrt{3}}{2} \\ &= \frac{2(1 \pm \sqrt{3})}{2} \\ &= 1 \pm \sqrt{3} \end{aligned}$$

The solutions are $1 + \sqrt{3}$ and $1 - \sqrt{3}$

22. $x^2 - 2x + 1 = 5$ can be rewritten as $x^2 - 2x - 4 = 0$

$$\begin{aligned} x &= \frac{-(-2) \pm \sqrt{(-2)^2 - 4(1)(-4)}}{2(1)} \\ &= \frac{2 \pm \sqrt{4+16}}{2} \\ &= \frac{2 \pm \sqrt{20}}{2} = \frac{2(1 \pm \sqrt{5})}{2} \\ &= 1 \pm \sqrt{5} \end{aligned}$$

The solutions are $1 + \sqrt{5}$ and $1 - \sqrt{5}$

23. Solve $3y^2 + 8y + 2 = 0$
Use the quadratic formula, with $a = 3$, $b = 8$, and $c = 2$, to solve for y.

$$\begin{aligned} y &= \frac{-b \pm \sqrt{b^2 - 4ac}}{2a} \\ y &= \frac{-8 \pm \sqrt{(8)^2 - 4(3)(2)}}{2(3)} \\ &= \frac{-8 \pm \sqrt{64 - 24}}{6} \\ &= \frac{-8 \pm \sqrt{40}}{6} \\ &= \frac{-8 \pm 2\sqrt{10}}{6} \\ &= \frac{2(-4 \pm \sqrt{10})}{6} \\ &= \frac{-4 \pm \sqrt{10}}{3} \end{aligned}$$

The solutions are $\frac{-4+\sqrt{10}}{3}$ and $\frac{-4-\sqrt{10}}{3}$

24. $2p^2 - 5p = 1$ can be rewritten as $2p^2 - 5p - 1$

$$\begin{aligned} p &= \frac{-(-5) \pm \sqrt{(-5)^2 - 4(2)(-1)}}{2(2)} \\ &= \frac{5 \pm \sqrt{25+8}}{4} \\ &= \frac{5 \pm \sqrt{33}}{4} \end{aligned}$$

The solutions are $\frac{5+\sqrt{33}}{4}$ and $\frac{5-\sqrt{33}}{4}$

25. Solve $x^2 - 2x + 10 = 0$
Using the quadratic formula with $a = 1$, $b = -2$, and $c = 10$

$$\begin{aligned} x &= \frac{-b \pm \sqrt{b^2 - 4ac}}{2a} \\ x &= \frac{-(-2) \pm \sqrt{(-2)^2 - 4(1)(10)}}{2(1)} \\ &= \frac{2 \pm \sqrt{4 - 40}}{2} \\ &= \frac{2 \pm \sqrt{-36}}{2} \\ &= \frac{2 \pm 6i}{2} \\ &= \frac{2(1 \pm 3i)}{2} \\ &= 1 \pm 3i \end{aligned}$$

The solutions are $1 + 3i$ and $1 - 3i$

26.

$$\begin{aligned} x &= \frac{-b \pm \sqrt{b^2 - 4ac}}{2a} \\ x &= \frac{-6 \pm \sqrt{(6)^2 - 4(1)(10)}}{2(1)} \\ &= \frac{-6 \pm \sqrt{36 - 40}}{2} \\ &= \frac{-6 \pm \sqrt{-4}}{2} \\ &= \frac{-6 \pm 2i}{2} \\ &= \frac{2(-3 \pm i)}{2} \\ &= -3 \pm i \end{aligned}$$

The solutions are $-3 + i$ and $-3 - i$

27. Solve $x^2 + 6x = 1$
Write the equation so that one side equals zero, that is $x^2 + 6x - 1 = 0$, then use the quadratic formula, with $a = 1$, $b = 6$, and $c = -1$, to solve for x.

$$\begin{aligned} x &= \frac{-b \pm \sqrt{b^2 - 4ac}}{2a} \\ x &= \frac{-6 \pm \sqrt{(6)^2 - 4(1)(-1)}}{2(1)} \\ &= \frac{-6 \pm \sqrt{36 + 4}}{2} \\ &= \frac{-6 \pm \sqrt{40}}{2} \\ &= \frac{-6 \pm 2\sqrt{10}}{2} \\ &= \frac{2(-3 \pm \sqrt{10})}{2} \\ &= -3 \pm \sqrt{10} \end{aligned}$$

The solutions are $-3 + \sqrt{10}$ and $-3 - \sqrt{10}$

28. $x^2 + 4x = 3$ can be rewritten as $x^2 + 4x - 3 = 0$

$$\begin{aligned} x &= \frac{-4 \pm \sqrt{(4)^2 - 4(1)(-3)}}{2(1)} \\ &= \frac{-4 \pm \sqrt{16 + 12}}{2} \\ &= \frac{-4 \pm \sqrt{28}}{2} = \frac{2(-2 \pm \sqrt{7})}{2} \\ &= -2 \pm \sqrt{7} \end{aligned}$$

The solutions are $-2 + \sqrt{7}$ and $-2 - \sqrt{7}$

29. Solve $x^2 + 4x + 8 = 0$
Using the quadratic formula with $a = 1$, $b = 4$, and $c = 8$

$$\begin{aligned} x &= \frac{-b \pm \sqrt{b^2 - 4ac}}{2a} \\ x &= \frac{-4 \pm \sqrt{(4)^2 - 4(1)(8)}}{2(1)} \\ &= \frac{-4 \pm \sqrt{16 - 32}}{2} \\ &= \frac{-4 \pm \sqrt{-16}}{2} \\ &= \frac{-4 \pm 4i}{2} \\ &= \frac{4(1 \pm i)}{2} \\ &= 2(1 \pm i) = 2 \pm 2i \end{aligned}$$

The solutions are $2 + 2i$ and $2 - 2i$

30.

$$\begin{aligned} x &= \frac{-10 \pm \sqrt{(10)^2 - 4(1)(27)}}{2(1)} \\ &= \frac{-10 \pm \sqrt{100 - 108}}{2} \\ &= \frac{-10 \pm \sqrt{-8}}{2} \\ &= \frac{-10 \pm 2i\sqrt{2}}{2} \\ &= \frac{2(-5 \pm i\sqrt{2})}{2} \\ &= -5 \pm i\sqrt{2} \end{aligned}$$

The solutions are $-5 + i\sqrt{2}$ and $-5 + i\sqrt{2}$

31. Solve $4x^2 = 4x - 1$
Write the equation so that one side equals zero, that is $4x^2 - 4x - 1 = 0$, then use the quadratic formula, with $a = 4$, $b = -4$, and $c = -1$, to solve for x.

$$\begin{aligned} x &= \frac{-b \pm \sqrt{b^2 - 4ac}}{2a} \\ x &= \frac{-(-4) \pm \sqrt{(-4)^2 - 4(4)(-1)}}{2(4)} \\ &= \frac{4 \pm \sqrt{16 + 16}}{8} \\ &= \frac{4 \pm \sqrt{32}}{8} \end{aligned}$$

$$= \frac{4 \pm 4\sqrt{2}}{8}$$
$$= \frac{4(1 \pm \sqrt{2})}{8}$$
$$= \frac{1 \pm \sqrt{2}}{2}$$

The solutions are $\frac{1+\sqrt{2}}{2}$ and $\frac{1-\sqrt{2}}{2}$

32. $-4x^2 = 4x - 1$ can be rewritten as $0 = 4x^2 + 4x - 1$

$$\begin{aligned} x &= \frac{-4 \pm \sqrt{(4)^2 - 4(4)(-1)}}{2(4)} \\ &= \frac{-4 \pm \sqrt{16+16}}{8} \\ &= \frac{-4 \pm \sqrt{32}}{8} = \frac{4(-1 \pm \sqrt{4})}{8} \\ &= \frac{-1 \pm \sqrt{2}}{2} \end{aligned}$$

The solutions are $\frac{-1+\sqrt{2}}{2}$ and $\frac{-1-\sqrt{2}}{2}$

33. Find $f(7)$, $f(10)$, and $f(12)$

$$\begin{aligned} f(7) &= \frac{1}{6}(7)^3 + \frac{1}{2}(7)^2 + \frac{1}{2}(7) \\ &= \frac{343}{6} + \frac{49}{2} + \frac{7}{2} \\ &= \frac{343}{6} + \frac{147}{6} + \frac{21}{6} \\ &= \frac{511}{6} \approx 85.1\overline{6} \approx 85 \text{ oranges} \\ f(10) &= \frac{1}{6}(10)^3 + \frac{1}{2}(10)^2 + \frac{1}{2}(10) \\ &= \frac{1000}{6} + 50 + 5 \\ &= \frac{500}{3} + \frac{150}{3} + \frac{15}{3} \\ &= \frac{665}{3} \approx 221.\overline{6} \approx 222 \text{ oranges} \\ f(12) &= \frac{1}{6}(12)^3 + \frac{1}{2}(12)^2 + \frac{1}{2}(12) \\ &= 288 + 72 + 6 \\ &= 366 \text{ oranges} \end{aligned}$$

34. **a)** $x = 2009 - 1985 = 24$

$$\begin{aligned} f(24) &= 4.8565 + 0.2841(24) + 0.1784(24)^2 \\ &= 4.8565 + 6.8184 + 102.7584 \\ &= 114.4333 \end{aligned}$$

The average payroll for 2009-10 is \$114.4333 million

b) Solve $100 = 4.8565 + 0.2841x + 0.1784x^2$. First, let us rewrite the equation as $0 = -95.1435 + 0.2841x + 0.1784x^2$ then we can use the quadratic formula to solve for x

$$\begin{aligned} x &= \frac{-0.2841 \pm \sqrt{0.2841^2 - 4(0.1784)(-95.1435)}}{2(0.1784)} \\ &= \frac{-0.2841 \pm \sqrt{0.0807 + 67.8944}}{0.3568} = \frac{-0.2841 \pm \sqrt{67.9751}}{0.3568} \\ &= \frac{-0.2841 \pm 8.2447}{0.3568} \\ &= \frac{-0.2841 + 8.2447}{0.3568} = 22.3111 \end{aligned}$$

Therefore, the average payroll will be \$100 million is the, $1985 + 23.3111 = 2007.3111$, 2007-08 season. **NOTE:** We could not choose the negative option of the quadratic formula since it would result in the result that is negative which corresponds to a year before 1985 and that does not make sense.

35. Solve $50 = 9.41 - 0.19x + 0.09x^2$. First, let us rewrite the equation as $0 = -40.59 - 0.19x + 0.09x^2$ then we can use the quadratic formula to solve for x

$$\begin{aligned} x &= \frac{-(-0.19) \pm \sqrt{(-0.19)^2 - 4(0.09)(-40.59)}}{2(0.09)} \\ &= \frac{0.19 \pm \sqrt{0.0361 + 14.6124}}{0.18} = \frac{0.19 \pm \sqrt{14.6485}}{0.18} \\ &= \frac{0.19 \pm 3.8273}{0.18} \\ &= \frac{0.19 + 3.8273}{0.18} = 22.3183 \end{aligned}$$

Therefore, the average price of a ticket will be \$50 will happen during the, $1990 + 22.3183 = 2012.3183$ 2012-13 season. **NOTE:** We could not choose the negative option of the quadratic formula since it would result in the result that is negative which corresponds to a year before 1990 and that does not make physical sense.

36. **a)**

$$\begin{aligned} w(72) &= 0.0728(72)^2 - 6.986(72) + 289 \\ &= 163.4032 \text{ pounds} \end{aligned}$$

b) Solve $170 = 0.0728h^2 - 6.986h + 289$, which can be written as $0.0728h^2 - 6.986h + 119 = 0$

$$\begin{aligned} h &= \frac{-(-6.986) \pm \sqrt{(-6.986)^2 - 4(0.0728)(119)}}{2(0.0728)} \\ &= \frac{6.986 \pm \sqrt{14.1514}}{0.1456} \\ &= \frac{6.986 \pm 3.7618)}{0.1456} \end{aligned}$$

The possible two answers are $\frac{6.986-3.7618}{0.1456} = 22.1440$ *in*, which is out side of the domain of the function, and $\frac{6.986+3.7618}{0.1456} = 73.8173$ *in*, which is in the domain interval of the function w. Therefore, the man is about 73.8 inches tall.

37. $f(x) = x^3 - x^2$

a) For large values of x, x^3 would be larger than x^2. $x^3 = x \cdot x \cdot x$ and $x^2 = x \cdot x$ so for very large values of x there is an extra factor of x in x^3 which causes x^3 to be larger than x^2.

b) As x gets very large the values of x^3 become much larger than those of x^2 and therefore we can "ignore" the effect of x^2 in the expression $x^3 - x^2$. Thus, we can approximate the function to look like x^3 for very large values of x.

c) Below is a graph of $x^3 - x^2$ and x^3 for $100 \leq x \leq 200$. It is hard to distinguish between the two graphs confirming the conclusion reached in part b).

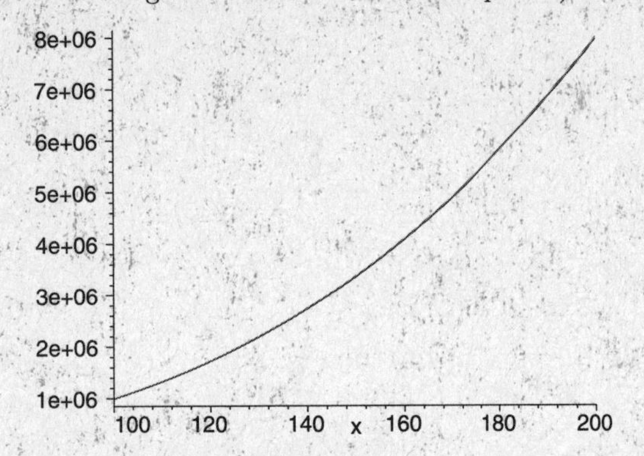

38. $f(x) = x^4 - 10x^3 + 3x^2 - 2x + 7$

a) For large values of x, x^4 will be larger than $| -10x^3 + 3x^2 - 2x + 7 |$ since the second term is a third degree polynomial (compared to a fourth degree polynomial) and has terms being subtracted.

b) Since the values of x^4 "dominate" the function for very large values of x the function will look like x^4 for very large values of x.

c) Below is a graph of $x^4 - 10x^3 + 3x^2 - 2x + 7$ and x^4 for $100 \leq x \leq 200$. The graphs are close to each other confirming our conclusion from part b).

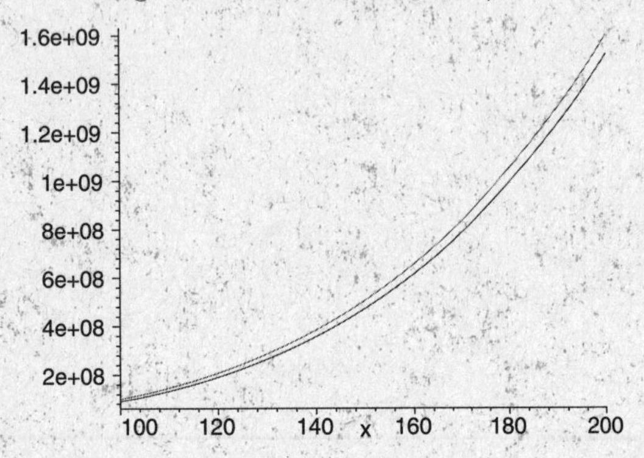

39. $f(x) = x^2 + x$

a) For values very close to 0, x is larger than x^2 since for values of x less than 1 $x^2 < x$.

b) For values of x very close to 0 $f(x)$ looks like x since the x^2 can be "ignored".

c) Below is a graph of $x^2 + x$ and x for $-0.01 \leq x \leq 0.01$. It is very hard to distinguish between the two graphs confirming our conclusion from part b).

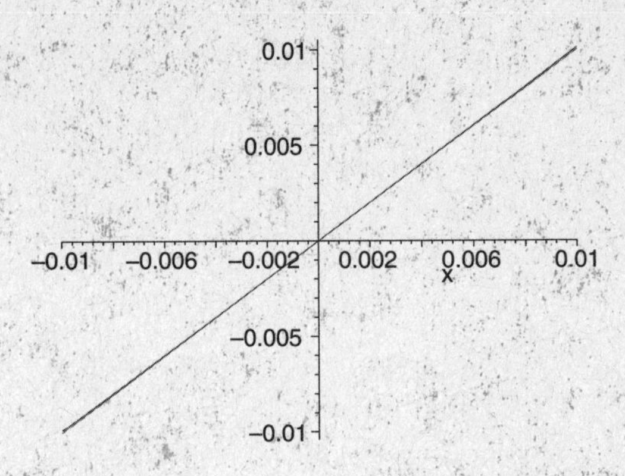

40. $f(x) = x^3 + 2x$

a) For x values very close to 0, $2x$ is larger than x^3 since for x values less than 1 the higher the degree the smaller the values of the term.

b) For x values very close to 0, the function will looks like $2x$ since the x^3 term may be "ignored".

c) Below is a graph of $x^3 + 2x$ and $2x$ for $-0.01 \leq x \leq 0.01$. It is very difficult to distinguish between the two graphs confirming our conclusion in part b).

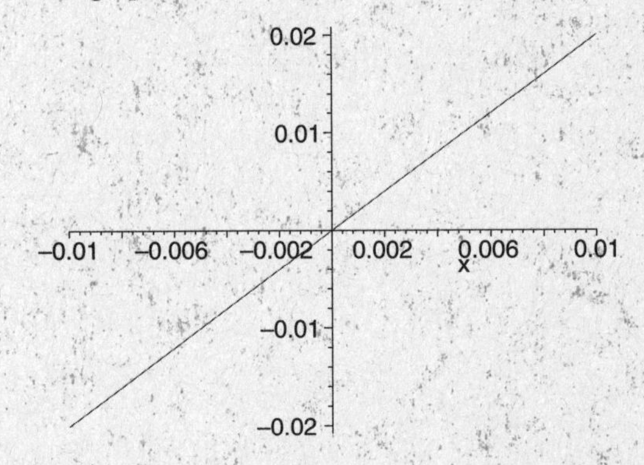

41. $f(x) = x^3 - x$

$$\begin{aligned} f(x) &= 0 \\ x^3 - x &= 0 \\ x(x^2 - 1) &= 0 \\ x(x-1)(x+1) &= 0 \\ x &= 0 \\ x &= 1 \\ x &= -1 \end{aligned}$$

42. $x = 2.359$

43. $x = -1.831$, $x = -0.856$, and $x = 3.188$

44. $x = 2.039$, and $x = 3.594$

45. $x = -10.153$, $x = -1.871$, $x = -0.821$, $x = -0.303$, $x = 0.098$, $x = 0.535$, $x = 1.219$, and $x = 3.297$

46. $y = 8.254x - 5.457$

47. $y = -0.279x + 4.036$

48. $y = 1.004x^2 + 1.904x - 0.601$

49. $y = 0.942x^2 - 2.651x - 27.943$

50. $y = 0.218x^3 + 0.188x^2 - 29.643x + 57.067$

51. $y = 0.237x^4 - 0.885x^3 - 29.224x^2 + 165.166x - 210.135$

Exercise Set 1.3

1. $y = |\,x\,|$ and $y = |\,x + 3\,|$

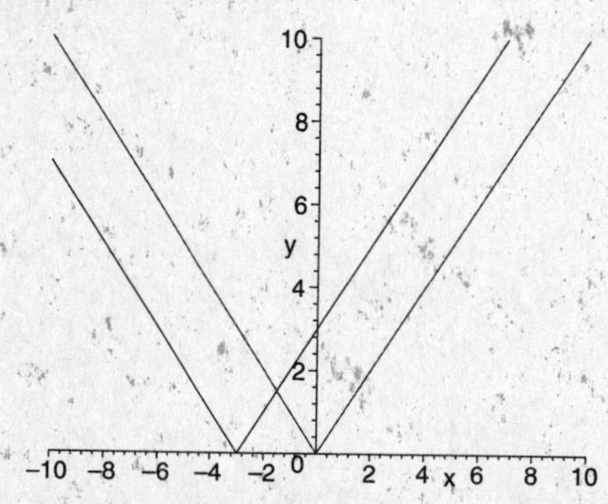

2. $y = |\,x\,|$ and $y = |\,x + 1\,|$

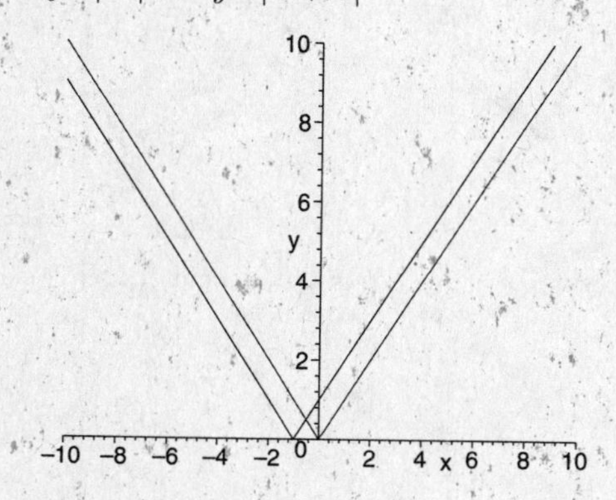

3. $y = \sqrt{x}$ and $y = \sqrt{x + 1}$

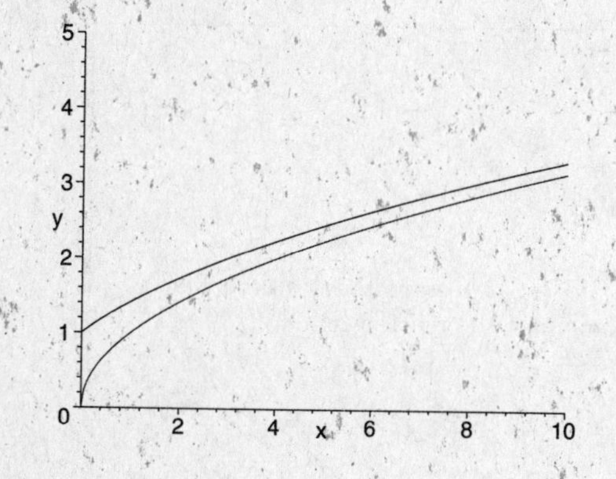

4. $y = \sqrt{x}$ and $y = \sqrt{x - 2}$

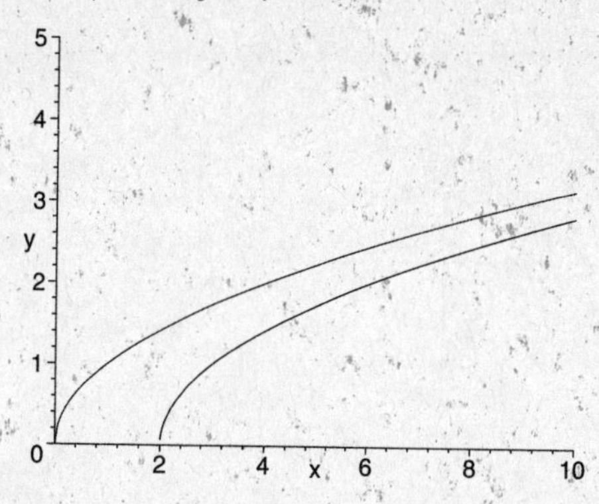

5. $y = \frac{2}{x}$

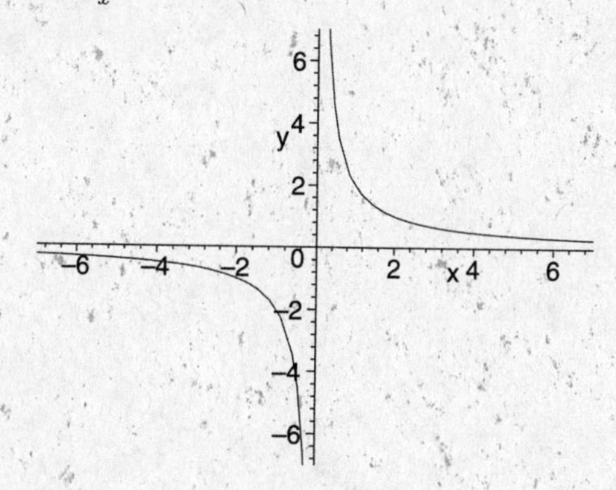

6. $y = \frac{3}{x}$

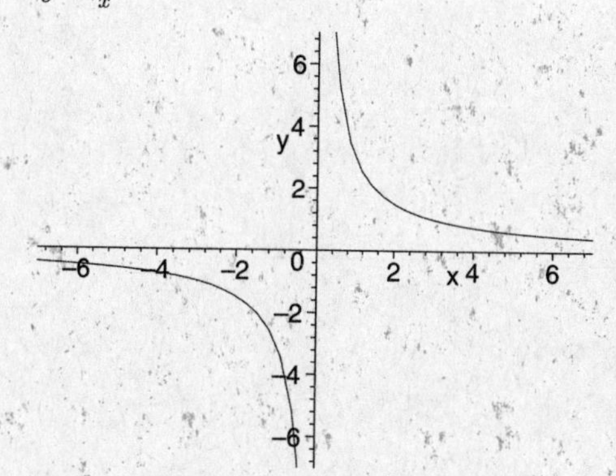

7. $y = \frac{-2}{x}$

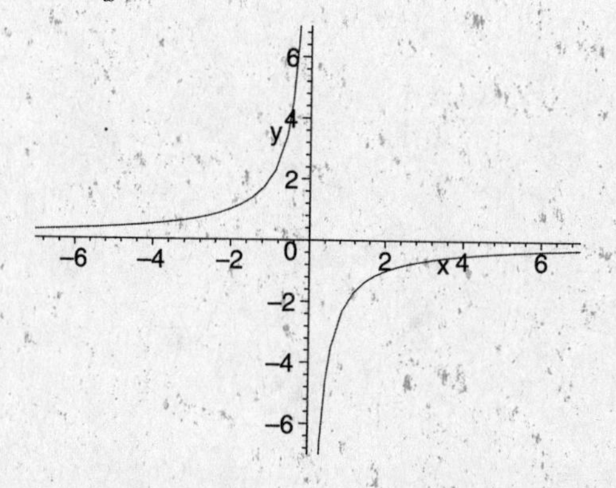

8. $y = \frac{-3}{x}$

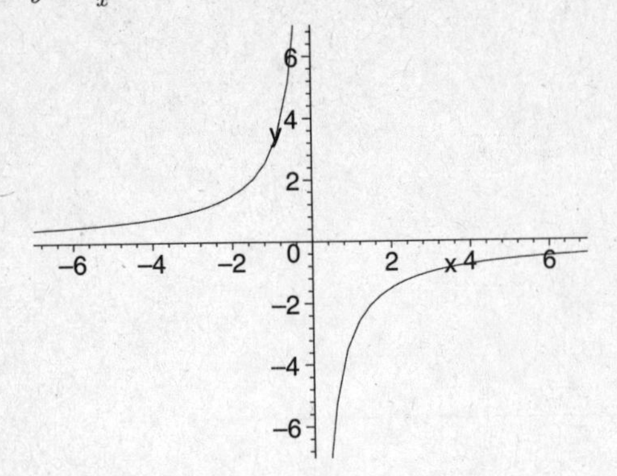

9. $y = \frac{1}{x^2}$

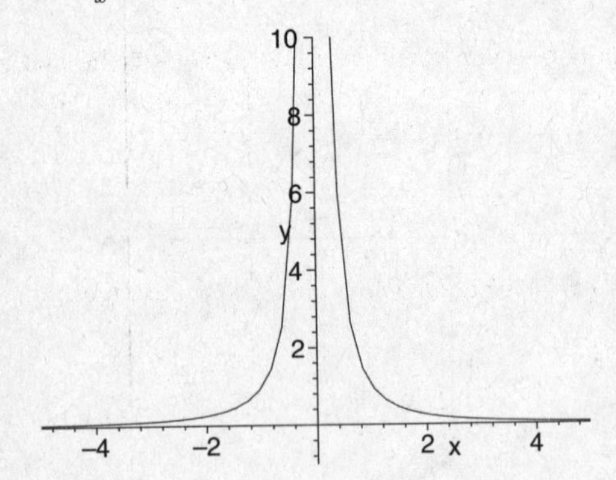

10. $y = \frac{1}{x-1}$

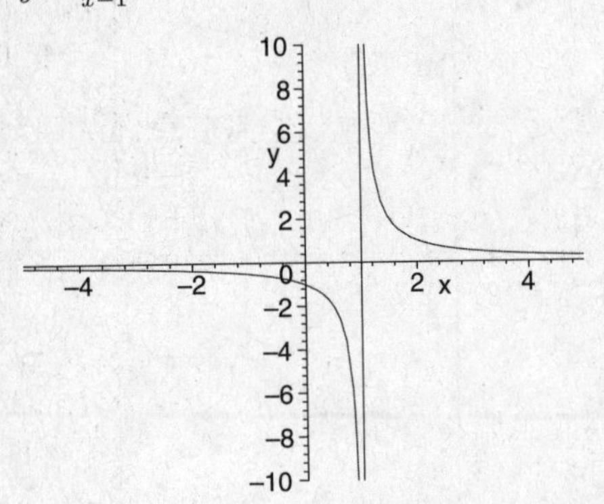

11. $y = \sqrt[3]{x}$

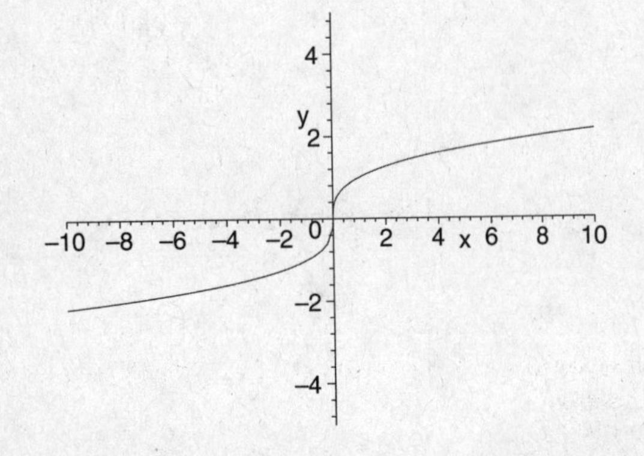

12. $y = \frac{1}{|x|}$

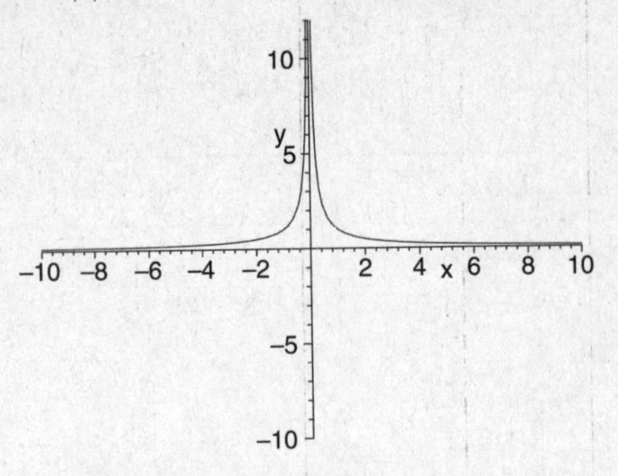

13. $y = \frac{x^2-9}{x+3}$. It is important to note here that $x = -3$ is not in the domain of the plotted function.

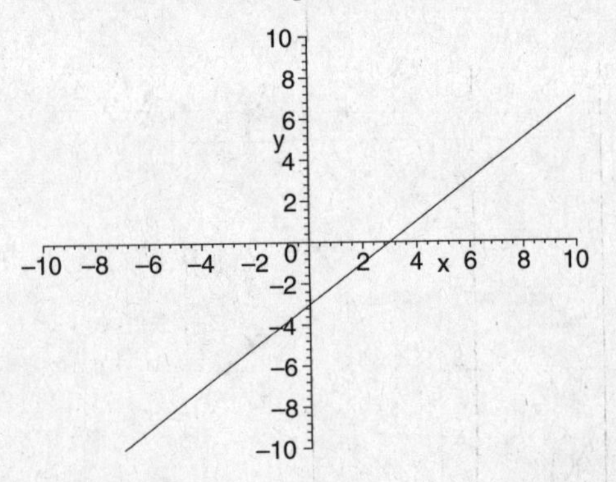

14. $y = \frac{x^2-4}{x-2}$. Note: $x = 2$ is not in the domain of the plotted function.

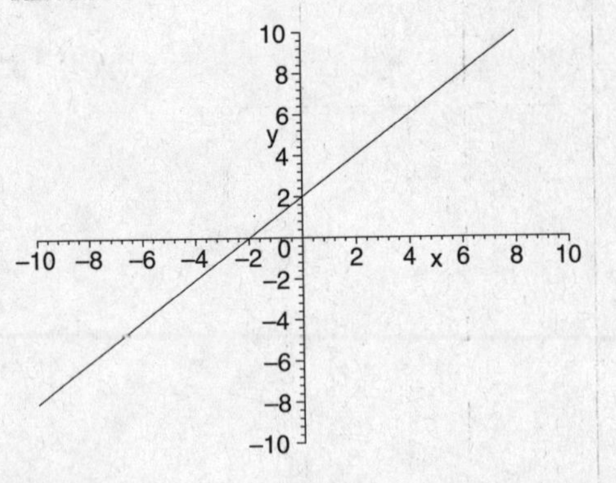

15. $y = \frac{x^1-1}{x-1}$. It is important to note here that $x = 1$ is not in the domain of the plotted function.

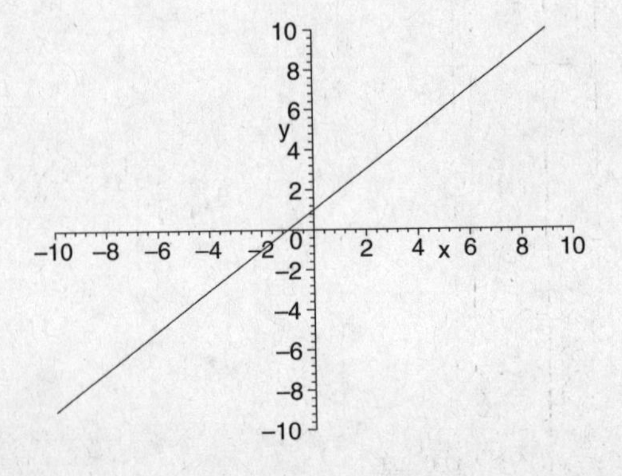

16. $y = \frac{x^2-25}{x+5}$. Note: $x = -5$ is not in the domain of the plotted function.

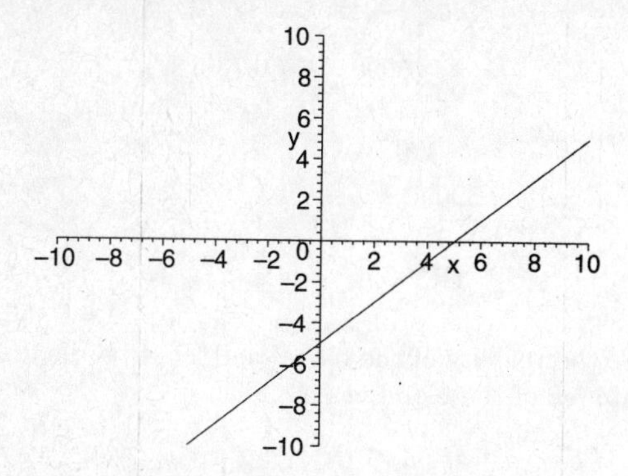

17. $\sqrt{x^3} = x^{\left(\frac{3}{2}\right)}$

18. $\sqrt{x^5} = x^{\left(\frac{5}{2}\right)}$

19. $\sqrt[5]{a^3} = a^{\left(\frac{3}{5}\right)}$

20. $\sqrt[4]{b^2} = b^{\left(\frac{2}{4}\right)} = b^{\left(\frac{1}{2}\right)}$

21. $\sqrt[7]{t} = t^{\left(\frac{1}{7}\right)}$

22. $\sqrt[8]{c} = c^{\left(\frac{1}{8}\right)}$

23. $\frac{1}{\sqrt[3]{t^4}} = \frac{1}{t^{\left(\frac{4}{3}\right)}} = t^{-\left(\frac{4}{3}\right)}$

24. $\frac{1}{\sqrt[5]{t^6}} = \frac{1}{t^{\left(\frac{6}{5}\right)}} = t^{-\left(\frac{6}{5}\right)}$

25. $\frac{1}{\sqrt{t}} = \frac{1}{t^{\left(\frac{1}{2}\right)}} = t^{-\left(\frac{1}{2}\right)}$

26. $\frac{1}{\sqrt{m}} = \frac{1}{m^{\left(\frac{1}{2}\right)}} = m^{-\left(\frac{1}{2}\right)}$

27. $\frac{1}{\sqrt{x^2+7}} = \frac{1}{(x^2+7)^{\left(\frac{1}{2}\right)}} = (x^2+7)^{-\left(\frac{1}{2}\right)}$

28. $\frac{1}{\sqrt{x^3+4}} = \frac{1}{(x^3+4)^{\left(\frac{1}{2}\right)}} = (x^3+4)^{-\left(\frac{1}{2}\right)}$

29. $x^{\frac{1}{5}} = \sqrt[5]{x}$

30. $t^{\frac{1}{7}} = \sqrt[7]{t}$

31. $y^{\frac{2}{3}} = \sqrt[3]{y^2}$

32. $t^{\frac{2}{5}} = \sqrt[5]{t^2}$

33. $t^{\frac{-2}{5}} = \frac{1}{t^{\frac{2}{5}}} = \frac{1}{\sqrt[5]{t^2}}$

34. $y^{\frac{-2}{3}} = \frac{1}{y^{\frac{2}{3}}} = \frac{1}{\sqrt[3]{y^2}}$

35. $b^{\frac{-1}{3}} = \frac{1}{b^{\frac{1}{3}}} = \frac{1}{\sqrt[3]{b^2}}$

36. $b^{\frac{-1}{5}} = \frac{1}{b^{\frac{1}{5}}} = \frac{1}{\sqrt[5]{b^2}}$

37. $e^{\frac{-17}{6}} = \frac{1}{e^{\frac{17}{6}}} = \frac{1}{\sqrt[6]{e^{17}}}$

38. $m^{\frac{-19}{6}} = \frac{1}{m^{\frac{19}{6}}} = \frac{1}{\sqrt[6]{m^{19}}}$

39. $(x-3)^{\frac{-1}{2}} = \frac{1}{(x-3)^{\frac{1}{2}}} = \frac{1}{\sqrt{x-3}}$

40. $(y+7)^{\frac{-1}{4}} = \frac{1}{(y+7)^{\frac{1}{4}}} = \frac{1}{\sqrt[4]{y+7}}$

41. $\frac{1}{t^{\frac{2}{3}}} = \frac{1}{\sqrt[3]{t^2}}$

42. $\frac{1}{w^{-\frac{4}{5}}} = w^{\frac{4}{5}} = \sqrt[5]{w^4}$

43. $9^{3/2} = (\sqrt{9})^3 = (3)^3 = 27$

44. $16^{5/2} = (\sqrt{16})^5 = (4)^5 = 1024$

45. $64^{2/3} = (\sqrt[3]{64})^3 = (4)^2 = 16$

46. $8^{2/3} = (\sqrt[3]{8})^2 = (2)^2 = 4$

47. $16^{3/4} = (\sqrt[4]{16})^3 = (2)^3 = 8$

48. $25^{5/2} = (\sqrt{25})^5 = (5)^5 = 3125$

49. The domain consists of all x-values such that the denominator does not equal 0, that is $x - 5 \neq 0$, which leads to $x \neq 5$. Therefore, the domain is $\{x | x \neq 5\}$

50. $x + 2 \neq 0$ leads to $x \neq -2$. Therefore, the domain is $(-\infty, -2) \cup (-2, \infty)$.

51. Solving for the values of the x in the denominator that make it 0.

$$\begin{aligned} x^2 - 5x + 6 &= 0 \\ (x-3)(x-2) &= 0 \\ \text{So} & \\ x &= 3 \text{ and} \\ x &= 2 \end{aligned}$$

Which means that the domain is the set of all x -values such that $x \neq 3$ or $x \neq 2$

52. Solving $x^2 + 6x + 5 = 0$ leads to $(x+3)(x+2) = 0$ which means the domain consists of all real numbers such that $x \neq -3$ and $x \neq -2$

53. The domain of a square root function is restricted by the value where the radicant is positive. Thus, the domain of $f(x) = \sqrt{5x+4}$ can be found by finding the solution to the inequality $5x + 4 \geq 0$.

$$\begin{aligned} 5x + 4 &\geq 0 \\ 5x &\geq -4 \\ x &\geq \frac{-4}{5} \end{aligned}$$

54. The domain is the solution to $2x - 6 \geq 0$.

$$\begin{aligned} 2x - 6 &\geq 0 \\ 2x &\geq 6 \\ x &\geq 3 \end{aligned}$$

55. To complete the table we will plug the given W values into the equation

$$\begin{aligned} T(20) &= (20)^{1.31} = 50.623 \approx 51 \\ T(30) &= (30)^{1.31} = 86.105 \approx 86 \\ T(40) &= (40)^{1.31} = 125.516 \approx 126 \end{aligned}$$

$$\begin{aligned} T(50) &= (50)^{1.31} = 168.132 \approx 168 \\ T(100) &= (100)^{1.31} = 416.869 \approx 417 \\ T(150) &= (150)^{1.31} = 709.054 \approx 709 \end{aligned}$$

Therefore the table is given by

W	0	10	20	30	40	50	100	150
T	0	20	51	86	126	168	417	709

Now the graph

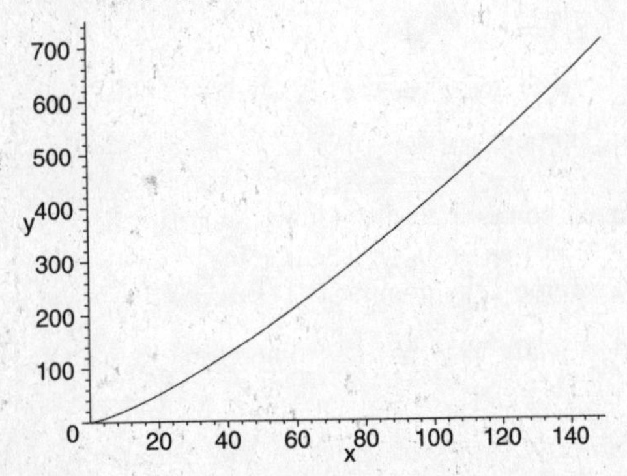

56. First find the constant of the variation. Let N represent the number of cities with a population greater than S.

$$\begin{aligned} N &= \frac{k}{S} \\ 48 &= \frac{k}{350000} \\ (48)(350000) &= k \\ 16800000 &= k \end{aligned}$$

So the variation equation is $N = \frac{16800000}{S}$. Now, we have to find N when $S = 200000$.

$$\begin{aligned} N &= \frac{16800000}{200000} \\ &= 84 \end{aligned}$$

57. **a)** $f(180) = 0.144(180)^{1/2} = 0.144(13.41640786) \approx 1.932\ m^2$.

b) $f(170) = 0.144(170)^{1/2} = 0.144(13.03840481) \approx 1.878\ m^2$.

c) The graph

58. **a)** $y(2.7) = 0.73(2.7)^{3.63} \approx 26.864\ kg$.

b) $y(2.7) = 0.73(7)^{3.63} \approx 853.156\ kg$.

c)

$$\begin{aligned} 5000 &= 0.73(x)^{3.63} \\ \frac{5000}{0.73} &= x^{3.63} \\ \left(\frac{5000}{0.73}\right)^{\frac{1}{3.63}} &= x \\ 11.393\ m &\approx x \end{aligned}$$

59. Let V be the velocity of the blood, and let A be the cross sectional area of the blood vessel. Then

$$V = \frac{k}{A}$$

Using $V = 30$ when $A = 3$ we can find k.

$$\begin{aligned} 30 &= \frac{k}{3} \\ (30)(3) &= k \\ 90 &= k \end{aligned}$$

Now we can write the proportial equation

$$V = \frac{90}{A}$$

we need to find A when $V = 0.026$

$$\begin{aligned} 0.026 &= \frac{90}{A} \\ 0.026A &= 90 \\ A &= \frac{90}{0.026} \\ &= 3461.538\ m^2 \end{aligned}$$

60. Let V be the velocity of the blood, and let A be the cross sectional area of the blood vessel. Then

$$V = \frac{k}{A}$$

Using $V = 28$ when $A = 2.8$ we can find k.

$$\begin{aligned} 28 &= \frac{k}{2.8} \\ (28)(2.8) &= k \\ 78.4 &= k \end{aligned}$$

Now we can write the proportial equation

$$V = \frac{78.4}{A}$$

we need to find A when $V = 0.025$

$$\begin{aligned} 0.025 &= \frac{78.4}{A} \\ 0.025A &= 78.4 \\ A &= \frac{78.4}{0.025} \\ &= 3136\ m^2 \end{aligned}$$

61.

$$\begin{aligned} x+7+\frac{9}{x} &= 0 \\ x(x+7+\frac{9}{x}) &= x(0) \\ x^2+7x+9 &= 0 \\ x &= \frac{-7\pm\sqrt{49-4(1)(9)}}{2} \\ &= \frac{-7\pm\sqrt{13}}{2} \\ x &= \frac{-7-\sqrt{13}}{2} \\ \text{and} & \\ x &= \frac{-7+\sqrt{13}}{2} \end{aligned}$$

62.

$$\begin{aligned} 1-\frac{1}{w} &= \frac{1}{w^2} \\ w^2-w &= 1 \\ w^2-w-1 &= 0 \\ w &= \frac{1\pm\sqrt{1+4}}{2} \\ &= \frac{1\pm\sqrt{5}}{2} \end{aligned}$$

63. $P = 1000t^{5/4} + 14000$

a) $t = 37$, $P = 1000(37)^{5/4} + 14000 = 105254.0514.$
$t = 40$, $P = 1000(40)^{5/4} + 14000 = 114594.6744$
$t = 50$, $P = 1000(50)^{5/4} + 14000 = 146957.3974$

b) Below is the graph of P for $0 \le t \le 50$.

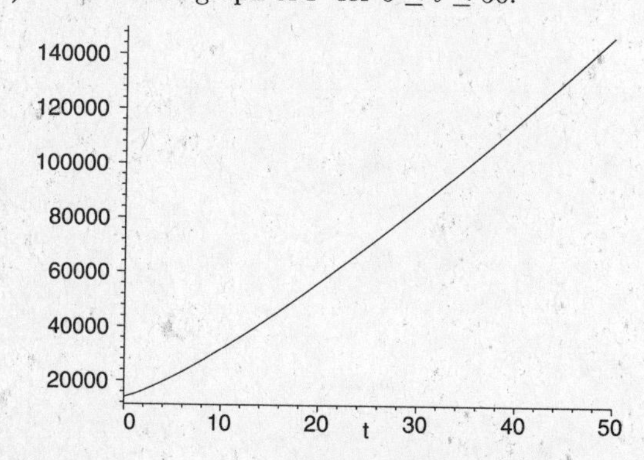

64. At most a function of degree n can have n y-intercepts. A polynomial of degree n can be factored into at most n linear terms and each of those linear terms leads to a y-intercept. This is sometimes called the Fundamental Theorem of Algebra

65. A rational function is a function given by the quotient of two polynomial functions while a polynomial function is a function that has the form $a_nx^n + a_{n-1}x^{n-1} + \cdots + a_1x + a_0$. Since every polynomial function can be written as a quotient of two other polynomial function then every polynomial function is a rational function.

66. $x = 1.5$ and $x = 9.5$

67. $x = 2.6458$ and $x = -2.6458$

68. $x = -2$ and $x = 3$

69. The function has no zeros

70. $x = 1$ and $x = 2$

Exercise Set 1.4

1. $(120^o)(\frac{\pi\ rad}{180^o}) = \frac{2\pi}{3}\ rad$

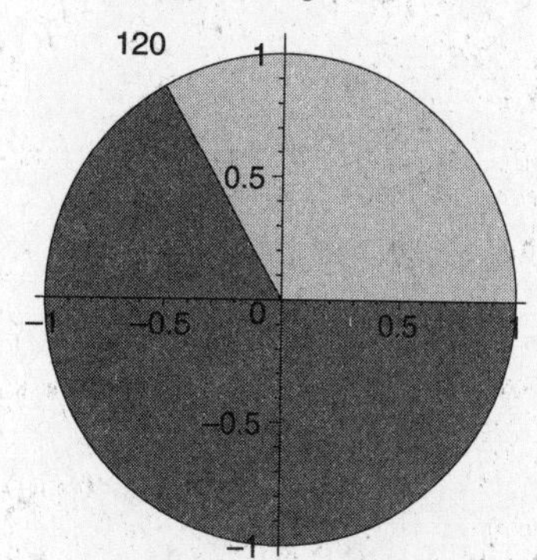

2. $(150^o)(\frac{\pi\ rad}{180^o}) = \frac{5\pi}{6}\ rad$

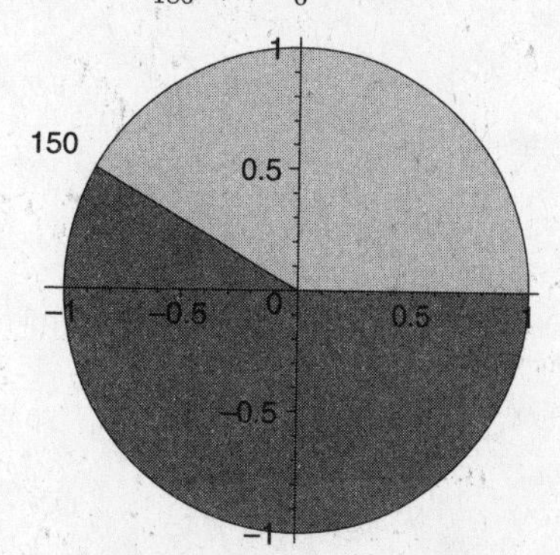

3. $(240^o)(\frac{\pi\ rad}{180^o}) = \frac{4\pi}{3}\ rad$

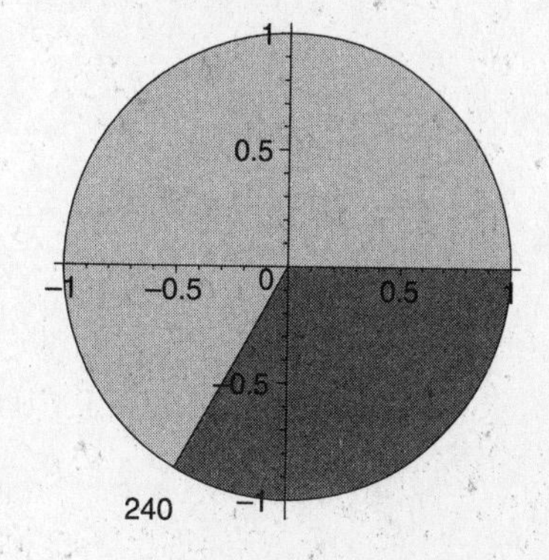

4. $(300^o)(\frac{\pi\ rad}{180^o}) = \frac{5\pi}{3}\ rad$

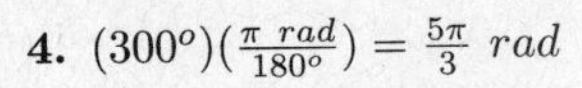

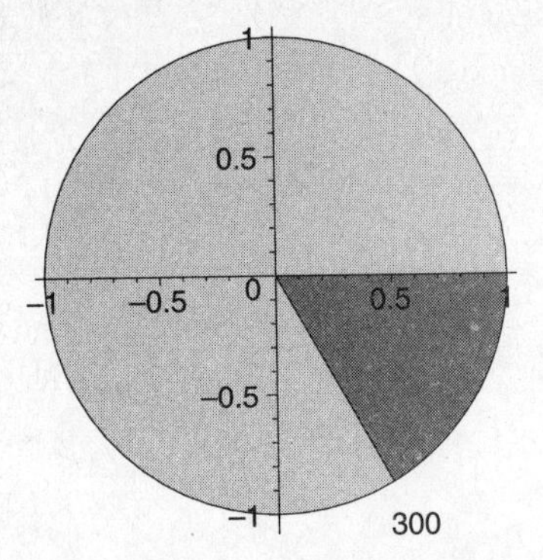

5. $(540^o)(\frac{\pi\ rad}{180^o}) = 3\pi\ rad$

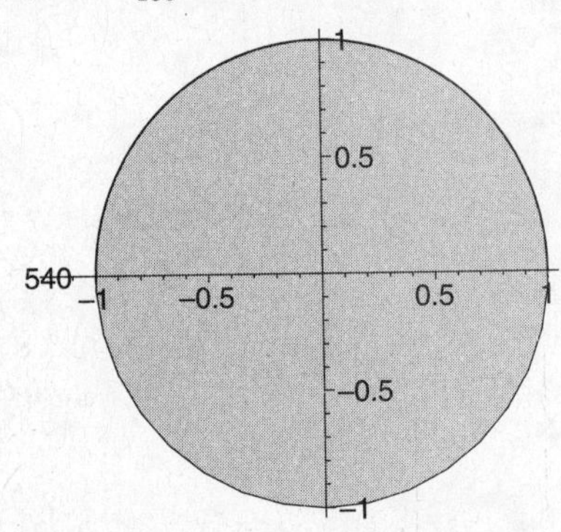

6. $(-450^o)(\frac{\pi\ rad}{180^o}) = \frac{-5\pi}{2}\ rad$

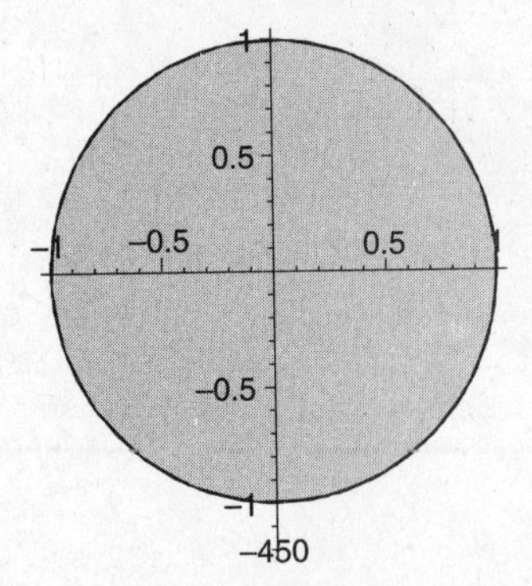

7. $(\frac{3\pi}{4})(\frac{180^o}{\pi\ rad}) = 135^o$

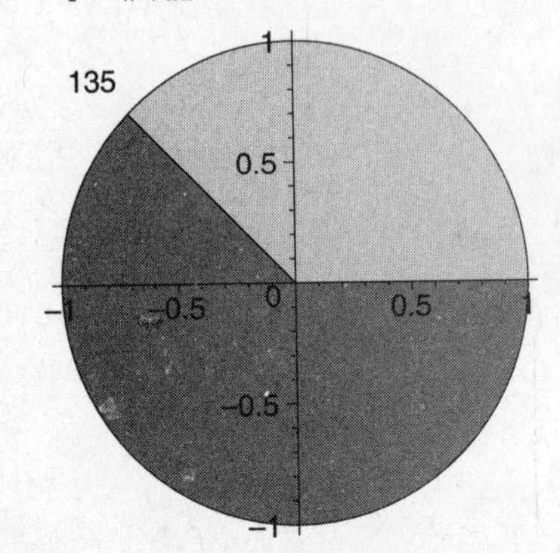

8. $(\frac{7\pi}{6})(\frac{180^o}{\pi\ rad}) = 210^o$

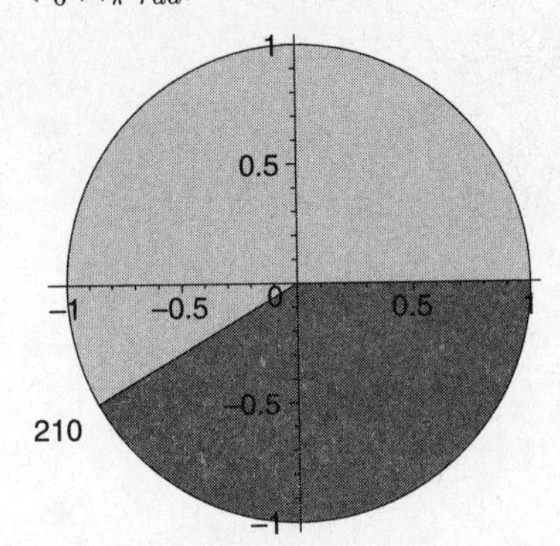

9. $(\frac{3\pi}{2})(\frac{180^o}{\pi\ rad}) = 270^o$

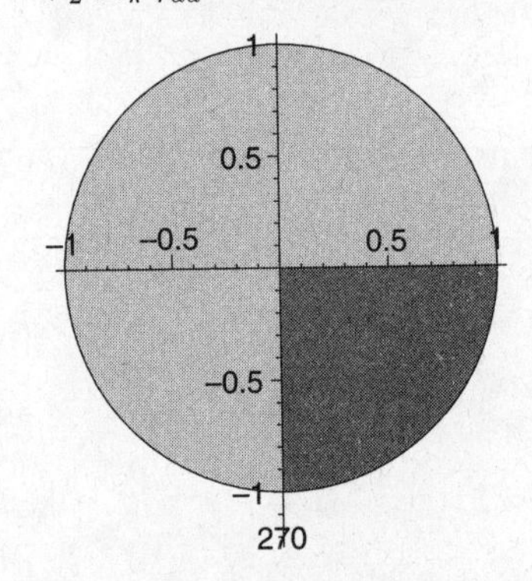

10. $(3\pi)(\frac{180^o}{\pi\ rad}) = 540^o$

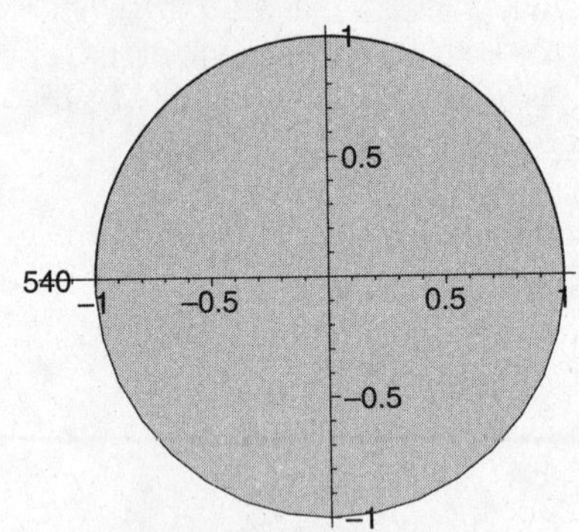

11. $(\frac{-\pi}{3})(\frac{180^o}{\pi\ rad}) = -60^o$

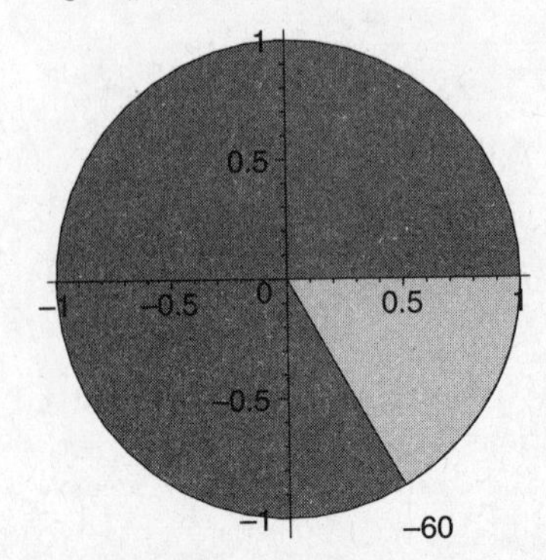

12. $(\frac{-11\pi}{15})(\frac{180^o}{\pi\ rad}) = -132^o$

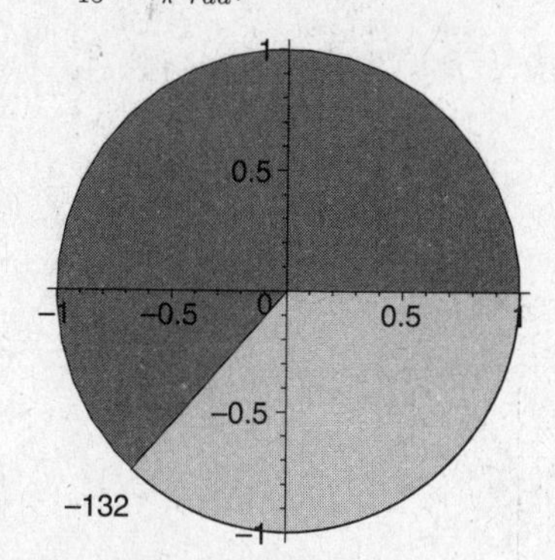

13. We need to solve $\theta_1 = \theta_2 + 360(k)$ for k. If the solution is an integer then the angles are coterminal otherwise they are not coterminal.

$$\begin{aligned} 395 &= 15 + 360(k) \\ 380 &= 360(k) \\ \frac{380}{360} &= k \\ 1.0\overline{5} &= k \end{aligned}$$

Since k is not an integer, we conclude that 15° and 395° are not coterminal.

14.

$$\begin{aligned} 225 &= -135 + 360(k) \\ 360 &= 360(k) \\ \frac{360}{360} &= k \\ 1 &= k \end{aligned}$$

Since k is an integer, we conclude that 225° and -135° are coterminal.

15. We need to solve $\theta_1 = \theta_2 + 360(k)$ for k. If the solution is an integer then the angles are coterminal otherwise they are not coterminal.

$$\begin{aligned} 107 &= -107 + 360(k) \\ 214 &= 360(k) \\ \frac{214}{360} &= k \\ 0.59\overline{4} &= k \end{aligned}$$

Since k is not an integer, we conclude that 15° and 395° are not coterminal.

16.

$$\begin{aligned} 140 &= 440 + 360(k) \\ -300 &= 360(k) \\ \frac{-300}{360} &= k \\ 1.6\overline{1} &= k \end{aligned}$$

Since k is not an integer, we conclude that 140° and 440° are not coterminal.

17. We need to solve $\theta_1 = \theta_2 + 2\pi(k)$ for k. If the solution is an integer then the angles are coterminal otherwise they are not coterminal.

$$\begin{aligned} \frac{\pi}{2} &= \frac{3\pi}{2} + 2\pi(k) \\ -\pi &= 2\pi(k) \\ \frac{-\pi}{2\pi} &= k \\ \frac{-1}{2} &= k \end{aligned}$$

Since k is not an integer, we conclude that $\frac{\pi}{2}$ and $\frac{3\pi}{2}$ are not coterminal.

18.

$$\begin{aligned} \frac{\pi}{2} &= -\frac{3\pi}{2} + 2\pi(k) \\ 2\pi &= 2\pi(k) \\ \frac{2\pi}{2\pi} &= k \\ 1 &= k \end{aligned}$$

Since k is an integer, we conclude that $\frac{\pi}{2}$ and $\frac{-3\pi}{2}$ are coterminal.

19. We need to solve $\theta_1 = \theta_2 + 2\pi(k)$ for k. If the solution is an integer then the angles are coterminal otherwise they are not coterminal.

$$\begin{aligned} \frac{7\pi}{6} &= \frac{-5\pi}{6} + 2\pi(k) \\ 2\pi &= 2\pi(k) \\ \frac{2\pi}{2\pi} &= k \\ 1 &= k \end{aligned}$$

Since k is an integer, we conclude that $\frac{7\pi}{6}$ and $\frac{-5\pi}{6}$ are coterminal.

20. We need to solve $\theta_1 = \theta_2 + 2\pi(k)$ for k. If the solution is an integer then the angles are coterminal otherwise they are not coterminal.

$$\begin{aligned} \frac{3\pi}{4} &= \frac{-\pi}{4} + 2\pi(k) \\ \pi &= 2\pi(k) \\ \frac{\pi}{2\pi} &= k \\ \frac{1}{2} &= k \end{aligned}$$

Since k is not an integer, we conclude that $\frac{3\pi}{4}$ and $\frac{-\pi}{4}$ are not coterminal.

21. $\sin 34^o = 0.5592$

22. $\sin 82^o = 0.9903$

23. $\cos 12^o = 0.9781$

24. $\cos 41^o = 0.7547$

25. $\tan 5^o = 0.0875$

26. $\tan 68^o = 2.4751$

27. $cot\ 34^o = \frac{1}{tan\ 34^o} = 1.4826$

28. $cot\ 56^o = \frac{1}{tan\ 56^o} = 0.6745$

29. $sec\ 23^o = \frac{1}{cos\ 23^o} = 1.0864$

30. $csc\ 72^o = \frac{1}{sin\ 72^o} = 1.0515$

31. $sin(\frac{\pi}{5}) = 0.5878$

32. $cos(\frac{2\pi}{5}) = 0.3090$

33. $tan(\frac{\pi}{7}) = 0.4816$

34. $cot(\frac{3\pi}{11}) = \frac{1}{tan(\frac{3\pi}{11})} = 0.8665$

35. $sec(\frac{3\pi}{8}) = \frac{1}{cos(\frac{3\pi}{8})} = 2.6131$

36. $csc(\frac{4\pi}{13}) = \frac{1}{sin(\frac{4\pi}{13})} = 1.2151$

37. $sin(2.3) = 0.7457$

38. $cos(0.81) = 0.6895$

39. $t = sin^{-1}(0.45) = 26.7437^o$

40. $t = sin^{-1}(0.87) = 60.4586^o$

41. $t = cos^{-1}(0.34) = 70.1231^o$

42. $t = cos^{-1}(0.72) = 43.9455^o$

43. $t = tan^{-1}(2.34) = 66.8605^o$

44. $t = tan^{-1}(0.84) = 40.0302^o$

45. $t = sin^{-1}(0.59) = 0.6311$

46. $t = sin^{-1}(0.26) = 0.2630$

47. $t = cos^{-1}(0.60) = 0.9273$

48. $t = cos^{-1}(0.78) = 0.6761$

49. $t = tan^{-1}(0.11) = 0.1096$

50. $t = tan^{-1}(1.26) = 0.8999$

51.

$$\begin{aligned} sin\ 57^o &= \frac{x}{40} \\ x &= 40sin\ 57^o \\ x &= 33.5468 \end{aligned}$$

52.

$$\begin{aligned} tan\ 20^o &= \frac{15}{x} \\ x &= \frac{15}{tan\ 20^o} \\ x &= 41.2122 \end{aligned}$$

53.

$$\begin{aligned} cos\ 50^o &= \frac{15}{x} \\ x &= \frac{15}{cos\ 50^o} \\ x &= 23.3359 \end{aligned}$$

54.

$$\begin{aligned} sin\ 25^o &= \frac{1.4}{x} \\ x &= \frac{1.4}{sin\ 25^o} \\ x &= 3.3127 \end{aligned}$$

55.

$$\begin{aligned} cos\ t &= \frac{40}{60} \\ t &= cos^{-1}(\frac{40}{60}) \\ t &= 48.1897^o \end{aligned}$$

56.

$$\begin{aligned} tan\ t &= \frac{20}{25} \\ t &= tan^{-1}(\frac{20}{25}) \\ t &= 38.6598^o \end{aligned}$$

57.

$$\begin{aligned} tan\ t &= \frac{18}{9.3} \\ t &= tan^{-1}(\frac{18}{9.3}) \\ t &= 62.6761^o \end{aligned}$$

58.

$$\begin{aligned} sin\ t &= \frac{30}{50} \\ t &= sin^{-1}\frac{30}{50}) \\ t &= 36.8699^o \end{aligned}$$

59. We can rewrite $75^o = 30^o + 45^o$ then use a sum identity

$$\begin{aligned} cos(A+B) &= cos\ Acos\ B - sin\ Asin\ B \\ cos\ 75^o &= cos(30^o + 45^o) \\ &= cos\ 30^o cos\ 45^o - sin\ 30^o sin\ 45^o \\ &= \frac{\sqrt{3}}{2}\cdot\frac{1}{\sqrt{2}} - \frac{1}{2}\cdot\frac{1}{\sqrt{2}} \\ &= \frac{\sqrt{3}}{2\sqrt{2}} - \frac{1}{2\sqrt{2}} \\ &= \frac{-1+\sqrt{3}}{2\sqrt{2}} \end{aligned}$$

60. The x coordinate can be found as follows

$$\begin{aligned} cos\ 20^o &= \frac{x}{200} \\ x &= 200c0s\ 20^o \\ &= 187.939 \end{aligned}$$

The y coordinate

$$\begin{aligned} sin\ 20^o &= \frac{y}{200} \\ y &= 200sin\ 20^o \\ &= 68.404 \end{aligned}$$

61. Five miles is the same as $5 \cdot 5280\ ft = 26400\ ft$. The difference in elevation, y, is

$$\begin{aligned} sin\ 4^o &= \frac{y}{26400} \\ y &= 26400 sin\ 4^o \\ &= 1841.57\ ft \end{aligned}$$

62. First a grade of 5% means that the ratio of the y coordinate to the x coordinate is 0.05 since $tan\ t = \frac{y}{x}$. This means that $x = \frac{y}{0.05} = 20y$ The distance from the base to the top is $6 \cdot 5280\ ft = 31680\ ft$. Using the pythagorean theorem

$$\begin{aligned} x^2 + y^2 &= 31680^2 \\ (20y)^2 + y^2 &= 1003622400 \\ 401y^2 &= 1003622400 \\ y &= \sqrt{\frac{1003622400}{401}} \\ &= 1582.02\ ft \end{aligned}$$

63. a)

$$\begin{aligned} cos\ 40^o &= \frac{x}{150} \\ x &= 150 cos\ 40^o \\ &= 114.907 \end{aligned}$$

b)

$$\begin{aligned} sin\ 40^o &= \frac{y}{150} \\ y &= 150 sin\ 40^o \\ &= 96.4181 \end{aligned}$$

c)

$$\begin{aligned} z^2 &= (x+180)^2 + y^2 \\ &= (114.907+180)^2 + (96.4181)^2 \\ z^2 &= 96266.58866 \\ z &= \sqrt{96266.58866} \\ &= 310.268 \end{aligned}$$

64.

$$\begin{aligned} v &= \frac{77000 \cdot 200 \cdot sec\ 60^o}{5000000} \\ &= \frac{15400000}{5000000 cos\ 60^o} \\ &= 6.16\ cm/sec \end{aligned}$$

65.

$$\begin{aligned} v &= \frac{77000 \cdot 100 \cdot sec\ 65^o}{4000000} \\ &= \frac{7700000}{4000000 cos\ 65^o} \\ &= 4.55494\ cm/sec \end{aligned}$$

66. a) $tan(67^o) = \frac{h}{x}$ so, $x = \frac{h}{tan(67^o)}$

b)

$$\begin{aligned} tan(24^o) &= \frac{h}{1012+x} \\ h &= tan(24^o)(1012+x) \\ h &= (1012)tan(24^o) + x\ tan(24^o) \\ h &= (1012)tan(24^o) + \frac{h}{tan(67^o)} tan(24^o) \\ h(1 - \frac{tan(24^o)}{tan(67^o)} &= (1012)tan(24^o) \\ h &= \frac{(1012)tan(24^o)}{1 - \frac{tan(24^o)}{tan(67^o)}} \\ &= 555.567\ ft \end{aligned}$$

67. a) When we consider the two triangles we have a new triangle that has three equal angles which is the definition of an equilateral triangle.

b) The short leg of each triangle is given by $2sin(30) = 2(\frac{1}{2}) = 1$

c) The long leg (L) is given by

$$\begin{aligned} 2^2 &= L^2 + 1^2 \\ 4 - 1 &= L^2 \\ \sqrt{3} &= L \end{aligned}$$

d) By considering all possible ratios between the long, short and hypotenuse of small triangles we obtain the trigonometric functions of $\frac{\pi}{6} = 30^o$ and $\frac{\pi}{3} = 60^o$

68. a) Since the triangle has two angles equal in magnitude it should have two sides that are equal as well. (It is an isosceles triangle)

b)

$$\begin{aligned} h^2 &= 1^1 + 1^1 \\ h^2 &= 2 \\ h &= \sqrt{2} \end{aligned}$$

c) Since the hypotenuse is known then we can use the figure to find the trigonometric functions of $\frac{\pi}{4} = 45^o$ using the ratios of the sides of the triangle.

69. a) The tangent of an angle is equal to the ratio of the opposite side to the adjacent side (of a right triangle), and for the small triangle that ratio is $\frac{5}{7}$.

b) For the large right triangle, the opposite side is 10 and the adjacent side is $7+7=14$. Thus the tangent is $\frac{10}{14}$

c) Because the trigonometric functions depend on the ratios of the sides and not the size of triangle. Note that the answer in part b) is equivalent to that in part a) even though the triangle in part b) was larger that that used in part a)

70. Let (x, y) be a non-origin point that defines the terminal side of an angle, t, and let $r = \sqrt{x^2+y^2}$ be the distance from the origin to the point (x, y). Then the trigonometric function are defined as follows:

$sin\ t = \frac{y}{r}$ and $csc\ t = \frac{r}{y}$ $(y \neq 0)$
$cos\ t = \frac{x}{r}$ and $sec\ t = \frac{r}{x}$ $(x \neq 0)$ and
$tan\ t = \frac{y}{x}$ $(x \neq 0)$ and $cot\ t = \frac{x}{y}$ $(y \neq 0)$
From the above definitions and recalling that the reciprocal of a non zero number x is given by $\frac{1}{x}$ we show that
$sin\ t = \frac{1}{csc\ t}$

$cos\ t = \frac{1}{sec\ t}$ and

$tan\ t = \frac{1}{cot\ t}$

71.

$$\begin{aligned} \frac{sin\ t}{cos\ t} &= \frac{y/r}{x/r} \\ &= \frac{y}{r} \div \frac{x}{r} \\ &= \frac{y}{r} \cdot \frac{r}{x} \\ &= \frac{y}{x} \\ &= tan\ t \end{aligned}$$

Thus

$$\frac{sin\ t}{cos\ t} = tan\ t$$

and

$$\begin{aligned} \frac{cos\ t}{sin\ t} &= \frac{x/r}{y/r} \\ &= \frac{x}{r} \div \frac{y}{r} \\ &= \frac{x}{r} \cdot \frac{r}{y} \\ &= \frac{x}{y} \\ &= cot\ t \end{aligned}$$

Thus

$$\frac{cos\ t}{sin\ t} = cot\ t$$

72. **a)** $sin(t) = \frac{u}{1} = u$

b) Consider the triangle made by the sides v, w, and y. The angle vw has a value of $90 - r$ (completes a straight angle). The sum of angles in any triangle is 180. Therefore

$$\begin{aligned} s + 90 + (90 - r) &= 180 \\ s + 180 - r &= 180 \\ s - r &= 0 \\ s &= r \end{aligned}$$

c) $sin(s) = \frac{w}{v}$ which means that $w = sin(s)v$. $cos(t) = \frac{v}{1} = v$
Thus, $w = sin(s)v = sin(s)\ cos(t)$

d) $sin(t) = \frac{u}{1} = u$ and $cos(r) = \frac{x}{u}$ which means that $x = u\ cos(r)$.
In part b) we showed that $r = s$ therefore $cos(r) = cos(s)$.
So, $x = u\ cos(r) = sin(t)cos(s)$

e) $sin(s+t) = \frac{(w+x)}{1} = w+x$. Using the results we have obtained from previous parts we can conclude
$sin(s+t) = w + x = sin(s)cos(t) + cos(s)sin(t)$

73. **a)** $sin(t) = \frac{u}{1} = u$, and $cos(t) = \frac{v}{1} = v/$

b) Consider the triangle made by the sides v, w, and y. The angle vw has a value of $90 - r$ (completes a straight angle). The sum of angles in any triangle is 180. Therefore

$$\begin{aligned} s + 90 + (90 - r) &= 180 \\ s + 180 - r &= 180 \\ s - r &= 0 \\ s &= r \end{aligned}$$

c) $cos(s) = \frac{y}{v}$, which means $y = cos(s)v$.
But from part a) $v = cos(t)$, therefore $y = cos(s)cos(t)$

d) $sin(r) = \frac{z}{u}$, which means $z = sin(r)u$.
Using results from part a) and part b) we get $sin(r) = sin(s)$ and $u = sin(t)$, therefore $z = sin(s)sin(t)$

e) $cos(s+t) = \frac{(y-z)}{1} = y - z$. Replacing ur results for y and z we get $cos(s+t) = cos(s)cos(t) - sin(s)sin(t)$

74. **a)** $cos(\frac{\pi}{2} - t) = \frac{u}{1} = u = sin(t)$

b) $sin(\frac{\pi}{2} - t) = \frac{v}{1} = v = cos(t)$

75. Use $cos^2 t + sin^2 t = 1$ as follows

$$\begin{aligned} cos^2 t + sin^2 t &= 1 \\ \frac{cos^2 t}{cos^2 t} + \frac{sin^2 t}{cos^2 t} &= \frac{1}{cos^2 t} \\ 1 + tan^t &= sec^2 t \end{aligned}$$

76.

$$\begin{aligned} cos^2 t + sin^2 t &= 1 \\ \frac{cos^2 t}{sin^2 t} + \frac{sin^2 t}{sin^2} &= \frac{1}{sin^2 t} \\ cot^2 t + 1 &= csc^2 t \end{aligned}$$

77. Let $2t = t + t$

$$\begin{aligned} sin(a+b) &= sin(a)cos(b) + cos(a)sin(b) \\ sin(2t) &= sin(t+t) \\ &= sin(t)cos(t) + cos(t)sin(t) \\ &= 2sin(t)cos(t) \end{aligned}$$

78. **a)**

$$\begin{aligned} cos(2t) &= cos(t+t) \\ &= cos(t)cos(t) - sin(t)sin(t) \\ &= cos^2(t) - sin^2(t) \end{aligned}$$

b)

$$\begin{aligned} cos(2t) &= cos^2(t) - sin^2(t) \\ &= cos^2(t) - (1 - cos^2(t)) \\ &= 2cos^2(t) - 1 \end{aligned}$$

c)

$$\begin{aligned} cos(2t) &= cos^2(t) - sin^2(t) \\ &= (1 - sin^2(t)) - sin^2(t) \\ &= 1 - 2sin^2(t) \end{aligned}$$

79. Using the result from Exercise 78 part (c)

$$\begin{aligned} cos(2t) &= 1 - 2sin^2(t) \\ cos(2t) - 1 &= -2sin^2(t) \\ \frac{cos(2t) - 1}{-2} &= sin^2(t) \\ \frac{1 - cos(2t)}{2} &= sin^2(t) \end{aligned}$$

80.

$$\begin{aligned} cos(2t) &= 2cos^2 - 1 \\ cos(2t) + 1 &= 2cos^2(t) \\ \frac{cos(2t) + 1}{2} &= cos^2(t) \end{aligned}$$

81. **a)** $V(0) = sin^p(0)sin^q(0)sin^r(0)sin^s(0) = 0$
$V(1) = sin^p(\frac{\pi}{2})sin^q(\frac{\pi}{2})sin^r(\frac{\pi}{2})sin^s(\frac{\pi}{2}) = 1$

b) When $h = 0$ the volume of the tree is zero since there is no height and therefore the proportion of volume under that height is zero. While at the top of the tree, $h = 1$, the proportion of volume under the tree is 1 since the entire tree volume falls below its height.

82. **a)**

$$\begin{aligned} V(0.5) &= sin^{-3.728}(\frac{\pi}{4})sin^{48.646}(\frac{\pi}{2\sqrt{2}}) \\ &\quad \times sin^{-123.208}(\frac{\pi}{2\sqrt[3]{2}})sin^{86.629}(\frac{\pi}{2\sqrt[4]{2}}) \\ &= 0.8208 \end{aligned}$$

b) $V(h)$

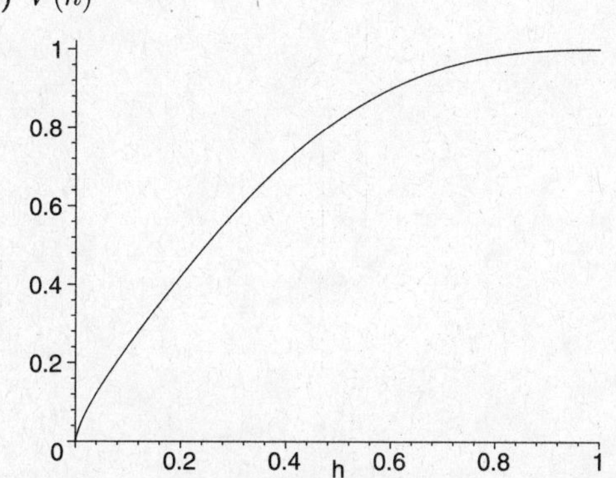

c) The result from part b) agrees with the definition of $V(h)$ since the values of $V(h)$ are limited between 0 and 1.

83.

$$\begin{aligned} V(\frac{1}{2}) &= sin^{-5.621}(\frac{\pi}{4})sin^{74.831}(\frac{\pi}{2\sqrt{2}}) \\ &\quad \times sin^{-195.644}(\frac{\pi}{2\sqrt[3]{2}})sin^{138.959}(\frac{\pi}{2\sqrt[4]{2}}) \\ &= 0.8219 \end{aligned}$$

Exercise Set 1.5

1. $5\pi/4$

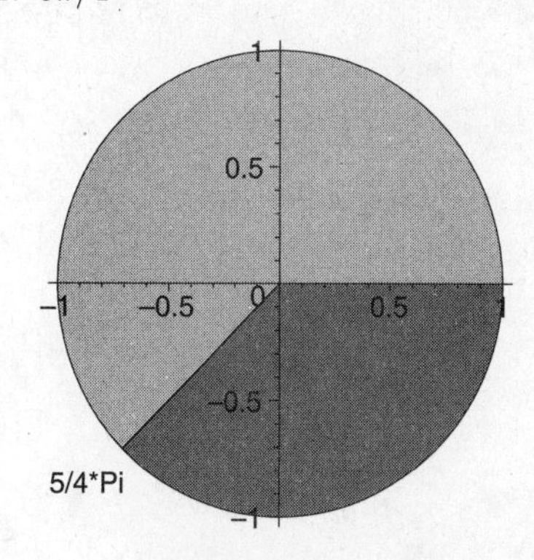

2. $-5\pi/6$

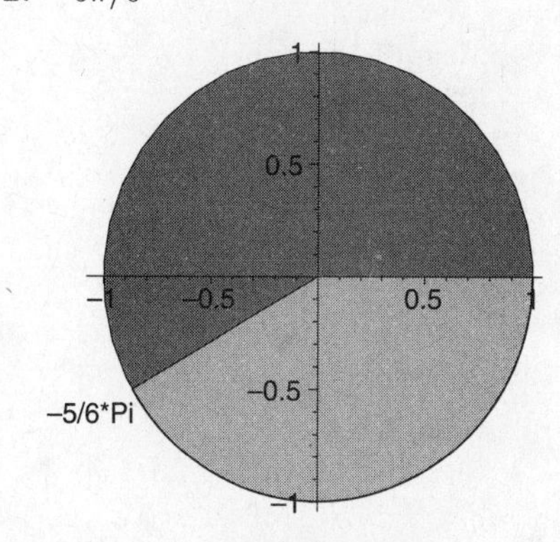

3. $-\pi$

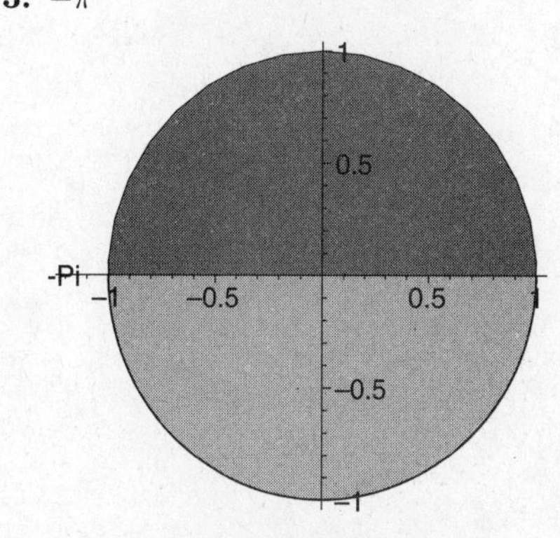

4. 2π

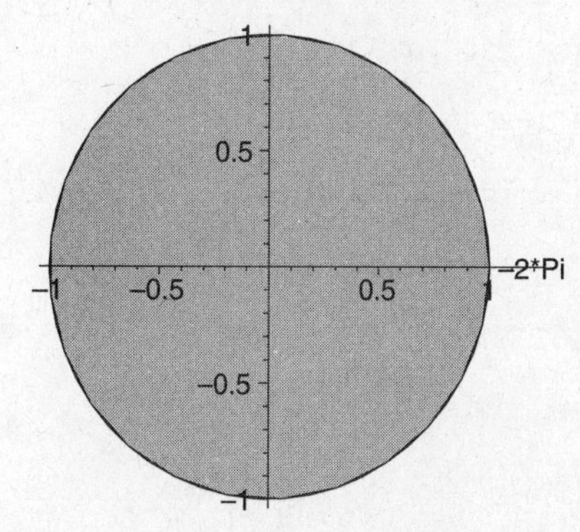

5. $13\pi/6$

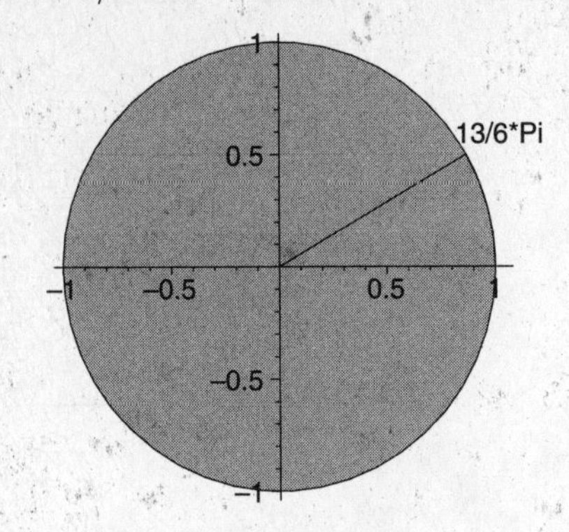

6. $-7\pi/4$

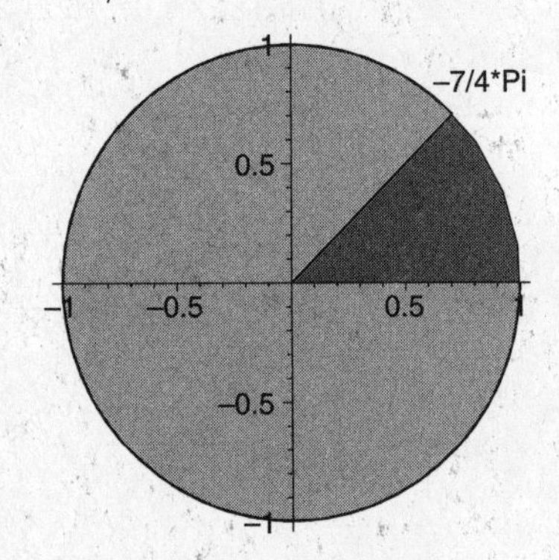

7. $cos(9\pi/2) = 0$

8. $sin(5\pi/4) = \frac{-1}{\sqrt{2}}$

9. $sin(-5\pi/6) = \frac{-1}{2}$

10. $cos(-5\pi/4) = \frac{-1}{\sqrt{2}}$

11. $cos(5\pi) = -1$

12. $sin(6\pi) = 0$

13. $tan(-4\pi/3) = -\sqrt{3}$

14. $tan(-7\pi/3) = -\sqrt{3}$

15. $cos\ 125^o = -0.5736$

16. $sin\ 164^o = 0.2756$

17. $tan(-220^o) = -0.8391$

18. $cos(-253^o) = -0.2924$

19. $sec\ 286^o = \frac{1}{cos\ 286^o} = 3.62796$

20. $csc\ 312^o = \frac{1}{sin\ 312^o} = -1.34563$

21. $sin(1.2\pi) = -0.587785$

22. $tan(-2.3\pi) = -1.37638$

23. $cos(-1.91) = -0.332736$

24. $sin(-2.04) = -0.891929$

25. $t = sin^{-1}(1/2) = \frac{\pi}{6} + 2n\pi$ and $\frac{5\pi}{6} + 2n\pi$

26. $t = sin^{-1}(-1) = \frac{3\pi}{2} + 2n\pi$

27. $2t = sin^{-1}(0) = n\pi$ so $t = \frac{n\pi}{2}$

28.

$$\begin{aligned} 2sin(t + \frac{\pi}{3}) &= -\sqrt{3} \\ sin(t + \frac{\pi}{3}) &= \frac{-\sqrt{3}}{2} \\ t + \frac{\pi}{3} &= sin^{-1}(\frac{-\sqrt{3}}{2}) \\ t &= \frac{-\pi}{3} - \frac{\pi}{3} + 2n\pi \\ &= \frac{-2\pi}{3} + 2n\pi \\ \text{and} & \\ t &= \frac{-\pi}{3} + \frac{4\pi}{3} + 2n\pi \\ &= \pi + 2n\pi \end{aligned}$$

29.

$$\begin{aligned} cos(3t + \frac{\pi}{4}) &= -\frac{1}{2} \\ 3t + \frac{\pi}{4} &= cos^{-1}(-\frac{1}{2}) \\ 3t &= -\frac{\pi}{4} + \frac{2\pi}{3} + 2n\pi \\ 3t &= \frac{5\pi}{12} 2n\pi \\ t &= \frac{5\pi}{36}\frac{2}{3}n\pi \\ \text{and} & \\ 3t &= -\frac{\pi}{4} + \frac{4\pi}{3} + 2n\pi \\ 3t &= \frac{13\pi}{12} + 2n\pi \\ t &= \frac{13}{36}\pi + \frac{2}{3}n\pi \end{aligned}$$

30.

$$\begin{aligned} cos(2t) &= 0 \\ 2t &= cos^{-1}(0) \\ 2t &= \frac{\pi}{2} + 2n\pi \\ t &= \frac{\pi}{4} + n\pi \\ \text{and} & \\ 2t &= \frac{3\pi}{2} + 2n\pi \\ t &= \frac{3\pi}{4} + n\pi \end{aligned}$$

31.

$$\begin{aligned} cos(3t) &= 1 \\ 3t &= cos^{-1}(1) \\ 3t &= 2n\pi \\ t &= \frac{2}{3}n\pi \end{aligned}$$

32.

$$\begin{aligned}
2cos(\frac{t}{2}) &= -\sqrt{3}\\
cos(\frac{t}{2}) &= \frac{-\sqrt{3}}{2}\\
\frac{t}{2} &= cos^{-1}(\frac{-\sqrt{3}}{2})\\
\frac{t}{2} &= \frac{5\pi}{6}+2n\pi\\
t &= \frac{5\pi}{3}+4n\pi\\
\text{and}&\\
\frac{t}{2} &= \frac{7\pi}{6}+2n\pi\\
t &= \frac{7\pi}{3}+4n\pi
\end{aligned}$$

33.

$$\begin{aligned}
2sin^2t-5sin\ t-3 &= 0\\
(2sin\ t+1)(sin\ t-3) &= 0\\
\text{The only solution comes from}&\\
(2sin\ t+1) &= 0\\
sin\ t &= -\frac{1}{2}\\
t &= sin^{-1}(-\frac{1}{2})\\
t &= \frac{7\pi}{6}+2n\pi\\
\text{and}&\\
t &= \frac{11\pi}{6}+2n\pi
\end{aligned}$$

34.

$$\begin{aligned}
cos^2x+5cos\ x &= 6\\
cos^2x+5cos\ x-6 &= 0\\
(cos\ x+6)(cos\ x-1) &= 0\\
\text{The only solution comes from}&\\
cos\ x-1 &= 0\\
x &= cos^{-1}(1)\\
x &= 2n\pi
\end{aligned}$$

35.

$$\begin{aligned}
cos^2x+5cos\ x &= -6\\
cos^2x+5cos\ x+6 &= 0\\
cos\ x &= \frac{-5\pm\sqrt{25-4(1)(6)}}{2}\\
&= \frac{-5\pm 1}{2}\\
&= \frac{-5-1}{2}=-3\\
\text{and}&\\
&= \frac{-5+1}{2}=-2
\end{aligned}$$

Since both values are larger than one, then the equation has no solutions.

36.

$$\begin{aligned}
sin^2t-2sin\ t-3 &= 0\\
(sin\ t-3)(sin\ t+1) &= 0\\
\text{The only solution comes from}&\\
sin\ t+1 &= 0\\
t &= sin^{-1}(-1)\\
&= \frac{3\pi}{2}+2n\pi
\end{aligned}$$

37. $y=2sin\ 2t+4$
amplitude $=2$, period $=\frac{2\pi}{2}=\pi$, mid-line $y=4$
maximum $=4+2=6$, minimum $=4-2=2$

38. $y=3cos\ 2t-3$
amplitude $=3$, period $=\frac{2\pi}{2}=\pi$, mid-line $y=-3$
maximum $=-3+3=0$, minimum $=-3-3=-6$

39. $y=5cos(t/2)+1$
amplitude $=5$, period $=\frac{2\pi}{\frac{1}{2}}=4\pi$, mid-line $y=1$
maximum $=1+5=6$, minimum $=1-5=-4$

40. $y=3sin(t/3)+2$
amplitude $=3$, period $=\frac{2\pi}{\frac{1}{3}}=6$, mid-line $y=2$
maximum $=2+3=5$, minimum $=2-3=-1$

41. $y=\frac{1}{2}sin(3t)-3$
amplitude $=\frac{1}{2}$, period $=\frac{2\pi}{3}$, mid-line $y=-3$
maximum $-3+\frac{1}{2}=\frac{-5}{2}$, minimum $=-3-\frac{1}{2}=\frac{-7}{2}$

42. $y=\frac{1}{2}cos(4t)+2$
amplitude $=\frac{1}{2}$, period $=\frac{2\pi}{4}=\frac{\pi}{2}$, mid-line $y=2$
maximum $=2+\frac{1}{2}=\frac{5}{2}$, minimum $=2-\frac{1}{2}=\frac{3}{2}$

43. $y=4sin(\pi t)+2$
amplitude $=4$, period $=\frac{2\pi}{\pi}=2$, mid-line $y=2$
maximum $=2+4=6$, minimum $=2-4=-2$

44. $y=3cos(3\pi t)-2$
amplitude $=3$, period $=\frac{2\pi}{3\pi}=\frac{2}{3}$, mid-line $y=-2$
maximum $=-2+3=1$, minimum $=-2-3=-5$

45. The maximum is 10 and the minimum is -4 so the amplitude is $\frac{10-(-4)}{2}=7$. The mid-line is $y=10-7=3$, and the period is 2π (the distance from one peak to the next one) which means that $b=\frac{2\pi}{2\pi}=1$. From the information above, and the graph, we conclude that the function is

$$y=7sin\ t+3$$

46. The maximum is 4 and the minimum is -1 so the amplitude is $\frac{4-(-1)}{2}=\frac{5}{2}$. The mid-line is $y=4-\frac{5}{2}=\frac{3}{2}$, and the period is 4π which means $b=\frac{\pi}{2}$. From the information above, and the graph, we conclude that the function is

$$y=\frac{5}{2}cos(t/2)+\frac{3}{2}$$

47. The maximum is 1 and the minimum is -3 so the amplitude is $\frac{1-(-3)}{2}=2$. The mid-line is $y=1-2=-1$, and the period is 4π which means $b=\frac{\pi}{2}$. From the information above, and the graph, we conclude that the function is

$$y=2cos(t/2)-1$$

48. The maximum is -0.5 and the minimum is -1.5 so the amplitude is $\frac{-0.5-(-1.5)}{2} = \frac{1}{2}$. The mid-line is $y = -0.5 - \frac{-1}{2} = -1$, and the period is 1 which means that $b = \frac{2\pi}{1} = 2\pi$. From the information above, and the graph, we conclude that the function is

$$y = \frac{1}{2}sin(2\pi t) - 1$$

49.

$$\begin{aligned} R &= 0.339 + 0.808cos\ 40^o\ cos\ 30^o \\ &\quad -0.196sin\ 40^o\ sin\ 30^o - 0.482cos\ 0^o\ cos\ 30^o \\ &= 0.571045\ megajoules/m^2 \end{aligned}$$

50.

$$\begin{aligned} R &= 0.339 + 0.808cos\ 30^o\ cos\ 20^o \\ &\quad -0.196sin\ 30^o\ sin\ 20^o - 0.482cos\ 180^o\ cos\ 20^o \\ &= 1.12788\ megajoules/m^2 \end{aligned}$$

51.

$$\begin{aligned} R &= 0.339 + 0.808cos\ 50^o\ cos\ 55^o \\ &\quad -0.196sin\ 50^o\ sin\ 55^o - 0.482cos\ 45^o\ cos\ 55^o \\ &= 0.234721\ megajoules/m^2 \end{aligned}$$

52.

$$\begin{aligned} R &= 0.339 + 0.808cos\ 50^o\ cos\ 0^o \\ &\quad -0.196sin\ 50^o\ sin\ 0^o - 0.482cos\ 0^o\ cos\ 0^o \\ &= 0.858372\ megajoules/m^2 \end{aligned}$$

53. Period is 5 so $b = \frac{2\pi}{5}$, $k = 2500$, $a = 250$. Therefore, the function is

$$V(t) = 250cos\ \frac{2\pi t}{5} + 2500$$

54. Period is 2 sp $b = \frac{2\pi}{2} = \pi$, $a = \frac{3400}{2} = 1700$, $k = 1700 + 1100 = 2800$. Therefore, the function is

$$V(t) = 1700cos\ \pi t + 2800$$

55. Since our lungs increase and decrease as we breathe then there is a maximum and minimum volume for the air capacity in our lungs. We have a regular period of time at which we breathe (inhale and exhale). These facotrs are reasons why the cosine model is appropriate for describing lung capacity.

56. The minimum is 35.33 and the maximum is 36.87 so the amplitude is $\frac{36.87-35.33}{2} = 0.77$. The period is 24 so $b = \frac{2\pi}{24} = \frac{\pi}{12}$, $k = 36.87 - 0.77 = 36.1$. Thus, the function is

$$T(t) = 0.77cos\ \frac{\pi}{12} + 36.1$$

57. The frequency is the reciprocal of the period. Therefore, $f = \frac{b}{2\pi} = \frac{880\pi}{2\pi} = 440\ Hz$

58. $f = \frac{440\pi}{2\pi} = 220\ Hz$

59. The amplitude is given as 5.3. $b = f \cdot 2\pi$ where f is the frequency, $b = 0.172 \cdot 2\pi = 1.08071$, $k = 143$. Therefore, the function is

$$p(t) = 5.3cos(1.08071t) + 143$$

60. $p(t) = 6.7cos(0.496372t) + 137$

61. $x = cos(140^o)$, $y = sin(140^o)$, $(-0.76604, 0.64279)$

62. $(-0.17365, -0.98481)$

63. $x = cos(\frac{9\pi}{5})$, $y = sin(\frac{9\pi}{5})$, $(0.80902, -0.58779)$

64. $(-0.22252, -0.97493)$

65. Rewrite $105^o = 45^o + 60^o$ and use a sum identity.

$$\begin{aligned} sin\ 105^o &= sin(45^o + 60^o) \\ &= sin\ 45^o\ cos\ 60^o + cos\ 45^o\ sin\ 60^o \\ &= \frac{1}{\sqrt{2}} \cdot \frac{1}{2} + \frac{1}{\sqrt{2}} \cdot \frac{\sqrt{3}}{2} \\ &= \frac{1}{2\sqrt{2}} + \frac{\sqrt{3}}{2\sqrt{2}} \\ &= \frac{1+\sqrt{3}}{2\sqrt{2}} \end{aligned}$$

66.

$$\begin{aligned} cos\ 165^o &= cos(120^o + 45^o) \\ &= cos\ 120^o\ cos\ 45^o - sin\ 120^o\ sin\ 45^o \\ &= \frac{-1}{2} \cdot \frac{1}{\sqrt{2}} - \frac{\sqrt{3}}{2} \cdot \frac{1}{\sqrt{2}} \\ &= -\frac{1}{2\sqrt{2}} - \frac{\sqrt{3}}{2\sqrt{2}} \\ &= -\frac{1+\sqrt{3}}{2\sqrt{2}} \end{aligned}$$

67. **a)** From the graph we can see that the point with angle t has an opposite x and y coordinate than the point with angle $t + \pi$. Since the x coordinate corresponds to the *cos* of the angle which the point makes and the y coordinate corresponds to the *sin* of the angle which the point makes it follows that $sin(t + \pi) = -sin(t)$ and $cos(t + \pi) = -cos(t)$.

b)

$$\begin{aligned} sin(t + \pi) &= sin\ tcos\ \pi + cos\ tsin\ \pi \\ &= sin\ t \cdot -1 + cos\ t \cdot 0 \\ &= -sin\ t \end{aligned}$$

and

$$\begin{aligned} cos(t + \pi) &= cos\ tcos\ \pi - sin\ tsin\ \pi \\ &= cos\ t \cdot -1 - sin\ t \cdot 0 \\ &= -cos\ t \end{aligned}$$

c)

$$tan(t + \pi) = \frac{sin(t + \pi)}{cos(t + \pi)}$$

$$\begin{aligned} &= \frac{-sin\ t}{-cos\ t} \\ &= \frac{sin\ t}{cos\ t} \\ &= tan\ t \end{aligned}$$

68. a) The amplitude could be thought of as half the difference between the maximum and minimum, $a = \frac{max-min}{2}$, which implies that $2a = max - min$. k is the average mean of the maximum and the minimum, $k = \frac{max+min}{2}$, which implies that $2k = max + min$. Solving the system of equations above for max and min gives the desired results.

b) The average mean of the maximum and minimum, using the results from part a), implies that the mid-line equation is $y = \frac{(k+a)+(k-a)}{2} = \frac{2k}{2} = k$.

c) Half the difference between the maximum and minimum, using the results from part a), implies that the amplitude is $\frac{(k+a)-(k-a)}{2} = \frac{2a}{2} = a$.

69. a) Since the radius of a unit circle is 1, the circumference of the unit circle is 2π. Therefore any point $t + 2\pi$ will have exactly the same terminal side as the point t, that is to say that the points t and $t+2\pi$ are coterminal on the unit circle. Therefore, $sin\ t = sin(t + 2\pi)$ for all numbers t.

b)

$$\begin{aligned} g(t + 2\pi/b) &= asin[b(t + 2\pi/b)] + k \\ &= asin(bt + 2\pi) + k \end{aligned}$$

from part a)

$$= asin(bt) + k$$

by definition

$$g(t + 2\pi/b) = g(t)$$

c) Since the function evaluated at $t+2\pi/b$ has the same value as the function evaluated at t and $2\pi/b \neq 0$ then $t + 2\pi/b$ is evaluated after t. Since we have a periodic function in $g(t)$ it follows that the period of the function is implied to be $2\pi/b$.

70. Since at the apex, L is large, T is small, and d is small, then the basilar membrane is affected mostly by low frequency sounds.

71. Since at the base, L is small, T is large, and d is large, then the basilar membrane is affected mostly by high frequency sounds.

72. $f = \frac{880\pi}{2\pi} = 440$

73. $f = \frac{880 \cdot 2^{-9/12}\pi}{2\pi} = 261.626$

74. From the equation, n has to be 12 in order for $\frac{880(2^{n/12})\pi}{2\pi}$ to equal 880. There are 12 notes above A above middle C.

75.

$$\begin{aligned} \frac{880(2^{n/12})\pi}{2\pi} &= 1760 \\ 2^{n/12-1} &= \frac{1760}{880} \\ 2^{n/12-1} &= 2 \end{aligned}$$

Comparing exponents we can conclude that

$$\begin{aligned} \frac{n}{12} - 1 &= 1 \\ \frac{n}{12} &= 2 \\ n &= 24 \end{aligned}$$

There are 24 notes above A above middle C.

76.

$$\begin{aligned} \frac{880(2^{n/12})\pi}{2\pi} &= 1320 \\ 2^{n/12-1} &= \frac{1320}{880} \\ 2^{n/12-1} &= 1.5 \\ (\frac{n}{12} - 1)ln(2) &= ln(1.5) \\ \frac{n}{12} - 1 &= \frac{ln(1.5)}{ln(2)} \\ \frac{n}{12} &= \frac{ln(1.5)}{ln(2)} + 1 \\ n &= 12\left(\frac{ln(1.5)}{ln(2)} + 1\right) \\ n &= 19.01955 \end{aligned}$$

There are 19 notes above A above middle C.

77.

$$\begin{aligned} \frac{880(2^{n/12})\pi}{2\pi} &= 2200 \\ 2^{n/12-1} &= \frac{2200}{880} \\ 2^{n/12-1} &= 2.5 \\ (\frac{n}{12} - 1)ln(2) &= ln(2.5) \\ \frac{n}{12} - 1 &= \frac{ln(2.5)}{ln(2)} \\ \frac{n}{12} &= \frac{ln(2.5)}{ln(2)} + 1 \\ n &= 12\left(\frac{ln(2.5)}{ln(2)} + 1\right) \\ n &= 27.86314 \end{aligned}$$

There are 28 notes above A above middle C

78. a) Left to the student

b) $y = \frac{1}{2}cos(2t) - \frac{1}{2}$

c) We use the double angle identity obtained in Exercise 79 of Section 1.4 and solve for $-sin^2(t)$ to obtain the model in part b).

79. a) Left to the student

b) $y = \frac{1}{2}cos(2t) + \frac{1}{2}$

c) We use the double angle identity obtained in Exercise 79 of Section 1.4 and solve for $cos^2(t)$ to obtain the model in part b).

80. a) Left to the student

b) $y = sin(2t)$

c) We use the double angle identity for $sin(2t)$ obtained in Exercise 77 of Section 1.4 to obtain the model in part b).

81. a) Left to the student

b) Left to the student

c) The horizontal shift moves every point of the original graph $\frac{\pi}{4}$ units to the right.

82. a) Left to the student

b) Left to the student

c) The horizontal shift moves every point of the original graph $\frac{\pi}{3}$ units to the left.

83. Left to the student

84. Left to the student

85. Left to the student

Chapter Review Exercises

1. a) 100 live births per 1000 women

b) 20 years old and 30 years old

2. $f(-2) = 2(-2)^2 - (-2) + 3 = 13$

3.

$$\begin{aligned} f(1+h) &= 2(1+h)^2 - (1+h) + 3 \\ &= 2(1+2h+h^2) - 1 - h + 3 \\ &= 2 + 4h + 2h^2 - 1 - h + 3 \\ &= 2h^2 + 3h + 4 \end{aligned}$$

4. $f(0) = 2(0)^2 - (0) + 3 = 3$

5. $f(-5) = (1-(-5))^2 = (1+5)^2 = 6^2 = 36$

6.

$$\begin{aligned} f(2-h) &= (1-(2-h))^2 \\ &= (h-1)^2 \\ &= h^2 - 2h + 1 \end{aligned}$$

7. $f(4) = (1-4)^2 = (-3)^2 = 9$

8. $f(x) = 2x^2 + 3x - 1$

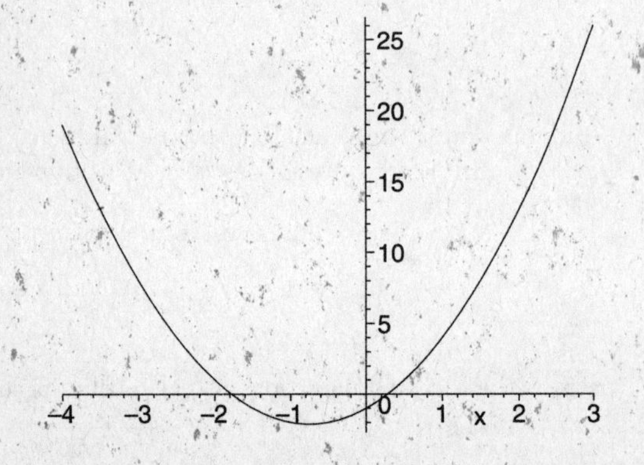

9. $y = 3x^3 - 6x + 1$

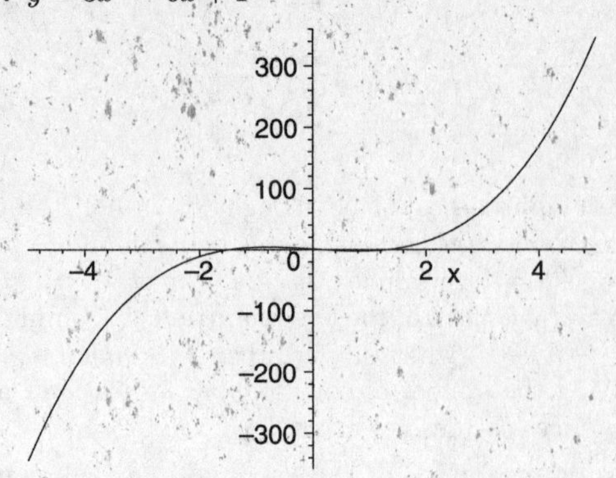

10. $y = | x + 1 |$

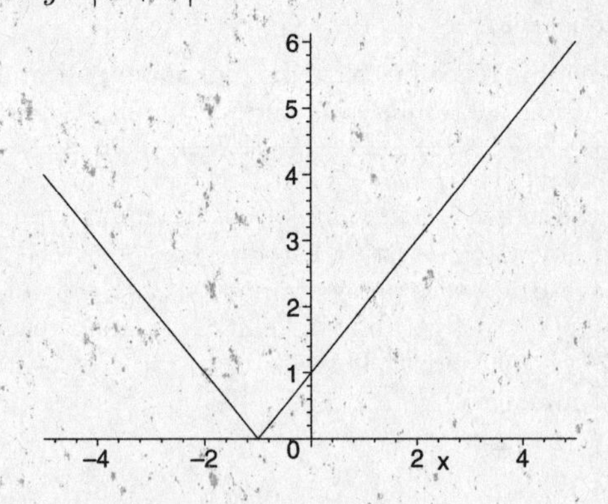

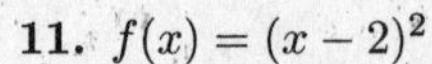

11. $f(x) = (x-2)^2$

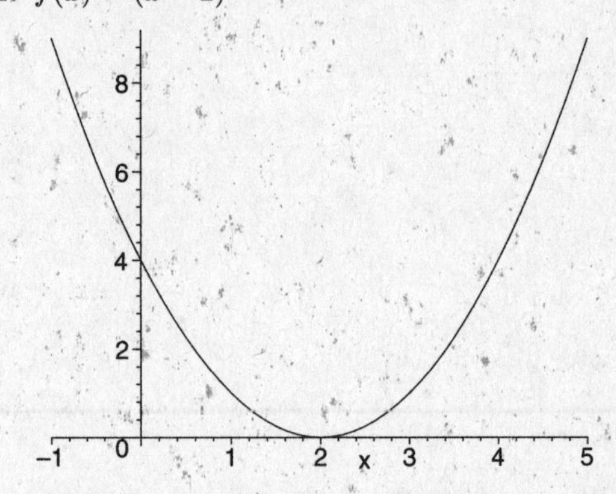

12. $f(x) = \frac{x^2-16}{x+4}$. It is important to note that $x = -4$ does not belong to the domain of the plotted function.

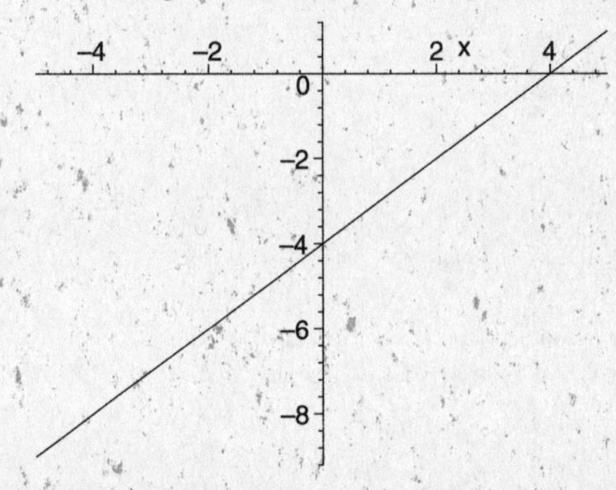

13. a) $f(2) = 1.2$

b) $x = -3$

14. $x = -2$

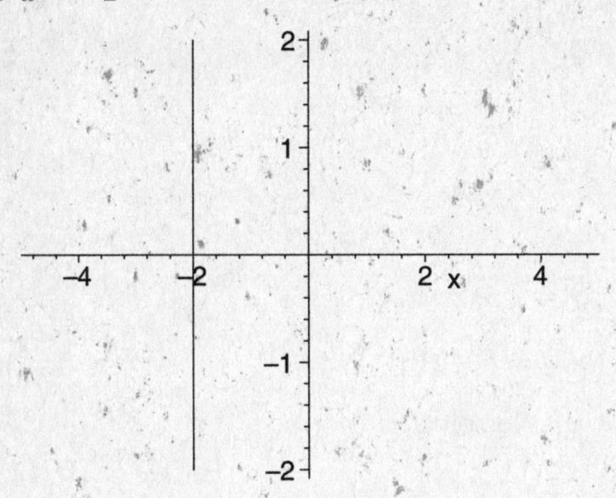

15. $y = 4 - 2x$

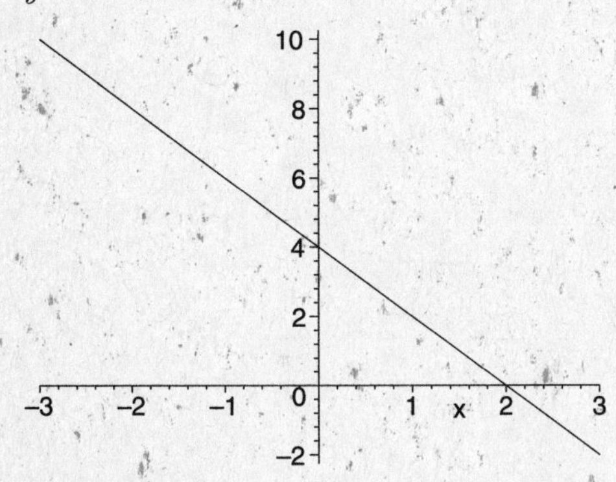

16. $m = \frac{-2-5}{4-(-7)} = \frac{-7}{11}$

$$\begin{aligned} y - y_1 &= m(x - x_1) \\ y - (-2) &= \frac{-7}{11}(x-4) \\ y &= -\frac{7x}{11} + \frac{28}{11} - 2 \\ y &= -\frac{7x}{11} + \frac{6}{11} \end{aligned}$$

17. Use the slope-point equation

$$\begin{aligned} y - y_1 &= m(x - x_1) \\ y - 11 &= 8(x - \frac{1}{2}) \\ y &= 8x - 4 + 11 \\ y &= 8x + 7 \end{aligned}$$

18. Slope $= -\frac{1}{6}$, y-intercept $(0, 3)$

19.

$$\begin{aligned} x^2 + 5x + 4 &= 0 \\ (x+1)(x+4) &= 0 \\ x + 1 &= 0 \\ x &= -1 \\ &Or \\ x + 4 &= 0 \\ x &= -4 \end{aligned}$$

20.

$$\begin{aligned} x^2 - 7x + 12 &= 0 \\ (x-3)(x-4) &= 0 \\ x - 3 &= 0 \\ x &= 3 \\ &Or \\ x - 4 &= 0 \\ x &= 4 \end{aligned}$$

21. $x^2 + 2x = 8$

$$\begin{aligned} x^2 + 2x - 8 &= 0 \\ (x+4)(x-2) &= 0 \\ x + 4 &= 0 \\ x &= -4 \\ &Or \\ x - 2 &= 0 \\ x &= 2 \end{aligned}$$

22. $x^2 + 6x = 20$

$$\begin{aligned} x^2 + 6x - 20 &= 0 \\ x &= \frac{-6 \pm \sqrt{36+80}}{2} \\ &= \frac{-6 \pm 2\sqrt{29}}{2} \\ &= -3 \pm \sqrt{29} \end{aligned}$$

23.

$$\begin{aligned} x^3 + 3x^2 - x - 3 &= 0 \\ x^2(x+3) - (x+3) &= 0 \\ (x+3)(x^2-1) &= 0 \\ (x+3)(x-1)(x+1) &= 0 \\ x + 3 &= 0 \\ x &= -3 \\ &Or \\ x - 1 &= 0 \\ x &= 1 \\ &Or \\ x + 1 &= 0 \\ x &= -1 \end{aligned}$$

24.

$$\begin{aligned} x^4 + 2x^3 - x - 2 &= 0 \\ x^3(x+2) - (x+2) &= 0 \\ (x+2)(x^3-1) &= 0 \\ x &= -2 \\ x &= -1 \end{aligned}$$

25. Using the points (one could use any two points on the line) $(0, 50$, and $(4, 350)$ the rate of change is

$$\frac{350-50}{4-0} = \frac{300}{4} = 75 \text{ pages per day}$$

26. The rate of change is

$$\frac{20-100}{12-0} = \frac{-80}{12} = \frac{-20}{3} \text{ meters per second}$$

27. The variation equation is $M = kW$, with k constant. When $W = 150$, $M = 60$ means

$$\begin{aligned} 60 &= k(150) \\ \frac{60}{150} &= k \\ \frac{2}{5} &= k \end{aligned}$$

Find M when $W = 210$

$$\begin{aligned} M &= \frac{2}{5}W \\ &= \frac{2}{5}(210) \\ &= 84\ lbs \end{aligned}$$

28. $5x^2 - x - 7 = 0$

$$\begin{aligned} x &= \frac{1 \pm \sqrt{1+140}}{10} \\ &= \frac{1 \pm \sqrt{141}}{10} \end{aligned}$$

29. $y^{1/6} = \sqrt[6]{y}$

30. $\sqrt[20]{x^3} = x^{3/20}$

31. $27^{2/3} = (\sqrt[3]{27})^2 = 3^2 = 9$

32.

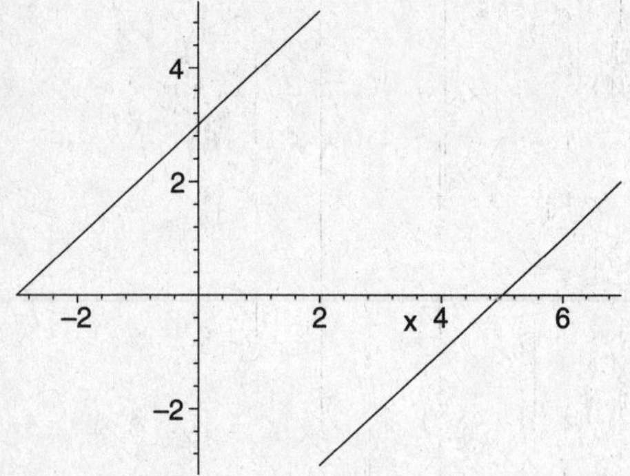

33. **a)** $m = \frac{92-74}{23-9} = \frac{18}{14} = \frac{9}{7}$

$$\begin{aligned} G - 74 &= \frac{9}{7}(x-9) \\ G &= \frac{9}{7}x - \frac{81}{7} + 74 \\ G &= \frac{9}{7}x + \frac{437}{7} \end{aligned}$$

b) $G(18) = \frac{9}{7}(18) + \frac{437}{7} = 85.6$
$G(25) = \frac{9}{7}(25) + \frac{437}{7} = 94.6$

34. $sin(2\pi/3) = \frac{\sqrt{3}}{2}$

35. $cos(-\pi) = -1$

36. $tan(7\pi/4) = -1$

37.

$$\begin{aligned} sin(70^o) &= \frac{x}{127} \\ x &= 127\ sin(70^o) \\ &= 119.341 \end{aligned}$$

38. $t = sin^{-1}(1) = \frac{\pi}{2} + 2n\pi$

39. $t = tan^{-1}(\sqrt{3}) = \frac{\pi}{3} + n\pi$

40. $2t = cos^{-1}(2)$, No solution

41.

$$\begin{aligned} 12cos^2(2t - \frac{\pi}{4}) &= 9 \\ cos^2(2t - \frac{\pi}{4}) &= \frac{9}{12} \\ cos^2(2t - \frac{\pi}{4}) &= \frac{3}{4} \\ cos(2t - \frac{\pi}{4}) &= \pm\sqrt{3/4} \end{aligned}$$

Two solutions

$$\begin{aligned} 2t - \frac{\pi}{4} &= cos^{-1}(\sqrt{3/4}) \\ 2t - \frac{\pi}{4} &= \frac{\pi}{6} + 2n\pi \\ 2t &= \frac{\pi}{6} + \frac{\pi}{4} + 2n\pi \\ 2t &= \frac{5\pi}{12} + 2n\pi \\ t &= \frac{5pi}{24} + n\pi \end{aligned}$$

and

$$\begin{aligned} 2t - \frac{\pi}{4} &= cos^{-1}(-\sqrt{3/4}) \\ 2t - \frac{\pi}{4} &= \frac{-\pi}{6} + 2n\pi \\ 2t &= \frac{-\pi}{6} + \frac{\pi}{4} + 2n\pi \\ 2t &= \frac{\pi}{12} + 2n\pi \\ t &= \frac{pi}{24} + n\pi \end{aligned}$$

42.

$$\begin{aligned} (2\ sin(t) - 1)(sin(t) + 4) &= 0 \\ sin(t) + 4 &= 0 \\ sin(t) &= -4 \text{ No solution} \\ 2\ sin(t) - 1 &= 0 \\ sin(t) &= \frac{1}{2} \\ t &= sin^{-1}(1/2) \\ t &= \frac{\pi}{6} + 2n\pi \\ t &= \frac{5\pi}{6} + 2n\pi \end{aligned}$$

43. $y = 2\ sin(t/3) - 4$
amplitude $= 2$, period $= \frac{2\pi}{(1/3)} = 6\pi$
mid-line $y = -4$, max $= -4 + 2 = 2$, min $= -4 - 2 = -6$

44. $y = \frac{1}{2}cos(2\pi t) + 3$
amplitude $= \frac{1}{2}$, period $= \frac{2\pi}{2\pi} = 1$
mid-line $y = 3$, max $= 3 + \frac{1}{2} = \frac{7}{2}$, min $= 3 - \frac{1}{2} = \frac{5}{2}$

45. Amplitude $= \frac{5-1}{2} = 2$, period $= \pi$, mid-line value $= 3$
$y = 2sin(2t) + 3$

46. Amplitude $= \frac{1-(-5)}{2} = 3$, period $= 2$, mid-line value $= -2$
$y = 3cos(\pi t) - 2$

47. **a)** Amplitude $= \frac{135-1}{2} = 67$, period $= 1/2$ means that $b = \frac{2\pi}{(1/2)} = 4\pi$, mid-line value $= 1 + 67 = 68$
Since the heel begins on the top of the eye we will use a cosine model
$h(t) = 67\ cos(4\pi t) + 68$

b) $h(10) = 67\ cos(40\pi) + 68 = 101.5\ m$

48. $(64^{5/3})^{-1/2} = 64^{-5/6} = \frac{1}{(\sqrt[6]{64})^5} = \frac{1}{32}$

49. $x = 0$, $x = -2$, and $x = 2$

50. $x = \pm\sqrt{10}$ and $x = \pm 2\sqrt{2}$

51. $(-1.8981, 0.7541)$, $(-0.2737, 1.0743)$, and $(2.0793, 0.6723)$

52. **a)** $G(x) = 0.6255x + 75.4766$

b) $G(18) = 0.6255(18) + 75.4766 = 86.7356$
$G(25) = 0.6255(25) + 75.4766 = 91.1141$

c) In Exercise 33, $G(18) = 85.6$ and $G(25) = 94.6$. The results obtained with the regression line are close to those obtained in Exercise 33.

53. **a)** $w(h) = 0.003968x^2 + 3.269048x - 76.428571$

b)

$$\begin{aligned} w(67) &= 0.003968(67)^2 + 3.269048(67) - 76.428571 \\ &= 160.415\ lbs \end{aligned}$$

Chapter 1 Test

1. **a)** Approximately 1150 minutes per month

b) About 62 years old

2. $f(x) = x^2 + 2$

a) $f(-3) = (-3)^2 + 2 = 11$

b) $f(x+h) = (x+h)^2 + 2 = x^2 + 2xh + h^2 + 2$

3. $f(x) = 2x^2 + 3$

a) $f(-2) = 2(-2)^2 + 3 = 8 + 3 = 11$

b) $f(x+h) = 2(x+h)^2 + 3 = 2(x^2 + 2xh + h^2) + 3 = 2x^2 + 4xh + 2h^2 + 3$

4. Slope $= -3$, y-intercept $(0, 2)$

5.

$$\begin{aligned} y - y_1 &= m(x - x_1) \\ y - (-5) &= \frac{1}{4}(x - 8) \\ y &= \frac{1}{4}x - 2 - 5 \\ y &= \frac{1}{4}x - 7 \end{aligned}$$

6. $m = \frac{10-(-5)}{-3-2} = 3$

7. Use the points $(0, 30)$ and $(3, 9)$
Average rate of change $= \frac{9-30}{3-0} = \frac{-21}{3} = -7$
The computer loses \$700 of its value each year.

8. Rate of change $= \frac{3-0}{6-0} = \frac{1}{2}$

9. Variation equation $F = kW$. Use $F = 120$ when $W = 180$ to find k

$$\begin{aligned} 120 &= k(180) \\ \frac{120}{180} &= k \\ \frac{2}{3} &= k \end{aligned}$$

The equation of variation is $F = \frac{2}{3}W$

10. **a)** $f(1) = -4$

b) $x = -3$ and $x = 3$

11.

$$\begin{aligned} x^2 + 4x - 2 &= 0 \\ x &= \frac{-4 \pm \sqrt{16+8}}{2} \\ &= \frac{-4 \pm \sqrt{24}}{2} \\ &= \frac{-4 \pm 2\sqrt{6}}{2} \\ &= -2 \pm \sqrt{6} \end{aligned}$$

12.

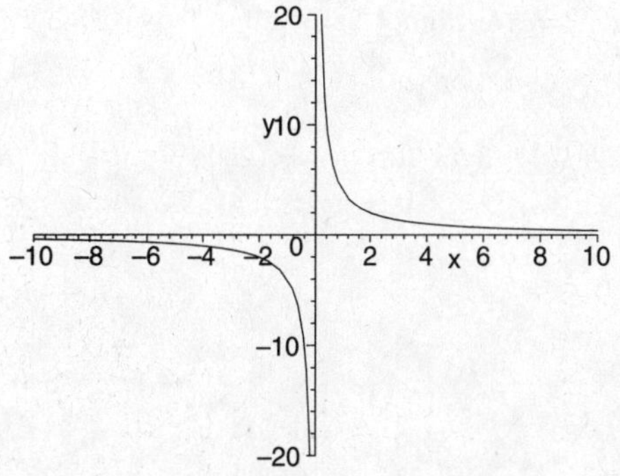

13. $1/\sqrt{t} = 1/t^{1/2} = t^{-1/2}$

14. $t^{-3/5} = 1/t^{3/5} = 1/\sqrt[5]{t^3}$

15. $f(x) = \frac{x^2-1}{x+1}$. It is important to note that $x = -1$ is not in the domain of the plotted function

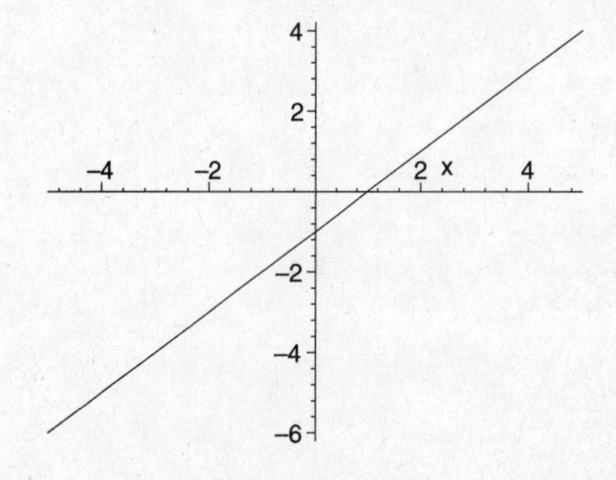

16. $sin(11\pi/6) = -\frac{1}{2}$

17. $cos(-3\pi/4) = \frac{sqrt2}{2}$

18. $tan(\pi) = 0$

19.

$$\begin{aligned} tan(40^o) &= \frac{3.28}{x} \\ x &= \frac{3.28}{tan(40^o)} \\ &= 3.909 \end{aligned}$$

20.

$$\begin{aligned} tan(t) &= \pm\sqrt{3} \\ t &= tan^{-1}(\sqrt{3}) \\ t &= \frac{\pi}{3} + 2n\pi \\ \text{and} & \\ t &= tan^{-1}(-\sqrt{3}) \\ t &= -\frac{\pi}{3} + 2n\pi \end{aligned}$$

21.

$$\begin{aligned} cos^2(t) &= 2 \\ cos(t) &= \pm\sqrt{2} \\ cos(t) &= 1.414 \end{aligned}$$

No solution, $cos(t)$ cannot have values larger than 1.

22.

$$\begin{aligned} 2sin^3(2t) - 3sin^2(2t) - 2sin(2t) &= 0 \\ sin(2t)(2sin(2t) - 1)(sin(2t) + 2) &= 0 \\ t &= \frac{n\pi}{2} \\ Or & \\ t &= \frac{-\pi}{12} + n\pi \\ Or & \\ t &= \frac{7\pi}{12} + n\pi \end{aligned}$$

23. Amplitude = 4, period = $\frac{2\pi}{2} = \pi$, mid-line $y = 4$
max = $4 + 4 = 8$, min = $4 - 4 = 0$

24. Amplitude = 6, period = $\frac{2\pi}{(1/3)} = 6\pi$, mid-line $y = -10$
max = $-10 + 6 = -4$, min = $-10 - 6 = -16$

25. Amplitude = $\frac{-0.5-(-1.5)}{2} = \frac{1}{2}$, period = $\frac{2\pi}{3}$,
$b = \frac{2\pi}{(2\pi/3)} = 3$, mid-line value is -1
Thus, equation of the line is $y = \frac{1}{2}cos(3t) - 1$

26. Amplitude = $\frac{4-1}{2} = \frac{3}{2}$, period = 1,
$b = \frac{2\pi}{1} = 2\pi$, mid-line value is 2.5
Thus, equation of the line is $y = \frac{3}{2}sin(2\pi t) + \frac{5}{2}$

27.

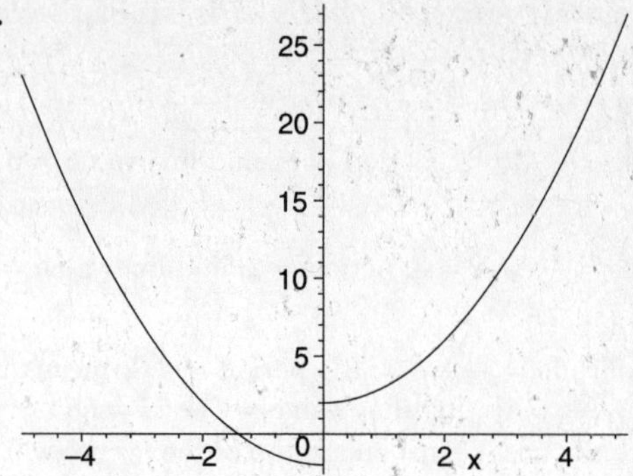

28.

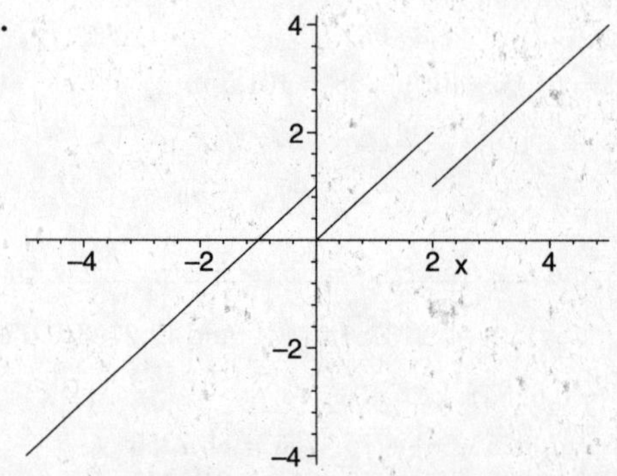

29. a) Find the slope $m = \frac{176-170}{80-50} = \frac{1}{5}$
Use slope-point equation

$$\begin{aligned} M - M_1 &= m(r - r_1) \\ M - 170 &= \frac{1}{5}(r - 50) \\ M &= \frac{1}{5}r - 10 + 170 \\ M &= \frac{1}{5}r + 160 \end{aligned}$$

b) $M(62) = \frac{1}{5}(62) + 160 = 172.4$
$M(75) = \frac{1}{5}(75) + 160 = 175$

30.

$$\begin{aligned} 3x + \frac{8}{x} - 1 &= 0 \\ 3x^2 - x + 8 &= 0 \\ x &= \frac{1 \pm \sqrt{1-96}}{6} \\ &= \frac{1 \pm i\sqrt{95}}{6} \\ &= \frac{1}{6} \pm \frac{i\sqrt{95}}{6} \end{aligned}$$

31. $x = --1.2543$

32. There are no real zeros for this function.

33. $(-1.21034, 2.36346)$

34. a) $M(r) = 0.2r + 160$

b) $M(62) = 172.4$
$M(75) = 174$

c) The results from the regression model are exactly the same as the result obtained in Exercise 29.

35. a) Linear Model: $y = 37.57614x + 294.47744$
Quadratic Model: $y = -0.59246x^2 + 74.60681x - 117.72472$
Cubic Model: $y = 0.02203x^3 - 2.60421x^2 + 125.71434x - 439.64751$
Quartic Model: $y = 0.00284x^4 - 0.32399x^3 + 11.45714x^2 - 88.51211x + 507.83874$

b)

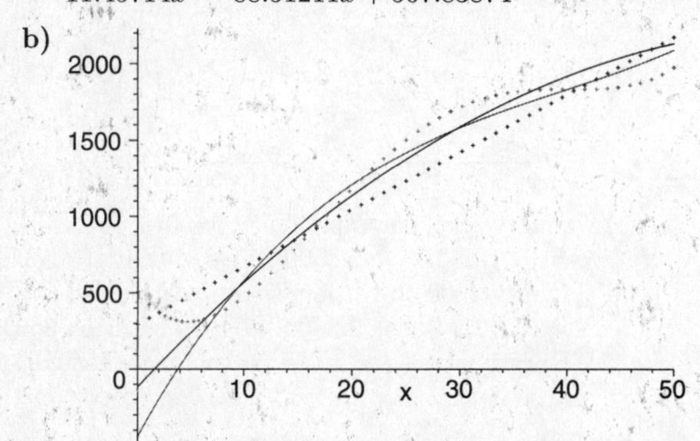

c) By consider the graph in part b) and the scatter plot of the data points, it seems like quartic model best fits the data. The reason for this conclusion is because the scatter plot and the quartic model have the least amount of deviation (sometimes called residue) between them compared to the other models.

d) Left to the student (answers vary).

Technology Connection

- **Page 5:**
Left to the student

- **Page 7:**

1. The line will look like a vertical line.
2. The line will look like a horizontal line.
3. The line will look like a vertical line.
4. The line will look like a horizontal line.

- **Page 10:**

1. Graphs are parallel
2. The function values differ by the constant value added.
3. Graphs are parallel

- **Page 19:**

1. $f(-5) = 6$, $f(-4.7) = 3.99$, $f(11) = 150$, $f(2/3) = -1.556$
2. $f(-5) = -21.3$, $f(-4.7) = -12.3$, $f(11) = -117.3$, $f(2/3) = 3.2556$
3. $f(-5) = -75$, $f(-4.7) = -45.6$, $f(11) = 420.6$, $f(2/3) = 1.6889$

- **Page 21:**

1.

-1	0	1	2	3	4	5	6	7
11	4	-1	-4	-5	-4	-1	4	11

2.

-3	-2	-1	0	1	2	3	4	5
-29	-15	-5	1	3	1	-5	-15	-29

- **Page 23:**

1. $x = 4.4149$
2. $x = -0.618034$ and $x = 1.618034$

- **Page 27:**

1. $y = -0.37393x + 1.02464$
2. $y = 0.46786x^2 - 3.36786x + 5.26429$
3. $y = 0.975x^3 - 6.031x^2 + 8.625x - 3.055$

- **Page 28:**

1. $x = 2$ and $x = -5$
2. $x = -4$ and $x = 6$
3. $x = -2$ and $x = 1$
4. $x = -1.414214$, $x = 0$, and $x == 1.414214$
5. $x = 0$ and $x = 700$
6. $x = -2.079356$, $x = 0.46295543$, and $x = 3.1164004$
7. $x = -3.095574$, $x = -0.6460838$, $x = 0.6460838$, and $x = 3.095574$
8. $x = -1$ and $x = 1$
9. $x = -2$, $x = 1.414214$, $x = 1$, and $x = 1.414214$
10. $x = -3$, $x = -1$, $x = 2$, and $x = 3$
11. $x = -0.3874259$ and $x = 1.7207592$
12. $x = 6.1329332$

- **Page 37:**

1. $[0, \infty)$
2. $[-2, \infty)$
3. $(-\infty, \infty)$
4. $(-\infty, \infty)$
5. $[1, \infty)$
6. $(-\infty, \infty)$
7. $[-3, \infty)$
8. $(-\infty, \infty)$
9. $(-\infty, \infty)$
10. $(-\infty, \infty)$
11. Not correct
12. Correct

- **Page 46:**

1. $t = 6.89210^o$
2. $t = 46.88639^o$
3. No solution
4. $t = 1.01599$
5. $t = 0.66874$
6. 0.46677

- **Page 56:**
Number 2 equation: shifts the $cos(\pi x)$ graph up by 1 unit
Number 3 equation: shifts the $cos(\pi x)$ graph up by 1 unit and shrinks the period by a factor of 2
Number 4 equation: shifts the $cos(\pi x)$ graph up by 1 unit, shrinks the period by a factor of 2, and increases the amplitude by a factor of 3
Number 5 equation: shifts the $cos(\pi x)$ graph up by 1 unit, shrinks the period by a factor of 2, increases the amplitude by a factor of 3, and shifts the graph to the right by 0.5 units

Extended Life Science Connection

1. **a)** $y = 1.343450619x + 311.3019556$

b)

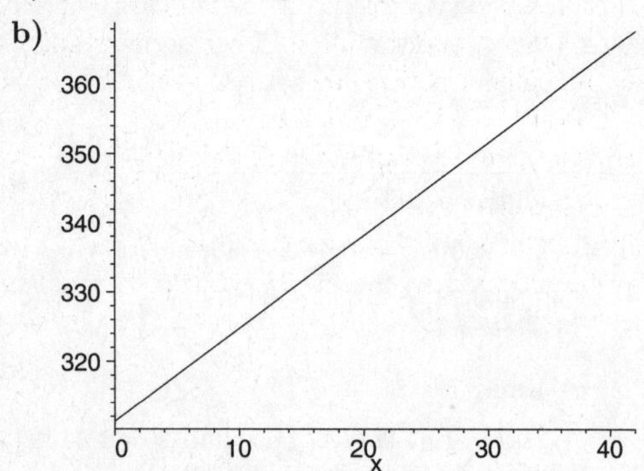

c) January 1990 corresponds to $t = 31$
$y = 1.343450619(31) + 311.3019556 = 352.95$
January 2000 corresponds to $t = 41$
$y = 1.343450619(41) + 311.3019556 = 366.38$
The estimates seem to be reasonable when compared to the data.

d) January 2010 corresponds to $t = 51$
$y = 1.343450619(51) + 311.3019556 = 379.82$
January 2050 corresponds to $t = 91$
$y = 1.343450619(91) + 311.3019556 = 433.56$
The estimates seem to be reasonable when compared to the data.

e) Find x when $y = 500$

$$\begin{aligned} y &= 1.34345x + 311.30196 \\ 500 &= 1.34345x + 311.30196 \\ 500 - 311.30196 &= 1.34345x \\ \frac{500 - 311.30196}{1.34345} &= x \\ 140.5 &\approx x \end{aligned}$$

The carbon dioxide concentration will reach 500 parts per million sometime in the year 2099.

2. **a)** $y = 0.0122244281x^2 + 0.8300246407x + 314.8103665$

b)

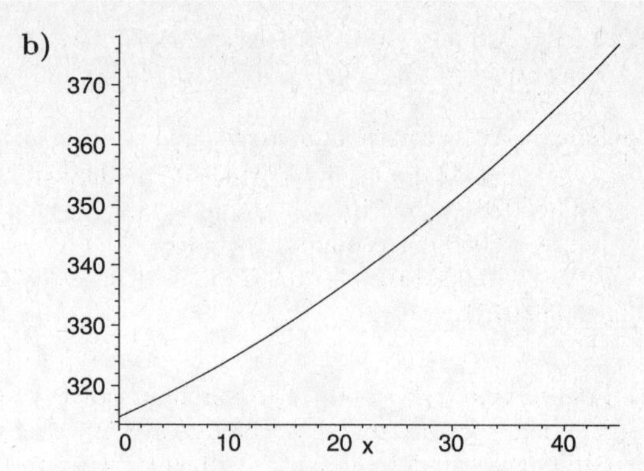

c) January 1990 corresponds to $t = 31$
$y = 0.0122(31)^2 + 0.8300(31) + 314.8104 = 352.29$
January 2000 corresponds to $t = 41$
$y = 0.0122(41)^2 + 0.8300(41) + 314.8104 = 369.39$
The estimates seem to be reasonable when compared to the data.

d) January 2010 corresponds to $t = 51$
$y = 0.0122(51)^2 + 0.8300(51) + 314.8104 = 388.94$
January 2050 corresponds to $t = 91$
$y = 0.0122(91)^2 + 0.8300(91) + 314.8104 = 491.57$

e) Find x when $y = 500$

$$\begin{aligned} 500 &= 0.0122x^2 + 0.8300x + \\ & 314.8104 \\ 0 &= 0.0122x^2 + 0.8300x - \\ & 185.1896 \\ x &= \frac{0.8300}{2(0.0122)} + \\ & \frac{\sqrt{(0.8300)^2 - 4(0.0122)(-185.1896)}}{2(0.0122)} \\ x &\approx 161.63 \end{aligned}$$

The carbon dioxide concentration will reach 500 parts per million sometime in the year 2120.

3. **a)** $y = -0.000307x^3 + 0.031536x^2 + 0.509387x + 315.8660781$

b)

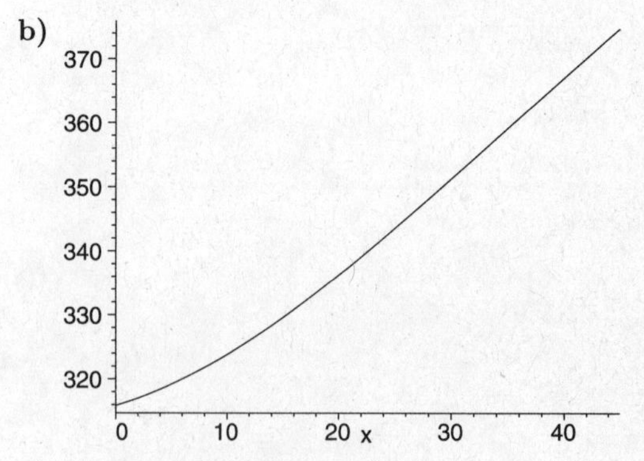

c) January 1990 corresponds to $t = 31$
$y = -0.000307x^3 + 0.031536x^2 + 0.509387x + 315.8660781 = 352.82$
January 2000 corresponds to $t = 41$
$y = -0.000307x^3 + 0.031536x^2 + 0.509387x + 315.8660781 = 368.60$

The estimates seem to be reasonable when compared to the data

d) January 2010 corresponds to $t = 51$
$y = -0.000307x^3 + 0.031536x^2 + 0.509387x + 315.8660781 = 383.15$
January 2050 corresponds to $t = 91$
$y = -0.000307x^3 + 0.031536x^2 + 0.509387x + 315.8660781 = 392.02$

e) Find x when $y = 500$. The maximum of the cubic function does not intersect the line $y = 500$ therefore under this model the carbon dioxide concentration will never reach 500 parts per million.

4. a)

380 370 360 350 340 330 320
0 10 20 t 30 40 50

b) The graph represents a steady increase in the concentration of carbon dioxide.

c)

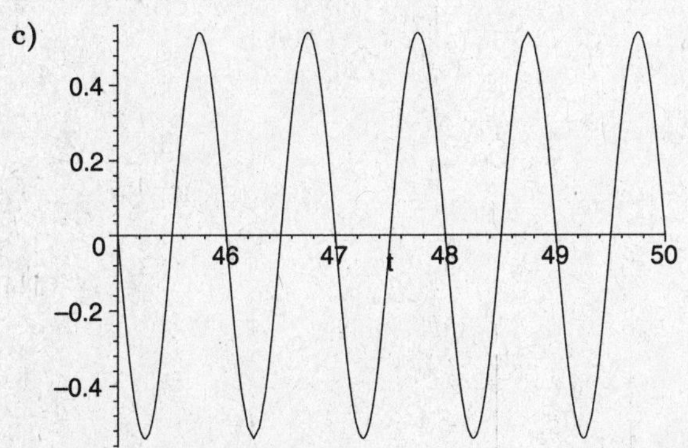

d) The graph shows an oscillating behavior for the concentration of carbon dioxide.

e)

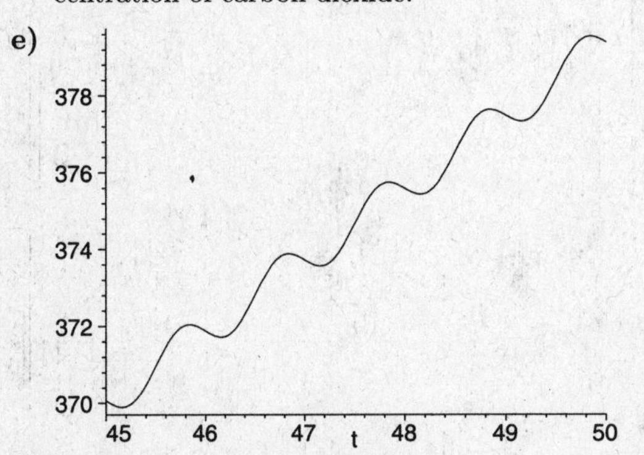

f) The graph behavior shows that there is a periodic fluctuation in the concentration of carbon dioxide.

5.

- **LINEAR MODEL:** This model is the easiest mathematically to compute and explain. It does resemble the scatter plot of the original data sets. Under this model, the concentration levels of carbon dioxide will increase with time indefinitely.
- **QUADRATIC MODEL:** This model also resembles the original data set's scatter plot. At relatively small values of t it allows for a longer time for the increase in the concentration of carbon dioxide since it is a parabola. As time increases though the level at which the concentration of carbon dioxide will increase will be quicker than the linear model.
- **CUBIC MODEL:** This modelalso resembled the original data set's scattor plot indicates that there is a level after which the concentration of carbon dioxide will not increase. It is the only model that did not allow the concentration level of carbon dioxide to reach 500 parts per million. This model suggests that as time increased the concentration of carbon dioxide will begin to decrease indefinitely.
- **PERIODIC MODEL:** This model, as the other, modeled the data set to a very good degree of accuracy. It was the only model that allowed for oscillating behavior in the future, which is more likely to happen than what the other models suggested.

Chapter 2

Differentiation

Exercise Set 2.1

1. The function is not continuous at $x = 1$ since the limit from the left of $x = 1$ is not equal to the limit from the right of $x = 1$ and therefore the limit of the function at $x = 1$ does not exist.

2. The function is not continuous at $x = -2$ since the limit from the left of $x = -2$ is not equal to the limit from the right of $x = -2$ and therefore the limit of the function at $x = -2$ does not exist.

3. The function is continuous at every point in the given plot. Note that the graph can be traced without a jump from one point to another.

4. The function is not continout at $x = -2$ since the value of the function at $x = -2$ is undefined.

5. a) As we approach the x-value of 1 from the right we notice that the y-value is approaching a value of -1. Thus, $\lim_{x\to1^+} f(x) = -1$. As we approach the x-value of 1 from the left we notice that the y-value is approaching a value of 2. Thus, $\lim_{x\to1^-} f(x) = 2$. Since $\lim_{x\to1^+} f(x) \neq \lim_{x\to1^-} f(x)$ then $\lim_{x\to1} f(x)$ does not exist.

   b) Reading the value from the graph $f(1) = -1$.

   c) Since the $\lim_{x\to1^-} f(x)$ does not exist, then $f(x)$ is not continuous at $x = 1$.

   d) As we approach the x-value of -2 from the right we notice that the y-value is approaching a value of 3. Thus, $\lim_{x\to-2^+} f(x) = 3$. As we approach the x-value of -2 from the left we notice that the y-value is approaching a value of 3. Thus, $\lim_{x\to-2^-} f(x) = 3$. Since $\lim_{x\to-2^+} f(x) = \lim_{x\to-2^-} f(x) = 3$ then $\lim_{x\to-2} f(x) = 3$.

   e) Reading the value from the graph $f(-2) = 3$.

   f) Since $\lim_{x\to-2} f(x) = 3$ and $f(-2) = 3$, then $f(x)$ is continuous at $x = -2$.

6. a) $\lim_{x\to1^+} g(x) = -2$, $\lim_{x\to1^-} g(x) = -2$, therefore $\lim_{x\to1} f(x) = -2$

   b) $g(1) = -2$

   c) Since $\lim_{x\to1} g(x) = -1 = g(1) = -2$, then $g(x)$ is continuous at $x = 1$

   d) $\lim_{x\to-2^+} g(x) = 4$, $\lim_{x\to-2^-} g(x) = -3$, thus $\lim_{x\to-2} g(x)$ does not exist

   e) $g(-2) = -3$

   f) Since $\lim_{x\to-2} g(x) \neq g(-2)$, then the function is not continuous at $x = -2$

7. a) As we approach the x-value of 1 from the right we notice that the y-value is approaching a value of 2. Thus, $\lim_{x\to1^+} h(x) = 2$. As we approach the x-value of 1 from the left we notice that the y-value is approaching a value of 2. Thus, $\lim_{x\to1^-} h(x) = 2$. Since $\lim_{x\to1^+} h(x) = \lim_{x\to1^-} h(x) = 2$ then $\lim_{x\to1} h(x) = 2$.

   b) Reading the value from the graph $h(1) = 2$.

   c) Since the $\lim_{x\to1} h(x) = 2$ and $= h(1) = 2$ then $h(x)$ is continuous at $x = 1$.

   d) As we approach the x-value of -2 from the right we notice that the y-value is approaching a value of 3. Thus, $\lim_{x\to-2^+} h(x) = 0$. As we approach the x-value of -2 from the left we notice that the y-value is approaching a value of 0. Thus, $\lim_{x\to-2^-} h(x) = 0$. Since $\lim_{x\to-2^+} h(x) = \lim_{x\to-2^-} h(x) = 0$ then $\lim_{x\to-2} h(x) = 0$.

   e) Reading the value from the graph $h(-2) = 0$.

   f) Since $\lim_{x\to-2} h(x) = 0$ and $h(-2) = 0$, then $h(x)$ is continuous at $x = -2$.

8. a) $\lim_{x\to1^+} t(x) \approx 0.25$, $\lim_{x\to1^-} t(x) \approx 0.25$, therefore $\lim_{x\to1} t(x) \approx 0.25$

   b) $t(1) =\approx 0.25$

   c) Since $\lim_{x\to1} t(x) = t(1) \approx 0.25$, then $t(x)$ is continuous at $x = 1$

   d) $\lim_{x\to-2^+} t(x) =$ undefined, $\lim_{x\to-2^-} t(x) =$ undefined, thus $\lim_{x\to-2} t(x)$ does not exists

   e) $t(-2) =$ undefined

   f) Since $\lim_{x\to-2} t(x)$ does not exist, then the function is not continuous at $x = -2$

9. a) As we approach the x value of 1 from the right we find that the y value is approching 3. Thus $\lim_{x\to1^+} f(x) = 3$

   b) As we approach the x value of 1 from the left, we find that the y value is approching 3. Thus $\lim_{x\to1^-} f(x) = 3$

   c) Since $\lim_{x\to1^+} f(x) = 3$ and $\lim_{x\to1^-} f(x) = 3$ then $\lim_{x\to1} f(x) = 3$

   d) From the given conditions $f(1) = 2$

   e) $f(x)$ is not continuous at $x = 1$ since $\lim_{x\to3} f(x) \neq f(1)$

   f) $f(x)$ is continuous at $x = 2$ since $\lim_{x\to2^+} f(x) = \lim_{x\to2^-} f(x) = 2 = f(2)$

10. a) $\lim_{x\to -2^+} f(x) = 0$

b) $\lim_{x\to -2^+} f(x) = 0$

c) $\lim_{x\to -2} f(x) = 0$

d) $f(-2) = 3$

e) $f(x)$ is not continuous at $x = -2$

f) $f(x)$ is continuous at $x = 1$

11. a) True. The values of y as we approch $x = 0$ from the right is the same as the value of the function at $x = 0$, which is 0

b) True. The values of y as we approch $x = 0$ from the left is the same as the value of the function at $x = 0$, which is 0

c) True. Since $\lim_{x\to 0^+} f(x) = 0$ and $\lim_{x\to 0^-} f(x) = 0$

d) False. Since $\lim_{x\to 3^+} f(x) = 3$ and $\lim_{x\to 3^-} f(x) = 1$

e) True. Since $\lim_{x\to 0^+} f(x) = \lim_{x\to 0^-} f(x) = 0$

f) False. Since $\lim_{x\to 3^+} f(x) \neq \lim_{x\to 3^-} f(x)$

g) True. Since $\lim_{x\to 0} f(x) = 0 = f(0)$

h) False. Since $\lim_{x\to 3} f(x)$ does not exist

12. a) True. Since $\lim_{x\to 2^+} g(x) = 1 = g(2)$

b) False. Since $\lim_{x\to 2^-} g(x) = -1 \neq 1 = g(2)$

c) False. Since $\lim_{x\to 2^+} g(x) = 1$ and $\lim_{x\to 2^-} g(x) = -1$

d) False. Since $\lim_{x\to 2^+} g(x) \neq \lim_{x\to 2^-} g(x)$

e) False. Since $\lim_{x\to 2} g(x)$ does not exist

13. a) False. As we approach the x value of -2 from the right we find that the y value is approching 2.

b) True. As we approach the x value of -2 from the left, we find that the y value is approching 0.

c) False. Since $\lim_{x\to -2^+} f(x) = 1$ and $\lim_{x\to 1^-} f(x) = 0$

d) False. Since $\lim_{x\to -2^+} f(x) \neq \lim_{x\to -2^-} f(x)$

e) False. Since $\lim_{x\to -2} f(x)$ does not exist

f) True. Since $\lim_{x\to 0^+} f(x) = \lim_{x\to 0^-} f(x) = 0$

g) True. The graph indicate a point (solid dot) at $(0, 2)$

h) False. Since $\lim_{x\to -2^+} f(x) \neq \lim_{x\to -2^-} f(x)$

i) False. Since $\lim_{x\to 0} f(x) \neq f(0)$

j) True. Since $\lim_{x\to -1^+} f(x) = \lim_{x\to -1^-} f(x) = f(-1)$

14. a) False. Since $\lim_{x\to 2^-} f(x) = 0$

b) False. Since $\lim_{x\to 2^+} f(x) = 3$

c) False.

d) False. Since $\lim_{x\to 2^+} f(x) \neq \lim_{x\to 2^-} f(x)$

e) True. Since $\lim_{x\to 4^+} f(x) \neq \lim_{x\to 4^-} f(x) = 3$

f) False. Since $\lim_{x\to 4} f(x) = 3$ and $f(4) = -1$

g) False. Since $\lim_{x\to 4} f(x) \neq f(4)$

h) True. Since $\lim_{x\to 0} f(x) = 4$ and $f(0) = 4$

i) True. Since $\lim_{x\to 3} f(x) = 3 = \lim_{x\to 5} f(x)$

j) False. Since $\lim_{x\to 2} f(x)$ does not exist

15. a) True. As we approach the x value of 0 from the right we find that the y value is approching 0, which is the value of the function at $x = 0$.

b) False. As we approach the x value of 0 from the left, we find that the y value is approching 2 instead of 0.

c) False. Since $\lim_{x\to 0^+} f(x) = 0$ and $\lim_{x\to 0^-} f(x) = 2$

d) True. Since $\lim_{x\to 2^+} f(x) = 4 = \lim_{x\to 2^-} f(x)$

e) False. Since $\lim_{x\to 0^-} f(x) \neq \lim_{x\to 0^+} f(x)$

f) True. Since $\lim_{x\to 2^+} f(x) = \lim_{x\to 2^-} f(x) = 4$

g) False. Since $\lim_{x\to 0} f(x)$ does not exist

h) True. Since $\lim_{x\to 2} f(x) = 4 = f(2)$

16. a) True. $\lim_{x\to 0^+} g(x) = 0 = g(0)$

b) True. $\lim_{x\to 0^-} g(x) = 0 = g(0)$

c) True.

d) True. $\lim_{x\to 0} g(x) = 0$

e) True. $\lim_{x\to 0} g(x) = 0 = g(0)$

17. The function p is not continuous at $x = 1$ since the $\lim_{x\to 1} p(x)$ does not exist. p is continuous at $x = 1.5$ since $\lim_{x\to 1.5} p(x) = 0.6 = p(1.5)$. p is not continuous at $x = 1$ since the $\lim_{x\to 2} p(x)$ does not exist. p is continuous at $x = 2.01$ since $\lim_{x\to 2.01} p(x) = 0.8 = p(2.01)$.

18. p is continuous at $x = 2.99$ since $\lim_{x\to 2.99} p(x) = 0.8 = p(2.99)$. p is not continuous at $x = 3$ since the $\lim_{x\to 3} p(x)$ does not exist. p is continuous at $x = 3.04$ since $\lim_{x\to 3.04} p(x) = 1 = p(3.04)$. p is not continuous at $x = 4$ since the $\lim_{x\to 4} p(x)$ does not exist.

19. $\lim_{x\to 1^-} p(x) = 0.4$, $\lim_{x\to 1^+} p(x) = 0.6$, therefore $\lim_{x\to 1} p(x)$ does not exist

20. $\lim_{x\to 1^-} p(x) = 0.6$, $\lim_{x\to 2^+} p(x) = 0.8$, therefore $\lim_{x\to 2} p(x)$ does not exist

21. $\lim_{x\to 2.6^-} p(x) = 0.8$, $\lim_{x\to 2.6^+} p(x) = 0.8$, therefore $\lim_{x\to 2.6} p(x) = 0.8$

22. $\lim_{x\to 3} p(x)$ does not exist since $\lim_{x\to 3^-} p(x) = 0.8$ and $\lim_{x\to 3^+} p(x) = 1$

23. $\lim_{x\to 3.4} p(x) = 1$ since $\lim_{x\to 3.4^-} p(x) = 1$, $\lim_{x\to 3.4^+} p(x) = 1$

24. C is continuous at $x = 0.1$ since $\lim_{x\to 0.1} C(x) = 2.3 = C(0.1)$. C is not continuous at $x = 0.2$ since the $\lim_{x\to 0.2} C(x)$ does not exist. C is continuous at $x = 0.25$ since $\lim_{x\to 0.25} C(x) = 3 = C(0.25)$. C is continuous at $x = 0.267$ since $\lim_{x\to 0.267} C(x) = 3 = C(0.267)$.

25. If we continue the pattern used for the taxi fare function, we see that for $x = 2.3$, which falls in the range of 2.2 and 2.4 miles, the fare will be \$5.60, for $x = 5$, which falls between the range 2.4 and 2.6 miles, the fare is \$5.90. For $x = 2.6$ and $x = 3$ we need to be careful since they act as a boundary of two possible fares. Therefore C is continuous at $x = 2.3$ since $\lim_{x\to 2.3} C(x) = 5.60 = C(2.3)$. C is continuous at $x = 2.5$ since $\lim_{x\to 2.5} C(x) = 5.90 = C(2.5)$. C is not continuous at $x = 2.6$ since $\lim_{x\to 2.6^-} C(x) = 5.90$ and $\lim_{x\to 2.6^+} C(x) = 6.20$ thus, $\lim_{x\to 2.6} C(x)$ does not exist. C is not continuous at $x = 3$ since $\lim_{x\to 3^-} C(x)6.50$ and $\lim_{x\to 3^+} C(x) = 6.80$ thus, $\lim_{x\to 3} C(x)$ does not exist.

26. $\lim_{x\to 1/4^-} C(x) = \$2.60$, $\lim_{x\to 1/4^+} C(x) = \$2.60$, therefore $\lim_{x\to 1/4} C(x) = \$2.60$

27. $\lim_{x\to 0.2^-} C(x) = \$2.30$, $\lim_{x\to 0.2^+} C(x) = \$2.60$, therefore $\lim_{x\to 0.2} C(x)$ does not exist

28. $\lim_{x\to 0.6^-} C(x) = \$2.90$, $\lim_{x\to 0.6^+} C(x) = \$3.20$, therefore $\lim_{x\to 0.6} C(x)$ does not exist

29. $\lim_{x\to 0.5^-} C(x) = \$2.90$, $\lim_{x\to 0.5^+} C(x) = \$2.90$, therefore $\lim_{x\to 0.5} C(x) = \$2.90$

30. $\lim_{x\to 0.4^-} C(x) = \$2.60$, $\lim_{x\to 0.4^+} C(x) = \$2.90$, therefore $\lim_{x\to 0.4} C(x)$ does not exist

31. The population function, $p(t)$, is discontinuous at $t^* = 0.5$, $t^* = 0.75$, $t^* = 1.25$, $t^* = 1.5$, and at $t^* = 1.75$ since at these points the population function has a "jump" which means that the $\lim_{t\to t^*} p(t)$ does not exist

32. There was a jump in the population at $t = 0.5$, $t = 0.75$, $t = 1.5$, and $t = 1.75$ due to births. While there was a decline in the populaion at $t = 1.25$ due to deaths.

33. $\lim_{t\to 1.5^+} p(t) = 12$

34. $\lim_{t\to 1.5^-} p(t) = 11$

35. The population function, $p(t)$, is discontinuous at $t^* = 0.1$, $t^* = 0.3$, $t^* = 0.4$, $t^* = 0.5$, $t^* = 0.6$, and at $t^* = 0.8$ since at these points the population function has a "jump" which means that the $\lim_{t\to t^*} p(t)$ does not exist

36. There was a jump in the population at $t = 0.1$, $t = 0.4$, $t = 0.5$, and $t = 0.6$ due to births. While there was a decline in the populaion at $t = 0.3$ and $t = 0.8$ due to deaths.

37. $\lim_{t\to 0.6^+} p(t) = 35$

38. $\lim_{t\to 0.6^-} p(t) = 33$

39. From the graph, the "I've got it" experience seems to occur after spending 20 hours on the task.

40. Once the task mastered one should get 100 correct trials out of 100 trials.

41. $\lim_{t\to 20^+} N(t) = 100$, $\lim_{t\to 20^-} N(t) = 30$, therefore $\lim_{t\to} N(t)$ does not exist

42. $\lim_{t\to 30^-} N(t) = 100$, $\lim_{t\to 30^+} N(t) = 100$, therefore $\lim_{t\to 30} N(t) = 100$

43. $N(t)$ is discontinuous at $t = 20$ since $\lim_{t\to 20} N(t)$ does not exist. $N(t)$ is continuous at $t = 30$ since $\lim_{t\to 30} N(t) = 100 = N(30)$

44. $N(t)$ is discontinuous at $t = 10$ since $\lim_{t\to 10} N(t)$ does not exist. $N(t)$ is continuous at $t = 26$ since $\lim_{t\to 26} N(t) = 100 = N(26)$

45. A function may not be continuous if the function is not defined at one of the points in the domain, it also may not be continuous if the limit at a point does not exist, it also may not be continuous if the limit at a point is different than the value of the function at that point.
NOTE: See the graphs on page 77.

46. $f(x)$ is continuous by C1 and C2, $\lim_{x\to 3} f(x) = 19$

47. $f(x)$ is continuous by C1 and C2, $\lim_{x\to 1} f(x) = 0$

48. $g(x)$ is continuous by C4, $\lim_{x\to -1} g(x) = \dfrac{1}{2}$

49. $g(x)$ is continuous by C4, $\lim_{x\to 1} g(x) = 1$

50. *tan* x is continuous by C5, $\lim_{x\to \frac{\pi}{4}} \tan x = 1$

51. *cot* x is continuous by C5, $\lim_{x\to \frac{\pi}{3}} \cot x = \dfrac{1}{\sqrt{3}}$

52. *sec* x is continuous by C5, $\lim_{x\to \frac{\pi}{6}} \sec x = \dfrac{2}{\sqrt{3}}$

53. *csc* x is continuous by C5, $\lim_{x\to \frac{\pi}{4}} \csc x = \sqrt{2}$

54. $f(x)$ is continuous by C5, $\lim_{x\to 3} f(x) = \sqrt{19}$

55. $f(x)$ is continuous by C5, $\lim_{x\to \frac{\pi}{3}} f(x) = \dfrac{\sqrt[4]{3}}{\sqrt{2}}$

56. $g(x)$ is continuous by C3, $\lim_{x\to \frac{\pi}{4}} f(x) = \dfrac{1}{2}$

57. $g(x)$ is continuous by C5, $ds\lim_{x\to \frac{\pi}{6}} g(x) = -\frac{1}{2}$

58. Limit approaches 0.92857

59. Limit approaches 0

60. Limit approaches 0

61. Limit approaches 1

62. Limit approaches 0

63. Limit does not exist

Exercise Set 2.2

1. $x^2 - 3$ is a continuous function (it is a polynomial). Therefore, we can use direct substitution

$$\begin{aligned}\lim_{x\to 1}(x^2-3) &= (1)^2 - 3\\ &= 1-3\\ &= -2\end{aligned}$$

2. $\lim_{x\to 1}(x^2+4) = (1)^2 + 4 = 1 + 4 = 5$

3. The function $f(x) = \frac{3}{x}$ is not continuous at $x - 0$ since the denominator equals zero. There are no algebraic simplifications that can be done to the function. To find the limit, we can either plug points that are approaching 0 from the right and the left and detrmine the limit from each side, or we can use the graph of the function to determine the limit (if it exists). Looking at the graph, we see that as x approches 0 from the left the y values are becoming more and more negative, and as x approches 0 from the right, the y values are becoming more and more positive. Therefore, since $\lim_{x\to 0^+}\frac{3}{x} \neq \lim_{x\to 0^-}\frac{3}{x}$ then $\lim_{x\to 0}\frac{3}{x}$ does not exist.

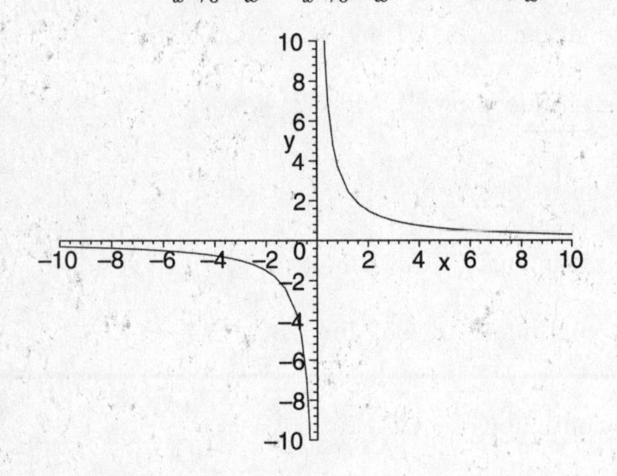

4. $\lim_{x\to 0^+}\frac{-4}{x}$ does not exist since the limit from the left of $x = 0$ does not equal the limit from the right of $x = 0$

5. $2x+5$ is a continuous function (it is a polynomial). Therefore, we can use direct substitution

$$\begin{aligned}\lim_{x\to 3}(2x+5) &= 2(3)+5\\ &= 6+5\\ &= 11\end{aligned}$$

6. $\lim_{x\to 4}(5-3x) = 5 - 3(4) = 5 - 12 = -7$

7. The function $\frac{x^2-25}{x+5}$ is discontinuous at $x = -5$, but it can be simplified algebraically.

$$\frac{x^2-25}{x+5} = \frac{(x-5)(x+5)}{x+5} = x-5$$

Therfore, $\lim_{x\to -5}\frac{x^2-25}{x+5} = \lim_{x\to -5}(x-5) = -5-5 = -10$

8. $\lim_{x\to -4}\frac{x^2-16}{x+4} = \frac{(x-4)(x+4)}{x+4} = \lim_{x\to -4}(x-4)$
$= -4 - 4 = -8$

9. Since $\frac{5}{x}$ is continuous at $x = -2$ we can use direct substitution.
$\lim_{x\to -2}\frac{5}{x} = \frac{5}{-2} = -\frac{5}{2}$

10. $\lim_{x\to -5}\frac{-2}{x} = \frac{-2}{-5} = \frac{2}{5}$

11. The function $\frac{x^2+x-6}{x-2}$ is discontinuous at $x = 2$, but it can be simplified algebraically. The limit is then computed as follows:

$$\begin{aligned}\lim_{x\to 2}\frac{x^2+x-6}{x-2} &= \lim_{x\to 2}\frac{(x-2)(x+3)}{x-2}\\ &= \lim_{x\to 2}(x+3)\\ &= 2+3\\ &= 5\end{aligned}$$

12.

$$\begin{aligned}\lim_{x\to -4}\frac{x^2-x-20}{x+4} &= \lim_{x\to -4}\frac{(x-5)(x+4)}{x+4}\\ &= \lim_{x\to -4}(x-5)\\ &= -4-5\\ &= -9\end{aligned}$$

13. Since $\sqrt[3]{x^2-17}$ is continuous at $x = 5$ we can use direct substitution

$$\begin{aligned}\lim_{x\to 5}\sqrt[3]{x^2-17} &= \sqrt[3]{5^2-17}\\ &= \sqrt[3]{25-17}\\ &= \sqrt[3]{8}\\ &= 2\end{aligned}$$

14. $\lim_{x\to 2}\sqrt{x^2+5} = \sqrt{2^2+5} = \sqrt{4+5} = \sqrt{9} = 3$

15. $\lim_{x\to \frac{\pi}{4}}(x + sin\ x) = \frac{\pi}{4} + sin\ \frac{\pi}{4} = \frac{\pi}{4} + \frac{1}{\sqrt{2}}$

16. $\lim_{x\to \frac{\pi}{6}}(cos\ x + tan\ x) = \frac{\sqrt{3}}{2} + \frac{1}{\sqrt{3}}$

17. $\lim_{x\to 0}\frac{1+sin\ x}{1-sin\ x} = \frac{1+0}{1-0} = 1$

18. $\lim_{x\to 0}\frac{1+cos\ x}{cos\ x} = \frac{1+1}{1} = 2$

19. Using the graph of $\frac{1}{x-2}$ we find that the limit as x approaches 2 does not exist since the limit from the left of $x = 2$ does not equal the limit from the right of $x = 2$

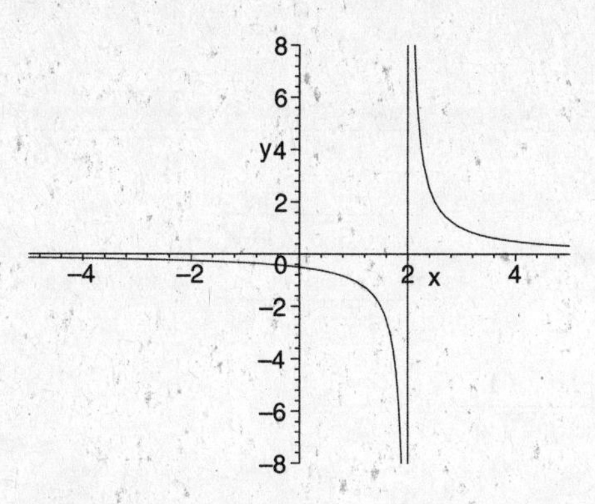

20. The limit to the left of $x = 1$ is ∞ and the limit to the right of $x = 1$ is ∞. Thus, $\lim_{x\to 1} \frac{1}{(x-1)^2}$ does not exist

21. Since $\frac{3x^2-4x+2}{7x^2-5x+3}$ is continuous at $x = 2$ we can use direct substitution

$$\begin{aligned}\lim_{x\to 2} \frac{3x^2-4x+2}{7x^2-5x+3} &= \frac{3(2)^2-4(2)+2}{7(2)^2-5(2)+3} \\ &= \frac{12-8+2}{28-10+3} \\ &= \frac{6}{21} \\ &= \frac{2}{7}\end{aligned}$$

22.

$$\begin{aligned}\lim_{x\to -1} \frac{4x^2+5x-7}{3x^2-2x+1} &= \frac{4(-1)^2+5(-1)-7}{3(-1)^2-2(-1)+1} \\ &= \frac{4-5-7}{3+2+1} \\ &= \frac{-8}{6} \\ &= -\frac{4}{3}\end{aligned}$$

23. The function $\frac{x^2+x-6}{x^2-4}$ is discontinuous at $x = 2$. But we can simplify it algebraically first then find the limit as follows

$$\begin{aligned}\lim_{x\to 2} \frac{x^2+x-6}{x^2-4} &= \lim_{x\to 2} \frac{(x-2)(x+3)}{(x-2)(x+2)} \\ &= \lim_{x\to 2} \frac{(x+3)}{(x+2)} \\ &= \frac{2+3}{2+2} \\ &= \frac{5}{4}\end{aligned}$$

24.

$$\begin{aligned}\lim_{x\to 4} \frac{x^2-16}{x^2-x-12} &= \lim_{x\to 4} \frac{(x-4)(x+4)}{(x-4)(x+3)} \\ &= \lim_{x\to 2} \frac{(x+4)}{(x+3)} \\ &= \frac{4+4}{4+3} \\ &= \frac{8}{7}\end{aligned}$$

25. Since we have a limit in terms of h, we can treat x as a constant. To evaluate the limit we can use direct substitution (we have a polynomial in h, which is continuous for all values of h).
$\lim_{h\to 0}(6x^2 + 6xh + 2h^2)$
$= 6x^2 + 6x(0) + 2(0)^2$
$= 6x^2 + 0 + 0$
$= 6x^2$

26. $\lim_{h\to 0}(10x + 5h) = 10x + 5(0) = 10x$

27. Since we have a limit in terms of h, we can treat x as a constant. Since $\frac{-2x-h}{x^2(x+h)^2}$ is continuous at $h = 0$ we can use direct substitution

$$\begin{aligned}\lim_{h\to 0} \frac{-2x-h}{x^2(x+h)^2} &= \frac{-2x-0}{x^2(x+0)^2} \\ &= \frac{-2x}{x^2(x)^2} \\ &= \frac{-2x}{x^4} \\ &= \frac{-2}{x^3}\end{aligned}$$

28. $\lim_{h\to 0} \frac{-5}{x(x+h)} = \frac{-5}{x(x+0)} = \frac{-5}{x^2}$

29. $\lim_{x\to 0} \frac{\tan x}{x} = \lim_{x\to 0} \frac{\sin x}{x} \cos x = 1 \cdot 1 = 1$ Recall that $\lim_{x\to 0} \frac{\sin x}{x} = 1$.

30. $\lim_{x\to 0} x \csc x = \lim_{x\to 0} \frac{x}{\sin x} = \frac{1}{1} = 1$

31. $\lim h \to 0 \frac{\sin x \sin h}{h} = \sin x \lim_{h\to 0} \frac{\sin h}{h}$
$= \sin x \cdot 1 = \sin x$

32. $\lim_{h\to 0} \frac{\sin x(\cos h - 1)}{h} = \sin x \lim_{h\to 0} \frac{\cos h - 1}{h} = \sin x \cdot 0 = 0$

33.

$$\begin{aligned}\lim_{x\to 0} \frac{x^2+3x}{x-2x^4} &= \lim_{x\to 0} \frac{x(x+3)}{x(1-2x^3)} \\ &= \lim_{x\to 0} \frac{(x+3)}{(1-2x)} \\ &= \frac{(0+3)}{(1-0)} \\ &= \frac{3}{1} = 3\end{aligned}$$

34. $\lim_{x\to 0} \frac{x^2-2x}{x^2+3x} = \lim_{x\to 0} \frac{x(1-2x^2)}{(x+3)} = \frac{0}{3} = 0$

35.

$$\begin{aligned}\lim_{x\to 0} \frac{x\sqrt{x}}{x+x^2} &= \lim_{x\to 0} \frac{x\sqrt{x}}{x(1+x)} \\ &= \lim_{x\to 0} \frac{\sqrt{x}}{(1+x)} \\ &= \frac{\sqrt{0}}{(1+0)} = 0\end{aligned}$$

36. $\lim_{x\to 0}\frac{x+x^2}{x\sqrt{x}}=\lim_{x\to 0}\frac{x(1+x)}{x\sqrt{x}}$
$=\lim_{x\to 0}\frac{1+x}{\sqrt{x}}=\frac{1}{\sqrt{x}}+\sqrt{x}$ The limit does not exist

37.

$$\begin{aligned}\lim_{x\to 2}\frac{x-2}{x^2-x-2} &= \lim_{x\to 2}\frac{x-2}{(x-2)(x+1)}\\ &= \lim_{x\to 2}\frac{1}{(x+1)}\\ &= \frac{1}{(2+1)}=\frac{1}{3}\end{aligned}$$

38.

$$\begin{aligned}\lim_{x\to -1}\frac{x^2-1}{x+1} &= \lim_{x\to 1}\frac{(x-1)(x+1)}{x+1}\\ &= \lim_{x\to -1}(x-1)\\ &= (-1-1)=-2\end{aligned}$$

39.

$$\begin{aligned}\lim_{x\to 3}\frac{x^2-9}{2x-6} &= \lim_{x\to 3}\frac{(x-3)(x+3)}{2(x-3)}\\ &= \lim_{x\to 3}\frac{(x+3)}{2}\\ &= \frac{(3+3)}{2}\\ &= \frac{6}{2}=3\end{aligned}$$

40.

$$\begin{aligned}\lim_{x\to -2}\frac{3x^2+5x-2}{x^2-3x-10} &= \lim_{x\to -2}\frac{(x+2)(3x-1)}{(x+2)(x-5)}\\ &= \lim_{x\to -2}\frac{(3x-1)}{(x-5)}\\ &= \frac{(3(-2)-1)}{(-2-5)}\\ &= \frac{-7}{-7}=1\end{aligned}$$

41.

$$\begin{aligned}\frac{a^2-4}{\sqrt{a^2+5}-3} &= \frac{a^2-4}{\sqrt{a^2+5}-3}\cdot\frac{\sqrt{a^2+5}+3}{\sqrt{a^2+5}+3}\\ &= \frac{(a^2-4)(\sqrt{a^2+5}+3)}{(a^2+5-9)}\\ &= \frac{(a^2-4)(\sqrt{a^2+5}+3)}{(a^2-4)}\\ &= \sqrt{a^2+5}+3\end{aligned}$$

Thus, $\lim_{a\to -2}(\sqrt{a^2+5}+3)=6$

42.

$$\begin{aligned}\frac{\sqrt{x}-1}{x-1} &= \frac{\sqrt{x}-1}{(\sqrt{x}-1)(\sqrt{x}+1)}\\ &= \frac{1}{\sqrt{x}+1}\end{aligned}$$

Thus, $\lim_{x\to 1}\frac{1}{\sqrt{x}+1}=\frac{1}{\sqrt{2}}$

43.

$$\begin{aligned}\frac{\sqrt{3-x}-\sqrt{3}}{x} &= \frac{\sqrt{3-x}-\sqrt{3}}{x}\cdot\frac{\sqrt{3-x}+\sqrt{3}}{\sqrt{3-x}+\sqrt{3}}\\ &= \frac{3-x-3}{x(\sqrt{3-x}+\sqrt{3})}\\ &= \frac{-1}{\sqrt{3-x}+\sqrt{3}}\end{aligned}$$

Thus, $\lim_{x\to 0}\frac{-1}{\sqrt{3-x}+\sqrt{3}}=\frac{-1}{2\sqrt{3}}$

44.

$$\begin{aligned}\frac{\sqrt{4+x}-\sqrt{4-x}}{x} &= \frac{\sqrt{4+x}-\sqrt{4-x}}{x}\cdot\frac{\sqrt{4+x}+\sqrt{4-x}}{\sqrt{4+x}+\sqrt{4-x}}\\ &= \frac{4+x-(4-x)}{x(\sqrt{4+x}+\sqrt{4-x})}\\ &= \frac{2}{\sqrt{4+x}+\sqrt{4-x}}\end{aligned}$$

Thus, $\lim_{x\to 0}\frac{2}{\sqrt{4+x}+\sqrt{4-x}}=\frac{1}{2}$

45. Limit approaches $\frac{3}{4}$

46.

$$\begin{aligned}\frac{\sqrt{7+2x}-\sqrt{7}}{x} &= \frac{\sqrt{7+2x}-\sqrt{7}}{x}\cdot\frac{\sqrt{7+2x}+\sqrt{7}}{\sqrt{7+2x}+\sqrt{7}}\\ &= \frac{7+2x-7}{x(\sqrt{7=2x}+\sqrt{7})}\\ &= \frac{2}{\sqrt{3-x}+\sqrt{3}}\end{aligned}$$

Thus, $\lim_{x\to 0}\frac{2}{\sqrt{7+2x}+\sqrt{7}}=\frac{1}{\sqrt{7}}$

47.

$$\begin{aligned}\frac{2-\sqrt{x}}{4-x} &= \frac{2-\sqrt{x}}{4-x}\cdot\frac{2+\sqrt{x}}{2+\sqrt{x}}\\ &= \frac{4-x}{(4-x)(2+\sqrt{x})}\\ &= \frac{1}{2+\sqrt{x}}\end{aligned}$$

Thus, $\lim_{x\to 4}\frac{1}{2+\sqrt{x}}=\frac{1}{4}$

48.

$$\begin{aligned}\frac{7-\sqrt{49-x^2}}{x} &= \frac{7-\sqrt{49-x^2}}{x}\cdot\frac{7+\sqrt{49-x^2}}{7+\sqrt{49-x^2}}\\ &= \frac{49-(49-x^2)}{x(7+\sqrt{49-x^2})}\\ &= \frac{x}{7+\sqrt{49-x^2}}\end{aligned}$$

Thus, $\lim_{x\to 0}\frac{x}{7+\sqrt{49-x^2}}=0$

Exercise Set 2.3

1. **a)** First we obtain the expression for $f(x+h)$ with $f(x) = 7x^2$

$$\begin{aligned} f(x+h) &= 7(x+h)^2 \\ &= 7(x^2+2xh+h^2) \\ &= 7x^2+14xh+7h^2 \end{aligned}$$

Then

$$\begin{aligned} \frac{f(x+h)-f(x)}{h} &= \frac{(7x^2+14xh+7h^2)-7x^2}{h} \\ &= \frac{14xh+7h^2}{h} \\ &= \frac{h(14x+7h)}{h} \\ &= 14x+7h \end{aligned}$$

b) For $x = 4$ and $h = 2$,

$$14x+7h = 14(4)+7(2) = 56+14 = 70$$

For $x = 4$ and $h = 1$,

$$14x+7h = 14(4)+7(1) = 56+7 = 63$$

For $x = 4$ and $h = 0.1$,

$$14x+7h = 14(4)+7(0.1) = 56+0.7 = 56.7$$

For $x = 4$ and $h = 0.01$,

$$14x+7h = 14(4)+7(0.01) = 56+0.07 = 56.07$$

2. **a)**

$$\begin{aligned} \frac{f(x+h)-f(x)}{h} &= \frac{(5x^2+10xh+5h^2)-5x^2}{h} \\ &= \frac{10xh+5h^2}{h} \\ &= \frac{h(10x+5h)}{h} \\ &= 10x+5h \end{aligned}$$

b) For $x = 4$ and $h = 2$,

$$10x+5h = 10(4)+5(2) = 40+10 = 50$$

For $x = 4$ and $h = 1$,

$$10x+5h = 10(4)+5(1) = 40+5 = 45$$

For $x = 4$ and $h = 0.1$,

$$10x+5h = 10(4)+5(0.1) = 40+0.5 = 40.5$$

For $x = 4$ and $h = 0.01$,

$$10x+5h = 10(4)+5(0.01) = 40+0.05 = 40.05$$

3. **a)** First we obtain the expression for $f(x+h)$ with $f(x) = -7x^2$

$$\begin{aligned} f(x+h) &= -7(x+h)^2 \\ &= -7(x^2+2xh+h^2) \\ &= -7x^2-14xh-7h^2 \end{aligned}$$

Then

$$\begin{aligned} \frac{f(x+h)-f(x)}{h} &= \frac{(-7x^2-14xh-7h^2)-(-7x^2)}{h} \\ &= \frac{-14xh-7h^2}{h} \\ &= \frac{h(-14x-7h)}{h} \\ &= -14x-7h \end{aligned}$$

b) For $x = 4$ and $h = 2$,

$$-14x-7h = -14(4)-7(2) = -56-14 = -70$$

For $x = 4$ and $h = 1$,

$$-14x-7h = -14(4)-7(1) = -56-7 = -63$$

For $x = 4$ and $h = 0.1$,

$$-14x-7h = -14(4)-7(0.1) = -56-0.7 = -56.7$$

For $x = 4$ and $h = 0.01$,

$$-14x-7h = -14(4)-7(0.01) = -56-0.07 = -56.07$$

4. **a)**

$$\begin{aligned} \frac{f(x+h)-f(x)}{h} &= \frac{(-5x^2-10xh-5h^2)-(-5x^2)}{h} \\ &= \frac{-10xh-5h^2}{h} \\ &= \frac{h(-10x-5h)}{h} \\ &= -10x-5h \end{aligned}$$

b) For $x = 4$ and $h = 2$,

$$-10x-5h = -10(4)-5(2) = -40-10 = -50$$

For $x = 4$ and $h = 1$,

$$-10x-5h = -10(4)-5(1) = -40-5 = -45$$

For $x = 4$ and $h = 0.1$,

$$-10x-5h = -10(4)-5(0.1) = -40-0.5 = -40.5$$

For $x = 4$ and $h = 0.01$,

$$-10x-5h = -10(4)-5(0.01) = -40-0.05 = -40.05$$

5. **a)** First we obtain the expression for $f(x+h)$ with $f(x) = 7x^3$

$$\begin{aligned} f(x+h) &= 7(x+h)^3 \\ &= 7(x^3+3x^2h+3xh^2+h^3) \\ &= 7x^3+21x^2h+21xh^2+7h^3 \end{aligned}$$

Then

$$\begin{aligned}\frac{f(x+h)-f(x)}{h} &= \frac{(7x^3+21x^2h+21xh^2+7h^3)-7x^3}{h}\\ &= \frac{21x^2h+21xh^2+7h^3}{h}\\ &= \frac{h(21x^2+21xh+7h^2)}{h}\\ &= 21x^2+21xh+7h^2\end{aligned}$$

b) For $x=4$ and $h=2$,

$$\begin{aligned}21x^2+21xh+7h^2 &= 21(4)^2+21(4)(2)+7(2)^2\\ &= 336+168+28\\ &= 532\end{aligned}$$

For $x=4$ and $h=1$,

$$\begin{aligned}21x^2+21xh+7h^2 &= 21(4)^2+21(4)(1)+7(1)^2\\ &= 336+84+7\\ &= 427\end{aligned}$$

For $x=4$ and $h=0.1$,

$$\begin{aligned}21x^2+21xh+7h^2 &= 21(4)^2+21(4)(0.1)^2\\ &= 336+8.4+0.07\\ &= 344.47\end{aligned}$$

For $x=4$ and $h=0.01$,

$$\begin{aligned}21x^2+21xh+7h^2 &= 21(4)^2+21(4)(0.01)+7(0.01)^2\\ &= 336+0.84+0.0007\\ &= 336.8407\end{aligned}$$

6. a)

$$\begin{aligned}\frac{f(x+h)-f(x)}{h} &= \frac{(5x^3+15x^2h+15xh^2+5h^3)-5x^3}{h}\\ &= \frac{15x^2h+15xh^2+5h^3}{h}\\ &= \frac{h(15x^2+15xh+5h^2)}{h}\\ &= 15x^2+15xh+5h^2\end{aligned}$$

b) For $x=4$ and $h=2$,

$$\begin{aligned}15x^2+15xh+5h^2 &= 15(4)^2+15(4)(2)+5(2)^2\\ &= 240+120+20\\ &= 380\end{aligned}$$

For $x=4$ and $h=1$,

$$\begin{aligned}15x^2+15xh+5h^2 &= 15(4)^2+15(4)(1)+5(1)^2\\ &= 240+60+5\\ &= 305\end{aligned}$$

For $x=4$ and $h=0.1$,

$$\begin{aligned}15x^2+15xh+5h^2 &= 15(4)^2+15(4)(0.1)+5(0.1)^2\\ &= 240+6+0.05\\ &= 246.05\end{aligned}$$

For $x=4$ and $h=0.01$,

$$\begin{aligned}15x^2+15xh+5h^2 &= 15(4)^2+15(4)(0.01)+5(0.01)^2\\ &= 240+.6+0.0005\\ &= 240.6005\end{aligned}$$

7. a) First we obtain the expression for $f(x+h)$ with $f(x)=\frac{5}{x}$

$$f(x+h) = \frac{5}{(x+h)}$$

Then

$$\begin{aligned}\frac{f(x+h)-f(x)}{h} &= \frac{\frac{5}{(x+h)}-\frac{5}{x}}{h}\\ &= \frac{\frac{5}{(x+h)}\cdot x(x+h)-\frac{5}{x}\cdot x(x+h)}{\frac{h}{1}\cdot x(x+h)}\\ &= \frac{5x-5(x+h)}{hx(x+h)}\\ &= \frac{-5h}{hx(x+h)}\\ &= \frac{-5}{x(x+h)}\end{aligned}$$

b) For $x=4$ and $h=2$,

$$\frac{-5}{x(x+h)} = \frac{-5}{4(4+2)} = \frac{-5}{26} \approx -0.208$$

For $x=4$ and $h=1$,

$$\frac{-5}{x(x+h)} = \frac{-5}{4(4+1)} = \frac{-5}{20} \approx -0.25$$

For $x=4$ and $h=0.1$,

$$\frac{-5}{x(x+h)} = \frac{-5}{4(4+0.1)} = \frac{-5}{16.4} \approx -0.305$$

For $x=4$ and $h=0.01$,

$$\frac{-5}{x(x+h)} = \frac{-5}{4(4+0.01)} = \frac{-5}{16.04} \approx -0.312$$

8. a)

$$\begin{aligned}\frac{f(x+h)-f(x)}{h} &= \frac{\frac{4}{(x+h)}-\frac{4}{x}}{h}\\ &= \frac{\frac{4}{(x+h)}\cdot x(x+h)-\frac{4}{x}\cdot x(x+h)}{\frac{h}{1}\cdot x(x+h)}\\ &= \frac{4x-4(x+h)}{hx(x+h)}\\ &= \frac{-4h}{hx(x+h)}\\ &= \frac{-4}{x(x+h)}\end{aligned}$$

b) For $x = 4$ and $h = 2$,

$$\frac{-4}{x(x+h)} = \frac{-4}{4(4+2)} = \frac{-1}{6} \approx -0.167$$

For $x = 4$ and $h = 1$,

$$\frac{-4}{x(x+h)} = \frac{-4}{4(4+1)} = \frac{-1}{5} \approx -0.2$$

For $x = 4$ and $h = 0.1$,

$$\frac{-4}{x(x+h)} = \frac{-4}{4(4+0.1)} = \frac{-1}{4.1} \approx -0.244$$

For $x = 4$ and $h = 0.01$,

$$\frac{-4}{x(x+h)} = \frac{-4}{4(4+0.01)} = \frac{-4}{4.04} \approx -0.249$$

9. **a)** First we obtain the expression for $f(x+h)$ with $f(x) = -2x+5$

$$\begin{aligned} f(x+h) &= -2(x+h)+5 \\ &= -2x-2h+5 \end{aligned}$$

Then

$$\begin{aligned} \frac{f(x+h)-f(x)}{h} &= \frac{(-2x-2h+5)-(-2x+5)}{h} \\ &= \frac{-2h}{h} \\ &= -2 \end{aligned}$$

b) Since the difference quotient is a constant, then the value of the difference quotient will be -2 for all the values of x and h.

10. **a)**

$$\begin{aligned} \frac{f(x+h)-f(x)}{h} &= \frac{(2x+2h+3)-(2x+3)}{h} \\ &= \frac{2h}{h} \\ &= 2 \end{aligned}$$

b) Since the difference quotient is a constant, then the value of the difference quotient will be 2 for all the values of x and h.

11. **a)** First we obtain the expression for $f(x+h)$ with $f(x) = x^2 - x$

$$\begin{aligned} f(x+h) &= (x+h)^2 - (x+h) \\ &= x^2 = 2xh + h^2 - x - h \end{aligned}$$

Then

$$\begin{aligned} \frac{f(x+h)-f(x)}{h} &= \frac{(x^2+2xh+h^2-x-h)-(x^2-x)}{h} \\ &= \frac{2xh+h^2-h}{h} \\ &= \frac{h(2x+h-1)}{h} \\ &= 2x+h-1 \end{aligned}$$

b) For $x = 4$ and $h = 2$,

$$2x+h-1 = 2(4)+2-1 = 8+2-1 = 9$$

For $x = 4$ and $h = 1$,

$$2x+h-1 = 2(4)+2-1 = 8+1-1 = 8$$

For $x = 4$ and $h = 0.1$,

$$2x+h-1 = 2(4)+0.1-1 = 8+0.1-1 = 7.1$$

For $x = 4$ and $h = 0.01$,

$$2x+h-1 = 2(4)+2-1 = 8+0.01-1 = 7.01$$

12. **a)**

$$\begin{aligned} \frac{f(x+h)-f(x)}{h} &= \frac{(x^2+2xh+h^2+x+h)-(x^2+x)}{h} \\ &= \frac{2xh+h^2+h}{h} \\ &= \frac{h(2x+h+1)}{h} \\ &= 2x+h-1 \end{aligned}$$

b) For $x = 4$ and $h = 2$,

$$2x+h+1 = 2(4)+2+1 = 8+2+1 = 11$$

For $x = 4$ and $h = 1$,

$$2x+h+1 = 2(4)+2+1 = 8+1+1 = 10$$

For $x = 4$ and $h = 0.1$,

$$2x+h+1 = 2(4)+0.1+1 = 8+0.1+1 = 9.1$$

For $x = 4$ and $h = 0.01$,

$$2x+h+1 = 2(4)+2+1 = 8+0.01+1 = 9.01$$

13. **a)** For the average growth rate during the first year we use the points $(0, 7.9)$ and $(12, 22.4)$

$$\begin{aligned} \frac{y_2-y_1}{x_2-x_1} &= \frac{22.4-7.9}{12-0} \\ &= \frac{14.5}{12} \\ &\approx 1.20834 \text{ pounds per month} \end{aligned}$$

b) For the average growth rate during the second year we use the points $(12, 22.4)$ and $(24, 27.8)$

$$\begin{aligned} \frac{y_2-y_1}{x_2-x_1} &= \frac{27.8-22.4}{24-12} \\ &= \frac{5.4}{12} \\ &= 0.45 \text{ pounds per month} \end{aligned}$$

c) For the average growth rate during the third year we use the points $(24, 27.8)$ and $(36, 31.5)$

$$\begin{aligned} \frac{y_2-y_1}{x_2-x_1} &= \frac{31.5-27.8}{36-24} \\ &= \frac{3.7}{12} \\ &\approx 0.30834 \text{ pounds per month} \end{aligned}$$

d) For the average growth rate during his first three years we use the points $(0, 7.9)$ and $(36, 31.5)$

$$\begin{aligned}\frac{y_2 - y_1}{x_2 - x_1} &= \frac{31.5 - 7.9}{36 - 0} \\ &= \frac{23.6}{36} \\ &\approx 1.967 \text{ pounds per month}\end{aligned}$$

e) The graph indicates that the highest growth rate out of the first three years of a boy's life happens at birth (that is were the graph is the steepest).

14. **a)** $\frac{20.5-7.9}{9-0} = 1.4$ pounds per month

b) $\frac{17.4-7.9}{6-0} \approx 1.583$ pounds per month

c) $\frac{13.2-7.9}{3-0} \approx 1.767$ pounds per month

d) Based on the the calculated average growth rates above, we can estimate the average growth rate to be larger than 1.767 for the first few weeks of a typical boy's life.

15. **a)** For the average growth rate between ages 12 and 18 months

$$\begin{aligned}\frac{y_2 - y_1}{x_2 - x_1} &= \frac{25.9 - 22.4}{18 - 12} \\ &= \frac{3.5}{6} \\ &\approx 0.583 \text{ pounds per month}\end{aligned}$$

b) For the average growth rate between ages 12 and 14 (we use the point at 15 months)

$$\begin{aligned}\frac{y_2 - y_1}{x_2 - x_1} &= \frac{24.5 - 22.4}{15 - 12} \\ &= \frac{2.1}{3} \\ &= 0.7 \text{ pounds per month}\end{aligned}$$

c) For the average growth rate between ages 12 and 13 (we can approximate the value of y when $x = 13$ by reading it from the graph)

$$\begin{aligned}\frac{y_2 - y_1}{x_2 - x_1} &\approx \frac{23.23 - 22.4}{15 - 12} \\ &\approx \frac{0.83}{1} \\ &\approx 0.83 \text{ pounds per month}\end{aligned}$$

d) The average growth of a typical boy when he is 12 months old is about 0.9 pounds per month

16. **a)** $\frac{98.6-98.6}{10-1} = 0$ degrees per day. Using this rate of change we would not be able to conclude the person was sick.

b) $\frac{99.5-98.6}{2-1} = 0.9$ degrees per day
$\frac{101-99.5}{3-2} = 1.5$ degrees per day
$\frac{102-101}{4-3} = 1$ degrees per day
$\frac{102.5-102}{5-4} = 0.5$ degrees per day
$\frac{102.5-102.5}{6-5} = 0$ degrees per day
$\frac{102.4-102.5}{7-6} = -0.1$ degrees per day
$\frac{102-102.4}{8-7} = -0.4$ degrees per day
$\frac{100-102}{9-8} = -2$ degrees per day
$\frac{98.6-100}{10-9} = -1.4$ degrees per day
$\frac{98.6-98.6}{11-10} = 0$ degrees per day

c) The temperature began to rise on day 1, reached the peak on day 5, began to subside on day 6, and was back to normal on day 10

d) By examining the graph, we notice an increase in temperaure after day 1 which reaches a maximum value of 102.5^o on day 5 and continues at the max temp through day 6 at which time the temperature begins to drop until it reaches the normal level of 98.6^o on day 10.

17. **a)** Average rate of change from $t = 0$ to $t = 8$

$$\begin{aligned}\frac{N_2 - N_1}{t_2 - t_1} &= \frac{10 - 0}{8 - 0} \\ &= \frac{10}{8} = 1.25 \text{ words per minute}\end{aligned}$$

Average rate of change from $t = 8$ to $t = 16$

$$\begin{aligned}\frac{N_2 - N_1}{t_2 - t_1} &= \frac{20 - 10}{16 - 8} \\ &= \frac{10}{8} = 1.25 \text{ words per minute}\end{aligned}$$

Average rate of change from $t = 16$ to $t = 24$

$$\begin{aligned}\frac{N_2 - N_1}{t_2 - t_1} &= \frac{25 - 20}{24 - 16} \\ &= \frac{5}{8} = 0.625 \text{ words per minute}\end{aligned}$$

Average rate of change from $t = 24$ to $t = 32$

$$\begin{aligned}\frac{N_2 - N_1}{t_2 - t_1} &= \frac{25 - 25}{32 - 24} \\ &= \frac{0}{8} = 0 \text{ words per minute}\end{aligned}$$

Average rate of change from $t = 32$ to $t = 36$

$$\begin{aligned}\frac{N_2 - N_1}{t_2 - t_1} &= \frac{25 - 25}{36 - 32} \\ &= \frac{0}{4} = 0 \text{ words per minute}\end{aligned}$$

b) The rate of change becomes 0 after 24 minutes because the number of words memorized does not change and remains at 25 words, that means that there is no change in the number of words memorized after 24 minutes.

18. **a)** $s(2) = 10(2)^2 = 10(4) = 40$ miles.
$s(5) = 10(5)^2 = 10(25) = 250$ miles

b) $s(5) - s(2) = 250 - 40 = 210$ miles. This represents the distance travels between $t = 2$ and $t = 5$ seconds.

c) $\frac{250-40}{5-2} = \frac{210}{3} = 70$ miles per hour

19. **a)** When $t = 3$, $s = 16(3)^2 = 16(9) = 144$ feet

b) When $t = 5$, $s = 16(5)^2 = 16(25) = 400$ feet

c) Average velocity $= \frac{400-144}{5-3} = \frac{256}{2} = 128$ feet per second

20. a) $\frac{30970-30680}{20} = 14.5$ miles per gallon
b) $\frac{30970-30680}{20} = 14.5$ miles per gallon

21. a) Population A: The average growth rate $= \frac{500-0}{4-0} = \frac{500}{4} = 125$ million per year
Population B: The average growth rate $= \frac{500-0}{4-0} =$ 125 million per year

b) We would not detect the fact that the population grow at different rates. The calculation shows the populations growing at the same average growth rate, since for either population we used the same points to calculate the average growth rate $(0, 0)$, and $(4, 500)$.

c) Population A:
Between $t = 0$ and $t = 1$, Average Growth Rate $= \frac{290-0}{1-0} = 290$ million people per year
Between $t = 1$ and $t = 2$, Average Growth Rate $= \frac{250-290}{2-1} = -40$ million people per year
Between $t = 2$ and $t = 3$, Average Growth Rate $= \frac{200-250}{3-2} = -50$ million people per year
Between $t = 3$ and $t = 4$, Average Growth Rate $= \frac{500-200}{4-3} = 300$ million people per year

Population B:
Between $t = 0$ and $t = 1$, Average Growth Rate $= \frac{125-0}{1-0} = 125$ million people per year
Between $t = 1$ and $t = 2$, Average Growth Rate $= \frac{250-125}{2-1} = 125$ million people per year
Between $t = 2$ and $t = 3$, Average Growth Rate $= \frac{375-250}{3-2} = 125$ million people per year
Between $t = 3$ and $t = 4$, Average Growth Rate $= \frac{400-375}{4-3} = 125$ million people per year

d) It is clear from part (c) that the first population has different growing rates depending on which interval of time we choose. Therefore, the statement "the population grew by 125 million each year" does not convey how population went through periods were the population increased and periods were the population decreased.

22. In the period between 1850 and 1860 the deer population is decreasing which corresponds to a negative rate of change. In the period between 1890 and 1960 the deer population is increasing which corresponds to a positive rate of change.

23. The rate of change in the period between 1800 and 1850 is similar to that of 1930 to 1950 is the sense that they both exhibit steady increase in the population. The drastic drop in the population shortly after 1850 is similar to the drop in population seen near 1975.

24.

$$\begin{aligned}\frac{f(x+h)-f(x)}{h} &= \frac{m(x+h)+b-(mx+b)}{h}\\ &= \frac{mx+mh+b-mx-b}{h}\\ &= \frac{mh}{h}\\ &= m\end{aligned}$$

25.

$$\begin{aligned}\frac{f(x+h)-f(x)}{h} &= \frac{a(x+h)^2+b(x+h)+c}{h} - \\ &\quad \frac{(ax^2+bx+c)}{h}\\ &= \frac{ax^2+2axh+ah^2+bx+bh+c}{h}\\ &\quad \frac{-ax^2+bx+c}{h}\\ &= \frac{2axh+ah^2+bh}{h}\\ &= 2ax+ah+b\\ &= a(2x+h)+b\end{aligned}$$

26.

$$\begin{aligned}\frac{f(x+h)-f(x)}{h} &= \frac{a(x+h)^3+b(x+h)^2-(ax^3+bx^2)}{h}\\ &= \frac{ax^3+3ax^2h+3axh^2+h^3}{h} + \\ &\quad \frac{bx^2+2bxh+bh^2-ax^3-bx^2}{h}\\ &= \frac{3ax^2h+3axh^2+h^3+2bxh+bh^2}{h}\\ &= 3ax^2+3axh+h^2+2bx+bh\\ &= (3ax^2+2bx)+h(3ax+b)\end{aligned}$$

27.

$$\begin{aligned}\frac{f(x+h)-f(x)}{h} &= \frac{\sqrt{x+h}-\sqrt{x}}{h}\\ &= \frac{\sqrt{x+h}-\sqrt{x}}{h}\cdot\frac{\sqrt{x+h}+\sqrt{x}}{\sqrt{x+h}+\sqrt{x}}\\ &= \frac{x+h-x}{h\sqrt{x+h}+\sqrt{x}}\\ &= \frac{1}{\sqrt{x+h}+\sqrt{x}}\end{aligned}$$

28.

$$\begin{aligned}\frac{f(x+h)-f(x)}{h} &= \frac{(x+h)^4-x^4}{h}\\ &= \frac{x^4+4x^3h+6x^2h^2+4xh^3+h^4-x^4}{h}\\ &= \frac{4x^3h+6x^2h^2+4xh^3+h^4}{h}\\ &= 4x^3+6x^2h+4xh^2+h^3\end{aligned}$$

29.

$$\begin{aligned}\frac{f(x+h)-f(x)}{h} &= \frac{\frac{1}{(x+h)^2}-\frac{1}{x^2}}{h}\\ &= \frac{x^2-(x+h)^2}{hx^2(x+h)^2}\\ &= \frac{x^2-x^2-2xh-h^2}{hx^2(x+h)^2}\\ &= \frac{-2x-h}{x^2(x+h)^2}\\ &= -\frac{2x+h}{x^2(x+h)^2}\end{aligned}$$

30.

$$\begin{aligned}\frac{f(x+h)-f(x)}{h} &= \frac{\frac{1}{1-(x+h)}-\frac{1}{1-x}}{h}\\ &= \frac{(1-x)-(1-(x+h))}{h(1-x)(1-(x+h))}\\ &= \frac{1-x-1+x+h}{h(1-x)(1-(x+h))}\\ &= \frac{1}{(1-x)(1-(x+h))}\end{aligned}$$

31.

$$\begin{aligned}\frac{f(x+h)-f(x)}{h} &= \frac{\frac{(x+h)}{(1+x+h)}-\frac{x}{1+x}}{h}\\ &= \frac{(x+h)(1+x)-x(1+x+h)}{h(1+x)(1+x+h)}\\ &= \frac{x^2+x+xh+h-x-x^2-xh}{h(1+x)(1+x+h)}\\ &= \frac{1}{(1+x)(1+x+h)}\end{aligned}$$

32.

$$\begin{aligned}\frac{f(x+h)-f(x)}{h} &= \frac{\sqrt{3-2(x+h)}-\sqrt{3-2x}}{h}\\ &= \frac{3-2(x+h)-(3-2x)}{h(\sqrt{3-2(x+h)}+\sqrt{3-2x})}\\ &= \frac{-2}{\sqrt{3-2(x+h)}+\sqrt{3-2x}}\end{aligned}$$

33.

$$\begin{aligned}\frac{f(x+h)-f(x)}{h} &= \frac{\frac{1}{\sqrt{x+h}}-\frac{1}{\sqrt{x}}}{h}\\ &= \frac{\sqrt{x}-\sqrt{x+h}}{h\sqrt{x}\sqrt{x+h}}\\ &= \frac{\sqrt{x}-\sqrt{x+h}}{h\sqrt{x}\sqrt{x+h}}\,\frac{\sqrt{x}+\sqrt{x+h}}{\sqrt{x}+\sqrt{x+h}}\\ &= \frac{x-(x+h)}{h\sqrt{x}\sqrt{x+h}(\sqrt{x}+\sqrt{x+h})}\\ &= \frac{-1}{\sqrt{x}\sqrt{x+h}(\sqrt{x}+\sqrt{x+h})}\end{aligned}$$

34.

$$\begin{aligned}\frac{f(x+h)-f(x)}{h} &= \frac{\frac{2(x+h)}{x+h-1}-\frac{2x}{x-1}}{h}\\ &= \frac{2(x+h)(x-1)-2x(x+h-1)}{h(x-1)(x+h-1)}\\ &= \frac{2x^2-2x+2xh-2h-2x^2-2xh+2x}{h(x-1)(x+h-1)}\\ &= \frac{-2}{(x-1)(x+h-1)}\end{aligned}$$

Exercise Set 2.4

1. a-b) $f(x)=5x^2$

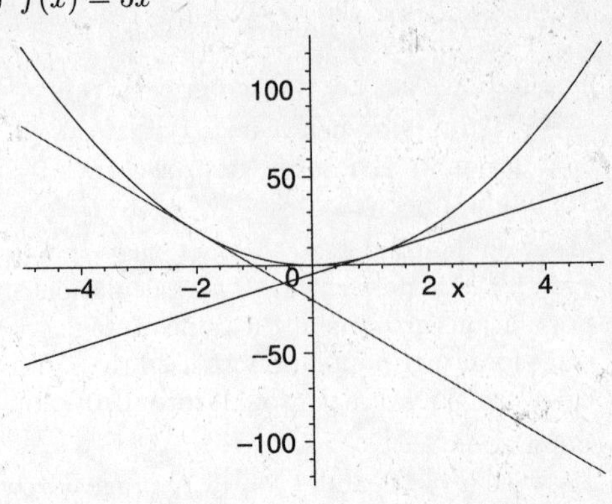

c)

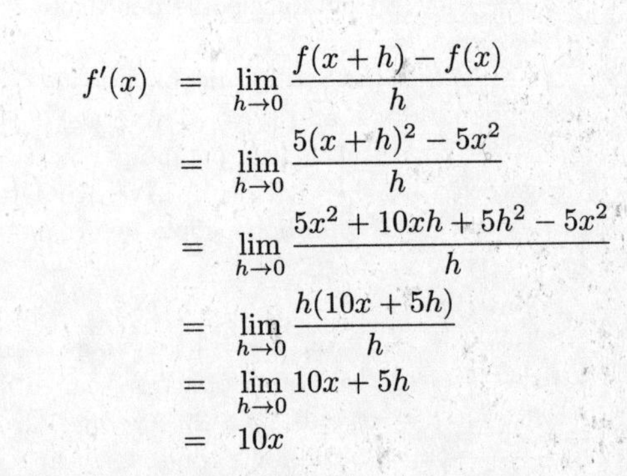

$$\begin{aligned}f'(x) &= \lim_{h\to 0}\frac{f(x+h)-f(x)}{h}\\ &= \lim_{h\to 0}\frac{5(x+h)^2-5x^2}{h}\\ &= \lim_{h\to 0}\frac{5x^2+10xh+5h^2-5x^2}{h}\\ &= \lim_{h\to 0}\frac{h(10x+5h)}{h}\\ &= \lim_{h\to 0}10x+5h\\ &= 10x\end{aligned}$$

d) $f'(-2)=10(-2)=-20$
$f'(0)=10(0)=0$
$f'(1)=10(1)=10$. These slopes are in agreement with the slopes of the tangent lines drawn in part (b).

2. a-b) $f(x)=7x^2$

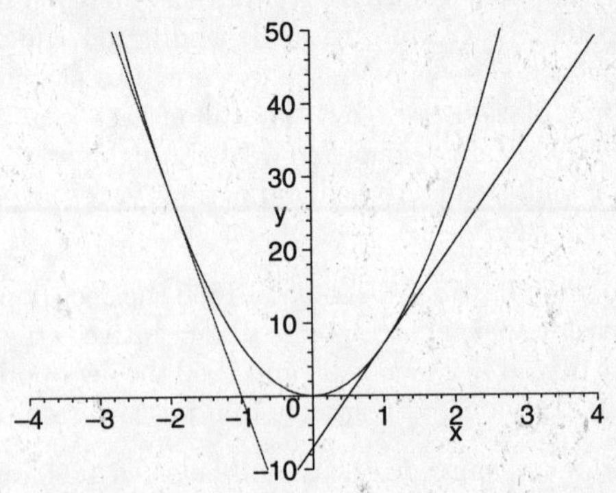

c)

$$\begin{aligned}f'(x) &= \lim_{h\to 0}\frac{f(x+h)-f(x)}{h}\\ &= \lim_{h\to 0}\frac{7(x+h)^2-7x^2}{h}\\ &= \lim_{h\to 0}\frac{7x^2+14xh+7h^2-7x^2}{h}\\ &= \lim_{h\to 0}\frac{h(14x+7h)}{h}\\ &= \lim_{h\to 0}14x+7h\\ &= 14x\end{aligned}$$

d) $f'(-2) = 14(-2) = -28$
$f'(0) = 14(0) = 0$
$f'(1) = 14(1) = 14$. These slopes are in agreement with the slopes of the tangent lines drawn in part (b).

3. a-b) $f(x) = -5x^2$

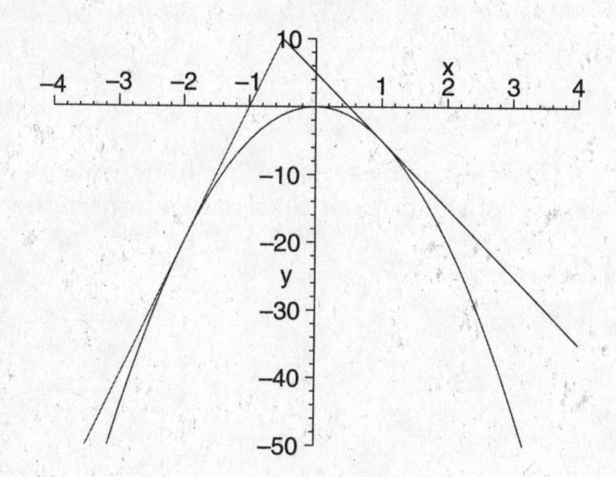

c)

$$\begin{aligned} f'(x) &= \lim_{h\to 0} \frac{f(x+h)-f(x)}{h} \\ &= \lim_{h\to 0} \frac{-5(x+h)^2-(-5x^2)}{h} \\ &= \lim_{h\to 0} \frac{-5x^2-10xh-5h^2-(-5x^2)}{h} \\ &= \lim_{h\to 0} \frac{h(-10x-5h)}{h} \\ &= \lim_{h\to 0} -10x-5h \\ &= -10x \end{aligned}$$

d) $f'(-2) = -10(-2) = 20$
$f'(0) = -10(0) = 0$
$f'(1) = -10(1) = -10$. These slopes are in agreement with the slopes of the tangent lines drawn in part (b).

4. a-b) $f(x) = -7x^2$

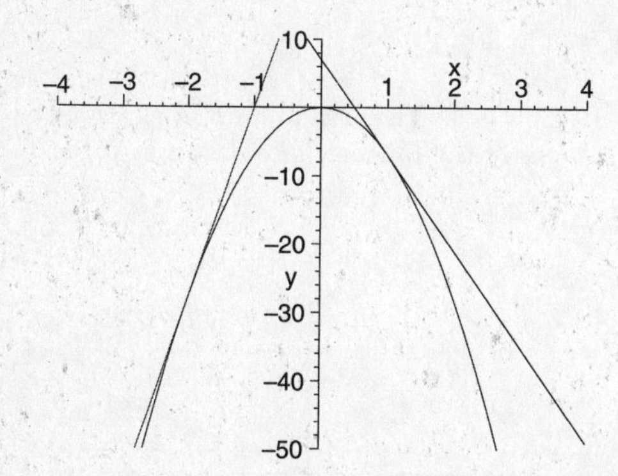

c)

$$\begin{aligned} f'(x) &= \lim_{h\to 0} \frac{f(x+h)-f(x)}{h} \\ &= \lim_{h\to 0} \frac{-7(x+h)^2-(-7x^2)}{h} \\ &= \lim_{h\to 0} \frac{-7x^2-14xh-7h^2-(-7x^2)}{h} \\ &= \lim_{h\to 0} \frac{h(-14x-7h)}{h} \\ &= \lim_{h\to 0} -14x-7h \\ &= -14x \end{aligned}$$

d) $f'(-2) = -14(-2) = 28$
$f'(0) = -14(0) = 0$
$f'(1) = -14(1) = -14$. These slopes are in agreement with the slopes of the tangent lines drawn in part (b).

5. a-b) $f(x) = x^3$

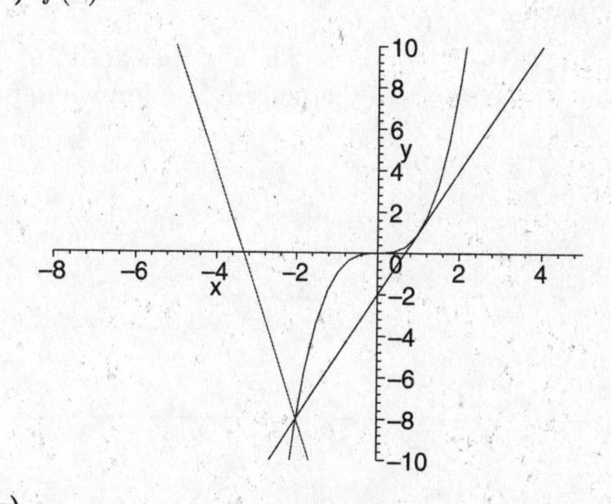

c)

$$\begin{aligned} f'(x) &= \lim_{h\to 0} \frac{f(x+h)-f(x)}{h} \\ &= \lim_{h\to 0} \frac{(x+h)^3-x^3}{h} \\ &= \lim_{h\to 0} \frac{x^3+3x^2h+3xh^2+h^3-x^3}{h} \\ &= \lim_{h\to 0} \frac{h(3x^2+3xh+h^2)}{h} \\ &= \lim_{h\to 0} 3x^2+3xh+h^2 \\ &= 3x^2 \end{aligned}$$

d) $f'(-2) = 3(-2)^2 = 12$
$f'(0) = 3(0)^2 = 0$
$f'(1) = 3(1)^2 = 3$. These slopes are in agreement with the slopes of the tangent lines drawn in part (b).

6. a-b) $f(x) = -x^3$

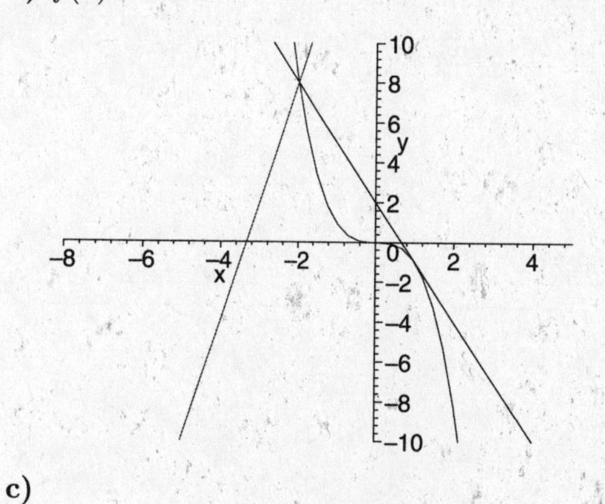

c)

$$\begin{aligned} f'(x) &= \lim_{h\to 0} \frac{f(x+h)-f(x)}{h} \\ &= \lim_{h\to 0} \frac{-(x+h)^3-(-x^3)}{h} \end{aligned}$$

$$= \lim_{h\to 0} \frac{-x^3 - 3x^2h - 3xh^2 - h^3 - (-x^3)}{h}$$
$$= \lim_{h\to 0} \frac{h(-3x^2 - 3xh - h^2)}{h}$$
$$= \lim_{h\to 0} -3x^2 - 3xh - h^2$$
$$= -3x^2$$

d) $f'(-2) = -3(-2)^2 = -12$
$f'(0) = -3(0)^2 = 0$
$f'(1) = -3(1)^2 = -3$. These slopes are in agreement with the slopes of the tangent lines drawn in part (b).

7. a-b) $f(x) = 2x + 3$

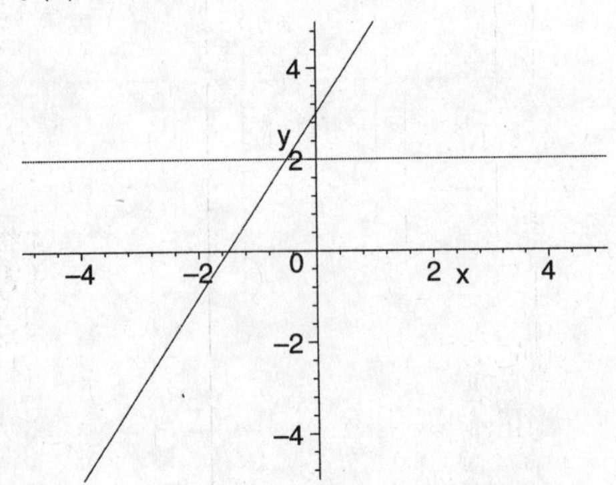

c)

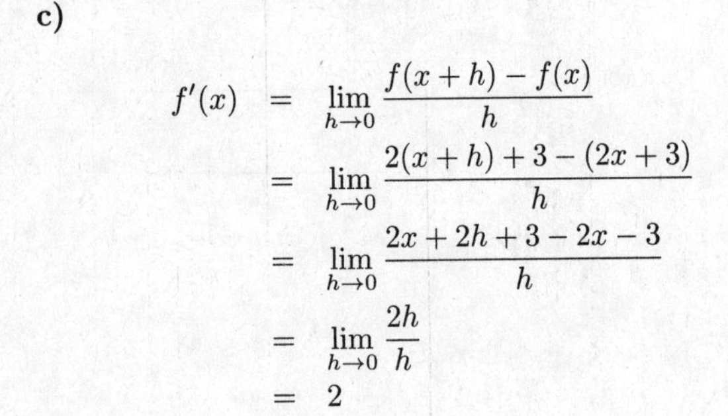

$$f'(x) = \lim_{h\to 0} \frac{f(x+h) - f(x)}{h}$$
$$= \lim_{h\to 0} \frac{2(x+h) + 3 - (2x+3)}{h}$$
$$= \lim_{h\to 0} \frac{2x + 2h + 3 - 2x - 3}{h}$$
$$= \lim_{h\to 0} \frac{2h}{h}$$
$$= 2$$

d) $f'(-2) = 2$
$f'(0) = 2$
$f'(1) = 2$. These slopes are in agreement with the slopes of the tangent lines drawn in part (b).

8. a-b) $f(x) = -2x + 5$

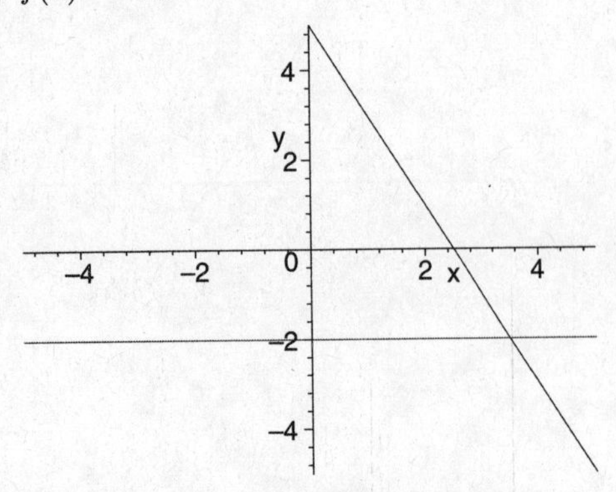

c)

$$f'(x) = \lim_{h\to 0} \frac{f(x+h) - f(x)}{h}$$
$$= \lim_{h\to 0} \frac{-2(x+h) + 5 - (-2x+5)}{h}$$
$$= \lim_{h\to 0} \frac{-2x - 2h + 5 + 2x - 5}{h}$$
$$= \lim_{h\to 0} \frac{-2h}{h}$$
$$= -2$$

d) $f'(-2) = -2$
$f'(0) = -2$
$f'(1) = -2$. These slopes are in agreement with the slopes of the tangent lines drawn in part (b).

9. a-b) $f(x) = -4x$

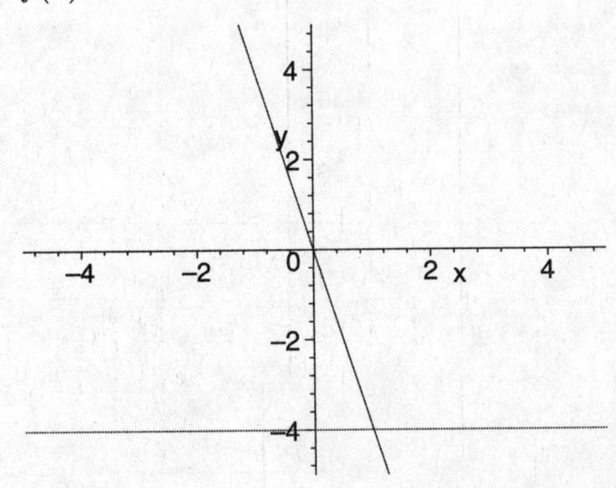

c)

$$f'(x) = \lim_{h\to 0} \frac{f(x+h) - f(x)}{h}$$
$$= \lim_{h\to 0} \frac{-4(x+h) - (-4x)}{h}$$
$$= \lim_{h\to 0} \frac{-4x - 4h + 4x}{h}$$
$$= \lim_{h\to 0} \frac{-4h}{h}$$
$$= -4$$

d) $f'(-2) = -4$
$f'(0) = -4$
$f'(1) = -4$. These slopes are in agreement with the slopes of the tangent lines drawn in part (b).

10. a-b) $f(x) = \frac{1}{2}x$

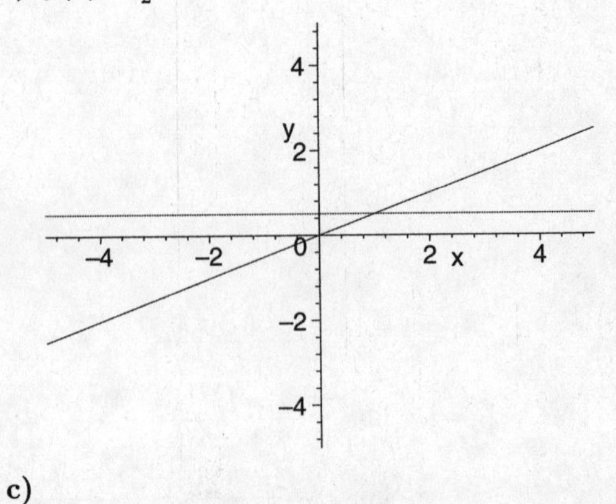

c)

$$f'(x) = \lim_{h\to 0} \frac{f(x+h) - f(x)}{h}$$

$$
\begin{aligned}
&= \lim_{h\to 0} \frac{\frac{1}{2}(x+h) - \frac{1}{2}x}{h} \\
&= \lim_{h\to 0} \frac{\frac{1}{2}x - \frac{1}{2}h - \frac{1}{2}x}{h} \\
&= \lim_{h\to 0} \frac{\frac{1}{2}h}{h} \\
&= \frac{1}{2}
\end{aligned}
$$

d) $f'(-2) = \frac{1}{2}$
$f'(0) = \frac{1}{2}$
$f'(1) = \frac{1}{2}$. These slopes are in agreement with the slopes of the tangent lines drawn in part (b).

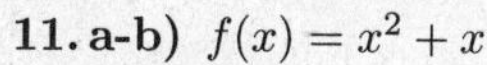

11. a-b) $f(x) = x^2 + x$

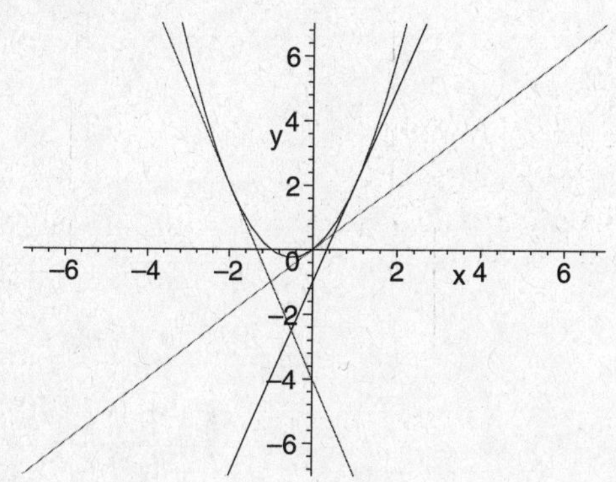

c)

$$
\begin{aligned}
f'(x) &= \lim_{h\to 0} \frac{f(x+h) - f(x)}{h} \\
&= \lim_{h\to 0} \frac{(x+h)^2 + (x+h) - (x^2 + x)}{h} \\
&= \lim_{h\to 0} \frac{x^2 + 2xh + h^2 + x + h - x^2 - x}{h} \\
&= \lim_{h\to 0} \frac{2xh + h^2 + h}{h} \\
&= \lim_{h\to 0} \frac{h(2x + h + 1)}{h} \\
&= 2x + 1
\end{aligned}
$$

d) $f'(-2) = 2(-2) + 1 = -3$
$f'(0) = 2(0) + 1 = 1$
$f'(1) = 2(1) + 1 = 3$. These slopes are in agreement with the slopes of the tangent lines drawn in part (b).

12. a-b) $f(x) = x^2 - x$

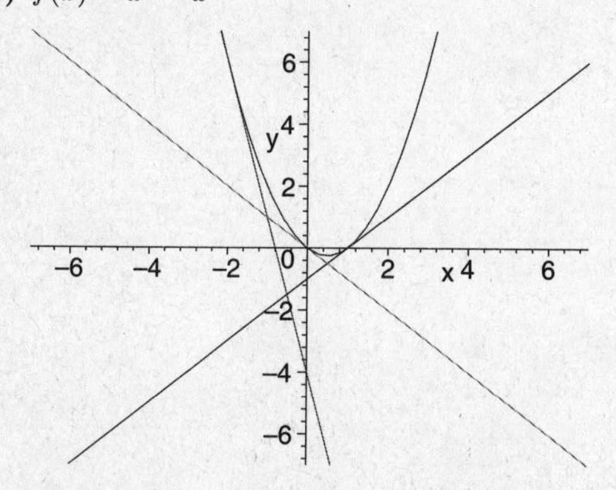

c)

$$
\begin{aligned}
f'(x) &= \lim_{h\to 0} \frac{f(x+h) - f(x)}{h} \\
&= \lim_{h\to 0} \frac{(x+h)^2 - (x+h) - (x^2 - x)}{h} \\
&= \lim_{h\to 0} \frac{x^2 + 2xh + h^2 - x - h - x^2 + x}{h} \\
&= \lim_{h\to 0} \frac{2xh + h^2 - h}{h} \\
&= \lim_{h\to 0} \frac{h(2x + h - 1)}{h} \\
&= 2x - 1
\end{aligned}
$$

d) $f'(-2) = 2(-2) - 1 = -5$
$f'(0) = 2(0) - 1 = -1$
$f'(1) = 2(1) - 1 = 1$. These slopes are in agreement with the slopes of the tangent lines drawn in part (b).

13. a-b) $f(x) = 2x^2 + 3x - 2$

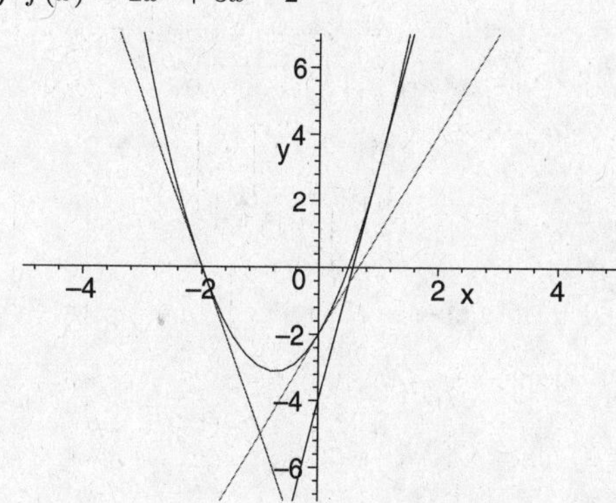

c)

$$
\begin{aligned}
f'(x) &= \lim_{h\to 0} \frac{f(x+h) - f(x)}{h} \\
&= \lim_{h\to 0} \frac{2(x+h)^2 + 3(x+h) - 2 - (2x^2 + 3x - 2)}{h} \\
&= \lim_{h\to 0} \frac{4xh + 2h^2 + 3h}{h} \\
&= \lim_{h\to 0} \frac{h(4x + 3)}{h} \\
&= 4x + 3
\end{aligned}
$$

d) $f'(-2) = 4(-2) + 3 = -5$
$f'(0) = 4(0) + 3 = 3$
$f'(1) = 4(1) + 3 = 7$. These slopes are in agreement with the slopes of the tangent lines drawn in part (b).

14. a-b) $f(x) = 5x^2 - 2x + 7$

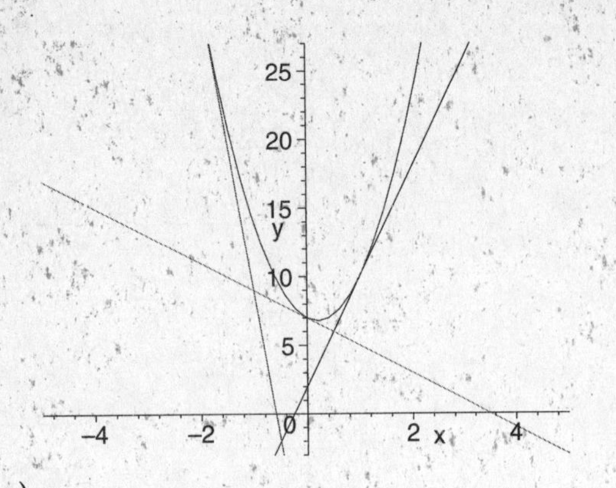

c)

$$\begin{aligned} f'(x) &= \lim_{h\to 0} \frac{f(x+h)-f(x)}{h} \\ &= \lim_{h\to 0} \frac{5(x+h)^2 - 2(x+h) + 7 - (5x^2 - 2x + 7)}{h} \\ &= \lim_{h\to 0} \frac{10xh + 5h^2 - 2h}{h} \\ &= \lim_{h\to 0} \frac{h(10x + 5h - 2)}{h} \\ &= 10x - 2 \end{aligned}$$

d) $f'(-2) = 10(-2) - 2 = -22$
$f'(0) = 10(0) - 2 = -2$
$f'(1) = 10(1) - 2 = 8$. These slopes are in agreement with the slopes of the tangent lines drawn in part (b).

15. a-b) $f(x) = \frac{1}{x}$

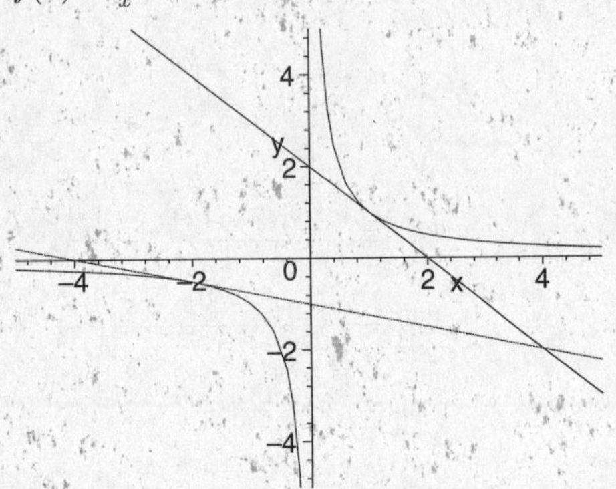

c)

$$\begin{aligned} f'(x) &= \lim_{h\to 0} \frac{f(x+h)-f(x)}{h} \\ &= \lim_{h\to 0} \frac{\frac{1}{(x+h)} - \frac{1}{x}}{h} \\ &= \lim_{h\to 0} \frac{\frac{1}{(x+h)} \cdot x(x+h) - \frac{1}{x} \cdot x(x+h)}{h \cdot x(x+h)} \\ &= \lim_{h\to 0} \frac{x - (x+h)}{hx(x+h)} \\ &= \lim_{h\to 0} \frac{-h)}{hx(x+h)} \\ &= \lim_{h\to 0} \frac{-1}{x(x+h)} \\ &= -\frac{1}{x^2} \end{aligned}$$

d) $f'(-2) = -\frac{1}{(-2)^2} = -\frac{1}{4}$
$f'(0) =$ does not exist
$f'(1) = -\frac{1}{(1)^2} = -1$. These slopes are in agreement with the slopes of the tangent lines drawn in part (b).

16. a-b) $f(x) = \frac{5}{x}$

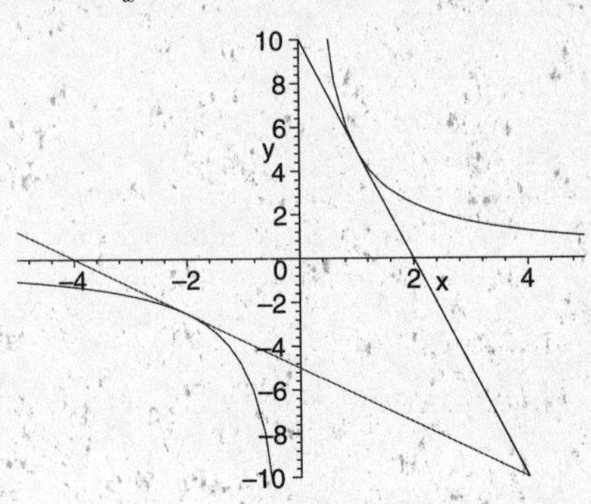

c)

$$\begin{aligned} f'(x) &= \lim_{h\to 0} \frac{f(x+h)-f(x)}{h} \\ &= \lim_{h\to 0} \frac{\frac{5}{(x+h)} - \frac{5}{x}}{h} \\ &= \lim_{h\to 0} \frac{\frac{5}{(x+h)} \cdot x(x+h) - \frac{5}{x} \cdot x(x+h)}{h \cdot x(x+h)} \\ &= \lim_{h\to 0} \frac{5x - 5(x+h)}{hx(x+h)} \\ &= \lim_{h\to 0} \frac{-5h)}{hx(x+h)} \\ &= \lim_{h\to 0} \frac{-5}{x(x+h)} \\ &= -\frac{5}{x^2} \end{aligned}$$

d) $f'(-2) = -\frac{5}{(-2)^2} = -\frac{5}{4}$
$f'(0) =$ does not exist
$f'(1) = -\frac{5}{(1)^2} = -5$. These slopes are in agreement with the slopes of the tangent lines drawn in part (b).

17. $f(x) = mx$

$$\begin{aligned} f'(x) &= \lim_{h\to 0} \frac{f(x+h)-f(x)}{h} \\ &= \lim_{h\to 0} \frac{m(x+h) - mx}{h} \\ &= \lim_{h\to 0} \frac{mx + mh - mx}{h} \\ &= \lim_{h\to 0} \frac{mh}{h} \\ &= m \end{aligned}$$

18. $f(x) = ax^2 + bx + c$

$$\begin{aligned} f'(x) &= \lim_{h\to 0} \frac{f(x+h)-f(x)}{h} \\ &= \lim_{h\to 0} \frac{a(x+h)^2 + b(x+h) + c - (ax^2 + bx + c)}{h} \end{aligned}$$

$$\begin{aligned}
&= \lim_{h\to 0} \frac{ax^2+2axh+ah^2+bx+bh-ax^2-bx-c}{h} \\
&= \lim_{h\to 0} \frac{2axh+ah^2+bh}{h} \\
&= \lim_{h\to 0} \frac{h(2ax+ah+b)}{h} \\
&= \lim_{h\to 0} 2ax+ah+b \\
&= 2ax+b
\end{aligned}$$

19. $f(x)=x^2$. From Example 3, $f'(x)=2x$. For the point $(3,9)$ we have $f'(3)=2(3)=6=m$. So the equation of the tangent line is

$$\begin{aligned}
y-y_1 &= m(x-x_1) \\
y-9 &= 6(x-3) \\
y &= 6x-18+9 \\
y &= 6x-9
\end{aligned}$$

For the point $(-1,1)$ we have $f'(-1)=2(-1)=-2$. So the equation of the tangent line is

$$\begin{aligned}
y-y_1 &= m(x-x_1) \\
y-1 &= -2(x-(-1)) \\
y-1 &= -2x-2 \\
y &= -2x-2+1 \\
y &= -2x-1
\end{aligned}$$

For the point $(10,100)$ we have $f'(10)=2(10)=20$. So the equation of the tangent line is

$$\begin{aligned}
y-y_1 &= m(x-x_1) \\
y-100 &= 20(x-10) \\
y &= 20x-200+100 \\
y &= 20x-100
\end{aligned}$$

20. $f(x)=x^3$. From Example 4, $f'(x)=3x^2$. For the point $(-2,-8)$ we have $f'(-2)=12=m$. So the equation of the tangent line is

$$\begin{aligned}
y-y_1 &= m(x-x_1) \\
y+8 &= 12(x+2) \\
y &= 12x+24-8 \\
y &= 12x+16
\end{aligned}$$

For the point $(0,0)$ we have $f'(0)=0$. So the equation of the tangent line is

$$\begin{aligned}
y-y_1 &= m(x-x_1) \\
y-0 &= 0(x-0) \\
y &= 0
\end{aligned}$$

For the point $(4,64)$ we have $f'(4)=48$. So the equation of the tangent line is

$$\begin{aligned}
y-y_1 &= m(x-x_1) \\
y-64 &= 48(x-4) \\
y &= 48x-192+64 \\
y &= 48x-128
\end{aligned}$$

21. From Exercise 14, $f'(x)=-\frac{5}{x^2}$. For the point $(1,5)$ we have $f'(1)=-\frac{5}{1^2}=-5=m$. So the equation of the tangent line is

$$\begin{aligned}
y-y_1 &= m(x-x_1) \\
y-5 &= -5(x-1) \\
y &= -5x+5+5 \\
y &= -5x+10
\end{aligned}$$

For the point $(-1,-5)$ we have $f'(-1)=-\frac{5}{(-1)^2}=-5$. So the equation of the tangent line is

$$\begin{aligned}
y-y_1 &= m(x-x_1) \\
y-(-5) &= -5(x-(-1)) \\
y+5 &= -5(x+1) \\
y &= -5x-5-5 \\
y &= -5x-10
\end{aligned}$$

For the point $(100,0.05)$, which can be rewritten as $(100,\frac{1}{20})$, we have $f'(10)=-\frac{5}{100^2}=-\frac{1}{2000}$. So the equation of the tangent line is

$$\begin{aligned}
y-y_1 &= m(x-x_1) \\
y-\frac{1}{20} &= -\frac{1}{2000}(x-100) \\
y &= -\frac{x}{2000}+\frac{1}{20}+\frac{1}{20} \\
y &= -\frac{x}{2000}+\frac{2}{20} \\
y &= -\frac{x}{2000}+\frac{1}{10}
\end{aligned}$$

22. $f(x)=\frac{2}{x}$. Using the difference quotient we find

$$\begin{aligned}
f'(x) &= \lim_{h\to 0} \frac{f(x+h)-f(x)}{h} \\
&= \lim_{h\to 0} \frac{\frac{2}{(x+h)}-\frac{2}{x}}{h} \\
&= \lim_{h\to 0} \frac{\frac{2}{(x+h)}\cdot x(x+h)-\frac{2}{x}\cdot x(x+h)}{h\cdot x(x+h)} \\
&= \lim_{h\to 0} \frac{2x-2(x+h)}{hx(x+h)} \\
&= \lim_{h\to 0} \frac{-2h)}{hx(x+h)} \\
&= \lim_{h\to 0} \frac{-2}{x(x+h)} \\
&= -\frac{2}{x^2}
\end{aligned}$$

For the point $(-1,-2)$ we have $f'(-1)=-\frac{2}{(-1)^2}=-2=m$. So the equation of the tangent line is

$$\begin{aligned}
y-y_1 &= m(x-x_1) \\
y-(-2) &= -2(x-(-1)) \\
y+2 &= -2(x+1) \\
y &= -2x-2-2 \\
y &= -2x-4
\end{aligned}$$

For the point $(2,1)$ we have $f'(2) = -\frac{2}{2^2} = -\frac{1}{2}$. So the equation of the tangent line is

$$\begin{aligned} y - y_1 &= m(x - x_1) \\ y - 1 &= -\frac{1}{2}(x-2) \\ y &= -\frac{x}{2} + 1 + 1 \\ y &= -\frac{x}{2} + 2 \end{aligned}$$

For the point $(10, \frac{1}{5})$ we have $f'(10) = -\frac{2}{10^2} = -\frac{1}{50}$. So the equation of the tangent line is

$$\begin{aligned} y - y_1 &= m(x - x_1) \\ y - \frac{1}{5} &= -\frac{1}{50}(x - 10) \\ y &= -\frac{x}{50} + \frac{1}{5} + \frac{1}{5} \\ y &= -\frac{x}{50} + \frac{2}{5} \end{aligned}$$

23. First, let us find the expression for $f'(x)$.

$$\begin{aligned} f'(x) &= \lim_{h\to 0} \frac{f(x+h) - f(x)}{h} \\ &= \lim_{h\to 0} \frac{4 - (x+h)^2 - (4 - x^2)}{h} \\ &= \lim_{h\to 0} \frac{4 - x^2 - 2xh - h^2 - 4 + x^2}{h} \\ &= \lim_{h\to 0} \frac{-2xh - h^2}{h} \\ &= \lim_{h\to 0} \frac{h(-2x - h)}{h} \\ &= \lim_{h\to 0} (-2x - h) \\ &= -2x \end{aligned}$$

For the point $(-1,3)$ we have $f'(-1) = -2(-1) = 2 = m$. So the equation of the tangent line is

$$\begin{aligned} y - y_1 &= m(x - x_1) \\ y - 3 &= 2(x - (-1)) \\ y - 3 &= -2x + 2 \\ y &= 2x + 2 + 3 \\ y &= 2x + 5 \end{aligned}$$

For the point $(0,4)$ we have $f'(0) = -2(0) = 0$. So the equation of the tangent line is

$$\begin{aligned} y - y_1 &= m(x - x_1) \\ y - 4 &= 0(x - 0) \\ y &= 0 + 4 \\ y &= 4 \end{aligned}$$

For the point $(5,-21)$ we have $f'(5) = -2(5) = -10$. So the equation of the tangent line is

$$\begin{aligned} y - y_1 &= m(x - x_1) \\ y - (-21) &= -10(x - 5) \\ y + 21 &= -10x + 50 \\ y &= -10x + 50 - 21 \\ y &= -10x + 29 \end{aligned}$$

24.

$$\begin{aligned} f'(x) &= \lim_{h\to 0} \frac{f(x+h) - f(x)}{h} \\ &= \lim_{h\to 0} \frac{(x+h)^2 - 2(x+h) - (x^2 - 2x)}{h} \\ &= \lim_{h\to 0} \frac{x^2 + 2xh + h^2 - 2x - 2h - x^2 + 2x}{h} \\ &= \lim_{h\to 0} \frac{2xh + h^2 - 2h}{h} \\ &= \lim_{h\to 0} \frac{h(2x + h - 2)}{h} \\ &= \lim_{h\to 0} (2x + h - 2) \\ &= 2x - 2 \end{aligned}$$

For the point $(-2,8)$ we have $f'(-2) = -6 = m$. So the equation of the tangent line is

$$\begin{aligned} y - y_1 &= m(x - x_1) \\ y - 8 &= -6(x - (-2)) \\ y - 8 &= -6x - 12 \\ y &= -6x - 12 + 8 \\ y &= -6x - 4 \end{aligned}$$

For the point $(1,-1)$ we have $f'(1) = 0$. So the equation of the tangent line is

$$\begin{aligned} y - y_1 &= m(x - x_1) \\ y - (-1) &= 0(x - 0) \\ y &= 0 - 1 \\ y &= -1 \end{aligned}$$

For the point $(4,8)$ we have $f'(4) = 6$. So the equation of the tangent line is

$$\begin{aligned} y - y_1 &= m(x - x_1) \\ y - 8 &= 6(x - 4) \\ y &= 6x - 24 + 8 \\ y &= 6x - 16 \end{aligned}$$

25. The function is not differentiable at x_0 since it is discontinuous, x_3 since it has a corner, x_4 since it has a corner, x_6 since it has a corner, and x_{12} since it has a vertical tangent.

26. The function is not differentiable at x_2, x_4, x_5, x_7 and x_8.

27. The function is not differentiable at integer values of x since the function is not continuous at integer values of x.

28. The function is not differentiable at values of x of the form $x = 0.2n$ where $n = 1, 2, 3, \cdots$.

29. The function is differentiable for all values in the domain.

30. The function is not differentiable in the year 1850, 1860, 1865, 1880, 1910, 1960, 1975, and 1980.

31. The function is differentiable for all values in the domain.

32. L_4 and L_6 are the only lines that appear to be tangent since the others intersect the curve and are no "tangent" to it.

33. As the points Q get closer to P the secant lines are getting closer to the tangent line at point P.

34.

$$\begin{aligned} f'(x) &= \lim_{h\to 0} \frac{f(x+h)-f(x)}{h} \\ &= \lim_{h\to 0} \frac{(x+h)^4 - x^4}{h} \\ &= \lim_{h\to 0} \frac{4x^3h + 6x^2h^2 + 4xh^3 + h^4}{h} \\ &= \lim_{h\to 0} 4x^3 + 6x^2h + 4xh^2 + h^3 \\ &= 4x^3 \end{aligned}$$

35.

$$\begin{aligned} f'(x) &= \lim_{h\to 0} \frac{f(x+h)-f(x)}{h} \\ &= \lim_{h\to 0} \frac{\frac{1}{x^2+2xh+h^2} - \frac{1}{x^2}}{h} \\ &= \lim_{h\to 0} \frac{x^2 - x^2 - 2xh - h^2}{hx^2(x^2+2xh+h^2)} \\ &= \lim_{h\to 0} \frac{-h(2x-h)}{hx^2(x^2+2xh+h^2)} \\ &= \lim_{h\to 0} \frac{-2x+h}{x^2(x^2+2xh+h^2)} \\ &= \frac{-2x}{x^2(x^2)} \\ &= \frac{-2}{x^3} \end{aligned}$$

36.

$$\begin{aligned} f'(x) &= \lim_{h\to 0} \frac{f(x+h)-f(x)}{h} \\ &= \lim_{h\to 0} \frac{\frac{1}{1-x-h} - \frac{1}{1-x}}{h} \\ &= \lim_{h\to 0} \frac{1-x-1+x+h}{h(1-x)(1-x-h)} \\ &= \lim_{h\to 0} \frac{1}{(1-x)(1-x-h)} \\ &= \frac{1}{(1-x)^2} \end{aligned}$$

37.

$$\begin{aligned} f'(x) &= \lim_{h\to 0} \frac{f(x+h)-f(x)}{h} \\ &= \lim_{h\to 0} \frac{\frac{x+h}{1+x+h} - \frac{x}{1+x}}{h} \\ &= \lim_{h\to 0} \frac{(x+h)(1+x) - x(1+x+h)}{h(1+x)(1+x+h)} \\ &= \lim_{h\to 0} \frac{x + x^2 + h + hx - x - x^2 - xh}{h(1+x)(1+x+h)} \\ &= \lim_{h\to 0} \frac{h}{h(1+x)(1+x+h)} \\ &= \lim_{h\to 0} \frac{1}{(1+x)(1+x+h)} \\ &= \frac{1}{(1+x)^2} \end{aligned}$$

38.

$$\begin{aligned} f'(x) &= \lim_{h\to 0} \frac{f(x+h)-f(x)}{h} \\ &= \lim_{h\to 0} \frac{\sqrt{x+h}-\sqrt{x}}{h} \\ &= \lim_{h\to 0} \frac{\sqrt{x+h}-\sqrt{x}}{h} \cdot \frac{\sqrt{x+h}+\sqrt{x}}{\sqrt{x+h}+\sqrt{x}} \\ &= \lim_{h\to 0} \frac{x+h-x}{h(\sqrt{x+h}+\sqrt{x})} \\ &= \lim_{h\to 0} \frac{h}{h(\sqrt{x+h}+\sqrt{x})} \\ &= \lim_{h\to 0} \frac{1}{(\sqrt{x+h}+\sqrt{x})} \\ &= \frac{1}{2\sqrt{x}} \end{aligned}$$

39.

$$\begin{aligned} f'(x) &= \lim_{h\to 0} \frac{f(x+h)-f(x)}{h} \\ &= \lim_{h\to 0} \frac{1}{\sqrt{x+h}} - \frac{1}{\sqrt{x}} \\ &= \lim_{h\to 0} \frac{\sqrt{x}-\sqrt{x+h}}{h\sqrt{x}\sqrt{x+h}} \\ &= \lim_{h\to 0} \frac{\sqrt{x}-\sqrt{x+h}}{h\sqrt{x}\sqrt{x+h}} \cdot \frac{\sqrt{x}+\sqrt{x+h}}{\sqrt{x}+\sqrt{x+h}} \\ &= \lim_{h\to 0} \frac{x-x-h}{h\sqrt{x}\sqrt{x+h}(\sqrt{x}+\sqrt{x+h})} \\ &= \lim_{h\to 0} \frac{-1}{\sqrt{x}\sqrt{x+h}(\sqrt{x}+\sqrt{x+h})} \\ &= \frac{-1}{2\sqrt{x^3}} \end{aligned}$$

40.

$$\begin{aligned} f'(x) &= \lim_{h\to 0} \frac{f(x+h)-f(x)}{h} \\ &= \lim_{h\to 0} \frac{\frac{3(x+h)}{x+h+5} - \frac{3x}{x+5}}{h} \\ &= \lim_{h\to 0} \frac{(3x+3h)(x+5) - 3x(x+h+5)}{h(x+5)(x+h+5)} \\ &= \lim_{h\to 0} \frac{3x^2 + 15x + 3xh + 15h - 3x^2 - 3xh - 15x}{h(x+5)(x+h+5)} \\ &= \lim_{h\to 0} \frac{15h}{h(x+5)(x+h+5)} \\ &= \lim_{h\to 0} \frac{15}{(x+5)(x+h+5)} \\ &= \frac{15}{(x+5)^2} \end{aligned}$$

41. The function $f(x)$ will is not differentiable at $x = -3$.

42. $f'(3) = 6$, $f'(-1) = -2$, $f'(10) = 20$

43. $f'(-2) = 12$, $f'(0) = 0$, $f'(4) = 48$

44. $f'(1) = -5$, $f'(-1) = -5$, $f'(100) = \frac{-1}{2000}$

45. $f'(-1) = -2$, $f'(2) = \frac{-1}{2}$, $f'(10) = \frac{-2}{100}$

46. $f'(-1) = 2$, $f'(0) = 0$, $f'(5) = -10$

47. $f'(-2) = -6$, $f'(1) = 0$, $f'(4) = 6$

48. **a)**

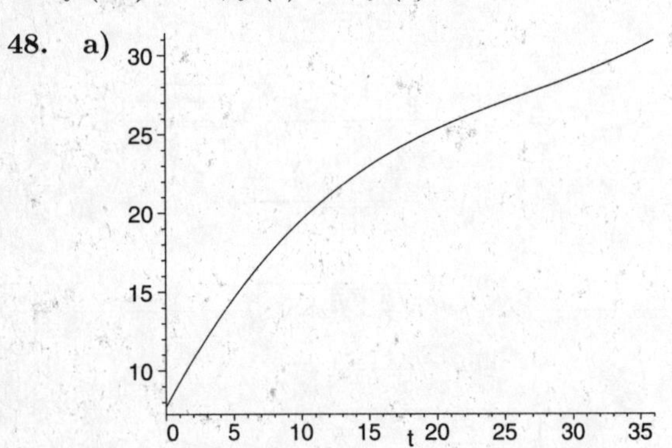

b) $y = 0.41x + 16.067$

c) Rate of change is 0.41

d)
- $(12, w(12)$ and $(24, w(24))$:
  $y = 0.472416x + 15.318$, Rate of change $= 0.47241$
- $(12, w(12)$ and $(18, w(18))$:
  $y = 0.5685x + 14.165$, Rate of change $= 0.0.5685$
- $(12, w(12)$ and $(15, w(15))$:
  $y = 0.6326x + 13.395$, Rate of change $= 0.0.6326$

e) The slope of the tangent line at $(12, w(12))$ appears to be approaching 0.71

Exercise Set 2.5

1. $\frac{dy}{dx} = 7x^{7-1} = 7x^6$

2. $\frac{dy}{dx} = 8x^7$

3. $\frac{dy}{dx} = 3 \cdot 2x^{2-1} = 6x$

4. $\frac{dy}{dx} = 20x^3$

5. $\frac{dy}{dx} = 4 \cdot 3x^{3-1} = 12x^2$

6. $\frac{dy}{dx} = 10cos(x)$

7. $\frac{dy}{dx} = 3 \cdot \frac{2}{3}x^{2/3-1} = 2x^{-1/3}$

8. $\frac{dy}{dx} = -2 \cdot \frac{1}{2}x^{-1/2} = -x^{-1/2}$

9. Rewrite as $y = x^{3/4}$, $\frac{dy}{dx} = \frac{3}{4}x^{3/4-1} = \frac{3}{4}x^{-1/4}$

10. Rewrite as $y = x^{4/7}$, $\frac{dy}{dx} = \frac{4}{7}x^{-3/7}$

11. $\frac{dy}{dx} = 4cos\ x$

12. $\frac{dy}{dx} = -3sin\ x$

13. $\frac{dy}{dx} = cos\ x - 1x^{-2} = cos\ x - \frac{1}{x^2}$

14. $\frac{dy}{dx} = \frac{d}{dx}(x^{1/2} - x^{-1/2}) = \frac{1}{2}x^{1/2-1} + \frac{1}{2}x^{-1/2-1}$
$= \frac{1}{2}x^{-1/2} + \frac{1}{2}x^{-3/2}$ or $\frac{1}{2\sqrt{x}} + \frac{1}{2x^{3/2}}$

15. $\frac{dy}{dx} = 2(2x+1)^{2-1}(2) = 4(2x+1)$

16. $\frac{dy}{dx} = 2(3x-2)(3) = 6(3x-2)$

17. $f'(x) = 0.25(3.2x^{3.2-1}) = 0.8x^{2.2}$

18. $f'(x) = 0.32(12.5x^{11.5} = 4x^{11.5}$

19. $f'(x) = 10cos\ x + 12sin\ x$

20. $f'(x) = \sqrt{5}\ cos\ x$

21. $f'(x) = -\sqrt[3]{9}\ sin\ x$

22. $f'(x) = \frac{-2}{x^2} - \frac{1}{2}$

23. Rewrite as $f(x) = 5x^{-1} + \frac{x}{5}$, $f'(x) = -5x^{-2} + \frac{1}{5}$

24. $f'(x) = -2x^{-5/3} + \frac{3}{4}x^{-1/4} + \frac{84}{5}x^{1/5} - \frac{8}{3}x^{-4}$

25. $f'(x) = -x^{-1/2} - x^{-3/4} - \frac{1}{2}x^{-5/4} + \frac{7}{2}x^{-3/2}$

26. $f(x) = x^{7/2}$
$f'(x) = \frac{7}{2}x^{5/2}$

27. $f(x) = x^{18/5} - x^{13/5}$
$f'(x) = \frac{18}{5}x^{13/5} - \frac{13}{5}x^{8/5}$

28. $f(x) = 2 + x^{-1}$
$f'(x) = -x^{-2}$

29. $f(x) = 3x^{-1} - 4x^{-2}$
$f'(x) = -3x^{-2} + 8x^{-3}$

30. $f(x) = x + 2 + 3x^{-1} + 4x^{-2}$
$f'(x) = 1 - 3x^{-2} - 8x^{-3}$

31. $f(x) = 4x + 3 - 2x^{-1}$
$f'(x) = 4 + 2x^{-2}$

32. $g(x) = 2x^{-1} - 4x^{-2} + 6x^{-3}$
$g'(x) = -2x^{-2} + 8x^{-3} - 18x^{-4}$

33. $p(x) = 6x^{-4} - 2x^{-3}$
$p'(x) = -24x^{-5} + 6x^{-4}$

34. $r(x) = 12cos\ x + 8sin\ x - 5$
$r'(x) = -12sin\ x + 8cos\ x$

35. $s(x) = 3\sqrt{2}cos\ x - 2\sqrt{2}sin\ x$
$s'(x) = -3\sqrt{2}sin\ x - 2\sqrt{2}cos\ x$

36. $f'(x) = \frac{3}{2}cos\ x + \frac{5}{8}sin\ x$

37. $q'(x) = 8\left(\frac{\sqrt{5}}{3}\right) cos\ x + 8\left(\frac{\sqrt[3]{5}}{7}\right) sin\ x$

38. $W(x) = \frac{x}{3} + \frac{2}{3} - x^{-1} + \frac{4}{3}cos\ x$
$W'(x) = \frac{1}{3} + x^{-2} - \frac{4}{3}sin\ x$

39. $U(x) = sin\ x - 2x^{1/2} + 3x^{-1/2}$
$U'(x) = cos\ x - x^{-1/2} - \frac{3}{2}x^{-3/2}$

40. $f'(x) = 2cos\ x$
The slops at $(0,0)$ is $f'(0) = cos(0) = 1$.
So the equation of the tangent lilne is

$$\begin{aligned} y - y_1 &= m(x - x_1) \\ y - 0 &= 1(x - 0) \\ y &= x \end{aligned}$$

41. $f'(x) = -3sin\ x$
The slope at $(0,4)$ is $f'(0) = 0$.
So the equation of the tangent line is

$$\begin{aligned} y - y_1 &= m(x - x_1) \\ y - 4 &= 0(x - 0) \\ y - 4 &= 0 \\ y &= 4 \end{aligned}$$

42. $f'(x) = \frac{3}{2}x^{1/2} + \frac{1}{2}x^{-1/2}$
The slope at $(4,10)$ is
$f'(4) = \frac{3}{2}(4)^{1/2} + \frac{1}{2}(4)^{-1/2} = 3.25$
So the equation of the line is

$$\begin{aligned} y - 10 &= 3.25(x - 4) \\ y - 10 &= = 3.25x - 13 \\ y &= 3.25x - 3 \end{aligned}$$

43. $f'(x) = \frac{4}{3}x^{1/3} + \frac{2}{3}x^{-2/3}$
The slope at $(8,20)$ is
$f'(8) = \frac{4}{3}(8)^{1/3} + \frac{2}{3}(8)^{-2/3} = \frac{17}{6}$
So the equation of the line is

$$\begin{aligned} y - y_1 &= m(x - x_1) \\ y - 20 &= \frac{17}{6}(x - 8) \\ y &= \frac{17}{6}x - \frac{68}{3} + 20 \\ y &= \frac{17}{6}x - \frac{8}{3} \end{aligned}$$

44. $y = x^{1/12}$
$\frac{dy}{dx} = \frac{1}{12}x^{-11/12}$

45. Rewrite $y = \left(\frac{x^{2/3}}{x^{3/2}}\right)^{1/3} = x^{-5/18}$
$\frac{dy}{dx} = \frac{-5}{18}x^{-23/8}$

46. $4sin\ x$
$\frac{dy}{dx} = 4cos\ x$

47. Use the trigonometric identity $1+tan^2x = sec^2x$ to rewrite $y = 3sec^2x\ cos^3x = 3cos\ x$
$\frac{dy}{dx} = -3sin\ x$

48. $y = \frac{1}{\sqrt{2}}sin\ x + \frac{1}{\sqrt{2}}cos\ x$
$\frac{dy}{dx} = \frac{1}{\sqrt{2}}(cos\ x - sin\ x)$

49. Use the sum identity for cosine to rewrite $y = \frac{\sqrt{3}}{2}cos\ x - \frac{1}{2}sin\ x$
$\frac{dy}{dx} = \frac{-\sqrt{3}}{2}sin\ x - \frac{1}{2}cos\ x$

50. $\frac{dy}{dx} = 2\left(x + \frac{1}{x}\right)\left(1 - x^{-2}\right)$
The slope at $(2, \frac{25}{4})$ is $\frac{dy}{dx}|_{x=2} = \frac{25}{4}$
The equation of the line is

$$\begin{aligned} y - \frac{25}{4} &= \frac{25}{4}(x - 2) \\ y &= \frac{25}{4}x - \frac{25}{4} \end{aligned}$$

51. $\frac{dy}{dx} = 2\left(\sqrt{x} - \frac{1}{\sqrt{x}}\right)\left(\frac{1}{2\sqrt{x}} + \frac{1}{2\sqrt{x^3}}\right)$
The slope at $(1,0)$ is $\frac{dy}{dx}|_{x=1} = 0$
The equation of the line is

$$\begin{aligned} y - y_1 &= m(x - x_1) \\ y - 0 &= 0(x - 1) \\ y &= 0 \end{aligned}$$

52. Let $D(x) = f(x) - g(x)$. Then

$$\begin{aligned} \frac{D(x+h) - D(x)}{h} &= \frac{[f(x+h) - g(x+h)] - [f(x) - g(x)]}{h} \\ &= \frac{f(x+h) - f(x)}{h} - \frac{g(x+h) - g(x)}{h} \end{aligned}$$

$$\begin{aligned} D'(x) &= \lim_{h\to 0} \frac{D(x+h) - D(x)}{h} \\ &= \lim_{h\to 0} \left[\frac{f(x+h) - f(x)}{h} + \frac{g(x+h) - g(x)}{h}\right] \\ &= \lim_{h\to 0} \frac{f(x+h) - f(x)}{h} + \lim_{h\to 0} \frac{g(x+h) - g(x)}{h} \\ &= f'(x) - g'(x) \end{aligned}$$

53. Let $f(x) = cos\ x$ then

$$\begin{aligned} f'(x) &= lim_{h\to 0}\frac{f(x+h) - f(x)}{h} \\ &= \lim_{h\to 0} \frac{cos(x+h) - cos(x)}{h} \\ &= \lim_{h\to 0} \frac{cos\ xcos\ h - sin\ xsin\ h - cos\ x}{h} \\ &= \lim_{h\to 0} cos\ x\left(\frac{cos\ h - 1}{h}\right) - sin\ x\left(\frac{sin\ h}{h}\right) \\ &= cos\ x \cdot \lim_{h\to 0} \frac{cos\ h - 1}{h} - sin\ x \cdot \lim_{h\to 0} \frac{sin\ h}{h} \\ &= cos\ x \cdot 0 - sin\ x \cdot 1 \\ &= -sin\ x \end{aligned}$$

54. The value of the slope of the tangent lines of the sine function correspond to whether the cosine function is has positive or negative values. For a positive slope of the tangent line means the cosine has a positive values at the aprticular x value, and vice versa.

55. Left to the student

56. Left to the student

57. The tangent line is horizontal at $x = \pm\ 0.6922$

58. The tangent line is horizontal at $x = 0$

59. The tangent line is horizontal at $x = -0.3456$ and at $x = 1.9289$

60. The tangent line is horizontal at $x = 1.9384$, $x = 3.6613$ and at $x = 5.6498$

Exercise Set 2.6

1. **a)** $v(t) = \frac{ds(t)}{dt} = 3t^2 + 1$
b) $a(t) = \frac{dv(t)}{dt} = 6t$
c) $v(4) = 3(4)^2 + 1 = 48 + 1 = 49$ feet per second
$a(4) = 6(4) = 24$ feet per squared seconds

2. **a)** $v(t) = \frac{ds}{dt} = 3$ Note that the velocity is constant and does not depent on time
b) $a(t) = \frac{dv}{dt} = 0$ Note that the velocity is constant and does not depent on time
c) $v(2) = 3$ miles per hour
$a(2) = 0$ miles per squared hours
d) Uniform motion means that the velocity and acceleration are constants namely, the velocity is the slope of the linear distance function and the acceleration is zero.

3. **a)** $v(t) = \frac{ds(t)}{dt} = -20t + 2$
b) $a(t) = \frac{dv(t)}{dt} = -20$
c) $v(1) = -20(1) + 2 = -20 + 2 = 18$ feet per second
$a(1) = -20$ feet per squared seconds

4. **a)** $v(t) = \frac{ds}{dt} = 2t - \frac{1}{2}$
b) $a(t) = \frac{dv}{dt} = 2$
c) $v(1) = 2(1) - \frac{1}{2} = 2 - \frac{1}{2} = \frac{3}{2} = 1.5$ feet per second
$a(1) = 2$ feet per squared seconds

5. **a)** $v(t) = \frac{ds(t)}{dt} = 5 + 2cos\ t$
b) $a(t) = \frac{dv(t)}{dt} = -2sin\ t$
c) Find $v(\frac{\pi}{4})$ and $a(\frac{\pi}{4})$

$$\begin{aligned} v(\frac{\pi}{4}) &= 5 + 2cos(\frac{\pi}{4}) \\ &= 5 + 2 \cdot \frac{\sqrt{2}}{2} \\ &= 5 + \sqrt{2} \\ &\approx 6.414\ m/sec \end{aligned}$$

$$\begin{aligned} a(\frac{\pi}{4}) &= -2sin(\frac{\pi}{4}) \\ &= 2 \cdot \frac{\sqrt{2}}{2} \\ &= \sqrt{2} \\ &\approx 1.414\ m/sec^2 \end{aligned}$$

d) Find t when $v(t) = 3$

$$\begin{aligned} 3 &= 5 + 2cos\ t \\ -2 &= 2cos\ t \\ -1 &= cos\ t \\ t &= cos^{-1}(-1) \\ &= \pi + 2n\pi \end{aligned}$$

6. **a)** $v(t) = 3 + sin\ t$
b) $a(t) = cos\ t$
c) Find $v(\frac{\pi}{3})$ and $a(\frac{\pi}{3})$

$$\begin{aligned} v(\frac{\pi}{3}) &= 3 + sin(\frac{\pi}{3}) \\ &= 3 + \frac{\sqrt{3}}{2} \\ &\approx 3.866\ m/sec \end{aligned}$$

$$\begin{aligned} a(\frac{\pi}{3}) &= cos(\frac{\pi}{3}) \\ &= \frac{1}{2} = 0.5\ m/sec^2 \end{aligned}$$

d) Find t when $v(t) = 2$

$$\begin{aligned} 2 &= 3 + sin\ t \\ -1 &= sin\ t \\ t &= sin^{-1}(-1) \\ &= \frac{3\pi}{2} + 2n\pi \end{aligned}$$

7. **a)** $\frac{dN}{da} = -2a + 300$
b) Since a is counted in thousands we need to find

$$\begin{aligned} N(10) &= -(10)^2 + 300(10) + 6 \\ &= -100 + 3000 + 6 \\ &= 2996 \end{aligned}$$

There will be 2996 units sold after spending \$10000 on advertsing
c) At $a = 10$, $\frac{DN}{da} = -2(10) + 300 = -20 + 300 = 280$ units per thousand dollars spent on advertising
d) The rate of change of the number of units sold depends on the amount spent on advertising according to the equation $\frac{dN}{da} = -2a + 300$ which means that for every a thousand dollars spent on advertising, the change in the units solds is $-2a+300$ units. If \$10000 is spent on advertising, then there will be a 280 unit increase in the number of units sold.

8. **a)** $\frac{dw}{dt} = 1.82 - 0.1192t + 0.002274t^2$
b) $w(0) = 8.15 + 1.82(0) - 0.0596(0)^2 + 0.000758(0)^3 =$ 8.15 pounds
c) $\frac{dw}{dt}|_{t-0} = 1.82 - 0.1192(0) + 0.002274(0)^2 = 1.82$ pounds per month
d) $w(12) = 8.15 + 1.82(12) - 0.0596(12)^2 + 0.000758(12)^3$ $= 22.72$ pounds
e) $\frac{dw}{dt}|_{t=12} = 1.82 - 0.1192(12) + 0.002274(12)^2 = 0.717$ pounds per month
f) Average rate of change $= \frac{w(12)-w(0)}{12} = \frac{22.72-8.15}{12} =$ 1.21 pounds per month
g)

$$\begin{aligned} 1.82 - 0.1192t + 0.002274t^2 &= 1.21 \\ 0.002274t^2 - 0.1192t + 0.61 &= 0 \end{aligned}$$

Using the quadratic formula we get that $t = 5.747$ months.

9. a) $\frac{dw}{dt} = 1.61 - 0.0968t + 0.0018t^2$

b) $w(0) = 7.60+1.61(0)-0.0484(0)^2+0.0006(0)^3 = 7.60$ pounds

c) $\frac{dw}{dt}|_{t=0} = 1.61-0.0968(0)+0.0018(0)^2 = 1.61$ pounds per month

d) $w(12) = 7.60+1.61(12)-0.0484(12)^2+0.0006(12)^3 = 20.987$ pounds

e) $\frac{dw}{dt}|_{12} = 1.61 - 0.0968(12) + 0.0018(12)^2 = 0.708$ pounds per month

f) Average rate of change $= \frac{w(12)-w(0)}{12} = \frac{20.987-7.6}{12} = 1.1156$ pounds per month

g)

$$\begin{aligned} 1.61 - 0.0968t + 0.0018t^2 &= 1.1156 \\ 0.0018t^2 - 0.0968t + 0.4944 &= 0 \\ \frac{0.0968 + \sqrt{(0.0968)^2 - 4(0.0018)(0.4944)}}{2(0.0018)} &= t \\ 5.7147 &= t \end{aligned}$$

10. a) $\frac{dD}{dF} = 2$ feet per degrees

b) For every increase of one degree there will be an increase of 2 feet in the stopping distance of an object on glare ice

11. a) $\frac{dC}{dr} = 2\pi$

b) For every increase of one centimeter of the radius the healing wound circumference increases by 2π centimeters

12. a) $\frac{dA}{dr} = 2\pi r$

b) For every increase of one centimeter of the radius the circular area of a healing wound increases by $2\pi r$ squared centimeters

13. a) $\frac{dT}{dt} = -0.2t + 1.2$

b) At $t = 1.5$

$$\begin{aligned} T(1.5) &= -0.1(1.5)^2 + 1.2(1.5) + 98.6 \\ &= 100.2 \text{ degrees} \end{aligned}$$

c) $\frac{dT}{dt}|_{t=1.5} = -0.2(1.5) + 1.2 = 0.9$ degrees per day

d) The sign of $T'(t)$ is significant because it indicates whether the rate of change in temperature is an increase (if positive) or a decrease (if negative). That is, whether the fever is increasing or decreasing during the illness

14. a) $\frac{dB}{dx} = 0.1x - 0.9x^2$

b) The sensitivity is affected by the following rule: for every increase in the dosage of x cubic centimeters the blood pressure is changed by an amount equal to $0.1x - 0.9x^2$

15. a) $\frac{dT}{dW} = 1.31W^{0.31}$

b) For every increase of W in body weight the territorial area of an animal increases by an amount equal to $1.31W^{0.31}$

16. a) $\frac{dH}{dW} = 1.41W^{0.41}$

b) For every increase of W in body weight the home range of an animal increases by an amount equal to $1.41W^{0.41}$

17. First rewrite $R(Q)$ as follows $R(Q) = \frac{k}{2}Q^2 - \frac{1}{3}Q^3$

a) $\frac{dR}{dQ} = \frac{k}{2}(2Q) - \frac{1}{3}(3Q^2) = kQ - Q^2$

b) For every increase Q in the dosage there will be a change in the reaction of the body to that dosage change equal to the amount $kQ - Q^2$

18. a) $\frac{dP}{dt} = 4000t$

b) $P(10) = 100000 + 2000(10)^2 = 300000$ people

c) $\frac{dP}{dt}|_{t=10} = 4000(10) = 40000$ people per year

d) The population of the city will reach 300000 people after 10 years.
The population growth rate at 10 years will be 40000 people per year, that is, the population will increase by 40000 for the next year

19. a) $\frac{dA}{dt} = 0.08$

b) The rate of change for the median age of women at first marriage is constant at 0.08. That is, each year the median age of women at first marriage in increasing by 0.08 years

20. a) $\frac{dV}{dh} = 1.22(\frac{1}{2}h^{-1/2}) = \frac{0.61}{\sqrt{h}}$

b) $V(40000) = 1.22\sqrt{40000} = 244$ miles

c) $\frac{dV}{dh}|_{h=40000} = \frac{0.61}{\sqrt{40000}} = 0.00305$ miles per feet

d) At a height of 40000 feet the view is 244 miles.
At a height of 40000, for every one foot increase in height corresponds to an increase in the view of 0.00305 miles

21. The average rate of change of a function is the value of the difference quotient evaluated over a period of time, while the instantaneous rate of change is the value of the slope of the tangent line at that particular instant.

22. The derivative of a function at a point x can be thought of as the slope of the tangent line at that point. It could also be thought of as the instantaneous rate of change of teh function at that point.

23.

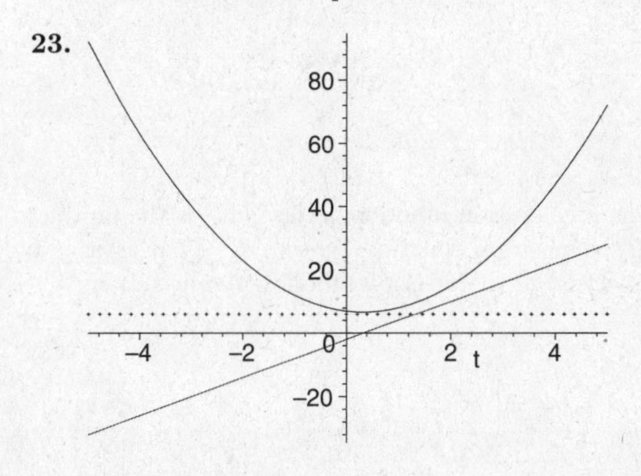

24.

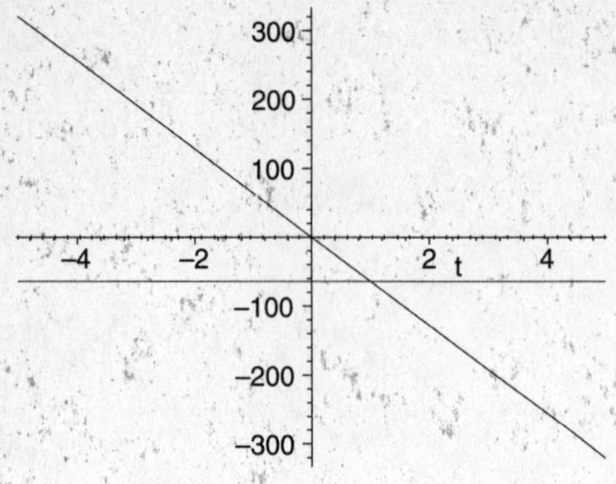

25.

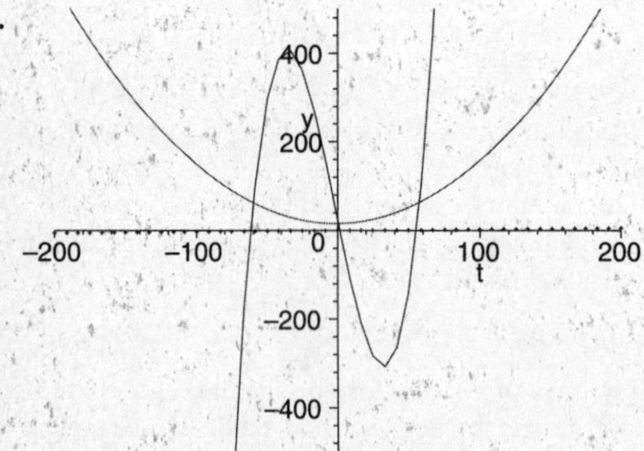

26.

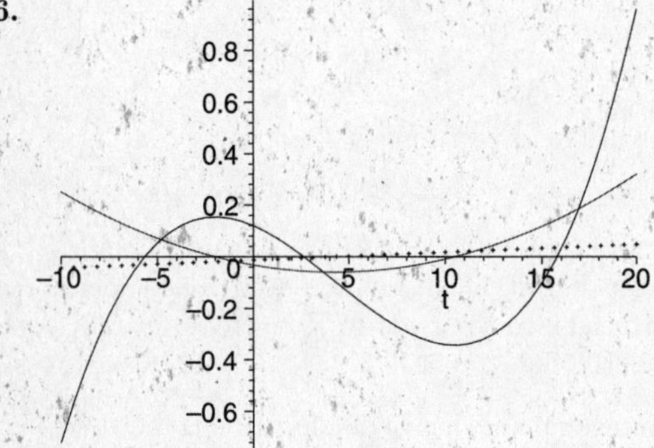

27. a)

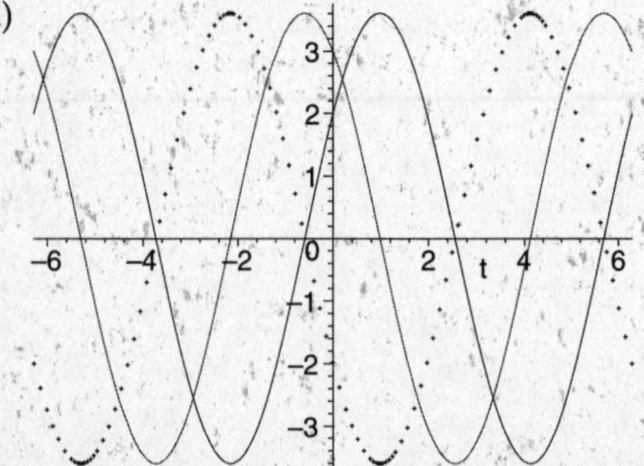

b) The acceleration function is the same as the product of the position function times -1. That is it is a reflection of the position function about the t axis.

28. a)

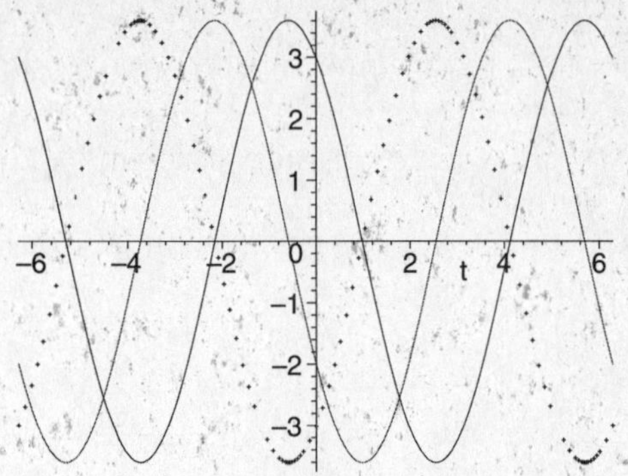

b) The acceleration function is the same as the product of the position function times -1. That is it is a reflection of the position function about the t axis.

Exercise Set 2.7

1. Method One: $x^3 \cdot x^8 = x^{3+8} = x^{11}$. $\frac{dy}{dx} = 11x^{10}$
Method Two (product rule):

$$\begin{aligned} \frac{dy}{dx} &= x^3 \cdot 8x^7 + 3x^2 \cdot x^8 \\ &= 8x^{10} + 3x^{10} \\ &= 11x^{10} \end{aligned}$$

2. Method One: $x^4 \cdot x^4 = x^{4+9} = x^{13}$. $\frac{dy}{dx} = 13x^{12}$
Method Two (product rule):

$$\begin{aligned} \frac{dy}{dx} &= x^4 \cdot 9x^8 + 4x^3 \cdot x^9 \\ &= 9x^{12} + 4x^{12} \\ &= 13x^{12} \end{aligned}$$

3. Method One: $x\sqrt{x} = x^{3/2}$, $\frac{dy}{dx} = \frac{3}{2}x^{1/2}$
Method Two (product rule):

$$\begin{aligned} \frac{dy}{dx} &= x(\frac{1}{2}x^{-1/2}) + (1)x^{1/2} \\ &= \frac{1}{2}x^{1/2} + x^{1/2} \\ &= \frac{3}{2}x^{1/2} \end{aligned}$$

4. Method One: $y = x^{8/3}$, $\frac{dy}{dx} = \frac{8}{3}x^{5/3}$
Method Two (product rule)

$$\begin{aligned} \frac{dy}{dx} &= x^2(\frac{2}{3}x^{-1/3} + 2x(x^{2/3}) \\ &= \frac{2}{3}x^{5/3} + 2x^{5/3} \\ &= \frac{8}{3}x^{5/3} \end{aligned}$$

5. Method One: $\frac{x^8}{x^5} = x^3$, $\frac{dy}{dx} = 3x^2$
Method Two (quotient rule):

$$\begin{aligned} \frac{dy}{dx} &= \frac{x^5(8x^7) - 5x^4(x^8)}{(x^5)^2} \\ &= \frac{8x^{12} - 5x^{12}}{x^{10}} \\ &= \frac{3x^{12}}{x^{10}} \\ &= 3x^2 \end{aligned}$$

6. Method One: $\frac{x^5}{x^8} = x^{-3}$, $\frac{dy}{dx} = -3x^{-4}$
Method Two (quotient rule):

$$\begin{aligned} \frac{dy}{dx} &= \frac{x^8(5x^4) - x^5(8x^7)}{(x^8)^2} \\ &= \frac{5x^{12} - 8x^{12}}{x^{16}} \\ &= \frac{-3x^{12}}{x^{16}} \\ &= -3x^{-4} \end{aligned}$$

7. Method One: $y = x^2 - 25$, $\frac{dy}{dx} = 2x$
Method Two (product rule):

$$\begin{aligned} \frac{dy}{dx} &= (x+5)(1) + (1)(x-5) \\ &= x + 5 + x - 5 \\ &= 2x \end{aligned}$$

8. Method One: $f(x) = \frac{(x+3)(x-3)}{x+3} = (x-3)$, $f'(x) = 1$
Methods Two (quotient rule):

$$\begin{aligned} f'(x) &= \frac{(x+3)(2x) - (x^2-9)(1)}{(x+3)^2} \\ &= \frac{2x^2 + 6x - x^2 + 9}{x^2 + 6x + 9} \\ &= \frac{x^2 + 6x + 9}{x^2 + 6x + 9} \\ &= 1 \end{aligned}$$

9.

$$\begin{aligned} y &= (8x^5 - 3x^2 + 20)(8x^4 - 3x^{1/2}) \\ \frac{dy}{dx} &= (8x^5 - 3x^2 + 20)(32x^3 - \frac{3}{2}x^{-1/2}) + \\ & (40x^4 - 6x)(8x^4 - 3x^{1/2}) \\ &= (8x^5 - 3x^2 + 20)(32x^3 - \frac{3}{2\sqrt{x}}) + \\ & (40x^4 - 6x)(8x^4 - 3\sqrt{x}) \end{aligned}$$

10.

$$\begin{aligned} f(x) &= (7x^6 + 4x^3 - 50)(9x^{10} - 7x^{1/2}) \\ f'(x) &= (7x^6 + 4x^3 - 50)(90x^9 - \frac{7}{2}x^{-1/2}) + \\ & (42x^5 + 12x^2)(9x^{10} - 7x^{1/2}) \\ &= (7x^6 + 4x^3 - 50)(90x^9 - \frac{7}{2\sqrt{x}}) + \\ & (42x^5 + 12x^2)(9x^{10} - 7\sqrt{x}) \end{aligned}$$

11. $f(x) = (x^{1/2} - x^{1/3})(2x+3)$

$$\begin{aligned} f'(x) &= (\frac{1}{2}x^{-1/2} - \frac{1}{3}x^{-2/3})(2x+3) + (\sqrt{x} - \sqrt[3]{x})(2) \\ &= 2(\sqrt{x} - \sqrt[3]{x}) + (2x+3)\left(\frac{1}{2\sqrt{x}} - \frac{1}{3\sqrt[3]{x^2}}\right) \end{aligned}$$

12.

$$\begin{aligned} f'(x) &= (\sqrt[3]{x} + 2x)sec^2x + tan\ x\left(\frac{1}{3}x^{-2/3} + 2\right) \\ &= (\sqrt[3]{x} + 2x)sec^2x + tan\ x\left(\frac{1}{3\sqrt[3]{x^2}} + 2\right) \end{aligned}$$

13. $f(x) = x^{1/2} tan\ x$

$$\begin{aligned} f'(x) &= x^{1/2} sec^2 x + tan\ x\left(\frac{1}{2}x^{-1/2}\right) \\ &= \sqrt{x}\ sec^2x + \left(\frac{1}{2\sqrt{x}}\right) tan\ x \end{aligned}$$

14.

$$\begin{aligned} g'(x) &= \sqrt[3]{x}\ sec\ x\ tan\ x + sec\ x\left(\frac{1}{3}x^{-2/3}\right) \\ &= \sqrt[3]{x}\ sec\ x\ tan\ x + \frac{sec\ x}{3\sqrt[3]{x^2}} \end{aligned}$$

15. $(2t+3)^2 = (2t+3)(2t+3) = 4t^2 + 12t + 9$
$f'(t) = 8t + 12$

16. $r(t) = (5t-4)^2 = (5t-4)(5t-4)$

$$\begin{aligned} r'(t) &= (5t-4)(5) + (5)(5t-4) \\ &= 25t - 20 + 25t - 20 \\ &= 50t - 40 \end{aligned}$$

17. $g(x) = (0.02x^2 + 1.3x - 11.7)(4.1x + 11.3)$

$$\begin{aligned} g'(x) &= (0.02x^2 + 1.3x - 11.7)(4.1) + \\ & (0.04x + 1.3)(4.1x + 11.3) \end{aligned}$$

18. $g(x) = (3.12x^2 + 10.2x - 5.01)(2.9x^2 + 4.3x - 2.1)$

$$\begin{aligned} g'(x) &= (3.12x^2 + 10.2x - 5.01)(5.8x + 4.3) + \\ & (6.24x + 10.2)(2.9x^2 + 4.3x - 2.1) \end{aligned}$$

19. $g(x) = sec\ x\ csc\ x$

$$\begin{aligned} g'(x) &= sec\ x(-csc\ x\ cot\ x) + csc\ x(sec\ x\ tan\ x) \\ &= \frac{-1}{sin^2x} + \frac{1}{cos^2x} \\ &= sec^2x - csc^2x \end{aligned}$$

20.

$$\begin{aligned} p'(x) &= cot\ x(-csc\ x\ cot\ x) + csc\ x(-csc^2x) \\ &= -csc\ x\ cot^2x - csc^3x \\ &= -csc\ x(cot^2x + csc^2x) \end{aligned}$$

21. $q(x) = \frac{sin\ x}{1 + cos\ x}$

$$\begin{aligned} q'(x) &= \frac{(1 + cos\ x)(cos\ x) - sin\ x(-sin\ x)}{(1 + cos\ x)^2} \\ &= \frac{cos\ x + cos^2x + sin^2x}{(1 + cos\ x)^2} \\ &= \frac{cos\ x + 1}{(1 + cos\ x)^2} \\ &= \frac{1}{1 + cos\ x} \end{aligned}$$

22.

$$\begin{aligned} f'(x) &= \frac{\sin x(\sin x) - (1-\cos x)(\cos x)}{\sin^2 x} \\ &= \frac{\sin^2 x - \cos x + \cos^2 x}{\sin^2 x} \\ &= \frac{1-\cos x}{1-\cos^2 x} \\ &= \frac{1}{1+\cos x} \end{aligned}$$

23. $s(t) = \tan^2 t$
$s'(t) = 2\tan t \sec^2 t$

24.

$$\begin{aligned} s'(t) &= 2\sec t(\sec t \tan t) \\ &= 2\sec^2 t \tan t \end{aligned}$$

25. $f(x) = (x + 2x^{-1})(x^2 - 3)$

$$\begin{aligned} f'(x) &= (x+2x^{-1})(2x) + (1-2x^{-2})(x^2-3) \\ &= 2x^2 + 4 + x^2 - 3 - 2 + 6x^{-2} \\ &= 3x^2 - 1 + 6x^{-2} \\ &= 3x^2 - 1 + \frac{6}{x^2} \end{aligned}$$

26. $y = 4x^4 - x^3 - 20x^2 + 5x$
$\frac{dy}{dx} = 16x^3 - 3x^2 - 40x + 5$

27. You could use the quotient rule, but a better technique to use would be to rewrite the function as follows:$q(x) = \frac{3}{5}x^2 - \frac{6}{5}x + \frac{4}{5}$ then

$$q'(x) = \frac{6}{5}x - \frac{6}{5}$$

28. $q(x) = \frac{4}{3}x^3 + \frac{2}{3}x^2 - \frac{5}{3}x + \frac{13}{3}$

$$q'(x) = 4x^2 + \frac{4}{3}x - \frac{5}{3}$$

29. $y = \frac{x^2+3x-4}{2x-1}$

$$\begin{aligned} \frac{dy}{dx} &= \frac{(2x-1)(2x+3) - (x^2+3x-4)(2)}{(2x-1)^2} \\ &= \frac{2x^2+6x-2x-3-2x^2-6x+8}{(2x-1)^2} \\ &= \frac{-2x+5}{(2x-1)^2} \end{aligned}$$

30. $y = \frac{(x-1)(x-3)}{(x-1)(x+1)} = \frac{x-3}{x+1}$

$$\begin{aligned} \frac{dy}{dx} &= \frac{(x+1)(1) - (x-3)(1)}{(x+1)^2} \\ &= \frac{x+1-x+3}{(x+1)^2} \end{aligned}$$

31. $w = \frac{3t-1}{t^2-2t+6}$

$$\begin{aligned} \frac{dw}{dt} &= \frac{(t^2-2t+6)(3) - (3t-1)(2t-2)}{(t^2-2t+6)^2} \\ &= \frac{3t^2-6t+18-[6t^2-6t-2t+2]}{(t^2-2t+6)^2} \\ &= \frac{-3t^2+2t+16}{(t^2-2t+6)^2} \end{aligned}$$

32. $q = \frac{t^2+7t-1}{2t^2-3t-7}$

$$\begin{aligned} \frac{dq}{dt} &= \frac{(2t^2-3t-7)(2t+7) - (t^2+7t-1)(4t-3)}{(2t^2-3t-7)^2} \\ &= \frac{11t^2-38t-52}{(2t^2-3t-7)^2} \end{aligned}$$

33. $f(x) = \frac{x}{\frac{1}{x}+1} = \frac{x^2}{1+x}$

$$\begin{aligned} f'(x) &= \frac{(1+x)(2x) - (x^2)(1)}{(1+x)^2} \\ &= \frac{2x+2x^2-x^2}{(1+x)^2} \\ &= \frac{2x+x^2}{(1+x)^2} \end{aligned}$$

34. $f(x) = \frac{\frac{1}{x}}{x+\frac{1}{x}} = \frac{1}{x^2+1}$

$$\begin{aligned} f'(x) &= \frac{(x^2+1)(0) - 1(2x)}{(x^2+1)^2} \\ &= \frac{-2x}{(x^2+1)^2} \end{aligned}$$

35. $y = \frac{\tan t}{1+\sec t}$ Which can be rewritten as $y = \frac{\frac{\sin t}{\cos t}}{1+\frac{1}{\cos t}}$ multiplying every term by $\cos t$ to clear the fractions gives $y = \frac{\sin t}{\cos t+1}$ which has a derivative of $\frac{dy}{dx} = \frac{1}{1+\cos t}$ (see problem 21 for details)

36. Rewrite $y = \frac{\cos t}{\sin t+1}$

$$\begin{aligned} \frac{dy}{dx} &= \frac{(\sin t+1)(-\sin t) - \cos t(\cos t)}{(\sin t+1)^2} \\ &= \frac{-\sin^2 t - \sin t - \cos^2 t}{(\sin t+1)^2} \\ &= \frac{-(\sin t+1)}{(\sin t+1)^2} \\ &= \frac{-1}{\sin t+1} \end{aligned}$$

37. $w = \frac{\tan x + x\sin x}{\sqrt{x}}$

$$\begin{aligned} \frac{dw}{dx} &= \frac{\sqrt{x}(\sec^2 x + x\cos x + \sin x) - \frac{(\tan x + x\sin x)}{2\sqrt{x}}}{(\sqrt{x})^2} \\ &= \frac{2x(\sec^2 x + x\cos x + \sin x) - \tan x - x\sin x}{2x^{3/2}} \\ &= \frac{2x\sec^2 x + 2x^2\cos x + x\sin x - \tan x}{2x^{3/2}} \end{aligned}$$

38.

$$\begin{aligned}\frac{dw}{dx} &= \frac{\sqrt{x}(-sinx + x^2cscxcotx - 2xcscx) - \frac{cosx - x^2cscx}{2\sqrt{x}}}{(\sqrt{x})^2} \\ &= \frac{-sin\ x}{\sqrt{x}} - x^{3/2}csc\ x\ cot\ x + 2x^{5/2}csc\ x \\ &\quad -\frac{cos\ x}{2x^{3/2}} - \frac{\sqrt{x}csc\ x}{2}\end{aligned}$$

39. $y = \frac{1+t^{1/2}}{1-t^{1/2}}$

$$\begin{aligned}\frac{dy}{dx} &= \frac{(1-t^{1/2})(\frac{1}{2t^{1/2}}) - (1+t^{1/2})(\frac{-1}{2t^{1/2}})}{(1-t^{1/2})^2} \\ &= \frac{1 - t^{1/2} + 1 + t^{1/2}}{2t^{1/2}(1-\sqrt{t})^2} \\ &= \frac{2}{2\sqrt{t}(1-\sqrt{t})^2} \\ &= \frac{1}{\sqrt{t}(1-\sqrt{t})^2}\end{aligned}$$

40. Rewrite $y = \frac{2x-3x^2}{15x^2+9}$

$$\begin{aligned}\frac{dy}{dx} &= \frac{(15x^2+9)(2-6x) - (2x-3x^2)(30x)}{(15x^2+9)^2} \\ &= \frac{30x^2 - 90x^3 + 18 - 54x - 60x^2 - 90x^3}{(15x^2+9)^2} \\ &= \frac{-150x^3 - 60x^2 - 54x + 18}{(15x^2+9)^2}\end{aligned}$$

41. $f(t) = t\ sin\ t\ tan\ t$

$$\begin{aligned}f'(t) &= (t\ sin\ t)(sec^2t) + (tan\ t)(t\ cos\ t + sin\ t) \\ &= t\ sin\ t\ sec^2t + t\ tan\ t\ cos\ t + sin\ t\ tan\ t \\ &= t\ sin\ t\ \frac{1}{cos^2t} + t\ \frac{sin\ t}{cos\ t}cos\ t + sin\ t\ tan\ t \\ &= t\ tan\ t\ sec\ t + t\ sin\ t + sin\ t\ tan\ t\end{aligned}$$

42.

$$\begin{aligned}g'(t) &= (t\ csc\ t)(-sin\ t) + (1 + cos\ t)(-t\ csc\ t\ cot\ t + csc\ t) \\ &= -t - t\ csc\ t\ cot\ t + csc\ t - t\ cot^2t + cot\ t\end{aligned}$$

43. - 84. Left to the student

85. **a)** $f(x) = \frac{x}{x+1}$

$$\begin{aligned}f'(x) &= \frac{(x+1)(1) - x(1)}{(x+1)^2} \\ &= \frac{x+1-x}{(x+1)^2} \\ &= \frac{1}{(x+1)^2}\end{aligned}$$

b) $g(x) = \frac{-1}{x+1}$

$$\begin{aligned}g'(x) &= \frac{(x+1)(0) - (-1)(1)}{(x+1)^2} \\ &= \frac{1}{(x+1)^2} \\ &= \frac{1}{(x+1)^2}\end{aligned}$$

c) Since the graphs of both functions are similar then the average rate of change for the functions will be the same (that is why the answers in part (a) and part (b) are equal).

86. **a)** $f(x) = \frac{x^2}{x^2-1}$

$$\begin{aligned}f'(x) &= \frac{(x^2-1)(2x) - x^2(2x)}{(x^2-1)^2} \\ &= \frac{2x^3 - 2x - 2x^3}{(x^2-1)^2} \\ &= \frac{-2x}{(x^2-1)^2}\end{aligned}$$

b) $g(x) = \frac{1}{x^2-1}$

$$\begin{aligned}g'(x) &= \frac{(x^2-1)(0) - (1)(2x)}{(x^2-1)^2} \\ &= \frac{0 - 2x}{(x^2-1)^2} \\ &= \frac{-2x}{(x^2-1)^2}\end{aligned}$$

c) Since the graphs of both functions are similar then the average rate of change for the functions will be the same (that is why the answers in part (a) and part (b) are equal).

87. $f(x) = sin^2x + cos^2x$

a)

$$\begin{aligned}f'(x) &= 2sin\ x\ cos\ x + 2cos\ x(-sin\ x) \\ &= 2sin\ x\ cos\ x - 2sin\ x\ cos\ x \\ &= 0\end{aligned}$$

b) Since $sin^2x + cos^2x = 1$ (fundamental trigonometric identity) we would expect the derivative to be zero since we are taking a derivative of a constant, which is always zero.

88. **a)** $f'(x) = 2tan\ x\ sec^2x$

b) $g'(x) = 2sec\ x\ (sec\ x\ tan\ x) = 2tan\ xsec^2x$

c) The answers in parts a) and b) are the same. The reason is that $tan^2x + 1 = sec^2x$ and therefore when we take the derivative we get the same result since the derivative of 1 is zero.

89. $y = \frac{8}{x^2+4}$

$$\begin{aligned}\frac{dy}{dx} &= \frac{(x^2+4)(0) - 8(2x)}{(x^2+4)^2} \\ &= \frac{0 - 16x}{(x^2+4)^2} \\ &= \frac{-16x}{(x^2+4)^2}\end{aligned}$$

For the point $(0, 2)$, $\frac{dy}{dx}|_{x=0} = m = \frac{-16(0)}{(0^2+4)^2} = 0$. The tangent line is

$$\begin{aligned}y - y_1 &= m(x - x_1) \\ y - 2 &= 0(x - 0) \\ y - 2 &= 0 \\ y &= 2\end{aligned}$$

For the point $(-2,-1)$, $\frac{dy}{dx}|_{x=-2} = m = \frac{-16(-2)}{((-2)^2+4)^2} = \frac{32}{8} = 4$. The tangent line is

$$\begin{aligned} y - y_1 &= m(x - x_1) \\ y - 1) &= 4(x - (-2)) \\ y - 1 &= 4x + 8 \\ y &= 4x + +8 + 1 \\ y &= 4x + 9 \end{aligned}$$

90. $y = \frac{4x}{1+x^2}$

$$\begin{aligned} \frac{dy}{dx} &= \frac{(1+x^2)(4) - 4x(2x)}{(1+x^2)^2} \\ &= \frac{4 + 4x^2 - 4x^2}{(1+x^2)^2} \\ &= \frac{4}{(1+x^2)^2} \end{aligned}$$

For the point $(0,0)$, $\frac{dy}{dx}|_{x=0} = m = 4$. The tangent line is

$$\begin{aligned} y - y_1 &= m(x - x_1) \\ y - 0 &= 4(x - 0) \\ y &= 4 \end{aligned}$$

For the point $(-1,-2)$, $\frac{dy}{dx}|_{x=-1} = 1$. The tangent line is

$$\begin{aligned} y - y_1 &= m(x - x_1) \\ y - (-2) &= 1(x - (-1)) \\ y + 2 &= x + 1 \\ y &= x - 1 \end{aligned}$$

91. $y = \frac{\sqrt{x}}{x+1} = \frac{x^{1/2}}{x+1}$

$$\begin{aligned} \frac{dy}{dx} &= \frac{(x+1)(\frac{1}{2}x^{-1/2}) - x^{1/2}(1)}{(x+1)^2} \\ &= \frac{\frac{1}{2}x^{1/2} + \frac{1}{2}x^{-1/2} - x^{1/2}}{(x+1)^2} \\ &= \frac{\frac{1}{2}x^{-1/2}}{(x+1)^2} \\ &= \frac{1}{2\sqrt{x}(x+1)^2} \end{aligned}$$

When $x = 1$, $y = \frac{\sqrt{1}}{1+1} = \frac{1}{2}$, and $\frac{dy}{dx}_{x=1} = m = \frac{1}{2\sqrt{1}(1+1)^2} = \frac{1}{8}$. The tangent line is

$$\begin{aligned} y - y_1 &= m(x - x_1) \\ y - \frac{1}{2} &= \frac{1}{8}(x - 1) \\ y - \frac{1}{2} &= \frac{1}{8}x - \frac{1}{8} \\ y &= \frac{1}{8}x - \frac{1}{8} + \frac{1}{2} \\ y &= \frac{1}{8}x - \frac{1}{8} + \frac{4}{8} \\ y &= \frac{1}{8}x + \frac{3}{8} \end{aligned}$$

When $x = \frac{1}{4}$, $y = \frac{\sqrt{\frac{1}{4}}}{1+\frac{1}{4}} = \frac{2}{5}$, and $\frac{dy}{dx}|_{x=\frac{1}{4}} = \frac{1}{2\sqrt{\frac{1}{4}}(1+\frac{1}{4})^2} = \frac{16}{25}$. The tangent line is

$$\begin{aligned} y - y_1 &= m(x - x_1) \\ y - \frac{2}{5} &= \frac{16}{25}(x - \frac{1}{4}) \\ y - \frac{2}{5} &= \frac{16}{25}x - \frac{4}{25} \\ y &= \frac{16}{25}x - \frac{4}{25} + \frac{2}{5} \\ y &= \frac{16}{25}x - \frac{4}{25} + \frac{10}{25} \\ y &= \frac{16}{25}x + \frac{6}{25} \end{aligned}$$

92. $y = \frac{x^2+3}{x-1}$

$$\begin{aligned} \frac{dy}{dx} &= \frac{(x-1)(2x) - (x^2+3)(1)}{(x-1)^2} \\ &= \frac{2x^2 - 2x - x^2 - 3}{(x-1)^2} \\ &= \frac{(x-3)(x+1)}{(x-1)^2} \end{aligned}$$

When $x = 2$, $y = \frac{2^2+3}{2-1} = 7$, and $\frac{dy}{dx}|_{x=2} = -1$. The tangent line is

$$\begin{aligned} y - y_1 &= m(x - x_1) \\ y - 7 &= -1(x - 2) \\ y - 7 &= -x + 2 \\ y &= -x + 9 \end{aligned}$$

When $x = 3$, $y = \frac{3^2+3}{3-1} = 6$ and $\frac{dy}{dx}|_3 = 0$. The tangent line is

$$\begin{aligned} y - y_1 &= m(x - x_1) \\ y - 6 &= 0(x - 3) \\ y - 6 &= 0 \\ y &= 6 \end{aligned}$$

93. $y = x\ sin\ x$

$$\frac{dy}{dx} = x\ cos\ x + sin\ x$$

When $x = \frac{\pi}{4}$

$$\begin{aligned} \frac{dy}{dx}|_{x=\frac{\pi}{4}} &= m \\ &= \frac{\pi}{4}cos(\frac{\pi}{4}) + sin(\frac{\pi}{4}) \\ &= \frac{\pi}{4}\frac{\sqrt{2}}{2} + \frac{\sqrt{2}}{2} \\ &= \frac{\sqrt{2}(\pi + 4)}{8} \end{aligned}$$

The tangent line is

$$y - y_1 = m(x - x_1)$$

$$\begin{aligned} y-\frac{\sqrt{2}\pi}{8} &= \sqrt{2}\left(\frac{\pi+4}{8}\right)(x-\frac{\pi}{4}) \\ y &= \sqrt{2}\left(\frac{\pi+4}{8}\right)x-\frac{\sqrt{2}\pi}{4}\left(\frac{\pi+4}{8}\right)+\frac{\sqrt{2}\pi}{8} \\ y &= \sqrt{2}\left(\frac{\pi+4}{8}\right)x-\frac{\pi^2}{16\sqrt{2}}-\frac{\sqrt{2}\pi}{8}+\frac{\sqrt{2}\pi}{8} \\ y &= \sqrt{2}\left(\frac{\pi+4}{8}\right)x-\frac{\pi^2}{16\sqrt{2}} \end{aligned}$$

94. $y = x\ tan\ x$

$$\frac{dy}{dx} = x\ sec^2x + tan\ x$$

When $x = \frac{\pi}{4}$

$$\begin{aligned} \frac{dy}{dx}\Big|_{x=\frac{\pi}{4}} &= m \\ &= \frac{\pi}{4}sec^2(\frac{\pi}{4}) + tan(\frac{\pi}{4}) \\ &= \frac{\pi}{4}(\frac{1}{2}) + 1 \\ &= \frac{\pi+8}{8} \end{aligned}$$

The tangent line is

$$\begin{aligned} y-y_1 &= m(x-x_1) \\ y-\frac{\pi}{4} &= \left(\frac{\pi+8}{8}\right)(x-\frac{\pi}{4}) \\ y &= \left(\frac{\pi+8}{8}\right)x-\frac{\pi}{4}\left(\frac{\pi+8}{8}\right)+\frac{\pi}{4} \\ y &= \left(\frac{\pi+8}{8}\right)x-\frac{\pi^2}{32} \end{aligned}$$

95. **a)** $T(t) = \frac{4t}{t^2+1} + 98.6$

$$\begin{aligned} \frac{dT}{dt} &= \frac{(t^2+1)(4)-4t(2t)}{(t^2+1)^2} + 0 \\ &= \frac{4t^2+4-8t^2}{(t^2+1)^2} \\ &= \frac{-4t^2+4}{(t^2+1)^2} \end{aligned}$$

b) When $t = 2$ hours

$$\begin{aligned} T &= \frac{4(2)}{2^2+1} + 98.6 \\ &= \frac{8}{5} + 98.6 \\ &= 100.2 \text{ degrees} \end{aligned}$$

c) When $t = 2$ hours

$$\begin{aligned} \frac{dT}{dt} &= \frac{-4(2)^2+4}{(2^2+1)^2} \\ &= \frac{-12}{5} \\ &= -2.4 \text{ degrees per hour} \end{aligned}$$

96. **a)** $P(t) = 10 + \frac{50t}{2t^2+9}$

$$\begin{aligned} \frac{dP}{dt} &= \frac{(2t^2+9)(50)-(50t)(4t)}{(2t^2+9)^2} \\ &= \frac{-100t^2+450}{(2t^2+9)^2} \end{aligned}$$

b) When $t = 8$ years

$$\begin{aligned} P &= 10 + \frac{50(8)}{2(8)^2+9} \\ &= 10 + \frac{400}{137} \\ &\approx 12.920 \end{aligned}$$

The population will reach approximately 12920 people after 8 years

c) When $t = 12$

$$\begin{aligned} \frac{dP}{dt} &= \frac{-100(12)^2+450}{(2(12)^2+9)^2} \\ &= \frac{-14400}{88209} \approx -0.163 \\ &\approx -163 \text{ people per year} \end{aligned}$$

97. **a)** Since $\frac{s(t)}{100} = tan\ t$ then $s(t) = 100\ tan\ t$

b) $\frac{d\ s(t)}{dt} = 100sec^2t$

c)

$$\begin{aligned} 100sec^2t &= 200 \\ sec^2t &= 2 \\ \frac{1}{cos^2t} &= 2 \\ cos^2t &= \frac{1}{2} \\ cos\ t &= \pm\frac{1}{\sqrt{2}} \\ t &= \pm\frac{\pi}{4}2n\pi \\ &= \frac{\pi}{4} + \frac{n\pi}{2} \end{aligned}$$

98. **a)**

$$\begin{aligned} v(t) &= s'(t) \\ &= \frac{(\sqrt{t}+1)(-3\ sin\ t)-(3\ cos\ t)(\frac{1}{2\sqrt{t}})}{(\sqrt{t}+1)^2} \end{aligned}$$

b)

$$\begin{aligned} v(1) &= \frac{(\sqrt{1}+1)(-3\ sin\ 1)-(3\ cos\ 1)(\frac{1}{2\sqrt{1}})}{(\sqrt{1}+1)^2} \\ &= -1.465\ in/sec \end{aligned}$$

NOTE: the negative sign indicates a downward movement.

c)

$$\begin{aligned} v(1) &= \frac{(\sqrt{\frac{\pi}{3}}+1)(-3\ sin\ \frac{\pi}{3}) - (3\ cos\ \frac{\pi}{3})(\frac{1}{2\sqrt{\frac{\pi}{3}}})}{(\sqrt{\frac{\pi}{3}}+1)^2} \\ &= 1.487\ in/sec \end{aligned}$$

99. $g(x) = \frac{(x^2+1)tan\ x}{(x^2-1)}$

$$\begin{aligned} g'(x) &= \frac{(x^2-1)sec^2x(x^2+1)+2x\ tan\ x}{(x^2-1)^2} - \\ &\quad \frac{2x(x^2+1)tan\ x}{(x^2-1)^2} \\ &= \frac{(x^4-1)sec^2x}{(x^2-1)^2} + \frac{2x\ tan\ x}{(x^2-1)^2} - \\ &\quad \frac{2x\ tan\ x}{(x^2-1)^2} - \frac{2x^3 tan\ x}{(x^2-1)^2} \\ &= \frac{(x^4-1)sec^2x}{(x^2-1)^2} - \frac{2x^3\ tan\ x}{(x^2-1)^2} \\ &= \frac{sec^2x(x^4-2x^3\ sin\ x\ cos\ x - 1)}{(x^2-1)^2} \\ &= \frac{sec^2x(x^4 - x^3\ sin(2x) - 1)}{(x^2-1)^2} \end{aligned}$$

100.

$$\begin{aligned} g'(t) &= \frac{(t^3+1)[(-csc\ x\ cot\ x)(t^3-1)+csc\ x(3t^2)]}{(t^3+1)^2} - \\ &\quad \frac{3t^2(t^3-1)csc\ x}{(t^3+1)^2} \\ &= \frac{[(-csc\ x\ cot\ x)(t^3-1)+csc\ x(3t^2)]}{(t^3+1)} - \\ &\quad \frac{3t^2(t^3-1)csc\ x}{(t^3+1)^2} \end{aligned}$$

101. $s(t) = \frac{tan\ t}{t\ cos\ t}$

$$\begin{aligned} s'(t) &= \frac{t\ cos\ tsec^2t - tan\ t(-tsin\ t + cos\ t)}{(t\ cos\ t)^2} \\ &= \frac{t\ sec\ t + t\ sin\ t\ tan\ t - sin\ t}{t^2\ cos^2 t} \\ &= \frac{sec^2t(t\ sec\ t + t\ sin\ t\ tan\ t - sin\ t}{t^2} \\ &= \frac{t\ sec^3t + t\ sin\ t\ tan\ tsec^2t - sin\ t\ sec^2t}{t^2} \\ &= \frac{sec\ t(t\ sec^2t + t\ sin\ t\ tan\ t\ sec\ t - sin\ t\ sec\ t)}{t^2} \\ &= \frac{sec\ t(t\ sec^2t + t\ tan^2\ t - tan\ t)}{t^2} \\ &= \frac{sec\ t(t\ sec^2t + tan\ t(t\ tan\ t - 1))}{t^2} \end{aligned}$$

102. Rewrite $f(x) = \frac{x\ sin\ x - cos\ x}{x\ sin\ x + cos\ x}$,

$$\begin{aligned} f'(x) &= \frac{(x\ sin\ x + cos\ x)(x\ cos\ x + sin\ x + sin\ x)}{(x\ sin\ x + cos\ x)^2} \\ &\quad - \frac{(x\ sin\ x - cos\ x)(x\ cos\ x + sin\ x - sin\ x)}{(x\ sin\ x + cos\ x)^2} \\ &= \frac{2x\ sin^2x + x\ cos^2x}{(x\ sin\ x + cos\ x)^2} \end{aligned}$$

103.

$$\begin{aligned} g'(x) &= \frac{(x+cosx)[-xsinx + cosx(xcosx+sinx) - sec^2x]}{(x+cos\ x)^2} \\ &\quad - \frac{(1-sinx)[xsinxcosx - tanx]}{(x+cosx)^2} \\ &= \frac{-x^2sin^2x + x^2cos^2x + xcosxsinx - xsec^2x}{(x+cosx)^2} \\ &\quad - \frac{xcosxsin^2x + xcos^3x + cos^2xsinx - secx}{(x+cosx)^2} \\ &\quad - \frac{xsinxcosx + tanx + xsin^2xcosx - sinxtanx}{(x+cosx)^2} \\ &= \frac{x\ cos^2x + (x^2 + sin\ x)cos^2x - x\ sec^2x}{(x+cos\ x)^2} \\ &\quad + \frac{-x^2sin^2x - sec\ x - sin\ x\ tan\ x + tan\ x}{(x+cos\ x)^2} \end{aligned}$$

104. $f(x) = \frac{x^{1/2}sin\ x - x^{3/2}cos\ x}{x^2+2x+3}$

$$\begin{aligned} f'(x) &= \frac{(x^2+2x+3)[x^{1/2}cos\ x + \frac{sin\ x}{2\sqrt{x}}]}{(x^2+2x+3)^2} \\ &\quad - \frac{(-x^{3/2}sin\ x + \frac{3}{2}x^{1/2}cos\ x)}{(x^2+2x+3)^2} \\ &\quad \frac{(\sqrt{x}sin\ x - x^{3/2}cos\ x)(2x+2)}{(x^2+2x+3)^2} \end{aligned}$$

105. Let $y = (x-1)(x-2)(x-3)$

a)

$$\begin{aligned} \frac{dy}{dx} &= (x-1)[(x-2)(1)+(x-3)(1)] + [(x-2)(x-3)](1) \\ &= (x-1)(x-2) + (x-1)(x-3) + (x-2)(x-3) \\ &= 3x^2 - 12x + 12 \end{aligned}$$

b) $y = (2x+1)(3x-5)(-x+3)$

$$\begin{aligned} \frac{dy}{dx} &= (2x+1)(3x-5)(-1) + (2x+3)(-x+3)(3) \\ &\quad +(3x-5)(-x+3)(2) \\ &= -(2x+1)(3x-5) + 3(2x+1)(-x+3) \\ &\quad +2(3x-5)(-x+3) \\ &= -18x^2 + 50x - 16 \end{aligned}$$

c) The derivative of a product of three functions is the sum of all possible combinations consisting of the product of two functions and the derivative of the third function.

d) The derivative of more than three function is the sum of all possible combinations consisting of the product of three of the functions and the derivative of the fourth function.
Let $y = x(x+1)(2x+3)(-x+1) = -2x^4 - 3x^3 + 2x^2 + 3x$ which has a derivative $y' = -8x^3 - 9x^2 + 4x + 3$. Let us use the rule to find the derivative of y:

$$\begin{aligned} y' &= -x(x+1)(2x+3) + x(2x+3)(-x+1) + \\ &\quad 2x(x+1)(-x+1) + (x+1)(2x+3)(-x+1) \\ &= -2x^3 - 5x^2 - 3x - 2x^3 - x^2 + 3x \\ &\quad -2x^3 + 2x - 2x^3 - 3x^2 + 2x + 3 \\ &= -8x^3 - 9x^2 + 4x + 3 \end{aligned}$$

106. $(\pm\, 1.414, -4)$

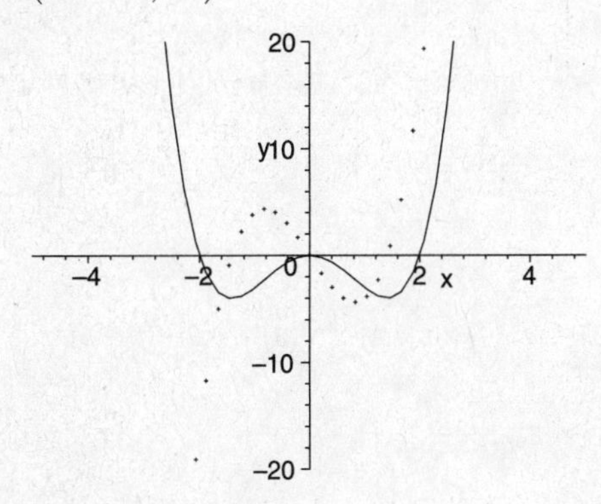

107. No horizontal tangent lines

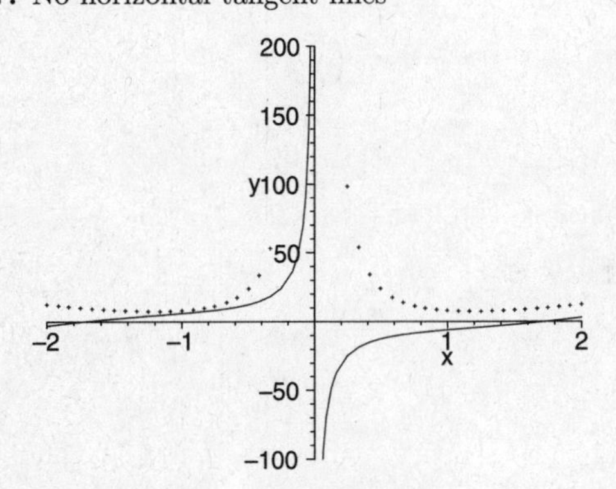

108. $(0, -1)$

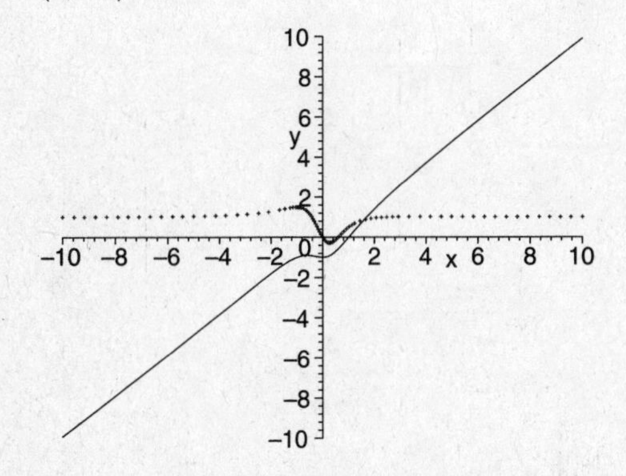

109. $(0.2, 0.75)$ and $(-0.2, -0.75)$

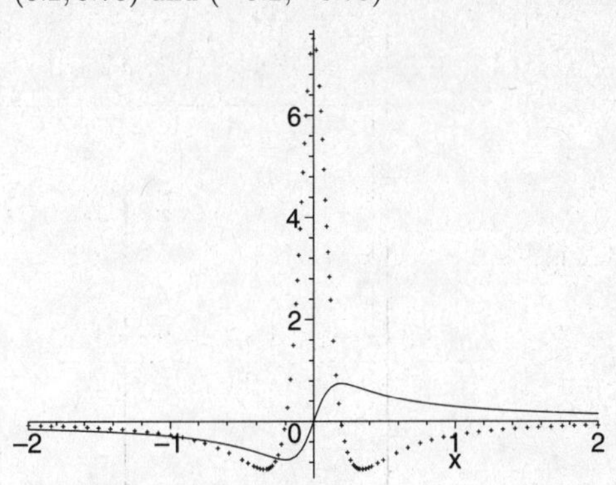

110. $(0, 0)$ and $(\pm\, 0.213, 0.0164)$

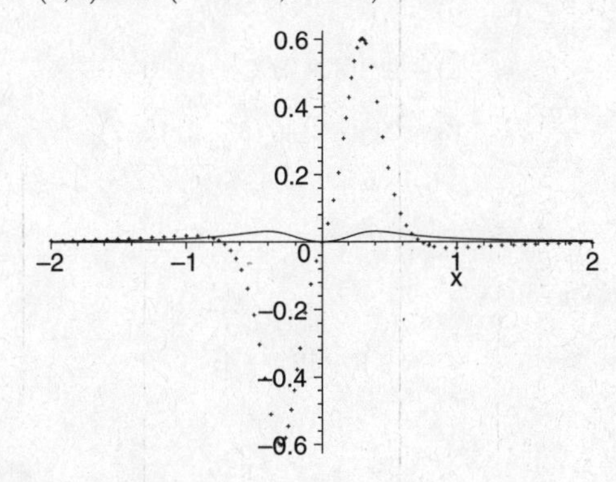

111. $(1, 2)$ and $(-1, -2)$

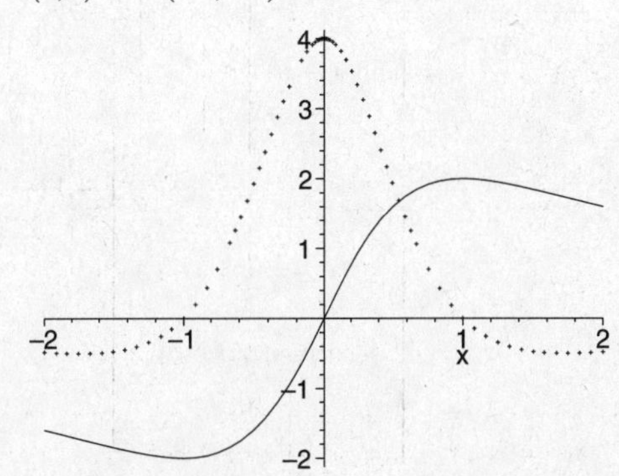

112. $y = \frac{4-4x^2}{(x^2+1)^2}$ seems to be the correct derivative of $y = \frac{4x}{x^2+1}$

Exercise Set 2.8

1. $y = (2x+1)^2$
Method One (chain rule):

$$\begin{aligned} \frac{dy}{dx} &= 2(2x+1)(2) \\ &= 4(2x+1) \\ &= 8x+4 \end{aligned}$$

Method Two (product rule): $y = (2x+1)(2x+1)$

$$\begin{aligned} \frac{dy}{dx} &= (2x+1)(2) + (2x+1)(2) \\ &= 4x+2+4x+2 \\ &= 8x+4 \end{aligned}$$

Method Three (expand first):

$$\begin{aligned} y &= 4x^2 + 4x + 1 \\ \frac{dy}{dx} &= 4(2x) + 4 \\ &= 8x + 4 \end{aligned}$$

2. $y = (3 - 2x)^2$
Method One:

$$\begin{aligned} \frac{dy}{dx} &= 2(3 - 2x)(-2) \\ &= -4(3 - 2x) \\ &= -12 + 8x \end{aligned}$$

Method Two:

$$\begin{aligned} y &= (3 - 2x)(3 - 2x) \\ \frac{dy}{dx} &= (3 - 2x)(-2) + (3 - 2x)(-2) \\ &= -6 + 4x - 6 + 4x \\ &= -12 + 8x \end{aligned}$$

Method Three:

$$\begin{aligned} y &= 9 - 12x + 4x^2 \\ \frac{dy}{dx} &= -12 + 4x \end{aligned}$$

3. $y = (1 - x)^{55}$

$$\begin{aligned} \frac{dy}{dx} &= 55(1 - x)^{54}(-1) \\ &= -55(1 - x)^{54} \end{aligned}$$

4. $y = tan^2 x$, $\frac{dy}{dx} = 2\ tan\ x\ sec^2 x$

5. $sec^2 x$

$$\begin{aligned} \frac{dy}{dx} &= 2\ sec\ x\ sec\ x\ tan\ x \\ &= 2\ tan\ x\ sec^2 x \end{aligned}$$

6. $y = (2x + 3)^{10}$
$\frac{dy}{dx} = 10(2x + 3)^9(2) = 20(2x + 3)^9$

7. $y = \sqrt{1 - 3x} = (1 - 3x)^{1/2}$

$$\begin{aligned} \frac{dy}{dx} &= \frac{1}{2}(1 - 3x)^{-1/2}(-3) \\ &= \frac{-3}{2\sqrt{1 - 3x}} \end{aligned}$$

8. $y = \sqrt[3]{x^2 + 1} = (x^2 + 1)^{1/3}$

$$\begin{aligned} \frac{dy}{dx} &= \frac{1}{3}(x^2 + 1)^{-2/3}(2x) \\ &= \frac{2x}{3\sqrt[3]{(x^2 + 1)^2}} \end{aligned}$$

9. $y = \frac{2}{3x^2+1} = 2(3x^2 + 1)^{-1}$

$$\begin{aligned} \frac{dy}{dx} &= 2(-1)(3x^2 + 1)^{-2}(6x) \\ &= \frac{-12x}{(3x^2 + 1)^2} \end{aligned}$$

10. $y = 3(2x + 4)^{-1/2}$

$$\begin{aligned} \frac{dy}{dx} &= 3(-\frac{1}{2}(2x + 4)^{-3/2}(2)) \\ &= \frac{-3}{\sqrt{(2x + 4)^3}} \end{aligned}$$

11. $s(t) = t(2t + 3)^{1/2}$

$$\begin{aligned} s'(t) &= t\left(\frac{1}{2}(2t + 3)^{-1/2}(2)\right) + (2t + 3)^{1/2}(1) \\ &= \frac{t}{\sqrt{2t + 3}} + \sqrt{2t + 3} \end{aligned}$$

12.

$$\begin{aligned} s'(t) &= t^2\left(\frac{1}{3}(3t + 4)^{-2/3}(3)\right) + 2t(3t + 4)^{1/3} \\ &= \frac{t^2}{\sqrt[3]{3t + 4}} + 2t\ \sqrt[3]{3t + 4} \end{aligned}$$

13. $s(t) = sin\left(\frac{\pi}{6}\ t + \frac{\pi}{3}\right)$

$$\begin{aligned} s'(t) &= cos\left(\frac{\pi}{6}\ t + \frac{\pi}{3}\right) \cdot \frac{\pi}{6} \\ &= \frac{\pi}{6}\ cos\left(\frac{\pi}{6}\ t + \frac{\pi}{3}\right) \end{aligned}$$

14. $s'(t) = -sin(3t - 4)(3) = -3sin(3t - 4)$

15. $g(x) = (1 + x^3)^3 - (1 + x^3)^4$

$$\begin{aligned} g'(x) &= 3(1 + x^3)^2(3x^2) - 4(1 + x^3)^3(3x^2) \\ &= 9x^2(1 + x^3)^2 - 12x^2(1 + x^3)^3 \end{aligned}$$

16.

$$\begin{aligned} \frac{dy}{dx} &= \frac{1}{2}(1 + sec\ x)^{-1/2}(sec\ x\ tan\ x) \\ &= \frac{sec\ x\ tan\ x}{2\sqrt{1 + sec\ x}} \end{aligned}$$

17. $y = \sqrt{1 - csc\ x} = (1 - csc\ x)^{1/2}$

$$\begin{aligned} \frac{dy}{dx} &= \frac{1}{2}(1 - csc\ x)^{-1/2}(-(-csc\ x\ cot\ x)) \\ &= \frac{csc\ x\ cot\ x}{2\sqrt{1 - csc\ x}} \end{aligned}$$

18.

$$\begin{aligned} g'(x) &= \frac{1}{2}x^{-1/2} + 3(x - 3)^2 \\ &= \frac{1}{2\sqrt{x}} + 3(x - 3)^2 \end{aligned}$$

19. $g(x) = (2x-1)^{1/3} + (4-x)^2$

$$\begin{aligned} g'(x) &= \frac{1}{3}(2x-1)^{-2/3}(2) + 2(4-x)(-1) \\ &= \frac{2}{3\sqrt[3]{(2x-1)^2}} - 2(4-x) \end{aligned}$$

20.

$$\begin{aligned} \frac{dy}{dx} &= x^2(-\sin x) + 2x \cos x - 2x \cos x \\ &\quad +2\sin x - 2(-\csc^2 x) \\ &= -x^2 \sin x + 2\sin x + 2\csc^2 x \end{aligned}$$

21. $y = x^3 \sin x + 5x \cos x + 4\sec x$

$$\begin{aligned} \frac{dy}{dx} &= x^3 \cos x + 3x^2 \sin x + 5x(-\sin x) \\ &\quad +5\cos x + 4\sec x \tan x \\ &= x^3 \cos x + 3x^2 \sin x - 5x \sin x \\ &\quad +5\cos x + 4\sec x \tan x \end{aligned}$$

22.

$$\begin{aligned} f'(x) &= (2x+3)^{1/2}(2x+3) + \frac{1}{2}(2x+3)^{-1/2}(2)(x^2+3x+1) \\ &= (2x+3)^{3/2} + \frac{x^2+3x+1}{\sqrt{2x+3}} \end{aligned}$$

23. $y = (x^2+x^3)^{1/2}(2x^2+3x+5)$

$$\begin{aligned} \frac{dy}{dx} &= (x^2+x^3)^{1/2}(4x+3) \\ &\quad +\frac{1}{2}(x^2+x^3)^{-1/2}(2x+3x^2)(2x^2+3x+5) \\ &= (4x+3)\sqrt{x^2+x^3} + \frac{(2x+3x^2)(2x^2+3x+5)}{2\sqrt{x^2+x^3}} \end{aligned}$$

24. $\frac{dy}{dx} = -\csc(1/x)(-1/x^2) = \frac{\csc(1/x)}{x^2}$

25. $f(t) = \cos\sqrt{t}$

$$\begin{aligned} f'(t) &= -\sin\sqrt{t}\left(\frac{1}{2}t^{-1/2}\right) \\ &= \frac{-\sin\sqrt{t}}{2\sqrt{t}} \end{aligned}$$

26.

$$\begin{aligned} f'(t) &= \cos\sqrt{t}\left(\frac{1}{2}t^{-1/2}\right) \\ &= \frac{\cos\sqrt{t}}{2\sqrt{t}} \end{aligned}$$

27. $f(x) + (3x+2)(2x+5)^{1/2}$

$$\begin{aligned} f'(x) &= (3x+2)\left(\frac{1}{2}(2x+5)^{-1/2}(2)\right) + 3(2x+5)^{1/2} \\ &= \frac{3x+2}{\sqrt{2x+5}} + 3\sqrt{2x+5} \end{aligned}$$

28.

$$\begin{aligned} f'(x) &= (5x-2)\left(\frac{1}{2}(3x+4)^{-1/2}(3)\right) + 5(3x+4)^{1/2} \\ &= \frac{3(5x-2)}{2\sqrt{3x+4}} + 5\sqrt{3x+5} \end{aligned}$$

29.

$$\begin{aligned} \frac{dy}{dx} &= \cos(\cos x)(-\sin x) \\ &= -\sin x \cos(\cos x) \end{aligned}$$

30. $\frac{dy}{dx} = \sec^2(\sin x)(\cos x)$

31. $y = (\cos(4t))^{1/2}$

$$\begin{aligned} \frac{dy}{dx} &= \frac{1}{2}(\cos(4t))^{-1/2}(-\sin(4t)(4)) \\ &= \frac{-2\sin(4t)}{\sqrt{\cos(4t)}} \end{aligned}$$

32.

$$\begin{aligned} f'(x) &= \frac{(x^3-1)^5[4(x^2+3)^3(2x)]}{(x^3-1)^{10}} - \\ &\quad \frac{(x^2+3)^4[5(x^3-1)^4(3x^2)]}{(x^3-1)^{10}} \\ &= \frac{8x(x^3-1)^5(x^2+3)^3 - 15x^2(x^2+3)^4(x^3-1)^4}{(x^3-1)^{10}} \end{aligned}$$

33. $f(x) = \frac{(x^3+2x^2+3x-1)^3}{(2x^4+1)^2}$

$$\begin{aligned} f'(x) &= \frac{(2x^4+1)^2[3(x^3+2x+3x-1)^2(3x^2+4x+3)]}{(2x^4+1)^4} - \\ &\quad \frac{(x^3+2x^2+3x-1)^3[2(2x^4+1)(8x^3)]}{(2x^4+1)^4} \\ &= \frac{3(x^3+2x^2+3x-1)^2(2x^4+1)^2(3x^2+4x+3)}{(2x^4+1)^4} - \\ &\quad \frac{16x^3(x^3+2x^2+3x-1)^3(2x^4+1)}{(2x^4+1)^4} \end{aligned}$$

34. $y = \left(\frac{2x+3}{3x-5}\right)^{1/3}$

$$\begin{aligned} \frac{dy}{dx} &= \frac{1}{3}\left(\frac{2x+3}{3x-5}\right)^{-2/3}\left[\frac{(3x-5)(2)-(2x+3)(3)}{(3x-5)^2}\right] \\ &= \frac{6x-10-6x-9}{3(3x-5)^{4/3}\sqrt[3]{(3x-5)^2}} \\ &= \frac{-19}{3(3x-5)^{4/3}\sqrt[3]{(3x-5)^2}} \end{aligned}$$

35. $y = \left(\frac{3x-4}{5x+3}\right)^{1/2}$

$$\begin{aligned} \frac{dy}{dx} &= \frac{1}{2}\left(\frac{3x-4}{5x+3}\right)^{-1/2}\left[\frac{(5x+3)(3)-(3x-4)(5)}{(5x+3)^2}\right] \\ &= \frac{15x+9-15x+20}{2\sqrt{3x-4}(5x+3)^{3/2}} \\ &= \frac{29}{2\sqrt{3x-4}(5x+3)^{3/2}} \end{aligned}$$

36.

$$\begin{aligned} f'(x) &= sec^2(x\sqrt{x-1})\left(x[\frac{1}{2}(x-1)^{-1/2}]+(x-1)^{1/2}(1)\right) \\ &= \frac{x\ sec^2(x\sqrt{x-1})}{2\sqrt{x-1}}+\sqrt{x-1}\ sec^2(x\sqrt{x-1}) \end{aligned}$$

37. $r(x) = x(0.01x^2 + 2.391x - 8.51)^5$

$$\begin{aligned} r'(x) &= x[5(0.01x^2+2.391x-8.51)^4(0.02x+2.391)]+ \\ &\quad (0.01x^2+2.391x-8.51)^5(1) \\ &= 5x(0.02x+2.391)(0.01x^2+2.391x-8.51)^4+ \\ &\quad (0.01x^2+2.391x-8.51)^5 \\ &= (0.01x^2+2.391x-8.51)^4(5x(0.02x+2.391)+ \\ &\quad (0.01x^2+2.391x-8.51)) \\ &= (0.01x^2+2.391x-8.51)^4(0.1x^2+11.955x+ \\ &\quad 0.01x^2+2.391x-8.51) \\ &= (0.01x^2+2.391x-8.51)^4(0.11x^2+ \\ &\quad 14.346x-8.51) \end{aligned}$$

38. $r(x) = (3.21x - 5.87)^3(2.36x - 5.45)^5$

$$\begin{aligned} r'(x) &= (3.21x-5.87)^3[5(2.36x-5.45)^4(2.36)]+ \\ &\quad (2.36x-5.45)^5[3(3.21x-5.87)^2(3.21)] \\ &= (2.36x-5.45)^4(3.21x-5.87)^2(11.8) \\ &\quad (3.21x-5.87)+9.63(2.36x-5.45)) \\ &= (2.36x-5.45)^4(3.21x-5.87)^2(60.6x-121.75) \end{aligned}$$

39. $y = (cot\ 5x - cos\ 5x)^{1/5}$

$$\begin{aligned} \frac{dy}{dx} &= \frac{1}{5}(cot\ 5x-cos\ 5x)^{-4/5}(-csc^2(5x)(5)+sin(5x)(5)) \\ &= \frac{sin\ 5x-csc\ 5x}{(cot\ 5x-cos\ 5x)^{4/5}} \end{aligned}$$

40.

$$\begin{aligned} \frac{dy}{dx} &= -csc\ x-cos(cos^2x)(2cos\ x(-sin\ x)) \\ &= -csc\ x+2sin\ x\ cos\ x\ cos(cos^2x) \end{aligned}$$

41. $y = sin(sec^4(x^2))$

$$\begin{aligned} \frac{dy}{dx} &= cos(sec^4(x^2))\cdot 4sec^3(x^2)(sec\ x^2\ tan\ x^2)\cdot 2x \\ &= 8x\ sec^4(x^2)\ tan(x^2)\ cos(sec^4(x^2)) \end{aligned}$$

42. $y = ((2x + 3)^{1/2} + 1)^{1/2}$

$$\begin{aligned} \frac{dy}{dx} &= \frac{1}{2}((2x+3)^{1/2}+1)^{-1/2}\left(\frac{1}{2}(2x+3)^{-1/2}(2)\right) \\ &= \frac{1}{2\sqrt{\sqrt{2x+3}+1}\sqrt{2x+3}} \end{aligned}$$

43. $y = ((x^2 + 2)^{1/4} + 1)^{1/3}$

$$\begin{aligned} \frac{dy}{dx} &= \frac{1}{3}((x^2+2)^{1/4}+1)^{-2/3}\left(\frac{1}{4}(x^2+2)^{-3/4}(2x)\right) \\ &= \frac{x}{6\sqrt[4]{(x^2+2)^3}\sqrt[3]{(\sqrt[4]{x^2+2}+1)^2}} \end{aligned}$$

44.

$$\begin{aligned} \frac{dy}{dx} &= \frac{(x+sinx)^2(1)-x[2(x+xsinx)(1+xcosx+sinx)]}{(x+sinx)^4} \\ &= \frac{(x+sinx)^2+(-2x^2-2x^2sinx)(1+xcosx+sinx)}{(x+sin\ x)^4} \end{aligned}$$

45. $y = \frac{sin^3 x}{x^2+5}$

$$\begin{aligned} \frac{dy}{dx} &= \frac{(x^2+5)3sin^2x\ cos\ x-sin^3x(2x)}{(x^2+5)^2} \\ &= \frac{3x^2\ sin^2x\ cos\ x+15sin^2x\ cos\ x-2x\ sin^3x}{(x^2+5)^2} \\ &= \frac{sin^2x(3x^2\ cos\ x-2x\ sin\ x+15\ cos\ x}{(x^2+5)^2} \end{aligned}$$

46.

$$\begin{aligned} \frac{dy}{dx} &= 2\ tan(\sqrt{t+2})sec^2(\sqrt{t+2})\cdot\frac{1}{2\sqrt{t+2}} \\ &= \frac{tan(\sqrt{t+2})\ sec^2(\sqrt{t+2})}{\sqrt{t+2}} \end{aligned}$$

47. $f(x) = cot^3(x\ sin(2x + 4))$

$$\begin{aligned} f'(x) &= 3cot^2(x\ sin(2x+4))(-csc^2(x\ sin(2x+4))\times \\ &\quad [x\ cos(2x+4)(2)+sin(2x+4)] \\ &= -3cot^2(x\ sin(2x+4))csc^2(x\ sin(2x+4))\times \\ &\quad [2x\ cos(2x+4)+sin(2x+4)] \end{aligned}$$

48.

$$\begin{aligned} \frac{dy}{dx} &= \frac{1}{2\sqrt{2+cos^2t}}\cdot 2\ cos\ t\ (-sin\ t) \\ &= \frac{-cos\ t\ sin\ t}{\sqrt{2+cos^2t}} \end{aligned}$$

49. $y = \sqrt{sec^4x + x}$

$$\begin{aligned} \frac{dy}{dx} &= \frac{1}{2\sqrt{sec^4x+x}}\cdot(4sec^3x(sec\ x\ tan\ x)+1) \\ &= \frac{2\ sec^4\ x\ tan\ x+1}{\sqrt{sec^4x+x}} \end{aligned}$$

50.

$$\begin{aligned} \frac{dy}{dx} &= \frac{1}{2\sqrt{x+csc\ x}}\cdot(1-csc\ x\ cot\ x) \\ &= \frac{1-csc\ x\ cot\ x}{2\sqrt{x+csc\ x}} \end{aligned}$$

51. $y = \sqrt{u} = u^{1/2}, \quad u = x^2 - 1$

$$\frac{dy}{du} = \frac{1}{2}u^{1/2-1} = \frac{1}{2}u^{-1/2} = \frac{1}{2\sqrt{u}}$$

$$\frac{du}{dx} = 2x$$

$$\begin{aligned} \frac{dy}{dx} = \frac{dy}{du}\cdot\frac{du}{dx} &= \frac{1}{2\sqrt{u}}\cdot 2x \\ &= \frac{2x}{2\sqrt{x^2-1}} \\ &= \frac{x}{\sqrt{x^2-1}} \end{aligned}$$

52. $y = \frac{15}{u^3} = 15u^{-3}$, $u = 2x+1$

$\frac{dy}{du} = -45u^{-4}$, or $\frac{-45}{u^4}$

$\frac{du}{dx} = 2$

$\frac{dy}{dx} = \frac{dy}{du} \cdot \frac{du}{dx} = \frac{-45}{u^4} \cdot 2 = \frac{-90}{(2x+1)^4}$

53. $y = u^{50}$, $u = 4x^3 - 2x^2$

$\frac{dy}{du} = 50u^{49}$

$\frac{du}{dx} = 12x^2 - 4x$

$\frac{dy}{dx} = \frac{dy}{du} \cdot \frac{du}{dx}$

$= 50u^{49}(12x^2 - 4x)$

$= 50(4x^3 - 2x^2)^{49}(12x^2 - 4x)$

54. $y = \frac{u+1}{u-1}$, $u = 1 + \sqrt{x} = 1 + x^{1/2}$

$\frac{dy}{du} = \frac{(u-1)(1) - 1 \cdot (u+1)}{(u-1)^2}$

$= \frac{u - 1 - u - 1}{(u-1)^2}$

$= \frac{-2}{(u-1)^2}$

$\frac{du}{dx} = \frac{1}{2}x^{-1/2}$

$\frac{dy}{dx} = \frac{dy}{du} \cdot \frac{du}{dx} = \frac{-2}{(u-1)^2} \cdot \frac{1}{2}x^{-1/2}$

$= \frac{-2}{(1 + x^{1/2} - 1)^2} \cdot \frac{1}{2}x^{-1/2}$

$= \frac{-1}{(x^{1/2})^2} \cdot x^{-1/2}$

$= \frac{-1}{x} \cdot x^{-1/2}$

$= -x^{-3/2}$

55. $y = u(u+1)$, $u = x^3 - 2x$

$\frac{dy}{du} = u \cdot 1 + 1 \cdot (u+1)$

$= u + u + 1$

$= 2u + 1$

$\frac{du}{dx} = 3x^2 - 2$

$\frac{dy}{dx} = \frac{dy}{du} \cdot \frac{du}{dx} = (2u+1)(3x^2 - 2)$

$= [2(x^3 - 2x) + 1](3x^2 - 2)$

$= (2x^3 - 4x + 1)(3x^2 - 2)$

56. $y = (u+1)(u-1)$, $u = x^3 + 1$

$\frac{dy}{du} = (u+1)(1) + 1 \cdot (u-1)$

$= u + 1 + u - 1$

$= 2u$

$\frac{du}{dx} = 3x^2$

$\frac{dy}{dx} = \frac{dy}{du} \cdot \frac{du}{dx} = 2u \cdot 3x^2$

$= 2(x^3 + 1)(3x^2)$

$= 6x^2(x^3 + 1)$

57. $y = \sqrt{x^2 + 3x} = (x^2 + 3x)^{1/2}$

$\frac{dy}{dx} = \frac{1}{2}(x^2 + 3x)^{-1/2}(2x + 3)$

$= \frac{2x+3}{2\sqrt{x^2 + 3x}}$

When $x = 1$, $\frac{dy}{dx} = \frac{2 \cdot 1 + 3}{2\sqrt{1^2 + 3 \cdot 1}}$

$= \frac{2+3}{2\sqrt{4}}$

$= \frac{5}{2 \cdot 2}$

$= \frac{5}{4}$

Thus, at $(1, 2)$, $m = \frac{5}{4}$. We use point-slope equation.

$y - y_1 = m(x - x_1)$

$y - 2 = \frac{5}{4}(x - 1)$

$y - 2 = \frac{5}{4}x - \frac{5}{4}$

$y = \frac{5}{4}x + \frac{3}{4}$

58. $y = (x^3 - 4x)^{10}$

$\frac{dy}{dx} = 10(x^3 - 4x)^9(3x^2 - 4)$

When $x = 2$, $\frac{dy}{dx} = 10(2^3 - 4 \cdot 2)^9(3 \cdot 2^2 - 4)$

$= 10(8 - 8)^9(12 - 4)$

$= 10 \cdot 0 \cdot 8$

$= 0$

Use the point-slope equation:

$y - 0 = 0(x - 2)$

$y = 0$

59. $y = x\sqrt{2x+3} = x(2x+3)^{1/2}$

$\frac{dy}{dx} = x \cdot \frac{1}{2}(2x+3)^{-1/2}(2) + 1 \cdot (2x+3)^{1/2}$

$= \frac{x}{\sqrt{2x+3}} + \sqrt{2x+3}$

When $x = 3$, $\frac{dy}{dx} = \frac{3}{\sqrt{2\cdot 3+3}} + \sqrt{2\cdot 3+3}$
$= \frac{3}{\sqrt{9}} + \sqrt{9}$
$= \frac{3}{3} + 3$
$= 1 + 3 = 4$

Thus, at $(3, 9)$, $m = 4$. We use point-slope equation.

$y - y_1 = m(x - x_1)$
$y - 9 = 4(x - 3)$
$y - 9 = 4x - 12$
$y = 4x - 3$

60. $y - \left(\frac{2x+3}{x-1}\right)^3$

$$\frac{dy}{dx} = 3\left(\frac{2x+3}{x-1}\right)^2\left[\frac{(x-1)(2) - 1\cdot(2x+3)}{(x-1)^2}\right]$$
$$= 3\left(\frac{2x+3}{x-1}\right)^2\left(\frac{2x-2-2x-3}{(x-1)^2}\right)$$
$$= 3\left(\frac{2x+3}{x-1}\right)^2\left(\frac{-5}{(x-1)^2}\right)$$

When $x = 2, \frac{dy}{dx} = 3\left(\frac{2\cdot 2+3}{2-1}\right)^2\left(\frac{-5}{(2-1)^2}\right)$
$= 3(7^2)(-5) = -735$

Use the point-slope equation:

$y - 343 = -735(x - 2)$
$y - 343 = -735x + 1470$
$y = -735x + 1813$

61. $f(x) = sin^2x$

$\frac{dy}{dx} = 2\ sin\ x\ cos\ x$

When $x = -\frac{\pi}{6}$, $\frac{dy}{dx} = 2sin(\frac{-\pi}{6})cos(\frac{-\pi}{6}) = \frac{-\sqrt{3}}{2}$

Use the point-slope equation:

$y - \frac{1}{4} = \frac{-\sqrt{3}}{2}(x - (-\frac{\pi}{6}))$
$y - \frac{1}{4} = -\frac{\sqrt{3}}{2}x - \frac{\sqrt{3}}{12}$
$y = -\frac{\sqrt{3}}{2}x - \frac{\sqrt{3}}{12} + \frac{1}{4}$
$y = \frac{1}{12}(-6\sqrt{3}x - \sqrt{3}\pi + 3)$

62. $f(x) = x\ sin\ 2x$

$\frac{dy}{dx} = 2x\ cos\ 2x + sin\ 2x$

When $x = \pi$, $\frac{dy}{dx} = 2\pi\ cos\ 2\pi + sin\ 2\pi = 2\pi$

Use the point-slope equation:

$y - 0 = 2\pi(x - \pi)$
$y = 2\pi x - 2\pi^2$

63. $f(x) = \frac{x^2}{(1+x)^5}$

a)

$$f'(x) = \frac{(1+x)^5(2x) - x^2(5(1+x)^4)}{(1+x)^{10}}$$
$$= \frac{(1+x)^4[2x + 2x^2 - 5x^2]}{(1+x)^{10}}$$
$$= \frac{2x - 3x^2}{(1+x)^6}$$

b)

$$f'(x) = x^2[-5(1+x)^{-6}(1)] + 2x(1+x)^{-5}$$
$$= \frac{-5x^2}{(1+x)^6} + \frac{2x}{(1+x)^5}$$
$$= \frac{-5x^2 + 2x(1+x)}{(1+x)^6}$$
$$= \frac{2x - 3x^2}{(1+x)^6}$$

c) The results in the previous parts are the same.

64. a) $g(x) = (x^3 + 5x)^2$
$g'(x) = 2(x^3 + 5x)(3x^2 + 5)$
$= 2(3x^5 + 20x^3 + 25x)$
$= 6x^5 + 40x^3 + 50x$

b) $g(x) = x^6 + 10x^4 + 25x^2$
$g'(x) = 6x^5 + 40x^3 + 50x$

c) The answers are the same

65. Using the Chain Rule:

Let $y = f(u)$. Then

$\frac{dy}{dx} = \frac{dy}{du}\cdot\frac{du}{dx}$
$= 3u^2(8x^3)$
$= 3(2x^4 + 1)^2(8x^3)$ Substituting $2x^4 + 1$ for u

When $x = -1$, $\frac{dy}{dx} = 3[2(-1)^4 + 1]^2[8(-1)^3]$
$= 3(2 + 1)^2(-8)$
$= 3\cdot 3^2(-8)$
$= -216$

Finding $f(g(x))$:

$f \circ g(x) = f(g(x)) = f(2x^4 + 1) = (2x^4 + 1)^3$

Then $(f \circ g)'(x) = 3(2x^4 + 1)^2(8x^3)$ and

$(f \circ g)'(-1) = -216$ as above.

66. Using the Chain Rule:

Let $y = f(u)$.

$$\frac{dy}{dx} = \frac{dy}{du} \cdot \frac{du}{dx}$$
$$= \frac{(u-1)(1)-(1)(u+1)}{(u-1)^2} \cdot \frac{1}{2}x^{-1/2}$$
$$= \frac{u-1-u-1}{(u-1)^2} \cdot \frac{1}{2}x^{-1/2}$$
$$= \frac{-2}{(u-1)^2} \cdot \frac{1}{2\sqrt{x}}$$
$$= \frac{-1}{\sqrt{x}(\sqrt{x}-1)^2}$$

When $x = 4$, $\frac{dy}{dx} = \frac{-1}{\sqrt{4}(\sqrt{4}-1)^2} = \frac{-1}{2 \cdot 1^2} = -\frac{1}{2}$.

Finding $f(g(x))$:

$$f \circ g(x) = f(g(x)) = f(\sqrt{x}) = \frac{\sqrt{x}+1}{\sqrt{x}-1} = \frac{x^{1/2}+1}{x^{1/2}-1}$$

Then $(f \circ g)'(x) = \dfrac{(x^{1/2}-1)\left(\frac{1}{2}x^{-1/2}\right) - \frac{1}{2}x^{-1/2}(x^{1/2}+1)}{(x^{1/2}-1)^2}$

$$= \frac{\frac{1}{2} - \frac{1}{2}x^{-1/2} - \frac{1}{2} - \frac{1}{2}x^{-1/2}}{(x^{1/2}-1)^2}$$
$$= \frac{-x^{-1/2}}{(x^{1/2}-1)^2}$$
$$= \frac{-1}{\sqrt{x}(\sqrt{x}-1)^2}$$

and $(f \circ g)'(4) = -\frac{1}{2}$ as above.

67. Using the Chain Rule:

Let $y = f(u) = \sqrt[3]{u} = u^{1/3}$. Then

$$\frac{dy}{dx} = \frac{dy}{du} \cdot \frac{du}{dx}$$
$$= \frac{1}{3}u^{-2/3} \cdot (-6x)$$
$$= -2x \cdot u^{-2/3}$$
$$= -2x(1-3x^2)^{-2/3} \quad \text{Substituting } 1-3x^2 \text{ for } u$$

When $x = 2$, $\frac{dy}{dx} = 2 \cdot 2(1 - 3 \cdot 2^2)^{-2/3}$
$$= -4(-11)^{-2/3} \approx -0.8087$$

Finding $f(g(x))$:

$f \circ g(x) = f(g(x)) = f(1-3x^2) = \sqrt[3]{1-3x^2}$, or $(1-3x^2)^{1/3}$

Then $(f \circ g)'(x) = \frac{1}{3}(1-3x^2)^{-2/3}(-6x) = -2x(1-3x^2)^{-2/3}$ and

$(f \circ g)'(2) = -4(-11)^{-2/3} \approx -0.8087$ as above.

68. Using the Chain Rule:

Let $y = f(u)$.

$$\frac{dy}{dx} = \frac{dy}{du} \cdot \frac{du}{dx}$$
$$= 10u^4 \cdot \frac{(4+x)(-1)-(1)(3-x)}{(4+x)^2}$$
$$= 10u^4 \cdot \frac{-4-x-3+x}{(4+x)^2}$$
$$= 10\left(\frac{3-x}{4+x}\right)^4 \cdot \frac{-7}{(4+x)^2}$$

When $x = -10$, $\frac{dy}{dx} = 10\left(\frac{3+10}{4-10}\right)^4 \cdot \frac{-7}{(4-10)^2}$
$$= 10\left(\frac{13}{-6}\right)^4\left(\frac{-7}{(-6)^2}\right)$$
$$= \frac{-70 \cdot 13^4}{6^6} \approx -42.8513$$

Finding $f(g(x))$:

$$f \circ g(x) = f(g(x)) = 2\left(\frac{3-x}{4+x}\right)^5$$

Then $(f \circ g)'(x) = 2 \cdot 5\left(\frac{3-x}{4+x}\right)^4 \cdot \frac{4+x(-1)-(1)(3-x)}{(4+x)^2}$
$$= 10\left(\frac{3-x}{4+x}\right)^4 \cdot \frac{-4-x-3+x}{(4+x)^2}$$
$$= 10\left(\frac{3-x}{4+x}\right)^4 \cdot \frac{-7}{(4+x)^2}$$

and $(f \circ g)'(-10) = 10\left(\frac{3+10}{4-10}\right)^4 \cdot \frac{-7}{(4-10)^2} = \frac{-70 \cdot 13^4}{6^6} \approx -42.8513$ as above.

69. $A = 1000(1+i)^3$

a)
$$\frac{dA}{di} = 1000(3(1+i)^2) = 3000(1+i)^2$$

b) $\frac{dA}{di}$ represents the rate at which the amount of investment is changing with respect to an annual interest rate i.

70. **a)**
$$\frac{dA}{di} = 1000\left(20(1+\frac{i}{4})^{19}(\frac{1}{4})\right) = 5000(1+\frac{i}{4})^{19}$$

b) $\frac{dA}{di}$ represents the rate at which the amount of investment is changing with respect to an quarterly componded interest rate i.

71. $D = 0.85A(c+25)$, $c = (140-y)\frac{w}{72x}$

a) To find D as a function of c, we substitute 5 for A in the formula for D.

$$D = 0.85A(c+25)$$

$$\begin{aligned} &= 0.85(5)(c+25) \\ &= 4.25(c+25) \\ &= 4.25(c+25) \\ &= 4.25c + 106.25 \end{aligned}$$

To find c as a function of w, we substitute 45 for y and 0.6 for x in the formula for c.

$$\begin{aligned} c &= (140-45)\frac{w}{72(0.6)} \\ &= 95 \cdot \frac{w}{43.2} \\ &\approx 2.199w \end{aligned}$$

b) $\frac{dD}{dc} = 4.25$

c) $\frac{dc}{dw} = 2.199$

d) First we find $D \circ c(w)$.

$$\begin{aligned} D \circ c(w) &= D(c(w)) \\ &= 4.25(2.199w) + 106.25 \\ &= 9.34575w + 106.25 \end{aligned}$$

Then we have

$$\frac{dD}{dw} = 9.34575 \approx 9.346.$$

e) $\frac{dD}{dw}$ represents the rate of change of the dosage with respect to the patient's weight. For each additional kilogram of weight, the dosage is increased by about 9.35 mg.

72. $D = A(c+25)$, $c = (140-y)\frac{w}{72x}$

a) To find D as a function of c, we substitute 5 for A in the formula for D.

$$\begin{aligned} D &= A(c+25) \\ &= 5(c+25) \\ &= 5c + 125 \end{aligned}$$

To find c as a function of w, we substitute 45 for y and 0.6 for x in the formula for c.

$$\begin{aligned} c &= (140-45)\frac{w}{72(0.6)} \\ &= 95 \cdot \frac{w}{43.2} \\ &\approx 2.199w \end{aligned}$$

b) $\frac{dD}{dc} = 5$

c) $\frac{dc}{dw} = 2.199$

d) First we find $D \circ c(w)$.

$$\begin{aligned} D \circ c(w) &= D(c(w)) \\ &= 5(2.199w) + 125 \\ &= 10.995w + 125 \end{aligned}$$

Then we have

$$\frac{dD}{dw} = 10.995.$$

e) $\frac{dD}{dw}$ represents the rate of change of the dosage with respect to the patient's weight. For each additional kilogram of weight, the dosage is increased by about 11 mg.

73. **a)** January 2009 corresponds to $t = 52$

$$\begin{aligned} C'(t) &= 0.74 + 0.02376t - 1.0814\pi\ cos(2\pi t) \\ C'(52) &= 0.74 + 0.02376(52) - 1.0814\pi cos(104\pi) \\ &= -1.4218\ ppmv/yr \end{aligned}$$

b) July 2009 corresponds to $t = 52.5$

$$\begin{aligned} C'(52.5) &= 0.74 + 0.02376(52.5) - 1.0814\pi cos(105\pi) \\ &= 5.3847\ ppmv/yr \end{aligned}$$

74. $T'(t) = \frac{0.82\pi}{12}\ cos\left(\frac{\pi}{12}[t+2]\right)$

$$\begin{aligned} T'(8) &= \frac{0.82\pi}{12}\ cos\left(\frac{\pi}{12}[8+2]\right) \\ &= -0.1859\ degrees/hr \end{aligned}$$

75. $y = ((x^2+4)^8 + 3\sqrt{x})^4$

$$\begin{aligned} \frac{dy}{dx} &= 4((x^2+4)^8 + 3\sqrt{x})^3[8(x^2+4)^7(2x) + \frac{3}{2\sqrt{x}}] \\ &= 4((x^2+4)^8 + 3\sqrt{x})^3\left(16x(x^2+4)^7 + \frac{3}{2\sqrt{x}}\right) \end{aligned}$$

76.

$$\begin{aligned} \frac{dy}{dx} &= \frac{1}{2}((x^5+x+1)^3 + 7\ sec^2 x)^{-1/2} \times \\ &\quad [3(x^5+x+1)^2(5x^4+1) + 14\ sec\ x\ sec\ x\ tan\ x] \\ &= \frac{3(x^5+x+1)^2(5x^4+1) + 14\ sec^2 x\ tan\ x}{2\sqrt{(x^5+x+1)^3 + 7\ sec^2 x}} \end{aligned}$$

77. Let $y = sin(sin(sin\ x))$ then

$$\begin{aligned} \frac{dy}{dx} &= cos(sin(sin\ x)) \cdot cos(sin\ x) \cdot cos\ x \\ &= cos\ x\ cos(sin\ x)\ cos(sin(sin\ x)) \end{aligned}$$

78. Let $y = cos(sec(sin\ 2x))$ then

$$\begin{aligned} \frac{dy}{dx} &= -sin(sec(sin\ 2x)) \cdot sec(sin\ 2x)\ tan(sin\ 2x) \cdot 2cos\ 2x \\ &= -2cos\ 2x sec(sin\ 2x)\ tan(sin\ 2x)\ sin(sec(sin\ 2x)) \end{aligned}$$

79. Let $y = tan(cot(sec\ 3x))$ then

$$\begin{aligned} \frac{dy}{dx} &= sec^2(cot(sec\ 3x)) \cdot -csc^2(sec\ 3x) \cdot 3\ sec\ 3x\ tan\ 3x \\ &= -3\ sec\ 3x\ tan\ 3x csc^2(sec\ 3x)\ sec^2(cot(sex\ 3x)) \end{aligned}$$

80. Let $y = csc(cos(sec(\sqrt{x^2+1})))$ then

$$\begin{aligned} \frac{dy}{dx} &= -csc(cos(sec(\sqrt{x^2+1})))\ cot(cos(sex(\sqrt{x^2+1}))) \times \\ &\quad -sin(sec(\sqrt{x^2+1})) \cdot sec(\sqrt{x^2+1})\ tan(\sqrt{x^2+1}) \times \\ &\quad \frac{2x}{2\sqrt{x^2+1}} \end{aligned}$$

81. $y = \left(\sin\left(\frac{3\pi}{2} + 3\right)\right)^{1/5}$ is a constant, which means $\frac{dy}{dx} = 0$.

82. The derivative of a constant is 0

83.

$$\begin{aligned}
\sin(a+x) &= \sin a \cos x + \cos a \sin x \\
\frac{d}{dx}(\sin(a+x)) &= \frac{d}{dx}(\sin a \cos x + \cos a \sin x) \\
&= -\sin a \sin x + \cos a \cos x \\
&= \cos a \cos x - \sin a \sin x \\
&= \cos(a+x)
\end{aligned}$$

84.

$$\begin{aligned}
\cos(a+x) &= \cos a \cos x - \sin a \sin x \\
\frac{d}{dx}(\cos(a+x)) &= \frac{d}{dx}(\cos a \cos x - \sin a \sin x) \\
&= -\cos a \sin x + \sin a \cos x \\
&= \sin a \cos x - \cos a \sin x \\
&= \sin(a+x)
\end{aligned}$$

85. Let $Q(x) = \frac{N(x)}{D(x)}$. Then we can write

$$Q(x) = N(x) \cdot [D(x)]^{-1}$$

using the property of negative exponents. Now we use the product differentiation rule

$$\begin{aligned}
Q'(x) &= N(x) \cdot -1[D(x)]^{-2} \cdot D'(x) + [D(x)]^{-1} \cdot N'(x) \\
&= \frac{-N(x) \cdot D'(x)}{[D(x)]^2} + \frac{N'(x)}{D(x)} \\
&= \frac{-N(x) \cdot D'(x)}{[D(x)]^2} + \frac{N'(x) \cdot D(x)}{[D(x)]^2} \\
&= \frac{N'(x) \cdot D(x) - N(x) \cdot D'(x)}{[D(x)]^2}
\end{aligned}$$

86. One might use the following: Composition of functions could be expressed as a process through which one function is evaluated using another function that falls in the original function's domain.

87. $(-2.145, -7.728)$ and $(2.145, 7.728)$

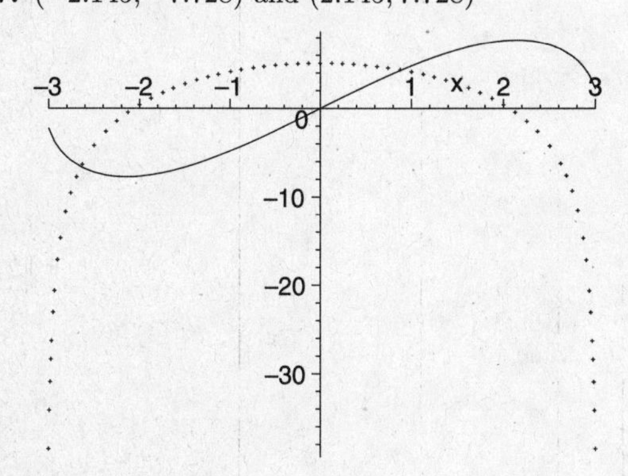

88. $(-1.475, 9.488)$

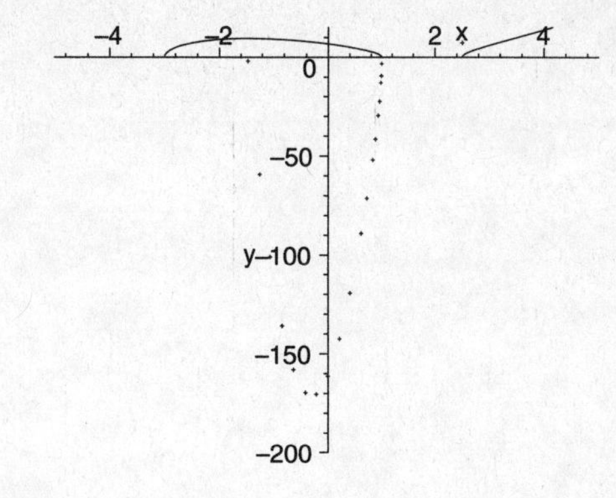

89.

$$\begin{aligned}
f'(x) &= x \cdot \frac{-2x}{2\sqrt{4-x^2}} + \sqrt{4-x^2} \\
&= \frac{-x^2}{\sqrt{4-x^2}} + \frac{4-x^2}{\sqrt{4-x^2}} \\
&= \frac{4-2x^2}{\sqrt{4-x^2}}
\end{aligned}$$

90.

$$\begin{aligned}
f'(x) &= \frac{4 \cdot \sqrt{x-10} - 4x \cdot \frac{1}{2\sqrt{x-10}}}{x-10} \\
&= \frac{4(x-10) - 2x}{(x-10)^{3/2}} \\
&= \frac{2x-40}{(x+10)^{3/2}}
\end{aligned}$$

Exercise Set 2.9

1. $y = 3x + 5$

$$\begin{aligned}
\frac{dy}{dx} &= 3 \\
\frac{d^2y}{dx^2} &= 0
\end{aligned}$$

2. $y = -4x + 7$

$$\begin{aligned}
\frac{dy}{dx} &= -4 \\
\frac{d^2y}{dx^2} &= 0
\end{aligned}$$

3. $y = -3(2x+2)^{-1}$

$$\begin{aligned}
\frac{dy}{dx} &= 3(2x+2)^{-2}(2) = 6(2x+2)^{-2} = \frac{6}{(2x+2)^2} \\
\frac{d^2y}{dx^2} &= -12(2x+2)^{-3}(2) = \frac{-24}{(2x+2)^3}
\end{aligned}$$

4. $y = -(3x-4)^{-1}$

$$\begin{aligned}
\frac{dy}{dx} &= (3x-4)^{-2}(3) = \frac{3}{(3x-4)^2} \\
\frac{d^2y}{dx^2} &= 3(-2(3x-4)^{-3}(3)) = \frac{-18}{(3x-4)^3}
\end{aligned}$$

5. $y = (2x+1)^{1/3}$

$$\begin{aligned} \frac{dy}{dx} &= \frac{1}{3}(2x+1)^{-2/3}(2) = \frac{2}{3(2x+1)^{2/3}} \\ \frac{d^2y}{dx^2} &= \frac{2}{3}(-\frac{2}{3}(2x+1)^{-5/3}(2)) = \frac{-8}{9(2x+1)^{5/2}} \end{aligned}$$

6. $f(x) = (3x+2)^{-3}$

$$\begin{aligned} f'(x) &= -3(3x+2)^{-4}(3) = -9(3x+2)^{-4} \\ f''(x) &= 36(3x+2)^{-5}(3) = \frac{108}{(3x+2)^5} \end{aligned}$$

7. $f(x) = (4-3x)^{-4}$

$$\begin{aligned} f'(x) &= -4(4-3x)^{-5}(-3) = 12(4-3x)^{-5} \\ f''(x) &= -60(4-3x)^{-6}(-3) = \frac{180}{(4-3x)^6} \end{aligned}$$

8. $y = \sqrt{x-1} = (x-1)^{1/2}$

$$\begin{aligned} \frac{dy}{dx} &= \frac{1}{2}(x-1)^{1/2-1} \cdot 1 \\ &= \frac{1}{2}(x-1)^{-1/2} \\ \frac{d^2y}{dx^2} &= \frac{1}{2} \cdot \left(-\frac{1}{2}\right)(x-1)^{-1/2-1} \cdot 1 \\ &= -\frac{1}{4}(x-1)^{-3/2} \\ &= -\frac{1}{4(x-1)^{3/2}} \\ &= -\frac{1}{4\sqrt{(x-1)^3}} \end{aligned}$$

9. $y = \sqrt{x+1} = (x+1)^{1/2}$

$$\begin{aligned} \frac{dy}{dx} &= \frac{1}{2}(x+1)^{-1/2} \cdot 1 \\ &= \frac{1}{2}(x+1)^{-1/2} \\ \frac{d^2y}{dx^2} &= -\frac{1}{4}(x+1)^{-3/2} \cdot 1 \\ &= -\frac{1}{4}(x+1)^{-3/2} \\ &= -\frac{1}{4(x+1)^{3/2}} \\ &= -\frac{1}{4\sqrt{(x+1)^3}} \end{aligned}$$

10. $f(x) = (3x+2)^{10}$

$$\begin{aligned} f'(x) &= 10(3x+2)^9(3) = 30(3x+2)^9 \\ f''(x) &= 270(3x+2)^8(3) = 810(3x+2)^8 \end{aligned}$$

11. $f(x) = (2x+9)^{16}$

$$\begin{aligned} f'(x) &= 16(2x+9)^{15}(2) = 32(2x+9)^{15} \\ f''(x) &= 480(2x+9)^{14}(2) = 960(2x+9)^{14} \end{aligned}$$

12. $g(x) = tan(2x)$

$$\begin{aligned} g'(x) &= sec^2(2x)(2) = 2\ sec^2(2x) \\ g''(x) &= 2(2\ sec(2x)\ sec(2x)\ tan(2x)(2)) \\ &= 8\ sec^2(2x)\ tan(2x) \end{aligned}$$

13. $g(x) = sec(3x+1)$

$$\begin{aligned} g'(x) &= sec(3x+1)\ tan(3x+1)(3) \\ &= 3\ sec(3x+1)\ tan(3x+1) \\ g''(x) &= 3\ sec(3x+1)\ sec(3x+1)(3) + \\ &\quad tan(3x+1)(3\ sec(3x+1)\ tan(3x+1)(3) \\ &= 9\ sec^3(3x+1) + 9\ sec(3x+1)\ tan^2(3x+1) \\ &= 9\ sec(3x+1)[sec^2(3x+1) + tan^2(3x+1)] \end{aligned}$$

14. $f(x) = 13x^2 + 2x + 7 - csc\ x$

$$\begin{aligned} f'(x) &= 26x + 2 - (-csc\ x\ cot\ x) \\ &= 26x + 2 + csc\ x\ cot\ x \\ f''(x) &= 26 + csc\ x(-csc^2 x) + cot\ x(-csc\ x\ cot\ x) \\ &= 26 - csc^3 x - csc\ x\ cot^2 x \end{aligned}$$

15. $f(x) = sec(2x+3) + 4x^2 + 3x - 7$

$$\begin{aligned} f'(x) &= sec(2x+3)\ tan(2x+3)(2) + 8x + 3 \\ &= 2\ sec(2x+3)\ tan(2x+3) + 8x + 3 \\ f''(x) &= 2\ sec(2x+3)\ sec^2(2x+3)(2) + \\ &\quad tan(2x+3)\ sec(2x+3)\ tan(2x+3)(2) + 8 \\ &= 4\ sec^3(2x+3) + 2\ sec(2x+3)\ tan^2(2x+3) + 8 \end{aligned}$$

16. $g(x) = mx + b$

$$\begin{aligned} g'(x) &= m \\ g''(x) &= 0 \end{aligned}$$

17. $y = ax^2 + bx + c$

$$\begin{aligned} \frac{dy}{dx} &= a \cdot 2x + b + 0 \\ &= 2ax + b \\ \frac{d^2y}{dx^2} &= 2a + 0 \\ &= 2a \end{aligned}$$

18. $y = \sqrt[3]{2x+4} = (2x+4)^{1/3}$

$$\begin{aligned} \frac{dy}{dx} &= \frac{1}{3}(2x+4)^{-2/3}(2) \\ &= \frac{2}{3}(2x+4)^{-2/3} \\ \frac{d^2y}{dx^2} &= -\frac{2}{3} \cdot \frac{2}{3}(2x+4)^{-5/3}(2) \\ &= -\frac{8}{9}(2x+4)^{-5/3} \\ &= -\frac{8}{9(2x+4)^{5/3}} \\ &= -\frac{8}{9\sqrt[3]{(2x+4)^5}} \end{aligned}$$

19. $y = \sqrt[4]{(x^2+1)^3} = (x^2+1)^{3/4}$

$$\begin{aligned}\frac{dy}{dx} &= \frac{3}{4}(x^2+1)^{-1/4}(2x)\\ &= \frac{3}{2}x(x^2+1)^{-1/4}\\ \frac{d^2y}{dx^2} &= \frac{3}{2}\left[x\left(-\frac{1}{4}\right)(x^2+1)^{-5/4}(2x) + 1\cdot(x^2+1)^{-1/4}\right]\\ &= -\frac{3x^2}{4(x^2+1)^{5/4}} + \frac{3}{2(x^2+1)^{1/4}}\\ &= -\frac{3x^2}{4\sqrt[4]{(x^2+1)^5}} + \frac{3}{2\sqrt[4]{x^2+1}}\end{aligned}$$

20. $y = 13x^{3.2} + 12x^{1.2} - 5x^{-0.25}$

$$\begin{aligned}\frac{dy}{dx} &= 41.6x^{2.2} + 14.4x^{0.2} - 1.25x^{-1.25}\\ \frac{d^2y}{dx^2} &= 91.52x^{1.2} + 2.88x^{-0.8} + 1.5625x^{2.25}\end{aligned}$$

21. $f(x) = (4x+3)\ cos\ x$

$$\begin{aligned}f'(x) &= (4x+3)\ (-sin\ x) + 4\ cos\ x\\ &= -(4x+3)\ sin\ x + 4\ cos\ x\end{aligned}$$

22. $s(t) = sin(at)$

$$\begin{aligned}s'(t) &= cos(at)\ a = a\ cos(at)\\ s''(t) &= a(-sin(at)\ a)\\ &= -a^2\ sin(at)\end{aligned}$$

23. $s(t) = cos(at+b)$

$$\begin{aligned}s'(t) &= -sin(at+b)\ a = -a\ sin(at+b)\\ s''(t) &= -a(cos(at+b)\ a)\\ &= -a^2\ cos(at+b)\end{aligned}$$

24. $y = x^{5/2} + x^{3/2} - x^{1/2}$

$$\begin{aligned}\frac{dy}{dx} &= \frac{5}{2}x^{3/2} + \frac{3}{2}x^{1/2} - \frac{1}{2}x^{-1/2}\\ \frac{d^2y}{dx^2} &= \frac{15}{4}x^{1/2} + \frac{3}{4}x^{-1/2} + \frac{1}{4}x^{-3/2}\end{aligned}$$

25. $y = \frac{(t^2+3)^{1/2}}{7} + (3t^2+1)^{1/3}$

$$\begin{aligned}\frac{dy}{dx} &= \frac{1}{7}\cdot\frac{1}{2}(t^2+3)^{-1/2}(2t) + \frac{1}{3}(3t^2+1)^{-2/3}(6t)\\ &= \frac{t}{7}(t^2+3)^{-1/2} + 2t(3t^2+1)^{-2/3}\\ \frac{d^2y}{dx^2} &= \frac{t}{7}(\frac{-1}{2}(t^2+3)^{-3/2}(2t)) + (2)(t^2+3)^{-1/2} +\\ & \quad 2t(\frac{-2}{3}(3t^2+1)^{-5/3}(6t)) + (2)(3t^2+1)^{-2/3}\\ &= \frac{-t^2}{7(t^2+3)^{3/2}} + \frac{2}{(t^2+3)^{1/2}}\\ & \quad -\frac{8t^2}{(3t^2+1)^{5/3}} + \frac{2}{3t^2+1}\end{aligned}$$

26. $y = \frac{1}{\pi^2}\ tan\ \pi t + \frac{4}{\pi^2}\ sec\ \pi t$

$$\begin{aligned}\frac{dy}{dx} &= \frac{1}{\pi^2}sec^2\pi t\cdot\pi + \frac{4}{\pi^2}\ sec\ \pi t\ tan\ \pi t\cdot\pi\\ &= \frac{1}{\pi}\ sec^2\pi t + \frac{4}{\pi}\ sec\ \pi t\ tan\ \pi t\\ \frac{d^2y}{dx^2} &= \frac{1}{\pi}(2\ sec\ \pi t\ sec\ \pi t\ tan\ \pi t\cdot\pi) +\\ & \quad \frac{4}{\pi}(sec\ \pi t\ sec^2\pi t\cdot\pi + sec\ \pi t\ tan\ \pi t\ tan\ \pi t\cdot\pi)\\ &= 2\ sec^2\pi t\ tan\ \pi t + 4\ sec^3\pi t + 4\ sec\ \pi t\ tan^2\pi t\end{aligned}$$

27. $y = x^4$

$$\begin{aligned}\frac{dy}{dx} &= 4x^3\\ \frac{d^2y}{dx^2} &= 4\cdot 3x^2\\ &= 12x^2\\ \frac{d^3y}{dx^3} &= 12\cdot 2x\\ &= 24x\\ \frac{d^4y}{dy^4} &= 24\end{aligned}$$

28. $y = x^5$

$$\begin{aligned}\frac{dy}{dx} &= 5x^4\\ \frac{d^2y}{dx^2} &= 20x^3\\ \frac{d^3y}{dx^3} &= 60x^2\\ \frac{d^4y}{dy^4} &= 120x\end{aligned}$$

29. $y = x^6 - x^3 + 2x$

$$\begin{aligned}\frac{dy}{dx} &= 6x^5 - 3x^2 + 2\\ \frac{d^2y}{dx^2} &= 30x^4 - 6x\\ \frac{d^3y}{dx^3} &= 120x^3 - 6\\ \frac{d^4y}{dx^4} &= 360x^2\\ \frac{d^5y}{dx^5} &= 720x\end{aligned}$$

30. $y = x^7 - 8x^2 + 2$

$$\begin{aligned}\frac{dy}{dx} &= 7x^6 - 16x\\ \frac{d^2y}{dx^2} &= 42x^5 - 16\\ \frac{d^3y}{dx^3} &= 210x^4\\ \frac{d^4y}{dx^4} &= 840x^3\\ \frac{d^5y}{dx^5} &= 2520x^2\\ \frac{d^6y}{dx^6} &= 5040x\end{aligned}$$

31. $y = (x^2-5)^{10}$

$$\begin{aligned}\frac{dy}{dx} &= 10(x^2-5)^9 \cdot 2x \\ &= 20x(x^2-5)^9 \\ \frac{d^2y}{dx^2} &= 20x \cdot 9(x^2-5)^8 \cdot 2x + 20(x^2-5)^9 \\ &= 360x^2(x^2-5)^8 + 20(x^2-5)^9 \\ &= 20(x^2-5)^8[18x^2 + (x^2-5)] \\ &= 20(x^2-5)^8(19x^2-5)\end{aligned}$$

32. $y = x^k$

$$\begin{aligned}\frac{dy}{dx} &= kx^{k-1} \\ \frac{d^2y}{dx^2} &= k(k-1)x^{k-2} \\ \frac{d^3y}{dx^3} &= k(k-1)(k-2)x^{k-3} \\ \frac{d^4y}{dx^4} &= k(k-1)(k-2)(k-3)x^{k-4} \\ \frac{d^5y}{dx^5} &= k(k-1)(k-2)(k-3)(k-4)x^{k-5}\end{aligned}$$

33. $y = sec(2x+3)$

$$\begin{aligned}\frac{dy}{dx} &= sec(2x+3)\ tan(2x+3)\ (2) \\ &= 2\ sec(2x+3)\ tan(2x+3) \\ \frac{d^y}{dx^2} &= 2\ sec(2x+3)\ sec^2(2x+3)(2) + \\ &\quad [2\ sec(2x+3)\ tan(2x+3)(2)]\ tan(2x+3) \\ &= 4\ sec^3(2x+3) + 4\ sec(2x+3)\ tan^2(2x+3) \\ \frac{d^3y}{dx^3} &= 4(3sec^2(2x+3)\ sec(2x+3)\ tan(2x+3)(2)) + \\ &\quad 4\ sec(2x+3)[2\ tan(2x+3)\ sec^2(2x+3)(2)] + \\ &\quad sec(2x+3)\ tan(2x+3)\ (2)\ tan^2(2x+3) \\ &= 40\ sec^3(2x+3)\ tan(2x+3) + \\ &\quad 8\ sec(2x+3)\ tan^3(2x+3)\end{aligned}$$

34. $y = cot(3x-1)$

$$\begin{aligned}\frac{dy}{dx} &= -3csc^2(3x-1) \\ \frac{d^2y}{dx^2} &= -3[2\ csc(3x-1)\ (-csc(3x-1)\ cot(3x-1)\ (3))] \\ &= 18\ csc^2(3x-1)\ cot(3x-1) \\ \frac{d^3y}{dx^3} &= 18\ csc^2(3x-1)(-csc^2(3x-1)(3)) + \\ &\quad 18[2\ csc^2(3x-1)cot(3x-1)(3))]cot(3x-1) \\ &= -54\ csc^4(3x-1) - 108\ csc^2(3x-1)\ cot^2(3x-1)\end{aligned}$$

35. $s(t)10\ cos(3t+2) - 4\ sin(3t+2)$

$$\begin{aligned}v(t) &= 10[-sin(3t+2)(3)] - 4[cos(3t+2)(3) \\ &= -30\ sin(3t+2) - 12\ cos(3t+2) \\ a(t) &= -30[cos(3t+2)(3)] - 12[-sin(3t+2)(3)] \\ &= -90\ cos(3t+2) - 36\ sin(3t+2) \\ &= 9[10\ cos(3t+2) - 4\ sin(3t+2)] \\ &= 9\ s(t)\end{aligned}$$

36. $s(t) = 6.8\ tan(2.6t-1)$

$$\begin{aligned}v(t) &= 6.8\ sec^2(2.6t-1)(2.6) = 17.68\ sec^2(2.6t-1) \\ a(t) &= 17.68[2\ sec(2.6t-1)\ sec(2.6t-1)\ tan(2.6t-1)(2.6)] \\ &= 91.936\ sec^2(2.6t-1)\ tan(2.6t-1)\end{aligned}$$

37. $s(t) = t^3 + t^2 + 2t$

$$\begin{aligned}v(t) &= s'(t) = 3t^2 + 2t + 2 \\ a(t) &= s''(t) = 6t + 2\end{aligned}$$

38. $s(t) = t^4 + t^2 + 3t$

$$\begin{aligned}v(t) &= s'(t) = 4t^3 + 2t + 3 \\ a(t) &= s''(t) = 12t^2 + 2\end{aligned}$$

39. $w(t) = 0.000758t^3 - 0.0596t^2 - 1.82t + 8.15$
The acceleration of a function that depends on time is the second derivative of the function with respect to time.

$$\begin{aligned}w'(t) &= 0.002274t^2 - 0.1192t + 1.82 \\ w''(t) &= 0.004548t - 0.1192\end{aligned}$$

40. $w(t) = 0.0006t^3 - 0.0484t^2 - 1.61t + 7.60$

$$\begin{aligned}w'(t) &= 0.0018t^2 - 0.0968t + 1.61 \\ w''(t) &= 0.0036t - 0.068\end{aligned}$$

41. $P(t)100000(1 + 0.6t + t^2)$

$$\begin{aligned}P'(t) &= 100000(0.6 + 2t) \\ P''(t) &= 100000(2) \\ &= 200000\end{aligned}$$

42. $P(t)100000(1 + 0.4t + t^2)$

$$\begin{aligned}P'(t) &= 100000(0.4 + 2t) \\ P''(t) &= 100000(2) \\ &= 200000\end{aligned}$$

43. $y = \frac{x}{(x-1)^{1/2}}$

$$\begin{aligned}y' &= \frac{\sqrt{x-1}(1) - x \cdot \frac{1}{2\sqrt{x-1}}}{x-1} \\ &= \frac{2(x-1) - x}{2(x-1)\sqrt{x-1}} \\ &= \frac{x-2}{2(x-1)^{3/2}}\end{aligned}$$

$$\begin{aligned}y'' &= \frac{2(x-1)^{3/2}(1) - (x-2)\left[2 \cdot \frac{3}{2}(x-1)^{1/2}\right]}{4(x-1)^3} \\ &= \frac{2(x-1)^{3/2} - 3(x-2)(x-1)^{1/2}}{4(x-1)^3} \\ &= \frac{(x-1)^{1/2}[2(x-1) - 3(x-2)]}{(x-1)^3} \\ &= \frac{4-x}{(x-1)^{5/2}}\end{aligned}$$

$$\begin{aligned} y''' &= \frac{4(x-1)^{5/2}(-1)-(4-x)\left[4\cdot\frac{5}{2}(x-1)^{3/2}\right]}{16(x-1)^5} \\ &= \frac{(x-1)^{3/2}[4(x-1)-10(4-x)]}{16(x-1)^5} \\ &= \frac{4x-4-40+10x}{16(x-1)^{7/2}} \\ &= \frac{3x-18}{16(x-1)^{7/2}} \end{aligned}$$

44. $= \frac{\sqrt{x}-1}{\sqrt{x}+1}$

$$\begin{aligned} y' &= \frac{(\sqrt{x}+1)(\frac{1}{2\sqrt{x}})-(\sqrt{x}-1)(\frac{1}{2\sqrt{x}})}{(\sqrt{x}+1)^2} \\ &= \frac{1}{\sqrt{x}(\sqrt{x}+1)^2} \\ &= [\sqrt{x}(\sqrt{x}+1)^2]^{-1} \end{aligned}$$

$$\begin{aligned} y'' &= -1[\sqrt{x}(\sqrt{x}+1)^2]^{-2}[\sqrt{x}[2(\sqrt{x}+1)(\frac{1}{2\sqrt{x}})]+ \\ &\quad -1[\sqrt{x}(\sqrt{x}+1)^2]^{-2}\frac{1}{2\sqrt{x}}(\sqrt{x}+1)^2] \\ &= -[\sqrt{x}(\sqrt{x}+1)^2]^{-2}\left(\frac{3}{2}\sqrt{x}+2+\frac{1}{2\sqrt{x}}\right) \end{aligned}$$

$$\begin{aligned} y''' &= -[\sqrt{x}(\sqrt{x}+1)^2]^{-2}[\frac{3}{2}x^{-1/2}+\frac{1}{4}x^{-3/2}]+ \\ &\quad 2[\sqrt{x}(\sqrt{x}+1)^2]^{-3}[\sqrt{x}[2(\sqrt{x}+1)(\frac{1}{2\sqrt{x}})]+ \\ &\quad -1[\sqrt{x}(\sqrt{x}+1)^2]^{-2}\frac{1}{2\sqrt{x}}] \\ &= -[\sqrt{x}(\sqrt{x}+1)^2]^{-2}[\frac{3}{2}x^{-1/2}+\frac{1}{4}x^{-3/2}]+ \\ &\quad 2[\sqrt{x}(\sqrt{x}+1)^2]^{-3}[\sqrt{x}(x+1)]-[\sqrt{x}(\sqrt{x}+1)^2]^{-2}\frac{1}{2\sqrt{x}}] \end{aligned}$$

45. $f(x) = \frac{x}{x-1}$

$$\begin{aligned} f'(x) &= \frac{(x-1)(1)-x(1)}{(x-1)^2} \\ &= \frac{-1}{(x-1)^2} \\ &= -(x-1)^{-2} \end{aligned}$$

$$\begin{aligned} f''(x) &= -(-2(x-1)^{-3}) \\ &= \frac{2}{(x-1)^3} \end{aligned}$$

46. $f(x) = (1+x^2)^{-1}$

$$\begin{aligned} f'(x) &= -1(1+x^2)^{-2}(2x) \\ &= -2x(1+x^2)^{-2} \end{aligned}$$

$$\begin{aligned} f''(x) &= -2x[-2(1+x^2)^{-3}(2x)]+(-2)(1+x^2)^{-2} \\ &= \frac{-8x^2}{(1+x^2)^3}+\frac{4}{(1+x^2)^2} \end{aligned}$$

47. $y = \sin x$

a) $\frac{dy}{dx} = \cos x$

b) $\frac{d^2y}{dx^2} = -\sin x$

c) $\frac{d^3y}{dx^3} = -\cos x$

d) $\frac{d^4y}{dx^4} = \sin x$

e) $\frac{d^8y}{dx^8} = \sin x$

f) $\frac{d^{10}y}{dx^{10}} = -\sin x$

g) $\frac{d^{837}y}{dx^{837}} = \cos x$

48. $y = \cos x$

a) $\frac{dy}{dx} = -\sin x$

b) $\frac{d^2y}{dx^2} = -\cos x$

c) $\frac{d^3y}{dx^3} = \sin x$

d) $\frac{d^4y}{dx^4} = \cos x$

e) $\frac{d^8y}{dx^8} = \cos x$

f) $\frac{d^{11}y}{dx^{10}} = \sin x$

g) $\frac{d^{523}y}{dx^{523}} = \sin x$

49. Functions that have the form $f(x) = A\sin x + B\cos x$ where A and B are constants, will satisfy the condition of their second derivative being the negative of the original function.

50. $f(x) = \frac{x-1}{x+2}$

$$\begin{aligned} f'(x) &= \frac{(x+2)(1)-(x-1)(1)}{(x+2)^2} \\ &= \frac{3}{(x+2)^2} = 3(x+2)^{-2} \\ f''(x) &= -6(x+2)^{-3} = \frac{-6}{(x+2)^3} \\ f'''(x) &= 18(x+2)^{-4} = \frac{18}{(x+2)^4} \\ f^{(4)(x)} &= -72(x+2)^{-5} = \frac{-72}{(x+2)^5} \\ f^{(5)(x)} &= 360(x+2)^{-6} = \frac{360}{(x+2)^6} \end{aligned}$$

51. $f(x) = \frac{x+3}{x-2}$

$$\begin{aligned} f'(x) &= \frac{(x-2)(1)-(x+3)(1)}{(x-2)^2} \\ &= \frac{-5}{(x-2)^2} = -5(x-2)^{-2} \\ f''(x) &= 10(x-2)^{-3} = \frac{10}{(x-2)^3} \\ f'''(x) &= -30(x-2)^{-4} = \frac{-30}{(x-2)^4} \\ f^{(4)}(x) &= 120(x-2)^{-5} = \frac{120}{(x-2)^5} \\ f^{(5)}(x) &= -600(x-2)^{-6} = \frac{-600}{(x-2)^6} \end{aligned}$$

52.

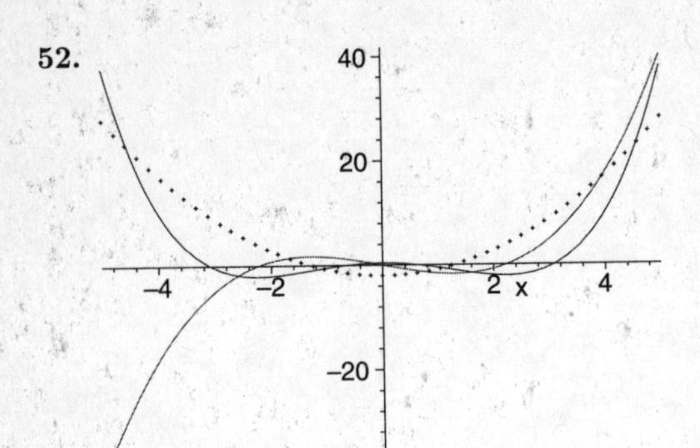

53.

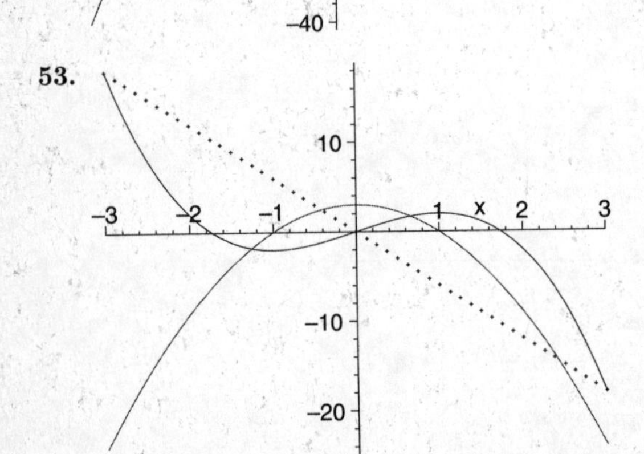

54.

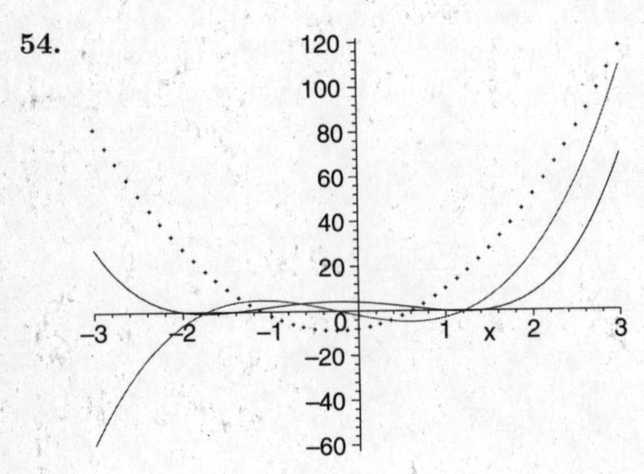

55.

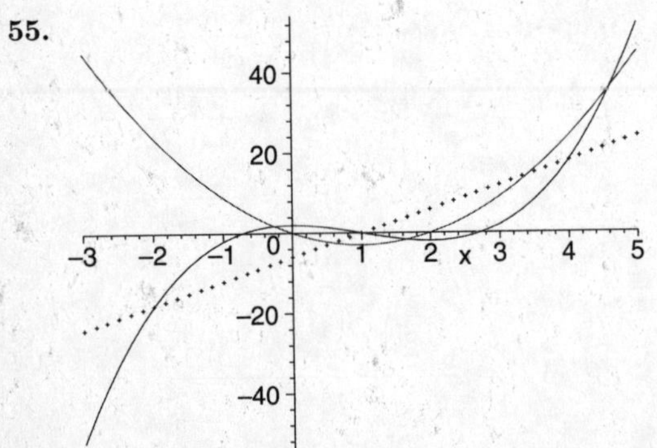

56.

57.

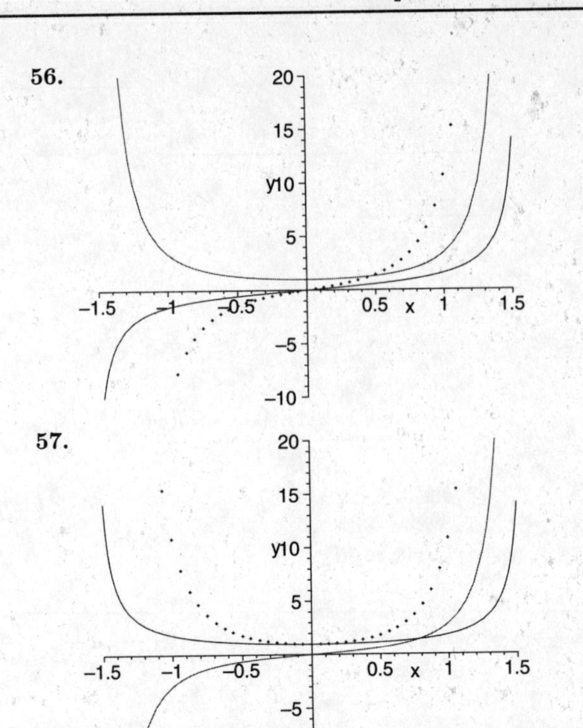

Chapter Review Exercises

1. a)

x	f(x)
-8	-11
-7.5	-10.5
-7.1	-10.1
-7.01	-10.01
-7.001	-10.001
-7.0001	-10.0001
-6	-9
-6.5	-9.5
-6.9	-9.9
-6.99	-9.99
-6.999	-9.999
-6.9999	-9.9999

b) $\lim_{x\to -7^-} f(x) = -10$

$\lim_{x\to -7^+} f(x) = -10$

$\lim_{x\to -7} f(x) = -10$

2. From the graph below, we can see that $\lim_{x\to -7} f(x) = -10$

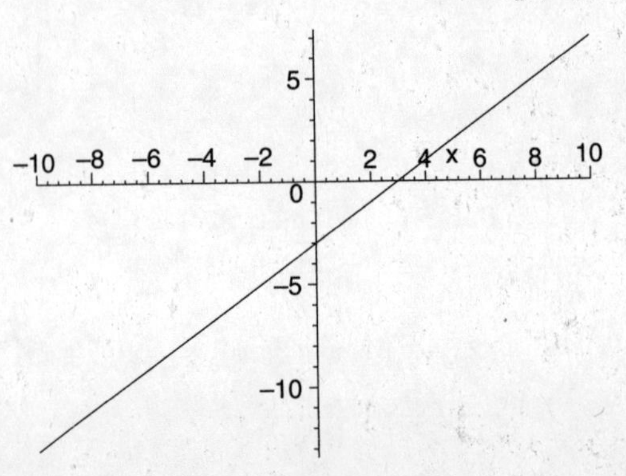

3. $f(x) = \frac{(x+7)(x-3)}{(x+7)} = (x-3)$, so $\lim_{x \to -7} f(x) = -10$

4. $\lim_{x \to -2} \frac{8}{x} = -4$

5. $\lim_{x \to 1} (2x^4 - 3x^2 + x + 4) = 2(1) - 3(1) + (1) + 4 = 4$

6. $\lim_{x \to 6} \frac{(x-6)(x+11)}{(x-6)} = 17$

7. $\lim_{x \to 4} \sqrt{x^2+9} = \sqrt{4^2+9} = \sqrt{25} = 5$

8. The function is not continuous at $x = -2$

9. The function is continuous

10. $\lim_{x \to 1} g(x) = -4$

11. $g(1) = -4$

12. Yes, $g(x)$ is continuoous at $x = 1$

13. $\lim_{x \to -2} g(x)$ does not exist

14. $g(-2) = -2$

15. No, $g(x)$ is not continuous at $x = -2$

16. $f(2) = 8$, $f(-1) = 2)$, average rate of change is

$$\frac{8-2}{2-(-1)} = \frac{6}{3} = 2$$

17.

$$\begin{aligned} \frac{f(x+h) - f(x)}{h} &= \frac{-3(x+h) + 5 - (-3x+5)}{h} \\ &= \frac{-3x - 3h + 5 + 3x - 5}{h} \\ &= \frac{-3h}{h} \\ &= -3 \end{aligned}$$

18.

$$\begin{aligned} \frac{f(x+h) - f(x)}{h} &= \frac{2(x+h)^2 - 3 - (2x^2 - 3)}{h} \\ &= \frac{2x^2 + 4xh + 2h^2 - 3 - 2x^2 + 3}{h} \\ &= \frac{4xh + 2h^2}{h} \\ &= 4x + 2h \end{aligned}$$

19. $\frac{dy}{dx} = 2x + 3$
The slope at $(-1, -2)$ is

$$\begin{aligned} m &= \frac{dy}{dx}\Big|_{x=-1} \\ &= 2(-1) + 3 \\ &= 1 \end{aligned}$$

Now we use the slope-point equation

$$\begin{aligned} y - (-2) &= 1(x - (-1)) \\ y &= x + 1 - 2 \\ y &= x - 1 \end{aligned}$$

20. $\frac{dy}{dx} = -2x + 8$

$$\begin{aligned} -2x + 8 &= 0 \\ -2x &= -8 \\ x &= 4 \end{aligned}$$

$$\begin{aligned} y &= -(4)^2 + 8(4) - 11 \\ &= -16 + 32 - 11 \\ &= 5 \end{aligned}$$

The slope of the tangent line is horizontal at $(4, 5)$

21. $\frac{dy}{dx} = 10x - 49$

$$\begin{aligned} 10x - 49 &= 1 \\ 10x &= 50 \\ x &= 5 \end{aligned}$$

$$\begin{aligned} y &= 5(5)^2 - 49(5) + 12 \\ &= 125 - 245 + 12 \\ &= -108 \end{aligned}$$

The slope of the tangent line is 1 at $(5, -108)$

22. $\frac{dy}{dx} = 4 \cdot 5x^4 = 20x^4$

23. $y = 3x^{1/3}$
$\frac{dy}{dx} = 3 \cdot \frac{1}{3} x^{-2/3} = \frac{1}{x^{2/3}}$

24. $y = -8x^{-8}$
$\frac{dy}{dx} = -64x^{-9} = \frac{-64}{x^9}$

25. $y = 21x^{4/3}$
$\frac{dy}{dx} = 21 \cdot \frac{4}{3} x^{1/3} = 28x^{1/3}$

26. $y = sec\ 5x$
$\frac{dy}{dx} = 5\ sec\ 5x\ tan\ 5x$

27. $y = x\ cot(x^2)$

$$\begin{aligned} \frac{dy}{dx} &= x\left(-csc^2(x^2)(2x)\right) + cot(x^2) \\ &= -2x^2\ csc^2(x^2) + cot(x^2) \end{aligned}$$

28. $y = 2.3\sqrt{0.4x + 5.3} + 0.01\ sin(0.17x - 0.31)$

$$\begin{aligned} \frac{dy}{dx} &= 2.3\frac{1}{2\sqrt{0.4x+5.3}}(0.4) + 0.01\ cos(0.17x - 0.31)(0.17) \\ &= \frac{0.46}{\sqrt{0.4x+5.3}} + 0.0017\ cos(0.17x - 0.31) \end{aligned}$$

29. $f(x) = \frac{1}{6}x^6 + 8x^4 - 5x$

$$\begin{aligned} f'(x) &= \frac{1}{6} \cdot 6x^5 + 8 \cdot 4x^3 - 5 \\ &= x^5 + 32x^3 - 5 \end{aligned}$$

30. $y = x^2 + 1$
$\frac{dy}{dx} = 2x$

31. $y = \frac{x^2+8}{8-x}$

$$\begin{aligned}\frac{dy}{dx} &= \frac{(8-x)(2x)-(x^2+8)(-1)}{(8-x)^2}\\ &= \frac{16x-2x^2+x^2+8}{(8-x)^2}\\ &= \frac{-x^2+16x+8}{(8-x)^2}\end{aligned}$$

32. $y = \frac{tan\ x}{x}$

$$\begin{aligned}\frac{dy}{dx} &= \frac{x\cdot sec^2x - tan\ x\cdot(1)}{x^2}\\ &= \frac{x\ sec^2x - tan\ x}{x^2}\end{aligned}$$

33. $y = (sin^2x + x)^{1/2}$

$$\begin{aligned}\frac{dy}{dx} &= \frac{1}{2}(sin^2x+x)^{-1/2}(2\ sin\ x\ cos\ x+1)\\ &= \frac{2\ sin\ x\ cos\ x+1}{2\sqrt{sin^2x+x}}\end{aligned}$$

34. $f(x) = tan^2(x\ cos\ x)$

$$\begin{aligned}f'(x) &= 2\ tan(x\ cos\ x)\ sec^2(x\ cos\ x)(x(-sin\ x)+cos\ x(1))\\ &= 2\ sec^2(x\ cos\ x)\ tan(x\ cos\ x)(cos\ xx\ sin\ x)\end{aligned}$$

35. $f(x) = cot(x - cos\ x)$

$$\begin{aligned}f'(x) &= -csc^2(x-cos\ x)(1-(-sin\ x))\\ &= -csc^2(x-cos\ x)(1+sin\ x)\end{aligned}$$

36. $f(x) = x^2(4x+3)^{3/4}$

$$\begin{aligned}f'(x) &= x^2[\frac{3}{4}(4x+3)^{-1/4}(4)] + 2x(4x+3)^{3/4}\\ &= \frac{3x^2}{(4x+3)^{1/4}} + 2x(4x+3)^{3/4}\end{aligned}$$

37. $y = x^3 - 2x^{-1}$

$$\begin{aligned}\frac{dy}{dx} &= 3x^2+2x^{-2}\\ \frac{d^2y}{dx^2} &= 6x-4x^{-3}\\ \frac{d^3y}{dx^3} &= 6+12x^{-4}\\ \frac{d^4y}{dx^4} &= -48x^{-5}\\ \frac{d^5y}{dx^5} &= 240x^{-6} = \frac{240}{x^6}\end{aligned}$$

38. $y = \frac{1}{42}x^7 - 10x^3 + 13x^2 + 28x - 5 + cos\ x$

$$\begin{aligned}\frac{dy}{dx} &= \frac{1}{6}x^6 - 30x^2+26x+28-sin\ x\\ \frac{d^2y}{dx^2} &= x^5-60x+26-cos\ x\\ \frac{d^3y}{dx^3} &= 5x^4-60+sin\ x\\ \frac{d^4y}{dx^4} &= 20x^3+cos\ x\end{aligned}$$

39. $s(t) = t + t^4$

a) $v(t) = s'(t) = 1 + 4t^3$

b) $a(t) = v'(t) = 12t^2$

c) $v(2) = 1 + 4(2)^3 = 1 + 4(8) = 33$
$a(2) = 12(2)^2 = 12(4) = 48$

40. **a)** $v(t) = 0.036t^2 - 1.85 - \frac{0.002\pi}{15}\ sin(\frac{\pi}{15}\ t)$

b) $a(t) = 0.072t - \frac{0.002\pi^2}{225}\ cos(\frac{\pi}{15}\ t)$

c) $v(2.5) = -1.62521$
$a(2.5) = 0.17992$

41. **a)** Amplitude $= \frac{135-1}{2} = 67$, period $= \frac{1}{2} = \frac{2\pi}{b} \to b = 4\pi$, mid-line$= 67+1 = 68$ and the capsule begins at the top means we are using a cosine model. Therefore $h(t) = 67\ cos(4\pi t) + 68$

b) Five minutes after reaching the bottom corrspond to $t = 20\ min = \frac{1}{3}$ hours since it is five minutes after half a period.

$$\begin{aligned}h\left(\frac{1}{12}\right) &= 67\ cos\left(4\pi\frac{1}{3}\right)+68\\ &= 34.5\ ft\end{aligned}$$

c) $h'(t) = -67\ sin(4\pi t)(4\pi) = -268\pi\ sin(4\pi t)$

$$\begin{aligned}h'\left(\frac{1}{3}\right) &= -268\pi\ sin\left(\frac{4\pi}{3}\right)\\ &= 729.1473\ ft/hr\end{aligned}$$

42. **a)** $P'(t) = 100t$

b) $P(20) = 10000 + 50(400) = 30000$ people

c) $P'(20) = 100(20) = 2000$ people/year

43.

$$\begin{aligned}(f\circ g)(x) &= (1-2x)^2+5\\ &= 1-4x+4x^2+5\\ &= 4x^2-4x+6\end{aligned}$$

$$\begin{aligned}(g\circ f)(x) &= 1-2(x^2+5)\\ &= 1-2x^2-10\\ &= -2x^2-9\end{aligned}$$

44.

$$\begin{aligned}\frac{dy}{dx} &= \frac{(1+x^3)[(x)(\frac{3}{2\sqrt{1+3x}})+\sqrt{1+3x}(1)] - x\sqrt{1+3x}(3x^2)}{(1+x^3)^2}\\ &= \frac{3x(1+x^3)+2(1+3x)(1+x^3)-6x^3(1+3x)}{2\sqrt{1+3x}(1+x^3)^2}\\ &= \frac{3x+3x^4+2+2x^3+6x+6x^4-6x^3-18x^4}{2\sqrt{1+3x}(1+x^3)^2}\\ &= \frac{-9x^4-4x^3+9x+2}{2\sqrt{1+3x}(1+x^3)^2}\end{aligned}$$

45. Limit approaches $-\frac{1}{4}$

46. Limit approaches $\frac{1}{6}$

47.

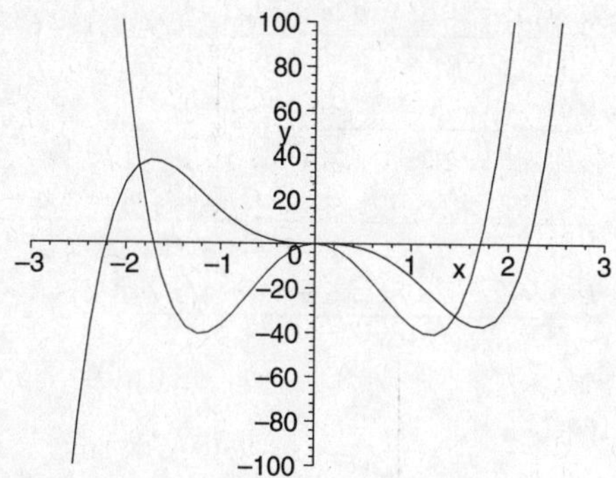

Chapter 2 Test

1. a)

x	f(x)
5	11
5.7	11.7
5.9	11.9
5.99	11.99
5.999	11.999
5.9999	11.9999
7	13
6.5	12.5
6.1	12.1
6.01	12.01
6.001	12.001
6.0001	12.0001
6.00001	12.00001

b) $\lim_{x\to 6^-} f(x) = 12$

$\lim_{x\to 6^+} f(x) = 12$

$\lim_{x\to 6} f(x) = 12$

2. From the graph below, we can see that $\lim_{x\to 6} f(x) = 12$

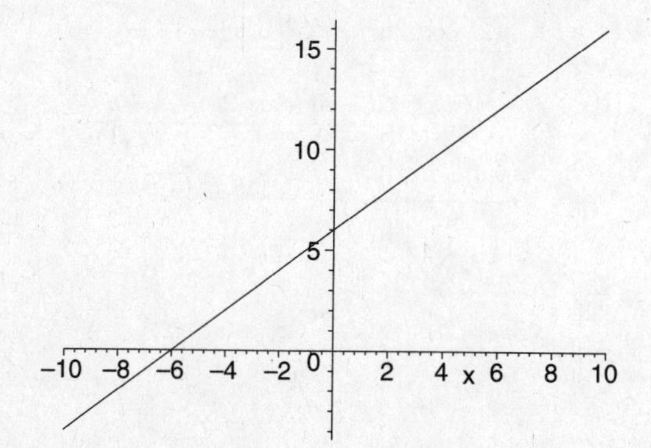

3. $f(x) = \frac{(x-6)(x+6)}{(x-6)} = (x+6)$, so $\lim_{x\to 6} f(x) = 12$

4. $\lim_{x\to -5} f(x)$ does not exist

5. $\lim_{x\to -4} f(x) = 0$

6. $\lim_{x\to -3} f(x)$ does not exist

7. $\lim_{x\to -2} f(x) = 2$

8. $\lim_{x\to -1} f(x) = 4$

9. $\lim_{x\to 1} f(x) = 1$

10. $\lim_{x\to 2} f(x) = 1$

11. $\lim_{x\to 3} f(x) = 1$

12. Function is continuous

13. Function is not continuous at $x = 3$

14. $\lim_{x\to 3} f(x)$ does not exist

15. $f(3) = 1$

16. No, the function is not continuous at $x = 3$

17. $\lim_{x\to 4} f(x) = 3$

18. $f(4) = 3$

19. Yes, the function is continupus at $x = 4$

20. $\lim_{x\to 1} (3x^4 - 2x^2 + 5 = 3(1) - 2(1) + 5 = 6$

21. $\lim_{x\to 2^+} \frac{x-2}{x(x-2)(x+2)} =$

$\lim_{x\to 2^+} \frac{1}{x(x+2)} = \frac{1}{8}$

22. $\lim_{x\to 0} \frac{7}{x}$ does not exist

23. $f(x) = 2x^2 + 3x - 9$, $\frac{f(x+h)-f(x)}{h}$

$$\begin{aligned} &= \frac{2(x+h)^2 + 3(x+h) - 9 - (2x^2 + 3x - 9)}{h} \\ &= \frac{2x^2 + 4xh + 2h^2 + 3x + 3h - 9 - 2x^2 - 3x + 9}{h} \\ &= \frac{4xh + 2h^2 + 3h}{h} \\ &= 4x + 3 + 2h \end{aligned}$$

24. First find $\frac{dy}{dx}$

$$\begin{aligned} \frac{dy}{dx} &= 1 + 4(-1x^{-2}) \\ &= 1 - \frac{4}{x^2} \end{aligned}$$

Next find the slope at $(4, 5)$

$$\begin{aligned} m &= 1 - \frac{4}{(4)^2} \\ &= 1 - \frac{1}{4} \\ &= \frac{3}{4} \end{aligned}$$

Finally use the point-slope equation

$$\begin{aligned} y - 5 &= \frac{3}{4}(x - 4) \\ y &= \frac{3}{4}x - 3 + 5 \\ y &= \frac{3}{4}x + 2 \end{aligned}$$

25. $\frac{dy}{dx} = 3x^2 - 6x$. The tangent line is horizontal when $\frac{dy}{dx} = 0$ so

$$\begin{aligned} 3x^2 - 6x &= 0 \\ 3x(x-2) &= 0 \\ &\text{either} \\ 3x &= 0 \\ x &= 0 \\ &\text{or} \\ x - 2 &= 0 \\ x &= 2 \end{aligned}$$

When $x = 0$, $y = 0^3 - 3(0)^2 = 0$, we have the point $(0, 0)$
When $x = 2$, $y = 2^3 - 3(2)^2 = 8 - 12 = -4$, we have the point $(2, -4)$

26. $\frac{dy}{dx} = 10 \cdot \frac{1}{2\sqrt{x}} = \frac{5}{\sqrt{x}}$

27. $y = -10x^{-1}$
$\frac{dy}{dx} = -10 \cdot -1x^{-2} = \frac{10}{x^2}$

28. $\frac{dy}{dx} = \frac{5}{4}x^{1/4}$

29. $y = -0.5x^2 + 0.61x + 90$
$\frac{dy}{dx} = -0.5(2x) + 0.61 = -x + 0.61$

30. $\frac{dy}{dx} = (sec^2\ 2x)(2) = 2\ sec^2\ 2x$

31. $f(x) = \frac{x}{5-x}$

$$\begin{aligned} f'(x) &= \frac{(5-x)(1) - x(-1)}{(5-x)^2} \\ &= \frac{5 - x + x}{(5-x)^2} \\ &= \frac{5}{(5-x)^2} \end{aligned}$$

32. $\frac{dy}{dx} = -5(x^5 - 4x^3 + x)^{-6}(5x^4 - 12x^2 + 1)$
$= \frac{-5(5x^4 - 12x^2 + 1)}{(x^5 - 4x^3 + x)^6}$

33. $f(x) = x\sqrt{x^2 + 5}$

$$\begin{aligned} f'(x) &= x \cdot \frac{2x}{2\sqrt{x^2+5}} + \sqrt{x^2+5} \\ &= \frac{x^2}{\sqrt{x^2+5}} + \sqrt{x^2+5} \cdot \frac{\sqrt{x^2+5}}{\sqrt{x^2+5}} \\ &= \frac{x^2}{\sqrt{x^2+5}} + \frac{x^2+5}{\sqrt{x^2+5}} \\ &= \frac{2x^2+5}{\sqrt{x^2+5}} \end{aligned}$$

34. $\frac{dy}{dx} = -sin(\sqrt{x}) \cdot \frac{1}{2\sqrt{x}} = \frac{-sin(\sqrt{x})}{2\sqrt{x}}$

35. $f(x) = tan\ 2x\ sec\ 3x$

$$\begin{aligned} f'(x) &= tan\ 2x[(sec\ 3x\ tan\ 3x)(3)] + [(sec^2\ 2x)(2)]sec\ 3x \\ &= sec\ 3x(3\ tan\ 2x\ tan\ 3x + 2\ sec^2\ 2x) \end{aligned}$$

36.

$$\begin{aligned} f'(x) &= \frac{(cos^2x - x)(2\ tan\ x\ sec^2x)}{(cos^2x - x)^2} - \\ &\quad \frac{(tan^2x)(-2cos\ x\ sin\ x - 1)}{(cos^2x - x)^2} \\ &= \frac{2\ tan\ x - 2x\ tan\ x sec^2x + 2sin^2x\ tan\ x + tan^2x}{(cos^2x - x)^2} \\ &= \frac{tan\ x(2 - 2x\ sec^2x + 2sin^2x + tan\ x)}{(cos^2x - x)^2} \end{aligned}$$

37. $f(x) = sin(cos(x^2))$

$$\begin{aligned} f'(x) &= cos(cos(x^2)) \cdot -sin(x^2) \cdot 2x \\ &= -2x\ sin(x^2)\ cos(cos(x^2)) \end{aligned}$$

38.

$$\begin{aligned} f'(x) &= 2x + cos(2x\sqrt{x+2})\left(\cdot 2x \cdot \frac{1}{2\sqrt{x+2}} + (2)\sqrt{x+2}\right) \\ &= 2x + \left(\frac{x}{\sqrt{x+2}} + 2\sqrt{x+2}\right) cos(2x\sqrt{x+2}) \end{aligned}$$

39. $y = x^4 - 3x^2$

$$\begin{aligned} \frac{dy}{dx} &= 4x^3 - 6x \\ \frac{d^2y}{dx^2} &= 12x^2 - 6 \\ \frac{d^3y}{dx^3} &= 24x \end{aligned}$$

40. **a)** $v(t) = \frac{2t\ cos\ 2t - sin\ 2t}{t^2}$

b)

$$\begin{aligned} a(t) &= \frac{t^2[2t(-sin\ 2t)(2) + 2\ cos\ 2t - (cos\ 2t)(2)]}{t^4} - \\ &\quad \frac{[(2t\ cos\ 2t - sin\ 2t)(2t)]}{t^4} \\ &= \frac{-4t^3\ sin\ 2t - 4t^2\ cos\ 2t - 2t\ sin\ 2t}{t^4} \\ &= \frac{2(1 - 2t^2)sin\ 2t - 4t\ cos\ 2t}{t^4} \end{aligned}$$

c) $s(\frac{7\pi}{6}) = \frac{sin(\frac{7\pi}{3})}{\frac{7\pi}{6}} = \frac{3\sqrt{3}}{7\pi}$
$v(\frac{7\pi}{6}) = \frac{\frac{7\pi}{3}cos(\frac{7\pi}{3}) - sin(\frac{7\pi}{3})}{\frac{49\pi^2}{36}} = \frac{6(3\sqrt{3} - 7\pi)}{49\pi^2}$
$a(\frac{7\pi}{6}) = \frac{291 - 2(\frac{7\pi}{6})^2 sin(\frac{7\pi}{3}) - \frac{14\pi}{3}cos(\frac{7\pi}{3})}{(\frac{7\pi}{6})^4} = \frac{-12(-18\sqrt{3} + 42\pi + 49\sqrt{3}\pi^2)}{343\pi^3}$

41. $M = -0.001t^3 + 0.1t^2$

a) $M'(t) = -0.003t^2 + 0.2t$

b) $M(10) = 0.001(10)^3 + 0.1(10)^2 = -1 + 10 = 9$ words

c) $M'(10) = -0.003(10)^2 + 0.2(10) = -0.3 + 2 = 1.7$ words/minutes

42.

$$\begin{aligned}(f \circ g)(x) &= (2x^3)^2 - (2x^3) \\ &= 4x^6 - 2x^3\end{aligned}$$

$$\begin{aligned}(g \circ f)(x) &= 2(x^2 - x)^3 \\ &= 2(x^6 - 3x^5 + 3x^4 - x^3) \\ &= 2x^6 - 6x^5 + 6x^4 - 2x^3\end{aligned}$$

43. $y = (1-3x)^{2/3}(1+3x)^{1/3}$

$$\begin{aligned}\frac{dy}{dx} &= (1-3x)^{2/3}\left(\frac{1}{3}(1+3x)^{-2/3}(3)\right) + \\ &\quad \left(\frac{2}{3}(1-3x)^{-1/3}(-3)\right)(1+3x)^{1/3} \\ &= \left(\frac{1-3x}{1+3x}\right)^{2/3} - 2\left(\frac{1+3x}{1-3x}\right)^{1/3}\end{aligned}$$

44. $\lim\limits_{x\to 3} \dfrac{(x-3)(x^2+3x+9)}{x-3} = \lim\limits_{x\to 3}(x^2+3x+9) = 27$

45. $\lim\limits_{t\to 0} \dfrac{\tan t + \sin t}{t} = \lim\limits_{t\to 0}\left(\dfrac{\tan t}{t} + \dfrac{\sin t}{t}\right) =$
$\lim\limits_{t\to 0}\left(\dfrac{\tan t}{t}\right) + \lim\limits_{t\to 0}\left(\dfrac{\sin t}{t}\right) = 1 + 1 = 2$

46. $(1.0836, 25.1029)$ and $(2.9503, 8.6247)$

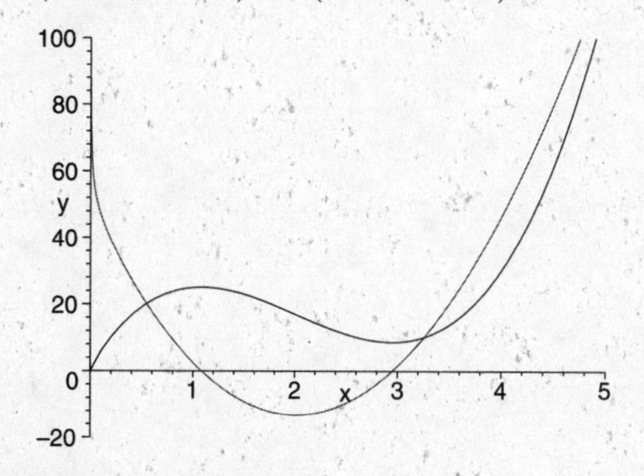

47. Limit approaches 0

Technology Connection

- **Page 73:**
  1. 3
  2. -0.2
- **Page 87:**
  1. 53
  2. 2.8284
  3. Limit does not exist
- **Page 95:**
  1. $f(x+h) = 64, \;\; f(x) + h = 38$
  2. $f(x+h) = 218.78, \;\; f(x) + h = 208.1$
- **Page 109:**
  1. $f'(x) = -\frac{3}{x^2}$, $f'(-2) = \frac{-3}{4}$, $f'(-1/2) = -12$
  2. $y = -\frac{3}{4}x - 3, \;\; y = -12x - 12$
- **112:**
  1. $f'(20) = 60, \;\; f'(37) = 26, \;\; f'(50) = 0,$ $f'(90) = -80$
  2. $f'(-5) = -96, \;\; f'(0) = -11, \;\; f'(7) = 24,$ $f'(12) = -11, \;\; f'(15) = -56$
  3. $f'(-2) = -36, \;\; f'(0) = 0, \;\; f'(2) = 12,$ $f'(4) = 0, \;\; f'(6.3) = -43.47$
  4. $f'(-2)$ does not exist, $f'(-1.3) = 0.4079,$ $f'(-0.5) = 1.8074, \;\; f'(0) = 2, \;\; f'(1) = 1.1547,$ $f'(2)$ does not exist
  5. $x = 3$ is not in the domain of the function $x\sqrt{4-x}$
- **Page 118:**
  1. $f'(24) = 152, \;\; f'(138) = -76, \;\; f'(150) = -100,$ $f'(190) = -180$
  2. $f'(-5) = -96, \;\; f'(0) = -11, \;\; f'(7) = 24,$ $f'(12) = -11, \;\; f'(15) = -56$
  3. $f'(-2) = -36, \;\; f'(0) = 0, \;\; f'(2) = 12,$ $f'(4) = 0, \;\; f'(6.3) = -43.47$
  4. $f'(0) = 1.61, \;\; f'(12) = 0.7076, \;\; f'(24) = 0.3236,$ $f'(6.3) = 0.458$
- **Page 137:**
  1. Part "c" is the correct derivative
  2. Left to the student
  3. Left to the student

Extended Life Science Connection

1. $f(x) = \sqrt{4+x}$

$$\begin{aligned}f'(x) &= \frac{1}{2\sqrt{4+x}} \\ f'(0) &= \frac{1}{2\sqrt{4+0}} \\ &= \frac{1}{2\sqrt{4}} \\ &= 0.25\end{aligned}$$

2. **a)**

h	$\frac{1}{2}$	$\frac{1}{4}$	$\frac{1}{8}$	$\frac{1}{16}$
$F_+(h)$	0.242641	0.24621	0.24808	0.24903

b)

h	$\frac{1}{2}$	$\frac{1}{4}$	$\frac{1}{8}$	$\frac{1}{16}$
$F_-(h)$	0.258342	0.25403	0.25198	0.25098

3. **a)** Forward difference:: $\frac{350-264}{(2/7)} = 301$ Backward difference: $\frac{264-167}{(2/7)} = 339.5$

b) Forward difference:: $\frac{306-264}{(1/7)} = 294$ Backward difference: $\frac{264-219}{(1/7)} = 315$ It seems that the answers from part a) are more accurate since they use points closer to the point under consideration (March 19) and therefore represent the slope of the tangent line more closely than points away from the point corresponding to March 19.

4.

$$\begin{aligned} F_c &= \frac{1}{2}[F_+(h) + F_-(h)] \\ &= \frac{1}{2}\left(\frac{f(h) - f(0)}{h} + \frac{f(0) - f(-h)}{h}\right) \\ &= \frac{1}{2}\left(f(h) - f(0) + f(0) - f(-h)\right) \\ &= \frac{1}{2}\left(\frac{f(h) - f(-h)}{h}\right) \\ &= \frac{f(h) - f(-h)}{2h} \end{aligned}$$

5. Since $F_c(h)$ cover more of the function domain containing $x = 0$ than does $F_+(h)$ or $F_-(h)$, then it makes sense that $F_c(h)$ be closer to $f'(0)$ than either of them.

6. Using the results from question 2 parts a) and b) and the definition of F_c we get

h	$\frac{1}{2}$	$\frac{1}{4}$	$\frac{1}{8}$	$\frac{1}{16}$
$F_c(h)$	0.25049	0.25012	0.25003	0.250005

7. The answers in Exercise 6 are closer to $f'(0) = 0.25$ than those from Exercise 2. The value of $h = \frac{1}{16}$ using the central difference quotient gives the closest value to 0.25. This is expected since the central difference quotient is the most accurate out of the three difference quotients and $\frac{1}{16}$ is the smallest value of h to be considered.

8. **a)** $F_c(1/7) = \frac{f(1/7) - f(-1/7)}{2(1/7)} = \frac{306-219}{(2/7)} = 304.5$

b) $F_c(2/7) = \frac{f(2/7) - f(-2/7)}{2(2/7)} = \frac{350-167}{(4/7)} = 320.25$

c) The answer from part a) gives the more accurate rate of change.

9. $F_c(4/7) = \frac{f(4/7) - f(-4/7)}{2(4/7)} = \frac{7739-7053}{(8/7)} = 600.25$
$F_c(4/7) = \frac{f(2/7) - f(-2/7)}{2(2/7)} = \frac{7628-7296}{(4/7)} = 581$

10. **a)** $f(x) = x^3 + x$
$f'(x) = 3x^2 + 1$, $f'(0) = 1$
Close to answer given by nDeriv, which is 1.000001

b) $f(x) = 1000x^3 + x$
$f'(x) = 3000x^2$, $f'(0) = 1$
Close to answer given by nDeriv, which is 1.001

c) $f(x) = 1000000x^3 + x$
$f'(x) = 3000000x^2 + 1$, $f'(0) = 1$
Does not match the answer from nDeriv, which is 2.

d) $f(x) = 1000000000x^3 + x$
$f'(x) = 3000000000x^2 + 1$, $f'(0) = 1$
Does not match the answer from nDeriv, which is 1001.

e) $f(x) = \mid x \mid$
$f'(x)$ does not exist at $x = 0$
Does not match the answer given by nDeriv, which is 0.

Chapter 3

Application of Differentiation

Exercise Set 3.1

1. $f(x) = x^2 - 4x + 5$. First, find the critical points (values of x at which the derivative is zero or undefined).

$f'(x) = 2x - 4$

$f'(x)$ exists for all real numbers. We solve $f'(x) = 0$:

$$\begin{aligned} 2x - 4 &= 0 \\ 2x &= 4 \\ x &= 2 \end{aligned}$$

The only critical point is at $x = 2$. We use 2 to divide the real number line into two intervals, A: $(-\infty, 2)$ and B: $(2, \infty)$:

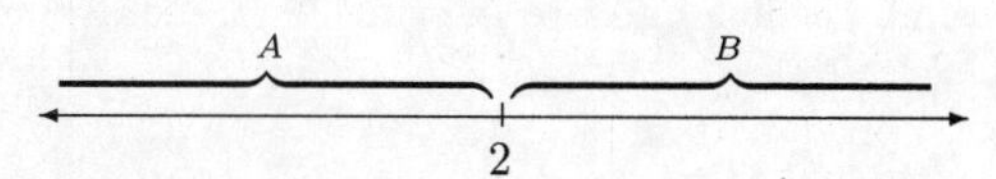

We use a test value in each interval to determine the sign of the derivative in each interval.

A: Test 0, $f'(0) = 2 \cdot 0 - 4 = -4 < 0$

B: Test 3, $f'(3) = 2 \cdot 3 - 4 = 2 > 0$

We see that $f(x)$ is decreasing on $(-\infty, 2)$ and increasing on $(2, \infty)$, and the change from decreasing to increasing indicates that a relative minimum occurs at $x = 2$. We substitute into the original equation to find $f(2)$:

$f(2) = 2^2 - 4 \cdot 2 + 5 = 1$

Thus, there is a relative minimum at $(2, 1)$. We use the information obtained to sketch the graph.

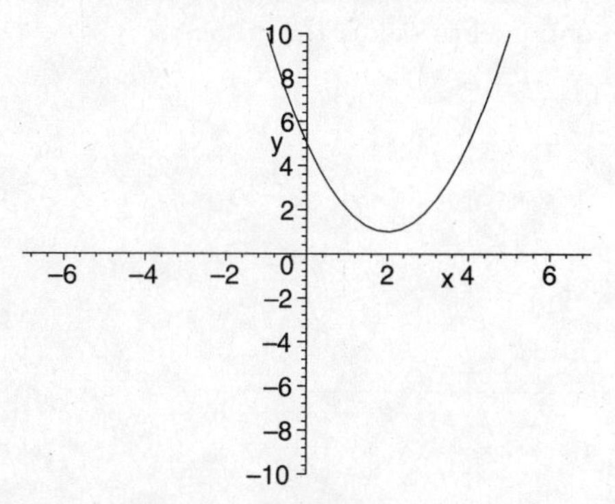

2. $f(x) = x^2 - 6x - 3$

$f'(x) = 2x - 6$

$f'(x)$ exists for all real numbers. Solve $f'(x) = 0$:

$$\begin{aligned} 2x - 6 &= 0 \\ x &= 3 \end{aligned}$$

The only critical point is at $x = 3$. Use it to divide the real number line into two intervals, A: $(-\infty, 3)$ and B: $(3, \infty)$.

A: Test 0, $f'(0) = 2 \cdot 0 - 6 = -6 < 0$

B: Test 4, $f'(4) = 2 \cdot 4 - 6 = 2 > 0$

Since $f(x)$ is decreasing on $(-\infty, 3)$ and increasing on $(3, \infty)$, there is a relative minimum at $x = 3$.

$f(3) = 3^2 - 6 \cdot 3 - 3 = -12$

Thus, there is a relative minimum at $(3, -12)$. We sketch the graph.

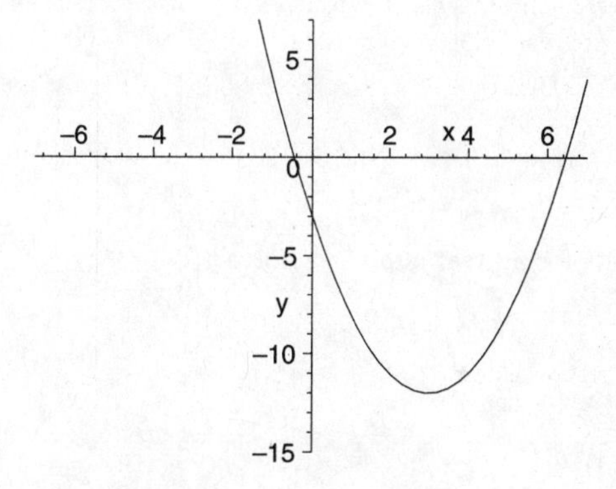

3. $f(x) = 5 + x - x^2$

First, find the critical points,

$f'(x) = 1 - 2x$

$f'(x)$ exists for all real numbers. We solve $f'(x) = 0$:

$$\begin{aligned} 1 - 2x &= 0 \\ 1 &= 2x \\ \frac{1}{2} &= x \end{aligned}$$

The only critical point is at $x = \frac{1}{2}$. We use $\frac{1}{2}$ to divide the real number line into two intervals, A: $\left(-\infty, \frac{1}{2}\right)$ and B: $\left(\frac{1}{2}, \infty\right)$:

A B

$\frac{1}{2}$

We use a test value in each interval to determine the sign of the derivative in each interval.

A: Test 0, $f'(0) = 1 - 2 \cdot 0 = 1 > 0$

B: Test 1, $f'(1) = 1 - 2 \cdot 1 = -1 < 0$

We see that $f(x)$ is increasing on $\left(-\infty, \frac{1}{2}\right)$ and decreasing on $\left(\frac{1}{2}, \infty\right)$, so there is a relative maximum at $x = \frac{1}{2}$. We find $f\left(\frac{1}{2}\right)$:

$$f\left(\frac{1}{2}\right) = 5 + \frac{1}{2} - \left(\frac{1}{2}\right)^2 = \frac{21}{4}$$

Thus, there is a relative maximum at $\left(\frac{1}{2}, \frac{21}{4}\right)$. We use the information obtained to sketch the graph.

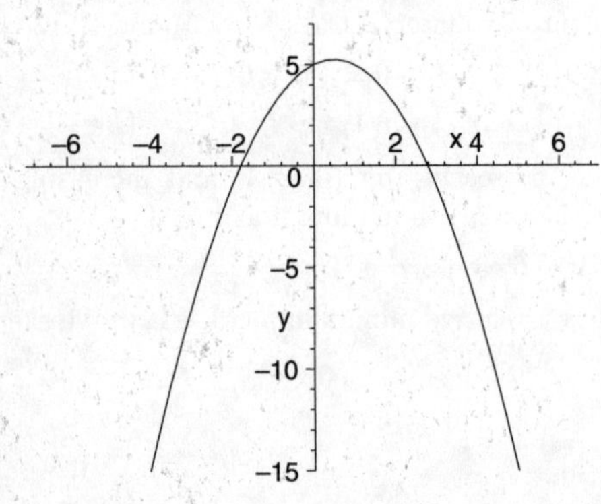

4. $f(x) = 2 - 3x - 2x^2$

$f'(x) = -3 - 4x$

$f'(x)$ exists for all real numbers. Solve $f'(x) = 0$:

$$\begin{aligned} -3 - 4x &= 0 \\ -\frac{3}{4} &= x \end{aligned}$$

The only critical point is $-\frac{3}{4}$. Use it to divide the real number line into two intervals, A: $\left(-\infty, -\frac{3}{4}\right)$ and B: $\left(-\frac{3}{4}, \infty\right)$.

A: Test -1, $f'(-1) = -3 - 4(-1) = 1 > 0$

B: Test 0, $\quad f'(0) = -3 - 4 \cdot 0 = -3 < 0$

Since $f(x)$ is increasing on $\left(-\infty, -\frac{3}{4}\right)$ and decreasing on $\left(-\frac{3}{4}, \infty\right)$, there is a relative maximum at $x = -\frac{3}{4}$.

$$f\left(-\frac{3}{4}\right) = 2 - 3\left(-\frac{3}{4}\right) - 2\left(-\frac{3}{4}\right)^2 = \frac{25}{8}$$

Thus, there is a relative maximum at $x = \left(-\frac{3}{4}, \frac{25}{8}\right)$. We sketch the graph.

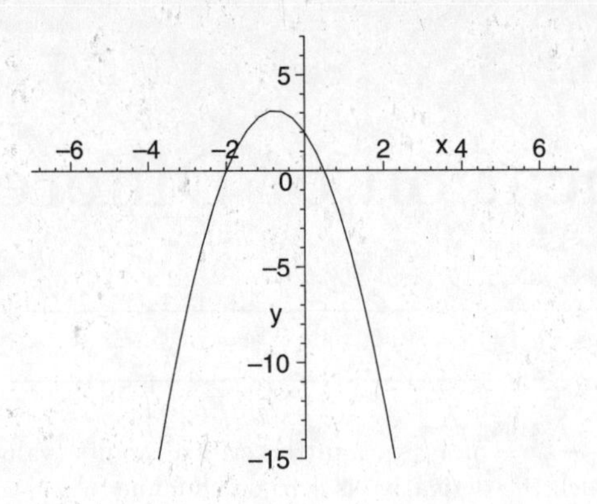

5. $f(x) = 1 + 6x + 3x^2$

First, find the critical points.

$f'(x) = 6 + 6x$

$f'(x)$ exists for all real numbers. We solve $f'(x) = 0$:

$$\begin{aligned} 6 + 6x &= 0 \\ 6x &= -6 \\ x &= -1 \end{aligned}$$

The only critical point is at $x = -1$. We use -1 to divide the real number line into two intervals, A: $(-\infty, -1)$ and B: $(-1, \infty)$:

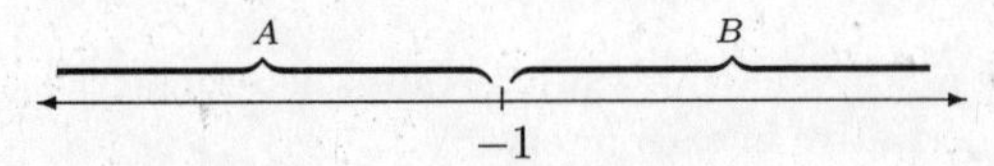

We use a test value in each interval to determine the sign of the derivative in each interval.

A: Test -2, $f'(-2) = 6 + 6(-2) = -6 < 0$

B: Test 0, $\quad f'(0) = 6 + 6 \cdot 0 = 6 > 0$

We see that $f(x)$ is decreasing on $(-\infty, -1)$ and increasing on $(-1, \infty)$, so there is a relative minimum at $x = -1$. We find $f(-1)$:

$$f(-1) = 1 + 6(-1) + 3(-1)^2 = -2$$

Thus, there is a relative minimum at $(-1, -2)$. We use the information obtained to sketch the graph.

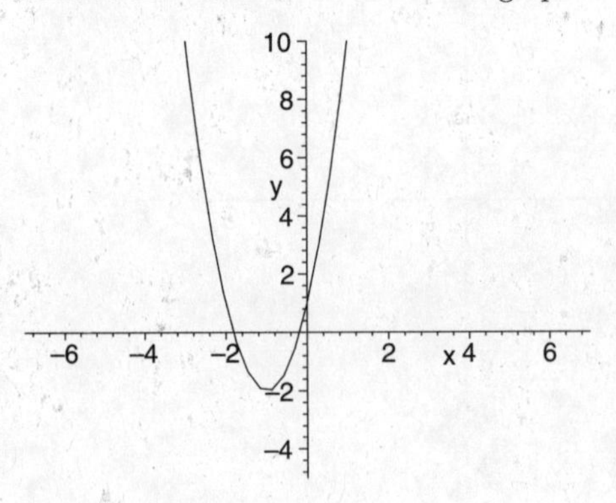

6. $f(x) = 0.5x^2 - 2x - 11$

$f'(x) = x - 2$

$f'(x)$ exists for all real numbers. Solve $f'(x) = 0$.

$$\begin{aligned} x-2 &= 0 \\ x &= 2 \end{aligned}$$

The only critical point is at $x = 2$. Use it to divide the real number line into two intervals, A: $(-\infty, 2)$ and B: $(2, \infty)$.

A: Test 0, $f'(0) = 0 - 2 = -2 < 0$

B: Test 3, $f'(3) = 3 - 2 = 1 > 0$

Since $f(x)$ is decreasing on $(-\infty, 2)$ and increasing on $(2, \infty)$, there is a relative minimum at $x = 2$.

$f(2) = 0.5(2)^2 - 2 \cdot 2 - 11 = -13$

Thus, there is a relative minimum at $(2, -13)$. We sketch the graph.

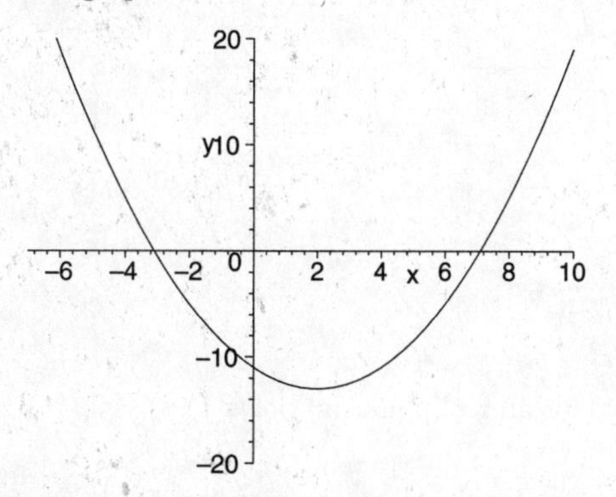

7. $f(x) = x^3 - x^2 - x + 2$

First, find the critical points.

$f'(x) = 3x^2 - 2x - 1$

$f'(x)$ exists for all real numbers. We solve $f'(x) = 0$:

$$\begin{aligned} 3x^2 - 2x - 1 &= 0 \\ (3x+1)(x-1) &= 0 \\ 3x + 1 &= 0 \\ 3x &= -1 \\ x &= -\frac{1}{3} \\ \text{Or} & \\ x - 1 &= 0 \\ x &= 1 \end{aligned}$$

The critical points are at $x = -\dfrac{1}{3}$ and $x = 1$. We use them to divide the real number line into three intervals, A: $\left(-\infty, -\dfrac{1}{3}\right)$, B: $\left(-\dfrac{1}{3}, 1\right)$, and C: $(1, \infty)$.

A B C

$-\dfrac{1}{3}$ 1

We use a test value in each interval to determine the sign of the derivative in each interval.

A: Test -1, $f'(-1) = 3(-1)^2 - 2(-1) - 1 = 3 + 2 - 1 = 4 > 0$

B: Test 0, $f'(0) = 3(0)^2 - 2(0) - 1 = -1 < 0$

C: Test 2, $f'(2) = 3(2)^2 - 2(2) - 1 = 12 - 4 - 1 = 7 > 0$

We see that $f(x)$ is increasing on $\left(-\infty, -\dfrac{1}{3}\right)$, decreasing on $\left(-\dfrac{1}{3}, 1\right)$, and increasing again on $(1, \infty)$, so there is a relative maximum at $x = -\dfrac{1}{3}$ and a relative minimum at $x = 1$. We find $f\left(-\dfrac{1}{3}\right)$:

$$\begin{aligned} f\left(-\frac{1}{3}\right) &= \left(-\frac{1}{3}\right)^3 - \left(-\frac{1}{3}\right)^2 - \left(-\frac{1}{3}\right) + 2 \\ &= -\frac{1}{27} - \frac{1}{9} + \frac{1}{3} + 2 \\ &= \frac{59}{27} \end{aligned}$$

Then we find $f(1)$:

$$\begin{aligned} f(1) &= 1^3 - 1^2 - 1 + 2 \\ &= 1 - 1 - 1 + 2 \\ &= 1 \end{aligned}$$

There is a relative maximum at $\left(-\dfrac{1}{3}, \dfrac{59}{27}\right)$, and there is a relative minimum at $(1, 1)$. We use the information obtained to sketch the graph.

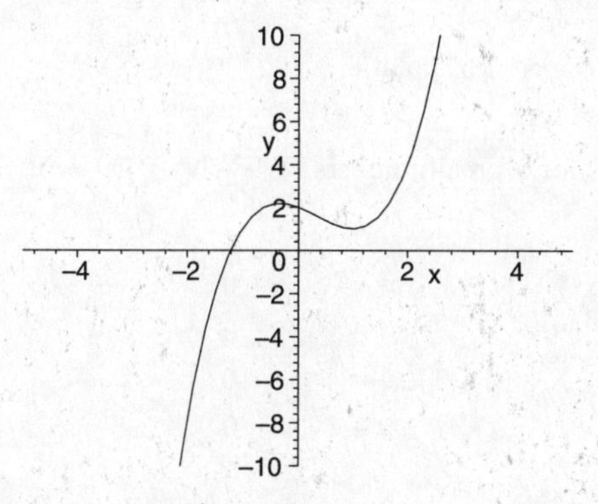

8. $f(x) = x^3 + \dfrac{1}{2}x^2 - 2x + 5$

$f'(x) = 3x^2 + x - 2$

$f'(x)$ exists for all real numbers. Solve $f'(x) = 0$.

$$\begin{aligned} 3x^2 + x - 2 &= 0 \\ (3x-2)(x+1) &= 0 \\ x &= \frac{2}{3} \\ \text{Or} & \\ x &= -1 \end{aligned}$$

The critical points are at $x = -1$ and $x = \dfrac{2}{3}$. Use them to divide the real number line into three intervals, A: $(-\infty, -1)$, B: $\left(-1, \dfrac{2}{3}\right)$, and C: $\left(\dfrac{2}{3}, \infty\right)$.

A: Test -2, $f'(-2) = 3(-2)^2 + (-2) - 2 = 8 > 0$

B: Test 0, $f'(0) = 3 \cdot 0^2 + 0 - 2 = -2 < 0$

C: Test 1, $f'(1) = 3 \cdot 1^2 + 1 - 2 = 2 > 0$

Since $f(x)$ is increasing on $(-\infty, -1)$, decreasing on $\left(-1, \frac{2}{3}\right)$, and increasing again on $\left(\frac{2}{3}, \infty\right)$, there is a relative maximum at $x = -1$ and a relative minimum at $x = \frac{2}{3}$.

$$f(-1) = (-1)^3 + \frac{1}{2}(-1)^2 - 2(-1) + 5 = \frac{13}{2}$$

$$f\left(\frac{2}{3}\right) = \left(\frac{2}{3}\right)^3 + \frac{1}{2}\left(\frac{2}{3}\right)^2 - 2\left(\frac{2}{3}\right) + 5 = \frac{113}{27}$$

There is a relative maximum at $\left(-1, \frac{13}{2}\right)$, and there is a relative minimum at $\left(\frac{2}{3}, \frac{113}{27}\right)$. We sketch the graph.

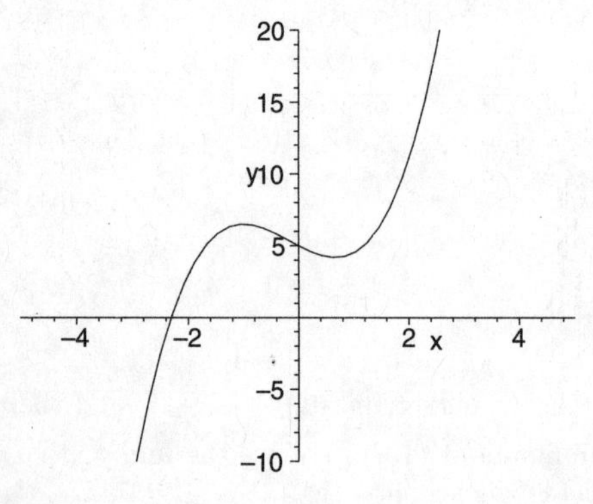

9. $f(x) = x^3 - 3x + 6$

First, find the critical points.

$f'(x) = 3x^2 - 3$

$f'(x)$ exists for all real numbers. We solve $f'(x) = 0$:

$$\begin{aligned} 3x^2 - 3 &= 0 \\ x^2 - 1 &= 0 \\ (x+1)(x-1) &= 0 \\ x + 1 &= 0 \\ x &= -1 \\ &\text{Or} \\ x - 1 &= = 0 \\ x &= 1 \end{aligned}$$

The critical points are at $x = -1$ and $x = 1$. We use them to divide the real number line into three intervals, A: $(-\infty, -1)$, B: $(-1, 1)$, and C: $(1, \infty)$.

A B C

-1 1

We use a test value in each interval to determine the sign of the derivative in each interval.

A: Test -2, $f'(-2) = 3(-2)^2 - 3 = 12 - 3 = 9 > 0$

B: Test 0, $f'(0) = 3 \cdot 0^2 - 3 = 0 - 3 = -3 < 0$

C: Test 2, $f'(2) = 3 \cdot 2^2 - 3 = 12 - 3 = 9 > 0$

We see that $f(x)$ is increasing on $(-\infty, -1)$, decreasing on $(-1, 1)$, and increasing again on $(1, \infty)$, so there is a relative maximum at $x = -1$ and a relative minimum at $x = 1$. We find $f(-1)$:

$$f(-1) = (-1)^3 - 3(-1) + 6 = -1 + 3 + 6 = 8$$

Then we find $f(1)$:

$$f(1) = 1^3 - 3 \cdot 1 + 6 = 1 - 3 + 6 = 4$$

There is a relative maximum at $(-1, 8)$, and there is a relative minimum at $(1, 4)$. We use the information obtained to sketch the graph.

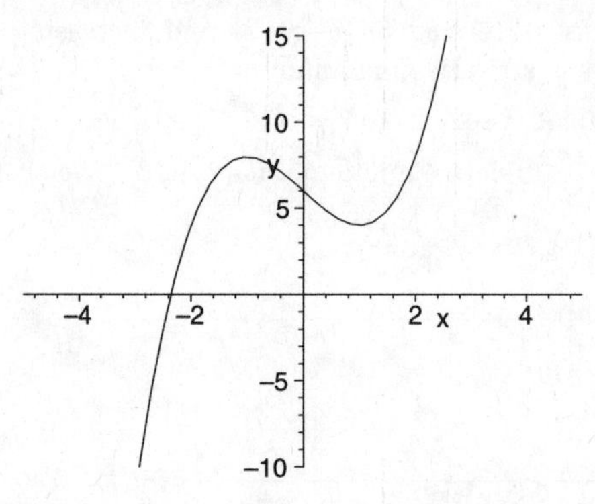

10. $f(x) = x^3 - 3x^2$

$f'(x) = 3x^2 - 6x$

$f'(x)$ exists for all real numbers. Solve $f'(x) = 0$.

$$\begin{aligned} 3x^2 - 6x &= 0 \\ x^2 - 2x &= 0 \\ x(x-2) &= 0 \\ x &= 0 \\ &\text{Or} \\ x &= 2 \end{aligned}$$

The critical points are at $x = 0$ and $x = 2$. Use them to divide the real number line into three intervals, A: $(-\infty, 0)$, B: $(0, 2)$, and C: $(2, \infty)$.

A: Test -1, $f'(-1) = 3(-1)^2 - 6(-1) = 9 > 0$

B: Test 1, $f'(1) = 3 \cdot 1^2 - 6 \cdot 1 = -3 < 0$

C: Test 3, $f'(3) = 3 \cdot 3^2 - 6 \cdot 3 = 9 > 0$

Since $f(x)$ is increasing on $(-\infty, 0)$, decreasing on $(0, 2)$, and increasing again on $(2, \infty)$, there is a relative maximum at $x = 0$ and a relative minimum at $x = 2$.

$f(0) = 0^3 - 3 \cdot 0^2 = 0$

$f(2) = 2^3 - 3 \cdot 2^2 = -4$

There is a relative maximum at $(0, 0)$, and there is a relative minimum at $(2, -4)$. We sketch the graph.

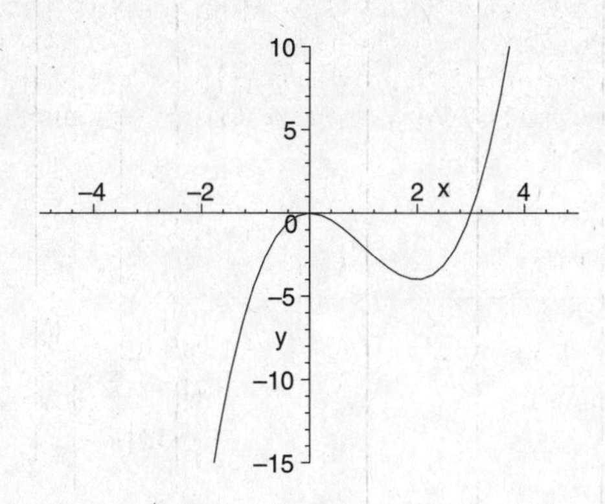

11. $f(x) = 2x^3$

First, find the critical points.

$f'(x) = 6x^2$

$f'(x)$ exists for all real numbers. We solve $f'(x) = 0$:

$$\begin{aligned} 6x^2 &= 0 \\ x^2 &= 0 \\ x &= 0 \end{aligned}$$

The only critical point is at $x = 0$. We use 0 to divide the real number line into two intervals, A: $(-\infty, 0)$ and B: $(0, \infty)$:

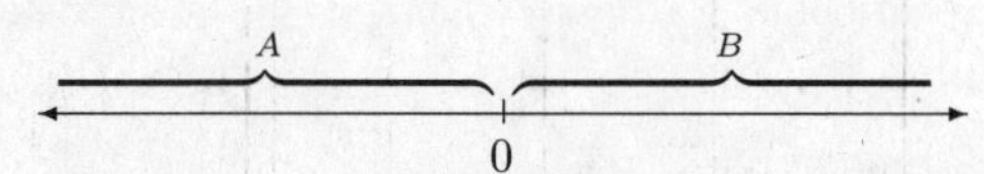

We use a test value in each interval to determine the sign of the derivative in each interval.

A: Test -1, $f'(-1) = 6(-1)^2 = 6 > 0$

B: Test 1, $f'(1) = 6(1)^2 = 6 > 0$

We see that $f(x)$ is increasing on $(-\infty, 0)$ and increasing on $(0, \infty)$, therefore there is no change from decreasing to increasing or from increasing to decreasing. Therefore the function does not has a relative extrema.

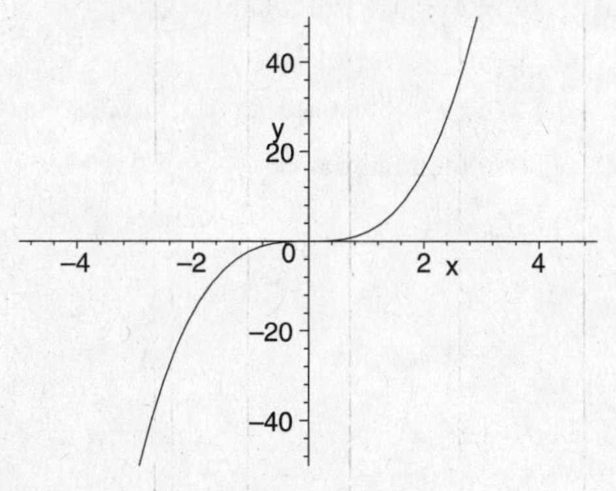

12. $f(x) = 1 - x^3$

First, find the critical points.

$f'(x) = -3x^2$

$f'(x)$ exists for all real numbers. We solve $f'(x) = 0$:

$$\begin{aligned} -3x^2 &= 0 \\ x^2 &= 0 \\ x &= 0 \end{aligned}$$

The only critical point is at $x = 0$. We use 0 to divide the real number line into two intervals, A: $(-\infty, 0)$ and B: $(0, \infty)$:

A B

0

We use a test value in each interval to determine the sign of the derivative in each interval.

A: Test -1, $f'(-1) = -3(-1)^2 = -3 < 0$

B: Test 1, $f'(1) = -3(1)^2 = -3 < 0$

We see that $f(x)$ is decreasing on $(-\infty, 0)$ and decreasing on $(0, \infty)$, therefore there is no change from decreasing to increasing or from increasing to decreasing. Therefore the function does not has a relative extrema.

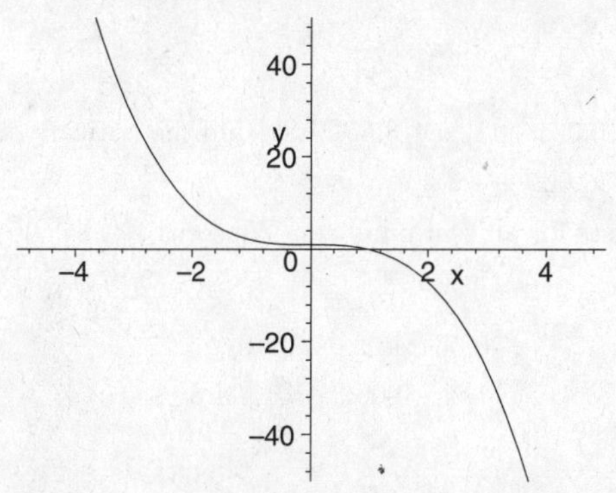

13. $f(x) = 0.02x^2 + 1.3x + 2.31$ First, find the critical points.

$f'(x) = 0.04x + 1.3$

$f'(x)$ exists for all real numbers. We solve $f'(x) = 0$:

$$\begin{aligned} 0.04x + 1.3 &= 0 \\ 0.04x &= -1.3 \\ x &= \frac{-1.3}{0.04} \\ x &= -32.5 \end{aligned}$$

The only critical point is at $x = -32.5$. We use -32.5 to divide the real number line into two intervals, A: $(-\infty, -32.5)$ and B: $(-32.5, \infty)$:

A B

-32.5

We use a test value in each interval to determine the sign of the derivative in each interval.

A: Test 0, $f'(0) = 0.04 \cdot 0 + 1.3 = 1.3 > 0$

B: Test -100, $f'(-100) = 0.04 \cdot (-100) + 1.3 = -2.7 < 0$

We see that $f(x)$ is decreasing on $(-\infty, -32.5)$ and increasing on $(-32.5, \infty)$, and the change from decreasing to increasing indicates that a relative minimum occurs at $x = -32.5$. We substitute into the original equation to find $f(-32.5)$:

$$\begin{aligned} f(-32.5) &= 0.02(-32.5)^2 + 1.3(-32.5) + 2.31 \\ &= 21.125 - 42.25 + 2.31 \\ &= -18.815 \end{aligned}$$

Thus, there is a relative minimum at $(-32.5, -18.815)$. We use the information obtained to sketch the graph.

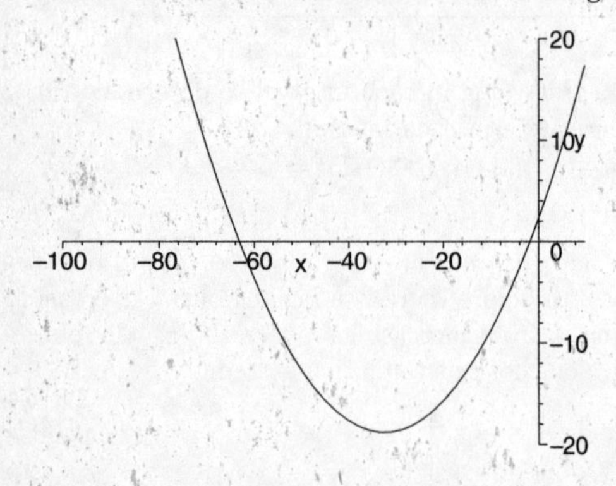

14. $f(x) = -0.03x^2 + 1.8x - 3.45$ First, find the critical points.
$f'(x) = -0.06x + 1.8$
$f'(x)$ exists for all real numbers. We solve $f'(x) = 0$:

$$\begin{aligned} -0.06x + 1.8 &= 0 \\ -0.06x &= -1.8 \\ x &= \frac{-1.8}{-0.06} \\ x &= 30 \end{aligned}$$

The only critical point is at $x = 30$. We use 30 to divide the real number line into two intervals, A: $(-\infty, 30)$ and B: $(30, \infty)$:

A B
30

We use a test value in each interval to determine the sign of the derivative in each interval.
A: Test 0, $f'(0) = -0.06 \cdot 0 + 1.8 = 1.8 > 0$
B: Test 100, $f'(-100) = -0.06 \cdot (100) + 1.3 = -4.7 < 0$
We see that $f(x)$ is increasing on $(-\infty, 30)$ and decreasing on $(30, \infty)$, and the change from increasing to decreasing indicates that a relative maximum occurs at $x = 30$. We substitute into the original equation to find $f(30)$:

$$\begin{aligned} f(30) &= -0.03(30)^2 + 1.8(30) - 3.45 \\ &= -27 + 54 - 3.45 \\ &= 23.55 \end{aligned}$$

Thus, there is a relative maximum at $(30, 23.55)$. We use the information obtained to sketch the graph.

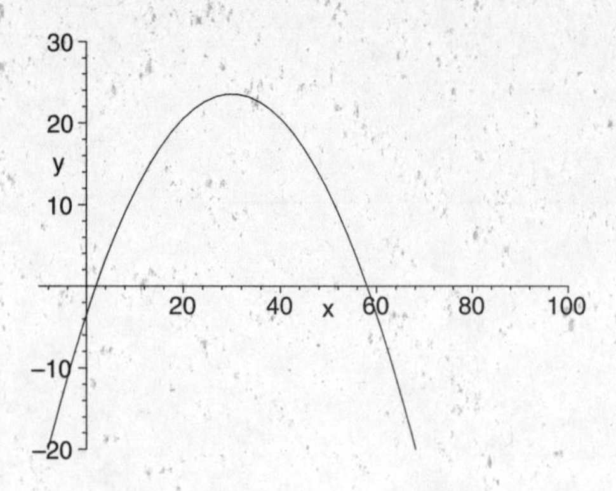

15. $f(x) = x^4 - 2x^3$
$f'(x) = 4x^3 - 6x^2$
$f'(x)$ exists for all real numbers. Solve $f'(x) = 0$.

$$\begin{aligned} 4x^3 - 6x^2 &= 0 \\ 2x^3 - 3x^2 &= 0 \\ x^2(2x - 3) &= 0 \\ x &= 0 \\ \text{Or} & \\ x &= \frac{3}{2} \end{aligned}$$

The critical points are at $x = 0$ and $x = \frac{3}{2}$. Use them to divide the real number line into three intervals, A: $(-\infty, 0)$, B: $\left(0, \frac{3}{2}\right)$, and C: $\left(\frac{3}{2}, \infty\right)$.
A: Test -1, $f'(-1) = 4(-1)^3 - 6(-1)^2 = -10 < 0$
B: Test 1, $f'(1) = 4 \cdot 1^3 - 6 \cdot 1^2 = -2 < 0$
C: Test 2, $f'(2) = 4 \cdot 2^3 - 6 \cdot 2^2 = 8 > 0$

Since $f(x)$ is decreasing on both $(-\infty, 0)$ and $\left(0, \frac{3}{2}\right)$ and is increasing on $\left(\frac{3}{2}, \infty\right)$, there is no relative extremum at $x = 0$ but there is a relative minimum at $x = \frac{3}{2}$.

$$f\left(\frac{3}{2}\right) = \left(\frac{3}{2}\right)^4 - 2\left(\frac{3}{2}\right)^3 = \frac{81}{16} - \frac{27}{4} = -\frac{27}{16}$$

There is a realtive minimum at $\left(\frac{3}{2}, -\frac{27}{16}\right)$. We sketch the graph.

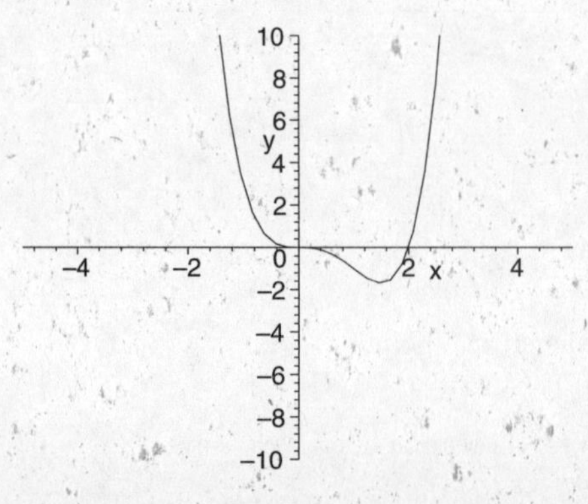

16. $f(x) = x^4 - 8x^2 + 3$

First, find the critical points.

$f'(x) = 4x^3 - 16x$

$f'(x)$ exists for all real numbers. We solve $f'(x) = 0$:

$$\begin{aligned} 4x^3 - 16x &= 0 \\ x^3 - 4x &= 0 \\ x(x^2 - 4) &= 0 \\ x(x+2)(x-2) &= 0 \\ x &= 0 \\ &\text{Or} \\ x + 2 &= 0 \\ x &= -2 \\ &\text{Or} \\ x - 2 &= 0 \\ x &= 2 \end{aligned}$$

The critical points are at $x = -2$, $x = 0$, and $x = 2$. We use them to divide the real number line into four intervals, A: $(-\infty, -2)$, B: $(-2, 0)$, C: $(0, 2)$, and D: $(2, \infty)$.

A B C D
−2 0 2

We use a test value in each interval to determine the sign of the derivative in each interval.

A: Test -3, $f'(-3) = 4(-3)^3 - 16(-3) = -108 + 48 = -60 < 0$

B: Test -1, $f'(-1) = 4(-1)^3 - 16(-1) = -4 + 16 = 12 > 0$

C: Test 1, $f'(1) = 4 \cdot 1^3 - 16 \cdot 1 = 4 - 16 = -12 < 0$

D: Test 3, $f'(3) = 4 \cdot 3^3 - 16 \cdot 3 = 108 - 48 = 60 > 0$

We see that $f(x)$ is decreasing on $(-\infty, -2)$, increasing on $(-2, 0)$, decreasing again on $(0, 2)$, and increasing again on $(2, \infty)$. Thus, there is a relative minimum at $x = -2$, a relative maximum at $x = 0$, and another relative minimum at $x = 2$.

We find $f(-2)$:

$f(-2) = (-2)^4 - 8(-2)^2 + 3 = 16 - 32 + 3 = -13$

Then we find $f(0)$:

$f(0) = 0^4 - 8 \cdot 0^2 + 3 = 0 - 0 + 3 = 3$

Finally, we find $f(2)$:

$f(2) = 2^4 - 8 \cdot 2^2 + 3 = 16 - 32 + 3 = -13$

There are relative minima at $(-2, -13)$ and $(2, -13)$, and there is a relative maximum at $(0, 3)$. We use the information obtained to sketch the graph.

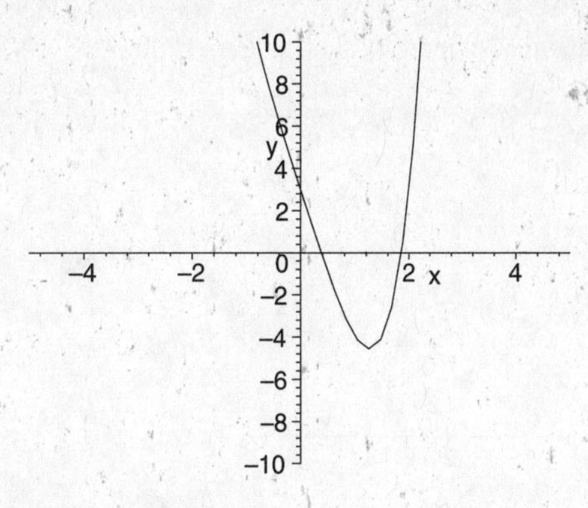

17. $f(x) = x\sqrt{8 - x^2}$

The domain of this function is between $[-\sqrt{8}, \sqrt{8}]$

First, find the critical points.

$$\begin{aligned} f'(x) &= x\left(\frac{1}{2}(8 - x^2)^{-1/2}(-2x)\right) + (8 - x^2)^{1/2}(1) \\ &= \frac{-x^2}{(8 - x^2)^{1/2}} + (8 - x^2)^{1/2} \end{aligned}$$

$f'(x)$ does not exist for $x = \pm\sqrt{8}$. We solve $f'(x) = 0$:

$$\begin{aligned} \frac{-x^2}{(8 - x^2)^{1/2}} + (8 - x^2)^{1/2} &= 0 \\ 8 - x^2 &= -x^2 \\ 8 &= 2x^2 \\ 4 &= x^2 \\ \pm 2 &= x \end{aligned}$$

The critical points are at $x = \pm\sqrt{8}$, $x = -2$ and $x = 2$. We use them to divide the real number line into three intervals, A: $(-\sqrt{8}, -2)$, B: $(-2, 2)$, C: $(2, \sqrt{8})$.

A B C
−2 2

We use a test value in each interval to determine the sign of the derivative in each interval.

A: Test -2.5, $f'(-2.5) = \dfrac{-(-2.5)^2}{(8 - (-2.5)^2)^{1/2}} + (8 - (-2.5)^2)^{1/2} < 0$

B: Test 0, $f'(0) = \dfrac{-(0)^2}{(8 - 0^2)^{1/2}} + (8 - 0^2)^{1/2} > 0$

C: Test 2.5, $f'(2.5) = \dfrac{-(2.5)^2}{(8 - (2.5)^2)^{1/2}} + (8 - (2.5)^2)^{1/2} < 0$

We see that $f(x)$ is decreasing on $(-\sqrt{8}, -2)$, and on $(2, \sqrt{8})$, and increasing on $(-2, 2)$,so there is a relative minimum at $x = -2$ and a relative maximum at $x = 2$.

We find $f(-2)$:

$f(-2) = (-2)\sqrt{8 - (-2)^2} = -4$

Then we find $f(2)$:

$f(2) = (2)\sqrt{8 - (2)^2} = 4$ There is a relative minimum at $(-2, -4)$, and there is a relative maximum at $(2, 4)$. We use the information obtained to sketch the graph.

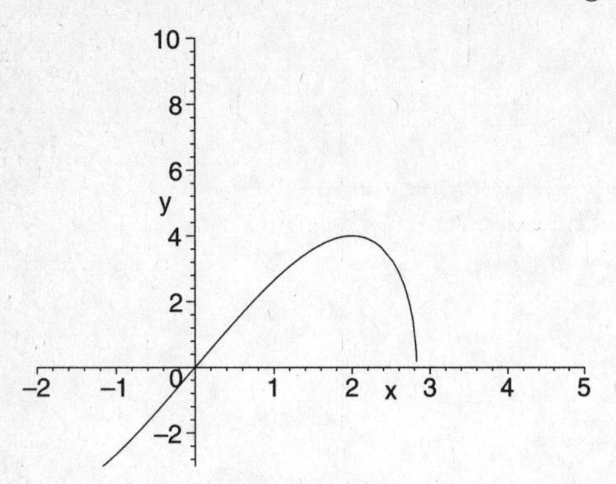

18. $f(x) = x\sqrt{16 - x^2}$

The domain of this function is between $[-4, 4]$

First, find the critical points.

$$\begin{aligned} f'(x) &= x\left(\frac{1}{2}(16 - x^2)^{-1/2}(-2x)\right) + (16 - x^2)^{1/2}(1) \\ &= \frac{-x^2}{(16 - x^2)^{1/2}} + (16 - x^2)^{1/2} \end{aligned}$$

$f'(x)$ does not exist for $x = \pm 4$. We solve $f'(x) = 0$:

$$\begin{aligned} \frac{-x^2}{(16 - x^2)^{1/2}} + (16 - x^2)^{1/2} &= 0 \\ 16 - x^2 &= -x^2 \\ 16 &= 2x^2 \\ 8 &= x^2 \\ \pm\sqrt{8} &= x \end{aligned}$$

The critical points are at $x = \pm 4$, $x = -\sqrt{8}$ and $x = \sqrt{8}$. We use them to divide the real number line into three intervals, A: $(-4, -\sqrt{8})$, B: $(-\sqrt{8}, \sqrt{8})$, C: $(\sqrt{8}, 4)$.

A B C

$-\sqrt{8}$ $\sqrt{8}$

We use a test value in each interval to determine the sign of the derivative in each interval.

A: Test -3, $f'(-3) = \dfrac{-(-3)^2}{(16 - (-3)^2)^{1/2}} + (16 - (-3)^2)^{1/2} < 0$

B: Test 0, $f'(0) = \dfrac{-(0)^2}{(16 - 0^2)^{1/2}} + (16 - 0^2)^{1/2} > 0$

C: Test 3, $f'(3) = \dfrac{-(3)^2}{(16 - 3^2)^{1/2}} + (16 - 3^2)^{1/2} < 0$

We see that $f(x)$ is decreasing on $(-4, -\sqrt{8})$, and on $(\sqrt{8}, 4)$, and increasing on $(-\sqrt{8}, \sqrt{8})$,so there is a relative minimum at $x = -\sqrt{8}$ and a relative maximum at $x = \sqrt{8}$.

We find $f(-\sqrt{8})$:

$f(-\sqrt{8}) = (-\sqrt{8})\sqrt{16 - (-\sqrt{8})^2} = -8$

Then we find $f(\sqrt{8})$:

$f(2) = (\sqrt{8})\sqrt{16 - (\sqrt{8})^2} = 8$ There is a relative minimum at $(-\sqrt{8}, -8)$, and there is a relative maximum at $(\sqrt{8}, 8)$. We use the information obtained to sketch the graph.

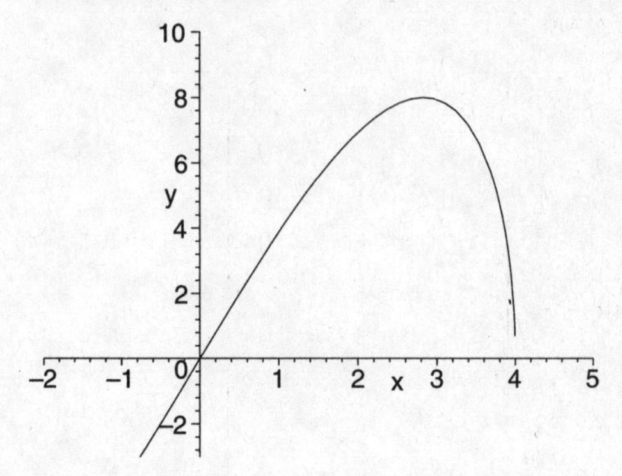

19. $f(x) = 1 - x^{2/3}$

First, find the critical points.

$f'(x) = -\dfrac{2}{3}x^{-1/3} = -\dfrac{2}{3\sqrt[3]{x}}$

$f'(x)$ does not exist for $x = 0$. The equation $f'(x) = 0$ has no solution, so the only critical point is at $x = 0$. We use it to divide the real number line into two intervals: A: $(-\infty, 0)$ and B: $(0, \infty)$.

A B

0

We use a test value in each interval to determine the sign of the derivative in each interval.

A: Test -1, $f'(-1) = -\dfrac{2}{3\sqrt[3]{-1}} = -\dfrac{2}{3(-1)} = \dfrac{2}{3} > 0$

B: Test 1, $f'(1) = -\dfrac{2}{3\sqrt[3]{1}} = -\dfrac{2}{3 \cdot 1} = -\dfrac{2}{3} < 0$

We see that $f(x)$ is increasing on $(-\infty, 0)$ and decreasing on $(0, \infty)$, so there is a relative maximum at $x = 0$.

We find $f(0)$:

$f(0) = 1 - 0^{2/3} = 1 - 0 = 1$

There is a relative maximum at $(0, 1)$. We use the information obtained to sketch the graph.

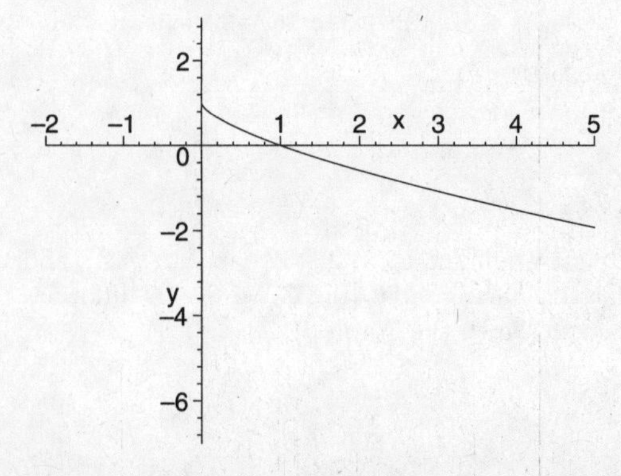

20. $f(x) = (x+3)^{2/3} - 5$

$f'(x) = \frac{2}{3}(x+3)^{-1/3} = \frac{2}{3\sqrt[3]{x+3}}$

$f'(x)$ does not exist for $x = -3$. The equation $f'(x) = 0$ has no solution, so the only critical point is at $x = -3$. Use it to divide the real number line into two intervals, A: $(-\infty, -3)$ and B: $(-3, \infty)$.

A: Test -4, $f'(-4) = \frac{2}{3\sqrt[3]{-4+3}} = -\frac{2}{3} < 0$

B: Test -2, $f'(-2) = \frac{2}{3\sqrt[3]{-2+3}} = \frac{2}{3} > 0$

Since $f(x)$ is decreasing on $(-\infty, -3)$ and increasing on $(-3, \infty)$, there is a relative minimum at $x = -3$.

$f(-3) = (-3+3)^{2/3} - 5 = -5$

There is a relative minimum at $(-3, -5)$. We sketch the graph.

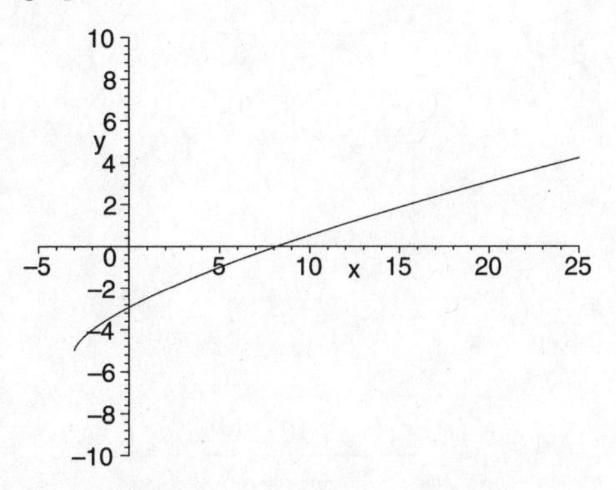

21. $f(x) = \frac{-8}{x^2+1} = -8(x^2+1)^{-1}$

First, find the critical points.

$f'(x) = -8(-1)(x^2+1)^{-2}(2x)$

$= 16x(x^2+1)^{-2}$

$= \frac{16x}{(x^2+1)^2}$

$f'(x)$ exists for all real numbers. We solve $f'(x) = 0$:

$$\begin{aligned} \frac{16x}{(x^2+1)^2} &= 0 \\ 16x &= 0 \\ x &= 0 \end{aligned}$$

The only critical point is at $x = 0$. We use it to divide the real number line into two intervals, A: $(-\infty, 0)$ and B: $(0, \infty)$.

A B

0

We use a test value in each interval to determine the sign of the derivative in each interval.

A: Test -1, $f'(-1) = \frac{16(-1)}{[(-1)^2+1]^2} = \frac{-16}{4} = -4 < 0$

B: Test 1, $f'(1) = \frac{16 \cdot 1}{(1^2+1)^2} = \frac{16}{4} = 4 > 0$

We see that $f(x)$ is decreasing on $(-\infty, 0)$ and increasing on $(0, \infty)$, so there is a relative minimum at $x = 0$.

We find $f(0)$:

$f(0) = \frac{-8}{0^2+1} = \frac{-8}{1} = -8$

There is a relative minimum at $(0, -8)$. We use the information obtained to sketch the graph.

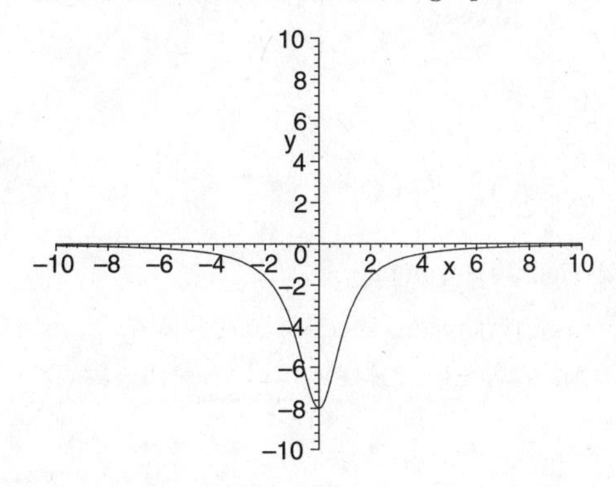

22. $f(x) = \frac{5}{x^2+1} = 5(x^2+1)^{-1}$

$$\begin{aligned} f'(x) &= 5(-1)(x^2+1)^{-2}(2x) \\ &= -10x(x^2+1)^{-2} \\ &= \frac{-10x}{(x^2+1)^2} \end{aligned}$$

$f'(x)$ exists for all real numbers. Solve $f'(x) = 0$.

$$\begin{aligned} \frac{-10x}{(x^2+1)^2} &= 0 \\ -10x &= 0 \\ x &= 0 \end{aligned}$$

The only critical point is at $x = 0$. Use it to divide the real number line into two intervals, A: $(-\infty, 0)$ and B: $(0, \infty)$.

A: Test -1, $f'(-1) = \frac{-10(-1)}{[(-1)^2+1]^2} = \frac{5}{2} > 0$

B: Test 1, $f'(1) = \frac{-10 \cdot 1}{(1^2+1)^2} = -\frac{5}{2} < 0$

Since $f(x)$ is increasing on $(-\infty, 0)$ and decreasing on $(0, \infty)$, there is a relative minimum at $x = 0$.

$f(0) = \frac{5}{0^2+1} = 5$

There is a relative maximum at $(0, 5)$. We sketch the graph.

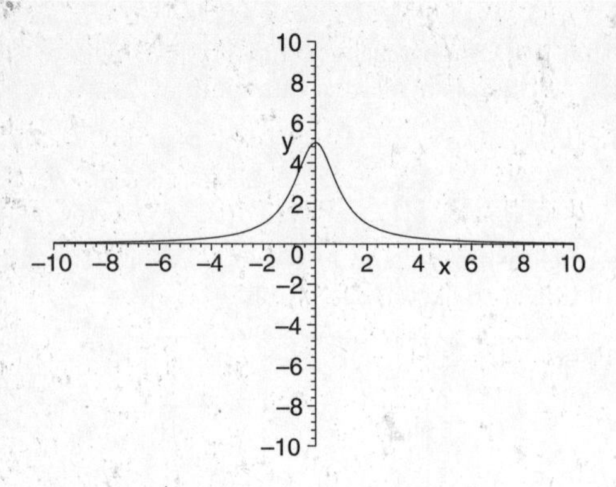

23. $f(x) = \dfrac{4x}{x^2+1}$

First, find the critical points.

$$\begin{aligned} f'(x) &= \frac{(x^2+1)(4) - 2x(4x)}{(x^2+1)^2} \\ &= \frac{4x^2+4-8x^2}{(x^2+1)^2} \\ &= \frac{4-4x^2}{(x^2+1)^2} \end{aligned}$$

$f'(x)$ exists for all real numbers. We solve $f'(x) = 0$:

$$\begin{aligned} \frac{4-4x^2}{(x^2+1)^2} &= 0 \\ 4-4x^2 &= 0 \\ 1-x^2 &= 0 \\ (1-x)(1+x) &= 0 \\ 1-x &= 0 \\ 1 &= x \\ \text{Or} \\ 1+x &= 0 \\ x &= -1 \end{aligned}$$

The critical points are at $x = -1$ and $x = 1$. We use them to divide the real number line into three intervals, A: $(-\infty, -1)$, B: $(-1, 1)$, and C: $(1, \infty)$.

A B C

−1 1

We use a test value in each interval to determine the sign of the derivative in each interval.

A: Test -2, $f'(-2) = \dfrac{4-4(-2)^2}{[(-2)^2+1]^2} = \dfrac{-12}{25} < 0$

B: Test 0, $f'(0) = \dfrac{4-4\cdot 0^2}{(0^2+1)^2} = \dfrac{4}{1} = 4 > 0$

C: Test 2, $f'(2) = \dfrac{4-4\cdot 2^2}{(2^2+1)^2} = \dfrac{-12}{25} < 0$

We see that $f(x)$ is decreasing on $(-\infty, -1)$, increasing on $(-1, 1)$, and decreasing again on $(1, \infty)$, so there is a relative minimum at $x = -1$ and a relative maximum at $x = 1$.

We find $f(-1)$:

$$f(-1) = \frac{4(-1)}{(-1)^2+1} = \frac{-4}{2} = -2$$

Then we find $f(1)$:

$$f(1) = \frac{4\cdot 1}{1^2+1} = \frac{4}{2} = 2$$

There is a relative minimum at $(-1, -2)$, and there is a relative maximum at $(1, 2)$. We use the information obtained to sketch the graph.

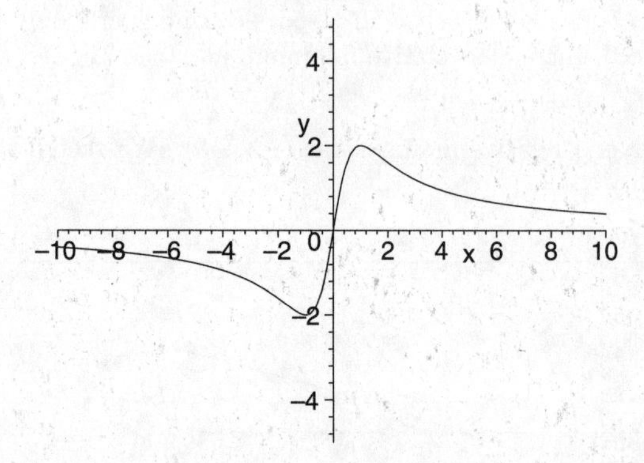

24. $f(x) = \dfrac{x^2}{x^2+1}$

$$\begin{aligned} f'(x) &= \frac{(x^2+1)(2x) - 2x(x^2)}{(x^2+1)^2} \\ &= \frac{2x^3+2x-2x^3}{(x^2+1)^2} \\ &= \frac{2x}{(x^2+1)^2} \end{aligned}$$

$f'(x)$ exists for all real numbers. Solve $f'(x) = 0$.

$$\begin{aligned} \frac{2x}{(x^2+1)^2} &= 0 \\ 2x &= 0 \\ x &= 0 \end{aligned}$$

The only critical point is at $x = 0$. Use it to divide the real number line into two intervals, A: $(-\infty, 0)$ and B: $(0, \infty)$.

A: Test -1, $f'(-1) = \dfrac{2(-1)}{[(-1)^2+1]^2} = \dfrac{-2}{4} = -\dfrac{1}{2} < 0$

B: Test 1, $f'(1) = \dfrac{2\cdot 1}{(1^2+1)^2} = \dfrac{2}{4} = \dfrac{1}{2} > 0$

Since $f(x)$ is decreasing on $(-\infty, 0)$ and increasing on $(0, \infty)$, there is a relative minimum at $x = 0$.

$$f(0) = \frac{0^2}{0^2+1} = 0$$

There is a relative minimum at $(0, 0)$. We sketch the graph.

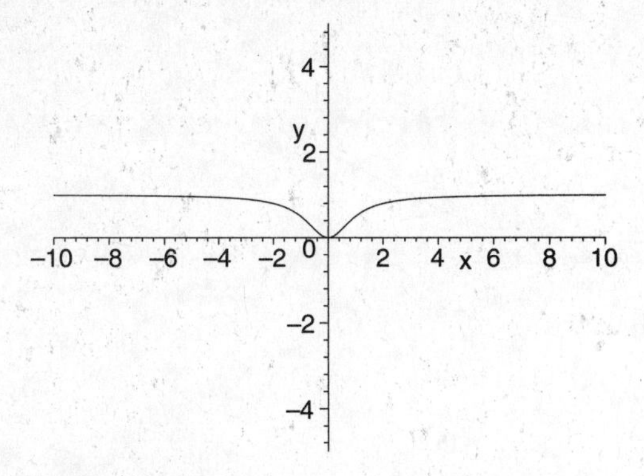

25. $f(x) = \sqrt[3]{x} = x^{1/3}$

First, find the critical points.

$f'(x) = \dfrac{1}{3}x^{-2/3} = \dfrac{1}{3\sqrt[3]{x^2}}$

$f'(x)$ does not exist for $x = 0$. The equation $f'(x) = 0$ has no solution, so the only critical point is at $x = 0$. We use it to divide the real number line into two intervals, A: $(-\infty, 0)$, and B: $(0, \infty)$.

A B
0

We use a test value in each interval to determine the sign of the derivative in each interval.

A: Test -1, $f'(-1) = \dfrac{1}{3\sqrt[3]{(-1)^2}} = \dfrac{1}{3 \cdot 1} = \dfrac{1}{3} > 0$

B: Test 1, $f'(1) = \dfrac{1}{3\sqrt[3]{1^2}} = \dfrac{1}{3 \cdot 1} = \dfrac{1}{3} > 0$

We see that $f(x)$ is increasing on both intervals, so the function has no relative extrema. We use the information obtained to sketch the graph.

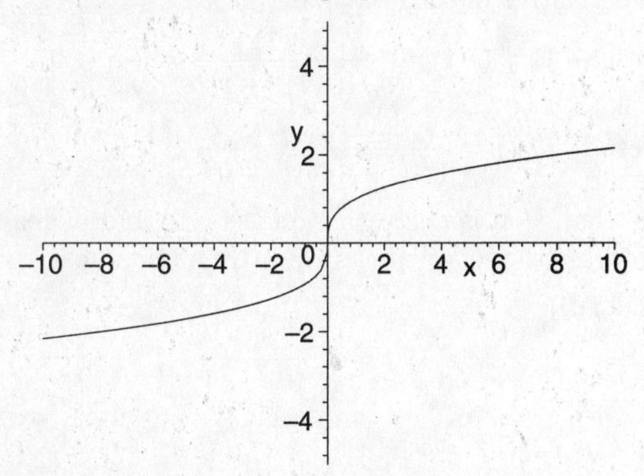

26. $f(x) = (x+1)^{1/3}$

$f'(x) = \dfrac{1}{3}(x+1)^{-2/3} = \dfrac{1}{3\sqrt[3]{(x+1)^2}}$

$f'(x)$ does not exist for $x = -1$. The equation $f'(x) = 0$ has no solution, so the only critical point is at $x = -1$. We use it to divide the real number line into two intervals, A: $(-\infty, -1)$ and B: $(-1, \infty)$.

A: Test -2, $f'(-2) = \dfrac{1}{3\sqrt[3]{(-2+1)^2}} = \dfrac{1}{3} > 0$

B: Test 0, $\quad f'(0) = \dfrac{1}{3\sqrt[3]{(0+1)^2}} = \dfrac{1}{3} > 0$

Since $f(x)$ is increasing on both intervals, there are no relative extrema. We sketch the graph.

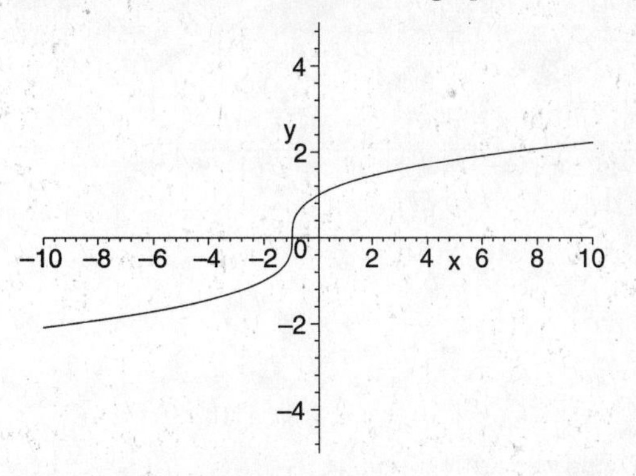

27. $f(x) = \sqrt{x^2 + 2x + 5} = (x^2 + 2x + 5)^{1/2}$

First, find the critical points.

$f'(x) = \dfrac{1}{2}(x^2 + 2x + 5)^{-1/2}(2x + 2) = \dfrac{x+1}{\sqrt{x^2 + 2x + 5}}$

$f'(x)$ exists for all x values. We solve $f'(x) = 0$,

$$\begin{aligned} f'(x) &= 0 \\ \frac{x+1}{\sqrt{x^2+2x+5}} &= 0 \\ x+1 &= 0 \\ x &= -1 \end{aligned}$$

So the only critical point is at $x = -1$. We use it to divide the real number line into two intervals, A: $(-\infty, -1)$, and B: $(-1, \infty)$.

A B
-1

We use a test value in each interval to determine the sign of the derivative in each interval.

A: Test -2, $f'(-2) = \dfrac{(-2)+1}{\sqrt{(-2)^2 + 2(-2) + 5}} = \dfrac{-1}{\sqrt{5}} < 0$

B: Test 0, $f'(0) = \dfrac{0+1}{\sqrt{(0)^2 + 2(0) + 5}} = \dfrac{1}{\sqrt{5}} > 0$

We see that $f(x)$ is decreasing on $(-\infty, -1)$ and increasing on $(-1, \infty)$, so the function has a relative minimum at $x = -1$.

We find $f(-1)$:

$$\begin{aligned} f(-1) &= \sqrt{(-1)^2 + 2(-1) + 5} \\ &= \sqrt{1 - 2 + 5} \\ &= \sqrt{4} = 2 \end{aligned}$$

There is a relative minimum at $(-1, 2)$. We use the information obtained to sketch the graph.

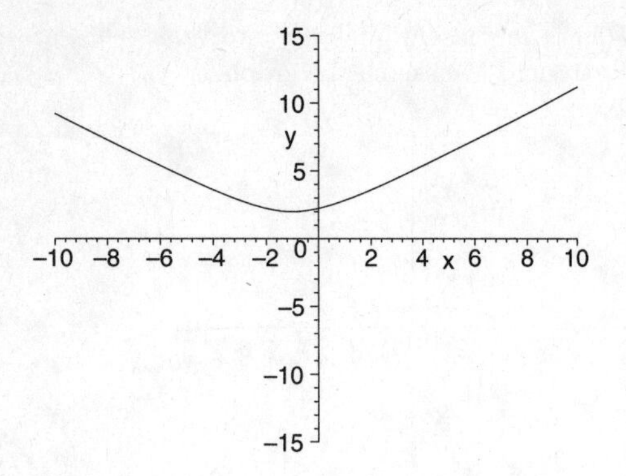

28. $f(x) = \sqrt{x^2 + 6x + 10} = (x^2 + 6x + 10)^{1/2}$

First, find the critical points.

$f'(x) = \frac{1}{2}(x^2 + 6x + 10)^{-1/2}(2x + 6) = \frac{2x + 6}{\sqrt{x^2 + 6x + 10}}$

$f'(x)$ exists for all x values. We solve $f'(x) = 0$,

$$\begin{aligned} f'(x) &= 0 \\ \frac{2x+6}{\sqrt{x^2 + 6x + 10}} &= 0 \\ 2x + 6 &= 0 \\ 2x &= -6 \\ x &= -3 \end{aligned}$$

So the only critical point is at $x = -3$. We use it to divide the real number line into two intervals, A: $(-\infty, -3)$, and B: $(-3, \infty)$.

A B

−3

We use a test value in each interval to determine the sign of the derivative in each interval.

A: Test -4, $f'(-4) = \frac{2(-4) + 6}{\sqrt{(-2)^2 + 6(-2) + 10}} = \frac{-2}{\sqrt{2}} < 0$

B: Test 0, $f'(0) = \frac{2(0) + 6}{\sqrt{(0)^2 + 6(0) + 10}} = \frac{6}{\sqrt{10}} > 0$

We see that $f(x)$ is decreasing on $(-\infty, -3)$ and increasing on $(-3, \infty)$, so the function has a relative minimum at $x = -3$.

We find $f(-3)$:

$$\begin{aligned} f(-3) &= \sqrt{(-3)^2 + 6(-3) + 10} \\ &= \sqrt{9 - 18 + 10} \\ &= \sqrt{1} = 1 \end{aligned}$$

There is a relative minimum at $(-3, 1)$. We use the information obtained to sketch the graph.

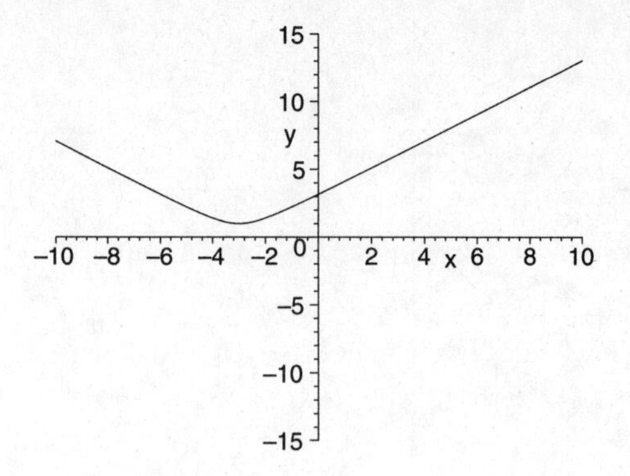

29. $f(x) = \frac{1}{\sqrt{x^2+1}} = (x^2 + 1)^{-1/2}$

First, find the critical points.

$f'(x) = \frac{-1}{2}(x^2 + 1)^{-3/2}(2x) = \frac{2x}{\sqrt{(x^2 + 1)^3}}$

$f'(x)$ exists for all x values. We solve $f'(x) = 0$,

$$\begin{aligned} f'(x) &= 0 \\ \frac{2x}{\sqrt{(x^2 + 1)^3}} &= 0 \\ 2x &= 0 \\ x &= 0 \end{aligned}$$

So the only critical point is at $x = 0$. We use it to divide the real number line into two intervals, A: $(-\infty, 0)$, and B: $(0, \infty)$.

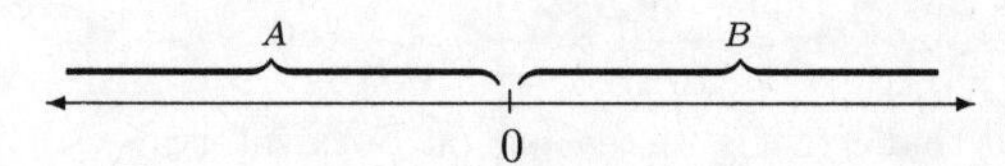

We use a test value in each interval to determine the sign of the derivative in each interval.

A: Test -1, $f'(-1) = \frac{2(-1)}{\sqrt{((-1)^2 + 1)^3}} = \frac{-2}{\sqrt{8}} < 0$

B: Test 1, $f'(1) = \frac{2(1)}{\sqrt{((1)^2 + 1)^3}} = \frac{2}{\sqrt{8}} > 0$

We see that $f(x)$ is decreasing on $(-\infty, 0)$ and increasing on $(0, \infty)$, so the function has a relative minimum at $x = 0$.

We find $f(0)$:

$$\begin{aligned} f(0) &= \sqrt{(0)^2 + 1} \\ &= \sqrt{0 + 1} \\ &= \sqrt{1} = 1 \end{aligned}$$

There is a relative minimum at $(0, 1)$. We use the information obtained to sketch the graph.

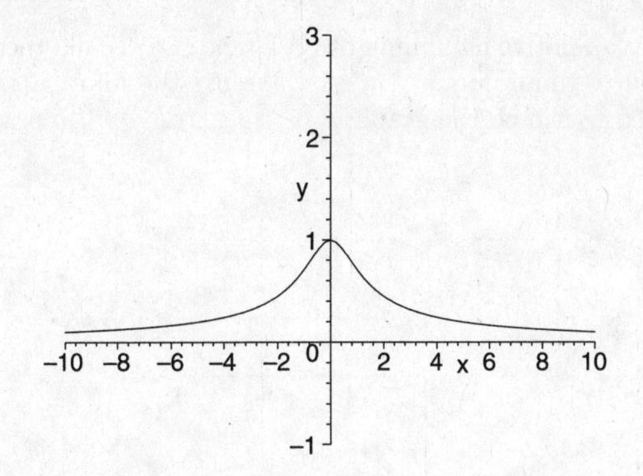

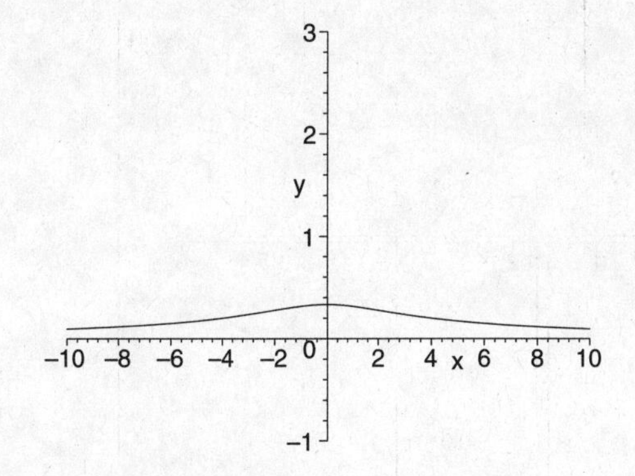

30. $f(x) = \frac{1}{\sqrt{x^2+9}} = (x^2+9)^{-1/2}$

First, find the critical points.

$f'(x) = \frac{-1}{2}(x^2+9)^{-3/2}(2x) = \frac{2x}{\sqrt{(x^2+9)^3}}$

$f'(x)$ exists for all x values. We solve $f'(x) = 0$,

$$\begin{aligned} f'(x) &= 0 \\ \frac{2x}{\sqrt{(x^2+9)^3}} &= 0 \\ 2x &= 0 \\ x &= 0 \end{aligned}$$

So the only critical point is at $x = 0$. We use it to divide the real number line into two intervals, A: $(-\infty, 0)$, and B: $(0, \infty)$.

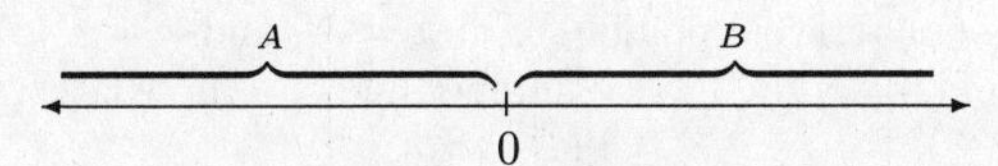

We use a test value in each interval to determine the sign of the derivative in each interval.

A: Test -1, $f'(-1) = \frac{2(-1)}{\sqrt{((-1)^2+9)^3}} = \frac{-2}{\sqrt{10}} < 0$

B: Test 1, $f'(1) = \frac{2(1)}{\sqrt{((1)^2+9)^3}} = \frac{2}{\sqrt{10}} > 0$

We see that $f(x)$ is decreasing on $(-\infty, 0)$ and increasing on $(0, \infty)$, so the function has a relative minimum at $x = 0$.

We find $f(0)$:

$$\begin{aligned} f(0) &= \sqrt{(0)^2+9} \\ &= \sqrt{0+9} \\ &= \sqrt{9} = 3 \end{aligned}$$

There is a relative minimum at $(0, 3)$. We use the information obtained to sketch the graph.

31. $f(x) = sin\ x$

First, find the critical points.

$f'(x) = cos\ x$

$f'(x)$ exists for all x values. We solve $f'(x) = 0$,

$$\begin{aligned} f'(x) &= 0 \\ cos\ x &= 0 \\ x &= \frac{\pi}{2} \\ \text{and} & \\ x &= \frac{3\pi}{2} \end{aligned}$$

So the only critical points are at $x = \frac{\pi}{2}$ and $x = \frac{3\pi}{2}$ and there might be extrema points at the end points $x = 0$ and $x = 2\pi$. We use them to divide the real number line into three intervals, A: $[0, \frac{\pi}{2})$, B: $(\frac{\pi}{2}, \frac{3\pi}{2})$, and C: $(\frac{3\pi}{2}, 2\pi]$.

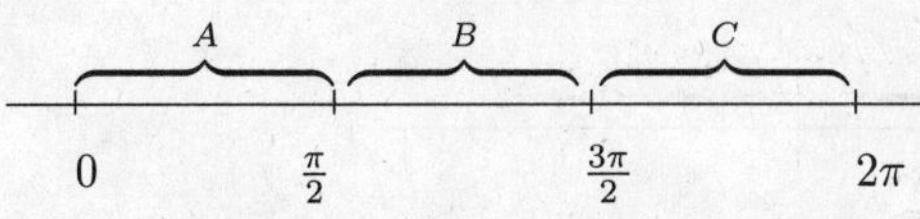

We use a test value in each interval to determine the sign of the derivative in each interval.

A: Test 0, $f'(0) = cos(0) = 1 > 0$

B: Test π, $f'(\pi) = cos(\pi) = -1 < 0$

C: Test 2π, $f'(2\pi) = cos(2\pi) = 1 > 0$

We see that $f(x)$ is decreasing on $(-\infty, \frac{\pi}{2})$ and increasing on $(\frac{\pi}{2}, \frac{3\pi}{2})$, we also see that $f(x)$ is increasing on $(\frac{3\pi}{2}, \infty)$ so the function has a relative maximum at $x = \frac{\pi}{2}$ and a relative minimum at $x = \frac{3\pi}{2}$.

We find $f(\frac{\pi}{2})$:

$$\begin{aligned} f(\frac{\pi}{2}) &= sin(\frac{\pi}{2}) \\ &= 1 \end{aligned}$$

We find $f(\frac{3\pi}{2})$:

$$\begin{aligned} f(\frac{3\pi}{2}) &= sin(\frac{3\pi}{2}) \\ &= -1 \end{aligned}$$

There is a relative maximum at $(\frac{\pi}{2}, 1)$ and there is a relative minimum at $(\frac{3\pi}{2}, -1)$. We use the information obtained to sketch the graph.

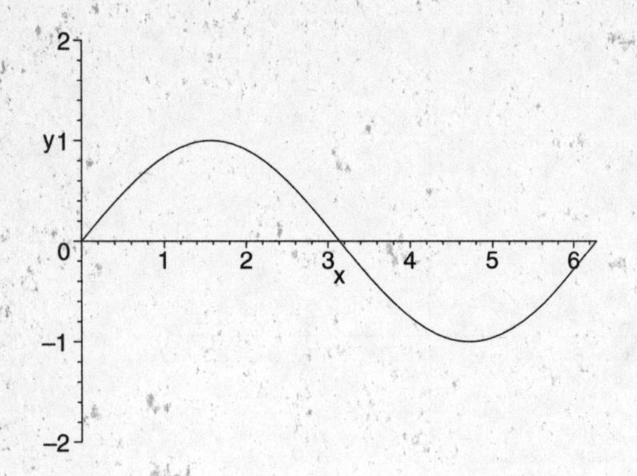

32. $f(x) = cos\ x$ First, find the critical points.

$f'(x) = -sin\ x$

$f'(x)$ exists for all x values. We solve $f'(x) = 0$,

$$\begin{aligned} f'(x) &= 0 \\ -sin\ x &= 0 \\ x &= 0 \\ \text{and} \\ x &= \pi \\ \text{and} \\ x &= 2\pi \end{aligned}$$

So the only critical points are at $x = 0$, $x = \pi$ and $x = 2\pi$. There might be extrema points at the end points $x = 0$ and $x = 2\pi$. We use them to divide the real number line into two intervals, A: $[0, \pi)$, B: $(\pi, 2\pi]$.

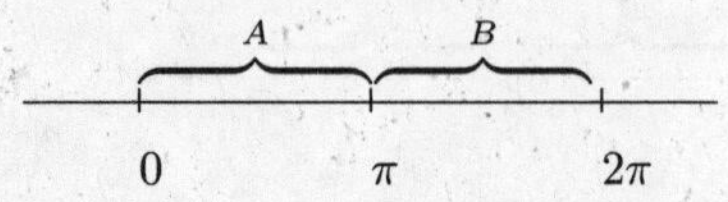

$0 \qquad \pi \qquad 2\pi$

We use a test value in each interval to determine the sign of the derivative in each interval.

A: Test $\frac{\pi}{2}$, $f'(\frac{\pi}{2}) = -sin(\frac{\pi}{2}) = -1 < 0$

B: Test $\frac{3\pi}{2}$, $f'(\frac{3\pi}{2}) = -sin(\frac{3\pi}{2}) = 1 > 0$

We see that $f(x)$ is decreasing on $[0, \pi)$ and increasing on $(\pi, 2\pi]$ so the function has a relative minimum at $x = \pi$ and the function has a relative maximum at the endpoints $x = 0$ and $x = 2\pi$.

We find $f(0)$:

$$\begin{aligned} f(0) &= cos(0) \\ &= 1 \end{aligned}$$

We find $f(\pi)$:

$$\begin{aligned} f(\pi) &= cos(\pi) \\ &= -1 \end{aligned}$$

We find $f(2\pi)$:

$$\begin{aligned} f(2\pi) &= cos(2\pi) \\ &= 1 \end{aligned}$$

There is a relative maximum at $(0, 1)$ and $(2\pi, 1)$ and there is a relative minimum at $(\pi, -1)$. We use the information obtained to sketch the graph.

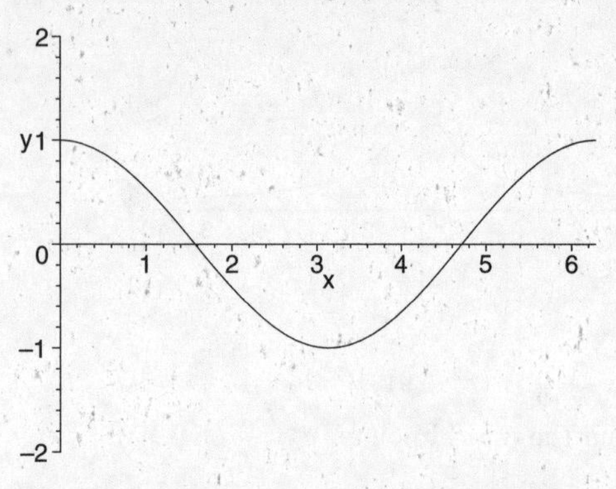

33. $f(x) = sin\ x - cos\ x$ First, find the critical points.

$f'(x) = cos\ x + sin\ x$

$f'(x)$ exists for all x values. We solve $f'(x) = 0$,

$$\begin{aligned} f'(x) &= 0 \\ cos\ x + sin\ x &= 0 \\ cos\ x &= -sin\ x \\ x &= \frac{3\pi}{4} \\ \text{and} \\ x &= \frac{7\pi}{4} \end{aligned}$$

So the only critical points are at $x = \frac{3\pi}{4}$ and $x = \frac{7\pi}{4}$. We use them to divide the real number line into three intervals, A: $[0, \frac{3\pi}{4})$, B: $(\frac{3\pi}{4}, \frac{7\pi}{4})$, C: $(\frac{7\pi}{4}, 2\pi]$.

A B C

$0 \qquad \frac{3\pi}{4} \qquad \frac{7\pi}{4} \qquad 2\pi$

We use a test value in each interval to determine the sign of the derivative in each interval.

A: Test 0, $f'(0) = cos(0) + sin(0) = 1 > 0$

B: Test $\frac{5\pi}{4}$, $f'(\frac{5\pi}{2}) = cos(\frac{5\pi}{2}) + sin(\frac{5\pi}{2}) = -1.414 < 0$

C: Test 2π, $f'(2\pi) = cos(2\pi) + sin(2\pi) = 1 > 0$

We see that $f(x)$ is increasing on $[0, \frac{3\pi}{4})$ and on $(\frac{7\pi}{4}, 2\pi]$ and decreasing on $(\frac{3\pi}{4}, \frac{7\pi}{4})$ so the function has a relative maximum at $x = \frac{3\pi}{4}$ and a relative minimum at $x = \frac{7\pi}{4}$.

We find $f(\frac{3\pi}{4})$:

$$\begin{aligned} f(\frac{3\pi}{4}) &= sin(\frac{3\pi}{4}) - cos(\frac{3\pi}{4}) \\ &= \frac{1}{\sqrt{2}} - (-\frac{1}{\sqrt{2}}) \\ &= \frac{2}{\sqrt{2}} \approx 1.414 \end{aligned}$$

We find $f(\frac{7\pi}{4})$:

$$f(\frac{7\pi}{4}) = sin(\frac{7\pi}{4}) - cos(\frac{7\pi}{4})$$

$$= -\frac{1}{\sqrt{2}} - (\frac{1}{\sqrt{2}})$$
$$= -\frac{2}{\sqrt{2}} \approx -1.414$$

There is a relative maximum at $(\frac{3\pi}{4}, \frac{2}{\sqrt{2}})$ and there is a relative minimum at $\frac{7\pi}{4}, \frac{2}{\sqrt{2}}$. We use the information obtained to sketch the graph.

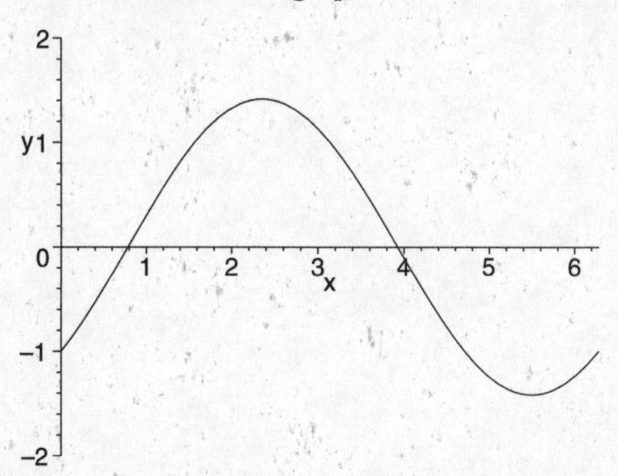

34. $f(x) = sin\ 3x$ First, find the critical points.

$f'(x) = 3cos\ 3x$

$f'(x)$ exists for all x values. We solve $f'(x) = 0$,

$$\begin{aligned} f'(x) &= 0 \\ 3cos\ 3x &= 0 \\ 3x &= \frac{\pi}{2} \\ x &= \frac{\pi}{6} \\ \text{and} \\ x &= \frac{\pi}{2} \\ \text{and} \\ x &= \frac{5\pi}{6} \\ \text{and} \\ x &= \frac{7\pi}{6} \\ \text{and} \\ x &= \frac{9\pi}{6} \\ \text{and} \\ x &= \frac{11\pi}{6} \end{aligned}$$

So the critical points are at $x = \frac{\pi}{6}$, $x = \frac{\pi}{2}$, $x = \frac{5\pi}{6}$, $x = \frac{7\pi}{6}$, $x = \frac{9\pi}{6}$, and $x = \frac{11\pi}{6}$

We use them to divide the real number line into seven intervals, A: $(0, \frac{\pi}{6})$, B: $(\frac{\pi}{6}, \frac{\pi}{2})$, C: $(\frac{\pi}{2}, \frac{5\pi}{6})$, D: $(\frac{5\pi}{6}, \frac{7\pi}{6})$, E: $(\frac{7\pi}{6}, \frac{9\pi}{6})$, F: $(\frac{9\pi}{6}, \frac{11\pi}{6})$, and G: $(\frac{11\pi}{6}, 2\pi)$.

A B C D E F G

$0 \quad \frac{\pi}{6} \quad \frac{\pi}{2} \quad \frac{5\pi}{6} \quad \frac{7\pi}{6} \quad \frac{9\pi}{6} \quad \frac{11\pi}{6} \quad 2\pi$

We use a test value in each interval to determine the sign of the derivative in each interval.

A: Test 0, $f'(0) = 3cos(3 \cdot 0) = 3 > 0$

B: Test $\frac{\pi}{3}$, $f'(\frac{\pi}{3}) = 3cos(3 \cdot \frac{\pi}{3}) = -3 < 0$

C: Test $\frac{2\pi}{3}$, $f'(\frac{2\pi}{3}) = 3cos(3 \cdot \frac{2\pi}{3}) = 3 > 0$

D: Test π, $f'(\pi) = 3cos(3 \cdot \pi) = -3 < 0$

E: Test $\frac{4\pi}{3}$, $f'(\frac{4\pi}{3}) = 3cos(3 \cdot \frac{4\pi}{3}) = 3 > 0$

F: Test $\frac{5\pi}{3}$, $f'(\pi) = 3cos(3 \cdot \frac{5\pi}{3}) = -3 < 0$

G: Test 2π, $f'(2\pi) = 3cos(3 \cdot 2\pi) = 3 > 0$

We see that $f(x)$ is increasing on $(0, \frac{\pi}{6})$, $(\frac{\pi}{2}, \frac{5\pi}{6})$, $(\frac{7\pi}{6}, \frac{9\pi}{6})$, and $(\frac{11\pi}{6}, 2\pi)$ and decreasing on $(\frac{\pi}{6}, \frac{\pi}{2})$, $(\frac{5\pi}{6}, \frac{7\pi}{6})$, $(\frac{9\pi}{6}, \frac{11\pi}{6})$, so the function has a relative maximum at $x = \frac{\pi}{6}$, $x = \frac{5\pi}{6}$, and $x = \frac{9\pi}{6}$ and the function has a relative minimum at $x = \frac{\pi}{2}$, $x = \frac{7\pi}{6}$ and $\frac{11\pi}{6}$.

We find $f(\frac{\pi}{6})$:

$$\begin{aligned} f(\frac{\pi}{6}) &= sin(3 \cdot \frac{\pi}{6}) \\ &= sin(\frac{\pi}{2}) \\ &= 1 \end{aligned}$$

We find $f(\frac{\pi}{2})$:

$$\begin{aligned} f(\frac{\pi}{2}) &= sin(3 \cdot \frac{\pi}{2}) \\ &= sin(3\frac{\pi}{2}) \\ &= -1 \end{aligned}$$

We find $f(\frac{5\pi}{6})$:

$$\begin{aligned} f(\frac{5\pi}{6}) &= sin(3 \cdot \frac{5\pi}{6}) \\ &= sin(5\frac{\pi}{2}) \\ &= 1 \end{aligned}$$

We find $f(\frac{7\pi}{6})$:

$$\begin{aligned} f(\frac{7\pi}{6}) &= sin(3 \cdot \frac{7\pi}{6}) \\ &= sin(7\frac{\pi}{2}) \\ &= -1 \end{aligned}$$

We find $f(\frac{9\pi}{6})$:

$$\begin{aligned} f(\frac{9\pi}{6}) &= sin(3 \cdot \frac{9\pi}{6}) \\ &= sin(3\frac{\pi}{2}) \\ &= 1 \end{aligned}$$

We find $f(\frac{11\pi}{6})$:

$$\begin{aligned} f(\frac{11\pi}{6}) &= sin(3 \cdot \frac{11\pi}{6}) \\ &= sin(11\frac{\pi}{2}) \\ &= -1 \end{aligned}$$

There is a relative maximum at $(\frac{\pi}{6}, 1)$, $(\frac{5\pi}{6}, 1)$, and $(\frac{9\pi}{6}, 1)$ and there is a relative minimum at $(\frac{\pi}{2}, -1)$, $(\frac{7\pi}{6}, -1)$, and $(\frac{11\pi}{6}, -1)$. We use the information obtained to sketch the graph.

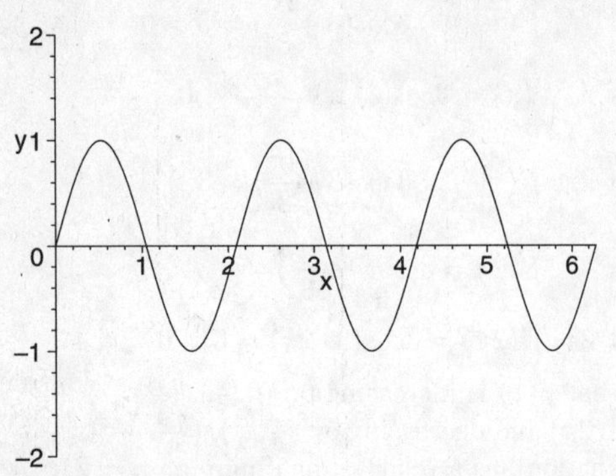

35. $f(x) = cos\ 2x$

First, find the critical points.

$f'(x) = -2sin\ 2x$

$f'(x)$ exists for all x values. We solve $f'(x) = 0$,

$$\begin{aligned} f'(x) &= 0 \\ -2sin\ 2x &= 0 \\ 2x &= 0 \\ x &= 0 \\ \text{and} & \\ x &= \frac{\pi}{2} \\ \text{and} & \\ x &= \pi \\ \text{and} & \\ x &= \frac{3\pi}{2} \\ \text{and} & \\ x &= 2\pi \end{aligned}$$

So the only critical points are at $x = 0$, $x = \frac{\pi}{2}$, $x = \pi$, $x = \frac{3\pi}{2}$ and $x = 2\pi$. We use them to divide the real number line into four intervals, A: $(0, \frac{\pi}{2})$, B: $(\frac{\pi}{2}, \pi)$, C: $(\pi, \frac{3\pi}{2})$ and D: $(\frac{3\pi}{2}, 2\pi)$.

A B C D

0 $\frac{\pi}{2}$ π $\frac{3\pi}{2}$ 2π

We use a test value in each interval to determine the sign of the derivative in each interval.

A: Test $\pi over 4$, $f'(\pi over4) = -2sin(\pi over2) = -2 < 0$

B: Test $\frac{3\pi}{4}$, $f'(\frac{3\pi}{4}) = -2sin(\frac{3\pi}{2}) = 2 > 0$

C: Test $\frac{5\pi}{4}$, $f'(\frac{5\pi}{4}) = -2sin(\frac{3\pi}{2}) = -2 < 0$

D: Test $\frac{7\pi}{4}$, $f'(\frac{7\pi}{4}) = -2sin(\frac{5\pi}{2}) = 2 > 0$

We see that $f(x)$ is decreasing on $(0, \frac{\pi}{2})$ and $(\pi, \frac{3\pi}{2})$ and the function is increasing on $(\frac{\pi}{2}, \pi)$ and $(\frac{3\pi}{2}, 2\pi)$ so the function has a relative maximum at $x = 0$ (end point), $x = \pi$ and $x = 2\pi$ (end point) and a relative minimum at $x = \frac{\pi}{2}$ and $x = \frac{3\pi}{2}$.

We find$f(0)$:

$$\begin{aligned} f(0) &= cos(2 \cdot 0) \\ &= cos(0) \\ &= 1 \end{aligned}$$

We find $f(\pi)$:

$$\begin{aligned} f(\pi) &= cos(2 \cdot \pi) \\ &= 1 \end{aligned}$$

We find $f(2\pi)$:

$$\begin{aligned} f(2\pi) &= cos(2 \cdot 2\pi) \\ &= 1 \end{aligned}$$

We find $f(\frac{\pi}{2})$:

$$\begin{aligned} f(2\pi) &= cos(2 \cdot \frac{\pi}{2}) \\ &= cos(\pi) \\ &= -1 \end{aligned}$$

We find $f(\frac{3\pi}{2})$:

$$\begin{aligned} f(2\pi) &= cos(2 \cdot \frac{3\pi}{2}) \\ &= cos(3\pi) \\ &= -1 \end{aligned}$$

There is a relative maximum at $(0, 1)$, $(\pi, 1)$ and $(2\pi, 1)$ and there is a relative minimum at $(\frac{\pi}{2}, -1)$ and $(\frac{3\pi}{2}, -1)$. We use the information obtained to sketch the graph.

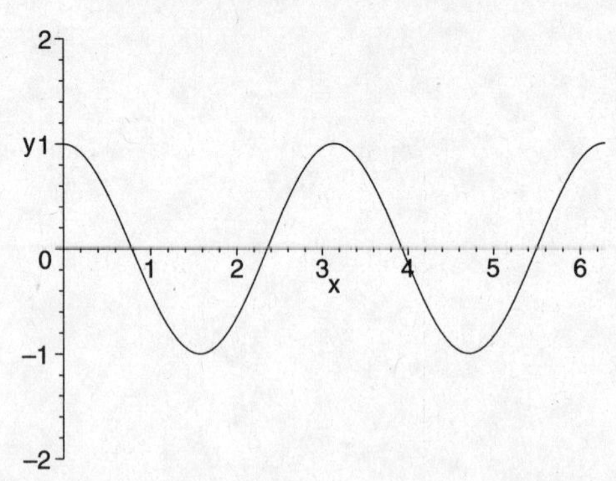

36. $f(x) = x + 2\ sin\ x$ First, find the critical points.

$f'(x) = 1 + 2cos\ x$

$f'(x)$ exists for all x values. We solve $f'(x) = 0$,

$$\begin{aligned} f'(x) &= 0 \\ 1 + 2cos\ x &= 0 \\ cos\ x &= -\frac{1}{2} \\ x &= \frac{2\pi}{3} \\ \text{and} & \\ x &= \frac{4\pi}{3} \end{aligned}$$

So the only critical points are at $x = \frac{2\pi}{3}$ and $x = \frac{4\pi}{3}$ and there might be extrema points at the end points $x = 0$ and $x = 2\pi$. We use them to divide the real number line into three intervals, A: $[0, \frac{2\pi}{3})$, B: $(\frac{2\pi}{3}, \frac{4\pi}{3})$, and C: $(\frac{4\pi}{3}, 2\pi]$.

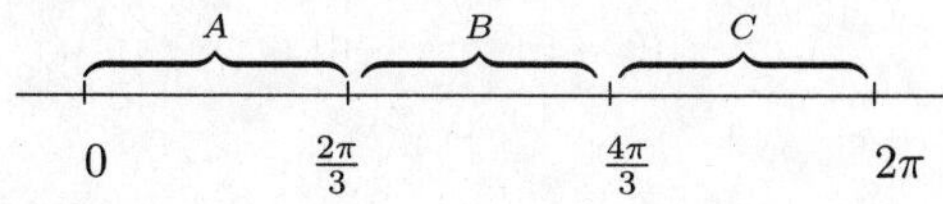

We use a test value in each interval to determine the sign of the derivative in each interval.

A: Test 0, $f'(0) = 1 + 2cos(0) = 3 > 0$

B: Test π, $f'(\pi) = 1 + 2cos(\pi) = -1 < 0$

C: Test 2π, $f'(2\pi) = 1 + 2cos(2\pi) = 3 > 0$

We see that $f(x)$ is increasing on $[0, \frac{2\pi}{3})$ and $(\frac{4\pi}{3}, 2\pi]$ and decreasing on $(\frac{2\pi}{3}, \frac{4\pi}{3})$ so the function has a relative maximum at $x = \frac{2\pi}{3}$ and 2π and a relative minimum at $x = 0$ and $x = \frac{4\pi}{3}$.

We find$f(0)$:

$$\begin{aligned} f(0) &= 0 + 2sin(0) \\ &= 0 \end{aligned}$$

We find $f(\frac{2\pi}{3})$:

$$\begin{aligned} f(\frac{2\pi}{3}) &= \frac{2\pi}{3} + 2sin(\frac{2\pi}{3}) \\ &\approx 3.826 \end{aligned}$$

We find $f(\frac{4\pi}{3})$:

$$\begin{aligned} f(\frac{4\pi}{3}) &= \frac{4\pi}{3} + 2sin(\frac{4\pi}{3}) \\ &\approx 2.457 \end{aligned}$$

We find $f(2\pi)$:

$$\begin{aligned} f(2\pi) &= 2\pi + 2sin(2\pi) \\ &= 2\pi \end{aligned}$$

There is a relative maximum at $(\frac{2\pi}{3}, 3.826)$ and $(2\pi, 2\pi)$ and there is a relative minimum at $(\frac{4\pi}{3}, 2.457)$. We use the information obtained to sketch the graph.

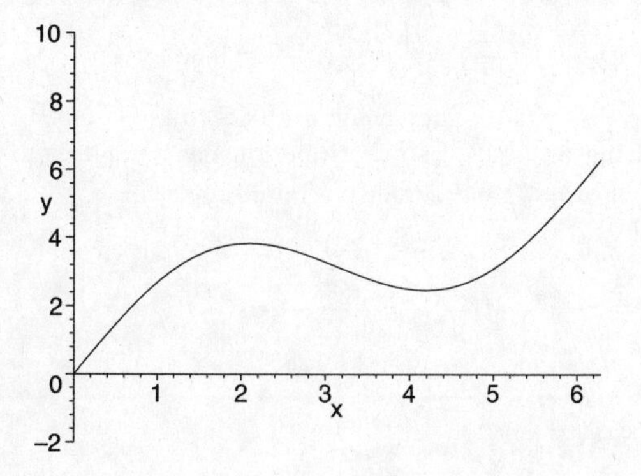

37. $f(x) = x + cos\ 2x$ First, find the critical points.

$f'(x) = 1 - 2sin\ 2x$

$f'(x)$ exists for all x values. We solve $f'(x) = 0$,

$$\begin{aligned} f'(x) &= 0 \\ 1 - 2sin\ 2x &= 0 \\ sin\ 2x &= \frac{1}{2} \\ x &= \frac{\pi}{12} \\ \text{and} \\ x &= \frac{5\pi}{12} \\ \text{and} \\ x &= \frac{13\pi}{12} \\ \text{and} \\ x &= \frac{17\pi}{12} \end{aligned}$$

So the only critical points are at $x = \frac{\pi}{12}$, $x = \frac{5\pi}{12}$, $x = \frac{13\pi}{12}$, and $x = \frac{17\pi}{12}$ and there might be extrema points at the end points $x = 0$ and $x = 2\pi$. We use them to divide the real number line into five intervals, A: $[0, \frac{\pi}{12})$, B: $(\frac{\pi}{12}, \frac{5\pi}{12})$, C: $(\frac{5\pi}{12}, \frac{13\pi}{12})$, D: $(\frac{13\pi}{12}, \frac{17\pi}{12})$, and E: $(\frac{17\pi}{12}, 2\pi]$.

A	B	C	D	E

$0 \quad \frac{\pi}{12} \quad \frac{5\pi}{12} \quad \frac{13\pi}{12} \quad \frac{17\pi}{12} \quad 2\pi$

We use a test value in each interval to determine the sign of the derivative in each interval.

A: Test 0, $f'(0) = 1 - 2sin(2 \cdot 0) = 1 > 0$

B: Test $\frac{\pi}{4}$, $f'(\frac{\pi}{4}) = 1 - 2sin(2 \cdot \frac{\pi}{4}) = -1 < 0$

C: Test π, $f'(\pi) = 1 - 2sin(2 \cdot \pi) = 1 > 0$

D: Test $\frac{15\pi}{12}$, $f'(\frac{15\pi}{12}) = 1 - 2sin(2 \cdot \frac{15\pi}{12}) = -1 < 0$

E: Test $2\pi f'(2\pi) = 1 - 2sin(2 \cdot 2\pi) = 1 > 0$

We see that $f(x)$ is increasing on $[0, \frac{\pi}{12})$, $(\frac{5\pi}{12}, \frac{13\pi}{12})$, and $(\frac{17\pi}{12}, 2\pi]$ and decreasing on $(\frac{\pi}{12}, \frac{5\pi}{12})$, $(\frac{13\pi}{12} and \frac{17\pi}{12})$ so the function has a relative maximum at $x = \frac{\pi}{12}$ and $x = \frac{13\pi}{12}$ and a relative minimum at $x = \frac{5\pi}{12}$, $x = \frac{17\pi}{12}$, and $x = 2\pi$.

We find $f(\frac{\pi}{12})$:

$$\begin{aligned} f(\frac{\pi}{12}) &= \frac{\pi}{12} + cos(2 \cdot \frac{\pi}{12}) \\ &\approx 1.128 \end{aligned}$$

We find $f(\frac{5\pi}{12})$:

$$\begin{aligned} f(\frac{5\pi}{12}) &= \frac{5\pi}{12} + cos(2 \cdot \frac{5\pi}{12}) \\ &\approx 0.443 \end{aligned}$$

We find $f(\frac{13\pi}{12})$:

$$\begin{aligned} f(\frac{13\pi}{12}) &= \frac{13\pi}{12} + cos(2 \cdot \frac{13\pi}{12}) \\ &\approx 4.269 \end{aligned}$$

We find $f(\frac{17\pi}{12})$:

$$\begin{aligned} f(\frac{17\pi}{12}) &= \frac{17\pi}{12} + cos(2 \cdot \frac{17\pi}{12}) \\ &\approx 3.585 \end{aligned}$$

We find $f(2\pi)$:

$$\begin{aligned} f(2pi) &= 2\pi + cos(2\cdot 2\pi) \\ &\approx 7.2832 \end{aligned}$$

There is a relative maximum at $(\frac{\pi}{12}, 1.128)$ and $(\frac{13\pi}{12}, 4.269)$ and a relative minimum at $(\frac{5\pi}{12}, 0.443)$, $(\frac{17\pi}{12}, 3.585)$, and $(2\pi, 7.283)$. We use the information obtained to sketch the graph.

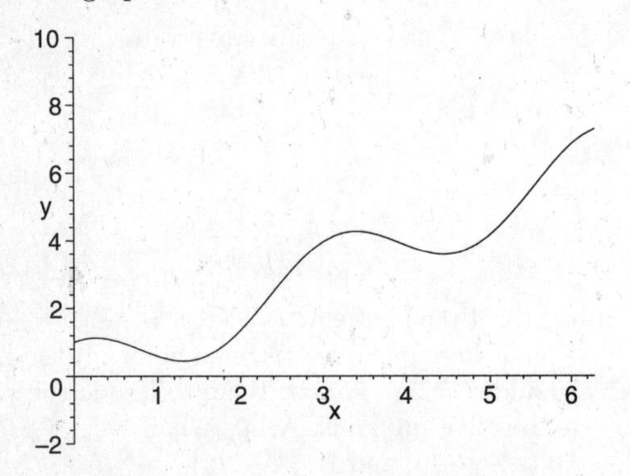

38. $f(x) = \frac{x}{4} + sin\ \frac{x}{2}$. First, find the critical points. $f'(x) = \frac{1}{4} + \frac{1}{2}cos\ \frac{x}{2}$. We solve $f'(x) = 0$

$$\begin{aligned} \frac{1}{4} + \frac{1}{2}cos\ \frac{x}{2} &= 0 \\ \frac{1}{2}cos\ \frac{x}{2} &= -\frac{1}{4} \\ cos\ \frac{x}{2} &= -\frac{1}{2} \\ x &= \frac{2\pi}{3} \end{aligned}$$

and

$$x = \frac{4\pi}{3}$$

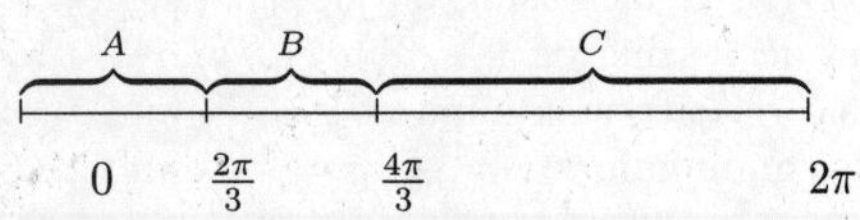

We use a test value in each interval to determine the sign of the derivative in each interval.

A: Test $\frac{\pi}{3}$, $f'(\frac{\pi}{3}) = \frac{1}{4} + \frac{1}{2}cos\ \frac{\pi}{6} = 0.683 > 0$

B: Test π, $f'(\pi) = \frac{1}{4} + \frac{1}{2}cos\frac{\pi}{2} = 0.25 > 0$

C: Test $\frac{5\pi}{3}$, $f'(\frac{5\pi}{3}) = \frac{1}{4} + \frac{1}{2}cos\frac{5\pi}{6} = -0.183 > 0$

We see that $f(x)$ is increasing on $[0, \frac{4\pi}{3})$, and decreasing on $(\frac{4\pi}{3}, 2\pi]$ so the function has a relative maximum at $x = \frac{4\pi}{3}$ We find $f(\frac{4\pi}{3})$

$$\begin{aligned} f(\frac{4\pi}{3}) &= \frac{1}{4}(\frac{4\pi}{3}) + sin(\frac{2\pi}{3}) \\ &= 0.1812 \end{aligned}$$

There is a relative maximum at $(\frac{4\pi}{3}, 0.1812)$. We sketch the graph.

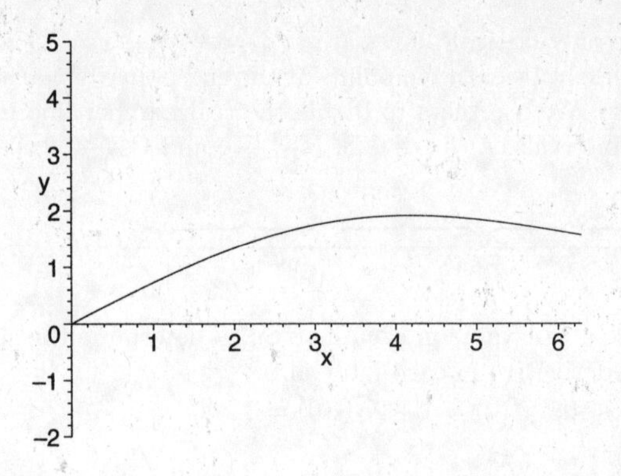

39. $f(x) = \frac{x}{3} + cos\frac{2x}{3}$. Find the critical values $f'(x) = \frac{1}{3} - \frac{2}{3}\ sin\ \frac{2x}{3}$ We solve $f'(x) = 0$

$$\begin{aligned} \frac{1}{3} - \frac{2}{3}\ sin\ \frac{2x}{3} &= 0 \\ \frac{2}{3}sin\ \frac{2x}{3} &= \frac{1}{3} \\ sin\ \frac{2x}{3} &= \frac{1}{2} \\ \frac{2x}{3} &= sin^{-1}(\frac{1}{2}) \\ \frac{2x}{3} &= \frac{pi}{6} \\ x &= \frac{\pi}{4} \end{aligned}$$

and

$$\begin{aligned} \frac{2x}{3} &= \frac{5\pi}{6} \\ x &= \frac{5\pi}{4} \end{aligned}$$

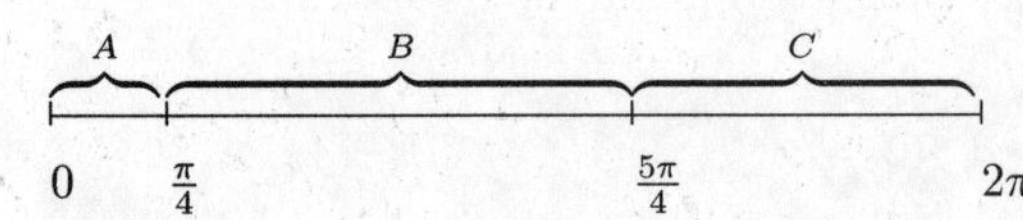

We use a test value in each interval to determine the sign of the derivative in each interval.

A: Test $\frac{\pi}{6}$, $f'(\frac{\pi}{6}) = \frac{1}{3} - \frac{2}{3}sin\ \frac{\pi}{9} = 0.105 > 0$

B: Test π, $f'(\pi) = \frac{1}{3} - \frac{2}{3}sin\frac{2\pi}{3} = -0.323 < 0$

C: Test $\frac{3\pi}{2}$, $f'(\frac{5\pi}{3}) = \frac{1}{3} - \frac{2}{3}sin\ \pi = 0.333 > 0$

We see that $f(x)$ is increasing on $[0, \frac{\pi}{4})$ and $(\frac{5\pi}{4}, 2\pi])$ and decreasing on $(\frac{\pi}{4}, \frac{5\pi}{4})$ so the function has a relative maximum at $x = \frac{\pi}{4}$ and a relative minimum at $x = \frac{5pi}{4}$. We find $f(\frac{\pi}{4})$

$$\begin{aligned} f(\frac{\pi}{4}) &= \frac{\pi}{12} + cos(\frac{\pi}{6}) \\ &= 1.128 \end{aligned}$$

We find $f(\frac{5\pi}{4})$

$$\begin{aligned} f(\frac{5\pi}{4}) &= \frac{5\pi}{12} + cos(\frac{5\pi}{12}) \\ &= 0.443 \end{aligned}$$

There is a relative maximum at $(\frac{\pi}{4}, 1.128)$ and a relative minimum at $(\frac{5\pi}{4}, 0.443)$. We sktch the graph

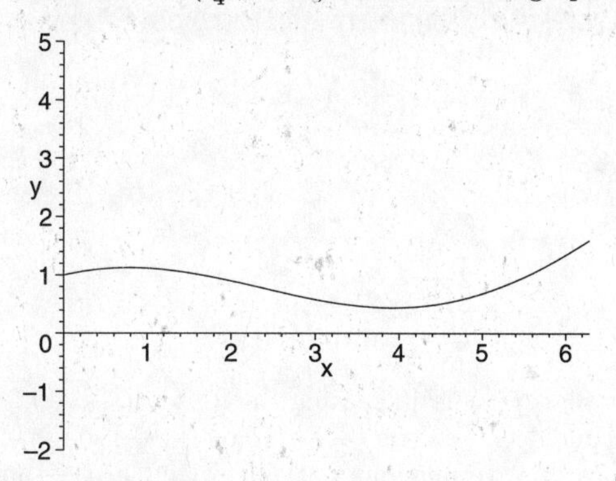

40. $f(x) = \frac{sin\ x}{2+cos\ x}$

$$\begin{aligned} f'(x) &= \frac{(2+cos\ x)(cos\ x) - sin\ x(-sin\ x)}{(2+cos\ x)^2} \\ &= \frac{2\ cos\ x + cos^2 x + sin^2 x}{(2+cos\ x)^2} \\ &= \frac{2\ cos\ x + 1}{(2+cos\ x)^2} \end{aligned}$$

We solve $f'(x) = 0$

$$\begin{aligned} \frac{2\ cos\ x+1}{(2+cos\ x)^2} &= 0 \\ 2\ cos\ x + 1 &= 0 \\ cos\ x &= -\frac{1}{2} \\ x &= \frac{2\pi}{3} \\ \text{and} \\ x &= \frac{4\pi}{3} \end{aligned}$$

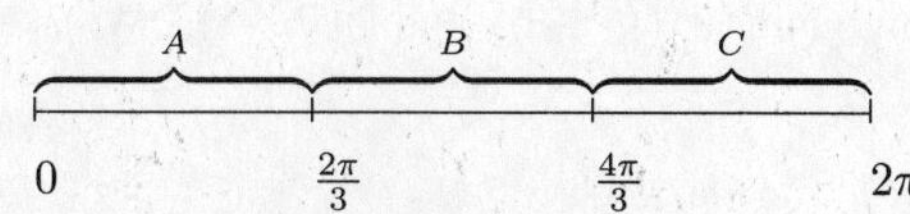

We use a test value in each interval to determine the sign of the derivative in each interval.

A: Test $\frac{\pi}{4}$, $f'(\frac{\pi}{4}) = 0.329 > 0$

B: Test π, $f'(\pi) = -1 < 0$

C: Test $\frac{3\pi}{2}$, $f'(\frac{3\pi}{2}) = 0.25 > 0$

We see that $f(x)$ is increasing on $[0, \frac{2\pi}{3})$ and $(\frac{4\pi}{3}, 2\pi])$ and decreasing on $(\frac{2\pi}{3}, \frac{4\pi}{3})$ so the function has a relative maximum at $x = \frac{2\pi}{3}$ and a relative minimum at $x = \frac{4\pi}{3}$.

We find $f(\frac{2\pi}{3})$

$$\begin{aligned} f(\frac{2\pi}{3}) &= \frac{sin\ \frac{2\pi}{3}}{2+cos\ \frac{2\pi}{3}} \\ &= 0.5774 \end{aligned}$$

We find $f(\frac{4\pi}{3})$

$$\begin{aligned} f(\frac{4\pi}{3}) &= \frac{sin\ \frac{4\pi}{3}}{2+cos\ \frac{4\pi}{3}} \\ &= -0.5774 \end{aligned}$$

There is a relative maximum at $(\frac{2\pi}{3}, 0.5774)$ and a relative minimum at $(\frac{4\pi}{3}, -0.5774)$. We sketch the graph

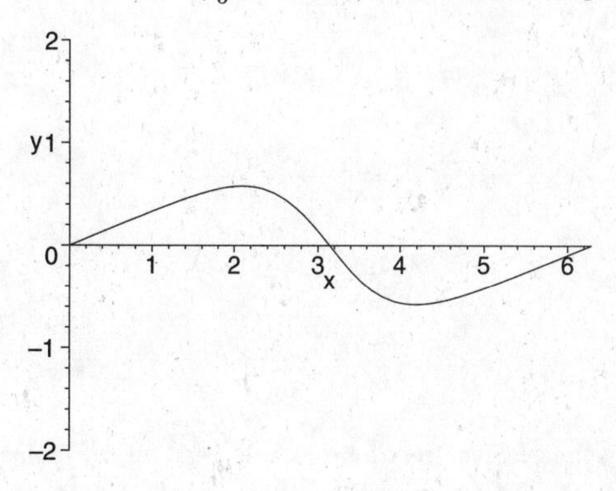

41. $f(x) = \frac{cos\ x}{2-sin\ x}$

Find the critical values

$$\begin{aligned} f'(x) &= \frac{(2-sin\ x)(-sin\ x) - cos\ x(-cos\ x)}{(2-sin\ x)^2} \\ &= \frac{-2\ sin\ x + sin^2 x + cos^2 x}{(2-sin\ x)^2} \\ &= \frac{1-2\ sin\ x}{(2-sin\ x)^2} \end{aligned}$$

We solve $f'(x) = 0$

$$\begin{aligned} \frac{1-2\ sin\ x}{(2-sin\ x)^2} &= 0 \\ 1-2\ sin\ x &= 0 \\ sin\ x &= \frac{1}{2} \\ x &= \frac{\pi}{6} \\ \text{and} \\ x &= \frac{5\pi}{6} \end{aligned}$$

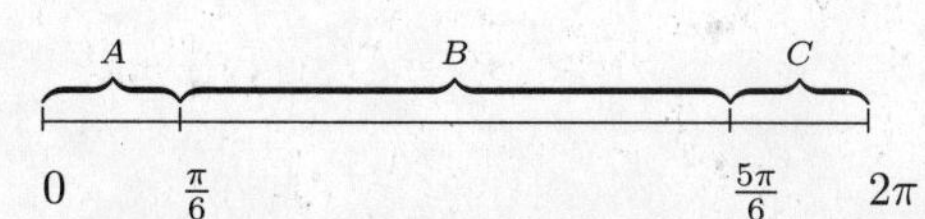

We use a test value in each interval to determine the sign of the derivative in each interval.

A: Test $\frac{\pi}{18}$, $f'(\frac{\pi}{18}) = 0.196 > 0$

B: Test $\frac{\pi}{4}$, $f'(\frac{\pi}{4}) = -0.248 < 0$

C: Test $\frac{11\pi}{12}$, $f'(\frac{11\pi}{12}) = 0.159 > 0$

We see that $f(x)$ is increasing on $[0, \frac{\pi}{6})$ and $(\frac{5\pi}{6}, 2\pi])$ and decreasing on $(\frac{\pi}{6}, \frac{5\pi}{6})$ so the function has a relative maximum at $x = \frac{\pi}{3}$ and a relative minimum at $x = \frac{5\pi}{6}$.

We find $f(\frac{\pi}{6})$

$$\begin{aligned} f(\frac{\pi}{6}) &= \frac{cos(\frac{\pi}{6})}{2 - sin(\frac{\pi}{6})} \\ &= \frac{\frac{\sqrt{3}}{2}}{2 - \frac{1}{2}} \\ &= \frac{\sqrt{3}}{3} \end{aligned}$$

We find $f(\frac{5\pi}{6})$

$$\begin{aligned} f(\frac{5\pi}{6}) &= \frac{cos(\frac{5\pi}{6})}{2 - sin(\frac{5\pi}{6})} \\ &= \frac{\frac{-\sqrt{3}}{2}}{2 - \frac{1}{2}} \\ &= \frac{-\sqrt{3}}{3} \end{aligned}$$

There is a relative maximum at $(\frac{\pi}{6}, \frac{\sqrt{3}}{3})$ and a relative minimum at $(\frac{5\pi}{6}, -\frac{\sqrt{3}}{3})$. We sketch the graph

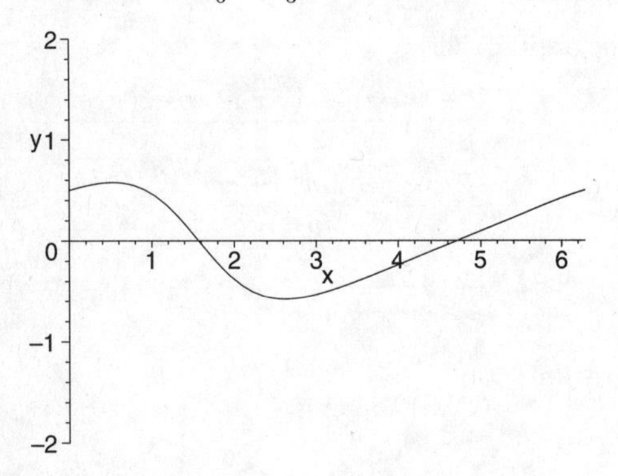

42. $f(x) = sin\ x\ cos\ x$

$f'(x) = -sin^2x + cos^2x = 1 - 2\ sin^2x$

We solve $f'(x) = 0$

$$\begin{aligned} 1 - 2\ sin^2x &= 0 \\ sin^2x &= \frac{1}{2} \\ sin\ x &= \pm\frac{1}{\sqrt{2}} \\ x &= \frac{\pi}{4} \\ &\text{and} \\ x &= \frac{3\pi}{4} \\ &\text{and} \\ x &= \frac{5\pi}{4} \\ &\text{and} \\ x &= \frac{7\pi}{4} \end{aligned}$$

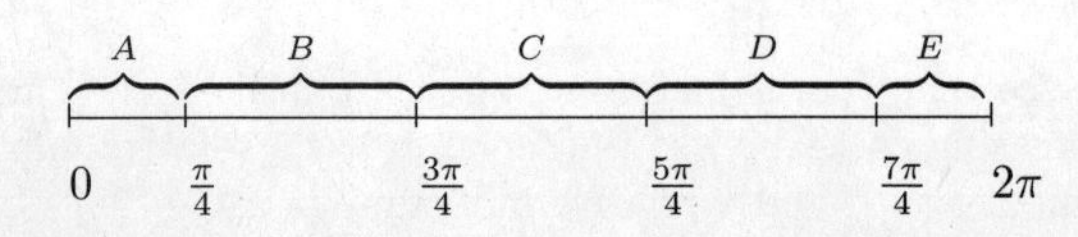

We use a test value in each interval to determine the sign of the derivative in each interval.

A: Test $\frac{\pi}{6}$, $f'(\frac{\pi}{6})$ =0.5 > 0

B: Test $\frac{\pi}{2}$, $f'(\frac{\pi}{2}) = -1 < 0$

C: Test π, $f'(\pi)$ =1 > 0

D: Test $\frac{3\pi}{2}$, $f'(\frac{3\pi}{2}) = -1 > 0$

E: Test $\frac{15\pi}{8}$, $f'(\frac{15\pi}{8})$ =0.707 > 0

We see that $f(x)$ is increasing on $[0, \frac{\pi}{4})$, $(\frac{3\pi}{4}, \frac{5\pi}{4})$, and $(\frac{7\pi}{4}, 2\pi]$ and decreasing on $(\frac{\pi}{4}, \frac{3\pi}{4})$ and $(\frac{5\pi}{4}, \frac{7\pi}{4})$ so the function has a relative maximum at $x = \frac{\pi}{4}$ and $x = \frac{5\pi}{4}$ and a relative minimum at $x = \frac{3\pi}{4}$ and $x = \frac{7\pi}{4}$. We find $f(\frac{\pi}{4})$

$$\begin{aligned} f(\frac{\pi}{4}) &= sin(\frac{\pi}{4})\ cos(\frac{\pi}{4}) \\ &= \frac{1}{2} \end{aligned}$$

We find $f(\frac{3\pi}{4})$

$$\begin{aligned} f(\frac{3\pi}{4}) &= sin(\frac{3\pi}{4})\ cos(\frac{3\pi}{4}) \\ &= -\frac{1}{2} \end{aligned}$$

We find $f(\frac{5\pi}{4})$

$$\begin{aligned} f(\frac{5\pi}{4}) &= sin(\frac{5\pi}{4})\ cos(\frac{5\pi}{4}) \\ &= \frac{1}{2} \end{aligned}$$

We find $f(\frac{7\pi}{4})$

$$\begin{aligned} f(\frac{7\pi}{4}) &= sin(\frac{7\pi}{4})\ cos(\frac{7\pi}{4}) \\ &= -\frac{1}{2} \end{aligned}$$

There is a relative maximum at $(\frac{\pi}{4}, \frac{1}{2})$ and $(\frac{5\pi}{4}, \frac{1}{2})$ and a relative minimum at $(\frac{3\pi}{4}, -\frac{1}{2})$ and $(\frac{7\pi}{4}, -\frac{1}{2})$. We sketch the graph

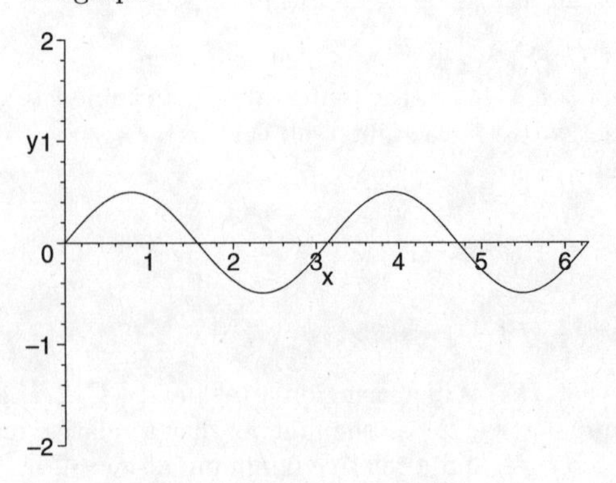

43. $f(x) = sin\ x - sin^2\ x$

Find the critical values

$$\begin{aligned} f'(x) &= cos\ x - 2\ sin\ x\ cos\ x \\ &= cos\ x(1 - 2\ sin\ x) \end{aligned}$$

We solve $f'(x) = 0$

$$
\begin{aligned}
\cos x(1 - 2 \sin x) &= 0 \\
\cos x &= 0 \\
x &= \frac{\pi}{2} \\
&\text{and} \\
x &= \frac{3\pi}{2} \\
1 - 2 \sin x &= 0 \\
\sin x &= \frac{1}{2} \\
x &= \frac{\pi}{6} \\
&\text{and} \\
x &= \frac{5\pi}{6}
\end{aligned}
$$

A B C D E

0 $\frac{\pi}{6}$ $\frac{\pi}{2}$ $\frac{5\pi}{6}$ $\frac{3\pi}{2}$ 2π

We use a test value in each interval to determine the sign of the derivative in each interval.

A: Test $\frac{\pi}{18}$, $f'(\frac{\pi}{18}) = 0.643 > 0$

B: Test $\frac{\pi}{4}$, $f'(\frac{\pi}{4}) = -0.293 < 0$

C: Test $\frac{2\pi}{3}$, $f'(\frac{2\pi}{3}) = 0.366 > 0$

D: Test π, $f'(\pi) = -1 < 0$

E: Test $\frac{7\pi}{4}$, $f'(\frac{7\pi}{4}) = 1.707 < 0$

We see that $f(x)$ is increasing on $[0, \frac{\pi}{6})$, $(\frac{\pi}{2}, \frac{5\pi}{6})$, and $(\frac{3\pi}{2}, 2\pi]$ and decreasing on $(\frac{\pi}{6}, \frac{\pi}{2})$ and $(\frac{5\pi}{6}, \frac{3\pi}{2})$ so the function has a relative maximum at $x = \frac{\pi}{6}$ and $x = \frac{5\pi}{6}$ and a relative minimum at $x = \frac{\pi}{2}$ and $x = \frac{3\pi}{2}$.

We find $f(\frac{\pi}{6})$

$$
\begin{aligned}
f(\frac{\pi}{6}) &= \sin(\frac{\pi}{6}) - \sin^2(\frac{\pi}{6}) \\
&= \frac{1}{2} - \frac{1}{4} \\
&= \frac{1}{4}
\end{aligned}
$$

We find $f(\frac{5\pi}{6})$

$$
\begin{aligned}
f(\frac{5\pi}{6}) &= \sin(\frac{5\pi}{6}) - \sin^2(\frac{5\pi}{6}) \\
&= \frac{1}{2} - \frac{1}{4} \\
&= \frac{1}{4}
\end{aligned}
$$

We find $f(\frac{\pi}{2})$

$$
\begin{aligned}
f(\frac{\pi}{2}) &= \sin(\frac{\pi}{2}) - \sin^2(\frac{\pi}{2}) \\
&= 1 - 1 \\
&= 0
\end{aligned}
$$

We find $f(\frac{3\pi}{2})$

$$
\begin{aligned}
f(\frac{3\pi}{2}) &= \sin(\frac{3\pi}{2}) - \sin^2(\frac{3\pi}{2}) \\
&= -1 - 1 \\
&= -2
\end{aligned}
$$

There is a relative maximum at $(\frac{\pi}{6}, \frac{1}{4})$ and $(\frac{5\pi}{6}, \frac{1}{4})$ and a relative minimum at $(\frac{\pi}{2}, 0)$ and $(\frac{3\pi}{2}, -2)$. We sketch the graph

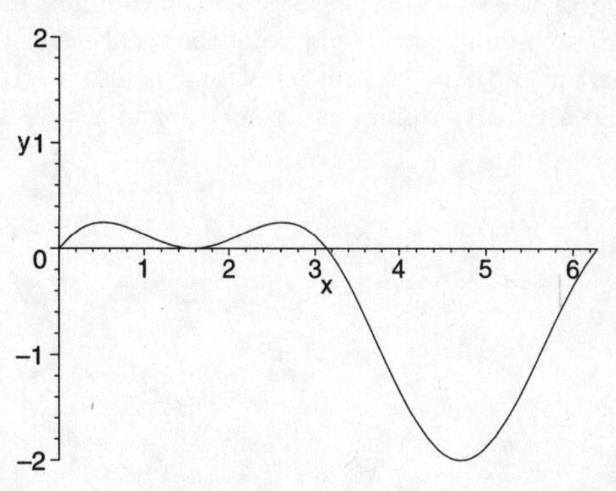

44. $f(x) = \cos x + \cos^2 x$

$$
\begin{aligned}
f'(x) &= -\sin x + 2 \cos x\,(-\sin x) \\
&= -\sin x(1 + 2 \cos x)
\end{aligned}
$$

We solve $f'(x) = 0$

$$
\begin{aligned}
-\sin x(1 + 2 \cos x) &= 0 \\
-\sin x &= 0 \\
x &= 0 \\
&\text{and} \\
x &= \pi \\
&\text{and} \\
x &= 2\pi \\
1 + 2 \sin x &= 0 \\
\cos x &= -\frac{1}{2} \\
x &= \frac{2\pi}{3} \\
&\text{and} \\
x &= \frac{4\pi}{3}
\end{aligned}
$$

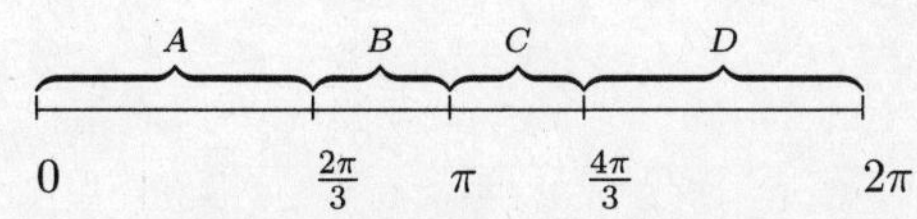

We use a test value in each interval to determine the sign of the derivative in each interval.

A: Test $\frac{\pi}{3}$, $f'(\frac{\pi}{3}) = -1.732 < 0$

B: Test $\frac{5\pi}{6}$, $f'(\frac{5\pi}{6}) = 0.366 > 0$

C: Test $\frac{7\pi}{6}$, $f'(\frac{7\pi}{6}) = -0.366 < 0$

D: Test $\frac{5\pi}{3}$, $f'(\frac{5\pi}{3}) = 1.732 > 0$

We see that $f(x)$ is decreasing on $[0, \frac{2\pi}{3})$ and $(\pi, \frac{4\pi}{3})$, and increasing on $(\frac{2\pi}{3}, \pi)$ and $(\frac{4\pi}{3}, 2\pi])$ so the function has a relative maximum at $x = 0$ (end point that is also a critical value), $x = \pi$ and $x = 2\pi$ (end point that is also a critical value) and a relative minimum at $x = \frac{2\pi}{3}$ and $x = \frac{4\pi}{3}$.

We find $f(0)$

$$\begin{aligned} f(0) &= cos(0) + cos^2(0) \\ &= 2 \end{aligned}$$

We find π

$$\begin{aligned} f(\pi) &= cos(\pi) + cos^2(\pi) \\ &= 0 \end{aligned}$$

We find $f(2\pi)$

$$\begin{aligned} f(2\pi) &= cos(2\pi) + cos^2(2\pi) \\ &= 2 \end{aligned}$$

We find $f(\frac{2\pi}{3})$

$$\begin{aligned} f(\frac{2\pi}{3}) &= cos(\frac{2\pi}{3} + cos^2(\frac{2\pi}{3}) \\ &= -\frac{1}{4} \end{aligned}$$

We find $f(\frac{4\pi}{3})$

$$\begin{aligned} f(\frac{4\pi}{3}) &= cos(\frac{4\pi}{3} + cos^2(\frac{4\pi}{3}) \\ &= -\frac{1}{4} \end{aligned}$$

There is a relative maximum at $(0, 2)$,$(\pi, 0)$ and $(2\pi, 2)$ and a relative minimum at $(\frac{2\pi}{3}, -\frac{1}{4})$ and $(\frac{4\pi}{3}, -\frac{1}{4})$. We sketch the graph

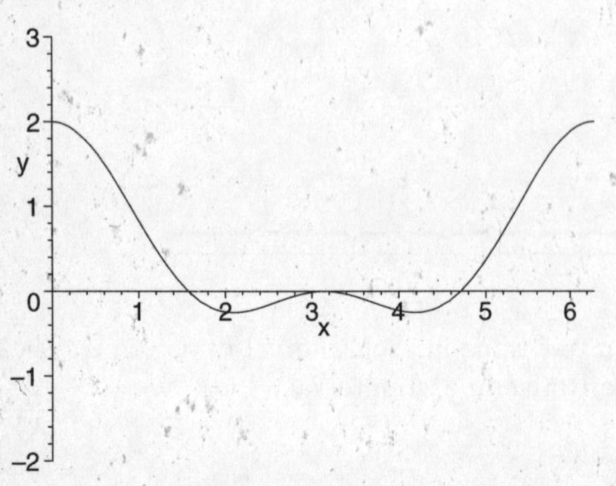

45. $f(x) = 9sin\ x - 4\ sin^3\ x$

Find the critical values

$$\begin{aligned} f'(x) &= 9\ cos\ x - 12\ sin^2 x\ cos\ x \\ &= 3\ cos\ x(3 - 4\ sin^2 x) \end{aligned}$$

We solve $f'(x) = 0$

$$\begin{aligned} 3\ cos\ x(3 - 4\ sin^2 x) &= 0 \\ cos\ x &= 0 \\ x &= \frac{\pi}{2} \\ \text{and} \\ x &= \frac{3\pi}{2} \\ 3 - 4\ sin^2 x &= 0 \\ sin^2 x &= \frac{3}{4} \\ sin\ x &= \pm\frac{\sqrt{3}}{2} \\ x &= \frac{\pi}{3} \\ \text{and} \\ x &= \frac{2\pi}{3} \\ \text{and} \\ x &= \frac{5\pi}{3} \end{aligned}$$

A B C D E F G

0 $\frac{\pi}{3}$ $\frac{\pi}{2}$ $\frac{2\pi}{3}$ $\frac{4\pi}{3}$ $\frac{3\pi}{2}$ $\frac{5\pi}{3}$ 2π

We use a test value in each interval to determine the sign of the derivative in each interval.

A: Test $\frac{\pi}{4}$, $f'(\frac{\pi}{4}) = 2.121 > 0$

B: Test $\frac{5\pi}{12}$, $f'(\frac{5\pi}{12}) = -0.568 < 0$

C: Test $\frac{7\pi}{12}$, $f'(\frac{7\pi}{12}) = 0.568 > 0$

D: Test π, $f'(\pi) = -9 < 0$

E: Test $\frac{17\pi}{12}$, $f'(\frac{17\pi}{12}) = 0.568 > 0$

F: Test $\frac{19\pi}{12}$, $f'(\frac{19\pi}{12}) = -0.568 < 0$

G: Test $\frac{11\pi}{6}$, $f'(\frac{11\pi}{6}) = 5.760 > 0$

We see that $f(x)$ is increasing on $[0, \frac{\pi}{3})$, $(\frac{\pi}{2}, \frac{2\pi}{3})$, $(\frac{4\pi}{3}, \frac{3\pi}{2})$ and $(\frac{5\pi}{3}, 2\pi]$ and decreasing on $(\frac{\pi}{3}, \frac{\pi}{2})$, $(\frac{2\pi}{3}, \frac{4\pi}{3})$ and $(\frac{3\pi}{2}, \frac{5\pi}{3})$ so the function has a relative maximum at $x = \frac{\pi}{3}$, $x = \frac{2\pi}{3}$ and $x = \frac{3\pi}{2}$ and a relative minimum at $x = \frac{\pi}{2}$, $x = \frac{4\pi}{3}$ and $x = \frac{5\pi}{3}$.

We find $f(\frac{\pi}{3})$

$$f(\frac{\pi}{3}) = 9\ sin(\frac{\pi}{3}) - 4\ sin^3(\frac{\pi}{3})$$

$$= \frac{9\sqrt{3}}{2} - \frac{12\sqrt{3}}{8}$$
$$= 3\sqrt{3}$$

We find $f(\frac{2\pi}{3})$

$$\begin{aligned} f(\frac{2\pi}{3}) &= 9\ sin(\frac{2\pi}{3}) - 4\ sin^3(\frac{2\pi}{3}) \\ &= \frac{9\sqrt{3}}{2} - \frac{12\sqrt{3}}{8} \\ &= 3\sqrt{3} \end{aligned}$$

We find $f(\frac{3\pi}{2})$

$$\begin{aligned} f(\frac{3\pi}{2}) &= 9\ sin(\frac{3\pi}{2}) - 4\ sin^3(\frac{3\pi}{2}) \\ &= -9 - 4(-1) \\ &= -5 \end{aligned}$$

We find $f(\frac{\pi}{2})$

$$\begin{aligned} f(\frac{\pi}{2}) &= 9\ sin(\frac{\pi}{2}) - 4\ sin^3(\frac{\pi}{2}) \\ &= 9 - 4 \\ &= 5 \end{aligned}$$

We find $f(\frac{4\pi}{3})$

$$\begin{aligned} f(\frac{4\pi}{3}) &= 9\ sin(\frac{4\pi}{3}) - 4\ sin^3(\frac{4\pi}{3}) \\ &= -\frac{9\sqrt{3}}{2} + \frac{12\sqrt{3}}{8} \\ &= -3\sqrt{3} \end{aligned}$$

We find $f(\frac{5\pi}{3})$

$$\begin{aligned} f(\frac{5\pi}{3}) &= 9\ sin(\frac{5\pi}{3}) - 4\ sin^3(\frac{5\pi}{3}) \\ &= -\frac{9\sqrt{3}}{2} + \frac{12\sqrt{3}}{8} \\ &= -3\sqrt{3} \end{aligned}$$

There is a relative maximum at $(\frac{\pi}{3}, 3\sqrt{3})$, $(\frac{2\pi}{3}, 3\sqrt{3})$ and $(\frac{3\pi}{2}, -5)$ and a relative minimum at $(\frac{\pi}{2}, 5)$,$(\frac{4\pi}{3}, -3\sqrt{3})$, and $(\frac{5\pi}{3}, -3\sqrt{3})$. We sketch the graph

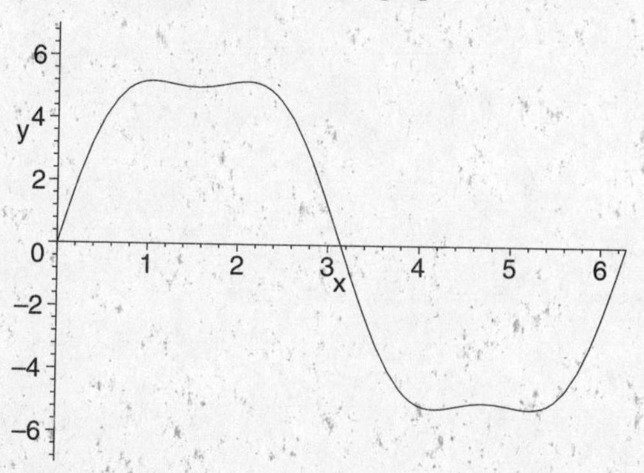

46. $f(x) = 3cos\ x - 2\ cos^3\ x$

$$\begin{aligned} f'(x) &= -3\ sin\ x + 6\ cos^2 x\ sin\ x \\ &= -3\ sin\ x(1 - 2cos^2 x) \end{aligned}$$

We solve $f'(x) = 0$

$$\begin{aligned} -3\ sin\ x(1 - 2cos^2 x) &= 0 \\ sin\ x &= 0 \\ x &= 0 \\ \text{and} & \\ x &= \pi \\ \text{and} & \\ x &= 2\pi \\ 1 - 2\ cos^2 x &= 0 \\ cos^2 x &= \frac{1}{2} \\ cos\ x &= \pm\frac{1}{\sqrt{2}} \\ x &= \frac{\pi}{4} \\ \text{and} & \\ x &= \frac{3\pi}{4} \\ \text{and} & \\ x &= \frac{5\pi}{4} \\ \text{and} & \\ x &= \frac{7\pi}{4} \end{aligned}$$

A	B	C	D	E	F
0 to $\frac{\pi}{4}$	$\frac{\pi}{4}$ to $\frac{3\pi}{4}$	$\frac{3\pi}{4}$ to π	π to $\frac{5\pi}{4}$	$\frac{5\pi}{4}$ to $\frac{7\pi}{4}$	$\frac{7\pi}{4}$ to 2π

We use a test value in each interval to determine the sign of the derivative in each interval.

A: Test $\frac{\pi}{6}$, $f'(\frac{\pi}{6}) = 0.75 > 0$

B: Test $\frac{\pi}{2}$, $f'(\frac{\pi}{2}) = -3 < 0$

C: Test $\frac{5\pi}{6}$, $f'(\frac{5\pi}{6}) = 0.75 > 0$

D: Test $\frac{7\pi}{6}$, $f'(\frac{7\pi}{6}) = -0.75 < 0$

E: Test $\frac{9\pi}{6}$, $f'(\frac{9\pi}{6}) = 3 > 0$

F: Test $\frac{11\pi}{6}$, $f'(\frac{11\pi}{6}) = -0.75 < 0$

We see that $f(x)$ is increasing on $[0, \frac{\pi}{4})$, $(\frac{3\pi}{4}, \pi)$, and $(\frac{5\pi}{4}, \frac{7\pi}{4})$ and decreasing on $(\frac{\pi}{4}, \frac{3\pi}{4})$, $(\pi, \frac{5\pi}{4})$ and $(\frac{7\pi}{4}, 2\pi]$ so the function has a relative maximum at $x = \frac{\pi}{4}$, $x = \pi$ and $x = \frac{7\pi}{4}$ and a relative minimum at $x = \frac{3\pi}{4}$ and $x = \frac{5\pi}{4}$.

We find $f(\frac{\pi}{4})$

$$f(\frac{\pi}{4}) = 3\ cos(\frac{\pi}{4}) - 2cos^3(\frac{\pi}{4})$$

$$= \sqrt{2}$$

We find $f(\frac{3\pi}{4})$

$$\begin{aligned} f(\frac{3\pi}{4}) &= 3\ cos(\frac{3\pi}{4}) - 2cos^3(\frac{3\pi}{4}) \\ &= -\sqrt{2} \end{aligned}$$

We find $f(\pi)$

$$\begin{aligned} f(\pi) &= 3\ cos(\pi) - 2cos^3(\frac{3\pi}{4}) \\ &= -1 \end{aligned}$$

We find $f(\frac{5\pi}{4})$

$$\begin{aligned} f(\frac{5\pi}{4}) &= 3\ cos(\frac{5\pi}{4}) - 2cos^3(\frac{5\pi}{4}) \\ &= -\sqrt{2} \end{aligned}$$

We find $f(\frac{7\pi}{4})$

$$\begin{aligned} f(\frac{7\pi}{4}) &= 3\ cos(\frac{7\pi}{4}) - 2cos^3(\frac{7\pi}{4}) \\ &= \sqrt{2} \end{aligned}$$

There is a relative maximum at $(\frac{\pi}{4}, \sqrt{2})$, $(\pi, -1)$ and $(\frac{7\pi}{4}, \sqrt{2})$ and a relative minimum at $(\frac{3\pi}{4}, -\sqrt{2})$ and $(\frac{5\pi}{4}, -\sqrt{2})$. We sketch the graph

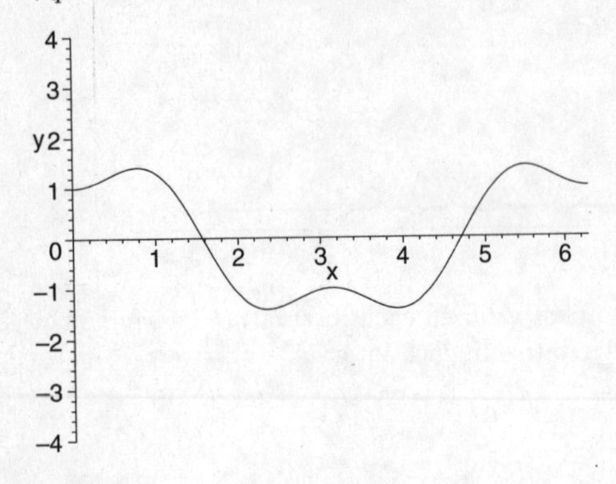

47. - 92. Left to the student.

93. $h(d) = -0.002d^2 + 0.8d + 6.6$

$h'(d) = -0.004d + 0.8$

Solve $h'(d) = 0$.

$$\begin{aligned} -0.004d + 0.8 &= 0 \\ -0.004d &= -0.8 \\ d &= 20 \end{aligned}$$

A: Test 100, $f'(50) = 0.4 > 0$

B: Test 300, $f'(50) = -0.4 < 0$

The function is increasing on to the left of $d = 200$ and decreasing to the right of $d = 200$ therefore there is a relative maximum at $d = 200$. We find $h(200)$

$$\begin{aligned} h(200) &= -0.002(200)^2 + 0.8(200) + 6.6 \\ &= -80 + 160 + 6.6 \\ &= 86.6 \end{aligned}$$

There is a relative maximum at $(200, 86.6)$

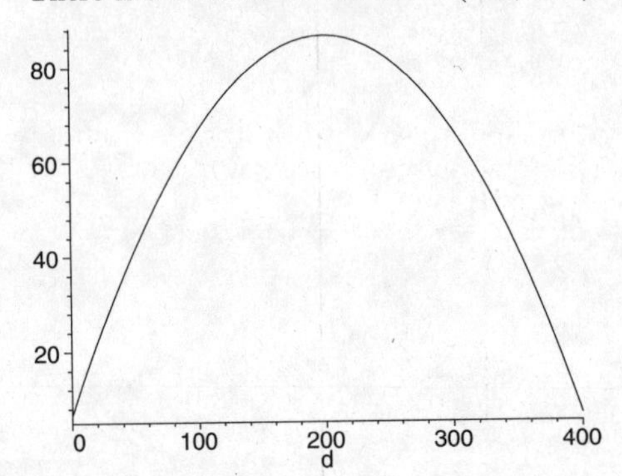

94. Answers vary.

95. $T(t) = -0.1t^2 + 1.2t + 98.6,\ 0 \le t \le 12$

$T'(t) = -0.2t + 1.2$

$T'(t)$ exists for all real numbers. Solve $T'(t) = 0$.

$$\begin{aligned} -0.2t + 1.2 &= 0 \\ -0.2t &= -1.2 \\ t &= 6 \end{aligned}$$

The only critical point is at $t = 6$. We use it to divide the interval $[0, 12]$ (the domain of $T(t)$) into two intervals, A: $[0, 6)$ and B: $(6, 12]$.

A: Test 0, $T'(0) = -0.2(0) + 1.2 = 1.2 > 0$

B: Test 7, $T'(7) = -0.2(7) + 1.2 = -0.2 < 0$

Since $T(t)$ is increasing on $[0, 6)$ and decreasing on $(6, 12]$, there is a relative maximum at $x = 6$.

$$T(6) = -0.1(6)^2 + 1.2(6) + 98.6 = 102.2$$

There is a relative maximum at $(6, 102.2°)$. We sketch the graph.

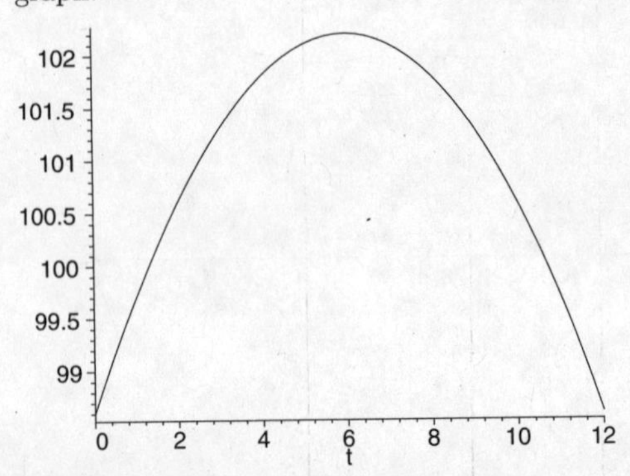

96. $f(x) = \sqrt{138.1 - 5.025x + 0.2902x^2}$
$= (138.1 - 5.025x + 0.2902x^2)^{1/2}$

First, find the critical points.

$$\begin{aligned} f'(x) &= \frac{1}{2}(138.1 - 5.025x + 0.2902x^2)^{-1/2} \times \\ &\quad (-5.025 + 0.5804x) \\ &= \frac{-5.025 + 0.5804x}{2\sqrt{138.1 - 5.025x + 0.2902x^2}} \end{aligned}$$

The equation $138.1 - 5.025x + 0.2902x^2 = 0$ has no real-number solutions, so there are no values of x for which $f'(x)$ does not exist.

We solve $f'(x) = 0$:

$$\begin{aligned} \frac{-5.025 + 0.5804x}{2\sqrt{138.1 - 5.025x + 0.2902x^2}} &= 0 \\ -5.025 + 0.5804x &= 0 \\ 0.5804x &= 5.025 \\ x &\approx 8.658 \end{aligned}$$

The only critical point is at x about 8.658. We use it to divide the number line into two intervals, A: $(-\infty, 8.658)$ and B: $(8.658, \infty)$.

We will use a test value in each interval to determine the sign of the derivative in each interval.

A: Test 8, $f'(8) = \dfrac{-5.025 + 0.5804(8)}{2\sqrt{138.1 - 5.025(8) + 0.2902(8)^2}} < 0$

B: Test 9, $f'(9) = \dfrac{-5.025 + 0.5804(9)}{2\sqrt{138.1 - 5.025(9) + 0.2902(9)^2}} > 0$

We see that $f(x)$ is decreasing on $(-\infty, 8.658)$ and increasing on $(8.658, \infty)$, so there is a relative minimum at $x \approx 8.658$.

We find $f(8.658)$:

$f(8.658) = \sqrt{138.1 - 5.025(8.658) + 0.2902(8.658)^2} \approx 10.786$

There is a relative minimum at about $(8.658, 10.786)$. Thus, the point farthest north on the path of the center of the eclipse is at a latitude of about 8.658 degrees East and at a longitude of about 10.786 degrees South.

We use the information obtained above to sketch the graph.

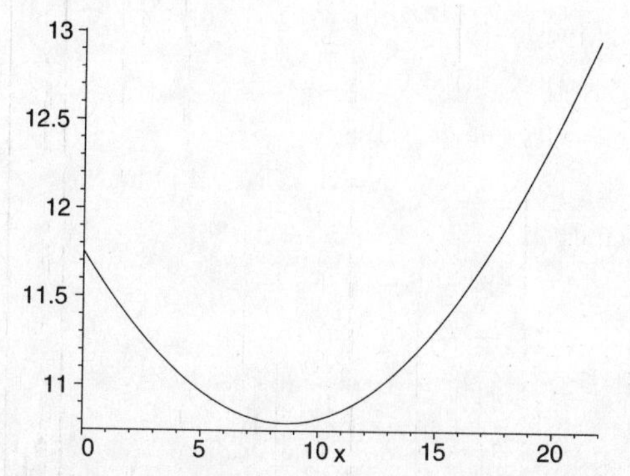

97. Let us consider the intervals $[a, b]$, $[c, d]$ and consider the sign of the slope of the tangent line. Next, consider the intervals $[b, c]$ and $[d, e]$ and again consider the sign of the slope of the tangent line. The sign of the slope of the tangent line at a point, which is the value of the derivative of the function at that point, informs us whether the function is increasing or decreasing.

98. The crtical values occur when the slope of the tangent line at the point is either zero or undefined (which usually indicates a maximum or a minimum).
$(1, 5)$ a relative maximum
$(3, 1)$ a relative minimum
$(4, 5)$ a relative maximum
$(5, 2)$ a relative minimum

99. The crtical values occur when the slope of the tangent line at the point is either zero or undefined (which usually indicates a maximum or a minimum).
$(2, 1)$ a relative minimum
$(4, 7)$ a relative maximum

100. The crtical values occur when the slope of the tangent line at the point is either zero or undefined (which usually indicates a maximum or a minimum).
$(1, 1)$ a relative minimum
$(5, 4)$ a relative maximum
$(6, 2)$ a relative minimum

101. Relative minimum at $(-5, 425)$ and $(4, -304)$ and a relative maximum at $(-2, 560)$.

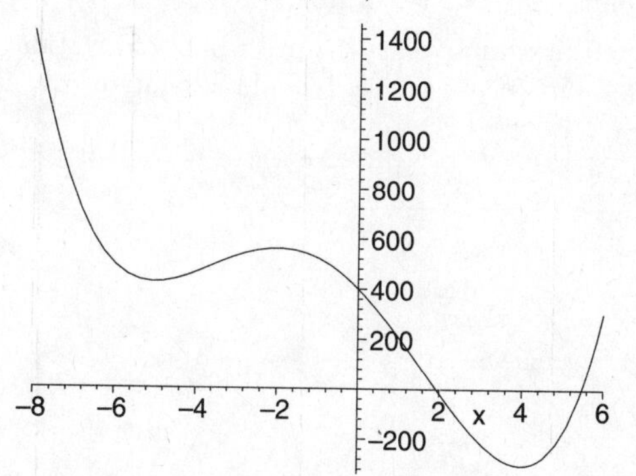

102. Relative maximum at $(-6.262, 3213.80)$, $(-0.559, 1440.06)$, and $(5.054, 6674.12)$ and a relative minimum at $(-3.683, -2288.03)$ and $(2.116, -1083.08)$.

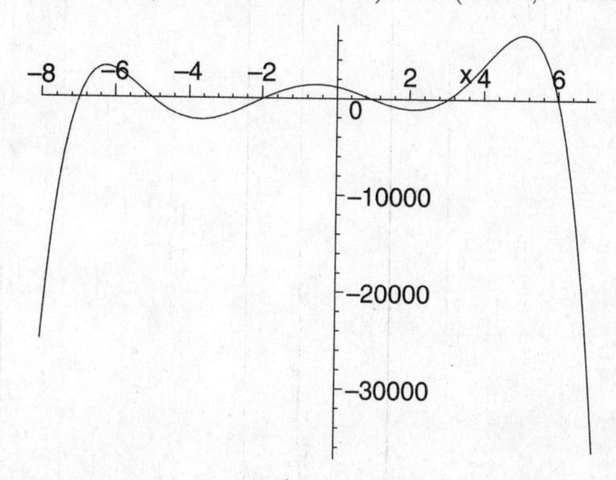

Exercise Set 3.2

1. $f(x) = 2 - x^2$

a) Find $f'(x)$ and $f''(x)$.

$f'(x) = -2x$

$f''(x) = -2$

b) Find the critical points.

Since $f'(x)$ exists for all values of x, the only critical points are where $-2x = 0$.

$-2x = 0$

$x = 0$

$f(0) = 2 - 0^2 = 2$

This gives the point $(0, 2)$ on the graph.

c) Use the Second-Derivative Test:

$f''(x) = -2$

$f''(0) = -2 < 0$

This tells us that $(0, 2)$ is a relative maximum. Then we can deduce that $f(x)$ is increasing on $(-\infty, 0)$ and decreasing on $(0, \infty)$.

The second derivative, $f''(x)$, exists and is -2 for all real numbers. Note that $f''(x)$ is never 0. Thus, there are no possible inflection points.

Since $f''(x)$ is always negative ($f''(x) = -2$), f is concave down on the interval $(-\infty, \infty)$.

d) Sketch the graph using the preceding information. By solving $2 - x^2 = 0$ we can easily find the x-intercepts. They are $(-\sqrt{2}, 0)$ and $(\sqrt{2}, 0)$.

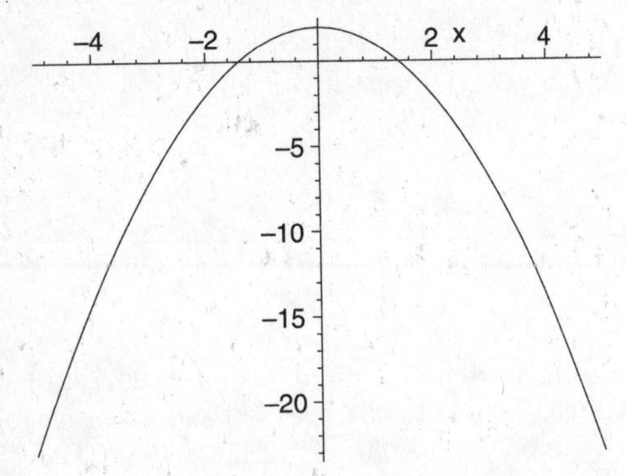

2. $f(x) = x^2 + x - 1$

a) Find $f'(x)$ and $f''(x)$.

$f'(x) = 2x + 1$

$f''(x) = 2$

b) Find the critical points.

Since $f'(x) = 2x + 1$ exists for all values of x, the only critical points are where $2x + 1 = 0$.

$2x + 1 = 0$

$2x = -1$

$x = -\frac{1}{2}$ Critical point

$$f\left(-\frac{1}{2}\right) = \left(-\frac{1}{2}\right)^2 + \left(-\frac{1}{2}\right) - 1 = \frac{1}{4} - \frac{2}{4} - \frac{4}{4} = -\frac{5}{4}$$

This gives the point $\left(-\frac{1}{2}, -\frac{5}{4}\right)$ on the graph.

$f''(x) = 2$

$f''\left(-\frac{1}{2}\right) = 2 > 0$

This tells us that $\left(-\frac{1}{2}, -\frac{5}{4}\right)$ is a relative minimum. Then we can deduce that $f(x)$ is decreasing on $\left(-\infty, -\frac{1}{2}\right)$ and increasing on $\left(-\frac{1}{2}, \infty\right)$.

c) Find the possible inflection points.

The second derivative, $f''(x)$, exists and is 2 for all real numbers. Note that $f''(x)$ is never 0. Thus, there are no possible inflection points.

Note that $f''(x)$ is always positive, $f''(x) = 2$. Thus, f is concave up on the interval $(-\infty, \infty)$.

d) Sketch the graph using the preceding information.

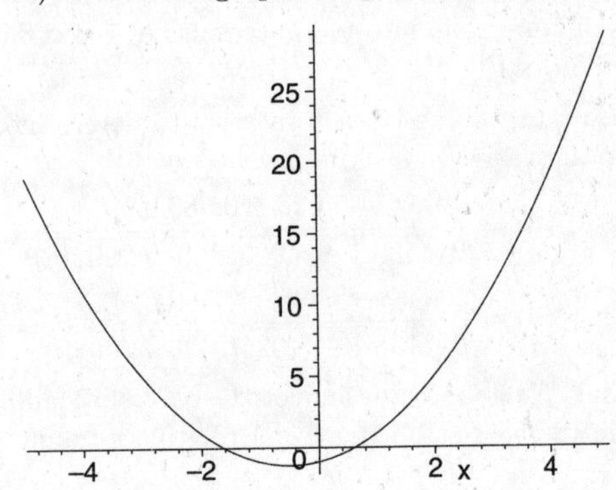

3. $f(x) = 2x^3 - 3x^2 - 36x + 28$

a) Find $f'(x)$ and $f''(x)$.

$f'(x) = 6x^2 - 6x - 36$

$f''(x) = 12x - 6$

b) Find the critical points of f.

Since $f'(x)$ exists for all values of x, the only critical points are where $6x^2 - 6x - 36 = 0$.

$6x^2 - 6x - 36 = 0$

$6(x + 2)(x - 3) = 0$

$x + 2 = 0$ *or* $x - 3 = 0$

$x = -2$ *or* $x = 3$ Critical points

Then $f(-2) = 2(-2)^3 - 3(-2)^2 - 36(-2) + 28$
$= -16 - 12 + 72 + 28$
$= 72,$

and $f(3) = 2 \cdot 3^3 - 3 \cdot 3^2 - 36 \cdot 3 + 28$
$= 54 - 27 - 108 + 28$
$= -53.$

These give the points $(-2, 72)$ and $(3, -53)$ on the graph.

c) Use the Second-Derivative Test:

$f''(-2) = 12(-2) - 6 = -30 < 0$, so $(-2, 72)$ is a relative maximum.

$f''(3) = 12 \cdot 3 - 6 = 30 > 0$, so $(3, -53)$ is a relative minimum.

Then if we use the points -2 and 3 to divide the real number line into three intervals, $(-\infty, -2)$, $(-2, 3)$, and $(3, \infty)$, we know that f is increasing on $(-\infty, -2)$, decreasing on $(-2, 3)$, and increasing again on $(3, \infty)$.

d) Find the possible inflection points.

$f''(x)$ exists for all valus of x, so we solve $f''(x) = 0$.

$12x - 6 = 0$

$12x = 6$

$x = \frac{1}{2}$ Possible inflection point

Then $f\left(\frac{1}{2}\right) = 2\left(\frac{1}{2}\right)^3 - 3\left(\frac{1}{2}\right)^2 - 36\left(\frac{1}{2}\right) + 28$

$= \frac{1}{4} - \frac{3}{4} - 18 + 28$

$= \frac{19}{2}.$

This gives the point $\left(\frac{1}{2}, \frac{19}{2}\right)$ on the graph.

e) To determine the concavity we use the possible inflection point, $\frac{1}{2}$, to divide the real number line into two invervals, A: $\left(-\infty, \frac{1}{2}\right)$ and B: $\left(\frac{1}{2}, \infty\right)$. Test a point in each interval.

A: Test 0, $f''(0) = 12 \cdot 0 - 6 = -6 < 0$

B: Test 1, $f''(1) = 12 \cdot 1 - 6 = 6 > 0$

Then f is concave down on $\left(-\infty, \frac{1}{2}\right)$ and concave up on $\left(\frac{1}{2}, \infty\right)$, so $\left(\frac{1}{2}, \frac{19}{2}\right)$ is an inflection point.

f) Sketch the graph using the preceding information.

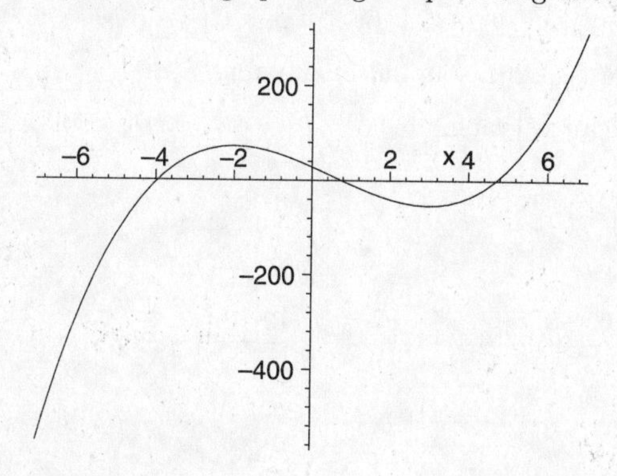

4. $f(x) = 3x^3 - 36x - 3$

a) $f'(x) = 9x^2 - 36$

$f''(x) = 18x$

b) $f'(x)$ exists for all real numbers. Solve:

$9x^2 - 36 = 0$

$9(x+2)(x-2) = 0$

$x = -2$ *or* $x = 2$ Critical points

$f(-2) = 3(-2)^3 - 36(-2) - 3 = 45$ and

$f(2) = 3 \cdot 2^3 - 36 \cdot 2 - 3 = -51$, so $(-2, 45)$ and $(2, -51)$ are on the graph.

c) $f''(-2) = 18(-2) = -36$, so $f(-2, 45)$ is a relative maximum.

$f''(2) = 18 \cdot 2 = 36$, so $(2, -51)$ is a relative minimum.

Then $f(x)$ is increasing on $(-\infty, -2)$, decreasing on $(-2, 2)$, and increasing again on $(2, \infty)$.

d) $f''(x)$ exists for all real numbers. Solve:

$18x = 0$

$x = 0$ Possible inflection point

$f(0) = 3 \cdot 0^3 - 36 \cdot 0 - 3 = -3$, so $(0, -3)$ is another point on the graph.

e) Use 0 to divide the real number line into two intervals, A: $(-\infty, 0)$ and B: $(0, \infty)$.

A: Test -1, $f''(-1) = 18(-1) = -18 < 0$

B: Test 1, $f''(1) = 18 \cdot 1 = 18 > 0$

Then $(0, -3)$ is an inflection point.

f) Sketch the graph.

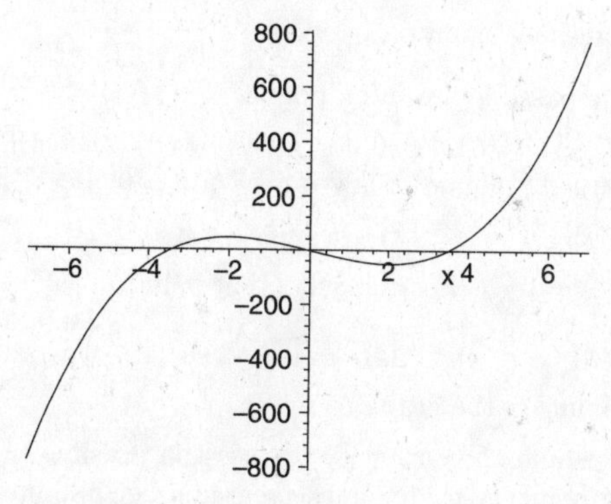

5. $f(x) = \frac{8}{3}x^3 - 2x + \frac{1}{3}$

a) Find $f'(x)$ and $f''(x)$.

$f'(x) = 8x^2 - 2$

$f''(x) = 16x$

b) Find the critical points of f.

Now $f'(x) = 8x^2 - 2$ exists for all values of x, so the only critical points of f are where $8x^2 - 2 = 0$.

$8x^2 - 2 = 0$

$8x^2 = 2$

$x^2 = \frac{2}{8}$

$x^2 = \frac{1}{4}$

$x = \pm\frac{1}{2}$ Critical points

Then $f\left(-\frac{1}{2}\right) = \frac{8}{3}\left(-\frac{1}{2}\right)^3 - 2\left(-\frac{1}{2}\right) + \frac{1}{3}$

$= -\frac{1}{3} + 1 + \frac{1}{3}$

$= 1,$

and $f\left(\frac{1}{2}\right) = \frac{8}{3}\left(\frac{1}{2}\right)^3 - 2\left(\frac{1}{2}\right) + \frac{1}{3}$

$= \frac{1}{3} - 1 + \frac{1}{3}$

$= -\frac{1}{3}$

These give the points $\left(-\frac{1}{2}, 1\right)$ and $\left(\frac{1}{2}, -\frac{1}{3}\right)$ on the graph.

c) Use the Second-Derivative Test:

$f''\left(-\frac{1}{2}\right) = 16\left(-\frac{1}{2}\right) - 8 < 0$, so $\left(-\frac{1}{2}, 1\right)$ is a relative maximum.

$f''\left(\frac{1}{2}\right) = 16 \cdot \frac{1}{2} = 8 > 0$, so $\left(\frac{1}{2}, -\frac{1}{3}\right)$ is a relative minimum.

Then if we use the points $-\frac{1}{2}$ and $\frac{1}{2}$ to divide the real number line into three intervals, A: $\left(-\infty, -\frac{1}{2}\right)$, B: $\left(-\frac{1}{2}, \frac{1}{2}\right)$, and C: $\left(\frac{1}{2}, \infty\right)$, we know that f is increasing on $\left(-\infty, -\frac{1}{2}\right)$, decreasing on $\left(-\frac{1}{2}, \frac{1}{2}\right)$, and increasing again on $\left(\frac{1}{2}, \infty\right)$.

d) Find the possible inflection points.

Now $f''(x) = 16x$ exists for all values of x, so the only critical points of f' are where $16x = 0$.

$16x = 0$

$x = 0$ Possible inflection point

Then $f(0) = \frac{8}{3} \cdot 0^3 - 2 \cdot 0 + \frac{1}{3} = \frac{1}{3}$, so $\left(0, \frac{1}{3}\right)$ is another point on the graph.

e) To determine the concavity we use the possible inflection point, 0, to divide the real number line into two invervals, A: $(-\infty, 0)$ and B: $(0, \infty)$. Test a point in each interval.

A: Test -1, $f''(-1) = 16(-1) = -16 < 0$

B: Test 1, $f''(1) = 16 \cdot 1 = 16 > 0$

Then $f'(x)$ is concave down on $(-\infty, 0)$ and concave up on $(0, \infty)$, so $\left(0, \frac{1}{3}\right)$ is an inflection point.

f) Sketch the graph using the preceding information.

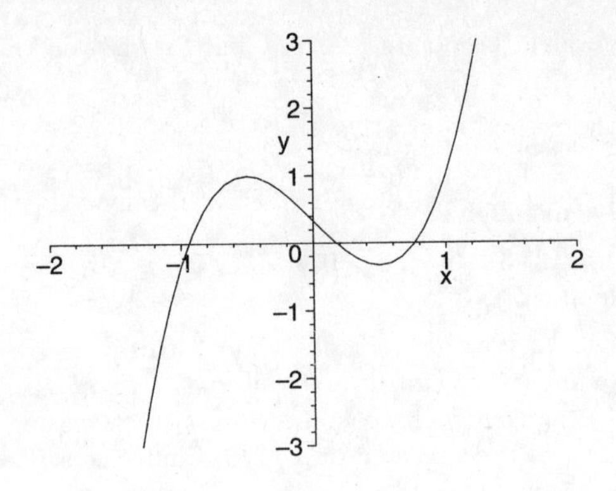

6. $f(x) = 80 - 9x^2 - x^3$

a) $f'(x) = -18x - 3x^2$

$f''(x) = -18 - 6x$

b) $f'(x)$ exists for all real numbers. Solve:

$-18x - 3x^2 = 0$

$-3x(6 + x) = 0$

$x = 0$ *or* $x = -6$ Critical points

$f(0) = 80 - 9 \cdot 0^2 - 0^3 = 80$ and

$f(-6) = 80 - 9(-6)^2 - (-6)^3 = -28$, so $(0, 80)$ and $(-6, -28)$ are on the graph.

c) $f''(0) = -18 - 6 \cdot 0 = -18$, so $(0, 80)$ is a relative maximum.

$f''(-6) = -18 - 6(-6) = 18$, so $(-6, -28)$ is a relative minimum.

Then $f(x)$ is decreasing on $(-\infty, -6)$, increasing on $(-6, 0)$, and decreasing again on $(0, \infty)$.

d) $f''(x)$ exists for all real numbers. Solve:

$-18 - 6x = 0$

$x = -3$ Possible inflection point

$f(-3) = 80 - 9(-3)^2 - (-3)^3 = 26$, so $(-3, 26)$ is another point on the graph.

e) Use -3 to divide the real number line into two intervals, A: $(-\infty, -3)$ and B: $(-3, \infty)$.

A: Test -4, $f''(-4) = -18 - 6(-4) = 6 > 0$

B: Test 0, $f''(0) = -18 - 6 \cdot 0 = -18 < 0$

Then $(-3, 26)$ is an inflection point.

f) Sketch the graph.

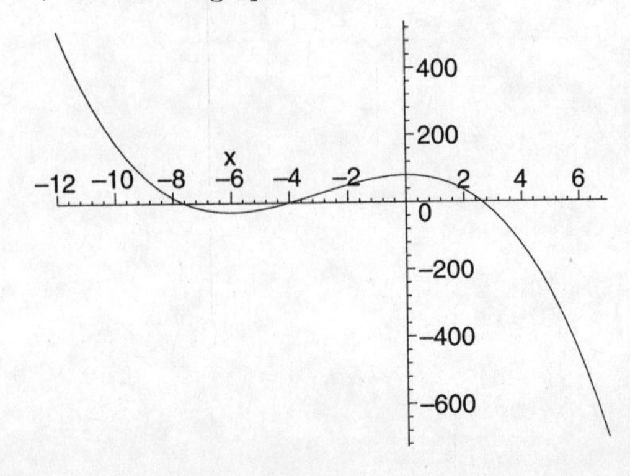

7. $f(x) = 3x^4 - 16x^3 + 18x^2$

a) $f'(x) = 12x^3 - 48x^2 + 36x$

$f''(x) = 36x^2 - 96x + 36$

b) Since $f'(x)$ exists for all values of x, the only critical points are where $f'(x) = 0$.

$12x^3 - 48x^2 + 36x = 0$

$12x(x^2 - 4x + 3) = 0$

$12x(x-1)(x-3) = 0$

$12x = 0$ *or* $x - 1 = 0$ *or* $x - 3 = 0$

$x = 0$ *or* $x = 1$ *or* $x = 3$

Then $f(0) = 3 \cdot 0^4 - 16 \cdot 0^3 + 18 \cdot 0^2 = 0$,

$f(1) = 3 \cdot 1^4 - 16 \cdot 1^3 + 18 \cdot 1^2 = 5$,

and $f(3) = 3 \cdot 3^4 - 16 \cdot 3^3 + 18 \cdot 3^2 = -27$.

These give the points $(0,0)$, $(1,5)$, and $(3,-27)$ on the graph.

c) Use the Second-Derivative Test:

$f''(0) = 36 \cdot 0^2 - 96 \cdot 0 + 36 = 36 > 0$, so $(0,0)$ is a relative minimum.

$f''(1) = 36 \cdot 1^2 - 96 \cdot 1 + 36 = -24 < 0$, so $(1,5)$ is a relative maximum.

$f''(3) = 36 \cdot 3^2 - 96 \cdot 3 + 36 = 72 > 0$, so $(3,-27)$ is a relative minimum.

Then if we use the points 0, 1, and 3 to divide the real number line into four intervals, $(-\infty,0)$, $(0,1)$, $(1,3)$, and $(3,\infty)$, we know that f is decreasing on $(-\infty,0)$ and on $(1,3)$ and is increasing on $(0,1)$ and $(3,\infty)$.

d) $f''(x)$ exists for all values of x, so the only possible inflection points are where $f''(x) = 0$.

$36x^2 - 96x + 36 = 0$

$12(3x^2 - 8x + 3) = 0$

$3x^2 - 8x + 3 = 0$

Using the quadratic formula, we find $x = \dfrac{4 \pm \sqrt{7}}{3}$, so $x \approx 0.45$ or $x \approx 2.22$ are possible inflection points.

Then $f(0.45) \approx 2.31$ and $f(2.22) \approx -13.48$, so $(0.45, 2.31)$ and $(2.22, -13.48)$ are two more points on the graph.

e) To determine the concavity we use the points 0.45 and 2.22 to divide the real number line into three invervals, A: $(-\infty, 0.45)$, B: $(0.45, 2.22)$, and C: $(2.22, \infty)$. Test a point in each interval.

A: Test 0, $f''(0) = 36 \cdot 0^2 - 96 \cdot 0 + 36 = 36 > 0$

B: Test 1, $f''(1) = 36 \cdot 1^2 - 96 \cdot 1 + 36 =$

$-24 < 0$

C: Test 3, $f''(3) = 36 \cdot 3^2 - 96 \cdot 3 + 36 = 72 > 0$

Then f is concave up on $(-\infty, 0.45)$, concave down on $(0.45, 2.22)$, and concave up on $(2.22, \infty)$, so $(0.45, 2.31)$ and $(2.22, -13.48)$ are inflection points.

f) Sketch the graph.

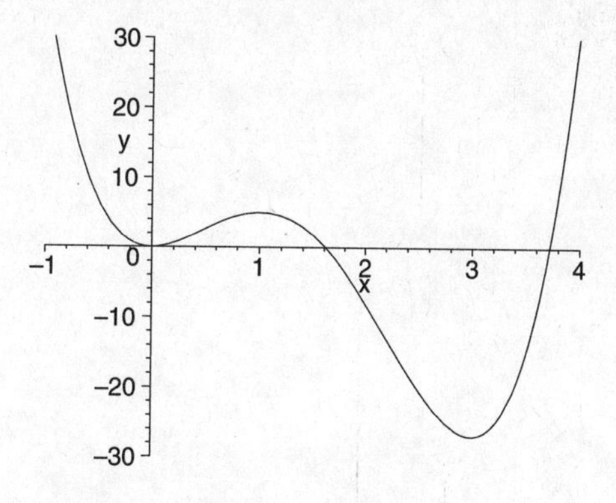

8. $f(x) = 3x^4 + 4x^3 - 12x^2 + 5$

a) $f'(x) = 12x^3 + 12x^2 - 24x$

$f''(x) = 36x^2 + 24x - 24$

b) $f'(x)$ exists for all real numbers. Solve:

$12x^3 + 12x^2 - 24x = 0$

$12x(x^2 + x - 2) = 0$

$12x(x+2)(x-1) = 0$

$x = 0$ *or* $x = -2$ *or* $x = 1$ Critical points

$f(-2) = -27$, $f(0) = 5$, and $f(1) = 0$, so $(-2,-27)$, $(0,5)$, and $(1,0)$ are on the graph.

c) $f''(-2) = 72$, so $(-2,-27)$ is a relative minimum.

$f''(0) = -24$, so $(0,5)$ is a relative maximum.

$f''(1) = 36$, so $(1,0)$ is a relative minimum.

Then $f(x)$ is decreasing on $(-\infty,-2)$ and on $(0,1)$ and is increasing on $(-2,0)$ and on $(1,\infty)$.

d) $f''(x)$ exists for all real numbers. Solve:

$36^2 + 24x - 24 = 0$

$12(3x^2 + 2x - 2) = 0$

$x = \dfrac{-1 \pm \sqrt{7}}{3}$, so $x \approx -1.22$ or $x \approx 0.55$ are possible inflection points.

$f(-1.22) \approx -13.48$ and $f(0.55) \approx 2.31$, so $(-1.22, -13.48)$ and $(0.55, 2.31)$ are two more points on the graph.

e) Use -1.22 and 0.55 to divide the real number line into three intervals, A: $(-\infty, -1.22)$, B: $(-1.22, 0.55)$, and C: $(0.55, \infty)$.

A: Test -2, $f''(-2) = 72 > 0$

B: Test 0, $f''(0) = -24 < 0$

C: Test 1, $f''(1) = 36 > 0$

Then $(-1.22, -13.48)$ and $(0.55, 2.31)$ are both inflection points.

f) Sketch the graph.

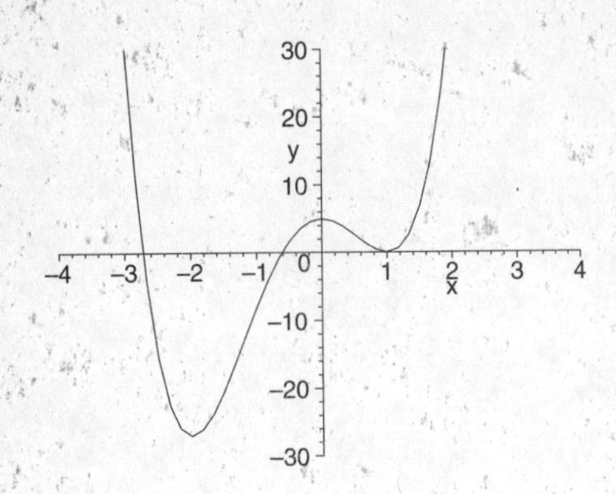

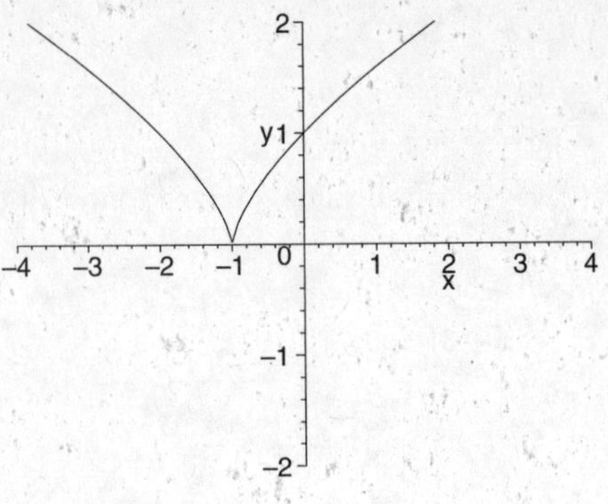

9. $f(x) = (x+1)^{2/3}$

a) $f'(x) = \dfrac{2}{3}(x+1)^{-1/3} = \dfrac{2}{3\sqrt[3]{x+1}}$

$f''(x) = -\dfrac{2}{9}(x+1)^{-4/3} = -\dfrac{2}{9\sqrt[3]{(x+1)^4}}$

b) Since $f'(-1)$ does not exist, -1 is a critical point. The equation $f'(x) = 0$ has no solution, so the only critical point is -1.

Now $f(-1) = (-1+1)^{2/3} = 0^{2/3} = 0$.

This gives the point $(-1, 0)$ on the graph.

c) We cannot use the Second-Derivative Test, because $f''(-1)$ is not defined. We will use the First-Derivative Test. Use -1 to divide the real number line into two intervals, A: $(-\infty, -1)$ and B: $(-1, \infty)$. Test a point in each interval.

A: Test -2, $f'(-2) = \dfrac{2}{3\sqrt[3]{-2+1}} = -\dfrac{2}{3} < 0$

B: Test 0, $f'(0) = \dfrac{2}{3\sqrt[3]{0+1}} = \dfrac{2}{3} > 0$

Since $f(x)$ is decreasing on $(-\infty, -1)$ and increasing on $(-1, \infty)$, there is a relative minimum at $(-1, 0)$.

d) Since $f''(-1)$ does not exist, -1 is a possible inflection point. The equation $f''(x) = 0$ has no solution, so the only possible inflection point is -1. We have already found $f(-1)$ in step (b).

e) To determine the concavity we use -1 to divide the real number line into two invervals as in step (c). Test a point in each interval.

A: Test -2, $f''(-2) = -\dfrac{2}{9\sqrt[3]{(-2+1)^4}} = -\dfrac{2}{9} < 0$

B: Test 0, $f''(0) = -\dfrac{2}{9\sqrt[3]{(0+1)^4}} = -\dfrac{2}{9} < 0$

Then f is concave down on both intervals, so there is no inflection point.

f) Sketch the graph using the preceding information.

10. $f(x) = (x-1)^{2/3}$

a) $f'(x) = \dfrac{2}{3}(x-1)^{-1/3} = \dfrac{2}{3\sqrt[3]{x-1}}$

$f''(x) = -\dfrac{2}{9}(x-1)^{-4/3} = -\dfrac{2}{9\sqrt[3]{(x-1)^4}}$

b) $f'(x)$ does not exist for $x = 1$. The equation $f'(x) = 0$ has no solution, so 1 is the only critical point.

$f(1) = (1-1)^{2/3} = 0$, so $(1, 0)$ is on the graph.

c) We use the First-Derivative Test, since $f''(1)$ is not defined. Use 1 to divide the real number line into two intervals, A: $(-\infty, 1)$ and B: $(1, \infty)$.

A: Test 0, $f'(0) = \dfrac{2}{3\sqrt[3]{0-1}} = -\dfrac{2}{3} < 0$

B: Test 2, $f'(2) = \dfrac{2}{3\sqrt[3]{2-1}} = \dfrac{2}{3} > 0$

$f(x)$ is decreasing on $(-\infty, 1)$ and increasing on $(1, \infty)$ and $(1, 0)$ is a relative minimum.

d) $f''(x)$ does not exist for $x = 1$. The equation $f''(x) = 0$ has no solution, so the only possible inflection point is 1. We found $f(1)$ in step (b).

e) Use 1 to divide the real number line as in step (c).

A: Test 0, $f''(0) = -\dfrac{2}{9\sqrt[3]{(0-1)^4}} = -\dfrac{2}{9} < 0$

B: Test 2, $f''(2) = -\dfrac{2}{9\sqrt[3]{(2-1)^4}} = -\dfrac{2}{9} < 0$

$f(x)$ is concave down on both intervals, so there is no inflection point.

f) Sketch the graph.

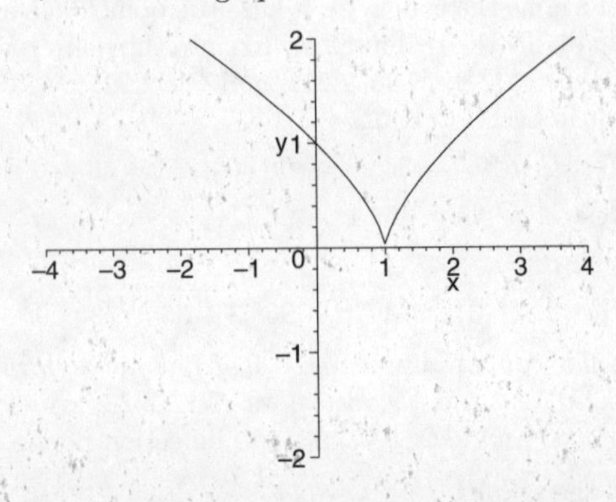

11. $f(x) = x^4 - 6x^2$

a) $f'(x) = 4x^3 - 12x$

$f''(x) = 12x^2 - 12$

b) Since $f'(x)$ exists for all values of x, the only critical points are where $4x^3 - 12x = 0$.

$$4x^3 - 12x = 0$$
$$4x(x^2 - 3) = 0$$
$$4x = 0 \quad or \quad x^2 - 3 = 0$$
$$x = 0 \quad or \quad x^2 = 3$$
$$x = 0 \quad or \quad x = \pm\sqrt{3}$$

The critical points are $-\sqrt{3}$, 0, and $\sqrt{3}$.

$$f(-\sqrt{3}) = (-\sqrt{3})^4 - 6(-\sqrt{3})^2$$
$$= 9 - 6 \cdot 3 = 9 - 18 = -9$$
$$f(0) = 0^4 - 6 \cdot 0^2 = 0 - 0 = 0$$
$$f(\sqrt{3}) = (\sqrt{3})^4 - 6(\sqrt{3})^2$$
$$= 9 - 6 \cdot 3 = 9 - 18 = -9$$

These give the points $(-\sqrt{3}, -9)$, $(0, 0)$, and $(\sqrt{3}, -9)$ on the graph.

c) Use the Second-Derivative Test:

$f''(-\sqrt{3}) = 12(-\sqrt{3})^2 - 12 = 12 \cdot 3 - 12 =$ $24 > 0$, so $(-\sqrt{3}, -9)$ is a relative minimum.

$f''(0) = 12 \cdot 0^2 - 12 = -12 < 0$, so $(0, 0)$ is a relative maximum.

$f''(\sqrt{3}) = 12(\sqrt{3})^2 - 12 = 12 \cdot 3 - 12 = 24 > 0$, so $(\sqrt{3}, -9)$ is a relative minimum.

Then if we use the points $-\sqrt{3}$, 0, and $\sqrt{3}$ to divide the real number line into four intervals, $(-\infty, -\sqrt{3})$, $(-\sqrt{3}, 0)$, $(0, \sqrt{3})$, and $(\sqrt{3}, \infty)$, we know that f is decreasing on $(-\infty, -\sqrt{3})$ and on $(0, \sqrt{3})$ and is increasing on $(-\sqrt{3}, 0)$ and on $(\sqrt{3}, \infty)$.

d) Since $f''(x)$ exists for all values of x, the only possible inflection points are where $12x^2 - 12 = 0$.

$$12x^2 - 12 = 0$$
$$x^2 - 1 = 0$$
$$(x + 1)(x - 1) = 0$$
$$x + 1 = 0 \quad \text{or} \quad x - 1 = 0$$
$$x = -1 \quad \text{or} \quad x = 1$$

The possible inflection points are -1 and 1.

$$f(-1) = (-1)^4 - 6(-1)^2 = 1 - 6 \cdot 1 = -5$$
$$f(1) = 1^4 - 6 \cdot 1^2 = 1 - 6 \cdot 1 = -5$$

These give the points $(-1, -5)$ and $(1, -5)$ on the graph.

e) To determine the concavity we use the points -1 and 1 to divide the real number line into three invervals, A: $(-\infty, -1)$, B: $(-1, 1)$, and $(1, \infty)$. Test a point in each interval.

A: Test -2, $f''(-2) = 12(-2)^2 - 12 = 36 > 0$

B: Test 0, $f''(0) = 12 \cdot 0^2 - 12 = -12 < 0$

C: Test 2, $f''(2) = 12 \cdot 2^2 - 12 = 36 > 0$

We see that f is concave up on the intervals $(-\infty, -1)$ and $(1, \infty)$ and concave down on the interval $(-1, 1)$, so $(-1, -5)$ and $(1, -5)$ are inflection points.

f) Sketch the graph using the preceding information. By solving $x^4 - 6x^2 = 0$ we can find the x-intercepts. They are helpful in graphing.

$$x^4 - 6x^2 = 0$$
$$x^2(x^2 - 6) = 0$$
$$x^2 = 0 \quad \text{or} \quad x^2 - 6 = 0$$
$$x = 0 \quad \text{or} \quad x^2 = 6$$
$$x = 0 \quad \text{or} \quad x = \pm\sqrt{6}$$

The x-intercepts are $(0, 0)$, $(-\sqrt{6}, 0)$, and $(\sqrt{6}, 0)$.

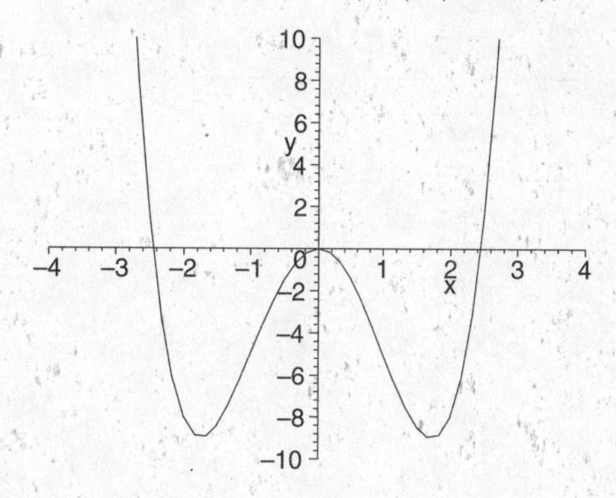

12. $f(x) = 2x^2 - x^4$

a) $f'(x) = 4x - 4x^3$

$f''(x) = 4 - 12x^2$

b) $f'(x)$ exists for all real numbers. Solve:

$$4x - 4x^3 = 0$$
$$4x(1 - x^2) = 0$$
$$x = 0 \ or \ x = \pm 1 \quad \text{Critical points}$$

$f(-1) = 1$, $f(0) = 0$, and $f(1) = 1$, so $(-1, 1)$, $(0, 0)$, and $(1, 1)$ are on the graph.

c) $f''(-1) = -8$, so $(-1, 1)$ is a relative maximum.

$f''(0) = 4$, so $(0, 0)$ is a relative minimum.

$f''(1) = -8$, so $(1, 1)$ is a relative maximum.

Then $f(x)$ is increasing on $(-\infty, -1)$ and on $(0, 1)$ and is decreasing on $(-1, 0)$ and on $(1, \infty)$.

d) $f''(x)$ exists for all real numbers. Solve:

$$4 - 12x^2 = 0$$
$$x^2 = \frac{1}{3}$$
$$x = \pm\frac{1}{\sqrt{3}} \quad \text{Possible inflection points}$$

$f\left(-\frac{1}{\sqrt{3}}\right) = \frac{5}{9}$ and $f\left(\frac{1}{\sqrt{3}}\right) = \frac{5}{9}$, so

$\left(-\frac{1}{\sqrt{3}}, \frac{5}{9}\right)$ and $\left(\frac{1}{\sqrt{3}}, \frac{5}{9}\right)$ are on the graph.

e) Use $-\frac{1}{\sqrt{3}}$ and $\frac{1}{\sqrt{3}}$ to divide the real number line into three intervals, A: $\left(-\infty, -\frac{1}{\sqrt{3}}\right)$, B: $\left(-\frac{1}{\sqrt{3}}, \frac{1}{\sqrt{3}}\right)$, and C: $\left(-\frac{1}{\sqrt{3}}, \infty\right)$.

A: Test -1, $f''(-1) = -8 < 0$

B: Test 0, $f''(0) = 4 > 0$

C: Test 1, $f''(1) = -8 < 0$

Then $\left(-\frac{1}{\sqrt{3}}, \frac{5}{9}\right)$ and $\left(\frac{1}{\sqrt{3}}, \frac{5}{9}\right)$ are both inflection points.

f) Sketch the graph.

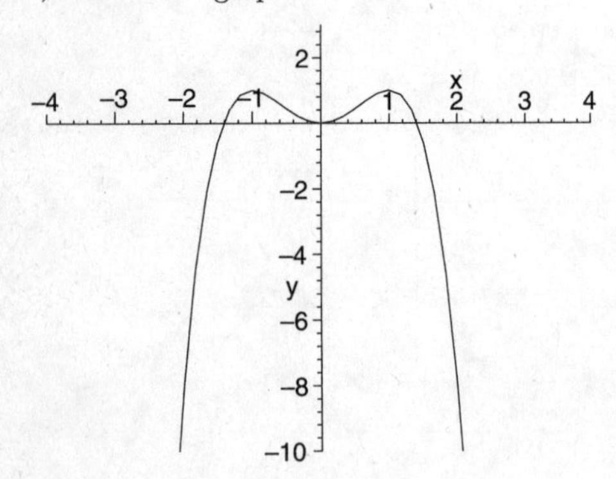

13. $f(x) = 3x^4 + 4x^3$

a) $f'(x) = 12x^3 + 12x^2$

$f''(x) = 36x^2 + 24x$

b) Since $f'(x)$ exists for all values of x, the only critical points of f are where $12x^3 + 12x^2 = 0$.

$12x^3 + 12x^2 = 0$

$12x^2(x+1) = 0$

$12x^2 = 0$ *or* $x + 1 = 0$

$x = 0$ *or* $x = -1$

The critical points are 0 and -1.

$f(0) = 3 \cdot 0^4 + 4 \cdot 0^3 = 0 + 0 = 0$

$f(-1) = 3(-1)^4 + 4(-1)^3 = 3 \cdot 1 + 4(-1)$
$= 3 - 4$
$= -1$

These give the points $(0, 0)$ and $(-1, -1)$ on the graph.

c) Use the Second-Derivative Test:

$f''(-1) = 36(-1)^2 + 24(-1) = 36 - 24 = 12 > 0$, so $(-1, -1)$ is a relative minimum.

$f''(0) = 36 \cdot 0^2 + 24 \cdot 0 = 0$, so this test fails. We will use the First-Derivative Test. Use 0 to divide the interval $(-1, \infty)$ into two intervals: A: $(-1, 0)$ and B: $(0, \infty)$. Test a point in each interval.

A: Test $-\frac{1}{2}$, $f'\left(-\frac{1}{2}\right) = 12\left(-\frac{1}{12}\right)^3 + 12\left(\frac{1}{2}\right)^2 = \frac{3}{2} > 0$

B: Test 2, $f'(2) = 12 \cdot 2^3 + 12 \cdot 2^2 = 144 > 0$

Since f is increasing on both intervals, $(0, 0)$ is not a relative extremum. Since $(-1, -1)$ is a relative minimum, we know that f is decreasing on $(-\infty, -1)$.

d) Now $f''(x)$ exists for all values of x, so the only possible inflection points are where $36x^2 + 24x = 0$.

$36x^2 + 24x = 0$

$12x(3x + 2) = 0$

$12x = 0$ or $3x + 2 = 0$

$x = 0$ or $x = -\frac{2}{3}$

The possible inflection points are 0 and $-\frac{2}{3}$.

$f(0) = 3 \cdot 0^4 + 4 \cdot 0^3 = 0$ Already found in step (b)

$f\left(-\frac{2}{3}\right) = 3\left(-\frac{2}{3}\right)^4 + 4\left(-\frac{2}{3}\right)^3$
$= 3 \cdot \frac{16}{81} + 4 \cdot \left(-\frac{8}{27}\right)$
$= \frac{16}{27} - \frac{32}{27}$
$= -\frac{16}{27}$

This gives one additional point $\left(-\frac{2}{3}, -\frac{16}{27}\right)$ on the graph.

e) To determine the concavity we use $-\frac{2}{3}$ and 0 to divide the real number line into three invervals, A: $\left(-\infty, -\frac{2}{3}\right)$, B: $\left(-\frac{2}{3}, 0\right)$, and C: $(0, \infty)$. Test a point in each interval.

A: Test -1, $f''(-1) = 36(-1)^2 + 24(-1) = 12 > 0$

B: Test $-\frac{1}{2}$, $f''\left(-\frac{1}{2}\right) = 36\left(-\frac{1}{2}\right)^2 + 24\left(-\frac{1}{2}\right) = -3 < 0$

C: Test 1, $f''(1) = 36 \cdot 1^2 + 24 \cdot 1 = 60 > 0$

We see that f is concave up on the intervals $\left(-\infty, -\frac{2}{3}\right)$ and $(0, \infty)$ and concave down on the interval $\left(-\frac{2}{3}, 0\right)$, so $\left(-\frac{2}{3}, -\frac{16}{27}\right)$ and $(0, 0)$ are both inflection points.

f) Sketch the graph using the preceding information. By solving $3x^4 + 4x^3 = 0$ we can find x-intercepts. They are helpful in graphing.

$3x^4 + 4x^3 = 0$

$x^3(3x + 4) = 0$

$x^3 = 0$ *or* $3x + 4 = 0$

$x = 0$ *or* $x = -\frac{4}{3}$

The intercepts are $(0, 0)$ and $\left(-\frac{4}{3}, 0\right)$.

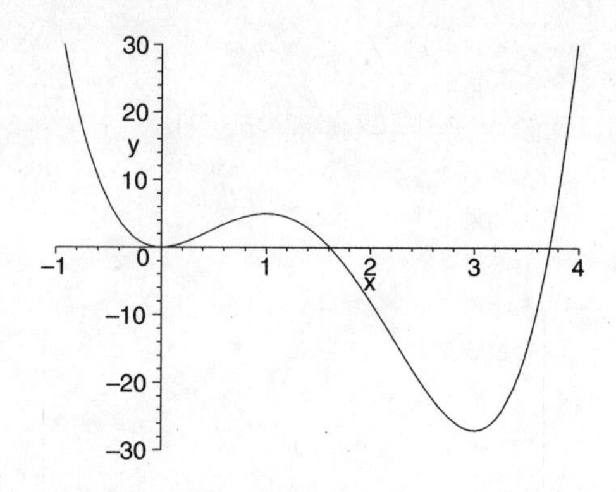

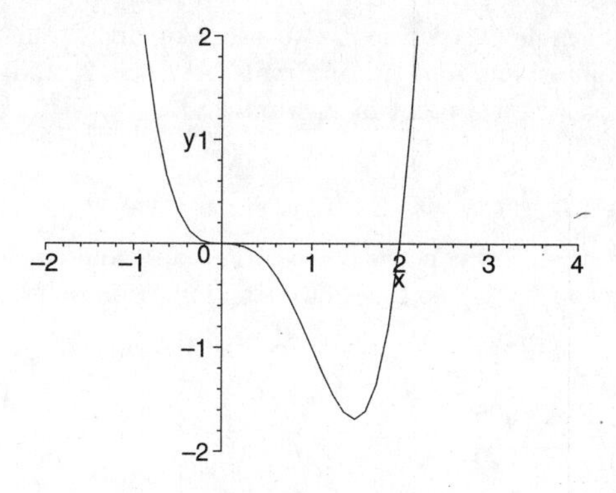

14. $f(x) = x^4 - 2x^3$

a) $f'(x) = 4x^3 - 6x^2$

$f''(x) = 12x^2 - 12x$

b) $f'(x)$ exists for all real numbers. Solve:

$$4x^3 - 6x^2 = 0$$
$$2x^2(2x - 3) = 0$$
$$x = 0 \text{ or } x = \frac{3}{2} \quad \text{Critical points}$$

$f(0) = 0$ and $f\left(\frac{3}{2}\right) = -\frac{27}{16}$, so $(0,0)$ and $\left(\frac{3}{2}, -\frac{27}{16}\right)$ are on the graph.

c) $f''(0) = 0$, so the Second-Derivative Test fails.

$f''\left(\frac{3}{2}\right) = 9 > 0$, so $\left(\frac{3}{2}, -\frac{27}{16}\right)$ is a relative minimum.

Use the First-Derivative Test for $x = 0$. Divide $\left(-\infty, \frac{3}{2}\right)$ into two intervals, A: $(-\infty, 0)$ and B: $\left(0, \frac{3}{2}\right)$.

A: Test -1, $f'(-1) = -10 < 0$

B: Test 1, $f'(1) = -2 < 0$

Since f is decreasing on both intervals, $(0,0)$ is not a relative extremum. We also know that f is increasing on $\left(\frac{3}{2}, \infty\right)$, since $\left(\frac{3}{2}, -\frac{27}{16}\right)$ is a relative minimum.

d) $f''(x)$ exists for all real numbers. Solve:

$$12x^2 - 12x = 0$$
$$12x(x - 1) = 0$$
$$x = 0 \text{ or } x = 1 \quad \text{Possible inflection points}$$

From step (b) we know $f(0) = 0$. Now $f(1) = -1$, so $(1, -1)$ is another point on the graph.

e) Use 0 and 1 to divide the real number line into three intervals, A: $(-\infty, 0)$, B: $(0, 1)$, and C: $(1, \infty)$.

A: Test -1, $f''(-1) = 24 > 0$

B: Test $\frac{1}{2}$, $f''\left(\frac{1}{2}\right) = -3 < 0$

C: Test 2, $f''(2) = 24 > 0$

Then $(0,0)$ and $(1,-1)$ are both inflection points.

f) Sketch the graph.

15. $f(x) = x^3 - 6x^2 - 135x$

a) $f'(x) = 3x^2 - 12x - 135$

$f''(x) = 6x - 12$

b) Since $f'(x)$ exists for all values of x, the only critical points of f are where $3x^2 - 12x - 135 = 0$.

$$3x^2 - 12x - 135 = 0$$
$$x^2 - 4x - 45 = 0$$
$$(x - 9)(x + 5) = 0$$
$$x - 9 = 0 \text{ or } x + 5 = 0$$
$$x = 9 \text{ or } x = -5$$

The critical points are 9 and -5.

$$f(9) = 9^3 - 6 \cdot 9^2 - 135 \cdot 9$$
$$= 729 - 486 - 1215$$
$$= -972$$
$$f(-5) = (-5)^3 - 6(-5)^2 - 135(-5)$$
$$= -125 - 150 + 675$$
$$= 400$$

These give the points $(9, -972)$ and $(-5, 400)$ on the graph.

c) Use the Second-Derivative Test:

$f''(-5) = 6(-5) - 12 = -30 - 12 = -42 < 0$, so $(-5, 400)$ is a relative maximum.

$f''(9) = 6 \cdot 9 - 12 = 54 - 12 = 42 > 0$, so $(9, -972)$ is a relative minimum.

Then if we use the points -5 and 9 to divide the real number line into three intervals, $(-\infty, -5)$, $(-5, 9)$, and $(9, \infty)$, we know that f is increasing on $(-\infty, -5)$ and on $(9, \infty)$ and is decreasing on $(-5, 9)$.

d) Now $f''(x)$ exists for all values of x, so the only possible inflection points are where $6x - 12 = 0$.

$$6x - 12 = 0$$
$$6x = 12$$
$$x = 2 \quad \text{Possible inflection point}$$
$$f(2) = 2^3 - 6 \cdot 2^2 - 135 \cdot 2$$
$$= 8 - 24 - 270$$
$$= -286$$

This gives another point $(2, -286)$ on the graph.

e) To determine the concavity we use 2 to divide the real number line into two invervals, A: $(-\infty, 2)$ and B: $(2, \infty)$. Test a point in each interval.

A: Test 0, $f''(0) = 6 \cdot 0 - 12 = -12 < 0$

B: Test 3, $f''(3) = 6 \cdot 3 - 12 = 6 > 0$

We see that f is concave down on $(-\infty, 2)$ and concave up on $(2, \infty)$, so $(2, -286)$ is an inflection point.

f) Sketch the graph using the preceding information.

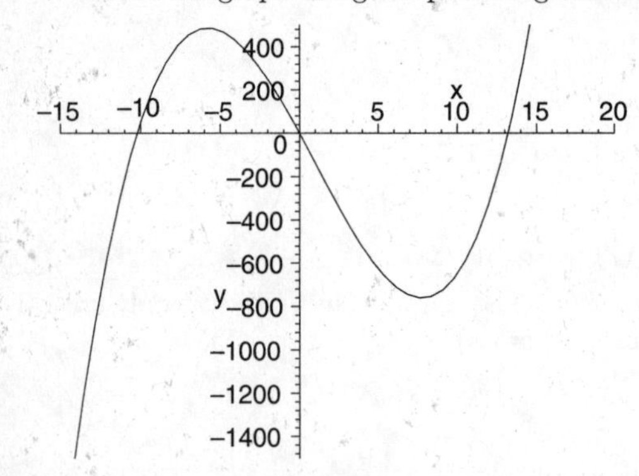

16. $f(x) = x^3 - 3x^2 - 144x - 140$

a) $f'(x) = 3x^2 - 6x - 144$

$f''(x) = 6x - 6$

b) $f'(x)$ exists for all real numbers. Solve:

$3x^2 - 6x - 144 = 0$

$3(x^2 - 2x - 48) = 0$

$3(x + 6)(x - 8) = 0$

$x = -6$ *or* $x = 8$ Critical points

$f(-6) = 400$, $f(8) = -972$, so $(-6, 400)$ and $(8, -972)$ are on the graph.

c) $f''(-6) = -42 < 0$, so $(-6, 400)$ is a relative maximum.

$f''(8) = 42 > 0$, so $(8, -972)$ is a relative minimum.

Then $f(x)$ is increasing on $(-\infty, -6)$ and on $(8, \infty)$ and is decreasing on $(-6, 8)$.

d) $f''(x)$ exists for all real numbers. Solve:

$6x - 6 = 0$

$x = 1$ Possible inflection point

$f(1) = -286$, so $(1, -286)$ is on the graph.

e) Use 1 to divide the real number line into two intervals, A: $(-\infty, 1)$, and B: $(1, \infty)$.

f) A: Test 0, $f''(0) = -6 < 0$

B: Test 2, $f''(2) = 6 > 0$

Then $(1, -286)$ is an inflection point.

Sketch the graph.

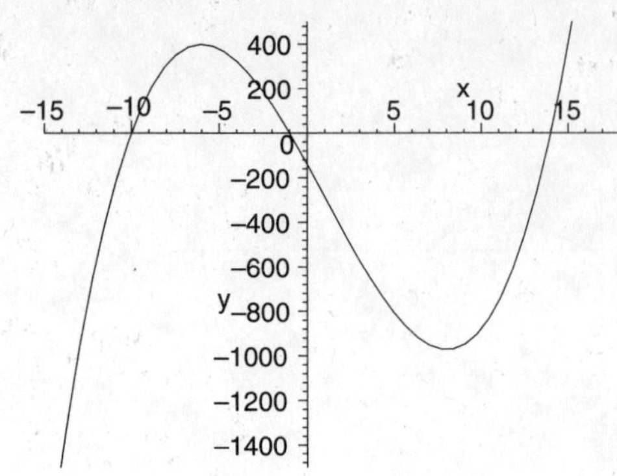

17. $f(x) = \dfrac{x}{x^2 + 1}$

a) $f'(x) = \dfrac{(x^2+1)(1) - 2x \cdot x}{(x^2+1)^2}$ Quotient Rule

$= \dfrac{x^2 + 1 - 2x^2}{(x^2+1)^2}$

$= \dfrac{1 - x^2}{(x^2+1)^2}$

$f''(x) = \dfrac{(x^2+1)^2(-2x) - 2(x^2+1)(2x)(1-x^2)}{[(x^2+1)^2]^2}$

Quotient Rule

$= \dfrac{(x^2+1)[(x^2+1)(-2x) - 4x(1-x^2)]}{(x^2+1)^4}$

$= \dfrac{-2x^3 - 2x - 4x + 4x^3}{(x^2+1)^3}$

$= \dfrac{2x^3 - 6x}{(x^2+1)^3}$

b) Since $f'(x)$ exists for all real numbers, the only critical points are where $f'(x) = 0$.

$\dfrac{1 - x^2}{(x^2+1)^2} = 0$

$1 - x^2 = 0$ Multiplying by $(x^2+1)^2$

$(1 + x)(1 - x) = 0$

$1 + x = 0$ *or* $1 - x = 0$

$x = -1$ *or* $1 = x$ Critical points

Then $f(-1) = \dfrac{-1}{(-1)^2 + 1} = -\dfrac{1}{2}$ and $f(1) = \dfrac{1}{1^2 + 1} = \dfrac{1}{2}$, so $\left(-1, -\dfrac{1}{2}\right)$ and $\left(1, \dfrac{1}{2}\right)$ are on the graph.

c) Use the Second-Derivative Test:

$f''(-1) = \dfrac{2(-1)^3 - 6(-1)}{[(-1)^2 + 1]^3}$

$= \dfrac{-2 + 6}{2^3}$

$= \dfrac{4}{8} = \dfrac{1}{2} > 0$, so $\left(-1, -\dfrac{1}{2}\right)$ is a relative minimum.

$$f''(1) = \frac{2 \cdot 1^3 - 6 \cdot 1}{[(1)^2 + 1]^3}$$
$$= \frac{2-6}{2^3}$$
$$= \frac{-4}{8} = -\frac{1}{2} < 0, \text{ so } \left(1, \frac{1}{2}\right) \text{ is a}$$
relative maximum.

Then if we use -1 and 1 to divide the real number line into three intervals, $(-\infty, -1)$, $(-1, 1)$, and $(1, \infty)$, we know that f is decreasing on $(-\infty, -1)$ and on $(1, \infty)$ and is increasing on $(-1, 1)$.

d) $f''(x)$ exists for all real numbers, so the only possible inflection points are where $f''(x) = 0$.

$$\frac{2x^3 - 6x}{(x^2+1)^3} = 0$$
$$2x(x^2 - 3) = 0$$
$$2x = 0 \text{ or } x^2 - 3 = 0$$
$$x = 0 \text{ or } \quad x^2 = 3$$
$$x = 0 \text{ or } \quad x = \pm\sqrt{3}$$
Possible inflection points

$$f(-\sqrt{3}) = \frac{-\sqrt{3}}{(-\sqrt{3})^2 + 1} = -\frac{\sqrt{3}}{4}$$
$$f(0) = \frac{0}{0^2+1} = 0$$
$$f(\sqrt{3}) = \frac{\sqrt{3}}{(\sqrt{3})^2 + 1} = \frac{\sqrt{3}}{4}$$

These give the points $\left(-\sqrt{3}, -\frac{\sqrt{3}}{4}\right)$, $(0, 0)$, and $\left(\sqrt{3}, \frac{\sqrt{3}}{4}\right)$ on the graph.

e) To determine the concavity we use $-\sqrt{3}$, 0, and $\sqrt{3}$ to divide the real number line into four intervals, A: $(-\infty, -\sqrt{3})$, B: $(-\sqrt{3}, 0)$, C: $(0, \sqrt{3})$, and D: $(\sqrt{3}, \infty)$. Test a point in each interval.

A: Test -2, $f''(-2) = \frac{-4}{125} < 0$

B: Test -1, $f''(-1) = \frac{1}{2} > 0$

C: Test 1, $f''(1) = \frac{-1}{2} < 0$

D: Test 2, $f''(2) = \frac{4}{125} > 0$

Then f is concave down on $(-\infty, -\sqrt{3})$ and on $(0, \sqrt{3})$ and is concave up on $(-\sqrt{3}, 0)$ and on $(\sqrt{3}, \infty)$, so $\left(-\sqrt{3}, -\frac{\sqrt{3}}{4}\right)$, $(0, 0)$, and $\left(\sqrt{3}, \frac{\sqrt{3}}{4}\right)$ are all inflection points.

f) Sketch the graph using the preceding information.

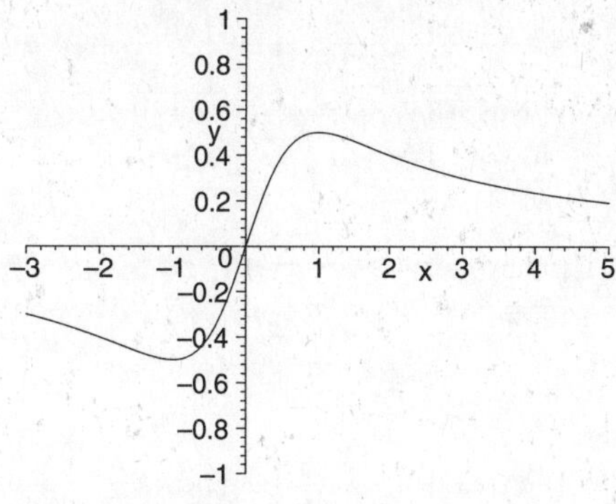

18. $f(x) = \frac{3}{x^2+1} = 3(x^2+1)^{-1}$

a) $f'(x) = 3(-1)(x^2+1)^{-2}(2x)$
$$= -6x(x^2+1)^{-2}, \text{ or } \frac{-6x}{(x^2+1)^2}$$
$$f''(x) = \frac{(x^2+1)^2(-6) - 2(x^2+1)(2x)(-6x)}{[(x^2+1)^2]^2}$$
$$= \frac{(x^2+1)[(x^2+1)(-6) - 2(2x)(-6x)]}{(x^2+1)^4}$$
$$= \frac{-6x^2 - 6 + 24x^2}{(x^2+1)^3}$$
$$= \frac{18x^2 - 6}{(x^2+1)^3}$$

b) Since $f'(x)$ exists for all real numbers, the only critical points are where $f'(x) = 0$.

$$\frac{-6x}{(x^2+1)^2} = 0$$
$$-6x = 0 \quad \text{Multiplying by } (x^2+1)^2$$
$$x = 0 \quad \text{Critical point}$$

Then $f(0) = \frac{3}{0^2+1} = 3$, so $(0, 3)$ is on the graph.

c) Use the Second-Derivative Test:

$$f''(0) = \frac{18 \cdot 0^2 - 6}{(0^2+1)^3} = -6 < 0, \text{ so } (0, 3) \text{ is a}$$
relative maximum.

Then if we use 0 to divide the real number line into two intervals, $(-\infty, 0)$ and $(0, \infty)$, we know that f is increasing on $(-\infty, 0)$ and decreasing on $(0, \infty)$.

d) $f''(x)$ exists for all real numbers, so the only possible inflection points are where $f''(x) = 0$.

$$\frac{18x^2 - 6}{(x^2+1)^3} = 0$$
$$18x^2 - 6 = 0 \quad \text{Multiplying by } (x^2+1)^3$$
$$18x^2 = 6$$
$$x^2 = \frac{1}{3}$$
$$x = \pm\frac{1}{\sqrt{3}} \quad \text{Possible inflection points}$$

$$f\left(-\frac{1}{\sqrt{3}}\right) = \frac{3}{\left(-\frac{1}{\sqrt{3}}\right)^2 + 1} = \frac{3}{\frac{1}{3}+1} = \frac{3}{\frac{4}{3}} = \frac{3}{1}\cdot\frac{3}{4} = \frac{9}{4}$$

$$f\left(\frac{1}{\sqrt{3}}\right) = \frac{3}{\left(\frac{1}{\sqrt{3}}\right)^2 + 1} = \frac{3}{\frac{1}{3}+1} = \frac{3}{\frac{4}{3}} = \frac{3}{1}\cdot\frac{3}{4} = \frac{9}{4}$$

These give the points $\left(-\frac{1}{\sqrt{3}}, \frac{9}{4}\right)$ and $\left(\frac{1}{\sqrt{3}}, \frac{9}{4}\right)$ on the graph.

e) To determine the concavity we use $-\frac{1}{\sqrt{3}}$ and $\frac{1}{\sqrt{3}}$ to divide the real number line into three intervals, A: $\left(-\infty, -\frac{1}{\sqrt{3}}\right)$, B: $\left(-\frac{1}{\sqrt{3}}, \frac{1}{\sqrt{3}}\right)$, and C: $\left(\frac{1}{\sqrt{3}}, \infty\right)$. Test a point in each interval.

A: Test -1, $f''(-1) = \frac{18(-1)^2 - 6}{[(-1)^2+1]^3} = \frac{12}{8} = \frac{3}{2} > 0$

C: Test 0, $f''(0) = \frac{18\cdot 0^2 - 6}{(0^2+1)^3} = -6 < 0$

D: Test 1, $f''(1) = \frac{18\cdot 1^2 - 6}{(1^2+1)^3} = \frac{12}{8} = \frac{3}{2} > 0$

Then f is concave up on $\left(-\infty, -\frac{1}{\sqrt{3}}\right)$, and on $\left(\frac{1}{\sqrt{3}}, \infty\right)$ and is concave down on $\left(-\frac{1}{\sqrt{3}}, \frac{1}{\sqrt{3}}\right)$, so $\left(-\frac{1}{\sqrt{3}}, \frac{9}{4}\right)$ and $\left(\frac{1}{\sqrt{3}}, \frac{9}{4}\right)$ are both inflection points.

f) Sketch the graph using the preceding information.

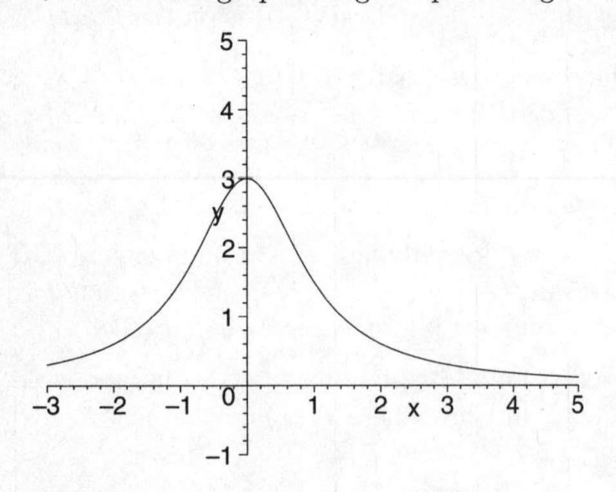

19. $f(x) = (x-1)^3$

a) $f'(x) = 3(x-1)^2(1) = 3(x-1)^2$

$f''(x) = 3\cdot 2(x-1)(1) = 6(x-1)$

b) Since $f'(x)$ exists for all real numbers, the only critical points are where $f'(x) = 0$.

$$3(x-1)^2 = 0$$
$$(x-1)^2 = 0$$
$$x - 1 = 0$$
$$x = 1 \quad \text{Critical point}$$

Then $f(1) = (1-1)^3 = 0^3 = 0$, so $(1, 0)$ is on the graph.

c) The Second-Derivative Test fails since $f''(1) = 0$, so we use the First-Derivative Test. Use 1 to divide the real number line into two intervals, A: $(-\infty, 1)$ and B: $(1, \infty)$. Test a point in each interval.

A: Test 0, $f'(0) = 3(0-1)^2 = 3 > 0$

B: Test 2, $f'(2) = 3(2-1)^2 = 3 > 0$

Since f is increasing on both intervals, $(1, 0)$ is not a relative extremum.

d) $f''(x)$ exists for all real numbers, so the only possible inflection points are where $f''(x) = 0$.

$$6(x-1) = 0$$
$$x - 1 = 0$$
$$x = 1 \quad \text{Possible inflection point}$$

From step (b), we know that $(1, 0)$ is on the graph.

e) To determine the concavity we use 1 to divide the real number line as in step (c).

A: Test 0, $f''(0) = 6(0-1) = -6 < 0$

B: Test 2, $f''(2) = 6(2-1) = 6 > 0$

Then f is concave down on $(-\infty, 1)$ and concave up on $(1, \infty)$, so $(1, 0)$ is an inflection point.

f) Sketch the graph using the preceding information.

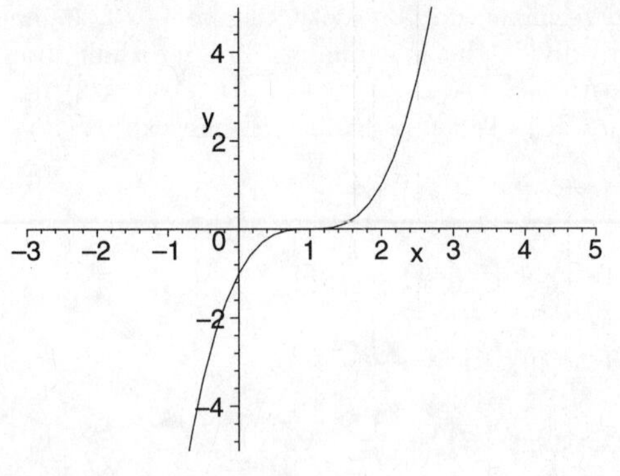

20. $f(x) = (x+2)^3$

a) $f'(x) = 3(x+2)^2$

$f''(x) = 6(x+2)$

b) $f'(x)$ exists for all real numbers. Solve:

$$3(x+2)^2 = 0$$
$$x + 2 = 0$$
$$x = -2 \quad \text{Critical point}$$

$f(-2) = (-2+2)^3 = 0$, so $(-2, 0)$ is on the graph.

c) $f''(-2) = 0$, so we use the First-Derivative Test. Use -2 to divide the real number line into two intervals, A: $(-\infty, -2)$ and B: $(-2, \infty)$.

A: Test -3, $f'(-3) = 3 > 0$

B: Test 0, $f'(0) = 12 > 0$

Since f is increasing on both intervals, $(-2, 0)$ is not a relative extremum.

d) $f''(x)$ exists for all real numbers. Solve:

$$6(x+2) = 0$$
$$x = -2 \quad \text{Possible inflection point}$$

We know $(-2, 0)$ is on the graph from step (b).

e) Use -2 to divide the real number line as in step (c).

A: Test -3, $f''(-3) = -6 < 0$

B: Test 0, $f''(0) = 12 > 0$

Then $(-2, 0)$ is an inflection point.

f) Sketch the graph.

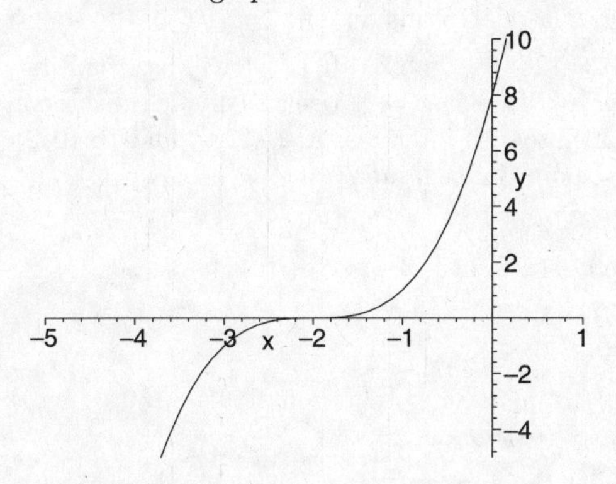

21. $f(x) = x^2(1-x)^2$
$$= x^2(1 - 2x + x^2)$$
$$= x^2 - 2x^3 + x^4$$

a) $f'(x) = 2x - 6x^2 + 4x^3$

$f''(x) = 2 - 12x + 12x^2$

b) Since $f'(x)$ exists for all real numbers, the only critical points are where $f'(x) = 0$.

$$2x - 6x^2 + 4x^3 = 0$$
$$2x(1 - 3x + 2x^2) = 0$$
$$2x(1-x)(1-2x) = 0$$
$$2x = 0 \quad or \quad 1 - x = 0 \quad or \quad 1 - 2x = 0$$
$$x = 0 \quad or \quad 1 = x \quad or \quad 1 = 2x$$
$$x = 0 \quad or \quad 1 = x \quad or \quad \frac{1}{2} = x$$

Critical points

$$f(0) = 0^2(1-0)^2 = 0$$
$$f(1) = 1^2(1-1)^2 = 0$$
$$f\left(\frac{1}{2}\right) = \left(\frac{1}{2}\right)^2\left(1 - \frac{1}{2}\right)^2 = \frac{1}{4} \cdot \frac{1}{4} = \frac{1}{16}$$

Thus, $(0, 0)$, $(1, 0)$, and $\left(\frac{1}{2}, \frac{1}{16}\right)$ are on the graph.

c) Use the Second-Derivative Test:

$f''(0) = 2 - 12 \cdot 0 + 12 \cdot 0^2 = 2 > 0$, so $(0, 0)$ is a relative minimum.

$f''(\frac{1}{2}) = 2 - 12 \cdot \frac{1}{2} + 12\left(\frac{1}{2}\right)^2 = 2 - 6 + 3 =$

$-1 < 0$, so $\left(\frac{1}{2}, \frac{1}{16}\right)$ is a relative maximum.

$f''(1) = 2 - 12 \cdot 1 + 12 \cdot 1^2 = 2 > 0$, so $(1, 0)$ is a relative minimum.

Then if we use the points 0, $\frac{1}{2}$, and 1 to divide the real number line into four intervals, $(-\infty, 0)$, $\left(0, \frac{1}{2}\right)$, $\left(\frac{1}{2}, 1\right)$ and $(1, \infty)$, we know that f is decreasing on $(-\infty, 0)$ and on $\left(\frac{1}{2}, 1\right)$ and is increasing on $\left(0, \frac{1}{2}\right)$ and on $(1, \infty)$.

d) $f''(x)$ exists for all real numbers, so the only possible inflection points are where $f''(x) = 0$.

$$2 - 12x + 12x^2 = 0$$
$$2(1 - 6x + 6x^2) = 0$$

Using the quadratic formula we find

$$x = \frac{3 \pm \sqrt{3}}{6}$$

$x \approx 0.21$ or $x \approx 0.79$ Possible inflection points

$f(0.21) \approx 0.03$ and $f(0.79) \approx 0.03$, so $(0.21, 0.03)$ and $(0.79, 0.03)$ are on the graph.

e) To determine the concavity we use 0.21 and 0.79 to divide the real number line into three intervals, A: $(-\infty, 0.21)$, B: $(0.21, 0.79)$, and C: $(0.79, \infty)$.

A: Test 0, $f''(0) = 2 - 12 \cdot 0 + 12 \cdot 0^2 = 2 > 0$

B: Test 0.5, $f''(0.5) = 2 - 12(0.5) + 12(0.5)^2 = -1 < 0$

C: Test 1, $f''(1) = 2 - 12 \cdot 1 + 12 \cdot 1^2 = 2 > 0$

Then f is concave up on $(-\infty, 0.21)$ and on $(0.79, \infty)$ and is concave down on $(0.21, 0.79)$, so $(0.21, 0.03)$ and $(0.79, 0.03)$ are both inflection points.

f) Sketch the graph using the preceding information.

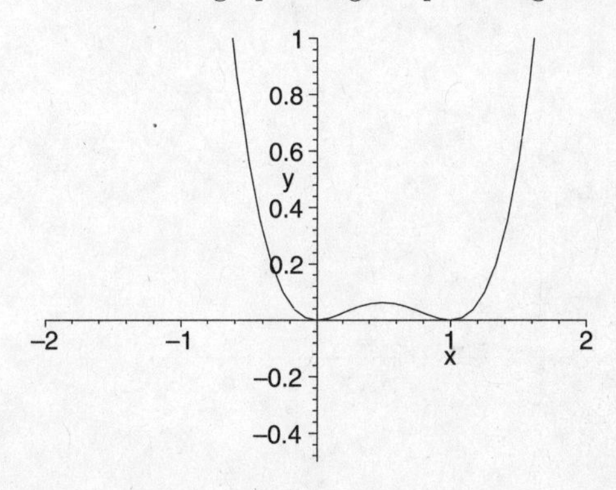

22. $f(x) = x^2(3-x)^2 = 9x^2 - 6x^3 + x^4$

a) $f'(x) = 18x - 18x^2 + 4x^3$

$f''(x) = 18 - 36x + 12x^2$

b) $f'(x)$ exists for all real numbers. Solve:

$$18x - 18x^2 + 4x^3 = 0$$
$$2x(9 - 9x + 2x^2) = 0$$
$$2x(3 - 2x)(3 - x) = 0$$

$x = 0$ *or* $x = \frac{3}{2}$ *or* $x = 3$ Critical points

$f(0) = 0$, $f\left(\frac{3}{2}\right) = \frac{81}{16}$, and $f(3) = 0$, so $(0,0)$ $\left(\frac{3}{2}, \frac{81}{16}\right)$, and $(3,0)$ are on the graph.

c) $f''(0) = 18 > 0$, so $(0,0)$ is a relative minimum.

$f''\left(\frac{3}{2}\right) = -9 < 0$, so $\left(\frac{3}{2}, \frac{81}{16}\right)$ is a relative maximum.

$f''(3) = 18 > 0$, so $(3,0)$ is a relative minimum.

Then f is decreasing on $(-\infty, 0)$ and on $\left(\frac{3}{2}, 3\right)$ and is increasing on $\left(0, \frac{3}{2}\right)$, and on $(3, \infty)$.

d) $f''(x)$ exists for all real numbers. Solve:

$$18 - 36x + 12x^2 = 0$$
$$6(3 - 6x + 2x^2) = 0$$
$$x = \frac{3 \pm \sqrt{3}}{2}$$

$x \approx 0.63$ or $x \approx 2.37$ Possible inflection points

$f(0.63) \approx 2.23$ and $f(2.37) \approx 2.23$, so $(0.63, 2.23)$ and $(2.37, 2.23)$ are on the graph.

e) Use 0.63 and 2.37 to divide the real number line into three intervals, A: $(-\infty, 0.63)$, B: $(0.63, 2.37)$, and C: $(2.37, \infty)$.

A: Test 0, $f''(0) = 18 > 0$

B: Test 1, $f''(1) = -6 < 0$

C: Test 3, $f''(3) = 18 > 0$

Then $(0.63, 2.23)$ and $(2.37, 2.23)$ are both inflection points.

f) Sketch the graph.

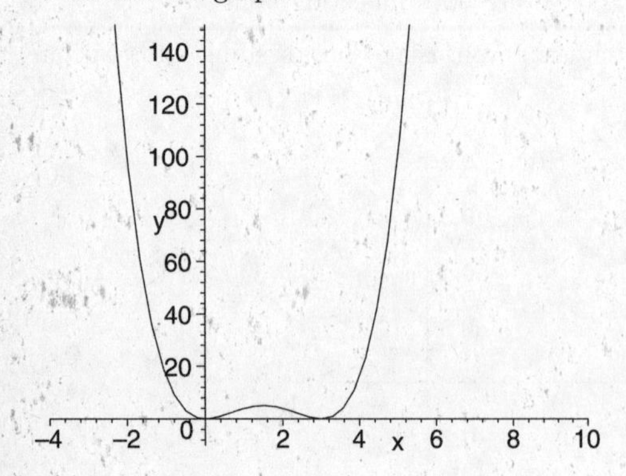

23. $f(x) = 20x^3 - 3x^5$

a) $f'(x) = 60x^2 - 15x^4$

$f''(x) = 120x - 60x^3$

b) Since $f'(x)$ exists for all real numbers, the only critical points are where $f'(x) = 0$.

$$60x^2 - 15x^4 = 0$$
$$15x^2(4 - x^2) = 0$$
$$15x^2(2 + x)(2 - x) = 0$$
$$15x^2 = 0 \quad or \quad 2 + x = 0 \quad or \quad 2 - x = 0$$
$$x = 0 \quad or \quad x = -2 \quad or \quad 2 = x$$

Critical points

$f(0) = 20 \cdot 0^3 - 3 \cdot 0^5 = 0$

$f(-2) = 20(-2)^3 - 3(-2)^5 = -160 + 96 = -64$

$f(2) = 20 \cdot 2^3 - 3 \cdot 2^5 = 160 - 96 = 64$

Thus, $(0,0)$, $(-2,-64)$, and $(2,64)$ are on the graph.

c) Use the Second-Derivative Test:

$f''(-2) = 120(-2) - 60(-2)^3 = -240 + 480 = 240 > 0$, so $(-2,-64)$ is a relative minimum.

$f''(2) = 120 \cdot 2 - 60 \cdot 2^3 = 240 - 480 = -240 < 0$, so $(2,64)$ is a relative maximum.

$f''(0) = 120 \cdot 0 - 60 \cdot 0^3 = 0$, so we will use the First-Derivative Test on $x = 0$. Use 0 to divide the interval $(-2,2)$ into two intervals, A: $(-2,0)$, and B: $(0,2)$. Test a point in each interval.

A: Test -1, $f'(-1) = 60(-1)^2 - 15(-1)^4 = 60 - 15 = 45 > 0$

B: Test 1, $f'(1) = 60 \cdot 1^2 - 15 \cdot 1^4 = 60 - 15 = 45 > 0$

Then f is increasing on both intervals, so $(0,0)$ is not a relative extremum.

If we use the points -2 and 2 to divide the real number line into three intervals, $(-\infty,-2)$, $(-2,2)$, and $(2,\infty)$, we know that f is decreasing on $(-\infty,-2)$ and on $(2,\infty)$ and is increasing on $(-2,2)$.

d) $f''(x)$ exists for all real numbers, so the only possible inflection points are where $f''(x) = 0$.

$$120x - 60x^3 = 0$$
$$60x(2 - x^2) = 0$$
$$60x = 0 \quad \text{or} \quad 2 - x^2 = 0$$
$$x = 0 \quad \text{or} \quad 2 = x^2$$
$$x = 0 \quad \text{or} \quad \pm\sqrt{2} = x$$

Possible inflection points

$f(-\sqrt{2}) = 20(-\sqrt{2})^3 - 3(-\sqrt{2})^5 = -40\sqrt{2} + 12\sqrt{2} = -28\sqrt{2}$

$f(0) = 0$ from step (b)

$f(\sqrt{2}) = 20(\sqrt{2})^3 - 3(\sqrt{2})^5 = 40\sqrt{2} - 12\sqrt{2} = 28\sqrt{2}$

Thus, $(-\sqrt{2}, -28\sqrt{2})$ and $(\sqrt{2}, 28\sqrt{2})$ are also on the graph.

e) To determine the concavity we use $-\sqrt{2}$, 0, and $\sqrt{2}$ to divide the real number line into four intervals, A: $(-\infty, -\sqrt{2})$, B: $(-\sqrt{2}, 0)$, C: $(0, \sqrt{2})$, and D: $(\sqrt{2}, \infty)$.

A: Test -2, $f''(-2) = 120(-2) - 60(-2)^3 = 240 > 0$

B: Test -1, $f''(-1) = 120(-1) - 60(-1)^3 = -60 < 0$

C: Test 1, $f''(1) = 120 \cdot 1 - 60 \cdot 1^3 = 60 > 0$

D: Test 2, $f''(2) = 120 \cdot 2 - 60 \cdot 2^3 = -240 < 0$

Then f is concave up on $(-\infty, -\sqrt{2})$ and on $(0, \sqrt{2})$ and is concave down on $(-\sqrt{2}, 0)$ and on $(\sqrt{2}, \infty)$, so $(-\sqrt{2}, -28\sqrt{2})$, $(0, 0)$, and $(\sqrt{2}, 28\sqrt{2})$ are all inflection points.

f) Sketch the graph using the preceding information.

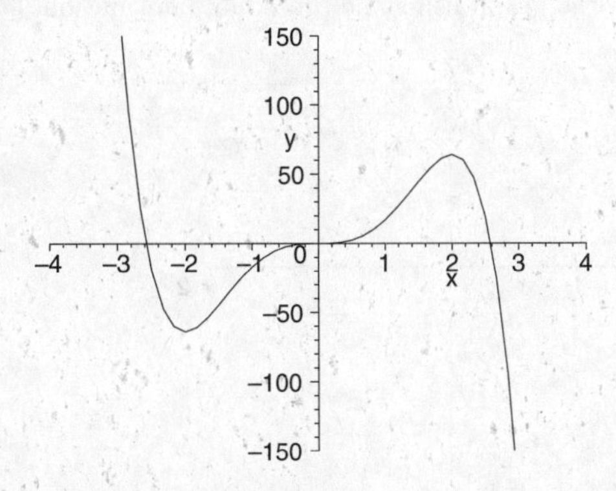

24. $f(x) = 5x^3 - 3x^5$

a) $f'(x) = 15x^2 - 15x^4$

$f''(x) = 30x - 60x^3$

b) $f'(x)$ exists for all real numbers. Solve:

$$15x^2 - 15x^4 = 0$$
$$15x^2(1 - x^2) = 0$$
$$15x^2(1 + x)(1 - x) = 0$$

$x = 0$ *or* $x = -1$ *or* $x = 1$ Critical points

$f(0) = 0$, $f(-1) = -2$, and $f(1) = 2$, so $(0, 0)$ $(-1, -2)$, and $(1, 2)$ are on the graph.

c) $f''(-1) = 30 > 0$, so $(-1, -2)$ is a relative minimum.

$f''(1) = -30 < 0$, so $(1, 2)$ is a relative maximum.

$f''(0) = 0$, so we use the First-Derivative Test on $x = 0$. Use 0 to divide the interval $(-1, 1)$ into two intervals: A: $(-1, 0)$ and B: $(0, 1)$.

A: Test $-\frac{1}{2}$, $f'\left(-\frac{1}{2}\right) = \frac{45}{16} > 0$

B: Test $\frac{1}{2}$, $f'\left(\frac{1}{2}\right) = \frac{45}{16} > 0$

Then f is increasing on both $(-1, 0)$ and $(0, 1)$, so $(0, 0)$ is not a relative extremum.

Then we know that f is decreasing on $(-\infty, -1)$ and $(1, \infty)$ and is increasing on $(-1, 1)$.

d) $f''(x)$ exists for all real numbers. Solve:

$$30x - 60x^3 = 0$$
$$30x(1 - 2x^2) = 0$$
$$x = 0 \text{ or } x = \pm\frac{1}{\sqrt{2}}$$

$x = 0$ *or* $x \approx \pm 0.71$ Possible inflection points

$f(0) = 0$ Found in step (b)

$f(-0.71) \approx -1.25$ and $f(0.71) \approx 1.25$ so $(-0.71, -1.25)$ and $(0.71, 1.25)$ are also on the graph.

e) Use -0.71, 0, and 0.71 to divide the real number line into four intervals, A: $(-\infty, -0.71)$, B: $(-0.71, 0)$, C: $(0, 0.71)$, and D: $(0.71, \infty)$.

A: Test -1, $f''(-1) = 30 > 0$

B: Test -0.5, $f''(-0.5) = -7.5 < 0$

C: Test 0.5, $f''(0.5) = 7.5 > 0$

D: Test 1, $f''(1) = -30 < 0$

Then $(-0.71, -1.25)$, $(0, 0)$, and $(0.71, 1.25)$ are all inflection points.

f) Sketch the graph.

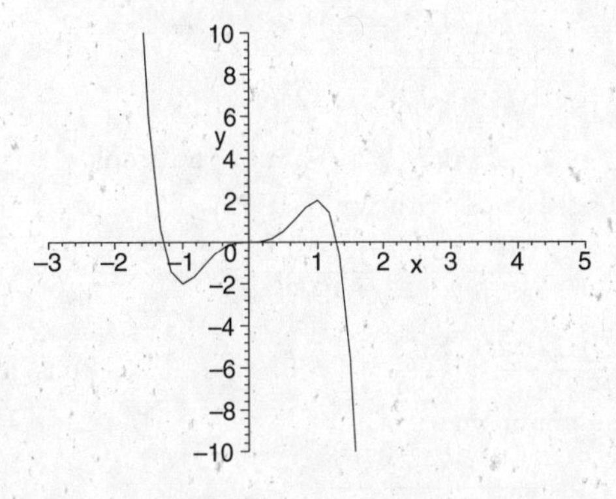

25. $f(x) = x\sqrt{4 - x^2} = x(4 - x^2)^{1/2}$

a) $f'(x) = x \cdot \frac{1}{2}(4 - x^2)^{-1/2}(-2x) + 1 \cdot (4 - x^2)^{1/2}$

$$= \frac{-x^2}{\sqrt{4 - x^2}} + \sqrt{4 - x^2}$$
$$= \frac{-x^2 + 4 - x^2}{\sqrt{4 - x^2}}$$
$$= \frac{4 - 2x^2}{\sqrt{4 - x^2}}, \text{ or } (4 - 2x^2)(4 - x^2)^{-1/2}$$

$f''(x) = (4 - 2x^2)\left(-\frac{1}{2}\right)(4 - x^2)^{-3/2}(-2x) + (-4x)(4 - x^2)^{-1/2}$

$$= \frac{x(4 - 2x^2)}{(4 - x^2)^{3/2}} - \frac{4x}{(4 - x^2)^{1/2}}$$
$$= \frac{x(4 - 2x^2) - 4x(4 - x^2)}{(4 - x^2)^{3/2}}$$
$$= \frac{4x - 2x^3 - 16x + 4x^3}{(4 - x^2)^{3/2}}$$
$$= \frac{2x^3 - 12x}{(4 - x^2)^{3/2}}$$

b) $f'(x)$ does not exist where $4-x^2=0$. Solve:

$$4-x^2=0$$
$$(2+x)(2-x)=0$$
$$2+x=0 \quad or \quad 2-x=0$$
$$x=-2 \quad or \quad 2=x$$

Note that $f(x)$ is not defined for $x<-2$ or $x>2$. (For these values $4-x^2<0$.) Therefore, relative extrema canot occur at $x=-2$ or $x=2$, because there is no open interval containing -2 or 2 on which the function is defined. For this reason, we do not consider -2 and 2 further in our discussion of relative extrema.

Critical points occur where $f'(x)=0$. Solve:

$$\frac{4-2x^2}{\sqrt{4-x^2}}=0$$
$$4-2x^2=0$$
$$4=2x^2$$
$$2=x^2$$
$$\pm\sqrt{2}=x \quad \text{Critical points}$$

$$f(-\sqrt{2})=-\sqrt{2}\sqrt{4-(-\sqrt{2})^2}=-\sqrt{2}\cdot\sqrt{2}=-2$$

$$f(\sqrt{2})=\sqrt{2}\sqrt{4-(\sqrt{2})^2}=\sqrt{2}\cdot\sqrt{2}=2$$

Then $(-\sqrt{2},-2)$ and $(\sqrt{2},2)$ are on the graph.

c) Use the Second-Derivative Test:

$$f''(-\sqrt{2})=\frac{2(-\sqrt{2})^3-12(-\sqrt{2})}{[4-(-\sqrt{2})^2]^{3/2}}=$$

$\dfrac{-4\sqrt{2}+12\sqrt{2}}{2^{3/2}}=\dfrac{8\sqrt{2}}{2\sqrt{2}}=4>0$, so $(-\sqrt{2},-2)$ is a relative minimum.

$$f''(\sqrt{2})=\frac{2(\sqrt{2})^3-12(\sqrt{2})}{[4-(\sqrt{2})^2]^{3/2}}=$$

$\dfrac{4\sqrt{2}-12\sqrt{2}}{2^{3/2}}=\dfrac{-8\sqrt{2}}{2\sqrt{2}}=-4<0$, so $(\sqrt{2},2)$ is a relative maximum.

If we use the points $-\sqrt{2}$ and $\sqrt{2}$ to divide the interval $(-2,2)$ into three intervals, $(-2,-\sqrt{2})$, $(-\sqrt{2},\sqrt{2})$, and $(\sqrt{2},2)$, we know that f is decreasing on $(-2,-\sqrt{2})$ and on $(\sqrt{2},2)$ and is increasing on $(-\sqrt{2},\sqrt{2})$.

d) $f''(x)$ does not exist where $4-x^2=0$. From step (b) we know that this occurs at $x=-2$ and at $x=2$. However, just as relative extrema cannot occur at $(-2,0)$ and $(2,0)$, they cannot be inflection points either. Inflection points could occur where $f''(x)=0$.

$$\frac{2x^3-12x}{(4-x^2)^{3/2}}=0$$
$$2x^3-12x=0$$
$$2x(x^2-6)=0$$
$$2x=0 \text{ or } x^2-6=0$$
$$x=0 \text{ or } x^2=6$$
$$x=0 \text{ or } x=\pm\sqrt{6}$$

Note that $f(x)$ is not defined for $x=\pm\sqrt{6}$. Therefore, the only possible inflection point is $x=0$.

$f(0)=0\sqrt{4-0^2}=0\cdot 2=0$

Then $(0,0)$ is on the graph.

e) To determine the concavity we use 0 to divide the interval $(-2,2)$ into two intervals, A: $(-2,0)$ and B: $(0,2)$.

A: Test -1, $f''(-1)=\dfrac{2(-1)^3-12(-1)}{[4-(-1)^2]^{3/2}}=$

$\dfrac{10}{3^{3/2}}>0$

B: Test 1, $f''(1)=\dfrac{2\cdot 1^3-12\cdot 1}{(4-1^2)^{3/2}}=\dfrac{-10}{3^{3/2}}<0$

Then f is concave up on $(-2,0)$ and concave down on $(0,2)$, so $(0,0)$ is an inflection point.

f) Sketch the graph using the preceding information.

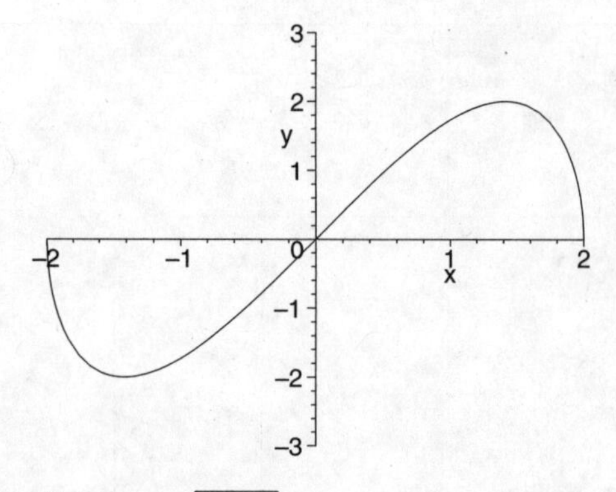

26. $f(x)=-x\sqrt{1-x^2}$

a) $f'(x)=\dfrac{2x^2-1}{\sqrt{1-x^2}}$

$f''(x)=\dfrac{-2x^3+3x}{(1-x^2)^{3/2}}$

b) $f'(x)$ does not exist for $x=\pm1$, but relative extrema cannot occur at $x=-1$ or at $x=1$, because $f(x)$ is not defined on $(-\infty,-1)$ or on $(1,\infty)$. Solve:

$$\frac{2x^2-1}{\sqrt{1-x^2}}=0$$
$$2x^2-1=0$$
$$x=\pm\frac{1}{\sqrt{2}}=\pm\frac{\sqrt{2}}{2} \quad \text{Critical points}$$

$f\left(-\dfrac{\sqrt{2}}{2}\right)=\dfrac{1}{2}$ and $f\left(\dfrac{\sqrt{2}}{2}\right)=-\dfrac{1}{2}$, so $\left(-\dfrac{\sqrt{2}}{2},\dfrac{1}{2}\right)$ and $\left(\dfrac{\sqrt{2}}{2},-\dfrac{1}{2}\right)$ are on the graph.

c) $f''\left(-\dfrac{\sqrt{2}}{2}\right)=-4<0$, so $\left(-\dfrac{\sqrt{2}}{2},\dfrac{1}{2}\right)$ is a relative maximum.

$f''\left(\dfrac{\sqrt{2}}{2}\right)=4>0$, so $\left(\dfrac{\sqrt{2}}{2},-\dfrac{1}{2}\right)$ is a relative minimum.

Then f is increasing on $\left(-1,-\dfrac{\sqrt{2}}{2}\right)$ and on $\left(\dfrac{\sqrt{2}}{2},1\right)$ and is decreasing on $\left(-\dfrac{\sqrt{2}}{2},\dfrac{\sqrt{2}}{2}\right)$.

d) $f''(x)$ does not exist for $x = \pm 1$, but inflection points cannot occur at those values because the domain of the function is $[-1, 1]$. Solve:

$$\frac{-2x^3 + 3x}{(1 - x^2)^{3/2}} = 0$$

$$-2x^3 + 3x = 0$$

$$x(-2x^2 + 3) = 0$$

$$x = 0 \text{ or } x = \pm\frac{\sqrt{6}}{2}$$

Note that only $x = 0$ is in the domain of f, so $x = 0$ is the only possible inflection point.

$f(0) = 0$, so $(0, 0)$ is on the graph.

e) Use 0 to divide the interval $(-1, 1)$ into two intervals, A: $(-1, 0)$ and B: $(0, 1)$.

A: Test $-\frac{1}{2}$, $f''\left(-\frac{1}{2}\right) = \frac{-10}{3^{3/2}} < 0$

B: Test $\frac{1}{2}$, $f''\left(\frac{1}{2}\right) = \frac{10}{3^{3/2}} > 0$

Then $(0, 0)$ is an inflection point.

f) Sketch the graph.

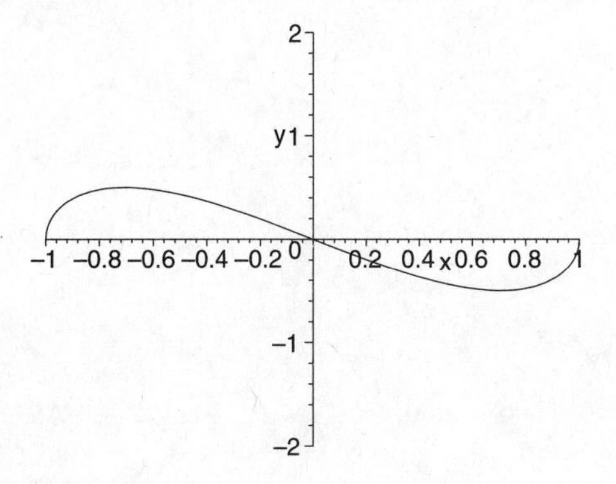

27. $f(x) = (x - 1)^{1/3} - 1$

a) $f'(x) = \frac{1}{3}(x - 1)^{-2/3}$, or $\frac{1}{3(x - 1)^{2/3}}$

$$f''(x) = \frac{1}{3}\left(-\frac{2}{3}\right)(x - 1)^{-5/3}$$

$$= -\frac{2}{9}(x - 1)^{-5/3}, \text{ or } -\frac{2}{9(x - 1)^{5/3}}$$

b) $f'(x)$ does not exist for $x = 1$. The equation $f'(x) = 0$ has no solution, so $x = 1$ is the only critical point. $f(1) = (1 - 1)^{1/3} - 1 = 0 - 1 = -1$, so $(1, -1)$ is on the graph.

c) Use the First-Derivative Test: Use 1 to divide the real number line into two intervals, A: $(-\infty, 1)$ and B: $(1, \infty)$. Test a point in each interval.

A: Test 0, $f'(0) = \frac{1}{3(0 - 1)^{2/3}} = \frac{1}{3 \cdot 1} = \frac{1}{3} > 0$

B: Test 2, $f'(2) = \frac{1}{3(2 - 1)^{2/3}} = \frac{1}{3 \cdot 1} = \frac{1}{3} > 0$

Then f is increasing on both intervals, so $(1, -1)$ is not a relative extremum.

d) $f''(x)$ does not exist for $x = 1$. The equation $f''(x) = 0$ has no solution, so $x = 1$ is the only possible inflection point. From step (b) we know $(1, -1)$ is on the graph.

e) To determine the concavity we use 1 to divide the real number line as in step (c). Test a point in each interval.

A: Test 0, $f''(0) = -\frac{2}{9(0 - 1)^{5/3}} = -\frac{2}{9(-1)} = \frac{2}{9} > 0$

B: Test 2, $f''(2) = -\frac{2}{9(2 - 1)^{5/3}} = -\frac{2}{9 \cdot 1} = -\frac{2}{9} < 0$

Then f is concave up on $(-\infty, 1)$ and concave down on $(1, \infty)$, so $(1, -1)$ is an inflection point.

f) Sketch the graph using the preceding information.

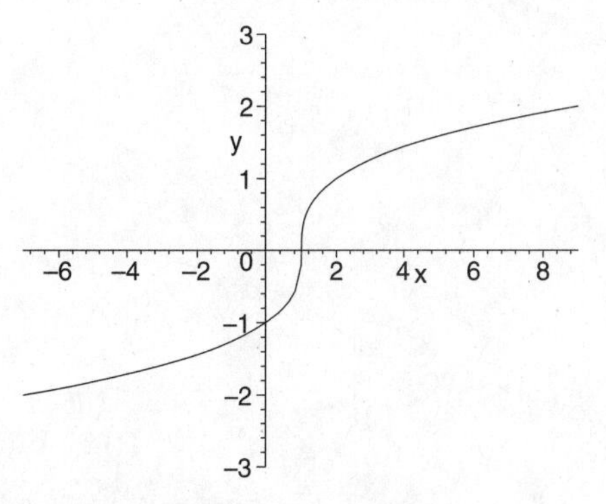

28. $f(x) = 2 - x^{1/3}$

a) $f'(x) = -\frac{1}{3}x^{-2/3}$, or $-\frac{1}{3x^{2/3}}$

$$f''(x) = \frac{2}{9}x^{-5/3}, \text{ or } \frac{2}{9x^{5/3}}$$

b) $f'(x)$ does not exist for $x = 0$. The equation $-\frac{1}{3x^{2/3}} = 0$ has no solution, so $x = 0$ is the only critical point.

$f(0) = 2$, so $(0, 2)$ is on the graph.

c) Use the First-Derivative Test. Use 0 to divide the real number line into two intervals, A: $(-\infty, 0)$ and B: $(0, \infty)$.

A: Test -1, $f'(-1) = -\frac{1}{3} < 0$

B: Test 1, $f'(1) = -\frac{1}{3} < 0$

f is decreasing on both intervals, so $(0, 2)$ is not a relative extremum.

d) $f''(x)$ does not exist for $x = 0$. The equation $\frac{2}{9x^{5/3}} = 0$ has no solution, so $x = 0$ is the only possible inflection point. From step (b) we know $(0, 2)$ is on the graph.

e) Use 0 to divide the real number line as in step (c).

A: Test -1, $f''(-1) = \dfrac{2}{9(-1)^{5/3}} = -\dfrac{2}{9} < 0$

B: Test 1, $f''(1) = \dfrac{2}{9(1)^{5/3}} = \dfrac{2}{9} > 0$

Then $(0, 2)$ is an inflection point.

f) Sketch the graph.

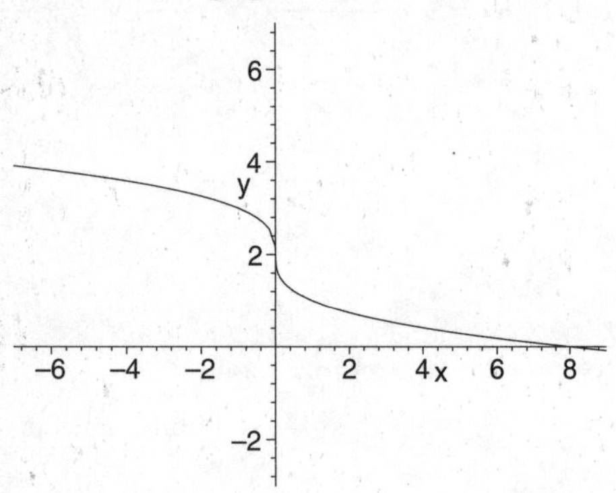

29. $f(x) = x + cos\ 2x$

$f'(x) = 1 - 2\ sin\ 2x$
We solve $f'(x) = 0$

$$\begin{aligned} 1 - 2\ sin\ 2x &= 0 \\ sin\ 2x &= \frac{1}{2} \\ 2x &= \frac{\pi}{6} \\ x &= \frac{\pi}{12} \end{aligned}$$

and

$$\begin{aligned} 2x &= \frac{5\pi}{6} \\ x &= \frac{5\pi}{12} \end{aligned}$$

and

$$2x = \frac{13\pi}{6}$$

$$x = \frac{13\pi}{12}$$

and

$$2x = \frac{17\pi}{6}$$

$$x = \frac{17\pi}{12}$$

The values above divide the number line into five intervals. We apply test points to check the sign of the first derivative in each of those intervals.

$f'(0) = 1 - 2\ sin(2(0)) = 1 > 0$

$f'(\frac{\pi}{3}) = 1 - 2\ sin(\frac{2\pi}{3}) = -0.7321 < 0$

$f'(\pi) = 1 - 2\ sin(2\pi) = 1 > 0$

$f'(\frac{5\pi}{4}) = 1 - 2\ sin(\frac{5\pi}{2}) = -1 < 0$

$f'(\frac{5\pi}{3}) = 1 - 2\ sin(\frac{10\pi}{3}) = 2.7321 > 0$

Thus, there is a relative maximum at $x = \frac{\pi}{12}$ and $x = \frac{13\pi}{12}$ and a relative minimum at $x = \frac{5\pi}{12}$ and $x = \frac{17\pi}{12}$.

$f(\frac{\pi}{12}) = \frac{\pi}{12} + cos\ \frac{\pi}{6} = 1.128$

$f(\frac{13\pi}{12}) = \frac{13\pi}{12} + cos(\frac{13\pi}{6}) = 4.269$

$f(\frac{5\pi}{12}) = \frac{5\pi}{12} + cos(\frac{5\pi}{6}) = 0.443$

$f(\frac{17\pi}{12}) = \frac{17\pi}{12} + cos(\frac{17\pi}{6}) = 3.585$

$$\begin{aligned} f''(x) &= -4\ cos\ 2x \\ f''(x) &= 0 \\ -4\ cos\ 2x &= 0 \\ cos\ 2x &= 0 \\ 2x &= \frac{\pi}{2} \\ x &= \frac{\pi}{4} \end{aligned}$$

and

$$\begin{aligned} 2x &= \frac{3\pi}{2} \\ x &= \frac{3\pi}{4} \end{aligned}$$

and

$$\begin{aligned} 2x &= \frac{5\pi}{2} \\ x &= \frac{5\pi}{4} \end{aligned}$$

and

$$\begin{aligned} 2x &= \frac{7\pi}{2} \\ x &= \frac{7\pi}{4} \end{aligned}$$

The second derivative zeros divide the number line into five intervals, we determine the sign of the second derivative in each of those intervals:

$f''(\frac{\pi}{6}) = -4\ cos(\frac{\pi}{3}) = -2 < 0$
$f''(\frac{\pi}{2}) = -4\ cos(\pi) = 4 > 0$ $f''(\pi) = -4\ cos(2\pi) = -4 < 0$
$f''(\frac{3\pi}{2}) = -4\ cos(3\pi) = 4 > 0$ $f''(\frac{15\pi}{8}) = -4\ cos(\frac{15\pi}{4}) = -2.828 < 0$

This means that we have inflection points at $x = \frac{\pi}{4}$, $x = \frac{\pi}{4}$, $x = \frac{5\pi}{4}$ and $x = \frac{7\pi}{4}$

$f(\frac{\pi}{4}) = \frac{\pi}{4} + cos(\frac{\pi}{2}) = 0.785$

$f(\frac{3\pi}{4}) = \frac{3\pi}{4} + cos(\frac{3\pi}{2}) = 2.356$

$f(\frac{\pi}{4}) = \frac{5\pi}{4} + cos(\frac{5\pi}{2}) = 3.927$

$f(\frac{\pi}{4}) = \frac{7\pi}{4} + cos(\frac{7\pi}{2}) = 5.498$

we sketch the graph

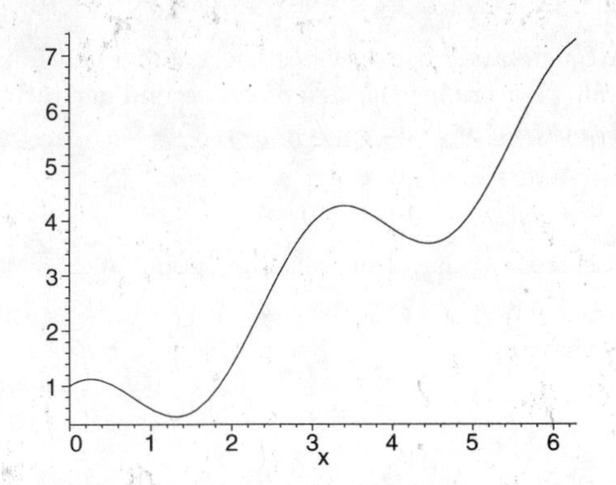

30. $f(x) = x - 2\ sin\ x$

$f'(x) = 1 - 2\ cos\ x$
We solve $f'(x) = 0$

$$\begin{aligned} 1 - 2\ cos\ x &= 0 \\ cos\ x &= \frac{1}{2} \\ x &= \frac{\pi}{3} \\ \text{and} & \\ 2x &= \frac{5\pi}{3} \end{aligned}$$

The values above divide the number line into three intervals. We apply test points to check the sign of the first derivative in each of those intervals.

$f'(0) = 1 - 2\ cos((0)) = -1 < 0$
$f'(\pi) = 1 - 2\ cos(\pi) = 3 > 0$
$f'(\pi) = 1 - 2\ cos(\frac{11\pi}{6}) = -0.732 < 0$

Thus, there is a relative maximum at $x = \frac{5\pi}{3}$ and a relative minimum at $x = \frac{\pi}{3}$.

$f(\frac{\pi}{3}) = \frac{\pi}{3} - 2\ sin\ \frac{\pi}{3} = -0.6849$
$f(\frac{5\pi}{3}) = \frac{5\pi}{3} - 2\ sin\ (\frac{5\pi}{3}) = 6.968$

$$\begin{aligned} f''(x) &= 2\ sin\ x \\ f''(x) &= 0 \\ 2\ sin\ x &= 0 \\ sin\ x &= 0 \\ x &= \pi \end{aligned}$$

The second derivative zero divides the number line into two intervals, we determine the sign of the second derivative in each of those intervals:

$f''(\frac{\pi}{4}) = 2\ sin(\frac{\pi}{4}) = 1.414 > 0$
$f''(\frac{3\pi}{2}) = 2\ sin(\frac{3\pi}{2}) = -2 < 0$

This means that we have inflection points at $x = \pi$.

$f(\pi) = \pi - 2\ sin(\pi) = 0.785$

we sketch the graph

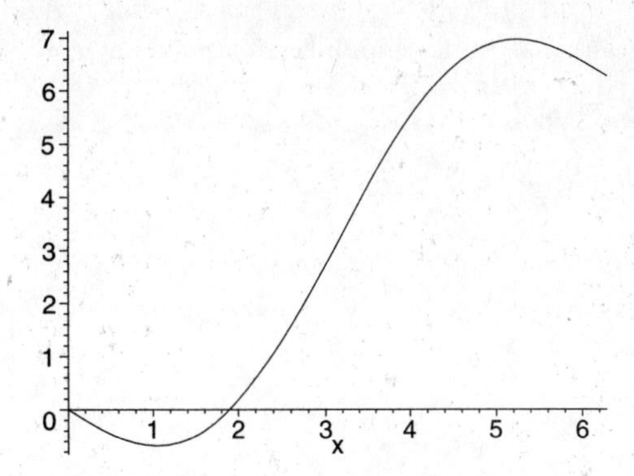

31. $f(x) = \frac{x}{3} - sin\ \frac{2x}{3}$

$f'(x) = \frac{1}{3} - \frac{2}{3}\ cos\ \frac{2x}{3}$
We solve $f'(x) = 0$

$$\begin{aligned} \frac{1}{3} - \frac{2}{3}\ cos\ \frac{2x}{3} &= 0 \\ cos\ \frac{2x}{3} &= \frac{1}{2} \\ \frac{2x}{3} &= \frac{\pi}{3} \\ x &= \frac{\pi}{2} \end{aligned}$$

The zero of the first derivative divides the number line into two intervals. We apply test points to check the sign of the first derivative in each of those intervals.

$f'(\frac{\pi}{4}) = \frac{1}{3} - \frac{2}{3}\ sin\ (\frac{\pi}{12}) = -0.244 < 0$

$f'(\pi) = \frac{1}{3} - \frac{2}{3}\ sin\ (\frac{2\pi}{3}) = 0.667 > 0$

Thus, there is a relative minimum at $x = \frac{\pi}{2}$.

$f(\frac{\pi}{2}) = \frac{\pi}{2} - sin\ (\frac{\pi}{3}) = -0.342$

$$\begin{aligned} f''(x) &= \frac{4}{9}\ sin\ \frac{2x}{3} \\ f''(x) &= 0 \\ \frac{4}{9}\ sin\ \frac{2x}{3} &= 0 \\ sin\ \frac{2x}{3} &= 0 \\ \frac{2x}{3} &= \pi \\ x &= \frac{3\pi}{2} \end{aligned}$$

The second derivative zero divides the number line into two intervals, we determine the sign of the second derivative in each of those intervals:

$f''(\pi) = \frac{4}{9}\ sin\ (\frac{2\pi}{3}) = 0.385 > 0$

$f''(2\pi) = \frac{4}{9}\ sin\ (\frac{4\pi}{3}) = -0.385 < 0$

This means that we have an inflection points at $x = \frac{3\pi}{2}$.
$f(\frac{3\pi}{2}) = \frac{\pi}{2} - sin\ (\pi) = \frac{\pi}{2} = 1.571$

We sketch the graph

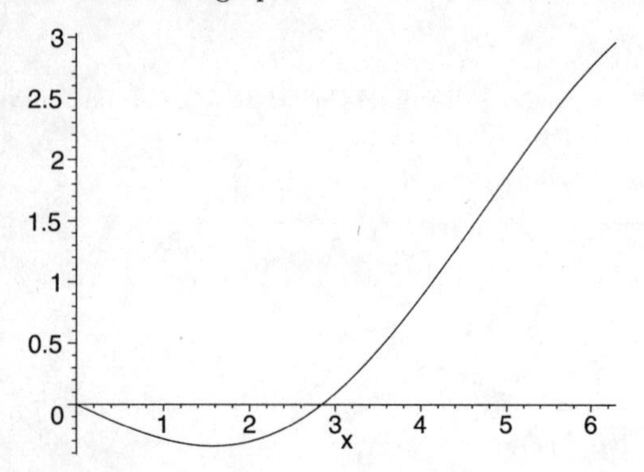

32. $f(x) = \frac{x}{4} + cos\ \frac{x}{2}$

$f'(x) = \frac{1}{4} - \frac{1}{2}\ sin\ \frac{x}{2}$
We solve $f'(x) = 0$

$$\begin{aligned} \frac{1}{4} - \frac{1}{2}\ sin\ \frac{x}{2} &= 0 \\ sin\ \frac{x}{2} &= \frac{1}{2} \\ \frac{x}{2} &= \frac{\pi}{6} \\ x &= \frac{\pi}{3} \\ \text{and} & \\ \frac{x}{2} &= \frac{5\pi}{6} \\ x &= \frac{5\pi}{3} \end{aligned}$$

The zeros of the first derivative divide the number line into three intervals. We apply test points to check the sign of the first derivative in each of those intervals.
$f'(0) = \frac{1}{4} - \frac{1}{2}\ sin(0) = 0.25 > 0$

$f'(\pi) = \frac{1}{4} - \frac{1}{2}\ sin(\frac{\pi}{2}) = -0.25 < 0$

$f'(\frac{11\pi}{6}) = \frac{1}{4} - \frac{1}{2}\ sin(\frac{11\pi}{3}) = 0.121 > 0$

Thus, there is a relative maximum at $x = \frac{\pi}{3}$ and a relative minimum at $x = \frac{5\pi}{3}$

$$\begin{aligned} f''(x) &= \frac{-1}{4}\ cos\ \frac{x}{2} \\ f''(x) &= 0 \\ \frac{-1}{4}\ cos\ \frac{x}{2} &= 0 \\ cos\ \frac{x}{2} &= 0 \\ \frac{x}{2} &= \frac{\pi}{2} \\ x &= \pi \end{aligned}$$

The second derivative zero divides the number line into two interval, we dtermine the sign of the second derivative in each of those intervals:
$f''(0) = \frac{-1}{4}\ cos(0) = -0.25 < 0$
$f''(\frac{3\pi}{2}) = \frac{-1}{4}\ cos(\frac{3\pi}{4}) = 0.177 > 0$
This means that we have an inflection point at $x = \pi$.
$f(\pi) = \frac{\pi}{4} + cos(\frac{\pi}{2}) = 0.785$
We sketch the graph

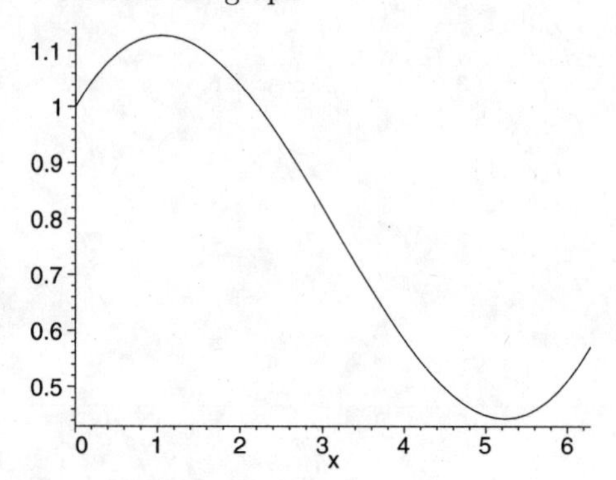

33. $f(x) = sin\ x + cos\ x$

$f'(x) = cos\ x - sin\ x$
We solve$f'(x) = 0$

$$\begin{aligned} cos\ x - sin\ x &= 0 \\ cos\ x &= sin\ x \\ x &= \frac{\pi}{4} \\ \text{and} & \\ x &= \frac{5\pi}{4} \end{aligned}$$

The zeros of the first derivative divide the number line into three intervals. We apply test points to check the sign of the first derivative in each of those intervals.
$f'(0) = cos(0) - sin(0) = 1 > 0$

$f'(\pi) = cos(\pi) - sin(\pi) = -1 < 0$

$f'(\frac{3\pi}{2}) = cos(\frac{3\pi}{2}) - sin(\frac{3\pi}{2}) = 1 > 0$
Thus, there is a relative maximum at $x = \frac{\pi}{4}$ and a relative minimum at $x = \frac{5\pi}{4}$.
$f(\frac{\pi}{4}) = sin(\frac{\pi}{4}) + cos(\frac{\pi}{4}) = 1.414$
$f(\frac{5\pi}{4}) = sin\frac{5\pi}{4}) + cos(\frac{5\pi}{4}) = -1.414$

$$\begin{aligned} f''(x) &= -sin\ x - cos\ x \\ f''(x) &= 0 \\ -sin\ x - cos\ x &= 0 \\ -sin\ x &= cos\ x \\ x &= \frac{3\pi}{4} \\ \text{and} & \\ x &= \frac{7\pi}{4} \end{aligned}$$

The second derivative zero divides the number line into three interval, we dtermine the sign of the second derivative in each of those intervals:

$f''(0) = -sin(0) - cos(0) = -1 < 0$
$f''(\pi) = -sin(\pi) - cos(\pi) = 1 > 0$
$f''(\frac{11\pi}{6}) = -sin(\frac{11\pi}{6}) - cos(\frac{11\pi}{6}) = -0.366 < 0$

This means that we have an inflection point at $x = \frac{3\pi}{4}$ and at $x = \frac{7\pi}{4}$.

$f(\frac{3\pi}{4}) = sin(\frac{3\pi}{4}) + cos(\frac{3\pi}{4}) = 0$

$f(\frac{3\pi}{4}) = sin(\frac{7\pi}{4}) + cos(\frac{7\pi}{4}) = 0$

We sketch the graph

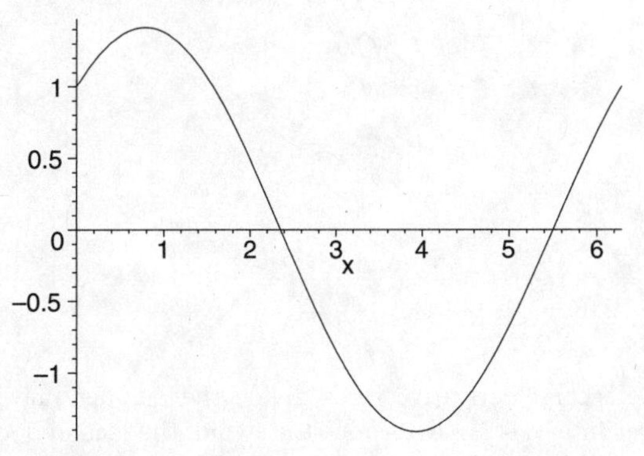

34. $f(x) = sin\ x - cos\ x$

$f'(x) = cos\ x + sin\ x$
We solve$f'(x) = 0$

$$\begin{aligned} cos\ x + sin\ x &= 0 \\ cos\ x &= -sin\ x \\ x &= \frac{3\pi}{4} \\ \text{and} \\ x &= \frac{7\pi}{4} \end{aligned}$$

The zeros of the first derivative divide the number line into three intervals. We apply test points to check the sign of the first derivative in each of those intervals.

$f'(0) = cos(0) + sin(0) = 1 > 0$

$f'(\pi) = cos(\pi) + sin(\pi) = -1 < 0$

$f'(\frac{11\pi}{6}) = cos(\frac{11\pi}{6}) - sin(\frac{11\pi}{6}) = 0.366 > 0$

Thus, there is a relative maximum at $x = \frac{3\pi}{4}$ and a relative minimum at $x = \frac{7\pi}{4}$.

$f(\frac{3\pi}{4}) = sin(\frac{3\pi}{4}) - cos(\frac{3\pi}{4}) = 1.414$

$f(\frac{7\pi}{4}) = sin\frac{7\pi}{4}) - cos(\frac{7\pi}{4}) = -1.414$

$$\begin{aligned} f''(x) &= -sin\ x + cos\ x \\ f''(x) &= 0 \\ -sin\ x + cos\ x &= 0 \\ sin\ x &= cos\ x \\ x &= \frac{\pi}{4} \\ \text{and} \\ x &= \frac{5\pi}{4} \end{aligned}$$

The second derivative zero divides the number line into three interval, we dtermine the sign of the second derivative in each of those intervals:

$f''(0) = -sin(0) + cos(0) = 1 > 0$

$f''(\pi) = -sin(\pi) + cos(\pi) = -1 < 0$

$f''(\frac{11\pi}{6}) = -sin(\frac{11\pi}{6}) + cos(\frac{11\pi}{6}) = 1.366 > 0$

This means that we have an inflection point at $x = \frac{\pi}{4}$ and at $x = \frac{5\pi}{4}$.

$f(\frac{\pi}{4}) = sin(\frac{\pi}{4}) - cos(\frac{\pi}{4}) = 0$

$f(\frac{5\pi}{4}) = sin(\frac{5\pi}{4}) - cos(\frac{5\pi}{4}) = 0$

We sketch the graph

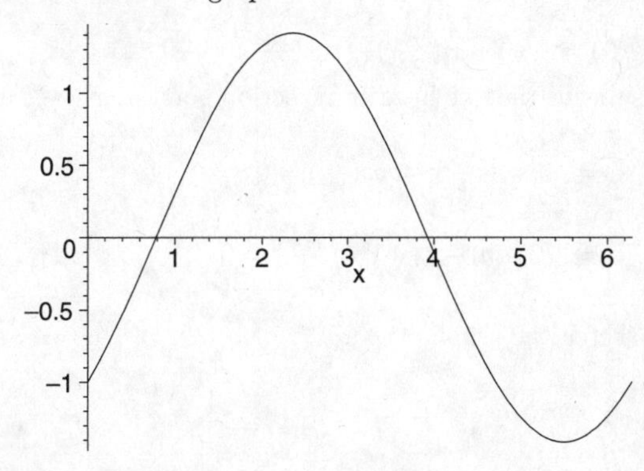

35. $f(x) = \sqrt{3}\ sin\ x + cos\ x$

$f'(x) = \sqrt{3}\ cos\ x - sin\ x$
We solve $f'(x) = 0$

$$\begin{aligned} \sqrt{3}\ cos\ x - sin\ x &= 0 \\ \sqrt{3}\ cos\ x &= sin\ x \\ tan\ x &= \sqrt{3} \\ x &= \frac{\pi}{3} \\ \text{and} \\ x &= \frac{4\pi}{3} \end{aligned}$$

The zeros of the first derivative divide the number line into three intervals. We apply test points to check the sign of the first derivative in each of those intervals. $f'(0) = \sqrt{3}\ cos(0) + sin(0) = 1.732 > 0$
$f'(\pi) = \sqrt{3}\ cos(\pi) + sin(\pi) = -1.732 < 0$
$f'(\frac{5\pi}{3}) = \sqrt{3}\ cos(\frac{5\pi}{3}) + sin(\frac{5\pi}{3}) = 1.732$

Thus, we have a relative maximum at $x = \frac{\pi}{3}$ and a relative minimum at $x = \frac{4\pi}{3}$.

$f(\frac{\pi}{3}) = \sqrt{3}\ sin(\frac{\pi}{3}) + cos(\frac{\pi}{3}) = 2$

$f(\frac{4\pi}{3}) = \sqrt{3}\ sin(\frac{4\pi}{3}) + cos(\frac{4\pi}{3}) = -2$

$$\begin{aligned} f''(x) &= -\sqrt{3}\ sin\ x - cos\ x \\ f''(x) &= 0 \\ -\sqrt{3}\ sin\ x - cos\ x &= 0 \\ tan\ x &= \frac{-1}{\sqrt{3}} \\ x &= \frac{5\pi}{6} \\ \text{and} \\ x &= \frac{11\pi}{6} \end{aligned}$$

The second derivative zeros divide the number line into three intervals, we determine the sign of the second derivative in each of those intervals:

$f''(0) = -\sqrt{3}\ sin(0) - cos(0) = -1 < 0$

$f''(\pi) = -\sqrt{3}\ sin(\pi) - cos(\pi) = 1 < 0$

$f''(\frac{35\pi}{18}) = -\sqrt{3}\ sin(\frac{35\pi}{18}) - cos(\frac{35\pi}{18}) = -0.845 < 0$

This means that we have an inflection point at $x = \frac{5\pi}{6}$ and at $x = \frac{11\pi}{6}$.

$f(\frac{5\pi}{6}) = \sqrt{3}\ sin(\frac{5\pi}{6}) + cos(\frac{5\pi}{6}) = 0$

$f(\frac{11\pi}{6}) = \sqrt{3}\ sin(\frac{11\pi}{6}) + cos(\frac{11\pi}{6}) = 0$

We sketch the graph

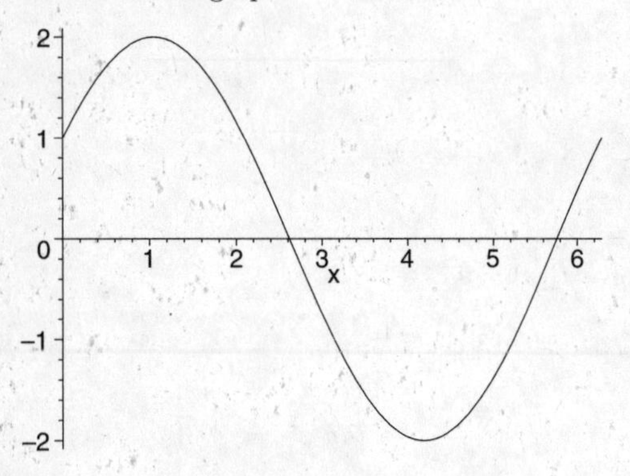

36. $f(x) = sin\ x - \sqrt{3}\ cos\ x$

$f'(x) = cos\ x + \sqrt{3}\ sin\ x$

We solve $f'(x) = 0$

$$\begin{aligned} cos\ x + \sqrt{3}\ sin\ x &= 0 \\ \sqrt{3}\ sin\ x &= -cos\ x \\ tan\ x &= \frac{1}{\sqrt{3}} \\ x &= \frac{5\pi}{6} \\ \text{and} \\ x &= \frac{11\pi}{6} \end{aligned}$$

The zeros of the first derivative divide the number line into three intervals. We apply test points to check the sign of the first derivative in each of those intervals. $f'(0) = cos(0) + \sqrt{3}\ sin(0) = 1 > 0$

$f'(\pi) = cos(\pi) + \sqrt{3}\ sin(\pi) = -1 < 0$

$f'(2\pi) = cos(2\pi) + \sqrt{3}\ sin(2\pi) = 1$

Thus, we have a relative maximum at $x = \frac{5\pi}{6}$ and a relative minimum at $x = \frac{11\pi}{6}$.

$f(\frac{5\pi}{6}) = sin(\frac{5\pi}{6}) - \sqrt{3}\ cos(\frac{5\pi}{6}) = 2$

$f(\frac{4\pi}{3}) = sin(\frac{11\pi}{6}) - \sqrt{3}\ cos(\frac{11\pi}{6}) = -2$

$$\begin{aligned} f''(x) &= -sin\ x + \sqrt{3}\ cos\ x \\ f''(x) &= 0 \\ -sin\ x + \sqrt{3}\ cos\ x &= 0 \\ tan\ x &= \frac{1}{\sqrt{3}} \\ x &= \frac{\pi}{6} \\ \text{and} \\ x &= \frac{7\pi}{6} \end{aligned}$$

The second derivative zeros divide the number line into three intervals, we determine the sign of the second derivative in each of those intervals:

$f''(0) = -sin(0) + \sqrt{3}\ cos(0) = 1.732 > 0$

$f''(\pi) = -sin(\pi) + \sqrt{3}\ cos(\pi) = -1.732 < 0$

$f''(2\pi) = -sin(2\pi) + \sqrt{3}\ cos(2\pi) = 1.732 > 0$

This means that we have an inflection point at $x = \frac{\pi}{6}$ and at $x = \frac{7\pi}{6}$.

$f(\frac{\pi}{6}) = sin(\frac{\pi}{6}) - \sqrt{3}\ cos(\frac{\pi}{6}) = -1$

$f(\frac{11\pi}{6}) = \sqrt{3}\ sin(\frac{11\pi}{6}) + cos(\frac{11\pi}{6}) = 1$

We sketch the graph

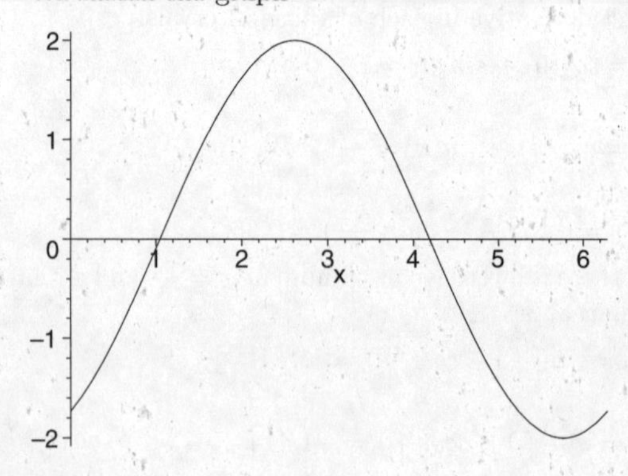

37. $f(x) = \frac{sin\ x}{2 - cos\ x}$

$f'(x) = \frac{2\ cos\ x - 1}{(cos\ x - 2)^2}$

We solve $f'(x) = 0$

$$\frac{2\ cos\ x - 1}{(cos\ x - 2)^2} = 0$$

$$\begin{aligned} \cos x &= \frac{1}{2} \\ x &= \frac{\pi}{3} \\ \text{and} \\ x &= \frac{5\pi}{3} \end{aligned}$$

The zeros of the first derivative divide the number line into three intervals. We apply test points to check the sign of the first derivative in each of those intervals. $f'(0) = \frac{2\ \cos(0)-1}{(\cos(0)-2)^2} = 1 > 0$
$f'(\pi) = \frac{2\ \cos(\pi)-1}{(\cos(\pi)-2)^2} = -\frac{1}{3} < 0$
$f'(2\pi) = \frac{2\ \cos(2\pi)-1}{(\cos(2\pi)-2)^2} = 1$

Thus, we have a relative maximum at $x = \frac{\pi}{3}$ and a relative minimum at $x = \frac{5\pi}{3}$.
$f(\frac{\pi}{3}) = \frac{\sin(\frac{\pi}{3})}{2-\cos(\frac{\pi}{3})} = 0.577$

$f(\frac{5\pi}{3}) = \frac{\sin(\frac{5\pi}{3})}{2-\cos(\frac{5\pi}{3})} = 0.174$

$$\begin{aligned} f''(x) &= \frac{2\ \sin x(1+\cos x)}{(\cos(x)-2)^3} \\ f''(x) &= 0 \\ \frac{2\ \sin x(1+\cos x)}{(\cos(x)-2)^3} &= 0 \\ \text{either} \\ \sin x &= 0 \\ x &= 0 \\ x &= \pi \\ x &= 2\pi \\ \text{or} \\ 1+\cos x &= 0 \\ \cos x &= -1 \\ x &= \pi \end{aligned}$$

The second derivative zeros divide the number line into two intervals (with the end points at 0 and 2π as potential inflection points), we determine the sign of the second derivative in each of those intervals:
$f''(\frac{\pi}{4}) = \frac{2\ \sin x(1+\cos(\frac{\pi}{4}))}{(\cos(\frac{\pi}{4})-2)^3} = -1.117 < 0$

$f''(\frac{3\pi}{2}) = \frac{2\ \sin x(1+\cos(\frac{3\pi}{2}))}{(\cos(\frac{3\pi}{2})-2)^3} = \frac{1}{4} > 0$

Note: checking a test point smaller than 0 will yield a positive value and testing a point larger than 2π will yield a negative value for the second derivative.

This means that we have an inflection point at $x = 0$, $x = \pi$ and at $x = 2\pi$.

$f(0) = \frac{\sin(0)}{2-\cos(0)} = 0$

$f(\pi) = \frac{\sin(\pi)}{2-\cos(\pi)} = 0$

$f(2\pi) = \frac{\sin(2\pi)}{2-\cos(2\pi)} = 0$

We sketch the graph

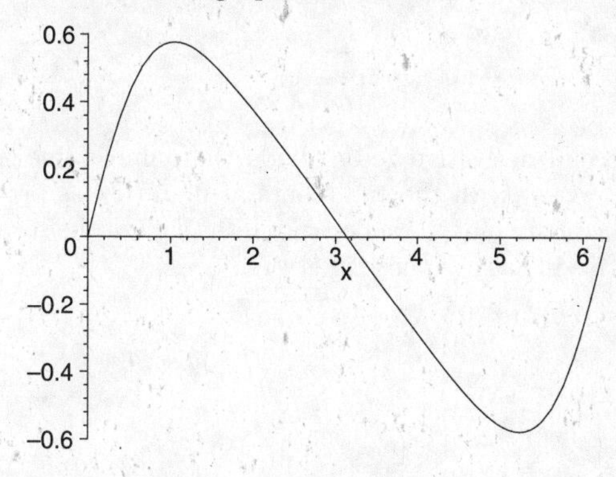

38. $f(x) = \frac{\cos x}{2+\sin x}$

$f'(x) = -\frac{\sin x}{2+\sin x} - \frac{\cos^2 x}{(2+\sin x)^2}$
We solve $f'(x) = 0$

$$\begin{aligned} -\frac{\sin x}{2+\sin x} - \frac{\cos^2 x}{(2+\sin x)^2} &= 0 \\ \sin x(2+\sin x) &= \cos x \\ \sin x &= \frac{-1}{2} \\ x &= \frac{7\pi}{6} \\ \text{and} \\ x &= \frac{11\pi}{6} \end{aligned}$$

The zeros of the first derivative divide the number line into three intervals. We apply test points to check the sign of the first derivative in each of those intervals. $f'(0) = -\frac{\sin(0)}{2+\sin(0)} - \frac{\cos^2(0)}{(2+\sin(0))^2} = -0.25 < 0$
$f'(\pi) = -\frac{\sin(\frac{3\pi}{2})}{2+\sin(\frac{3\pi}{2})} - \frac{\cos^2(\frac{3\pi}{2})}{(2+\sin(\frac{3\pi}{2}))^2} = 1 > 0$
$f'(2\pi) = -\frac{\sin(2\pi)}{2+\sin(2\pi)} - \frac{\cos^2(2\pi)}{(2+\sin(2\pi))^2} = -0.25$

Thus, we have a relative maximum at $x = \frac{11\pi}{6}$ and a relative minimum at $x = \frac{7\pi}{6}$.
$f(\frac{7\pi}{6}) = -\frac{\sin(\frac{7\pi}{6})}{2+\sin(\frac{7\pi}{6})} - \frac{\cos^2(\frac{7\pi}{6})}{(2+\sin(\frac{7\pi}{6}))^2} = -0.577$

$f(\frac{11\pi}{6}) = -\frac{\sin(\frac{11\pi}{6})}{2+\sin(\frac{11\pi}{6})} - \frac{\cos^2(\frac{11\pi}{6})}{(2+\sin(\frac{11\pi}{6}))^2} = 0.577$

$$\begin{aligned} f''(x) &= -\frac{\cos x}{2+\sin x} + \frac{3\ \sin x\ \cos x}{(2+\sin x)^2} + \\ &\quad \frac{2\cos^2 x}{(2+\sin x)^3} \\ f''(x) &= 0 \\ -2\ \cos x(\sin x - 1) &= 0 \\ \text{either} \\ \cos x &= 0 \\ x &= \frac{\pi}{2} \\ x &= \frac{3\pi}{2} \end{aligned}$$

or

$$\begin{aligned} \sin x - 1 &= 0 \\ \sin x &= 1 \\ x &= \frac{\pi}{2} \end{aligned}$$

The second derivative zeros divide the number line into two intervals (with the end points at 0 and 2π as potential inflection points), we determine the sign of the second derivative in each of those intervals:

$f''(0) = -0.25 < 0$

$f''(\pi) = 0.25 > 0$

$f''(2\pi) = -0.25 < 0$

This means that we have an inflection point at $x = \frac{\pi}{2}$ and at $x = \frac{3\pi}{2}$.

$f(\frac{\pi}{2}) = \frac{\cos(\frac{\pi}{2})}{2+\sin(\frac{\pi}{2})} = 0$

$f(\frac{3\pi}{2}) = \frac{\cos(\frac{3\pi}{2})}{2+\sin(\frac{3\pi}{2})} = 0$

We sketch the graph

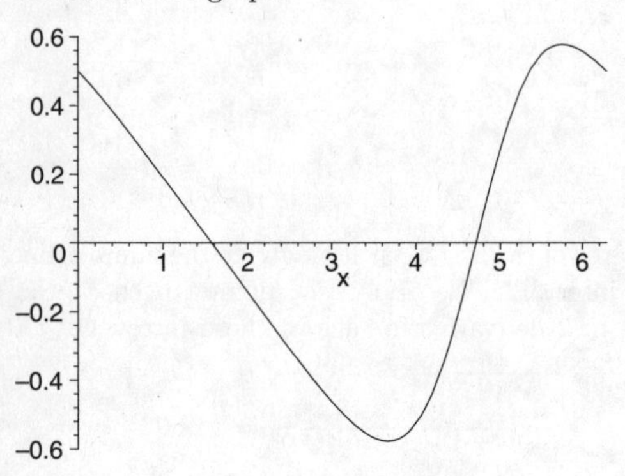

39. $f(x) = \cos^2 x$

$f'(x) = -2 \cos x \sin x = -\sin 2x$

We solve $f'(x) = 0$

$$\begin{aligned} -\sin 2x &= 0 \\ x &= 0 \\ x &= \frac{\pi}{2} \\ x &= \pi \\ x &= \frac{3\pi}{2} \\ x &= 2\pi \end{aligned}$$

The zeros of the first derivative divide the number line into four intervals, plus the two endpoints. We apply test points to check the sign of the first derivative in each of those intervals. $f'(\frac{\pi}{4}) = -\sin(\frac{\pi}{2}) = -1 < 0$

$f'(\frac{3\pi}{4}) = -\sin(\frac{3\pi}{2}) = 1 > 0$

$f'(\frac{5\pi}{4}) = -\sin(\frac{5\pi}{2}) = -1 < 0$

$f'(\frac{7\pi}{4}) = -\sin(\frac{7\pi}{2}) = 1 > 0$

Thus, we have a relative maximum at $x = \pi$ and a relative minimum at $x = \frac{\pi}{2}$, and $x = \frac{3\pi}{2}$.

$f(\pi) = \cos^2(\pi) = 1$

$f(\frac{\pi}{2}) = \cos^2(\frac{\pi}{2}) = 0$

$f(\frac{3\pi}{2}) = \cos^2(\frac{3\pi}{2}) = 0$

$f(0) = \cos^2(0) = 1$ and $f(2\pi) = \cos^2(2\pi) = 1$ which means that there is a relative maximum at $x = 0$ and $x = 2\pi$ as well.

$$\begin{aligned} f''(x) &= -2 \cos 2x \\ f''(x) &= 0 \\ -2 \cos 2x &= 0 \\ x &= \frac{\pi}{4} \\ x &= \frac{3\pi}{4} \\ x &= \frac{5\pi}{4} \\ x &= \frac{7\pi}{4} \end{aligned}$$

The second derivative zeros divide the number line into five intervals, we determine the sign of the second derivative in each of those intervals:

$f''(0) = -2 \cos(0) = -2 < 0$

$f''(\frac{\pi}{2}) = -2 \cos(\pi) = 2 > 0$

$f''(\pi) = -2 \cos(2\pi) = -2 < 0$

$f''(\frac{3\pi}{2}) = -2 \cos(3\pi) = 2 > 0$

$f''(2\pi) = -2 \cos(4\pi) = -2 < 0$

This means that we have an inflection point at $x = \frac{\pi}{4}$, $x = \frac{3\pi}{4}$, $x = \frac{5\pi}{4}$, and at $x = \frac{7\pi}{4}$.

$f(\frac{\pi}{4}) = -2 \cos^2(\frac{\pi}{4}) = 0.5$

$f(\frac{3\pi}{4}) = -2 \cos^2(\frac{3\pi}{4}) = 0.5$

$f(\frac{5\pi}{4}) = -2 \cos^2(\frac{5\pi}{4}) = 0.5$

$f(\frac{7\pi}{4}) = -2 \cos^2(\frac{7\pi}{4}) = 0.5$

We sketch the graph

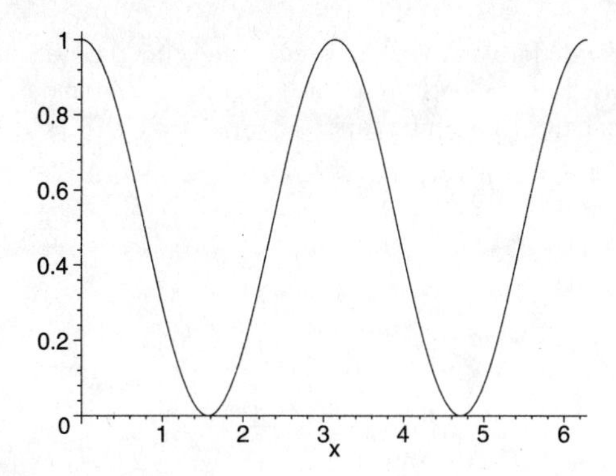

40. $f(x) = sin^4\ x$

$f'(x) = 4\ sin^3\ x\ cos\ x$
We solve $f'(x) = 0$

$$\begin{aligned} 4\ sin^3\ x\ cos\ x &= 0 \\ x &= 0 \\ x &= \frac{\pi}{2} \\ x &= \pi \\ x &= \frac{3\pi}{2} \\ x &= 2\pi \end{aligned}$$

The zeros of the first derivative divide the number line into four intervals, plus the two endpoints. We apply test points to check the sign of the first derivative in each of those intervals. $f'(\frac{\pi}{4}) = 4\ sin^3(\frac{\pi}{4})\ cos(\frac{\pi}{4}) = 1 > 0$

$f'(\frac{3\pi}{4}) = 4\ sin^3(\frac{3\pi}{4})\ cos(\frac{3\pi}{4}) = -1 < 0$

$f'(\frac{5\pi}{4}) = 4\ sin^3(\frac{5\pi}{4})\ cos(\frac{5\pi}{4}) = 1 > 0$

$f'(\frac{7\pi}{4}) = 4\ sin^3(\frac{7\pi}{4})\ cos(\frac{7\pi}{4}) = -1 < 0$

Thus, we have a relative maximum at $x = \frac{\pi}{2}$, and at $x = \frac{3\pi}{2}$ and a relative minimum at $x = \pi$
$f(\frac{\pi}{2}) = sin^4(\frac{\pi}{2}) = 1$

$f(\frac{3\pi}{2}) = sin^4(\frac{3\pi}{2}) = 1$

$f(\pi) = sin^4(\pi) = 0$

$f(0) = sin^4(0) = 0$ and $f(2\pi) = sin^4(2\pi) = 0$ which means that there is a relative minimum at $x = 0$ and $x = 2\pi$ as well.

$$\begin{aligned} f''(x) &= -4\ sin^4\ x + 12\ sin^2\ x\ cos^2\ x \\ &= -4\ sin^2\ x(sin^2\ x - 3\ cos^2\ x) \\ f''(x) &= 0 \end{aligned}$$

either

$$\begin{aligned} -4\ sin^2\ x &= 0 \\ x &= 0 \\ x &= \frac{\pi}{2} \end{aligned}$$

or

$$\begin{aligned} sin^2\ x - 3\ cos^2\ x &= 0 \\ tan^2\ x &= 3 \\ x &= \frac{\pi}{3} \\ x &= \frac{2\pi}{3} \\ x &= \pi \\ x &= \frac{4\pi}{3} \\ x &= \frac{5\pi}{3} \end{aligned}$$

The second derivative zeros divide the number line into six intervals, we determine the sign of the second derivative in each of those intervals:
$f''(\frac{\pi}{4}) = 2 > 0$

$f''(\frac{\pi}{2}) = -4 < 0$

$f''(\frac{3\pi}{4}) = 2 > 0$

$f''(\frac{5\pi}{4}) = 2 > 0$

$f''(\frac{3\pi}{2}) = -4 < 0$

$f''(\frac{7\pi}{4}) = 2 > 0$

This means that we have an inflection point at $x = \frac{\pi}{3}$, $x = \frac{2\pi}{3}$, $x = \frac{4\pi}{3}$, and at $x = \frac{5\pi}{3}$.

$f(\frac{\pi}{3}) = sin^4(\frac{\pi}{3}) = 0.5625$

$f(\frac{2\pi}{3}) = sin^4(\frac{2\pi}{3}) = 0.5625$

$f(\frac{4\pi}{3}) = sin^4(\frac{4\pi}{3}) = 0.5625$

$f(\frac{5\pi}{3}) = sin^4(\frac{5\pi}{3}) = 0.5625$

We sketch the graph

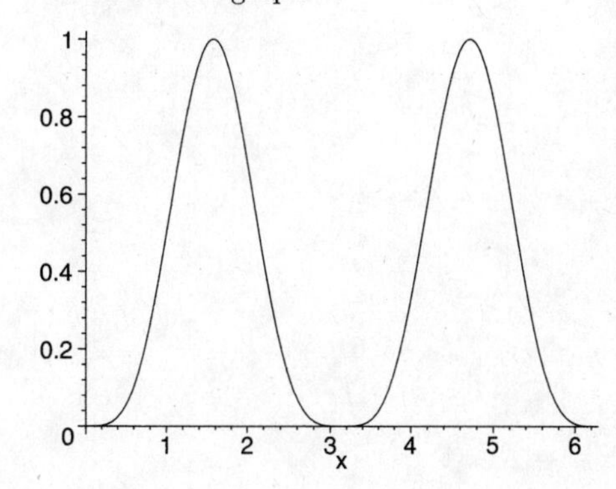

41. $f(x) = cos^4\ x$

$f'(x) = -4\ cos^3\ x\ sin\ x$
We solve $f'(x) = 0$

$$\begin{aligned} -4\ cos^3\ x\ sin\ x &= 0 \\ x &= 0 \\ x &= \frac{\pi}{2} \\ x &= \pi \\ x &= \frac{3\pi}{2} \\ x &= 2\pi \end{aligned}$$

The zeros of the first derivative divide the number line into four intervals, plus the two endpoints. We apply test points to check the sign of the first derivative in each of those intervals. $f'(\frac{\pi}{4}) = -4\ cos^3(\frac{\pi}{4})\ sin(\frac{\pi}{4}) = -1 < 0$

$f'(\frac{3\pi}{4}) = -4\ cos^3(\frac{3\pi}{4})\ sin(\frac{3\pi}{4}) = 1 > 0$

$f'(\frac{5\pi}{4}) = -4\ cos^3(\frac{5\pi}{4})\ sin(\frac{5\pi}{4}) = -1 < 0$

$f'(\frac{7\pi}{4}) = -4\ cos^3(\frac{7\pi}{4})\ cos(\frac{7\pi}{4}) = 1 > 0$

Thus, we have a relative minimum at $x = \frac{\pi}{2}$, and at $x = \frac{3\pi}{2}$ and a relative maximum at $x = \pi$

$f(\frac{\pi}{2}) = cos^4(\frac{\pi}{2}) = 0$

$f(\frac{3\pi}{2}) = cos^4(\frac{3\pi}{2}) = 0$

$f(\pi) = cos^4(\pi) = 1$

$f(0) = cos^4(0) = 1$ and $f(2\pi) = cos^4(2\pi) = 1$ which means that there is a relative maximum at $x = 0$ and $x = 2\pi$ as well.

$$\begin{aligned} f''(x) &= 12sin^2\ x\ cos^2\ x - 4\ cos^4\ x \\ &= 4\ cos^2\ x(3sin^2\ x -\ cos^2\ x) \\ f''(x) &= 0 \\ \text{either} & \\ 4\ cos^2\ x &= 0 \\ x &= \frac{\pi}{2} \\ x &= \frac{3\pi}{2} \\ \text{or} & \\ 3sin^2\ x -\ cos^2\ x &= 0 \\ tan^2\ x &= \frac{1}{3} \\ x &= \frac{\pi}{6} \\ x &= \frac{5\pi}{6} \\ x &= \frac{7\pi}{6} \\ x &= \frac{5\pi}{3} \\ x &= \frac{11\pi}{6} \end{aligned}$$

The second derivative zeros divide the number line into seven intervals, we determine the sign of the second derivative in each of those intervals:

$f''(0) = 4\ cos^2(0)(3\ sin^2(0) - cos^2(0)) = -4 < 0$

$f''(\frac{\pi}{4}) = 4\ cos^2(\frac{\pi}{4})(3\ sin^2(\frac{\pi}{4}) - cos^2(\frac{\pi}{4})) = 2 > 0$

$f''(\frac{3\pi}{4}) = 4\ cos^2(\frac{3\pi}{4})(3\ sin^2(\frac{3\pi}{4}) - cos^2(\frac{3\pi}{4})) = 2 > 0$

$f''(\pi) = 4\ cos^2(\pi)(3\ sin^2(\pi) - cos^2(\pi)) = -4 < 0$

$f''(\frac{5\pi}{4}) = 4\ cos^2(\frac{5\pi}{4})(3\ sin^2(\frac{5\pi}{4}) - cos^2(\frac{5\pi}{4})) = -4 < 0$

$f''(\frac{7\pi}{4}) = 4\ cos^2(\frac{7\pi}{4})(3\ sin^2(\frac{7\pi}{4}) - cos^2(\frac{7\pi}{4})) = -4 < 0$

$f''(2\pi) = 4\ cos^2(2\pi)(3\ sin^2(2\pi) - cos^2(2\pi)) = -4 < 0$

This means that we have an inflection point at $x = \frac{\pi}{6}$, $x = \frac{5\pi}{6}$, $x = \frac{7\pi}{6}$, and at $x = \frac{11\pi}{6}$.

$f(\frac{\pi}{6}) = cos^4(\frac{\pi}{6}) = 0.5625$

$f(\frac{5\pi}{6}) = cos^4(\frac{5\pi}{6}) = 0.5625$

$f(\frac{7\pi}{6}) = cos^4(\frac{7\pi}{3}) = 0.5625$

$f(\frac{11\pi}{3}) = cos^4(\frac{11\pi}{3}) = 0.5625$

We sketch the graph

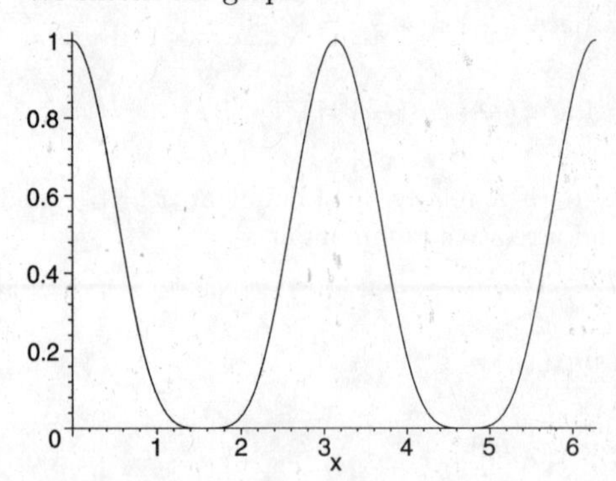

42. $f(x) = 3\ sin\ x - sin^3\ x$

$f'(x) = 3\ cos\ x - 3\ sin^2\ x\ cos\ x = 3\ cos^3\ x$
We solve $f'(x) = 0$

$$\begin{aligned} 3\ cos^3\ x &= 0 \\ x &= \frac{\pi}{2} \\ x &= \frac{3\pi}{2} \end{aligned}$$

The zeros of the first derivative divide the numler line into three intervals. We apply test points to check the sign of the first derivative in each of those intervals.

$f'(\frac{\pi}{4}) = 1.061 > 0$

$f'(\pi) = -3 < 0$

$f'(\frac{7\pi}{4}) = 1.061 > 0$

Thus, we have a relative maximum at $x = \frac{\pi}{2}$ and a relative minimum at $x = \frac{3\pi}{2}$.

$f(\frac{\pi}{2}) = 3\ sin(\frac{\pi}{2}) - sin^3(\frac{\pi}{2}) = 2$

$f(\frac{3\pi}{2}) = 3\ sin(\frac{3\pi}{2}) - sin^3(\frac{3\pi}{2}) = -2$

$$\begin{aligned} f''(x) &= -3\ sin\ x - 6\ sin\ x\ cos^2\ x + 3\ sin^3\ x \\ &= -9\ sin\ x\ cos^2\ x \\ f''(x) &= 0 \\ \text{either} \\ -9\ sin\ x &= 0 \\ x &= 0 \\ x &= \pi \\ x &= 2\pi \\ \text{or} \\ cos\ x &= 0 \\ x &= \frac{\pi}{2} \\ x &= \frac{3\pi}{2} \end{aligned}$$

The second derivative zeros divide the number line into four intervals, we determine the sign of the second derivative in each of those intervals:

$f''(\frac{\pi}{4}) = -3.182 < 0$

$f''(\frac{3\pi}{4}) = -3.182 < 0$

$f''(\frac{5\pi}{4}) = 3.182 > 0$

$f''(\frac{7\pi}{4}) = 3.182 > 0$

This means that we have an inflection point at $x = \pi$.

$f'(\pi) = 3\ sin(\pi) - sin^3(\pi) = 0$

We sketch the graph

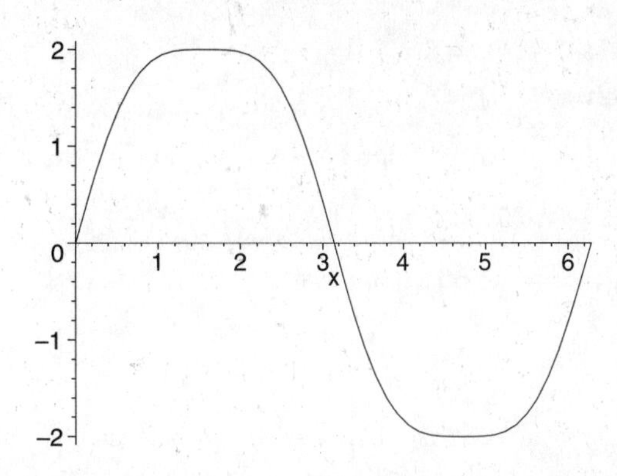

43. $f(x) = x^3 + 3x + 1$

$f'(x) = 3x^2 + 3$

$f''(x) = 6x$

$f''(x)$ exists for all values of x, so we solve $f''(x) = 0$.

$6x = 0$

$x = 0$ Possible inflection point

We use the possible inflection point, 0, to divide the real number line into two intervals, A: $(-\infty, 0)$ and B: $(0, \infty)$. Test a point in each interval.

A: Test -1, $f''(-1) = 6(-1) = -6 < 0$

B: Test 1, $f''(1) = 6 \cdot 1 = 6 > 0$

Then f is concave down on $(-\infty, 0)$ and concave up on $(0, \infty)$. We find that $f(0) = 0^3 + 3 \cdot 0 + 1 = 1$, so $(0, 1)$ is an inflection point.

44. $f(x) = x^3 - 6x^2 + 12x - 6$

$f'(x) = 3x^2 - 12x + 12$

$f''(x) = 6x - 12$

$f''(x)$ exists for all real numbers. Solve:

$6x - 12 = 0$

$x = 2$ Possible inflection point

Use 2 to divide the real number line into two intervals, A: $(-\infty, 2)$ and B: $(2, \infty)$.

A: Test 0, $f''(0) = 6 \cdot 0 - 12 = -12 < 0$

B: Test 3, $f''(3) = 6 \cdot 3 - 12 = 6 > 0$

Note that $f(2) = 2^3 - 6 \cdot 2^2 + 12 \cdot 2 - 6 = 2$. Then $(2, 2)$ is an inflection point.

45. $f(x) = \frac{4}{3}x^3 - 2x^2 + x$

$f'(x) = 4x^2 - 4x + 1$

$f''(x) = 8x - 4$

$f''(x)$ exists for all values of x, so we solve $f''(x) = 0$.

$8x - 4 = 0$

$8x = 4$

$x = \frac{1}{2}$ Possible inflection point

We use the possible inflection point, $\frac{1}{2}$, to divide the real number line into two intervals, A: $\left(-\infty, \frac{1}{2}\right)$ and B: $\left(\frac{1}{2}, \infty\right)$. Test a point in each interval.

A: Test 0, $f''(0) = 8 \cdot 0 - 4 = -4 < 0$
B: Test 1, $f''(1) = 8 \cdot 1 - 4 = 4 > 0$

Then f is concave down on $\left(-\infty, \frac{1}{2}\right)$ and concave up on $\left(\frac{1}{2}, \infty\right)$. We find that $f\left(\frac{1}{2}\right) = \frac{4}{3}\left(\frac{1}{2}\right)^3 - 2\left(\frac{1}{2}\right)^2 + \frac{1}{2} = \frac{1}{6}$, so $\left(\frac{1}{2}, \frac{1}{6}\right)$ is an inflection point.

46. $f(x) = x^4 - 4x^3 + 10$

$f'(x) = 4x^3 - 12x^2$

$f''(x) = 12x^2 - 24x$

$f''(x)$ exists for all real numbers. Solve:

$12x^2 - 24x = 0$

$12x(x-2) = 0$

$x = 0$ *or* $x = 2$ Possible inflection points

Use 0 and 2 to divide the real number line into three intervals, A: $(-\infty, 0)$, B: $(0, 2)$, and C: $(2, \infty)$.

A: Test -1, $f''(-1) = 12(-1)^2 - 24(-1) = 36 > 0$

B: Test 1, $f''(1) = 12 \cdot 1^2 - 24 \cdot 1 = -12 < 0$

C: Test 3, $f''(3) = 12 \cdot 3^2 - 24 \cdot 3 = 36 > 0$

Note that $f(0) = 0^4 - 4 \cdot 0^3 + 10 = 10$ and $f(2) = 2^4 - 4 \cdot 2^3 + 10 = -6$. Then $(0, 10)$ and $(2, -6)$ are inflection points.

47. $f(x) = x - \sin x$

$f'(x) = 1 - \cos x$

$f''(x) = \sin x$

$f''(x)$ exists for all real numbers. Solve:

$\sin x = 0$

$x = n\,\pi$ Possible inflection points

Use -2π, $-\pi$, 0, π, and 2π to divide the real number line,

A: Test $-\frac{3\pi}{2}$, $f''(\frac{-3\pi}{2}) = \sin(\frac{-3\pi}{2}) = 1 > 0$

B: Test $-\frac{\pi}{2}$, $f''(-\frac{\pi}{2}) = \sin(\frac{-\pi}{2}) = -1 < 0$

C: Test $\frac{\pi}{2}$, $f''(\frac{\pi}{2}) = \sin(\frac{\pi}{2}) = 1 > 0$

D: Test $\frac{3\pi}{2}$, $f''(\frac{3\pi}{2}) = \sin(\frac{3\pi}{2}) = -1 < 0$

Note that since $\sin/x$ is a periodic function this patern of changing signs will continue indefinitely. Therefore, the inflection points will occur at $n\pi$.

48. $f(x) = 2x + 1 + \cos 2x$

$f'(x) = 2 - 2\sin 2x$
$f''(x) = -4\cos 2x$

$f''(x)$ exists for all values of x.

$$\begin{aligned} f''(x) &= 0 \\ -4\cos 2x &= 0 \\ 2x &= \frac{\pi}{2} + n\pi \\ x &= \frac{\pi}{4} + \frac{n\pi}{2} \end{aligned}$$

Use $\frac{-\pi}{2}$, $\frac{-\pi}{4}$, $\frac{\pi}{4}$, and $\frac{\pi}{2}$ to divide the real number line,

A: Test $-\pi$, $f''(\pi) = -4\cos(2\pi) = -4 < 0$
B: Test $-\frac{\pi}{3}$, $f''(-\frac{\pi}{3}) = -4\cos(\frac{-2\pi}{3}) = 2 > 0$
C: Test 0, $f''(0) = -4\cos(0) = -4 < 0$
D: Test $\frac{\pi}{4}$, $f''(\frac{\pi}{4}) = -4\cos(\frac{\pi}{2}) = 2 < 0$
E: Test $\frac{\pi}{2}$, $f''(\frac{\pi}{2}) = -4\cos(\pi) = -4 < 0$
Note that since $\cos/2x$ is a periodic function this patern of changing signs will continue indefinitely. Therefore, the inflection points will occur at $\frac{\pi}{4} + \frac{n\pi}{2}$.

49. $f(x) = \tan x$

$f'(x) = \sec^2 x$
$f''(x) = 2\sec^2 x \tan x$

$f''(x)$ does not exist for $x = \frac{\pi}{2} + n\pi$.

$$\begin{aligned} f''(x) &= 0 \\ 2\sec^2 x \tan x &= 0 \\ x &= n\pi \end{aligned}$$

Use -2π, $-\pi$, 0, π, and 2π to divide the real number line,
A: Test $-\frac{5\pi}{4}$, $f''(\frac{-5\pi}{4}) = 2\sec^2(\frac{-5\pi}{4})\tan(\frac{-5\pi}{4}) = 4 > 0$
B: Test $-\frac{\pi}{4}$, $f''(-\frac{\pi}{4}) = 2\sec^2(\frac{-\pi}{4})\tan(\frac{-\pi}{4}) = -4 < 0$
C: Test $\frac{\pi}{4}$, $f''(\frac{\pi}{4}) = 2\sec^2(\frac{\pi}{4})\tan(\frac{\pi}{4}) = 4 > 0$
D: Test $\frac{5\pi}{4}$, $f''(\frac{5\pi}{4}) = 2\sec^2(\frac{5\pi}{4})\tan(\frac{5\pi}{4}) = -4 < 0$
Note that since $2\sec^2\ \tan x$ is a periodic function this patern of changing signs will continue indefinitely. Therefore, the inflection points will occur at $n\pi$.

50. $f(x) = \cot x$

$f'(x) = -\csc^2 x$
$f''(x) = 2\csc^2 x \cot x$

$f''(x)$ does not exist for $x = n\pi$.

$$\begin{aligned} f''(x) &= 0 \\ 2\csc^2 x \cot x &= 0 \\ x &= \frac{\pi}{2} + n\pi \end{aligned}$$

Use $\frac{-3\pi}{2}$, $\frac{-\pi}{2}$, $\frac{\pi}{2}$, and $\frac{3\pi}{2}$ to divide the real number line,
A: Test $-\frac{5\pi}{4}$, $f''(\frac{-5\pi}{4}) = 2\csc^2(\frac{-5\pi}{4})\cot(\frac{-5\pi}{4}) = 4 > 0$
B: Test $-\frac{\pi}{4}$, $f''(-\frac{\pi}{4}) = 2\csc^2(\frac{-\pi}{4})\cot(\frac{-\pi}{4}) = -4 < 0$
C: Test $\frac{\pi}{4}$, $f''(\frac{\pi}{4}) = 2\csc^2(\frac{\pi}{4})\cot(\frac{\pi}{4}) = 4 > 0$
D: Test $\frac{5\pi}{4}$, $f''(\frac{5\pi}{4}) = 2\csc^2(\frac{5\pi}{4})\cot(\frac{5\pi}{4}) = -4 < 0$
Note that since $2\csc^2 x \cot x$ is a periodic function this patern of changing signs will continue indefinitely. Therefore, the inflection points will occur at $\frac{\pi}{2} + n\pi$.

51. $f(x) = \tan x + \sec x$

$f'(x) = \sec^2 x + \sec x \tan x$

$$\begin{aligned} f''(x) &= 2\sec^2 x \tan x + \sec^3 x + \tan^2 x \sec x \\ &= \frac{-\cos^2 x + 2\sin x + 2}{\cos^3 x} \\ &= \frac{\sin^2 x - 1 + 2\sin x + 2}{\cos^3 x} \\ &= \frac{(\sin x + 1)^2}{\cos^3 x} \end{aligned}$$

The values that make $f''(x) = 0$ are not in the domain of the second derivative, therefore there are no inflection points.

52. $f(x) = cot\ x + csc\ x$

$f'(x) = -csc^2\ x - csc\ x\ cot\ x$

$$\begin{aligned} f''(x) &= 2\ csc^2\ x\ cot\ x + csc^3\ x + cot^2\ x\ csc\ x \\ &= \frac{cos^2\ x + 2\ cos\ x + 1}{sin^3\ x} \\ &= \frac{(cos\ x + 1)^2}{sin^3\ x} \end{aligned}$$

The values that make $f''(x) = 0$ are not in the domain of the second derivative, therefore there are no inflection points.

53. - 104 Left to the student.

105. $V(r) = k(20r^2 - r^3) = 20kr^2 - kr^3$, $0 \le r \le 20$
The maximum occurs at the critical value of the function, since the endpoints result in a value of zero.

$$\begin{aligned} V'(r) &= 40kr - 3kr^2 \\ V'(r) &= 0 \\ 40kr - 3kr^2 &= 0 \\ kr(40 - 3r) &= 0 \\ r &= 0 \text{ not acceptable} \\ 40 - 3r &= 0 \\ r &= \frac{40}{3} \end{aligned}$$

106. $T(x) = 0.0027(x^3 - 34x + 240)$, $0 \le x \le 24$
The minimum occurs at the critical value.

$$\begin{aligned} T'(x) &= 0.0027(3x^2 - 34) \\ T'(x) &= 0 \\ 0.0027(3x^2 - 34) &= 0 \\ x &= -\sqrt{\frac{34}{3}} \text{ not acceptable} \\ x &= \sqrt{\frac{34}{3}} \\ &\approx 3.3665 \end{aligned}$$

The minimum occurs about 3.4 hours after midnight and the minimum value is

$$\begin{aligned} T(3.4) &= 0.0027([3.4]^3 - 34[3.4] + 240 \\ &= 0.442 \end{aligned}$$

107. $T(x) = 0.0338x^4 - 0.996x^3 + 8.57x^2 - 18.4x + 43.5$

a)

$$\begin{aligned} T'(x) &= 0.1352x^3 - 2.988x^2 + 17.14x - 18.4 \\ T''(x) &= 0.4056x^2 - 5.976x + 17.14 \end{aligned}$$

Using the quadratic formula we solve $T''(x) = 0$ to get

$$\begin{aligned} x &= 10.833 \approx 11 \\ x &= 3.901 \approx 4 \end{aligned}$$

The zeros of the second derivative (middle od April and middle of November) divide the number line into three intervals. We next use test points to check the sign of the second derivative at each of those intervals.
$T''(0) = 0.4056(0)^2 - 5.976(0) + 17.14 = 17.14 > 0$
$T''(10) = 0.4056(10)^2 - 5.976(10) + 17.14 = -2.06 < 0$
$T''(12) = 0.4056(12)^2 - 5.976(12) + 17.14 = 3.83 > 0$
This means that both zeros of the second derivative are inflection points.

b) The inflection points represent the values at which the rate of change of the temperature is changing the fastest or slowest.

108. $N(x) = 748x^3 - 6820x^2 - 5520x + 916000$

a)

$$\begin{aligned} N'(x) &= 2244x^2 - 13640x - 5520 \\ N'(x) &= 0 \\ 2244x^2 - 13640x - 5520 &= 0 \\ x &= -0.381 \text{ not acceptable} \\ x &= 6.459 \end{aligned}$$

The zero of the first derivative divides the number line into two intervals. We next test the sign of the first derivative in each of those intervals.
$N'(0) = -5520 < 0$
$N'(10) = 82480 > 0$ Therefore, there was a relative minimum after 6.5 years (middle of 1996) that has a value of

$$\begin{aligned} N(6.5) &= 748(6.5)^3 - 6820(6.5)^2 - 5520(6.5) + 916000 \\ &= 7977394.5 \end{aligned}$$

b)

$$\begin{aligned} N''(x) &= 4488x - 13640 \\ N''(x) &= 0 \\ 4488x - 13640 &= 0 \\ x &= 3.039 \end{aligned}$$

The zero of the second derivative divides the number line into two intervals. We next test the sign of the second derivative in each of those intervals.
$N''(0) = -13640 < 0$
$N''(10) = 31240 > 0$
Thus, there was an inflection point shortly after 1993 that has a value of

$$\begin{aligned} N(3.039) &= 748(3.039)^3 - 6820(3.039)^2 - \\ &\quad 5520(3.039) + 916000 \\ &= 857232.4 \end{aligned}$$

c) 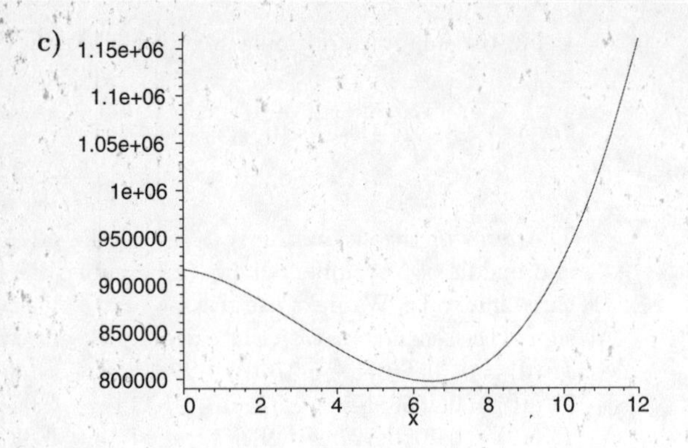

109. $N(x) = -0.00006x^3 + 0.006x^2 - 0.1x + 1.9$

a)

$$\begin{aligned} -0.00018x^2 + 0.012x - 0.1 &= N'(x) \\ N'(x) &= 0 \\ -0.00018x^2 + 0.012x - 0.1 &= 0 \end{aligned}$$

Using the quadratic formula

$$\begin{aligned} x &= 9.763 \\ x &= 56.904 \end{aligned}$$

The zeros of the first derivative (ages 9.763 and 56.904) divide the number line into three intervals. We next check the sign of the first derivative in each of those intervals.
$N'(0) = -0.00018(0)^2 + 0.012(0) - 0.1 = -0.1 < 0$
$N'(10) = -0.00018(10)^2 + 0.012(10) - 0.1 = 0.002 > 0$
$N'(80) = -0.00018(80)^2 + 0.012(80) - 0.1 = -0.292 < 0$
Thus, we have a relative maximum at $x = 56.904$ and a relative minim at $x = 9.763$.

$$\begin{aligned} N(9.763) &= -0.00006(9.763)^3 + 0.006(9.763)^2 - \\ &\quad 0.1(9.763) + 1.9 \\ &= 1.440 \\ N(56.904) &= -0.00006(56.904)^3 + 0.006(56.904)^2 - \\ &\quad 0.1(56.904) + 1.9 \\ &= 4.582 \end{aligned}$$

b)

$$\begin{aligned} N''(x) &= -0.00036x + 0.012 \\ N''(x) &= 0 \\ -0.00036x + 0.012 &= 0 \\ x &= \frac{100}{3} \end{aligned}$$

The zero of the second derivative divides the number line into two intervals. We check the sign of the second derivative in each of those of intervals.
$N''(10) = -0.00036(10) + 0.012 = 0.0084 > 0$
$N''(50) = -0.00036(50) + 0.012 = -0.006 < 0$
Thus, there is an inflection point at $x = \frac{100}{3}$ that has a value of

$$\begin{aligned} N\left(\frac{100}{3}\right) &= -0.00006\left(\frac{100}{3}\right)^3 + 0.006\left(\frac{100}{3}\right)^2 - \\ &\quad 0.1\left(\frac{100}{3}\right) + 1.9 \\ &= 3.011 \end{aligned}$$

c) 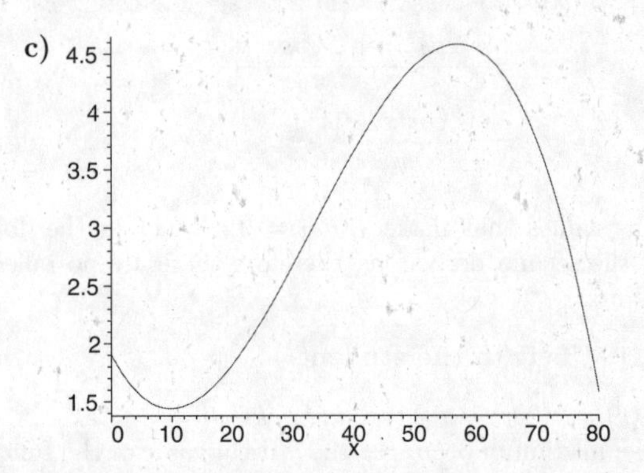

110. g is the graph of the derivative of h since the maximums and minumums occur where g is zero.

111. h is the graph of the derivative of g since the maximums and minumums occur where h is zero.

112. **a)** Relative maximum at $(1, 5)$ and $(5, 5)$
b) Relative mu=inimum at $(3, 1)$
c) Inflection points at $(2, 3)$ and $(4, 3)$
d) Increasing on $(0, 1)$ and $(3, 5)$
e) Decreasing on $(1, 3)$ and $(5, 6)$
f) Concave up on $(2, 4)$
g) Concave down on $(0, 2)$ and $(4, 6)$

113. **a)** Relative maximum at approximately $(2, 7)$
b) Relative mu=inimum at $(8, 0)$
c) Inflection points at $(4, 1)$
d) Increasing on $(0, 2)$ and $(8, 12)$
e) Decreasing on $(2, 8)$
f) Concave up on $(4, 12)$
g) Concave down on $(0, 4)$

114. Left to the student.

115. Left to the student. (Answers vary)

116. **a)** Applying the point $(0, 0)$ we get

$$\begin{aligned} 0 &= a\ cos(0) + k \\ a &= -k \end{aligned}$$

Thus

$$y = a\ cos\ bx - a$$

Applying the points $(\pi/4, 1/2)$ and $(3\pi/4, 1/2)$ we get

$$\begin{aligned} \frac{1}{2} &= a\ cos(b\pi/4) - a \\ \frac{1}{2} &= a\ cos(3b\pi/4) - a \end{aligned}$$

Thus

$$\begin{aligned} cos(b\pi/4) &= cos(3b\pi/4) \\ b &= 2 \end{aligned}$$

Now,

$$y = a\ cos\ 2x - a$$

Using the point $(\pi/2, 1)$ we get

$$\begin{aligned} 1 &= a\ cos(\pi) - a \\ 1 &= -2a \\ a &= \frac{-1}{2} \end{aligned}$$

Therefore,

$$y = -\frac{1}{2}\ cos\ 2x + \frac{1}{2}$$

b) Left to the student

c)

$$\begin{aligned} f(x) &= sin^2\ x \\ f'(x) &= 2\ sin\ x\ cos\ x \\ y &= -\frac{1}{2}\ cos\ 2x + \frac{1}{2} \\ \frac{dy}{dx} &= -\frac{1}{2} \cdot -2\ sin\ 2x \\ &= sin\ 2x \end{aligned}$$

Thus,

$$sin^2\ x = -\frac{1}{2}\ cos\ 2x + \frac{1}{2}$$

and

$$sin\ 2x = 2\ sin\ x\ cos\ x$$

117. Relative minimum at $(0,0)$, and a relative maximum at $(1,1)$

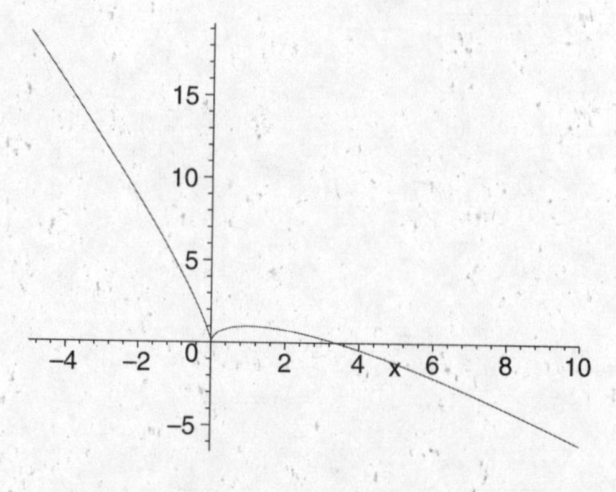

118. Relative maximum at $(0,0)$ and a relative minimum at $(1,-2)$

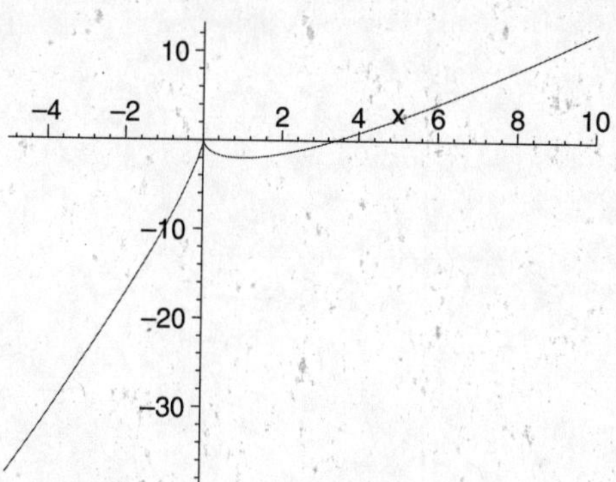

119. Relative maximum at $(0,0)$ and a relative minimum at $(0.8,-1.1)$

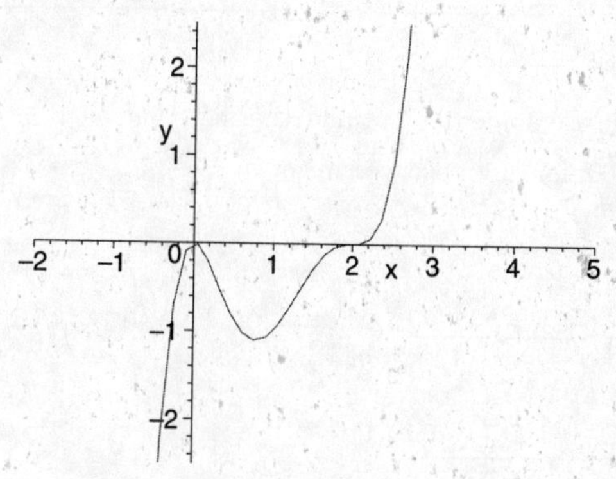

120. No relative extrema

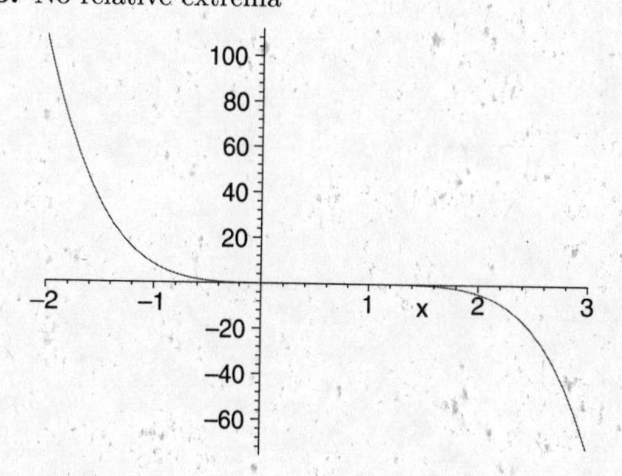

121. Relative minimum at $(0.25,-0.25)$

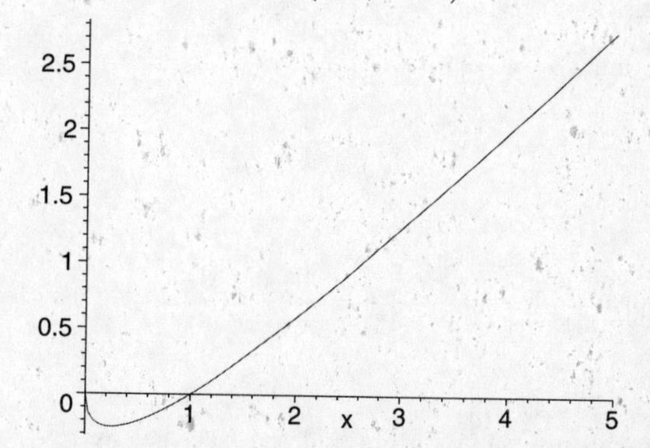

122. Relative maximum at $(-1, 1.414)$ and a relative minimum at $(1, -1.414)$

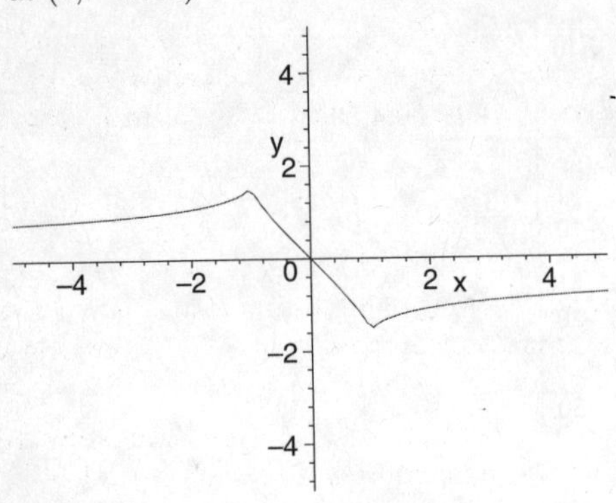

Exercise Set 3.3

1. Find $\lim_{x\to\infty} \frac{2x-4}{5x}$.

We will use some algebra and the fact that as $x \to \infty$, $\frac{b}{ax^n} \to 0$, for any positive integer n.

$$\lim_{x\to\infty} \frac{2x-4}{5x}$$

$$= \lim_{x\to\infty} \frac{2x-4}{5x} \cdot \frac{(1/x)}{(1/x)} \quad \text{Multiplying by a form of 1}$$

$$= \lim_{x\to\infty} \frac{2x \cdot \frac{1}{x} - 4 \cdot \frac{1}{x}}{5x \cdot \frac{1}{x}}$$

$$= \lim_{x\to\infty} \frac{2 - \frac{4}{x}}{5}$$

$$= \frac{2-0}{5} \quad \text{As } x \to \infty, \frac{4}{x} \to 0.$$

$$= \frac{2}{5}$$

2. $\lim_{h\to 0} \frac{-5}{x(x+h)} = \frac{-5}{x^2}$

3. Find $\lim_{x\to\infty} \left(5 - \frac{2}{x}\right)$.

We will use the fact that as $x \to \infty$, $\frac{b}{ax^n} \to 0$, for any positive integer n.

$$\lim_{x\to\infty} \left(5 - \frac{2}{x}\right)$$

$$= 5 - 0 \quad \text{As } x \to \infty, \frac{2}{x} \to 0.$$

$$= 5$$

4. $\lim_{x\to\infty} \frac{3x+1}{4x} = \lim_{x\to\infty} \frac{3+\frac{1}{x}}{4} = \frac{3+0}{4} = \frac{3}{4}$

5. Find $\lim_{x\to\infty} \frac{2x-5}{4x+3}$.

We will use some algebra and the fact that as $x \to \infty$, $\frac{b}{ax^n} \to 0$, for any positive integer n.

$$\lim_{x\to\infty} \frac{2x-5}{4x+3}$$

$$= \lim_{x\to\infty} \frac{2x-5}{4x+3} \cdot \frac{(1/x)}{(1/x)} \quad \text{Multiplying by a form of 1}$$

$$= \lim_{x\to\infty} \frac{2x \cdot \frac{1}{x} - 5 \cdot \frac{1}{x}}{4x \cdot \frac{1}{x} + 3 \cdot \frac{1}{x}}$$

$$= \lim_{x\to\infty} \frac{2 - \frac{5}{x}}{4 + \frac{3}{x}}$$

$$= \frac{2-0}{4+0} \quad \text{As } x \to \infty, \frac{5}{x} \to 0 \text{ and } \frac{3}{x} \to 0.$$

$$= \frac{2}{4} = \frac{1}{2}$$

6. $\lim_{x\to\infty} \left(7 + \frac{3}{x}\right) = 7 + 0 = 7$

7. Find $\lim_{x\to\infty} \frac{2x^2-5}{3x^2-x+7}$.

We will use some algebra and the fact that as $x \to \infty$, $\frac{b}{ax^n} \to 0$, for any positive integer n.

$$\lim_{x\to\infty} \frac{2x^2-5}{3x^2-x+7}$$

$$= \lim_{x\to\infty} \frac{2x^2-5}{3x^2-x+7} \cdot \frac{(1/x^2)}{(1/x^2)} \quad \text{Multiplying by a form of 1}$$

$$= \lim_{x\to\infty} \frac{2x^2 \cdot \frac{1}{x^2} - 5 \cdot \frac{1}{x^2}}{3x^2 \cdot \frac{1}{x^2} - x \cdot \frac{1}{x^2} + 7 \cdot \frac{1}{x^2}}$$

$$= \lim_{x\to\infty} \frac{2 - \frac{5}{x^2}}{3 - \frac{1}{x} + \frac{7}{x^2}}$$

$$= \frac{2-0}{3-0+0} \quad \text{As } x \to \infty, \frac{5}{x^2} \to 0, \frac{1}{x} \to 0, \text{ and } \frac{7}{x^2} \to 0$$

$$= \frac{2}{3}$$

8. $\lim_{x\to\infty} \frac{6x+1}{5x-2} = \lim_{x\to\infty} \frac{6+\frac{1}{x}}{5-\frac{2}{x}} = \frac{6+0}{5-0} = \frac{6}{5}$

9. Find $\lim_{x\to\infty} \frac{4-3x}{5-2x^2}$.

We divide the numerator and the denominator by x^2, the highest power of x in the denominator.

$$\lim_{x\to\infty} \frac{4-3x}{5-2x^2} = \lim_{x\to\infty} \frac{\frac{4}{x^2} - \frac{3}{x}}{\frac{5}{x^2} - 2}$$

$$= \frac{0-0}{0-2}$$

$$= 0$$

10. $\lim_{x\to\infty} \dfrac{4-3x-12x^2}{1+5x+3x^2} = \lim_{x\to\infty} \dfrac{\dfrac{4}{x^2}-\dfrac{3}{x}-12}{\dfrac{1}{x^2}+\dfrac{5}{x}+3} =$

$\dfrac{0-0-12}{0+0+3} = \dfrac{-12}{3} = -4$

11. Find $\lim_{x\to\infty} \dfrac{8x^4-3x^2}{5x^2+6x}$.

We divide the numerator and the denominator by x^2, the highest power of x in the denominator.

$$\lim_{x\to\infty} \frac{8x^4-3x^2}{5x^2+6x} = \lim_{x\to\infty} \frac{8x^2-3}{5+\dfrac{6}{x}}$$

$$= \frac{\lim_{x\to\infty} 8x^2-3}{5+0} = \infty$$

12. $\lim_{x\to\infty} \dfrac{6x^2-x}{4x^4-3x^3} = \lim_{x\to\infty} \dfrac{\dfrac{6}{x^2}-\dfrac{1}{x^3}}{4-\dfrac{3}{x}} = \dfrac{0-0}{4-0} = 0$

13. Find $\lim_{x\to\infty} \dfrac{6x^4-5x^2+7}{8x^6+4x^3-8x}$.

We divide the numerator and the denominator by x^6, the highest power of x in the denominator.

$$\lim_{x\to\infty} \frac{6x^4-5x^4+7}{8x^6+4x^3-8x} = \lim_{x\to\infty} \frac{\dfrac{6}{x^2}-\dfrac{5}{x^4}+\dfrac{7}{x^6}}{8+\dfrac{4}{x^3}-\dfrac{8}{x^5}}$$

$$= \frac{0-0+0}{8+0-0}$$

$$= 0$$

14. $\lim_{x\to\infty} \dfrac{6x^5-x^3}{4x^2-3x^3} = \lim_{x\to\infty} \dfrac{6x^2-1}{\dfrac{4}{x}-3} =$

$\dfrac{\lim_{x\to\infty} 6x^2-1}{0-3} = -\infty$

15. Find $\lim_{x\to\infty} \dfrac{11x^5+4x^3-6x^2+2}{6x^3+5x^2+3x-1}$.

We divide the numerator and the denominator by x^3, the highest power of x in the denominator.

$$\lim_{x\to\infty} \frac{11x^5+4x^3-6x+2}{6x^3+5x^2+3x-1} = \lim_{x\to\infty} \frac{11x^2+4-\dfrac{6}{x^2}+\dfrac{2}{x^3}}{6+\dfrac{5}{x}+\dfrac{3}{x^2}-\dfrac{1}{x^3}}$$

$$= \frac{\lim_{x\to\infty} 11x^2+4-0+0}{6+0+0-0}$$

$$= \infty$$

16. $\lim_{x\to\infty} \dfrac{7x^9-6x^3+2x^2-10}{2x^6+4x^2-x+23} = \lim_{x\to\infty} \dfrac{7x^3-\dfrac{6}{x^3}+\dfrac{2}{x^4}-\dfrac{10}{x^6}}{2+\dfrac{4}{x^4}-\dfrac{1}{x^5}+\dfrac{23}{x^6}}$

$$= \frac{\lim_{x\to\infty} 7x^3-0+0-0}{2+0-0+0}$$

$$= \infty$$

17. $f(x) = \dfrac{4}{x}$, or $4x^{-1}$

a) *Intercepts.* Since the numerator is the constant 4, there are no x-intercepts. The number 0 is not in the domain of the function, so there are no y-intercepts.

b) *Asymptotes.*

Vertical. The denominator is 0 for $x=0$, so the line $x=0$ is a vertical asymptote.

Horizontal. The degree of the numerator is less than the degree of the denominator, so $y=0$ is a horizontal asymptote.

Oblique. There is no oblique asymptote since the degree of the numerator is not one more than the degree of the denominator.

c) *Derivatives.*

$$f'(x) = -4x^{-2} = -\frac{4}{x^2}$$

$$f''(x) = 8x^{-3} = \frac{8}{x^3}$$

d) *Critical points.* The number 0 is not in the domain of f. Now $f'(x)$ exists for all values of x except 0. The equation $f'(x)=0$ has no solution, so there are no critical points.

e) *Increasing, decreasing, relative extrema.* Use 0 to divide the real number line into two intervals, A: $(-\infty,0)$ and B: $(0,\infty)$. Test a point in each interval.

A: Test -1, $f'(-1) = -\dfrac{4}{(-1)^2} = -4 < 0$

B: Test 1, $f'(1) = -\dfrac{4}{1^2} = -4 < 0$

Then f is decreasing on both intervals. Since there are no critical points, there are no relative extrema.

f) *Inflection points.* $f''(0)$ does not exist, but because $f(0)$ does not exist there cannot be an inflection point at 0. The equation $f''(x)=0$ has no solution, so there are no inflection points.

g) *Concavity.* Use 0 to divide the real number line as in step (e). Note that for any $x<0$, $x^3<0$, so

$$f''(x) = \frac{8}{x^3} < 0$$

and for any $x>0$, $x^3>0$, so

$$f''(x) = \frac{8}{x^3} > 0.$$

Then f is concave down on $(-\infty,0)$ and concave up on $(0,\infty)$.

h) *Sketch.* Use the preceding information to sketch the graph. Compute function values as needed.

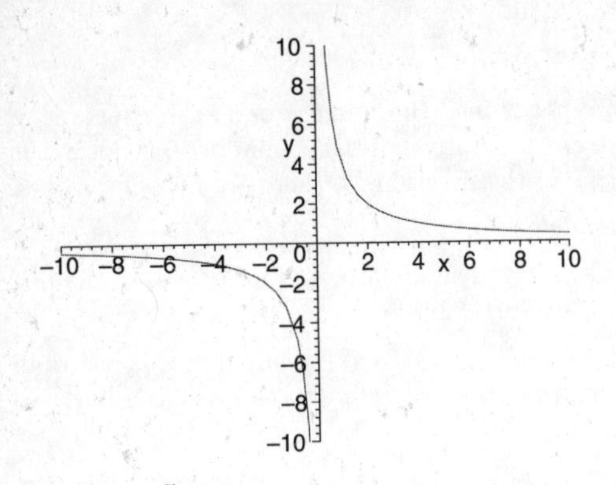

h) Sketch.

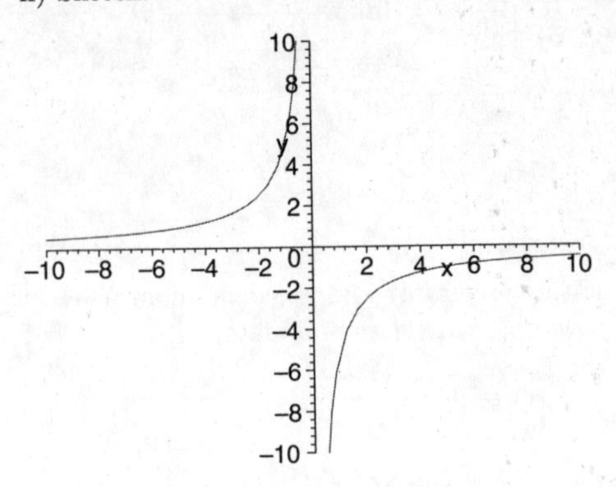

18. $f(x) = -\dfrac{5}{x}$

a) *Intercepts.* Since the numerator is the constant 5, there are no x-intercepts. There are no y-intercepts, because 0 is not in the domain of the function.

b) *Asymptotes.*

Vertical. The denominator is 0 for $x = 0$, so $x = 0$ is a vertical asymptote.

Horizontal. The degree of the numerator is less than the degree of the denominator, so $y = 0$ is a horizontal asymptote.

Oblique. There is no oblique asymptote since the degree of the numerator is not one more than the degree of the denominator.

c) *Derivatives.*

$$f'(x) = 5x^{-2} = \frac{5}{x^2}$$

$$f''(x) = -10x^{-3} = -\frac{10}{x^3}$$

d) *Critical points.* $f'(x)$ exists for all values of x except 0, but 0 is not in the domain of the function. The equation $f'(x) = 0$ has no solution, so there are no critical points.

e) *Increasing, decreasing, relative extrema.* Use 0 to divide the real number line into two intervals, A: $(-\infty, 0)$ and B: $(0, \infty)$.

A: Test -1, $f'(-1) = 5 > 0$

B: Test 1, $f'(1) = 5 > 0$

Then f is increasing on both intervals. Since there are no critical points, there are no relative extrema.

f) *Inflection points.* $f''(0)$ does not exist, but because $f(0)$ does not exist there cannot be an inflection point at 0. The equation $f''(x) = 0$ has no solution, so there are no inflection points.

g) *Concavity.* Use 0 to divide the real number line as in step (e). Note that for any $x < 0$, $x^3 < 0$, so

$$f''(x) = -\frac{10}{x^3} > 0$$

and for any $x > 0$, $x^3 > 0$, so

$$f''(x) = -\frac{10}{x^3} < 0.$$

Then f is concave up on $(-\infty, 0)$ and concave down on $(0, \infty)$.

19. $f(x) = \dfrac{-2}{x-5}$

a) *Intercepts.* Since the numerator is the constant -2, there are no x-intercepts. To find the y-intercepts we compute $f(0)$:

$$f(0) = \frac{-2}{0-5} = \frac{-2}{-5} = \frac{2}{5}$$

Then $\left(0, \dfrac{2}{5}\right)$ is the y-intercept.

b) *Asymptotes.*

Vertical. The denominator is 0 for $x = 5$, so the line $x = 5$ is a vertical asymptote.

Horizontal. The degree of the numerator is less than the degree of the denominator, so $y = 0$ is a horizontal asymptote.

Oblique. There is no oblique asymptote since the degree of the numerator is not one more than the degree of the denominator.

c) *Derivatives.*

$$f'(x) = 2(x-5)^{-2} = \frac{2}{(x-5)^2}$$

$$f''(x) = -4(x-5)^{-3} = -\frac{4}{(x-5)^3}$$

d) *Critical points.* $f'(5)$ does not exist, but because $f(5)$ does not exist, $x = 5$ is not a critical point. The equation $f'(x) = 0$ has no solution, so there are no critical points.

e) *Increasing, decreasing, relative extrema.* Use 5 to divide the real number line into two intervals, A: $(-\infty, 5)$ and B: $(5, \infty)$. Test a point in each interval.

A: Test 0, $f'(0) = \dfrac{2}{(0-5)^2} = \dfrac{2}{25} > 0$

B: Test 6, $f'(6) = \dfrac{2}{(6-5)^2} = 2 > 0$

Then f is increasing on both intervals. Since there are no critical points, there are no relative extrema.

f) *Inflection points.* $f''(5)$ does not exist, but because $f(5)$ does not exist there cannot be an inflection point at 5. The equation $f''(x) = 0$ has no solution, so there are no inflection points.

g) *Concavity.* Use 5 to divide the real number line as in step (e). Note that for any $x < 5$, $(x-5)^3 < 0$, so

$$f''(x) = -\frac{4}{(x-5)^3} > 0$$

and for any $x > 5$, $(x-5)^3 > 0$, so

$$f''(x) = -\frac{4}{(x-5)^3} < 0.$$

Then f is concave up on $(-\infty, 5)$ and concave down on $(5, \infty)$.

h) *Sketch.* Use the preceding information to sketch the graph. Compute function values as needed.

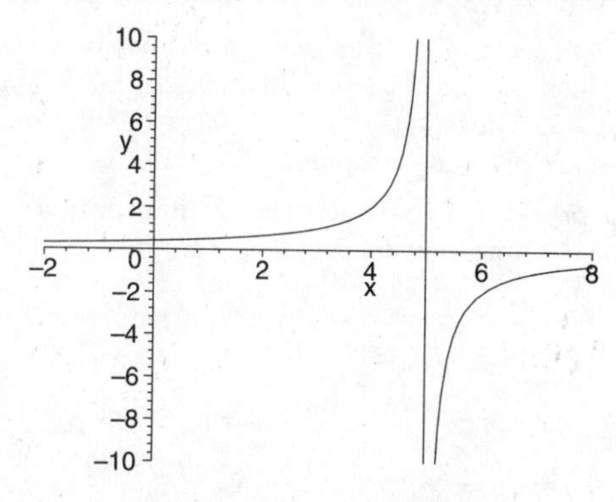

20. $f(x) = \dfrac{1}{x-5}$

a) *Intercepts.* Since the numerator is the constant 1, there are no x-intercepts.

$f(0) = -\frac{1}{5}$, so $\left(0, -\frac{1}{5}\right)$ is the y-intercept.

b) *Asymptotes.*

Vertical. The denominator is 0 for $x = 5$, so $x = 5$ is a vertical asymptote.

Horizontal. The degree of the numerator is less than the degree of the denominator, so $y = 0$ is a horizontal asymptote.

Oblique. There is no oblique asymptote since the degree of the numerator is not one more than the degree of the denominator.

c) *Derivatives.*

$$f'(x) = -(x-5)^{-2} = \frac{-1}{(x-5)^2}$$

$$f''(x) = 2(x-5)^{-3} = \frac{2}{(x-5)^3}$$

d) *Critical points.* $f'(5)$ does not exist, but because $f(5)$ does not exist $x = 5$ is not a critical point. The equation $f'(x) = 0$ has no solution, so there are no critical points.

e) *Increasing, decreasing, relative extrema.* Use 5 to divide the real number line into two intervals, A: $(-\infty, 5)$ and B: $(5, \infty)$.

A: Test 0, $f'(0) = -\frac{1}{25} < 0$

B: Test 6, $f'(6) = -1 < 0$

Then f is decreasing on both intervals. Since there are no critical points, there are no relative extrema.

f) *Inflection points.* $f''(5)$ does not exist, but because $f(5)$ does not exist there cannot be an inflection point at 5. The equation $f''(x) = 0$ has no solution, so there are no inflection points.

g) *Concavity.* Use 5 to divide the real number line as in step (e). Note that for any $x < 5$, $(x-5)^3 < 0$, so

$$f''(x) = \frac{2}{(x-5)^3} < 0$$

and for any $x > 5$, $(x-5)^3 > 0$, so

$$f''(x) = \frac{2}{(x-5)^3} > 0.$$

Then f is concave down on $(-\infty, 5)$ and concave up on $(5, \infty)$.

h) Sketch.

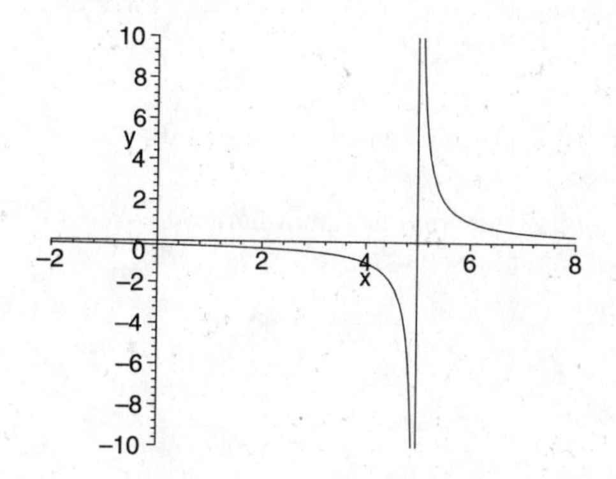

21. $f(x) = \dfrac{1}{x-3}$

a) *Intercepts.* Since the numerator is the constant 1, there are no x-intercepts.

$f(0) = \frac{1}{0-3} = -\frac{1}{3}$, so $\left(0, -\frac{1}{3}\right)$ is the y-intercept.

b) *Asymptotes.*

Vertical. The denominator is 0 for $x = 3$, so the line $x = 3$ is a vertical asymptote.

Horizontal. The degree of the numerator is less than the degree of the denominator, so $y = 0$ is a horizontal asymptote.

Oblique. There is no oblique asymptote since the degree of the numerator is not one more than the degree of the denominator.

c) *Derivatives.*

$$f'(x) = -(x-3)^{-2} = -\frac{1}{(x-3)^2}$$

$$f''(x) = 2(x-3)^{-3} = \frac{2}{(x-3)^3}$$

d) *Critical points.* $f'(3)$ does not exist, but because $f(3)$ does not exist, $x = 3$ is not a critical point. The equation $f'(x) = 0$ has no solution, so there are no critical points.

e) *Increasing, decreasing, relative extrema.* Use 3 to divide the real number line into two intervals, A: $(-\infty, 3)$ and B: $(3, \infty)$. Test a point in each interval.

A: Test 0, $f'(0) = -\dfrac{1}{(0-3)^2} = -\dfrac{1}{9} < 0$

B: Test 4, $f'(4) = -\dfrac{1}{(4-3)^2} = -1 < 0$

Then f is decreasing on both intervals. Since there are no critical points, there are no relative extrema.

f) *Inflection points.* $f''(3)$ does not exist, but because $f(3)$ does not exist there cannot be an inflection point at 3. The equation $f''(x) = 0$ has no solution, so there are no inflection points.

g) *Concavity.* Use 3 to divide the real number line as in step (e). Note that for any $x < 3$, $(x-3)^3 < 0$, so

$$f''(x) = \frac{2}{(x-3)^3} < 0$$

and for any $x > 3$, $(x-3)^3 > 0$, so

$$f''(x) = \frac{2}{(x-3)^3} > 0.$$

Then f is concave down on $(-\infty, 3)$ and concave up on $(3, \infty)$.

h) *Sketch.* Use the preceding information to sketch the graph. Compute function values as needed.

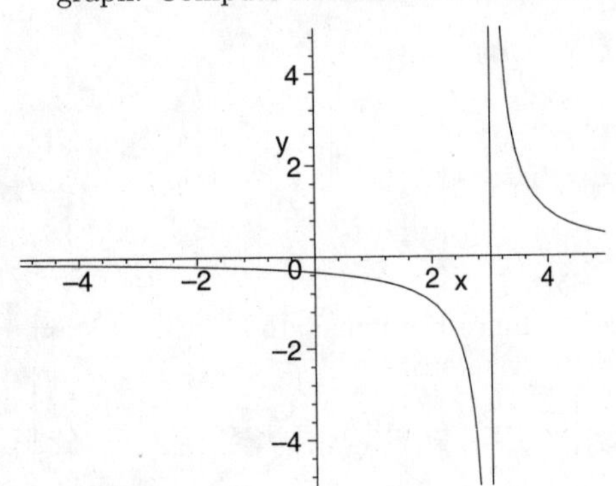

22. $f(x) = \dfrac{1}{x+2}$

a) *Intercepts.* Since the numerator is the constant 1, there are no x-intercepts.

$f(0) = \dfrac{1}{2}$, so $\left(0, \dfrac{1}{2}\right)$ is the y-intercept.

b) *Asymptotes.*

Vertical. The denominator is 0 for $x = -2$, so $x = -$ 2 is a vertical asymptote.

Horizontal. The degree of the numerator is less than the degree of the denominator, so $y = 0$ is a horizontal asymptote.

Oblique. There are no oblique asymptotes since the degree of the numerator is not one more than the degree of the denominator.

c) *Derivatives.*

$$f'(x) = -(x+2)^{-2} = \frac{-1}{(x+2)^2}$$

$$f''(x) = 2(x+3)^{-3} = \frac{2}{(x+2)^3}$$

d) *Critical points.* $f'(-2)$ does not exist, but because $f(-2)$ does not exist $x = -2$ is not a critical point. The equation $f'(x) = 0$ has no solution, so there are no critical points.

e) *Increasing, decreasing, relative extrema.* Use -2 to divide the real number line into two intervals, A: $(-\infty, -2)$ and B: $(-2, \infty)$.

A: Test -3, $f'(-3) = -1 < 0$

B: Test 0, $f'(0) = -\dfrac{1}{4} < 0$

Then f is decreasing on both intervals. Since there are no critical points, there are no relative extrema.

f) *Inflection points.* $f''(-2)$ does not exist, but because $f(-2)$ does not exist there cannot be an inflection point at -2. The equation $f''(x) = 0$ has no solution, so there are no inflection points.

g) *Concavity.* Use -2 to divide the real number line as in step (e). Note that for any $x < -2$, $(x+2)^3 < 0$, so

$$f''(x) = \frac{2}{(x+2)^3} < 0$$

and for any $x > -2$, $(x+2)^3 > 0$, so

$$f''(x) = \frac{2}{(x+2)^3} > 0.$$

Then f is concave down on $(-\infty, -2)$ and concave up on $(-2, \infty)$.

h) Sketch.

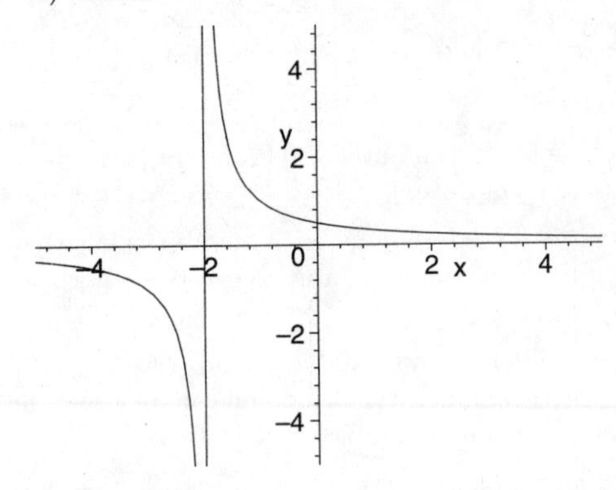

23. $f(x) = \dfrac{-2}{x+5}$

a) *Intercepts.* Since the numerator is the constant -2, there are no x-intercepts.

$f(0) = \dfrac{-2}{0+5} = -\dfrac{2}{5}$, so $\left(0, -\dfrac{2}{5}\right)$ is the y-intercept.

b) *Asymptotes.*

Vertical. The denominator is 0 for $x = -5$, so the line $x = -5$ is a vertical asymptote.

Horizontal. The degree of the numerator is less than the degree of the denominator, so $y = 0$ is a horizontal asymptote.

Oblique. There is no oblique asymptote since the degree of the numerator is not one more than the degree of the denominator.

c) *Derivatives.*

$$f'(x) = 2(x+5)^{-2} = \frac{2}{(x+5)^2}$$

$$f''(x) = -4(x+5)^{-3} = \frac{-4}{(x+5)^3}$$

d) *Critical points.* $f'(-5)$ does not exist, but because $f(-5)$ does not exist, $x = -5$ is not a critical point. The equation $f'(x) = 0$ has no solution, so there are no critical points.

e) *Increasing, decreasing, relative extrema.* Use -5 to divide the real number line into two intervals, A: $(-\infty, -5)$ and B: $(-5, \infty)$. Test a point in each interval.

A: Test -6, $f'(0) = \dfrac{2}{(-6+5)^2} = 2 > 0$

B: Test 0, $f'(0) = \dfrac{2}{(0+5)^2} = \dfrac{2}{25} > 0$

Then f is increasing on both intervals. Since there are no critical points, there are no relative extrema.

f) *Inflection points.* $f''(-5)$ does not exist, but because $f(-5)$ does not exist there cannot be an inflection point at -5. The equation $f''(x) = 0$ has no solution, so there are no inflection points.

g) *Concavity.* Use -5 to divide the real number line as in step (e). Test a point in each interval.

A: Test -6, $f''(-6) = \dfrac{-4}{(-6+5)^3} = 4 > 0$

B: Test 0, $f''(0) = \dfrac{-4}{(0+5)^3} = -\dfrac{4}{125} < 0$

Then f is concave up on $(-\infty, -5)$ and concave down on $(-5, \infty)$.

h) *Sketch.* Use the preceding information to sketch the graph. Compute function values as needed.

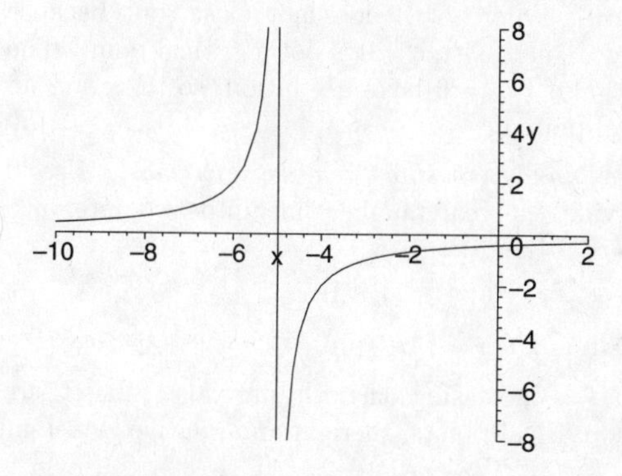

24. $f(x) = \dfrac{-3}{x-3}$

a) *Intercepts.* Since the numerator is the constant -3, there are no x-intercepts.

$f(0) = 1$, so $(0, 1)$ is the y-intercept.

b) *Asymptotes.*

Vertical. The denominator is 0 for $x = 3$, so $x = 3$ is a vertical asymptote.

Horizontal. The degree of the numerator is less than the degree of the denominator, so $y = 0$ is a horizontal asymptote.

Oblique. There are no oblique asymptotes since the degree of the numerator is not one more than the degree of the denominator.

c) *Derivatives.*

$$f'(x) = 3(x-3)^{-2} = \frac{3}{(x-3)^2}$$

$$f''(x) = -6(x-3)^{-3} = \frac{-6}{(x-3)^3}$$

d) *Critical points.* $f'(3)$ does not exist, but because $f(3)$ does not exist $x = 3$ is not a critical point. The equation $f'(x) = 0$ has no solution, so there are no critical points.

e) *Increasing, decreasing, relative extrema.* Use 3 to divide the real number line into two intervals, A: $(-\infty, 3)$ and B: $(3, \infty)$.

A: Test 0, $f'(0) = \dfrac{1}{3} > 0$

B: Test 4, $f'(4) = 3 > 0$

Then f is increasing on both intervals. Since there are no critical points, there are no relative extrema.

f) *Inflection points.* $f''(3)$ does not exist, but because $f(3)$ does not exist there cannot be an inflection point at 3. The equation $f''(x) = 0$ has no solution, so there are no inflection points.

g) *Concavity.* Use 3 to divide the real number line as in step (e).

A: Test 0, $f''(0) = \dfrac{2}{9} > 0$

B: Test 4, $f''(4) = -6 < 0$.

Then f is concave up on $(-\infty, 3)$ and concave down on $(3, \infty)$.

h) Sketch.

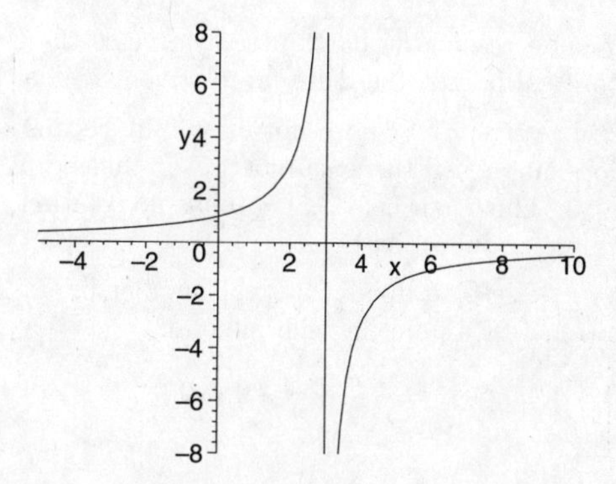

25. $f(x) = \dfrac{2x+1}{x}$

a) *Intercepts.* To find the x-intercepts, solve $f(x) = 0$.

$$\frac{2x+1}{x} = 0$$
$$2x + 1 = 0$$
$$2x = -1$$
$$x = -\frac{1}{2}$$

Since $x = -\frac{1}{2}$ does not make the denominator 0, the x-intercept is $\left(-\frac{1}{2}, 0\right)$. The number 0 is not in the domain of f, so there are no y-intercepts.

b) *Asymptotes.*

Vertical. The denominator is 0 for $x = 0$, so the line $x = 0$ is a vertical asymptote.

Horizontal. The numerator and denominator have the same degree, so $y = \frac{2}{1}$, or $y = 2$, is a horizontal asymptote.

Oblique. There is no oblique asymptote since the degree of the numerator is not one more than the degree of the denominator.

c) *Derivatives.*

$$f'(x) = -\frac{1}{x^2}$$

$$f''(x) = 2x^{-3}, \text{ or } \frac{2}{x^3}$$

d) *Critical points.* $f'(0)$ does not exist, but because $f(0)$ does not exist $x = 0$ is not a critical point. The equation $f'(x) = 0$ has no solution, so there are no critical points.

e) *Increasing, decreasing, relative extrema.* Use 0 to divide the real number line into two intervals, A: $(-\infty, 0)$ and B: $(0, \infty)$. Test a point in each interval.

A: Test -1, $f'(-1) = -\dfrac{1}{(-1)^2} = -1 < 0$

B: Test 1, $f'(1) = -\dfrac{1}{1^2} = -1 < 0$

Then f is decreasing on both intervals. Since there are no critical points, there are no relative extrema.

f) *Inflection points.* $f''(0)$ does not exist, but because $f(0)$ does not exist there cannot be an inflection point at 0. The equation $f''(x) = 0$ has no solution, so there are no inflection points.

g) *Concavity.* Use 0 to divide the real number line as in step (e). Test a point in each interval.

A: Test -1, $f''(-1) = \dfrac{2}{(-1)^3} = -2 < 0$

B: Test 1, $f''(1) = \dfrac{2}{1^3} = 2 > 0$

Then f is concave down on $(-\infty, 0)$ and concave up on $(0, \infty)$.

h) *Sketch.* Use the preceding information to sketch the graph. Compute function values as needed.

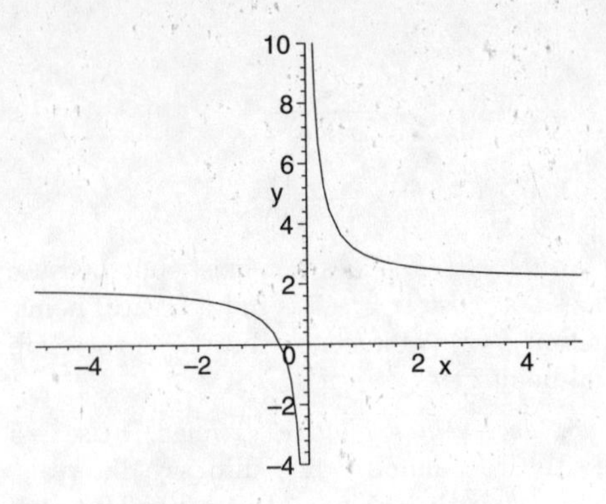

26. $f(x) = \dfrac{3x-1}{x}$

a) *Intercepts.* $f(x) = 0$ for $x = \frac{1}{3}$ and this value does not make the denominator 0, so $\left(\frac{1}{3}, 0\right)$ is the x-intercept. The number 0 is not in the domain of the function, so there are no y-intercepts.

b) *Asymptotes.*

Vertical. The denominator is 0 for $x = 0$, so $x = 0$ is a vertical asymptote.

Horizontal. The numerator and denominator have the same degree, so $y = \frac{3}{1}$, or $y = 3$, is a horizontal asymptote.

Oblique. There are no oblique asymptotes since the degree of the numerator is not one more than the degree of the denominator.

c) *Derivatives.*

$$f'(x) = \frac{1}{x^2}$$

$$f''(x) = -2x^{-3}, \text{ or } -\frac{2}{x^3}$$

d) *Critical points.* $f'(0)$ does not exist, but because $f(0)$ does not exist $x = 0$ is not a critical point. The equation $f'(x) = 0$ has no solution, so there are no critical points.

e) *Increasing, decreasing, relative extrema.* Use 0 to divide the real number line into two intervals, A: $(-\infty, 0)$ and B: $(0, \infty)$.

A: Test -1, $f'(-1) = 1 > 0$

B: Test 1, $f'(1) = 1 > 0$

Then f is increasing on both intervals. Since there are no critical points, there are no relative extrema.

f) *Inflection points.* $f''(0)$ does not exist, but because $f(0)$ does not exist there cannot be an inflection point at 0. The equation $f''(x) = 0$ has no solution, so there are no inflection points.

g) *Concavity.* Use 0 to divide the real number line as in step e).

A: Test -1, $f''(-1) = 2 > 0$

B: Test 1, $f''(1) = -2 < 0$.

Then f is concave up on $(-\infty, 0)$ and concave down on $(0, \infty)$.

h) Sketch.

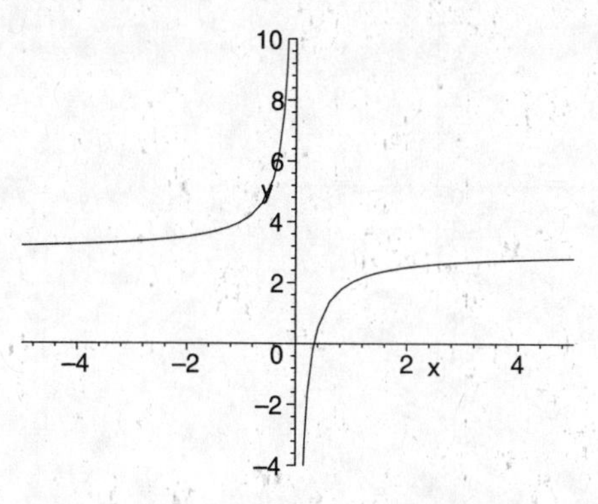

27. $f(x) = x + \frac{9}{x} = \frac{x^2+9}{x}$

a) *Intercepts.* The equation $f(x) = 0$ has no real number solution, so there are no x-intercepts. The number 0 is not in the domain of the function, so there are no y-intercepts.

b) *Asymptotes.*

Vertical. The denominator is 0 for $x = 0$, so the line $x = 0$ is a vertical asymptote.

Horizontal. The degree of the numerator is greater than the degree of the denominator, so there are no horizontal asymptotes.

Oblique. As $|x|$ gets very large, $f(x) = x + \frac{9}{x}$ approaches x, so $y = x$ is an oblique asymptote.

c) *Derivatives.*

$$f'(x) = 1 - 9x^{-2} = 1 - \frac{9}{x^2}$$

$$f''(x) = 18x^{-3} = \frac{18}{x^3}$$

d) *Critical points.* $f'(0)$ does not exist, but because $f(0)$ does not exist $x = 0$ is not a critical point. Solve $f'(x) = 0$.

$$1 - \frac{9}{x^2} = 1$$

$$1 = \frac{9}{x^2}$$

$$x^2 = 9$$

$$x = \pm 3$$

Thus, -3 and 3 are critical points. $f(-3) = -6$ and $f(3) = 6$, so $(-3, -6)$ and $(3, 6)$ are on the graph.

e) *Increasing, decreasing, relative extrema.* Use -3, 0, and 3 to divide the real number line into four intervals, A: $(-\infty, -3)$, B: $(-3, 0)$, C: $(0, 3)$, and D: $(3, \infty)$. Test a point in each interval.

A: Test -4, $f'(-4) = 1 - \frac{9}{(-4)^2} = \frac{7}{16} > 0$

B: Test -1, $f'(-1) = 1 - \frac{9}{(-1)^2} = -8 < 0$

C: Test 1, $f'(1) = 1 - \frac{9}{1^2} = -8 < 0$

D: Test 4, $f'(4) = 1 - \frac{9}{4^2} = \frac{7}{16} > 0$

Then f is increasing on $(-\infty, -3)$ and on $(3, \infty)$ and is decreasing on $(-3, 0)$ and on $(0, 3)$. Thus, there is a relative maximum at $(-3, -6)$ and a relative minimum at $(3, 6)$.

f) *Inflection points.* $f''(0)$ does not exist, but because $f(0)$ does not exist there cannot be an inflection point at 0. The equation $f''(x) = 0$ has no solution, so there are no inflection points.

g) *Concavity.* Use 0 to divide the real number line into two intervals, A: $(-\infty, 0)$ and B: $(0, \infty)$. Test a point in each interval.

A: Test -1, $f''(-1) = \frac{18}{(-1)^3} = -18 < 0$

B: Test 1, $f''(1) = \frac{18}{1^3} = 18 > 0$

Then f is concave down on $(-\infty, 0)$ and concave up on $(0, \infty)$.

h) *Sketch.* Use the preceding information to sketch the graph. Compute other function values as needed.

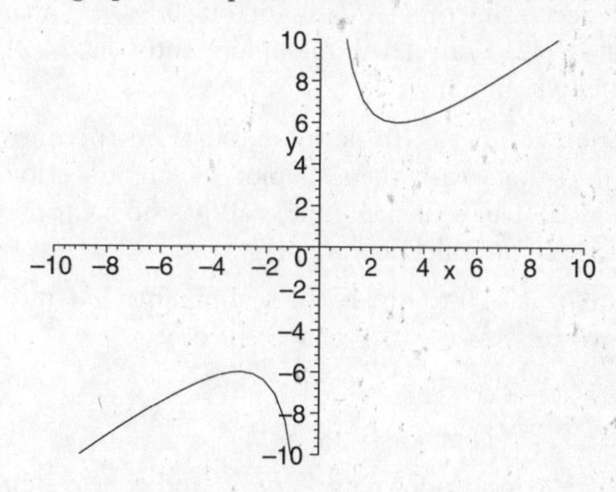

28. $f(x) = x + \frac{2}{x} = \frac{x^2+2}{x}$

a) *Intercepts.* The equation $f(x) = 0$ has no real number solutions, so there are no x-intercepts. The number 0 is not in the domain of the function, so there are no y-intercepts.

b) *Asymptotes.*

Vertical. The denominator is 0 for $x = 0$, so the line $x = 0$ is a vertical asymptote.

Horizontal. The degree of the numerator is greater than the degree of the denominator, so there are no horizontal asymptotes.

Oblique. As $|x|$ gets very large, $f(x) = x + \frac{2}{x}$ approaches x, so $y = x$ is an oblique asymptote.

c) *Derivatives.*

$$f'(x) = 1 - 2x^{-2} = 1 - \frac{2}{x^2}$$

$$f''(x) = 4x^{-3} = \frac{4}{x^3}$$

d) *Critical points.* $f'(0)$ does not exist, but because $f(0)$ does not exist $x = 0$ is not a critical point. $f'(x) = 0$ for $x = \pm\sqrt{2}$, so $-\sqrt{2}$ and $\sqrt{2}$ are critical points. $f(-\sqrt{2}) = -2\sqrt{2}$ and $f(\sqrt{2}) = 2\sqrt{2}$, so $(-\sqrt{2}, -2\sqrt{2})$ and $(\sqrt{2}, 2\sqrt{2})$ are on the graph.

e) *Increasing, decreasing, relative extrema.* Use $-\sqrt{2}$, 0, and $\sqrt{2}$ to divide the real number line into four intervals, A: $(-\infty, -\sqrt{2})$, B: $(-\sqrt{2}, 0)$, C: $(0, \sqrt{2})$, and D: $(\sqrt{2}, \infty)$.

A: Test -2, $f'(-2) = \frac{1}{2} > 0$

B: Test -1, $f'(-1) = -1 < 0$

C: Test 1, $f'(1) = -1 < 0$

D: Test 2, $f'(2) = \frac{1}{2} > 0$

Then f is increasing on $(-\infty, -\sqrt{2})$ and on $(\sqrt{2}, \infty)$ and is decreasing on $(-\sqrt{2}, 0)$ and on $(0, \sqrt{2})$. Thus, $(-\sqrt{2}, -2\sqrt{2})$ is a relative maximum and $(\sqrt{2}, 2\sqrt{2})$ is a relative minimum.

f) *Inflection points.* $f''(0)$ does not exist, but because $f(0)$ does not exist there cannot be an inflection point at 0. The equation $f''(x) = 0$ has no solution, so there are no inflection points.

g) *Concavity.* Use 0 to divide the real number line into two intervals, A: $(-\infty, 0)$ and B: $(0, \infty)$.

A: Test -1, $f''(-1) = -4 < 0$

B: Test 1, $f''(1) = 4 > 0$

Then f is concave down on $(-\infty, 0)$ and concave up on $(0, \infty)$.

h) Sketch.

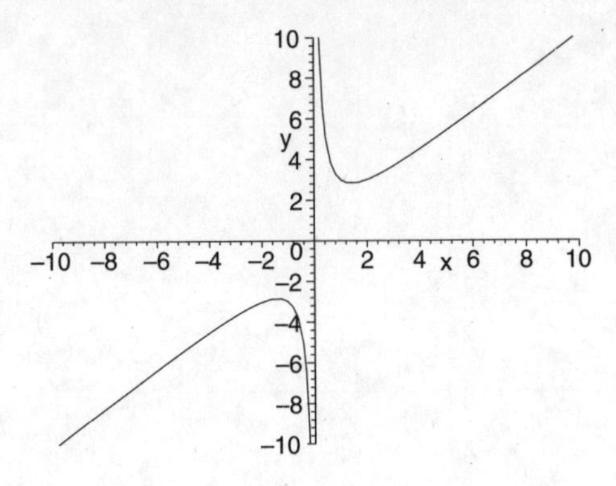

29. $f(x) = \frac{2}{x^2}$

a) *Intercepts.* Since the numerator is the constant 2, there are no x-intercepts. The number 0 is not in the domain of the function, so there are no y-intercepts.

b) *Asymptotes.*

Vertical. The denominator is 0 for $x = 0$, so the line $x = 0$ is a vertical asymptote.

Horizontal. The degree of the numerator is less than the degree of the denominator, so $y = 0$ is a horizontal asymptote.

Oblique. There is no oblique asymptote since the degree of the numerator is not one more than the degree of the denominator.

c) *Derivatives.*

$$f'(x) = -4x^{-3} = -\frac{4}{x^3}$$

$$f''(x) = 12x^{-4} = \frac{12}{x^4}$$

d) *Critical points.* $f'(0)$ does not exist, but because $f(0)$ does not exist $x = 0$ is not a critical point. The equation $f'(x) = 0$ has no solution, so there are no critical points.

e) *Increasing, decreasing, relative extrema.* Use 0 to divide the real number line into two intervals, A: $(-\infty, 0)$ and B: $(0, \infty)$.

A: Test -1, $f'(-1) = -\frac{4}{(-1)^3} = 4 > 0$

B: Test 1, $f'(1) = -\frac{4}{1^3} = -4 < 0$

Then f is increasing on $(-\infty, 0)$ and is decreasing on $(0, \infty)$. Since there are no critical points, there are no relative extrema.

f) *Inflection points.* $f''(0)$ does not exist, but because $f(0)$ does not exist there cannot be an inflection point at 0. The equation $f''(x) = 0$ has no solution, so there are no inflection points.

g) *Concavity.* Use 0 to divide the real number line as in step (e). Test a point in each interval.

A: Test -1, $f''(-1) = \frac{12}{(-1)^4} = 12 > 0$

B: Test 1, $f''(1) = \frac{12}{1^4} = 12 > 0$

Then f is concave up on both intervals.

h) *Sketch.* Use the preceding information to sketch the graph. Compute function values as needed.

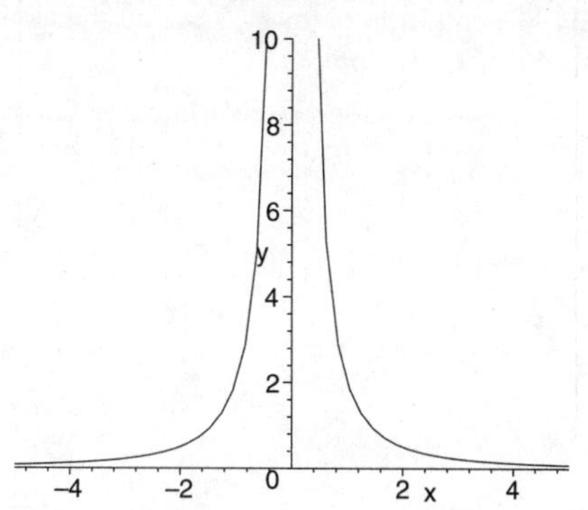

30. $f(x) = \dfrac{-1}{x^2}$

a) *Intercepts.* Since the numerator is the constant -1, there are no x-intercepts. The number 0 is not in the domain of the function, so there are no y-intercepts.

b) *Asymptotes.*

Vertical. The denominator is 0 for $x = 0$, so the line $x = 0$ is a vertical asymptote.

Horizontal. The degree of the numerator is less than the degree of the denominator, so $y = 0$ is a horizontal asymptote.

Oblique. There is no oblique asymptote since the degree of the numerator is not one more than the degree of the denominator.

c) *Derivatives.*

$$f'(x) = 2x^{-3} = \frac{2}{x^3}$$

$$f''(x) = -6x^{-4} = -\frac{6}{x^4}$$

d) *Critical points.* $f'(0)$ does not exist, but because $f(0)$ does not exist $x = 0$ is not a critical point. The equation $f'(x) = 0$ has no solution, so there are no critical points.

e) *Increasing, decreasing, relative extrema.* Use 0 to divide the real number line into two intervals, A: $(-\infty, 0)$ and B: $(0, \infty)$.

A: Test -1, $f'(-1) = -2 < 0$

B: Test 1, $f'(1) = 2 > 0$

Then f is decreasing on $(-\infty, 0)$ and is increasing on $(0, \infty)$. Since there are no critical points, there are no relative extrema.

f) *Inflection points.* $f''(0)$ does not exist, but because $f(0)$ does not exist there cannot be an inflection point at 0. The equation $f''(x) = 0$ has no solution, so there are no inflection points.

g) *Concavity.* Use 0 to divide the real number line as in step (e).

A: Test -1, $f''(-1) = -6 < 0$

B: Test 1, $f''(1) = -6 < 0$

Then f is concave down on both intervals.

h) Sketch.

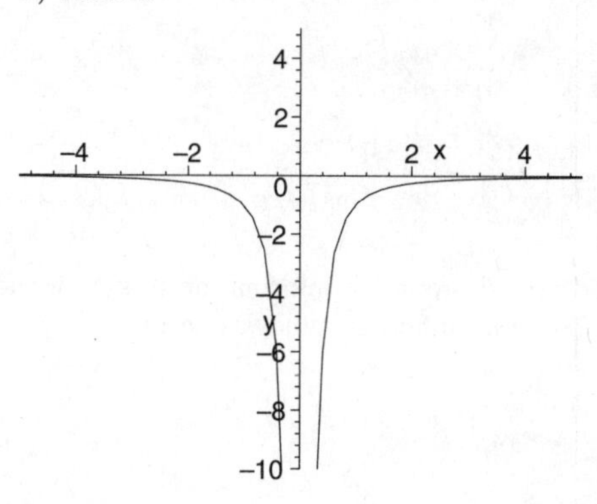

31. $f(x) = \dfrac{x}{x-3}$

a) *Intercepts.* The numerator is 0 for $x = 0$ and this value of x does not make the denominator 0, so $(0, 0)$ is the x-intercept. $f(0) = \dfrac{0}{0-3} = 0$, so the y-intercept is the x-intercept $(0, 0)$.

b) *Asymptotes.*

Vertical. The denominator is 0 for $x = 3$, so the line $x = 3$ is a vertical asymptote.

Horizontal. The numerator and the denominator have the same degree, so $y = \dfrac{1}{1}$, or $y = 1$, is a horizontal asymptote.

Oblique. There is no oblique asymptote since the degree of the numerator is not one more than the degree of the denominator.

c) *Derivatives.*

$$f'(x) = -\frac{3}{(x-3)^2}$$

$$f''(x) = 6(x-3)^{-3} = \frac{6}{(x-3)^3}$$

d) *Critical points.* $f'(3)$ does not exist, but because $f(3)$ does not exist $x = 3$ is not a critical point. The equation $f'(x) = 0$ has no solution, so there are no critical points.

e) *Increasing, decreasing, relative extrema.* Use 3 to divide the real number line into two intervals, A: $(-\infty, 3)$ and B: $(3, \infty)$. Test a point in each interval.

A: Test 0, $f'(0) = -\dfrac{1}{3} < 0$

B: Test 4, $f'(4) = -3 < 0$

Then f is decreasing on both intervals. Since there are no critical points, there are no relative extrema.

f) *Inflection points.* $f''(3)$ does not exist, but because $f(3)$ does not exist there cannot be an inflection point at 3. The equation $f''(x) = 0$ has no solution, so there are no inflection points.

g) *Concavity.* Use 3 to divide the real number line as in step (e).

A: Test 0, $f''(0) = -\frac{2}{9} < 0$

B: Test 4, $f''(4) = 6 > 0$

Then f is concave down on $(-\infty, 3)$ and concave up on $(3, \infty)$.

h) *Sketch.* Use the preceding information to sketch the graph. Compute function values as needed.

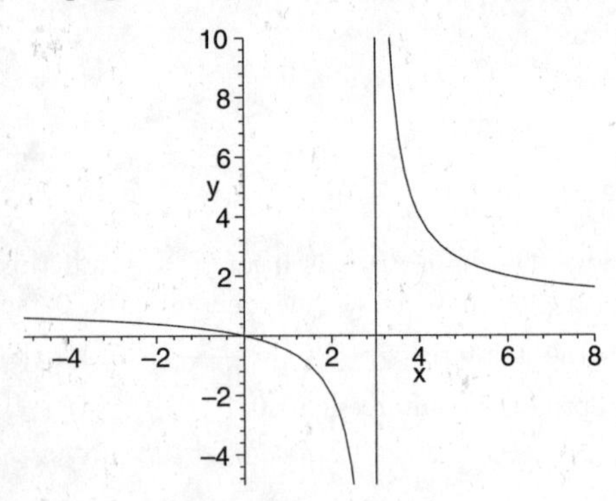

32. $f(x) = \dfrac{x}{x+2}$

a) *Intercepts.* The numerator is 0 for $x = 0$ and this value of x does not make the denominator 0, so $(0, 0)$ is the x-intercept. $f(0) = 0$, so the y-intercept is the x-intercept $(0, 0)$.

b) *Asymptotes.*

Vertical. The denominator is 0 for $x = -2$, so the line $x = -2$ is a vertical asymptote.

Horizontal. The degree of the numerator and denominator have the same degree, so $y = \frac{1}{1}$, or $y = 1$, is the horizontal asymptote.

Oblique. There is no oblique asymptote since the degree of the numerator is not one more than the degree of the denominator.

c) *Derivatives.*

$$f'(x) = \frac{2}{(x+2)^2}$$

$$f''(x) = -4(x+2)^{-3} = -\frac{4}{(x+2)^3}$$

d) *Critical points.* $f'(-2)$ does not exist, but because $f(-2)$ does not exist $x = -2$ is not a critical point. The equation $f'(x) = 0$ has no solution, so there are no critical points.

e) *Increasing, decreasing, relative extrema.* Use -2 to divide the real number line into two intervals, A: $(-\infty, -2)$ and B: $(-2, \infty)$.

A: Test -3, $f'(-3) = 2 > 0$

B: Test -1, $f'(-1) = 2 > 0$

Then f is increasing on both intervals. Since there are no critical points, there are no relative extrema.

f) *Inflection points.* $f''(-2)$ does not exist, but because $f(-2)$ does not exist there cannot be an inflection point at -2. The equation $f''(x) = 0$ has no solution, so there are no inflection points.

g) *Concavity.* Use -2 to divide the real number line as in step (e).

A: Test -3, $f''(-3) = 4 > 0$

B: Test -1, $f''(-1) = -4 < 0$

Then f is concave up on $(-\infty, -2)$ and concave down on $(-2, \infty)$.

h) Sketch.

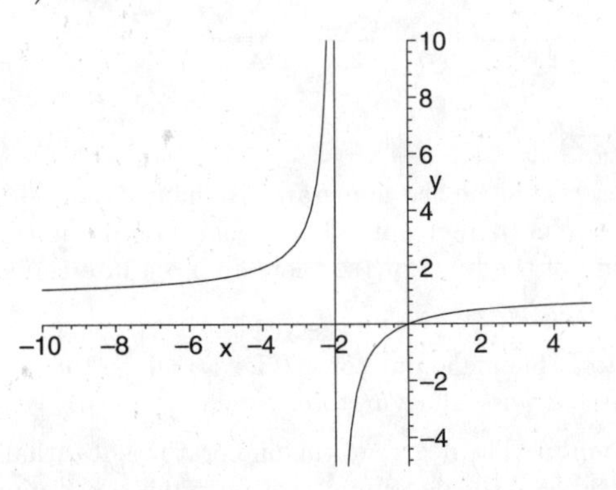

33. $f(x) = \dfrac{1}{x^2+3}$

a) *Intercepts.* Since the numerator is the constant 1, there are no x-intercepts.

$f(0) = \dfrac{1}{0^2+3} = \dfrac{1}{3}$, so $\left(0, \dfrac{1}{3}\right)$ is the y-intercept.

b) *Asymptotes.*

Vertical. $x^2 + 3 = 0$ has no real number solutions, so there are no vertical asymptotes.

Horizontal. The degree of the numerator is less than the degree of the denominator, so $y = 0$ is a horizontal asymptote.

Oblique. There is no oblique asymptote since the degree of the numerator is not one more than the degree of the denominator.

c) *Derivatives.*

$$f'(x) = -\frac{2x}{(x^2+3)^2}$$

$$f''(x) = \frac{6x^2-6}{(x^2+3)^3}$$

d) *Critical points.* $f'(x)$ exists for all real numbers. Solve $f'(x) = 0$.

$$-\frac{2x}{(x^2+3)^2} = 0$$

$$-2x = 0$$

$$x = 0 \quad \text{Critical point}$$

From step (a) we already know $\left(0, \dfrac{1}{3}\right)$ is on the graph.

e) *Increasing, decreasing, relative extrema.* Use 0 to divide the real number line into two intervals, A: $(-\infty, 0)$ and B: $(0, \infty)$. Test a point in each interval.

A: Test -1, $f'(-1) = \frac{1}{8} > 0$

B: Test 1, $f'(1) = -\frac{1}{8} < 0$

Then f is increasing on $(-\infty, 0)$ and decreasing on $(0, \infty)$. Thus, $\left(0, \frac{1}{3}\right)$ is a relative maximum.

f) *Inflection points.* $f''(x)$ exists for all real numbers. Solve $f''(x) = 0$.

$$\frac{6x^2-6}{(x^2+3)^3} = 0$$
$$6x^2 - 6 = 0$$
$$6(x+1)(x-1) = 0$$
$$x = -1 \text{ or } x = 1 \quad \text{Possible inflection points}$$

$f(-1) = \frac{1}{4}$ and $f(1) = \frac{1}{4}$, so $\left(-1, \frac{1}{4}\right)$ and $\left(1, \frac{1}{4}\right)$ are on the graph.

g) *Concavity.* Use -1 and 1 to divide the real number line into three intervals, A: $(-\infty, -1)$, B: $(-1, 1)$, and C: $(1, \infty)$. Test a point in each interval.

A: Test -2, $f''(-2) = \frac{18}{343} > 0$

B: Test 0, $f''(0) = -\frac{2}{9} < 0$

C: Test 2, $f''(2) = \frac{18}{343} > 0$

Then f is concave up on $(-\infty, -1)$ and on $(1, \infty)$ and is concave down on $(-1, 1)$. Thus, $\left(-1, \frac{1}{4}\right)$ and $\left(1, \frac{1}{4}\right)$ are both inflection points.

h) *Sketch.* Use the preceding information to sketch the graph. Compute other function values as needed.

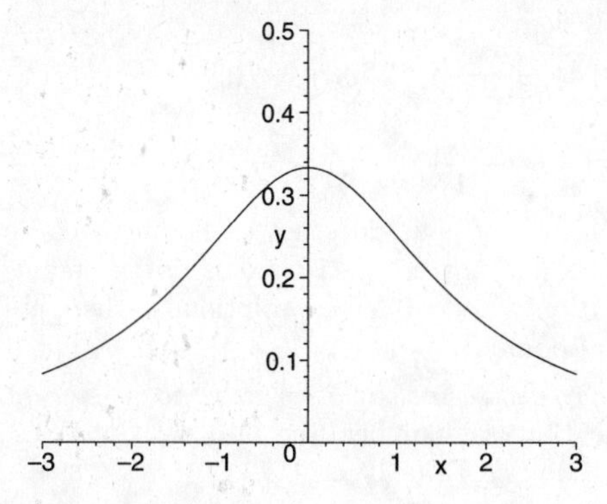

34. $f(x) = \frac{-1}{x^2+2}$

a) *Intercepts.* Since the numerator is the constant -1, there are no x-intercepts.

$f(0) = -\frac{1}{2}$, so $\left(0, -\frac{1}{2}\right)$ is the y-intercept.

b) *Asymptotes.*

Vertical. $x^2 + 2 = 0$ has no real number solutions, so there are no vertical asymptotes.

Horizontal. The degree of the numerator is less than the degree of the denominator, so $y = 0$ is a horizontal asymptote.

Oblique. There is no oblique asymptote since the degree of the numerator is not one more than the degree of the denominator.

c) *Derivatives.*

$$f'(x) = \frac{2x}{(x^2+2)^2}$$
$$f''(x) = \frac{-6x^2+4}{(x^2+2)^3}$$

d) *Critical points.* $f'(x)$ exists for all real numbers. $f'(x) = 0$ for $x = 0$, so 0 is a critical point. From step (a) we already know $\left(0, -\frac{1}{2}\right)$ is on the graph.

e) *Increasing, decreasing, relative extrema.* Use 0 to divide the real number line into two intervals, A: $(-\infty, 0)$ and B: $(0, \infty)$.

A: Test -1, $f'(-1) = -\frac{2}{9} < 0$

B: Test 1, $f'(1) = \frac{2}{9} > 0$

Then f is decreasing on $(-\infty, 0)$ and increasing on $(0, \infty)$. Thus, $\left(0, -\frac{1}{2}\right)$ is a relative minimum.

f) *Inflection points.* $f''(x)$ exists for all real numbers. $f''(x) = 0$ for $x = \pm\frac{\sqrt{6}}{3}$, so $-\frac{\sqrt{6}}{3}$ and $\frac{\sqrt{6}}{3}$ are possible inflection points. $f\left(-\frac{\sqrt{6}}{3}\right) = -\frac{3}{8}$ and $f\left(\frac{\sqrt{6}}{3}\right) = -\frac{3}{8}$, so $\left(-\frac{\sqrt{6}}{3}, -\frac{3}{8}\right)$ and $\left(\frac{\sqrt{6}}{3}, -\frac{3}{8}\right)$ are on the graph.

g) *Concavity.* Use $-\frac{\sqrt{6}}{3}$ and $\frac{\sqrt{6}}{3}$ to divide the real number line into three intervals, A: $\left(-\infty, -\frac{\sqrt{6}}{3}\right)$, B: $\left(-\frac{\sqrt{6}}{3}, \frac{\sqrt{6}}{3}\right)$, and C: $\left(\frac{\sqrt{6}}{3}, \infty\right)$.

A: Test -1, $f''(-1) = -\frac{2}{27} < 0$

B: Test 0, $f''(0) = \frac{1}{2} > 0$

C: Test 1, $f''(1) = -\frac{2}{27} < 0$

Then f is concave down on $\left(-\infty, -\frac{\sqrt{6}}{3}\right)$ and on $\left(\frac{\sqrt{6}}{3}, \infty\right)$ and is concave up on $\left(-\frac{\sqrt{6}}{3}, \frac{\sqrt{6}}{3}\right)$. Thus, $\left(-\frac{\sqrt{6}}{3}, -\frac{3}{8}\right)$ and $\left(\frac{\sqrt{6}}{3}, -\frac{3}{8}\right)$ are both inflection points.

h) Sketch.

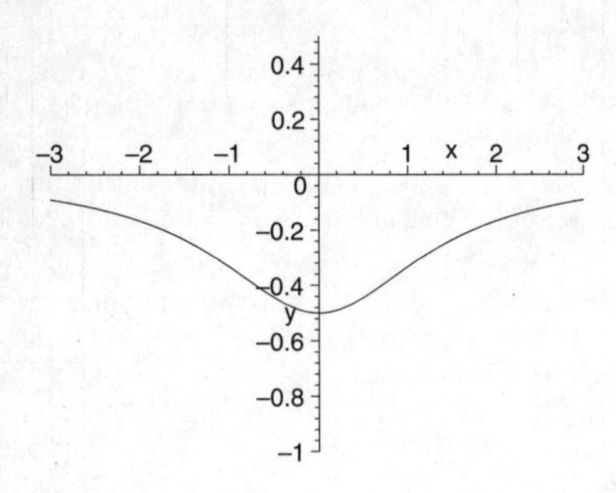

35. $f(x) = \dfrac{x-1}{x+2}$

a) *Intercepts.* The numerator is 0 for $x = 1$ and this value of x does not make the denominator 0, so $(1, 0)$ is the x-intercept.

$f(0) = \dfrac{0-1}{0+2} = -\dfrac{1}{2}$, so $\left(0, -\dfrac{1}{2}\right)$ is the y-intercept.

b) *Asymptotes.*

Vertical. The denominator is 0 for $x = -2$, so the line $x = -2$ is a vertical asymptote.

Horizontal. The numerator and the denominator have the same degree, so $y = \dfrac{1}{1}$, or $y = 1$, is a horizontal asymptote.

Oblique. There is no oblique asymptote since the degree of the numerator is not one more than the degree of the denominator.

c) *Derivatives.*

$$f'(x) = \frac{3}{(x+2)^2}$$

$$f''(x) = \frac{-6}{(x+2)^3}$$

d) *Critical points.* $f'(-2)$ does not exist, but because $f(-2)$ does not exist $x = -2$ is not a critical point. The equation $f'(x) = 0$ has no solution, so there are no critical points.

e) *Increasing, decreasing, relative extrema.* Use -2 to divide the real number line into two intervals, A: $(-\infty, -2)$ and B: $(-2, \infty)$. Test a point in each interval.

A: Test -3, $f'(-3) = 3 > 0$

B: Test -1, $f'(-1) = 3 > 0$

Then f is increasing on both intervals. Since there are no critical points, there are relative extrema.

f) *Inflection points.* $f''(-2)$ does not exist, but because $f(-2)$ does not exist there cannot be an inflection point at -2. The equation $f''(x) = 0$ has no solution, so there are no inflection points.

g) *Concavity.* Use -2 to divide the real number line as in step (e). Test a point in each interval.

A: Test -3, $f''(-3) = 6 > 0$

B: Test -1, $f''(-1) = -6 < 0$

Then f is concave up on $(-\infty, -2)$ and concave down on $(-2, \infty)$.

h) *Sketch.* Use the preceding information to sketch the graph. Compute function values as needed.

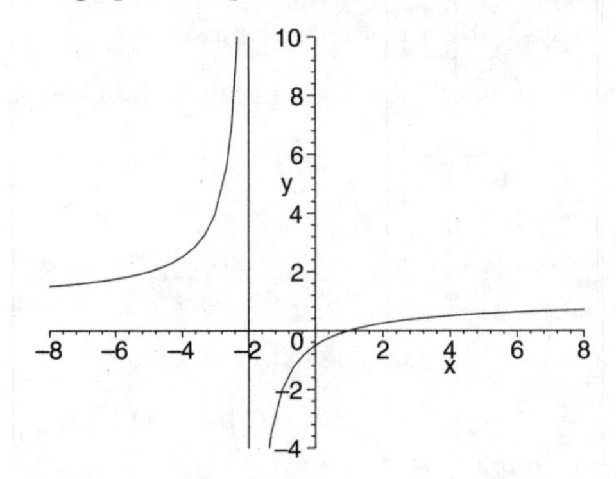

36. $f(x) = \dfrac{x-2}{x+1}$

a) *Intercepts.* The numerator is 0 for $x = 2$ and this value of x does not make the denominator 0, so $(2, 0)$ is the x-intercept. $f(0) = -2$, so $(0, 2)$ is the y-intercept.

b) *Asymptotes.*

Vertical. The denominator is 0 for $x = -1$, so the line $x = -1$ is a vertical asymptote.

Horizontal. The numerator and denominator have the same degree, so $y = \dfrac{1}{1}$, or $y = 1$, is a horizontal asymptote.

Oblique. There is no oblique asymptote since the degree of the numerator is not one more than the degree of the denominator.

c) *Derivatives.*

$$f'(x) = \frac{3}{(x+1)^2}$$

$$f''(x) = \frac{-6}{(x+1)^3}$$

d) *Critical points.* $f'(-1)$ does not exist, but because $f(-1)$ does not exist $x = -1$ is not a critical point. The equation $f'(x) = 0$ has no solution, so there are no critical points.

e) *Increasing, decreasing, relative extrema.* Use -1 to divide the real number line into two intervals, A: $(-\infty, -1)$ and B: $(-1, \infty)$.

A: Test -2, $f'(-2) = 3 > 0$

B: Test 0, $f'(0) = 3 > 0$

Then f is increasing on both intervals. Since there are no critical points, there are relative extrema.

f) *Inflection points.* $f''(-1)$ does not exist, but because $f(-1)$ does not exist there cannot be an inflection point at -1. The equation $f''(x) = 0$ has no solution, so there are no inflection points.

g) *Concavity.* Use -1 to divide the real number line as in step (e).

A: Test -2, $f''(-2) = 6 > 0$

B: Test 0, $f''(0) = -6 < 0$

Then f is concave up on $(-\infty, -1)$ and concave down on $(-1, \infty)$.

h) Sketch.

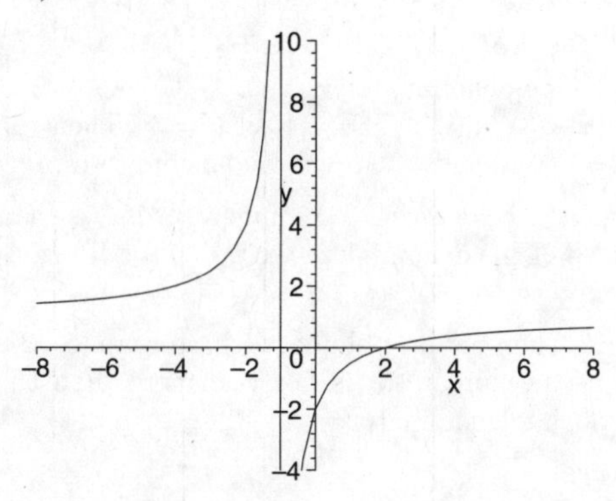

37. $f(x) = \dfrac{x^2 - 4}{x + 3}$

a) *Intercepts.* The numerator $x^2 - 4 = (x+2)(x-2)$ is 0 for $x = -2$ and or $x = 2$, and neither of these values makes the denominator 0. Thus, the x-intercepts are $(-2, 0)$ and $(2, 0)$.

$f(0) = \dfrac{0^2 - 4}{0 + 3} = -\dfrac{4}{3}$, so $\left(0, -\dfrac{4}{3}\right)$ is the y-intercept.

b) *Asymptotes.*

Vertical. The denominator is 0 for $x = -3$, so the line $x = -3$ is a vertical asymptote.

Horizontal. The degree of the numerator is greater than the degree of the denominator, so there are no horizontal asymptotes.

Oblique.

$$f(x) = x - 3 + \frac{5}{x+3}$$

$$\begin{array}{r|l} & x - 3 \\ \hline x + 3 & x^2 \qquad - 4 \\ & \underline{x^2 + 3x} \\ & -3x - 4 \\ & \underline{-3x - 9} \\ & 5 \end{array}$$

As $|x|$ gets very large, $f(x)$ approaches $x - 3$, so $y = x - 3$ is an oblique asymptote.

c) *Derivatives.*

$$f'(x) = \frac{x^2 + 6x + 4}{(x+3)^2}$$

$$f''(x) = \frac{10}{(x+3)^3}$$

d) *Critical points.* $f'(-3)$ does not exist, but because $f(-3)$ does not exist $x = -3$ is not a critical point. Solve $f'(x) = 0$.

$$\frac{x^2 + 6x + 4}{(x+3)^2} = 0$$

$$x^2 + 6x + 4 = 0$$

$x = -3 \pm \sqrt{5}$ Using the quadratic formula

$x \approx -5.24$ or $x \approx -0.76$ Critical points

$f(-5.24) \approx -10.47$ and $f(-0.76) \approx -1.53$, so $(-5.24, -10.47)$ and $(-0.76, -1.53)$ are on the graph.

e) *Increasing, decreasing, relative extrema.* Use -5.24, -3, and -0.76 to divide the real number line into four intervals, A: $(-\infty, -5.24)$, B: $(-5.24, -3)$, C: $(-3, -0.76)$, and D: $(-0.76, \infty)$. Test a point in each interval.

A: Test -6, $f'(-6) = \dfrac{4}{9} > 0$

B: Test -4, $f'(-4) = -4 < 0$

C: Test -2, $f'(-2) = -4 < 0$

D: Test 0, $f'(0) = \dfrac{4}{9} > 0$

Then f is increasing on $(-\infty, -5.24)$ and on $(-0.76, \infty)$ and is decreasing on $(-5.24, -3)$ and on $(-3, -0.76)$. Thus, $(-5.24, -10.47)$ is a relative maximum and $(-0.76, -1.53)$ is a relative minimum.

f) *Inflection points.* $f''(-3)$ does not exist, but because $f(-3)$ does not exist there cannot be an inflection point at -3. The equation $f''(x) = 0$ has no solution, so there are no inflection points.

g) *Concavity.* Use -3 to divide the real number line into two intervals, A: $(-\infty, -3)$ and B: $(-3, \infty)$. Test a point in each interval.

A: Test -4, $f''(-4) = -10 < 0$

B: Test -2, $f''(-2) = 10 > 0$

Then f is concave down on $(-\infty, -3)$ and concave up on $(-3, \infty)$.

h) *Sketch.* Use the preceding information to sketch the graph. Compute other function values as needed.

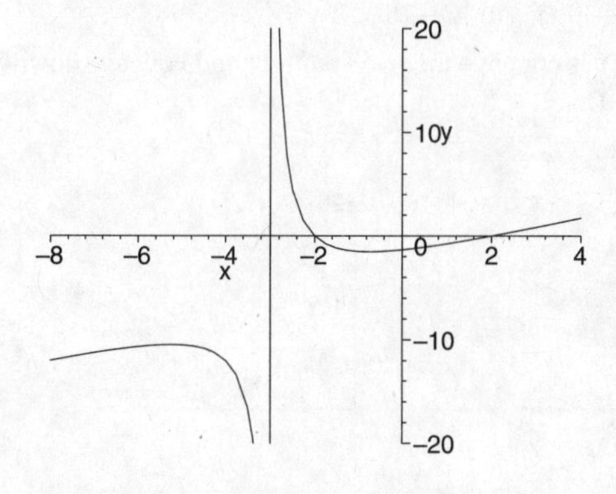

38. $f(x) = \dfrac{x^2-9}{x+1}$

a) *Intercepts.* The numerator is 0 for $x = -3$ or $x = 3$, and neither of these values makes the denominator 0. Thus, the x-intercepts are $(-3, 0)$ and $(3, 0)$.

$f(0) = -9$, so $(0, -9)$ is the y-intercept.

b) *Asymptotes.*

Vertical. The denominator is 0 for $x = -1$, so the line $x = -1$ is a vertical asymptote.

Horizontal. The degree of the numerator is greater than the degree of the denominator, so there are no horizontal asymptotes.

Oblique. By dividing, we get $f(x) = x - 1 - \dfrac{8}{x+1}$. As $|x|$ gets very large, $f(x)$ approaches $x - 1$, so $y = x - 1$ is an oblique asymptote.

c) *Derivatives.*

$$f'(x) = \frac{x^2+2x+9}{(x+1)^2}$$

$$f''(x) = \frac{-16}{(x+1)^3}$$

d) *Critical points.* $f'(-1)$ does not exist, but because $f(-1)$ does not exist $x = -1$ is not a critical point. The equation $f'(x) = 0$ has no real number solutions, so there are no critical points.

e) *Increasing, decreasing, relative extrema.* Use -1 to divide the real number line into two intervals, A: $(-\infty, -1)$ and B: $(-1, \infty)$.

A: Test -2, $f'(-2) = 9 > 0$

B: Test 0, $f'(0) = 9 > 0$

Then f is increasing on both intervals. Since there are no critical points, there are no relative extrema.

f) *Inflection points.* $f''(-1)$ does not exist, but because $f(-1)$ does not exist there cannot be an inflection point at -1. The equation $f''(x) = 0$ has no solution, so there are no inflection points.

g) *Concavity.* Use -1 to divide the real number line as in step (e).

A: Test -2, $f''(-2) = 16 > 0$

B: Test 0, $f''(0) = -16 < 0$

Then f is concave up on $(-\infty, -1)$ and concave down on $(-1, \infty)$.

h) Sketch.

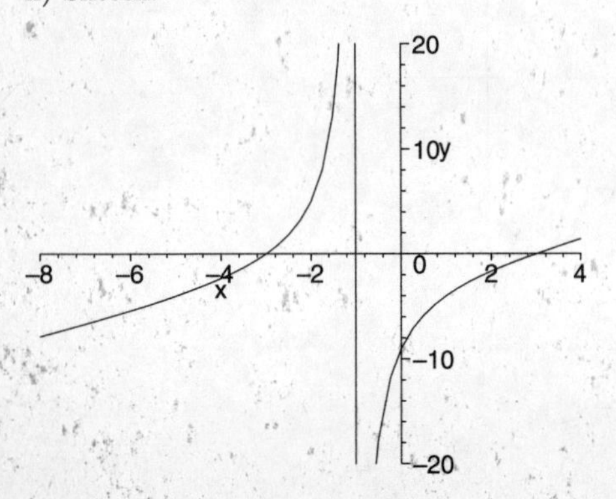

39. $f(x) = \dfrac{x-1}{x^2-2x-3}$

a) *Intercepts.* The numerator is 0 for $x = 1$, and this value of x does not make the denominator 0. Then $(1, 0)$ is the x-intercept.

$f(0) = \dfrac{0-1}{0^2 - 2\cdot 0 - 3} = \dfrac{1}{3}$, so $\left(0, \dfrac{1}{3}\right)$ is the y-intercept.

b) *Asymptotes.*

Vertical. The denominator $x^2 - 2x - 3 = (x+1)(x-3)$ is 0 for $x = -1$ or $x = 3$. Then the lines $x = -1$ and $x = 3$ are vertical asymptotes.

Horizontal. The degree of the numerator is less than the degree of the denominator, so $y = 0$ is a horizontal asymptote.

Oblique. There is no oblique asymptote since the degree of the numerator is not one more than the degree of the denominator.

c) *Derivatives.*

$$f'(x) = \frac{-x^2+2x-5}{(x^2-2x-3)^2}$$

$$f''(x) = \frac{2x^3-6x^2+30x-26}{(x^2-2x-3)^3}$$

d) *Critical points.* $f'(-1)$ and $f'(3)$ do not exist, but because $f(-1)$ and $f(3)$ do not exist $x = -1$ and $x = 3$ are not critical points. The equation $f'(x) = 0$ has no real number solution, so there are no critical points.

e) *Increasing, decreasing, relative extrema.* Use -1 and 3 to divide the real number line into three intervals, A: $(-\infty, -1)$, B: $(-1, 3)$, and C: $(3, \infty)$. Test a point in each interval.

A: Test -2, $f'(-2) = -\dfrac{13}{25} < 0$

B: Test 0, $f'(0) = -\dfrac{5}{9} < 0$

C: Test 4, $f'(4) = -\dfrac{13}{25} < 0$

Then f is decreasing on all three intervals. Since there are no critical points, there are no relative extrema.

f) *Inflection points.* $f''(-1)$ and $f''(3)$ do not exist, but because $f(-1)$ and $f(3)$ do not exist there cannot be an inflection point at -1 or at 3. Solve $f''(x) = 0$.

$$\frac{2x^3-6x^2+30x-26}{(x^2-2x-3)^3} = 0$$

$$2x^3 - 6x^2 + 30x - 26 = 0$$

$$(x-1)(2x^2-4x+26) = 0$$

$$x - 1 = 0 \quad\text{or}\quad 2x^2 - 4x + 26 = 0$$

$$x = 1 \qquad \text{No real number solution}$$

$f(1) = 0$, so $(1, 0)$ is on the graph and is a possible inflection point.

g) *Concavity.* Use -1, 1, and 3 to divide the real number line into four intervals, A: $(-\infty, -1)$, B: $(-1, 1)$, C: $(1, 3)$, and D $(3, \infty)$. Test a point in each interval.

A: Test -2, $f''(-2) = -\frac{126}{125} < 0$

B: Test 0, $f''(0) = \frac{26}{27} > 0$

C: Test 2, $f''(2) = -\frac{26}{27} < 0$

D: Test 4, $f''(4) = \frac{126}{125} > 0$

Then f is concave down on $(-\infty, -1)$ and on $(1, 3)$ and is concave up on $(-1, 1)$ and on $(3, \infty)$. Thus, $(1, 0)$ is an inflection point.

h) *Sketch.* Use the preceding information to sketch the graph. Compute other function values as needed.

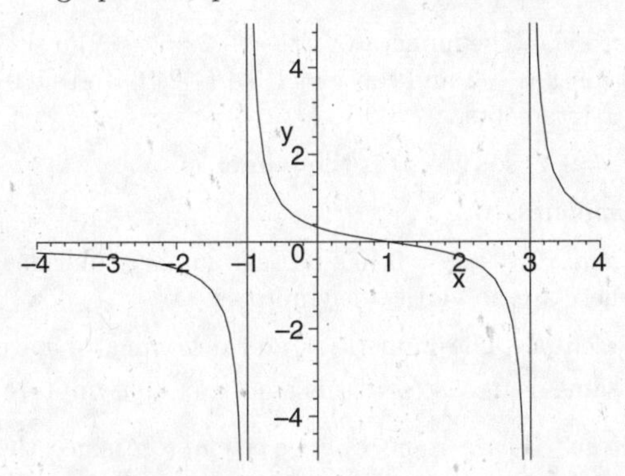

40. $f(x) = \dfrac{x+2}{x^2 + 2x - 15}$

a) *Intercepts.* The numerator is 0 for $x = -2$, and this value of x does not make the denominator 0. Then $(-2, 0)$ is the x-intercept.

$f(0) = -\frac{2}{15}$, so $\left(0, -\frac{2}{15}\right)$ is the y-intercept.

b) *Asymptotes.*

Vertical. The denominator $x^2 + 2x - 15 = (x+5)(x-3)$ is 0 for $x = -5$ or $x = 3$, so the lines $x = -5$ and $x = 3$ are vertical asymptotes.

Horizontal. The degree of the numerator is less than the degree of the denominator, so $y = 0$ is a horizontal asymptote.

Oblique. There is no oblique asymptote since the degree of the numerator is not one more than the degree of the denominator.

c) *Derivatives.*

$$f'(x) = \frac{-x^2 - 4x - 19}{(x^2 + 2x - 15)^2}$$

$$f''(x) = \frac{2x^3 + 12x^2 + 114x + 136}{(x^2 + 2x - 15)^3}$$

d) *Critical points.* $f'(-5)$ and $f'(3)$ do not exist, but because $f(-5)$ and $f(3)$ do not exist $x = -5$ and $x = 3$ are not critical points. The equation $f'(x) = 0$ has no real number solution, so there are no critical points.

e) *Increasing, decreasing, relative extrema.* Use -5 and 3 to divide the real number line into three intervals, A: $(-\infty, -5)$, B: $(-5, 3)$, and C: $(3, \infty)$.

A: Test -6, $f'(-6) = -\frac{31}{81} < 0$

B: Test 0, $f'(0) = -\frac{19}{225} < 0$

C: Test 4, $f'(4) = -\frac{51}{81} < 0$

Then f is decreasing on all three intervals. Since there are no critical points, there are no relative extrema.

f) *Inflection points.* $f''(-5)$ and $f''(3)$ do not exist, but because $f(-5)$ and $f(3)$ do not exist there cannot be an inflection point at -5 or 3. The equation $f''(x) = 0$ has no rational solutions. Since $f''(-1.5) < 0$ and $f''(-1) > 0$, we know that there is a solution of $f''(x) = 0$ and hence a possible inflection point between -1.5 and -1.

g) *Concavity.* We will use -1.5 as an estimate for the possible inflection point. Then use -5, -1.5, and 3 to divide the real number line into four intervals, A: $(-\infty, -5)$, B: $(-5, -1.5)$, C: $(-1.5, 3)$, and D: $(3, \infty)$.

A: Test -6, $f''(-6) = -\frac{548}{729} < 0$

B: Test -4, $f''(-4) = \frac{256}{343} > 0$

C: Test 2, $f''(2) = -\frac{428}{343} < 0$

D: Test 4, $f''(4) = \frac{912}{729} > 0$

Then f is concave down on $(-\infty, -5)$ and on $(1.5, 3)$ approximately and is concave up on $(-5, -1.5)$ approximately and on $(3, \infty)$. Thus, there is an inflection point between -1.5 and -1.

h) Sketch.

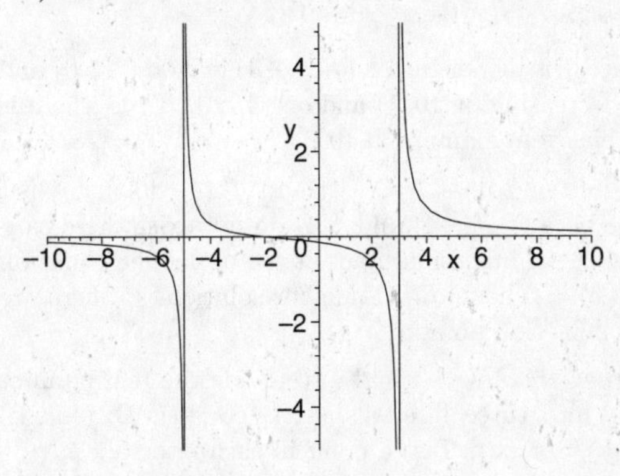

41. $f(x) = \dfrac{2x^2}{x^2 - 16}$

a) *Intercepts.* The numerator is 0 for $x = 0$, and this value of x does not make the denominator 0, so $(0, 0)$ is the x-intercept.

$f(0) = 0$, so the y-intercept is the x-intercept $(0, 0)$.

b) *Asymptotes.*

Vertical. The denominator $x^2 - 16 = (x+4)(x-4)$ is 0 for $x = -4$ or $x = 4$, so the lines $x = -4$ and $x = 4$ are vertical asymptotes.

Horizontal. The numerator and denominator have the same degree, so $y = \frac{2}{1}$, or $y = 2$, is a horizontal asymptote.

Oblique. There is no oblique asymptote since the degree of the numerator is not one more than the degree of the denominator.

c) *Derivatives.*

$$f'(x) = \frac{-64x}{(x^2-16)^2}$$

$$f''(x) = \frac{192x^2 + 1024}{(x^2-16)^3}$$

d) *Critical points.* $f'(-4)$ and $f'(4)$ do not exist, but because $f(-4)$ and $f(4)$ do not exist $x = -4$ and $x = 4$ are not critical points. Solve $f'(x) = 0$.

$$\frac{-64x}{(x^2-16)^2} = 0$$

$$-64x = 0$$

$$x = 0 \quad \text{Critical point}$$

From step (a) we already know that $(0,0)$ is on the graph.

e) *Increasing, decreasing, relative extrema.* Use -4, 0, and 4 to divide the real number line into four intervals, A: $(-\infty,-4)$, B: $(-4,0)$, C: $(0,4)$, and D: $(4,\infty)$.

A: Test -5, $f'(-5) = \frac{320}{81} > 0$

B: Test -1, $f'(-1) = \frac{64}{225} > 0$

C: Test 1, $f'(1) = -\frac{64}{225} < 0$

D: Test 5, $f'(5) = -\frac{320}{81} < 0$

Then f is increasing on $(-\infty,-4)$ and on $(-4,0)$ and is decreasing on $(0,4)$ and on $(4,\infty)$. Thus, there is a relative maximum at $(0,0)$.

f) *Inflection points.* $f''(-4)$ and $f''(4)$ do not exist, but because $f(-4)$ and $f(4)$ do not exist there cannot be an inflection point at -4 or 4. The equation $f''(x) = 0$ has no real-number solution, so there are no inflection points.

g) *Concavity.* Use -4 and 4 to divide the real number line into three intervals, A: $(-\infty,-4)$, B: $(-4,4)$, and C: $(4,\infty)$. Test a point in each interval.

A: Test -5, $f''(-5) = \frac{5824}{729} > 0$

B: Test 0, $f''(0) = -\frac{1}{4} < 0$

C: Test 5, $f''(5) = -\frac{5824}{729} > 0$

Then f is concave up on $(-\infty,-4)$ and on $(4,\infty)$ and is concave down on $(-4,4)$.

h) *Sketch.* Use the preceding information to sketch the graph. Compute other function values as needed.

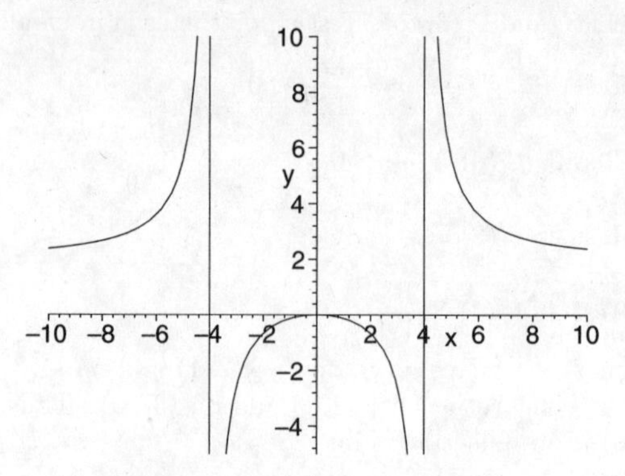

42. $f(x) = \frac{x^2 + x - 2}{2x^2 + 1}$

a) *Intercepts.* The numerator $x^2+x-2 = (x+2)(x-1)$ is 0 for $x = -2$ and for $x = 1$, so $(-2,0)$ and $(1,0)$ are x-intercepts.

$f(0) = -2$, so $(0,-2)$ is the y-intercept.

b) *Asymptotes.*

Vertical. $2x^2 + 1 = 0$ has no real-number solutions, so there are no vertical asymptotes.

Horizontal. The numerator and denominator have the same degree, so $y = \frac{1}{2}$ is a horizontal asymptote.

Oblique. There is no oblique asymptote since the degree of the numerator is not one more than the degree of the denominator.

c) *Derivatives.*

$$f'(x) = \frac{-2x^2 + 10x + 1}{(2x^2+1)^2}$$

$$f''(x) = \frac{8x^3 - 60x^2 - 12x + 10}{(2x^2+1)^3}$$

d) *Critical points.* $f'(x)$ exists for all real values of x.

$f'(x) = 0$ for $x = \frac{5 - 3\sqrt{3}}{2} \approx -0.10$ or

$x = \frac{5 + 3\sqrt{3}}{2} \approx 5.10$

$f(-0.10) \approx -2.05$ and $f(5.10) \approx 0.55$, so $(-0.10,-2.05)$ and $(5.10,0.55)$ are on the graph.

e) *Increasing, decreasing, relative extrema.* Use -0.10 and 5.10 to divide the real number line into three intervals, A: $(-\infty,-0.10)$, B: $(-0.10,5.10)$, and C: $(5.10,\infty)$.

A: Test -1, $f'(-1) = -\frac{11}{9} < 0$

B: Test 0, $f'(0) = 1 > 0$

C: Test 6, $f'(6) = -\frac{11}{5329} < 0$

Then f is decreasing on $(-\infty,-0.10)$ and on $(5.10,\infty)$ and is increasing on $(-0.10,5.10)$. Thus, $(-0.10,-2.05)$ is a relative minimum and $(5.10,0.55)$ is a relative maximum.

f) *Inflection points.* $f''(x)$ exists for all real values of x. The equation $f''(x) = 0$ is difficult to solve, but we can determine that $f''(7.5) < 0$ and $f''(7.75) > 0$ so there is a solution and hence a possible inflection point in the interval $(7.5, 7.75)$.

g) *Concavity.* From step (f) we know that f is concave down on $(-\infty, 7.5)$ and concave up on $(7.75, \infty)$ and that the concavity changes in the interval $(7.5, 7.75)$. Thus, there is an inflection point in $(7.5, 7.75)$.

h) Sketch.

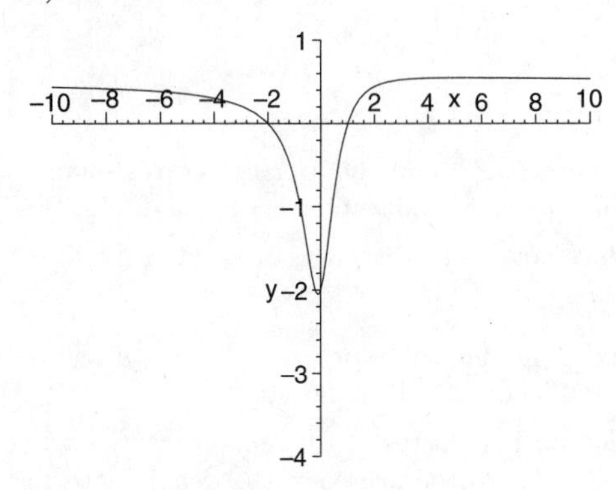

43. $f(x) = \dfrac{1}{x^2 - 1}$

a) *Intercepts.* Since the numerator is the constant 1, there are no x-intercepts.

$f(0) = \dfrac{1}{0^2 - 1} = -1$, so the y-intercept is $(0, -1)$.

b) *Asymptotes.*

Vertical. The denominator $x^2 - 1 = (x+1)(x-1)$ is 0 for $x = -1$ or $x = 1$, so the lines $x = -1$ and $x = 1$ are vertical asymptotes.

Horizontal. The degree of the numerator is less than the degree of the denominator, so $y = 0$ is a horizontal asymptote.

Oblique. There is no oblique asymptote since the degree of the numerator is not one more than the degree of the denominator.

c) *Derivatives.*

$$f'(x) = \frac{-2x}{(x^2 - 1)^2}$$

$$f''(x) = \frac{2(3x^2 + 1)}{(x^2 - 1)^3}$$

d) *Critical points.* $f'(-1)$ and $f'(1)$ do not exist, but because $f(-1)$ and $f(1)$ do not exist $x = -1$ and $x = 1$ are not critical points. Solve $f'(x) = 0$.

$$\frac{-2x}{(x^2 - 1)^2} = 0$$

$$-2x = 0$$

$$x = 0 \quad \text{Critical point}$$

From step (a) we already know that $(0, -1)$ is on the graph.

e) *Increasing, decreasing, relative extrema.* Use -1, 0, and 1 to divide the real number line into four intervals, A: $(-\infty, -1)$, B: $(-1, 0)$, C: $(0, 1)$, and D: $(1, \infty)$. Test a point in each interval.

A: Test -2, $f'(-2) = \dfrac{4}{9} > 0$

B: Test $-\dfrac{1}{2}$, $f'\left(-\dfrac{1}{2}\right) = \dfrac{16}{9} > 0$

C: Test $\dfrac{1}{2}$, $f'\left(\dfrac{1}{2}\right) = -\dfrac{16}{9} < 0$

D: Test 2, $f'(2) = -\dfrac{4}{9} < 0$

Then f is increasing on $(-\infty, -1)$ and on $(-1, 0)$ and is decreasing on $(0, 1)$ and on $(1, \infty)$. Thus, there is a relative maximum at $(0, -1)$.

f) *Inflection points.* $f''(-1)$ and $f''(1)$ do not exist, but because $f(-1)$ and $f(1)$ do not exist there cannot be an inflection point at -1 or at 1. The equation $f''(x) = 0$ has no real-number solution, so there are no inflection points.

g) *Concavity.* Use -1 and 1 to divide the real number line into three intervals, A: $(-\infty, -1)$, B: $(-1, 1)$, and C: $(1, \infty)$. Test a point in each interval.

A: Test -2, $f''(-2) = \dfrac{26}{9} > 0$

B: Test 0, $f''(0) = -2 < 0$

C: Test 2, $f''(2) = \dfrac{26}{9} > 0$

Then f is concave up on $(-\infty, -1)$ and on $(1, \infty)$ and is concave down on $(-1, 1)$.

h) *Sketch.* Use the preceding information to sketch the graph. Compute other function values as needed.

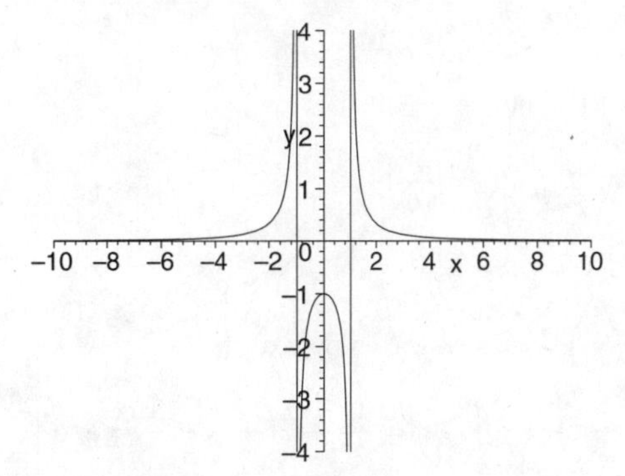

44. $f(x) = \dfrac{10}{x^2 + 4}$

a) *Intercepts.* Since the numerator is the constant 10, there are no x-intercepts.

$f(0) = 2.5$, so $(0, 2.5)$ is the y-intercept.

b) *Asymptotes.*

Vertical. $x^2 + 4 = 0$ has no real-number solutions, so there are no vertical asymptotes.

Horizontal. The degree of the numerator is less than the degree of the denominator, so $y = 0$ is a horizontal asymptote.

Oblique. There is no oblique asymptote since the degree of the numerator is not one more than the degree of the denominator.

c) *Derivatives.*

$$f'(x) = \frac{-20x}{(x^2+4)^2}$$

$$f''(x) = \frac{20(3x^2-4)}{(x^2+4)^3}$$

d) *Critical points.* $f'(x)$ exists for all real values of x.

$f'(x) = 0$ for $x = 0$

From step (a) we know that $(0, 2.5)$ is on the graph.

e) *Increasing, decreasing, relative extrema.* Use 0 to divide the real number line into two intervals, A: $(-\infty, 0)$ and B: $(0, \infty)$.

A: Test -1, $f'(-1) = \frac{4}{5} > 0$

B: Test 1, $f'(1) = -\frac{4}{5} < 0$

Then f is increasing on $(-\infty, 0)$ and is decreasing on $(0, \infty)$. Thus, $(0, 2.5)$ is a relative maximum.

f) *Inflection points.* $f''(x)$ exists for all real values of x. $f''(x) = 0$ for $x = \pm\frac{2}{\sqrt{3}}$, so there are possible inflection points at $x = -\frac{2}{\sqrt{3}}$ and at $x = \frac{2}{\sqrt{3}}$.

g) *Concavity.* Use $-\frac{2}{\sqrt{3}}$ and $\frac{2}{\sqrt{3}}$ to divide the real number line into three intervals, A: $\left(-\infty, -\frac{2}{\sqrt{3}}\right)$, B: $\left(-\frac{2}{\sqrt{3}}, \frac{2}{\sqrt{3}}\right)$, and C: $\left(\frac{2}{\sqrt{3}}, \infty\right)$.

A: Test -2, $f''(-2) = \frac{5}{16} > 0$

B: Test 0, $f''(0) = -\frac{5}{4} < 0$

C: Test 2, $f''(2) = \frac{5}{16} > 0$

Then f is concave up on $\left(-\infty, -\frac{2}{\sqrt{3}}\right)$ and on $\left(\frac{2}{\sqrt{3}}, \infty\right)$ and is concave down on $\left(-\frac{2}{\sqrt{3}}, \frac{2}{\sqrt{3}}\right)$.

$f\left(-\frac{2}{\sqrt{3}}\right) = f\left(\frac{2}{\sqrt{3}}\right) = \frac{15}{8}$, so $\left(-\frac{2}{\sqrt{3}}, \frac{15}{8}\right)$ and $\left(\frac{2}{\sqrt{3}}, \frac{15}{8}\right)$ are inflection points.

h) *Sketch.* Use the preceding information to sketch the graph. Compute other function values as needed.

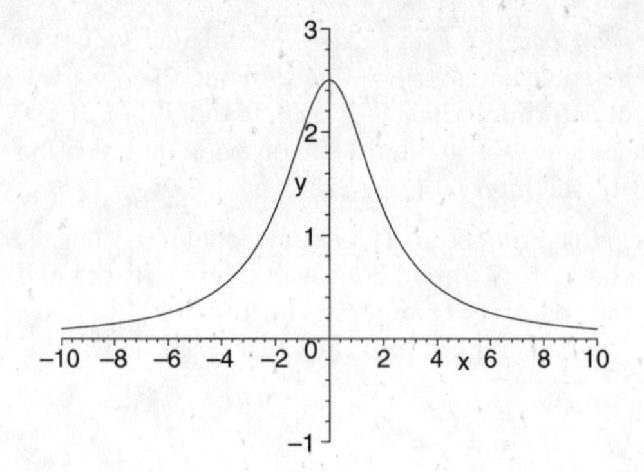

45. $f(x) = \frac{x^2+1}{x}$

a) *Intercepts.* Since the numerator has no real-number solutions, there are no x-intercepts.

$f(0)$ does not exist, so there is no y-intercept.

b) *Asymptotes.*

Vertical. The denominator x is 0 for $x = 0$, so the line $x = 0$ is a vertical asymptote.

Horizontal. The degree of the numerator is not less than or equal to the degree of the denominator, so there is no horizontal asymptote.

Oblique. The degree of the numerator is one more than the degree of the denominator, so there is an oblique asymptote. When we divide $x^2 + 1$ by x we have $f(x) = \frac{x^2+1}{x} = x + \frac{1}{x}$. As $|x|$ gets very large, $\frac{1}{x}$ approaches 0. Thus, $y = x$ is an oblique asymptote.

c) *Derivatives.*

$$f'(x) = \frac{x^2-1}{x^2}$$

$$f''(x) = \frac{2}{x^3}$$

d) *Critical points.* $f'(0)$ does not exist, but because $f(0)$ does not exist 0 is not a critical point. Solve $f'(x) = 0$.

$$\frac{x^2-1}{x^2} = 0$$

$$x^2 - 1 = 0$$

$$(x+1)(x-1) = 0$$

$x = -1$ or $x = 1$ Critical points

$f(-1) = -2$ and $f(1) = 2$, so $(-1, -2)$ and $(1, 2)$ are on the graph.

e) *Increasing, decreasing, relative extrema.* Use -1, 0, and 1 to divide the real number line into four intervals, A: $(-\infty, -1)$, B: $(-1, 0)$, C: $(0, 1)$, and D: $(1, \infty)$. Test a point in each interval.

A: Test -2, $f'(-2) = \frac{3}{4} > 0$

B: Test $-\frac{1}{2}$, $f'\left(-\frac{1}{2}\right) = -3 < 0$

C: Test $\frac{1}{2}$, $f'\left(\frac{1}{2}\right) = -3 < 0$

D: Test 2, $f'(2) = \frac{3}{4} > 0$

Then f is increasing on $(-\infty, -1)$ and on $(1, \infty)$ and is decreasing on $(-1, 0)$ and $(0, 1)$. Thus, there is a relative maximum at $(-1, -2)$ and a relative minimum at $(1, 2)$.

f) *Inflection points.* $f''(0)$ does not exist, but because $f(0)$ does not exist there cannot be an inflection point at 0. The equation $f''(x) = 0$ has no solution, so there are no inflection points.

g) *Concavity.* Use 0 to divide the real number line into two intervals, A: $(-\infty, 0)$ and B: $(0, \infty)$. Test a point in each interval.

A: Test -1, $f''(-1) = -2 < 0$

B: Test 1, $f''(1) = 2 > 0$

Then f is concave down on $(-\infty, 0)$ and is concave up on $(0, \infty)$.

h) *Sketch.* Use the preceding information to sketch the graph. Compute other function values as needed.

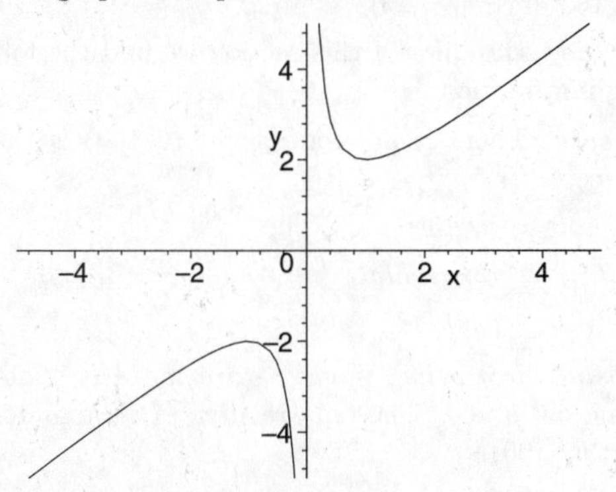

46. $f(x) = \dfrac{x^3}{x^2 - 1}$

a) *Intercepts.* $f(x) = 0$ when $x = 0$, so the x-intercept is $(0, 0)$. Note that this is also the y-intercept.

b) *Asymptotes.*

Vertical. $x^2 - 1 = 0$ when $x = -1$ or $x = 1$, so the lines $x = -1$ and $x = 1$ are vertical asymptotes.

Horizontal. The degree of the numerator is not less than or equal to the degree of the numerator, so there is no horizontal asymptote.

Oblique. $f(x) = x + \dfrac{x}{x^2 - 1}$, so $y = x$ is an oblique asymptote.

c) *Derivatives.*

$$f'(x) = \frac{x^4 - 3x^2}{(x^2 - 1)^2}$$

$$f''(x) = \frac{8x^5 - 22x^3 + 6x}{(x^2 - 1)^3}$$

d) *Critical points.* $f'(-1)$ and $f'(1)$ do not exist, but because $f(-1)$ and $f(1)$ do not exist -1 and 1 are not critical points. $f'(x) = 0$ for $x = -\sqrt{3}$, $x = 0$, or $x = \sqrt{3}$.

e) *Increasing, decreasing, relative extrema.* Use $-\sqrt{3}$, -1, 0, 1, and $\sqrt{3}$ to divide the real number line into six intervals, A: $(-\infty, -\sqrt{3})$, B: $(-\sqrt{3}, -1)$, C: $(-1, 0)$, D: $(0, 1)$, E: $(1, \sqrt{3})$, and F: $(\sqrt{3}, \infty)$.

A: Test -2, $f'(-2) = \frac{4}{9} > 0$

B: Test $-\frac{3}{2}$, $f'\left(-\frac{3}{2}\right) = -\frac{27}{25} < 0$

C: Test $-\frac{1}{2}$, $f'\left(-\frac{1}{2}\right) = -\frac{11}{16} < 0$

D: Test $\frac{1}{2}$, $f'\left(\frac{1}{2}\right) = -\frac{11}{6} < 0$

E: Test $\frac{3}{2}$, $f'\left(\frac{3}{2}\right) = -\frac{27}{25} < 0$

F: Test 2, $f'(2) = \frac{4}{9} > 0$

Then f is increasing on $(-\infty, -\sqrt{3})$ and on $(\sqrt{3}, \infty)$ and is decreasing on $(-\sqrt{3}, -1)$, on $(-1, 0)$, on $(0, 1)$ and on $(1, \sqrt{3})$. Thus, $\left(-\sqrt{3}, -\frac{3\sqrt{3}}{2}\right)$ is a relative maximum and $\left(\sqrt{3}, \frac{3\sqrt{3}}{2}\right)$ is a relative minimum.

f) *Inflection points.* $f''(-1)$ and $f''(1)$ do not exist, but because $f(-1)$ and $f(1)$ do not exist there cannot be an inflection point at either -1 or 1. $f''(x) = 0$ when $x = \dfrac{11 - \sqrt{73}}{8}$ or $x = 0$ or $x = \dfrac{11 + \sqrt{73}}{8}$, so there are possible inflection points at these points.

g) *Concavity.* Use -1, 0, $\dfrac{11 - \sqrt{73}}{8}$, 1, and $\dfrac{11 + \sqrt{73}}{8}$ to divide the real number line into six intervals: A: $(-\infty, -1)$, B: $(-1, 0)$, C: $\left(0, \frac{11 - \sqrt{73}}{8}\right)$, D: $\left(\frac{11 - \sqrt{73}}{8}, 1\right)$, E: $\left(1, \frac{11 + \sqrt{73}}{8}\right)$, and F: $\left(\frac{11 + \sqrt{73}}{8}, \infty\right)$. Testing a number in each interval, we find that in A, $f''(x) < 0$; in B, $f''(x) > 0$; in C, $f''(x) < 0$; in D, $f''(x) < 0$; in E, $f''(x) > 0$; and in F, $f''(x) > 0$. Thus, f is concave down on intervals A, C, and D and is concave up on intervals B, E, and F. There is an inflection point at $(0, 0)$.

h) *Sketch.* Use the preceding information to sketch the graph. Compute other function values as needed.

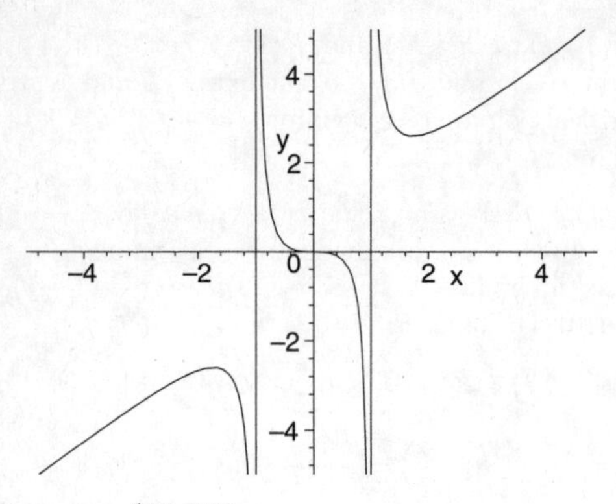

47. $C(p) = \dfrac{\$48,000}{100-p}$

We will only consider the interval $[0, 100)$ since it is not possible to remove less than 0% or more than 100% of the pollutants and $C(p)$ is not defined for $p = 100$.

a) $C(0) = \dfrac{\$48,000}{100-0} = \$480$

$C(20) = \dfrac{\$48,000}{100-20} = \$600$

$C(80) = \dfrac{\$48,000}{100-80} = \$2400$

$C(90) = \dfrac{\$48,000}{100-90} = \$4800$

b) $\lim\limits_{p\to 100^-} C(p) = \lim\limits_{p\to 100^-} \dfrac{\$48,000}{100-p} = \infty$

c) The cost of removing 100% of the pollutants is infinitely high.

d) Using the techniques of this section we find the following additional information.

Intercepts. No p-intercept; $(0, 480)$ is the C-intercept.

Asymptotes. *Vertical.* $p = 100$

Horizontal. $C = 0$

Oblique. None

Increasing, decreasing, relative extrema. $C(p)$ is increasing on $[0, 100)$. There are no relative extrema.

Inflection points, concavity. $C(p)$ is concave up on $[0, 100)$. There is no inflection point.

We use this information and compute other function values as needed to sketch the graph.

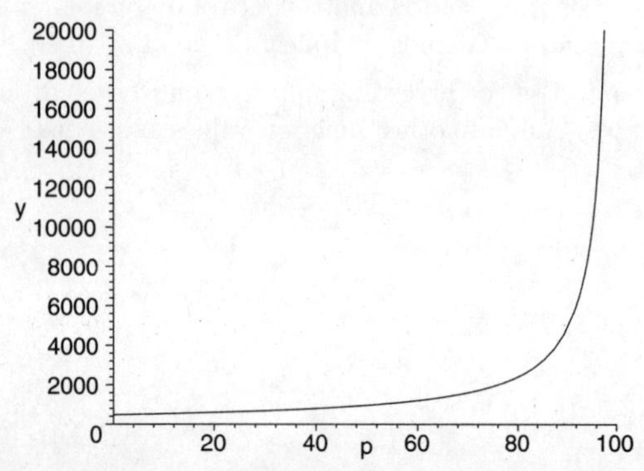

e) From the result in part (b), we see that the company cannot afford to remove 100% of the pollutants.

48. $A(t) = \dfrac{100}{t^2+1}$

a) $A(0) = \dfrac{100}{0^2+1} = 100$ cc

$A(1) = \dfrac{100}{1^2+1} = 50$ cc

$A(2) = \dfrac{100}{2^2+1} = 20$ cc

$A(7) = \dfrac{100}{7^2+1} = 2$ cc

$A(10) = \dfrac{100}{10^2+1} \approx 0.99$ cc

b) $\lim\limits_{t\to\infty} A(t) = \lim\limits_{t\to\infty} \dfrac{100}{t^2+1} = 0$

c) $A'(t) = -\dfrac{200t}{(t^2+1)^2}$

$A''(t) = \dfrac{600t^2 - 200}{(t^2+1)^3}$

$A'(t) = 0$ when $t = 0$

$A''(0) = -200 < 0$ so there is a relative maximum at $t = 0$.

$A(0) = 100$, so the relative maximum value over $[0, \infty)$ is 100 cc at $t = 0$.

d) Using the techniques of this section we find the following information.

Intercepts. There is no x-intercept; $(0, 100)$ is the y-intercept.

Asymptotes. *Vertical.* None

Horizontal. $y = 0$

Oblique. None

Increasing, decreasing, relative extrema. $A(t)$ is decreasing on $[0, \infty)$. The only relative extremum occurs at $(0, 100)$.

Inflection points, concavity. $A(t)$ is concave down on $\left(0, \dfrac{1}{\sqrt{3}}\right)$ and concave up on $\left(\dfrac{1}{\sqrt{3}}, \infty\right)$. The point $\left(\dfrac{1}{\sqrt{3}}, 75\right)$ is an inflection point.

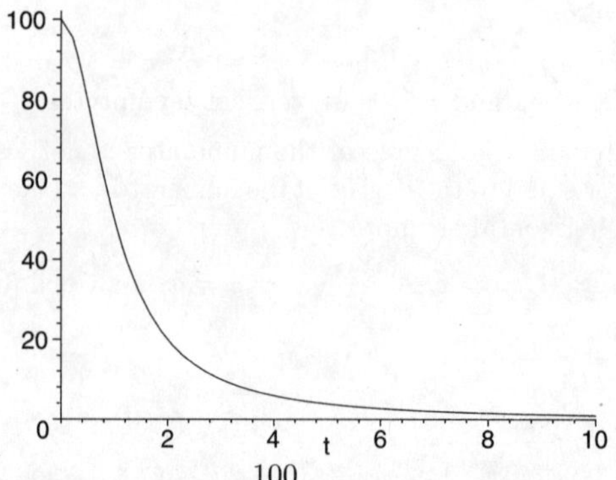

e) The equation $\dfrac{100}{t^2+1} = 0$ has no solution, so according to the given function the medication never completely leaves the bloodstream.

49. $T(t) = \dfrac{6t}{t^2+1} + 98.6$

a)

$$\begin{aligned} T(0) &= \frac{6(0)}{0+1} + 98.6 = 98.6 \\ T(1) &= \frac{6(1)}{1+1} + 98.6 = 101.6 \\ T(2) &= \frac{6(2)}{4+1} + 98.6 = 101 \\ T(5) &= \frac{6(5)}{25+1} + 98.6 = 99.8 \\ T(10) &= \frac{6(10)}{100+1} + 98.6 = 99.2 \end{aligned}$$

b)

$$\begin{aligned} \lim_{t\to\infty} T(t) &= \lim_{t\to\infty} \frac{6t}{t^2+1} + 98.6 \\ &= \lim_{t\to\infty} \frac{\frac{6}{t}}{1+\frac{1}{t^2}} + 98.6 \\ &= 0 + 98.6 \\ &= 98.6 \end{aligned}$$

c) The maximum occurs at the critical value of the function

$$\begin{aligned} T'(t) &= \frac{6(t^2+1) - 6t(2t)}{(t^2+1)^2} \\ &= \frac{6-6t^2}{(t^2+1)^2} \\ T'(t) &= 0 \\ \frac{6-6t^2}{(t^2+1)^2} &= 0 \\ 6-6t^2 &= 0 \\ t &= 1 \end{aligned}$$

Note we ignore the negative option since the question involves the interval $[0,\infty)$.
From part a) we know that $T(1) = 101.6$

d)

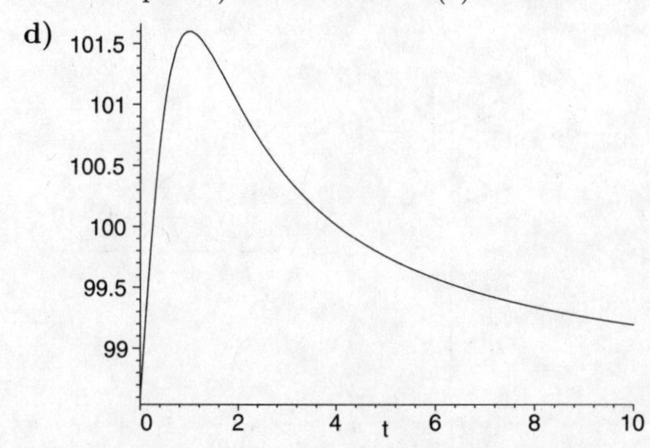

e) According to this model the temperature does not return to 98.6^o since that temperature is reaches only when t approaches infinity.

50. $P(t) = \dfrac{250t}{t^2+4} + 500$

a)

$$\begin{aligned} P(0) &= \frac{250(0)}{0+4} + 500 = 500 \\ P(1) &= \frac{250(1)}{1+4} + 500 = 550 \\ P(2) &= \frac{250(2)}{4+4} + 500 = 562.5 \\ P(5) &= \frac{250(5)}{25+4} + 500 = 543.1 \\ P(10) &= \frac{250(10)}{100+4} + 500 = 524.04 \end{aligned}$$

b)

$$\begin{aligned} \lim_{t\to\infty} P(t) &= \lim_{t\to\infty} \frac{250t}{t^2+4} + 500 \\ &= \lim_{t\to\infty} \frac{\frac{250}{t}}{1+\frac{4}{t^2}} + 98.6 \\ &= 0 + 500 \\ &= 500 \end{aligned}$$

c) The maximum occurs at the critical value of the function

$$\begin{aligned} T'(t) &= \frac{250(t^2+4) - 250t(2t)}{(t^2+4)^2} \\ &= \frac{1000-250t^2}{(t^2+4)^2} \\ T'(t) &= 0 \\ \frac{1000-250t^2}{(t^2+4)^2} &= 0 \\ 1000-250t^2 &= 0 \\ t &= 2 \end{aligned}$$

Note we ignore the negative option since the question involves the interval $[0,\infty)$.
From part a) we know that $P(2) = 562.5$

d)

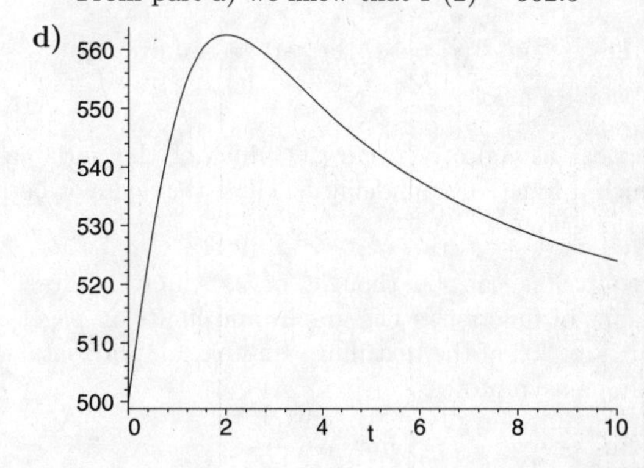

51. a) $E(9) = 9 \cdot \dfrac{4}{9} = 4.00$

$E(8) = 9 \cdot \dfrac{4}{8} = 4.50$

$E(7) = 9 \cdot \dfrac{4}{7} \approx 5.14$

$E(6) = 9 \cdot \dfrac{4}{6} = 6.00$

$E(5) = 9 \cdot \frac{4}{5} = 7.20$

$E(4) = 9 \cdot \frac{4}{4} = 9.00$

$E(3) = 9 \cdot \frac{4}{3} = 12.00$

$E(2) = 9 \cdot \frac{4}{2} = 18.00$

$E(1) = 9 \cdot \frac{4}{1} = 36.00$

$E\left(\frac{2}{3}\right) = 9 \cdot \frac{4}{\frac{2}{3}} = 9 \cdot \left(4 \cdot \frac{3}{2}\right) = 54.00$

$E\left(\frac{1}{3}\right) = 9 \cdot \frac{4}{\frac{1}{3}} = 9 \cdot \left(4 \cdot \frac{3}{1}\right) = 108.00$

We complete the table.

Innings pitched (i)	Earned-run average (E)
9	4.00
8	4.50
7	5.14
6	6.00
5	7.20
4	9.00
3	12.00
2	18.00
1	36.00
$\frac{2}{3}$	54.00
$\frac{1}{3}$	108.00

b) As i approaches 0 from the right, the values of $E(i)$ increase without bound, so $\lim_{i \to 0^+} E(i) = \infty$.

c) Since $\lim_{i \to 0^+} E(i) = \infty$, the earned run average would be ∞.

52. Vertical asymptotes occur at values of the variable for which a function is undefined. Thus, they cannot be part of the graph.

53. Asymptotes can be thought of as "limiting lines" for graphs of functions. The graphs and limits on pages 201, 202, and 203 of the text illustrate vertical, horizontal, and oblique asymptotes.

54. $\lim_{x \to -\infty} \frac{-3x^2+5}{2-x} = \lim_{x \to -\infty} \frac{-3x + \frac{5}{x}}{\frac{2}{x} - 1} =$

$\frac{\lim_{x \to -\infty} -3x + 0}{0 - 1} = -\lim_{x \to -\infty} -3x = -\infty$

55.

$$\lim_{x \to 5} \frac{x^2-6x+5}{x^2-3x-10} = \lim_{x \to 5} \frac{(x-1)(x-5)}{(x+2)(x-5)}$$

$$= \lim_{x \to 5} \frac{(x-1)}{(x+2)}$$

$$= \frac{5-1}{5+2}$$

$$= \frac{4}{7}$$

56. $\lim_{x \to -2} \frac{x^3+8}{x^2-4} = \lim_{x \to -2} \frac{(x+2)(x^2-2x+4)}{(x+2)(x-2)}$

$= \lim_{x \to -2} \frac{x^2-2x+4}{x-2}$ Assuming $x \neq -2$

$= \frac{(-2)^2 - 2(-2) + 4}{-2-2}$

$= \frac{4+4+4}{-4}$

$= -3$

57. We divide the numerator and the denominator by x^2, the highest power of x in the denominator.

$\lim_{x \to \infty} \frac{-6x^3+7x}{2x^2-3x-10} = \lim_{x \to \infty} \frac{-6x + \frac{7}{x}}{2 - \frac{3}{x} - \frac{10}{x^2}}$

$= \frac{\lim_{x \to \infty} -6x + 0}{2 - 0 - 0}$

$= -\infty$

(The numerator increases without bound negatively while the denominator approaches 2.)

58. $\lim_{x \to -\infty} \frac{-6x^3+7x}{2x^2-3x-10} = \lim_{x \to -\infty} \frac{-6x + \frac{7}{x}}{2 - \frac{3}{x} - \frac{10}{x^2}}$

$= \frac{\lim_{x \to -\infty} -6x + 0}{2 - 0 - 0}$

$= \infty$

59. $\lim_{x \to 1} \frac{x^3-1}{x^2-1} = \lim_{x \to 1} \frac{(x-1)(x^2+x+1)}{(x-1)(x+1)} =$

$\lim_{x \to 1} \frac{x^2+x+1}{x+1} = \frac{1+1+1}{1+1} = \frac{3}{2}$

60.

$$\lim_{x \to -\infty} \frac{7x^5+x-9}{6x+x^3} = \lim_{x \to -\infty} \frac{\frac{7x^2}{6} + \frac{1}{x^2} - \frac{9}{x^3}}{\frac{6}{x^2} + 1}$$

$$= \frac{\lim_{x \to -\infty} \frac{7x^2}{6} + 0 - 0}{0+1}$$

$$= \infty$$

61.

$$\lim_{x \to -\infty} \frac{2x^4+x}{x+1} = \lim_{x \to -\infty} \frac{2x^3+1}{1+\frac{1}{x}}$$

$$= \frac{\lim_{x \to -\infty} 2x^3+1}{1+0}$$

$$= -\infty$$

62. Undefined

63. Undefined

64. Since the numerator is bounded by ± 1 and the denominator grows indefinitely, then

$$\lim_{x\to\infty} \frac{\sin x}{x} = 0$$

see figure below

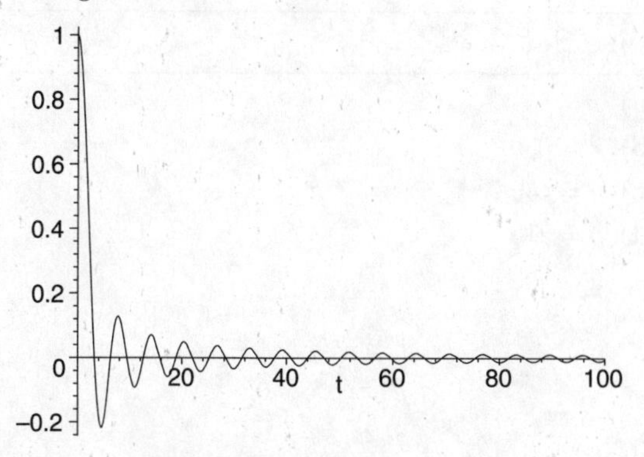

65. Since the numerator is bounded by ± 1 and the denominator grows indefinitely, then

$$\lim_{x\to\infty} \frac{\cos x}{x} = 0$$

see figure below

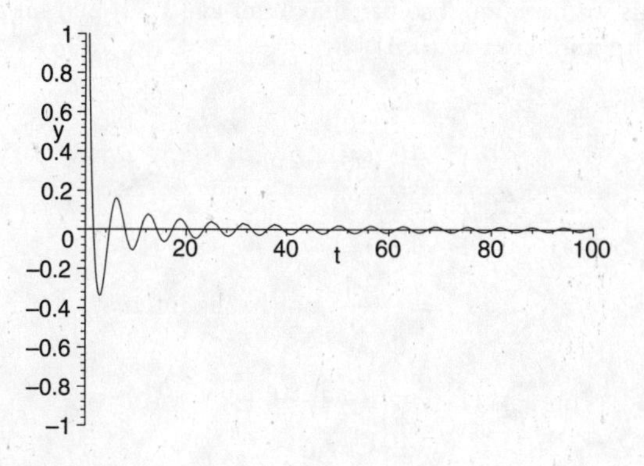

66. $\lim_{x\to 3^+} f(x) = -\infty$

67. $\lim_{x\to 3^-} f(x) = \infty$

68. $\lim_{x\to -3^+} f(x) = \infty$

69. $\lim_{x\to -3^-} f(x) = -\infty$

70. $\lim_{x\to\infty} f(x) = 1$

71. $\lim_{x\to -\infty} f(x) = 1$

72. $\lim_{x\to 2^+} f(x) = \infty$

73. $\lim_{x\to 2^-} f(x) = -\infty$

74. $\lim_{x\to -1^+} f(x) = \infty$

75. $\lim_{x\to -1^-} f(x) = -\infty$

76. $\lim_{x\to -\infty} f(x) = 0$

77. $\lim_{x\to -\infty} f(x) = 0$

78.

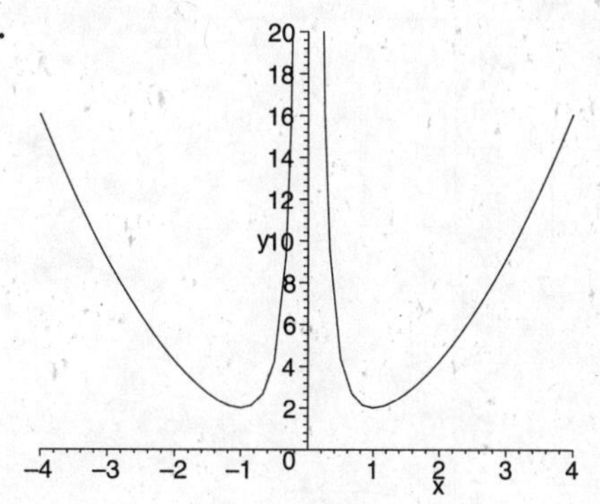

79.

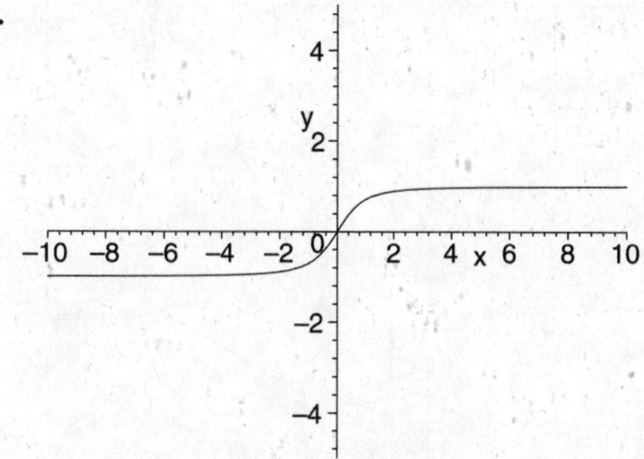

80.

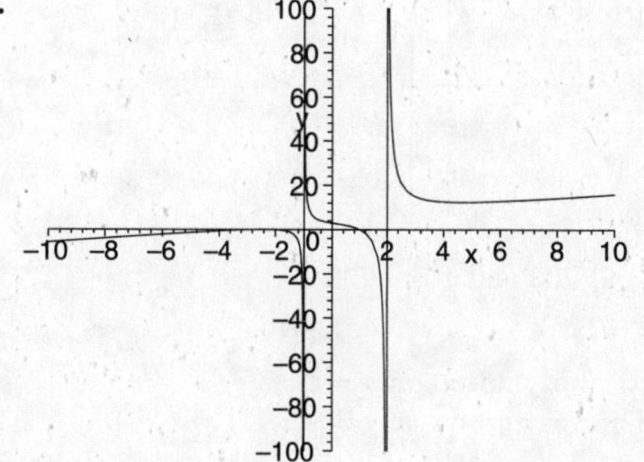

81.

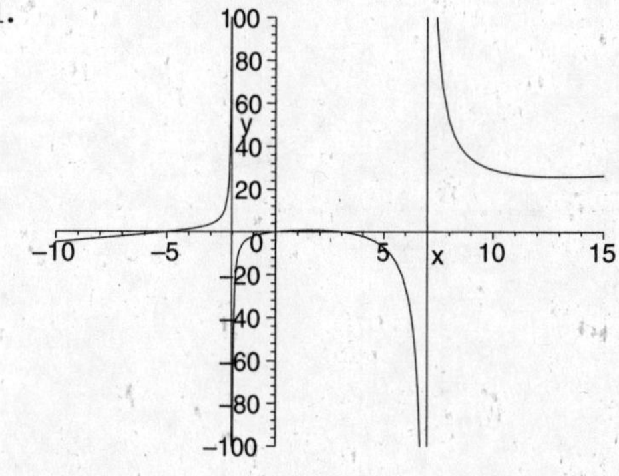

82.

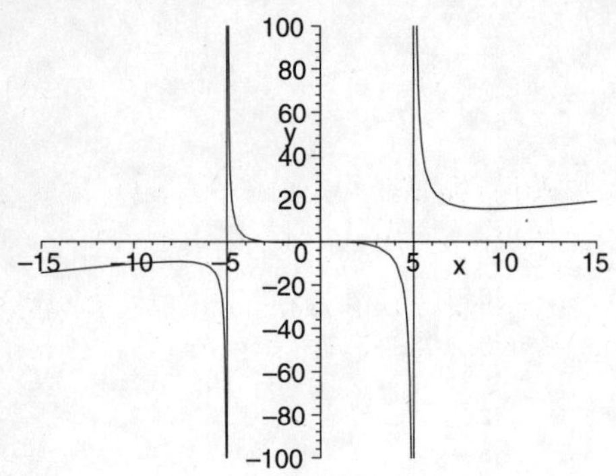

83.

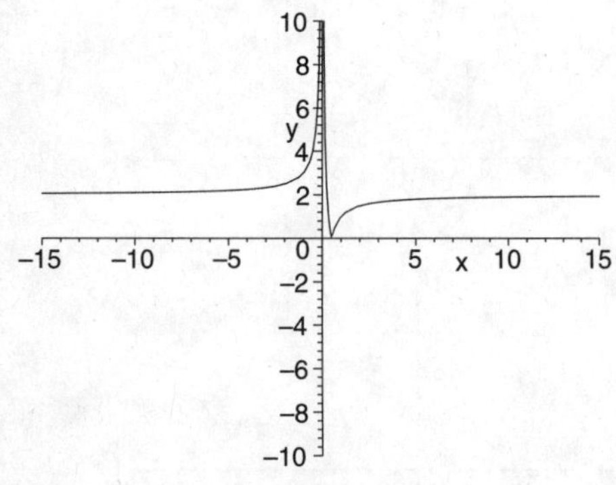

84.

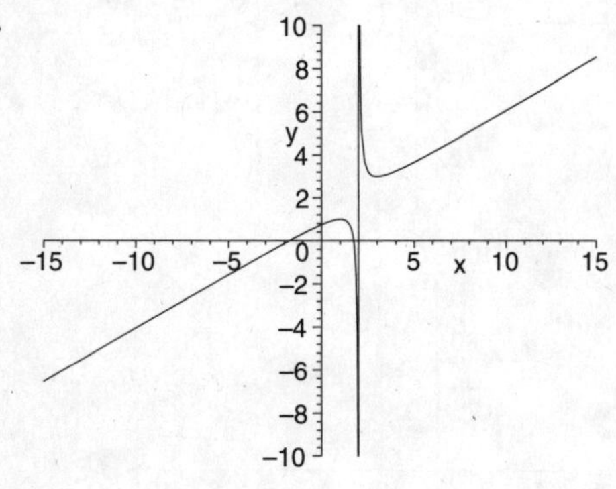

a) $x = -1.732$ and $x = 1.732$

b) $y = 0.75$

c) Vertical asymptote at $x = 2$
Oblique asymptote at $y = \frac{1}{2}x + 1$

85.

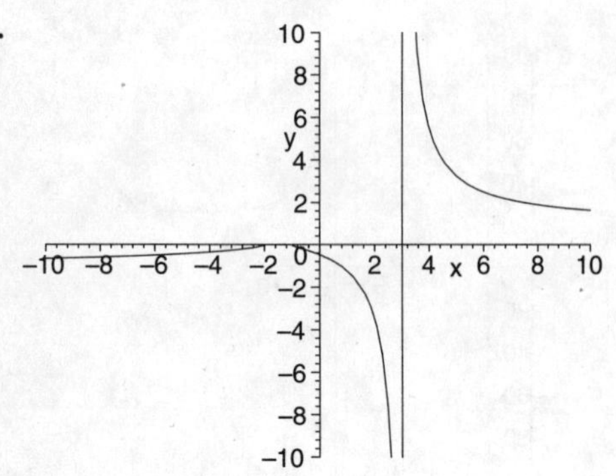

a) $\lim_{x \to \infty} f(x) = 1$
$\lim_{x \to -\infty} f(x) = -1$

b) The function is undefined over the interval $(-2, -1)$.

c) The domain of the function is given by

$$x < -2, \ -1 < x < 3, \text{ and } x > 3$$

The function is undefined on the interval $(-2, -1)$ since in that interval the radicant is negative.

d) $\lim_{x \to -2^+} f(x) = 0$
$\lim_{x \to -2^-} f(x) = 0$

Exercise Set 3.4

1. **a)** $x = 85$
b) $x = 0$
c) $y = 150$
d) $y = 210$

2. $f(x) = 4 + x + x^2$

$$\begin{aligned} f'(x) &= 1 + 2x \\ f'(x) &= 0 \\ 1 + 2x &= 0 \\ x &= -\frac{1}{2} \end{aligned}$$

The critical value does not belong in the given interval. Next, we check the endpoints of the interval
$f(0) = 4 + 0 + 0 = 4$
$f(2) = 4 + 2 + 4 = 10$
Thus we have an absolute maximim at $(2, 10)$ and an absolute minimum at $(0, 4)$.

3. $f(x) = x^3 - x^2 - x + 2$

$$\begin{aligned} f'(x) &= 3x^2 - 2x - 1 = (3x + 1)(x - 1) \\ f'(x) &= 0 \\ (3x + 1)(x + 1) &= 0 \\ x &= \frac{-1}{3} \quad \text{not in the interval} \\ x &= 1 \\ f(1) &= 1 - 1 - 1 + 2 \\ &= 1 \end{aligned}$$

Next, we check the endpoints of the interval
$f(0) = 0 - 0 - 0 + 2 = 2$
$f(2) = 8 - 4 - 2 + 2 = 4$
Thus we have an absolute maximim at $(2, 4)$ and an absolute minimum at $(1, 1)$.

4. $f(x) = x^3 + \frac{1}{2}x^2 - 2x + 5$

$$\begin{aligned} f'(x) &= 3x^2 + x - 2 = (3x - 2)(x + 1) \\ f'(x) &= 0 \\ (3x + 1)(x + 1) &= 0 \\ x &= \frac{2}{3} \\ x &= -1 \quad \text{not in the interval} \\ f(\tfrac{2}{3}) &= \left(\frac{2}{3}\right)^3 + \frac{1}{2}\left(\frac{2}{3}\right)^2 - 2\left(\frac{2}{3}\right) + 5 \\ &= \frac{113}{27} = 4.185 \end{aligned}$$

Next, we check the endpoints of the interval
$f(0) = 0 + 0 - 0 + 5 = 5$
$f(2) = 1 + \frac{1}{2} - 2 + 5 = 4.5$
Thus we have an absolute maximim at $(0, 5)$ and an absolute minimum at $\left(\frac{2}{3}, \frac{113}{27}\right)$.

5. $f(x) = 3x - 2$ $f'(x) = 3$, which means the function does not have critical values. Next, we check the endpoints of the interval
$f(-1) = -3 - 2 = -5$
$f(1) = 3 - 2 = 1$
Thus we have an absolute maximim at $(1, 1)$ and an absolute minimum at $(-1, -5)$.

6. $f(x) = 2x + 4$ $f'(x) = 2$, which means the function does not have critical values. Next, we check the endpoints of the interval
$f(-1) = -2 + 4 = 2$
$f(1) = 2 + 4 = 6$
Thus, we have an absolute maximim at $(1, 6)$ and an absolute minimum at $(-1, 2)$.

7. $f(x) = 3 - 2x - 5x^2$

$$\begin{aligned} f'(x) &= -2 - 10x \\ f'(x) &= 0 \\ -2 - 10x &= 0 \\ x &= \frac{-1}{5} \\ f\left(\frac{-1}{5}\right) &= 3 + \frac{2}{5} - \frac{1}{5} \\ &= \frac{16}{5} \end{aligned}$$

Next, we check the endpoints of the interval
$f(-3) = 3 + 6 - 45 = -36$
$f(3) = 3 - 6 - 45 = -48$
Thus, we have an absolute maximum at $\left(\frac{-1}{5}, \frac{16}{5}\right)$ and an absolute minimum at $(3, -48)$.

8. $f(x) = 1 + 6x - 3x^2$

$$\begin{aligned} f'(x) &= 6 - 6x \\ f'(x) &= 0 \\ 6 - 6x &= 0 \\ x &= 1 \\ f(1) &= 1 + 6 - 3 \\ &= 4 \end{aligned}$$

Next, we check the endpoints of the interval
$f(0) = 1 + 0 - 0 = 1$
$f(4) = 1 + 24 - 48 = -23$
Thus, we have an absolute maximum at $(1, 4)$ and an absolute minimum at $(4, -23)$.

9. $f(x) = 1 - x^3$

$$\begin{aligned} f'(x) &= -3x^2 \\ f'(x) &= 0 \\ -3x^2 &= 0 \\ x &= 0 \\ f(0) &= 1 - 0 \\ &= 1 \end{aligned}$$

Next, we check the endpoints of the interval
$f(-8) = 1 - (-512) = 513$
$f(8) = 1 - 512 = -511$
Thus, we have an absolute maximum at $(-8, 513)$ and an absolute minimum at $(8, -511)$.

10. $f(x) = 2x^3$

$$\begin{aligned} f'(x) &= 6x^2 \\ f'(x) &= 0 \\ 6x^2 &= 0 \\ x &= 0 \\ f(0) &= 0 \end{aligned}$$

Next, we check the endpoints of the interval
$f(-10) = 2(1000) = 2000$
$f(10) = 2(-1000) = -2000$
Thus, we have an absolute maximum at $(10, 2000)$ and an absolute minimum at $(-10, -2000)$.

11. $f(x) = 12 + 9x - 3x^2 - x^3$

$$\begin{aligned} f'(x) &= 9 - 6x - 3x^2 \\ &= 3(3 + x)(1 - x) \\ f'(x) &= 0 \\ 3(3 + x)(1 - x) &= 0 \\ x &= -3 \\ x &= 1 \\ f(-3) &= 12 - 27 - 27 + 27 = -15 \\ f(1) &= 12 + 9 - 3 - 1 = 17 \end{aligned}$$

Note that the critical values are the same as the endpoints of the interval.
Thus, we have an absolute maximum at $(1, 17)$ and an absolute minimum at $(-3, -15)$.

12. $f(x) = x^3 - 6x^2 + 10$

$$\begin{aligned} f'(x) &= 3x^2 - 12x \\ &= 3x(x - 4) \\ f'(x) &= 0 \\ 3x(x - 4) &= 0 \\ x &= 0 \\ x &= 4 \\ f(0) &= 0 - 0 + 10 = 10 \\ f(4) &= 64 - 96 + 10 = -22 \end{aligned}$$

Note that the critical values are the same as the endpoints of the interval.
Thus, we have an absolute maximum at $(0, 10)$ and an absolute minimum at $(4, -22)$.

13. $f(x) = x^4 - 2x^3$

$$\begin{aligned} f'(x) &= 4x^3 - 6x^2 \\ &= 2x^2(2x-3) \\ f'(x) &= 0 \\ 2x^2(2x-3) &= 0 \\ x &= 0 \\ x &= \frac{3}{2} \\ f(0) &= 0 - 0 = 0 \\ f\left(\frac{3}{2}\right) &= \frac{81}{16} - \frac{27}{4} = -\frac{27}{16} \end{aligned}$$

Next, we check the endpoints of the interval
$f(-2) = 16 - 2(-8) = 32$
$f(2) = 16 - 2(8) = 0$
Thus, we have an absolute maximum at $(-2, 32)$ and an absolute minimum at $\left(\frac{3}{2}, -\frac{27}{16}\right)$.

14. $f(x) = x^3 - x^4$

$$\begin{aligned} f'(x) &= 3x^2 - 4x^3 \\ &= x^2(3 - 4x^3) \\ f'(x) &= 0 \\ x^2(3-4x) &= 0 \\ x &= 0 \\ x &= \frac{3}{4} \\ f(0) &= 0 - 0 = 0 \\ f\left(\frac{3}{4}\right) &= \frac{27}{64} - \frac{81}{256} = \frac{27}{256} \end{aligned}$$

Next, we check the endpoints of the interval
$f(-1) = -1 - 1 = -2$
$f(2) = 1 - 1 = 0$
Thus, we have an absolute maximum at $\left(\frac{3}{4}, \frac{27}{256}\right)$ and an absolute minimum at $(-1, -2)$.

15. $f(x) = x^4 - 2x^2 + 5$

$$\begin{aligned} f'(x) &= 4x^3 - 4x \\ &= 4x(x^2 - 1) \\ &= 4x(x-1)(x+1) \\ f'(x) &= 0 \\ 4x(x-1)(x+1) &= 0 \\ x &= 0 \\ x &= 1 \\ x &= -1 \\ f(0) &= 0 - 0 + 5 = 5 \\ f(1) &= 1 - 2 + 5 = 4 \\ f(-1) &= 1 - 2 + 5 = 4 \end{aligned}$$

Next, we check the endpoints of the interval
$f(-2) = 16 - 8 + 5 = 13$
$f(2) = 16 - 8 + 5 = 13$
Thus, we have an absolute maximum at $(-2, 13)$ and at $(-2, 13)$ and an absolute minimum at $(-1, 4)$ and at $(1, 4)$.

16. $f(x) = x^4 - 8x^2 + 3$

$$\begin{aligned} f'(x) &= 4x^3 - 16x \\ &= 4x(x^2 - 4) \\ &= 4x(x-2)(x+2) \\ f'(x) &= 0 \\ 4x(x-2)(x+2) &= 0 \\ x &= 0 \\ x &= 2 \\ x &= -2 \\ f(0) &= 0 - 0 + 3 = 3 \\ f(-2) &= 16 - 32 + 3 = -13 \\ f(2) &= 16 - 32 + 3 = -13 \end{aligned}$$

Next, we check the endpoints of the interval
$f(-3) = 81 - 72 + 3 = 12$
$f(3) = 81 - 72 + 3 = 12$
Thus, we have an absolute maximum at $(-3, 12)$ and at $(3, 12)$ and an absolute minimum at $(-2, -13)$ and at $(2, -13)$.

17. $f(x) = (x+3)^{2/3} - 5$
$f'(x) = \frac{2}{3}(x+3)^{-1/3}$, which is undefined at $x = -3$ (the only critical value).
$f'(-3) = 0 - 5 = -5$
Next, we check the endpoints of the interval
$f(-4) = 1 - 5 = -4$
$f(5) = 4 - 5 = -1$
Thus, we have an absolute maximum at $(5, -1)$ and an absolute minimum at $(-3, -5)$.

18. $f(x) = 1 - x^{2/3}$
$f'(x) = -\frac{2}{3}x^{-1/3}$, which is undefined at $x = 0$ (the only critical value).
$f'(0) = 1 - 0 = 1$
Next, we check the endpoints of the interval
$f(-8) = 1 - 4 = -3$
$f(8) = 1 - 4 = -3$
Thus, we have an absolute maximum at $(0, 1)$ and an absolute minimum at $(-8, -3)$ and at $(8, -3)$.

19. $f(x) = x + \dfrac{1}{x}$

$$\begin{aligned} f'(x) &= 1 - \frac{1}{x^2} \\ f'(x) &= 0 \\ 1 - \frac{1}{x^2} &= 0 \\ x^2 - 1 &= 0 \\ x &= -1 \text{ not acceptable} \\ x &= 1 \\ f(1) &= 1 + 1 = 2 \end{aligned}$$

Next, we check the endpoint of the interval
$f(20) = 20 + 0.05 = 20.05$
Thus, we have an absolute maximum at $(20, 20.05)$ and an absolute minimum at $(1, 2)$.

20. $f(x) = x + \frac{4}{x}$

$$\begin{aligned} f'(x) &= 1 - \frac{4}{x^2} \\ f'(x) &= 0 \\ 1 - \frac{4}{x^2} &= 0 \\ x^2 - 4 &= 0 \\ x &= 2 \text{ not acceptable} \\ x &= -2 \\ f(-2) &= -2 - 2 = -4 \end{aligned}$$

Next, we check the endpoints of the interval
$f(-8) = -8 - 0.5 = -8.5$
$f(-1) = -1 - 4 = -5$
Thus, we have an absolute maximum at $(-2, -4)$ and an absolute minimum at $(-8, -8.5)$.

21. $f(x) = \frac{x^2}{x^2 + 1}$

$$\begin{aligned} f'(x) &= \frac{2x(x^2+1) - 2x(x^2)}{(x^2+1)^2} \\ &= \frac{2x}{(x^2+1)^2} \\ f'(x) &= 0 \\ \frac{2x}{(x^2+1)^2} &= 0 \\ x &= 0 \\ f(0) &= \frac{0}{1} = 0 \end{aligned}$$

Next, we check the endpoints of the interval
$f(-2) = \frac{4}{4+1} = 0.8$
$f(2) = \frac{4}{4+1} = 0.8$
Thus, we have an absolute maximum at $(-2, 0.8)$ and at $(2, 0.8)$ and an absolute minimum at $(0, 0)$.

22. $f(x) = \frac{4x}{x^2 + 1}$

$$\begin{aligned} f'(x) &= \frac{4(x^2+1) - 4x(2x)}{(x^2+1)^2} \\ &= \frac{4 - 4x^2}{(x^2+1)^2} \\ f'(x) &= 0 \\ \frac{4 - 4x^2}{(x^2+1)^2} &= 0 \\ x &= -1 \\ x &= 1 \\ f(1) &= \frac{4}{1+1} = 2 \\ f(-1) &= \frac{-4}{1+1} = -2 \end{aligned}$$

Next, we check the endpoints of the interval
$f(-3) = \frac{-12}{9+1} = -1.2$
$f(3) = \frac{12}{9+1} = 1.2$
Thus, we have an absolute maximum at $(1, 2)$ and an absolute minimum at $(-1, -2)$.

23. $f(x) = (x+1)^{1/3}$
$f'(x) = \frac{1}{3}(x+1)^{-2/3}$, which is undefined at $x = -1$ (the only critical value).
$f(-1) = 0$
Next, we check the endpoints of the interval
$f(-2) = (-1)^{-1/3} = -1$
$f(26) = (27)^{1/3} = 3$
Thus, we have an absolute maximum at $(26, 3)$ and an absolute minimum at $(-2, -1)$.

24. $f(x) = x^{1/3}$
$f'(x) = \frac{1}{3}x^{-2/3}$, which is undefined at $x = 0$ a value outside the given interval.
Next we check the endpoints of the interval
$f(8) = 8^{1/3} = 2$
$f(64) = 64^{1/3} = 4$
Thus, we have an absolute maximum at $(64, 4)$ and an absolute minimum at $(8, 2)$.

25. $f(x) = \frac{x+2}{x^2+5}$

$$\begin{aligned} f'(x) &= \frac{(x^2+5) - 2x(x+2)}{(x^2+5)^2} \\ &= \frac{5 - 4x - x^2}{(x^2+5)^2} \\ &= \frac{(5+x)(1-x)}{(x^2+5)^2} \\ f'(x) &= 0 \\ \frac{(5+x)(1-x)}{(x^2+5)^2} &= 0 \\ x &= -5 \\ x &= 1 \\ f(-5) &= \frac{-3}{25+5} = -0.1 \\ f(1) &= \frac{3}{1+5} = 0.5 \end{aligned}$$

Next, we check the endpoints of the interval
$f(-6) = \frac{-4}{36+5} = -0.0976$
$f(6) = \frac{8}{36+5} = 0.19512$
Thus, we have an absolute maximum at $(1, 0.5)$ and an absolute minimum at $(-5, -0.1)$.

26. $f(x) = \frac{x+2}{x^2+3x+3}$

$$\begin{aligned} f'(x) &= \frac{(x^2+3x+3) - (x+2)(2x+3)}{(x^2+3x+3)^2} \\ &= \frac{-x^2 - 4x - 3}{(x^2+3x+3)^2} \\ &= \frac{-(x+1)(x+3)}{(x^2+5)^2} \end{aligned}$$

$$\begin{aligned} f'(x) &= 0 \\ \frac{-(x+1)(x+3)}{(x^2+3x+3)^2} &= 0 \\ x &= -1 \\ x &= -3 \\ f(-3) &= \frac{-1}{9-9+3} = \frac{1}{3} \\ f(-1) &= \frac{1}{1-3+3} = 1 \end{aligned}$$

Next, we check the endpoints of the interval
$f(-4) = \frac{-2}{16-12+3} = -0.2857$
$f(4) = \frac{6}{16+12+3} = 0.1936$
Thus, we have an absolute maximum at $(-1, 1)$ and an absolute minimum at $(-3, -\frac{1}{3})$.

27. $f(x) = x(x - x^2)^{1/2}$

$$\begin{aligned} f'(x) &= (x-x^2)^{1/2} + \frac{x(1-2x)}{2(x-x^2)^{1/2}} \\ f'(x) &= 0 \\ &\rightarrow \\ 2(x-x^2) &= x(2x-1) \\ 4x^2 - 3x &= 0 \\ x &= 0 \\ x &= \frac{3}{4} \\ f(0) &= 0 \\ f\left(\frac{3}{4}\right) &= \frac{3}{4}\left(\frac{3}{4} - \frac{9}{16}\right)^{1/2} = 0.32476 \end{aligned}$$

Next, we check the end point of the interval
$f(1) = 1(0) = 0$
Thus, we have an absolute maximum at $\left(\frac{3}{4}, 0.32476\right)$, and an absolute minimum at $(0, 0)$ and at $(1, 0)$.

28. $f(x) = x(8x - x^2)^{1/2}$

$$\begin{aligned} f'(x) &= (8x-x^2)^{1/2} + \frac{x(8-2x)}{2(8x-x^2)^{1/2}} \\ f'(x) &= 0 \\ &\rightarrow \\ 2(8x-x^2) &= x(2x-8) \\ 2x^2 - 12x &= 0 \\ x &= 0 \\ x &= 6 \\ f(0) &= 0 \\ f(6) &= 6(48-36)^{1/2} = 20.785 \end{aligned}$$

Next, we check the end point of the interval
$f(8) = 8(64 - 64) = 0$
Thus, we have an absolute maximum at $(6, 20.785)$, and an absolute minimum at $(0, 0)$ and at $(8, 0)$.

29. $f(x) = x(x + 3)^{1/2}$

$$\begin{aligned} f'(x) &= (x+3)^{1/2} + \frac{x}{2(x+3)^{1/2}} \\ f'(x) &= 0 \\ &\rightarrow \\ 2(x+3) &= -x \\ 3x + 6 &= 0 \\ x &= -2 \\ f(-2) &= -2(-2+3)^{1/2} = -2 \end{aligned}$$

Next, we check the end points of the interval
$f(-3) = -3(0) = 0$
$f(6) = 6(9)^{1/2} = 18$
Thus, we have an absolute maximum at $(6, 18)$, and an absolute minimum at $(-2, -2)$.

30. $f(x) = x(6 - x)^{1/2}$

$$\begin{aligned} f'(x) &= (6-x)^{1/2} - \frac{x}{2(6-x)^{1/2}} \\ f'(x) &= 0 \\ &\rightarrow \\ 2(6-x) &= x \\ 12 - 3x &= 0 \\ x &= 4 \\ f(4) &= 4(2)^{1/2} = 5.6569 \end{aligned}$$

Next, we check the end points of the interval
$f(0) = 0(6)^{1/2} = 0$
$f(6) = 6(0) = 0$
Thus, we have an absolute maximum at $(4, 5.6569)$, and an absolute minimum at $(0, 0)$ and at $(6, 0)$.

31. $f(x) = x + 2\ sin\ x$

$$\begin{aligned} f'(x) &= 1 + 2\ cos\ x \\ f'(x) &= 0 \\ 1 + 2\ cos\ x &= 0 \\ cos\ x &= \frac{-1}{2} \\ x &= \frac{2\pi}{3} \\ x &= \frac{4\pi}{3} \\ f\left(\frac{2\pi}{3}\right) &= \frac{2\pi}{3} + 2\ sin\left(\frac{2\pi}{3}\right) = 3.8264 \\ f\left(\frac{4\pi}{3}\right) &= \frac{4\pi}{3} + 2\ sin\left(\frac{4\pi}{3}\right) = 2.4567 \end{aligned}$$

Next, we check the endpoints of the interval
$f(0) = 0 + 0 = 0$
$f(2\pi) = 2\pi + 0 = 2\pi$
Thus, we have an absolute maximum at $(2\pi, 2\pi)$ and an absolute minimum at $(0, 0)$.

32. $f(x) = x - \cos 2x$

$$\begin{aligned} f'(x) &= 1 + 2\sin 2x \\ f'(x) &= 0 \\ 1 + 2\sin 2x &= 0 \\ \sin 2x &= \frac{-1}{2} \\ 2x &= \frac{7\pi}{6} \\ x &= \frac{7\pi}{12} \\ 2x &= \frac{11\pi}{6} \\ x &= \frac{11\pi}{12} \\ f\left(\frac{7\pi}{12}\right) &= \frac{7\pi}{12} - \cos\left(\frac{7\pi}{6}\right) = 2.6986 \\ f\left(\frac{11\pi}{12}\right) &= \frac{11\pi}{12} - \cos\left(\frac{11\pi}{6}\right) = 2.0138 \end{aligned}$$

Next, we check the endpoints of the interval
$f(0) = 0 - 1 = -1$
$f(2\pi) = 2\pi - 1 = 5.2832$
Thus, we have an absolute maximum at $(2\pi, 5.2832)$ and an absolute minimum at $(0, -1)$.

33. $f(x) = \dfrac{\sin x}{2 + \sin x}$

$$\begin{aligned} f'(x) &= \frac{(2+\sin x)\cos x - \sin x\cos x}{(2+\sin x)^2} \\ &= \frac{2\cos x}{(2+\sin x)^2} \\ f'(x) &= 0 \\ \frac{2\cos x}{(2+\sin x)^2} &= 0 \\ 2\cos x &= 0 \\ x &= \frac{\pi}{2} \\ x &= \frac{3\pi}{3} \\ f\left(\frac{\pi}{2}\right) &= \frac{1}{2+1} = \frac{1}{3} \\ f\left(\frac{3\pi}{2}\right) &= \frac{-1}{2-1} = -1 \end{aligned}$$

Next, we check the endpoints of the interval
$f(0) = \frac{0}{2} = 0$
$f(2\pi) = \frac{0}{2} = 0$

Thus, we have an absolute maximum at $\left(\dfrac{\pi}{2}, \dfrac{1}{3}\right)$ and an absolute minimum at $\left(\dfrac{3\pi}{2}, -1\right)$.

34. $f(x) = \dfrac{\sin x}{3 - \sin x}$

$$\begin{aligned} f'(x) &= \frac{(3-\sin x)\cos x + \sin x\cos x}{(3-\sin x)^2} \\ &= \frac{3\cos x}{(3-\sin x)^2} \\ f'(x) &= 0 \\ \frac{3\cos x}{(3-\sin x)^2} &= 0 \\ 3\cos x &= 0 \\ x &= \frac{\pi}{2} \\ x &= \frac{3\pi}{3} \\ f\left(\frac{\pi}{2}\right) &= \frac{1}{3-1} = \frac{1}{2} \\ f\left(\frac{3\pi}{2}\right) &= \frac{-1}{3+1} = -\frac{1}{4} \end{aligned}$$

Next, we check the endpoints of the interval
$f(0) = \frac{0}{2} = 0$
$f(2\pi) = \frac{0}{2} = 0$

Thus, we have an absolute maximum at $\left(\dfrac{\pi}{2}, \dfrac{1}{2}\right)$ and an absolute minimum at $\left(\dfrac{3\pi}{2}, -\dfrac{1}{4}\right)$.

35. $f(x) = \dfrac{\sin x}{(1 + \sin x)^2}$

$$\begin{aligned} \frac{(1+\sin x)^2\cos x - 2\sin x(1+\sin x)\cos x}{(1+\sin x)^4} &= f'(x) \\ \frac{\cos x}{(1+\sin x)^4} &= \\ f'(x) &= 0 \\ \frac{\cos x}{(1+\sin x)^4} &= 0 \\ \cos x &= 0 \\ x &= \frac{\pi}{2} \\ f\left(\frac{\pi}{2}\right) &= \frac{1}{(1+1)^2} = \frac{1}{4} \end{aligned}$$

Note that $f'(x)$ is undefined at $x = -\frac{\pi}{2}$ but the value does not belong to the given interval.
Next, we check the endpoints of the interval
$f(0) = \frac{0}{1} = 0$
$f(\pi) = \frac{0}{1} = 0$

Thus, we have an absolute maximum at $(\left(\dfrac{\pi}{2}, \dfrac{1}{4}\right)$ and an absolute minimum at $(0,0)$ and at $(\pi, 0)$.

36. $f(x) = \dfrac{\cos^2 x}{(2 + \cos x)}$

$$\begin{aligned} \frac{(2+\cos x)(-2\cos x\sin x) + \cos^2 x\sin x}{(2+\cos x)^2} &= f'(x) \\ \frac{\cos x\sin x(\cos x - 4)}{(2+\cos x)^2} &= \\ f'(x) &= 0 \\ \frac{\cos x\sin x(\cos x - 4)}{(2+\cos x)^2} &= 0 \\ \cos x &= 0 \end{aligned}$$

or

$$\begin{aligned} \sin x &= 0 \\ x &= 0 \\ x &= \frac{\pi}{2} \\ x &= \pi \\ x &= \frac{3\pi}{2} \\ x &= 2\pi \\ f(0) &= \frac{1}{2+1} = \frac{1}{3} \\ f\left(\frac{\pi}{2}\right) &= \frac{0}{(2+0)} = 0 \\ f(\pi) &= \frac{1}{(2-1)} = 1 \\ f\left(\frac{3\pi}{2}\right) &= \frac{0}{(2+0)} = 0 \\ f(2\pi) &= \frac{1}{2+1} = \frac{1}{3} \end{aligned}$$

Thus, we have an absolute maximum at $((\pi, 1)$ and an absolute minimum at $\left(\frac{\pi}{2}, 0\right)$ and at $\left(\frac{3\pi}{2}, 0\right)$.

37. $f(x) = 2x - \tan x$

$$\begin{aligned} f'(x) &= 2 - \sec^2 x \\ &= 2 - \frac{1}{\cos^2 x} \\ f'(x) &= 0 \\ 2 - \frac{1}{\cos^2 x} &= 0 \\ \cos^2 x &= \frac{1}{2} \\ \cos x &= \frac{-1}{\sqrt{2}} \\ \cos x &= \frac{1}{\sqrt{2}} \\ x &= \frac{\pi}{4} \\ f\left(\frac{\pi}{4}\right) &= \frac{\pi}{2} - 1 = 0.5708 \end{aligned}$$

Next, we check the endpoints of the interval
$f(0) = 0 - 0 = 0$
$f\left(\frac{\pi}{3}\right) = \frac{2\pi}{3} - 1.73205 = 0.36234$
Thus, we have an absolute maximum at $\left(\frac{\pi}{4}, 0.5708\right)$ and an absolute minimum at $(0, 0)$.

38. $f(x) = x + \cot x$

$$\begin{aligned} f'(x) &= 1 - \csc^2 x \\ &= 1 - \frac{1}{\sin^2 x} \\ f'(x) &= 0 \\ 1 - \frac{1}{\sin^2 x} &= 0 \\ \sin^2 x &= 1 \\ \sin x &= -1 \\ \sin x &= 1 \end{aligned}$$

$$\begin{aligned} x &= \frac{\pi}{2} \\ f\left(\frac{\pi}{2}\right) &= 1.5708 \end{aligned}$$

Next, we check the endpoints of the interval
$f\left(\frac{\pi}{6}\right) = 2.2556$
$f\left(\frac{5\pi}{6}\right) = 0.8859$
Thus, we have an absolute maximum at $\left(\frac{\pi}{6}, 2.2556\right)$ and an absolute minimum at $\left(\frac{5\pi}{6}, 0.8859\right)$.

39. $f(x) = 3 \sin x - 2 \sin^3 x$

$$\begin{aligned} f'(x) &= 3 \cos x - 6 \sin^2 x \cos x \\ &= 3 \cos x(1 - 2 \sin^2 x) \\ f'(x) &= 0 \\ 3 \cos x(1 - 2 \sin^2 x) &= 0 \\ \cos x &= 0 \\ &\text{or} \\ 1 - 2 \sin^2 x &= 0 \\ \sin^2 x &= \frac{1}{2} \\ \sin x &= \pm\frac{1}{\sqrt{2}} \\ x &= \frac{\pi}{2} \\ x &= \frac{3\pi}{2} \\ x &= \frac{\pi}{4} \\ x &= \frac{3\pi}{4} \\ x &= \frac{5\pi}{4} \\ x &= \frac{7\pi}{4} \\ f\left(\frac{\pi}{2}\right) &= 3 - 2 = 1 \\ f\left(\frac{3\pi}{2}\right) &= -3 + 2 = -1 \\ f\left(\frac{\pi}{4}\right) &= = 2.1213 - 0.7071 = 1.4142 \\ f\left(\frac{3\pi}{4}\right) &= = 2.1213 - 0.7071 = 1.4142 \\ f\left(\frac{5\pi}{4}\right) &= = -2.1213 + 0.7071 = -1.4142 \\ f\left(\frac{7\pi}{4}\right) &= = -2.1213 + 0.7071 = -1.4142 \end{aligned}$$

Next, we check the endpoints of the interval
$f(0) = 0 - 0 = 0$
$f(2\pi) = 0 - 0 = 0$
Thus, we have an absolute maximum at $\left(\frac{\pi}{4}, 1.4142\right)$ and at $\left(\frac{3\pi}{4}, 1.4142\right)$ and an absolute minimum at $\left(\frac{5\pi}{4}, -1.4142\right)$ and at $\left(\frac{7\pi}{4}, -1.4142\right)$.

40. $f(x) = 2\cos^3 x - 3\cos x$

$$\begin{aligned}
f'(x) &= -6\cos^2 x \sin x + 3\sin x \\
&= -3\sin x(2\cos^2 x - 1) \\
f'(x) &= 0 \\
-3\sin x(2\cos^2 x - 1) &= 0 \\
\sin x &= 0 \\
&\text{or} \\
2\cos^2 x - 1 &= 0 \\
\cos^2 x &= \frac{1}{2} \\
\cos x &= \pm\frac{1}{\sqrt{2}} \\
x &= 0 \\
x &= \pi \\
x &= 2\pi \\
x &= \frac{\pi}{4} \\
x &= \frac{\pi}{2} \\
x &= \frac{3\pi}{4} \\
x &= \frac{5\pi}{4} \\
x &= \frac{7\pi}{4} \\
f(0) &= 2 - 3 = -1 \\
f(\pi) &= -2 + 3 = 1 \\
f(2\pi) &= 2 - 3 = -1 \\
f\left(\frac{\pi}{4}\right) &= = 0.7071 - 2.1213 = -1.4142 \\
f\left(\frac{3\pi}{4}\right) &= = -0.7071 + 2.1213 = 1.4142 \\
f\left(\frac{5\pi}{4}\right) &= = -0.7071 + 2.1213 = 1.4142 \\
f\left(\frac{7\pi}{4}\right) &= = 0.7071 - 2.1213 = -1.4142
\end{aligned}$$

Thus, we have an absolute maximum at $\left(\frac{3\pi}{4}, 1.4142\right)$ and at $\left(\frac{5\pi}{4}, 1.4142\right)$ and an absolute minimum at $\left(\frac{\pi}{4}, -1.4142\right)$ and at $\left(\frac{7\pi}{4}, -1.4142\right)$.

41. $f(x) = x - \frac{4}{3}x^3$

$$\begin{aligned}
f'(x) &= 1 - x^2 \\
f'(x) &= 0 \\
1 - x^2 &= 0 \\
x &= -1 \text{ not acceptable} \\
x &= 1 \\
f(1) &= 1 - \frac{4}{3} = -\frac{1}{3}
\end{aligned}$$

$$f''(x) = -2x$$
$$f''(1) = -2 < 0$$

The function has an absolute maximum of $\frac{-1}{3}$, no absolute minimum.

42. $f(x) = 16x - \frac{4}{3}x^3$

$$\begin{aligned}
f'(x) &= 16 - x^2 \\
f'(x) &= 0 \\
16 - x^2 &= 0 \\
x &= -4 \text{ not acceptable} \\
x &= 4 \\
f(4) &= 64 - \frac{4}{3}(64) = -\frac{64}{3}
\end{aligned}$$

$$f''(x) = -2x$$
$$f''(4) = -8 < 0$$

The function has an absolute maximum of $\frac{-64}{3}$, no absolute minimum.

43. $f(x) = -0.001x^2 + 4.8x - 60$

$$\begin{aligned}
f'(x) &= -0.002x + 4.8 \\
f'(x) &= 0 \\
-0.002x + 4.8 &= 0 \\
x &= 2400 \\
f(2400) &= -5760 + 11520 - 60 = 5700 \\
f''(x) &= -0.002 < 0
\end{aligned}$$

The fnction has an absolute maximum of 5700, no absolute minimum.

44. $f(x) = -0.01x^2 + 1.4x - 30$

$$\begin{aligned}
f'(x) &= -0.02x + 1.4 \\
f'(x) &= 0 \\
-0.02x + 1.4 &= 0 \\
x &= 70 \\
f(70) &= = 19 \\
f''(x) &= -0.02 < 0
\end{aligned}$$

The fnction has an absolute maximum of 19, no absolute minimum.

45. $f(x) = 2x + \frac{72}{x}$

$$\begin{aligned}
f'(x) &= 2 - \frac{72}{x^2} \\
f'(x) &= 0 \\
2 - \frac{72}{x^2} &= 0 \\
x^2 &= 36 \\
x &= 6 \\
f(6) &= 12 + 12 = 24 \\
f''(x) &= \frac{72}{x^3} \\
f''(6) &= \frac{1}{3} > 0
\end{aligned}$$

The function has an absolute minimum of 24, no absolute maximum.

46. $f(x) = x + \dfrac{3600}{x}$

$$\begin{aligned} f'(x) &= 1 - \frac{3600}{x^2} \\ f'(x) &= 0 \\ 1 - \frac{3600}{x^2} &= 0 \\ x^2 &= 3600 \\ x &= 60 \\ f(60) &= 60 + 60 = 120 \\ f''(x) &= \frac{3600}{x^3} \\ f''(60) &= \frac{1}{60} > 0 \end{aligned}$$

The function has an absolute minimum of 120, no absolute maximum.

47. $f(x) = x^2 + \dfrac{432}{x}$

$$\begin{aligned} f'(x) &= 2x - \frac{432}{x^2} \\ f'(x) &= 0 \\ 2x - \frac{432}{x^2} &= 0 \\ x^3 &= 216 \\ x &= 6 \\ f(6) &= 36 + 72 = 108 \\ f''(x) &= 2 + \frac{432}{x^3} \\ f''(6) &= 2 + 2 = 4 > 0 \end{aligned}$$

The function has an absolute minimum of 108, no absolute maximum.

48. $f(x) = x^2 + \dfrac{250}{x}$

$$\begin{aligned} f'(x) &= 2x - \frac{250}{x^2} \\ f'(x) &= 0 \\ 2x - \frac{250}{x^2} &= 0 \\ x^3 &= 125 \\ x &= 5 \\ f(5) &= 25 + 50 = 75 \\ f''(x) &= 2 + \frac{250}{x^3} \\ f''(5) &= 2 + 2 = 4 > 0 \end{aligned}$$

The function has an absolute minimum of 75, no absolute maximum.

49. $f(x) = (x+1)^3$

$$\begin{aligned} f'(x) &= 3(x+1)^2 \\ f'(x) &= 0 \\ 3(x+1)^2 &= 0 \\ x &= -1 \\ f(-1) &= 0 \\ f''(x) &= 6(x+1) \\ f''(-1) &= 0 \end{aligned}$$

The function has no absolute maximum or minimum.

50. $f(x) = (x-1)^3$

$$\begin{aligned} f'(x) &= 3(x-1)^2 \\ f'(x) &= 0 \\ 3(x-1)^2 &= 0 \\ x &= 1 \\ f(1) &= 0 \\ f''(x) &= 6(x-1) \\ f''(-1) &= 0 \end{aligned}$$

The function has no absolute maximum or minimum.

51. $f(x) = 2x - 3$
The function is linear, which means that on the interval $(-\infty, \infty)$ the function has no absolute maximum or minimum.

52. $f(x) = 9 - 5x$
The function is linear, which means that on the interval $(-\infty, \infty)$ the function has no absolute maximum or minimum.

53. $f(x) = x^{2/3}$

$$f'(x) = \frac{2}{3}x^{-1/3}$$

$f''(x) = -\frac{2}{9}x^{-4/3}$ $f'(x)$ and $f''(x)$ are undefined at $x = 0$

$$\begin{aligned} f(-1) &= (-1)^{2/3} = 1 \\ f(0) &= (0)^{2/3} = 0 \\ f(1) &= (1)^{2/3} = 1 \\ f''(1) &= \frac{-2}{9} \\ f''(-1) &= \frac{-2}{9} \end{aligned}$$

The function has an absolute maximum of 1 and an absolute minimum of 0.

54. $f(x) = x^{2/3}$

$$\begin{aligned} f'(x) &= \frac{2}{3}x^{-1/3} \\ f''(x) &= -\frac{2}{9}x^{-4/3} \end{aligned}$$

$f'(x)$ and $f''(x)$ are undefined at $x = 0$
$f(0) = 0$ The function has an absolute minimum of 0, no absolute maximum.

55. $f(x) = \frac{1}{3}x^3 - x + \frac{2}{3}$
The function grows indefinitely over $(-\infty, \infty)$, which means it has no absolute maximum or minimum.

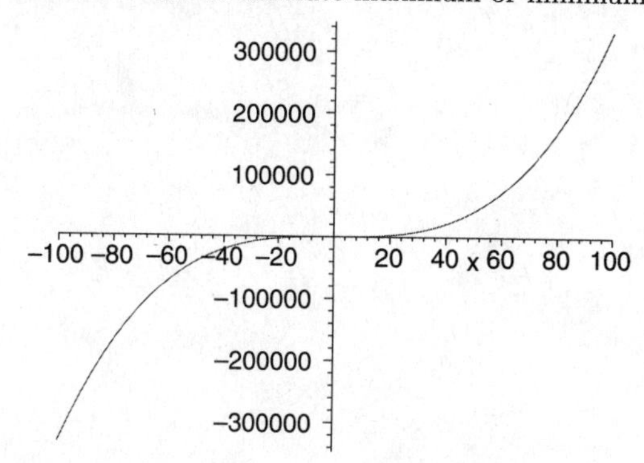

56. $f(x) = \frac{1}{3}x^3 - \frac{1}{2}x^2 - 2x + 1$
The function grows indefinitely over $(-\infty, \infty)$, which means it has no absolute maximum or minimum.

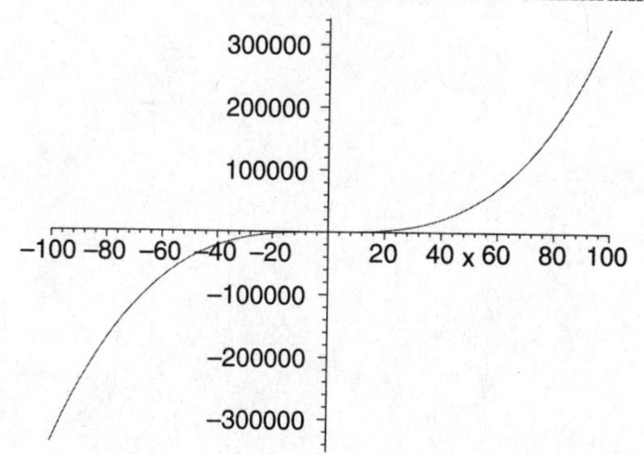

57. $f(x) = x^4 - 2x^2$

$$\begin{aligned}
f'(x) &= 4x^3 - 4x \\
&= 4x(x^2 - 1) \\
f'(x) &= 0 \\
4x(x^2 - 1) &= 0 \\
x &= 0 \\
x &= \pm 1 \\
f(-1) &= 1 - 2 = -1 \\
f(0) &= 0 - 0 = 0 \\
f(1) &= 1 - 2 = -1
\end{aligned}$$

The function has an absolute minimum of -1 and no absolute maximum.

58. $f(x) = 2x^4 - 4x^2 + 2$

$$\begin{aligned}
f'(x) &= 8x^3 - 8x \\
&= 8x(x^2 - 1) \\
f'(x) &= 0 \\
8x(x^2 - 1) &= 0 \\
x &= 0 \\
x &= \pm 1
\end{aligned}$$

$$\begin{aligned}
f(-1) &= 2 - 4 + 2 = 0 \\
f(0) &= 0 - 0 + 2 = 2 \\
f(1) &= 2 - 4 + 2 = 0
\end{aligned}$$

The function has an absolute minimum of -1 and no absolute maximum.

59. $f(x) = \tan x + \cot x$

$$\begin{aligned}
f'(x) &= \sec^2 x - \csc^2 x \\
&= \frac{\sin^2 x - \cos^2 x}{\cos^2 x \sin^2 x} \\
f'(x) &= 0 \\
\frac{\sin^2 x - \cos^2 x}{\cos^2 x \sin^2 x} &= 0 \\
\sin^2 x - \cos^2 x &= 0 \\
\sin x &= \pm\cos x \\
x &= \frac{\pi}{4} \\
f\left(\frac{\pi}{4}\right) &= 1 + 1 = 2
\end{aligned}$$

The function is undefined at $x = 0$ and $x = \frac{\pi}{2}$
$f''(x) = 2\sec^2 x \tan x + 2\csc^2 x \cot x$
$f''\left(\frac{\pi}{4}\right) = 1 + 1 = 2 > 0$ The function has an absolute minimum of 2 and no absolute maximum.

60. $f(x) = \dfrac{\sin x}{1 + \sin x}$

$$\begin{aligned}
f'(x) &= \frac{(1 + \sin x)\cos x - \sin x \cos x}{(1 + \sin x)^2} \\
&= \frac{\cos x}{(1 + \sin x)^2} \\
f'(x) &= 0 \\
\cos x &= 0 \\
x &= \frac{\pi}{2} \\
f\left(\frac{\pi}{2}\right) &= \frac{1}{2} \\
f''(x) &= \frac{-(1 + \sin x)^2 \sin x - 2(1 + \sin x)\cos^2 x}{(1 + \sin x)^4} \\
f''\left(\frac{\pi}{2}\right) &= \frac{-1}{8} < 0
\end{aligned}$$

The function has an absolute maximum of 2 and no absolute minimum.

61. $f(x) = \dfrac{1}{\sin x + \cos x}$

$$\begin{aligned}
f'(x) &= \frac{\cos x - \sin x}{(\sin x + \cos x)^2} \\
f'(x) &= 0 \\
\frac{\cos x - \sin x}{(\sin x + \cos x)^2} &= 0 \\
\cos x &= \sin x \\
x &= \frac{\pi}{4}
\end{aligned}$$

$$f\left(\frac{\pi}{4}\right) = \frac{1}{\frac{1}{\sqrt{2}}+\frac{1}{\sqrt{2}}} = \frac{\sqrt{2}}{2}$$

$$f''(x) = \frac{2\ sin\ x\ cos\ x - 1}{(sin\ x + cos\ x)^4}$$

$$f''\left(\frac{\pi}{4}\right) = \frac{1}{(\frac{1}{\sqrt{2}}+\frac{1}{\sqrt{2}})^4} > 0$$

The function has an absolute minimum of $\frac{\sqrt{2}}{2}$ and no absolute maximum.

62. $f(x) = \dfrac{1}{x + 2\ cos\ x}$

$$f'(x) = \frac{2\ sin\ x - 1}{(x+2\ cos\ x)^2}$$
$$f'(x) = 0$$
$$\frac{2\ sin\ x - 1}{(x+2\ cos\ x)^2} = 0$$
$$sin\ x = \frac{1}{2}$$
$$x = \frac{\pi}{6}$$
$$x = \frac{5\pi}{6}$$
$$f\left(\frac{\pi}{6}\right) = 0.4433$$
$$f\left(\frac{5\pi}{6}\right) = 1.1287$$
$$f(0) = 0.5$$
$$f(\pi) = 0.8760$$

The function has an absolute maximum of 1.1287 and an absolute minimum of 0.4433.

63. $f(x) = \dfrac{1}{x - 2\ sin\ x}$

$$f'(x) = \frac{2\ cos\ x - 1}{(x-2\ sin\ x)^2}$$
$$f'(x) = 0$$
$$\frac{2\ cos\ x - 1}{(x-2\ sin\ x)^2} = 0$$
$$cos\ x = \frac{1}{2}$$
$$x = \frac{\pi}{3}$$
$$f\left(\frac{\pi}{3}\right) = -1.46$$
$$f''(x) = \frac{-2\ sin\ x(x-2\ sin\ x)}{(x-2\ sin\ x)^3} - \frac{2(2\ cos\ x - 1)(1-cos\ x)}{(x-2\ sin\ x)^3}$$
$$f''\left(\frac{\pi}{3}\right) = -3.693 < 0$$

The function has an absolute maximum of -1.46 and no absolute minimum.

64. $f(x) = 2\ csc\ x + cot\ x$

$$f'(x) = -2\ csc\ x\ cot\ x - csc^2\ x$$
$$f'(x) = 0$$
$$-2\ csc\ x\ cot\ x - csc^2\ x = 0$$
$$x = \frac{2\pi}{3}$$
$$f\left(\frac{2\pi}{3}\right) = 1.7321$$
$$f''(x) = -2\ csc\ xcot^2\ x + 2\ csc^3\ x + 2\ csc^2\ x\ cot\ x$$
$$f''\left(\frac{2\pi}{3}\right) = 0.7698 > 0$$

The function has an absolute minimum of 1.7321 and no absolute maximum.

65. $f(x) = 2\ csc\ x + cot\ x$

$$f'(x) = -2\ csc\ x\ cot\ x - csc^2\ x$$
$$f'(x) = 0$$
$$-2\ csc\ x\ cot\ x - csc^2\ x = 0$$
$$x = \frac{4\pi}{3}$$
$$f\left(\frac{4\pi}{3}\right) = -1.7321$$
$$f''(x) = -2\ csc\ xcot^2\ x + 2\ csc^3\ x + 2\ csc^2\ x\ cot\ x$$
$$f''\left(\frac{4\pi}{3}\right) = -0.7698 < 0$$

The function has an absolute maximum of -1.7321 and no absolute minimum.

66. $f(x) = tan\ x - 2\ sec\ x$

$$f'(x) = sec^2\ x - 2\ sec\ x\ tan\ x$$
$$f'(x) = 0$$
$$sec^2\ x - 2\ sec\ x\ tan\ x = 0$$
$$x = \frac{\pi}{6}$$
$$f\left(\frac{\pi}{6}\right) = -1.7321$$
$$f''(x) = 2\ tan\ x\ sec^2\ x - 2\ sec^3\ x - 2\ sec\ x\ tan^2\ x$$
$$f\left(\frac{\pi}{6}\right) = -2.309 < 0$$

The function has an absolute maximum of -1.7321 and no absolute minimum.

67. $f(x) = tan\ x - 2\ sec\ x$

$$f'(x) = sec^2\ x - 2\ sec\ x\ tan\ x$$

$$\begin{aligned} f'(x) &= 0 \\ sec^2\ x - 2\ sec\ x\ tan\ x &= 0 \\ x &= \frac{5\pi}{6} \\ f\left(\frac{5\pi}{6}\right) &= 1.7321 \\ f''(x) &= 2\ tan\ x\ sec^2\ x - 2\ sec^3\ x - \\ & \quad 2\ sec\ x\ tan^2\ x \\ f\left(\frac{5\pi}{6}\right) &= 2.309 > 0 \end{aligned}$$

The function has an absolute minimum of 1.7321 and no absolute maximum.

68. $f(x) = \dfrac{1}{1 + cos\ x}$

$$\begin{aligned} f'(x) &= \frac{sin\ x}{(1 + cos\ x)^2} \\ f'(x) &= 0 \\ sin\ x &= 0 \\ x &= 0 \\ f(0) &= \frac{1}{2} \\ f''(x) &= \frac{2 - cos\ x}{(1 + cos\ x)^2} \\ f''(0) &= \frac{1}{2} > 0 \end{aligned}$$

The function has an absolute minimum of $\frac{1}{2}$ and no absolute maximum.

69. $f(x) = \dfrac{1}{1 - 2\ sin\ x}$

$$\begin{aligned} f'(x) &= \frac{2\ cos\ x}{(1 - 2\ sin\ x)^2} \\ f'(x) &= 0 \\ cos\ x &= 0 \\ x &= \frac{\pi}{2} \\ f\left(\frac{\pi}{2}\right) &= -1 \\ f''(x) &= \frac{8\ cos^2\ x}{(1 - 2\ sin\ x)^3} - \frac{2\ sin\ x}{(1 - 2\ sin\ x)^2} \\ f''\left(\frac{\pi}{2}\right) &= -2 < 0 \end{aligned}$$

The function has an absolute maximum of -1 and no absolute minimum.

70. $B(x) = 0.05x^2 - 0.3x^3$

$$\begin{aligned} B'(x) &= 0.1x - 0.9x^2 \\ &= 0.1x(1 - 9x) \\ B'(x) &= 0 \\ 0.1x(1 - 9x) &= 0 \\ x &= 0 \\ x &= \frac{1}{9} \approx 0.1111 \end{aligned}$$

$$\begin{aligned} B(0) &= 0 \\ B(0.1111) &= 0.05(0.1111)^2 - 0.3(0.1111)^3 \\ &= 0.00021 \end{aligned}$$

We check the endpoint of the given interval $B(0.16) = 0.05(0.16)^2 - 0.3(0.16)^3 = 0.000052$ Thus, the maximum blood pressure of 0.00021 occurs at a dosage of 0.1111

71. $y = -6.1x^2 + 752x + 22620$

$$\begin{aligned} y' &= -12.2x + 752 \\ -12.2x + 752 &= 0 \\ x &= \frac{752}{12.2} = 61.64 \\ y'' &= -12.2 < 0 \end{aligned}$$

The maximum number of accidents occurs at $x = 61.64$ mph.

72. $T(t) = -0.1t^2 + 1.2t + 98.6$

$$\begin{aligned} T'(t) &= -0.2t + 1.2 \\ -0.2t + 1.2 &= 0 \\ t &= 6 \\ T(6) &= -0.1(6)^2 + 1.2(6) + 98.6 \\ &= 102.2 \\ T''(t) &= -0.2 < 0 \end{aligned}$$

The maximum temperature value is 102.2^o and occurs after 6 days.

73. $r(x) = 104.5x^2 - 1501.5x + 6016$

$$\begin{aligned} r'(x) &= 209x - 1501.5 \\ 209x - 1501.5 &= 0 \\ x &= \frac{1501.5}{209} = 7.18 \\ r(7.18) &= 104.5(7.18)^2 - 1501.5(7.18) + 6016 \\ &= 15434.75 \\ r''(x) &= 209 > 0 \end{aligned}$$

The death rate is minimized at 7.18 hours of sleep per night.

74. **a)** $D(h) = 0.139443\ \sqrt{h} - 0.043705\ h^{0.725}$

b) $D(h) = 0.139443\ \sqrt{h} - 0.043705\ h^{0.725}$

$$\begin{aligned} D'(h) &= \frac{0.697215}{\sqrt{h}} - \frac{0.031686125}{h^{0.275}} \\ \frac{0.697215}{\sqrt{h}} - \frac{0.031686125}{h^{0.275}} &= 0 \\ h^{0.775} &= 0.0220920416 \\ h &= 33.282 \\ D(33.282) &= 0.139443\ \sqrt{33.282} - \\ & \quad 0.043705\ (33.282)^{0.725} \\ &= 0.2497 \end{aligned}$$

We check the endpoints of the interval $D(0) = 0.139443\ \sqrt{0} - 0.043705(0)^{0.725} = 0$

$D(215) = 0.139443\ \sqrt{215} - 0.043705(215)^{0.725} = -0.1010$

The function has an absolute maximum of 0.2497 and an absolute minimum of -0.1010.

75. **a)** $D(h) = 0.139443\ \sqrt{h} - 0.238382\ h^{0.3964}$

b) $D(h) = 0.139443\ \sqrt{h} - 0.238382\ h^{0.3964}$

$$\begin{aligned}
D'(h) &= \frac{0.697215}{\sqrt{h}} - \frac{0.0944946248}{h^{0.6036}} \\
\frac{0.697215}{\sqrt{h}} - \frac{0.0944946248}{h^{0.6036}} &= 0 \\
h^{1.1036} &= 0.0658830698 \\
h &= 18.816 \\
D(18.816) &= 0.139443\ \sqrt{18.816} - \\
&\quad 0.0944946\ (18.816)^{0.3964} \\
&= -0.1581
\end{aligned}$$

We check the endpoints of the interval
$D(0) = 0.139443\ \sqrt{0} - 0.0944946(0)^{0.3964} = 0$
$D(215) = 0.139443\ \sqrt{215} - 0.0944946(215)^{0.3964} = 0.0408$

The function has an absolute maximum of 0.0408 and an absolute minimum of -0.1581.

76. $y = (x-a)^2 + (x-b)^2$

$$\begin{aligned}
y' &= 2(x-a) + 2(x-b) \\
&= 4x - 2a - 2b \\
y'' &= 4 > 0 \\
4x - 2a - 2b &= 0 \\
4x &= 2(a-b) \\
x &= \frac{a-b}{4}
\end{aligned}$$

77. Left to the student (answers vary).

78. Left to the student (answers vary).

79. **a)** $cos\ \theta_0 = \frac{\text{Adj}}{\text{Hyp}} = \frac{1}{\sqrt{3}}$

b) $sin\ \theta_0 = \frac{\text{Opp}}{\text{Hyp}} = \frac{\sqrt{2}}{\sqrt{3}}$

c) From Example 4

$$\begin{aligned}
S''(\theta) &= \frac{3}{2}a^2\left[\frac{(\sqrt{3}sin(\theta))(sin^2(\theta))}{sin^4(\theta)}\right] - \\
&\quad \frac{3}{2}a^2\left[\frac{(1-\sqrt{3}cos(\theta))(2sin(\theta)cos(\theta))}{sin^4(\theta)}\right] \\
S''(\theta_0) &= \frac{3}{2}a^2\left[\frac{\sqrt{3}(\frac{\sqrt{2}}{\sqrt{3}})(\frac{\sqrt{2}}{\sqrt{3}})^2}{(\frac{\sqrt{2}}{\sqrt{3}})^4}\right] - \\
&\quad \frac{3}{2}a^2\left[\frac{(1-\sqrt{3}\frac{1}{\sqrt{3}})2(\frac{\sqrt{2}}{\sqrt{3}})(\frac{1}{\sqrt{3}})}{(\frac{\sqrt{2}}{\sqrt{3}})^4}\right] \\
&= \frac{3}{2}a^2\left[\frac{\frac{2\sqrt{2}}{3}}{\frac{4}{9}} - \frac{0}{\frac{4}{9}}\right] \\
&= \frac{9\sqrt{2}a^2}{4}
\end{aligned}$$

80. Absolute maximum at $x = 0$ and absolute minimum at $x = 3$

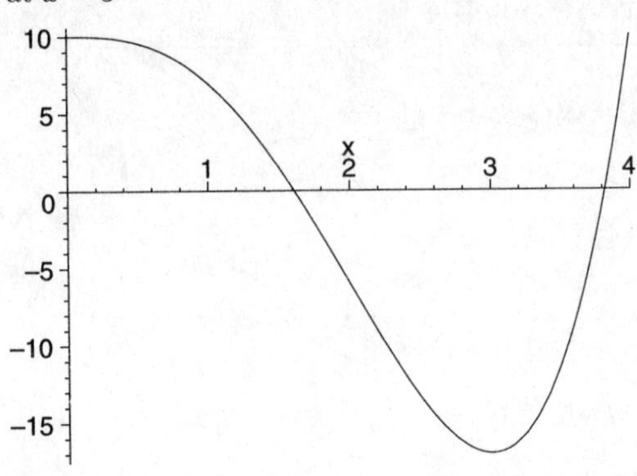

81. Absolute maximum at $x = 4$ and absolute minimum at $x = 2$

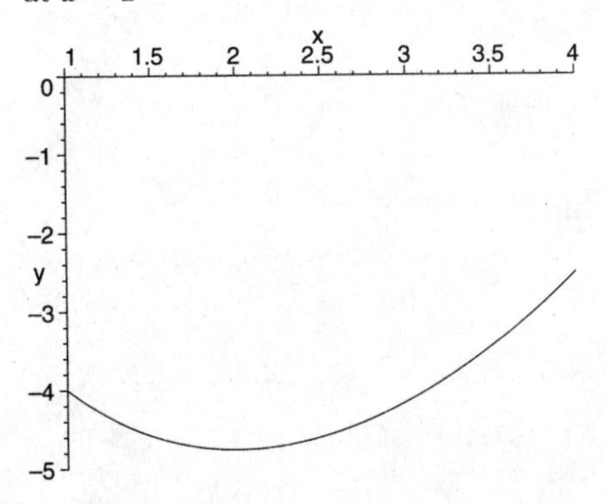

82. Absolute minimum at $x = \pm 1$, no absolute maximum

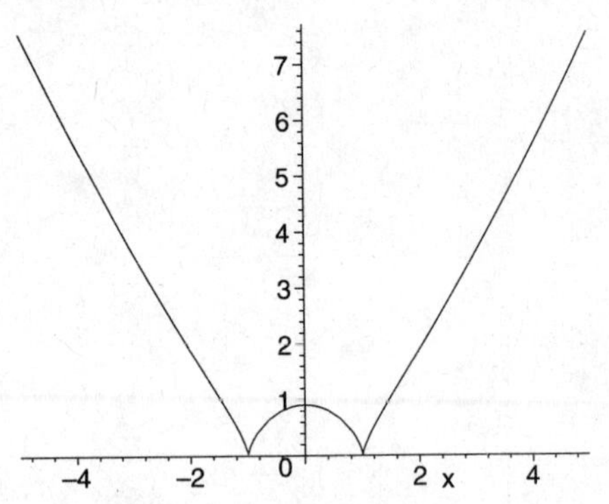

83. No absolute maximum and no absolute minimum

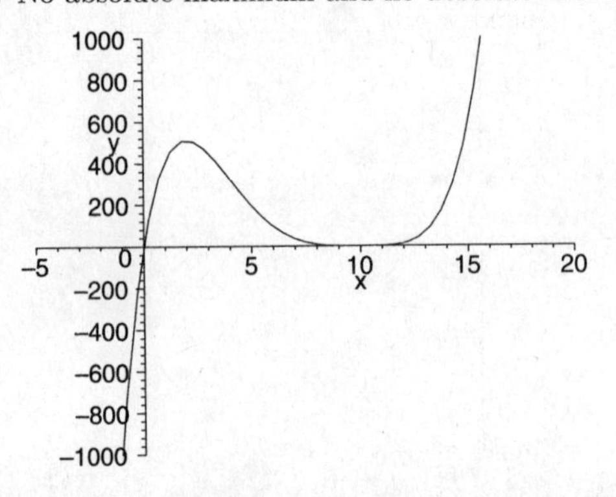

Exercise Set 3.5

1. $Q = xy$ and $x + y = 50$

$$\begin{aligned} Q &= x(50 - x) \\ &= 50x - x^2 \\ Q' &= 50 - 2x \\ 50 - 2x &= 0 \\ x &= 25 \\ y &= 50 - 25 = 25 \end{aligned}$$

The two numbers are $x = 25$ and $y = 25$

2. $Q = xy$ and $x + y = 70$

$$\begin{aligned} Q &= x(70 - x) \\ &= 70x - x^2 \\ Q' &= 70 - 2x \\ 70 - 2x &= 0 \\ x &= 35 \\ y &= 70 - 35 = 35 \end{aligned}$$

The two numbers are $x = 35$ and $y = 35$

3. There cannot be a minimum product since the there is only one critical value for the function and the second derivative is positive for values of x.

4. There cannot be a minimum product since the there is only one critical value for the function and the second derivative is positive for values of x.

5. $Q = xy$ and $y - x = 4$

$$\begin{aligned} Q &= x(x + 4) \\ &= x^2 + 4x \\ Q' &= 2x + 4 \\ 2x + 4 &= 0 \\ x &= -2 \\ y &= -2 + 4 = 2 \end{aligned}$$

The two number are $x = -2$ and $y = 2$

6. $Q = xy$ and $y - x = 6$

$$\begin{aligned} Q &= x(x + 6) \\ &= x^2 + 6x \\ Q' &= 2x + 6 \\ 2x + 6 &= 0 \\ x &= -3 \\ y &= -3 + 6 = 3 \end{aligned}$$

The two number are $x = -3$ and $y = 3$

7. $Q = xy^2$ and $x + y^2 = 1$

$$\begin{aligned} Q &= x(1 - x) \\ &= x - x^2 \\ Q' &= 1 - 2x \\ 1 - 2x &= 0 \\ x &= \frac{1}{2} \\ y^2 &= 1 - \frac{1}{2} = \frac{1}{2} \\ y &= \frac{1}{\sqrt{2}} \end{aligned}$$

When $x = \frac{1}{2}$ and $y = \frac{1}{\sqrt{2}}$

$$\begin{aligned} Q &= \frac{1}{2}\left(\frac{1}{\sqrt{2}}\right)^2 \\ &= \frac{1}{4} \end{aligned}$$

8. $Q = xy^2$ and $x + y^2 = 4$

$$\begin{aligned} Q &= x(4 - x) \\ &= 4x - x^2 \\ Q' &= 4 - 2x \\ 4 - 2x &= 0 \\ x &= 2 \\ y^2 &= 4 - 2 = 2 \\ y &= \sqrt{2} \end{aligned}$$

When $x = 2$ and $y = \sqrt{2}$

$$\begin{aligned} Q &= 2(\sqrt{2})^2 \\ &= 4 \end{aligned}$$

9. $Q = 2x^2 + 3y^2$ and $x + y = 5$

$$\begin{aligned} Q &= 2x^2 + 3(5 - x)^2 \\ &= 2x^2 + 75 - 30x + 3x^2 \\ &= 5x^2 - 30x + 75 \\ Q' &= 10x - 30 \\ 10x - 30 &= 0 \\ x &= 3 \\ y &= 5 - 3 = 2 \end{aligned}$$

When $x = 3$ and $y = 2$

$$\begin{aligned} Q &= 2(3)^2 + 3(2)^2 \\ &= 18 + 12 \\ &= 30 \end{aligned}$$

10. $Q = x^2 + 2y^2$ and $x + y = 3$

$$\begin{aligned} Q &= x^2 + 2(3 - x)^2 \\ &= x^2 + 18 - 6x + 2x^2 \\ &= 3x^2 - 6x + 18 \\ Q' &= 6x - 6 \\ 6x - 6 &= 0 \\ x &= 1 \\ y &= 3 - 1 = 2 \end{aligned}$$

When $x = 1$ and $y = 2$

$$\begin{aligned} Q &= (1)^2 + 2(2)^2 \\ &= 1 + 8 \\ &= 9 \end{aligned}$$

11. $Q = x^2 + y^2$ and $x + y = 20$

$$\begin{aligned} Q &= x^2 + (20 - x)^2 \\ &= x^2 + 400 - 40x + x^2 \\ &= 2x^2 - 40x + 400 \\ Q' &= 4x - 40 \\ 4x - 40 &= 0 \\ x &= 10 \\ y &= 20 - 10 = 10 \end{aligned}$$

When $x = 10$ and $y = 10$

$$\begin{aligned} Q &= (10)^2 + (10)^2 \\ &= 100 + 100 \\ &= 200 \end{aligned}$$

12. $Q = x^2 + y^2$ and $x + y = 10$

$$\begin{aligned} Q &= x^2 + (10 - x)^2 \\ &= x^2 + 100 - 20x + x^2 \\ &= 2x^2 - 20x + 100 \\ Q' &= 4x - 20 \\ 4x - 20 &= 0 \\ x &= 5 \\ y &= 105 = 5 \end{aligned}$$

When $x = 5$ and $y = 5$

$$\begin{aligned} Q &= (5)^2 + (5)^2 \\ &= 25 + 25 \\ &= 25 \end{aligned}$$

13. $Q = xy$ and $\frac{4}{3}x^2 + y = 16$

$$\begin{aligned} Q &= x(16 - \frac{4}{3}x^2) \\ &= 16x - \frac{4}{3}x^3 \\ Q' &= 16 - 4x^2 \\ &= 4(4 - x^2) \\ 4(4 - x^2) &= 0 \\ x &= 2 \\ y &= 16 - \frac{4}{3}(4) = \frac{32}{3} \end{aligned}$$

and

When $x = 2$ and $y = \frac{32}{3}$

$$\begin{aligned} Q &= 2\left(\frac{32}{3}\right) \\ &= \frac{64}{3} \end{aligned}$$

14. $Q = xy$ and $x + \frac{4}{3}y^2 = 1$

$$\begin{aligned} Q &= (1 - \frac{4}{3}y^2)y \\ &= y - \frac{4}{3}y^3 \\ Q' &= 1 - 4y^2 \\ 1 - 4y^2 &= 0 \\ y &= \frac{1}{2} \\ x &= 1 - \frac{4}{3}(\frac{1}{4} = \frac{2}{3} \end{aligned}$$

When $x = \frac{2}{3}$ and $y = \frac{1}{2}$

$$\begin{aligned} Q &= \frac{2}{3} \cdot \frac{1}{2} \\ &= \frac{1}{3} \end{aligned}$$

15. $Q = \sqrt{x} + \sqrt{y}$ and $x + y = 1$

$$\begin{aligned} Q &= \sqrt{x} + \sqrt{1 - x} \\ &= x^{1/2} + (1 - x)^{1/2} \\ Q' &= \frac{1}{2\sqrt{x}} - \frac{1}{2\sqrt{1 - x}} \\ \frac{1}{2\sqrt{x}} - \frac{1}{2\sqrt{1 - x}} &= 0 \\ \sqrt{x} &= \sqrt{1 - x} \\ x &= 1 - x \\ x &= \frac{1}{2} \\ y &= 1 - \frac{1}{2} = \frac{1}{2} \end{aligned}$$

When $x = \frac{1}{2}$ and $y = \frac{1}{2}$

$$\begin{aligned} Q &= \sqrt{\frac{1}{2}} + \sqrt{\frac{1}{2}} \\ &= \frac{2}{\sqrt{2}} \\ &= \sqrt{2} \end{aligned}$$

16. $Q = \sqrt{x} + \sqrt{y}$ and $2x + y = 6$

$$\begin{aligned} Q &= \sqrt{x} + \sqrt{6 - 2x} \\ &= x^{1/2} + (6 - 2x)^{1/2} \\ Q' &= \frac{1}{2\sqrt{x}} - \frac{1}{\sqrt{6 - 2x}} \\ \frac{1}{2\sqrt{x}} - \frac{1}{\sqrt{6 - 2x}} &= 0 \\ 2\sqrt{x} &= \sqrt{6 - 2x} \\ 4x &= 6 - 2x \\ x &= 1 \\ y &= 6 - 2(1) = 4 \end{aligned}$$

When $x = 1$ and $y = 4$

$$\begin{aligned} Q &= \sqrt{1} + \sqrt{4} \\ &= 1 + 2 \\ &= 3 \end{aligned}$$

17. $A = lw$ and $2l + w = 20$

$$\begin{aligned} A &= l(20-2l) \\ &= 20l - 2l^2 \\ A' &= 20 - 4l \\ 20 - 4l &= 0 \\ l &= 5 \\ w &= 20 - 2(5) = 10 \end{aligned}$$

When $l = 5$ and $w = 10$

$$\begin{aligned} A &= 5(10) \\ &= 50 \end{aligned}$$

The rectangular fence is 5 yards by 10 yards with a maximum area of 50 squared yards.

18.

$$\begin{aligned} A &= x(240 - 3x) \\ &= 240x - 3x^2 \\ A' &= 240 - 6x \\ 240 - 6x &= 0 \\ x &= 40 \\ A &= 40[240 - 3(40)] \\ A &= 480 \end{aligned}$$

The maximum area that can be enclosed is 480 square yards.

19. $A = lw$ and $2l + 2w = 54 \to l + w = 27$

$$\begin{aligned} A &= l(27 - l) \\ &= 27l - l^2 \\ A' &= 27 - 2l \\ 27 - 2l &= 0 \\ l &= 13.5 \\ w &= 27 - 13.5 = 13.5 \end{aligned}$$

When $l = 13.5$ and $w = 13.5$

$$\begin{aligned} A &= 13.5(13.5) \\ &= 182.25 \end{aligned}$$

The room is 13.5 feet by 13.5 feet yards with a maximum area of 182.25 squared feet.

20. $A = lw$ and $2l + 2w = 34 \to l + w = 17$

$$\begin{aligned} A &= l(17 - l) \\ &= 17l - l^2 \\ A' &= 17 - 2l \\ 17 - 2l &= 0 \\ l &= 8.5 \\ w &= 17 - 8.5 = 8.5 \end{aligned}$$

When $l = 8.5$ and $w = 8.5$

$$\begin{aligned} A &= 8.5(8.5) \\ &= 72.25 \end{aligned}$$

The rectangle is 8.5 feet by 8.5 feet yards with a maximum area of 72.25 squared feet.

21. Length: $l = 30 - 2x$, Width: $30 - 2x$, and $Height : h = x$

$$\begin{aligned} V &= (30 - 2x)(30 - 2x)x \\ &= 900x - 120x^2 + 4x^3 \\ V' &= 900 - 240x + 12x^2 \\ &= 12(15 - x)(5 - x) \\ 12(15 - x)(5 - x) &= 0 \\ x &= 15 \text{ not acceptable} \\ x &= 5 \\ l &= 30 - 2(5) = 20 \\ w &= 30 - 2(5) = 20 \\ h &= 5 \\ V &= 20(20)5 \\ &= 2000 \end{aligned}$$

The box has a length and width of 20 inches and a height of 5 inches with a maximum volume of 2000 squared inches.

22. Length: $l = 20 - 2x$, Width: $20 - 2x$, and $Height : h = x$

$$\begin{aligned} V &= (20 - 2x)(20 - 2x)x \\ &= 400x - 80x^2 + 4x^3 \\ V' &= 400 - 160x + 12x^2 \\ &= 4(10 - 3x)(10 - x) \\ 4(10 - 3x)(10 - x) &= 0 \\ x &= 10 \text{ not acceptable} \\ x &= \frac{10}{3} \\ l &= 20 - 2\left(\frac{10}{3}\right) = \frac{40}{3} \\ w &= 20 - 2\left(\frac{10}{3}\right) = \frac{40}{3} \\ h &= \frac{10}{3} \\ V &= \frac{10}{3} \\ &= 2000 \end{aligned}$$

The box has a length and width of 20 inches and a height of 5 inches with a maximum volume of 2000 squared inches.

23. Length: x, Width: x, and Height: y

$SA = x^2 + 4xy$ and $x^2y = 62.5 \to y = \dfrac{62.5}{x^2}$

$$\begin{aligned} SA &= x^2 + 4x\left(\frac{62.5}{x^2}\right) \\ &= x^2 + \frac{250}{x} \\ SA' &= 2x - \frac{250}{x^2} \end{aligned}$$

$$\begin{aligned} 2x - \frac{250}{x^2} &= 0 \\ 2x^3 &= 250 \\ x &= 5 \\ y &= \frac{62.5}{5^2} = 2.5 \end{aligned}$$

When $x = 5$ and $y = 2.5$

$$\begin{aligned} SA &= 5^2 + 4(5)(2.5) \\ &= 75 \end{aligned}$$

The sqaure based box has dimensions $5 \times 5 \times 2.5$ inches and a minimum surface area of 75 squared inches.

24. Length: x, Width: x, and Height: y
$SA = x^2 + 4xy$ and $x^2y = 32 \rightarrow y = \frac{32}{x^2}$

$$\begin{aligned} SA &= x^2 + 4x\left(\frac{32}{x^2}\right) \\ &= x^2 + \frac{128}{x} \\ SA' &= 2x - \frac{128}{x^2} \\ 2x - \frac{128}{x^2} &= 0 \\ 2x^3 &= 128 \\ x &= 4 \\ y &= \frac{32}{4^2} = 2 \end{aligned}$$

When $x = 4$ and $y = 2$

$$\begin{aligned} SA &= 4^2 + 4(4)(2) \\ &= 48 \end{aligned}$$

The sqaure based box has dimensions $4 \times 4 \times 2$ inches and a minimum surface area of 48 squared inches.

25. $A = \frac{1}{2}(2\ sin\frac{\theta}{2})(cos\frac{\theta}{2})$

$$\begin{aligned} A &= sin\ \frac{\theta}{2}\ cos\ \frac{\theta}{2} \\ A' &= -sin^2\ \theta + cos^2\ \theta \\ -sin^2\ \frac{\theta}{2} + cos^2\ \frac{\theta}{2} &= 0 \\ sin\ \frac{\theta}{2} &= cos\ \frac{\theta}{2} \\ \frac{\theta}{2} &= \frac{\pi}{4} \\ \theta &= \frac{\pi}{2} \\ A &= sin\frac{\pi}{4}\ cos\frac{\pi}{4} \\ &= \frac{1}{2} \end{aligned}$$

The angle that maximizes the triangle is $\frac{\pi}{2}$ and the maximum area is $\frac{1}{2}$ square unit of length.

26.

$$\begin{aligned} \frac{1}{2}(cos\ \theta + 1)(sin\ \theta) &= A \\ \frac{1}{2}\left[(cos\ \theta + 1)(cos\ \theta) - sin^2\ \theta\right] &= A' \\ \frac{1}{2}\left[2\ cos^2\ \theta + cos\ \theta - 1\right] &= \\ \frac{1}{2}\left[(2\ cos\ \theta - 1)(cos\ \theta + 1)\right] &= \\ \frac{1}{2}\left[(2\ cos\ \theta - 1)(cos\ \theta + 1)\right] &= 0 \\ cos\ \theta &= -1 \\ \theta &= \pi \text{ not acceptable} \\ cos\ \theta &= \frac{1}{2} \\ \theta &= \frac{\pi}{3} \\ A &= \frac{1}{2}\left[cos\ \frac{\pi}{3} + 1\right]\left[sin\ \frac{\pi}{3}\right] \\ &= \frac{3\sqrt{3}}{8} = 0.6495 \end{aligned}$$

The maximum area possible is 0.6495 squared units of length.

27. Price of ticket: x
Number of people at the game: $N = -10000x + 130000$

$$\begin{aligned} x(-10000x + 130000) + 1.50(-10000x + 130000) &= R \\ -10000x^2 + 130000x - 15000x + 19500 &= \\ -10000x^2 + 115000x + 19500 &= \\ -20000x + 115000 &= R' \\ -20000x + 115000 &= 0 \\ x &= 5.75 \\ -10000(5.75) + 130000 &= N \\ 72500 &= \end{aligned}$$

The maximum revenue occurs when the price of the ticket is \$5.75 and 72500 people attend the game.

28. Price of the room: x
Number of rented rooms: $N = -x + 90$

$$\begin{aligned} P &= x(-x + 90) - 6(-x + 90) \\ &= -x^2 + 90x + 6x + 450 \\ &= -x^2 + 96x + 450 \\ P' &= -2x + 96 \\ -2x + 96 &= 0 \\ x &= 48 \end{aligned}$$

The maximum profit occurs when the price of the room is \$48.

29. Number of trees per acre: N
Number of bushels per tree: $B = -N + 50$

$$\begin{aligned} Y &= N(-N + 50) \\ &= -N^2 + 50N \\ Y' &= -2N + 50 \\ -2N + 50 &= 0 \\ N &= 25 \end{aligned}$$

The farmer should plant 25 trees per acre to maximize bushel yields per tree.

30. Price of ticket: x
Number of people at the theater: $N = -10x + 130$

$$\begin{aligned} x(-10x+130) &= R \\ -10x^2 + 130x &= \\ -20x + 130 &= R' \\ -20x + 130 &= 0 \\ x &= 6.5 \end{aligned}$$

The maximum revenue occurs when the price of the ticket is \$6.5.

31. $x^2y = 320 \rightarrow y = \dfrac{320}{x^2}$

$$\begin{aligned} C &= 0.15x^2 + 0.10x^2 + 0.025(4xy) \\ &= 0.25x^2 + 0.1x\left(\frac{320}{x^2}\right) \\ &= 0.25x^2 + \frac{32}{x} \\ C' &= 0.5x - \frac{32}{x^2} \\ 0.5x - \frac{32}{x^2} &= 0 \\ x^3 &= 64 \\ x &= 4 \\ y &= \frac{320}{16} \\ &= 20 \end{aligned}$$

The dimensions of the box that minimize the cost are $4 \times 4 \times 20$ feet.

32. The area of the 7×13 sign does not have as large an area as the 10×10 sign. The customer could refuse this offer if they are concerned with maximizing the size of the sign. (Answers may vary)

33. Amount invested: x, Rate: r and $x = kr$

$$\begin{aligned} P &= 0.18kr - kr^2 \\ P' &= 0.18k - 2kr \\ 0.18k - 2kr &= 0 \\ r &= 0.09 \end{aligned}$$

Interest rate to maximize profit should be 9%

34. $xy = 73.125 \rightarrow y = \dfrac{73.125}{x}$
Length: $l = x - 1.5$, Width: $w = y - 1$

$$\begin{aligned} A &= (x-1.5)(y-1) \\ &= (x-1.5)\left(\frac{73.125}{x} - 1\right) \\ &= 74.625 - x - \frac{109.6875}{x} \\ A' &= -1 + \frac{109.6875}{x^2} \\ -1 + \frac{109.6875}{x^2} &= 0 \\ x &= 10.4732 \\ y &= \frac{73.125}{10.4732} \\ &= 6.9821 \end{aligned}$$

The outside dimensions should be 10.7432×6.9821 inches. The actual page measured 10×7.5 inches.

35. Price: p, Percentage: r, Ordered pair (p, r)
$(25, 2.13)$ and $(26, 2.09)$

a)

$$\begin{aligned} r - 2.13 &= -0.04(p-25) \\ r &= -0.04p + 3.13 \end{aligned}$$

b) $R = p(100r)$, the 100 was needed to get the number of people with personalized license plates.

$$\begin{aligned} R &= p(-4p+313) \\ &= -4p^2 + 313p \\ R' &= -8p + 313 \\ -8p + 313 &= 0 \\ p &= 39.13 \end{aligned}$$

The price that maximizes revenue is \$39.13.

36.

$$\begin{aligned} w'(t) &= 0.002274t^2 - 0.1192t + 1.82 \\ w''(t) &= 0.004548t - 0.1192 \\ 0.004548t - 0.1192 &= 0 \\ t &= 26.2093 \end{aligned}$$

The boy's growing rate is slowest after 26.2093 months.

37. Length: y, gerth: 4x, $4x + y = 84$

$$\begin{aligned} V &= x^2y \\ &= x^2(84-4x) \\ &= 84x^2 - 4x^3 \\ V' &= 168x - 12x^2 \\ 168x - 12x^2 &= 0 \\ 12x(14-x) &= 0 \\ x &= 0 \text{ not acceptable} \\ x &= 14 \\ y &= 84 - 4(14) = 28 \end{aligned}$$

The maximum box has dimensions $14 \times 14 \times 28$ inches.

38. $xy = 48 \rightarrow y = \dfrac{48}{x}$, rate per yard: k

$$\begin{aligned} C &= 2xk + 1.5yk \\ &= 2xk + 1.5k\left(\frac{48}{x}\right) \\ &= 2xk + \frac{72k}{x} \end{aligned}$$

$$\begin{aligned} C' &= 2k - \frac{72k}{x^2} \\ 2k - \frac{72k}{x^2} &= 0 \\ x &= 6 \\ y &= 8 \end{aligned}$$

The fence should be 6×8 yards.

39. $2y + 2x + \pi x = 24$

$$\begin{aligned} A &= 2xy + \frac{\pi}{2}x \\ &= 2x(-x - \frac{\pi}{2}x + 12) + \frac{\pi}{2}x^2 \\ &= -2x^2 - \frac{\pi}{2}x^2 + 24x \\ A' &= -4x - \pi x + 24 \\ -4x - \pi x + 24 &= 0 \\ x(4+\pi) &= 24 \\ x &= \frac{24}{4+\pi} \\ y &= -\frac{24}{4+\pi} - \frac{\pi}{2}\left(\frac{24}{4+\pi}\right) + 12 \\ &= \frac{24}{4+\pi} \end{aligned}$$

40. Left to the student

41. Let x be the number

$$\begin{aligned} S &= \frac{1}{x} + 5x^2 \\ S' &= -\frac{1}{x^2} + 10x \\ -\frac{1}{x^2} + 10x &= 0 \\ x^3 &= \frac{1}{10} \\ x &= \sqrt[3]{\frac{1}{10}} \end{aligned}$$

42. Let x be the number

$$\begin{aligned} S &= \frac{1}{x} + 4x^2 \\ S' &= -\frac{1}{x^2} + 8x \\ -\frac{1}{x^2} + 8x &= 0 \\ x^3 &= \frac{1}{8} \\ x &= \sqrt[3]{\frac{1}{8}} \\ &= \frac{1}{2} \end{aligned}$$

43.

$$\begin{aligned} S &= \pi x^2 + (24-x)^2 \\ &= \pi x^2 + 576 - 48x + x^2 \\ &= (\pi+1)x^2 - 48x + 576 \\ S' &= 2(\pi+1)x - 48 \\ 2(\pi+1)x - 48 &= 0 \\ x &= \frac{24}{\pi+1} \\ 24 - x &= 24 - \frac{24}{\pi+1} \\ &= \frac{24\pi}{\pi+1} \end{aligned}$$

There is no maximum if the string is cut. A maximum could occur if the string is not cut and the whole string is used to make the circle.

44.

$$\begin{aligned} C &= 5000\sqrt{x^2+1} + 3000(4-x) \\ C' &= \frac{5000x}{\sqrt{x^2+1}} - 3000 \\ \frac{5000x}{\sqrt{x^2+1}} - 3000 &= 0 \\ \sqrt{x^2+1} &= \frac{3x}{5} \\ x^2 + 1 &= \frac{9}{25}x^2 \\ x^2 &= \frac{25}{16} \\ x &= 1.25 \end{aligned}$$

The power line should be constructed so that it is 1.25 miles up shore from point B.

45. left to the student

46.

$$\begin{aligned} \sqrt{x^2+b^2} + r + \sqrt{(p-x)^2+a^2} &= D \\ \frac{x}{\sqrt{x^2+b^2}} - \frac{(p-x)}{\sqrt{(p-x)^2+a^2}} &= D' \\ \frac{x}{\sqrt{x^2+b^2}} - \frac{(p-x)}{\sqrt{(p-x)^2+a^2}} &= 0 \\ x^2[(p-x)^2+a^2] &= (p-x)^2(x^2+b^2) \\ (a^2-b^2)x^2 - 2pb^2x + p^2b^2 &= 0 \\ \frac{2pb^2 \pm \sqrt{4p^2b^4 - 4(a^2-b^2)p^2b^2}}{2(a^2-b^2)} &= \\ \frac{2pb^2 \pm \sqrt{8p^2b^4 - 4a^2p^2b^2}}{2(a^2-b^2)} &= \\ \frac{pb^2 \pm pb\sqrt{2b^2-a^2}}{(a^2-b^2)} &= \end{aligned}$$

47. $Q = x^3 + 2y^3$ and $x + y = 1$

$$\begin{aligned} Q &= (1-y)^3 + 2y^3 \\ &= 1 - 3y + 3y^2 - y^3 + 2y^3 \\ &= y^3 + 3y^2 - 3y + 1 \\ Q' &= 3y^2 + 6y - 3 \\ 3y^2 + 6y - 3 &= 0 \end{aligned}$$

$$\begin{aligned} y &= \frac{-6 \pm \sqrt{36+36}}{6} \\ &= -1+\sqrt{2} \\ x &= 1-(-1 \pm +\sqrt{2}) \\ &= 2-\sqrt{2} \end{aligned}$$

When $x = 2 - \sqrt{2}$ and $y = -1 + \sqrt{2}$

$$\begin{aligned} Q &= (2-\sqrt{2})^3 + 2(-1+\sqrt{2})^3 \\ &= 8 - 12\sqrt{2} + 12 - 2\sqrt{2} + \\ &\quad 2(-1 + 3\sqrt{2} - 6 + 2\sqrt{2}) \\ &= 6 - 4\sqrt{2} \end{aligned}$$

48. $Q = 3x + y^3$ and $x^2 + y^2 = 2$

$$\begin{aligned} Q &= 3\sqrt{2-y^2} + y^3 \\ Q' &= \frac{-3y}{\sqrt{2-y^2}} + 3y^2 \\ \frac{-3y}{\sqrt{2-y^2}} + 3y^2 &= 0 \\ \sqrt{2-y^2} &= \frac{1}{y} \\ 2-y^2 &= \frac{1}{y^2} \\ 2y^2 - y^4 &= 1 \\ y^4 - 2y^2 + 1 &= 0 \\ (y^2-1)^2 &= 0 \\ y &= 1 \\ x &= \pm 1 \\ y &= -1 \\ x &= \pm 1 \end{aligned}$$

When $x = 1$ and $y = 1$, $Q = 4$
When $x = 1$ and $y = -1$, $Q = 2$
When $x = -1$ and $y = 1$, $Q = -2$
When $x = -1$ and $y = -1$, $Q = -4$
Thus, Q reaches a minimum when $x = -1$ and $y = -1$.

Exercise Set 3.6

1. $f(x) = x^2$, $a = 3$
$f'(x) = 2x$

$$\begin{aligned} L(x) &= f(3) + f'(3)(x-3) \\ &= 9 + 6(x-3) \\ &= 6x - 9 \end{aligned}$$

2. $f(x) = x^3$, $a = 2$
$f'(x) = 3x^2$

$$\begin{aligned} L(x) &= 8 + 12(x-2) \\ &= 12x - 16 \end{aligned}$$

3. $f(x) = \frac{1}{x}$, $a = 4$
$f'(x) = \frac{-1}{x^2}$

$$\begin{aligned} L(x) &= f(4) + f'(4)(x-4) \\ &= \frac{1}{4} - \frac{1}{16}(x-4) \\ &= -\frac{1}{16}x + \frac{1}{2} \end{aligned}$$

4. $f(x) = \frac{1}{x^2}$, $a = 0.5$
$f'(x) = \frac{-2}{x^3}$

$$\begin{aligned} L(x) &= 4 - 16(x-0.5) \\ &= -16x + 22 \end{aligned}$$

5. $f(x) = x^{3/2}$, $a = 4$
$f'(x) = \frac{3}{2}x^{1/2}$

$$\begin{aligned} L(x) &= f(4) + f'(4)(x-4) \\ &= 8 + 3(x-4) \\ &= 3x - 4 \end{aligned}$$

6. $f(x) = \sqrt{x}$, $a = 100$
$f'(x) = \frac{1}{2\sqrt{x}}$

$$\begin{aligned} L(x) &= 10 + \frac{1}{20}(x-100) \\ &= \frac{1}{20}x + 5 \end{aligned}$$

7. $f(x) = \cos x$, $a = 0$
$f'(x) = -\sin x$

$$\begin{aligned} L(x) &= f(0) + f'(0)(x-0) \\ &= 1 + 0(x-0) \\ &= 1 \end{aligned}$$

8. $f(x) = \tan x$, $a = 0$
$f'(x) = \sec^2 x$

$$\begin{aligned} L(x) &= 0 + 1(x-0) \\ &= x \end{aligned}$$

9. $f(x) = x \cos x$, $a = 0$
$f'(x) = -x \sin x + \cos x$

$$\begin{aligned} L(x) &= f(0) + f'(0)(x-0) \\ &= 0 + 1(x-0) \\ &= x \end{aligned}$$

10. $f(x) = x \sec x$, $a = 0$
$f'(x) = x \sec x \tan x + \sec x$

$$\begin{aligned} L(x) &= 0 + 1(x-0) \\ &= x \end{aligned}$$

11. $f(x) = \sqrt{x}$, $a = 16$
$f'(x) = \dfrac{1}{2\sqrt{x}}$

$$\begin{aligned} L(19) &= f(16) + f'(16)(19-16) \\ &= 4 + \frac{1}{8}(3) \\ &= 4.375 \end{aligned}$$

12. $f(x) = \sqrt{x}$, $a = 24$
$f'(x) = \dfrac{1}{2\sqrt{x}}$

$$\begin{aligned} L(24) &= f(25) + f'(25)(24-25) \\ &= 5 + \frac{1}{10}(-1) \\ &= 4.9 \end{aligned}$$

13. $f(x) = \sqrt{x}$, $a = 100$
$f'(x) = \dfrac{1}{2\sqrt{x}}$

$$\begin{aligned} L(99.1) &= f(100) + f'(100)(99.1-100) \\ &= 10 + \frac{1}{20}(-0.9) \\ &= 9.955 \end{aligned}$$

14. $f(x) = \sqrt{x}$, $a = 100$
$f'(x) = \dfrac{1}{2\sqrt{x}}$

$$\begin{aligned} L(103.4) &= f(100) + f'(100)(103.4-100) \\ &= 10 + \frac{1}{20}(3.4) \\ &= 10.17 \end{aligned}$$

15. $f(x) = \sqrt[3]{x}$, $a = 8$
$f'(x) = \dfrac{1}{3\sqrt[3]{x^2}}$

$$\begin{aligned} L(10) &= f(8) + f'(8)(10-8) \\ &= 2 + \frac{1}{12}(2) \\ &= 2.16667 \end{aligned}$$

16. $f(x) = \sqrt[3]{x}$, $a = 1$
$f'(x) = \dfrac{1}{3\sqrt[3]{x^2}}$

$$\begin{aligned} L(0.91) &= f(1) + f'(1)(0.91-1) \\ &= 1 + \frac{1}{3}(-0.09) \\ &= 0.97 \end{aligned}$$

17. $f(x) = \sqrt{x}$, $a = 100$
$f'(x) = \dfrac{1}{2\sqrt{x}}$

$$\begin{aligned} L(97) &= f(100) + f'(100)(97-100) \\ &= 10 + \frac{1}{20}(-3) \\ &= 9.85 \end{aligned}$$

18. $f(x) = \sqrt[3]{x}$, $a = 1728$
$f'(x) = \dfrac{1}{3\sqrt[3]{x^2}}$

$$\begin{aligned} L(1729.03) &= f(1728) + f'(1728)(1729.03-1728) \\ &= 12 + \frac{1}{36}(1.03) \\ &= 12.0286 \end{aligned}$$

19. $f(x) = sin\ x$, $a = 0$
$f'(x) = cos\ x$

$$\begin{aligned} L(0.1) &= f(0) + f'(0)(0.1-0) \\ &= 0 + 1(0.1) \\ &= 0.1 \end{aligned}$$

20. $f(x) = sin\ x$, $a = 0$
$f'(x) = cos\ x$

$$\begin{aligned} L(-0.05) &= f(0) + f'(0)(-0.05-0) \\ &= 0 + 1(-0.05) \\ &= -0.05 \end{aligned}$$

21. $f(x) = tan\ x$, $a = 0$
$f'(x) = sec^2\ x$

$$\begin{aligned} L(-0.04) &= f(0) + f'(0)(-0.04-0) \\ &= 0 + 1(-0.04-0) \\ &= -0.04 \end{aligned}$$

22. $f(x) = tan\ x$, $a = 0$
$f'(x) = sec^2\ x$

$$\begin{aligned} L(0.02) &= f(0) + f'(0)(0.02-0) \\ &= 0 + 1(0.02-0) \\ &= 0.02 \end{aligned}$$

23. $f(x) = \frac{1}{3}x^2 - x + 1over3$
$f'(x) = \frac{2}{3}x - 1$

$$\begin{aligned} x_{n+1} &= x_n - \frac{f(x_n)}{f'(x_n)} \\ x_1 &= 0 \\ x_2 &= 0 - \frac{0.33333}{-1} = 0.33333 \\ x_3 &= 0.33333 - \frac{0.03704}{-0.77778} = 0.38095 \\ x_4 &= 0.38095 - \frac{0.00076}{-0.74603} = 0.38197 \\ x_5 &= 0.38197 - \frac{0}{-7.454} = 0.38197 \end{aligned}$$

$x = 0.38197$

24. $f(x) = \frac{1}{3}x^2 - x + 1over3$
$f'(x) = \frac{2}{3}x - 1$

$$x_{n+1} = x_n - \frac{f(x_n)}{f'(x_n)}$$

$$\begin{aligned} x_1 &= 3 \\ x_2 &= 3 - \frac{0.33333}{1} = 2.66667 \\ x_3 &= 2.66667 - \frac{0.03704}{0.77778} = 2.61904 \\ x_4 &= 2.61904 - \frac{0.00075}{0.74603} = 2.61804 \\ x_5 &= 2.61804 - \frac{0}{0.74536} = 2.61804 \end{aligned}$$

$x = 2.61804$

25. $f(x) = x^3 - 3x + 3$
$f'(x) = 3x^2 - 3$

$$\begin{aligned} x_{n+1} &= x_n - \frac{f(x_n)}{f'(x_n)} \\ x_1 &= -3 \\ x_2 &= -3 - \frac{-15}{24} = -2.375 \\ x_3 &= -2.375 - \frac{-3.27148}{13.92188} = -2.14001 \\ x_4 &= -2.14001 - \frac{-0.38045}{10.73892} = -2.10458 \\ x_5 &= -2.10458 - \frac{-0.00799}{10.28777} = -2.10381 \\ x_6 &= -2.10381 - \frac{-0.00007}{10.27805} = -2.10380 \\ x_7 &= -2.10380 - \frac{-0.00004}{10.27805} = -2.10380 \end{aligned}$$

$x = -2.10380$

26. $f(x) = x^4 - 4x + 1$
$f'(x) = 4x^3 - 4$

$$\begin{aligned} x_{n+1} &= x_n - \frac{f(x_n)}{f'(x_n)} \\ x_1 &= 0 \\ x_2 &= 0 - \frac{1}{-4} = 0.25 \\ x_3 &= 0.25 - \frac{0.00391}{-3.9375} = 0.25099 \\ x_4 &= 0.25099 - \frac{0}{-3.93675} = 0.25099 \end{aligned}$$

$x = 0.25099$

27. $f(x) = x\sqrt{x+1} - 4$
$f'(x) = \dfrac{x}{\sqrt{x+1}} + \sqrt{x+1}$

$$\begin{aligned} x_{n+1} &= x_n - \frac{f(x_n)}{f'(x_n)} \\ x_1 &= 2 \\ x_2 &= 2 - \frac{-0.5359}{2.5774} = 2.2079 \\ x_3 &= 2.2079 - \frac{-0.0455}{2.6677} = 2.2250 \\ x_4 &= 2.2250 - \frac{-0.0043}{2.675} = 2.2266 \\ x_5 &= 2.2266 - \frac{-0.0004}{2.6757} = 2.2267 \\ x_6 &= 2.2267 - \frac{-0.0002}{2.6757} = 2.2268 \\ x_7 &= 2.2268 - \frac{0}{2.6757} = 2.2268 \end{aligned}$$

$x = 2.2268$

28. $f(x) = x\sqrt{x+2} - 8$
$f'(x) = \dfrac{x}{\sqrt{x+2}} + \sqrt{x+2}$

$$\begin{aligned} x_{n+1} &= x_n - \frac{f(x_n)}{f'(x_n)} \\ x_1 &= 2 \\ x_2 &= 2 - \frac{-4}{2.5} = 3.6 \\ x_3 &= 3.6 - \frac{0.519515}{3.1271} = 3.43398 \\ x_4 &= 3.43398 - \frac{0.00493}{3.0677} = 3.43238 \\ x_5 &= 3.43238 - \frac{0}{3.0671} = 3.43238 \end{aligned}$$

$x = 3.43238$

29. $f(x) = \cos 2x - x$
$f'(x) = -2 \sin 2x - 1$

$$\begin{aligned} x_{n+1} &= x_n - \frac{f(x_n)}{f'(x_n)} \\ x_1 &= 0 \\ x_2 &= 0 - \frac{1}{-1} = 1 \\ x_3 &= 1 - \frac{-1.416}{-2.819} = 0.49769 \\ x_4 &= 0.49769 - \frac{0.04648}{-2.678} = 0.51505 \\ x_5 &= 0.51505 - \frac{-0.0003}{-2.715} = 0.51494 \\ x_6 &= 0.51494 - \frac{0}{-2.714} = 0.51494 \end{aligned}$$

$x = 0.51494$

30. $f(x) = \sin x + x - 1$
$f'(x) = \cos x + 1$

$$\begin{aligned} x_{n+1} &= x_n - \frac{f(x_n)}{f'(x_n)} \\ x_1 &= 0 \\ x_2 &= 0 - \frac{-1}{2} = 0.5 \\ x_3 &= 0.5 - \frac{-0.0206}{1.8776} = 0.51097 \\ x_4 &= 0.51097 - \frac{0}{1.8723} = 0.51097 \end{aligned}$$

$x = 0.51097$

31. $f(x) = \sin x - \cos x + x$
$f'(x) = \cos x + \sin x + 1$

$$x_{n+1} = x_n - \frac{f(x_n)}{f'(x_n)}$$

$$\begin{aligned} x_1 &= 0 \\ x_2 &= 0 - \frac{-1}{2} = 0.5 \\ x_3 &= 0.5 - \frac{0.10184}{2.357} = 0.45679 \\ x_4 &= 0.45679 - \frac{0.00039}{2.3385} = 0.45663 \\ x_5 &= 0.45663 - \frac{0}{2.3385} = 0.45663 \end{aligned}$$

$x = 0.45663$

32. $f(x) = 2x - sin\ x - cos\ x^2$
$f'(x) = 2 - cos\ x + 2x\ sin\ x^2$

$$\begin{aligned} x_{n+1} &= x_n - \frac{f(x_n)}{f'(x_n)} \\ x_1 &= \frac{\pi}{4} \\ x_2 &= \frac{\pi}{4} - \frac{0.04799}{2.2015} = 0.7636 \\ x_3 &= 0.7636 - \frac{0.0009}{2.1185} = 0.76317 \\ x_4 &= 0.76317 - \frac{0}{2.1169} = 0.76317 \end{aligned}$$

$x = 0.76317$

33. Initial guess $x = -1.5$ leads to solution $x = -1.142495$
Initial guess $x = 0.25$ leads to solution $x = 0.176245$
Initial guess $x = 4.9$ leads to solution $x = 4.96625$

34. Initial guess $x = -3$ leads to solution $x = -2.95728$
Initial guess $x = -0.25$ leads to solution $x = -0.15237$
Initial guess $x = 1$ leads to solution $x = 1.10964$

35. Initial guess $x = -3$ leads to solution $x = -2.86640$
Initial guess $x = -0.5$ leads to solution $x = -0.56682$
Initial guess $x = 0.5$ leads to solution $x = 0.40865$

36. Initial guess $x = -1.5$ leads to solution $x = -1.4613$
Initial guess $x = 0.6$ leads to solution $x = 0.57437$
Initial guess $x = 1.25$ leads to solution $x = 1.24907$

37. Use Newton's method on

$$0.05x^2 - 0.3x^3 - 0.0001$$

Initial guess $x = 0$ leads to solution $x = 0.05452$
Initial guess $x = 0.1$ leads to solution $x = 0.15229$
The two dosages are $x = 0.055$ and $x = 0.15$ cubic centimeters.

38. Use Newton's method on

$$-0.00006x^3 + 0.006x^2 - 0.1x - 1.1$$

Initial guess $x = 32$ leads to solution $x = 33.22222$
Initial guess $x = 73$ leads to solution $x = 74.21361$
The ages for which there are 3 million hearing impaired Americans are $x = 33.2$ years and 74.2 years.

39. Use Newton's method on

$$-6.85 + 1.82t - 0.0596t^2 + 0.000758t^3$$

Initial guess $x = 4$ leads to solution $x = 4.34880$
The age at which the median weight of boys is 15 pounds is $t = 4.35$ months.

40. Use Newton's method on

$$-16.85 + 1.82t - 0.0596t^2 + 0.000758t^3$$

Initial guess $x = 15.5$ leads to solution $x = 15.76586$
The age at which the median weight of boys is 25 pounds is $t = 15.77$ months.

41. Use Newton's method on

$$-0.000775x^3 + 0.0696x^2 - 0.209x - 35.32$$

Initial guess $x = 30$ leads to solution $x = 29.93332$
The rate is 40 per 100000 in the end of 1959.

42. Use Newton's method on

$$0.0338x^4 - 0.996x^3 + 8.57x^2 - 18.4x - 16.5$$

Initial guess $x = 5$ leads to solution $x = 4.75755$
Initial guess $x = 10$ leads to solution $x = 9.85639$
The average temperature in NY is 60° on middle of May and middle of October.

43. Use Newton's method on

$$-0.000054x^4 + 0.0067x^3 - 0.0997x^2 - 0.84x - 300.25$$

Initial guess $x = 60$ leads to solution $x = 57.09821$
Initial guess $x = 95$ leads to solution $x = 97.50401$
At the age of 57 years and 97.5 years 300 out of 100000 woemn will have breast cancer.

44. $f(r) = r^2 - 0.94r - 0.0192\left[1 - \left(\frac{0.94}{r}\right)^{24}\right]$

$f'(r) = 2r - 0.94 + \left(\frac{0.94^{23}}{r^{25}}\right)$

$$\begin{aligned} x_{n+1} &= x_n - \frac{f(x_n)}{f'(x_n)} \\ x_1 &= 1 \\ x_2 &= 1 - \frac{0.04515}{1.1644} = 0.96122 \\ x_3 &= 0.96122 - \frac{0.01244}{1.263} = 0.95138 \\ x_4 &= 0.95138 - \frac{0.00601}{1.3256} = 0.94684 \\ x_5 &= 0.94684 - \frac{0.00341}{1.3626} = 0.94434 \\ x_6 &= 0.94434 - \frac{0.00209}{1.3856} = 0.94283 \\ x_7 &= 0.94283 - \frac{0.00133}{1.4004} = 0.94188 \\ x_8 &= 0.94188 - \frac{0.00087}{1.4101} = 0.94126 \\ x_9 &= 0.94126 - \frac{0.00058}{1.4166} = 0.94085 \\ x_{10} &= 0.94085 - \frac{0.00039}{1.4209} = 0.94057 \\ x_{11} &= 0.94057 - \frac{0.00026}{1.4239} = 0.94039 \\ x_{12} &= 0.94039 - \frac{0.00018}{1.4259} = 0.94027 \\ x_{13} &= 0.94027 - \frac{0.00012}{1.4272} = 0.94019 \\ x_{14} &= 0.94019 - \frac{0}{1.4281} = 0.94019 \end{aligned}$$

$x = 0.94019$

45. $f(r) = r^{15} - 0.99r^{14} - 0.0858\left[1 - \left(\frac{0.99}{r}\right)^{36}\right]$

$f'(r) = 15r^{14} - 13.86r^{13} + \left(\frac{2.151081148}{r^{37}}\right)$

$$\begin{aligned} x_{n+1} &= x_n - \frac{f(x_n)}{f'(x_n)} \\ x_1 &= 1.1 \\ x_2 &= 1.1 - \frac{0.33386}{9.0507} = 1.0631 \\ x_3 &= 1.0631 - \frac{0.09303}{4.4002} = 1.042 \\ x_4 &= 1.0420 - \frac{0.02021}{2.5499} = 1.0340 \\ x_5 &= 1.0340 - \frac{0.00248}{1.9274} = 1.03276 \\ x_6 &= 1.03276 - \frac{0}{1.8279} = 1.03276 \end{aligned}$$

$r = 1.03276$

46. $f(r) = r^3 - 0.99r^2 - 0.3168\left[1 - \left(\frac{0.99}{r}\right)^{16}\right]$

$f'(r) = 3r^2 - 1.98r - \frac{4.315869150}{r^{17}}$

After 36 iterations of Newton's method we reach $r = 1.19989$

47. $f(r) = r^4 - 0.98r^3 - 0.1764\left[1 - \left(\frac{0.98}{r}\right)^{15}\right]$

$f'(r) = 4r^3 - 2.94r^2 - \frac{1.954253846}{r^{16}}$

After 7 iterations of Newton's method we reach $r = 1.08846$

48. **a)**

$$\begin{aligned} v &= \frac{77000 \cdot 150 \cdot sec\ 60^o}{5000000} \\ &= 4.62\ \frac{cm}{s} \end{aligned}$$

b) $f(t) = 2.31\ sec\ t$, $f'(t) = 2.31\ sec\ t\ tan\ t$

$$\begin{aligned} L(0.01) &= f(\frac{\pi}{3}) + f'(\frac{\pi}{3})(0.01) \\ &= 4.7 \end{aligned}$$

Difference in measurement $= 4.7 - 4.62 = 0.08\ \frac{cm}{s}$

49. **a)**

$$\begin{aligned} v &= \frac{77000 \cdot 100 \cdot sec\ \frac{4\pi}{9}}{4000000} \\ &= 11.086\ \frac{cm}{s} \end{aligned}$$

b) $f(t) = 1.925\ sec\ t$, $f'(t) = 11.086\ sec\ t\ tan\ t$

$$\begin{aligned} L(0.01) &= f(\frac{4\pi}{9}) + f'(\frac{4\pi}{9})(0.01) \\ &= 11.7147 \end{aligned}$$

Difference in measurement $= 11.7147 - 11.086 = 0.6287\ \frac{cm}{s}$

c) The difference in measurement is more sensitive to angle measurement when the frequency f gets smaller.

50. **a)** Newton's method does not converge to a solution.

b) Newton's method does not converge to a solution for $f(x) = x^{1/3}$ since the derivative at $x_1 = 0$ is not defined.

51. **a)** $x_1 = 3$
$x_2 = 3 - \frac{-3}{-1} = 0$
$x_3 = 0 - \frac{0}{8} = 0$
$x = 0$

b) The tangent line at $x_1 = 3$ (which intersects x_{n+1}) intersected $x = 0$ instead of the closer solutions.

52. **a)**

$$\begin{aligned} \frac{1}{A_m} tan(L_m x) - \frac{1}{A_t} cot(L_t x) &= 0 \\ \frac{1}{10A_t} tan(8x) - \frac{1}{A_t} cot(9.7x) &= 0 \\ tan(8x) - 10\ cot(9.7x) &= 0 \end{aligned}$$

b) $x = 0.14$, $x = 0.22$, and $x = 0.28$

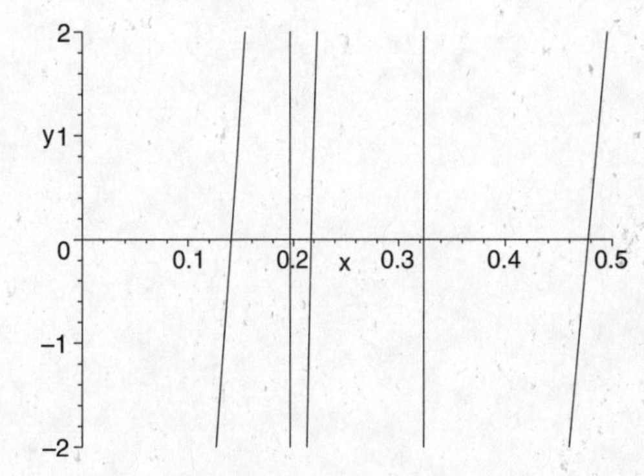

c) $x = 0.14065$, $x = 0.21719$, and $x = 0.47752$

d) For $x = 0.14065$, $f = 792.43\ Hz$
For $x = 0.21719$, $f = 1223.67\ Hz$
For $x = 0.47752$, $f = 2690.39\ Hz$

53. For $x = 0.04609$, $f = 259.674\ Hz$
For $x = 0.34184$, $f = 1925.96\ Hz$
For $x = 0.52360$, $f = 2950.01\ Hz$

54. For $x = 0.15381$, $f = 866.58\ Hz$
For $x = 0.27909$, $f = 1572.42\ Hz$
For $x = 0.52360$, $f = 2950.01\ Hz$

55. For $x = 0.11602$, $f = 653.67\ Hz$
For $x = 0.32838$, $f = 1850.12\ Hz$
For $x = 0.42340$, $f = 2385.47\ Hz$

Exercise Set 3.7

1. $xy - x + 2y = 3$

$$\begin{aligned} x\frac{dy}{dx} + y - 1 + 2\frac{dy}{dx} &= 0 \\ (x+2)\frac{dy}{dx} &= 1 - y \\ \frac{dy}{dx} &= \frac{1-y}{x+2} \end{aligned}$$

For $\left(-5, \frac{2}{3}\right)$

$$\begin{aligned}\frac{dy}{dx} &= \frac{1-\frac{2}{3}}{-5+2}\\ &= \frac{1}{9}\end{aligned}$$

2. $xy + y^2 - 2x = 0$

$$\begin{aligned}x\frac{dy}{dx} + y - 2y\frac{dy}{dx} - 2 &= 0\\ (x-2y)\frac{dy}{dx} &= 2-y\\ \frac{dy}{dx} &= \frac{2-y}{x-2y}\end{aligned}$$

For $(1, -2)$

$$\begin{aligned}\frac{dy}{dx} &= \frac{2-(-2)}{1-2(-2)}\\ &= \frac{4}{5}\end{aligned}$$

3. $x^2 + y^2 = 1$

$$\begin{aligned}2x + 2y\frac{dy}{dx} &= 0\\ \frac{dy}{dx} &= \frac{-x}{y}\end{aligned}$$

For $\left(\frac{1}{2}, \frac{\sqrt{3}}{2}\right)$

$$\begin{aligned}\frac{dy}{dx} &= -\frac{\frac{1}{2}}{\frac{\sqrt{3}}{2}}\\ &= -\frac{1}{\sqrt{3}}\end{aligned}$$

4. $x^2 - y^2 = 1$

$$\begin{aligned}2x - 2y\frac{dy}{dx} &= 0\\ \frac{dy}{dx} &= \frac{x}{y}\end{aligned}$$

For $(\sqrt{3}, \sqrt{2})$

$$\frac{dy}{dx} = \frac{\sqrt{3}}{\sqrt{2}}$$

5. $x^2y - 2x^3 - y^3 + 1 = 0$

$$\begin{aligned}x^2\frac{dy}{dx} + 2xy - 6x^2 - 3y^2\frac{dy}{dx} &= 0\\ (x^2-3y^2)\frac{dy}{dx} &= 6x^2 - 2xy\\ \frac{dy}{dx} &= \frac{6x^2-2xy}{x^2-3y^2}\end{aligned}$$

For $(2, -3)$

$$\begin{aligned}\frac{dy}{dx} &= \frac{6(2)^2 - 2(2)(-3)}{(2)^2 - 3(-3)^2}\\ &= -\frac{36}{23}\end{aligned}$$

6. $4x^3 - y^4 - 3y + 5x + 1 = 0$

$$\begin{aligned}12x^2 - 4y^3\frac{dy}{dx} - 3\frac{dy}{dx} + 5 &= 0\\ (4y^3+3)\frac{dy}{dx} &= 12x^2 + 5\\ \frac{dy}{dx} &= \frac{12x^2+5}{4y^3+3}\end{aligned}$$

For $(1, -2)$

$$\begin{aligned}\frac{dy}{dx} &= \frac{12(1)^2+5}{4(-2)^3+3}\\ &= -\frac{17}{29}\end{aligned}$$

7. $sin\ y + x^2 = cos\ y$

$$\begin{aligned}\frac{dy}{dx}\ cos\ y + 2x &= -\frac{dy}{dx}\ sin\ x\\ (cos\ y + sin\ y)\frac{dy}{dx} &= -2x\\ \frac{dy}{dx} &= \frac{-2x}{cos\ y + sin\ y}\end{aligned}$$

For $(1, 2\pi)$

$$\begin{aligned}\frac{dy}{dx} &= \frac{-2(1)}{cos\ 2\pi + sin\ 2\pi}\\ &= -2\end{aligned}$$

8. $tan^2 x = \frac{2}{3}\ cos\ y$

$$\begin{aligned}2\ tan\ x\ sec^2\ x &= -\frac{2}{3}\ sin\ y\frac{dy}{dx}\\ \frac{dy}{dx} &= -\frac{3\ tan\ x\ sec^2\ x}{sin\ y}\end{aligned}$$

For $\left(\frac{\pi}{6}, \frac{\pi}{3}\right)$

$$\begin{aligned}\frac{dy}{dx} &= -\frac{3\ tan\ \frac{\pi}{6}\ sec^2\ \frac{\pi}{6}}{sin\ \frac{\pi}{3}}\\ &= \frac{8}{3}\end{aligned}$$

9. $x\ sin\ x = y(1 + cos\ y)$

$$\begin{aligned}x\ cos\ x + sin\ x &= y(-sin\ y)\frac{dy}{dx} + (1 + cos\ y)\frac{dy}{dx}\\ x\ cos\ x + sin\ x &= (-y\ sin\ y + cos\ y + 1)\frac{dy}{dx}\\ \frac{dy}{dx} &= \frac{x\ cos\ x + sin\ x}{-y\ sin\ y + cos\ y + 1}\end{aligned}$$

For $\left(\frac{\pi}{2}, \frac{\pi}{3}\right)$

$$\begin{aligned}\frac{dy}{dx} &= \frac{\frac{\pi}{2} cos\ \frac{\pi}{2} + sin\ \frac{\pi}{2}}{-\frac{\pi}{3}\ sin\ \frac{\pi}{3} + cos\ \frac{\pi}{3} + 1}\\ &= \frac{6}{9 - \pi\sqrt{3}}\end{aligned}$$

10. $sin^2\ x + 3\ cos^2\ y = 1$

$$\begin{aligned} 2\ sin\ x cos\ x - 6\ cos\ y\ sin\ y\frac{dy}{dx} &= 0 \\ sin\ 2x - 3\ sin\ 2y\frac{dy}{dx} &= 0 \\ \frac{dy}{dx} &= \frac{sin\ 2x}{3\ sin\ 2y} \end{aligned}$$

For $\left(\frac{\pi}{6}, \frac{2\pi}{3}\right)$

$$\begin{aligned} \frac{dy}{dx} &= \frac{sin\ \frac{\pi}{3}}{3\ sin\ \frac{4\pi}{3}} \\ &= -\frac{1}{3} \end{aligned}$$

11. $2xy + 3 = 0$

$$\begin{aligned} 2x\frac{dy}{dx} + 2y &= 0 \\ \frac{dy}{dx} &= \frac{-y}{x} \end{aligned}$$

12. $x^2 + 2xy = 3y^2$

$$\begin{aligned} 2x + 2x\frac{dy}{dx} + 2y &= 6y\frac{dy}{dx} \\ 2x + 2y &= (6y - 2x)\frac{dy}{dx} \\ \frac{dy}{dx} &= \frac{x + y}{3y - x} \end{aligned}$$

13. $x^2 - y^2 = 16$

$$\begin{aligned} 2x - 2y\frac{dy}{dx} &= 0 \\ \frac{dy}{dx} &= \frac{x}{y} \end{aligned}$$

14. $x^2 + y^2 = 25$

$$\begin{aligned} 2x + 2y\frac{dy}{dx} &= 0 \\ \frac{dy}{dx} &= -\frac{x}{y} \end{aligned}$$

15. $y^5 = x^3$

$$\begin{aligned} 5y^4\frac{dy}{dx} &= 3x^2 \\ \frac{dy}{dx} &= \frac{3x^2}{5y^4} \end{aligned}$$

16. $y^5 = x^3$

$$\begin{aligned} 3y^2\frac{dy}{dx} &= 5x^4 \\ \frac{dy}{dx} &= \frac{5x^4}{3y^2} \end{aligned}$$

17. $x^2y^3 + x^3y^4 = 11$

$$\begin{aligned} 3x^2y^2\frac{dy}{dx} + 2xy^3 + 4x^3y^3\frac{dy}{dx} + 3x^2y^4 &= 0 \\ (3x^2y^2 + 4x^3y^3)\frac{dy}{dx} &= -2xy^3 - 3x^2y^4 \\ \frac{dy}{dx} &= \frac{-2xy^3 - 3x^2y^4}{3x^2y^2 + 4x^3y^3} \end{aligned}$$

18. $x^3y^2 - x^5y^3 = -19$

$$\begin{aligned} 2x^3y\frac{dy}{dx} + 3x^2y^2 - 3x^5y^2\frac{dy}{dx} - 5x^4y^3 &= 0 \\ (2x^3y - 3x^5y^2)\frac{dy}{dx} &= -3x^2y^2 - 5x^4y^3 \\ \frac{dy}{dx} &= \frac{5x^4y^3 - 3x^2y^2}{2x^3y - 3x^5y^2} \end{aligned}$$

19. $\sqrt{x} + \sqrt{y} = 1$

$$\begin{aligned} \frac{1}{2\sqrt{x}} + \frac{1}{2\sqrt{y}}\frac{dy}{dx} &= 0 \\ \frac{dy}{dx} &= -\frac{\sqrt{y}}{\sqrt{x}} \end{aligned}$$

20. $\frac{1}{x^2} + \frac{1}{y^2} = 5$

$$\begin{aligned} -\frac{2}{x^3} - \frac{2}{y^3}\frac{dy}{dx} &= 0 \\ \frac{dy}{dx} &= -\frac{y^3}{x^3} \end{aligned}$$

21. $y^3 = \frac{x - 1}{x + 1}$

$$\begin{aligned} 3y^2\frac{dy}{dx} &= \frac{x + 1 - (x - 1)}{(x + 1)^2} \\ \frac{dy}{dx} &= \frac{2}{3(x + 1)^2y^2} \end{aligned}$$

22. $y^2 = \frac{x^2 - 1}{x^2 + 1}$

$$\begin{aligned} 2y\frac{dy}{dx} &= \frac{2x(x^2 + 1) - 2x(x^2 - 1)}{(x^2 + 1)^2} \\ \frac{dy}{dx} &= \frac{2x}{(x^2 + 1)^2y} \end{aligned}$$

23. $x^{3/2} + y^{2/3} = 1$

$$\begin{aligned} \frac{3}{2}x^{1/2} + \frac{2}{3y^{1/3}}\frac{dy}{dx} &= 0 \\ \frac{dy}{dx} &= -\frac{\sqrt{x}}{\sqrt[3]{y}} \end{aligned}$$

24. $(x - y)^3 + (x + y)^3 = x^5 + y^5$

$$3(x - y)^2\left(1 - \frac{dy}{dx}\right) + 3(x + y)^2\left(1 + \frac{dy}{dx}\right)$$

$$\begin{aligned} &= 5x^4 + 5y^4\frac{dy}{dx} \\ \frac{dy}{dx} &= \frac{-5x^4 + 3(x-y)^2 + 3(x+y)^2}{5y^4 + 3(x-y)^2 - 3(x+y)^2} \\ &= \frac{6x^2 - 5x^4 + 6y^2}{5y^4 - 12xy} \end{aligned}$$

25. $\frac{x^2y + xy + 1}{2x + y} = 1 \rightarrow x^2y + xy + 1 = 2x + y$

$$\begin{aligned} x^2\frac{dy}{dx} + 2xy + x\frac{dy}{dx} + y &= 2 + \frac{dy}{dx} \\ (x^2 + x - 1)\frac{dy}{dx} &= 2 - 2xy - y \\ \frac{dy}{dx} &= \frac{2 - 2xy - y}{x^2 + x - 1} \end{aligned}$$

26. $\frac{xy}{x + y} = 2 \rightarrow xy = 2x + 2y$

$$\begin{aligned} x\frac{dy}{dx} + y &= 2 + 2\frac{dy}{dx} \\ (x-2)\frac{dy}{dx} &= 2 - y \\ \frac{dy}{dx} &= \frac{2-y}{x-2} \end{aligned}$$

27. $4\ sin\ x\ \ cos\ y = 3$

$$\begin{aligned} -4\ sin\ x\ sin\ y\frac{dy}{dx} + 4\ cos\ x\ cos\ y &= 0 \\ \frac{dy}{dx} &= \frac{cos\ x\ cos\ y}{sin\ x\ sin\ y} \\ &= cot\ x\ cot\ y \end{aligned}$$

28. $x\ tan^2y = y^3$

$$\begin{aligned} 2x\ tan\ y\ sec^2y\frac{dy}{dx} + tan^2y &= 3y^2\frac{dy}{dx} \\ (3y^2 - 2x\ tan\ y\ sec^2y)\frac{dy}{dx} &= tan^2y \\ \frac{dy}{dx} &= \frac{tan^2y}{2y^2 - 2x\ tan\ y\ sec^2y} \end{aligned}$$

29. $x + y = sin(\sqrt{y - x})$

$$\begin{aligned} 1 + \frac{dy}{dx} &= cos(\sqrt{y-x})\frac{\frac{dy}{dx} - 1}{2\sqrt{y-x}} \\ 2\sqrt{y-x} + 2\sqrt{y-x}\frac{dy}{dx} &= cos(\sqrt{y-x})\frac{dy}{dx} - \\ & \quad cos(\sqrt{y-x}) \\ (2\sqrt{y-x} - cos(\sqrt{y-x}))\frac{dy}{dx} &= -2\sqrt{y-x} + cos(\sqrt{y-x}) \\ \frac{dy}{dx} &= -\frac{2\sqrt{y-x} + cos(\sqrt{y-x})}{2\sqrt{y-x} - cos(\sqrt{y-x})} \end{aligned}$$

30. $sin(xy) = cos\ y$

$$\begin{aligned} cos(xy)(x\frac{dy}{dx} + y) &= -sin\ y\ \frac{dy}{dx} \\ (sin\ y + x\ cos(xy))\frac{dy}{dx} &= -y\ cos(xy) \\ \frac{dy}{dx} &= -\frac{y\ cos(xy)}{sin\ y + x\ cos(xy)} \end{aligned}$$

31. $A^3 + B^3 = 9$ When $A = 2$, $B = \sqrt[3]{9 - 8} = 1$

$$\begin{aligned} 3A^2\frac{dA}{dt} + 3B^2\frac{dB}{dt} &= 0 \\ 3(2)^2\frac{dA}{dt} + 3(1)^2(3) &= 0 \\ \frac{dA}{dt} &= \frac{-9}{12} \\ &= -\frac{3}{4} \end{aligned}$$

32. $G^2 + H^2 = 25$
When $G = 0$, $H = \sqrt{25 - 0} = 5$
When $G = 1$, $H = \sqrt{25 - 1} = \sqrt{24}$
When $G = 3$, $H = \sqrt{25 - 9} = 4$

$$\begin{aligned} 2G\frac{dG}{dt} + 2H\frac{dH}{dt} &= 0 \\ \frac{dH}{dt} &= -\frac{G}{H}\frac{dG}{dt} \end{aligned}$$

When $G = 0$, $\frac{dH}{dt} = \frac{-0}{5}3 = 0$
When $G = 1$, $\frac{dH}{dt} = \frac{-1}{\sqrt{24}}3 = -\frac{3}{\sqrt{24}}$
When $G = 3$, $\frac{dH}{dt} = \frac{-3}{4}3 = -\frac{9}{4}$

33. $V = \frac{4}{3}\pi r^3$

$$\begin{aligned} \frac{dV}{dt} &= 4\ \pi\ r^2\frac{dr}{dt} \\ &= 4\ \pi(1.2)^2(0.03) \\ &= 0.54287\ \frac{cm}{day} \end{aligned}$$

34. $A = \pi r^2$

$$\begin{aligned} \frac{dA}{dt} &= 2\ \pi\ r\frac{dr}{dt} \\ &= 2\ \pi(25)(-1) \\ &= -50\ \frac{mm}{day} \end{aligned}$$

35. $V = \frac{p}{4Lv}\ (R^2 - r^2)$

a)

$$\begin{aligned} \frac{dV}{dt} &= \frac{2Rp}{4Lv}\frac{dR}{dt} \\ &= \frac{2R(100)}{4(1)(0.05)}\ \frac{dR}{dt} \\ &= 1000R\ \frac{dR}{dt} \end{aligned}$$

b)

$$\begin{aligned} \frac{dV}{dt} &= 1000R\ \frac{dR}{dt} \\ &= 1000(0.0075)(-0.0015) \\ &= -0.01125\ \frac{mm^3}{min} \end{aligned}$$

36. a)

$$\begin{aligned}\frac{dV}{dt} &= \frac{2Rp}{4Lv}\frac{dR}{dt}\\ &= \frac{2R(100)}{4(1)(0.05)}\frac{dR}{dt}\\ &= 1000R\,\frac{dR}{dt}\end{aligned}$$

b)

$$\begin{aligned}\frac{dV}{dt} &= 1000R\,\frac{dR}{dt}\\ &= 1000(0.02)(0.0025)\\ &= 0.05\ \frac{mm^3}{min}\end{aligned}$$

37. $S = \frac{\sqrt{hw}}{60}$

$$\begin{aligned}\frac{dS}{dt} &= \frac{h}{120\sqrt{hw}}\frac{dw}{dt}\\ &= \frac{180}{120\sqrt{(180)(85)}}(-4)\\ &= -0.0485\ \frac{m^2}{month}\end{aligned}$$

38.

$$\begin{aligned}\frac{dS}{dt} &= \frac{h}{120\sqrt{hw}}\frac{dw}{dt}\\ &= \frac{160}{120\sqrt{(160)(60)}}(-3)\\ &= -0.0408\ \frac{m^2}{month}\end{aligned}$$

39. $D^2 = x^2 + y^2$ After one hour, $x = 25$, $y = 60$
$D = \sqrt{25^2 + 60^2} = 65$

$$\begin{aligned}2D\frac{dD}{dt} &= 2x\frac{dx}{dt} + 2y\frac{dy}{dt}\\ \frac{dD}{dt} &= \frac{x\frac{dx}{dt} + y\frac{dy}{dt}}{D}\\ &= \frac{25(25) + 60(60)}{65}\\ &= 65\ mph\end{aligned}$$

40. $26^2 = x^2 + y^2$ When $x = 10$, $y = \sqrt{26^2 - 10^2} = 24$

$$\begin{aligned}0 &= 2x\frac{dx}{dt} + 2y\frac{dy}{dt}\\ \frac{dy}{dt} &= -\frac{x}{y}\frac{dx}{dt}\\ &= -\frac{10}{24}(5)\\ &= 2.083\ \frac{ft}{sec}\end{aligned}$$

41. $tan\theta = \frac{h}{100}$

$$\begin{aligned}sec^2\,\theta\,\frac{d\theta}{dt} &= \frac{1}{100}\frac{dh}{dt}\\ \frac{dh}{dt} &= 100\ sec^2\,\theta\,\frac{d\theta}{dt}\\ &= 100\ sec^2(\frac{\pi}{6})(0.1)\\ &= 13.3333\ \frac{m}{min}\end{aligned}$$

42. $17^2 = x^2 + y^2$

a) When $x = 8$, $y = \sqrt{17^2 - 8^2} = 15$

$$\begin{aligned}0 &= 2x\frac{dx}{dt} + 2y\frac{dy}{dt}\\ \frac{dy}{dt} &= -\frac{x}{y}\frac{dx}{dt}\\ &= -\frac{8}{15}(2)\\ &= 1.06667\ \frac{ft}{sec}\end{aligned}$$

b) $cos\ \theta = \frac{x}{17}$
When $x = 8$, $\theta = cos^{-1}\frac{8}{17} = 1.0808\ rad$

$$\begin{aligned}-sin\ \theta\frac{d\theta}{dt} &= \frac{1}{17}\frac{dx}{dt}\\ \frac{d\theta}{dt} &= -\frac{1}{17\ sin\ \theta}\frac{dx}{dt}\\ &= -\frac{1}{17\ sin(1.0808)}(2)\\ &= 0.1333\ \frac{rad}{sec}\end{aligned}$$

43. **a)** $R^2 = x^2 + 9$
When $R = 10$, $x = \sqrt{100 - 9} = \sqrt{91}$

$$\begin{aligned}2R\,\frac{dR}{dt} &= 2x\,\frac{dx}{dt}\\ \frac{dx}{dt} &= \frac{R}{x}\frac{dR}{dt}\\ &= \frac{10}{\sqrt{91}}(2)\\ &= 2.0966\ \frac{ft}{sec}\end{aligned}$$

b) $sin\ \theta = \frac{3}{R}$
When $R = 10$, $\theta = sin^{-1}(\frac{3}{10}) = 0.30469\ rad$

$$\begin{aligned}cos\ \theta\frac{d\theta}{dt} &= -\frac{3}{R^2}\frac{dR}{dt}\\ \frac{d\theta}{dt} &= -\frac{3}{R^2\ cos\ \theta}\frac{dR}{dt}\\ &= -\frac{3}{100\ cos(0.30469)}(2)\\ &= -\frac{3}{5\ \sqrt{91}}\\ &= -0.0629\ \frac{rad}{sec}\end{aligned}$$

44. When the car passes the police car $\theta = 0$

$$\begin{aligned} tan\ \theta &= \frac{x}{20} \\ sec^2\ \theta \frac{d\theta}{dt} &= \frac{1}{20}\frac{dx}{dt} \\ \frac{d\theta}{dt} &= \frac{cos^2\ \theta}{20}\frac{dx}{dt} \\ &= \frac{1}{20}(75) \\ &= 3.75\ \frac{rad}{hour} \end{aligned}$$

45. $xy + x - 2y = 4$

$$\begin{aligned} x\frac{dy}{dx} + y + 1 - 2\frac{dy}{dx} = 0 & \\ (x-2)\frac{dy}{dx} &= -1 - y \\ \frac{dy}{dx} &= \frac{y+1}{2-x} \\ \frac{d^2y}{dx^2} &= \frac{(2-x)(\frac{dy}{dx}) + (y+1)}{(2-x)^2} \\ &= \frac{(2-x)\frac{y+1}{2-x} + y + 1}{(2-x)^2} \\ &= \frac{(y+1)+(y+1)}{(2-x)^2} \\ &= \frac{2y+2}{(2-x)^2} \end{aligned}$$

46. $y^2 - xy + x^2 = 5$

$$\begin{aligned} 2y\frac{dy}{dx} - x\frac{dy}{dx} - y + 2x &= 0 \\ (2y-x)\frac{dy}{dx} &= y - 2x \\ \frac{dy}{dx} &= \frac{y-2x}{2y-x} \\ \frac{(2y-x)(\frac{dy}{dx}-2) - (y-2x)(2(\frac{dy}{dx})-1)}{(2y-x)^2} &= \frac{d^2y}{dx^2} \\ &= \frac{3x\frac{dy}{dx} - 3y}{(2y-x)^2} \\ &= -\frac{6x^2+6y^2}{(2y-x)^3} \end{aligned}$$

47. $x^2 - y^2 = 5$

$$\begin{aligned} 2x - 2y\frac{dy}{dx} &= 0 \\ \frac{dy}{dx} &= \frac{x}{y} \\ \frac{d^2y}{dx^2} &= \frac{y - x\frac{dy}{dx}}{y^2} \\ &= \frac{y - \frac{x^2}{y}}{y^2} \\ &= \frac{y^2-x^2}{y^3} \end{aligned}$$

48. $x^3 - y^3 = 8$

$$\begin{aligned} 3x^2 - 3y^2\frac{dy}{dx} &= 0 \\ \frac{dy}{dx} &= \frac{x^2}{y^2} \\ \frac{d^2y}{dx^2} &= \frac{2xy^2 - 2x^2y\frac{dy}{dx}}{y^4} \\ &= \frac{2xy^3 - 2x^4}{y^5} \end{aligned}$$

49. Left to the student (answers vary)

50. Left to the student (answers vary)

51.

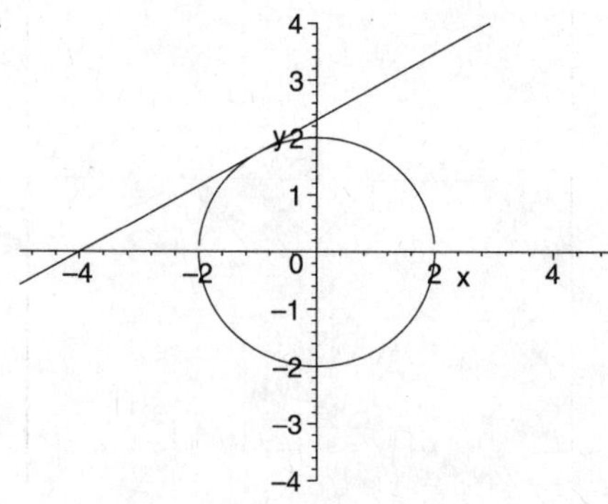

52.

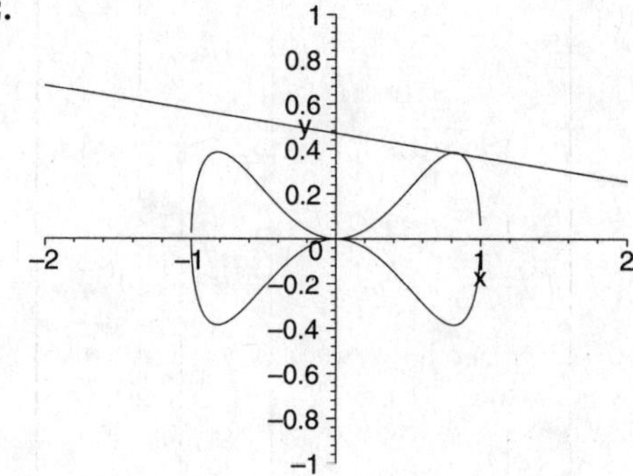

53.

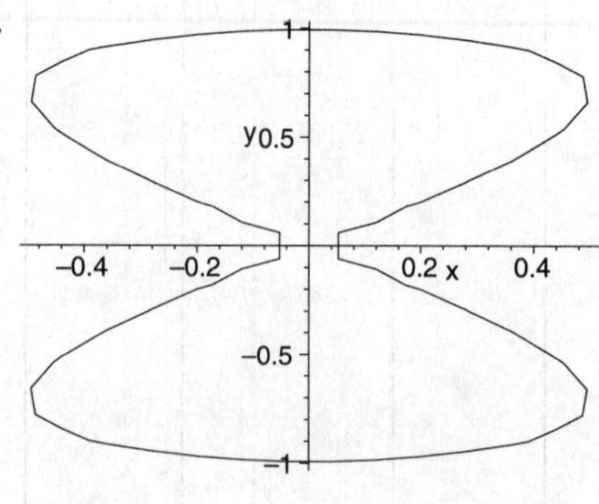

54.

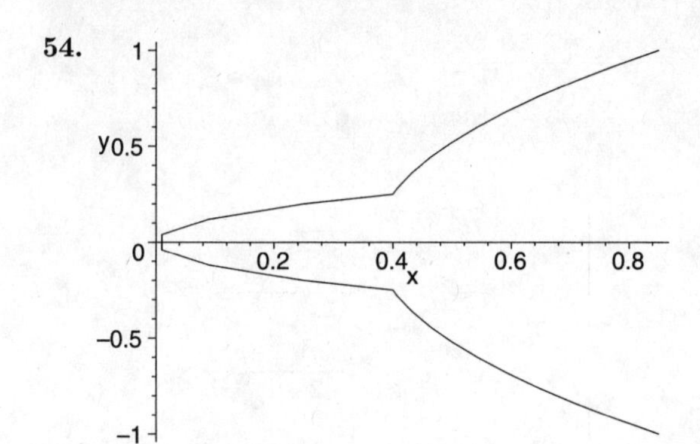

55.

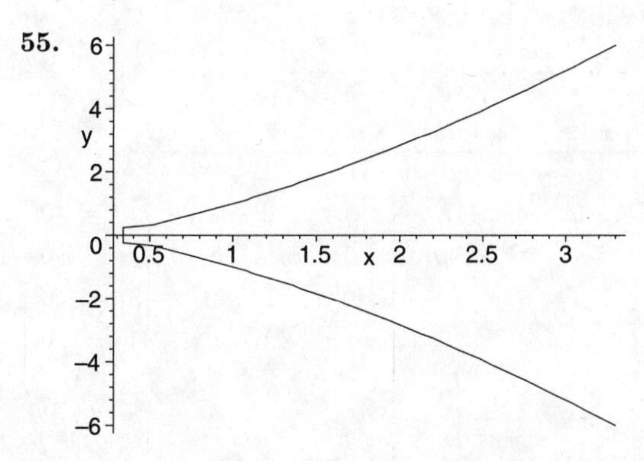

Chapter Review Exercises

1. $f(x) = 3 - 2x - x^2$

$$\begin{aligned} f'(x) &= -2 - 2x \\ -2 - 2x &= = 0 \\ x &= -1 \\ f(-1) &= 3 + 2 - 1 = 4 \\ f''(x) &= -2 \end{aligned}$$

The function has a relative maximum at $(-1, 4)$

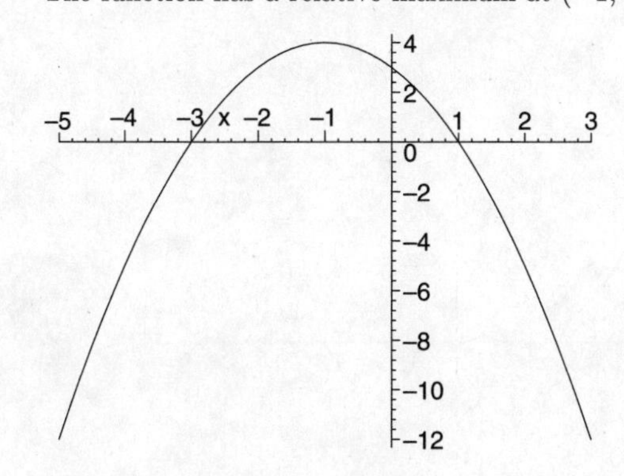

2. $f(x) = x^4 - 2x^2 + 3$

$$f'(x) = 4x^3 - 4x$$

$$\begin{aligned} 4x^3 - 4x &= 0 \\ 4x(x^2 - 1) &= 0 \\ x &= 0 \\ f(0) &= 0 - 0 + 3 = 3 \\ x &= 1 \\ f(1) &= 1 - 2 + 3 = 2 \\ x &= -1 \\ f(-1) &= 1 - 2 + 3 = 2 \\ f''(x) &= 12x^2 - 4 \\ 12x^2 - 4 &= 0 \\ x &= \frac{1}{\sqrt{3}} \end{aligned}$$

$$f(\frac{1}{\sqrt{3}})7 = \tfrac{1}{9} - \tfrac{2}{3} + 3 = \tfrac{22}{9}$$

$$x = -\frac{1}{\sqrt{3}}$$

$$f(-\frac{1}{\sqrt{3}})7 = \tfrac{1}{9} - \tfrac{2}{3} + 3 = \tfrac{22}{9}$$

The function has relative minima at $(1, 2)$ and $(-1, 2)$ and a relative maximum at $(0, 3)$

The function has inflection points at $\left(\frac{1}{\sqrt{3}}, \frac{22}{9}\right)$ and $\left(-\frac{1}{\sqrt{3}}, \frac{22}{9}\right)$

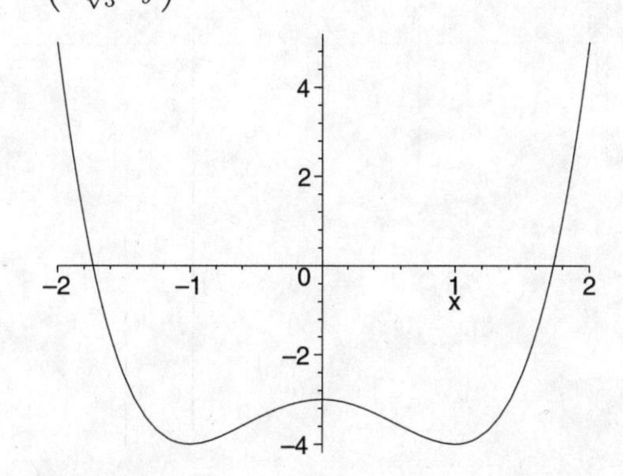

3. $f(x)\dfrac{-8x}{x^2 + 1}$

$$\begin{aligned} f'(x) &= \frac{-8(x^2 + 1) + 16x^2}{(x^2 + 1)^2} \\ &= \frac{8x^2 - 8}{(x^2 + 1)^2} \\ \frac{8x^2 - 8}{(x^2 + 1)^2} &= 0 \\ 8x^2 - 8 &= 0 \\ x^2 &= 1 \\ x &= 1 \\ f(1) &= \frac{-8}{2} = -4 \\ x &= -1 \\ f(-1) &= \frac{(8}{2) = 4} \\ f''(x) &= \frac{(x^2 + 1)^2(16x) - 8x^2(2x)}{(x^2 + 1)^4} \\ &= \frac{-16x(x^2 - 3)}{(x^2 + 1)^3} \\ \frac{-16x(x^2 - 3)}{(x^2 + 1)^3} &= 0 \\ -16x(x^2 - 3) &= 0 \end{aligned}$$

$$\begin{aligned} x &= 0 \\ f(0) &= 0 \\ x &= \sqrt{3} \\ f(\sqrt{3}) &= \frac{-8\sqrt{3}}{4} = -2\sqrt{3} \\ x &= -\sqrt{3} \\ f(-\sqrt{3}) &= \frac{8\sqrt{3}}{4} = 2\sqrt{3} \end{aligned}$$

The function has relative maximum at $(-1, 4)$ and a relative mainimum at $(1, -4)$
The function has inflection points at $(-\sqrt{3}, 2\sqrt{3})$, $(0, 0)$, and $(\sqrt{3}, -2\sqrt{3})$

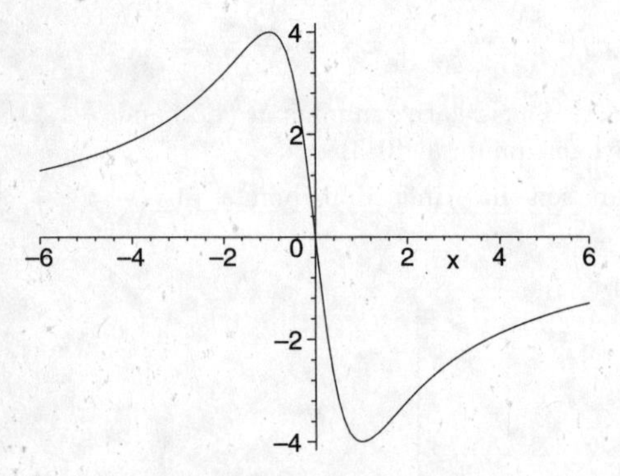

4. $f(x) = 4 + (x-1)^3$

$$\begin{aligned} f'(x) &= 3(x-1)^2 \\ 3(x-1)^2 &= 0 \\ x &= 1 \\ f(1) &= 4 + 0 = 4 \end{aligned}$$

Note that the first derivative is always positive, which means that the function does not have relative extrema points.

$$\begin{aligned} f''(x) &= 6x - 6 \\ 6x - 6 &= 0 \\ x &= 1 \end{aligned}$$

The function has an inflection point at $(1, 4)$

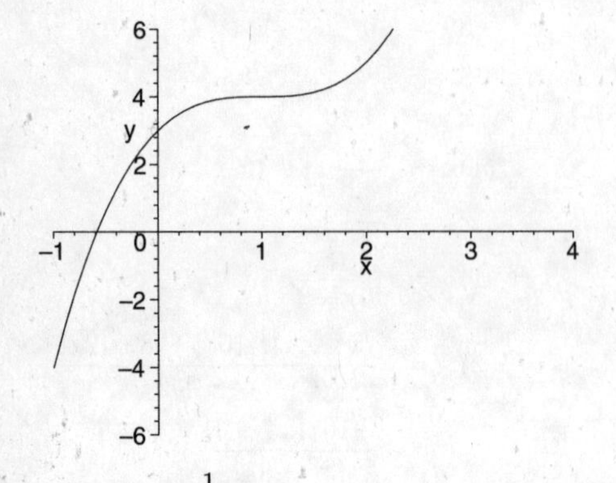

5. $f(x) = \dfrac{1}{2 \sin x + 7}$

$$f'(x) = \frac{2 \cos x}{(2 \sin x + 7)^2}$$

$$\begin{aligned} \frac{2 \cos x}{(2 \sin x + 7)^2} &= 0 \\ \cos x &= 0 \\ x &= \frac{\pi}{2} \\ f(\frac{\pi}{2}) &= \frac{1}{2+7} = \frac{1}{9} \\ x &= \frac{3\pi}{2} \\ f(\frac{3\pi}{2}) &= \frac{1}{-2+7} = \frac{1}{5} \end{aligned}$$

$$\begin{aligned} \frac{-2\sin x\,(2 \sin x + 7)^2 - 8 \cos^2 x(2 \sin x + 7)}{(2 \sin x + 7)^4} &= f''(x) \\ \frac{-2 \sin x\,(2 \sin x + 7)^2 - 8 \cos^2 x(2 \sin x + 7)}{(2 \sin x + 7)^4} &= 0 \\ -2 \sin x\,(2 \sin x + 7)^2 - 8 \cos^2 x(2 \sin x + 7) &= 0 \\ -4 \sin^2 x - 14 \sin x - 8 \cos^2 x &= 0 \\ 4 \sin^2 x - 14 \sin x - 8 &= 0 \\ (2 \sin x + 1)(\sin x - 4) &= 0 \\ \sin x &= \frac{-1}{2} \\ x &= \frac{7\pi}{6} \\ \frac{1}{-1+7} = \frac{1}{6} &= f(\frac{7\pi}{6}) \\ x &= \frac{11\pi}{6} \\ \frac{1}{-1+7} = \frac{1}{6} &= f(\frac{11\pi}{6}) \end{aligned}$$

The function has a relative maximum at $\left(\frac{3\pi}{2}, \frac{1}{5}\right)$ and a relative minimum at $\left(\frac{\pi}{2}, \frac{1}{9}\right)$
The function has an inflection point at $\left(\frac{7\pi}{6}, \frac{1}{6}\right)$ and $\left(\frac{11\pi}{6}, \frac{1}{6}\right)$

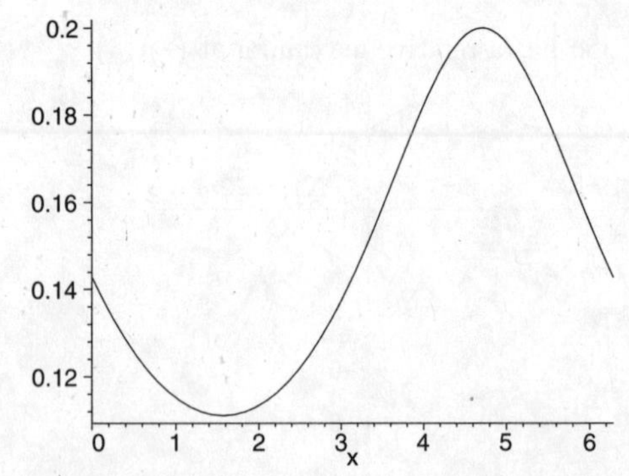

6. $f(x) = x^{2/3}$

$$\begin{aligned} f'(x) &= \frac{2}{3x^{1/3}} \\ f''(x) &= -\frac{2}{9x^{4/3}} \end{aligned}$$

The function has an inflection point at $(0, 0)$

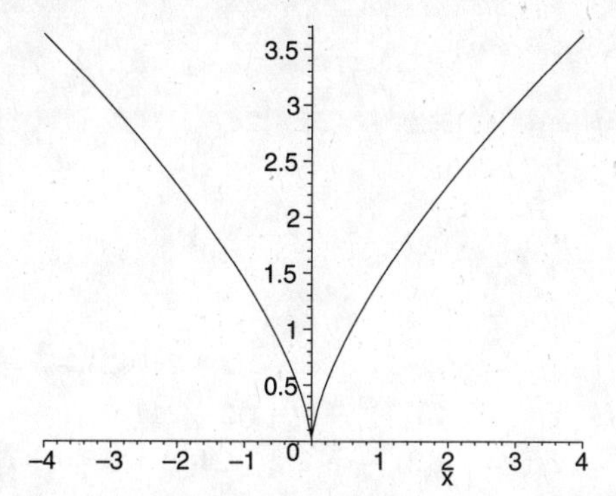

7. $f(x) = sin^2 x + 2\ sin\ x$

$$
\begin{aligned}
f'(x) &= 2\ sin\ x\ cos\ x + 2\ cos\ x \\
&= 2\ cos\ x(sin\ x + 1) \\
2\ cos\ x(sin\ x + 1) &= 0 \\
cos\ x &= 0 \\
x &= \frac{\pi}{2} \\
f(\frac{\pi}{2}) &= 1 + 2 = 3 \\
x &= \frac{3\pi}{2} \\
f(\frac{3\pi}{2}) &= 1 - 2 = -1 \\
sin\ x &= -1 \\
x\ \frac{3\pi}{2} f''(x) &= 2\ cos^x - 2\ sin^2\ x - 2\ sin\ x \\
2(1 - 2sin^2 x) - 2\ sin\ x &= 0 \\
4sin^2 x + 2\ sin\ x - 2 &= 0 \\
(2\ sin\ x - 1)^2 &= 0 \\
sin\ x &= \frac{1}{2} \\
x &= \frac{\pi}{6} \\
f(\frac{\pi}{6}) &= \frac{1}{4} + 1 = \frac{5}{4} \\
x &= \frac{5\pi}{6} \\
f(\frac{\pi}{6}) &= \frac{1}{4} + 1 = \frac{5}{4}
\end{aligned}
$$

The function has a relative maximum at $\left(\frac{\pi}{2}, 3\right)$ and a relative minimum at $\left(\frac{3\pi}{2}, -1\right)$ The function has an inflection points at $\left(\frac{\pi}{6}, \frac{5}{4}\right)$ and $\left(\frac{5\pi}{6}, \frac{5}{4}\right)$

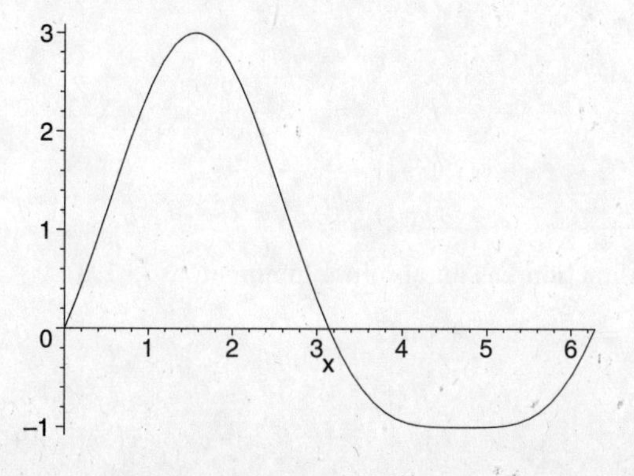

8. $f(x) = 2x + cos\ 2x$

$$
\begin{aligned}
f'(x) &= 2 - 2\ sin\ 2x \\
2 - 2\ sin\ 2x &= 0 \\
sin\ 2x &= 1 \\
2x &= \frac{\pi}{2} \\
x &= \frac{\pi}{4} \\
x &= \frac{5\pi}{4} \\
f(\frac{\pi}{4}) &= \frac{\pi}{2} + 0 = \frac{\pi}{2} \\
f(\frac{5\pi}{4}) &= \frac{5\pi}{2} + 0 = \frac{5\pi}{2} \\
f''(x) &= -4\ cos\ 2x \\
-4\ cos\ 2x &= 0 \\
2x &= \frac{\pi}{2} \\
x &= \frac{\pi}{4} \\
x &= \frac{3\pi}{4} \\
x &= \frac{5\pi}{4} \\
x &= \frac{7\pi}{4} \\
f(\frac{3\pi}{4}) &= \frac{3\pi}{2} + 0 = \frac{3\pi}{2} \\
f(\frac{7\pi}{4}) &= \frac{7\pi}{2} + 0 = \frac{7\pi}{2}
\end{aligned}
$$

The function has an inflection point at $\left(\frac{\pi}{4}, \frac{\pi}{2}\right)$, $\left(\frac{3\pi}{4}, \frac{3\pi}{2}\right)$, $\left(\frac{5\pi}{4}, \frac{5\pi}{2}\right)$, and $\left(\frac{7\pi}{4}, \frac{7\pi}{2}\right)$

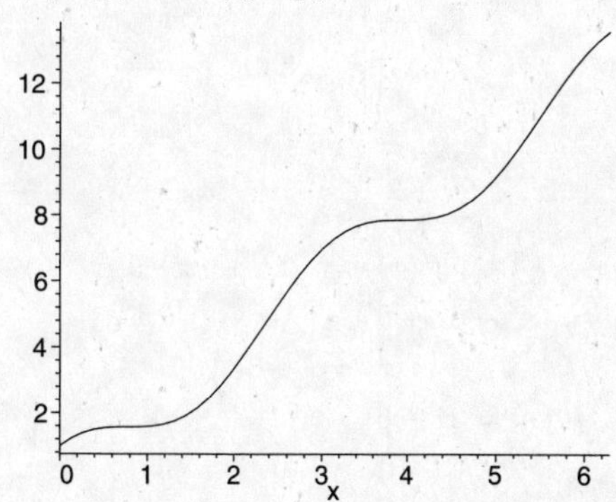

9.

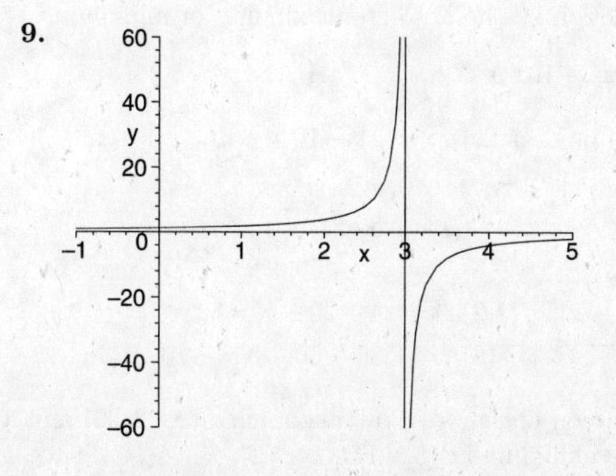

10.

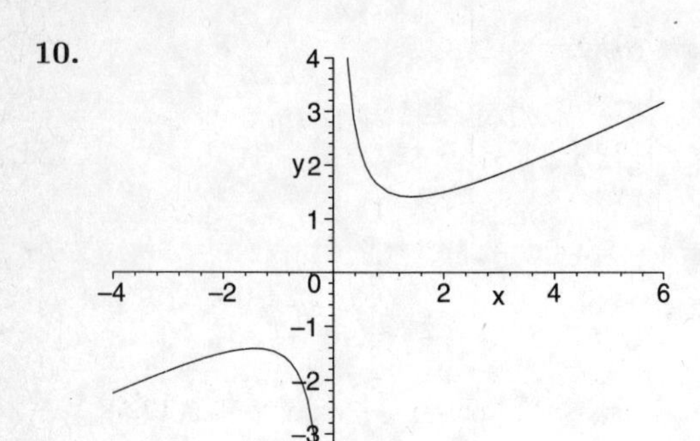

11.

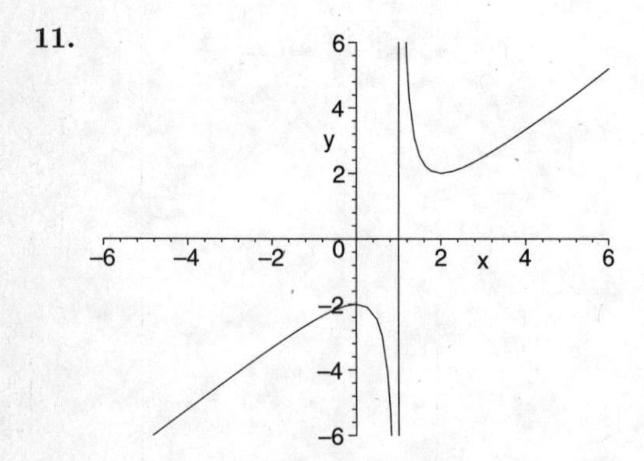

12.

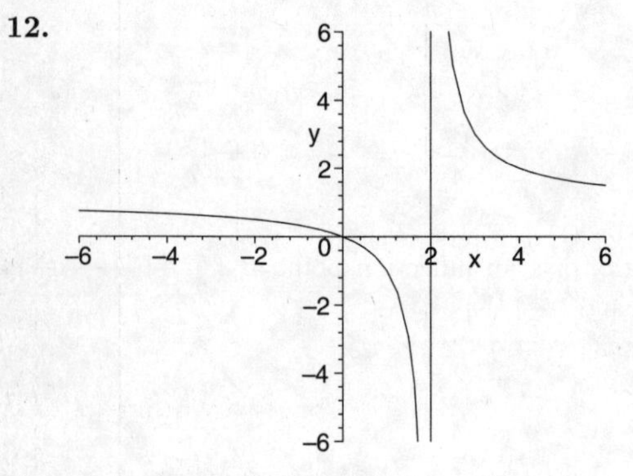

13. $f(x) = \frac{1}{3}x^3 + 3x^2 + 9x + 2$

$$\begin{aligned} f'(x) &= x^2 + 6x + 9 \\ x^2 + 6x + 9 &= 0 \\ x &= -3 \\ f''(x) = 2x + 6 & \\ f''(-3) &= 0 \end{aligned}$$

The function has no absolute maximums or minimums.

14. $f(x) = x^2 - 10x + 8$

$$\begin{aligned} f'(x) &= 2x - 10 \\ 2x - 10 &= 0 \\ x &= 5 \\ f(5) &= 25 - 50 + 8 = -17 \\ f(-2) &= 4 + 20 + 8 = 32 \\ f(6) &= 36 - 60 + 8 = -16 \end{aligned}$$

The function has an absolute maximum at $(-2, 32)$ and a absolute minimum at $(5, -17)$

15. $f(x) = 4x^3 - 6x^2 - 24x + 5$

$$\begin{aligned} f'(x) &= 12x^2 - 12x - 24 \\ 12x^2 - 12x - 24 &= 0 \\ x^2 - x - 2 &= 0 \\ x &= -1 \\ x &= 2 \\ f(-2) &= 4(-2)^3 - 6(-2)2 - 24(-2) + 5 = -3 \\ f(-1) &= 4(-1)^3 - 6(-1)2 - 24(-1) + 5 = 19 \\ f(2) &= 4(2)^3 - 6(2)2 - 24(2) + 5 = -35 \\ f(3) &= 4(3)^3 - 6(3)2 - 24(3) + 5 = -13 \end{aligned}$$

The function has an absolute maximum at $(-1, 19)$ and a absolute minimum at $(2, -35)$

16. $f(x) = \dfrac{sin\ x}{2 + sin\ x + cos\ x}$

$$\begin{aligned} \frac{cos\ x(2 + sin\ x + cos\ x) - sin\ x(cos\ x - sin\ x)}{(2 + sin\ x + cos\ x)^2} &= f'(x) \\ \frac{2cos\ x + 1}{(2 + sin\ x + cos\ x)^2} &= \\ \frac{2cos\ x + 1}{(2 + sin\ x + cos\ x)^2} &= 0 \\ cos\ x &= -\frac{1}{2} \\ x &= \frac{2\pi}{3} \\ x &= \frac{4\pi}{3} \\ f(0) &= 0 \\ f(\frac{2\pi}{3}) &= 0.366 \\ f(\frac{4\pi}{3}) &= -1.366 \\ f(2\pi) &= 0 \end{aligned}$$

The function has an absolute maximum at $\left(\frac{2\pi}{3}, 0.366\right)$ and an absolute minimum at $\left(\frac{4\pi}{3}, -1.366\right)$

17. $f(x) = x^2 - \dfrac{2}{x}$

$$\begin{aligned} f'(x) &= 2x + \frac{2}{x^2} \\ 2x + \frac{2}{x^2} &= 0 \\ 2x^3 &= -2 \\ x &= -1 \\ f''(x) &= 2 - \frac{2}{x^3} \\ f''(-1) &= 4 > 0 \\ f(-1) &= 1 + 2 = 3 \end{aligned}$$

The function has an absolute minimum at $(-1, 3)$

18. $f(x) = 4\ sin^3x + 3\ sin^2x$

$$\begin{aligned} f'(x) &= 12\ sin^2x\ cos\ x + 6\ sin\ xcos\ x \\ &= 6\ sin\ x\ cos\ x(2sin\ x + 1) \end{aligned}$$

$$\begin{aligned} 6\,sin\ x\ cos\ x(2sin\ x+1) &= 0 \\ x &= 0 \to f=0 \\ x &= \frac{\pi}{2} \to f=7 \\ x &= \pi \to f=0 \\ x &= \frac{7\pi}{6} \to f=0.25 \\ x &= \frac{3\pi}{2} \to f=0.25 \\ x &= 2\pi \to f=0 \end{aligned}$$

The function has an absolute maximum at $(\frac{\pi}{2}, 7)$ and an absolute minimum at $(\frac{3\pi}{2}, -1)$

19. $f(x) = 9\ tan^2x + cot^2x$

$$\begin{aligned} 18\ tan\ x\ sec^2x - 2\ cot\ x\ csc^2\ x &= f'(x) \\ 18\frac{sin\ x}{cos^3x} - 2\frac{cos\ x}{sin^3\ x} &= \\ 18\frac{sin\ x}{cos^3x} - 2\frac{cos\ x}{sin^3\ x} &= 0 \\ 9\ sin^4\ x - cos^4\ x &= 0 \\ (3\ sin^2x - cos^2x)(3\ sin^2 + cos^2x) &= 0 \\ x &= \frac{\pi}{6} \\ f''(\frac{\pi}{6}) &= 0.125 > 0 \\ f(\frac{\pi}{6}) &= 3+3=6 \end{aligned}$$

The function has an absolute minimum at $(\frac{\pi}{6}, 6)$

20. $f(x) = 5x^2 + \frac{5}{x^2}$

$$\begin{aligned} f'(x) &= 10x - \frac{10}{x^3} \\ 10x - \frac{10}{x^3} &= 0 \\ 10x^4 - 10 &= 0 \\ x &= 1 \\ f''(x) &= 10 + \frac{30}{x^4} \\ f''(1) &= 40 > 0 \\ f(1) &= 5+5=10 \end{aligned}$$

The function has an absolute minimun at $(1, 10)$

21. $f(x) = cot\ x - 2\ csc\ x$

$$\begin{aligned} -csc^2x + 2\ csc\ x\ cot\ x &= f'(x) \\ -csc\ x(csc\ x - 2\ cot\ x) &= \\ -csc^2x(csc\ x - 2\ cot\ x) &= 0 \\ x &= \frac{\pi}{3} \\ 2\ csc^2x\ cot\ x - 2\ csc^3x - 2\ csc\ x\ cot^2x &= f''(x) \\ -1.925 &= f''(\frac{\pi}{3}) < 0 \\ f(\frac{\pi}{3}) &= \frac{1}{\sqrt{3}} - 2\frac{2}{\sqrt{3}} \\ &= -\sqrt{3} \end{aligned}$$

The function has an absolute minimum at $(\frac{\pi}{3}, -\sqrt{3})$

22. $f(x) = -x^2 + 5x + 7$

$$\begin{aligned} f'(x) &= -2x+5 \\ -2x+5 &= 0 \\ x &= 2.5 \\ f''(x) &= -2 < 0 \\ f(2.5) &= -6.25 + 12.25 + 7 \\ &= 13.25 \end{aligned}$$

The function has an absolute maximum at $(2.5, 13.5)$

23. $Q = xy$ and $x + y = 60$

$$\begin{aligned} Q &= x(60-x) \\ &= 60x - x^2 \\ Q' &= 60 - 2x \\ 60-2x &= 0 \\ x &= 30 \\ y &= 60 - 30 = 30 \end{aligned}$$

The two number are 30 and 30

24. $Q = x^2 - 2y^2$ and $x - 2y = 1$

$$\begin{aligned} Q &= (2y+1)^2 - 2y^2 \\ &= 4y^2 + 4y + 1 - 2y^2 \\ &= 2y^2 + 4y + 1 \\ Q' &= 4y + 4 \\ 4y+4 &= 0 \\ y &= -1 \\ x &= 2(-1) + 1 = -1 \\ Q &= 1 - 2(1) = -1 \end{aligned}$$

The minimum value is -1

25. $x^2y = 2500$ and $C = 2x^2 + 3x^2 + 4xy = 5x^2 + 4xy$

$$\begin{aligned} C &= 5x^2 + 4x\left(\frac{2500}{x^2}\right) \\ &= 5x^2 + \frac{10000}{x} \\ C' &= 10x - \frac{10000}{x^2} \\ 10x + \frac{10000}{x} &= 0 \\ 10x^3 + 10000 &= 0 \\ x &= 10 \\ y &= \frac{2500}{100} = 25 \end{aligned}$$

The dimensions of the box are $10 \times 10 \times 25$ feet

26. $f(x) = \frac{x}{x-1}$

$$\begin{aligned} f'(x) &= \frac{x-1-x}{(x-1)^2} \\ &= \frac{-1}{(x-1)^2} \\ f(2) &= \frac{2}{1} = 2 \end{aligned}$$

$$\begin{aligned} f'(2) &= \frac{-1}{1} = -1 \\ L(x) &= f(2) + f'(2)(x-2) \\ &= 2 - (x-2) \\ &= 4 - x \end{aligned}$$

27. $f(x) = \dfrac{sin\ x}{1 + cos\ x}$

$$\begin{aligned} f'(x) &= \frac{cos\ x(1 + cos\ x) + sin^2 x}{(1 + cos\ x)^2} \\ f(0) &= 0 \\ f'(0) &= \frac{2+0}{4} = \frac{1}{2} \\ L(x) &= f(0) + f'(0)(x-0) \\ &= 0\frac{1}{2}x \\ &= \frac{x}{2} \end{aligned}$$

28. $f(x) = \sqrt{x}$ and $f'(x) = \dfrac{1}{2\sqrt{x}}$

$$\begin{aligned} L(63) &= f(64) + f'(64)(63-64) \\ &= 8 + \frac{1}{16}(-1) \\ &= 7.9375 \end{aligned}$$

29. $f(x) = x^3 - 5x + 3$ and $f'(x) = 3x^2 - 5$

$$\begin{aligned} x_1 &= 1 \\ x_2 &= 1 - \frac{-1}{-2} = 0.5 \\ x_3 &= 0.5 - \frac{0.625}{-4.25} = 0.6471 \\ x_4 &= 0.6471 - \frac{0.03562}{-3.744} = 0.6566 \\ x_5 &= 0.6566 - \frac{0.00018}{-3.707} = 0.6566 \\ x_6 &= 0.6566 - \frac{0}{-3.707} = 0.6566 \end{aligned}$$

$x = 0.6566$

30.

$$\begin{aligned} x_1 &= 3 \\ x_2 &= 3 - \frac{-0.7589}{-1.59} = 2.5227 \\ x_3 &= 2.5227 - \frac{-0.0563}{-1.319} = 2.4800 \\ x_4 &= 2.4800 - \frac{-0.0007}{-1.285} = 2.4796 \\ x_5 &= 2.4976 - \frac{-0.0002}{-1.285} = 2.4795 \\ x_6 &= 2.4795 - \frac{0}{-1.285} = 2.4795 \end{aligned}$$

$x = 2.4795$

31. $sin(y^2) + x = \sqrt{y}$

$$\begin{aligned} 2y\ cos(y^2)\frac{dy}{dx} + 1 &= \frac{1}{2\sqrt{y}}\frac{dy}{dx} \\ \left(2y\ cos(y^2) - \frac{1}{2\sqrt{y}}\right)\frac{dy}{dx} &= -1 \\ \left(\frac{4y^{3/2}\ cos(y^2) - 1}{2\sqrt{y}}\right)\frac{dy}{dx} &= -1 \\ \frac{dy}{dx} &= -\frac{2\sqrt{y}}{4y^{3/2}\ cos(y^2)} \end{aligned}$$

32. $2x^3 + 2y^3 = -9xy$

$$\begin{aligned} 6x^2 + 6y^2\frac{dy}{dx} &= -9x\frac{dy}{dx} - 9y \\ (6y^2 + 9x)\frac{dy}{dx} &= -(9y + 6x^2) \\ \frac{dy}{dx} &= -\frac{9y + 6x^2}{6y^2 + 9x} \\ &= -\frac{2x^2 + 3y}{3x + 2y^2} \end{aligned}$$

When $x = -1$ and $y = -2$

$$\begin{aligned} \frac{dy}{dx} &= -\frac{2(-1)^2 + 3(-2)}{3(-1) + 2(-2)^2} \\ &= \frac{4}{5} \end{aligned}$$

33. $x^2 + y^2 = 25^2 = 625$
When $x = 7$, $y = \sqrt{625 - 49} = 24$

$$\begin{aligned} 2x\frac{dx}{dt} + 2y\frac{dy}{dt} &= 0 \\ 2(7)(6) + 2(24)\frac{dy}{dt} &= 0 \\ \frac{dy}{dt} &= -\frac{84}{48} \\ &= -1.75\ \frac{ft}{sec} \end{aligned}$$

34. $f(x) = (x-3)^{2/5}$

$$f'(x) = \frac{2}{5}(x-3)^{-3/5}$$

$x = 3$ is a critical value of the function

$$\begin{aligned} f(3) &= 0 \\ f''(x) &= -\frac{6}{5}(x-3)^{-8/5} \end{aligned}$$

Since the second derivative is always positive for values in the domain, the function has an absolute minimum at $(3, 0)$

35. $(x-y)^4 + (x+y)^4 = x^6 + y^6$

$$\begin{aligned} 4(x-y)^3(1 - \frac{dy}{dx}) + 4(x+y)^3(1 + \frac{dy}{dx}) &= 6x^5 + 6y^5 \\ 8x^3 + 24xy^2 + 24x^2y\frac{dy}{dx} + 8y^3\frac{dy}{dx} &= 6x^5 + 6y^5\frac{dy}{dx} \end{aligned}$$

$$\begin{aligned}(24x^2y + 8y^3 - 6y^5)\frac{dy}{dx} &= 6x^5 - 8x^3 + 24xy^2 \\ \frac{dy}{dx} &= \frac{6x^5 - 8x^3 + 24xy^2}{24x^2y + 8y^3 - 6y^5} \\ &= \frac{3x^5 - 4x^3 + 12xy^2}{12x^2y + 4y^3 - 3y^5}\end{aligned}$$

36. $y = x^4 - 8x^3 - 270x^2$

$$\begin{aligned}\frac{dy}{dx} &= 4x^3 - 24x^2 + 540x \\ &= 4x(x-25)(x+9) \\ 4x(x-15)(x+9) &= 0 \\ x &= 0 \rightarrow y = 0 \\ x &= 15 \rightarrow y = -37125 \\ x &= -9 \rightarrow y = -9477\end{aligned}$$

The function has a relative maximum at $(0, 0)$ and a relative minimum at $(15, -37125)$

37. The function has a relative maximum at $(-1.714, 37.445)$ and a relative minimum at $(1.714, -37.445)$

38. The function has a relative maximum at $(0, 1.08)$ and a relative minimum at $(-3, -1)$ and at $(3, -1)$

39. **a)**
- LINEAR: $y = -124.6181 + 6.9982x$
- QUADRATIC: $-53.5148 + 2.8812x + 0.0439x^2$
- CUBIC: $5.2770 - 11.3593x + 0.4796x^2 - 0.0033x^3$
- QUARTIC: $-0.246 - 0.841x - 0.997x^2 + 0.0067x^3 - 0.00006x^4$

b) Left to the student

c) Left to the student

d) The quartic function has a maximum of 466.325 at age close to 79

Chapter 3 Test

1. $f(x) = x^2 - 4x - 5$

$$\begin{aligned}f'(x) &= 2x - 4 \\ 2x - 4 &= 0 \\ x &= 2 \\ f''(x) &= 2 > 0 \\ f(2) &= 4 - 8 - 5 = -9\end{aligned}$$

The function has a relative minimum at $(2, -9)$

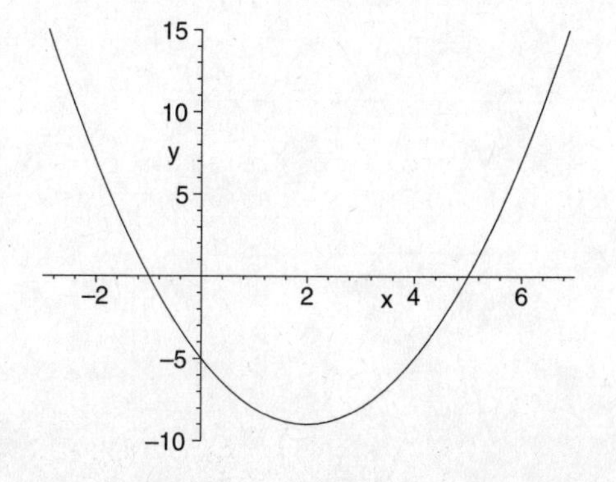

2. $f(x) = 2x^4 - 4x^2 + 1$

$$\begin{aligned}f'(x) &= 8x^3 - 8x \\ 8x^3 - 8x &= 0 \\ x &= 0 \rightarrow f = 1 \\ x &= -1 \rightarrow f = -1 \\ x &= 1 \rightarrow f = -1 \\ f''(x) &= 24x^2 - 8 \\ 24x^2 - 8 &= 0 \\ x &= -\frac{1}{\sqrt{3}} \rightarrow f = \frac{7}{9} \\ x &= \frac{1}{\sqrt{3}} \rightarrow f = \frac{7}{9}\end{aligned}$$

The function has a relative maximun at $(0, 1)$ and relative minimums at $(-1, -1)$ and $(1, -1)$

The function has inflection points at $\left(-\frac{1}{\sqrt{3}}, \frac{7}{9}\right)$ and $\left(\frac{1}{\sqrt{3}}, \frac{7}{9}\right)$

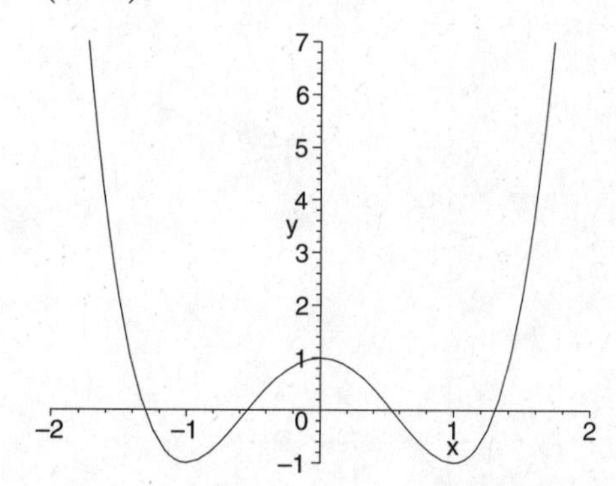

3. $f(x) = (x-2)^{2/3} - 4$

$$\begin{aligned}f'(x) &= \frac{2}{3}(x-2)^{-1/3} \\ f''(x) &= -\frac{2}{9}(x-2)^{-4/3}\end{aligned}$$

$f'(x)$ and $f''(x)$ are undefined at $x = 2$, the first derivative changes signs around the critical value. The function has a relative minimum at $(2, -4)$

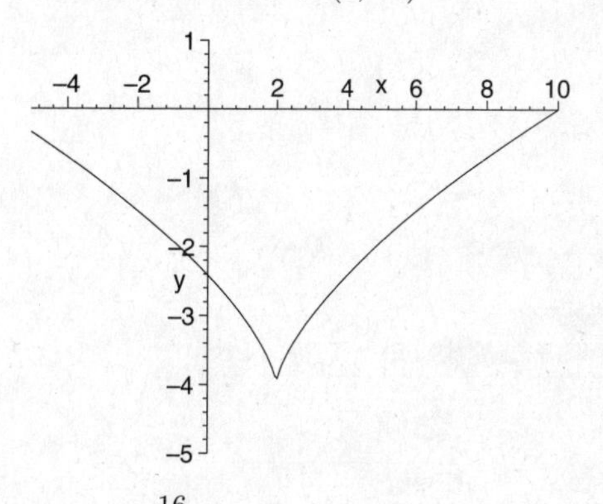

4. $f(x) = \dfrac{16}{x^2 + 4}$

$$f'(x) = \frac{32x}{(x^2+4)^2}$$

$$\begin{aligned}
\frac{32x}{(x^2+4)^2} &= 0\\
x &= 0\\
f(0) &= \frac{16}{0+4} = 4\\
f''(x) &= \frac{32(x^2+4)^2 - 64x(x^2+4)(2x)}{(x^2+4)^4}\\
&= \frac{-96x^4+512}{(x^2+4)^4}\\
\frac{-96x^4+512}{(x^2+4)^4} &= 0\\
x &= \pm\frac{2}{\sqrt{3}}\\
f''(\frac{2}{\sqrt{3}}) &= \frac{16}{\frac{16}{3}} = 3\\
f''(-\frac{2}{\sqrt{3}}) &= \frac{16}{\frac{16}{3}} = 3
\end{aligned}$$

The function has a relative maximum at $(0,4)$ and inflections points at $\left(\frac{2}{\sqrt{3}},3\right)$ and $\left(-\frac{2}{\sqrt{3}},3\right)$

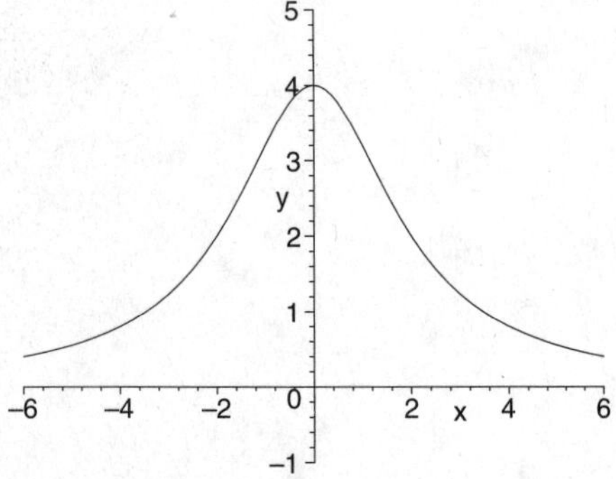

5. $f(x) = \dfrac{sin\ x}{2\ sin\ x - 7}$

$$\begin{aligned}
\frac{2\ sin\ x\ cos\ x - 7\ cos\ x - 2\ sin\ x\ cos\ x}{(2\ sin\ x-7)^2} &= f'(x)\\
\frac{-7\ cos\ x}{(2\ sin\ x-7)^2} &=\\
\frac{-7\ cos\ x}{(2\ sin\ x-7)^2} &= 0\\
x &= \frac{\pi}{2}\\
\frac{1}{2-7} = \frac{-1}{5} &= f(\frac{\pi}{2})\\
x &= \frac{3\pi}{2}\\
\frac{-1}{-2-7} = \frac{1}{9} &= f(\frac{3\pi}{2})\\
\frac{14\ sin^2x - 49\ sin\ x + 28\ cos^2x}{(2\ sin\ x-7)^3} &= f''(x)\\
\frac{-7(2\ sin^2x + 7\ sin\ x - 4)}{(2\ sin\ x-7)^3} &=\\
\frac{-7(2\ sin^2x + 7\ sin\ x - 4)}{(2\ sin\ x-7)^3} &= 0\\
sin\ x &= \frac{1}{2}
\end{aligned}$$

$$\begin{aligned}
x &= \frac{\pi}{6}\\
f(\frac{\pi}{6}) &= \frac{(0.5}{1-7) = \frac{-1}{12}}\\
x &= \frac{5\pi}{6}\\
f(\frac{5\pi}{6}) &= \frac{(0.5}{1-7) = \frac{-1}{12}}
\end{aligned}$$

The function has a relative maximum at $\left(\frac{3\pi}{2},\frac{1}{9}\right)$ and a relative minimum at $\left(\frac{\pi}{2},\frac{-1}{5}\right)$
The function has inflection points at $\left(\frac{\pi}{6},\frac{-1}{12}\right)$ and $\left(\frac{5\pi}{6},\frac{-1}{12}\right)$

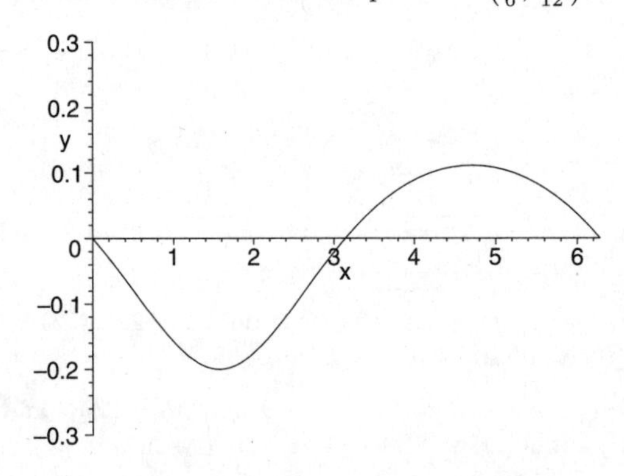

6. $f(x) = cos^2x - 2\ sin\ x$

$$\begin{aligned}
f'(x) &= -2\ cos\ x\ sin\ x - 2\ cos\ x\\
&= -2\ cos\ x(sin\ x - 1)\\
-2\ cos\ x(sin\ x-1) &= 0\\
x &= \frac{\pi}{2} \to f = -2\\
x &= \frac{3\pi}{2} \to f = 2\\
f''(x) &= 2\ sin^2x - 2\ cos^2\ x + 2\ sin\ x\\
&= 2(2\ sin\ x - 1)^2\\
2\ sin\ x - 1 &= 0\\
x &= \frac{\pi}{6} \to f = \frac{-1}{4}\\
x &= \frac{5\pi}{6} \to f = \frac{-1}{4}
\end{aligned}$$

The function has inflection points at $\left(\frac{\pi}{6},\frac{-1}{4}\right)$ and $\left(\frac{5\pi}{6},\frac{-1}{4}\right)$

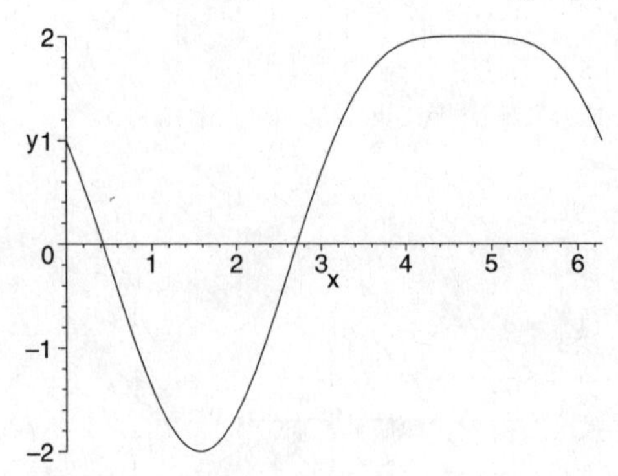

7. $f(x) = (x+2)^3$

$$f'(x) = 3(x+2)2 \geq 0$$

$$\begin{aligned} 3(x+2)^2 &= 0 \\ x &= -2 \\ f(-2) &= 0 \\ f''(x) &= 6(x+2) \\ 6(x+2) &= 0 \\ x &= -2 \end{aligned}$$

The function has an inflection point at $(-2, 0)$

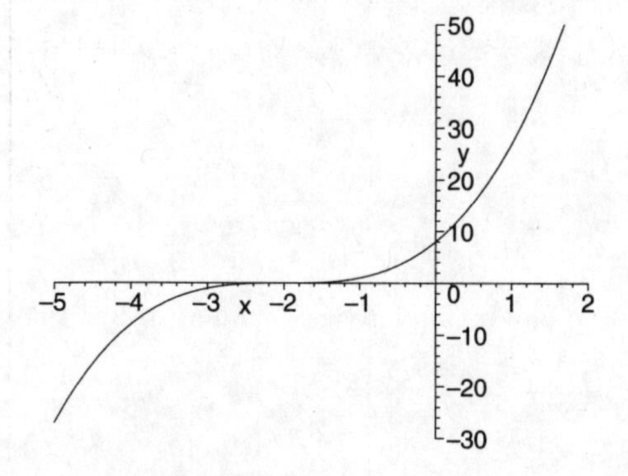

8. $f(x) = x\sqrt{9-x^2}$

$$\begin{aligned} f'(x) &= \frac{x^2}{\sqrt{9-x^2}} + \sqrt{9-x^2} \\ \frac{x^2}{\sqrt{9-x^2}} + \sqrt{9-x^2} &= 0 \\ x^2 &= 9 - x^2 \\ x &= \frac{3}{\sqrt{2}} \to f = \frac{9}{2} \\ x^2 &= 9 - x^2 \\ x &= \frac{-3}{\sqrt{2}} \to f = \frac{-9}{2} \\ f''(x) &= \frac{x}{\sqrt{9-x^2}} + \frac{x^3}{(9-x^2)^{3/2}} \\ \frac{x}{\sqrt{9-x^2}} + \frac{x^3}{(9-x^2)^{3/2}} &= 0 \\ x &= 0 \to f = 0 \end{aligned}$$

The function has a relative maximim at $\left(\frac{3}{\sqrt{2}}, \frac{9}{2}\right)$ and a relative minimum at $\left(\frac{-3}{\sqrt{2}}, \frac{-9}{2}\right)$

The function has an inflection point at $(0, 0)$

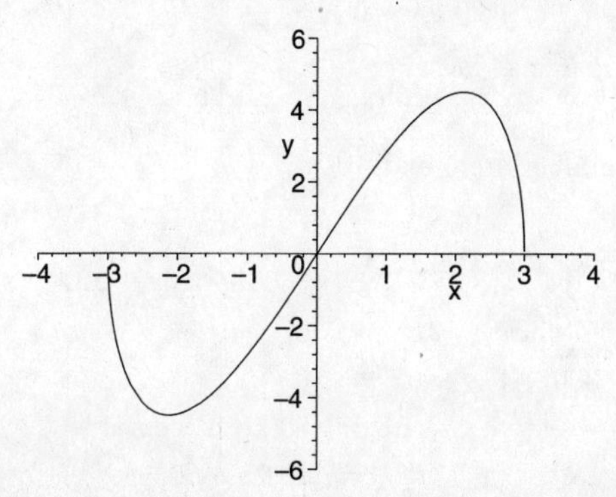

9.

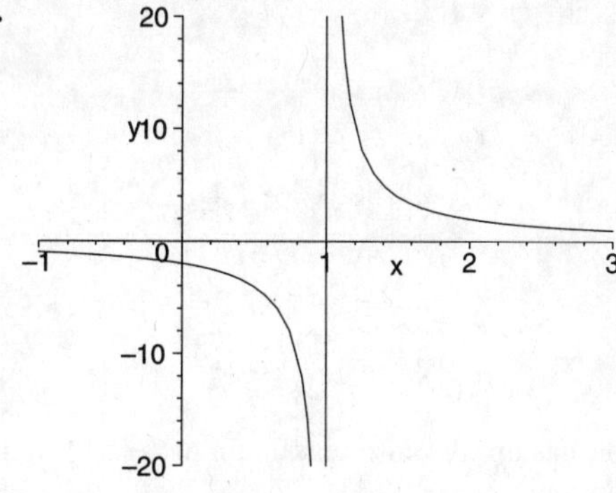

10.

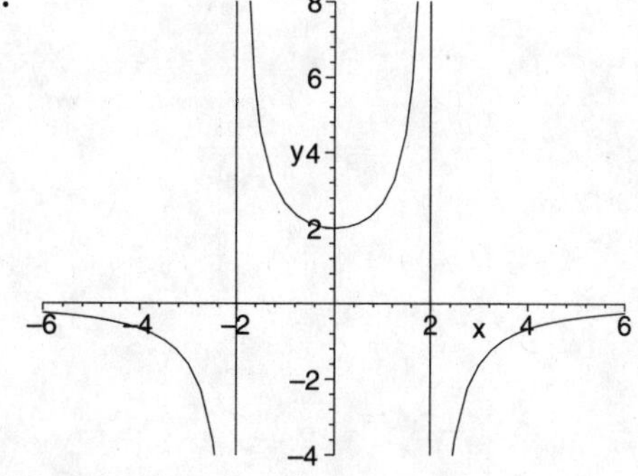

11.

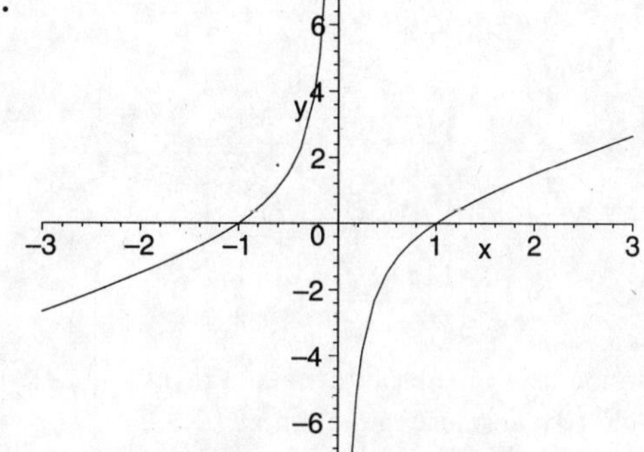

12.

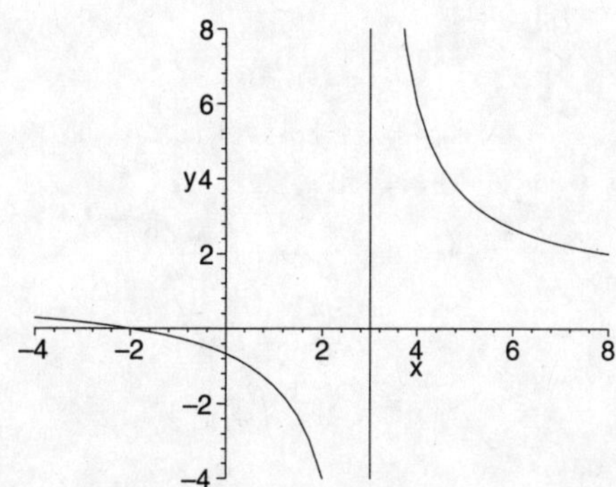

13. $f(x) = x(6-x) = 6x - x^2$

$$\begin{aligned} f'(x) &= 6 - 2x \\ 6 - 2x &= 0 \\ x &= 3 \\ f(3) &= 3(3) = 9 \end{aligned}$$

$$f''(x) = -2 < 0$$

The function has an absolute maximum at $(3, 9)$

14. $f(x) = x^3 + x^2 - x + 1$

$$\begin{aligned} f'(x) &= 3x^2 + 2x - 1 \\ 3x^2 + 2x - 1 &= 0 \\ x &= \frac{1}{3} \to f = \frac{22}{27} \\ x &= -1 \to f = 2 \\ x &= -2 \to f = -1 \\ x &= \frac{1}{2} \to = \frac{7}{8} \end{aligned}$$

The function has an absolute maximum at $(-1, 2)$ and an absolute minimum at $(-2, -1)$

15. $f(x) = cos^2x - sin\ x$

$$\begin{aligned} f'(x) &= -2\ cos\ x\ sin\ x - cos\ x \\ -2\ cos\ x\ sin\ x - cos\ x &= 0 \\ -cos\ x(2\ sin\ x + 1) &= 0 \\ x &= \frac{\pi}{2} \end{aligned}$$

$f(\frac{\pi}{2}) = 0 - 1 = -1$

$$\begin{aligned} x &= \frac{3\pi}{2}) \\ f(\frac{3\pi}{2}) &= 0 + 1 = 1 \\ x &= \frac{7\pi}{6} \\ f(\frac{7\pi}{6} &= \frac{3}{4} + \frac{1}{2} = \frac{5}{4} \\ x &= \frac{11\pi}{6} \\ f(\frac{11\pi}{6} &= \frac{3}{4} + \frac{1}{2} = \frac{5}{4} \end{aligned}$$

The function has an absolute maximim at $(\frac{7\pi}{6}, \frac{5}{4})$ and $(\frac{11\pi}{6}, \frac{5}{4})$ and an absolute minimum at $(\frac{\pi}{2}, -1)$

16. $f(x) = sin\ x(1 + cos\ x)$

$$\begin{aligned} f'(x) &= -sin^2x + cos\ x + cos^2x \\ &= 2\ cos^2x + cos\ x - 1 \\ 2\ cos^2x + cos\ x - 1 &= 0 \\ x &= \frac{\pi}{3} \to f = \frac{3\sqrt{3}}{4} \\ x &= \frac{5\pi}{3} \to f = -\frac{3\sqrt{3}}{4} \\ x &= \pi \to f = 0 \end{aligned}$$

The function an absolute maximum at $\left(\frac{\pi}{3}, \frac{3\sqrt{3}}{4}\right)$ and an absolute minimum at $\left(\frac{5\pi}{3}, -\frac{3\sqrt{3}}{4}\right)$

17. $f(x) = 9\ sec^2x + csc^2x$

$$\begin{aligned} 18\ sec^2x\ tan\ x - 2\ csc^2x\ cot\ x &= f'(x) \\ 18\frac{sin\ x}{cos^3x} - 2\frac{cos\ x}{sin^3x} &= \\ 18\frac{sin\ x}{cos^3x} - 2\frac{cos\ x}{sin^3x} &= 0 \\ 9\ sin^4x - cos^4x &= 0 \\ (3\ sin^2x - cos^2x)(3\ sin^2x + cos^2x) &= 0 \\ tan\ x &= \frac{1}{\sqrt{3}} \\ x &= \frac{\pi}{6} \\ f(\frac{\pi}{6}) &= 12 + 4 = 16 \end{aligned}$$

The function has an absolute minimum at $(\frac{\pi}{6}, 16)$

18. $f(x) = sin^2x - tan^2x$

$$\begin{aligned} 2\ sin\ x\ cos\ x - 2\ tan\ x\ sec^2\ x &= f'(x) \\ 2\ sin\ x(cos\ x - \frac{1}{cos^3x}) &= \\ 2\ sin\ x(cos\ x - \frac{1}{cos^3x}) &= 0 \\ x &= 0 \to f = 0 \end{aligned}$$

The function has an absolute maximum at $(0, 0)$

19. $f(x) = x^2 + \frac{128}{x}$

$$\begin{aligned} f'(x) &= 2x - \frac{128}{x^2} \\ 2x - \frac{128}{x^2} &= 0 \\ x^3 &= 64 \\ x &= 4 \\ f(4) &= 16 + 32 = 48 \end{aligned}$$

The function has an absolute minimum at $(4, 48)$

20. $Q = xy$ and $y - x = 8$

$$\begin{aligned} Q &= x(8 - x) = 8x - x^2 \\ Q' &= 8 - 2x \\ 8 - 2x &= 0 \\ x &= 4 \\ y &= 8 - 4 = 4 \end{aligned}$$

The two number are 4 and 4

21. $Q = x^2 + y^2$ and $x - y = 10$

$$\begin{aligned} Q &= (y + 10)^2 + y^2 \\ &= 2y^2 + 20y + 100 \\ Q' &= 4y + 20 \\ 4y + 20 &= 0 \\ y &= -5 \\ x &= 5 + 10 = 5 \end{aligned}$$

The two number are 5 and -5

22. $V = x(60 - 2x)^2 = 3600x - 240x^2 + 4x^3$

$$\begin{aligned} V' &= 3600 - 480x + 12x^2 \\ 3600 - 480x + 12x^2 &= 0 \\ 300 - 40x + x^2 &= 0 \\ x &= 30 \text{ not acceptable} \\ x &= 10 \end{aligned}$$

The dimensions of the box are 40 × 40 × 10 inches and it has a maximum volume of 16000 cubic inches

23. $f(x) = x\sqrt{x+1}$, $f'(x) = \dfrac{x}{2\sqrt{x+1}} + \sqrt{x+1}$

$$\begin{aligned} L(x) &= f(8) + f'(8)(x-8) \\ &= 24 + \frac{13}{3}(x-8) \\ &= \frac{13}{3}x - \frac{32}{3} \end{aligned}$$

24. $f(x) = (\sin x + 1)^2$, $f'(x) = 2\cos x\ (\sin x + 1)$

$$\begin{aligned} L(x) &= f(0) + f'(0)(x-0) \\ &= 1 + 2(x-0) \\ &= 2x+1 \end{aligned}$$

25. $f(x) = \sqrt{x}$, $f'(x) = \dfrac{1}{2\sqrt{x}}$

$$\begin{aligned} L(104) &= f(100) + f'(100)(104-100) \\ &= 10 + \frac{1}{20}(4) \\ &= 10.2 \end{aligned}$$

26. $f(x) = x^3 + 3x + 5$, $f'(x) = 3x^2 + 3$

$$\begin{aligned} x_1 &= 0 \\ x_2 &= 0 - \frac{5}{3} = -1.6667 \\ x_3 &= 1.6667 - \frac{-4.63}{11.3333} = -1.2581 \\ x_4 &= -1.2581 - \frac{-0.7659}{7.7487} = -1.1593 \\ x_5 &= -1.1593 - \frac{-0.0359}{7.0319} = -1.1542 \\ x_6 &= -1.1542 - \frac{-0.0001}{6.9965} = -1.1542 \\ x_7 &= -1.1542 - \frac{0}{6.9964} = -1.1542 \end{aligned}$$

$x = -1.1542$

27. $f(x) = \cos x - \sin x - x$, $f'(x) = -\sin x - \cos x - 1$

$$\begin{aligned} x_1 &= 1 \\ x_2 &= 1 - \frac{-1.301}{-2.382} = 0.4538 \\ x_3 &= 0.4538 - \frac{0.00656}{-2.337} = 0.4566 \\ x_4 &= 0.4566 - \frac{0}{-2.338} = 0.4566 \end{aligned}$$

$x = 0.4566$

28. $x^3 + y^3 = 9$

$$\begin{aligned} 3x^2 + 3y^2\frac{dy}{dx} &= 0 \\ \frac{dy}{dx} &= \frac{-x^2}{y^2} \end{aligned}$$

When $x = 1$ and $y = 2$

$$\begin{aligned} \frac{dy}{dx} &= \frac{-(1)^2}{(2)^2} \\ &= -\frac{1}{4} \end{aligned}$$

29. $x^2 + y^2 = 169$
When $x = 12$, $y = \sqrt{169 - 144} = 5$

$$\begin{aligned} 2x\frac{dx}{dt} + 2y\frac{dy}{dt} &= 0 \\ \frac{dy}{dt} &= -\frac{x}{y}\frac{dx}{dt} \\ &= -\frac{12}{5}(0.4) \\ &= 0.96\ \frac{ft}{sec} \end{aligned}$$

30. $f(x) = \dfrac{x^2}{1+x^3}$

$$\begin{aligned} f'(x) &= \frac{2x(1+x^3) - 3x^2(x^2)}{(1+x^3)^2} \\ &= \frac{2x - x^4}{(1+x^3)^2} \\ \frac{2x - x^4}{(1+x^3)^2} &= 0 \\ x &= 0 \\ x &= \sqrt[3]{2} \\ f(0) &= 0 \\ f(\sqrt[3]{2}) &= \frac{\sqrt[3]{4}}{1+2} = \frac{\sqrt[3]{4}}{3} \end{aligned}$$

The function has an absolute maximum at $(\sqrt[3]{2}, \frac{1}{3}\sqrt[3]{4})$ and an absolute minimum at $(0,0)$

31. The function has a relative maximum at $(1.09, 25.1)$ and a relative minimum at $(2.97, 8.6)$

Technology Connection

- **Page 171**

  1. Relative minimum at $(0,0)$
  2. Relative maximum at $(1.675, 2.361)$
  3. Relative maximum at $(-1, 21.167)$ and a relative minimum at $(2, -2.334)$
  4. Relative maximum at $(0, 22)$ and relative minimum at $(-3.1997, -0.0119)$ and $(-3.1997, -0.0119)$

- **Page 175**

1-2.

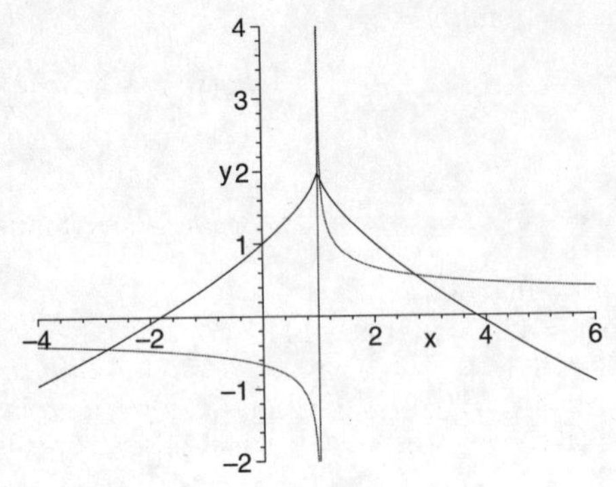

3. Relative maximum at $(1, 2)$

- **Page 188** Relative minimum at $(1, -1)$

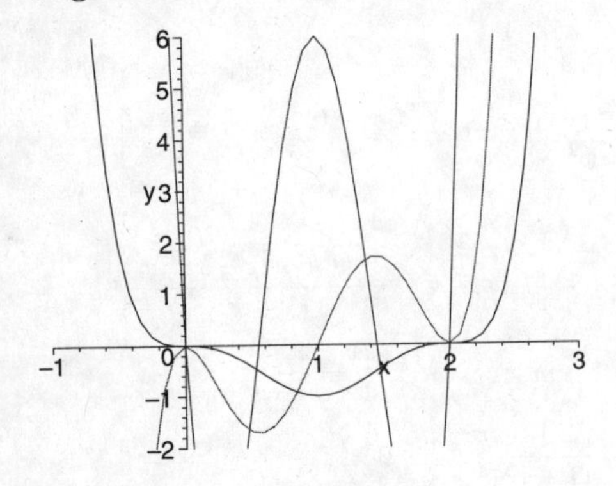

- **Page 196**

1. $\lim_{x\to\infty} \frac{2x+5}{x} = 2$

2.

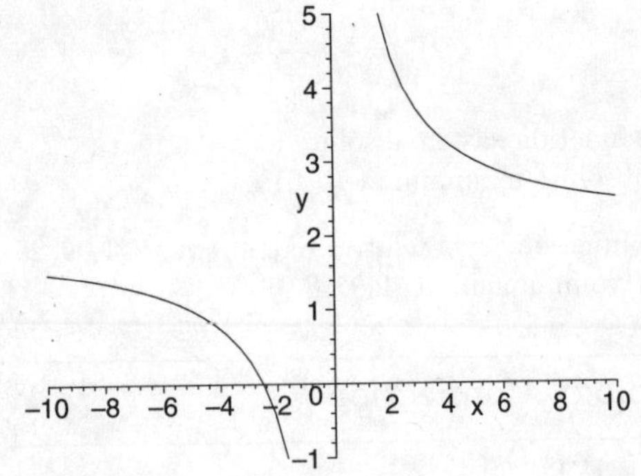

- **Page 198**

1. $\lim_{x\to\infty} \frac{2x^2+x-7}{3x^2-4x+1} = \frac{2}{3}$

2. $\lim_{x\to\infty} \frac{5x+4}{2x^3-3} = 0$

3. $\lim_{x\to\infty} \frac{5x^2-2}{4x+5} = \infty$

4. $\lim_{x\to\infty} \frac{x^2-1}{x^2+x-6} = 1$

- **Page 201**

1. Vertical asymptote at $x = -7$ and $x = 4$

2. Vertical asymptote at $x = -2$, $x = 0$, and $x = 3$

- **Page 202**

1. Horizontal asymptote at $y = 0$

2. Horizontal asymptote at $y = 3$

3. Horizontal asymptote at $y = 0$

4. Horizontal asymptote at $y = \frac{1}{2}$

- **Page 203**

1. Oblique asymptote at $y = 3x + 1$

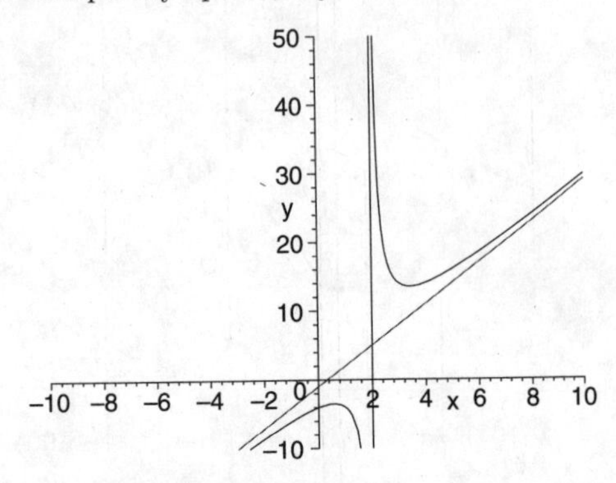

2. Oblique asymptote at $y = 5x$

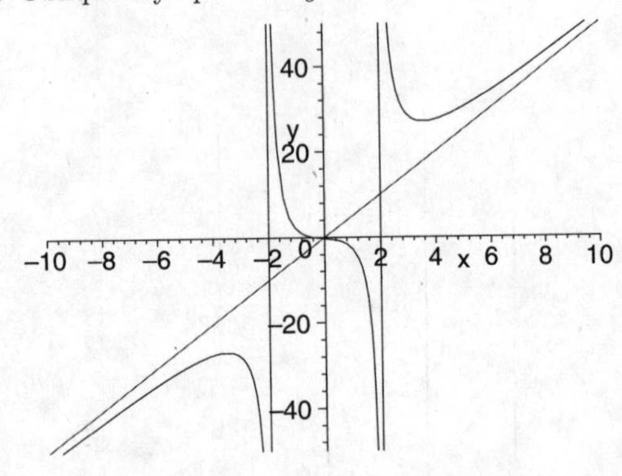

- **Page 204**

1. x-intercepts at $(0, 0)$, $(3, 0)$, and $(-5, 0)$
y-intercept at $(0, 0)$

2. x-intercepts at $(0, 0)$, $(-3, 0)$, and $(1, 0)$
y-intercept at $(0, 0)$

- **Page217**
On $[-2, 1]$: Absolute maximum at $(-0.5, 2.125)$ On $[-1, 2]$: Absolute maximum at $(-0.5, 2.125)$ and absolute minimum at $(1, 1)$

- **Page219** Absolute minimum at $(2, -4)$

- **Page224** Absolute minimum at $(0.316, 6.235)$

Extended Life Science Connection

1.

$$\begin{aligned} w &= p^2(1-a) + 2pq(1) + q^2(1-b) \\ &= p^2 - ap^2 + 2p(1-p) + (1-p)^2 - q^2b \\ &= p^2 - ap^2 + 2p - 2p^2 + 1 - 2p + p^2 - q^2b \\ &= 1 - ap^2 - q^2b \end{aligned}$$

2.

$$\begin{aligned} f(p) &= \frac{2p^2(1-a)+2pq}{2w} \\ &= \frac{p^2-2p^2a+pq}{1-ap^2-q^2b} \\ &= \frac{p(p(1-a)+q)}{1-ap^2-q^2b} \\ &= 2p\frac{p(1-a)+q}{1-ap^2-q^2b} \end{aligned}$$

3.

$$\begin{aligned} p &= p\frac{p(1-a)+q}{1-p^2a-q^2b} \\ p &= p\frac{p(1-a)+(1-p)}{1-p^2a-(1-p)^2b} \\ p &= 0 \rightarrow \\ 0 &= 0\frac{0(1-a)+(1-0)}{1-0a-b} \\ 0 &= 0 \\ p &= 1 \rightarrow \\ 1 &= 1\frac{1(1-a)+(1-1)}{1-a-0b} \\ 1 &= \frac{1-a}{1-a} \\ 1 &= 1 \end{aligned}$$

Therefore, $p=0$ and $p=1$ are solutions to equation (2). When $p=0$ the population will consist of B alleles only and when $p=1$ the population in not polymorphic. When $p=1$ the population consists only of A alleles (not polymorphic)

4.

$$\begin{aligned} p &= p\frac{p(1-a)+q}{1-p^2a-q^2b} \\ p(1-p^2a-q^2b) &= p[(p(1-a)+q] \\ p-p^3a-p(1-p)^2b &= p^2-pa+p(1-p) \\ -(a+b)p^3+(2b+a)p^2-pb &= 0 \\ -p[(a+b)p^2-(2b+a)p+b] &= 0 \\ p &= 0 \text{ as expected} \\ \frac{(2b-a)\pm\sqrt{(2b+a)^2-4(a+b)b}}{2(a+b)} &= p \\ \frac{(2b+a)\pm\sqrt{4b^2+4ab+a^2-4ab-4b^2}}{2(a+b)} &= \\ \frac{(2b+a)\pm\sqrt{a^2}}{2(a+b)} &= \\ \frac{2b+a\pm a}{2(a+b)} &= \\ \frac{2b+a+a}{2(a+b)} &= \\ \frac{2(a+b)}{2(a+b)} &= \\ &= 1 \text{ as expected} \\ p &= \frac{2b+a-a}{2(a+b)} \\ &= \frac{2b}{2(a+b)} \\ &= \frac{b}{a+b} \end{aligned}$$

For $p=\frac{b}{a+b}$ the population would be polymorphic

5. **a)** Since $p=\frac{b}{a+b}$, if a and b are positive then $a+b$ is positive and the quotient of two positive numbers is positive. If a and b are negative then $a+b$ is negative and the quotient of two negative numbers is positive.

b) If a and b have different signs then p would be negative which would mean that that proportion of A alleles is negative which means there is a negative number of alleles in the population, and that cannot happen.

6. **a)** $p=0$, $p=1$, and $p=\frac{0.5}{0.5+0.5}=0.5$

b) $p=0$, $p=1$

c) $p=0$, $p=1$, and $p=\frac{-0.5}{-0.5-0.5}=0.5$

7.

$$\begin{aligned} f(p) &= \frac{2p^2(1-a)+2(1-p)p}{2(1-ap^2-(1-p)^2b} \\ &= \frac{-ap^2+p}{2(1-ap^2-(1-p)^2b} \\ f'(p) &= \frac{[-(a+b)p^2+2bp-b+1][-2ap+1]}{[-(a+b)p^2+2bp-b+1]^2} - \\ &\quad \frac{[-ap^2+p][-2(a+b)p+2b]}{[-(a+b)p^2+2bp-b+1]^2} \\ L(p) &= f(0)+f'(0)p \\ &= 0+\frac{1}{1-b}p \\ &= \frac{1}{1-b}\,p \end{aligned}$$

8. If b is negative then $\frac{1}{1-b}<1$. This means that for the next generation p is multiplied by a number smaller than 1, that is, $f(p)<p$. Thus, if b is negative then the equilibrium at $p=0$ is stable.

9.

$$\begin{aligned} f(p) &= \frac{2p^2(1-a)+2(1-p)p}{2(1-ap^2-(1-p)^2b} \\ &= \frac{-ap^2+p}{2(1-ap^2-(1-p)^2b} \\ f'(p) &= \frac{[-(a+b)p^2+2bp-b+1][-2ap+1]}{[-(a+b)p^2+2bp-b+1]^2} - \\ &\quad \frac{[-ap^2+p][-2(a+b)p+2b]}{[-(a+b)p^2+2bp-b+1]^2} \\ L(p) &= f(1)+f'(1)(p-1) \\ &= 0+\frac{1}{1-a}(p-1) \\ &= \frac{1}{1-b}\,(p-1) \end{aligned}$$

10. Similar arguments as those for Exercise 8.

11. Left to the student

12.

$$\begin{aligned}
f(p) &= \frac{2p^2(1-a)+2(1-p)p}{2(1-ap^2-(1-p)^2b} \\
&= \frac{-ap^2+p}{2(1-ap^2-(1-p)^2b} \\
f'(p) &= \frac{[-(a+b)p^2+2bp-b+1][-2ap+1]}{[-(a+b)p^2+2bp-b+1]^2} - \\
&\quad \frac{[-ap^2+p][-2(a+b)p+2b]}{[-(a+b)p^2+2bp-b+1]^2} \\
f\left(\frac{b}{a+b}\right) &= \frac{b}{a+b} \\
f'\left(\frac{b}{a+b}\right) &= 1-\frac{ab}{a+b-ab} \\
L(p) &= f\left(\frac{b}{a+b}\right)+f'\left(\frac{b}{a+b}\right)\left(p-\frac{b}{a+b}\right) \\
&= \left(\frac{b}{a+b}\right)+\left(1-\frac{ab}{a+b-ab}\right)\left(p-\frac{1}{1-b}\right)
\end{aligned}$$

13. If a and b are positive then $w = 1-ap^2-bq^2$ is less than 1 and therefore $f(p) > p$ thus unstable, on the other hand, if a and b are negative then w is bigger than 1 and $f(p) < p$ thus stable.

Chapter 4

Exponential and Logarithmic Functions

Exercise Set 4.1

1. Graph: $y = 4^x$

First we find some function values.

Note: For

$x = -2$, $y = 4^{-2} = \dfrac{1}{4^2} = \dfrac{1}{16} = 0.0625$

$x = -1$, $y = 4^{-1} = \dfrac{1}{4} = 0.25$

$x = 0$, $y = 4^0 = 1$

$x = 1$, $y = 4^1 = 4$

$x = 2$, $y = 4^2 = 16$

x	y
-2	0.0625
-1	0.25
0	1
1	4
2	16

Plot these points and connect them with a smooth curve.

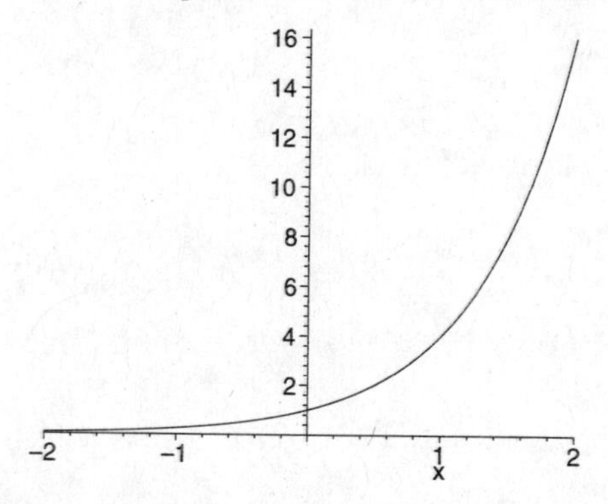

2.

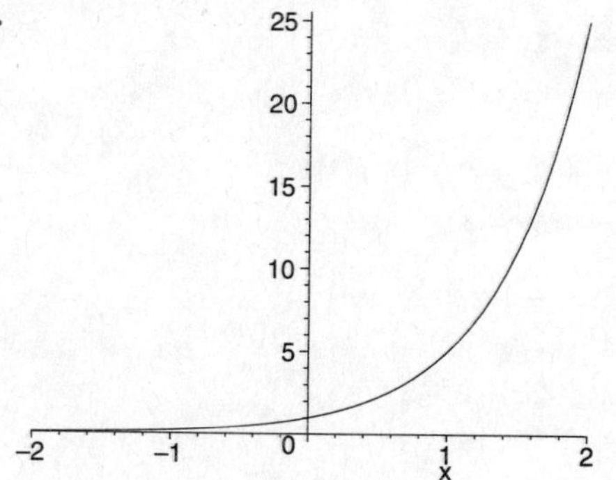

3. Graph: $y = (0.4)^x$

First we find some function values.

Note: For

$x = -2$, $y = (0.4)^{-2} = \dfrac{1}{(0.4)^2} = 6.25$

$x = -1$, $y = (0.4)^{-1} = \dfrac{1}{0.4} = 2.5$

$x = 0$, $y = (0.4)^0 = 1$

$x = 1$, $y = (0.4)^1 = 0.4$

$x = 2$, $y = (0.4)^2 = 0.16$

x	y
-2	6.25
-1	2.5
0	1
1	0.4
2	0.16

Plot these points and connect them with a smooth curve.

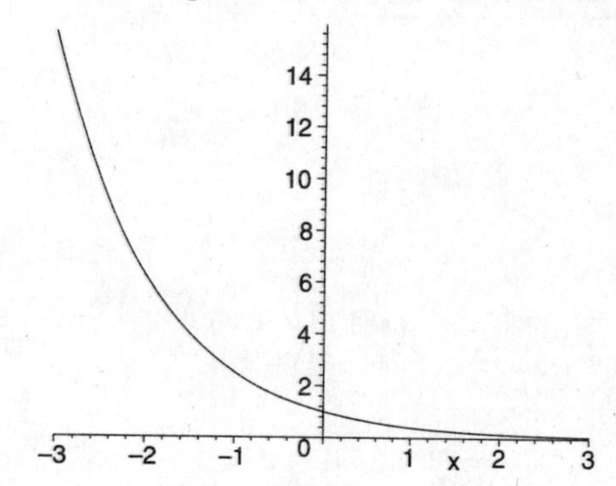

4.

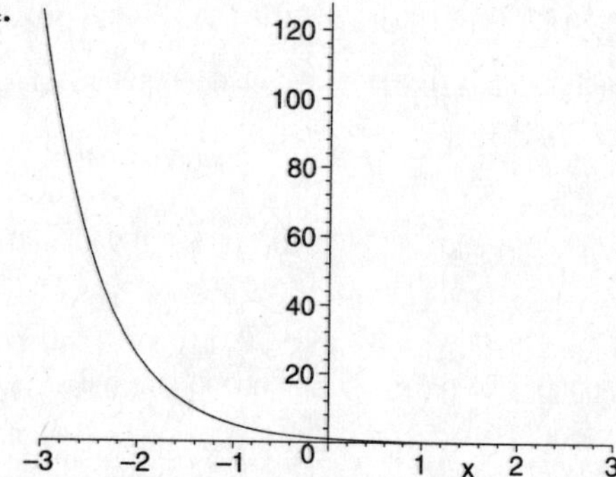

5. Graph: $x = 4^y$

First we find some function values.

Note: For

$y = -2$, $x = 4^{-2} = \dfrac{1}{4^2} = \dfrac{1}{16}$

$y = -1$, $x = 4^{-1} = \dfrac{1}{4}$

$y = 0$, $x = 4^0 = 1$

$y = 1$, $x = 4^1 = 4$

$y = 2$, $x = 4^2 = 16$

x	y
$\frac{1}{16}$	-2
$\frac{1}{4}$	-1
1	0
4	1
16	2

Plot these points and connect them with a smooth curve.

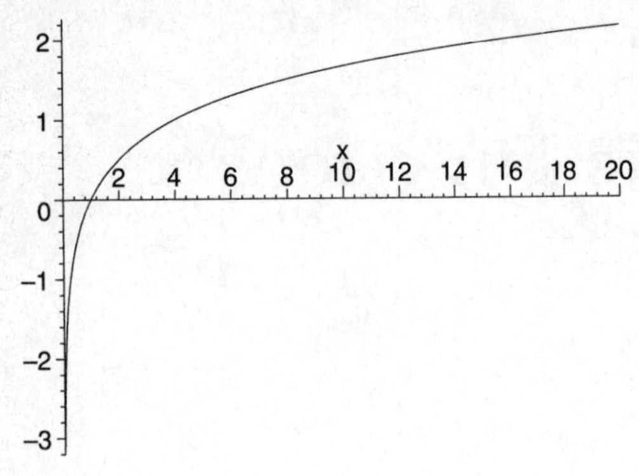

6.

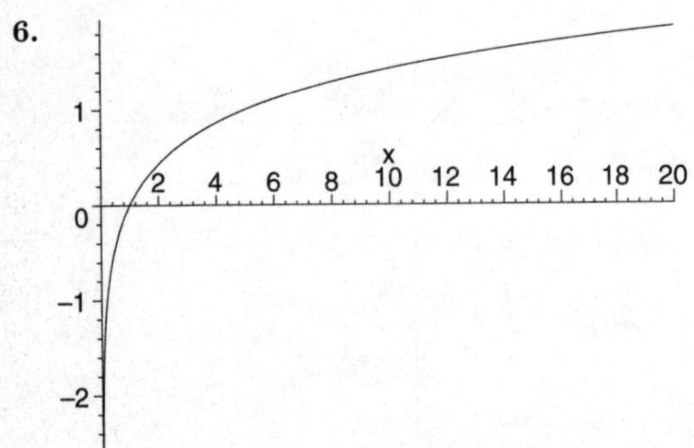

7. $N = 1000(1+0.20)^1 = 1000(1.2) = 1200$
$N = 1000(1+0.20)^2 = 1000(1.44) = 1440$
$N = 1000(1+0.20)^5 = 1000(2.488) = 2488$

8. $N = 50(1+0.05)^1 = 1000(1.05) = 1050$
$N = 1000(1+0.05)^2 = 1000(1.103) = 1103$
$N = 1000(1+0.05)^5 = 1000(1.276) = 1276$

9. $N = 286000000(1+0.021)^1 = 286000000(1.021) = 291006000$
$N = 286000000(1+0.021)^2 = 286000000(1.042441) = 298138126$
$N = 286000000(1+0.021)^5 = 286000000(1.109503586) = 3173188026$

10. $N = 50000000(1+0.019)^1 = 50000000(1.019) = 50950000$
$N = 50000000(1+0.019)^2 = 50000000(1.038361) = 51918050$
$N = 50000000(1+0.019)^5 = 50000000(1.098679244) = 54933962$

11. $f(x) = e^{3x}$

$f'(x) = 3e^{3x}$ $\left[\frac{d}{dx}e^{f(x)} = f'(x)e^{f(x)}\right]$

12. $f(x) = e^{2x}$
$f'(x) = 2e^{2x}$

13. $f(x) = 5e^{-2x}$

$f'(x) = 5 \cdot \underbrace{(-2)e^{-2x}}$ $\left[\frac{d}{dx}[c \cdot f(x)] = c \cdot f'(x)\right]$
$\left[\frac{d}{dx}e^{f(x)} = f'(x)e^{f(x)}\right]$

$= -10e^{-2x}$

14. $f(x) = 4e^{-3x}$
$f'(x) = 4(-3)e^{-3x} = -12e^{-3x}$

15. $f(x) = 3 - e^{-x}$
$f'(x) = 0 - \underbrace{(-1)e^{-x}}$ $\left[\frac{d}{dx}e^{f(x)} = f'(x)e^{f(x)}\right]$
$= e^{-x}$

16. $f(x) = 2 - e^{-x}$
$f'(x) = -(-1)e^{-x} = e^{-x}$

17. $f(x) = -7e^x$
$f'(x) = -7e^x$ $\left[\frac{d}{dx} - 7e^x = -7 \cdot \frac{d}{dx}e^x\right]$
$\left[\frac{d}{dx}e^x = e^x\right]$

18. $f(x) = -4e^x$
$f'(x) = -4e^x$

19. $f(x) = \frac{1}{2}e^{2x}$

$f'(x) = \frac{1}{2} \cdot \underbrace{2e^{2x}}$ $\left[\frac{d}{dx}[c \cdot f(x)] = c \cdot f'(x)\right]$
$\left[\frac{d}{dx}e^{f(x)} = f'(x)e^{f(x)}\right]$

$= e^{2x}$

20. $f(x) = \frac{1}{4}e^{4x}$

$f'(x) = \frac{1}{4} \cdot 4e^{4x} = e^{4x}$

21. $f(x) = x^4e^x$
$f'(x) = x^4 \cdot e^x + 4x^3 \cdot e^x$ Using the Product Rule
$= x^3e^x(x+4)$

22. $f(x) = x^5e^x$
$f'(x) = x^5e^x + 5x^4e^x$
$= x^4e^x(x+5)$

23. $f(x) = (x^2+3x-9)e^x$
$f'(x) = (x^2+3x-9)e^x + (2x+3)e^x$ Using the Product Rule
$= (x^2+3x-9+2x+3)e^x$
$= (x^2+5x-6)e^x$

24. $f(x) = (x^2-2x+2)e^x$
$f'(x) = (x^2-2x+2)e^x + (2x-2)e^x$
$= (x^2-2x+2+2x-2)e^x$
$= x^2e^x$

25. $f(x) = (\sin x)e^x$
$f'(x) = (\sin x)e^x + (\cos x)e^x$

26. $f(x) = (\cos x)e^x$
$f'(x) = (\cos x)e^x - (\sin x)e^x$

27. $f(x) = \dfrac{e^x}{x^4}$

$f'(x) = \dfrac{x^4 \cdot e^x - 4x^3 \cdot e^x}{x^8}$ Using the Quotient Rule

$= \dfrac{x^3 e^x(x-4)}{x^3 \cdot x^5}$ Factoring both numerator and denominator

$= \dfrac{x^3}{x^3} \cdot \dfrac{e^x(x-4)}{x^5}$

$= \dfrac{e^x(x-4)}{x^5}$ Simplifying

28. $f(x) = \dfrac{e^x}{x^5}$

$f'(x) = \dfrac{x^5 e^x - 5x^4 e^x}{x^{10}}$

$= \dfrac{x^4 e^x(x-5)}{x^{10}}$

$= \dfrac{e^x(x-5)}{x^6}$

29. $f(x) = e^{-x^2+7x}$

$f'(x) = (-2x+7)e^{-x^2+7x}$ $\left[\dfrac{d}{dx}e^{f(x)} = f'(x)e^{f(x)}\right]$

$[f(x) = -x^2 + 7x,$
$f'(x) = -2x + 7]$

30. $f(x) = e^{-x^2+8x}$

$f'(x) = (-2x+8)e^{-x^2+8x}$

31. $f(x) = e^{-x^2/2}$

$= e^{(-1/2)x^2}$

$f'(x) = \left(-\dfrac{1}{2} \cdot 2x\right)e^{(-1/2)x^2}$

$= -xe^{-x^2/2}$

32. $f(x) = e^{x^2/2}$

$f'(x) = \dfrac{2x}{2}e^{x^2/2} = xe^{x^2/2}$

33. $y = e^{\sqrt{x-7}}$

$= e^{(x-7)^{1/2}}$

$\dfrac{dy}{dx} = \dfrac{1}{2}(x-7)^{-1/2} \cdot e^{(x-7)^{1/2}}$

$= \dfrac{e^{(x-7)^{1/2}}}{2(x-7)^{1/2}}$

$= \dfrac{e^{\sqrt{x-7}}}{2\sqrt{x-7}}$

34. $y = e^{\sqrt{x-4}}$

$\dfrac{dy}{dx} = \dfrac{1}{2}(x-4)^{-1/2}e^{\sqrt{x-4}}$

$= \dfrac{e^{\sqrt{x-4}}}{2\sqrt{x-4}}$

35. $y = \sqrt{e^x - 1}$

$= (e^x - 1)^{1/2}$

$\dfrac{dy}{dx} = \dfrac{1}{2}(e^x - 1)^{-1/2} \cdot e^x$ [Extended Power Rule;
↑— $\dfrac{d}{dx}(e^x - 1) = e^x - 0 = e^x$]

$= \dfrac{e^x}{2\sqrt{e^x - 1}}$

36. $y = \sqrt{e^x + 1}$

$\dfrac{dy}{dx} = \dfrac{1}{2}(e^x + 1)^{-1/2}(e^x)$

$= \dfrac{e^x}{2\sqrt{e^x + 1}}$

37. $y = tan(e^x + 1)$

$\dfrac{dy}{dx} = sec^2(e^x + 1) \cdot e^x$

$= e^x \ sec^2(e^x + 1)$

38. $y = sin \ e^{2x}$

$\dfrac{dy}{dx} = cos \ e^{2x} \cdot 2e^{2x}$

$= 2e^{2x} \ cos \ e^{2x}$

39. $y = e^{tan \ x}$

$\dfrac{dy}{dx} = e^{tan \ x} \cdot sec^2 x$

40. $y = e^{cos \ x}$

$\dfrac{dy}{dx} = e^{cos \ x} \cdot -sin \ x$

$= -sin \ x \ e^{cos \ x}$

41. $y = (2x + cos \ x)e^{3x+1}$

$\dfrac{dy}{dx} = (2x + cos \ x) \cdot e^{3x+1}(1) + e^{3x+1}(2 - sin \ x)$

$= e^{3x+1}(2x + cos \ x - sin \ x + 2)$

42. $y = (sec \ x + tan \ x)e^x$

$\dfrac{dy}{dx} = (sec \ x + tan \ x)e^x + e^x(sec \ x \ tan \ x + sec^2 x)$

$= e^x(sec \ x + tan \ x + sec \ x \ tan \ x + sec^2 x)$

43. $y = xe^{-2x} + e^{-x} + x^3$

$\dfrac{dy}{dx} = x \cdot (-2) \cdot e^{-2x} + 1 \cdot e^{-2x} + (-1) \cdot e^{-x} + 3x^2$

$= -2xe^{-2x} + e^{-2x} - e^{-x} + 3x^2$

$= (1 - 2x)e^{-2x} - e^{-x} + 3x^2$

44. $y = e^x + x^3 - xe^x$

$\dfrac{dy}{dx} = e^x + 3x^2 - (xe^x + e^x)$

$= e^x + 3x^2 - xe^x - e^x$

$= 3x^2 - xe^x$

$= x(3x - e^x)$

45. $y = 1 - e^{-x}$

$$\frac{dy}{dx} = 0 - (-1)e^{-x} = e^{-x}$$

46. $y = 1 - e^{-3x}$

$$\frac{dy}{dx} = -(-3)e^{-3x} = 3e^{-3x}$$

47. $y = 1 - e^{-kx}$

$$\frac{dy}{dx} = 0 - (-k)e^{-kx} = ke^{-kx}$$

48. $y = 1 - e^{-mx}$

$$\frac{dy}{dx} = -(-m)e^{-mx} = me^{-mx}$$

49. $y = (e^{3x} + 1)^5$

$$\frac{dy}{dx} = 5(e^{3x} + 1)^4 \cdot 3e^{3x} = 15e^{3x}(e^{3x} + 1)^4$$

50. $y = (e^{x^2} - 2)^4$

$$\frac{dy}{dx} = 4(e^{x^2} - 2)^3(2xe^{x^2}) = 8xe^{x^2}(e^{x^2} - 2)^3$$

51. $y = \dfrac{e^{3t} - e^{7t}}{e^{4t}}$

$= \dfrac{e^{3t}(1 - e^{4t})}{e^{3t} \cdot e^{t}}$ Factoring

$= \dfrac{1 - e^{4t}}{e^{t}}$ Simplifying

$\dfrac{dy}{dt} = \dfrac{e^{t}(-4e^{4t}) - e^{t}(1 - e^{4t})}{(e^{t})^2}$ Using the Quotient Rule

$= \dfrac{e^{t}(-4e^{4t} - 1 + e^{4t})}{e^{t} \cdot e^{t}}$

$= \dfrac{-3e^{4t} - 1}{e^{t}}$

$= -3e^{3t} - e^{-t}$

52. $y = \sqrt[3]{e^{3t} + t}$

$$\frac{dy}{dx} = \frac{1}{3}(e^{3t} + t)^{-2/3}(3e^{3t} + 1)$$

53. $y = \dfrac{e^{x}}{x^2 + 1}$

$$\frac{dy}{dx} = \frac{(x^2 + 1)e^{x} - 2x \cdot e^{x}}{(x^2 + 1)^2} = \frac{e^{x}(x^2 - 2x + 1)}{(x^2 + 1)^2} = \frac{e^{x}(x - 1)^2}{(x^2 + 1)^2}$$

54. $y = \dfrac{e^{x}}{1 - e^{x}}$

$$\frac{dy}{dx} = \frac{(1 - e^{x})e^{x} - (-e^{x})e^{x}}{(1 - e^{x})^2} = \frac{e^{x} - e^{2x} + e^{2x}}{(1 - e^{x})^2} = \frac{e^{x}}{(1 - e^{x})^2}$$

55. $f(x) = e^{\sqrt{x}} + \sqrt{e^{x}}$

$= e^{x^{1/2}} + e^{x/2}$

$$f'(x) = \frac{1}{2}x^{-1/2}e^{x^{1/2}} + \frac{1}{2}e^{x/2} = \frac{e^{\sqrt{x}}}{2\sqrt{x}} + \frac{\sqrt{e^{x}}}{2}$$

56. $f(x) = \dfrac{1}{e^{x}} + e^{1/x} = e^{-x} + e^{x^{-1}}$

$$f'(x) = -e^{-x} + (-x^{-2})e^{x^{-1}} = -e^{-x} - x^{-2}e^{1/x}$$

57. $f(x) = e^{x/2} \cdot \sqrt{x - 1}$

$= e^{x/2} \cdot (x - 1)^{1/2}$

$f'(x) = e^{x/2} \cdot \dfrac{1}{2}(x - 1)^{-1/2} + \dfrac{1}{2}e^{x/2} \cdot (x - 1)^{1/2}$

$= \dfrac{1}{2}e^{x/2}\left((x - 1)^{-1/2} + (x - 1)^{1/2}\right)$

$= \dfrac{1}{2}e^{x/2}\left(\dfrac{1}{\sqrt{x - 1}} + \sqrt{x - 1}\right)$

$= \dfrac{1}{2}e^{x/2}\left(\dfrac{1}{\sqrt{x - 1}} + \sqrt{x - 1} \cdot \dfrac{\sqrt{x - 1}}{\sqrt{x - 1}}\right)$

$= \dfrac{1}{2}e^{x/2}\left(\dfrac{1}{\sqrt{x - 1}} + \dfrac{x - 1}{\sqrt{x - 1}}\right)$

$= \dfrac{1}{2}e^{x/2}\left(\dfrac{x}{\sqrt{x - 1}}\right)$

58. $f(x) = \dfrac{xe^{-x}}{1 + x^2}$

$f'(x) = \dfrac{(1 + x^2)[x(-e^{-x}) + e^{-x}] - 2x(xe^{-x})}{(1 + x^2)^2}$

$= \dfrac{-xe^{-x} + e^{-x} - x^3e^{-x} + x^2e^{-x} - 2x^2e^{-x}}{(1 + x^2)^2}$

$= \dfrac{-x^3e^{-x} - x^2e^{-x} - xe^{-x} + e^{-x}}{(1 + x^2)^2}$

$= \dfrac{e^{-x}(-x^3 - x^2 - x + 1)}{(1 + x^2)^2}$

59. $f(x) = \dfrac{e^x - e^{-x}}{e^x + e^{-x}}$

$$f'(x) = \frac{(e^x+e^{-x})(e^x+e^{-x})-(e^x-e^{-x})(e^x-e^{-x})}{(e^x+e^{-x})^2}$$
$$= \frac{(e^{2x}+e^0+e^0+e^{-2x})-(e^{2x}-e^0-e^0+e^{-2x})}{(e^x+e^{-x})^2}$$
$$= \frac{e^{2x}+1+1+e^{-2x}-e^{2x}+1+1-e^{-2x}}{(e^x+e^{-x})^2}$$
$$= \frac{4}{(e^x+e^{-x})^2}$$

60. $f(x) = e^{e^x}$

$f'(x) = e^x e^{e^x} = e^{x+e^x}$

61. Graph: $f(x) = e^{2x}$

Using a calculator we first find some function values.

Note: For

$x = -2,\ f(-2) = e^{2(-2)} = e^{-4} = 0.0183$

$x = -1,\ f(-1) = e^{2(-1)} = e^{-2} = 0.1353$

$x = \ \ 0,\ f(0) = e^{2\cdot 0} = e^0 = 1$

$x = \ \ 1,\ f(1) = e^{2\cdot 1} = e^2 = 7.3891$

$x = \ \ 2,\ f(2) = e^{2\cdot 2} = e^4 = 54.598$

x	$f(x)$
-2	0.0183
-1	0.1353
0	1
1	7.3891
2	54.598

Plot these points and connect them with a smooth curve.

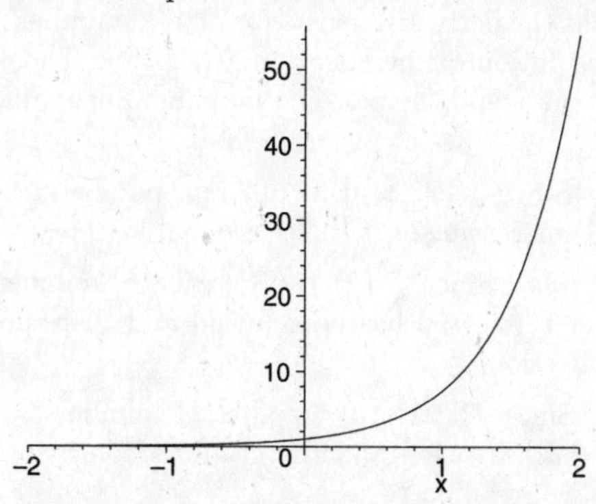

Derivatives. $f'(x) = 2e^{2x}$ and $f''(x) = 4e^{2x}$.

Critical points of f. Since $f'(x) > 0$ for all real numbers x, we know that the derivative exists for all real numbers and there is no solution of the equation $f'(x) = 0$. There are no critical points and therefore no maximum or minimum values.

Increasing. Since $f'(x) > 0$ for all real numbers x, the function f is increasing over the entire real line, $(-\infty, \infty)$.

Inflection points. Since $f''(x) > 0$ for all real numbers x, the equation $f''(x) = 0$ has no solution and there are no points of inflection.

Concavity. Since $f''(x) > 0$ for all real numbers x, the function f' is increasing and the graph is concave up over the entire real line.

62.

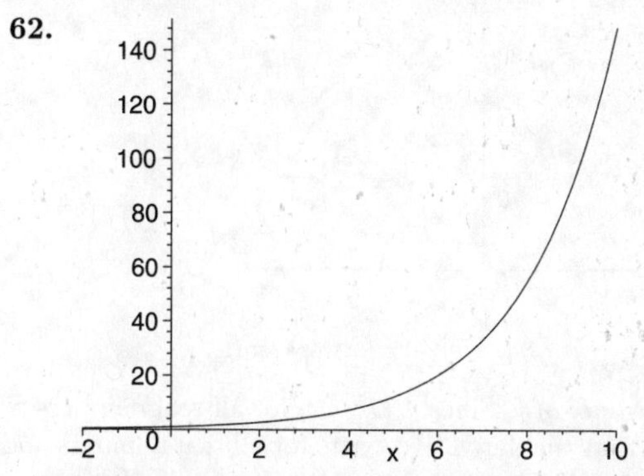

Derivatives. $f'(x) = \dfrac{1}{2}e^{(1/2)x}$ and $f''(x) = \dfrac{1}{4}e^{(1/2)x}$.

Critical points of f. Since $f'(x) > 0$ for all real numbers x, we know that the derivative exists for all real numbers and there is no solution of the equation $f'(x) = 0$. There are no critical points and therefore no maximum or minimum values.

Increasing. Since $f'(x) > 0$ for all real numbers x, the function f is increasing over the entire real line, $(-\infty, \infty)$.

Inflection points. Since $f''(x) > 0$ for all real numbers x, the equation $f''(x) = 0$ has no solution and there are no points of inflection.

Concavity. Since $f''(x) > 0$ for all real numbers x, the function f' is increasing and the graph is concave up over the entire real line.

63. Graph: $f(x) = e^{-2x}$

Using a calculator we first find some function values.

Note: For

$x = -2,\ f(-2) = e^{-2(-2)} = e^4 = 54.598$

$x = -1,\ f(-1) = e^{-2(-1)} = e^2 = 7.3891$

$x = \ \ 0,\ f(0) = e^{-2\cdot 0} = e^0 = 1$

$x = \ \ 1,\ f(1) = e^{-2\cdot 1} = e^{-2} = 0.1353$

$x = \ \ 2,\ f(2) = e^{-2\cdot 2} = e^{-4} = 0.0183$

x	$f(x)$
-2	54.598
-1	7.3891
0	1
1	0.1353
2	0.0183

Plot these points and connect them with a smooth curve.

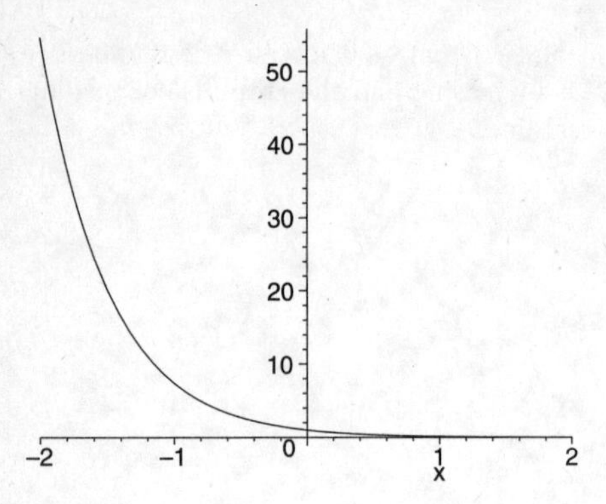

Derivatives. $f'(x) = -2e^{-2x}$ and $f''(x) = 4e^{-2x}$.

Critical points of f. Since $f'(x) < 0$ for all real numbers x, we know that the derivative exists for all real numbers and there is no solution of the equation $f'(x) = 0$. There are no critical points and therefore no maximum or minimum values.

Decreasing. Since $f'(x) < 0$ for all real numbers x, the function f is decreasing over the entire real line, $(-\infty, \infty)$.

Inflection points. Since $f''(x) > 0$ for all real numbers x, the equation $f''(x) = 0$ has no solution and there are no points of inflection.

Concavity. Since $f''(x) > 0$ for all real numbers x, the function f' is increasing and the graph is concave up over the entire real line.

64.

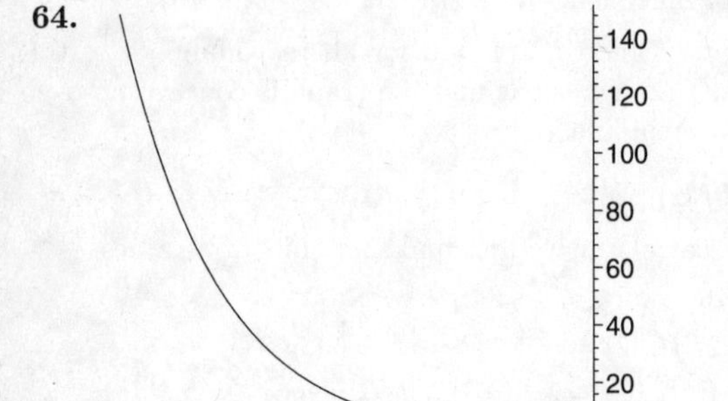

Derivatives. $f'(x) = -\frac{1}{2}e^{(-1/2)x}$ and $f''(x) = \frac{1}{4}e^{(-1/2)x}$.

Critical points of f. Since $f'(x) < 0$ for all real numbers x, we know that the derivative exists for all real numbers and there is no solution of the equation $f'(x) = 0$. There are no critical points and therefore no maximum or minimum values.

Decreasing. Since $f'(x) < 0$ for all real numbers x, the function f is decreasing over the entire real line, $(-\infty, \infty)$.

Inflection points. Since $f''(x) > 0$ for all real numbers x, the equation $f''(x) = 0$ has no solution and there are no points of inflection.

Concavity. Since $f''(x) > 0$ for all real numbers x, the function f' is increasing and the graph is concave up over the entire real line.

65. Graph: $f(x) = 3 - e^{-x}$, for nonnegative values of x.

Using a calculator we first find some function values.

Note: For

$x = 0,\ f(0) = 3 - e^{-0} = 3 - 1 = 2$

$x = 1,\ f(1) = 3 - e^{-1} = 3 - 0.3679 = 2.6321$

$x = 2,\ f(2) = 3 - e^{-2} = 3 - 0.1353 = 2.8647$

$x = 3,\ f(3) = 3 - e^{-3} = 3 - 0.0498 = 2.9502$

$x = 4,\ f(4) = 3 - e^{-4} = 3 - 0.0183 - 2.9817$

$x = 6,\ f(6) = 3 - e^{-6} = 3 - 0.0025 = 2.9975$

x	$f(x)$
0	2
1	2.6321
2	2.8647
3	2.9502
4	2.9817
6	2.9975

Plot these points and connect them with a smooth curve.

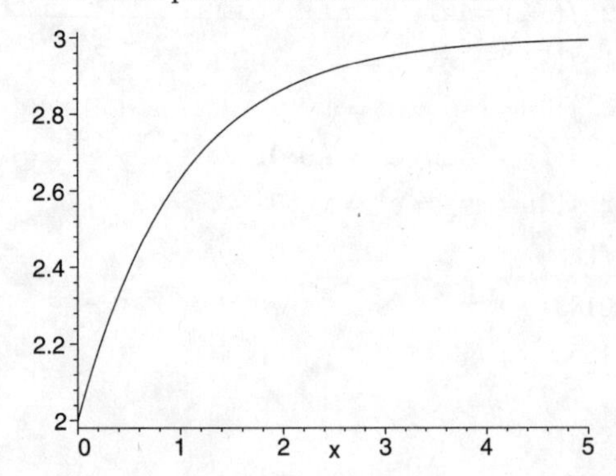

Derivatives. $f'(x) = e^{-x}$ and $f''(x) = -e^{-x}$.

Critical points of f. Since $f'(x) > 0$ for all real numbers x, we know that the derivative exists for all real numbers and there is no solution of the equation $f'(x) = 0$. There are no critical points and therefore no maximum or minimum values.

Increasing. Since $f'(x) > 0$ for all real numbers x, the function f is increasing over the entire real line, $(-\infty, \infty)$.

Inflection points. Since $f''(x) < 0$ for all real numbers x, the equation $f''(x) = 0$ has no solution and there are no points of inflection.

Concavity. Since $f''(x) < 0$ for all real numbers x, the function f' is decreasing and the graph is concave down over the entire real line.

66.

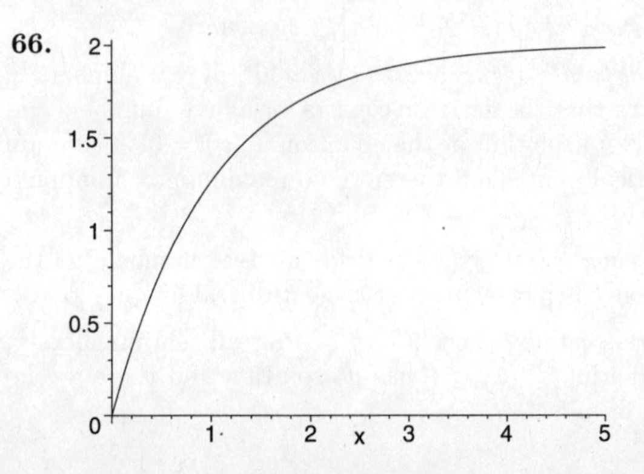

Derivatives. $f'(x) = 2e^{-x}$ and $f''(x) = -2e^{-x}$.

Critical points of f. Since $f'(x) > 0$ for all real numbers x, we know that the derivative exists for all real numbers and there is no solution of the equation $f'(x) = 0$. There are no critical points and therefore no maximum or minimum values.

Increasing. Since $f'(x) > 0$ for all real numbers x, the function f is increasing over the entire real line, $(-\infty, \infty)$.

Inflection points. Since $f''(x) < 0$ for all real numbers x, the equation $f''(x) = 0$ has no solution and there are no points of inflection.

Concavity. Since $f''(x) < 0$ for all real numbers x, the function f' is decreasing and the graph is concave down over the entire real line.

67. - 72. Left to the student

73. We first find the slope of the tangent line at $(0, 1)$, $f'(0)$:

$$f(x) = e^x$$
$$f'(x) = e^x$$
$$f'(0) = e^0 = 1$$

Then we find the equation of the line with slope 1 and containing the point $(0, 1)$:

$$y - y_1 = m(x - x_1) \quad \text{Point-slope equation}$$
$$y - 1 = 1(x - 0)$$
$$y - 1 = x$$
$$y = x + 1$$

74. $f(x) = 2e^{-3x}$

$$f'(x) = -6e^{-3x}$$
$$f'(0) = -6e^{-3\cdot 0} = -6$$
$$y - 2 = -6(x - 0)$$
$$y - 2 = -6x$$
$$y = -6x + 2$$

75. Left to the student

76. Left to the student

77. $C(t) = 10t^2e^{-t}$

a) $C(0) = 10 \cdot 0^2 \cdot e^{-0} = 0$ ppm

$$C(1) = 10 \cdot 1^2 \cdot e^{-1} \approx 10(0.367879) \approx 3.7 \text{ ppm}$$

$$C(2) = 10 \cdot 2^2 \cdot e^{-2} \approx 40(0.135335) \approx 5.4 \text{ ppm}$$

$$C(3) = 10 \cdot 3^2 \cdot e^{-3} \approx 90(0.049787) \approx 4.48 \text{ ppm}$$

$$C(10) = 10 \cdot 10^2 \cdot e^{-10} \approx 1000(0.000045) \approx 0.05 \text{ ppm}$$

b) We plot the points $(0, 0)$, $(1, 3.7)$, $(2, 5.4)$, $(3, 4.48)$, and $(10, 0.05)$ and other points as needed. Then we connect the points with a smooth curve.

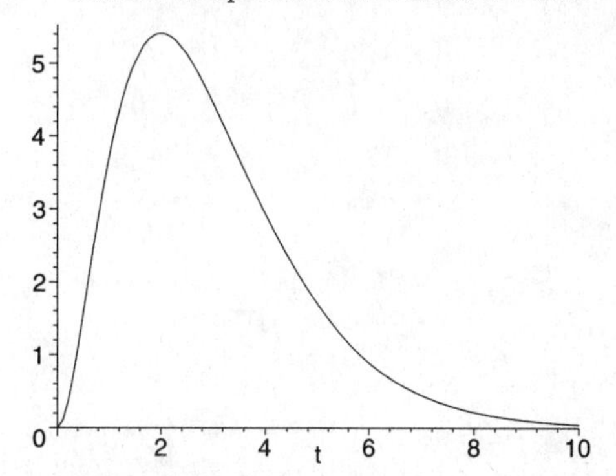

c) $C'(t) = 10t^2(-1)e^{-t} + 20te^{-t}$
$= -10te^{-t}(t - 2)$, or $10te^{-t}(2 - t)$

d) We find the maximum value of $C(t)$ on $[0, \infty)$.

$C'(t)$ exists for all t in $[0, \infty)$. We solve $C'(t) = 0$.

$$10te^{-t}(2 - t) = 0$$
$$t(2 - t) = 0 \quad (10e^t \neq 0)$$
$$t = 0 \text{ or } 2 - t = 0$$
$$t = 0 \text{ or } 2 = t$$

Since the function has two critical points and the interval is not closed, we must examine the graph to find the maximum value. We see that the maximum value of the concentration is about 5.4 ppm. It occurs at $t = 2$ hr.

e) The derivative represents the rate of change of the concentration of the medication with respect to the time t.

78. **a)** $P(0) = 40 + 60e^(-0.7(0)) = 100\%$
$P(0) = 40 + 60e^(-0.7(1)) = 69.795\%$
$P(0) = 40 + 60e^(-0.7(2)) = 54.796\%$
$P(0) = 40 + 60e^(-0.7(6)) = 40.9\%$
$P(0) = 40 + 60e^(-0.7(10)) = 40.055\%$

b) $\lim_{t\to\infty} P(t) = 40 + 60e^{-\infty} = 40\%$

c)

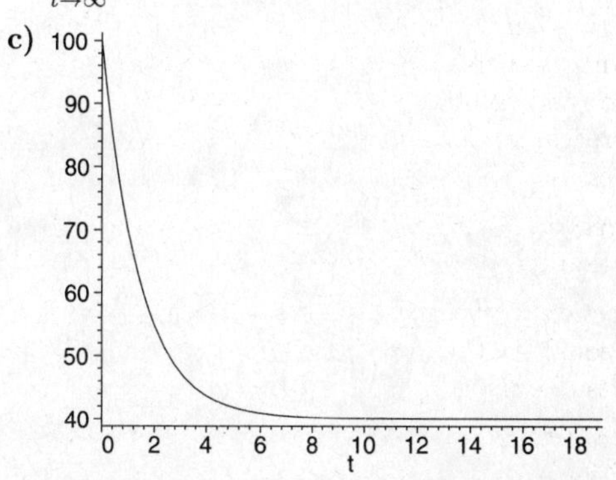

d) $P'(t) = 60 \cdot -0.7e^{-0.7t} = -42e^{-0.7t}$

e) The derivative represents the rate of change of the amount of knowledge retained by elephants with respect to time

79. a) $r(T) = 0.153\left(e^{0.141(T-9.5)} - e^{4.2159-0.153(39.4-T)}\right)$
$= 0.153\left(e^{0.141T-1.3395} - e^{0.153T-1.8123}\right)$

b)

$$\begin{aligned} r'(t) &= 0.153 \cdot \left(e^{0.141T-1.3395}\right)(0.141) - \\ &\quad 0.153 \cdot \left(e^{0.153T-1.8123}\right)(0.153) \\ &= 0.021573\left(e^{0.141T-1.3395}\right) - \\ &\quad 0.023409\left(e^{0.153T-1.8123}\right) \end{aligned}$$

80. a) $r(T) = 0.124\left(e^{0.129(T-9.5)} - e^{4.128-0.144(41.5-T)}\right)$
$= 0.124\left(e^{0.129T-1.2255} - e^{0.144T-1.848}\right)$

b)

$$\begin{aligned} r'(t) &= 0.124 \cdot \left(e^{0.129T-1.2255}\right)(0.129) - \\ &\quad 0.124 \cdot \left(e^{0.144T-1.848}\right)(0.144) \\ &= 0.015996\left(e^{0.129T-1.2255}\right) - \\ &\quad 0.017856\left(e^{0.144T-1.848}\right) \end{aligned}$$

81. $y = x^{3/2}\, e^{3x^3+2x-1}$

$$\begin{aligned} \frac{dy}{dx} &= x^{3/2}[e^{3x^3+2x-1}(9x^2+2)] + e^{3x^3+2x-1}(\frac{3}{2}x^{1/2}) \\ &= \frac{1}{2}\sqrt{x}\, e^{3x^3+2x-1}(2x(9x^2+2)+3) \\ &= \frac{1}{2}\sqrt{x}\, e^{3x^3+2x-1}(18x^3+4x^2+3) \end{aligned}$$

82. $y = x^{5/3}\, e^{x+1/x}$

$$\begin{aligned} \frac{dy}{dx} &= x^{5/3}[e^{x+1/x} \cdot (1 - \frac{1}{x^2})] + \\ &\quad e^{x+1/x} \cdot \frac{5}{3}x^{2/3} \\ &= \frac{1}{3}\, x^{2/3}\, e^{x+1/x}[x - \frac{1}{x} + 5] \\ &= \frac{1}{3}\, e^{x+1/x}\left(\frac{x^2+5x-1}{\sqrt[3]{x}}\right) \end{aligned}$$

83. $y = x^{1/2} + (e^x)^{1/2} + (xe^x)^{1/2}$

$$\begin{aligned} \frac{dy}{dx} &= \frac{1}{2\sqrt{x}} + \frac{1}{2\sqrt{e^x}} \cdot e^x + \\ &\quad \frac{1}{2\sqrt{xe^x}} \cdot (xe^x + e^x) \\ &= \frac{1}{2}\left(\frac{xe^x + e^x}{\sqrt{xe^x}} + \sqrt{e^x} + \frac{1}{\sqrt{x}}\right) \end{aligned}$$

84. $y = 3x^{-2} - e^{-2x} + 5\sqrt{x} - 2$

$$\begin{aligned} \frac{dy}{dx} &= -6x^{-3} + 2e^{-2x} + \frac{5}{2\sqrt{x}} \\ &= -6x^{-3} + \frac{2}{e^{2x}} + \frac{5}{2\sqrt{x}} \end{aligned}$$

85. $y = sin(cos\ e^x)$

$$\begin{aligned} \frac{dy}{dx} &= cos(cos\ e^x) \cdot -sin\ e^x \cdot e^x \\ &= -e^x\ sin(e^x)\ cos(cos\ e^x) \end{aligned}$$

86. $y = e^{sin\ e^x}$

$$\begin{aligned} \frac{dy}{dx} &= e^{sin\ e^x} \cdot cos\ e^x \cdot e^x \\ &= e^x\ cos\ e^x\ e^{sin\ e^x} \end{aligned}$$

87. $y = 1 + e^{1+e^{1+e^x}}$

$$\begin{aligned} \frac{dy}{dx} &= e^{1+e^{1+e^x}} \cdot e^{1+e^x} \cdot e^x \\ &= e^{x+1+e^x+1+e^{1+e^x}} \\ &= e^{x+2+e^x+e^{1+e^x}} \end{aligned}$$

88. $y = 1 + e^{1+e^{1+e}}$
Note that y is a constant therefore, $\frac{dy}{dx} = 0$

89. $f(t) = (1+t)^{1/t}$

$$\begin{aligned} f(1) &= 2^1 = 2 \\ f(0.5) &= 1.5^2 = 2.25 \\ f(0.2) &= 1.2^5 = 2.48832 \\ f(0.1) &= 1.1^{10} = 2.59374 \\ f(0.001) &= 1.001^{1000} = 2.71692 \end{aligned}$$

90. $g(t) = t^{1/(t-1)}$

$$\begin{aligned} g(0.5) &= 0.5^{-2} = 4 \\ g(0.9) &= 0.9^{-10} = 2.86797 \\ g(0.99) &= 0.99^{-100} = 2.732 \\ g(0.999) &= 0.999^{-1000} = 2.71964 \\ g(0.9998) &= 0.9998^{-5000} = 2.71855 \end{aligned}$$

91. $f(x) = x^2 e^{-x}$

$$\begin{aligned} f'(x) &= x^2(-e^{-x}) + 2x(e^{-x}) \\ &= e^{-x}(2x - x^2) \end{aligned}$$

Solve $f'(x) = 0$

$$\begin{aligned} e^{-x}(2x - x^2) &= 0 \\ 2x - x^2 &= 0 \\ x(2 - x) &= 0 \\ x &= 0 \\ \text{and} \\ x &= 2 \end{aligned}$$

We use test values to determine the nature of the critical values we found.
$f'(1) = e^{-1}(2-1) = e^{-1} > 0$
$f'(3) = e^{-3}(6-9) = -3e^{-3} < 0$
Therefore, the function has a maximum at $x = 2$. We find $f(2)$

$$\begin{aligned} f(2) &= 2^2 \cdot e^{-2} \\ &= \frac{4}{e^2} \end{aligned}$$

The maximum occurs at $(2, \frac{4}{e^2})$

92. $f(x) = xe^x$

$$\begin{aligned} f'(x) &= x(e^x) + e^x(1) \\ &= e^x(x+1) \end{aligned}$$

Solve $f'(x) = 0$

$$\begin{aligned} e^x(x+1) &= 0 \\ x+1 &= 0 \\ x &= -1 \end{aligned}$$

We use test values to determine the nature of the critical values we found.
$f'(-1/2) = e(-1/2)(-1/2+1) = \frac{1}{2}e^{-1/2} > 0$
$f'(-3/2) = e(-1/2)(-3/2+1) = \frac{-1}{2}e^{-1/2} < 0$
Therefore, the function has a minimum at $x = -1$. We find $f(-1)$

$$\begin{aligned} f(-1) &= -1 \cdot e^{-1} \\ &= \frac{-1}{e} \end{aligned}$$

The minimum occurs at $(-1, e)$

93. $F(t) = e^{-(9/(t-15)+0.56/(35-t))}$

a) $\lim_{t \to 15^+} F(t) = 0$

b) $\lim_{t \to 35^-} F(t) = 0$

c)

$$\begin{aligned} F'(t) &= e^{-(9/(t-15)+0.56/(35-t))} \cdot \\ & \quad -(-9/(t-15)^2 + 0.56/(35-t)^2 \end{aligned}$$

Solve for $F'(t) = 0$

$$\begin{aligned} \left(\frac{9}{(t-15)^2} - \frac{0.56}{(35-t)^2}\right) &= 0 \\ \frac{9}{(t-15)^2} &= \frac{0.56}{(35-t)^2} \\ 9(35-t)^2 - 0.56(t-15)^2 &= 0 \\ 11025 - 630t + 9t^2 - 0.56t^2 + 16.8t - 126 &= 0 \\ 8.44t^2 - 613.2t + 10899 &= 0 \end{aligned}$$

Using the quadratic formula we get: $x = 31.0071$. The other zero of the function falls outside the interval $15 < t < 35$. Using test values to determine the sign of the first derivative on either side of $x = 31.0071$ gives $F'(20) > 0$ and $F'(33) < 0$. Therefore $F(t)$ has a maximum at $x = 31.0071$. Find $F(31.0071)$

$$\begin{aligned} F(31.0071) &= e^{-(9/16.0071+0.56/3.9983)} \\ &= e^{-0.702499766} \\ &= 0.49535 \end{aligned}$$

94. $F(t) = e^{-(9/(t-15)+0.69/(31-t))}$

a) $\lim_{t \to 15^+} F(t) = 0$

b) $\lim_{t \to 31^-} F(t) = 0.56978$

c)

$$\begin{aligned} F'(t) &= e^{-(9/(t-15)+0.69/(31-t))} \cdot \\ & \quad -(-9/(t-15)^2 + 0.69/(31-t)^2 \end{aligned}$$

Solve for $F'(t) = 0$

$$\begin{aligned} \left(\frac{9}{(t-15)^2} - \frac{0.69}{(31-t)^2}\right) &= 0 \\ \frac{9}{(t-15)^2} &= \frac{0.69}{(31-t)^2} \\ 9(31-t)^2 - 0.69(t-15)^2 &= 0 \\ 8649 - 558t + 9t^2 - 0.69t^2 + 20.7t - 155.25 &= 0 \\ 8.31t^2 - 537.3t + 8493.75 &= 0 \end{aligned}$$

Using the quadratic formula we get: $x = 27.53047$. The other zero of the function falls outside the interval $15 < t < 31$. Using test values to determine the sign of the first derivative on either side of $x = 31.0071$ gives $F'(20) > 0$ and $F'(30) < 0$. Therefore $F(t)$ has a maximum at $x = 27.53047$. Find $F(27.53047)$

$$\begin{aligned} F(27.53047) &= e^{-(9/12.53047+0.69/3.46953)} \\ &= e^{-0.7182491958} \\ &= 0.39967 \end{aligned}$$

95. $D(t) = 34.4 - \frac{30.48}{1+29.44e^{-0.072t}}$

a) Find $D(110)$ and $D(150)$

$$\begin{aligned} D(110) &= 34.4 - \frac{30.48}{1+29.44e^{-0.072(110)}} \\ &= 34.4 - 30.157 \\ &= 4.253 \\ D(150) &= 34.4 - \frac{30.48}{1+29.44e^{-0.072(150)}} \\ &= 34.4 - 30.462 \\ &= 3.938 \end{aligned}$$

b) $\lim t \to \infty D(t) = 3.92$ Which is to say that the annual death rate in Mexico will not go below 3.92 per 1000 citizens.

96. a)

$$\begin{aligned} D'(t) &= -30.48(1+29.44e^{-0.072t})^{-2} \times \\ & \quad (29.44e^{-0.072t}(-0.072)) \\ &= \frac{64.608e^{-0.072t}}{(1+29.44e^{-0.072t})^2} \end{aligned}$$

The rate of change in the annual death rate per 1000 poulation per year is given by $\frac{64.608e^{-0.072t}}{(1+29.44e^{-0.072t})^2}$

b)

$$\begin{aligned} D'(108) - D'(107) &= \frac{64.608e^{-0.072(108)}}{(1+29.44e^{-0.072(108)})^2} - \\ & \quad \frac{64.608e^{-0.072(107)}}{(1+29.44e^{-0.072(107)})^2} \\ &= 0.02646 - 0.02838 \\ &= -0.00192 \end{aligned}$$

Between 2007 and 2008 the annual death rate will decline by 0.00192 deaths per 1000 population

97. Rewrite as follows $f(x) = 3.2e^{1.07x} - 5$, which gives $f'(x) = 3.424e^{1.07x}$, and now try to find the zero of $f(x)$ using Newton's method starting with a guess of $x_n = 1.5$

$$\begin{aligned} x_{n+1} &= x_n - f(x_n)/f'(x_n) \\ &= 1.5 - \frac{10.929}{17.044} = 0.8588 \\ &= 0.8588 - \frac{3.021}{8.5825} = 0.5068 \\ &= 0.5068 - \frac{0.50373}{5.889} = 0.4213 \\ &= 0.4213 - \frac{0.02257}{5.3742} = 0.4171 \\ &= 0.4171 - \frac{0.000049}{5.3501} = 0.417091 \\ &= 0.417091 - \frac{0.0000013}{5.35} = 0.417091 \end{aligned}$$

98. Rewrite as follows $f(x) = 2e^{0.04x} - 6.3x$, which gives $f'(x) = 0.08e^{0.04x} - 6.3$, and now try to find the zero of $f(x)$ using Newton's method starting with a guess of $x_n = 3$

$$\begin{aligned} x_{n+1} &= x_n - f(x_n)/f'(x_n) \\ &= 3 - \frac{-16.65}{-6.21} = 0.318841 \\ &= 0.318841 - \frac{0.01697}{-6.219} = 0.321570 \\ &= 0.321570 - \frac{0.00000076}{-6.219} = 0.321570 \end{aligned}$$

99. $P = 1 - \frac{1}{1+e^{-0.055-0.083T}}$

$$\begin{aligned} P(20) &= 1 - \frac{1}{1+e^{-0.055-0.083(20)}} \\ &= 1 - 0.84748 \\ &= 0.15252 \\ P(22) &= 1 - \frac{1}{1+e^{-0.055-0.083(22)}} \\ &= 1 - 0.86773 \\ &= 0.13227 \end{aligned}$$

The probability is declining by $0.15252 - 0.13227 = 0.02025$ per hour

100. $P = 1 - \frac{1}{1+e^{-5.297+31.669m}}$

$$\begin{aligned} P(0.2) &= 1 - \frac{1}{1+e^{-5.297+31.669(0.2)}} \\ &= 1 - 0.26177 \\ &= 0.73823 \\ P(0.3) &= 1 - \frac{1}{1+e^{-5.297+31.669(0.3)}} \\ &= 1 - 0.01472 \\ &= 0.98528 \end{aligned}$$

The probability is declining by $0.98528 - 0.73823 = 0.24705$ per week

101. The student use the power rule on an exponent that is not a constant, there is the error in the students work. The correct answer is $\frac{d}{dx}e^x = e^x$

102. 3^x only has positive values in its range while x^3 has all real numbers as its range. When $x < 0$, 3^x increases at a slower rate than x^3 but when $x > 0$, 3^x increases at a higher rate than that of x^3

103. Left to the student

104.

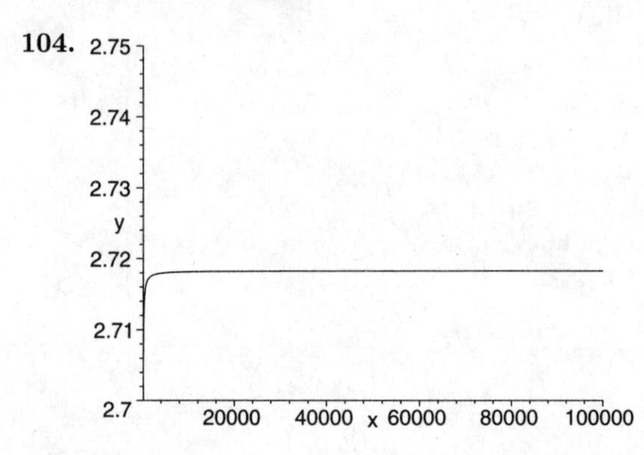

105. Relative minimum at $(0, 0)$ and a relative maximum at $(2, 0.5413)$

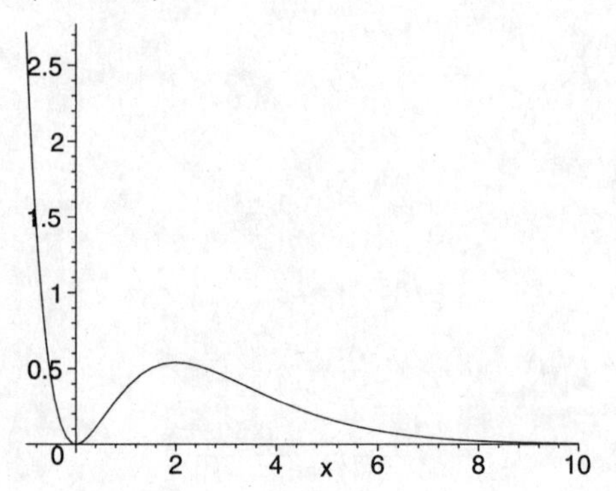

106. Relative maximum at $(0, 1)$

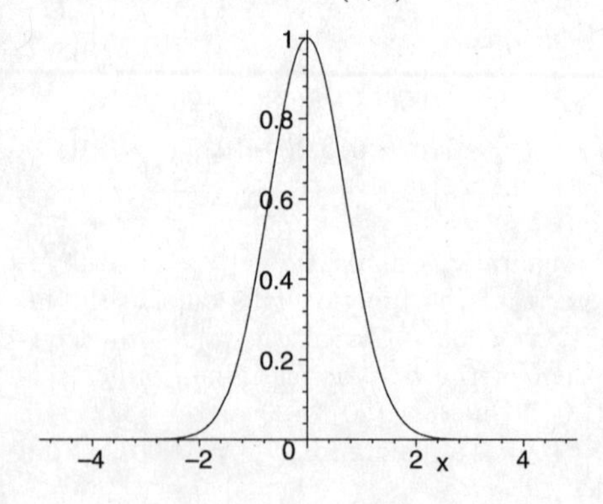

107.

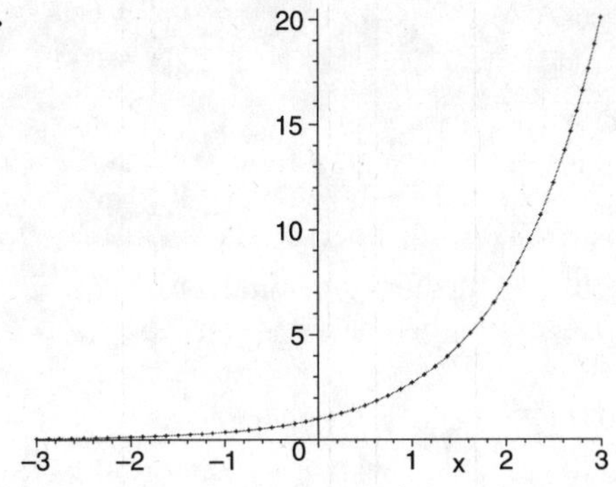

108.

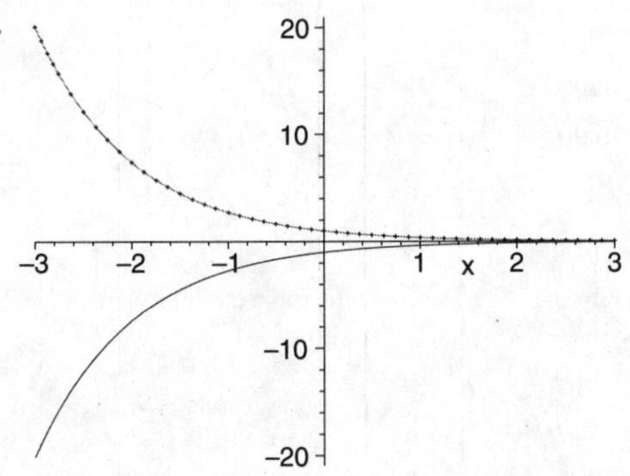

109.

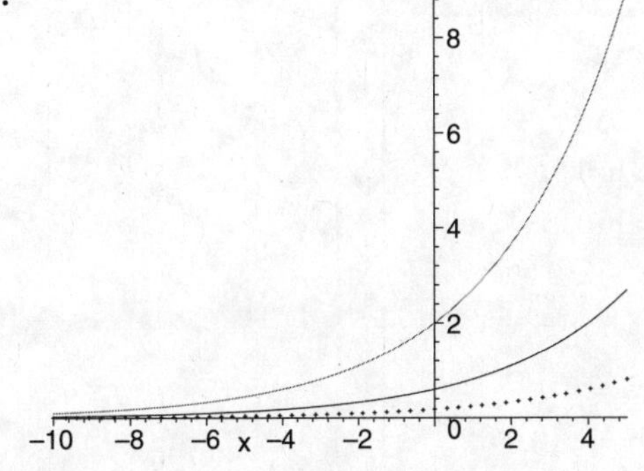

110.

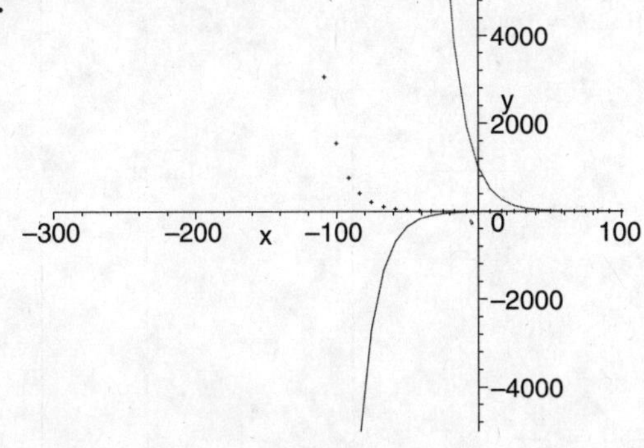

Exercise Set 4.2

1. $\log_2 8 = 3$ Logarithmic equation

$2^3 = 8$ Exponential equation;
2 is the base, 3 is the exponent

2. $3^4 = 81$

3. $\log_8 2 = \dfrac{1}{3}$ Logarithmic equation

$8^{1/3} = 2$ Exponential equation;
8 is the base, 1/3 is the exponent

4. $27^{1/3} = 3$

5. $\log_a K = J$ Logarithmic equation

$a^J = K$ Exponential equation;
a is the base, J is the exponent

6. $a^K = J$

7. $-\log_{10} h = p$ Logarithmic equation

$\log_{10} h = -p$ Multiplying by -1

$10^{-p} = h$ Exponential equation;
10 is the base, $-p$ is the exponent.

8. $-\log_b V = w$

$\log_b V = -w$

$b^{-w} = V$

9. $e^M = b$ Exponential equation;
e is the base, M is the exponent

$\log_e b = M$ Logarithmic equation

or $\ln b = M$ $\ln b$ is the abbreviation for $\log_e b$

10. $\log_e p = t$, or $\ln p = t$

11. $10^2 = 100$ Exponential equation; 10 is the base, 2 is the exponent

$\log_{10} 100 = 2$ Logarithmic equation

12. $\log_{10} 1000 = 3$

13. $10^{-1} = 0.1$ Exponential equation; 10 is the base, -1 is the exponent

$\log_{10} 0.1 = -1$ Logarithmic equation

14. $\log_{10} 0.01 = -2$

15. $M^p = V$ Exponential equation; M is the base, p is the exponent

$\log_M V = p$ Logarithmic equation

16. $\log_Q T = n$

17.
$$\begin{aligned}\log_b 15 &= \log_b 3 \cdot 5\\ &= \log_b 3 + \log_b 5 \quad \text{(P1)}\\ &= 1.099 + 1.609\\ &= 2.708\end{aligned}$$

18.
$$\begin{aligned}\log_b \frac{3}{5} &= \log_b 3 - \log_b 5 \quad \text{(P2)}\\ &= 1.099 - 1.609\\ &= -0.51\end{aligned}$$

19.
$$\begin{aligned}\log_b \frac{1}{5} &= \log_b 1 - \log_b 5 \quad \text{(P2)}\\ &= 0 - \log_b 5 \quad \text{(P6)}\\ &= -\log_b 5\\ &= -1.609\end{aligned}$$

20. $\log_b \sqrt{b^3} = \log_b b^{3/2} = \frac{3}{2}$

21. $\log_b 5b = \log_b 5 + \log_b b$ (P1)
$= 1.609 + 1$ (P4)
$= 2.609$

22. $\log_b 75 = \log_b 3 \cdot 5^2$
$= \log_b 3 + \log_b 5^2$ (P1)
$= \log_b 3 + 2\log_b 5$ (P3)
$= 1.099 + 2(1.609)$
$= 4.317$

23. $\ln 20 = \ln 4 \cdot 5$
$= \ln 4 + \ln 5$ (P1)
$= 1.3863 + 1.6094$
$= 2.9957$

24. $\ln \frac{5}{4} = \ln 5 - \ln 4$ (P2)
$= 1.6094 - 1.3863$
$= 0.2231$

25. $\ln \frac{1}{4} = \ln 1 - \ln 4$ (P2)
$= 0 - 1.3863$ (P6)
$= -1.3863$

26. $\ln 4e = \ln 4 + \ln e$ (P1)
$= 1.3863 + 1$ (P4)
$= 2.3863$

27. $\ln \sqrt{e^8} = \ln e^{8/2}$
$= \ln e^4$
$= 4$ (P5)

28. $\ln 100 = \ln 4 \cdot 5^2$
$= \ln 4 + \ln 5^2$ (P1)
$= \ln 4 + 2\ln 5$ (P3)
$= 1.3863 + 2(1.6094)$
$= 4.6051$

29. $\ln 3927 = 8.275631$ Using a calculator and rounding to six decimal places

30. 9.392662

31. $\ln 0.0182 = -4.006334$

32. -7.047017

33. $\ln 8100 = 8.999619$

34. -4.509860

35. $e^t = 100$
$\ln e^t = \ln 100$ Taking the natural logarithm on both sides
$t = \ln 100$ (P5)
$t = 4.605170$ Using a calculator
$t \approx 4.6$

36. $e^t = 1000$
$\ln e^t = \ln 1000$
$t = \ln 1000$
$t \approx 6.9$

37. $e^t = 60$
$\ln e^t = \ln 60$ Taking the natural logarithm on both sides
$t = \ln 60$ (P5)
$t = 4.094345$ Using a calculator
$t \approx 4.1$

38. $e^t = 90$
$\ln e^t = \ln 90$
$t = \ln 90$
$t \approx 4.5$

39. $e^{-t} = 0.1$
$\ln e^{-t} = \ln 0.1$ Taking the natural logarithm on both sides
$-t = \ln 0.1$ (P5)
$t = -\ln 0.1$
$t = -(-2.302585)$ Using a calculator
$t = 2.302585$
$t \approx 2.3$

40. $e^{-t} = 0.01$
$\ln e^{-t} = \ln 0.01$
$-t = \ln 0.01$
$t = -\ln 0.01$
$t \approx 4.6$

41. $e^{-0.02t} = 0.06$
$\ln e^{-0.02t} = \ln 0.06$ Taking the natural logarithm on both sides
$-0.02t = \ln 0.06$ (P5)
$t = \frac{\ln 0.06}{-0.02}$
$t = \frac{-2.813411}{-0.02}$ Using a calculator
$t \approx 141$

42. $e^{0.07t} = 2$
$\ln e^{0.07t} = \ln 2$
$0.07t = \ln 2$
$t = \frac{\ln 2}{0.07}$
$t \approx 9.9$

43. $y = -6\ln x$
$\frac{dy}{dx} = -6 \cdot \frac{1}{x}$ $\left[\frac{d}{dx}[c \cdot f(x)] = c \cdot f'(x)\right]$
$= -\frac{6}{x}$

44. $y = -4\ln x$

$\dfrac{dy}{dx} = -4 \cdot \dfrac{1}{x} = -\dfrac{4}{x}$

45. $y = x^4 \ln x - \dfrac{1}{2}x^2$

$\dfrac{dy}{dx} = \underbrace{x^4 \cdot \dfrac{1}{x} + 4x^3 \cdot \ln x}_{} - \dfrac{1}{2} \cdot 2x$ ← Product Rule on $x^4 \ln x$

$= x^3 + 4x^3 \ln x - x$

$= x^3(1 + 4\ln x) - x$

46. $y = x^5 \ln x - \dfrac{1}{4}x^4$

$\dfrac{dy}{dx} = x^5 \cdot \dfrac{1}{x} + 5x^4 \ln x - x^3$

$= x^4 + 5x^4 \ln x - x^3$

$= x^4(1 + 5\ln x) - x^3$

47. $y = \dfrac{\ln x}{x^4}$

$\dfrac{dy}{dx} = \dfrac{x^4 \cdot \frac{1}{x} - 4x^3 \cdot \ln x}{x^8}$ Using the Quotient Rule

$= \dfrac{x^3 - 4x^3 \ln x}{x^8}$

$= \dfrac{x^3(1 - 4\ln x)}{x^3 \cdot x^5}$ Factoring both numerator and denominator

$= \dfrac{x^3}{x^3} \cdot \dfrac{1 - 4\ln x}{x^5}$

$= \dfrac{1 - 4\ln x}{x^5}$ Simplifying

48. $y = \dfrac{\ln x}{x^5}$

$\dfrac{dy}{dx} = \dfrac{x^5 \cdot \frac{1}{x} - 5x^4 \ln x}{x^{10}}$

$= \dfrac{x^4 - 5x^4 \ln x}{x^{10}}$

$= \dfrac{x^4(1 - 5\ln x)}{x^{10}}$

$= \dfrac{1 - 5\ln x}{x^6}$

49. $y = \ln \dfrac{x}{4}$

$y = \ln x - \ln 4$ (P2)

$\dfrac{dy}{dx} = \dfrac{1}{x} - 0$

$= \dfrac{1}{x}$

50. $y = \ln \dfrac{x}{2} = \ln x - \ln 2$

$\dfrac{dy}{dx} = \dfrac{1}{x}$

51. $y = \ln\ cos\ x$

$\dfrac{dy}{dx} = \dfrac{-sin\ x}{cos\ x}$

$= -tan\ x$

52. $y = \ln\ tan\ x$

$\dfrac{dy}{dx} = \dfrac{sec^2\ x}{tan\ x}$

$= \dfrac{sin^3\ x}{cos^3\ x}$

53. $f(x) = \ln(\ln 4x)$

$f'(x) = 4 \cdot \dfrac{1}{4x} \cdot \dfrac{1}{\ln 4x}$ $\left[\dfrac{d}{dx} \ln g(x) = g'(x) \cdot \dfrac{1}{g(x)}\right]$

$\left[\dfrac{d}{dx} \ln 4x = 4 \cdot \dfrac{1}{4x}\right]$

$= \dfrac{1}{x} \cdot \dfrac{1}{\ln 4x}$

$= \dfrac{1}{x \ln 4x}$

54. $f(x) = \ln(\ln 3x)$

$f'(x) = 3 \cdot \dfrac{1}{3x} \cdot \dfrac{1}{\ln 3x} = \dfrac{1}{x \ln 3x}$

55. $f(x) = \ln\left(\dfrac{x^2 - 7}{x}\right)$

$f'(x) = \dfrac{x \cdot 2x - 1 \cdot (x^2 - 7)}{x^2} \cdot \dfrac{1}{\frac{x^2 - 7}{x}}$

$\left[\dfrac{d}{dx} \ln g(x) = g'(x) \cdot \dfrac{1}{g(x)}\right]$

(Using Quotient Rule to find $\dfrac{d}{dx}\dfrac{x^2 - 7}{x}$)

$= \dfrac{2x^2 - x^2 + 7}{x^2} \cdot \dfrac{x}{x^2 - 7}$ $\left[\dfrac{1}{\frac{x^2 - 7}{x}} = \dfrac{x}{x^2 - 7}\right]$

$= \dfrac{x^2 + 7}{x^2} \cdot \dfrac{x}{x^2 - 7}$

$= \dfrac{x}{x} \cdot \dfrac{x^2 + 7}{x(x^2 - 7)}$

$= \dfrac{x^2 + 7}{x(x^2 - 7)}$

We could also have done this another way using P2:

$f(x) = \ln\left(\dfrac{x^2 - 7}{x}\right)$

$= \ln(x^2 - 7) - \ln x$

$$f'(x) = 2x \cdot \frac{1}{x^2-7} - \frac{1}{x}$$
$$= \frac{2x}{x^2-7} \cdot \frac{x}{x} - \frac{1}{x} \cdot \frac{x^2-7}{x^2-7}$$ Multiplying by forms of 1
$$= \frac{2x^2}{x(x^2-7)} - \frac{x^2-7}{x(x^2-7)}$$
$$= \frac{2x^2-x^2+7}{x(x^2-7)}$$
$$= \frac{x^2+7}{x(x^2-7)}$$

56. $f(x) = \ln\left(\frac{x^2+5}{x}\right) = \ln(x^2+5) - \ln x$
$$f'(x) = 2x \cdot \frac{1}{x^2+5} - \frac{1}{x}$$
$$= \frac{2x}{x^2+5} \cdot \frac{x}{x} - \frac{1}{x} \cdot \frac{x^2+5}{x^2+5}$$
$$= \frac{2x^2-x^2-5}{x(x^2+5)}$$
$$= \frac{x^2-5}{x(x^2+5)}$$

57. $f(x) = e^x \ln x$
$$f'(x) = e^x \cdot \frac{1}{x} + e^x \cdot \ln x$$ Using the Product Rule
$$= e^x\left(\frac{1}{x} + \ln x\right)$$

58. $f(x) = e^{2x} \ln x$
$$f'(x) = e^{2x} \cdot \frac{1}{x} + 2e^{2x} \ln x$$
$$= e^{2x}\left(\frac{1}{x} + 2\ \ln x\right)$$

59. $f(x) = e^x sec x$
$$f'(x) = e^x \cdot sec\ x\ tan\ x + e^x sec\ x$$
$$= e^x\ sec\ x(tan\ x + 1)$$

60. $f(x) = (sin\ 2x)e^x$
$$f'(x) = (sin\ 2x)e^x + 2e^x cos\ 2x$$
$$= e^x(sin\ 2x + 2\ cos\ 2x)$$

61. $f(x) = \ln(e^x + 1)$
$$f'(x) = e^x \cdot \frac{1}{e^x+1} \qquad \left[\frac{d}{dx}(e^x+1) = e^x\right]$$
$$= \frac{e^x}{e^x+1}$$

62. $f(x) = \ln(e^x - 2)$
$$f'(x) = e^x \cdot \frac{1}{e^x-2} = \frac{e^x}{e^x-2}$$

63. $f(x) = (\ln x)^2$
$$f'(x) = 2(\ln x)^1 \cdot \frac{1}{x}$$ Extended Power Rule
$$= \frac{2\ln x}{x}$$

64. $f(x) = (\ln x)^3$
$$f'(x) = 3(\ln x)^2 \cdot \frac{1}{x} = \frac{3(\ln x)^2}{x}$$

65. $y = (\ln x)^{-4}$
$$\frac{dy}{dx} = -4(\ln x)^{-5} \cdot \frac{1}{x}$$
$$= \frac{-4(\ln x)^{-5}}{x}$$

66. $y = (\ln x)^n$
$$\frac{dy}{dx} = n(\ln x)^{n-1} \cdot \frac{1}{x} = \frac{n(\ln x)^{n-1}}{x}$$

67. $f(t) = \ln(t^3+1)^5$
$$f'(t) = 5(t^3+1)^4 \cdot 3t^2 \cdot \frac{1}{(t^3+1)^5}$$
$$= \frac{15t^2}{t^3+1}$$ Simplifying

68. $f(t) = \ln(t^2+t)^3$
$$f'(t) = 3(t^2+t)^2(2t+1) \cdot \frac{1}{(t^2+t)^3}$$
$$= \frac{3(2t+1)}{t^2+t}$$

69. $f(x) = [\ln(x+5)]^4$
$$f'(x) = 4[\ln(x+5)]^3 \cdot 1 \cdot \frac{1}{x+5}$$
$$= \frac{4[\ln(x+5)]^3}{x+5}$$

70. $f(x) = \ln[\ln(\ln 3x)]$
$$f'(x) = \frac{d}{dx}\ln(\ln 3x) \cdot \frac{1}{\ln(\ln 3x)}$$
$$= \frac{d}{dx}\ln 3x \cdot \frac{1}{\ln 3x} \cdot \frac{1}{\ln(\ln 3x)}$$
$$= 3 \cdot \frac{1}{3x} \cdot \frac{1}{\ln 3x} \cdot \frac{1}{\ln(\ln 3x)}$$
$$= \frac{1}{x \cdot \ln 3x \cdot \ln(\ln 3x)}$$

71. $f(t) = \ln[(t^3+3)(t^2-1)]$
$$f'(t) = [(t^3+3)(2t)+(3t^2)(t^2-1)] \cdot \frac{1}{(t^3+3)(t^2-1)}$$
$$= \frac{2t^4+6t+3t^4-3t^2}{(t^3+3)(t^2-1)}$$
$$= \frac{5t^4-3t^2+6t}{(t^3+3)(t^2-1)}$$

72. $f(t) = \ln\frac{1-t}{1+t} = \ln(1-t) - \ln(1+t)$
$$f'(t) = -1 \cdot \frac{1}{1-t} - \frac{1}{1+t} = \frac{-(1+t)-(1-t)}{(1-t)(1+t)} =$$
$$\frac{-1-t-1+t}{1-t^2} = \frac{-2}{1-t^2}$$

73. $y = \ln \dfrac{x^5}{(8x+5)^2}$

$y = \ln x^5 - \ln(8x+5)^2$ Property 2

$\dfrac{dy}{dx} = 5x^4 \cdot \dfrac{1}{x^5} - 2(8x+5)\cdot 8 \cdot \dfrac{1}{(8x+5)^2}$

$= \dfrac{5}{x} - \dfrac{16}{8x+5}$

$= \dfrac{5}{x}\cdot\dfrac{8x+5}{8x+5} - \dfrac{16}{8x+5}\cdot\dfrac{x}{x}$ Multiplying by forms of 1

$= \dfrac{40x+25-16x}{x(8x+5)}$

$= \dfrac{24x+25}{8x^2+5x}$

74. $y = \ln\sqrt{5+x^2} = \dfrac{1}{2}\ln(5+x^2)$

$\dfrac{dy}{dx} = \dfrac{1}{2}\cdot 2x \cdot \dfrac{1}{5+x^2} = \dfrac{x}{5+x^2}$

75. $y = \dfrac{\ln\ sin\ x}{sin\ x}$

$\dfrac{dy}{dx} = \dfrac{sin\ x\left(\frac{1}{sin\ x}\cdot cos\ x\right) - \ln\ sin\ x \cdot cos\ x}{sin^2\ x}$

$= \dfrac{cos\ x(1-\ln\ sin\ x)}{sin^2\ x}$

76. $f(x) = \dfrac{1}{5}x^5\left(\ln x - \dfrac{1}{5}\right)$

$f'(x) = \dfrac{1}{5}x^5\cdot\dfrac{1}{x} + x^4\left(\ln x - \dfrac{1}{5}\right)$

$= \dfrac{1}{5}x^4 + x^4\ln x - \dfrac{1}{5}x^4$

$= x^4 \ln x$

77. $y = \dfrac{x^{n+1}}{n+1}\left(\ln x - \dfrac{1}{n+1}\right)$

$\dfrac{dy}{dx} = \dfrac{x^{n+1}}{n+1}\left(\dfrac{1}{x} - 0\right) + x^n\left(\ln x - \dfrac{1}{n+1}\right)$

$\left[\dfrac{d}{dx}\dfrac{x^{n+1}}{n+1} = \dfrac{1}{n+1}\cdot(n+1)x^n = x^n\right]$

$= \dfrac{x^n}{n+1} + x^n\ln x - \dfrac{x^n}{n+1}$

$= x^n \ln x$

78. $y = \dfrac{x\ln x - x}{x^2+1}$

$\dfrac{dy}{dx} = \dfrac{(x^2+1)\left(x\cdot\dfrac{1}{x} + \ln x - 1\right) - 2x(x\ln x - x)}{(x^2+1)^2}$

$= \dfrac{(x^2+1)(\ln x) - 2x(x\ln x - x)}{(x^2+1)^2}$

$= \dfrac{x^2\ln x + \ln x - 2x^2\ln x + 2x^2}{(x^2+1)^2}$

$= \dfrac{-x^2\ln x + \ln x + 2x^2}{(x^2+1)^2}$

79. $y = \ln\left(t + \sqrt{1+t^2}\right) = \ln[t + (1+t^2)^{1/2}]$

$\dfrac{dy}{dx} = \left[1 + \dfrac{1}{2}(1+t^2)^{-1/2}\cdot 2t\right]\cdot\dfrac{1}{t+(1+t^2)^{1/2}}$

$= \left(1 + \dfrac{t}{(1+t^2)^{1/2}}\right)\cdot\dfrac{1}{t+(1+t^2)^{1/2}}$

$= \left(\dfrac{(1+t^2)^{1/2}}{(1+t^2)^{1/2}} + \dfrac{t}{(1+t^2)^{1/2}}\right)\cdot\dfrac{1}{t+(1+t^2)^{1/2}}$

$= \dfrac{t+(1+t^2)^{1/2}}{(1+t^2)^{1/2}}\cdot\dfrac{1}{t+(1+t^2)^{1/2}}$

$= \dfrac{1}{(1+t^2)^{1/2}}$

$= \dfrac{1}{\sqrt{1+t^2}}$

80. $f(x) = \ln\dfrac{1+\sqrt{x}}{1-\sqrt{x}} = \ln(1+\sqrt{x}) - \ln(1-\sqrt{x})$

$f'(x) = \dfrac{1}{2}x^{-1/2}\cdot\dfrac{1}{1+\sqrt{x}} + \dfrac{1}{2}x^{-1/2}\cdot\dfrac{1}{1-\sqrt{x}}$

$= \dfrac{1}{2\sqrt{x}(1+\sqrt{x})} + \dfrac{1}{2\sqrt{x}(1-\sqrt{x})}$

$= \dfrac{1-\sqrt{x}+1+\sqrt{x}}{2\sqrt{x}(1+\sqrt{x})(1-\sqrt{x})}$

$= \dfrac{2}{2\sqrt{x}(1-x)}$

$= \dfrac{1}{\sqrt{x} - x\sqrt{x}}$

$= \dfrac{\sqrt{x}}{x-x^2}$ Rationalizing the denominator

81. $y = sin(\ln\ x)$

$\dfrac{dy}{dx} = cos(\ln\ x)\cdot\dfrac{1}{x}$

$= \dfrac{cos(\ln\ x)}{x}$

82. $y = cot(2\ \ln\ x)$

$\dfrac{dy}{dx} = -cot(2\ \ln\ x)\ csc(2\ \ln\ x)\cdot\dfrac{1}{x}\cdot 2$

$= \dfrac{-2\ cot(2\ \ln\ x)\ csc(2\ \ln\ x)}{x}$

83. $y = (sin\ x)\ln\ (tan\ x)$

$\dfrac{dy}{dx} = (sin\ x)\left(\dfrac{sec^2\ x}{tan\ x}\right) + (cos\ x)\ \ln\ (tan\ x)$

$= cos\ x\ \ln\ (tan\ x) + sec\ x$

84. $y = (cos\ x)\ln\ (cot\ x)$

$\dfrac{dy}{dx} = (cos\ x)\left(\dfrac{-csc^2\ x}{cot\ x}\right) + (-sin\ x)\ \ln\ (cot\ x)$

$= -sin\ x\ \ln\ (cot\ x) - csc\ x$

85. $y = \ln(\sec 2x + \tan 2x)$

$$\frac{dy}{dx} = \frac{1}{\sec 2x + \tan 2x} \cdot (2\sec 2x \tan 2x + 2\sec^2 2x)$$
$$= \frac{2\sec 2x(\sec 2x + \tan 2x)}{\sec 2x + \tan 2x}$$
$$= 2\sec 2x$$

86. $y = \ln(\csc 3x + \cot 3x)$

$$\frac{dy}{dx} = \frac{-1}{\csc 3x + \cot 3x} \cdot (-3\csc 3x \cot 3x - 3\csc^2 3x)$$
$$= \frac{-3\csc 3x(\csc 3x + \cot 3x)}{\csc 3x + \cot 3x}$$
$$= -3\csc 3x$$

87. $y = \ln\left(\frac{e^x - e^{-x}}{e^x + e^{-x}}\right)$

$$\frac{dy}{dx} = \frac{e^x + e^{-x}}{e^x - e^{-x}} \cdot \frac{(e^x + e^{-x})(e^x + e^{-x})}{(e^x + e^{-x})^2} - \frac{e^x + e^{-x}}{e^x - e^{-x}} \cdot \frac{(e^x - e^{-x})(e^x - e^{-x})}{(e^x + e^{-x})^2}$$
$$= \frac{e^{2x} + 1 + 1 + e^{-2x} - e^{2x} + 1 + 1 - e^{-2x}}{(e^x - e^{-x})(e^x + e^{-x})}$$
$$= \frac{4}{e^{2x} - e^{-2x}}$$
$$= \frac{4e^{2x}}{e^{4x} - 1}$$

88. $y = \ln\left(\frac{e^x + e^{-x}}{e^x - e^{-x}}\right)$

$$\frac{dy}{dx} = \frac{e^x - e^{-x}}{e^x + e^{-x}} \cdot \frac{(e^x - e^{-x})(e^x - e^{-x})}{(e^x - e^{-x})^2} - \frac{e^x - e^{-x}}{e^x + e^{-x}} \cdot \frac{(e^x + e^{-x})(e^x + e^{-x})}{(e^x - e^{-x})^2}$$
$$= \frac{e^{2x} - 1 - 1 + e^{-2x} - e^{2x} - 1 - 1 - e^{-2x}}{(e^x - e^{-x})(e^x + e^{-x})}$$
$$= \frac{4}{e^{2x} - e^{-2x}}$$
$$= \frac{-4e^{2x}}{e^{4x} - 1}$$

89. Using the point $(10, 164)$ and $(70, 3045)$ we get

$$164 = A(10)^c$$
$$\rightarrow$$
$$A = \frac{164}{10^c}$$
$$3045 = A(70)^c$$
$$\rightarrow$$
$$3045 = \frac{164}{10^c} \cdot 70^c$$
$$\frac{3045}{164} = (7)^c$$
$$c = \log_7\left(\frac{3045}{164}\right)$$
$$= 1.501297$$
$$\rightarrow$$
$$A = \frac{164}{10^{1.501297}}$$
$$= 5.170666$$

Therefore, $y = 5.17x^{1.50}$

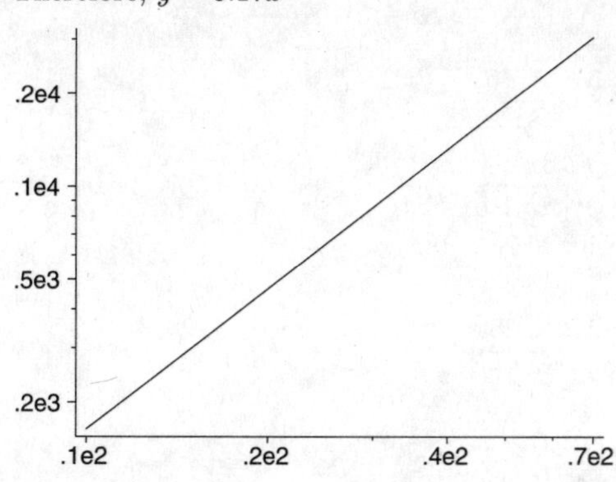

90.

$$0.3 = 100^c A$$
$$\rightarrow$$
$$A = \frac{0.3}{100^c}$$
$$0.0775 = 1500^c \cdot \frac{0.3}{100^c}$$
$$0.25834 = 15^c$$
$$c = \frac{\ln(0.25834)}{\ln(15)}$$
$$= -0.4998$$
$$\rightarrow$$
$$A = \frac{0.3}{100^{-0.4998}}$$
$$= 2.997$$

$y = 2.997x^{-0.4998}$

91. Using the point $(30, 25)$ and $(1000000000, 250)$ we get

$$250 = A1000000000^c$$
$$\rightarrow$$
$$A = \frac{250}{1000000000^c}$$
$$25 = A30^c$$
$$\rightarrow$$
$$25 = \frac{250}{1000000000^c} \cdot 30^c$$
$$\frac{25}{250} = \left(\frac{3}{100000000}\right)^c$$

$$\begin{aligned} c &= \frac{\log(\frac{1}{10})}{\log(\frac{3}{100000000})} \\ &= 0.133 \end{aligned}$$

$\rightarrow$

$$\begin{aligned} A &= \frac{250}{1000000000^{0.133}} \\ &= 15.907 \end{aligned}$$

Therefore, $y = 15.907x^{0.133}$

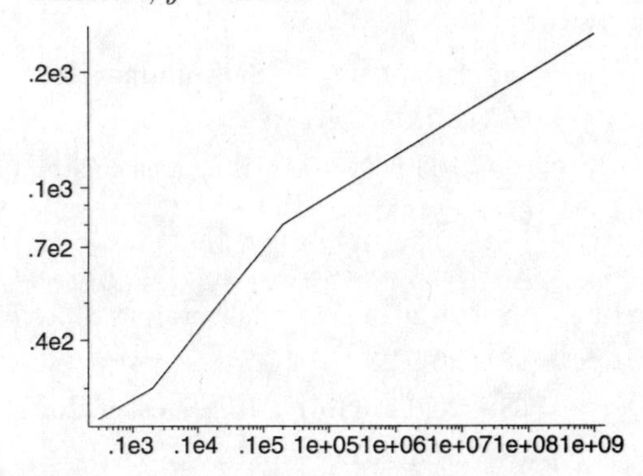

92.

$$\begin{aligned} 15 &= 2^c A \\ &\rightarrow \\ A &= \frac{15}{2^c} \\ 270 &= 2500^c \cdot \frac{15}{2^c} \\ 18 &= 1250^c \\ c &= \frac{\ln(18)}{\ln(1250)} \\ &= 0.40533 \\ &\rightarrow \\ A &= \frac{15}{2^{0.40533}} \\ &= 11.326 \end{aligned}$$

$y = 11.326x^{0.40533}$

.2e3
.1e3
.5e2
.2e2
.1e2 .1e3 .1e4

93. Using the point (25000, 1079) and (500000, 29) we get

$$\begin{aligned} 1079 &= A25000^c \\ &\rightarrow \\ A &= \frac{1079}{25000^c} \\ 29 &= A500000^c \\ &\rightarrow \\ 29 &= \frac{1079}{25000^c} \cdot 500000^c \\ \frac{29}{1079} &= (20)^c \\ c &= \frac{\log(\frac{29}{1079})}{\log(20)} \\ &= -1.207 \\ &\rightarrow \\ A &= \frac{1079}{25000^{-1.207}} \\ &= 219449722.1 \end{aligned}$$

Therefore, $y = 219449722.1x^{-1.207}$

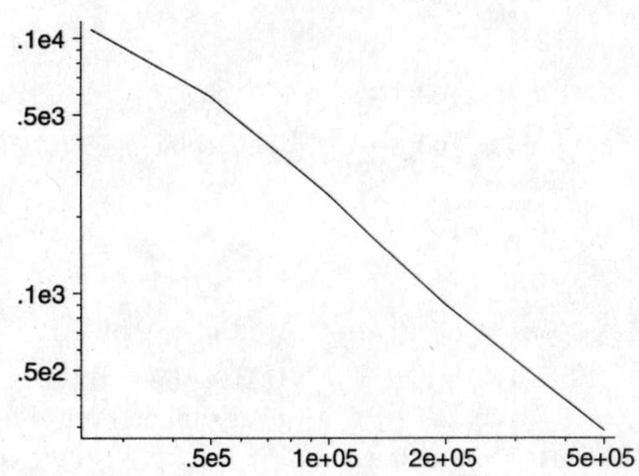

94.

$$\begin{aligned} 205 &= 10000^c A \\ &\rightarrow \\ A &= \frac{205}{10000^c} \\ 7 &= 300000^c \cdot \frac{205}{10000^c} \\ \frac{7}{205} &= 30^c \\ c &= \frac{\ln\left(\frac{7}{205}\right)}{\ln(30)} \\ &= -0.993 \\ &\rightarrow \\ A &= \frac{205}{10000^{-0.993}} \\ &= 1920497.654 \end{aligned}$$

$y = 1920497.654x^{-0.993}$

.1e3
.5e2
.1e2
.1e5 .5e5 1e+05

95. a) $P(t) = 100(1 - e^{-0.2t})$

$P(1) = 100(1 - e^{-0.2\cdot 1})$ Substituting 1 for t

$= 100(1 - e^{-0.2})$

$= 100(0.181269)$ Using a calculator

$\approx 18.1\%$

$P(6) = 100(1 - e^{-0.2\cdot 6})$

$= 100(1 - e^{-1.2})$

$= 100(0.698806)$

$\approx 69.9\%$

b) $P(t) = 100(1 - e^{-0.2t})$

$P'(t) = 100[-(-0.2)e^{-0.2t}]$

$= 20e^{-0.2t}$

c) $P(t) = 100(1 - e^{-0.2t})$

$90 = 100(1 - e^{-0.2t})$ Replacing $P(t)$ by 90

$0.9 = 1 - e^{-0.2t}$

$-0.1 = -e^{-0.2t}$ Adding -1

$0.1 = e^{-0.2t}$ Multiplying by -1

$\ln 0.1 = \ln e^{-0.2t}$ Taking the natural logarithm on both sides

$\ln 0.1 = -0.2t$

$\frac{\ln 0.1}{-0.2} = t$

$\frac{-2.302585}{-0.2} = t$ Using a calculator

$11.5 \approx t$

Thus it will take approximately 11.5 months for 90% of the doctors to become aware of the new medicine.

d) $\lim_{t\to\infty} P(t) = \lim_{t\to\infty} 100(1 - e^{-0.2t}) = 100.$

This indicates that 100% of doctors will eventually accept the new medicine.

96. $R = A\ln r - Br$

$\frac{dR}{dr} = \frac{A}{r} - B$

$\frac{dR}{dr}$ exists for all $r > 0$. Solve:

$\frac{A}{r} - B = 0$

$\frac{A}{r} = B$

$\frac{A}{B} = r$ Critical point

$\frac{d^2R}{dr^2} = -Ar^{-2} = -\frac{A}{r^2} < 0$ for all r. $(A > 0)$

Thus, we have a maximum at $r = \frac{A}{B}$. When $r = \frac{A}{B}$, $R = A\ln\left(\frac{A}{B}\right) - B\left(\frac{A}{B}\right) = A(\ln A - \ln B) - A =$ $A\ln A - A\ln B - A$. This is the maximum value of R.

97. a) $S(t) = 68 - 20\ln(t+1),\ t \ge 0$

$S(0) = 68 - 20\ln(0+1)$ Substituting 0 for t

$= 68 - 20\ln 1$

$= 68 - 20\cdot 0$

$= 68 - 0$

$= 68$

Thus the average score when they initially took the test was 68%.

b) $S(4) = 68 - 20\ln(4+1)$ Substituting 4 for t

$= 68 - 20\ln 5$

$= 68 - 20(1.609438)$ Using a calculator

$= 68 - 32.18876$

$\approx 36\%$

c) $S(24) = 68 - 20\ln(24+1)$ Substituting 24 for t

$= 68 - 20\ln 25$

$= 68 - 20(3.218876)$ Using a calculator

$= 68 - 64.37752$

$\approx 3.6\%$

d) First we reword the question:

3.6 (the average score after 24 months) is what percent of 68 (the average score when $t = 0$).

Then we translate and solve:

$3.6 = x \cdot 68$

$\frac{3.6}{68} = x$

$0.052941 = x$

$5\% \approx x$

e) $S(t) = 68 - 20\ln(t+1),\ t \ge 0$

$S'(t) = 0 - 20\cdot 1\cdot \frac{1}{t+1}$

$= -\frac{20}{t+1}$

f) $S'(t) < 0$ for all $t \ge 0$. Thus $S(t)$ is a decreasing function and has a maximum value of 68% when $t = 0$.

g) $\lim_{t\to\infty} S(t) = \lim_{t\to\infty} 68 - 20\ \ln\ (t+1) = -\infty.$

Clearly, the score cannot be less than 0, but this limit indicates that eventually everything will be forgotten.

98. $S(t) = 78 - 15\ln(t+1),\ t \ge 0$

a) $S(0) = 78 - 15\ln(0+1) = 78 - 15\ln 1 =$ $78 - 15\cdot 0 = 78\%$

b) $S(4) = 78 - 15\ln(4+1) = 78 - 15\ln 5 \approx$ 54%

c) $S(24) = 78 - 15\ln(24+1) = 78 - 15\ln 25 \approx$ 30%

d) $\frac{30}{78} \approx 38.5\%$

e) $S'(t) = -\frac{15}{t+1}$

f) $S'(t) < 0$ for all $t \geq 0$, so $S(t)$ is a decreasing function and has a maximum value of 78% when $t = 0$.

g) $\lim_{t\to\infty} S(t) = \lim_{t\to\infty} 78 - 15\ \ln(t+1) = -\infty.$

Clearly, the score cannot be less than 0, but this limit indicates that eventually everything will be forgotten.

99. $v(p) = 0.37 \ln p + 0.05$ p in thousands, v in ft per sec

a) $v(531) = 0.37 \ln 531 + 0.05$ Substituting 531 for p

$$= 0.37(6.274762) + 0.05$$
$$= 2.321662 + 0.05$$
$$= 2.371662$$
$$\approx 2.37 \text{ ft/sec}$$

b) $v(7900) = 0.37 \ln 7900 + 0.05$

$$= 0.37(8.974618) + 0.05$$
$$= 3.320609 + 0.05$$
$$= 3.370609$$
$$\approx 3.37 \text{ ft/sec}$$

c) $v'(p) = 0.37 \cdot \frac{1}{p} + 0$

$$= \frac{0.37}{p}$$

d) $v'(p)$ is the acceleration of the walker.

100. $W(t) = 100(1 - e^{-0.3t})$

a) $W(1) = 100(1 - e^{-0.3(1)}) \approx 26$ words per minute

$W(8) = 100(1 - e^{-0.3(8)}) \approx 91$ words per minute

b) $W'(t) = 100[-(-0.3)e^{-0.3t}]$

$$= 30e^{-0.3t}$$

c)
$$\begin{aligned} 95 &= 100(1 - e^{-0.3t}) \\ 0.95 &= 1 - e^{-0.3t} \\ e^{-0.3t} &= 0.05 \\ \ln e^{-0.3t} &= \ln 0.05 \\ -0.3t &= \ln 0.05 \\ t &= \frac{\ln 0.05}{-0.3} \\ t &\approx 10 \text{ weeks} \end{aligned}$$

d) $\lim_{t\to\infty} W(t) = \lim_{t\to\infty} 100(1 - e^{-0.3t}) = 100.$

This indicates that the typist's speed will eventually level off at 100 words per minute.

101. $f(x) = \ln[\ln\ x]^3$

$$\begin{aligned} f'(x) &= \frac{1}{[\ln\ x]^3} \cdot 3\ [\ln\ x]^2 \cdot \frac{1}{x} \\ &= \frac{3}{x\ \ln\ x} \end{aligned}$$

102. $f(x) = \frac{\ln\ x}{1 + (\ln\ x)^2}$

$$\begin{aligned} f'(x) &= \frac{(1 + [\ln\ x]^2) \cdot \frac{1}{x} - \ln\ x \cdot 2\ \ln\ x \cdot \frac{1}{x}}{(1 + (\ln\ x)^2)^2} \\ &= \frac{1 - [\ln\ x]^2}{x(1 + (\ln\ x)^2)^2} \end{aligned}$$

103. Using L'Hospital's Rule (taking the derivative of the numerator and denominator then applying the limit)

$$\begin{aligned} \lim_{h\to 0} \frac{\ln(1+h)}{h} &= \lim_{h\to 0} \frac{\frac{1}{(1+h)}}{1} \\ &= \lim_{h\to 0} \frac{1}{(1+h)} \\ &= \frac{1}{1+0} \\ &= 1 \end{aligned}$$

104.

$$\begin{aligned} P &= P_0 e^{-kt} \\ \frac{P}{P_0} &= e^{-kt} \\ \ln\left(\frac{P}{P_0}\right) &= -kt\ \ln(e) \\ \ln\left(\frac{P}{P_0}\right) &= -kt \\ -\frac{\ln\left(\frac{P}{P_0}\right)}{k} &= t \end{aligned}$$

105.

$$\begin{aligned} P &= P_0 e^{-kt} \\ \frac{P}{P_0} &= e^{kt} \\ \ln\left(\frac{P}{P_0}\right) &= kt\ \ln(e) \\ \ln\left(\frac{P}{P_0}\right) &= kt \\ \frac{\ln\left(\frac{P}{P_0}\right)}{k} &= t \end{aligned}$$

106. Left to the student

107. Left to the student

108. a) $r(T) = 0.153\left(e^{0.141(T-9.5)} - e^{4.2159 - 0.153(39.4-T)}\right)$

$$r(T) = 0.153\left(e^{0.141T - 1.3395} - e^{0.153T - 1.8123}\right)$$

b) $\frac{dr}{dT} = 0.153(0.141e^{0.141T-1.3395} - 0.153e^{0.153T-1.823})$

$$\frac{dr}{dT} = 0.021573e^{0.141T-1.3395} - 0.023409e^{0.153T-1.823})$$

c)

$$\begin{aligned}
\frac{dr}{dT} &= 0 \\
0.021573e^{0.141T-1.3395} &= 0.023409e^{0.153T-1.823} \\
\left(\frac{0.021573}{0.023409}\right) &= e^{0.153-1.823-(0.141T-1.3395)} \\
0.9215686275 &= e^{0.012T-0.4835} \\
\ln(0.9215686275) &= 0.012T - 0.4835 \\
T &= \frac{\ln(0.9215686275) + 0.4835}{0.012} \\
&= 33.485
\end{aligned}$$

109. **a)** $r(T) = 0.124\left(e^{0.129(T-9.5)} - e^{4.128-0.144(41.5-T)}\right)$

$r(T) = 0.124\left(e^{0.129T-1.2255} - e^{0.144T-1.848}\right)$

b) $\frac{dr}{dT} = 0.124(0.129e^{0.129T-1.2255} - 0.144e^{0.144T-1.848})$

$\frac{dr}{dT} = 0.015996e^{0.129T-1.2255} - 0.017856e^{0.144T-1.848})$

c)

$$\begin{aligned}
\frac{dr}{dT} &= 0 \\
0.015996e^{0.129T-1.2255} &= 0.017856e^{0.144T-1.848} \\
\left(\frac{0.015996}{0.017856}\right) &= e^{0.144-1.848-(0.129T-1.2255)} \\
0.890844286 &= e^{0.015T-0.6225} \\
\ln(0.890844286) &= 0.015T - 0.6225 \\
T &= \frac{\ln(0.890844286) + 0.6225}{0.015} \\
&= 34.17
\end{aligned}$$

110.

$$\frac{dr}{dT} = A\left(be^{b(T-T_L)} - ce^{b(T_U-T_L)+c(T_U-T)}\right)$$

Plugging the values of constants left to the student.

111.

$$\begin{aligned}
A\left(be^{b(T-T_L)} - ce^{b(T_U-T_L)-c(T_U-T)}\right) &= 0 \\
ce^{b(T_U-T_L)-c(T_U-T)} &= be^{b(T-T_L)} \\
e^{b(T_U-T_L)-c(T_U-T)-b(T-T_L)} &= \frac{b}{c} \\
e^{b(T_U)-c(T_U)-cT-bT} &= \frac{b}{c} \\
b(T_U) - c(T_U) - cT - bT &= \ln(\frac{b}{c}) \\
T_U(b-c) - T(b-c) &= \ln(\frac{b}{c}) \\
\ln(\frac{b}{c}) - T_U(b-c) &= -T(b-c) \\
T_U - \frac{1}{b-c}\ln(\frac{b}{c}) &= T
\end{aligned}$$

Plugging the values of the constants left to the student.

112.

a - b)

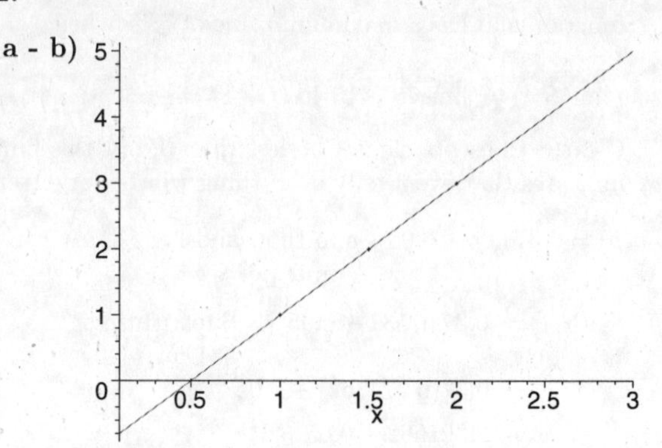

c) The line in the graph in part b) is $2x - 1$ since $\log(10^{2x-1}) = 2x - 1$

113.

a - b)

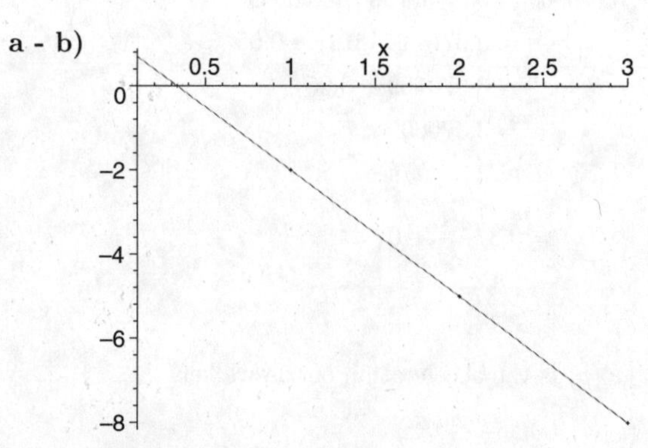

c) The line in the graph in part b) is $-3x + 1$ since $\log(10^{-3x+1}) = -3x + 1$

114. **a)**

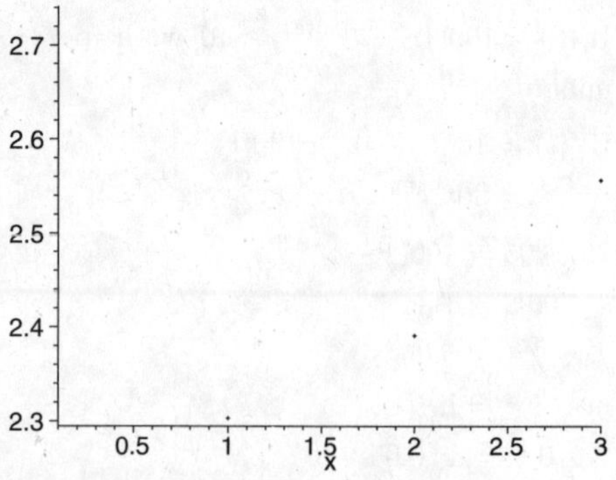

b) $m = \frac{\log(540)-\log(201)}{5-1} = 0.10730$

$$\begin{aligned}
y &= 0.10730x + b \\
\log(201) &= 0.10730(1) + b \\
b &= log(201) - 0.10730 \\
&= 2.19589
\end{aligned}$$

c) $N = 10^{0.10730t+2.19589}$

115. **a)**

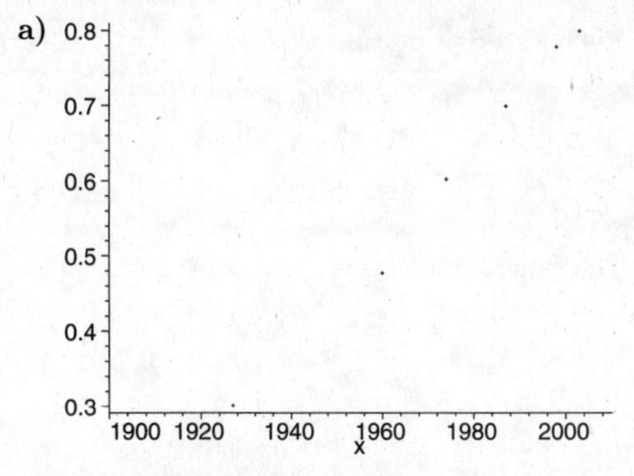

b) $m = \frac{\log(6.3)-\log(2)-}{2003-1927} = 0.00656$

$$\begin{aligned} y &= 0.00656x + b \\ \log(2) &= 0.00656(1927) + b \\ b &= -12.34 \end{aligned}$$

c) $N = 10^{0.00656t-12.34}$

116. The graph of $y = \ln(x)$ can be obtained by taking a reflection around the line $y = x$ of the graph of $y = e^x$

117. The statement means that the graph of the tangent line of the function $y = \ln(4x)$ is given by the graph of $y = \frac{1}{x}$

118. π^e is larger

119. $\sqrt[e]{e} = e^{1/e} = 1.44467$ which is larger than $\sqrt[x]{x} = x^{1/x}$ for all $x > 0$

120. $\lim_{x\to 1} \ln x = 0$

121. $\lim_{x\to\infty} \ln x = \infty$

122.

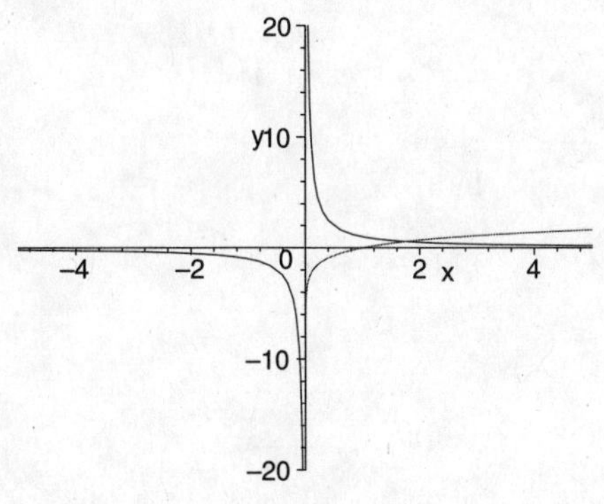

123.

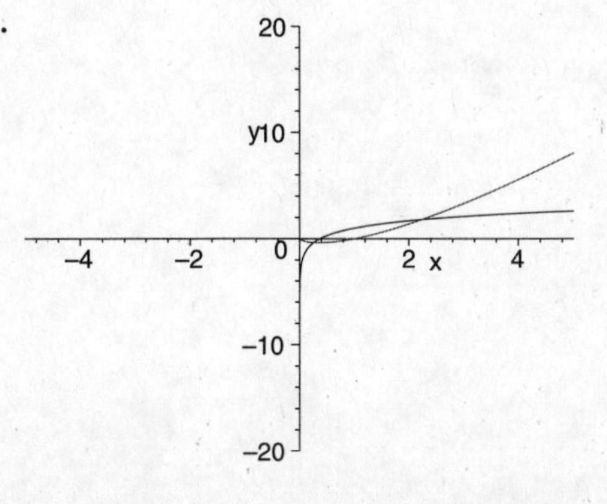

124.

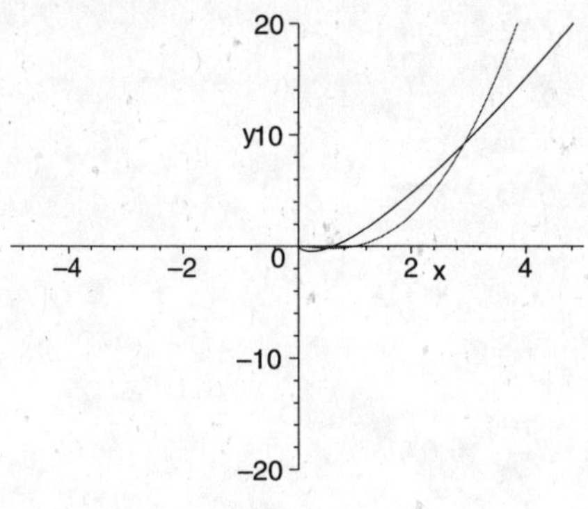

125.

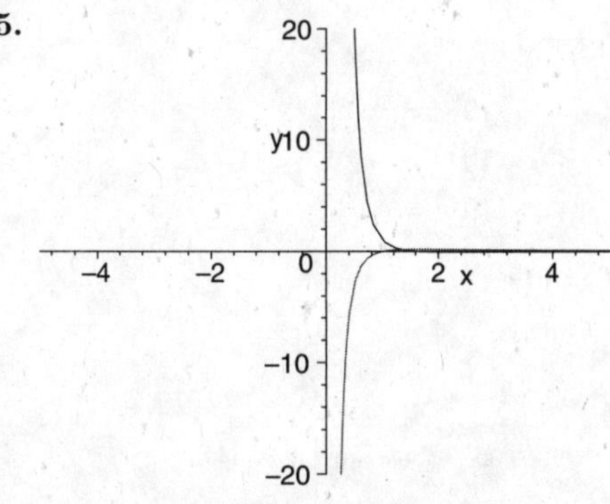

126. The function has a minimum at $(1/e, -1/e)$

127. The function has a minimum at $(1/\sqrt{e}, -1/2e)$

128. $y = 16.485x^{0.136}$

129. $y = 21.5x^{0.364}$

130. $y = 337219504.1x^{-1.237}$

131. $y = 1316788.481x^{-0.958}$

132. $x = 4063.76$ and $x = 20000$

133. $x = 203.19$

134. $x = 40.88$ and $x = 1000$

135. $x = 204.41$

136. $x = 143.32$ and $x = 10993.19$

137. $x = 83.82$ and $x = 22444.11$

138. $x = 143.32$ and $x = 10993.19$

139. $x = 85.68$ and$x = 9503.27$

Exercise Set 4.3

1. The solution of $\frac{dQ}{dt} = kQ$ is $Q(t) = ce^{kt}$, where t is the time. At $t = 0$, we have some "initial" population $Q(0)$ that we will represent by Q_0. Thus $Q_0 = Q(0) = ce^{k\cdot 0} = ce^0 = c \cdot 1 = c$.

Thus, $Q_0 = c$, so we can express $Q(t)$ as $Q(t) = Q_0\, e^{kt}$.

2. $R(t) = R_0\, e^{kt}$

3. The solution of $\frac{dy}{dt} = 2y$ is
$y = c\,e^{2t}$, $y = 5$
when $t = 0$ gives

$$\begin{aligned} 5 &= c\,e^{2(0)} \\ 5 &= c \cdot 1 \\ 5 &= c \end{aligned}$$

Therefore, $y = 5\,e^{2t}$

4. The solution of $\frac{dH}{dt} = 0.5y$ is
$H = c\,e^{0.5t}$
$H = 10$ when $t = 0$ gives

$$\begin{aligned} 10 &= c\,e^{0.5(0)} \\ 10 &= c \cdot 1 \\ 10 &= c \end{aligned}$$

Therefore, $H = 10\,e^{0.5t}$

5. **a)** $P = 1000\,e^{0.033t}$

b) when $t = 30$

$$\begin{aligned} P &= 1000\,e^{0.033(30)} \\ &= 1000\,e^{0.99} \\ &= 2691 \end{aligned}$$

c) when $t = 60$

$$\begin{aligned} P &= 1000\,e^{0.033(60)} \\ &= 1000\,e^{1.98} \\ &= 7243 \end{aligned}$$

d) when $t = 1440$

$$\begin{aligned} P &= 1000\,e^{0.033(1440)} \\ &= 1000\,e^{47.52} \\ &= 4.34 \times 10^{23} \end{aligned}$$

e) The generation time is

$$\begin{aligned} T &= \frac{\ln(2)}{k} \\ &= \frac{\ln(2)}{0.033} \\ &= 21 \end{aligned}$$

6. **a)** $P = 1000\,e^{0.000641t}$

b) when $t = 30$

$$\begin{aligned} P &= 1000\,e^{0.000461(30)} \\ &= 1000\,e^{0.01923} \\ &= 1019 \end{aligned}$$

c) when $t = 60$

$$\begin{aligned} P &= 1000\,e^{0.000641(60)} \\ &= 1000\,e^{0.03846} \\ &= 1039 \end{aligned}$$

d) when $t = 1440$

$$\begin{aligned} P &= 1000\,e^{0.000641(1440)} \\ &= 1000\,e^{0.92304} \\ &= 2517 \end{aligned}$$

e) The generation time is

$$\begin{aligned} T &= \frac{\ln(2)}{k} \\ &= \frac{\ln(2)}{0.000641} \\ &= 1081 \end{aligned}$$

7. **a)** $k = \frac{\ln(2)}{T}$

$$\begin{aligned} k &= \frac{\ln(2)}{0.47} \\ &= 1.4748 \end{aligned}$$

b) $P = 200\,e^{1.4748t}$ for $t = 3$

$$\begin{aligned} P &= 200\,e^{1.4748(3)} \\ &= 200\,e^{4.4244} \\ &= 16692 \end{aligned}$$

c) For $t = 24$

$$\begin{aligned} P &= 200\,e^{1.4748(24)} \\ &= 200\,e^{35.3952} \\ &= 4.7 \times 10^{17} \end{aligned}$$

d) Find t when $P = 3(200) = 600$

$$\begin{aligned} 600 &= 200\,e^{1.4748t} \\ 3 &= e^{1.4748t} \\ \ln(3) &= 1.4748t \\ \frac{\ln(3)}{1.4748} &= t \\ 0.7449 &= t \end{aligned}$$

8. **a)** $k = \frac{\ln(2)}{T}$

$$\begin{aligned} k &= \frac{\ln(2)}{2} \\ &= 0.3466 \end{aligned}$$

b) $P = 500\,e^{0.3466t}$ for $t = 3$

$$\begin{aligned} P &= 500\,e^{0.3466(3)} \\ &= 500\,e^{1.03398} \\ &= 1414 \end{aligned}$$

c) For $t = 24$

$$\begin{aligned} P &= 500\,e^{0.3466(24)} \\ &= 500\,e^{8.3184} \\ &= 2049299 \end{aligned}$$

d) Find t when $P = 3(500) = 1500$

$$\begin{aligned} 1500 &= 500\ e^{0.3466t} \\ 3 &= e^{0.3466t} \\ \ln(3) &= 0.3466t \\ \frac{\ln(3)}{0.3466} &= t \\ 3.1697 &= t \end{aligned}$$

9. Find P_o for $P = 20000$, $k = \frac{\ln(2)}{40}$, and $t = 120$

$$\begin{aligned} P &= P_o\ e^{kt} \\ 20000 &= P_o\ e^{\frac{\ln(2)}{40}(120)} \\ 20000 &= P_o e^{3\ln(2)} = 8P_o \\ \frac{20000}{8} &= P_o \\ 2500 &= P_o \end{aligned}$$

10. Find P_o for $P = 10000$, $k = \frac{\ln(2)}{0.47}$, and $t = 2.5$

$$\begin{aligned} P &= P_o\ e^{kt} \\ 10000 &= P_o\ e^{\frac{\ln(2)}{0.47}(2.5)} \\ 10000 &= P_o(3.686953088) \\ \frac{10000}{3.686953088} &= P_o \\ 2712 &= P_o \end{aligned}$$

11. **a)** $P = c\ e^{0.009t}$, when $t = 0$, $P = 281$ gives

$$\begin{aligned} 281 &= c\ e^{0.009(0)} \\ 281 &= c \cdot 1 \\ c &= 281 \end{aligned}$$

Thus, $P = 281\ e^{0.009t}$

b) Find P when $t = 15$

$$\begin{aligned} P &= 281\ e^{0.009(15)} \\ &= 281\ e^{0.135} \\ &= 321.6 \text{ million} \end{aligned}$$

c) $T = \frac{\ln(2)}{0.009} = 77.02$ years

12. **a)** $P = c\ e^{0.0021t}$, when $t = 0$, $P = 59.8$ gives

$$\begin{aligned} 59.8 &= c\ e^{0.0021(0)} \\ 59.8 &= c \cdot 1 \\ c &= 59.8 \end{aligned}$$

Thus, $P = 59.8\ e^{0.0021t}$

b) Find P when $t = 13$

$$\begin{aligned} P &= 59.8\ e^{0.0021(13)} \\ &= 59.8\ e^{0.0273} \\ &= 61.455 \text{ million} \end{aligned}$$

c) $T = \frac{\ln(2)}{0.0021} = 330.07$ years

13. The balance grows at the rate given by

$$\frac{dP}{dt} = 0.065P.$$

a) $P(t) = P_0\ e^{0.065t}$

b) $P(1) = 1000e^{0.065 \cdot 1}$ Substituting 1000 for P_0 and 1 for t

$$\begin{aligned} &= 1000e^{0.065} \\ &= 1000(1.067159) \\ &\approx 1067.16 \end{aligned}$$

The balance after 1 year is \$1067.16.

$P(2) = 1000e^{0.065 \cdot 2}$ Substituting 1000 for P_0 and 2 for t

$$\begin{aligned} &= 1000e^{0.13} \\ &= 1000(1.138828) \\ &\approx 1138.83 \end{aligned}$$

The balance after 2 years is \$1138.83.

c) $T = \frac{\ln 2}{k}$

$= \frac{0.693147}{0.065}$ Substituting 0.693147 for $\ln 2$ and 0.065 for k

≈ 10.7

An investment of \$1000 will double itself in 10.7 years.

14. a) $P(t) = P_0\ e^{0.08t}$

b) $P(1) = 20,000e^{0.08 \cdot 1} \approx \$21,665.74$

$P(2) = 20,000e^{0.08 \cdot 2} \approx \$23,470.22$

c) $T = \frac{\ln 2}{0.08} \approx 8.7$ yr

15. $k = \frac{\ln 2}{T}$

$= \frac{0.693147}{10}$ Substituting 0.693147 for $\ln 2$ and 10 for T

$= 0.0693147$

$\approx 6.9\%$

The annual interest rate is 6.9%.

16. $k = \frac{\ln 2}{12} \approx 0.0577 \approx 5.8\%$

17. $T = \frac{\ln 2}{k} = \frac{\ln 2}{0.035} \approx 19.8$ yr

18. $k = \frac{\ln 2}{69.31} \approx 0.01 \approx 1\%$

19. $k = \frac{\ln 2}{T} = \frac{\ln 2}{6.931} \approx 0.10 \approx 10\%$

20. $k = \frac{\ln 2}{17.3} \approx 0.04 \approx 4\%$

21. $T = \frac{\ln 2}{k} = \frac{\ln 2}{0.02794} \approx 24.8$ yr

22. $k = \frac{\ln 2}{19.8} \approx 0.035 \approx 3.5\%$

23. $R(b) = e^{21.4b}$ See Example 7

$80 = e^{21.4b}$ Substituting 80 for $R(b)$

$\ln 80 = \ln e^{21.4b}$

$\ln 80 = 21.4b$

$\dfrac{\ln 80}{21.4} = b$

$0.20 \approx b$ Rounding to the nearest hundredth

Thus when the blood alcohol level is 0.20%, the risk of having an accident is 80%.

24. $R(b) = e^{21.4b}$

$90 = e^{21.4b}$

$\ln 90 = \ln e^{21.4b}$

$\ln 90 = 21.4b$

$\dfrac{\ln 90}{21.4} = b$

$0.21\% \approx b$

25. a)

$$P(t) = P_0\, e^{kt}$$
$$216,000,000 = 2,508,000\, e^{k\cdot 200}$$
$$\frac{216,000,000}{2,508,000} = e^{200k}$$
$$\ln\frac{216,000}{2508} = \ln e^{200k}$$
$$\frac{\ln\dfrac{216,000}{2508}}{200} = k$$
$$0.022 \approx k$$

The growth rate was approximately 2.2%.

b) It is reasonable to assume that the population of the United States grew exponentially between 1776 and 1976 rather than linearly or in some other pattern.

26. **a)** $y = 45811.9 \cdot 1.0587^t$

b) $y = 4511.9 \cdot e^{0.057t}$

c) When $t = 50$, $y = 795268$

d) $t = \dfrac{\ln\left(\frac{1000000}{45811.9}\right)}{\ln(1.0587)} = 54$

Which corresponds to the year 2014

e) $T = \dfrac{\ln(2)}{0.057} = 12.16$

27. $(0, 47432)$ and $(40, 432976)$

a) From the point $(0, 47432)$ we get

$$\begin{aligned} y &= a \cdot b^t \\ 47432 &= a \cdot b^0 \\ 47432 &= a \end{aligned}$$

From the point $(40, 432976)$ we get

$$\begin{aligned} 432976 &= 47432 \cdot b^{40} \\ \frac{432976}{47432} &= b^{40} \\ \ln\left(\frac{432976}{47432}\right) &= 40\ln(b) \\ \frac{\ln\left(\frac{432976}{47432}\right)}{40} &= \ln(b) \\ 0.0552846 &= \ln(b) \end{aligned}$$

So, $y = 47432\, e^{0.0552846t}$

b) When $t = 50$

$$\begin{aligned} y &= 47432\, e^{0.0552846(50)} \\ &= 752595 \end{aligned}$$

c) When $y = 1000000$

$$\begin{aligned} 1000000 &= 47432\, e^{0.0552846t} \\ \frac{1000000}{47432} &= e^{0.0552846t} \\ \ln\left(\frac{1000000}{47432}\right) &= 0.0552846t \\ \frac{\ln\left(\frac{1000000}{47432}\right)}{0.0552846} &= t \\ 55.14 &= t \end{aligned}$$

Which corresponds to the year 2015

d) $T = \dfrac{\ln(2)}{0.0552846} = 12.54$ years

e) As expected the answers in this exercise are close to those of Exercise 26.

28. **a)** $y = 85.886796 \cdot 1.061898^t$. $\ln(b) = \ln(1.061898) = 0.0601$ represents the growth rate.

b) $y = 85.886796\, e^{0.0601t}$

c) When $t = 5$, $y = 116$

d) $t = \dfrac{\ln\left(\frac{1000}{85.886796}\right)}{\ln(1.061898)} = 40.87$

e) $T = \dfrac{\ln(2)}{0.0601} = 11.53$

29. $(0, 85)$ and $(4, 109)$

a) From the point $(0, 85)$ we get

$$\begin{aligned} y &= a \cdot b^t \\ 85 &= a \cdot b^0 \\ 85 &= a \end{aligned}$$

From the point $(4, 109)$ we get

$$\begin{aligned} 109 &= 85 \cdot b^4 \\ \frac{109}{85} &= b^4 \\ \ln\left(\frac{109}{85}\right) &= 4\ln(b) \\ \frac{\ln\left(\frac{109}{85}\right)}{4} &= \ln(b) \\ 0.062174 &= \ln(b) \end{aligned}$$

So, $y = 85\, e^{0.062174t}$

b) When $t = 5$

$$\begin{aligned} y &= 85\, e^{0.062174(5)} \\ &= 116 \end{aligned}$$

c) When $y = 1000000$

$$\begin{aligned} 1000 &= 85\, e^{0.062174t} \\ \frac{1000}{85} &= e^{0.062174t} \\ \ln\left(\frac{1000}{85}\right) &= 0.062174t \\ \frac{\ln\left(\frac{1000}{85}\right)}{0.062174} &= t \\ 39.65 &= t \end{aligned}$$

d) $T = \dfrac{\ln(2)}{0.062174} = 11.15$ years

e) As expected the answers in this exercise are close to those of Exercise 28.

30. a) $A(x) = 67367.603 - 24944.7753\ \ln(x)$

b)

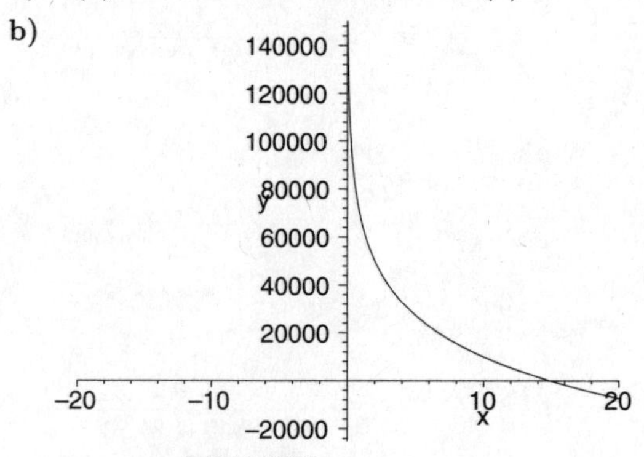

c) $A'(x) = \dfrac{-24944.7753}{x}$

d) The value of $A'(x)$ when x is close to 14 is close to -2000.

31. a) Using a graphing calculator $P(x) = 14.88 \cdot 0.9996^t$

b) $\ln(0.9996) = -0.0004$ So, $P(x) = 14.88\ e^{-0.0004t}$

c)

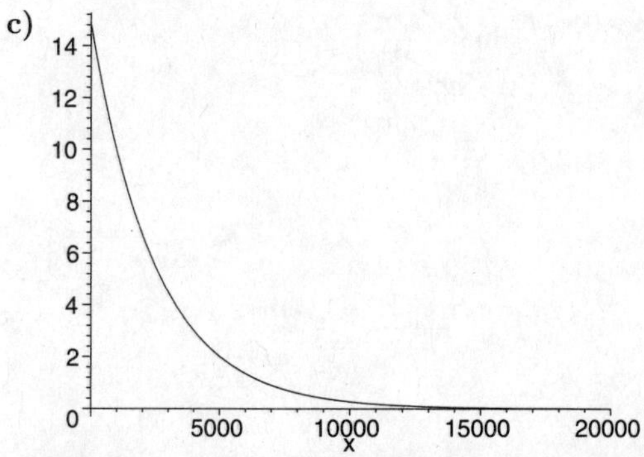

d) $P'(x) = 14.88 \cdot -0.0004\, e^{-0.0004t} = -0.00595\, e^{-0.0004t}$

e) When x gets closer to sea level the slope of the tangent line at $x = 0$ is about -0.005

f) $P(A(x)) = x = A(P(x))$ since the exponential function and the logrithmic functions are inverses of each other.

32. $P(t) = \dfrac{2500}{1 + 5.25\ e^{-0.32t}}$

a) $$\begin{aligned} P(0) &= \frac{2500}{1+5.25\ e^{-0.32(0)}} \\ &= \frac{2500}{1+5.25\ e^{0}} \\ &= \frac{2500}{1+5.25(1)} \\ &= 400 \end{aligned}$$

$$\begin{aligned} P(1) &= \frac{2500}{1+5.25\ e^{-0.32(1)}} \\ &= \frac{2500}{1+5.25\ e^{-0.32}} \\ &\approx 520 \end{aligned}$$

$$\begin{aligned} P(5) &= \frac{2500}{1+5.25\ e^{-0.32(5)}} \\ &= \frac{2500}{1+5.25\ e^{-1.6}} \\ &\approx 1214 \end{aligned}$$

$$\begin{aligned} P(10) &= \frac{2500}{1+5.25\ e^{-0.32(10)}} \\ &= \frac{2500}{1+5.25\ e^{-3.2}} \\ &\approx 2059 \end{aligned}$$

$$\begin{aligned} P(15) &= \frac{2500}{1+5.25\ e^{-0.32(15)}} \\ &= \frac{2500}{1+5.25\ e^{-4.8}} \\ &\approx 2396 \end{aligned}$$

$$\begin{aligned} P(20) &= \frac{2500}{1+5.25\ e^{-0.32(20)}} \\ &= \frac{2500}{1+5.25\ e^{-6.4}} \\ &\approx 2478 \end{aligned}$$

b) $P'(t) =$

$$\frac{(1+5.25e^{-0.32t})(0) - 5.25(-0.32)e^{-0.32t}(2500)}{(1+5.25\ e^{-0.32t})^2}$$

Quotient Rule

$$= \frac{4200\ e^{-0.32t}}{(1+5.25\ e^{-0.32t})^2}$$

c) The derivative $P'(t)$ exists for all real numbers. The equation $P'(t) = 0$ has no solution. Thus, the function has no critical points and hence no relative extrema. $P'(t) > 0$ for all real numbers, so $P(t)$ is increasing on $[0, \infty)$. The second derivative can be used to show that the graph has an inflection point at $(5.18, 1250)$. The function is concave up on $(0, 5.18)$ and concave down on $(5.18, \infty)$.

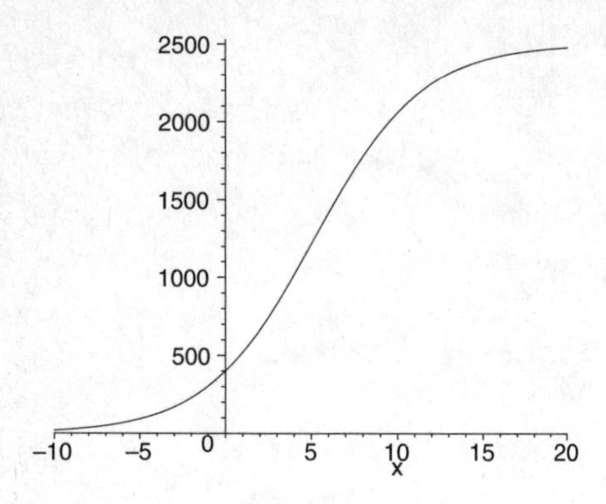

33. $G(x) = \dfrac{100}{1 + 43.3\ e^{-0.0425x}}$

a) When $x = 100$

$$\begin{aligned} G(100) &= \frac{100}{1 + 43.3\ e^{-0.0425(100)}} \\ &= \frac{100}{1 + 43.3\ e^{-4.25}} \\ &= 61.82\% \end{aligned}$$

When $x = 150$

$$\begin{aligned} G(150) &= \frac{100}{1 + 43.3\ e^{-0.0425(150)}} \\ &= \frac{100}{1 + 43.3\ e^{-6.375}} \\ &= 91.13\% \end{aligned}$$

When $x = 300$

$$\begin{aligned} G(300) &= \frac{100}{1 + 43.3\ e^{-0.0425(300)}} \\ &= \frac{100}{1 + 43.3\ e^{-12.75}} \\ &= 99.99\% \end{aligned}$$

b)

$$\begin{aligned} G'(x) &= 100 \cdot -1(1 + 43.3\ e^{-0.0425x})^{-2} \times \\ &\quad (43.3 \cdot -0.0425\ e^{-0.0425x}) \\ &= \frac{184.025\ e^{-0.0425x}}{(1 + 43.3\ e^{-0.0425x})^2} \end{aligned}$$

c)

$$G''(x) = \frac{\left(-184.025\ e^{-0.0425x}\right)\left[0.0425 - 1.84025\ e^{-0.0425x}\right]}{\left(1 + 43.3\ e^{-0.0425x}\right)^3}$$

The possible inflection point occurs when $G''(x) = 0$ or undefined, which means the only possible inflection points occurs at

$$\begin{aligned} 0.0425 - 1.84025\ e^{-0.0425x} &= 0 \\ e^{-0.0425x} &= \frac{0.0425}{1.84025} \\ -0.0425x &= \ln\left(\frac{0.0425}{1.84025}\right) \\ x &= \frac{\ln\left(\frac{0.0425}{1.84025}\right)}{-0.0425} \\ &= 88.66 \end{aligned}$$

d)

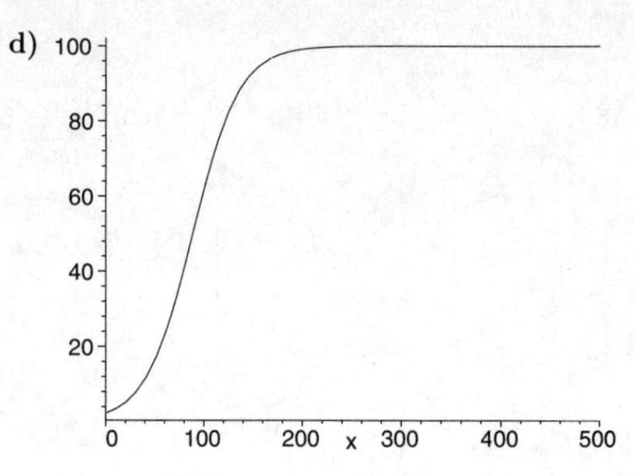

34. $C(t) = \dfrac{40.2}{1 + 335\ e^{-0.092x}}$

a) When $t = 20$

$$\begin{aligned} C(20) &= \frac{40.2}{1 + 335\ e^{-0.092(20)}} \\ &= 0.74164 \end{aligned}$$

When $t = 50$

$$\begin{aligned} C(50) &= \frac{40.2}{1 + 335\ e^{-0.092(50)}} \\ &= 9.2046 \end{aligned}$$

When $t = 100$

$$\begin{aligned} C(100) &= \frac{40.2}{1 + 335\ e^{-0.092(100)}} \\ &= 38.884 \end{aligned}$$

b)

$$\begin{aligned} 30 &= \frac{40.2}{1 + 335\ e^{-0.092(t)}} \\ \frac{40.2}{30} &= 1 + 335\ e^{-0.092t} \\ \frac{\ln\left(\frac{\frac{40.2}{30} - 1}{355}\right)}{-0.092} &= t \\ 74.923 &= t \end{aligned}$$

c)

$$\begin{aligned} C'(t) &= 40.2\left(-(1 + 335\ e^{-0.092t})^{-2}\right)(-30.82\ e^{-0.092t}) \\ &= \frac{1238.964\ e^{-0.092t}}{(1 + 335\ e^{-0.092t})^2} \end{aligned}$$

d)

$$\begin{aligned} C''(t) &= \frac{\left(e^{-0.092t}\right)\left[1 + 335\ e^{-0.092t}\right]}{\left(1 + 335\ e^{-0.092t}\right)^3} \\ &\quad \times \left(113.985 - 76369.741\ e^{-0.092t}\right) \end{aligned}$$

The possible inflection point occurs when $C''(t) = 0$ or undefined, which means the only possible inflection points occurs at

$$\begin{aligned} 113.985 - 76369.741\ e^{-0.092t} &= 0 \\ e^{-0.092t} &= \frac{113.985}{76369.741} \end{aligned}$$

$$\begin{aligned}
-0.092t &= \ln\left(\frac{113.985}{76369.741}\right) \\
t &= \frac{\ln\left(\frac{113.985}{73669.741}\right)}{-0.092} \\
&= 70.731
\end{aligned}$$

e) 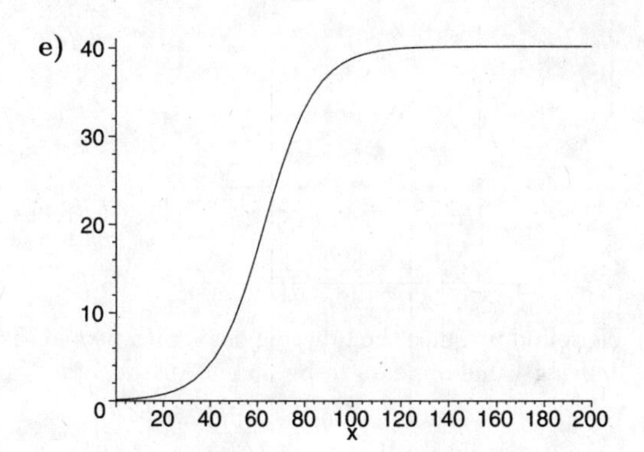

35. $A(t) = \dfrac{105}{1 + 32900\, e^{-0.04t}}$

a) When $t = 100$

$$\begin{aligned}
A(100) &= \frac{105}{1 + 32900\, e^{-0.04(100)}} \\
&= \frac{105}{1 + 32900\, e^{-4}} \\
&= 0.174
\end{aligned}$$

When $t = 150$

$$\begin{aligned}
A(150) &= \frac{105}{1 + 32900\, e^{-0.04(150)}} \\
&= \frac{105}{1 + 32900\, e^{-6}} \\
&= 1.272
\end{aligned}$$

When $t = 200$

$$\begin{aligned}
A(200) &= \frac{105}{1 + 32900\, e^{-0.04(200)}} \\
&= \frac{105}{1 + 32900\, e^{-8}} \\
&= 8.723
\end{aligned}$$

b)

$$\begin{aligned}
A'(x) &= 105 \cdot -1(1 + 32900\, e^{-0.04t})^{-2} \times \\
&\quad (32900 \cdot -0.04\, e^{-0.04t}) \\
&= \frac{138180\, e^{-0.04t}}{(1 + 32900\, e^{-0.04t})^2}
\end{aligned}$$

c)

$$\begin{aligned}
A''(t) &= \frac{\left(e^{-0.04t}\right)\left[1 + 32900\, e^{-0.04t}\right]}{(1 + 32900\, e^{-0.04t})^3} \times \\
&\quad \left[-5527.2\left(1 + 32900\, e^{-0.04t}\right) + 363689760\right]
\end{aligned}$$

The possible inflection point occurs when $A''(t) = 0$ or undefined, which means the possible inflection points occurs at

$$\begin{aligned}
0 &= -5527.2\left(1 + 32900\, e^{-0.04t}\right) + \\
&\quad 363689760 \\
1 + 32900\, e^{-0.04t} &= \frac{363689760}{5527.2} \\
\frac{\frac{363689760}{5527.2} - 1}{32900} &= e^{-0.04t} \\
\ln\left(\frac{\frac{363689760}{5527.2} - 1}{32900}\right) &= -0.04t \\
\frac{\ln\left(\frac{\frac{363689760}{5527.2} - 1}{32900}\right)}{-0.04} &= t \\
t &= -17.33
\end{aligned}$$

not an acceptable answer

or

$$\begin{aligned}
1 + 32900\, e^{-0.04t} &= 0 \\
e^{-0.04t} &= \frac{1}{32900} \\
-0.04t &= \ln\left(\frac{1}{32900}\right) \\
t &= \frac{\ln\left(\frac{1}{32900}\right)}{-0.04} \\
&= 260.031
\end{aligned}$$

We find $A(260.031)$

$$\begin{aligned}
A(200) &= \frac{105}{1 + 32900\, e^{-0.04(260.031)}} \\
&= \frac{105}{1 + 32900\, e^{-10.40124}} \\
&= 52.5
\end{aligned}$$

d) 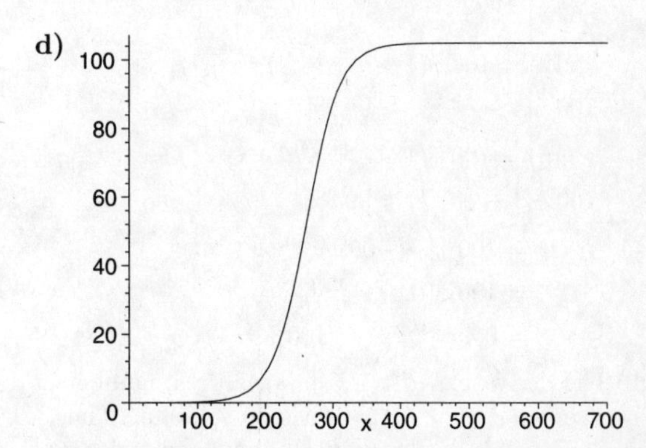

e) The maximum occurs at the limiting value which is 105

36. $p(t) = 1 - e^{-0.28t}$

a)
$p(1) = 1 - e^{-0.28(1)} \approx 0.244$
$p(2) = 1 - e^{-0.28(2)} \approx 0.429$
$p(5) = 1 - e^{-0.28(5)} \approx 0.753$
$p(11) = 1 - e^{-0.28(11)} \approx 0.954$
$p(16) = 1 - e^{-0.28(16)} \approx 0.989$
$p(20) = 1 - e^{-0.28(20)} \approx 0.996$

b) $p'(t) = 0.28\, e^{-0.28t}$

c) The derivative $p'(t)$ exists for all real numbers. The equation $p'(t) = 0$ has no solution. Thus, the function has no critical points and hence no relative extrema. $p'(t) > 0$ for all real numbers, so $p(t)$ is increasing on $[0, \infty)$. $p''(t) = -0.0784\, e^{-0.028t}$, so $p''(t) < 0$ for all real numbers and hence is concave down on $[0, \infty)$.

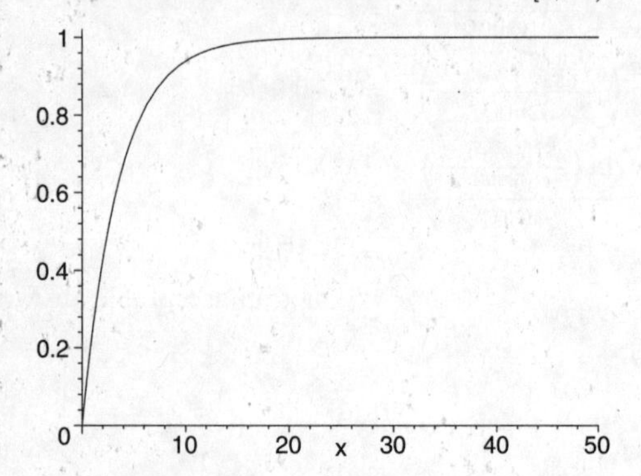

37. $P(t) = 100\%(1 - e^{-0.4t})$

a) $P(0) = 100\%(1 - e^{-0.4(0)}) = 100\%(1 - e^0) = 100\%(1 - 1) = 100\%(0) = 0\%$

$P(1) = 100\%(1 - e^{-0.4(1)}) = 100\%(1 - e^{-0.4}) \approx 33.0\%$

$P(2) = 100\%(1 - e^{-0.4(2)}) = 100\%(1 - e^{-0.8}) \approx 55.1\%$

$P(3) = 100\%(1 - e^{-0.4(3)}) = 100\%(1 - e^{-1.2}) \approx 69.9\%$

$P(5) = 100\%(1 - e^{-0.4(5)}) = 100\%(1 - e^{-2}) \approx 86.5\%$

$P(12) = 100\%(1 - e^{-0.4(12)}) = 100\%(1 - e^{-4.8}) \approx 99.2\%$

$P(16) = 100\%(1 - e^{-0.4(16)}) = 100\%(1 - e^{-6.4}) \approx 99.8\%$

b) $P'(t) = 100\%[-(-0.4)\, e^{-0.4t}]$
$= 100\%(0.4)\, e^{-0.4t}$
$= 0.4\, e^{-0.4t} \quad (100\% = 1)$

c) The derivative $P'(t)$ exists for all real numbers. The equation $P'(t) = 0$ has no solution. Thus, the function has no critical points and hence no relative extrema. $P'(t) > 0$ for all real numbers, so $P(t)$ is increasing on $[0, \infty)$. $P''(t) = -0.16\, e^{-0.4t}$, so $P''(t) < 0$ for all real numbers and hence is concave down on $[0, \infty)$.

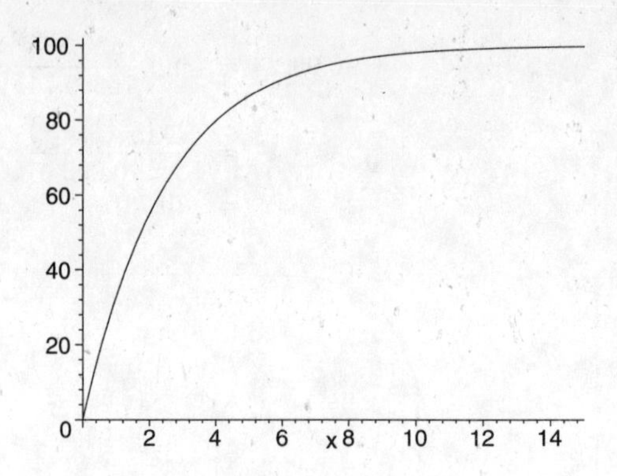

38. a) $N(t) = \dfrac{29.47232081}{1 + 79.56767122e^{-0.809743969t}}$

b) We round up since the function deals with people. The limiting value appears to be about 30 students.

c) We graph the function in the window $[-1, 20, -1, 35]$, Yscl $= 5$.

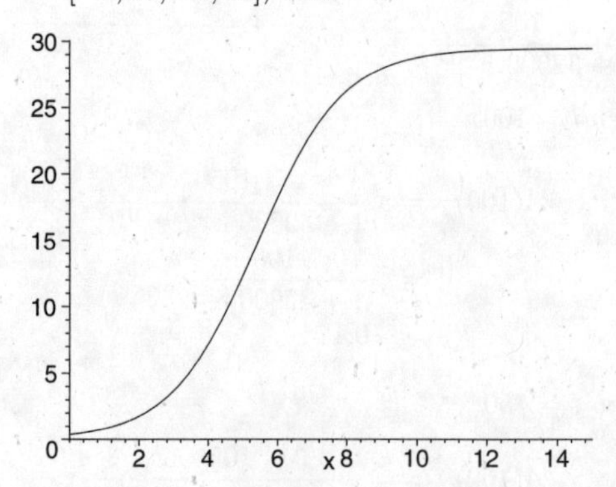

d) Use the quotient rule to find $N'(t)$.

$$N'(t) = \frac{1898.885181e^{-0.809743969t}}{(1 + 79.56767122e^{-0.809743969t})^2}$$

e) $\lim_{t \to \infty} N'(t) = 0$; this indicates that eventually the population does not change. That is, it reaches and remains at a limiting value.

39. $i = e^{0.073} - 1 = 0.0757 = 7.57\%$

40. $i = e^{0.08} - 1 = 0.0833 = 8.33\%$

41. $k = \ln(0.0924 + 1) = 0.0884 = 8.84\%$

42. $k = \ln(0.0661 + 1) = 0.064 = 6.4\%$

43.

$$\begin{aligned} 3P_0 &= P_0e^{kt} \\ 3 &= e^{kt} \\ \frac{\ln(3)}{k} &= t \end{aligned}$$

44.

$$\begin{aligned} 4P_0 &= P_0e^{kt} \\ 4 &= e^{kt} \\ \frac{\ln(4)}{k} &= t \end{aligned}$$

45. Answers vary

46. After the second year

47. $2 = e^{24k}$

$$k = \frac{ln(2)}{24} = 0.0289$$

48. $y = y_0\, e^{kt}$

$$\begin{aligned} y_2 &= y_0\, e^{kt_2} \\ y_1 &= y_0\, e^{kt_1} \\ \text{dividing} & \\ \frac{y_2}{y_1} &= e^{k(t_2-t_1)} \\ \ln\left(\frac{y_2}{y_1}\right) &= k(t_2 - t_1) \\ \frac{\ln\left(\frac{y_2}{y_1}\right)}{(t_2 - t_1)} &= k \end{aligned}$$

49. $k = \dfrac{\ln(2)}{T}$ and $T = \dfrac{\ln(2)}{k}$

k	1%	2%	4.621%	6.931%	14%
T	69.315	34.657	15	10	4.951

The graph of $T = \ln\, 2/k$ is not linear since the independent variable appears in the denominator

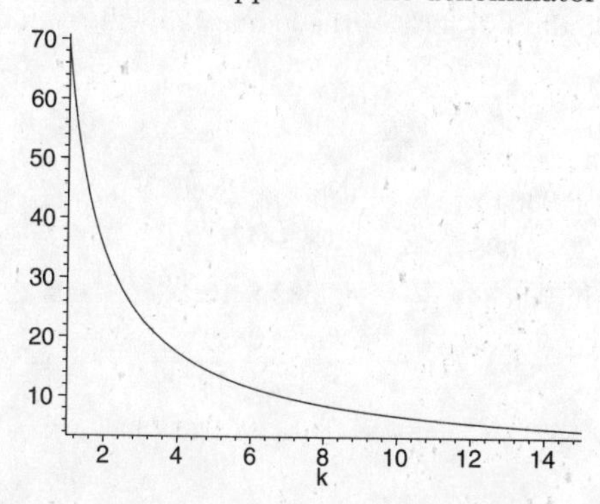

50. An exponential function grows indefinitely while a logistic function has a limiting growth value

51. The rule of 69 is used to approximate the doubling time (generation time) of growing things. The name comes from the approximation to $\ln(2) = 0.6931$ which is used to find the doubling time

Exercise Set 4.4

1. (d)

2. (f)

3. (a)

4. (e)

5. (c)

6. $k = \dfrac{\ln 2}{3} \approx 0.231 \approx 23.1\%$ per min

7. $k = \dfrac{\ln 2}{T} = \dfrac{\ln 2}{22} \approx 0.032 \approx 3.2\%$ per yr

8. $T = \dfrac{\ln 2}{0.096} \approx 7.2$ days

9. $k = \dfrac{\ln 2}{T} = \dfrac{\ln 2}{25} \approx 0.028 \approx 2.8\%$ per year

10. $k = \dfrac{\ln 2}{4560} \approx 0.00015 \approx 0.015\%$ per yr

11. $k = \dfrac{\ln 2}{T} = \dfrac{\ln 2}{23,105} \approx 0.00003 \approx 0.003\%$ per yr

12. $P(100) = 1000\, e^{-0.032(100)} \approx 40.8$ g

13.

$$\begin{aligned} P(t) &= P_0\, e^{-kt} \\ P(20) &= 1000\, e^{-0.231(20)} \quad \text{Substituting 1000 for } P_0, \text{ 0.231 for } k \text{ and 20 for } t \\ &= 1000\, e^{-4.62} \\ &\approx 9.9 \end{aligned}$$

Thus 9.9 grams of polonium will remain after 20 minutes.

14. The amount of carbon-14 present is 60% P_0.

$$\begin{aligned} 60\%\, P_0 &= P_0\, e^{-0.0001205t} \\ 0.6 &= e^{-0.0001205t} \\ \ln\, 0.6 &= \ln\, e^{-0.0001205t} \\ \ln\, 0.6 &= -0.0001205t \\ \frac{\ln\, 0.6}{-0.0001205} &= t \\ 4239 &\approx t \end{aligned}$$

The tusk is about 4239 years old.

15. $N(t) = N_0\, e^{-0.0001205t}$ See Example 3(b).

If a piece of wood has lost 90% of its carbon-14 from an initial amount P_0, then 10% P_0 is the amount present. To find the age of the wood, we solve the following equation for t:

$$10\%\ P_0 = P_0\, e^{-0.0001205t}$$
Substituting 10% P_0 for $P(t)$
$$0.1 = e^{-0.0001205t}$$
$$\ln 0.1 = \ln e^{-0.0001205t}$$
$$\ln 0.1 = -0.0001205t$$
$$-2.302585 = -0.0001205t \quad \text{Using a calculator}$$
$$\frac{-2.302585}{-0.0001205} = t$$
$$19{,}109 \approx t$$

Thus, the piece of wood is about 19,109 years old.

16. $A = A_0 e^{-kt}$

First find k. The half-life is 60.1 days.

$$k = \frac{\ln 2}{60.1} \approx 0.0115$$

75% of the Iodine-125 remains, so we have:

$$0.75 = 1 \cdot e^{-0.0115t}$$
$$\ln 0.75 = \ln e^{-0.0115t}$$
$$\ln 0.75 = -0.0115t$$
$$\frac{\ln 0.75}{-0.0115} = t$$
$$25 \approx t$$

The sample was on the shelf for about 25 days.

17. If an artifact has lost 60% of its carbon-14 from an initial amount P_0, then 40% P_0 is the amount present. To find the age t we solve the following equation for t.

$$40\% P_0 = P_0\, e^{-0.0001205t}$$
Substituting 40% P_0 for $P(t)$
(See Example 3(b).)
$$0.4 = e^{-0.0001205t}$$
$$\ln 0.4 = \ln e^{-0.0001205t}$$
$$\ln 0.4 = -0.0001205t$$
$$-0.916291 = -0.0001205t \quad \text{Using a calculator}$$
$$\frac{-0.916291}{-0.0001205} = t$$
$$7604 \approx t$$

The artifact is about 7604 years old.

18. For carbon-14 the decay rate is 0.0001205

a) The amount of carbon-14 that remains is

$$\begin{aligned} N &= N_0\, e^{-0.0001205 \cdot 5300} \\ &= 0.528\ N_0 \end{aligned}$$

Percentage lost is $100 - 52.8 = 47.2\%$

b)

$$\begin{aligned} N &= N_0\, e^{-0.0001205 \cdot 5309} \\ &= 0.527\ N_0 \end{aligned}$$

Percentage lost is $100 - 52.7 = 47.3\%$

19. For carbon-14 the decay rate is 0.0001205
The amount of carbon-14 that remains is

$$\begin{aligned} N &= N_0\, e^{-0.0001205 \cdot 10000} \\ &= N_0\, e^{-1.205} \\ &= 0.2997\ N_0 \end{aligned}$$

Amount lost is $100 - 29.97 = 70.03\%$

20. $k = \dfrac{\ln(1/2)}{60.1} = -0.011533$

a)

$$\begin{aligned} 0.75N_0 &= N_0\, e^{-0.011533t} \\ 0.75 &= e^{-0.011533t} \\ \ln(0.75) &= -0.011533t \\ \frac{\ln(0.75)}{-0.011533} &= t \\ t &= 24.94 \text{ days} \end{aligned}$$

b)

$$\begin{aligned} N &= N_0\, e^{-0.011533 \cdot 48} \\ &= N_0\, e^{-0.553584} \\ &= 0.5749N_0 \end{aligned}$$

After 48 days 57.49% of the iodine 125 will remain

21. $k = \dfrac{\ln(1/2)}{1.3 \times 10^9} = -5.332 \times 10^{-10}$

$$\begin{aligned} 0.0884N_0 &= N_0\, e^{-5.332\times 10^{-10}t} \\ 0.0884 &= e^{-5.332\times 10^{-10}t} \\ \ln(0.0884) &= -5.332 \times 10^{-10}t \\ t &= \frac{\ln(0.0884)}{-5.332 \times 10^{-10}} \\ &= 4.55 \times 10^9 \text{ years} \end{aligned}$$

22.

$$\begin{aligned} 0.302N_0 &= N_0\, e^{-5.332\times 10^{-10}t} \\ 0.302 &= e^{-5.332\times 10^{-10}t} \\ \ln(0.302) &= -5.332 \times 10^{-10}t \\ t &= \frac{\ln(0.302)}{-5.332 \times 10^{-10}} \\ &= 2.25 \times 10^9 \text{ years} \end{aligned}$$

23.

$$\begin{aligned} 0.653N_0 &= N_0\, e^{-5.332\times 10^{-10}t} \\ 0.653 &= e^{-5.332\times 10^{-10}t} \\ \ln(0.653) &= -5.332 \times 10^{-10}t \\ t &= \frac{\ln(0.653)}{-5.332 \times 10^{-10}} \\ &= 7.99 \times 10^8 \text{ years} \end{aligned}$$

24. a) $A = A_0\, e^{-kt}$

b) First find k. The half-life of A is 3.3 hr.

$$k = \frac{\ln 2}{3.3} \approx 0.21$$

$$\text{Then } 1 = 10\, e^{-0.21t}$$

$$0.1 = e^{-0.21t}$$

$$\ln 0.1 = \ln e^{-0.21t}$$

$$\ln 0.1 = -0.21t$$

$$\frac{\ln 0.1}{-0.21} = t$$

$$11 \approx t$$

There will be 1 g left after about 11 hr.

25. a) When A decomposes at a rate proportional to the amount of A present, we know that

$$\frac{dA}{dt} = -kA.$$

The solution of this equation is $A = A_0\, e^{-kt}$.

b) We first find k. The half-life of A is 3 hr.

$$k = \frac{\ln 2}{T}$$

$$k = \frac{0.693147}{3} \quad \text{Substituting 0.693147 for ln 2 and 3 for } T$$

$$\approx 0.23, \text{ or } 23\%$$

We now substitute 8 for A_0, 1 for A, and 0.23 for k and solve for t.

$$A = A_0\, e^{-kt}$$

$$1 = 8\, e^{-0.23t} \quad \text{Substituting}$$

$$\frac{1}{8} = e^{-0.23t}$$

$$0.125 = e^{-0.23t}$$

$$\ln 0.125 = \ln e^{-0.23t}$$

$$-2.079442 = -0.23t \quad \text{Using a calculator}$$

$$\frac{-2.079442}{-0.23} = t$$

$$9 \approx t$$

After 9 hr there will be 1 gram left.

26. $W = W_0\, e^{-0.009t}$

a) $k = 0.009$, or 0.9%

The animal loses 0.9% of its weight each day.

b) $W = W_0\, e^{-0.009(30)}$

$W \approx 0.763\, W_0$, or $76.3\%\, W_0$

76.3% of the initial weight remains after 30 days.

27. a) $W = W_0\, e^{-0.008t}$

$k = 0.008$, or 0.8%

The starving animal loses 0.8% of its weight each day.

b) $W = W_0\, e^{-0.008t}$

$$W = W_0\, e^{-0.008(30)} \quad \text{Substituting 30 for } t$$

$$W = W_0\, e^{-0.24}$$

$$W = 0.786628\, W_0 \quad \text{Using a calculator}$$

$$W \approx 78.7\%\, W_0$$

Thus, after 30 days, 78.7% of the initial weight remains.

28. $I = I_0\, e^{-0.01(100)}$

$I \approx 0.37\, I_0$, or $37\%\, I_0$

29. a) $I = I_0\, e^{-\mu x}$

$$I = I_0\, e^{-1.4(1)} \quad \text{Substituting 1.4 for } \mu \text{ and 1 for } x$$

$$I = I_0\, e^{-1.4}$$

$$I = I_0(0.246597) \quad \text{Using a calculator}$$

$$I \approx 25\%\, I_0$$

$$I = I_0\, e^{-1.4(2)} \quad \text{Substituting 1.4 for } \mu \text{ and 2 for } x$$

$$I = I_0\, e^{-2.8}$$

$$I = I_0(0.060810) \quad \text{Using a calculator}$$

$$I \approx 6.1\%\, I_0$$

$$I = I_0\, e^{-1.4(3)} \quad \text{Substituting 1.4 for } \mu \text{ and 3 for } x$$

$$I = I_0\, e^{-4.2}$$

$$I = I_0(0.014996) \quad \text{Using a calculator}$$

$$I \approx 1.5\%\, I_0$$

b) $I = I_0\, e^{-\mu x}$

$$I = I_0\, e^{-1.4(10)} \quad \text{Substituting 1.4 for } \mu \text{ and 10 for } x$$

$$I = I_0\, e^{-14}$$

$$I = I_0(0.00000083) \quad \text{Using a calculator}$$

$$I \approx 0.00008\%\, I_0$$

30. a) $T(t) = ae^{-kt} + C$

$$100 = ae^{-k \cdot 0} + 40$$

$$60 = a$$

b) At $t = 5$, $T = 90°$.

$$T(t) = 60\,e^{-kt} + 40$$
$$90 = 60\,e^{-k\cdot 5} + 40$$
$$50 = 60\,e^{-5k}$$
$$\frac{5}{6} = e^{-5k}$$
$$\ln \frac{5}{6} = \ln e^{-5k}$$
$$\ln \frac{5}{6} = -5k$$
$$\frac{\ln \frac{5}{6}}{-5} = k$$
$$0.04 \approx k$$
$$T(t) = 60\,e^{-0.04t} + 40$$

c) $T(10) = 60\,e^{-0.04(10)} + 40 \approx 80.2°$

d)
$$41 = 60\,e^{-0.04t} + 40$$
$$1 = 60\,e^{-0.04t}$$
$$\frac{1}{60} = e^{-0.04t}$$
$$\ln \frac{1}{60} = \ln e^{-0.04t}$$
$$\ln \frac{1}{60} = -0.04t$$
$$\frac{\ln \frac{1}{60}}{-0.04} = t$$
$$102.4 \text{ min} \approx t$$

e) $T'(t) = -2.4e^{-0.04t}$; the rate of change of the temperature is $-2.4e^{-0.04t}$ degrees per minute.

31. a) $T(t) = ae^{-kt} + C$ Newton's law of Cooling

At $t = 0$, $T = 100°$. We solve the following equation for a.

$$100 = ae^{-k\cdot 0} + 75 \quad \text{Substituting 100 for } T, \text{ 0 for } t, \text{ and 75 for } C$$
$$25 = ae^0$$
$$25 = a \qquad (e^0 = 1)$$

The value of the constant is 25°.

Thus, $T(t) = 25\,e^{-kt} + 75$.

b) Now we find k using the fact that at $t = 10$, $T = 90°$.

$$T(t) = 25\,e^{-kt} + 75$$
$$90 = 25\,e^{-k\cdot 10} + 75 \quad \text{Substituting 90 for } T \text{ and 10 for } t$$
$$15 = 25\,e^{-10k}$$
$$\frac{15}{25} = e^{-10k}$$
$$0.6 = e^{-10k}$$
$$\ln 0.6 = e^{-10k}$$
$$\ln 0.6 = -10k$$
$$\frac{\ln 0.6}{-10} = k$$
$$\frac{-0.510826}{-10} = k$$
$$0.05 \approx k$$

Thus, $T(t) = 25\,e^{-0.05t} + 75$.

c)
$$T(t) = 25\,e^{-0.05t} + 75$$
$$T(20) = 25\,e^{-0.05(20)} + 75 \quad \text{Substituting 20 for } t$$
$$= 25\,e^{-1} + 75$$
$$= 25(0.367879) + 75 \quad \text{Using a calculator}$$
$$= 9.196975 + 75$$
$$\approx 84.2$$

The temperature after 20 minutes is 84.2°.

d)
$$T(t) = 25\,e^{-0.05t} + 75$$
$$80 = 25\,e^{-0.05t} + 75 \quad \text{Substituting 80 for } T$$
$$5 = 25\,e^{-0.05t}$$
$$\frac{5}{25} = e^{-0.05t}$$
$$0.2 = e^{-0.05t}$$
$$\ln 0.2 = \ln e^{-0.05t}$$
$$\ln 0.2 = -0.05t$$
$$\frac{\ln 0.02}{-0.05} = t$$
$$\frac{-1.609438}{-0.05} = t$$
$$32 \approx t$$

It takes 32 minutes for the liquid to cool to 80°.

e) $T'(t) = -1.25e^{-0.05t}$; the rate of change of the temperature is $-1.25e^{-0.05t}$ degrees per minute.

32. Find a in $T(t) = ae^{-kT} + C$. Assume $T = 98.6°$ at $t = 0$.

$$98.6 = ae^{-k\cdot 0} + 10$$
$$88.6 = a$$
$$T(t) = 88.6\,e^{-kt} + 10$$

We want to find the number of hours N since the murder. We first use the two temperature readings to find k.

$61.6 = 88.6\,e^{-kN} + 10$, or $51.6 = 88.6\,e^{-kN}$

$57.2 = 88.6\,e^{-k(N+1)} + 10$, or $47.2 = 88.6\,e^{-k(N+1)}$

$$\frac{51.6}{47.2} = \frac{88.6\,e^{-kN}}{88.6\,e^{-k(N+1)}} = e^k$$

$$\ln\frac{51.6}{47.2} = \ln e^k$$

$$0.09 \approx k$$

Now we find N.

$$51.6 = 88.6\,e^{-0.09N} \qquad \text{Substituting}$$

$$\frac{51.6}{88.6} = e^{-0.09N}$$

$$\ln\frac{51.6}{88.6} = \ln e^{-0.09N}$$

$$\ln\frac{51.6}{88.6} = -0.09N$$

$$\frac{\ln\frac{51.6}{88.6}}{-0.09} = N$$

$$6 \approx N$$

The coroner arrived at 2 A.M., so the murder was committed at about 8 P.M.

33. We first find a in the equation $T(t) = ae^{-kt} + C$.

Assuming the temperature of the body was normal when the murder occurred, we have $T = 98.6°$at $t = 0$. Thus

$98.6° = ae^{-k\cdot 0} + 60°$ (Room temperature is 60°.)

so

$a = 38.6°$.

Thus T is given by $T(t) = 38.6\,e^{-kt} + 60$.

We want to find the number of hours N since the murder was committed. To find N we must first determine k. From the two temperature readings, we have

$$85.9 = 38.6\,e^{-kN} + 60, \text{ or } 25.9 = 38.6\,e^{-kN}$$

$$83.4 = 38.6\,e^{-k(N+1)} + 60, \text{ or } 23.4 = 38.6\,e^{-k(N+1)}$$

Dividing the first equation by the second, we get

$$\frac{25.9}{23.4} = \frac{38.6\,e^{-kN}}{38.6\,e^{-k(N+1)}} = e^{-kN+k(N+1)} = e^k.$$

We solve this equation for k:

$$\ln\frac{25.9}{23.4} = \ln e^k$$

$$\ln 1.106838 = k$$

$$0.10 \approx k$$

Now we substitute 0.10 for k in the equation $25.9 = 38.6\,e^{-kN}$ and solve for N.

$$25.9 = 38.6\,e^{-0.10N}$$

$$\frac{25.9}{38.6} = e^{-0.10N}$$

$$\ln\frac{25.9}{38.6} = \ln e^{-0.10N}$$

$$\ln 0.670984 = -0.10N$$

$$\frac{-0.399009}{-0.10} = N$$

$$4 \text{ hr} \approx N$$

The coroner arrived at 11 P.M., so the murder was committed about 7 P.M.

34. **a)** Let 1995 correspond to $t = 0$

$$\begin{aligned} 144 &= 150\,e^{7k} \\ \ln\left(\frac{144}{150}\right) &= 7k \\ \frac{\ln\left(\frac{144}{150}\right)}{7} &= k \\ -0.005832 &= k \end{aligned}$$

Thus, $N = 150\,e^{-0.005832t}$

b) In 2010, $t = 15$

$$\begin{aligned} N &= 150\,e^{-0.005832\cdot 15} \\ &= 150\cdot 0.9162411337 \\ &= 137.44 \end{aligned}$$

c)

$$\begin{aligned} 100 &= 150\,e^{-0.005832t} \\ \frac{100}{150} &= e^{-0.005832t} \\ \ln\left(\frac{2}{3}\right) &= -0.005832t \\ \frac{\ln\left(\frac{2}{3}\right)}{-0.005832} &= t \\ 69.5 &= t \end{aligned}$$

The population of Russia will be 100 million in the year 2065.

35. **a)** Let 1995 correspond to $t = 0$

$$\begin{aligned} 48.8 &= 51.9\,e^{6k} \\ \ln\left(\frac{48.8}{51.9}\right) &= 6k \\ \frac{\ln\left(\frac{48.8}{51.9}\right)}{6} &= k \\ 0.010265 &= k \end{aligned}$$

Thus, $N = 51.9\,e^{-0.010265t}$

b) In 2015, $t = 20$

$$\begin{aligned} N &= 51.9\,e^{-0.010265\cdot 20} \\ &= 51.9\cdot 0.8144029589 \\ &= 42.27 \end{aligned}$$

c)

$$\begin{aligned} 1 &= 51.9\,e^{-0.010265t} \\ \frac{1}{51.9} &= e^{-0.010265t} \\ \ln\left(\frac{1}{51.9}\right) &= -0.010265t \\ \frac{\ln\left(\frac{1}{51.9}\right)}{-0.010265} &= t \\ 384.7 &= t \end{aligned}$$

The population of Ukraine will be 1 in the year 2380.

36. $P = 14.7\, e^{-0.00005a}$

a) $P = 14.7\, e^{-0.00005(1000)} \approx 14.0 \text{ lb/in}^2$

b) $P = 14.7\, e^{-0.00005(20{,}000)} \approx 5.4 \text{ lb/in}^2$

c)
$$\begin{aligned}1.47 &= 14.7\, e^{-0.00005a}\\ 0.1 &= e^{-0.00005a}\\ \ln\, 0.1 &= \ln\, e^{0.00005a}\\ \ln\, 0.1 &= -0.00005a\\ \frac{\ln\, 0.1}{-0.00005} &= a\\ 46{,}052 \text{ ft} &\approx a\end{aligned}$$

d) $P' = -0.000735e^{-0.00005a}$; the rate of change of the pressure is $-0.000735e^{-0.00005a}$ lb/in² per unit of altitude.

37. a)
$$\begin{aligned}P(t) &= 50\, e^{-0.004t}\\ P(375) &= 50\, e^{-0.004(375)} \quad \text{Substituting 375 for } t\\ &= 50\, e^{-1.5}\\ &= 50(0.223130)\\ &\approx 11\end{aligned}$$

After 375 days, 11 watts will be available.

b)
$$\begin{aligned}T &= \frac{\ln 2}{k}\\ &= \frac{0.693147}{0.004} \quad \text{Substituting 0.693147 for } \ln 2 \text{ and 0.004 for } k\\ &\approx 173\end{aligned}$$

The half-life of the power supply is 173 days.

c)
$$\begin{aligned}P(t) &= 50\, e^{-0.004t}\\ 10 &= 50\, e^{-0.004t} \quad \text{Substituting 10 for } P(t)\\ \frac{10}{50} &= e^{-0.004t}\\ 0.2 &= e^{-0.004t}\\ \ln\, 0.2 &= \ln\, e^{-0.004t}\\ \ln\, 0.2 &= -0.004t\\ \frac{\ln\, 0.2}{-0.004} &= t\\ \frac{-1.609438}{-0.004} &= t \quad \text{Using a calculator}\\ 402 &\approx t\end{aligned}$$

The satellite can stay in operation 402 days.

d) When $t = 0$,
$$\begin{aligned}P &= 50\, e^{-0.004(0)} \quad \text{Substituting 0 for } t\\ &= 50\, e^0\\ &= 50 \cdot 1\\ &= 50\end{aligned}$$

At the beginning the power output was 50 watts.

e) $P'(t) = -0.2e^{-0.004t}$; the power is changing at a rate of $-0.2e^{-0.004t}$ watts per day.

38.
$$\begin{aligned}N(t) &= N_0\, e^{-0.0001205t}\\ N &= N_0\, e^{-0.0001205(4000)}\\ N &\approx N_0\,(0.62), \text{ or } 62\%\ N_0\end{aligned}$$

Thus, 62% of the carbon-14 remains, so 38% has been lost.

39. a) $y = 84.94353992 - 0.5412834098 \ln x$

b) When $x = 8$, $y = 83.8\%$;
when $x = 10$, $y = 83.7\%$;
when $x = 24$, $y = 83.2\%$;
when $x = 36$, $y = 83.09\%$

c)
$$\begin{aligned}82 &= 84.94353992 - 0.5412834098 \ln x\\ -2.94353992 &= -0.5412834098 \ln x\\ 5.438075261 &\approx \ln x\\ x &\approx e^{5.438075261}\\ x &\approx 230\end{aligned}$$

The test scores will fall below 82% after about 230 months.

d) $y' = -\dfrac{0.5412834098}{x}$; the test scores are changing at the rate of $-\dfrac{0.5412834098}{x}$ percent per month.

40. **a)** Let 1980 correspond to $t = 0$

$$\begin{aligned}64 &= 72\, e^{20k}\\ \ln\left(\frac{64}{72}\right) &= 20k\\ \frac{\ln\left(\frac{64}{72}\right)}{20} &= k\\ -0.005889 &= k\end{aligned}$$

Thus, $N = 72\, e^{-0.005889t}$

b) In 2010, $t = 30$

$$\begin{aligned}N &= 72\, e^{-0.005889 \cdot 30}\\ &= 150 \cdot 0.8380524814\\ &= 125.71\end{aligned}$$

c)

$$\begin{aligned}20 &= 72\, e^{-0.005889t}\\ \frac{20}{72} &= e^{-0.005889t}\\ \ln\left(\frac{20}{72}\right) &= -0.005889t\\ \frac{\ln\left(\frac{20}{72}\right)}{-0.005889} &= t\\ 217.5 &= t\end{aligned}$$

The population of Russia will be 100 million in the year 2198.

41. **a)** Let 1980 correspond to $t = 0$

$$\begin{aligned}48 &= 52\, e^{20k}\\ \ln\left(\frac{48}{52}\right) &= 20k\\ \frac{\ln\left(\frac{48}{52}\right)}{20} &= k\\ -0.004002 &= k\end{aligned}$$

Thus, $N = 52\, e^{-0.004002t}$

b) In 2010, $t = 30$

$$\begin{aligned} N &= 52\, e^{-0.004002 \cdot 30} \\ &= 52 \cdot 0.8868636211 \\ &= 46.12 \end{aligned}$$

c)

$$\begin{aligned} 10 &= 52\, e^{-0.004002t} \\ \frac{10}{52} &= e^{-0.004002t} \\ \ln\left(\frac{10}{52}\right) &= -0.004002t \\ \frac{\ln\left(\frac{10}{52}\right)}{-0.004002} &= t \\ 411.9 &= t \end{aligned}$$

The population of Russia will be 100 million in the year 2392.

42. Answers vary.

Exercise Set 4.5

1. $5^4 = e^{4 \cdot \ln 5}$ Theorem 13: $a^x = e^{x \cdot \ln a}$

$\approx e^{4(1.609438)}$ Using a calculator

$\approx e^{6.4378}$

2. $2^3 = e^{3 \cdot \ln 2} \approx e^{2.0794}$

3. $3.4^{10} = e^{10 \cdot \ln 3.4}$ Theorem 13: $a^x = e^{x \cdot \ln a}$

$\approx e^{10(1.223775)}$ Using a calculator

$\approx e^{12.238}$

4. $(1.04)^{24} = e^{24 \cdot \ln 1.04} \approx e^{0.941}$

5. $4^k = e^{k \cdot \ln 4}$ Theorem 13: $a^x = e^{x \cdot \ln a}$

6. $5^R = e^{R \cdot \ln 5}$

7. $8^{kT} = e^{kT \cdot \ln 8}$ Theorem 13: $a^x = e^{x \cdot \ln a}$

8. $10^{kR} = e^{kR \cdot \ln 10}$

9. $y = 6^x$

$\frac{dy}{dx} = (\ln 6)6^x$ Theorem 14: $\frac{dy}{dx}a^x = (\ln a)a^x$

10. $y = 7^x$

$\frac{dy}{dx} = (\ln 7)7^x$

11. $f(x) = 10^x$

$f'(x) = (\ln 10)10^x$ Theorem 14

12. $f(x) = 100^x$

$f'(x) = (\ln 100)100^x$

13. $f(x) = x(6.2)^x$

$f'(x) = x\left[\frac{d}{dx}(6.2)^x\right] + \left[\frac{d}{dx}x\right](6.2)^x$ Product Rule

$= x \cdot \underbrace{(\ln 6.2)(6.2)^x}_{\text{Theorem 14}} + 1 \cdot (6.2)^x$

$= (6.2)^x[x \ln 6.2 + 1]$

14. $f(x) = x(5.4)^x$

$f'(x) = x(\ln 5.4)(5.4)^x + (5.4)^x$

$= (5.4)^x(x \ln 5.4 + 1)$

15. $y = x^3 10^x$

$\frac{dy}{dx} = x^3 \cdot \underbrace{(\ln 10)10^x}_{\text{Theorem 14}} + 3x^2 \cdot 10^x$ Product Rule

$= 10^x x^2(x \ln 10 + 3)$

16. $y = x^4 5^x$

$\frac{dy}{dx} = x^4(\ln 5)5^x + 4x^3 5^x$

$= 5^x x^3(x \ln 5 + 4)$

17. $y = \log_4 x$

$\frac{dy}{dx} = \frac{1}{\ln 4} \cdot \frac{1}{x}$ Theorem 16: $\frac{d}{dx}\log_a x = \frac{1}{\ln a} \cdot \frac{1}{x}$

18. $y = \log_5 x$

$\frac{dy}{dx} = \frac{1}{\ln 5} \cdot \frac{1}{x}$

19. $f(x) = 2 \log x$

$f'(x) = 2 \cdot \frac{d}{dx} \log x$

$= 2 \cdot \frac{1}{\ln 10} \cdot \frac{1}{x}$ Theorem 16 ($\log x = \log_{10} x$)

$= \frac{2}{\ln 10} \cdot \frac{1}{x}$

20. $f(x) = 5 \log x$

$f'(x) = 5 \cdot \frac{1}{\ln 10} \cdot \frac{1}{x}$

$= \frac{5}{\ln 10} \cdot \frac{1}{x}$

21. $f(x) = \log \frac{x}{3}$

$f(x) = \log x - \log 3$ (P2)

$f'(x) = \frac{1}{\ln 10} \cdot \frac{1}{x} - 0$ Theorem 16 ($\log x = \log_{10} x$)

$= \frac{1}{\ln 10} \cdot \frac{1}{x}$

22. $f(x) = \log \frac{x}{5} = \log x - \log 5$

$f'(x) = \frac{1}{\ln 10} \cdot \frac{1}{x}$

23. $y = x^3 \log_8 x$

$\frac{dy}{dx} = x^3\underbrace{\left(\frac{1}{\ln 8} \cdot \frac{1}{x}\right)} + 3x^2 \cdot \log_8 x$ Product Rule

↑ Theorem 16

$= x^2 \cdot \frac{1}{\ln 8} + 3x^2 \cdot \log_8 x$

$= x^2\left(\frac{1}{\ln 8} + 3\log_8 x\right)$

24. $y = x \log_6 x$

$\frac{dy}{dx} = x \cdot \frac{1}{\ln 6} \cdot \frac{1}{x} + \log_6 x$

$= \frac{1}{\ln 6} + \log_6 x$

25. $y = \csc x \log_2 x$

$\frac{dy}{dx} = \csc x \cdot \frac{1}{\ln 2} \cdot \frac{1}{x} - \csc x \cot x \cdot \frac{\ln x}{\ln 2}$

$= \frac{\csc x - x \csc x \cot x \ln x}{x \ln 2}$

26. $y = \cot x \log_5 (2x)$

$\frac{dy}{dx} = \cot x \cdot \frac{1}{\ln 5} \cdot \frac{1}{x} - \csc^2 x \cdot \frac{\ln x}{\ln 5}$

$= \frac{\cot x - x \csc^2 x \ln x}{x \ln 5}$

27. $y = \log_{10} \sin x$

$\frac{dy}{dx} = \frac{1}{\ln 10} \cdot \frac{1}{\sin x} \cdot \cos x$

$= \frac{1}{\ln 10} \cdot \cot x$

28. $y = \log_2 \sec x$

$\frac{dy}{dx} = \frac{1}{\ln 2} \cdot \frac{1}{\sec x} \cdot \sec x \tan x$

$= \frac{1}{\ln 2} \cdot \tan x$

29. $y = \log_x 3$

$y = \frac{\ln 3}{\ln x}$

$\frac{dy}{dx} = -\ln 3 \cdot \frac{-1}{\ln^2 x} \cdot \frac{1}{x}$

$= \frac{-\ln 3}{\ln^2 x}$

30. $y = x \log_x 10$

$y = \frac{x \ln 10}{\ln x}$

$\frac{dy}{dx} = x \cdot \left(\ln 10 \cdot \frac{-1}{\ln^2 x} \cdot \frac{1}{x}\right) + \frac{\ln 10}{\ln x}$

$= \frac{\ln 10}{\ln x}\left(1 - \frac{1}{\ln x}\right)$

31. $g(x) = (\log_x 10)(\log_{10} x)$

$g'(x) = \log_x 10 \cdot \frac{1}{x \ln 10} + \log_{10} x \cdot \left(\ln 10 \cdot \frac{-1}{\ln^2 x} \cdot \frac{1}{x}\right)$

$= \frac{\ln 10}{\ln x} \cdot \frac{1}{x \ln 10} + \frac{\ln x}{\ln 10} \cdot \left(\ln 10 \cdot \frac{-1}{\ln^2 x} \cdot \frac{1}{x}\right)$

$= 0$

32. $g(x) = 2^x \log_2 x)$

$g'(x) = 2^x \cdot \frac{1}{x \ln 2} + \frac{\ln x}{\ln 2} \cdot 2^x \ln 2$

$= \frac{2^x}{x \ln 2} + \ln x$

33. a) $N(t) = 250,000\left(\frac{1}{4}\right)^t$

$N'(t) = 250,000\frac{d}{dx}\left(\frac{1}{4}\right)^t$

$= 250,000 \cdot \left(\ln \frac{1}{4}\right)\left(\frac{1}{4}\right)^t$ Theorem 14

$= 250,000(\ln 1 - \ln 4)\left(\frac{1}{4}\right)^t$ (P2)

$= -250,000(\ln 4)\left(\frac{1}{4}\right)^t$ $(\ln 1 = 0)$

b) The rate of change of the number of cans still in use, when 250,000 cans are initially distributed, is $-250,000(\ln 4)\left(\frac{1}{4}\right)^t$ cans per year.

34. a) $V(t) = \$5200(0.80)^t$

$V'(t) = \$5200(\ln 0.80)(0.80)^t$

b) The value of the office machine is changing at a rate of $\$5200(\ln 0.80)(0.80)^t$ per year.

35. $R = \log \frac{I}{I_0}$

$R = \log \frac{10^5 \cdot I_0}{I_0}$ Substituting $10^5 \cdot I_0$ for I

$= \log 10^5$

$= 5$ (P5)

The magnitude on the Richter scale is 5.

36. $R = \log \frac{10^{6.9} \cdot I_0}{I_0} = \log 10^{6.9} = 6.9$

37. a) $I = I_0 \cdot 10^R$

$I = I_0 \cdot 10^7$ Substituting 7 for R

$= 10^7 \cdot I_0$

b) $I = I_0 \cdot 10^R$

$I = I_0 \cdot 10^8$ Substituting 8 for R

$= 10^8 \cdot I_0$

c) The intensity in (b) is 10 times that in (a).

$10^8 I_0 = 10 \cdot 10^7 I_0$

d) $I = I_0 10^R$

$\frac{dI}{dR} = I_0 \cdot \frac{d}{dR} 10^R$ I_0 is a constant

$= I_0 \cdot (\ln 10)10^R$ Theorem 14

$= (I_0 \cdot \ln 10)10^R$

e) The intensity is changing at a rate of $(I_0 \cdot \ln 10)^R$.

38. $I = I_0 10^{0.1L}$

a) $I = I_0 10^{0.1(100)} = 10^{10} \cdot I_0$

b) $I = I_0 10^{0.1(10)} = 10 \cdot I_0$

c) The intensity in (a) is 10^9 times that in (b).

$10^{10}\ I_0 = 10^9 \cdot 10\ I_0$

d) $\frac{dI}{dL} = \frac{d}{dL} I_0\, e^{0.1L \cdot \ln 10} \quad (10^{0.1L} = e^{0.1L \cdot \ln 10})$

$= I_0 \cdot 0.1 \cdot \ln 10 \cdot e^{0.1L \cdot \ln 10}$

$= 0.1 \cdot I_0 \cdot \ln 10 \cdot 10^{0.1L}$

e) The intensity is changing at a rate of $0.1 \cdot I_0 \cdot \ln 10 \cdot 10^{0.1L}$.

39. a) $R = \log \frac{I}{I_0}$

$R = \log I - \log I_0$ (P2)

$\frac{dR}{dI} = \frac{1}{\ln 10} \cdot \frac{1}{I} - 0$ Theorem 16; I_0 is a constant.

$= \frac{1}{\ln 10} \cdot \frac{1}{I}$

b) The magnitude is changing at a rate of $\frac{1}{\ln 10} \cdot \frac{1}{I}$.

40. a) $L = 10 \log \frac{I}{I_0} = 10(\log I - \log I_0)$

$\frac{dL}{dI} = 10\Big(\frac{1}{\ln 10} \cdot \frac{1}{I} - 0\Big)$

$= \frac{10}{\ln 10} \cdot \frac{1}{I}$

b) The loudness is changing at a rate of $\frac{10}{\ln 10} \cdot \frac{1}{I}$.

41. a) $y = m \log x + b$

$\frac{dy}{dx} = m \cdot \frac{d}{dx} \log x + 0$ m and b are constants

$= m\Big(\frac{1}{\ln 10} \cdot \frac{1}{x}\Big)$ Theorem 16

$= \frac{m}{\ln 10} \cdot \frac{1}{x}$

b) The response is changing at a rate of $\frac{m}{\ln 10} \cdot \frac{1}{x}$.

42. a)

$$\begin{aligned} 7 &= -\log[H^+] \\ -7 &= \log[H^+] \\ 10^{-7} &= [H^+] \end{aligned}$$

b) $x = [H^+]$
Therefore, $[H^+] = 0.001t + 10^{-7}$

c)

$$\frac{d}{dt}\left[-\log\left(0.001t + 10^{-7}\right)\right] = \frac{-1}{\ln 10} \cdot \frac{0.001}{0.001t + 10^{-7}}$$

d) When $t = 0$, $x = 10^{-7}$ thus, the pH is $-\log(10^{-7}) = 7$

e) The pH is changing most rapidly at $t = 0$ which corresponds to a pH of 7

43. a) When $t = 0$ $[OH^-] = x = 10^{-7}$ moles/liter

b) When $t = 0$ $[H^+] = \frac{10^{-14}}{10^{-7}} = 10^{-7}$ moles/liter

c) When $t = 0$ $pH = -\log\left(10^{-7}\right) = 7$

d)

$$\begin{aligned} [H^+][OH^-] &= 10^{-14} \\ \log\left([H^+][OH^-]\right) &= \log(10^{-14} \\ \log\left([H^+][0.002t + 10^{-7}]\right) &= -14 \\ \log[H^+] + \log\left(0.002t + 10^{-7}\right) &= -14 \\ 14 + \log\left(0.002t + 10^{-7}\right) &= -\log[H^+] \end{aligned}$$

Thus, $pH = 14 + \log\left(0.002t + 10^{-7}\right)$

e) $\frac{d}{dt}\,[OH^-] = \frac{0.002}{\ln 10\ (0.002t + 10^{-7}}$

f) The pH is changing most rapidly at $t = 0$ which corresponds to a pH of 7

44. $y = 2^{x^4}$

$\frac{dy}{dx} = (\ln 2)2^{x^4} \cdot \frac{d}{dx} x^4$

$= 4x^3(\ln 2)2^{x^4}$

45. $y = x^x,\ x > 0$

$y = e^{x \ln x}$ Theorem 13: $a^x = e^{x \ln a}$

$\frac{dy}{dx} = \Big(x \cdot \frac{1}{x} + 1 \cdot \ln x\Big)e^{x \ln x}$

$= (1 + \ln x)x^x$ Substituting x^x for $e^{x \ln x}$

46. $y = \log_3 (x^2 + 1)$

$\frac{dy}{dx} = \frac{1}{\ln 3} \cdot \frac{1}{x^2 + 1} \cdot \frac{d}{dx} (x^2 + 1)$

$= \frac{2x}{(\ln 3)(x^2 + 1)}$

47. $f(x) = x^{e^x},\ x > 0$

$f(x) = e^{e^x \ln x}$ Theorem 13: $a^x = e^{x \ln a}$

$f'(x) = \Big(e^x \cdot \frac{1}{x} + e^x \ln x\Big)e^{e^x \ln x}$

$= e^x\Big(\frac{1}{x} + \ln x\Big)x^{e^x}$ Substituting x^{e^x} for $e^{e^x \ln x}$

$= e^x\, x^{e^x}\Big(\ln x + \frac{1}{x}\Big)$

48. $y = a^{f(x)}$

$\frac{dy}{dx} = f'(x)(\ln a)a^{f(x)}$

49. $y = \log_a f(x),\ f(x) > 0$

$a^y = f(x)$ Exponential equation

$e^{y \ln a} = f(x)$ Theorem 13

Differential implicitly to find dy/dx.

$$\frac{d}{dx}e^{y\ \ln\ a} = \frac{d}{dx}f(x)$$

$$\frac{dy}{dx}\cdot \ln\ a \cdot e^{y\ \ln\ a} = f'(x)$$

$$\frac{dy}{dx}\cdot \ln\ a \cdot f(x) = f'(x)$$ Substituting $f(x)$ for $e^{y\ \ln\ a}$

$$\frac{dy}{dx} = \frac{1}{\ln\ a}\cdot\frac{f'(x)}{f(x)}$$

50. $y = [f(x)]^{g(x)},\ f(x) > 0$

$$\ln\ y = \ln\ [f(x)]^{g(x)} = g(x)\ \ln\ f(x)$$

$$\frac{1}{y}\frac{dy}{dx} = g(x)\cdot\frac{1}{f(x)}\cdot f'(x) + g'(x)\ \ln\ f(x)$$

Differentiating implicity

$$\frac{dy}{dx} = y\Big[g(x)\cdot\frac{f'(x)}{f(x)} + g'(x)\ \ln\ f(x)\Big]$$

$$= f(x)^{g(x)}\Big[g(x)\cdot\frac{f'(x)}{f(x)} + g'(x)\ \ln\ f(x)\Big]$$

51. Since a^x can be written as $e^{x\cdot\ln\ a}$, we can find the derivative of $f(x) = a^x$ using the rule for differentiating an exponential function, base e.

52. If we first write $\log_a\ x$ as $a^{\log_a\ x}$ we have $a^{\log_a\ x} = x$. Then we can take the natural logarithm on both sides of this expression, use P3, and write $\log_a\ x = \frac{1}{\ln\ a}\cdot\ln\ x$. This allows us to find $\frac{d}{dx}\log_a\ x$ using the rules for differentiating $f(x) = c\cdot\ln\ x$.

Chapter Review Exercises

1. $y = \ln\ x$

$$\frac{dy}{dx} = \frac{1}{x}$$

2. $y = e^x$

$$\frac{dy}{dx} = e^x$$

3. $y = \ln\ (x^4 + 5)$

$$\frac{dy}{dx} = \frac{1}{x^4+5}\cdot 4x^3 = \frac{4x^3}{x^4+5}$$

4. $y = e^{2\sqrt{x}}$

$$\frac{dy}{dx} = e^{2\sqrt{x}}\cdot\frac{2}{2\sqrt{x}} = \frac{e^{2\sqrt{x}}}{\sqrt{x}}$$

5. $y = \ln\ (\sin\ x + x)$

$$\frac{dy}{dx} = \frac{1}{\sin\ x + x}\cdot \cos\ x + 1 = \frac{\cos\ x + 1}{\sin\ x + x}$$

6. $f(x) = e^{4x} + x^4$

$$\frac{dy}{dx} = 4\ e^{4x} + 4x^3$$

7. $f(x) = \frac{\ln\ x}{\tan\ x}$

$$f'(x) = \frac{\tan\ x\cdot\frac{1}{x} - \ln\ x\cdot\sec^2\ x}{\tan^2\ x} = \frac{\cot\ x}{x} - \csc^2\ x\ \ln\ x$$

8. $f(x) = e^{x^2}\cdot\ln\ 4x$

$$f'(x) = e^{x^2}\cdot\frac{4}{4x} + 2x\ e^{x^2}\cdot\ln\ 4x = \frac{e^{x^2}}{x} + 2x\ e^{x^2}\ \ln\ 4x = \frac{e^{x^2}(2x^2\ln\ 4x + 1)}{x}$$

9. $f(x) = e^{4x} - \ln\ \frac{x}{4}$

$$f'(x) = 4\ e^{4x} - \frac{1}{\frac{x}{4}}\cdot\frac{1}{4} = 4\ e^{4x} - \frac{1}{x}$$

10. $g(x) = x^8 - 8\ \ln\ x$

$$g'(x) = 8x^7 - \frac{8}{x} = \frac{8(x^8-1)}{x}$$

11. $y = \frac{\ln\ e^x}{e^x} = \frac{x}{e^x}$

$$\frac{dy}{dx} = \frac{e^x\cdot 1 - x\cdot e^x}{(e^x)^2} = \frac{e^x\ (1-x)}{(e^x)^2} = \frac{(1-x)}{e^x} = e^{-x}(1-x)$$

12.

$$\log_a\ 14 = \log_a\ (2\cdot 7) = \log_a\ 2 + \log_a\ 7 = 1.8301 + 5.0999 = 6.93$$

13.

$$\log_a\ \frac{2}{7} = \log_a\ 2 - \log_a\ 7 = 1.8301 - 5.0999 = -3.2698$$

14.

$$\log_a\ 28 = \log_a\ (2^2\cdot 7) = \log_a\ 2^2 + \log_a\ 7 = 2\log_a\ 2 + \log_a\ 7 = 2(1.8301) + 5.0999 = 8.7601$$

15.

$$\begin{aligned}\log_a 3.5 &= \log_a \frac{7}{2}\\ &= \log_a 7 - \log_a 2\\ &= 5.0999 - 1.8301\\ &= 3.2698\end{aligned}$$

16.

$$\begin{aligned}\log_a \sqrt{7} &= \log_a 7^{1/2}\\ &= \frac{1}{2}\log_a 7\\ &= \frac{1}{2}\cdot 5.0999\\ &= 2.54995\end{aligned}$$

17.

$$\begin{aligned}\log_a \frac{1}{4} &= \log_a 2^{-2}\\ &= -2\ \log_a 2\\ &= -2(1.8301)\\ &= -3.6602\end{aligned}$$

18.

$$\begin{aligned}\frac{dQ}{dt} &= kQ\\ \frac{dQ}{Q} &= k\ dt\\ \text{integrating} & \\ \ln Q &= kt + C\\ C \text{ constant} & \\ Q &= e^{kt+C}\\ Q &= e^{kt}\cdot e^{C}\\ \text{Let } e^C = Q_0 & \\ Q &= Q_0\ e^{kt}\end{aligned}$$

19. $k = \dfrac{\ln(2)}{16} = 0.043 = 4.3\%$

20. $T = \dfrac{\ln(2)}{0.068} = 10.2$

21. **a)** $C = 4.65\ e^{kt}$. When $t = 43$, $C = 25.38$

$$\begin{aligned}25.38 &= 4.65\ e^{43k}\\ \frac{25.38}{4.65} &= e^{43k}\\ \ln\left(\frac{25.38}{4.65}\right) &= 43k\\ \frac{\ln\left(\frac{25.38}{4.65}\right)}{43} &= k\\ 0.0395 &= k\end{aligned}$$

Therefore $C = 4.65\ e^{0.0395t}$

b) When $t = 53$, $y = 4.65\ e^{0.0395(53)} = 37.73$
When $t = 58$, $y = 4.65\ e^{0.0395(58)} = 45.96$

22. **a)**

$$\begin{aligned}\frac{dN}{dt} &= 0.12N\\ \frac{dN}{N} &= 0.12dt\\ \ln N &= 0.12t + C\\ N &= e^{0.12t+C}\\ N &= N_0\ e^{0.12t}\\ 60 &= N_0\ e^{0}\\ 60 &= N_0\end{aligned}$$

Thus, $N = 60\ e^{0.12t}$

b) When $t = 10$, $y = 60\ e^{1.2} = 199$

c) The time to double is $T = \dfrac{\ln(2)}{0.12} = 5.8$ years

23. $T = \dfrac{\ln(1/2)}{0.13} = 5.3$ years

24. $k = \dfrac{\ln(1/2)}{3.8} = 0.182 = 18.2\%$

25. **a)**

$$\begin{aligned}\frac{dA}{dt} &= -0.07A\\ \frac{dA}{A} &= -0.07dt\\ \ln A &= -0.07t + C\\ A &= e^{-0.07t+C}\\ A &= A_0\ e^{-0.07t}\\ 800 &= A_0\ e^{0}\\ 800 &= A_0\end{aligned}$$

Thus, $A = 800\ e^{-0.07t}$

b) When $t = 20$, $y = 800\ e^{-1.4} = 197$ grams

c) The half-life is $T = \dfrac{\ln(1/2)}{-0.07} = 9.9$ days

26. **a)** When $t = 1$, $y = 1 - 4^{-0.7(1)} = 0.5$
When $t = 2$, $y = 1 - 4^{-0.7(2)} = 0.75$
When $t = 5$, $y = 1 - 4^{-0.7(5)} = 0.97$
When $t = 10$, $y = 1 - 4^{-0.7(10)} = 0.9991$
When $t = 14$, $y = 1 - 4^{-0.7(14)} = 0.99994$

b) $p'(t) = 0.7\ e^{-0.7t}$

c) $p'(t)$ represents the rate of change of the probability of mastering a certain assembly line task with respect to the number of learning trials

d)

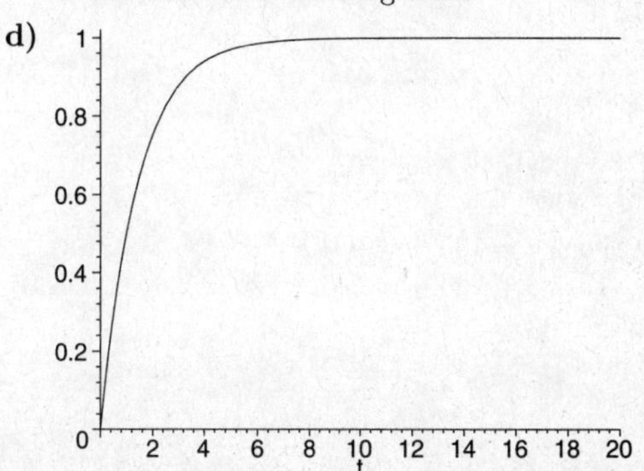

27. $\dfrac{dy}{dx} = 3^x \ \ln 3$

28. $\dfrac{dy}{dx} = \dfrac{1}{x \ln 5}$

29.

$$\begin{aligned}
\frac{dy}{dx} &= \frac{(e^{2x} - e^{-2x})(2e^{2x} - 2e^{-2x})}{(e^{2x} - e^{-2x})^2} - \\
&\quad \frac{(e^{2x} + e^{-2x})(2e^{2x} + 2e^{-2x})}{(e^{2x} - e^{-2x})^2} \\
&= \frac{2e^{4x} + 2e^{-4x} - 4 - 2e^{4x} - 2e^{-4x} - 4}{(e^{2x} - e^{-2x})^2} \\
&= \frac{-8}{(e^{2x} - e^{-2x})^2} \\
&= \frac{-8}{\left(e^{2x} - \frac{1}{2^{2x}}\right)^2} \\
&= \frac{-8e^{4x}}{(e^{4x} - 1)^2}
\end{aligned}$$

30.

$$\begin{aligned}
f'(x) &= x^4 \cdot \frac{1}{x} + 4x^3 \cdot \ln\, x \\
&= x^3(1 + 4\, ln\, x) \\
f'(x) &= 0 \\
x^3(1 + 4\, ln\, x) &= 0 \\
1 + 4\, ln\, x &= 0 \\
\ln\, x &= \frac{-1}{4} \\
x &= e^{-1/4}
\end{aligned}$$

When $x = e^{-1/4}$, $f(x) = \left(e^{-1/4}\right)^4 \ \ln\ e^{-1/4} = \frac{-1}{4\,e}$

31.

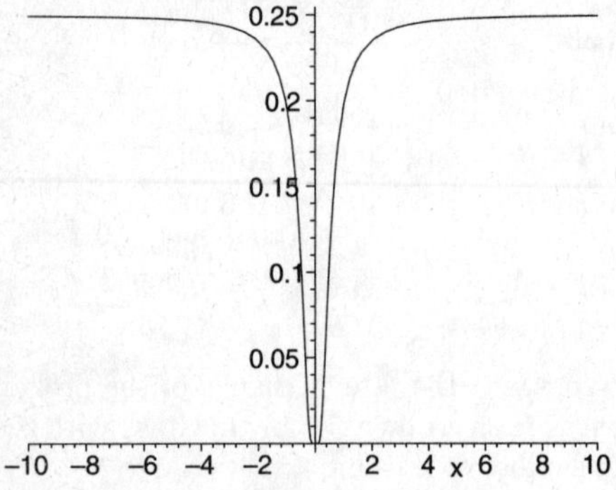

32. $\displaystyle\lim_{x \to 0} \frac{e^{1/x}}{(1 + e^{1/x})^2} = 0$

33. **a)** $y = 984(1.039)^t = 984\ e^{0.0387t}$
Exponential growth rate is $k = 3.87\%$

b) When $t = 100$, $y = 984\ e^{0.0387(100)} \approx 47110$
When $t = 200$, $y = 984\ e^{0.0387(200)} = 2254840$

c) $T = \dfrac{\ln(2)}{0.0387} = 17.9$ days

d)

$$\begin{aligned}
6200000000 &= 984\ e^{0.0387t} \\
\frac{6200000000}{984} &= e^{0.0387t} \\
\ln\left(\frac{6200000000}{984}\right) &= 0.0387t \\
\frac{\ln\left(\frac{6200000000}{984}\right)}{0.0387} &= t \\
404.6 &= t
\end{aligned}$$

e) The model obtained does not seem to be realistic. From the answer of part (d) we would expect 6.2 billion reported cases of SARS within 405 days after March 17, 2003 which corresponds to the middle of February of 2004, and we know that was not the case.

Chapter 4 Test

1. $y = e^x$

$\dfrac{dy}{dx} = e^x$

2. $y = (tan\ x)\ \ln\ x$

$$\begin{aligned}
\frac{dy}{dx} &= tan\ x \cdot \frac{1}{x} + sec^2\ x \cdot \ln\ x \\
&= \ln\ x\ sec^2\ x + \frac{tan\ x}{x}
\end{aligned}$$

3. $f(x) = e^{-x^2}$

$$\begin{aligned}
f'(x) &= e^{-x^2} \cdot -2x \\
&= -2x\ e^{-x^2}
\end{aligned}$$

4. $f(x) = \ln \dfrac{x}{7}$

$$\begin{aligned}
f'(x) &= \frac{7}{x} \cdot \frac{1}{7} \\
&= \frac{1}{x}
\end{aligned}$$

5. $f(x) = e^x - 5x^3$

$f'(x) = e^x - 15x^2$

6. $f(x) = 3e^x \ln\ x$

$$\begin{aligned}
f'(x) &= 3e^x \cdot \frac{1}{x} + 3e^x \cdot \ln\ x \\
&= \frac{3e^x(x \ln\ x + 1)}{x}
\end{aligned}$$

7. $y = \ln(e^x - sin\ x)$

$$\begin{aligned}
\frac{dy}{dx} &= \frac{1}{e^x - sin\ x} \cdot (e^x - cos\ x) \\
&= \frac{e^x - cos\ x}{e^x - sin\ x}
\end{aligned}$$

8. $y = \dfrac{\ln x}{e^x}$

$$\begin{aligned}\frac{dy}{dx} &= \frac{e^x \cdot \frac{1}{x} - \ln x \cdot e^x}{(e^x)^2} \\ &= \frac{(1 - x \ln x)}{x e^x}\end{aligned}$$

9.

$$\begin{aligned}\log_b 18 &= \log_b 2 \cdot 9 \\ &= \log_b 2 + \log_b 9 \\ &= 0.2560 + 0.8114 \\ &= 1.0674\end{aligned}$$

10.

$$\begin{aligned}\log_b 4.5 &= \log_b \frac{9}{2} \\ &= \log_b 9 - \log_b 2 \\ &= 0.8114 - 0.2560 \\ &= 0.5554\end{aligned}$$

11.

$$\begin{aligned}\log_b 3 &= \log_b \sqrt{9} \\ &= \log_b 9^{1/2} \\ &= \frac{1}{2} \cdot \log_b 9 \\ &= \frac{1}{2} \cdot 0.8114 \\ &= 0.4057\end{aligned}$$

12.

$$\begin{aligned}\frac{dM}{dt} &= kM \\ \frac{dM}{M} &= k\,dt\end{aligned}$$

integrating

$$\ln M = kt + C$$

C constant

$$\begin{aligned}M &= e^{kt+C} \\ M &= e^{kt} \cdot e^C\end{aligned}$$

Let $e^C = M_0$

$$M = M_0\, e^{kt}$$

13. $k = \dfrac{\ln(2)}{4} = 0.173 = 17.3\%$

14. $T = \dfrac{\ln(2)}{0.6931} = 10$ years

15. **a)** $C = 0.748\, e^{kt}$. When $t = 10$, $C = 1.042$

$$\begin{aligned}1.042 &= 0.748\, e^{10k} \\ \frac{1.042}{0.748} &= e^{10k} \\ \ln\left(\frac{1.042}{0.748}\right) &= 10k \\ \frac{\ln\left(\frac{1.042}{0.748}\right)}{10} &= k \\ 0.0331 &= k\end{aligned}$$

Therefore $C = 0.748\, e^{0.0331t}$

b) When $t = 17$, $y = 0.748\, e^{0.0331(17)} = \$1.31$
When $t = 27$, $y = 0.748\, e^{0.0331(27)} = \$1.83$

c) Answers vary

16. **a)**

$$\begin{aligned}\frac{dA}{dt} &= -0.1A \\ \frac{dA}{A} &= -0.1dt \\ \ln A &= -0.1t + C \\ A &= e^{-0.1t+C} \\ A &= A_0\, e^{-0.1t} \\ 3 &= A_0\, e^0 \\ 3 &= A_0\end{aligned}$$

Thus, $A = 3\, e^{-0.1t}$

b) When $t = 10$, $y = 3\, e^{-1} = 1.1$ cc

c) The half-life is $T = \dfrac{\ln(1/2)}{-0.1} = 6.93$ hours

17. $T = \dfrac{\ln(1/2)}{0.0083} = 83.5$ days

18. $k = \dfrac{\ln(1/2)}{2300000} = 0.0000003 = 0.00003\%$ per year

19. **a)** When $t = 0$

$$\begin{aligned}P(0) &= \frac{100}{1 + 24e^{(0)}} \\ &= \frac{100}{25} \\ &= 4\%\end{aligned}$$

b) When $t = 1$, $P(1) = \dfrac{100}{1 + 24e^{-0.28}} = 5.22\%$

When $t = 5$, $P(5) = \dfrac{100}{1 + 24e^{-0.28(5)}} = 14.45\%$

When $t = 10$, $P(10) = \dfrac{100}{1 + 24e^{-0.28(10)}} = 40.66\%$

When $t = 15$, $P(15) = \dfrac{100}{1 + 24e^{-0.28(15)}} = 73.54\%$

When $t = 20$, $P(20) = \dfrac{100}{1 + 24e^{-0.28(20)}} = 91.85\%$

When $t = 30$, $P(30) = \dfrac{100}{1 + 24e^{-0.28(30)}} = 99.46\%$

When $t = 35$, $P(35) = \dfrac{100}{1 + 24e^{-0.28(35)}} = 99.87\%$

c)

$$\begin{aligned}P'(t) &= -100\left(1 + 24e^{-0.28t}\right)^{-2} \cdot 24e^{-0.28t} \cdot -0.28 \\ &= \frac{672e^{-0.28t}}{\left(1 + 24e^{-0.28t}\right)^2}\end{aligned}$$

d) $P'(t)$ represents the rate of change of the percentage of people who buy products with respect to the number of times the ad is run

e)

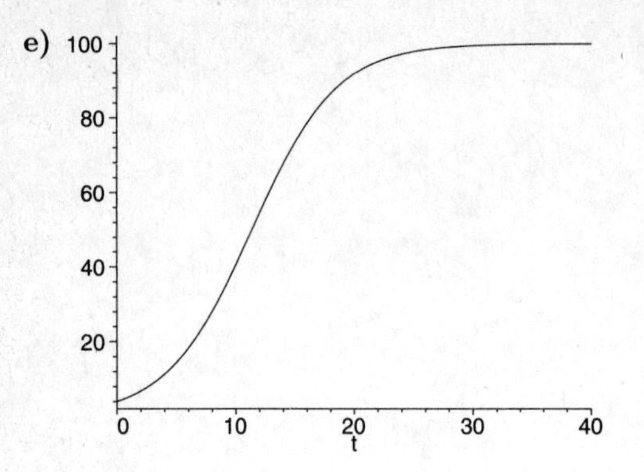

20. $f(x) = 20^x$

$f'(x) = 20^x \ln(20)$

21. $y = \log_{20} x$

$$\frac{dy}{dx} = \frac{1}{\ln(20)} \cdot \frac{1}{x} = \frac{1}{x \ln(20)}$$

22. $y = x(\ln x)^2 - 2x \ln x + 2x$

$$\begin{aligned} \frac{dy}{dx} &= (\ln x)^2 + 2x \ln x \cdot \frac{1}{x} \\ &\quad -2\ln x - 2x \cdot \frac{1}{x} + 2 \\ &= (\ln x)^2 \end{aligned}$$

23. $f(x) = x^4 e^{-x}$

$$\begin{aligned} f'(x) &= 4x^3 \cdot e^{-x} + x^4 \cdot -e^{-x} \\ &= e^{-x}(4x^3 - x^4) \end{aligned}$$

Solving $f'(x) = 0$ we get

$$\begin{aligned} 4x^3 - x^4 &= 0 \\ x^3(4 - x) &= 0 \\ x &= 0 \\ &\text{or} \\ x &= 4 \end{aligned}$$

Using test values to determine the sign of the first derivative we get: $f'(x) < 0$ on $(-\infty, 0]$
$f'(x) > 0$ on $[0, 4]$
$f'(x) < 0$ on $[4, \infty)$
Thus we have a relative minimum at $x = 0$ and

$$\begin{aligned} f(0) &= (0)^4 \cdot e^0 \\ &= 0 \end{aligned}$$

Relative minimum at $(0, 0)$ And a relative maximum at $x = 4$

$$\begin{aligned} f(4) &= (4)^4 \cdot e^{-4} \\ &= 256\, e^{-4} \\ &\approx 4.689 \end{aligned}$$

24.

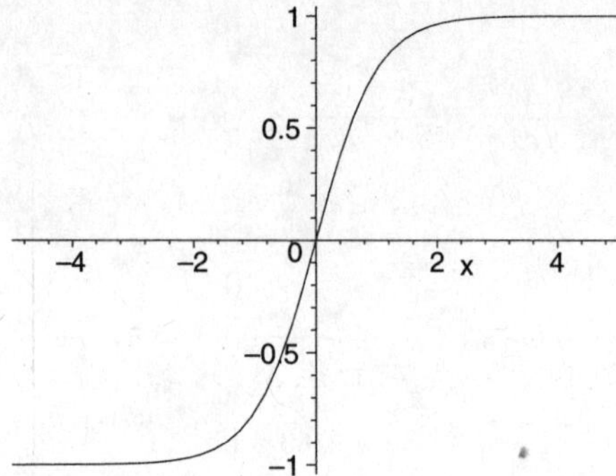

25. $\lim_{x \to 0} \frac{e^x - e^{-x}}{e^x + e^{-x}} = 0$

26. **a)** $y = 0.027(19.2)^t = 0.027\, e^{2.96t}$

b) When $t = 6$, $y = 1352605$
When $t = 7$, $y = 25970025$

c)

$$\begin{aligned} 6300000000 &= 0.027\, e^{2.96t} \\ \frac{6300000000}{0.027} &= e^{2.96t} \\ \ln\left(\frac{6300000000}{0.027}\right) &= 2.96t \\ \frac{\ln\left(\frac{6300000000}{0.027}\right)}{2.96} &= t \\ 8.86 &= t \end{aligned}$$

d) $T = \ln(2)/2.96 = 0.2342$ hours

Technology Connection

- **Page 265:**

  1. $5^\pi = 156.9925453$

  2. $5^{\sqrt{3}} = 16.24245082$

  3. $7^{-\sqrt{2}} = 0.0638044384$

  4. $18^{-\pi} = 0.000138793867$

- **Page 282:**
  $f(3) = 1000$
  $f(0.699) = 5.0003$
  $g(5) = 0.69897$
  $g(1000) = 3$

- **Page 285:**

  1. $t = 6.90776$

  2. $t = -4.09435$

  3. $t = 74.89331$

  4. $t = 64.21959$

  5. $t = 38.74448$

- **Page 303:**

  1. $y = 8.2432$

  2. $y = 9.638$

3. $y = 11.269$

4. $y = 13.176$

5. $y = 30187.92\,(1.04686554^x)$

$y = 30187.92\ e^{0.0458004996x}$

6. 75449, 149974, 298111

Extended Technology Application

1. **a)**

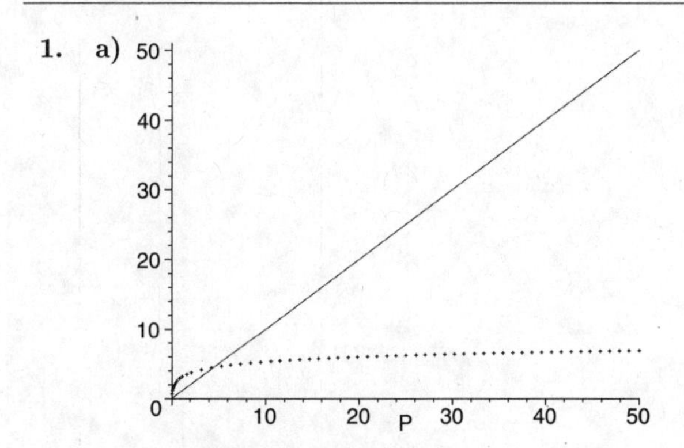

b - c) $f'(P) = \frac{1}{P} \to P_0 = 1$
$H(1) = \ln(20) - 1 = 1.9957$

2. **a)**

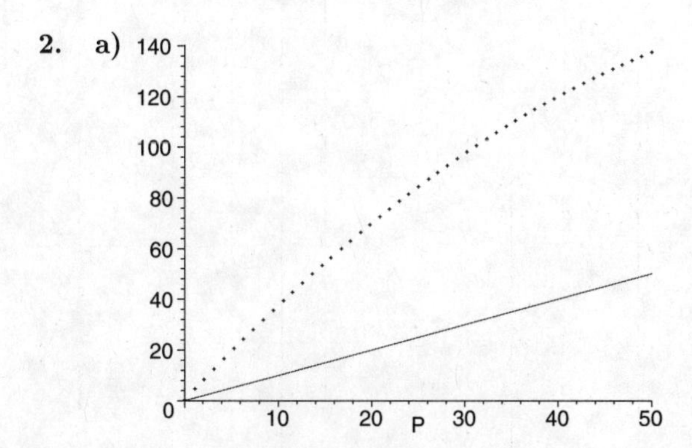

b - c) $f'(P) = -0.05P + 4 \to P_0 = 60$
$H(60) = -0.025(60)^2 + 4(60) - 60 = 90$

3. **a)**

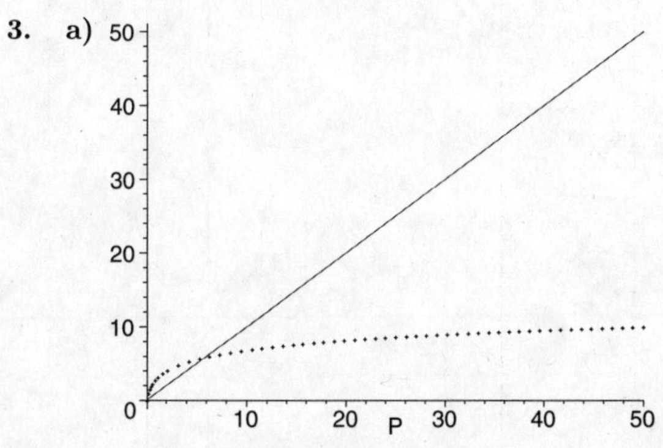

b - c) $f'(P) = \frac{2P+1}{P(P+1)} \to P_0 = \frac{1+\sqrt{5}}{2}$
$H\left(\frac{1+\sqrt{5}}{2}\right) = \ln\left([4+4\sqrt{5}][\frac{3+\sqrt{5}}{2}]\right) - \frac{1+\sqrt{5}}{2} = 1.905$

4. **a)**

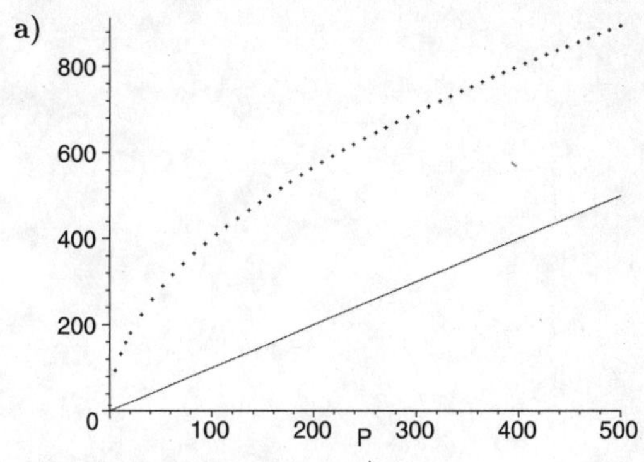

b - c) $P_0 = 400$, $H(400) = 400$

5. **a)**

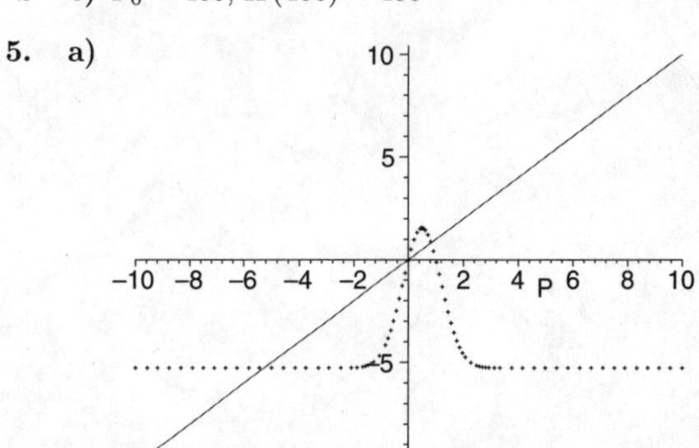

b - c) $P_0 = 0.428$, $H(0.428) = 1.0834$

6. **a)** $y = -0.0011083333x^3 + 0.0715357143x^2 - 0.0338095238x + 4$

b)

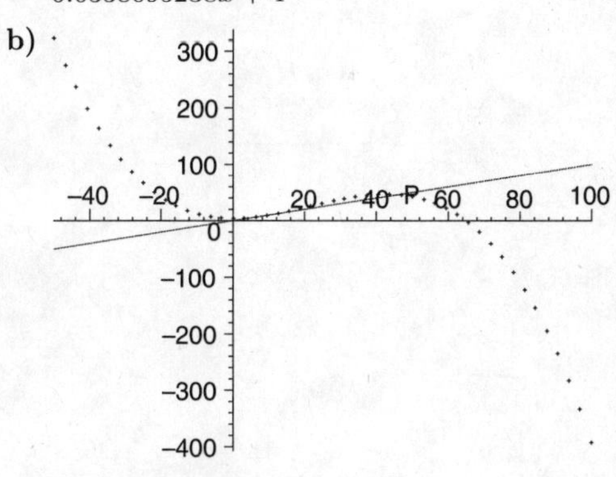

c) $P_0 = 33.841447$

Chapter 5

Integration

Exercise Set 5.1

1. $\int x^6\,dx$

$= \dfrac{x^{6+1}}{6+1} + C \quad \left(\int x^r\,dx = \dfrac{x^{r+1}}{r+1} + C\right)$

$= \dfrac{x^7}{7} + C$

2. $\displaystyle\int x^7\,dx = \frac{x^8}{8} + C$

3. $\int 2\,dx$

$= 2x + C$ (For k a constant, $\int k\,dx = kx + C$.)

4. $\int 4\,dx = 4x + C$

5. $\int x^{1/4}\,dx$

$= \dfrac{x^{1/4+1}}{\frac{1}{4}+1} + C \quad \left(\int x^r\,dx = \dfrac{x^{r+1}}{r+1} + C\right)$

$= \dfrac{x^{5/4}}{\frac{5}{4}} + C$

$= \dfrac{4}{5}x^{5/4} + C$

6. $\displaystyle\int x^{1/3}\,dx = \frac{x^{4/3}}{\frac{4}{3}} + C = \frac{3}{4}x^{4/3} + C$

7. $\int (x^2 + x - 1)\,dx$

$= \int x^2\,dx + \int x\,dx - \int 1\,dx$

The integral of a sum is the sum of the integrals.

$= \dfrac{x^3}{3} + \dfrac{x^2}{2} - x + C$ ← DON'T FORGET THE C!

$\left[\int x^r\,dx = \dfrac{x^{r+1}}{r+1} + C\right]$

(For k a constant, $\int k\,dx = kx + C$.)

8. $\displaystyle\int (x^2 - x + 2)\,dx = \frac{x^3}{3} - \frac{x^2}{2} + 2x + C$

9. $\int (t^2 - 2t + 3)\,dt$

$= \int t^2\,dt - \int 2t\,dt + \int 3\,dt$

The integral of a sum is the sum of the integrals.

$= \dfrac{t^3}{3} - 2\cdot\dfrac{t^2}{2} + 3t + C$

$\left[\int x^r\,dx = \dfrac{x^{r+1}}{r+1} + C\right]$

(For k a constant, $\int k\,dx = kx + C$.)

$= \dfrac{t^3}{3} + t^2 + 3t + C$

10. $\int (3t^2 - 4t + 7)\,dt = t^3 - 2t^2 + 7t + C$

11. $\int 5\,e^{8x}\,dx$

$= \dfrac{5}{8}e^{8x} + C \quad \left(\int be^{ax}\,dx = \dfrac{b}{a}e^{ax} + C\right)$

12. $\displaystyle\int 3e^{5x}\,dx = \frac{3}{5}e^{5x} + C$

13. $\int (w^3 - w^{8/7})\,dw$

$= \int x^3\,dx - \int x^{8/7}\,dx$

The integral of a sum is the sum of the integrals.

$= \dfrac{w^4}{4} - \dfrac{w^{8/7+1}}{\frac{8}{7}+1} + C \quad \left[\int x^r\,dx = \dfrac{x^{r+1}}{r+1} + C\right]$

$= \dfrac{w^4}{4} - \dfrac{w^{15/7}}{\frac{15}{7}} + C$

$= \dfrac{w^4}{4} - \dfrac{7}{15}w^{15/7} + C$

14. $\displaystyle\int (t^4 - t^{6/5})\,dt = \frac{t^5}{5} - \frac{t^{11/5}}{\frac{11}{5}} + C$

$= \dfrac{t^5}{5} - \dfrac{5}{11}t^{11/5} + C$

15. $\displaystyle\int \frac{1000}{r}\,dr$

$= 1000\displaystyle\int \frac{1}{r}\,dr$ The integral of a constant times a function is the constant times the integral.

$= 1000\ln r + C$ $\displaystyle\int \frac{1}{x}\,dx = \ln x + C,\ x > 0;$ we generally consider $r > 0$

16. $\displaystyle\int \frac{500}{x}\,dx = 500\ln x + C$

17. $\int \frac{dx}{x^2} = \int \frac{1}{x^2}\,dx = \int x^{-2}\,dx$

$= \frac{x^{-2+1}}{-2+1} + C \qquad \left[\int x^r\,dx = \frac{x^{r+1}}{r+1} + C\right]$

$= \frac{x^{-1}}{-1} + C$

$= -x^{-1} + C$, or $-\frac{1}{x} + C$

18. $\int \frac{dx}{x^3} = \int x^{-3}\,dx = \frac{x^{-2}}{-2} + C = -\frac{1}{2}x^{-2} + C$, or $-\frac{1}{2x^2} + C$

19. $\int \sqrt{s}\,ds = \int s^{1/2}\,ds$

$= \frac{s^{1/2+1}}{\frac{1}{2}+1} + C$

$= \frac{s^{3/2}}{\frac{3}{2}} + C$

$= \frac{2}{3}s^{3/2} + C$

20. $\int \sqrt[3]{x^2}\,dx = \int x^{2/3}\,dx = \frac{x^{5/3}}{\frac{5}{3}} + C = \frac{3}{5}\,x^{5/3} + C$

21. $\int \frac{-6}{\sqrt[3]{x^2}}\,dx = \int \frac{-6}{x^{2/3}}\,dx = \int -6x^{-2/3}\,dx$

$= -6\int x^{-2/3}\,dx$

$= -6 \cdot \frac{x^{-2/3+1}}{-\frac{2}{3}+1} + C$

$= -6 \cdot \frac{x^{1/3}}{\frac{1}{3}} + C$

$= -6 \cdot 3x^{1/3} + C$

$= -18x^{1/3} + C$

22. $\int \frac{20}{\sqrt[5]{x^4}}\,dx = \int 20\,x^{-4/5}\,dx$

$= 20 \cdot \frac{x^{1/5}}{\frac{1}{5}} + C$

$= 100x^{1/5} + C$

23. $\int 8\,e^{-2x}\,dx = \frac{8}{-2}\,e^{-2x} + C$

$= -4\,e^{-2x} + C$

24. $\int 7\,e^{-0.25x}\,dx = \frac{7}{-0.25}\,e^{-0.25x} + C$

$= -28\,e^{-0.25x} + C$

25. $\int \left(x^2 - \frac{3}{2}\sqrt{x} + x^{-4/3}\right) dx$

$= \int x^2\,dx - \frac{3}{2}\int x^{1/2}\,dx + \int x^{-4/3}\,dx$

$= \frac{x^{2+1}}{2+1} - \frac{3}{2} \cdot \frac{x^{1/2+1}}{\frac{1}{2}+1} + \frac{x^{-4/3+1}}{-\frac{4}{3}+1} + C$

$= \frac{x^3}{3} - \frac{3}{2} \cdot \frac{x^{3/2}}{\frac{3}{2}} + \frac{x^{-1/3}}{-\frac{1}{3}} + C$

$= \frac{x^3}{3} - x^{3/2} - 3x^{-1/3} + C$

26. $\int \left(x^4 + \frac{1}{8\sqrt{x}} - \frac{4}{5}x^{-2/5}\right) dx$

$= \int \left(x^4 + \frac{1}{8}x^{-1/2} - \frac{4}{5}x^{-2/5}\right) dx$

$= \frac{x^5}{5} + \frac{1}{8} \cdot \frac{x^{1/2}}{\frac{1}{2}} - \frac{4}{5} \cdot \frac{x^{3/5}}{\frac{3}{5}} + C$

$= \frac{x^5}{5} + \frac{1}{4}x^{1/2} - \frac{4}{3}x^{3/5} + C$

27. $\int 5\,sin\,2\theta\,d\theta = \frac{-5}{2}\,cos\,2\theta + C$

28. $\int \frac{1}{4}\,cos\,2\theta\,d\theta = \frac{1}{8}\,sin\,2\theta + C$

29. $\int \left(5\,sin\,5x - 4\,cos\,2x\right) dx$

$= -cos\,5x - 2\,sin\,2x + C$

30. $\int \left(3\,cos\,2\pi x - 8\,sin\,\pi x\right) dx = \frac{3}{2\pi}\,sin\,2\pi x + \frac{8}{\pi}cos\,\pi x + C$

31. $\int 3\,sec^2\,3x\,dx = tan\,3x + C$

32. $\int 5\,csc^2\,2x\,dx = \frac{-5}{2}cot\,2x + C$

33. $\int \frac{1}{3}\,sec\,\frac{x}{9}\,tan\,\frac{x}{9}\,dx = 3\,sec\,\frac{x}{9} + C$

34. $\int 4\,sec\,2x\,tan\,2x\,dx = 2\,sec\,2x + C$

35. $\int (sec\,x + tan\,x)sec\,x\,dx = \int (sec^2\,x + sec\,x\,tan\,x)\,dx$

$= tan\,x + sec\,x + C$

36. $\int (csc\,x + cot\,x)csc\,x\,dx = \int (csc^2\,x + csc\,x\,cot\,x)\,dx$

$= -cot\,x - cot\,x + C$

37. $\int \left[\frac{1}{t} + \frac{1}{t^2} - \frac{1}{e^t}\right] dt = \ln t - \frac{1}{t} + \frac{1}{e^t} + C$

38. $\int (4e^{7w} - w^{-2/5} + 7w^{100})\,dw = \frac{4}{7}e^{7w} + \frac{10}{3}w^{3/5} + \frac{7}{101}w^{101} + C$

39. Find the function f such that $f'(x) = x - 3$ and $f(2) = 9$.

We first find $f(x)$ by integrating.

$f(x) = \int (x-3)\,dx$
$= \int x\,dx - \int 3\,dx$
$= \frac{x^2}{2} - 3x + C$

The condition $f(2) = 9$ allows us to find C.

$f(x) = \frac{x^2}{2} - 3x + C$

$f(2) = \frac{2^2}{2} - 3 \cdot 2 + C = 9$ Substituting 2 for x and 9 for $f(2)$

$2 - 6 + C = 9$

$C = 13$

Thus, $f(x) = \frac{x^2}{2} - 3x + 13$.

40. $f(x) = \int (x-5)\,dx = \frac{x^2}{2} - 5x + C$

$f(1) = \frac{1^2}{2} - 5 \cdot 1 + C = 6$

$C = \frac{21}{2}$

$f(x) = \frac{x^2}{2} - 5x + \frac{21}{2}$

41. Find the function f such that $f'(x) = x^2 - 4$ and $f(0) = 7$.

We first find $f(x)$ by integrating.

$f(x) = \int (x^2 - 4)\,dx$
$= \int x^2\,dx - \int 4\,dx$
$= \frac{x^3}{3} - 4x + C$

The condition $f(0) = 7$ allows us to find C.

$f(x) = \frac{x^3}{3} - 4x + C$

$f(0) = \frac{0^3}{3} - 4 \cdot 0 + C = 7$ Substituting 0 for x and 7 for $f(0)$

Solving for C we get $C = 7$.

Thus, $f(x) = \frac{x^3}{3} - 4x + 7$.

42. $f(x) = \int (x^2+1)\,dx = \frac{x^3}{3} + x + C$

$f(0) = \frac{0^3}{3} + 0 + C = 8$

$C = 8$

$f(x) = \frac{x^3}{3} + x + 8$

43. $f(x) = \int 2\cos 3x\,dx = \frac{2}{3}\sin 3x + C$

$f(0) = \frac{2}{3}\sin(0) + C = 1$

$C = 1$

$f(x) = \frac{2}{3}\sin 3x + 1$

44. $f(x) = \int 3\sin 8x\,dx = \frac{-3}{8}\cos 8x + C$

$f(0) = \frac{-3}{8}\cos(0) + C = 3$

$-\frac{3}{8} + C = 1$

$C = \frac{11}{8}$

$f(x) = \frac{-3}{8}\cos 8x + \frac{11}{8}$

45. $f(x) = \int 5e^{2x}\,dx = \frac{5}{2}e^{2x} + C$

$f(0) = \frac{5}{2} + C = -10$

$C = \frac{-25}{2}$

$f(x) = \frac{5}{2}e^{2x} - \frac{25}{2}$

46. $f(x) = \int 3e^{-x}\,dx = -3e^{-x} + C$

$f(0) = -3 + C = 6$

$C = 9$

$f(x) = -3e^{-x} + 9$

47. $f(t) = \int 157t + 1000\,dt = \frac{157}{2}t^2 + 1000t + C$

$f(0) = 0 + 0 + C$

$156239 = C$

$f(t) = \frac{157}{2}t^2 + 1000t + 156239$

$f(2) = \frac{157}{2}(2)^2 + 1000(2) + 156239$

$= 158553$

48. $f(t) = \int 500 + 10t + 0.3t^3 = 500t + 5t^2 + 0.075t^4 + C$

$f(0) = 0 + 0 + C$

$452937 = C$

$f(t) = 452937 + 500t + 5t^2 + 0.075t^4$

$f(4) = 452937 + 500(4) + 5(4)^2 + 0.075(4)^4$

$= 455036$

49. $v(t) = 3t^2, \quad s(0) = 4$

We find $s(t)$ by integrating $v(t)$.

$s(t) = \int v(t)\,dt$
$= \int 3t^2\,dt$
$= 3 \cdot \frac{t^3}{3} + C$
$= t^3 + C$

The condition $s(0) = 4$ allows us to find C.

$s(0) = 0^3 + C = 4$ Substituting 0 for t and 4 for $s(0)$

Solving for C, we get $C = 4$.

Thus, $s(t) = t^3 + 4$.

50. $s(t) = \int 2t\,dt = t^2 + C$

$s(0) = 0^2 + C = 10$

$C = 10$

$s(t) = t^2 + 10$

51. $a(t) = 4t, \qquad v(0) = 20$

We find $v(t)$ by integrating $a(t)$.

$$\begin{aligned} v(t) &= \int a(t)\,dt \\ &= \int 4t\,dt \\ &= 4 \cdot \frac{t^2}{2} + C \\ &= 2t^2 + C \end{aligned}$$

The condition $v(0) = 20$ allows us to find C.

$v(0) = 2 \cdot 0^2 + C = 20$ Substituting 0 for t and 20 for $v(0)$

Solving for C, we get $C = 20$.

Thus, $v(t) = 2t^2 + 20$.

52. $v(t) = \int 6t\,dt = 3t^2 + C$

$v(0) = 3 \cdot 0^2 + C = 30$

$C = 30$

$v(t) = 3t^2 + 30$

53. $a(t) = -2t + 6, \qquad v(0) = 6 \qquad$ and $s(0) = 10$

We find $v(t)$ by integrating $a(t)$.

$$\begin{aligned} v(t) &= \int a(t)\,dt \\ &= \int (-2t + 6)\,dt \\ &= -t^2 + 6t + C_1 \end{aligned}$$

The condition $v(0) = 6$ allows us to find C_1.

$v(0) = -0^2 + 6 \cdot 0 + C_1 = 6$ Substituting 0 for t and 6 for $v(0)$

Solving for C_1, we get $C_1 = 6$.

Thus, $v(t) = -t^2 + 6t + 6$.

We find $s(t)$ by integrating $v(t)$.

$$\begin{aligned} s(t) &= \int v(t)\,dt \\ &= \int (-t^2 + 6t + 6)\,dt \\ &= -\frac{t^3}{3} + 3t^2 + 6t + C_2 \end{aligned}$$

The condition $s(0) = 10$ allows us to find C_2.

$s(0) = -\frac{0^3}{3} + 3 \cdot 0^2 + 6 \cdot 0 + C_2$ Substituting 0 for t and 10 for $s(0)$

Solving for C_2, we get $C_2 = 10$.

Thus, $s(t) = -\frac{1}{3}t^3 + 3t^2 + 6t + 10$.

54. $v(t) = \int (-6t + 7)\,dt = -3t^2 + 7t + C_1$

$v(0) = -3 \cdot 0^2 + 7 \cdot 0 + C_1 = 10$

$C_1 = 10$

$v(t) = -3t^2 + 7t + 10$

$$\begin{aligned} s(t) &= \int (-3t^2 + 7t + 10)\,dt \\ &= -t^3 + \frac{7}{2}t^2 + 10t + C_2 \end{aligned}$$

$s(0) = -0^3 + \frac{7}{2} \cdot 0^2 + 10 \cdot 0 + C_2 = 20$

$C_2 = 20$

$s(t) = -t^3 + \frac{7}{2}t^2 + 10t + 20$

55. $a(t) = -32$ ft/sec^2

$v(0) =$ initial velocity $= v_0$

$s(0) =$ initial height $= s_0$

We find $v(t)$ by integrating $a(t)$.

$$\begin{aligned} v(t) &= \int a(t)\,dt \\ &= \int (-32)\,dt \\ &= -32t + C_1 \end{aligned}$$

The condition $v(0) = v_0$ allows us to find C_1.

$v(0) = -32 \cdot 0 + C_1 = v_0$ Substituting 0 for t and v_0 for $v(0)$

$C_1 = v_0$

Thus, $v(t) = -32t + v_0$.

We find $s(t)$ by integrating $v(t)$.

$$\begin{aligned} s(t) &= \int v(t)\,dt \\ &= \int (-32t + v_0)\,dt \\ &= -16t^2 + v_0 t + C_2 \qquad v_0 \text{ is constant} \end{aligned}$$

The condition $s(0) = s_0$ allows us to find C_2.

$s(0) = -16 \cdot 0^2 + v_0 \cdot 0 + C_2 = s_0$ Substituting 0 for t and s_0 for $s(0)$

$C_2 = s_0$

Thus, $s(t) = -16t^2 + v_0 t + s_0$.

56. $v(t) = \int -32\,dt = -32t + C_1$

$v(0) = -32 \cdot 0 + C_1 = 80$

$C_1 = 80$

$v(t) = -32t + 80$

$s(t) = \int (-32t + 80)\,dt = -16t^2 + 80t + C_2$

$s(0) = -16t \cdot 0^2 + 80 \cdot 0 + C_2 = 10$

$C_2 = 10$

$s(t) = -16t^2 + 80t + 10$

Solve $-16t^2 + 80t + 10 = 0$.

$$t = \frac{10 \pm \sqrt{110}}{4}$$

Since $t > 0$, $t = \frac{10 + \sqrt{110}}{4} \approx 5.1$ sec.

57. $a(t) = k$ Constant acceleration

$v(t) = \int a(t)\,dt = \int k\,dt = kt \quad (v(0) = 0;\text{ thus } C = 0.)$

$$s(t) = \int v(t)\,dt = \int kt\,dt = k \cdot \frac{t^2}{2} = \frac{1}{2}kt^2$$

$(s(0) = 0;\text{ thus } C = 0.)$

We know that

$$a(t) = k = \frac{60\text{ mph}}{\frac{1}{2}\text{ min}}$$

and that

$$t = \frac{1}{2}\text{ min.}$$

Thus

$$s(t) = \frac{1}{2}kt^2$$

$$\begin{aligned} s\left(\frac{1}{2}\text{ min}\right) &= \frac{1}{2} \cdot \frac{60\text{ mph}}{\frac{1}{2}\text{ min}} \cdot \left(\frac{1}{2}\text{ min}\right)^2 \\ &= \frac{1}{2} \cdot \frac{60\text{ mi}}{\text{hr}} \cdot \frac{1}{2}\text{ min} \\ &= \frac{1}{2} \cdot \frac{60\text{ mi}}{\text{hr}} \cdot \frac{1}{120}\text{ hr} \\ &= \frac{60}{240}\text{ mi} \\ &= \frac{1}{4}\text{ mi} \end{aligned}$$

The car travels $\frac{1}{4}$ mi during that time.

58. $a(t) = -68.5$

$v(t) = \int a(t)\,dt = \int -68.5\,dt = -68.5t + C$

(v(0)=102.7)

$102.7 = 0 + C \to C = 102.7$

$$s(t) = \int v(t)\,dt = \int -68.5t + 102.7\,dt$$

$= -34.25t^2 + 102.7t + C$

$(s(0)=0; thus\ C = 0)$

$s(t) = -34.25t^2 + 102.7t$

The time it takes to go from $102.7 ft/sec$ to 0 is

$$\begin{aligned} -68.5t + 102.7 &= 0 \\ 68.5t &= 102.7 \\ t &= \frac{102.7}{68.5} \\ &= 1.499 \end{aligned}$$

Thus

$s(1.499) = -34.25(1.499)^2 + 102.7(1.499)$

$= 76.98$

The car travels almost 77 feet during that time.

59. $a(t) = -68.5$

$v(t) = \int a(t)\,dt = \int -68.5\,dt = -68.5t + C$

(v(0)=132)

$132 = 0 + C \to C = 102.7$

$$s(t) = \int v(t)\,dt = \int -68.5t + 132\,dt$$

$= -34.25t^2 + 132t + C$

$(s(0)=0; thus\ C = 0)$

$s(t) = -34.25t^2 + 132t$

The time it takes to go from $132 ft/sec$ to 0 is

$$\begin{aligned} -68.5t + 132 &= 0 \\ 68.5t &= 132 \\ t &= \frac{132}{68.5} \\ &= 1.927 \end{aligned}$$

Thus

$s(1.927) = -34.25(1.927)^2 + 132(1.927)$

$= 127.182$

The car travels almost 127.2 feet during that time.

60. $A'(t) = -43.4t^{-2}$

a)

$$\begin{aligned} A(t) &= 43.4t^{-1} + C \\ 39.7 &= 43.4(1)^{-1} + C \\ -3.7 &= C \end{aligned}$$

Thus, $A(t) = 43.4t^{-1} - 3.7$

b)

$$\begin{aligned} A(7) &= 43.4(7)^{-1} - 3.7 \\ &= 2.5 \end{aligned}$$

61. $M'(t) = 0.2t - 0.003t^2$

a) We integrate to find $M(t)$.

$M(t) = \int(0.2t - 0.003t^2)\,dt$

$= 0.1t^2 - 0.001t^3 + C$

We use $M(0) = 0$ to find C.

$M(0) = 0.1(0)^2 - 0.001(0)^3 + C = 0$

$C = 0$

$M(t) = 0.1t^2 - 0.001t^3$

b) $M(8) = 0.1(8)^2 - 0.001(8)^3$

$= 6.4 - 0.512$

$= 5.888$

≈ 6 words

62. $I'(t) = 3.389e^{0.1049t}$

a)

$$\begin{aligned} I(t) &= \int 3.389e^{0.1049t}\, dt \\ &= 32.307e^{0.1049t} + C \\ 0 &= 32.307 + C \\ -32.307 &= C \\ I(t) &= 32.307e^{0.1049t} - 32.307 \end{aligned}$$

b)

$$\begin{aligned} I(27) &= 32.307e^{0.1049(27)} - 32.307 \\ &= 516 \end{aligned}$$

c)

$$\begin{aligned} I(34) &= 32.307e^{0.1049(34)} - 32.307 \\ &= 1111 \end{aligned}$$

d) $1111 - 516 = 595$

63. $N'(t) = 38.2e^{0.0376t}$

a)

$$\begin{aligned} N(t) &= \int 38.2e^{0.0376t}\, dt \\ &= 1015.957e^{0.0376t} + C \\ 1622 &= 1015.957 + C \\ 606.043 &= C \\ N(t) &= 1015.957e^{0.0376t} + 606.043 \end{aligned}$$

b)

$$\begin{aligned} N(30) &= 1015.957e^{0.0376(30)} + 606.043 \\ &= 3742 \end{aligned}$$

c)

$$\begin{aligned} N(60) &= 1015.957e^{0.0376(60)} + 606.043 \\ &= 10294 \end{aligned}$$

64. $f'(t) = t^{1/2} + t^{-1/2}$

$$\begin{aligned} f(t) &= \frac{2}{3}t^{3/2} + 2t^{1/2} + C \\ 0 &= \frac{2}{3}(4)^{3/2} + 2(4)^{1/2} + C \\ -\frac{28}{3} &= C \\ f(t) &= \frac{2}{3}t^{3/2} + 2t^{1/2} - \frac{28}{3} \end{aligned}$$

65. $f'(t) = t^{\sqrt{3}}$

$$\begin{aligned} f(t) &= \int t^{\sqrt{3}}\, dt \\ &= \frac{t^{\sqrt{3}+1}}{\sqrt{3}+1} + C \\ 8 &= 0 + C \\ 8 &= C \\ f(t) &= \frac{t^{\sqrt{3}+1}}{\sqrt{3}+1} + 8 \end{aligned}$$

66.

$$\begin{aligned} \int (5t+4)^2\, dt &= \int (25t^2 + 40t + 16)\, dt \\ &= \frac{25}{3}t^3 + 20t^2 + 16t + C \end{aligned}$$

67.

$$\begin{aligned} \int (x-1)^2 x^3\, dx &= \int (x^5 - 2x^4 + x^3)\, dx \\ &= \frac{x^6}{6} - \frac{2}{5}x^5 + \frac{x^4}{4} + C \end{aligned}$$

68.

$$\begin{aligned} \int (1-t)\sqrt{t}\, dt &= \int (t^{1/2} - t^{3/2})\, dt \\ &= \frac{2}{3}t^{3/2} - \frac{2}{5}t^{5/2} + C \end{aligned}$$

69.

$$\begin{aligned} \int \frac{(t+3)^2}{\sqrt{t}}\, dt &= \int (t^{3/2} + 6t^{1/2} + 9t^{-1/2})\, dt \\ &= \frac{2}{5}t^{5/2} + 4t^{3/2} + 18t^{1/2} + C \end{aligned}$$

70.

$$\begin{aligned} \int \frac{x^4 - 6x^2 - 7}{x^3}\, dx &= \int (x - 6x^{-1} - 7x^{-3})dx \\ &= \frac{x^2}{2} - 6\ln(x) + \frac{7}{2}x^{-2} + C \end{aligned}$$

71.

$$\begin{aligned} \int (t+1)^3\, dt &= \int (t^3 + 3t^2 + 3t + 1)\, dt \\ &= \frac{t^4}{4} + t^3 + \frac{3}{2}t^2 + t + C \end{aligned}$$

72.

$$\begin{aligned} \int \frac{1}{\ln(10)}\frac{dx}{x} &= \frac{1}{\ln(10)}\int \frac{dx}{x} \\ &= \frac{\ln(x)}{\ln(10} + C \end{aligned}$$

73.

$$\begin{aligned} \int be^{ax}\, dx &= b\int e^{ax}\, dx \\ &= \frac{b}{a}e^{ax} + C \end{aligned}$$

74.

$$\begin{aligned} \int (3x-5)(2x+1)\, dx &= \int (6x^2 + 3x - 5)\, dx \\ &= 2x^3 + \frac{3}{2}x^2 - 5x + C \end{aligned}$$

75.

$$\begin{aligned}\int \sqrt[3]{64x^4}\, dx &= \sqrt[3]{64}\int x^{4/3}\, dx\\ &= 4\cdot\frac{3}{7}x^{7/3}+C\\ &= \frac{12}{7}x^{7/3}+C\end{aligned}$$

76.

$$\begin{aligned}\int \frac{x^2-1}{x+1}dx &= \int (x-1)\, dx\\ &= \frac{x^2}{2}-x+C\end{aligned}$$

77.

$$\begin{aligned}\int \frac{t^3+8}{t+2}dt &= \int \frac{(t+2)(t^2-2t+4)}{(t+2)}dt\\ &= \int (t^2-2t+4)dt\\ &= \frac{t^3}{3}-t^2+4t+C\end{aligned}$$

78.

$$\begin{aligned}\int cos\ x\ tan\ x\ dx &= \int sin\ x\ dx\\ &= -cos\ x+C\end{aligned}$$

79.

$$\begin{aligned}\int (cos^3x+cos\ x\ sin^2x)\ dx &= \int [cos\ x(cos^2x+sin^2x)]\ dx\\ &= \int cos\ x\ dx\\ &= sin\ x+C\end{aligned}$$

80.

$$\begin{aligned}\int tan^2 3x\ dx &= \int (sec^2 3x-1)\ dx\\ &= \frac{tan\ 3x}{3}-x+C\end{aligned}$$

81.

$$\begin{aligned}\int cot^2 2x\ dx &= \int (csc^2 2x-1)\ dx\\ &= -\frac{cot\ 2x}{2}-x+C\end{aligned}$$

82. Answers could vary. The integral is not unique since the student forgot to mention the constant of integration.

83. Answers could vary. The antiderivative of a function represents the area under the curve of that function.

Exercise Set 5.2

1. a) $f(x)=\dfrac{1}{x^2}$

In the drawing in the text the interval $[1,7]$ has been divided into 6 subintervals, each having width 1 $\left(\Delta x=\dfrac{7-1}{6}=1\right)$.

The heights of the rectangles shown are

$f(1)=\dfrac{1}{1^2}=1$

$f(2)=\dfrac{1}{2^2}=\dfrac{1}{4}=0.2500$

$f(3)=\dfrac{1}{3^2}=\dfrac{1}{9}\approx 0.1111$

$f(4)=\dfrac{1}{4^2}=\dfrac{1}{16}=0.0625$

$f(5)=\dfrac{1}{5^2}=\dfrac{1}{25}=0.0400$

$f(6)=\dfrac{1}{6^2}=\dfrac{1}{36}\approx 0.0278$

The area of the region under the curve over $[1,7]$ is approximately the sum of the areas of the 6 rectangles.

Area of each rectangle:

1st rectangle: $1\cdot 1=1$

$[f(1)=1 \text{ and } \Delta x=1]$

2nd rectangle: $0.2500\cdot 1=0.2500$

$[f(2)=0.2500 \text{ and } \Delta x=1]$

3rd rectangle: $0.1111\cdot 1=0.1111$

4th rectangle: $0.0625\cdot 1=0.0625$

5th rectangle: $0.0400\cdot 1=0.0400$

6th rectangle: $0.0278\cdot 1=0.0278$

The total area is $1+0.2500+0.1111+0.0625+0.0400+0.0278$, or 1.4914.

b) $f(x)=\dfrac{1}{x^2}$

The interval $[1,7]$ has been divided into 12 subintervals, each having width 0.5 $\left(\Delta x=\dfrac{7-1}{12}=\dfrac{6}{12}=0.5\right)$. The heights of six of the rectangles were computed in part (a). The others are computed below:

$$f(1.5) = \frac{1}{(1.5)^2} = \frac{1}{2.25} \approx 0.4444$$
$$f(2.5) = \frac{1}{(2.5)^2} = \frac{1}{6.25} \approx 0.1600$$
$$f(3.5) = \frac{1}{(3.5)^2} = \frac{1}{12.25} \approx 0.0816$$
$$f(4.5) = \frac{1}{(4.5)^2} = \frac{1}{20.25} \approx 0.0494$$
$$f(5.5) = \frac{1}{(5.5)^2} = \frac{1}{30.25} \approx 0.0331$$
$$f(6.5) = \frac{1}{(6.5)^2} = \frac{1}{42.25} \approx 0.0237$$

The area of the region under the curve over $[1, 7]$ is approximately the sum of the areas of the 12 rectangles.

Area of each rectangle:

1st rectangle: $1(0.5) = 0.5$

$[f(1) = 1 \text{ and } \Delta x = 0.5]$

2nd rectangle: $0.4444(0.5) = 0.2222$

$[f(1.5) \approx 0.4444 \text{ and } \Delta x = 0.5]$

3rd rectangle: $0.2500(0.5) = 0.1250$

4th rectangle: $0.1600(0.5) = 0.0800$

5th rectangle: $0.1111(0.5) \approx 0.0556$

6th rectangle: $0.0816(0.5) = 0.0408$

7th rectangle: $0.0625(0.5) \approx 0.0313$

8th rectangle: $0.0494(0.5) = 0.0247$

9th rectangle: $0.0400(0.5) = 0.0200$

10th rectangle: $0.0331(0.5) \approx 0.0166$

11th rectangle: $0.0278(0.5) = 0.0139$

12th rectangle: $0.0237(0.5) \approx 0.0119$

The total area is $0.5 + 0.2222 + 0.1250 + 0.0800 + 0.0556 + 0.0408 + 0.0313 + 0.0247 + 0.0200 + 0.0166 + 0.0139 + 0.0119$, or 1.1420. (Answers may vary slightly depending on when rounding was done.)

2. a) $\Delta x = 1$

$$\begin{aligned}\sum_{i=1}^{5} f(x_i)\Delta x &= f(0)\cdot 1 + f(1)\cdot 1 + f(2)\cdot 1 + \\ &\quad f(3)\cdot 1 + f(4)\cdot 1 \\ &= (0^2+1)\cdot 1 + (1^2+1)\cdot 1 + \\ &\quad (2^2+1)\cdot 1 + (3^2+1)\cdot 1 + \\ &\quad (4^2+1)\cdot 1 \\ &= 1+2+5+10+17 \\ &= 35\end{aligned}$$

b) $\Delta x = \dfrac{5-0}{10} = 0.5$

$$\begin{aligned}\sum_{i=1}^{10} f(x_i)\Delta x &= f(0)\cdot 0.5 + f(0.5)\cdot 0.5 + \\ &\quad f(1)\cdot 0.5 + f(1.5)\cdot 0.5 + f(2)\cdot 0.5 + \\ &\quad f(2.5)\cdot 0.5 + f(3)\cdot 0.5 + \\ &\quad f(3.5)\cdot 0.5 + f(4)\cdot 0.5 + \\ &\quad f(4.5)\cdot 0.5 + f(5)\cdot 0.5 \\ &= (0^2+1)\cdot 0.5 + (0.5^2+1)\cdot 0.5 + \\ &\quad (1^2+1)\cdot 0.5 + (1.5^2+1)\cdot 0.5 + \\ &\quad (2^2+1)\cdot 0.5 + (2.5^2+1)\cdot 0.5 + \\ &\quad (3^2+1)\cdot 0.5 + (3.5^2+1)\cdot 0.5 + \\ &\quad (4^2+1)\cdot 0.5 + (4.5^2+1)\cdot 0.5 \\ &= 40.625\end{aligned}$$

3. The shaded region represents an antiderivative. It also represents velocity, the antiderivative of acceleration.

4. The shaded region represents an antiderivative. It also represents distance, the antiderivative of velocity.

5. The shaded region represents an antiderivative. It also represents total energy used in time t.

6. The shaded region represents an antiderivative. It also represents the total number of divorces in time t.

7. The shaded region represents an antiderivative. It also represents the amount of the drug in the blood.

8. The shaded region represents an antiderivative. It also represents total sales.

9. The shaded region represents an antiderivative. It also represents the number of words memorized in time t.

10. The shaded region represents an antiderivative. It also represents the number of pages typed in t hours.

11. $\Delta x = \frac{2-0}{4} = \frac{1}{2}$

$$\begin{aligned}\int_0^2 x^2\,dx &= f(1/2)(1/2) + f(1)(1/2) + f(3/2)(1/2) + \\ &\quad f(2)(1/2) \\ &= (1/4)(1/2) + (1)(1/2) + (9/4)(1/2) + (4)(1/2) \\ &= 1/8 + 1/2 + 9/8 + 2 \\ &= 3.75\end{aligned}$$

12. $\Delta x = \frac{1-(-1)}{4} = \frac{1}{2}$

$$\begin{aligned}\int_{-1}^1 x^2\,dx &= f(-1/2)(1/2) + f(0)(1/2) + f(1/2)(1/2) + \\ &\quad f(1)(1/2) \\ &= (1/4)(1/2) + (0)(1/2) + (1/4)(1/2) + (1)(1/2) \\ &= 1/8 + 0 + 1/8 + 1/2 \\ &= 0.75\end{aligned}$$

13. $\Delta x = \frac{5-4}{6} = \frac{1}{6}$

$$\begin{aligned}\int_4^5 x\,dx &= f(25/6)(1/6) + f(26/6)(1/6) + f(27/6)(1/6) + \\ &\quad f(28/6)(1/6) + f(29/6)(1/6) + f(5)(1/6) \\ &= (25/6)(1/6) + (26/6)(1/6) + (27/6)(1/6) + \\ &\quad (28/6)(1/6) + (29/6)(1/6) + (5)(1/6) \\ &= 25/36 + 26/36 + 27/36 + 28/36 + 29/36 + 5/6 \\ &= 4.58333\end{aligned}$$

14. $\Delta x = \frac{1-0}{6} = \frac{1}{6}$

$$\begin{aligned}\int_4^5 x\,dx &= f(1/6)(1/6) + f(2/6)(1/6) + f(3/6)(1/6) + \\ &\quad f(4/6)(1/6) + f(5/6)(1/6) + f(1)(1/6) \\ &= (1/6)(1/6) + (2/6)(1/6) + (3/6)(1/6) + \\ &\quad (4/6)(1/6) + (5/6)(1/6) + (1)(1/6) \\ &= 1/36 + 2/36 + 3/36 + 4/36 + 5/36 + 1/6 \\ &= 0.58333\end{aligned}$$

15. $\Delta x = \frac{\pi-0}{4} = \frac{\pi}{4}$

$$\begin{aligned}\int_0^\pi sin\ x\,dx &= sin(\pi/4)(\pi/4) + sin(2\pi/4)(\pi/4) + \\ &\quad sin(3\pi/4)(\pi/4) + sin(\pi)(\pi/4) \\ &= (1/\sqrt{2})(\pi/4) + (1)(\pi/4) + (1/\sqrt{2})(\pi/4) + 0 \\ &= 1.89612\end{aligned}$$

16. $\Delta x = \frac{\pi-0}{4} = \frac{\pi}{4}$

$$\begin{aligned}\int_0^\pi cos\ x\,dx &= cos(\pi/4)(\pi/4) + cos(2\pi/4)(\pi/4) + \\ &\quad cos(3\pi/4)(\pi/4) + cos(\pi)(\pi/4) \\ &= (1/\sqrt{2})(\pi/4) + 0 + (-1/\sqrt{2})(\pi/4) + \\ &\quad (1)(\pi/4) \\ &= 1.11072\end{aligned}$$

17. a) $\int_a^b f(x)\,dx = 0$, because there is the same area above the x-axis as below. That is, the area is $A - A$, or 0.

b) $\int_a^b f(x)\,dx < 0$, because there is more area below the x-axis than above. The area is $A - 2A$, or $-A$.

18. a) $\int_a^b f(x)\,dx > 0$, because there is more area above the x-axis than below. The area is $3A - A$, or $2A$.

b) $\int_a^b f(x)\,dx < 0$, because the area is entirely below the x-axis. The area is $-3A$.

19. $P'(t) = 200e^{-t}$

$$\begin{aligned}\Delta t &= \frac{2-0}{6} = \frac{1}{3} \\ \int_0^2 P'(t)\,dt &= P'(1/3)(1/3) + P'(2/3)(1/3) + P'(1)(1/3) + \\ &\quad P'(4/3)(1/3) + P'(5/3)(1/3) + P'(2)(1/3) \\ &= 143.31(1/3) + 102.68(1/3) + 73.58(1/3) + \\ &\quad 52.72(1/3) + 37.78(1/3) + 27.07(1/3) \\ &= 145.71333 \approx 146\end{aligned}$$

20. $P'(t) = 10e^t$

$$\begin{aligned}\Delta t &= \frac{3-0}{6} = \frac{1}{2} \\ \int_0^2 P'(t)\,dt &= P'(1/2)(1/2) + P'(1)(1/2) + P'(3/2)(1/2) \\ &\quad + P'(2)(1/2) + P'(5/2)(1/2) + P'(3)(1/2) \\ &= 8.24 + 13.59 + 22.41 + 36.95 + \\ &\quad 60.92 + 100.43 \\ &= 242.54 \approx 243\end{aligned}$$

21. $P'(t) = -500(20 - t)$

$$\begin{aligned}\Delta t &= \frac{20-0}{5} = 4 \\ \int_0^{20} P'(t)\,dt &= P'(4)(4) + P'(8)4 + P'(12)(4) + \\ &\quad P'(16)(4) + P'(20)(4) \\ &= (-8000)(4) + (-6000)(4) + \\ &\quad (-4000)(4) + (-2000)(4) + 0 \\ &= -80000\end{aligned}$$

22. $P'(t) = -50t^2$

$$\begin{aligned}\Delta t &= \frac{10-0}{5} = 2 \\ \int_0^{10} P'(t)\,dt &= P'(2)(2) + P'(4)(2) + P'(6)(2) + \\ &\quad P'(8)(2) + P'(10)(2) \\ &= -400 - 1600 - 3600 - 6400 - 10000 \\ &= -22000\end{aligned}$$

23. $v(t) = 3t^2 + 2t$

$$\begin{aligned}\Delta t &= \frac{5-1}{4} = 1 \\ \int_1^5 v(t)\,dt &= v(1)(1) + v(2)(1) + v(3)(1) + v(4)(1) \\ &= 5 + 16 + 33 + 56 \\ &= 110\end{aligned}$$

24. $v(t) = 4t^3 + 2t$

$$\begin{aligned}\Delta t &= \frac{3-0}{6} = \frac{1}{2} \\ \int_1^5 v(t)\,dt &= v(1/2)(1/2) + v(1)(1/2) + v(3/2)(1/2) + \\ &\quad v(2)(1/2) + v(5/2)(1/2) + v(3)(1/2) \\ &= 0.75 + 3 + 8.25 + 18 + 33.75 + 57 \\ &= 120.75\end{aligned}$$

25. $f(x) = x$. Since we are integrating over $[0, 2]$ the length of the subintervals is given by

$$\Delta x = \frac{2-0}{n} = \frac{2}{n}$$

Now, we need an expression for x_i:
$x_0 = 0$

$x_1 = 0 + 2/n = 2/n$
$x_2 = 2/n + 2/n = 4/n = 2(2/n)$
$x_3 = 2(2/n) + 2/n = 3(2/n)$
In general, we can write $x_i = i \cdot (\frac{2}{n})$

$$\begin{aligned}
\sum_{i=1}^{n} f(x_i)\Delta x &= \sum_{i=1}^{n} f(i \cdot \frac{2}{n})\frac{2}{n} \\
&= \sum_{i=1}^{n} i \cdot \frac{2}{n} \cdot \frac{2}{n} \\
&= \sum_{i=1}^{n} i \cdot \frac{4}{n^2} \\
&= \frac{4}{n^2}\sum_{i=1}^{n} i \\
&= \frac{4}{n^2} \cdot \frac{n(n+1)}{2} \\
&= \frac{2(n+1)}{n} \\
\int_0^2 x dx &= \lim_{n\to\infty} \sum_{i=1}^{n} f(x_i)\Delta x \\
&= \lim_{n\to\infty} \frac{2(n+1)}{n} \\
&= \lim_{n\to\infty} 2 + \frac{2}{n} \\
&= 2
\end{aligned}$$

26. $f(x) = 2x$. Since we are integrating over $[0,1]$ the length of the subintervals is given by

$$\Delta x = \frac{1-0}{n} = \frac{1}{n}$$

Now, we need an expression for x_i:
$x_0 = 0$
$x_1 = 0 + 1/n = 1/n$
$x_2 = 1/n + 1/n = 1/n = 2(1/n)$
$x_3 = 2(1/n) + 1/n = 3(1/n)$
In general, we can write $x_i = i \cdot (\frac{1}{n})$

$$\begin{aligned}
\sum_{i=1}^{n} f(x_i)\Delta x &= \sum_{i=1}^{n} f(i \cdot \frac{1}{n})\frac{1}{n} \\
&= \sum_{i=1}^{n} 2i \cdot \frac{1}{n} \cdot \frac{1}{n} \\
&= \sum_{i=1}^{n} 2i \cdot \frac{1}{n^2} \\
&= \frac{2}{n^2}\sum_{i=1}^{n} i \\
&= \frac{2}{n^2} \cdot \frac{n(n+1)}{2} \\
&= \frac{(n+1)}{n} \\
\int_0^1 2x dx &= \lim_{n\to\infty} \sum_{i=1}^{n} f(x_i)\Delta x \\
&= \lim_{n\to\infty} \frac{(n+1)}{n} \\
&= \lim_{n\to\infty} 1 + \frac{1}{n} \\
&= 1
\end{aligned}$$

27. $f(x) = 3x^2$. Since we are integrating over $[0,1]$ the length of the subintervals is given by

$$\Delta x = \frac{1-0}{n} = \frac{1}{n}$$

Now, we need an expression for x_i:
$x_0 = 0$
$x_1 = 0 + 1/n = 1/n$
$x_2 = 1/n + 1/n = 1/n = 2(1/n)$
$x_3 = 2(1/n) + 1/n = 3(1/n)$
In general, we can write $x_i = i \cdot (\frac{1}{n})$

$$\begin{aligned}
\sum_{i=1}^{n} f(x_i)\Delta x &= \sum_{i=1}^{n} f(i \cdot \frac{1}{n})\frac{1}{n} \\
&= \sum_{i=1}^{n} i^2 \cdot 3(\frac{1}{n})^2 \cdot \frac{1}{n} \\
&= \sum_{i=1}^{n} i^2 \cdot \frac{3}{n^3} \\
&= \frac{3}{n^3}\sum_{i=1}^{n} i^2 \\
&= \frac{3}{n^3} \cdot \frac{n(n+1)(2n+1)}{6} \\
&= \frac{2n^3 + 3n^2 + n}{2n^3} \\
\int_0^1 3x^2 dx &= \lim_{n\to\infty} \sum_{i=1}^{n} f(x_i)\Delta x \\
&= \lim_{n\to\infty} \frac{2n^3 + 3n^2 + n}{2n^3} \\
&= \lim_{n\to\infty} 1 + \frac{3}{2n} + \frac{1}{2n^2} \\
&= 1
\end{aligned}$$

28. $f(x) = x^2$. Since we are integrating over $[0,3]$ the length of the subintervals is given by

$$\Delta x = \frac{3-0}{n} = \frac{3}{n}$$

Now, we need an expression for x_i:
$x_0 = 0$
$x_1 = 0 + 3/n = 3/n$
$x_2 = 3/n + 3/n = 6/n = 2(3/n)$
$x_3 = 2(3/n) + 3/n = 3(3/n)$
In general, we can write $x_i = i \cdot (\frac{3}{n})$

$$\begin{aligned}
\sum_{i=1}^{n} f(x_i)\Delta x &= \sum_{i=1}^{n} f(i \cdot \frac{3}{n})\frac{3}{n} \\
&= \sum_{i=1}^{n} i^2 \cdot (\frac{3}{n})^2 \cdot \frac{3}{n} \\
&= \sum_{i=1}^{n} i^2 \cdot \frac{27}{n^3}
\end{aligned}$$

$$
\begin{aligned}
&= \frac{27}{n^3}\sum_{i=1}^{n} i^2 \\
&= \frac{27}{n^3}\cdot\frac{n(n+1)(2n+1)}{6} \\
&= \frac{9(2n^3+3n^2+n)}{2n^3}
\end{aligned}
$$

$$
\begin{aligned}
\int_0^3 x^2dx &= \lim_{n\to\infty}\sum_{i=1}^{n} f(x_i)\Delta x \\
&= \lim_{n\to\infty} 9\cdot\frac{2n^3+3n^2+n}{2n^3} \\
&= \lim_{n\to\infty} 9+\frac{27}{2n}+\frac{9}{2n^2} \\
&= 9
\end{aligned}
$$

29. $f(x) = x^3$. Since we are integrating over $[0,4]$ the length of the subintervals is given by

$$\Delta x = \frac{4-0}{n} = \frac{4}{n}$$

Now, we need an expression for x_i:
$x_0 = 0$
$x_1 = 0 + 4/n = 4/n$
$x_2 = 4/n + 4/n = 8/n = 2(4/n)$
$x_3 = 2(4/n) + 4/n = 3(4/n)$
In general, we can write $x_i = i\cdot(\frac{4}{n})$

$$
\begin{aligned}
\sum_{i=1}^{n} f(x_i)\Delta x &= \sum_{i=1}^{n} f(i\cdot\frac{4}{n})\frac{4}{n} \\
&= \sum_{i=1}^{n} i^3\cdot(\frac{4}{n})^3\cdot\frac{4}{n} \\
&= \sum_{i=1}^{n} i^3\cdot\frac{256}{n^4} \\
&= \frac{256}{n^4}\sum_{i=1}^{n} i^3 \\
&= \frac{256}{n^4}\cdot\frac{n2(n+1)^2}{4} \\
&= \frac{64(n^4+2n^3+n^2)}{n^4}
\end{aligned}
$$

$$
\begin{aligned}
\int_0^4 x^3dx &= \lim_{n\to\infty}\sum_{i=1}^{n} f(x_i)\Delta x \\
&= \lim_{n\to\infty}\frac{64(n^4+2n^3+n^2)}{n^4} \\
&= \lim_{n\to\infty} 64+\frac{128}{n}+\frac{64}{n^2} \\
&= 64
\end{aligned}
$$

30. $f(x) = 4x^3$. Since we are integrating over $[0,1]$ the length of the subintervals is given by

$$\Delta x = \frac{1-0}{n} = \frac{1}{n}$$

Now, we need an expression for x_i:
$x_0 = 0$
$x_1 = 0 + 1/n = 1/n$
$x_2 = 1/n + 1/n = 2/n = 2(1/n)$
$x_3 = 2(1/n) + 1/n = 3(1/n)$
In general, we can write $x_i = i\cdot(\frac{1}{n})$

$$
\begin{aligned}
\sum_{i=1}^{n} f(x_i)\Delta x &= \sum_{i=1}^{n} f(i\cdot\frac{1}{n})\frac{1}{n} \\
&= \sum_{i=1}^{n} 4i^3\cdot(\frac{1}{n})^3\cdot\frac{1}{n} \\
&= \sum_{i=1}^{n} 4i^3\cdot\frac{1}{n^4} \\
&= \frac{4}{n^4}\sum_{i=1}^{n} i^3 \\
&= \frac{4}{n^4}\cdot\frac{n2(n+1)^2}{4} \\
&= \frac{(n^4+2n^3+n^2)}{n^4}
\end{aligned}
$$

$$
\begin{aligned}
\int_0^1 4x^3dx &= \lim_{n\to\infty}\sum_{i=1}^{n} f(x_i)\Delta x \\
&= \lim_{n\to\infty}\frac{(n^4+2n^3+n^2)}{n^4} \\
&= \lim_{n\to\infty} 1+\frac{2}{n}+\frac{1}{n^2} \\
&= 1
\end{aligned}
$$

31. $f(x) = x^2$. Since we are integrating over $[1,3]$ the length of the subintervals is given by

$$\Delta x = \frac{3-1}{n} = \frac{2}{n}$$

Now, we need an expression for x_i:
$x_0 = 1$
$x_1 = 1 + 2/n$
$x_2 = 1 + 2/n + 2/n = 1 + 2(2/n)$
$x_3 = 1 + 2(2/n) + 2/n = 1 + 3(2/n)$
In general, we can write $x_i = 1 + i(\frac{2}{n})$

$$
\begin{aligned}
\sum_{i=1}^{n} f(x_i)\Delta x &= \sum_{i=1}^{n} f(1+\frac{2i}{n})\frac{2}{n} \\
&= \sum_{i=1}^{n}(1+\frac{2i}{n})^2\cdot\frac{2}{n} \\
&= \sum_{i=1}^{n}\frac{2}{n}\left[1+\frac{4i}{n}+\frac{4i^2}{n^2}\right] \\
&= \frac{2}{n}\left[\sum_{i=1}^{n}1+\sum_{i=1}^{n}\frac{4i}{n}+\sum_{i=1}^{n}\frac{4i^2}{n^2}\right] \\
&= \frac{2}{n}\left[n+\frac{4}{n}\sum_{i=1}^{n}i+\frac{4}{n^2}\sum_{i=1}^{n}i^2\right] \\
&= \frac{2}{n}\left[n+\frac{4}{n}\cdot\frac{n(n+1)}{2}\right]+ \\
&\quad \frac{2}{n}\left[\frac{4}{n^2}\cdot\frac{n(n+1)(2n+1)}{6}\right] \\
&= 2+\frac{8}{n^2}\cdot\frac{n^2+n}{2}+\frac{8}{n^3}\cdot\frac{2n^3+3n^2+n}{6} \\
&= 2+4+\frac{4}{n}+\frac{8}{3}+\frac{4}{n}+\frac{4}{3n^2}
\end{aligned}
$$

$$= \frac{26}{3} + \frac{8}{n} + \frac{4}{3n^2}$$

$$\int_1^3 x^2 dx = \lim_{n\to\infty} \sum_{i=1}^{n} f(x_i)\Delta x$$

$$= \lim_{n\to\infty} \left[\frac{26}{3} + \frac{8}{n} + \frac{4}{3n^2}\right]$$

$$= \frac{26}{3}$$

32. $f(x) = x^3$. Since we are integrating over $[1, 4]$ the length of the subintervals is given by

$$\Delta x = \frac{4-1}{n} = \frac{3}{n}$$

Now, we need an expression for x_i:
$x_0 = 1$
$x_1 = 1 + 3/n$
$x_2 = 1 + 3/n + 3/n = 1 + 2(3/n)$
$x_3 = 1 + 2(3/n) + 2/n = 1 + 3(3/n)$
In general, we can write $x_i = 1 + i(\frac{3}{n})$

$$\sum_{i=1}^{n} f(x_i)\Delta x = \sum_{i=1}^{n} f(1 + \frac{3i}{n})\frac{3}{n}$$

$$= \sum_{i=1}^{n} (1 + \frac{3i}{n})^3 \cdot \frac{3}{n}$$

$$= \sum_{i=1}^{n} \frac{3}{n}\left[1 + \frac{9i}{n} + \frac{27i^2}{n^2} + \frac{27i^3}{n^3}\right]$$

$$= \frac{3}{n}\left[\sum_{i=1}^{n} 1 + \sum_{i=1}^{n} \frac{9i}{n} + \sum_{i=1}^{n} \frac{27i^2}{n^2} + \sum_{i=1}^{n} \frac{27i^3}{n^3}\right]$$

$$= \frac{3}{n}\left[n + \frac{9}{n}\sum_{i=1}^{n} i + \frac{27}{n^2}\sum_{i=1}^{n} i^2 + \frac{27}{n^3}\sum_{i=1}^{n} i^3\right]$$

$$= \frac{3}{n}\left[n + \frac{9}{n} \cdot \frac{n(n+1)}{2}\right] +$$

$$\frac{3}{n}\left[\frac{27}{n^2} \cdot \frac{n(n+1)(2n+1)}{6}\right] +$$

$$\frac{3}{n}\left[\frac{27}{n^3} \cdot \frac{n^2(n+1)^2}{4}\right]$$

$$= 3 + \frac{27}{2} + \frac{27}{2n} + 27 + \frac{81}{2n} +$$

$$\frac{81}{4} + \frac{81}{2n} + \frac{81}{4n^2}$$

$$= \frac{255}{5} + \frac{54}{n} + \frac{81}{n^2}$$

$$\int_1^4 x^3 dx = \lim_{n\to\infty} \sum_{i=1}^{n} f(x_i)\Delta x$$

$$= \lim_{n\to\infty} \left[\frac{255}{5} + \frac{54}{n} + \frac{81}{n^2}\right]$$

$$= \frac{255}{4}$$

33. $\int_0^2 (x + x^2) dx = \int_0^2 x\ dx + \int_0^2 x^2\ dx$
The first integral was evaluated in Exercise 25, and yielded a value of 2. The second integral is now computed: $\Delta x = \frac{2-0}{n} = \frac{2}{n}$

$x_0 = 0$
$x_1 = 0 + 2/n = 2/n$
$x_2 = 2/n + 2/n = 2(2/n)$
$x_3 = 2(2/n) + 2/n = 3(2/n)$
In general, we can write $x_i = i \cdot \frac{2}{n}$

$$\sum_{i=1}^{n} f(x_i)\Delta x = \sum_{i=1}^{n} f(i \cdot \frac{2}{n})\frac{2}{n}$$

$$= \sum_{i=1}^{n} i^2 \cdot (\frac{2}{n})^2 \frac{2}{n}$$

$$= \sum_{i=1}^{n} i^2 \cdot \frac{8}{n^3}$$

$$= \frac{8}{n^3}\sum_{i=1}^{n} i^2$$

$$= \frac{8}{n^3} \cdot \frac{n(n+1)(2n+1)}{6}$$

$$= \frac{4}{3}\frac{2n^3 + 3n^2 + n}{n^3}$$

$$= \frac{4}{3}\left[2 + \frac{3}{n} + \frac{1}{n^2}\right]$$

$$= \frac{8}{3} + \frac{4}{n} + \frac{4}{3n^2}$$

$$\int_0^2 x^2\ dx = \lim_{n\to\infty} \frac{8}{3} + \frac{4}{n} + \frac{4}{3n^2}$$

$$= \frac{8}{3}$$

Thus,

$$\int_0^2 (x + x^2)\ dx = 2 + \frac{8}{3} = \frac{14}{3}$$

34. - 42. Left to the student

43. $\int_0^\pi sin\ x\ dx = 2$

44. $\int_0^\pi cos\ x\ dx = 0$

45. $\int_0^4 \sqrt{x}\ dx = 5.33334$

46. $\int_0^3 \frac{2}{\sqrt{x+1}} dx = 4$

47. $\int_2^4 \ln(x)\ dx = 2.15888$

48. $\int_1^e \frac{1}{x} dx = 1$

Exercise Set 5.3

1. Find the area under the curve $y = 4$ on the interval $[1, 3]$.

$A(x) = \int 4\ dx$
$= 4x + C$

Since we know that $A(1) = 0$ (there is no area above the number 1), we can substitute for x and $A(x)$ to determine C.

$A(1) = 4 \cdot 1 + C = 0$ Substituting 1 for x and 0 for $A(1)$

Solving for C we get:

$4 + C = 0$

$C = -4$

Thus, $A(x) = 4x - 4$.

Then the area on the interval $[1, 3]$ is $A(3)$.

$A(3) = 4 \cdot 3 - 4$ Substituting 3 for x

$= 12 - 4$

$= 8$

2. $A(x) = \int 5\,dx = 5x + C$

Use $A(1) = 0$ to find C.

$A(1) = 5 \cdot 1 + C = 0$

$C = -5$

$A(x) = 5x - 5$

The area on $[1, 3]$ is $A(3)$.

$A(3) = 5 \cdot 3 - 5 = 10$

3. Find the area under the curve $y = 2x$ on the interval $[1, 3]$.

$A(x) = \int 2x\,dx$

$= x^2 + C$

Since we know that $A(1) = 0$ (there is no area above the number 1), we can substitute for x and $A(x)$ to determine C.

$A(1) = 1^2 + C = 0$ Substituting 1 for x and 0 for $A(1)$

Solving for C we get:

$1 + C = 0$

$C = -1$

Thus, $A(x) = x^2 - 1$.

Then the area on the interval $[1, 3]$ is $A(3)$.

$A(3) = 3^2 - 1$ Substituting 3 for x

$= 9 - 1$

$= 8$

4. $A(x) = \displaystyle\int x^2\,dx = \frac{x^3}{3} + C$

Use $A(0) = 0$ to find C.

$A(0) = \dfrac{0^3}{3} + C = 0$

$C = 0$

$A(x) = \dfrac{x^3}{3}$

The area on $[0, 3]$ is $A(3)$.

$A(3) = \dfrac{3^3}{3} = 9$

5. Find the area under the curve $y = x^2$ on the interval $[0, 5]$.

$A(x) = \int x^2\,dx$

$= \dfrac{x^3}{3} + C$

Since we know that $A(0) = 0$ (there is no area above the number 0), we can substitute for x and $A(x)$ to determine C.

$A(0) = \dfrac{0^3}{3} + C = 0$ Substituting 0 for x and 0 for $A(0)$

Solving for C, we get $C = 0$:

Thus, $A(x) = \dfrac{x^3}{3}$.

Then the area on the interval $[0, 5]$ is $A(5)$.

$A(5) = \dfrac{5^3}{3}$ Substituting 5 for x

$= \dfrac{125}{3}$, or $41\dfrac{2}{3}$

6. $A(x) = \displaystyle\int x^3\,dx = \frac{x^4}{4} + C$

Use $A(0) = 0$ to find C.

$A(0) = \dfrac{0^4}{4} + C = 0$

$C = 0$

$A(x) = \dfrac{x^4}{4}$

The area on $[0, 2]$ is $A(2)$.

$A(2) = \dfrac{2^4}{4} = 4$

7. Find the area under the curve $y = x^3$ on the interval $[0, 1]$.

$A(x) = \int x^3\,dx$

$= \dfrac{x^4}{4} + C$

Since we know that $A(0) = 0$, we can substitute for x and $A(x)$ to determine C.

$A(0) = \dfrac{0^4}{4} + C = 0$ Substituting 0 for x and 0 for $A(0)$

Solving for C, we get $C = 0$.

Thus, $A(x) = \dfrac{x^4}{4}$.

Then the area on the interval $[0, 1]$ is $A(1)$.

$A(1) = \dfrac{1^4}{4}$ Substituting 1 for x

$= \dfrac{1}{4}$

8. $A(x) = \displaystyle\int (1 - x^2)\,dx = x - \frac{x^3}{3} + C$

Use $A(-1) = 0$ to find C.

$A(-1) = -1 - \dfrac{(-1)^3}{3} + C = 0$

$-\dfrac{2}{3} + C = 0$

$C = \dfrac{2}{3}$

$A(x) = x - \dfrac{x^3}{3} + \dfrac{2}{3}$

The area on $[-1, 1]$ is $A(1)$.

$A(1) = 1 - \dfrac{1^3}{3} + \dfrac{2}{3} = \dfrac{4}{3}$, or $1\dfrac{1}{3}$

9. Find the area under the curve $y = 4 - x^2$ on the interval $[-2, 2]$.

$A(x) = \int (4 - x^2)\,dx$

$= 4x - \frac{x^3}{3} + C$

Since we know that $A(-2) = 0$, (there is no area above the number -2),we can substitute for x and $A(x)$ to determine C.

$A(-2) = 4(-2) - \frac{(-2)^3}{3} + C = 0$

Substituting -2 for x and 0 for $A(-2)$

Solving for C, we get:

$-8 + \frac{8}{3} + C = 0$

$-\frac{24}{8} + \frac{8}{3} + C = 0$

$-\frac{16}{3} + C = 0$

$C = \frac{16}{3}$

Thus, $A(x) = 4x - \frac{x^3}{3} + \frac{16}{3}$.

The area on the interval $[-2, 2]$ is $A(2)$.

$A(2) = 4 \cdot 2 - \frac{2^3}{3} + \frac{16}{3}$ Substituting 2 for x

$= 8 - \frac{8}{3} + \frac{16}{3}$

$= \frac{24}{3} - \frac{8}{3} + \frac{16}{3}$

$= \frac{32}{3}$, or $10\frac{2}{3}$

10. $A(x) = \int e^x\,dx = e^x + C$

Use $A(0) = 0$ to find C.

$A(0) = e^0 + C = 0$

$1 + C = 0$

$C = -1$

$A(x) = e^x - 1$

The area on $[0, 2]$ is $A(2)$.

$A(2) = e^2 - 1 \approx 6.3891$

11. Find the area under the curve $y = e^x$ on the interval $[0, 3]$.

$A(x) = \int e^x\,dx$

$= e^x + C$

Since we know that $A(0) = 0$, (there is no area above the number 0),we can substitute for x and $A(x)$ to determine C.

$A(0) = e^0 + C = 0$ Substituting 0 for x and 0 for $A(0)$

$1 + C = 0$ $(e^0 = 1)$

$C = -1$

Thus, $A(x) = e^x - 1$.

The area on the interval $[0, 3]$ is $A(3)$.

$A(3) = e^3 - 1$

$= 20.085537 - 1$ Using a calculator

≈ 19.086

12. $A(x) = \int \frac{2}{x}\,dx = 2\ln x + C$

Use $A(1) = 0$ to find C.

$A(1) = 2\ln 1 + C = 0$

$2 \cdot 0 + C = 0$

$C = 0$

$A(x) = 2\ln x$

The area on $[1, 4]$ is $A(4)$.

$A(4) = 2\ln 4 \approx 2.773$

13. Find the area under the curve $y = \frac{3}{x}$ on the interval $[1, 6]$.

$A(x) = \int \frac{3}{x}\,dx = 3\int \frac{1}{x}\,dx$

$= 3\ln x + C$

Since we know that $A(1) = 0$, (there is no area above the number 1), we can substitute for x and $A(x)$ to determine C.

$A(1) = 3\ln 1 + C = 0$ Substituting 1 for x and 0 for $A(1)$

$3 \cdot 0 + C = 0$ $(\ln 1 = 0)$

$C = 0$

Thus, $A(x) = 3\ln x$.

The area on the interval $[1, 6]$ is $A(6)$.

$A(6) = 3\ln 6$ Substituting 6 for x

≈ 5.375 Using a calculator

14. $A(x) = \int 3\,sin\,x\,dx = -3cos\,x + C$

Use $A(0) = 0$ to find C.

$A(0) = -3cos(0) + C = 0$

$-3 \cdot 0 + C = 0$

$C = 3$

$A(x) = -3cos\,x + 3$

The area on $[0, \frac{\pi}{4}]$ is $A(\frac{\pi}{4})$

$A(\frac{\pi}{4}) = -3cos(\frac{\pi}{4}) + 3$

≈ 0.8787 Using a calculator

15. $\int_0^{1.5} (x - x^2)\,dx$

$= \left[\frac{x^2}{2} - \frac{x^3}{3}\right]_0^{1.5}$

$= \left(\frac{(1.5)^2}{2} - \frac{(1.5)^3}{3}\right) - \left(\frac{0^2}{2} - \frac{0^3}{3}\right)$

Substituting 0 for x

Substituting 1.5 for x

$= \left(\frac{2.25}{2} - \frac{3.375}{3}\right) - (0 - 0)$

$= 1.125 - 1.125$

$= 0$

The area above the x-axis is equal to the area below the x-axis.

16. $\int_0^2 (x^2 - x)\,dx$

$= \left[\frac{x^3}{3} - \frac{x^2}{2}\right]_0^2$

$= \left(\frac{2^3}{3} - \frac{2^2}{2}\right) - \left(\frac{0^3}{3} - \frac{0^2}{2}\right)$

$= \frac{8}{3} - 2$

$= \frac{2}{3}$

The area above the x-axis is greater than the area below the x-axis.

17. $\int_0^{3\pi/2} \cos x\,dx$

$= \left[\sin x\right]_0^{3\pi/2}$

$= \sin(3\pi/2) - \sin(0)$

$= -1 - 0$

$= -1$

This means that the area between $[0, \frac{\pi}{2}]$ (area above the x-axis) is less than that area between $[\frac{\pi}{2}, \frac{3\pi}{2}]$ (area below the x-axis) by 1 unit of area.

18. $\int_0^b -2e^{3x} dx$

$= \left[-\frac{2}{3}e^{3x}\right]_0^b$

$= -\frac{2}{3}e^{3b} - \left(-\frac{2}{3}e^{3\cdot 0}\right)$

$= -\frac{2}{3}e^{3b} + \frac{2}{3}$, or $-\frac{2}{3}(e^{3b} - 1)$

The area is below the x-axis.

19. - 36.] Left to the student

37. $\int_a^b e^t\,dt$

$= [e^t]_a^b$

$= e^b - e^a$

38. $\int_0^a (ax - x^2)\,dx$

$= \left[\frac{ax^2}{2} - \frac{x^3}{3}\right]_0^a$

$= \left(\frac{a \cdot a^2}{2} - \frac{a^3}{3}\right) - \left(\frac{a \cdot 0^2}{2} - \frac{0^3}{3}\right)$

$= \frac{a^3}{2} - \frac{a^3}{3}$

$= \frac{a^3}{6}$

39. $\int_a^b 3t^2\,dt$

$= \left[3 \cdot \frac{t^3}{3}\right]_a^b$

$= [t^3]_a^b$

$= b^3 - a^3$

40. $\int_{-5}^2 4t^3\,dt$

$= [t^4]_{-5}^2$

$= 2^4 - (-5)^4$

$= 16 - 625$

$= -609$

41. $\int_1^e \left(x + \frac{1}{x}\right) dx$

$= \left[\frac{x^2}{2} + \ln x\right]_1^e$

$= \left(\frac{e^2}{2} + \ln e\right) - \left(\frac{1^2}{2} + \ln 1\right)$

$= \frac{e^2}{2} + 1 - \frac{1}{2}$ $\quad (\ln e = 1,\ \ln 1 = 0)$

$= \frac{e^2}{2} + \frac{1}{2}$

42. $\int_1^e \left(x - \frac{1}{x}\right) dx$

$= \left[\frac{x^2}{2} - \ln x\right]_1^e$

$= \left(\frac{e^2}{2} - \ln e\right) - \left(\frac{1^2}{2} - \ln 1\right)$

$= \frac{e^2}{2} - 1 - \frac{1}{2} + 0$

$= \frac{e^2}{2} - \frac{3}{2}$

43. $\int_0^{\pi/6} \frac{5}{2} \sin 2x\,dx$

$= \left[-5\,\frac{\cos 2x}{4}\right]_0^{\pi/6}$

$= -5\cos(\pi/6) - (-5\cos(0))$

$= -\frac{5}{8} - (-\frac{5}{4})$

$= \frac{5}{8}$

44. $\int_{\pi/3}^{2\pi/3} 4\cos\frac{1}{2}x\,dx$

$= \left[8\sin\frac{1}{2}x\right]_{\pi/3}^{2\pi/3}$

$= 8\sin(\pi/3) - 8\sin(\pi/6)$

$= 8 \cdot \frac{\sqrt{3}}{2} - 8 \cdot \frac{1}{2}$

$= 4\sqrt{3} - 4$

45. $\int_{-4}^{1} \frac{10}{17} t^3\, dt$

$= \frac{10}{17} \int_{-4}^{1} t^3\, dt$

$= \frac{10}{17} \left[\frac{t^4}{4}\right]_{-4}^{1}$

$= \frac{10}{17} \left(\frac{1^4}{4} - \frac{(-4)^4}{4}\right)$

$= \frac{10}{17} \left(\frac{1}{4} - 64\right)$

$= \frac{10}{17} \cdot \left(-\frac{255}{4}\right)$

$= -\frac{2550}{68}$

$= -\frac{1275}{34}$

46. $\int_{0}^{1} \frac{12}{13} t^2\, dt$

$= \frac{12}{13} \left[\frac{t^3}{3}\right]_{0}^{1}$

$= \frac{12}{13} \left(\frac{1^3}{3} - \frac{0^3}{3}\right)$

$= \frac{12}{13} \cdot \frac{1}{3}$

$= \frac{4}{13}$

47. Find the area under $y = x^3$ on $[0, 2]$.

$\int_0^2 x^3\, dx$

$= \left[\frac{x^4}{4}\right]_0^2$

$= \frac{2^4}{4} - \frac{0^4}{4}$

$= 4 - 0$

$= 4$

48. $\int_0^1 x^4\, dx$

$= \left[\frac{x^5}{5}\right]_0^1$

$= \frac{1^5}{5} - \frac{0^5}{5}$

$= \frac{1}{5}$

49. Find the area under $y = x^2 + x + 1$ on $[2, 3]$.

$\int_2^3 (x^2 + x + 1)\, dx$

$= \left[\frac{x^3}{3} + \frac{x^2}{2} + x\right]_2^3$

$= \left(\frac{3^3}{3} + \frac{3^2}{2} + 3\right) - \left(\frac{2^3}{3} + \frac{2^2}{2} + 2\right)$

$= \left(9 + \frac{9}{2} + 3\right) - \left(\frac{8}{3} + 2 + 2\right)$

$= 12 + \frac{9}{2} - \frac{8}{3} - 4$

$= 8 + \frac{27}{6} - \frac{16}{6}$

$= 8 + \frac{11}{6}$

$= 8 + 1\frac{5}{6}$

$= 9\frac{5}{6}$

50. $\int_{-2}^{1} (2 - x - x^2)\, dx$

$= \left[2x - \frac{x^2}{2} - \frac{x^3}{3}\right]_{-2}^{1}$

$= \left(2 \cdot 1 - \frac{1^2}{2} - \frac{1^3}{3}\right) - \left[2(-2) - \frac{(-2)^2}{2} - \frac{(-2)^3}{3}\right]$

$= 2 - \frac{1}{2} - \frac{1}{3} + 4 + 2 - \frac{8}{3}$

$= 4\frac{1}{2}$

51. Find the area under $y = 5 - x^2$ on $[-1, 2]$.

$\int_{-1}^{2} (5 - x^2)\, dx$

$= \left[5x - \frac{x^3}{3}\right]_{-1}^{2}$

$= \left(5 \cdot 2 - \frac{2^3}{3}\right) - \left[5(-1) - \frac{(-1)^3}{3}\right]$

$= \left(10 - \frac{8}{3}\right) - \left(-5 + \frac{1}{3}\right)$

$= \left(\frac{30}{3} - \frac{8}{3}\right) - \left(-\frac{15}{3} + \frac{1}{3}\right)$

$= \frac{22}{3} - \left(-\frac{14}{3}\right)$

$= \frac{22}{3} + \frac{14}{3}$

$= \frac{36}{3}$

$= 12$

52. $\int_{-2}^{3} e^x\, dx$

$= [e^x]_{-2}^{3}$

$= e^3 - e^{-2}$, or $e^3 - \frac{1}{e^2}$

53. Find the area under $y = e^x$ on $[-1, 5]$.

$\int_{-1}^{5} e^x\,dx$
$= [e^x]_{-1}^{5}$
$= e^5 - e^{-1}$
$= e^5 - \frac{1}{e}$

54. $\int_1^4 \left(2x + \frac{1}{x^2}\right) dx = \int_1^4 (2x + x^{-2})\,dx$

$$= \left[x^2 + \frac{x^{-1}}{-1}\right]_1^4$$
$$= \left[x^2 - \frac{1}{x}\right]_1^4$$
$$= \left(4^2 - \frac{1}{4}\right) - \left(1^2 - \frac{1}{1}\right)$$
$$= 16 - \frac{1}{4} - 1 + 1$$
$$= 15\frac{3}{4}$$

55.

$$\begin{aligned}\int_2^3 \frac{x^2-1}{x-1}dx &= \int_2^3 \frac{(x-1)(x+1)}{x-1}dx \\ &= \int_2^3 (x+1)\,dx \\ &= \left[\frac{1}{2}x^2 + x\right]_2^3 \\ &= \left[\frac{9}{2} + 3\right] - [2+2] \\ &= \frac{7}{2}\end{aligned}$$

56.

$$\begin{aligned}\int_1^5 \frac{x^5 - x^{-1}}{x^2}dx &= \int_1^5 (x^3 - x^{-3})\,dx \\ &= \left[\frac{x^4}{4} + \frac{1}{2x^2}\right]_1^5 \\ &= \left[\frac{625}{4} + \frac{1}{50}\right] - \\ &\quad \left[\frac{1}{4} + \frac{1}{2}\right] \\ &= \frac{3888}{25}\end{aligned}$$

57.

$$\begin{aligned}\int_4^{16} (x-1)\sqrt{x}\,dx &= \int_4^{16} (x^{3/2} - x^{1/2})\,dx \\ &= \left[\frac{2}{5}x^{5/2} - \frac{2}{3}x^{3/2}\right]_4^{16} \\ &= \left[\frac{2}{5}(16)^{5/2} - \frac{2}{3}(16)^{3/2}\right] - \\ &\quad \left[\frac{2}{5}(4)^{5/2} - \frac{2}{3}(4)^{3/2}\right] \\ &= \frac{2048}{5} - \frac{128}{3} - \frac{64}{5} + \frac{16}{3} \\ &= \frac{5392}{15}\end{aligned}$$

58.

$$\begin{aligned}\int_0^1 (x+2)^3\,dx &= \left[\frac{1}{4}(x+2)^4\right]_0^1 \\ &= \frac{1}{4}[81 - 0] \\ &= \frac{81}{4}\end{aligned}$$

59.

$$\begin{aligned}\int_1^8 \frac{\sqrt[3]{x^2} - 1}{\sqrt[3]{x}}dx &= \int_1^8 (x^{1/3} - x^{-1/3})\,dx \\ &= \left[\frac{3}{4}x^{4/3} - \frac{3}{2}x^{2/3}\right]_1^8 \\ &= \left[\frac{3}{4}8^{4/3} - \frac{3}{2}8^{2/3}\right] - \\ &\quad \left[\frac{3}{4}1^{4/3} - \frac{3}{2}1^{2/3}\right] \\ &= 12 - 6 - \frac{3}{4} + \frac{3}{2} \\ &= \frac{27}{4}\end{aligned}$$

60.

$$\begin{aligned}\int_0^1 \frac{x^3+8}{x+2}dx &= \int_0^1 (x^2 - 2x + 4)\,dx \\ &= \left[\frac{1}{3}x^3 - x^2 + 4x\right]_0^1 \\ &= \frac{1}{3} - 1 + 4 - 0 \\ &= \frac{10}{3}\end{aligned}$$

61.

$$\begin{aligned}\int_1^2 (4x+3)(5x-2)dx &= \int_1^2 (20x^2 + 7x - 6)\,dx \\ &= \left[\frac{20}{3}x^3 + \frac{7}{2}x^2 - 6x\right]_1^2 \\ &= \frac{160}{3} + 14 - 12 - \frac{20}{3} - \frac{7}{2} + 6 \\ &= \frac{307}{6}\end{aligned}$$

62.

$$\begin{aligned}\int_2^5 (t+\sqrt{3})(t-\sqrt{3})dt &= \int_2^5 (t^2 - 3)dt \\ &= \left[\frac{t^3}{3} - 3t\right]_2^5 \\ &= \frac{125}{3} - 15 - \frac{8}{3} + 6 \\ &= 30\end{aligned}$$

63.

$$\int_0^1 (t+1)^3 dt = \int_0^1 (t^3 + 3t^2 + 3t + 1)dt$$

$$= \left[\frac{t^4}{4} + t^3 + \frac{3t^2}{2} + t\right]_0^1$$
$$= \frac{1}{4} + 1 + \frac{3}{2} + 1 - 0$$
$$= \frac{15}{4}$$

64.

$$\begin{aligned}\int_1^3 \left(x - \frac{1}{x}\right)^2 dx &= \int_1^3 (x^2 - 2 + x^{-2})\, dx \\ &= \left[\frac{x^3}{3} - 2x - x^{-1}\right]_1^3 \\ &= 9 - 6 - \frac{1}{3} - \frac{1}{3} + 2 + 1 \\ &= \frac{16}{3}\end{aligned}$$

65.

$$\begin{aligned}\int_1^3 \frac{t^5 - t}{t^3} dt &= \int_1^3 (t^2 - t^{-2}) dt \\ &= \left[\frac{t^3}{3} + t^{-1}\right]_1^3 \\ &= 9 + \frac{1}{3} - \frac{1}{3} - 1 \\ &= 8\end{aligned}$$

66.

$$\begin{aligned}\int_4^9 \frac{t+1}{\sqrt{t}} dt &= \int_4^9 (t^{1/2} + t^{-1/2}) dt \\ &= \left[\frac{2t^{3/2}}{3} + 2t^{1/2}\right]_4^9 \\ &= 54 + 6 - 16 - 4 \\ &= 40\end{aligned}$$

67.

$$\begin{aligned}\int_3^5 \frac{x^2 - 4}{x - 2} dx &= \int_3^5 \frac{(x-2)(x+2)}{x-2} dx \\ &= \int_3^5 (x+2) dx \\ &= \left[\frac{x^2}{2} + 2x\right]_3^5 \\ &= \frac{25}{2} + 10 - \frac{9}{2} - 6 \\ &= 12\end{aligned}$$

68.

$$\begin{aligned}\int_0^1 \frac{t^3 + 1}{t + 1} dt &= \int_0^1 (t^2 - t + 1) dt \\ &= \left[\frac{t^3}{3} - \frac{t^2}{2} + t\right]_0^1 \\ &= \frac{1}{3} - \frac{1}{2} + 1 - 0 \\ &= \frac{5}{6}\end{aligned}$$

69. The average is given by

$$\begin{aligned}\frac{1}{1-(-1)} \int_{-1}^1 2x^3 dx &= \frac{1}{2} \int_{-1}^1 2x^3 dx \\ &= \frac{1}{2} \cdot \left[\frac{1}{2}x^4\right]_{-1}^1 \\ &= \frac{1}{4}(1^4 - (-1)^4) \\ &= 0\end{aligned}$$

70. The average is given by

$$\begin{aligned}\frac{1}{2-(-2)} \int_{-2}^2 (4 - x^2) dx &= \frac{1}{4} \int_{-2}^2 (4 - x^2) dx \\ &= \frac{1}{4}\left[4x - \frac{x^3}{3}\right]_{-2}^2 \\ &= \frac{1}{4}\left(8 - \frac{8}{3} + 8 - \frac{8}{3}\right) \\ &= \frac{8}{3}\end{aligned}$$

71. The average is given by

$$\begin{aligned}\frac{1}{1-0} \int_0^1 e^x dx &= \int_0^1 e^x dx \\ &= \left[e^x\right]_0^1 \\ &= e - 1\end{aligned}$$

72. The average is given by

$$\begin{aligned}\frac{1}{1-0} \int_0^1 e^{-x} dx &= \int_0^1 e^{-x} dx \\ &= \left[e^{-x}\right]_0^1 \\ &= e^{-1} - 1 \\ &= \frac{1}{e} - 1\end{aligned}$$

73. The average is given by

$$\begin{aligned}\frac{1}{2-0} \int_0^2 (x^2 - x + 1) dx &= \frac{1}{2} \int_0^2 (x^2 - x + 1) dx \\ &= \frac{1}{2}\left[\frac{x^3}{3} - \frac{x^2}{2} + x\right]_0^2 \\ &= \frac{1}{2}\left(\frac{8}{3} - 2 + 2 - 0\right) \\ &= \frac{4}{3}\end{aligned}$$

74. The average is given by

$$\begin{aligned}\frac{1}{4-0} \int_0^4 (x^2 + x - 2) dx &= \frac{1}{4} \int_0^4 (x^2 + x - 2) dx \\ &= \frac{1}{4}\left[\frac{x^3}{3} + \frac{x^2}{2} - 2x\right]_0^4 \\ &= \frac{1}{4}\left(\frac{64}{3} + 8 - 8\right) \\ &= \frac{16}{3}\end{aligned}$$

75. The average is given by

$$\begin{aligned}\frac{1}{6-2}\int_0^2 (3x+1)dx &= \frac{1}{4}\int_2^6 (3x+1)dx \\ &= \frac{1}{4}\left[\frac{3x^2}{2}+x\right]_2^6 \\ &= \frac{1}{4}(54+6-6-2) \\ &= 13\end{aligned}$$

76. The average is given by

$$\begin{aligned}\frac{1}{7-3}\int_0^2 (4x+1)dx &= \frac{1}{4}\int_3^7 (4x+1)dx \\ &= \frac{1}{4}\left[2x^2+x\right]_3^7 \\ &= \frac{1}{4}(98+7-18-3) \\ &= 21\end{aligned}$$

77. The average is given by

$$\begin{aligned}\frac{1}{1-0}\int_0^1 x^n dx &= \int_0^1 x^n dx \\ &= \left[\frac{x^{n+1}}{n+1}\right]_0^1 \\ &= \frac{1}{n+1}-0 \\ &= \frac{1}{n+1}\end{aligned}$$

78. The average is given by

$$\begin{aligned}\frac{1}{2-1}\int_1^2 x^n dx &= \int_1^2 x^n dx \\ &= \left[\frac{x^{n+1}}{n+1}\right]_1^2 \\ &= \frac{2^{n+1}}{n+1}-\frac{1}{n+1} \\ &= \frac{2^{n+1}-1}{n+1}\end{aligned}$$

79. The distance is given by

$$\begin{aligned}\int_1^5 (3t^2+2t)dt &= \left[t^3+t^2\right]_1^5 \\ &= 125+25-1-1 \\ &= 148\end{aligned}$$

80. The distance is given by

$$\begin{aligned}\int_0^3 (4t^3+2t)dt &= \left[t^4+t^2\right]_0^3 \\ &= 81+9 \\ &= 90\end{aligned}$$

81. The population increase is given by

$$\begin{aligned}\int_0^2 200e^{-t}dt &= \left[-200e^{-t}\right]_0^2 \\ &= -200e^{-2}+200 \\ &= 200\left(1-\frac{1}{e^2}\right)\end{aligned}$$

82. The population increase is given by

$$\begin{aligned}\int_0^3 10e^t dt &= \left[10e^t\right]_0^3 \\ &= 10e^3+10 \\ &= 10\left(e^3+1\right)\end{aligned}$$

83. The population decrease is given by

$$\begin{aligned}\int_0^{10} (-500(20-t)dt &= -500\left[20t-\frac{t^2}{2}\right]_0^{10} \\ &= -500(200-50) \\ &= -75000\end{aligned}$$

84. The population decrease is given by

$$\begin{aligned}\int_0^6 -50t^2\,dt &= -50\left[\frac{t^3}{3}\right]_0^6 \\ &= -50(72-0) \\ &= -3600\end{aligned}$$

85. The work done is given by

$$\begin{aligned}&\int_2^{11} 71.3x-4.15x^2+0.434x^3\,dx \\ &= \left[35.65x^2-1.383x^3+.1085x^4\right]_2^{11} \\ &= [4061-133.27] \\ &= 3927.73\ N\cdot mm\end{aligned}$$

86. The work done is given by

$$\begin{aligned}&\int_2^{11} 71.0x-10.3x^2+0.986x^3\,dx \\ &= \left[35x^2-3.433x^3+.2465x^4\right]_2^{11} \\ &= [3274.2-116.48] \\ &= 3157.72\ N\cdot mm\end{aligned}$$

87. **a)** The initial dosage is 42.03 $\mu g/ml$

b) The aberage amoount is given by

$$\begin{aligned}&\frac{1}{120-10}\int_{10}^{120} 42.03e^{-0.01050t}\,dt \\ &= \frac{1}{110}\left[\frac{42.03}{-0.01050}e^{-0.01050t}\right]_{10}^{120} \\ &= \frac{[-1135-(-3604)]}{110} \\ &= 22.45\ \mu g/ml\end{aligned}$$

88. The average temperature is given by

$$\frac{1}{12}\int_0^{12} 43.5 - 18.4x + 8.57x^2 - 0.996x^3 + 0.0338x^4\ dx$$
$$= \frac{1}{12}\left[43.5x - 9.2x^2 + 2.8567x^3 - 0.249x^4 + 0.25576x^5\right]_0^{12}$$
$$= \frac{[652.36 - 0]}{12}$$
$$= 54^o$$

89. a)

$$\int_0^b 2\pi r\ dr = 2\pi\left[\frac{r^2}{2}\right]_0^b = 2\pi\frac{b^2}{2} = \pi b^2$$

b) The circle could be thought of as a rectangle with length πb and width b

90. a)

$$\int_0^b 4\pi r^2\ dr = 4\pi\left[\frac{r^3}{3}\right]_0^b = 4\pi\frac{b^3}{3} = \frac{4}{3}\pi b^3$$

b) The volume could be thought of as a stacking of surface areas

91. The lower limit of the integral, $x = 1$, was not evaluated

92. $\int \ln x\ dx \neq \frac{1}{x}$

Exercise Set 5.4

1. First graph the system of equations and shade the region bounded by the graphs.

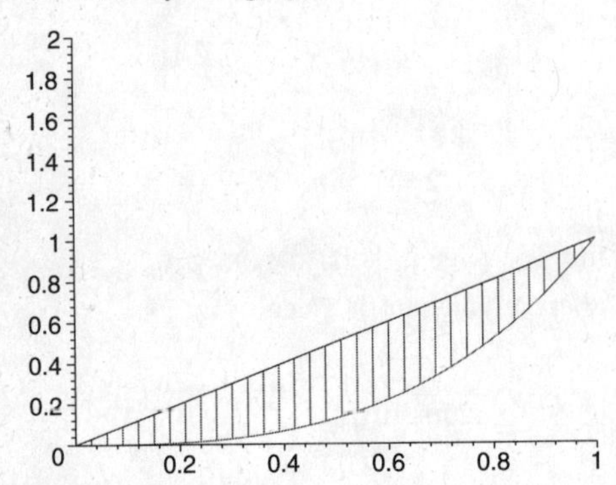

Here the boundaries are easily determined by looking at the graph. Note which is the upper graph. Here it is $x \geq x^3$ over the interval $[0, 1]$.

Compute the area as follows:

$$\int_0^1 (x - x^3)\ dx$$
$$= \left[\frac{x^2}{2} - \frac{x^4}{4}\right]_0^1$$
$$= \left(\frac{1^2}{2} - \frac{1^4}{4}\right) - \left(\frac{0^2}{2} - \frac{0^4}{4}\right)$$
$$= \frac{1}{2} - \frac{1}{4}$$
$$= \frac{1}{4}$$

2.

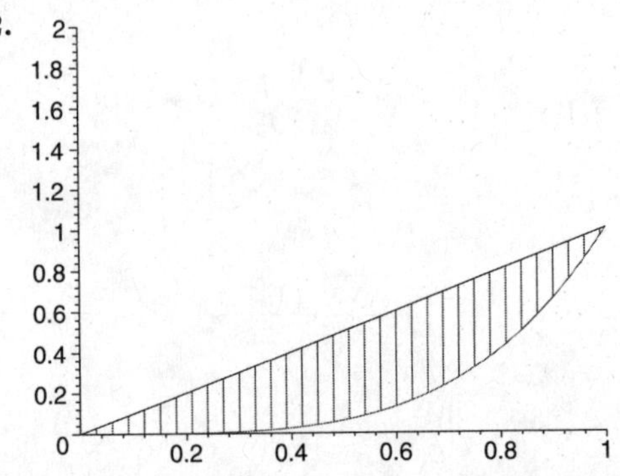

We find the boundaries by looking at the graph or by solving the following:

$$x^4 = x$$
$$x^4 - x = 0$$
$$x(x^3 - 1) = 0$$
$$x = 0 \text{ or } x = 1$$

$$\int_0^1 (x - x^4)\ dx$$
$$= \left[\frac{x^2}{2} - \frac{x^5}{5}\right]_0^1$$
$$= \left(\frac{1^2}{2} - \frac{1^5}{5}\right) - \left(\frac{0^2}{2} - \frac{0^5}{5}\right)$$
$$= \frac{3}{10}$$

3. First graph the system of equations and shade the region bounded by the graphs.

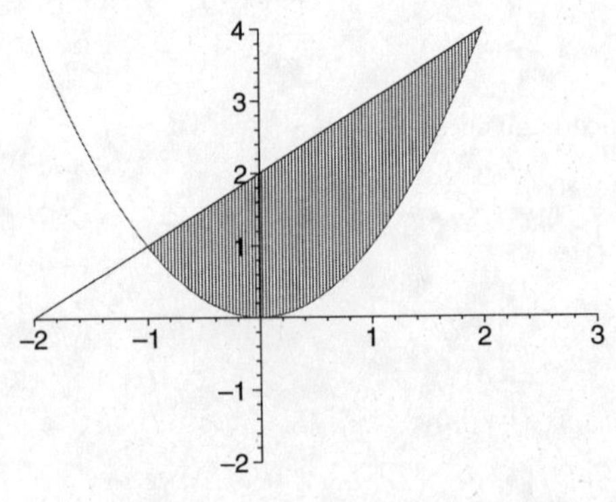

Here the boundaries are easily determined by the graph. Note which is the upper graph. Here it is $(x+2) \geq x^2$ over the interval $[-1, 2]$.

Compute the area as follows:

$\int_{-1}^{2}[(x+2)-x^2]\,dx$

$= \int_{-1}^{2}(-x^2+x+2)\,dx$

$= \left[-\frac{x^3}{3}+\frac{x^2}{2}+2x\right]_{-1}^{2}$

$= \left[-\frac{2^3}{3}+\frac{2^2}{2}+2\cdot 2\right]-\left[-\frac{(-1)^3}{3}+\frac{(-1)^2}{2}+2(-1)\right]$

$= \left(-\frac{8}{3}+2+4\right)-\left(\frac{1}{3}+\frac{1}{2}-2\right)$

$= -\frac{8}{3}+2+4-\frac{1}{3}-\frac{1}{2}+2$

$= -\frac{9}{3}-\frac{1}{2}+2+4+2=4\frac{1}{2}$, or $\frac{9}{2}$

4.

We find the boundaries by looking at the graph or by solving the following:

$x^2-2x=x$

$x^2-3x=0$

$x(x-3)=0$

$x=0$ or $x=3$

$\int_{0}^{3}[x-(x^2-2x)]\,dx$

$= \int_{0}^{3}(-x^2+3x)\,dx$

$= \left[-\frac{x^3}{3}+\frac{3x^2}{2}\right]_{0}^{3}$

$= \left(-\frac{3^3}{3}+\frac{3\cdot 3^2}{2}\right)-\left(-\frac{0^3}{3}+\frac{3\cdot 0^2}{2}\right)$

$= \frac{9}{2}$

5. First graph the system of equations and shade the region bounded by the graphs.

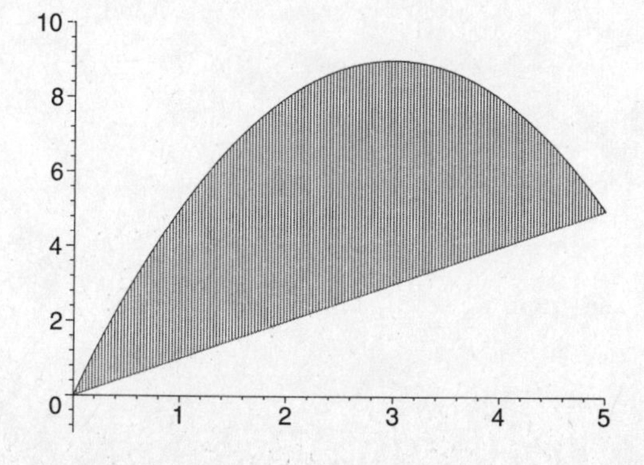

Here the boundaries are easily determined by the graph. Note which is the upper graph. Here it is $(6x-x^2)\geq x$ over the interval $[0,5]$.

Compute the area as follows:

$\int_{0}^{5}[(6x-x^2)-x]\,dx$

$= \int_{0}^{5}(-x^2+5x)\,dx$

$= \left[-\frac{x^3}{3}+5\cdot\frac{x^2}{2}\right]_{0}^{5}$

$= \left(-\frac{5^3}{3}+5\cdot\frac{5^2}{2}\right)-\left(-\frac{0^3}{3}+5\cdot\frac{0^2}{2}\right)$

$= \left(-\frac{125}{3}+\frac{125}{2}\right)-0$

$= -\frac{250}{6}+\frac{375}{6}$

$= \frac{125}{6}$

6.

We find the boundaries by looking at the graph or by solving the following:

$x^2-6x=-x$

$x^2-5x=0$

$x(x-5)=0$

$x=0$ or $x=5$

$\int_{0}^{5}[-x-(x^2+5x)\,dx$

$= \int_{0}^{5}(-x^2+5x)\,dx$

$= \left(-\frac{x^3}{3}+\frac{5x^2}{2}\right)_{0}^{5}$

$= \left(-\frac{5^3}{3}+\frac{5\cdot 5^2}{2}\right)-\left(-\frac{0^3}{3}+\frac{5\cdot 0^2}{2}\right)$

$= \frac{125}{6}$

7. First graph the system of equations and shade the region bounded by the graphs.

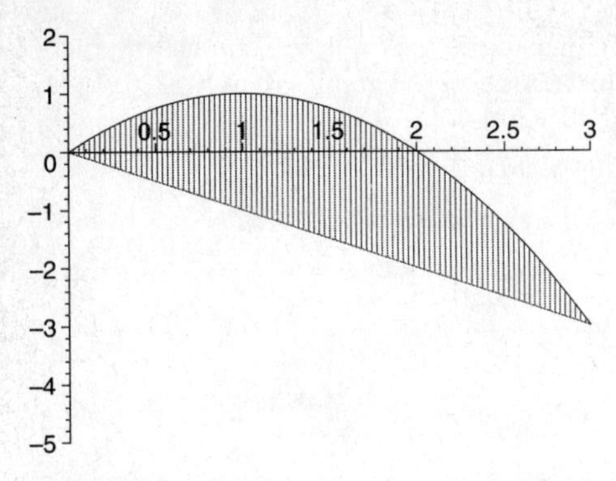

The boundaries are easily determined by looking at the graph. Note which is the upper graph over the shaded region. Here it is $(2x - x^2) \geq -x$ over the interval $[0, 3]$.

Compute the area as follows:

$$\int_0^3 [(2x - x^2) - (-x)]\, dx$$
$$= \int_0^3 (3x - x^2)\, dx$$
$$= \left[\frac{3}{2}x^2 - \frac{x^3}{3}\right]_0^3$$
$$= \left(\frac{3}{2} \cdot 3^2 - \frac{3^3}{3}\right) - \left(\frac{3}{2} \cdot 0^2 - \frac{0^3}{3}\right)$$
$$= \frac{27}{2} - \frac{27}{3}$$
$$= \frac{81}{6} - \frac{54}{6}$$
$$= \frac{27}{6}$$
$$= \frac{9}{2}$$

8.

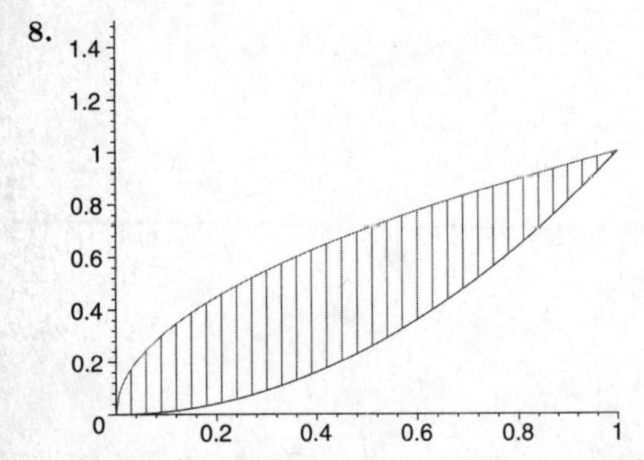

We find the boundaries by looking at the graph or by solving the following:

$$x^2 = \sqrt{x}$$
$$x^4 = x$$
$$x^4 - x = 0$$
$$x(x^3 - 1) = 0$$
$$x = 0 \text{ or } x = 1$$

$$\int_0^1 (\sqrt{x} - x^2)\, dx$$
$$= \left[\frac{2}{3}x^{3/2} - \frac{x^3}{3}\right]_0^1$$
$$= \left(\frac{2}{3} \cdot 1^{3/2} - \frac{1^3}{3}\right) - \left(\frac{2}{3} \cdot 0^{3/2} - \frac{0^3}{3}\right)$$
$$= \frac{1}{3}$$

9. First graph the system of equations and shade the region bounded by the graphs.

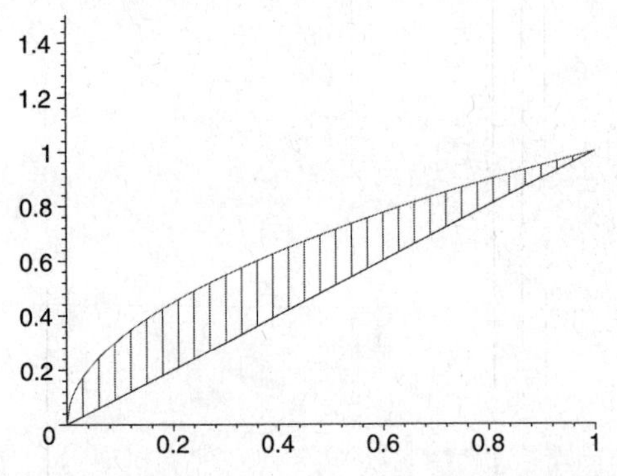

The boundaries are easily determined by looking at the graph. Note which is the upper graph over the shaded region. Here it is $\sqrt[4]{x} \geq x$ over the interval $[0, 1]$.

Compute the area as follows:

$$\int_0^1 (\sqrt[4]{x} - x)\, dx$$
$$= \int_0^1 (x^{1/4} - x)\, dx$$
$$= \left[\frac{4}{5}x^{5/4} - \frac{1}{2}x^2\right]_0^1$$
$$= \left(\frac{4}{5} \cdot 1^{5/4} - \frac{1}{2} \cdot 1^2\right) - \left(\frac{4}{5} \cdot 0^{5/4} - \frac{1}{2} \cdot 0^2\right)$$
$$= \frac{4}{5} - \frac{1}{2} - 0$$
$$= \frac{8}{10} - \frac{5}{10}$$
$$= \frac{3}{10}$$

10.

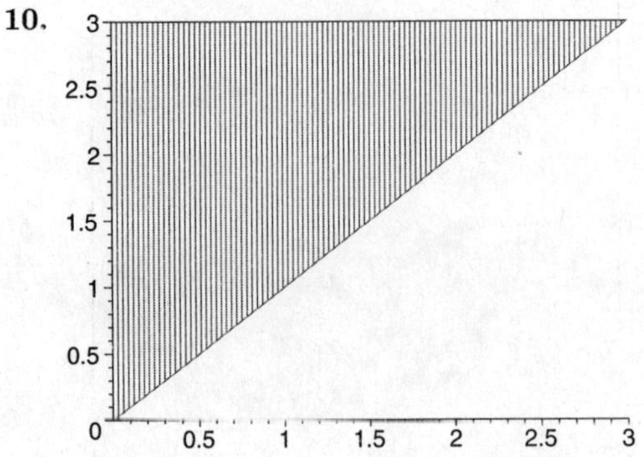

From the graph we see that the boundaries are 0 and 3.

$\int_0^3 (3-x)\,dx$

$= \left[3x - \frac{x^2}{2}\right]_0^3$

$= \left(3\cdot 3 - \frac{3^2}{2}\right) - \left(3\cdot 0 - \frac{0^2}{2}\right)$

$= \frac{9}{2}$

(Note that this is the area of a triangle with base 3 and height 3.)

11. Graph the system of equations and shade the region bounded by the graphs.

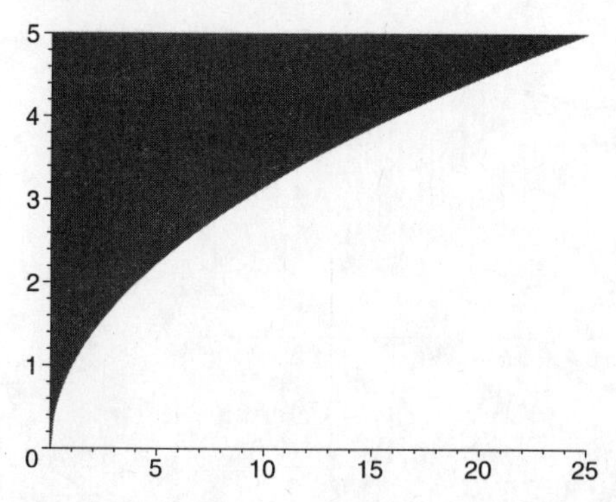

The boundaries are easily determined by looking at the graph. Here $5 \geq \sqrt{x}$ over the interval $[0, 25]$.

Compute the area as follows:

$\int_0^{25} (5-\sqrt{x})\,dx$

$= \int_0^{25} (5 - x^{1/2})\,dx$

$= \left[5x - \frac{x^{3/2}}{3/2}\right]_0^{25}$

$= \left[5x - \frac{2}{3}x^{3/2}\right]_0^{25}$

$= \left(5\cdot 25 - \frac{2}{3}\cdot 25^{3/2}\right) - \left(5\cdot 0 - \frac{2}{3}\cdot 0^{3/2}\right)$

$= 125 - \frac{250}{3} - 0 \qquad [25^{3/2} = (5^2)^{3/2} = 5^3 = 125]$

$= \frac{375}{3} - \frac{250}{3}$

$= \frac{125}{3}$, or $41\frac{2}{3}$

12.

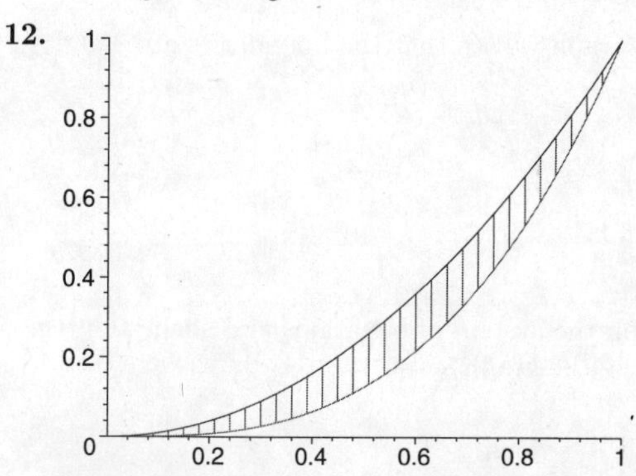

We find the boundaries by looking at the graph or by solving the following:

$x^2 = x^3$

$0 = x^3 - x^2$

$0 = x^2(x-1)$

$x = 0$ *or* $x = 1$

Note that $x^2 \geq x^3$ on $[0, 1]$.

$\int_0^1 (x^2 - x^3)\,dx$

$= \left[\frac{x^3}{3} - \frac{x^4}{4}\right]_0^1$

$= \left(\frac{1^3}{3} - \frac{1^4}{4}\right) - \left(\frac{0^3}{3} - \frac{0^4}{4}\right)$

$= \frac{1}{12}$

13. First graph the system of equations and shade the region bounded by the graphs.

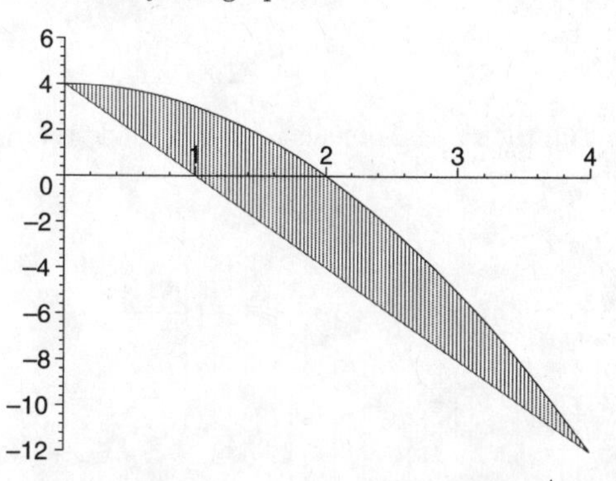

Then determine the first coordinates of possible points of intersection by solving a system of equations as follows. At the points of intersection, $y = 4 - x^2$ and $y = 4 - 4x$, so

$4 - x^2 = 4 - 4x$

$0 = x^2 - 4x$

$0 = x(x-4)$

$x = 0$ *or* $x = 4$

Thus the interval with which we are concerned is $[0, 4]$. Note that $4 - x^2 \geq 4 - 4x$ over the interval $[0, 4]$.

Compute the area as follows:

$\int_0^4 [(4-x^2) - (4-4x)]\,dx$

$= \int_0^4 (-x^2 + 4x)\,dx$

$= \left[-\frac{x^3}{3} + 2x^2\right]_0^4$

$= \left(-\frac{4^3}{3} + 2\cdot 4^2\right) - \left(-\frac{0^3}{3} + 2\cdot 0^2\right)$

$= -\frac{64}{3} + 32$

$= -\frac{64}{3} + \frac{96}{3}$

$= \frac{32}{3}$

14.

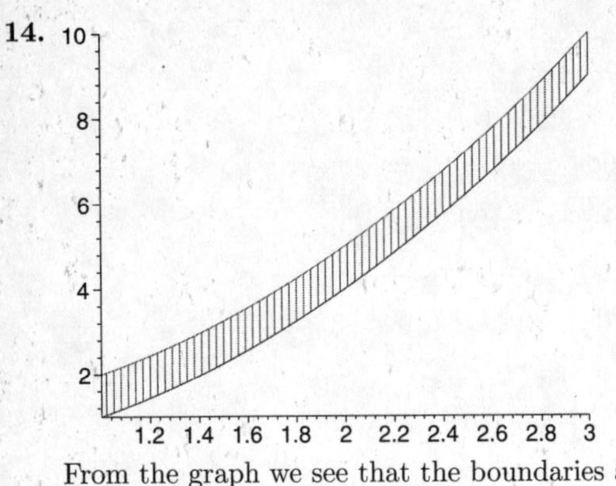

From the graph we see that the boundaries are 1 and 3.

$\int_1^3 [(x^2+1)-x^2]\,dx$
$= \int_1^3 dx$
$= [x]_1^3$
$= 3-1$
$= 2$

15. First graph the system of equations and shade the region bounded by the graphs.

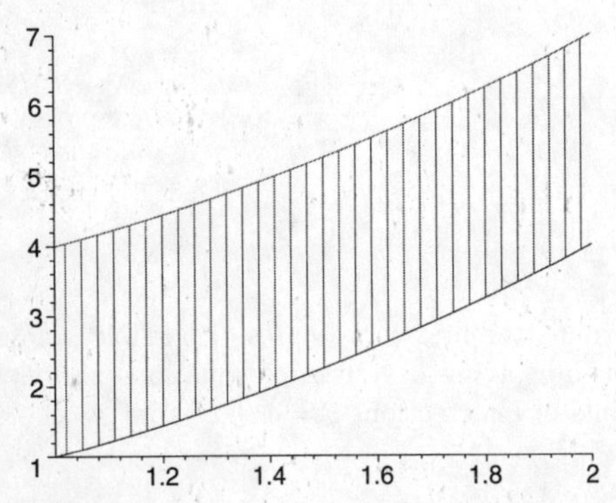

From the graph we can easily determine the interval with which we are concerned. Here $x^2+3 \geq x^2$ over the interval $[1, 2]$.

Compute the area as follows:

$\int_1^2 [(x^2+3)-x^2]\,dx$
$= \int_1^2 3\,dx$
$= [3x]_1^2$
$= 3\cdot 2 - 3\cdot 1$
$= 6-3$
$= 3$

16.

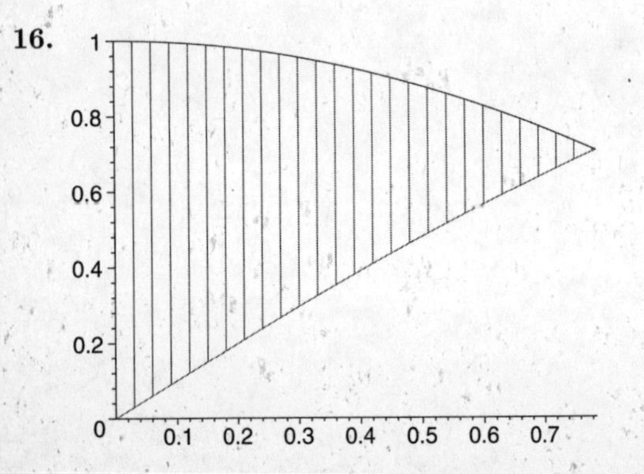

From the graph we see that the boundaries are 0 and $\frac{\pi}{4}$.

$\int_0^{\pi/4} [\cos x - \sin x]\,dx$
$= [\sin x + \cos x]_0^{\pi/4}$
$= \left[\frac{1}{\sqrt{2}} + \frac{1}{\sqrt{2}}\right] - [0+1]$
$= \sqrt{2} - 1$

17. First graph the system of equations and shade the region bounded by the graphs.

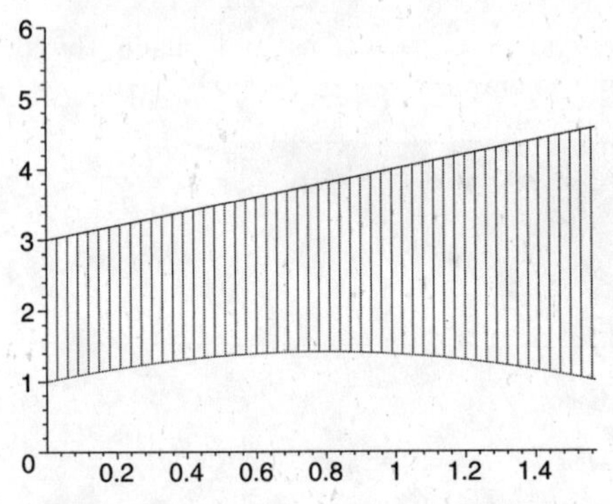

From the graph we can easily determine the interval with which we are concerned. Here $x+3 \geq \sin x + \cos x$ over the interval $[0, \frac{\pi}{2}]$.

Compute the area as follows:

$\int_0^{\frac{\pi}{2}} [x+3-\sin x - \cos x]\,dx$
$= \left[\frac{1}{2}x^2 + 3x + \cos x - \sin x\right]_0^{\frac{\pi}{2}}$
$= \left[\frac{\pi^2}{8} + \frac{3\pi}{2} - 1\right] - [1]$
$= \frac{\pi^2}{8} + \frac{3\pi}{2} - 2$

18.

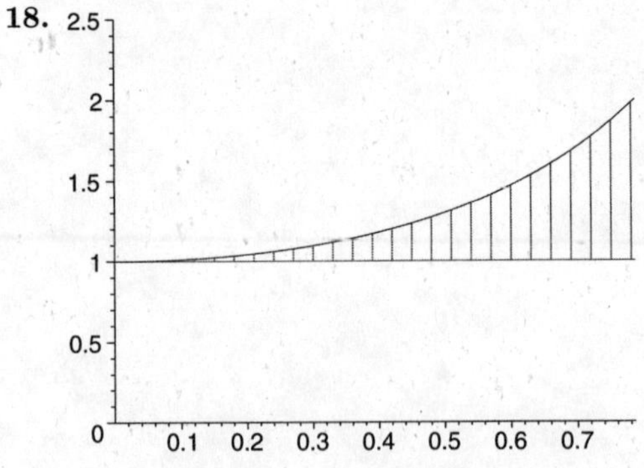

From the graph we see that the boundaries are 0 and $\frac{\pi}{4}$.

$\int_0^{\pi/4} [\sec^2 x - 1]\,dx$
$= [\tan x - x]_0^{\pi/4}$
$= \left[1 - \frac{\pi}{4}\right] - 0$
$= 1 - \frac{\pi}{4}$

19. First graph the system of equations and shade the region bounded by the graphs.

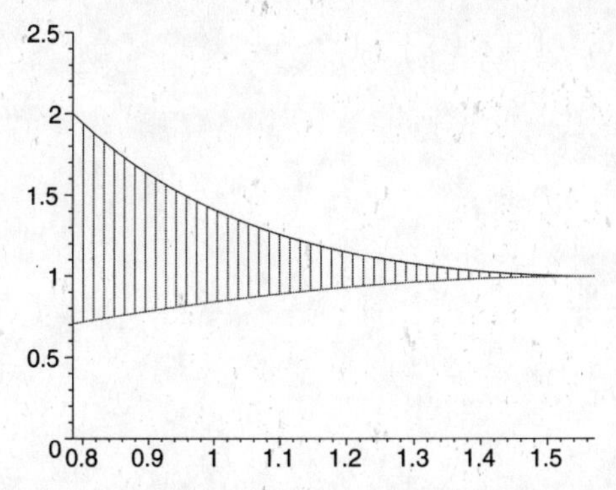

From the graph we can easily determine the interval with which we are concerned. Here $x+3 \geq sin\ x + cos\ x$ over the interval $[\frac{\pi}{4}, \frac{\pi}{2}]$.

Compute the area as follows:

$$\int_{\frac{\pi}{4}}^{\frac{\pi}{2}} [csc^2 x - sin\ x]\, dx$$
$$= \Big[-cot\ x + cos\ x\Big]_{\frac{\pi}{4}}^{\frac{\pi}{2}}$$
$$= [0] - \left[-1 + \tfrac{1}{\sqrt{2}}\right]$$
$$= 1 - \tfrac{1}{\sqrt{2}}$$

20.

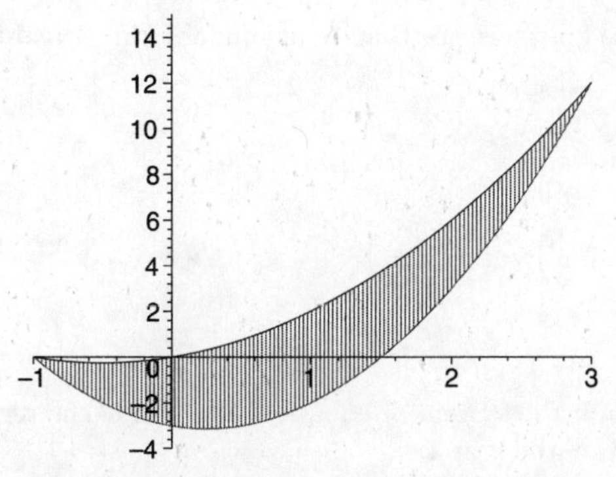

From the graph we see that the boundaries are -1 and 3.

$$\int_{-1}^{3} [x^2 + x - (2x^2 - x - 3)]\, dx$$
$$= \int_{-1}^{3} [-x^2 + 2x + 3]\, dx$$
$$= \Big[-\tfrac{1}{3}x^3 + x^2 + 3x\Big]_{-1}^{3}$$
$$= [-9 + 9 + 9] - [-\tfrac{1}{3} + 1 - 3]$$
$$= \tfrac{22}{3}$$

21. First graph the system of equations and shade the region bounded by the graphs.

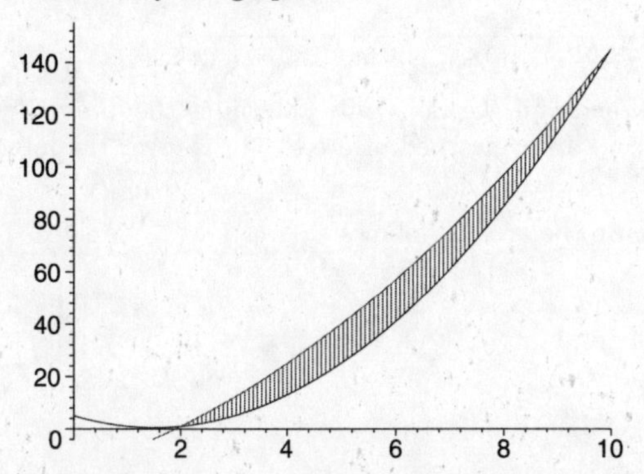

From the graph we can easily determine the interval with which we are concerned. Here $x^2 + 6x - 15 \geq 2x^2 - 6x + 5$ over the interval $[2, 10]$.

Compute the area as follows:

$$\int_{2}^{10} [x^2 + 6x - 15 - 2x^2 + 6x - 5]\, dx$$
$$= \int_{2}^{10} [-x^2 + 12x - 20]\, dx$$
$$= \Big[-\tfrac{1}{3}x^3 + 6x^2 - 20x\Big]_{2}^{10}$$
$$= [-\tfrac{1000}{3} + 600 - 200] - [-\tfrac{8}{3} + 24 - 40]$$
$$= \tfrac{256}{3}$$

22.

$$\int_{-1}^{0} [0 - (2x + x^2 - x^3)]\, dx + \int_{0}^{1} [(2x + x^2 - x^3) - 0]\, dx$$
$$= \int_{-1}^{0} (-2x - x^2 + x^3)\, dx + \int_{0}^{1} (2x + x^2 - x^3)\, dx$$
$$= \left[-x^2 - \frac{x^3}{3} + \frac{x^4}{4}\right]_{-1}^{0} + \left[x^2 + \frac{x^3}{3} - \frac{x^4}{4}\right]_{0}^{1}$$
$$= \left(-0^2 - \frac{0^3}{3} + \frac{0^4}{4}\right) - \left[-(-1)^2 - \frac{(-1)^3}{3} + \frac{(-1)^4}{4}\right] + \left(1^2 + \frac{1^3}{3} - \frac{1^4}{4}\right) - \left(0^2 + \frac{0^3}{3} - \frac{0^4}{4}\right)$$
$$= 0 + \frac{5}{12} + \frac{13}{12} - 0$$
$$= \frac{18}{12} = \frac{3}{2}$$

23. $f(x) \geq g(x)$ on $[-5, -1]$, and $g(x) \geq f(x)$ on $[-1, 3]$. We use two integrals to find the total area.

$$\int_{-5}^{-1} [(x^3 + 3x^2 - 9x - 12) - (4x + 3)]\, dx + \int_{-1}^{3} [(4x + 3) - (x^3 + 3x^2 - 9x - 12)]\, dx$$
$$= \int_{-5}^{-1} (x^3 + 3x^2 - 13x - 15)\, dx + \int_{-1}^{3} (-x^3 - 3x^2 + 13x + 15)\, dx$$
$$= \left[\frac{x^4}{4} + x^3 - \frac{13x^2}{2} - 15x\right]_{-5}^{-1} + \left[-\frac{x^4}{4} - x^3 + \frac{13x^2}{2} + 15x\right]_{-1}^{3}$$
$$= \left[\frac{(-1)^4}{4} + (-1)^3 - \frac{13(-1)^2}{2} - 15(-1)\right] - \left[\frac{(-5)^4}{4} + (-5)^3 - \frac{13(-5)^2}{2} - 15(-5)\right] + \left[-\frac{3^4}{4} - 3^3 + \frac{13(3)^2}{2} + 15 \cdot 3\right] - \left[-\frac{(-1)^4}{4} - (-1)^3 + \frac{13(-1)^2}{2} + 15(-1)\right]$$
$$= \left(\frac{1}{4} - 1 - \frac{13}{2} + 15\right) - \left(\frac{625}{4} - 125 - \frac{325}{2} + 75\right) + \left(-\frac{81}{4} - 27 + \frac{117}{2} + 45\right) - \left(-\frac{1}{4} + 1 + \frac{13}{2} - 15\right)$$
$$= \frac{31}{4} + \frac{225}{4} + \frac{225}{4} + \frac{31}{4}$$
$$= 128$$

24. $\int_{-1}^{4}[(x+28)-(x^4-8x^3+18x^2)]\,dx$
$= \int_{-1}^{4}(-x^4+8x^3-18x^2+x+28)\,dx$
$= \left[-\frac{x^5}{5}+2x^4-6x^3+\frac{x^2}{2}+28x\right]_{-1}^{4}$
$= \left(-\frac{4^5}{5}+2\cdot 4^4-6\cdot 4^3+\frac{4^2}{2}+28\cdot 4\right)-$
$\left[-\frac{(-1)^5}{5}+2(-1)^4-6(-1)^3+\frac{(-1)^2}{2}+28(-1)\right]$
$= \frac{216}{5}+\frac{193}{10}$
$= \frac{625}{10}=62\frac{1}{2}$

25. $f(x)\geq g(x)$ on $[1,4]$. We find the area.
$\int_{1}^{4}[(4x-x^2)-(x^2-6x+8)]\,dx$
$= \int_{1}^{4}(-2x^2+10x-8)\,dx$
$= \left[-\frac{2x^3}{3}+5x^2-8x\right]_{1}^{4}$
$= \left(-\frac{2\cdot 4^3}{3}+5\cdot 4^2-8\cdot 4\right)-\left(-\frac{2\cdot 1^3}{3}+5\cdot 1^2-8\cdot 1\right)$
$= \left(-\frac{128}{3}+80-32\right)-\left(-\frac{2}{3}+5-8\right)$
$= \frac{16}{3}+\frac{11}{3}=\frac{27}{3}$
$= 9$

26. $\int_{-2}^{1}x^2\,dx+\int_{1}^{3}1\cdot dx$
$= \left[\frac{x^3}{3}\right]_{-2}^{1}+[x]_{1}^{3}$
$= \left[\frac{1^3}{3}-\frac{(-2)^3}{3}\right]+(3-1)$
$= \frac{1}{3}+\frac{8}{3}+3-1$
$= 5$

27. Find the area under

$$f(x)=\begin{cases}4-x^2, & \text{if } x<0\\ 4, & \text{if } x\geq 0\end{cases} \quad \text{on } [-2,3]$$

We have to break the integral into two parts in order to complete this problem

$\int_{-2}^{3}f(x)\,dx$
$= \int_{-2}^{0}f(x)\,dx+\int_{0}^{3}f(x)\,dx$
$= \int_{-2}^{0}(4-x^2)\,dx+\int_{0}^{3}4\,dx$
$= \left[4x-\frac{x^3}{3}\right]_{-2}^{0}+[4x]_{0}^{3}$
$= \left\{\left[4\cdot 0-\frac{0^3}{3}\right]-\left[4(-2)-\frac{(-2)^3}{3}\right]\right\}+(4\cdot 3-4\cdot 0)$
$= (0-0)-\left(-8+\frac{8}{3}\right)+(12-0)$
$= -\left(-\frac{24}{3}+\frac{8}{3}\right)+12$
$= -\left(-\frac{16}{3}\right)+12$
$= \frac{16}{3}+\frac{36}{3}$
$= \frac{52}{3}$, or $17\frac{1}{3}$

28.

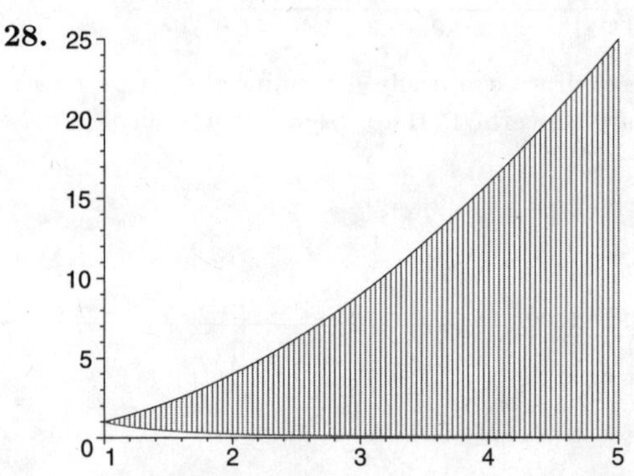

From the graph we see that the boundaries are 1 and 5.

$\int_{1}^{5}(x^2-x^{-2})\,dx$
$= \left[\frac{x^3}{3}+\frac{1}{x}\right]_{1}^{5}$
$= \left(\frac{5^3}{3}+\frac{1}{5}\right)-\left(\frac{1^3}{3}+\frac{1}{1}\right)$
$= \frac{608}{15}$

29. First graph the system of equations and shade the region bounded by the graph.

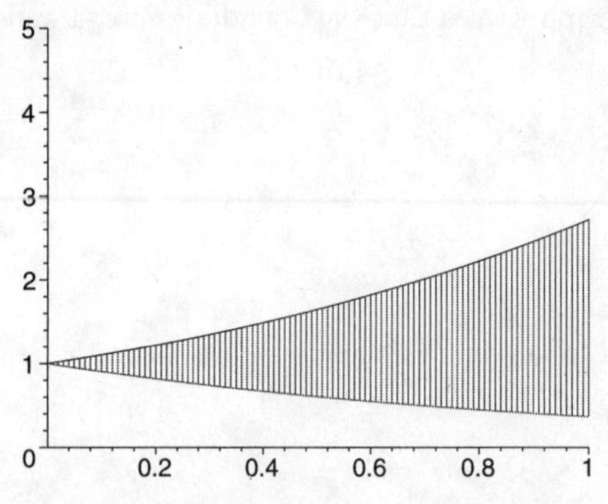

From the graph we can easily determine the interval with which we are concerned. Here $e^x\geq e^{-x}$ over the interval $[0,1]$.

Compute the area as follows:

$\int_0^1 (e^x - e^{-x})\, dx$

$= \left[e^x + e^{-x}\right]_0^1$

$= (e^1 + e^{-1}) - (e^0 + e^{-0})$

$= \left(e + \frac{1}{e}\right) - (1+1)$

$= e + \frac{1}{e} - 2$

$= \frac{e^2 - 2e + 1}{e}$

$= \frac{(e-1)^2}{e} \approx 1.086$

30.

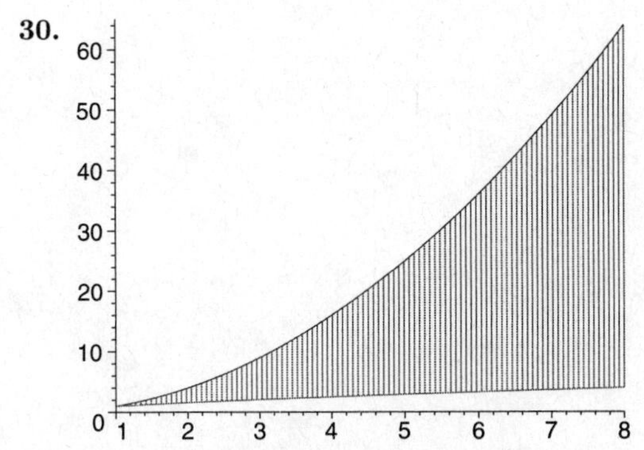

From the graph we see that the boundaries are 1 and 8.

$\int_1^8 (x^2 - \sqrt[3]{x^2})\, dx$

$= \left[\frac{x^3}{3} - \frac{3}{5}x^{5/3}\right]_1^8$

$= \left(\frac{8^3}{3} - \frac{3}{5}\cdot 8^{5/3}\right) - \left(\frac{1^3}{3} - \frac{3}{5}\cdot 1^{5/3}\right)$

$= \left(\frac{512}{3} - \frac{96}{5}\right) - \left(\frac{1}{3} - \frac{3}{5}\right)$

$= \frac{2276}{15}$

31. First graph the system of equations and shade the region bounded by the graph.

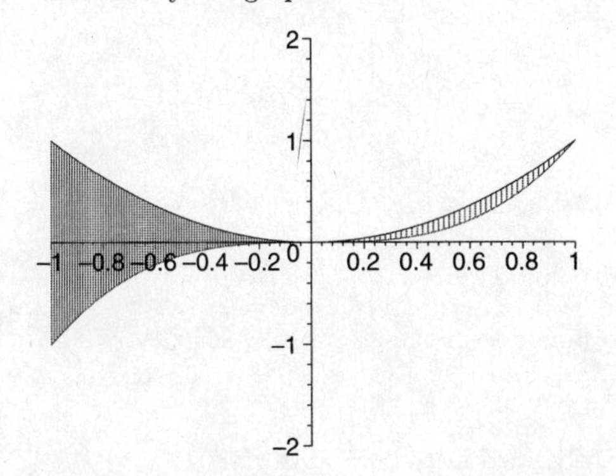

From the graph we can easily determine the interval with which we are concerned. Here $x^2 \geq x^3$ over the interval $[-1, 1]$.

Compute the area as follows:

$\int_{-1}^1 (x^2 - x^3)\, dx$

$= \left[\frac{x^3}{3} - \frac{x^4}{4}\right]_{-1}^1$

$= \left(\frac{1^3}{3} - \frac{1^4}{4}\right) - \left(\frac{(-1)^3}{3} - \frac{(-1)^4}{4}\right)$

$= \left(\frac{1}{3} - \frac{1}{4}\right) - \left(-\frac{1}{3} - \frac{1}{4}\right)$

$= \frac{1}{3} - \frac{1}{4} + \frac{1}{3} + \frac{1}{4} = \frac{2}{3}$

32.

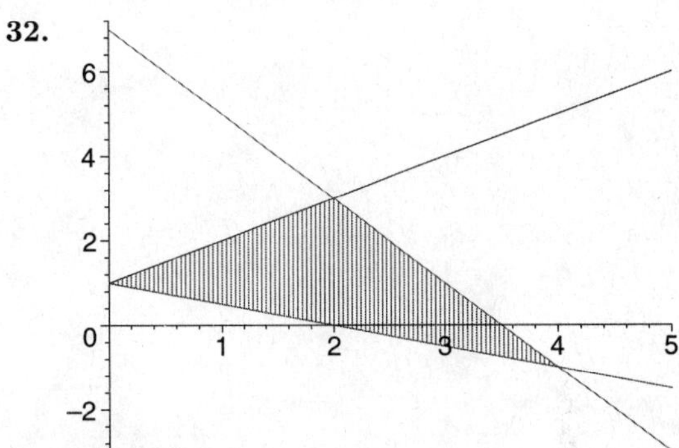

On $[0, 2]$, $x + 1 \geq -\frac{x}{2} + 1$, and on $[2, 4]$,

$-2x + 7 \geq -\frac{x}{2} + 1$.

$$\int_0^2 \left[(x+1) - \left(-\frac{x}{2} + 1\right)\right] dx + \int_2^4 \left[(-2x+7) - \left(-\frac{x}{2} + 1\right)\right] dx$$

$= \int_0^2 \frac{3}{2}x\, dx + \int_2^4 \left(-\frac{3}{2}x + 6\right) dx$

$= \left[\frac{3x^2}{4}\right]_0^2 + \left[-\frac{3x^2}{4} + 6x\right]_2^4$

$= \frac{3\cdot 2^2}{4} - \frac{3\cdot 0^2}{4} + \left(-\frac{3\cdot 4^2}{4} + 6\cdot 4\right) - \left(-\frac{3\cdot 2^2}{4} + 6\cdot 2\right)$

$= 3 - 0 + 12 - 9$

$= 6$

33. On $[0, 11]$, $f(x) \geq g(x)$.

$\int_0^{11} [f(x) - g(x)]\, dx$

$= \int_0^{11} [0.3x + 6.15x^2 - 0.552x^3]\, dx$

$= \left[0.15x^2 + \frac{14.45}{3}x^3 - 0.138x^4\right]_0^{11}$

$= [726.24 - 0]$

$= 726.24 \;\; N\cdot mm$

34. $T(t) = 5.76 + 24t - 16t^2$, $f(t) = 8$

$$\int_0^1 [T(t) - f(t)]\,dt$$
$$= \int_0^1 [-2.24 + 24t - 16t^2]\,dt$$
$$= \left[-2.24t + 12t^2 - \frac{16}{3}t^3\right]_0^1$$
$$= [-2.24 + 12 - \tfrac{16}{3} - 0]$$
$$= 4.43$$

35. $T(t) = 25 + 3e^{-t} - 20(t - 0.5)^2$, $f(t) = 8$

$$\int_0^1 [T(t) - f(t)]\,dt$$
$$= \int_0^1 [17 + 3e^{-t} - 20(t - 0.5)^2]\,dt$$
$$= \left[17t - 3e^{-t} - \frac{20}{3}(t - 0.5)^3\right]_0^1$$
$$= [17 - \tfrac{3}{e} - \tfrac{5}{6} - [-3 + \tfrac{5}{6}]$$
$$= 17.23 \text{ degree days}$$

36. a) $-0.003t^2 \geq -0.009t^2$

Thus, $M'(t) \geq m'(t)$, so subject B has the higher rate of memorization.

b) $\int_0^{10} [(-0.003t^2 + 0.2t) - (-0.009t^2 + 0.2t)]\,dt$
$$= \int_0^{10} 0.006t^2\,dt$$
$$= \left[0.002t^3\right]_0^{10}$$
$$= 0.002 \cdot 10^3 - 0.002 \cdot 0^3$$
$$= 2 - 0$$
$$= 2$$

Subject B memorizes 2 more words then subject A during the first 10 minutes.

37.

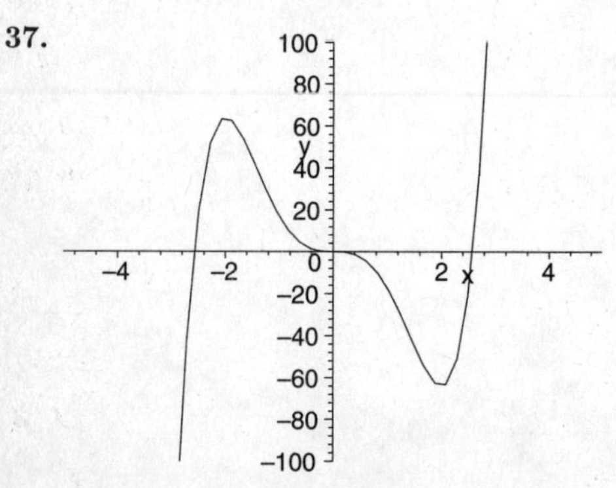

From the graph we see that the first coordinate of the relative maximum is $x = -2$ and the first coordinate of the relative minimum is $x = 2$

$$\int_{-2}^0 [3x^5 - 20x^3 - 0]\,dx + \int_0^2 [0 - 3x^5 + 20x^3]\,dx$$
$$= \left[\frac{1}{2}x^6 - 5x^4\right]_{-2}^0 + \left[-\frac{1}{2}x^6 + 5x^4\right]_0^2$$
$$= [-32 + 80] - [32 - 80]$$
$$= 96$$

38.

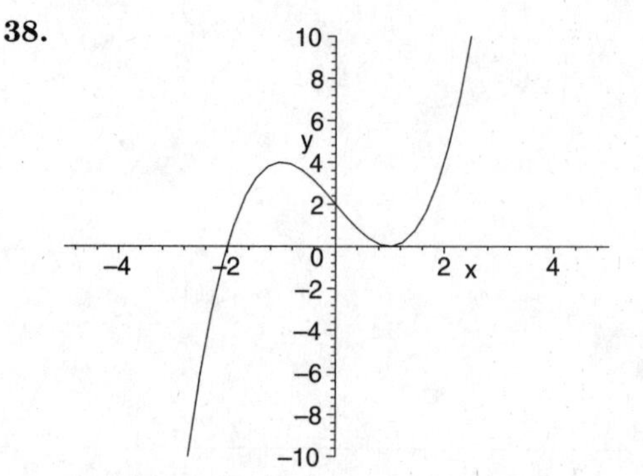

From the graph we see that the first coordinate of the relative maximum is $x = -1$ and the first coordinate of the relative minimum is $x = 1$

$$\int_{-1}^0 [x^3 - 3x + 2 - 0]\,dx + \int_0^1 [0 - x^3 + 3x - 2]\,dx$$
$$= \left[\frac{1}{4}x^4 - \frac{3}{2}x^2 + 2x\right]_{-1}^0 + \left[-\frac{1}{4}x^4 + \frac{3}{2}x^2 - 2x\right]_0^1$$
$$= [0 - \tfrac{1}{4} + \tfrac{3}{2} - 2] - [\tfrac{-1}{4} + \tfrac{3}{2} - 2]$$
$$= -\tfrac{3}{2}$$

39. $V = \dfrac{p}{4Lv}(R^2 - r^2)$

$$Q = \int_0^R \left[2\pi \cdot \frac{p}{4Lv}(R^2 - r^2)r\right] dr$$
$$= \frac{\pi p}{2Lv}\int_0^R [R^2 r - r^3]\,dr$$
$$= \frac{\pi p}{2Lv}\left[\frac{R^2}{2}r^2 - \frac{1}{4}r^4\right]_0^R$$
$$= \frac{\pi p}{2Lv}\left[\frac{R^4}{2} - \frac{1}{4}R^4\right]$$
$$= \frac{\pi p R^4}{8Lv}$$

40. The area bounded in the graph is 4

41. The area bounded in the graph is 24.961

42. The area bounded in the graph is 5.886

43. The area bounded in the graph is 416.708

44. The area bounded in the graph is 0.238

45. 

a)

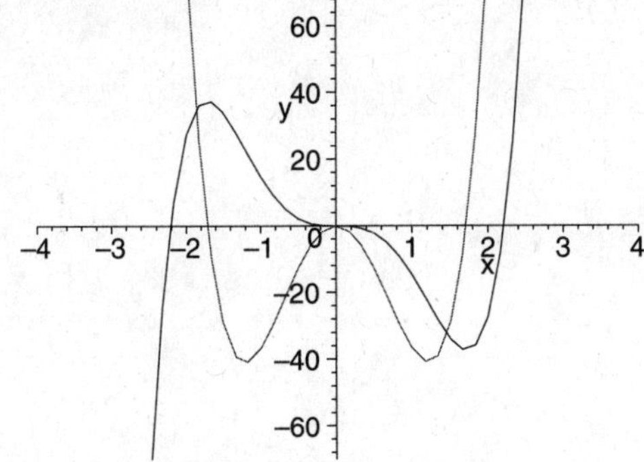

b) $x = -1.8623$, $x = 0$, and $x = 1.4594$

c) The area is 64.5239

d) The area is 17.683

Exercise Set 5.5

1. $\displaystyle\int \frac{3x^2\,dx}{7+x^3}$

Let $u = 7 + x^3$, then $du = 3x^2\,dx$.

$= \displaystyle\int \frac{du}{u}$ Substituting u for $7 + x^3$ and du for $3x^2\,dx$

$= \displaystyle\int \frac{1}{u}\,du$

$= \ln u + C$ Using Formula C

$= \ln(7 + x^3) + C$

2. $\displaystyle\int \frac{3x^2}{1+x^3}\,dx$

Let $u = 1 + x^3$, then $du = 3x^2\,dx$.

$= \displaystyle\int \frac{du}{u}$

$= \ln u + C$

$= \ln(1 + x^3) + C$

3. $\int e^{4x}\,dx$

Let $u = 4x$, then $du = 4\,dx$.

We do not have $4\,dx$. We only have dx and need to supply a 4. We do this by multiplying by $\frac{1}{4} \cdot 4$ as follows.

$\displaystyle\frac{1}{4} \cdot 4 \int e^{4x}\,dx$ Multiplying by 1

$= \displaystyle\frac{1}{4}\int 4e^{4x}\,dx$

$= \displaystyle\frac{1}{4}\int e^{4x}(4\,dx)$

$= \displaystyle\frac{1}{4}\int e^{u}\,du$ Substituting u for $4x$ and du for $4\,dx$

$= \displaystyle\frac{1}{4}e^{u} + C$ Using Formula B

$= \displaystyle\frac{1}{4}e^{4x} + C$

4. $\int e^{3x}\,dx$

Let $u = 3x$, then $du = 3\,dx$.

$= \displaystyle\frac{1}{3}\int e^{3x} \cdot 3\,dx$

$= \displaystyle\frac{1}{3}\int e^{u}\,du$

$= \displaystyle\frac{1}{3}e^{u} + C$

$= \displaystyle\frac{1}{3}e^{3x} + C$

5. $\int e^{x/2}\,dx = \int e^{(1/2)x}\,dx$

Let $u = \frac{1}{2}x$, then $du = \frac{1}{2}\,dx$.

We do not have $\frac{1}{2}\,dx$. We only have dx and need to supply a $\frac{1}{2}$ by multiplying by $2 \cdot \frac{1}{2}$ as follows.

$\displaystyle 2 \cdot \frac{1}{2}\int e^{x/2}\,dx$ Multiplying by 1

$= \displaystyle 2\int \frac{1}{2}e^{x/2}\,dx$

$= \displaystyle 2\int e^{x/2}\left(\frac{1}{2}\,dx\right)$

$= 2\int e^{u}\,du$ Substituting u for $x/2$ and du for $\frac{1}{2}\,dx$

$= 2\,e^{u} + C$ Using Formula B

$= 2\,e^{x/2} + C$

6. $\int e^{x/3}\,dx$

Let $u = \frac{x}{3}$, then $du = \frac{1}{3}\,dx$.

$= \displaystyle 3\int e^{x/3} \cdot \frac{1}{3}\,dx$

$= 3\int e^{u}\,du$

$= 3\,e^{u} + C$

$= 3\,e^{x/3} + C$

7. $\int x^3\,e^{x^4}\,dx$

Let $u = x^4$, then $du = 4x^3\,dx$.

We do not have $4x^3\,dx$. We only have $x^3\,dx$ and need to supply a 4. We do this by multiplying by $\frac{1}{4} \cdot 4$ as follows.

$\displaystyle\frac{1}{4} \cdot 4\int x^3\,e^{x^4}\,dx$ Multiplying by 1

$= \displaystyle\frac{1}{4}\int 4x^3\,e^{x^4}\,dx$

$= \displaystyle\frac{1}{4}\int e^{x^4}(4x^3\,dx)$

$= \displaystyle\frac{1}{4}\int e^{u}\,du$ Substituting u for x^4 and du for $4x^3\,dx$

$= \displaystyle\frac{1}{4} \cdot e^{u} + C$ Using Formula B

$= \displaystyle\frac{1}{4}e^{x^4} + C$

8. $\int x^4 e^{x^5}\,dx$

Let $u = x^5$, then $du = 5x^4\,dx$.

$= \frac{1}{5}\int e^{x^5}\cdot 5x^4\,dx$

$= \frac{1}{5}\int e^u\,du$

$= \frac{1}{5}e^u + C$

$= \frac{1}{5}e^{x^5} + C$

9. $\int t^2 e^{-t^3}\,dt$

Let $u = -t^3$, then $du = -3t^2\,dt$.

We do not have $-3t^2\,dt$. We only have $t^2\,dt$. We need to supply a -3 by multiplying by $-\frac{1}{3}\cdot(-3)$ as follows.

$-\frac{1}{3}\cdot(-3)\int t^2 e^{-t^3}\,dt$ Multiplying by 1

$= -\frac{1}{3}\int -3t^2 e^{-t^3}\,dt$

$= -\frac{1}{3}\int e^{-t^3}(-3t^2\,dt)$

$= -\frac{1}{3}\int e^u\,du$ Substituting u for $-t^3$ and du for $-3t^2\,dt$

$= -\frac{1}{3}e^u + C$ Using Formula B

$= -\frac{1}{3}e^{-t^3} + C$

10. $\int te^{-t^2}\,dt$

Let $u = -t^2$, then $du = -2t\,dt$.

$= -\frac{1}{2}\int e^{-t^2}\cdot(-2t)\,dt$

$= -\frac{1}{2}\int e^u\,du$

$= -\frac{1}{2}e^u + C$

$= -\frac{1}{2}e^{-t^2} + C$

11. $\int \frac{\ln 4x\,dx}{x}$

Let $u = \ln 4x$, then $du = \frac{1}{x}\,dx$.

$= \int \ln 4x\left(\frac{1}{x}\,dx\right)$

$= \int u\,du$ Substituting u for $\ln 4x$ and du for $\frac{1}{x}\,dx$

$= \frac{u^2}{2} + C$ Using Formula A

$= \frac{(\ln 4x)^2}{2} + C$

12. $\int \frac{\ln 5x\,dx}{x}$

Let $u = \ln 5x$, then $du = \frac{1}{x}\,dx$.

$= \int u\,du$

$= \frac{u^2}{2} + C$

$= \frac{(\ln 5x)^2}{2} + C$

13. $\int \frac{dx}{1+x}$

Let $u = 1 + x$, then $du = dx$.

$= \int \frac{du}{u}$ Substituting u for $1+x$ and du for dx

$= \int \frac{1}{u}\,du$

$= \ln u + C$ Using Formula C

$= \ln(1+x) + C$

14. $\int cos(2x+3)\,dx$

Let $u = 2x + 3$, then $du = 2\,dx$.

$= \frac{1}{2}\int cos(u)\,du$ Substituting u for $2x+3$ and $\frac{1}{2}du$ for dx

$= \frac{1}{2}sin(u) + C$

$= \frac{1}{2}sin(2x+3) + C$

15. $\int 3\,sin(4x+2)\,dx$

Let $u = 4x + 2$, then $du = 4\,dx$.

$= \frac{3}{4}\int sin(u)\,du$ Substituting u for $4x+2$ and $\frac{1}{4}du$ for dx

$= -\frac{3}{4}cos(u) + C$

$= -\frac{3}{4}cos(4x+2) + C$

16. $\int 2\,sec^2(2x+3)\,dx$

Let $u = 2x + 3$, then $du = 2\,dx$.

$= \frac{2}{2}\int sec^2(u)\,du$ Substituting u for $2x+3$ and $\frac{1}{2}du$ for dx

$= tan(u) + C$

$= tan(2x+3) + C$

17. $\int csc(2x+3)\,cot(2x+3)\,dx$

Let $u = 2x + 3$, then $du = 2\,dx$.

$= \frac{1}{2}\int csc(u)\,cot(u)\,du$ Substituting u for $2x+3$ and $\frac{1}{2}du$ for dx

$= -csc(u) + C$

$= -csc(2x+3) + C$

18. $\int cot^2(2-x)\,dx = \int csc^2(2-x) - 1$

Let $u = 2 - x$, then $du = -\,dx$.

$= \int [-csc^2(u) + 1]\,du$ Substituting u for $2 - x$ and $-du$ for dx

$= cot(u) + u + C$

$= cot(2-x) + 2 - x + C$

$= cot(2-x) - x + K$

19. $\int \frac{dx}{4-x}$

Let $u = 4 - x$, then $du = -dx$.

We do not have $-dx$. We only have dx and need to supply a -1 by multiplying by $-1 \cdot (-1)$ as follows.

$-1 \cdot (-1) \int \frac{dx}{4-x}$ Multiplying by 1

$= -1 \int -1 \cdot \frac{dx}{4-x}$

$= -\int \frac{1}{4-x}(-dx)$

$= -\int \frac{1}{u}\,du$ Substituting u for $4 - x$ and du for $-dx$

$= -\ln u + C$ Using Formula C

$= -\ln(4-x) + C$

20. $\int \frac{dx}{1-x}$

Let $u = 1 - x$, then $du = -dx$.

$= -\int \frac{1}{1-x}(-dx)$

$= -\int \frac{1}{u}\,dx$

$= -\ln u + C$

$= -\ln(1-x) + C$

21. $\int t^2(t^3-1)^7\,dt$

Let $u = t^3 - 1$, then $du = 3t^2\,dt$.

We do not have $3t^2\,dt$. We only have $t^2\,dt$. We need to supply a 3 by multiplying by $\frac{1}{3} \cdot 3$ as follows.

$\frac{1}{3} \cdot 3 \int t^2(t^3-1)^7\,dt$ Multiplying by 1

$= \frac{1}{3}\int 3t^2(t^3-1)^7\,dt$

$= \frac{1}{3}\int (t^3-1)^7\,3t^2\,dt$

$= \frac{1}{3}\int u^7\,du$ Substituting u for $t^3 - 1$ and du for $3t^2\,dt$

$= \frac{1}{3} \cdot \frac{u^8}{8} + C$ Using Formula A

$= \frac{1}{24}(t^3-1)^8 + C$

22. $\int t(t^2-1)^5\,dt$

Let $u = t^2 - 1$, then $du = 2t\,dt$.

$= \frac{1}{2}\int (t^2-1)^5 \cdot 2t\,dt$

$= \frac{1}{2}\int u^5\,du$

$= \frac{1}{2} \cdot \frac{u^6}{6} + C$

$= \frac{(t^2-1)^6}{12} + C$

23. $\int x\,sin\,x^2\,dx$

Let $u = x^2$, then $du = 2x\,dx$.

$= \frac{1}{2}\int sin\,u\,du$

$= -\frac{1}{2}cos\,u + C$

$= -\frac{1}{2}cos\,x^2 + C$

24. $\int x^2\,cos\,x^3\,dx$

Let $u = x^3$, then $du = 3x^2\,dx$.

$= \frac{1}{3}\int cos\,u\,du$

$= \frac{1}{3}sin\,u + C$

$= \frac{1}{3}sin\,x^3 + C$

25. $\int (x+1)\,sec^2(x^2+2x+3)\,dx$

Let $u = x^2 + 2x + 3$, then $du = 2x + 2\,dx$.

$= \frac{1}{2}\int sec^2\,u\,du$

$= \frac{1}{2}tan(u) + C$

$= \frac{1}{2}tan(x^2+2x+3) + C$

26. $\int (x^2+2x+2)\,csc(x^3+3x^2+6x-7)\,cot(x^3+3x^2+6x-7)\,dx$

Let $u = x^3 + 3x^2 + 6x - 7$, then $du = 3x^2 + 6x + 6\,dx$.

$= \frac{1}{3}csc(u)\,cot(u)\,du$

$= -\frac{1}{3}csc\,u + C$

$= -\frac{1}{3}csc(x^3+3x^2+6x-7) + C$

27. $\int \frac{e^x\,dx}{4+e^x}$

Let $u = 4 + e^x$, then $du = e^x\,dx$.

$= \int \frac{du}{u}$ Substituting u for $4 + e^x$ and du for $e^x\,dx$

$= \int \frac{1}{u}\,du$

$= \ln u + C$ Using Formula C

$= \ln(4 + e^x) + C$

28. $\displaystyle\int \frac{e^t\,dt}{3+e^t}$

Let $u = 3 + e^t$, then $du = e^t\,dt$.

$= \displaystyle\int \frac{du}{u}$

$= \ln u + C$

$= \ln(3 + e^t) + C$

29. $\displaystyle\int \frac{\ln x^2}{x}\,dx$

Let $u = \ln x^2$, then $du = \left(2x \cdot \frac{1}{x^2}\right) dx = \frac{2}{x}\,dx$.

We do not have $\frac{2}{x}\,dx$. We only have $\frac{1}{x}\,dx$ and need to supply a 2 by multiplying by $\frac{1}{2} \cdot 2$ as follows.

$\displaystyle\frac{1}{2} \cdot 2 \int \frac{\ln x^2}{x}\,dx$ Multiplying by 1

$= \displaystyle\frac{1}{2}\int 2 \cdot \frac{\ln x^2}{x}\,dx$

$= \displaystyle\frac{1}{2}\int \ln x^2 \cdot \frac{2}{x}\,dx$

$= \displaystyle\frac{1}{2}\int u\,du$ Substituting u for $\ln x^2$ and du for $\frac{2}{x}\,dx$

$= \displaystyle\frac{1}{2} \cdot \frac{u^2}{2} + C$ Using Formula A

$= \displaystyle\frac{u^2}{4} + C$

$= \displaystyle\frac{1}{4}(\ln x^2)^2 + C,$

or $\displaystyle\frac{1}{4}(2\ln x)^2 + C = \frac{1}{4} \cdot 4(\ln x)^2 + C$

$= (\ln x)^2 + C$

30. $\displaystyle\int \frac{(\ln x)^2}{x}\,dx$

Let $u = \ln x$, then $du = \frac{1}{x}\,dx$.

$= \int u^2\,du$

$= \displaystyle\frac{u^3}{3} + C$

$= \displaystyle\frac{(\ln x)^3}{3} + C$

31. $\displaystyle\int \frac{dx}{x\ln x}$

Let $u = \ln x$, then $du = \frac{1}{x}\,dx$.

$= \displaystyle\int \frac{1}{\ln x}\left(\frac{1}{x}\,dx\right)$

$= \displaystyle\int \frac{1}{u}\,du$ Substituting u for $\ln x$ and du for $\frac{1}{x}\,dx$

$= \ln u + C$ Using Formula C

$= \ln(\ln x) + C$

32. $\displaystyle\int \frac{dx}{x\ln x^2}$

Let $u = \ln x^2$, then $du = \frac{2}{x}\,dx$.

$= \displaystyle\frac{1}{2}\int \frac{1}{\ln x^2} \cdot \frac{2}{x}\,dx$

$= \displaystyle\frac{1}{2}\int \frac{1}{u}\,du$

$= \displaystyle\frac{1}{2}\ln u + C$

$= \displaystyle\frac{1}{2}\ln(\ln x^2) + C$

33. $\int \sqrt{ax+b}\,dx$, or $\int (ax+b)^{1/2}\,dx$

Let $u = ax + b$, then $du = a\,dx$.

We do not have $a\,dx$. We only have dx and need to supply an a by multiplying by $\frac{1}{a} \cdot a$ as follows.

$\displaystyle\frac{1}{a} \cdot a \int \sqrt{ax+b}\,dx$ Multiplying by 1

$= \displaystyle\frac{1}{a}\int a\sqrt{ax+b}\,dx$

$= \displaystyle\frac{1}{a}\int \sqrt{ax+b}\,(a\,dx)$

$= \displaystyle\frac{1}{a}\int \sqrt{u}\,du$ Substituting u for $ax + b$ and du for $a\,dx$

$= \displaystyle\frac{1}{a}\int u^{1/2}\,du$

$= \displaystyle\frac{1}{a} \cdot \frac{u^{3/2}}{\frac{3}{2}} + C$ Using Formula A

$= \displaystyle\frac{2}{3a} \cdot u^{3/2} + C$

$= \displaystyle\frac{2}{3a}(ax+b)^{3/2} + C$

34. $\int x\sqrt{ax^2+b}\,dx$

Let $u = ax^2 + b$, then $du = 2ax\,dx$.

$= \displaystyle\frac{1}{2a}\int \sqrt{ax^2+b} \cdot 2ax\,dx$

$= \displaystyle\frac{1}{2a}\int u^{1/2}\,du$

$= \displaystyle\frac{1}{2a} \cdot \frac{u^{3/2}}{\frac{3}{2}} + C$

$= \displaystyle\frac{1}{3a}(ax^2+b)^{3/2} + C$

35. $\int b\,e^{ax}\,dx$

$= b\int e^{ax}\,dx$

Let $u = ax$, then $du = a\,dx$.

We do not have $a\,dx$. We only have dx and need to supply an a by multiplying by $\frac{1}{a} \cdot a$ as follows.

$= b \cdot \frac{1}{a} \cdot a \int e^{ax}\, dx$

$= \frac{b}{a} \int a\, e^{ax}\, dx$

$= \frac{b}{a} \int e^{ax}\, (a\, dx)$

$= \frac{b}{a} \int e^{u}\, du$ Substituting u for ax and du for $a\, dx$

$= \frac{b}{a} e^{u} + C$ Using Formula B

$= \frac{b}{a} e^{ax} + C$

36. $\int P_0\, e^{kt}\, dt$

Let $u = kt$, then $du = k\, dt$.

$= \frac{1}{k} \int P_0\, e^{kt} \cdot k\, dt$

$= \frac{1}{k} \int P_0\, e^{u}\, du$

$= \frac{P_0}{k} e^{u} + C$

$= \frac{P_0}{k} e^{kt} + C$

37. $\int a\, sin(bx + c)\, dx$

Let $u = bx + c$, then $du = b\, dx$.

$= \frac{1}{b} \int sin\ u\, du$

$= -\frac{1}{b}\, cos(u) + D$

$= -\frac{1}{b}\, cos(bx + c) + D$

38. $\int a\, cos(bx + c)\, dx$

Let $u = bx + c$, then $du = b\, dx$.

$= \frac{1}{b} \int cos\ u\, du$

$= \frac{1}{b}\, sin(u) + D$

$= \frac{1}{b}\, sin(bx + c) + D$

39. $\int \frac{3x^2\, dx}{(1 + x^3)^5}$

Let $u = 1 + x^3$, then $du = 3x^2\, dx$.

$= \int \frac{1}{(1 + x^3)^5} \cdot 3x^2\, dx$

$= \int \frac{1}{u^5}\, du$ Substituting u for $1 + x^3$ and du for $3x^2\, dx$

$= \int u^{-5}\, du$

$= \frac{u^{-4}}{-4} + C$ Using Formula A

$= -\frac{1}{4u^4} + C$

$= -\frac{1}{4(1 + x^3)^4} + C$

40. $\int \frac{x^3\, dx}{(2 - x^4)^7}$

Let $u = 2 - x^4$, then $du = -4x^3\, dx$.

$= -\frac{1}{4} \int \frac{-4x^3\, dx}{(2 - x^4)^7}$

$= -\frac{1}{4} \int \frac{du}{u^7}$

$= -\frac{1}{4} \int u^{-7}\, du$

$= -\frac{1}{4} \cdot \frac{u^{-6}}{-6} + C$

$= \frac{1}{24\, u^6} + C$

$= \frac{1}{24(2 - x^4)^6} + C$

41. $\int cot\ x\, dx = \int \frac{cos\ x}{sin\ x}\, dx$

Let $u = sin\ x$, then $du = cos\ x\, dx$.

$= \int \frac{du}{u}$

$= \ln u + C$

$= \ln\, (sin\ x) + C$

42. $\int 12x \sqrt[5]{1 + 6x^2}\, dx$

Let $u = 1 + 6x^2$, then $du = 12x\, dx$.

$= \int u^{1/5}\, du$

$= \frac{5}{6} u^{6/5} + C$

$= \frac{5}{6} (1 + 6x^2)^{6/5} + C$

43. $\int 5t\sqrt{1 - 4t^2}\, dt$

Let $u = 1 - 4t^2$, then $du = -8t\, dt$.

$= -\frac{5}{8} \int u^{1/2}\, du$

$= -\frac{5}{12} u^{3/2} + C$

$= -\frac{5}{12} (1 - 4t^2)^{3/2} + C$

44. $\int e^t\ sin\ e^t\, dt$

Let $u = e^t$, then $du = e^t\, dt$.

$= \int sin\ u\, du$

$= -cos\ u + C$

$= -cos\ e^t + C$

45. $\int e^t\ sin\ e^t\, dt$

Let $u = \sqrt{w}$, then $du = \frac{dw}{2\sqrt{w}}$.

$= \frac{1}{2} \int e^{u}\, du$

$= \frac{1}{2} e^{u} + C$

$= \frac{1}{2} e^{\sqrt{w}} + C$

46. $\int \frac{x^2}{e^{x^3}}\, dx$

Let $u = x^3$, then $du = 3x^2\, dx$.

$= \frac{1}{3}\int e^{-u}\, du$

$= -\frac{1}{3}e^{-u} + C$

$= -\frac{1}{3}e^{-x^3} + C$

$= -\frac{1}{3e^{x^3}}$

47. $\int sin^2 x\ cos\ x\, dx$

Let $u = sin\ x$, then $du = cos\ x\, dx$.

$= \int u^2\, du$

$= \frac{1}{3}u^3 + C$

$= \frac{1}{3}sin^3 x + C$

48. $\int cos^2 x\ sin\ x\, dx$

Let $u = cos\ x$, then $du = -sin\ x\, dx$.

$= -\int u^2\, du$

$= -\frac{1}{3}u^3 + C$

$= -\frac{1}{3}cos^3 x + C$

49. $\int e^t\ sin\ e^t\, dt$

Let $u = -sin\ t$, then $du = -cos\ t\, dt$.

$= -\int e^u\, du$

$= -e^u + C$

$= -e^{sin\ t} + C$

50. $\int \frac{sin\ t}{e^{cos\ t}}\, dt$

Let $u = cos\ t$, then $du = -sin\ t\, dt$.

$= -\int e^{-u}\, du$

$= e^{-u} + C$

$= \frac{1}{e^{cos\ t}} + C$

51. $\int r^2\ sin(3r^3 + 7)\, dr$

Let $u = 3r^3$, then $du = 9r^2\, dr$.

$= \frac{1}{9}\int sin\ u\, du$

$= -\frac{1}{9}cos\ u + C$

$= -\frac{1}{9}cos(3r^3 + 7) + C$

52. $\frac{1}{x}\ sec^2\ \ln\ x\, dx$

Let $u = \ln\ x$, then $du = \frac{dx}{x}$.

$= \int sec^2 u\, du$

$= tan\ u + C$

$= tan\ \ln\ x + C$

53. $\int_0^1 2x\, e^{x^2}\, dx$

First find the indefinite integral.

$\int 2x\, e^{x^2}\, dx$

Let $u = x^2$, then $du = 2x\, dx$.

$= \int e^{x^2}(2x\, dx)$

$= \int e^u\, du$ — Substituting u for x^2 and du for $2x\, dx$

$= e^u + C$

$= e^{x^2} + C$

Then evaluate the definite integral on $[0, 1]$.

$\int_0^1 2x\, e^{x^2}\, dx$

$= \left[e^{x^2}\right]_0^1$

$= e^{1^2} - e^{0^2}$

$= e - 1$

54. $\int_0^1 3x^2\, e^{x^3}\, dx$

First find the indefinite integral.

$\int 3x^2\, e^{x^3}\, dx$

Let $u = x^3$, then $du = 3x^2\, dx$.

$= \int e^u\, du$

$= e^u + C$

$= e^{x^3} + C$

Then evaluate the definite integral on $[0, 1]$.

$\int_0^1 3x^2\, e^{x^3}\, dx$

$= \left[e^{x^3}\right]_0^1$

$= e^{1^3} - e^{0^3}$

$= e - 1$

55. $\int_0^1 x(x^2 + 1)^5\, dx$

First find the indefinite integral.

$\int x(x^2 + 1)^5\, dx$

Let $u = x^2 + 1$, then $du = 2x\, dx$.

We only have $x\, dx$ and need to supply a 2 by multiplying by $\frac{1}{2} \cdot 2$.

$\frac{1}{2} \cdot 2\int x(x^2 + 1)^5\, dx$ — Multiplying by 1

$= \frac{1}{2}\int 2x(x^2 + 1)^5\, dx$

$= \frac{1}{2}\int (x^2 + 1)^5 \cdot 2x\, dx$

$= \frac{1}{2}\int u^5\, du$ — Substituting u for $x^2 + 1$ and du for $2x\, dx$

$= \frac{1}{2} \cdot \frac{u^6}{6} + C$ Using Formula A

$= \frac{(x^2+1)^6}{12} + C$

Then evaluate the definite integral on $[0, 1]$.

$\int_0^1 x(x^2+1)^5\, dx$

$= \left[\frac{(x^2+1)^6}{12}\right]_0^1$

$= \frac{(1^2+1)^6}{12} - \frac{(0^2+1)^6}{12}$

$= \frac{64}{12} - \frac{1}{12}$

$= \frac{63}{12}$

$= \frac{21}{4}$

56. $\int_1^2 x(x^2-1)^7\, dx$

First find the indefinite integral.

$\int x(x^2-1)^7\, dx$

Let $u = x^2 - 1$, then $du = 2x\, dx$.

$= \frac{1}{2}\int (x^2-1)^7 \cdot 2x\, dx$

$= \frac{1}{2}\int u^7\, du$

$= \frac{1}{2} \cdot \frac{u^8}{8} + C$

$= \frac{(x^2-1)^8}{16} + C$

Then evaluate the definite integral on $[1, 2]$.

$\int_1^2 x(x^2-1)^7\, dx$

$= \left[\frac{(x^2-1)^8}{16}\right]_1^2$

$= \left[\frac{(2^2-1)^8}{16}\right] - \left[\frac{(1^2-1)^8}{16}\right]$

$= \frac{3^8}{16} = \frac{6561}{16}$

57. $\int_0^4 \frac{dt}{1+t}$

First find the indefinite integral.

$\int \frac{dt}{1+t}$

Let $u = 1 + t$, then $du = dt$.

$= \int \frac{du}{u}$ Substituting u for $1+t$ and du for dt

$= \int \frac{1}{u}\, du$

$= \ln u + C$ Using Formula C

$= \ln(1+t) + C$

Then evaluate the definite integral on $[0, 4]$.

$\int_0^4 \frac{dt}{1+t}$

$= [\ln(1+t)]_0^4$

$= \ln(1+4) - \ln(1+0)$

$= \ln 5 - \ln 1$

$= \ln 5 - 0$

$= \ln 5$

58. $\int_1^3 e^{2x}\, dx$

First find the indefinite integral.

$\int e^{2x}\, dx$

Let $u = 2x$, then $du = 2\, dx$.

$= \frac{1}{2}\int e^{2x} \cdot 2\, dx$

$= \frac{1}{2}\int e^u\, du$

$= \frac{1}{2}e^u + C$

$= \frac{1}{2}e^{4x} + C$

Then evaluate the definite integral on $[1, 3]$.

$\int_1^3 e^{2x}\, dx$

$= \left[\frac{1}{2}e^{2x}\right]_1^3$

$= \frac{1}{2}e^{2\cdot 2} - \frac{1}{2}e^{2\cdot 1}$

$= \frac{1}{2}e^4 - \frac{1}{2}e^1$, or $\frac{e^4 - e^1}{2}$

59. $\int_1^4 \frac{2x+1}{x^2+x-1}\, dx$

First find the indefinite integral.

$\int \frac{2x+1}{x^2+x-1}\, dx$

Let $u = x^2 + x = 1$, then $du = (2x+1)\, dx$.

$= \int \frac{1}{x^2+x-1}(2x+1)\, dx$

$= \int \frac{1}{u}\, du$ Substituting u for x^2+x-1 and du for $(2x+1)\, dx$

$= \ln u + C$

$= \ln(x^2+x-1) + C$

Then evaluate the definite integral on $[1, 4]$.

$\int_1^4 \frac{2x+1}{x^2+x-1}\, dx$

$= [\ln(x^2+x-1)]_1^4$

$= \ln(4^4+4-1) - \ln(1^2+1-1)$

$= \ln 19 - \ln 1$

$= \ln 19$ $(\ln 1 = 0)$

60. $\displaystyle\int_1^3 \frac{2x+3}{x^2+3x}\,dx$

First find the indefinite integral.

$\displaystyle\int \frac{2x+3}{x^2+3x}\,dx$

Let $u = x^2+3x$, then $du = 2x+3$.

$\displaystyle = \int \frac{du}{u}$

$= \ln u + C$

$= \ln(x^2+3x) + C$

Then evaluate the definite integral on $[1,3]$.

$\displaystyle\int_1^3 \frac{2x+3}{x^2+3x}\,dx$

$= \left[\ln(x^2+3x)\right]_1^3$

$= \ln(3^2+3\cdot 3) - \ln(1^2+3\cdot 1)$

$= \ln 18 - \ln 4$

$\displaystyle = \ln\frac{18}{4}$

$\displaystyle = \ln\frac{9}{2}$

61. $\int_0^b e^{-x}\,dx$

First find the indefinite integral.

$\int e^{-x}\,dx$

Let $u = -x$, then $du = -dx$.

We only have dx and need to supply a -1 by multiplying by $-1\cdot(-1)$.

$-1\cdot(-1)\int e^{-x}\,dx$

$= -\int -e^{-x}\,dx$

$= -\int e^{-x}\,(-dx)$

$= -\int e^u\,du$ Substituting u for $-x$ and du for $-dx$

$= -e^u + C$ Using Formula B

$= -e^{-x} + C$

Then evaluate the definite integral on $[0,b]$.

$\int_0^b e^{-x}\,dx$

$= \left[-e^{-x}\right]_0^b$

$= (-e^{-b}) - (-e^{-0})$

$= -e^{-b} + e^0$

$= -e^{-b} + 1$

$\displaystyle = 1 - \frac{1}{e^b}$

62. $\int_0^b 2e^{-2x}\,dx$

First find the indefinite integral.

$\int 2e^{-2x}\,dx$

Let $u = -2x$, then $du = -2\,dx$.

$= -\int e^{-2x}(-2\,dx)$

$= -\int e^u\,du$

$= -e^u + C$

$= -e^{-2x} + C$

Then evaluate the definite integral on $[0,b]$.

$\int_0^b 2e^{-2x}\,dx$

$= \left[-e^{-2x}\right]_0^b$

$= -e^{-2b} - (-e^{-2\cdot 0})$

$\displaystyle = -e^{-2b} + 1, \text{ or } 1 - \frac{1}{e^{2b}}$

63. $\int_0^b m\,e^{-mx}\,dx$

First find the indefinite integral.

$\int m\,e^{-mx}\,dx$ m is a constant

Let $u = -mx$, then $du = -m\,dx$.

We only have $m\,dx$ and need to supply a -1 by multiplying by $-1\cdot(-1)$.

$-1\cdot(-1)\int m\,e^{-mx}\,dx$

$= -\int -m\,e^{-mx}\,dx$

$= -\int e^{-mx}\,(-m\,dx)$

$= -\int e^u\,du$ Substituting u for $-mx$ and du for $-m\,dx$

$= -e^u + C$ Using Formula B

$= -e^{-mx} + C$

Then evaluate the definite integral on $[0,b]$.

$\int_0^b me^{-mx}\,dx$

$= \left[-e^{-mx}\right]_0^b$

$= (-e^{-mb}) - (-e^{-m\cdot 0})$

$= -e^{-mb} + 1$ $(e^{-m\cdot 0} = e^0 = 1)$

$= 1 - e^{-mb}$

$\displaystyle = 1 - \frac{1}{e^{mb}}$

64. $\int_0^b k\,e^{-kx}\,dx$

First find the indefinite integral.

$\int k\,e^{-kx}\,dx$

Let $u = -kx$, then $du = -k\,dx$.

$= -\int e^{-kx}\,(-k\,dx)$

$= -\int e^u\,du$

$= -e^u + C$

$= -e^{-kx} + C$

Then evaluate the definite integral on $[0,b]$.

$\int_0^b k\,e^{-kx}\,dx$

$= \left[-e^{-kx}\right]_0^b$

$= -e^{-kb} - (-e^{-k\cdot 0})$

$\displaystyle = -e^{-kb} + 1, \text{ or } 1 - \frac{1}{e^{kb}}$

65. $\int_0^4 (x-6)^2\, dx$

First find the indefinite integral.

$\int (x-6)^2\, dx$

Let $u = x-6$, then $du = dx$.

$= \int u^2\, du$ Substituting u for $x-6$ and du for dx

$= \frac{u^3}{3} + C$

$= \frac{(x-6)^3}{3} + C$

Then evaluate the definite integral on $[0, 4]$.

$\int_0^4 (x-6)^2\, dx$

$= \left[\frac{(x-6)^3}{3}\right]_0^4$

$= \frac{(4-6)^3}{3} - \frac{(0-6)^3}{3}$

$= -\frac{8}{3} - \left(-\frac{216}{3}\right)$

$= -\frac{8}{3} + \frac{216}{3}$

$= \frac{208}{3}$

66. $\int_0^3 (x-5)^2\, dx$

First find the indefinite integral.

$\int (x-5)^2\, dx$

Let $u = x-5$, then $du = dx$.

$= \int u^2\, du$

$= \frac{u^3}{3} + C$

$= \frac{(x-5)^3}{3} + C$

Then evaluate the definite integral on $[0, 3]$.

$\int_0^3 (x-5)^2\, dx$

$= \left[\frac{(x-5)^3}{3}\right]_0^3$

$= \frac{(3-5)^3}{3} - \frac{(0-5)^3}{3}$

$= -\frac{8}{3} + \frac{125}{3}$

$= \frac{117}{3} = 39$

67. $\int_{-1/3}^0 \cos(\pi x + \pi/3)\, dx$

First find the indefinite integral.

$\int \cos(\pi x + \pi/3)^2\, dx$

Let $u = \pi x + \pi/3$, then $du = \pi dx$.

$= \int \frac{1}{\pi} \cos u\, du$ Substituting u for $\pi x + \pi/3$ and $\frac{1}{\pi} du$ for dx

$= \frac{1}{\pi} \sin u + C$

$= \frac{1}{\pi} \sin(\pi x + \pi/3) + C$

Then evaluate the definite integral on $[-1/3, 0]$.

$\int_{-1/3}^0 \cos(\pi x + \pi/3)\, dx$

$= \left[\frac{1}{\pi} \sin(\pi x + \pi/3)\right]_{-1/3}^0$

$= \frac{\sqrt{3}}{2\pi} - 0$

$= \frac{\sqrt{3}}{2\pi}$

68. $\int_0^{\sqrt{\pi/6}} 2x \sin 3x^2\, dx$

First find the indefinite integral.

$\int 2x \sin 3x^2\, dx$

Let $u = 3x^2$, then $du = 6x\, dx$.

$= \int \frac{1}{3} \sin u\, du$ Substituting u for $3x^2$ and $\frac{1}{3} du$ for dx

$= -\frac{1}{3} \cos u + C$

$= -\frac{1}{\pi} \cos 3x^2 + C$

Then evaluate the definite integral on $[0, \sqrt{\pi/6}]$.

$\int_0^{\sqrt{\pi/6}} 2x \sin 3x^2\, dx$

$= \left[-\frac{1}{3} \cos 3x^2\right]_0^{\sqrt{\pi/6}}$

$= 0 - (-\frac{1}{3})$

$= \frac{1}{3}$

69. $\displaystyle\int_0^2 \frac{3x^2\,dx}{(1+x^3)^5}$

From Exercise 39 we know that the indefinite integral is

$$\int \frac{3x^2\,dx}{(1+x^3)^5} = -\frac{1}{4(1+x^3)^4} + C.$$

Now we evaluate the definite integral on $[0, 2]$.

$$\int_0^2 \frac{3x^2\,dx}{(1+x^3)^5}$$
$$= \left[-\frac{1}{4(1+x^3)^4}\right]_0^2$$
$$= \left[-\frac{1}{4(1+2^3)^4}\right] - \left[-\frac{1}{4(1+0^3)^4}\right]$$
$$= -\frac{1}{26,244} + \frac{1}{4}$$
$$= \frac{6560}{26,244}$$
$$= \frac{1640}{6561}$$

70. $\displaystyle\int_{-1}^0 \frac{x^3\,dx}{(2-x^4)^7}$

From Exercise 40 we know that

$$\int \frac{x^3\,dx}{(2-x^4)^7} = \frac{1}{24(2-x^4)^6} + C.$$

Then $\displaystyle\int_{-1}^0 \frac{x^3\,dx}{(2-x^4)^7}$

$$= \left[\frac{1}{24(2-x^4)^6}\right]_{-1}^0$$
$$= \frac{1}{24(2-0^4)^6} - \frac{1}{24[2-(-1)^4]^6}$$
$$= \frac{1}{1536} - \frac{1}{24}$$
$$= -\frac{63}{1536} = -\frac{21}{512}$$

71. $\int_0^{\sqrt{7}} 7x\sqrt[3]{1+x^2}\,dx = 7\int_0^{\sqrt{7}} x\sqrt[3]{1+x^2}\,dx$

First find the indefinite integral.

$7\int x\sqrt[3]{1+x^2}\,dx$

Let $u = 1+x^2$, then $du = 2x\,dx$.

$$= 7\cdot\frac{1}{2}\cdot 2\int x\sqrt[3]{1+x^2}\,dx$$
$$= 7\cdot\frac{1}{2}\int 2x\sqrt[3]{1+x^2}\,dx$$
$$= \frac{7}{2}\int \sqrt[3]{1+x^2}\,(2x\,dx)$$
$$= \frac{7}{2}\int \sqrt[3]{u}\,du$$
$$= \frac{7}{2}\int u^{1/3}\,du$$
$$= \frac{7}{2}\cdot\frac{u^{4/3}}{4/3} + C$$
$$= \frac{21}{8}u^{4/3} + C$$
$$= \frac{21}{8}(1+x^2)^{4/3} + C$$

Then evaluate the definite integral on $[0, \sqrt{7}]$.

$$7\int_0^{\sqrt{7}} x\sqrt[3]{1+x^2}\,dx$$
$$= \left[\frac{21}{8}(1+x^2)^{4/3}\right]_0^{\sqrt{7}}$$
$$= \frac{21}{8}(1+(\sqrt{7})^2)^{4/3} - \frac{21}{8}(1+0^2)^{4/3}$$
$$= \frac{21}{8}\cdot 8^{4/3} - \frac{21}{8}\cdot 1^{4/3}$$
$$= \frac{21}{8}\cdot 16 - \frac{21}{8}\cdot 1$$
$$= 42 - \frac{21}{8}$$
$$= \frac{315}{8}$$

72. $\int_0^1 12x\sqrt[5]{1-x^2}\,dx$

First find the indefinite integral.

$\int 12x\sqrt[5]{1-x^2}\,dx$

Let $u = 1-x^2$, then $du = -2x\,dx$.

$$= 12\left(-\frac{1}{2}\right)\int \sqrt[5]{1-x^2}\,(-2x\,dx)$$
$$= -6\int \sqrt[5]{u}\,du$$
$$= -6\int u^{1/5}\,du$$
$$= -6\frac{u^{6/5}}{6/5} + C$$
$$= -5\,u^{6/5} + C$$
$$= -5(1-x^2)^{6/5} + C$$

Then evaluate the definite integral on $[0, 1]$.

$$\int_0^1 12x\sqrt[5]{1-x^2}\,dx$$
$$= \left[-5(1-x^2)^{6/5}\right]_0^1$$
$$= -5(1-1^2)^{6/5} - \left[-5(1-0^2)^{6/5}\right]$$
$$= 5$$

73. **a)** $\displaystyle K\int_0^H (H-x)^{3/2}\,dx$

$$= \left[-\frac{2K}{5}(H-x)^{5/2}\right]_0^H$$
$$= 0 - \frac{-2K}{5}H^{5/2}$$
$$= \frac{2}{5}KH^{5/2}$$

b) $\displaystyle K\int_0^{H/2} (H-x)^{3/2}\,dx$

$$= \left[-\frac{2K}{5}(H-x)^{5/2}\right]_0^{H/2}$$
$$= \frac{-2K}{5}(H/2)^{5/2} - \frac{-2K}{5}H^{5/2}$$
$$= \frac{-\sqrt{2}KH^{5/2}}{20} + \frac{2KH^{5/2}}{5}$$
$$= \frac{1}{20}(8-\sqrt{2})KH^{5/2}$$

c) We divide the answers from the previous two parts

$$\frac{\frac{1}{20}(8-\sqrt{2})KH^{5/2}}{\frac{2}{5}KH^{5/2}} = \frac{(8-\sqrt{2})}{20}\cdot\frac{5}{2} = \frac{(8-\sqrt{2})}{8} = 0.8232$$

d) Upper half proportion is given by

$$1-\frac{(8-\sqrt{2})}{8} = \frac{\sqrt{2}}{8} = 0.1768$$

74. a) First find the indefinite integral.

$\int 100,000\, e^{0.025t}\, dt$

Let $u = 0.025t$, then $du = 0.025\, dt$.

$$= 100,000\left(\frac{1}{0.025}\right)\int e^{0.025t}\,(0.025)\,dt$$
$$= 4,000,000 \int e^u\, du$$
$$= 4,000,000\, e^u + C$$
$$= 4,000,000\, e^{0.025t} + C$$

Then evaluate the definite integral on $[0, 99]$.

$$4,000,000\left[e^{0.025t}\right]_0^{99}$$
$$= 4,000,000(e^{0.025(99)} - e^{0.025(0)})$$
$$= 4,000,000(e^{2.475} - e^0)$$
$$= 4,000,000(e^{2.475} - 1)$$
$$\approx 43,526,828$$

b) From part a) we know that

$\int D(t)\, dt = 4,000,000\, e^{0.025t} + C$

Now evaluate the definite integral on $[80, 99]$.

$$4,000,000\left[e^{0.025t}\right]_{80}^{99}$$
$$= 4,000,000(e^{0.025(99)} - e^{0.025(80)})$$
$$= 4,000,000(e^{2.475} - e^2)$$
$$\approx 17,970,604$$

75. We have to break the integral into two parts

$$\int_{-2}^{0}[-x\sqrt{4-x^2} - 0]\,dx + \int_0^2 [0-(-x\sqrt{4-x^2})]\,dx$$
$$= \left[\frac{1}{3}(4-x^2)^{3/2}\right]_{-2}^{0} + \left[-\frac{1}{3}(4-x^2)^{3/2}\right]_0^2$$
$$= \frac{8}{3} - 0 + 0 - \frac{-8}{3}$$
$$= \frac{16}{3}$$

76. We have to break the integral into two parts

$$\int_{-4}^{0}[0-(x\sqrt{16-x^2})]\,dx + \int_0^4 [x\sqrt{16-x^2} - 0]\,dx$$
$$= \left[\frac{1}{3}(16-x^2)^{3/2}\right]_{-4}^{0} + \left[-\frac{1}{3}(16-x^2)^{3/2}\right]_0^4$$
$$= \frac{64}{3} - 0 + 0 - \frac{-64}{3}$$
$$= \frac{128}{3}$$

77. Using the hint

$$\int \frac{t^2+2t}{(t+1)^2}\,dt$$
$$= \int 1 - \frac{1}{(t+1)^2}\,dt$$
$$= \int 1-(t+1)^{-2}\,dt$$
$$= t + \frac{1}{(t+1)} + C$$

78. Using the hint from the previous problem

$$\int \frac{x^2+6x}{(x+3)^2}\,dx$$
$$= \int 1 - \frac{9}{(x+3)^2}\,dx$$
$$= \int 1-(x+3)^{-2}\,dx$$
$$= x + \frac{1}{(x+3)} + C$$

79. Using the hint

$$\int \frac{x+3}{x+1}\,dx$$
$$= \int 1 + \frac{2}{(x+1)}\,dx$$
$$= x + 2\ln|x+1| + C$$

80. Using the hint from the previous problem

$$\int \frac{t-5}{t-4}\,dt$$
$$= \int 1 + \frac{-1}{t-4}\,dt$$
$$= t - \ln|t-4| + C$$

81. Let $u = \ln x$ then $du = \dfrac{dx}{x}$

$$\int \frac{dx}{x(\ln x)^n}$$
$$= \int u^{-n}\,du$$
$$= \frac{u^{-n+1}}{-n+1} + C$$
$$= \frac{(\ln x)^{-n+1}}{1-n} + C$$
$$= \frac{1}{(1-n)\,(\ln x)^{n-1}} + C$$

82. Using the hint

$$\int \frac{dx}{e^x+1}$$
$$= \int \frac{e^{-x}}{1+e^{-x}}\,dx$$
$$= \ln(1+e^{-x}) + C$$

83. Let $u = \ln(\ln x)$ then $du = \dfrac{dx}{x\ \ln x}$

$$\int \frac{dx}{x\ \ln x\,[\ln(\ln x)]}$$
$$= \int \frac{du}{u}$$
$$= \ln u + C$$
$$= \ln|\ln(\ln x)| + C$$

84. Let $u = 1 + ae^{-mx}$ then $du = -am\ e^{-mx}\ dx$

$$\int \frac{e^{-mx}}{1+ae^{-mx}}\,dx$$
$$= -\frac{1}{am}\int \frac{du}{u}$$
$$= -\frac{1}{am}\ln u + C$$
$$= -\frac{1}{am}\ln|1+ae^{-mx}| + C$$

85. Using the hint

$$\int sec\ x\,dx$$
$$= \int \frac{sec\ x\ tan\ x + sec^2 x}{tan\ x + sec\ x}\,dx$$

Let $u = tan\ x + sec\ x$ then $du = sec^2 x + sec\ x\ tan\ x\ dx$

$$= \int \frac{du}{u}$$
$$= \ln u + C$$
$$= \ln|\,sec\ x + tan\ x\,| + C$$

86. Using the hint from the previous problem

$$\int csc\ x\,dx$$
$$= \int \frac{csc\ x\ cot\ x + csc^2 x}{cot\ x + csc\ x}\,dx$$

Let $u = cot\ x + csc\ x$ then $du = -(csc^2 x + cot\ x\ csc\ x)\,dx$

$$= -\int \frac{du}{u}$$
$$= -\ln u + C$$
$$= -\ln|\,csc\ x + cot\ x\,| + C$$

87. Left to the student

Exercise Set 5.6

1. $\int 5x\,e^{5x}\,dx = \int x(5e^{5x}\,dx)$

Let

$u = x$ and $dv = 5e^{5x}\,dx$.

Then $du = dx$ and $v = e^{5x}$.

$$\underset{u}{\int x}(\underset{dv}{5e^{5x}\,dx}) = \underset{u}{x}\cdot \underset{v}{e^{5x}} - \int \underset{v}{e^{5x}}\cdot \underset{du}{dx}$$

Using Theorem 7:
$\int u\,dv = uv - \int v\,du$

$$= xe^{5x} - \frac{1}{5}e^{5x} + C$$

2. $\int 2x\,e^{2x}\,dx$

Let

$u = x$ and $dv = 2e^{2x}\,dx$.

Then

$du = dx$ and $v = e^{2x}$.

$$\int x(2e^{2x}\,dx) = xe^{2x} - \int e^{2x}\,dx$$
$$= xe^{2x} - \frac{1}{2}e^{2x} + C$$

3. $\int x\ sin\ x\,dx$

Let

$u = x$ and $dv = sin\ x\,dx$.

Then $du = dx$ and $v = -cos\ x$

$$\begin{aligned}\int x\ sin\ x\,dx &= x\cdot -cos\ x + \int cos\ x\,dx\\ &= -x\ cos\ x + sin\ x + C\\ &= sin\ x - x\ cos\ x + C\end{aligned}$$

4. $\int x\ cos\ x\,dx$

Let

$u = x$ and $dv = cos\ x\,dx$.

Then $du = dx$ and $v = sin\ x$

$$\begin{aligned}\int x\ cos\ x\,dx &= x\cdot sin\ x - \int sin\ x\,dx\\ &= x\ sin\ x - cos\ x + C\end{aligned}$$

5. $\int xe^{2x}\,dx$

Let

$u = x$ and $dv = e^{2x}\,dx$.

Then $du = dx$ and $v = \frac{1}{2}e^{2x}$

$$\begin{aligned}\int xe^{2x}\,dx &= x\cdot\frac{1}{2}e^{2x} - \int \frac{1}{2}e^{2x}dx\\ &= \frac{1}{2}xe^{2x} - \frac{1}{4}e^{2x}\end{aligned}$$

6. $\int xe^{3x}\,dx$

Let

$u = x$ and $dv = e^{3x}\,dx$.

Then $du = dx$ and $v = \frac{1}{3}e^{2x}$

$$\begin{aligned}\int xe^{3x}\,dx &= x\cdot\frac{1}{3}e^{3x} - \int \frac{1}{3}e^{3x}dx\\ &= \frac{1}{3}xe^{3x} - \frac{1}{9}e^{3x}\end{aligned}$$

7. $\int xe^{-2x}\,dx$

Let

$u = x$ and $dv = e^{-2x}\,dx$.

Then $du = dx$ and $v = -\frac{1}{2}e^{-2x}$.

$$\int x\,e^{-2x}\,dx = x\cdot\left(-\frac{1}{2}e^{-2x}\right) - \int\left(-\frac{1}{2}e^{-2x}\right)dx$$

$$= -\frac{1}{2}xe^{-2x} - \frac{-\frac{1}{2}}{-2}e^{-2x} + C$$

$$\left(\int be^{ax}\,dx = \frac{b}{a}e^{ax} + C\right)$$

$$= -\frac{1}{2}xe^{-2x} - \frac{1}{4}e^{-2x} + C$$

8. $\int xe^{-x}\,dx$

Let $u = x$ and $dv = e^{-x}\,dx$.

Then

$du = dx$ and $v = -e^{-x}$.

$$\int xe^{-x}\,dx = -xe^{-x} - \int -e^{-x}\,dx$$

$$= -xe^{-x} - e^{-x} + C$$

9. $\int x^2 \ln x\,dx = \int (\ln x)\,x^2\,dx$

Let

$u = \ln x$ and $dv = x^2\,dx$.

Then $du = \frac{1}{x}\,dx$ and $v = \frac{x^3}{3}$.

$$\underset{u}{\int (\ln x)}\ \underset{dv}{x^2\,dx} = \underset{u}{\ln x} \cdot \underset{v}{\frac{x^3}{3}} - \int \underset{v}{\frac{x^3}{3}} \cdot \underset{du}{\frac{1}{x}\,dx}$$

Integration by Parts

$$= \frac{x^3}{3}\ln x - \frac{1}{3}\int x^2\,dx$$

$$= \frac{x^3}{3}\ln x - \frac{1}{3}\cdot\frac{x^3}{3} + C$$

$$= \frac{x^3}{3}\ln x - \frac{x^3}{9} + C$$

10. $\int x^3 \ln x\,dx$

Let $u = \ln x$ and $dv = x^3\,dx$.

Then

$du = \frac{1}{x}\,dx$ and $v = \frac{x^4}{4}$.

$$\int x^3 \ln x\,dx = \frac{x^4 \ln x}{4} - \int \frac{x^4}{4}\cdot\frac{1}{x}\,dx$$

$$= \frac{x^4 \ln x}{4} - \int \frac{x^3}{4}\,dx$$

$$= \frac{x^4 \ln x}{4} - \frac{x^4}{16} + C$$

11. $\int x \ln x^2\,dx$

Let

$u = \ln x^2$ and $dv = x\,dx$.

Then $du = \frac{1}{x^2}\cdot 2x\,dx = \frac{2}{x}\,dx$ and $v = \frac{x^2}{2}$.

$$\int (\ln x^2)\,x\,dx = (\ln x^2)\cdot\frac{x^2}{2} - \int \frac{x^2}{2}\cdot\frac{2}{x}\,dx$$

$$= \frac{x^2}{2}\ln x^2 - \int x\,dx$$

$$= \frac{x^2}{2}\ln x^2 - \frac{x^2}{2} + C$$

12. $\int x^2 \ln x^3\,dx$

Let $u = \ln x^3$ and $dv = x^2\,dx$.

Then

$du = \frac{3}{x}\,dx$ and $v = \frac{x^3}{3}$.

$$\int x^2 \ln x^3\,dx = \frac{x^3 \ln x^3}{3} - \int \frac{x^3}{3}\cdot\frac{3}{x}\,dx$$

$$= \frac{x^3 \ln x^3}{3} - \int x^2\,dx$$

$$= \frac{x^3 \ln x^3}{3} - \frac{x^3}{3} + C, \text{ or}$$

$$\frac{x^3 \cdot 3 \ln x}{3} - \frac{x^3}{3} + C$$

$$= x^3 \ln x - \frac{x^3}{3} + C$$

13. $\int \ln(x+3)\,dx$

Let

$u = \ln(x+3)$ and $dv = dx$.

Then $du = \frac{1}{x+3}\,dx$ and $v = x+3$

$$\int \ln(x+3)\,dx = \ln(x+3)\cdot(x+3) - \int x+3\cdot\frac{dx}{x+3)}$$

$$= (x+3)\ln(x+3) - \int dx$$

$$= (x+3)\ln(x+3) - x + C$$

14. $\int \ln(x+1)\,dx$

Let $u = \ln(x+1)$ and $dv = dx$.

Then

$du = \frac{1}{x+1}\,dx$ and $v = x+1$. (Choosing $C = 1$)

$$\int \ln(x+1)\,dx = (x+1)\ln(x+1) - \int \frac{x+1}{x+1}\,dx$$

$$= (x+1)\ln(x+1) - \int dx$$

$$= (x+1)\ln(x+1) - x + C$$

15. $\int (x+2)\ln x\,dx = \int (\ln x)(x+2)\,dx$

Let

$u = \ln x$ and $dv = (x+2)\,dx$.

Then

$du = \frac{1}{x}\,dx$ and $v = \frac{(x+2)^2}{2}$.

$u = \ln x$ and $dv = (x+2)\,dx$.

Then

$du = \frac{1}{x}\,dx$ and $v = \frac{(x+2)^2}{2}$.

$\int (\ln x)(x+2)\,dx$

$$= (\ln x)\cdot\frac{(x+2)^2}{2} - \int \frac{(x+2)^2}{2}\cdot\frac{1}{x}\,dx$$

$$= \frac{x^2+4x+4}{2}\ln x - \int \frac{x^2+4x+4}{2x}\,dx$$

$$= \frac{x^2+4x+4}{2}\ln x - \int\left(\frac{x}{2}+2+\frac{2}{x}\right)dx$$
$$= \frac{x^2+4x+4}{2}\ln x - \frac{1}{2}\int x\,dx - 2\int dx - 2\int\frac{1}{x}\,dx$$
$$= \frac{x^2+4x+4}{2}\ln x - \frac{1}{2}\cdot\frac{x^2}{2} - 2\cdot x - 2\cdot\ln x + C$$
$$= \frac{x^2+4x+4}{2}\ln x - \frac{1}{2}\cdot\frac{x^2}{2} - 2\cdot x - 2\cdot\ln x + C$$
$$= \left(\frac{x^2+4x+4}{2} - 2\right)\ln x - \frac{x^2}{4} - 2x + C$$
$$= \left(\frac{x^2}{2} + 2x\right)\ln x - \frac{x^2}{4} - 2x + C$$

16. $\int (x+1)\ln x\,dx$

Let

$u = \ln x$ and $dv = (x+1)\,dx.$

Then

$du = \frac{1}{x}\,dx$ and $v = \frac{(x+1)^2}{2}.$

$\int (x+1)\ln x\,dx$
$$= \frac{(x+1)^2\ln x}{2} - \int\frac{(x+1)^2}{2}\cdot\frac{1}{x}\,dx$$
$$= \frac{(x^2+2x+1)\ln x}{2} - \int\frac{x^2+2x+1}{2x}\,dx$$
$$= \frac{(x^2+2x+1)\ln x}{2} - \int\left(\frac{x}{2}+1+\frac{1}{2x}\right)dx$$
$$= \frac{(x^2+2x+1)\ln x}{2} - \frac{x^2}{4} - x - \frac{1}{2}\ln x + C$$
$$= \left(\frac{x^2+2x+1-1}{2}\right)\ln x - \frac{x^2}{4} - x + C$$
$$= \left(\frac{x^2}{2}+x\right)\ln x - \frac{x^2}{4} - x + C$$

17. $\int (x-1)\sin x\,dx$ Let $u = x-1$ and $dv = \sin x\,dx.$

Then $du = dx$ and $v = -\cos x$

$$\int (x-1)\sin x\,dx = (x-1)\cdot -\cos x + \int \cos x\,dx = (1-x)\cos x + \sin x + C$$

18. $\int (x-2)\cos x\,dx$ Let $u = x-2$ and $dv = \cos x\,dx.$

Then $du = dx$ and $v = \sin x$

$$\int (x-2)\cos x\,dx = (x-2)\cdot\sin x - \int\sin x\,dx = (x-2)\sin x + \cos x + C$$

19. $\int x\sqrt{x+2}\,dx$

Let

$u = x$ and $dv = \sqrt{x+2}\,dx = (x+2)^{1/2}\,dx.$

Then

$du = dx$ and $v = \frac{(x+2)^{3/2}}{3/2} = \frac{2}{3}(x+2)^{3/2}.$

$\int \underset{u}{x}\,\underset{dv}{\sqrt{x+2}\,dx}$
$$= x\cdot\frac{2}{3}(x+2)^{3/2} - \int\frac{2}{3}(x+2)^{3/2}\,dx$$
$$= \frac{2}{3}x(x+2)^{3/2} - \frac{2}{3}\int(x+2)^{3/2}\,dx$$
$$= \frac{2}{3}x(x+2)^{3/2} - \frac{2}{3}\cdot\frac{(x+2)^{5/2}}{5/2} + C$$
$$= \frac{2}{3}x(x+2)^{3/2} - \frac{4}{15}(x+2)^{5/2} + C$$

20. $\int x\sqrt{x+4}\,dx$

Let

$u = x$ and $dv = \sqrt{x+4}\,dx.$

Then

$du = dx$ and $v = \frac{2}{3}(x+4)^{3/2}.$

$\int x\sqrt{x+4}\,dx$
$$= \frac{2}{3}x(x+4)^{3/2} - \int\frac{2}{3}(x+4)^{3/2}\,dx$$
$$= \frac{2}{3}x(x+4)^{3/2} - \frac{2}{3}\cdot\frac{2}{5}(x+4)^{5/2} + C$$
$$= \frac{2}{3}x(x+4)^{3/2} - \frac{4}{15}(x+4)^{5/2} + C$$

21. $\int x^3\ln 2x\,dx = \int(\ln 2x)(x^3\,dx)$

Let

$u = \ln 2x$ and $dv = x^3\,dx.$

Then $du = \frac{1}{x}\,dx$ and $v = \frac{x^4}{4}.$

$$\int(\ln 2x)(x^3\,dx) = (\ln 2x)\cdot\frac{x^4}{4} - \int\frac{x^4}{4}\cdot\frac{1}{x}\,dx$$
$$= \frac{x^4}{4}\ln 2x - \frac{1}{4}\int x^3\,dx$$
$$= \frac{x^4}{4}\ln 2x - \frac{1}{4}\cdot\frac{x^4}{4} + C$$
$$= \frac{x^4}{4}\ln 2x - \frac{x^4}{16} + C$$

22. $\int x^2\ln 5x\,dx$

Let

$u = \ln 5x$ and $dv = x^2\,dx.$

Then

$du = \frac{1}{x}\,dx$ and $v = \frac{x^3}{3}.$

$$\int x^2\ln 5x\,dx = \frac{x^3\ln 5x}{3} - \int\frac{x^2}{3}\,dx$$
$$= \frac{x^3\ln 5x}{3} - \frac{x^3}{9} + C$$

23. $\int x^2 e^x \, dx$

Let

$u = x^2$ and $dv = e^x \, dx$.

Then $du = 2x \, dx$ and $v = e^x$.

$$\overset{u}{}\quad \overset{dv}{}\quad \overset{u}{}\quad \overset{v}{}\quad \overset{v}{}\quad \overset{du}{}$$

$$\int x^2 \, e^x \, dx = x^2 \, e^x - \int e^x \cdot 2x \, dx$$

Integration by Parts

$$= x^2 \, e^x - \int 2xe^x \, dx$$

We evaluate $\int 2xe^x \, dx$ using the Integration by Parts formula.

$\int 2xe^x \, dx$

Let

$u = 2x$ and $dv = e^x \, dx$.

Then

$du = 2 \, dx$ and $v = e^x$.

$$\overset{u}{}\quad \overset{dv}{}\quad \overset{u}{}\quad \overset{v}{}\quad \overset{v\,du}{}$$

$$\int 2x \, e^x \, dx = 2x \cdot e^x - \int 2e^x \, dx$$

$$= 2xe^x - 2e^x + K$$

Thus,

$$\int x^2 \, e^x \, dx = x^2 e^x - (2xe^x - 2e^x + K)$$

$$= x^2 \, e^x - 2xe^x + 2e^x + C \quad (C = -K)$$

Since we have an integral $\int f(x)g(x) \, dx$ where $f(x)$, or x^2, can be differentiated repeatedly to a derivative that is eventually 0 and $g(x)$, or e^x, can be integrated repeatedly easily, we can use tabular integration.

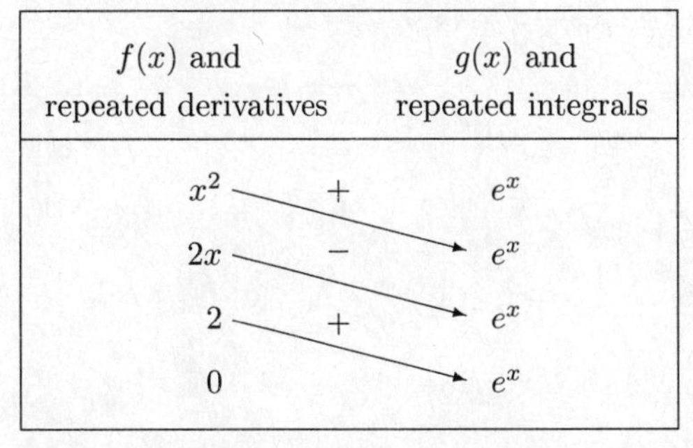

We add the products along the arrows, making the alternate sign changes.

$$\int x^2 \, e^x \, dx = x^2 e^x - 2xe^x + 2e^x + C$$

24. $\int (\ln x)^2 \, dx$

Let

$u = (\ln x)^2$ and $dv = dx$.

Then

$du = \dfrac{2 \ln x}{x} \, dx$ and $v = x$.

$$\int (\ln x)^2 \, dx = x(\ln x)^2 - \int 2 \ln x \, dx$$

$$= x(\ln x)^2 - 2 \int \ln x \, dx$$

$$= x(\ln x)^2 - 2(x \ln x - x + C)$$

See Example 1.

$$= x(\ln x)^2 - 2x \ln x + 2x + K$$

where $K = -2C$

25. $\int x^2 \sin 2x \, dx$

Let

$u = x^2$ and $dv = \sin 2x \, dx$.

Then $du = 2x \, dx$ and $v = -\dfrac{1}{2} \cos 2x$.

$$\int x^2 \sin 2x \, dx = x^2 \cdot -\frac{\cos 2x}{2} + \frac{1}{2} \int \cos 2x \cdot 2x \, dx$$

We evaluate $\int 2x \cos 2x \, dx$ using the Integration by Parts formula

Let

$u = x$ and $dv = 2 \cos 2x \, dx$.

Then

$du = dx$ and $v = \sin 2x$.

$$\int 2x\cos 2x \, dx = x \sin 2x - \int \sin 2x \, dx$$

$$= x \sin 2x + \frac{1}{2} \cos 2x + K$$

Thus,

$$\int x^2 \sin 2x \, dx = -\frac{1}{2} x^2 \cos 2x - \frac{1}{2} x \sin 2x - \frac{1}{4} \cos 2x - K$$

Since we have an integral $\int f(x)g(x) \, dx$ where $f(x)$, or x^2, can be differentiated repeatedly to a derivative that is eventually 0 and $g(x)$, or e^x, can be integrated repeatedly easily, we can use tabular integration.

$f(x)$ and repeated derivatives		$g(x)$ and repeated integrals
x^2	+	$\sin 2x$
$2x$	−	$-\frac{1}{2}\cos 2x$
2	+	$-\frac{1}{4}\sin 2x$
0		$\frac{1}{8}\cos 2x$

We add the products along the arrows, making the alternate sign changes.

$$\int x^2 \sin 2x \, dx = -\tfrac{1}{2} x^2 \cos 2x + \tfrac{1}{2} x \sin 2x + \tfrac{1}{4} \cos 2x + C$$

26. $\int x^{-5} \ln x \, dx$

Let

$u = \ln x$ and $dv = x^{-5} \, dx$.

Then

$du = \dfrac{1}{x} \, dx$ and $v = -\dfrac{x^{-4}}{4}$.

$$\int x^{-5} \ln x \, dx = -\frac{x^{-4}}{4} \ln x - \int -\frac{x^{-4}}{4} \cdot \frac{1}{x} \, dx$$

$$= -\frac{x^{-4}}{4} \ln x + \frac{1}{4} \int x^{-5} \, dx$$

$$= -\frac{x^{-4}}{4} \ln x - \frac{x^{-4}}{16} + C$$

27. $\int x^3 e^{-2x}\, dx$

We will use tabular integration.

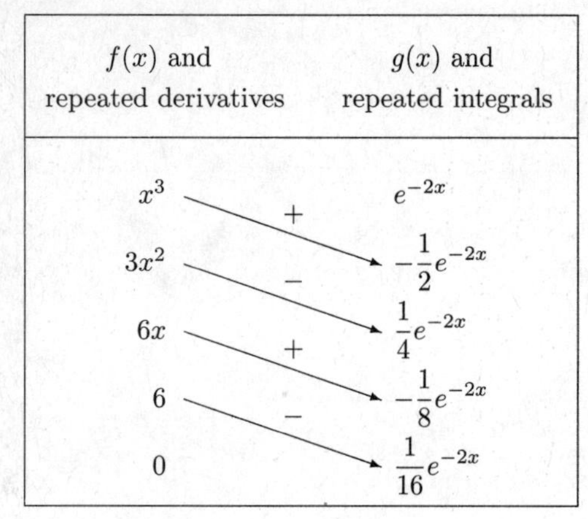

$f(x)$ and repeated derivatives		$g(x)$ and repeated integrals
x^3	$+$	e^{-2x}
$3x^2$	$-$	$-\frac{1}{2}e^{-2x}$
$6x$	$+$	$\frac{1}{4}e^{-2x}$
6	$-$	$-\frac{1}{8}e^{-2x}$
0		$\frac{1}{16}e^{-2x}$

$\int x^3 e^{-2x}\, dx$

$$= x^3\left(-\frac{1}{2}e^{-2x}\right) - 3x^2\left(\frac{1}{4}e^{-2x}\right) + 6x\left(-\frac{1}{8}e^{-2x}\right) - 6\left(\frac{1}{16}e^{-2x}\right) + C$$

$$= -\frac{1}{2}x^3 e^{-2x} - \frac{3}{4}x^2 e^{-2x} - \frac{3}{4}xe^{-2x} - \frac{3}{8}e^{-2x} + C$$

$$= e^{-2x}\left(-\frac{1}{2}x^3 - \frac{3}{4}x^2 - \frac{3}{4}x - \frac{3}{8}\right) + C$$

28. $x^5 \cos 4x\, dx$

We will use tabular integration.

$f(x)$ and repeated derivatives		$g(x)$ and repeated integrals
x^5	$+$	$\cos 4x$
$5x^4$	$-$	$\frac{1}{4}\sin 4x$
$20x^3$	$+$	$-\frac{1}{16}\cos 4x$
$60x^2$	$-$	$-\frac{1}{64}\sin 4x$
$120x$	$+$	$\frac{1}{256}\cos 4x$
120	$-$	$\frac{1}{1024}\sin 4x$
0		$-\frac{1}{4096}\cos 4x$

$$\int x^5 \cos 4x\, dx = \frac{1}{4}x^5 \sin 4x + \frac{5}{16}x^4 \cos 4x - \frac{5}{16}x^3 \sin 4x - \frac{15}{64}x^2 \cos 4x + \frac{15}{128}x \sin 4x + \frac{15}{512}\cos 4x$$

29. $\int x \sec^2 x\, dx$

Let

$u = x$ and $dv = \sec^2 x\, dx$.

Then

$du = dx$ and $v = \tan x$

$$\int x \sec^2 x\, dx = x \tan x - \int \tan x\, dx = x \tan x + \ln|\cos x| + C$$

30. $\int x \csc^2 x\, dx$

Let

$u = x$ and $dv = \csc^2 x\, dx$.

Then

$du = dx$ and $v = -\cot x$

$$\int x \csc^2 x\, dx = -x \cot x + \int \cot x\, dx = -x \cot x + \ln|\sin x| + C$$

31. $\int_1^2 x^2 \ln x\, dx$

In Exercise 9 above we found the indefinite integral.

$$\int x^2 \ln x\, dx = \frac{x^3}{3}\ln x - \frac{x^3}{9} + C$$

Evaluate the definite integral.

$$\int_1^2 x^2 \ln x\, dx = \left[\frac{x^3}{3}\ln x - \frac{x^3}{9}\right]_1^2$$

$$= \left(\frac{2^3}{3}\ln 2 - \frac{2^3}{9}\right) - \left(\frac{1^3}{3}\ln 1 - \frac{1^3}{9}\right)$$

$$= \left(\frac{8}{3}\ln 2 - \frac{8}{9}\right) - \left(\frac{1}{3}\ln 1 - \frac{1}{9}\right)$$

$$= \frac{8}{3}\ln 2 - \frac{8}{9} + \frac{1}{9} \qquad (\ln 1 = 0)$$

$$= \frac{8}{3}\ln 2 - \frac{7}{9}$$

32. From Exercise 10 we know

$$\int x^3 \ln x\, dx = \frac{x^4 \ln x}{4} - \frac{x^4}{16} + C$$

Evaluate the definite integral.

$$\int_1^2 x^3 \ln x\, dx = \left[\frac{x^4 \ln x}{4} - \frac{x^4}{16}\right]_1^2$$

$$= \left(\frac{2^4 \ln 2}{4} - \frac{2^4}{16}\right) - \left(\frac{1^4 \ln 1}{4} - \frac{1^4}{16}\right)$$

$$= (4 \ln 2 - 1) - \left(0 - \frac{1}{16}\right)$$

$$= 4 \ln 2 - \frac{15}{16}$$

33. $\int_2^6 \ln(x+3)\, dx$

In Exercise 13 above we found the indefinite integral.

$\int \ln(x+3)\, dx = (x+3)\ln(x+3) - x + C$

Evaluate the definite integral.

$\int_2^6 \ln x(x+3)\, dx$
$= \left[(x+3)\ \ln(x+3) - x\right]_2^6$
$= [(6+3)\ln(6+3) - 6] - [(2+3)\ln(2+3) - 2]$
$= (9\ \ln\ 9 - 6) - (5\ \ln\ 5 - 2)$
$= 9\ \ln\ 9 - 6 - 5\ \ln\ 5 + 2$
$= 9\ \ln\ 9 - 5\ \ln\ 5 - 4$

34. From Exercise 14 we know

$\int \ln(x+1)\, dx = (x+1)\ \ln(x+1) - x + C$

Evaluate the definite integral.

$\int_0^5 \ln(x+1)\, dx = \left[(x+1)\ \ln(x+1) - x\right]_0^5$
$= \left[(5+1)\ \ln(5+1) - 5\right] - \left[(0+1)\ \ln(0+1) - 0\right]$
$= 6\ \ln\ 6 - 5$

35. a) We first find the indefinite integral.

$\int xe^x\, dx$

Let

$u = x$ and $dv = e^x\, dx$.

Then

$du = dx$ and $v = e^x$.

$\int xe^x\, dx = xe^x - \int e^x\, dx$
$= xe^x - e^x + C$

b) Evaluate the definite integral.

$\int_0^1 xe^x\, dx = \left[xe^x - e^x\right]_0^1$
$= (1 \cdot e^1 - e^1) - (0 \cdot e^0 - e^0)$
$= (e - e) - (0 - 1)$
$= 0 - (-1)$
$= 1$

36. $\int_0^1 xe^{-x}\, dx$ Let

$u = x$ and $dv = e^{-x}\, dx$.

Then

$du = dx$ and $v = -e^{-x}$

$$\begin{aligned}\int xe^{-x}\, dx &= -xe^{-x} - \int -e^{-x}\, dx \\ &= -xe^{-x} - e^{-x} + C\end{aligned}$$

Thus,

$$\begin{aligned}\int_0^1 xe^{-x}\, dx &= \left[-xe^{-x} - e^{-x}\right]_0^1 \\ &= [-e^{-1} - e^{-1}] - [0 - 1] \\ &= 1\end{aligned}$$

37. $\int_0^{5\pi/6} 3x\ \cos\ x\, dx$ Let

$u = 3x$ and $dv = \cos\ x\, dx$.

Then

$du = 3\, dx$ and $v = \sin\ x$

$$\begin{aligned}\int 3x\ \cos\ x\, dx &= 3x\ \sin\ x - \int 3\ \sin\ x\, dx \\ &= 3x\ \sin\ x + 3\ \cos\ x + C\end{aligned}$$

Thus,

$$\begin{aligned}\int_0^{5\pi/6} 3x\ \cos\ x\, dx &= \left[3x\ \sin\ x + 3\ \cos\ x\right]_0^{5\pi/6} \\ &= \left[\frac{5\pi}{4} - \frac{3\sqrt{3}}{2}\right] - [0 + 3] \\ &= \frac{5\pi}{4} - 3 - \frac{3\sqrt{3}}{2}\end{aligned}$$

38. $\int_0^{4\pi/3} 5x\ \sin\ x\, dx$ Let

$u = 5x$ and $dv = \sin\ x\, dx$.

Then

$du = 5\, dx$ and $v = -\cos\ x$

$$\begin{aligned}\int 5x\ \sin\ x\, dx &= -5x\ \cos\ x - \int 5\ \cos\ x\, dx \\ &= -5x\ \cos\ x + 5\ \sin\ x + C\end{aligned}$$

Thus,

$$\begin{aligned}\int_0^{4\pi/3} 5x\ \sin\ x\, dx &= \left[-5x\ \cos\ x + 5\ \sin\ x\right]_0^{4\pi/3} \\ &= \left[\frac{10\pi}{3} - \frac{5\sqrt{3}}{2}\right] - [0 + 0] \\ &= \frac{10\pi}{3} - \frac{5\sqrt{3}}{2}\end{aligned}$$

39. $M'(t) = 10t\sqrt{t+15}$

Let $u = 10t$, and $dv = \sqrt{t+15}\, dt$

Then $du = 10dt$ and $v = \frac{2}{3}(t+15)^{3/2}$

$$\begin{aligned}M(t) &= \int 10t\sqrt{t+15}\, dt \\ &= \frac{20}{3}\ t(t+15)^{3/2} - \frac{20}{3}\int (t+15)^{3/2}\, dt \\ &= \frac{20}{3}\ t(t+15)^{3/2} - \frac{8}{3}(t+15)^{5/2} + C \\ 150000 &= 0 - \frac{8}{3} \cdot (15)^{5/2} + C \\ C &= 150000 + \frac{8(15)^{5/2}}{3} \\ M(t) &= \frac{20}{3}\ t(t+15)^{3/2} - \frac{8}{3}(t+15)^{5/2} + 150000 + \frac{8(15)^{5/2}}{3} \\ M(10) &= \frac{20}{3}\ 10(25)^{3/2} - \frac{8}{3}(25)^{3/2} + 150000 + \frac{8(15)^{5/2}}{3} \\ &= \frac{25000}{3} - \frac{25000}{3} + 150000 + \frac{8(15)^{5/2}}{3} \\ &= 152324\end{aligned}$$

40. $N'(t) = 1000t^2e^{-0.2t}$

$$\begin{aligned} N(t) &= \int 10t\sqrt{t+15}\,dt \\ &= -5000t^2e^{-0.2t} - 50000te^{-0.2t} - 250000e^{-0.2t} + C \\ 2000 &= -250000 + C \\ C &= 252000 \\ N(t) &= -5000t^2e^{-0.2t} - 50000te^{-0.2t} - 250000e^{-0.2t} \\ & \quad +252000 \\ N(60) &= -5000(60)^2e^{-0.2(60)} - 50000(60)e^{-0.2(60)} \\ & \quad -250000e^{-0.2(60)} + 252000 \\ &= 251869 \end{aligned}$$

41. a) We first find the indefinite integral.

$\int 10te^{-t}\,dt = 10\int te^{-t}\,dt$

Let

$u = t$ and $dv = e^{-t}\,dt$.

Then

$du = dt$ and $v = -e^{-t}$.

$$\begin{aligned} 10\int te^{-t}\,dt &= 10\big[t(-e^{-t}) - \textstyle\int -e^{-t}\,dt\big] \\ &= 10(-te^{-t} + \textstyle\int e^{-t}\,dt) \\ &= 10(-te^{-t} - e^{-t} + K) \\ &= -10te^{-t} - 10e^{-t} + C \quad (C = 10K) \end{aligned}$$

Then evaluate the definite integral.

$$\begin{aligned} \textstyle\int_0^T te^{-t}\,dt &= \big[-10te^{-t} - 10e^{-t}\big]_0^T \\ &= (-10Te^{-T} - 10e^{-T}) - \\ &\quad (-10\cdot 0\cdot e^{-0} - 10e^{-0}) \\ &= (-10Te^{-T} - 10e^{-T}) - (0 - 10) \\ &= -10Te^{-T} - 10e^{-T} + 10 \\ &= -10\big[e^{-T}(T+1) - 1\big], \text{ or} \\ &\quad 10\big[e^{-T}(-T-1) + 1\big] \end{aligned}$$

b) Substitute 4 for T.

$$\begin{aligned} \textstyle\int_0^4 te^{-t}\,dt &= -10\big[e^{-4}(4+1) - 1\big] \\ &= -50e^{-4} + 10 \\ &\approx -50(0.018316) + 10 \\ &\approx -0.915800 + 10 \\ &\approx 9.084 \end{aligned}$$

42. a) We first find the indefinite integral.

$\int te^{-kt}\,dt$

Let

$u = t$ and $dv = e^{-kt}\,dt$.

Then

$du = dt$ and $v = -\frac{1}{k}e^{-kt}$.

$$\begin{aligned} \int te^{-kt}\,dt &= -\frac{t}{k}e^{-kt} - \int -\frac{1}{k}e^{-kt}\,dt \\ &= -\frac{t}{k}e^{-kt} - \frac{1}{k^2}e^{-kt} + C \end{aligned}$$

Evaluate the definite integral.

$$\begin{aligned} \textstyle\int_0^T te^{-kt}\,dt &= \left[-\frac{t}{k}e^{-kt} - \frac{1}{k^2}e^{-kt}\right]_0^T \\ &= \left(-\frac{T}{k}e^{-kT} - \frac{1}{k^2}e^{-kT}\right) - \\ &\quad \left(-\frac{0}{k}e^{-k\cdot 0} - \frac{1}{k^2}e^{-k\cdot 0}\right) \\ &= -e^{-kT}\left(\frac{T}{k} + \frac{1}{k^2}\right) + \frac{1}{k^2} \end{aligned}$$

b) Substitute 10 for T and 0.2 for k.

$$\begin{aligned} \textstyle\int_0^{10} te^{-0.2t}\,dt &= -e^{-0.2(10)}\left[\frac{10}{0.2} + \frac{1}{(0.2)^2}\right] + \\ &\quad \frac{1}{(0.2)^2} \\ &= -e^{-2}(50 + 25) + 25 \\ &\approx 14.850 \text{ mg} \end{aligned}$$

43. Left to the student

44. **a)** Let $u = x + 2 \rightarrow x = u - 2$ then $du = dx$

$\int x\sqrt[3]{x+2}\,dx$

$$\begin{aligned} &= \int (u-2)u^{1/3}\,du \\ &= \int u^{4/3} - 2u^{1/3}\,du \\ &= \frac{3}{7}u^{7/3} - \frac{3}{2}u^{4/3} + C \\ &= \frac{3}{7}(x+2)^{7/3} - \frac{3}{2}(x+2)^{4/3} + C \end{aligned}$$

b) Let $u = x$ and $dv = (x+2)^{1/3}\,dx$

Then $du = dx$ and $v = \frac{3}{4}(x+2)^{4/3}$

$\int x\sqrt[3]{x+2}\,dx$

$$\begin{aligned} &= \frac{3}{4}x(x+2)^{4/3} - \int \frac{3}{4}(x+2)^{4/3}\,dx \\ &= \frac{3}{4}x(x+2)^{4/3} - \frac{9}{28}(x+2)^{7/3} + C \end{aligned}$$

c) $\frac{3}{7}(x+2)^{7/3} - \frac{3}{2}(x+2)^{4/3} + C$

$$\begin{aligned} &= \frac{3}{4}(x+2)^{4/3}\left[x - \frac{3}{7}(x+2)\right] + C \\ &= \frac{3}{4}(x+2)^{4/3}\left(\frac{4}{7}x - \frac{6}{7}\right) + C \\ &= 3(x+2)^{4/3}\left(\frac{x}{7} - \frac{2}{14}\right) + C \end{aligned}$$

$\frac{3}{4}x(x+2)^{4/3} - \frac{9}{28}(x+2)^{7/3} + C$

$$\begin{aligned} &= 3(x+2)^{4/3}\left[\frac{1}{7}(x+2) - \frac{1}{2}\right] + C \\ &= 3(x+2)^{4/3}\left(\frac{x}{7} - \frac{2}{14}\right) + C \end{aligned}$$

45. Let $u = \ln x$ and $dv = x^{1/2}$
Then $du = \dfrac{dx}{x}$ and $v = \dfrac{2}{3}x^{3/2}$

$$\begin{aligned}\int \sqrt{x}\ln x\,dx &= \frac{2}{3}x^{3/2}\ln x - \frac{2}{3}\int x^{1/2}\,dx \\ &= \frac{2}{3}x^{3/2}\ln x - \frac{4}{9}x^{3/2} + C\end{aligned}$$

46. Let $u = \ln x$ and $dv = x^n$
Then $du = \dfrac{dx}{x}$ and $v = \dfrac{1}{n+1}\,x^{n+1}$

$$\begin{aligned}\int x^n \ln x\,dx &= \frac{1}{n+1}\,x^{n+1}\ln x - \frac{1}{n+1}\int x^n\,dx \\ &= \frac{1}{n+1}\,x^{n+1}\ln x - \frac{1}{(n+1)^2}\,x^{n+1} + C\end{aligned}$$

47. Let $u = te^t$ and $dv = \dfrac{1}{(t+1)^2}\,dt$
Then $du = (te^t + e^t)\,dt$ and $v = \dfrac{-1}{t+1}$

$$\int \frac{te^t}{(t+1)^2}\,dt = \frac{-te^t}{t+1} + \int \frac{te^t}{t+1}\,dt + \int \frac{e^t}{t+1}\,dt$$

For the first integral on the right hand side
Let $u = \dfrac{t}{t+1}$ and $dv = e^t\,dt$
Then $du = \dfrac{dt}{t+1}$ and $v = e^t$
For the second integral on the right hand side
Let $u = \dfrac{1}{t+1}$ and $dv = e^t\,dt$
Then $du = \dfrac{-dt}{(t+1)^2}$ and $v = e^t$ $\displaystyle\int \frac{te^t}{(t+1)^2}\,dt$

$$\begin{aligned}&= \frac{-te^t}{t+1} + \int \frac{te^t}{t+1}\,dt + \int \frac{e^t}{t+1}\,dt \\ &= \frac{-te^t}{t+1} + \frac{te^t}{t+1} - \int \frac{e^t}{(t+1)^2}\,dt + \\ &\quad \frac{e^t}{t+1} + \int \frac{e^t}{(t+1)^2}\,dt \\ &= \frac{e^t}{t+1} + C\end{aligned}$$

48. Let $u = (\ln x)^2$ and $dv = x^2\,dx$
Then $du = \dfrac{2\,\ln x}{x}\,dx$ and $v = \dfrac{1}{3}x^3$

$$\int x^2(\ln x)^2\,dx = \frac{x^3}{3}(\ln x)^2 - \frac{2}{3}\int x^2\,\ln x\,dx$$

Let $u = \ln x$ and $dv = x^2$
Then $du = \dfrac{1}{x}$ and $dv = \dfrac{x^3}{3}$ $\int x^2(\ln x)^2\,dx$

$$\begin{aligned}&= \frac{x^3}{3}(\ln x)^2 - \frac{2}{3}\int x^2\,\ln x\,dx \\ &= \frac{x^3}{3}(\ln x)^2 - \frac{2x^3}{9}\,\ln x + \frac{2x^3}{27} + C\end{aligned}$$

49. Let $u = \ln x$ and $dv = x^{-1/2}$
Then $du = \dfrac{dx}{x}$ and $v = 2x^{\frac{1}{2}}$

$$\begin{aligned}\int \frac{\ln x}{x^{1/2}}\,dx &= 2\sqrt{x}\,\ln x - \int 2x^{-1/2}\,dx \\ &= 2\sqrt{x}\,\ln x - 4\sqrt{x} + C\end{aligned}$$

50. Let $u = (\ln x)^2$ and $dv = x^n\,dx$
Then $du = \dfrac{2\,\ln x}{x}$ and $v = \dfrac{x^{n+1}}{n+1}$

$$\int x^n(\ln x)^2\,dx = \frac{x^{n+1}}{n+1}\,(\ln x)^2 - \frac{2}{n+1}\int x^n\,\ln x\,dx$$

Let $u = \ln x$ and $dv = x^n\,dx$
Then $du = \dfrac{1}{x}$ and $v = \dfrac{x^{n+1}}{n+1}$
$\int x^n(\ln x)^2\,dx$

$$\begin{aligned}&= \frac{x^{n+1}}{n+1}\,(\ln x)^2 - \frac{2}{n+1}\int x^n\,\ln x\,dx \\ &= \frac{x^{n+1}\,(\ln x)^2}{n+1} - \frac{2x^{n+1}\,\ln x}{(n+1)^2} - \frac{2}{(n+1)^2}\int x^n\,dx \\ &= \frac{x^{n+1}\,(\ln x)^2}{n+1} - \frac{2x^{n+1}\,\ln x}{(n+1)^2} - \frac{2x^{n+1}}{(n+1)^3} + C\end{aligned}$$

51. Let $u = 13t^2 - 48$ and $dv = (4t+7)^{-1/5}$
Then $du = 26t\,dt$ and $v = \dfrac{5}{16}(4t+7)^{4/5}$

$$\begin{aligned}\int \frac{13t^2 - 48}{\sqrt[5]{4t+7}}\,dt &= \frac{5}{16}(13t^2 - 48)(4t+7)^{4/5} - \\ &\quad \int \frac{65}{8}t(4t+7)^{4/5}\,dt\end{aligned}$$

Let $u = \dfrac{65}{8}t$ and $dv = (4t+7)^{4/5}\,dt$
Then $du = 26t\,dt$ and $v = \dfrac{5}{36}(4t+7)^{9/5}$ $\displaystyle\int \frac{13t^2 - 48}{\sqrt[5]{4t+7}}\,dt$

$$\begin{aligned}&= \frac{5}{16}(13t^2 - 48)(4t+7)^{4/5} - \int \frac{65}{8}t(4t+7)^{4/5}\,dt \\ &= \frac{5}{16}(13t^2 - 48)(4t+7)^{4/5} - \frac{325t(4t+7)^{9/4}}{288} - \\ &\quad \frac{325}{288}\int (4t+7)^{9/5}\,dt \\ &= \frac{5(13t^2 - 48)(4t+7)^{4/5}}{16} - \frac{325t(4t+7)^{9/4}}{288} - \\ &\quad \frac{1625(4t+7)^{14/5}}{4032} + C\end{aligned}$$

52. Let $u = 27x^3 + 83x - 2$ and $dv = (3x+8)^{1/6}\,dx$
Then $du = 81x^2 + 82\,dx$ and $v = \dfrac{2}{7}(3x+8)^{7/6}$
$\int (27x^3 + 83x - 2)\sqrt[6]{3x+8}\,dx$

$$\begin{aligned}&= \frac{2}{7}\,(27x^3 + 83x - 2)(3x+8)^{7/6} - \\ &\quad \frac{2}{7}\int (81x^2 + 83)(3x+8)^{7/6}\,dx\end{aligned}$$

Let $u = 81x^2 + 83$ and $dv = (3x+8)^{7/6}\,dx$

Then $du = 162x\,dx$ and $v = \frac{2}{13}(3x+8)^{13/6}$

$$\int (27x^3 + 83x - 2)\sqrt[6]{3x+8}\,dx$$

$$= \frac{2}{7}(27x^3+83x-2)(3x+8)^{7/6} - \frac{4}{91}(81x^2+83)(3x+8)^{13/6} - \frac{648}{91}\int x(3x+8)^{13/6}\,dx$$

Let $u = x$ and $dv = (3x+8)^{13/6}\,dx$

Then $du = dx$ and $v = \frac{2}{19}(3x+8)^{19/6}$

$$\int (27x^3 + 83x - 2)\sqrt[6]{3x+8}\,dx$$

$$= \frac{2}{7}(27x^3+83x-2)(3x+8)^{7/6} - \frac{4}{91}(81x^2+83)(3x+8)^{13/6} - \frac{1296}{1729}x(3x+8)^{19/6} - \frac{1296}{1729}\int (3x+8)^{19/6}\,dx$$

$$= \frac{2}{7}(27x^3+83x-2)(3x+8)^{7/6} - \frac{4}{91}(81x^2+83)(3x+8)^{13/6} - \frac{1296}{1729}x(3x+8)^{19/6} - \frac{2592}{43225}(3x+8)^{25/6} + C$$

53. Left to the student

54. Left to the student

55. Left to the student

56. Left to the student

57. Left to the student

58. Left to the student

59. $\int_1^{10} x^5 \ln x\,dx = 355986.43$

Exercise Set 5.7

1. $\int xe^{-3x}\,dx$

This integral fits Formula 6 in Table 1.

$$\int xe^{ax}\,dx = \frac{1}{a^2}\cdot e^{ax}(ax-1) + C$$

In our integral $a = -3$, so we have, by the formula,

$$\int xe^{-3x}\,dx = \frac{1}{(-3)^2}\cdot e^{-3x}(-3x-1) + C$$

$$= \frac{1}{9}e^{-3x}(-3x-1) + C$$

$$\text{or } -\frac{1}{9}e^{-3x}(3x+1) + C$$

2. $\int xe^{4x}\,dx$

This integral fits Formula 6 in Table 1.

$$\int xe^{ax}\,dx = \frac{1}{a^2}\cdot e^{ax}(ax-1) + C$$

We have $a = 4$.

$$\int xe^{4x} = \frac{1}{4^2}e^{4x}(4x-1) + C$$

$$= \frac{1}{16}e^{4x}(4x-1) + C$$

3. $\int 5^x\,dx$

This integral fits Formula 11 in Table 1.

$$\int a^x\,dx = \frac{a^x}{\ln a} + C,\ a > 0,\ a \neq 1$$

In our integral $a = 5$, so we have, by the formula,

$$\int 5^x\,dx = \frac{5^x}{\ln 5} + C$$

4. $\int \frac{1}{\sqrt{x^2-9}}\,dx$

This integral fits Formula 25 in Table 1.

$$\int \frac{1}{\sqrt{x^2-a^2}}\,dx = \ln(x + \sqrt{x^2-a^2}) + C$$

We have $a^2 = 9$.

$$\int \frac{1}{\sqrt{x^2-9}}\,dx = \ln(x + \sqrt{x^2-9}) + C$$

5. $\int \frac{1}{16-x^2}\,dx$

This integral fits Formula 27 in Table 1.

$$\int \frac{1}{a^2-x^2}\,dx = \frac{1}{2a}\ln\left(\frac{a+x}{a-x}\right) + C$$

In our integral $a = 4$, so we have, by the formula,

$$\int \frac{1}{16-x^2}\,dx = \int \frac{1}{4^2-x^2}\,dx$$

$$= \frac{1}{2\cdot 4}\ln\frac{4+x}{4-x} + C$$

$$= \frac{1}{8}\ln\frac{4+x}{4-x} + C$$

6. $\int \frac{1}{x\sqrt{4+x^2}}\,dx$

This integral fits Formula 28 in Table 1.

$$\int \frac{1}{x\sqrt{a^2+x^2}}\,dx = -\frac{1}{a}\ln\left(\frac{a+\sqrt{a^2+x^2}}{x}\right) + C$$

We have $a^2 = 4$ and $a = 2$.

$$\int \frac{1}{x\sqrt{4+x^2}}\,dx = -\frac{1}{2}\ln\left(\frac{2+\sqrt{4+x^2}}{x}\right) + C$$

7. $\int \frac{x}{5-x}\,dx$

This integral fits Formula 30 in Table 1.

$$\int \frac{x}{ax+b}\,dx = \frac{b}{a^2} + \frac{x}{a} - \frac{b}{a^2}\ln(ax+b) + C$$

In our integral $a = -1$ and $b = 5$, so we have, by the formula,

$$\int \frac{x}{5-x}\,dx = \frac{5}{(-1)^2}+\frac{x}{(-1)}-\frac{5}{(-1)^2}\ \ln(-1\cdot x+5)+C$$
$$= 5 - x - 5\ \ln(5-x) + C$$

8. $\int \frac{x}{(1-x)^2}\,dx$

This integral fits Formula 31 in Table 1.

$$\int \frac{x}{(ax+b)^2}\,dx = \frac{b}{a^2(ax+b)} + \frac{1}{a^2}\ \ln(ax+b) + C$$

We have $a = -1$ and $b = 1$.

$$\int \frac{x}{(1-x)^2}\,dx = \frac{1}{(-1)^2(-1\cdot x+1)}+$$
$$\frac{1}{(-1)^2}\ \ln(-1\cdot x+1) + C$$
$$= \frac{1}{1-x} + \ln(1-x) + C$$

9. $\int \frac{1}{x(5-x)^2}\,dx$

This integral fits Formula 33 in Table 1.

$$\int \frac{1}{x(ax+b)^2}\,dx = \frac{1}{b(ax+b)} + \frac{1}{b^2}\ \ln\left(\frac{x}{ax+b}\right) + C$$

In our integral $a = -1$ and $b = 5$, so we have, by the formula,

$$\int \frac{1}{x(5-x)^2}\,dx = \int \frac{1}{x(-x+5)^2}\,dx$$
$$= \frac{1}{5(-x+5)} + \frac{1}{5^2}\ \ln\left(\frac{x}{-x+5}\right) + C$$
$$= \frac{1}{5(5-x)} + \frac{1}{25}\ \ln\left(\frac{x}{5-x}\right) + C$$

10. $\int \sqrt{x^2+9}\,dx$

This integral fits Formula 34 in Table 1.

$$\int \sqrt{x^2 \pm a^2}\,dx = \frac{1}{2}[x\sqrt{x^2\pm a^2}\pm$$
$$a^2\ \ln(x+\sqrt{x^2\pm a^2})] + C$$

We have $a^2 = 9$.

$$\int \sqrt{x^2+9}\,dx = \frac{1}{2}[x\sqrt{x^2+9}+9\ \ln(x+\sqrt{x^2+9})] + C$$

11. $\int \ln 3x\,dx$

$= \int (\ln 3 + \ln x)dx$

$= \int \ln 3\,dx + \int \ln x\,dx$

$= (\ln 3)\,x + \int \ln x\,dx$

The integral in the second term fits Formula 8 in Table 1.

$\int \ln x\,dx = x\ \ln x - x + C$

$\int \ln 3x\,dx = (\ln 3)x + \int \ln x\,dx$

$= (\ln 3)x + x\ln x - x + C$

12. $\int \ln \frac{4}{5}x\,dx$

$$= \int \left(\ln \frac{4}{5} + \ln x\right)dx$$
$$= \int \ln \frac{4}{5}\,dx + \int \ln x\,dx$$
$$= \left(\ln \frac{4}{5}\right)x + \int \ln x\,dx$$
$$= \left(\ln \frac{4}{5}\right)x + x\ \ln x - x + C \quad \text{Using Formula 8.}$$

This can also be expressed as $x\ \ln\left(\frac{4}{5}x\right) - x + C$.

13. $\int x^4 e^{5x}\,dx$

This integral first Formula 7 in Table 1.

$$\int x^n e^{ax}\,dx = \frac{x^n e^{ax}}{a} - \frac{n}{a}\int x^{n-1} e^{ax}\,dx$$

In our integral $n = 4$ and $a = 5$, so we have, by the formula,

$$\int x^4 e^{5x}\,dx$$
$$= \frac{x^4 e^{5x}}{5} - \frac{4}{5}\int x^3 e^{5x}\,dx$$

In the integral in the second term where $n = 3$ and $a = 5$, we again apply Formula 7.

$$= \frac{x^4 e^{5x}}{5} - \frac{4}{5}\left[\frac{x^3 e^{5x}}{5} - \frac{3}{5}\int x^2 e^{5x}\,dx\right]$$

We continue to apply Formula 7.

$$= \frac{x^4 e^{5x}}{5} - \frac{4}{25}x^3 e^{5x} + \frac{12}{25}\left[\frac{x^2 e^{5x}}{5} - \frac{2}{5}\int x\,e^{5x}\,dx\right]$$
$$= \frac{x^4 e^{5x}}{5} - \frac{4}{25}x^3 e^{5x} + \frac{12}{125}x^2 e^{5x} -$$
$$\frac{24}{125}\left[\frac{x\,e^{5x}}{5} - \frac{1}{5}\int x^0 e^{5x}\,dx\right]$$
$$= \frac{x^4 e^{5x}}{5} - \frac{4}{25}x^3 e^{5x} + \frac{12}{125}x^2 e^{5x} - \frac{24}{625}x\,e^{5x} +$$
$$\frac{24}{625}\int e^{5x}\,dx$$

We now apply Formula 5, $\int e^{ax}\,dx = \frac{1}{a}\cdot e^{ax} + C$.

$$= \frac{x^4 e^{5x}}{5} - \frac{4}{25}x^3 e^{5x} + \frac{12}{125}x^2 e^{5x} - \frac{24}{625}x\,e^{5x} +$$
$$\frac{24}{3125}e^{5x} + C$$

14. $\int x^3 e^{-2x}\,dx$

This integral fits Formula 7 in Table 1.

$$\int x^n e^{ax}\,dx = \frac{x^n e^{ax}}{a} - \frac{n}{a}\int x^{n-1} e^{ax}\,dx$$

We have $n = 3$ and $a = -2$.

$$\int x^3 e^{-2x}\,dx = \frac{x^3 e^{-2x}}{-2} - \frac{3}{-2}\int x^2 e^{-2x}\,dx$$

We continue to apply Formula 7.

$$= \frac{x^3 e^{-2x}}{-2} + \frac{3}{2}\left[\frac{x^2 e^{-2x}}{-2} - \frac{2}{-2}\int xe^{-2x}\,dx\right]$$
$$= \frac{x^3 e^{-2x}}{-2} - \frac{3}{4}x^2 e^{-2x} + \frac{3}{2}\left[\frac{xe^{-2x}}{-2} - \frac{1}{-2}\int e^{-2x}\,dx\right]$$

We now apply Formula 5, $\int e^{ax}\,dx = \frac{1}{a}\cdot e^{ax} + C$.

$$= \frac{x^3\,e^{-2x}}{-2} - \frac{3}{4}x^2\,e^{-2x} - \frac{3}{4}\,xe^{-2x} + \frac{3}{4}\cdot\frac{1}{-2}e^{-2x} + C$$

$$= -\frac{1}{2}x^3\,e^{-2x} - \frac{3}{4}x^2\,e^{-2x} - \frac{3}{4}xe^{-2x} - \frac{3}{8}e^{-2x} + C$$

15. $\int x^3 \sin x \ln x\,dx$

This integral fits Formula 22 in Table 1.

$$\int x^3 \sin x\,dx$$

$$= -x^3 \cos x + 3\int x^2 \cos x$$

$$= -x^3 \cos x + 3\left[x^2 \sin x - 2\int x \sin x\right]$$

$$= -x^3 \cos x + 3x^2 \sin x - 6\left[-x \cos x + \sin x\right] + C$$

$$= -x^3 \cos x + 3x^2 \sin x + 6x \cos x - 6 \sin x + C$$

16. $\int 5x^4 \ln x\,dx = 5\int x^4 \ln dx$

This integral fits Formula 10 in Table 1.

$$\int x^n \ln x\,dx = x^{n+1}\left[\frac{\ln x}{n+1} - \frac{1}{(n+1)^2}\right] + C,\ n \neq -1$$

We have $n = 4$.

$$5\int x^4 \ln x\,dx = 5x^{4+1}\left[\frac{\ln x}{4+1} - \frac{1}{(4+1)^2}\right] + C$$

$$= 5x^5\left(\frac{\ln x}{5} - \frac{1}{25}\right) + C$$

$$= x^5 \ln x - \frac{x^5}{5} + C$$

17. $\int \sec 2x\,dx$

This integral fits Formula 16 in Table 1.

$$\int \sec 2x\,dx = \frac{1}{2}\ln|\sec 2x + \tan 2x| + C$$

18. $\int \csc 3x\,dx$

This integral fits Formula 17 in Table 1.

$$\int \csc 3x\,dx = -\frac{1}{3}\ln|\csc 3x + \cot 3x| + C$$

19. $\int 2\tan(2x+1)\,dx$. Let $u = 2x+1$ then $du = 2\,dx$

This integral fits Formula 14 in Table 1.

$$\int \tan u\,du = -\ln|\cos u| + C$$

$$= -\ln|\cos(2x+1)| + C$$

20. $\int 4\cot(3x-2)\,dx$

Let $u = 3x - 2$ then $du = 3\,dx$

This integral fits Formula 15 in Table 1.

$$\frac{4}{3}\int \cot u\,du = \frac{4}{3}\ln|\sin u| + C$$

$$= \frac{4}{3}\ln|\sin(3x-2)| + C$$

21. $\int \frac{dx}{\sqrt{x^2+7}}$

This integral fits Formula 24 in Table 1.

$$\int \frac{1}{\sqrt{x^2+a^2}}\,dx = \ln(x + \sqrt{x^2+a^2}) + C$$

In our integral $a^2 = 7$, so we have, by the formula,

$$\int \frac{dx}{\sqrt{x^2+7}} = \ln(x + \sqrt{x^2+7}) + C$$

22. $\int \frac{3\,dx}{x\sqrt{1-x^2}} = 3\int \frac{dx}{x\sqrt{1-x^2}}$

This integral fits Formula 29 in Table 1.

$$\int \frac{1}{x\sqrt{a^2-x^2}}\,dx = -\frac{1}{a}\ln\left(\frac{a+\sqrt{a^2-x^2}}{a}\right) + C,$$

$$0 < x < a$$

We have $a^2 = 1$ and $a = 1$.

$$3\int \frac{dx}{x\sqrt{1-x^2}} = 3\left[-\frac{1}{1}\ln\left(\frac{1+\sqrt{1-x^2}}{x}\right)\right] + C$$

$$= -3\ln\left(\frac{1+\sqrt{1-x^2}}{x}\right) + C$$

23. $\int \frac{10\,dx}{x(5-7x)^2} = 10\int \frac{1}{x(-7x+5)^2}\,dx$

This integral fits Formula 33 in Table 1.

$$\int \frac{1}{x(ax+b)^2}\,dx = \frac{1}{b(ax+b)} + \frac{1}{b^2}\ln\left(\frac{x}{ax+b}\right) + C$$

In our integral $a = -7$ and $b = 5$, so we have, by the formula,

$$= 10\int \frac{1}{x(-7x+5)^2}\,dx$$

$$= 10\left[\frac{1}{5(-7x+5)} + \frac{1}{5^2}\ln\left(\frac{x}{-7x+5}\right)\right] + C$$

$$= \frac{2}{5-7x} + \frac{2}{5}\ln\left(\frac{x}{5-7x}\right) + C$$

24. $\int \frac{2}{5x(7x+2)}\,dx = \frac{2}{5}\int \frac{1}{x(7x+2)}\,dx$

This integral fits Formula 32 in Table 1.

$$\int \frac{1}{x(ax+b)}\,dx = \frac{1}{b}\ln\left(\frac{x}{ax+b}\right) + C$$

We have $a = 7$ and $b = 2$.

$$\frac{2}{5}\int \frac{1}{x(7x+2)}\,dx = \frac{2}{5}\cdot\frac{1}{2}\ln\left(\frac{x}{7x+2}\right) + C$$

$$= \frac{1}{5}\ln\left(\frac{x}{7x+2}\right) + C$$

25. $\int \frac{-5}{4x^2-1}\,dx = -5\int \frac{1}{4x^2-1}\,dx$

This integral almost fits Formula 26 in Table 1.

$$\int \frac{1}{x^2-a^2}\,dx = \frac{1}{2a}\ln\left(\frac{x-a}{x+a}\right) + C$$

But the x^2 coefficient needs to be 1. We factor out 4 as follows. Then we apply Formula 26.

$$-5\int \frac{1}{4x^2-1}\,dx = -5\int \frac{1}{4\left(x^2-\frac{1}{4}\right)}\,dx$$
$$= -\frac{5}{4}\int \frac{1}{x^2-\frac{1}{4}}\,dx \quad \left(a^2=\frac{1}{4},\ a=\frac{1}{2}\right)$$
$$= -\frac{5}{4}\left[\frac{1}{2\cdot\frac{1}{2}}\ \ln\left(\frac{x-\frac{1}{2}}{x+\frac{1}{2}}\right)\right]+C$$
$$= -\frac{5}{4}\ \ln\left(\frac{x-1/2}{x+1/2}\right)+C$$

26. $\int\sqrt{9t^2-1}\,dt = \int\sqrt{9\left(t^2-\frac{1}{9}\right)}\,dt$
$$= 3\int\sqrt{t^2-\frac{1}{9}}\,dt$$

This integral fits Formula 34 in Table 1.

$$\int\sqrt{x^2\pm a^2}\,dx = \frac{1}{2}\left[x\sqrt{x^2\pm a^2}\pm a^2\ \ln(x+\sqrt{x^2\pm a^2})\right]+C$$

We have $a^2=\frac{1}{9}$.

$$3\int\sqrt{t^2-\frac{1}{9}}\,dt$$
$$= 3\cdot\frac{1}{2}\left[t\sqrt{t^2-\frac{1}{9}}-\frac{1}{9}\ \ln\left(t+\sqrt{t^2-\frac{1}{9}}\right)\right]+C$$
$$= \frac{3}{2}\left[t\sqrt{t^2-\frac{1}{9}}-\frac{1}{9}\ \ln\left(t+\sqrt{t^2-\frac{1}{9}}\right)\right]+C$$

27. $\int\sqrt{4m^2+16}\,dm$

This integral almost fits Formula 34 in Table 1.

$$\int\sqrt{x^2+a^2}\,dx$$
$$= \frac{1}{2}\left[x\sqrt{x^2+a^2}+a^2\ \ln\left(x+\sqrt{x^2+a^2}\right)\right]+C$$

But the x^2 coefficient needs to be 1. We factor out 4 as follows. Then we apply Formula 34.

$$\int\sqrt{4m^2+16}\,dm$$
$$= \int\sqrt{4(m^2+4)}\,dm$$
$$= 2\int\sqrt{m^2+4}\,dm$$
$$= 2\cdot\frac{1}{2}\left[m\sqrt{m^2+4}+4\ \ln\left(m+\sqrt{m^2+4}\right)\right]+C$$
$$= m\sqrt{m^2+4}+4\ \ln\left(m+\sqrt{m^2+4}\right)+C$$

28. $\int\frac{3\ \ln x}{x^2}\,dx = 3\int x^{-2}\,\ln x\,dx$

This integral fits Formula 10 in Table 1.

$$\int x^n\ \ln x\,dx = x^{n+1}\left[\frac{\ln x}{n+1}-\frac{1}{(n+1)^2}\right]+C,\ n\pm -1$$

We have $n=-2$.

$$3\int x^{-2}\ \ln x\,dx = 3\cdot x^{-2+1}\left[\frac{\ln x}{-2+1}-\frac{1}{(-2+1)^2}\right]+C$$
$$= 3x^{-1}(-\ln x-1)+C,\ \text{or}$$
$$\frac{3}{x}(-\ln x-1)+C$$

29. $\int\frac{-5\ \ln x}{x^3}\,dx = -5\int x^{-3}\ \ln x\,dx$

This integral fits Formula 10 in Table 1.

$$\int x^n\ \ln x\,dx = x^{n+1}\left[\frac{\ln x}{n+1}-\frac{1}{(n+1)^2}\right]+C,\ n\neq -1$$

In our integral $n=-3$, so we have, by the formula,

$$-5\int x^{-3}\ \ln x\,dx$$
$$= -5\left[x^{-3+1}\left(\frac{\ln x}{-3+1}-\frac{1}{(-3+1)^2}\right)\right]+C$$
$$= -5\left[x^{-2}\left(\frac{\ln x}{-2}-\frac{1}{4}\right)\right]+C$$
$$= \frac{5\ \ln x}{2x^2}+\frac{5}{4x^2}+C$$

30. $\int(\ln x)^4\,dx$

This integral fits Formula 9 in Table 1.

$$\int(\ln x)^n\,dx = x(\ln x)^n-n\int(\ln x)^{n-1}\,dx,\quad n\neq -1$$

We have $n=4$.

$$\int(\ln x)^4\,dx$$
$$= x(\ln x)^4-4\int(\ln x)^3\,dx$$

We continue to apply Formula 9.

$$= x(\ln x)^4-4[x(\ln x)^3-3\int(\ln x)^2\,dx]$$
$$= x(\ln x)^4-4x(\ln x)^3+12[x(\ln x)^2-2\int\ln x\,dx]$$

Now we apply Formula 8, $\int\ln x\,dx = x\ \ln x-x+C$.

$$= x(\ln x)^4-4x(\ln x)^3+12x(\ln x)^2-24(x\ \ln x-x)+C$$
$$= x(\ln x)^4-4x(\ln x)^3+12x(\ln x)^2-24x\ \ln x+24x+C$$

31. $\int\frac{e^x}{x^{-3}}\,dx = \int x^3\,e^x\,dx$

This integral fits Formula 7 in Table 1.

$$\int x^n\,e^{ax}\,dx = \frac{x^n\,e^{ax}}{a}-\frac{n}{a}\int x^{n-1}\,e^{ax}\,dx$$

In our integral $n=3$ and $a=1$, so we have, by the formula,

$$\int x^3\,e^x\,dx$$
$$= x^3\,e^x-3\int x^2\,e^x\,dx$$

We continue to apply Formula 7.

$$= x^3\,e^x-3\left(x^2\,e^x-2\int x\,e^x\,dx\right)\quad (n=2,\ a=1)$$
$$= x^3\,e^x-3x^2\,e^x+6\int x\,e^x\,dx$$
$$= x^3\,e^x-3x^2\,e^x+6\left(x\,e^x-\int x^0\,e^x\,dx\right)\quad (n=1,\ a=1)$$
$$= x^3\,e^x-3x^2\,e^x+6x\,e^x-6\int e^x\,dx$$
$$= x^3\,e^x-3x^2\,e^x+6x\,e^x-6\,e^x+C$$

32. $\int\frac{3}{\sqrt{4x^2+100}}\,dx = \int\frac{3}{2\sqrt{x^2+25}}\,dx$
$$= \frac{3}{2}\int\frac{1}{\sqrt{x^2+25}}\,dx$$

This integral fits Formula 24 in Table 1.

$$\int \frac{1}{\sqrt{x^2+a^2}}\,dx = \ln(x+\sqrt{x^2+a^2})+C$$

We have $a^2 = 25$.

$$\frac{3}{2}\int \frac{1}{\sqrt{x^2+25}}\,dx = \frac{3}{2}\ln\left(x+\sqrt{x^2+25}\right)+C$$

33. $\int x\sqrt{1+2x}\,dx$

This integral fits Formula 35 in Table 1.

$$\int x\sqrt{a+bx}\,dx = \frac{2}{15b^2}(3bx-2a)(a+bx)^{3/2}+C$$

In our integral $a = 1$ and $b = 2$.

$$\begin{aligned}&\int x\sqrt{1+2x}\,dx\\ &= \frac{2}{15\cdot 2^2}(3\cdot 2x-2\cdot 1)(1+2x)^{3/2}+C\\ &= \frac{2}{60}(6x-2)(1+2x)^{3/2}+C\\ &= \frac{1}{30}\cdot 2(3x-1)(1+2x)^{3/2}+C\\ &= \frac{1}{15}(3x-1)(1+2x)^{3/2}+C\end{aligned}$$

34. $\int x\sqrt{2+3x}\,dx$

This integral fits Formula 35 in Table 1.

$$\int x\sqrt{a+bx}\,dx = \frac{2}{15b^2}(3bx-2a)(a+bx)^{3/2}+C$$

We have $a = 2$ and $b = 3$.

$$\begin{aligned}&\int x\sqrt{2+3x}\,dx\\ &= \frac{2}{15\cdot 3^2}(3\cdot 3x-2\cdot 2)(2+3x)^{3/2}+C\\ &= \frac{2}{135}(9x-4)(2+3x)^{3/2}+C\end{aligned}$$

35. $\int (\ln x)^4\,dx$

$$= x\ln^4 x - 4x\ln^3 x + 12x\ln^3 x - 24x\ln x + 24x + C$$

36. $\int x^3(\ln x)^2\,dx$

$$= \frac{1}{2}x^2\ln^2 x - \frac{1}{2}x^2\ln x + \frac{1}{4}x^2 + C$$

37. $\int \frac{1}{x(x^2-1)}\,dx$

$$= \frac{1}{2}\ln|x^2-1| - \ln|x| + C$$

38. $\int \frac{1}{\sqrt[3]{x}+\sqrt[4]{x}}\,dx$

$$\begin{aligned}= {} & \ln|\sqrt{x}-1| - \ln|\sqrt{x}+1| + 3\sqrt[3]{x}\\ & -4\sqrt[4]{x} + 2\sqrt{x} + \ln|\sqrt[3]{x}-\sqrt[6]{x}+1| -\\ & \ln|\sqrt[3]{x}+\sqrt[6]{x}+1| - 2\ln|\sqrt[6]{x}+1| +\\ & 2\ln|\sqrt[3]{x}-1| - \frac{12}{7}\sqrt[12]{x^7} -\\ & 2\ln|\sqrt[6]{x}-\sqrt[12]{x}+1| + 4\ln|\sqrt[12]{x}+1| +\\ & 2\ln|\sqrt[6]{x}+\sqrt[12]{x}+1| - 4\ln|\sqrt[12]{x}-1| -\\ & \ln|\sqrt[3]{x^2}+\sqrt[3]{x}+1| + 6\sqrt[6]{x} - 12\sqrt[12]{x} +\\ & \frac{3}{2}\sqrt[3]{x^2} + 2\ln|\sqrt[6]{x}-1| + \ln|x-1|\\ & -2\ln|\sqrt[4]{x}-1| + 2\ln|\sqrt[4]{x}+1| -\\ & \frac{12}{5}\sqrt[12]{x^5} + C\end{aligned}$$

39. $\int x^4 \sin 3x\,dx$

$$= -\frac{1}{3}x^4\cos 3x + \frac{4}{9}x^3\sin 3x + \frac{4}{9}x^2\cos 3x - \frac{8}{81}\cos 3x - \frac{8}{27}x\sin 3x + C$$

40. $\int x^6\cos 2x\,dx$

$$= \frac{1}{2}x^6\sin 2x + \frac{3}{2}x^5\cos 2x - \frac{15}{4}x^4\sin 2x - \frac{15}{2}x^3\cos 2x + \frac{45}{4}x^2\sin 2x - \frac{45}{8}\sin 2x + \frac{45}{4}x\cos 2x + C$$

41. $\int e^{2x}\sin 3x\,dx$

$$= \frac{-3}{13}e^{2x}\cos 3x + \frac{2}{13}e^{2x}\sin 3x + C$$

42. $\int e^{-x}\cos 2x\,dx$

$$= \frac{-1}{5}e^{-x}\cos 2x + \frac{2}{5}e^{-x}\sin 2x + C$$

43. **a)** Trapezoid: 0.742984098
b) Simpson: 0.74685538

44. **a)** Trapezoid: 18.21055563
b) Simpson: 17.75228722

45. **a)** Trapezoid: 0.449975605
b) Simpson: 0.447138991

46. **a)** Trapezoid: 0.784240767
b) Simpson: 0.785397945

47. **a)** Trapezoid: 1.503577487
b) Simpson: 1.505472368

48. **a)** Trapezoid: 0.754348074
b) Simpson: 0.748092562

49. **a)** Trapezoid: 0.270958739
b) Simpson: 0.270918581

50. **a)** Trapezoid: 0.127255793
b) Simpson: 0.127440729

51. $p'(t) = \frac{1}{t(2+t)^2}$ The integral follows formula number 33 in Table 1 with $a = 1$ and $b = 2$

$$\begin{aligned} p(t) &= \int p'(t)\,dt \\ &= \int \frac{1}{t(2+t)^2}\,dt \\ &= \frac{1}{2(2+t)} + \frac{1}{4}\ln\left|\frac{t}{2+t}\right| + C \end{aligned}$$

Applying the initial condition $p(2) = 0.8267$

$$\begin{aligned} 0.8267 &= \frac{1}{2(2+2)} + \frac{1}{4}\ln\left|\frac{2}{2+2}\right| + C \\ 0.8267 &= 0.125 - 0.1733 + C \\ C &= 0.5284 \\ p(t) &= \frac{1}{2(2+t)} + \frac{1}{4}\ln\left|\frac{2}{2+t}\right| + 0.5284 \end{aligned}$$

52. Using a spreadsheet and following Example 6, the number of degree days between 9:00 P.M. July 5 and 9:00 P.M. July 7, 2003 are 15.

53. Using a spreadsheet and following Example 6, the number of degree days between 9:00 P.M. July 5 and 9:00 P.M. July 7, 2003 are 17.

54. $\int \frac{8}{3x^2 - 2x}\,dx = 8\int \frac{1}{x(3x-2)}\,dx$
Using formula number in Table 1 with $a = 3$ and $b = -2$

$$\begin{aligned} 8\int \frac{1}{x(3x-2)}\,dx &= 8\left[\frac{1}{-2}\ln\left|\frac{x}{3x-2}\right| + C\right] \\ &= -4\ln\left|\frac{x}{3x-2}\right| + C \end{aligned}$$

55. $\int \frac{x\,dx}{4x^2 - 12x + 9} = \int \frac{x\,dx}{(2x-3)^2)}$
Using formula number in Table 1 with $a = 2$ and $b = -3$

$$\begin{aligned} \int \frac{x\,dx}{(2x-3)} &= \frac{-3}{4(2x-3)} + \frac{1}{4}\ln\left|\frac{x}{2x-3}\right| + C \\ &= \frac{3}{4(3-2x)} + \frac{1}{4}\ln\left|\frac{x}{2x-3}\right| + C \end{aligned}$$

56. $\int \frac{dx}{x^3 - 4x^2 + 4x} = \int \frac{dx}{x(x-2)^2)}$
Using formula number in Table 1 with $a = 1$ and $b = -2$

$$\begin{aligned} \int \frac{dx}{x(x-2)^2} &= \frac{1}{-2(x-2)} + \frac{1}{4}\ln\left|\frac{x}{x-2}\right| + C \\ &= \frac{1}{2(2-x)} + \frac{1}{4}\ln\left|\frac{x}{x-2}\right| + C \end{aligned}$$

57. $\int e^x\sqrt{e^{2x}+1}\,dx$
Let $y = e^x$ then $dy = e^x\,dx$
Using formula number in Table 1 with $a = 1$

$$\begin{aligned} \int e^x\sqrt{e^2x+1}\,dx &= \int \sqrt{y^2+1}\,dy \\ &= \ln\left|y + \sqrt{y^2+1}\right| + C \\ &= \ln\left|e^x + \sqrt{e^{2x}+1}\right| + C \end{aligned}$$

58. Let $y = e^{-x}$ then $dy = -e^{-x}$
Using formula number in Table 1 with $a = -1$ and $b = 3$

$$\int \frac{e^{-2x}}{9 - 6e^{-x} + e^{-2x}}\,dx$$

$$\begin{aligned} &= \int \frac{-y}{(3-y)^2}\,dy \\ &= \frac{-3}{(3-y)} - \ln|3-y| + C \\ &= \frac{3}{y-3} - \ln|3-y| + C \end{aligned}$$

59. Let $y = \ln x$ then $dy = \frac{dx}{x}$
Using formula number in Table 1 with $a = 7$

$$\int \frac{\sqrt{(\ln x)^2 + 49}}{2x}\,dx$$

$$\begin{aligned} &= \frac{1}{2}\int \sqrt{y^2+49}\,dy \\ &= \frac{1}{2}\ln\left|y + \sqrt{y^2+49}\right| + C \\ &= \frac{1}{2}\ln\left|\ln x + \sqrt{(\ln x)^2+49}\right| + C \end{aligned}$$

60. **a)** $\int_{-h}^{h} ax^2 + bx + c\,dx$

$$\begin{aligned} &= \left[\frac{ax^3}{3} + \frac{bx^2}{2} + cx\right]_{-h}^{h} \\ &= \frac{ah^3}{3} + \frac{bh^2}{2} + ch - \left(\frac{-ah^3}{3} + \frac{bh^2}{2} - ch\right) \\ &= \frac{2ah^3}{3} + 2ch \end{aligned}$$

b) $f(h) = ah^2 + bh + c$
$f(0) = c$
$f(-h) = ah^2 - bh + c$

c) $\int_{-h}^{h} ax^2 + bx + c\,dx$

$$\begin{aligned} &= \frac{2ah^3}{3} + 2ch \\ &= \frac{h}{3}\left[2ah^2 + 6c\right] \\ &= \frac{h}{3}\left[ah^2 - bh + c + 4c + c + bh + ah^2\right] \\ &= \frac{h}{3}\left[f(-h) + 4f(0) + f(h)\right] \end{aligned}$$

d) Left to the student (answers Vary)

61. **a)** $\int_{-h}^{h} ax^3 + bx^2 + cx + d\,dx$

$$= \left[\frac{ax^4}{4} + \frac{bx^3}{3} + \frac{cx^2}{2} + dx\right]_{-h}^{h}$$

$$= \frac{ah^4}{4} + \frac{bh^3}{3} + \frac{ch^2}{2} + dh - \left(\frac{ah^4}{4} - \frac{bh^3}{3} + \frac{ch^2}{2} - dh\right)$$

$$= \frac{2bh^3}{3} + 2dh$$

b) $f(h) = ah^3 + bh^2 + ch + d$
$f(0) = d$
$f(-h) = -ah^3 + bh^2 - ch + d$

c) $\int_{-h}^{h} ax^3 + bx^2 + cx + d\,dx$

$$= \frac{2bh^3}{3} + 2dh$$

$$= \frac{h}{3}\left[2bh^2 + 6d\right]$$

$$= \frac{h}{3}\left[-ah^3 + bh^2 - ch + d + 4d + d + ch + bh^2 + ah^3\right]$$

$$= \frac{h}{3}\left[f(-h) + 4f(0) + f(h)\right]$$

d) Left to the student (answers Vary)

Exercise Set 5.8

1. Find the volume of the solid of revolution generated by rotating about the x-axis the region under the graph of

$$y = x$$

from $x = 0$ to $x = 1$.

$V = \int_a^b \pi\left[f(x)\right]^2 dx$ Volume of a solid of revolution

$V = \int_0^1 \pi\, x^2\, dx$ Substituting 0 for a, 1 for b, and x for $f(x)$

$$= \left[\pi \cdot \frac{x^3}{3}\right]_0^1$$

$$= \frac{\pi}{3}\left[x^3\right]_0^1$$

$$= \frac{\pi}{3}(1^3 - 0^3)$$

$$= \frac{\pi}{3} \cdot 1$$

$$= \frac{\pi}{3}$$

2. $V = \int_0^2 \pi(\sqrt{x})^2\, dx$

$$= \int_0^2 \pi\, x\, dx$$

$$= \left[\frac{\pi\, x^2}{2}\right]_0^2$$

$$= \frac{\pi}{2}\left[x^2\right]_0^2$$

$$= \frac{\pi}{2}(2^2 - 0^2)$$

$$= \frac{\pi}{2} \cdot 4$$

$$= 2\pi$$

3. Find the volume of the solid of revolution generated by rotating about the x-axis the region under the graph of

$y = \sqrt{\sin x}$

from $x = 0$ to $x = \frac{\pi}{2}$.

$$V = \int_a^b \pi\left[f(x)\right]^2 dx$$

$$V = \int_0^{\pi/2} \pi \sin x\, dx$$

$$= \pi\left[-\cos x\right]_0^{\pi/2}$$

$$= \pi\left[0 - (-1)\right]$$

$$= \pi$$

4.

$y = \sqrt{\cos x}$

from $x = 0$ to $x = \frac{\pi}{4}$.

$$V = \int_a^b \pi\left[f(x)\right]^2 dx$$

$$V = \int_0^{\pi/4} \pi \cos x\, dx$$

$$= \pi\left[\sin x\right]_0^{\pi/4}$$

$$= \pi\left[\frac{1}{\sqrt{2}} - 0\right]$$

$$= \frac{\pi}{\sqrt{2}} = \frac{\sqrt{2}\,\pi}{2}$$

5. Find the volume of the solid of revolution generated by rotating about the x-axis the region under the graph of

$$y = e^x$$

from $x = -2$ to $x = 5$.

$V = \int_a^b \pi[f(x)]^2\,dx$ Volume of a solid of revolution

$V = \int_{-2}^5 \pi\,[e^x]^2\,dx$ Substituting -2 for a, 5 for b, and e^x for $f(x)$

$= \int_{-2}^5 \pi\,e^{2x}\,dx$

$= \left[\pi \cdot \frac{1}{2}e^{2x}\right]_{-2}^5$

$= \frac{\pi}{2}\left[e^{2x}\right]_{-2}^5$

$= \frac{\pi}{2}(e^{2\cdot 5} - e^{2(-2)})$

$= \frac{\pi}{2}(e^{10} - e^{-4})$

6. $V = \int_{-3}^2 \pi\,(e^x)^2\,dx$

$= \int_{-3}^2 \pi\,e^{2x}\,dx$

$= \left[\frac{\pi\,e^{2x}}{2}\right]_{-3}^2$

$= \frac{\pi}{2}\left[e^{2x}\right]_{-3}^2$

$= \frac{\pi}{2}(e^{2\cdot 2} - e^{2(-3)})$

$= \frac{\pi}{2}(e^4 - e^{-6})$

7. Find the volume of the solid of revolution generated by rotating about the x-axis the region under the graph of

$$y = \frac{1}{x}$$

from $x = 1$ to $x = 3$.

$V = \int_a^b \pi[f(x)]^2\,dx$ Volume of a solid of revolution

$V = \int_1^3 \pi\left[\frac{1}{x}\right]^2 dx$ Substituting 1 for a, 3 for b, and $\frac{1}{x}$ for $f(x)$

$= \int_1^3 \pi \cdot \frac{1}{x^2}\,dx$

$= \int_1^3 \pi\,x^{-2}\,dx$

$= \left[\pi\frac{x^{-1}}{-1}\right]_1^3$

$= -\pi\left[\frac{1}{x}\right]_1^3$

$= -\pi\left(\frac{1}{3} - \frac{1}{1}\right)$

$= -\pi \cdot \left(-\frac{2}{3}\right)$

$= \frac{2}{3}\pi$

8. $V = \int_1^4 \pi\left(\frac{1}{x}\right)^2 dx$

$= \int_1^4 \pi\,x^{-2}\,dx$

$= \left[\frac{\pi\,x^{-1}}{-1}\right]_1^4$

$= -\pi\left[\frac{1}{x}\right]_1^4$

$= -\pi\left(\frac{1}{4} - 1\right)$

$= \frac{3\pi}{4}$

9. Find the volume of the solid of revolution generated by rotating about the x-axis the region under the graph of

$$y = \frac{2}{\sqrt{x}}$$

from $x = 1$ to $x = 3$.

$V = \int_a^b \pi[f(x)]^2\,dx$ Volume of a solid of revolution

$V = \int_1^3 \pi\left[\frac{2}{\sqrt{x}}\right]^2 dx$ Substituting 1 for a, 3 for b, and $\frac{2}{\sqrt{x}}$ for $f(x)$

$= \int_1^3 \pi \cdot \frac{4}{x}\,dx$

$= 4\pi\left[\ln x\right]_1^3$

$= 4\pi(\ln 3 - \ln 1)$

$= 4\pi\ln 3$ $(\ln 1 = 0)$

10. $V = \int_1^4 \pi\left(\frac{1}{\sqrt{x}}\right)^2 dx$

$= \int_1^4 \pi\left(\frac{1}{x}\right) dx$

$= \left[\pi\ln x\right]_1^4$

$= \pi\left[\ln x\right]_1^4$

$= \pi\,(\ln 4 - \ln 1)$

$= \pi\ln 4$

11. Find the volume of the solid of revolution generated by rotating about the x-axis the region under the graph of

$$y = 4$$

from $x = 1$ to $x = 3$.

$V = \int_a^b \pi[f(x)]^2\,dx$ Volume of a solid of revolution

$V = \int_1^3 \pi\,[4]^2\,dx$ Substituting 1 for a, 3 for b, and 4 for $f(x)$

$= \int_1^3 16\pi\,dx$

$= 16\pi\left[x\right]_1^3$

$= 16\pi(3 - 1)$

$= 32\pi$

12. $V = \int_1^3 \pi\,(5)^2\,dx$
$= \int_1^3 25\pi\,dx$
$= \left[25\,\pi\,x\right]_1^3$
$= 25\pi\,(3-1)$
$= 50\pi$

13. Find the volume of the solid of revolution generated by rotating about the x-axis the region under the graph of

$y = x^2$

from $x = 0$ to $x = 2$.

$V = \int_a^b \pi\left[f(x)\right]^2 dx$ Volume of a solid of revolution

$V = \int_0^2 \pi\left[x^2\right]^2 dx$ Substituting 0 for a, 2 for b, and x^2 for $f(x)$

$= \int_0^2 \pi\,x^4\,dx$

$= \left[\pi\cdot\frac{x^5}{5}\right]_0^2$

$= \frac{\pi}{5}(2^5 - 0^5)$

$= \frac{32}{5}\pi$

14. $V = \int_{-1}^2 \pi\,(x+1)^2\,dx$

$= \left[\frac{\pi\,(x+1)^3}{3}\right]_{-1}^2$

$= \frac{\pi}{3}[(2+1)^3 - (-1+1)^3]$

$= \frac{\pi}{3}\cdot 27$

$= 9\pi$

15. Find the volume of the solid of revolution generated by rotating about the x-axis the region under the graph of

$y = cos\ x$

from $x = 0$ to $x = \frac{\pi}{2}$.

$$V = \int_a^b \pi\left[f(x)\right]^2 dx$$

$$V = \int_0^{\pi/2} \pi\ cos^2\ x\,dx$$

$$= \pi\left[\frac{1}{2}\,x + \frac{1}{2}\ sin\ x\ cos\ x\right]_0^{\pi/2}$$

$$= \frac{\pi}{2}\left[\frac{\pi}{2} + 0 - 0\right]$$

$$= \frac{\pi^2}{4}$$

16.

$y = sec\ 2x$

from $x = -\frac{\pi}{8}$ to $x = \frac{\pi}{8}$.

$$V = \int_a^b \pi\left[f(x)\right]^2 dx$$

$$V = \int_{-\pi/8}^{\pi/8} \pi\ sec^2\ 2x\,dx$$

$$= \pi\left[\frac{1}{2}\ tan\ 2x\right]_{\frac{-\pi}{8}}^{\pi/8}$$

$$= \frac{\pi}{2}\left[1 - (-1)\right]$$

$$= \pi$$

17. Find the volume of the solid of revolution generated by rotating about the x-axis the region under the graph of

$y = tan\ x$

from $x = 0$ to $x = \frac{\pi}{4}$.

$$V = \int_a^b \pi\left[f(x)\right]^2 dx$$

$$V = \int_0^{\pi/2} \pi\ tan^2\ x\,dx$$

$$= \pi\left[tan\ x + x\right]_0^{\pi/4}$$

$$= \pi\left[1 + \frac{\pi}{4} - 0\right]$$

$$= \pi + \frac{\pi^2}{4}$$

18.

$y = csc\ x$

from $x = \frac{\pi}{4}$ to $x = \frac{\pi}{2}$.

$$V = \int_a^b \pi\left[f(x)\right]^2 dx$$

$$V = \int_{\pi/4}^{\pi/2} \pi\ csc^2\ x\,dx$$

$$= \pi\left[-\ cot\ x\right]_{\frac{\pi}{4}}^{\pi/2}$$

$$= \pi\left[0 - (-1)\right]$$

$$= \pi$$

19. Find the volume of the solid of revolution generated by rotating about the x-axis the region under the graph of

$y = \sqrt{1+x}$

from $x = 2$ to $x = 10$.

$V = \int_a^b \pi\left[f(x)\right]^2 dx$ Volume of a solid of revolution

$V = \int_2^{10} \pi\left(\sqrt{1+x}\right)^2 dx$ Substituting 2 for a, 10 for b, and $\sqrt{1+x}$ for $f(x)$

$= \int_2^{10} \pi\,(1+x)\,dx$

$= \left[\pi\cdot\frac{(1+x)^2}{2}\right]_2^{10}$

$= \frac{\pi}{2}\left[(1+x)^2\right]_2^{10}$

$= \frac{\pi}{2}[(1+10)^2 - (1+2)^2]$

$= \frac{\pi}{2}(11^2 - 3^2)$

$= \frac{\pi}{2}(121 - 9)$

$= \frac{\pi}{2} \cdot 112$

$= 56\pi$

20. $V = \int_1^2 \pi\,(2\sqrt{x})^2\,dx$

$= \int_1^2 4\pi\,x\,dx$

$= \left[2\pi\,x^2\right]_1^2$

$= 2\pi\,(2^2 - 1^2)$

$= 6\pi$

21. Find the volume of the solid of revolution generated by rotating about the x-axis the region under the graph of

$$y = \sqrt{4 - x^2}$$

from $x = -2$ to $x = 2$.

$V = \int_a^b \pi[f(x)]^2\,dx$ Volume of a solid of revolution

$V = \int_{-2}^2 \pi\left(\sqrt{4-x^2}\right)^2 dx$ Substituting -2 for a, 2 for b, and $\sqrt{4-x^2}$ for $f(x)$

$= \int_{-2}^2 \pi(4-x^2)\,dx$

$= \pi\left[4x - \frac{x^3}{3}\right]_{-2}^2$

$= \pi\left[\left(4\cdot 2 - \frac{2^3}{3}\right) - \left(4\cdot(-2) - \frac{(-2)^3}{3}\right)\right]$

$= \pi\left(8 - \frac{8}{3} + 8 - \frac{8}{3}\right)$

$= \pi\left(16 - \frac{16}{3}\right)$

$= \pi\left(\frac{48}{3} - \frac{16}{3}\right)$

$= \frac{32}{3}\pi$

22.

$V = \int_{-r}^{r} \pi\,(\sqrt{r^2 - x^2})^2\,dx$

$= \int_{-r}^{r} \pi\,(r^2 - x^2)\,dx$

$= \left[\pi\left(r^2\,x - \frac{x^3}{3}\right)\right]_{-r}^{r}$

$= \pi\left[\left(r^2\cdot r - \frac{r^3}{3}\right) - \left(r^2(-r) - \frac{(-r)^3}{3}\right)\right]$

$= \pi\left[\left(r^3 - \frac{r^3}{3}\right) - \left(-r^3 + \frac{r^3}{3}\right)\right]$

$= \pi\left(\frac{2}{3}r^3 + \frac{2}{3}r^3\right)$

$= \frac{4}{3}\pi r^3$

23. Find the volume of the solid with cross-sectional area

$$A(x) = \frac{1}{2}\,x^2$$

from $x = 0$ to $x = 6$.

$V = \int_a^b A(x)\,dx$

$V = \int_0^6 \left(\frac{1}{2}\,x^2\right) dx$

$= \left[\frac{1}{6}\,x^3\right]_0^6$

$= \left(36 - 0\right)$

$= 36$

24.

$$A(x) = 1 - x^2$$

from $x = -1$ to $x = 1$.

$V = \int_a^b A(x)\,dx$

$V = \int_{-1}^1 \left(1 - x^2\right) dx$

$= \left[x - \frac{1}{3}\,x^3\right]_{-1}^1$

$= \left(1 - \frac{1}{3}\right) - \left(-1 + \frac{1}{3}\right)$

$= \frac{4}{3}$

25. Find the volume of the solid with cross-sectional area

$$A(x) = \frac{\sqrt{3}}{2}\,x^2$$

from $x = 0$ to $x = 9$.

$V = \int_a^b A(x)\,dx$

$V = \int_0^9 \left(\frac{\sqrt{3}}{2}\,x^2\right) dx$

$= \left[\frac{\sqrt{3}}{6}\,x^3\right]_0^9$

$= \left(\frac{243\,\sqrt{3}}{2} - 0\right)$

$= \frac{243\,\sqrt{3}}{2}$

26.

$$A(x) = \frac{1}{x^2}$$

from $x = 1$ to $x = 4$.

$$V = \int_a^b A(x)\,dx$$
$$V = \int_1^4 \left(\frac{1}{x^2}\right) dx$$
$$= \left[-\frac{1}{x}\right]_1^4$$
$$= -\frac{1}{4} - (-1)$$
$$= \frac{3}{4}$$

27. $A(x) = x^4$

from $x = 0$ to $x = 4$.

$$V = \int_a^b A(x)\,dx$$
$$V = \int_0^4 \left(x^4\right) dx$$
$$= \left[\frac{1}{5}\,x^5\right]_0^4$$
$$= \frac{1024}{5} - 0$$
$$= \frac{1024}{5}$$

28. $A(x) = 4x^4$

from $x = 0$ to $x = 3$.

$$V = \int_a^b A(x)\,dx$$
$$V = \int_0^3 \left(4x^4\right) dx$$
$$= \left[\frac{4}{5}\,x^5\right]_0^3$$
$$= \frac{972}{5} - 0$$
$$= \frac{972}{5}$$

29. $A(x) = 2x^2$

from $x = 0$ to $x = 5$.

$$V = \int_a^b A(x)\,dx$$
$$V = \int_0^5 \left(2x^2\right) dx$$
$$= \left[\frac{2}{3}\,x^3\right]_0^5$$
$$= \frac{250}{3} - 0$$
$$= \frac{250}{3}$$

30. $A(x) = \frac{1}{8}\,x^2$

from $x = 0$ to $x = 8$.

$$V = \int_a^b A(x)\,dx$$
$$V = \int_0^8 \left(\frac{1}{8}\,x^2\right) dx$$
$$= \left[\frac{1}{24}\,x^3\right]_0^8$$
$$= \frac{64}{3} - 0$$
$$= \frac{64}{3}$$

31. $10 = K \cdot 8 \to K = \frac{5}{4}$ (See Example 4 on page 410 as a reference)

$A(x) = \frac{25}{16}x^2$

from $x = 0$ to $x = 8$.

$$V = \int_a^b A(x)\,dx$$
$$V = \int_0^8 \left(\frac{25}{16}\,x^2\right) dx$$
$$= \left[\frac{25}{48}\,x^3\right]_0^8$$
$$= \frac{800}{3} - 0$$
$$= \frac{800}{3}$$

32. $5 = K \cdot 12 \to K = \frac{5}{12}$

$A(x) = \frac{25}{144}x^2$

from $x = 0$ to $x = 12$.

$$V = \int_a^b A(x)\,dx$$
$$V = \int_0^{12} \left(\frac{25}{144}\,x^2\right) dx$$
$$= \left[\frac{25}{432}\,x^3\right]_0^{12}$$
$$= 100 - 0$$
$$= 100$$

33. For $r = 1.5$ when $H = 75$ and $x = 0$, $K = 38.2285$

Thus, $r(x) = 0.05886(75 - x)^{3/4}$

$V = \int_a^b \pi \left[r(x)\right]^2 dx$

$V = \int_0^{75} \pi \left(0.05886(75 - x)^{3/4}\right)^2 dx$

$= \int_0^{75} \pi \left(0.003464(75 - x)^{3/2}\right) dx$

$= -0.003464\pi \left[\frac{2}{5}(75 - x)^{5/2}\right]_0^{75}$

$= -0.003464\pi \left(0 - \frac{-2}{5}\ 75^{5/2}\right)$

$= 212.058$

34. For $r = 1$ when $H = 50$ and $x = 0$, $K = 0.05318$
Thus, $r(x) = 0.05318(50 - x)^{3/4}$

$V = \int_a^b \pi \left[r(x)\right]^2 dx$

$V = \int_0^{50} \pi \left(0.05318(50 - x)^{3/4}\right)^2 dx$

$= \int_0^{50} \pi \left(0.002828(50 - x)^{3/2}\right) dx$

$= -0.002828\pi \left[\frac{2}{5}(50 - x)^{5/2}\right]_0^{50}$

$= -0.002828\pi \left(0 - \frac{-2}{5}\ 50^{5/2}\right)$

$= 62.832$

35. Find the volume of the solid of revolution generated by rotating about the x-axis the region under the graph of

$y = \sqrt{\ln x}$

from $x = e$ to $x = e^3$.

$V = \int_a^b \pi \left[f(x)\right]^2 dx$

$V = \int_e^{e^3} \pi \left[\sqrt{\ln x}\right]^2 dx$

$= \int_3^{e^3} \pi \ln x \, dx$

$= \pi \left[x \ln x - x\right]_e^{e^3}$

$= \pi[(e^3 \ln e^3 - e^3) - (e \ln e - e)]$

$= \pi[(3e^3 - e^3) - (e - e)]$

$= 2\pi e^3$

36. $V = \int_1^2 \pi \left(\sqrt{x}\, e^{-x}\right)^2 dx$

$= \int_1^2 \pi\, x\, e^{-x} dx$

$= \left[\pi \cdot \frac{1}{(-1)^2} e^{-x}(-x - 1)\right]_1^2$

$= \pi \left[\frac{-x - 1}{e^x}\right]_1^2$

$= \pi \left[\frac{-2 - 1}{e^2} - \frac{-1 - 1}{e^1}\right]$

$= \pi \left[\frac{-3}{e^2} + \frac{2}{3^2}\right]$

$= \pi \left(\frac{-3 + 2e}{e^2}\right)$

37. **a)** The resulting solid of revolution is a cone with a base radius of r and a height of h

b)

$$\begin{aligned} V &= \int_0^h \pi y^2 \, dx \\ &= \pi \int_0^h \frac{r^2}{h^2} x^2 \, dx \\ &= \frac{\pi r^2}{3h^2}\left[x^3\right]_0^h \\ &= \frac{\pi r^2}{3h^2}\left[h^3 - 0\right] \\ &= \frac{1}{3}\pi r^2 h \end{aligned}$$

38. **a)** Trapezoid: 53

b) Simpson: 52.06667

Exercise Set 5.9

1. $\displaystyle\int_2^\infty \frac{dx}{x^2}$

$= \lim_{b\to\infty} \int_2^b x^{-2} dx$

$= \lim_{b\to\infty} \left[\frac{x^{-1}}{-1}\right]_2^b$

$= \lim_{b\to\infty} \left[-\frac{1}{x}\right]_2^b$

$= \lim_{b\to\infty} \left(-\frac{1}{b} - \left(-\frac{1}{2}\right)\right)$

$= \frac{1}{2} \qquad \left(\text{As } b \to \infty,\ -\frac{1}{b} \to 0 \text{ and } -\frac{1}{b} + \frac{1}{2} \to \frac{1}{2}.\right)$

The limit does exist. Thus the improper integral is convergent.

2. $\displaystyle\int_2^\infty \frac{dx}{x} = \lim_{b\to\infty} \int_2^b \frac{1}{x} dx$

$= \lim_{b\to\infty} \left[\ln x\right]_2^b$

$= \lim_{b\to\infty} (\ln b - \ln 2)$

Note that $\ln b$ increases indefinitely as b increases. Therefore, the limit does not exist. If the limit does not exist, we say the improper integral is divergent.

3. $\displaystyle\int_4^\infty \frac{dx}{x} = \lim_{b\to\infty} \int_4^b \frac{1}{x} dx$

$= \lim_{b\to\infty} \left[\ln x\right]_4^b$

$= \lim_{b\to\infty} (\ln b - \ln 4)$

Note that $\ln b$ increases indefinitely as b increases. Therefore, the limit does not exist. If the limit does not exist, we say the improper integral is divergent.

4. $\int_0^\infty x^2\,dx = \lim_{b\to\infty}\int_0^b \frac{1}{3}x^3\,dx$

$= \lim_{b\to\infty}\frac{1}{3}[b^3 - 0]$

$= \lim_{b\to\infty}\frac{b^3}{3}$

Note that b^3 increases indefinitely as b increases. Therefore, the limit does not exist. If the limit does not exist, we say the improper integral is divergent.

5. $\int_{-\infty}^{-1}\frac{dt}{t^2}$

$= \lim_{b\to-\infty}\int_b^{-1} t^{-2}\,dt$

$= \lim_{b\to-\infty}\left[\frac{t^{-1}}{-1}\right]_b^{-1}$

$= \lim_{b\to-\infty}\left[-\frac{1}{t}\right]_b^{-1}$

$= \lim_{b\to-\infty}\left(1-\frac{1}{b}\right)$

$= 1 \quad \left(\text{As } b\to-\infty,\ -\frac{1}{b}\to 0 \text{ and } 1-\frac{1}{b}\to 1.\right)$

The limit does exist. Thus the improper integral is convergent.

6. $\int_{-\infty}^{-4}\frac{dx}{x}$

$= \lim_{b\to-\infty}\int_b^{-4}\frac{dx}{x}$

$= \lim_{b\to-\infty}\left[\ln x\right]_b^{-4}$

$= \lim_{b\to-\infty}\left[\ln -4 - \ln -\infty\right]$

The limit does not exist. If the limit does not exist, we say the improper integral is divergent.

7. $\int_0^\infty 4\,e^{-4x}\,dx = \lim_{b\to\infty}\int_0^b 4\,e^{-4x}\,dx$

$= \lim_{b\to\infty}\left[\frac{4}{-4}e^{-4x}\right]_0^b$

$= \lim_{b\to\infty}\left[-e^{-4x}\right]_0^b.$

$= \lim_{b\to\infty}[-e^{-4b} - (-e^{-4\cdot 0})]$

$= \lim_{b\to\infty}\left(-\frac{1}{e^{4b}}+1\right)$

$= 1$

The integral is convergent.

8. $\int_0^\infty \frac{du}{1+u} = \lim_{b\to\infty}\int_0^b \frac{1}{1+u}\,du$

$= \lim_{b\to\infty}\left[\ln(1+u)\right]_0^b$

$= \lim_{b\to\infty}[\ln(1+b) - \ln(1+0)]$

$= \lim_{b\to\infty}[\ln(1+b) - \ln 1]$

$= \lim_{b\to\infty}\ln(1+b)$

Note that $\ln(1+b)$ increases indefinitely as b increases. Therefore, the limit does not exist. Thus, the improper integral is divergent.

9. $\int_0^\infty e^x\,dx = \lim_{b\to\infty}\int_0^b e^x\,dx$

$= \lim_{b\to\infty}\left[e^x\right]_0^b$

$= \lim_{b\to\infty}(e^b - e^0)$

$= \lim_{b\to\infty}(e^b - 1)$

As $b\to\infty$, $e^b\to\infty$. Thus the limit does not exist. The improper integral is divergent.

10. $\int_0^\infty e^{2x}\,dx = \lim_{b\to\infty}\int_0^b e^{2x}\,dx$

$= \lim_{b\to\infty}\left[\frac{1}{2}e^{2x}\right]_0^b$

$= \lim_{b\to\infty}\left(\frac{1}{2}e^{2b} - \frac{1}{2}e^{2\cdot 0}\right)$

The limit does not exist. The integral is divergent.

11. $\int_{-\infty}^\infty e^{2x}\,dx = \lim_{b\to\infty}\int_{-b}^b e^{2x}\,dx$

$= \lim_{b\to\infty}\left[\frac{1}{2}e^{2x}\right]_{-b}^b$

$= \lim_{b\to\infty}(e^b - e^{-b})$

As $b\to\pm\infty$, $e^b\to\infty$. Thus the limit does not exist. The improper integral is divergent.

12. $\int_{-\infty}^\infty \frac{x\,dx}{(1+x^2)^2} = \lim_{b\to\infty}\int_{-b}^b \frac{x\,dx}{(1+x^2)^2}$

$= \lim_{b\to\infty}\left[\frac{-1}{2(1+x^2)}\right]_{-b}^b$

$= \lim_{b\to\infty}[0]$

$= 0$

The integral is convergent.

13. $\int_{-\infty}^\infty \frac{t\,dt}{(1+t^2)^3} = \lim_{b\to\infty}\int_{-b}^b \frac{t\,dt}{(1+t^2)^3}$

$= \lim_{b\to\infty}\left[\frac{-1}{4(1+t^2)^2}\right]_{-b}^b$

$= \lim_{b\to\infty}[0]$

$= 0$

The integral is convergent.

14. $\int_{-1}^\infty \frac{2x+3}{(x^2+3x+6)^4}\,dx = \lim_{b\to\infty}\int_{-1}^b \frac{2x+3\,dt}{(x^2+3x+6)^4}$

$= \lim_{b\to\infty}\left[\frac{-1}{3(x^2+3x+6)^3}\right]_{-1}^b$

$= \lim_{b\to\infty}\left[\frac{-1}{(b^2+3b+6)^3} - \frac{-1}{(4)^3}\right]$

$= 0 + \frac{1}{64}$

$= \frac{1}{64}$

The integral is convergent.

15. $\int_{-\infty}^\infty 2x\,e^{-3x^2}\,dx = \lim_{b\to\infty}\int_{-b}^b 2xe^{-3x^2}\,dx$

$= \lim_{b\to\infty}\left[\frac{-1}{3}e^{-3x^2}\right]_{-b}^b$

$= \lim_{b\to\infty}\left(\frac{-1}{3}e^{-3b^2} - \frac{-1}{3}e^{-3b^2}\right)$

As $b\to\pm\infty$, $e^b\to\infty$. Thus the limit does not exist. The improper integral is divergent.

16. $\int_{-\infty}^{\infty} x^2\, e^{-x^3}\, dx = \lim\limits_{b\to\infty} \int_{-b}^{b} x^2\, e^{-x^3}\, dx$

$= \lim\limits_{b\to\infty} \left[\frac{-1}{3} e^{-x^3}\right]_{-b}^{b}$

$= \lim\limits_{b\to\infty} \left(\frac{-1}{3} e^{-b^3} - \frac{-1}{3} e^{b^3}\right)$

As $b \to \pm\infty$, $e^b \to \infty$. Thus the limit does not exist. The improper integral is divergent.

17. $\int_0^{\infty} 2t^2\, e^{-2t}\, dt = \lim\limits_{b\to\infty} \int_0^b 2t^2\, e^{-2t}\, dt$

$= \lim\limits_{b\to\infty} \left[t^2\, e^{-2t} - t\, e^{-2t} + \frac{1}{2}\, e^{-2t}\right]_0^b$

$= \lim\limits_{b\to\infty} \left(-b^2\, e^{-2b} - b\, e^{-2b} - \frac{1}{2}\, e^{-2b} - \frac{-1}{2}\right)$

$= \frac{1}{2}$

The integral is convergent.

18. $\int_{-\infty}^{0} x\, e^x\, dx = \lim\limits_{b\to-\infty} \int_b^0 x\, e^x\, dx$

$= \lim\limits_{b\to-\infty} \left[x\, e^x - e^x\right]_b^0$

$= \lim\limits_{b\to-\infty} (-1 - b\, e^b + e^b)$

$= -1$

The integral is convergent.

19. $\int_{-\infty}^{1} 2x\, e^{3x}\, dx = \lim\limits_{b\to-\infty} \int_b^1 2x\, e^{3x}\, dx$

$= \lim\limits_{b\to-\infty} \left[\frac{2}{3} x\, e^{3x} - \frac{2}{9}\, e^{3x}\right]_b^1$

$= \lim\limits_{b\to-\infty} \left(\frac{2}{3} e^3 - \frac{2}{9} e^3 - \frac{2}{3} b\, e^{3b} + \frac{2}{9} e^{3b}\right)$

$= \frac{4}{9} e^3$

The integral is convergent.

20. $\int_0^{\infty} \frac{dt}{t^{2/3}} = \lim\limits_{b\to\infty} \int_0^b t^{-2/3}\, dt$

$= \lim\limits_{b\to\infty} \left[3t^{1/3}\right]_0^b$

$= \lim\limits_{b\to\infty} \left(3b^{1/3} - 0\right)$

$= \infty$

The integral is divergent.

21. $\int_1^{\infty} \frac{dx}{x^{1/2}} = \lim\limits_{b\to\infty} \int_1^b x^{-1/2}\, dx$

$= \lim\limits_{b\to\infty} \left[2x^{1/2}\right]_1^b$

$= \lim\limits_{b\to\infty} \left(2b^{1/2} - 2\right)$

$= \infty$

The integral is divergent.

22. $\int_0^{\infty} \frac{e^t}{(5+e^t)^2}\, dt = \lim\limits_{b\to\infty} \int_0^b e^t(5+e^t)^{-2}\, dt$

$= \lim\limits_{b\to\infty} \left[\frac{-1}{(5+e^t)}\right]_0^b$

$= \lim\limits_{b\to\infty} \left(\frac{-1}{5+e^b} + \frac{1}{5}\right)$

$= \frac{1}{5}$

The integral is convergent.

23. $\int_0^{\infty} \frac{1 - \sin x}{(x + \cos x)^2}\, dx = \lim\limits_{b\to\infty} \int_0^b (1 - \sin x)(x + \cos x)^{-2}\, dx$

$= \lim\limits_{b\to\infty} \left[\frac{-1}{(x + \cos x)}\right]_0^b$

$= \lim\limits_{b\to\infty} \left(\frac{-1}{b + \cos b} + 1\right)$

$= 1$

The integral is convergent.

24. $\int_0^{\infty} \sin\theta\, d\theta = \lim\limits_{b\to\infty} \int_0^b \sin\theta\, d\theta$

$= \lim\limits_{b\to\infty} \left[-\cos\theta\right]_0^b$

$= \lim\limits_{b\to\infty} (-\cos b + 1)$

The limit does not exist. The integral is divergent.

25. $\int_1^{\infty} \frac{1}{x(x+1)^2}\, dx = \lim\limits_{b\to\infty} \int_1^b \frac{1}{x^3 + 2x^2 + 3x}\, dx$

$= \lim\limits_{b\to\infty} \left[\ln x + \frac{1}{x+1} - \ln(x+1)\right]_1^b$

$= \lim\limits_{b\to\infty} \left[\ln \frac{x}{x+1} + \frac{1}{x+1}\right]_1^b$

$= \left(\ln 1 + 0 - \ln\frac{1}{2} - \frac{1}{2}\right)$

$= \ln 2 - \frac{1}{2}$

The integral is convergent.

26. $\int_5^{\infty} \frac{dt}{t^2 - 16} = \lim\limits_{b\to\infty} \int_5^b \frac{1}{8(t-4)} - \frac{1}{8(t+4)}\, dt$

$= \lim\limits_{b\to\infty} \left[\frac{1}{8}\ln(t-4) - \frac{1}{8}\ln(t+4)\right]_5^b$

$= \lim\limits_{b\to\infty} \left[\frac{1}{8(b-4)} - \frac{1}{8(b+4)} - \frac{\ln 9}{8}\right]$

$= \frac{\ln 9}{8}$

The integral is convergent.

27. $\int_1^{\infty} \frac{x}{\sqrt{x+2}}\, dx = \lim\limits_{b\to\infty} \int_1^b \frac{x}{\sqrt{x+2}}\, dx$

$= \lim\limits_{b\to\infty} \left[x - 2\ - ln(x+2)\right]_1^b$

$= \lim\limits_{b\to\infty} \left[b - 2\ln(b+2) - 1 + 2\ln(3)\right]$

The limit does not exist. The integral is divergent.

28. $$\int_e^\infty \frac{\ln x}{x^2}\,dx = \lim_{b\to\infty}\int_e^b \frac{\ln x}{x^2}\,dx$$
$$= \lim_{b\to\infty}\left[-\frac{\ln x}{x}-\frac{1}{x}\right]_e^b$$
$$= \lim_{b\to\infty}\left[-\frac{\ln b}{b}-\frac{1}{b}+\frac{1}{e}+\frac{1}{e}\right]$$
$$= \frac{2}{e}$$

The integral is convergent.

29. $$\int_e^\infty \frac{\ln x}{x}\,dx = \lim_{b\to\infty}\int_e^b \frac{\ln x}{x}\,dx$$
$$= \lim_{b\to\infty}\left[\ln^2 x\right]_e^b$$
$$= \lim_{b\to\infty}\left[\ln^2 b - 1\right]$$

The limit does not exist. The integral is divergent.

30. $$\int_2^\infty \frac{dt}{t\sqrt{t^2+1}} = \lim_{b\to\infty}\int_2^b \frac{dt}{t\sqrt{t^2+1}}$$
$$= \lim_{b\to\infty}\left[\mathit{arcsinh}\ t\right]_2^b$$
$$= \lim_{b\to\infty}\left[\mathit{arcsinh}\ b - \mathit{arcsinh}\ 2\right]$$

The limit does not exist. The integral is divergent.

31. $$\int_{-\infty}^\infty 3xe^{-x^2/2}\,dx = \lim_{b\to\infty}\int_{-b}^b 3xe^{x^2/2}\,dx$$
$$= \lim_{b\to\infty}\left[-3\,e^{-x^2/2}\right]_{-b}^b$$
$$= \lim_{b\to\infty}\left[-3\,e^{-b^2/2}+3\,e^{-b^2/2}\right]$$
$$= 0$$

The integral is convergent.

32. $$\int_{-\infty}^\infty 2xe^{-x^2}\,dx = \lim_{b\to\infty}\int_{-b}^b 2xe^{x^2}\,dx$$
$$= \lim_{b\to\infty}\left[-e^{-x^2}\right]_{-b}^b$$
$$= \lim_{b\to\infty}\left[-e^{-b^2}+e^{-b^2}\right]$$
$$= 0$$

The integral is convergent.

33. $\int_0^\infty m\,e^{-mx}\,dx,\ m>0$
$$= \lim_{b\to\infty}\int_0^b m\,e^{-mx}\,dx$$
$$= \lim_{b\to\infty}\left[\frac{m}{-m}e^{-mx}\right]_0^b$$
$$= \lim_{b\to\infty}\left[-e^{-mx}\right]_0^b$$
$$= \lim_{b\to\infty}[-e^{-mb}-(-e^{-m\cdot 0})]$$
$$= \lim_{b\to\infty}\left(1-\frac{1}{e^{mb}}\right)$$
$$= 1$$

The limit does exist. Thus the improper integral is convergent.

34. $\int_0^\infty A\,e^{-kt}\,dt,\ k>0$
$$= \lim_{b\to\infty}\int_0^b A\,e^{-kt}\,dt$$
$$= \lim_{b\to\infty}\left[\frac{A}{-k}e^{-kt}\right]_0^b$$
$$= \lim_{b\to\infty}\left[-\frac{A}{k}\,e^{-kt}\right]_0^b$$
$$= \lim_{b\to\infty}\left[-\frac{A}{k}\,e^{-kb}+\frac{A}{k}\right]$$
$$= \lim_{b\to\infty}\left(\frac{A}{k}-\frac{A}{k\,e^{mb}}\right)$$
$$= \frac{A}{k}$$

The limit does exist. Thus the improper integral is convergent.

35. The area is given by
$$\int_2^\infty \frac{1}{x^2}\,dx = \lim_{b\to\infty}\int_2^b x^{-2}\,dx$$
$$= \lim_{b\to\infty}\left[-x^{-1}\right]_2^b$$
$$= \lim_{b\to\infty}\left[-\frac{1}{x}\right]_2^b$$
$$= \lim_{b\to\infty}\left(-\frac{1}{b}-\left(-\frac{1}{2}\right)\right)$$
$$= \frac{1}{2}$$

The area of the region is $\frac{1}{2}$.

36. $$\int_2^\infty \frac{1}{x}\,dx = \lim_{b\to\infty}\int_2^b \frac{1}{x}\,dx$$
$$= \lim_{b\to\infty}\left[\ln x\right]_2^b$$
$$= \lim_{b\to\infty}(\ln b - \ln 2)$$

The limit does not exist. As b increases, $\ln b$ increases indefinitely. The area is infinite.

37. The area is given by
$$\int_0^\infty 2x\,e^{-x^2}\,dx$$
$$= \lim_{b\to\infty}\int_0^b 2x\,e^{-x^2}\,dx$$

(We use the substitution $u=-x^2$ to integrate.)
$$= \lim_{b\to\infty}\left[-e^{-x^2}\right]_0^b$$
$$= \lim_{b\to\infty}[-e^{-b^2}-(-e^{-0^2})]$$
$$= \lim_{b\to\infty}\left(-\frac{1}{e^{b^2}}+1\right)$$
$$= 1 \quad \left(\text{As } b\to\infty,\ -\frac{1}{e^{b^2}}\to 0 \text{ and } -\frac{1}{e^{b^2}}+1\to 1.\right)$$

The area of the region is 1.

38. $\int_6^\infty \frac{1}{\sqrt{(3x-2)^3}} = \lim_{b\to\infty}\int_6^b (3x-2)^{-3/2}\,dx$

$= \lim_{b\to\infty}\left[-\frac{2}{3}(3x-2)^{-1/2}\right]_6^b$

$= \lim_{b\to\infty}\left[\left(-\frac{2}{3}(3b-2)^{-1/2}\right)-\left(-\frac{2}{3}(3\cdot 6-2)^{-1/2}\right)\right]$

$= \lim_{b\to\infty}\left(-\frac{2}{3\sqrt{3b-2}}+\frac{2}{3\sqrt{16}}\right)$

$= \frac{1}{6}$

39. Note that 60.1 days $= \frac{60.1}{365}$ yr ≈ 0.164658.

a) $\frac{1}{2}P_0 = P_0\,e^{-k(0.164658)}$

$0.5 = e^{-0.164658k}$

$\ln 0.5 = \ln e^{-0.164658k}$

$\ln 0.5 = -0.164658k$

$\frac{\ln 0.5}{-0.164658} = k$

$4.20963 \approx k$

The decay rate is 420.963% per year.

b) The first month is $\frac{1}{12}$ yr.

$E = \int_0^{1/12} 10\,e^{-4.20963t}\,dt$

$= \frac{10}{-4.29063}\left[e^{-4.20963t}\right]_0^{1/12}$

$= \frac{10}{-4.20963}\left[e^{-4.20963(\frac{1}{12})} - e^{-4.20963(0)}\right]$

$= \frac{10}{-4.20963}(e^{-0.3508025} - 1)$

≈ 0.702858 rems

c) $E = \int_0^\infty 10\,e^{-4.20963t}\,dt$

$= \lim_{b\to\infty}\int_0^b 10\,e^{-4.20963t}\,dt$

$= \frac{10}{4.20963}$ $\quad\left[\int_0^\infty Pe^{-kt}\,dt = \frac{P}{k}\right]$

≈ 2.37551 rems

40. a) Note that 16.99 days ≈ 0.046548 yr.

Solve: $\frac{1}{2}P_0 = P_0\,e^{-k(0.046548)}$

$k \approx 14.891$, or 1489.1% per year

b) The first month is $\frac{1}{12}$ yr.

$E = \int_0^{1/12} 10\,e^{-14.891t}\,dt$

≈ 0.47739 rems

c) $E = \int_0^\infty 10\,e^{-14.891t}\,dt$

$= \frac{10}{14.891} \approx 0.67155$ rems

41. $\frac{P}{k} = \frac{1}{0.0000286}$

≈ 34965 lbs

42. $\frac{P}{k} = \frac{1}{0.023}$

≈ 43.5 lbs

43. $\int_1^\infty x^r\,dx = \lim_{b\to\infty}\int_1^b x^r\,dx$

$= \lim_{b\to\infty}\left[\frac{1}{r+1}\,x^{r+1}\right]_1^b$

In order for the limit to converge, the exponent $r+1$ must be negative, that is $r+1<0$.
Therefore, the integral is convergent for $r<-1$ and divergent otherwise.

44. $\int_e^\infty \frac{(\ln x)^r}{x}\,dx = \lim_{b\to\infty}\int_e^b \frac{(\ln x)^r}{x}\,dx$

$= \lim_{b\to\infty}\left[\frac{1}{r+1}\,(\ln x)^{r+1}\right]_e^b$

In order for the limit to converge, the exponent $r+1$ must be negative, that is $r+1<0$.
Therefore, the integral is convergent for $r<-1$ and divergent otherwise.

45. $\int_0^\infty te^{-kt}\,dt = \lim_{b\to\infty}\int_0^b te^{-kt}\,dt$

$= \lim_{b\to\infty}\left[\frac{-te^{-kt}}{k} - \frac{e^{-kt}}{k^2}\right]_0^b$

$= \lim_{b\to\infty}\left[\frac{-be^{-bk}}{k} - e^{-bk} + 0 + \frac{1}{k^2}\right]$

$= \frac{1}{k^2}$

The total amount of the drug dosage that goes through the body is $\frac{1}{k^2}$.

46. $100 = \frac{1}{k^2}$

$k = \frac{1}{10}$

47. If $y = \frac{1}{x^2}$ then $\int y\,dx = \frac{-1}{x}$

$\int_1^\infty y\,dx = \lim_{b\to\infty}\left[\frac{-1}{b}+1\right]$

$= 1$

If $y = \frac{1}{x}$ then $\int y\,dx = \ln x$

$\int_1^\infty y\,dx = \lim_{b\to\infty}\left[\ln b - 0\right]$

$= \infty$

The region under $y = \frac{1}{x^2}$ could be painted. Since the integral of that y is convergent while the other integral is divergent.

48.

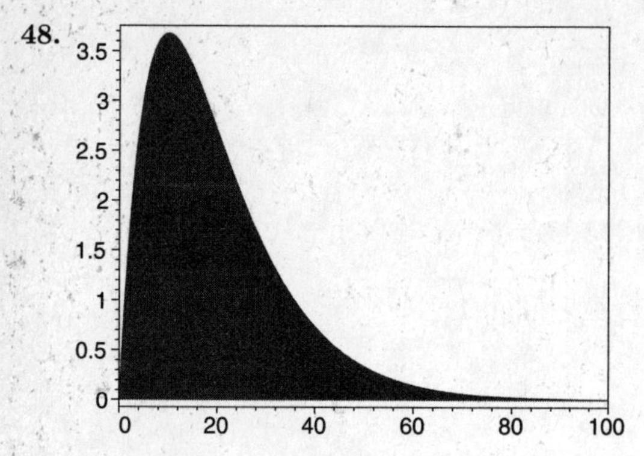

49. $\int_1^{\infty} \frac{4}{1+x^2}\,dx = \pi$

50. $\int_1^{\infty} \frac{6}{5+e^x}\,dx = \frac{6}{5}[\ln(e+5)-1] = 1.252310134$

Chapter Review Exercises

1. $\int 8x^4\,dx = \frac{8}{5}\,x^5 + C$

2. $\int (3x^2 + 2\,\sin\,x + 3)\,dx = x^3 - 2\,\cos\,x + 3x + C$

3. $\int \left(3t^2 + 7t + \frac{1}{t}\right)\,dt = t^3 + \frac{7}{2}\,t^2 + \ln t + C$

4.

$$\begin{aligned}\int_{-2}^{1} 4 - x^2\,dx &= \left[4x - \frac{1}{3}x^3\right]_{-2}^{1} \\ &= 4 - \frac{1}{3} - \left(-8 + \frac{8}{3}\right) \\ &= 9\end{aligned}$$

5.

$$\begin{aligned}\int_0^{\pi/4} cos\ x\,dx &= [sin\ x]_0^{\pi/4} \\ &= \frac{1}{\sqrt{2}} - 0 \\ &= \frac{1}{\sqrt{2}}\end{aligned}$$

6. Left to the student

7. Left to the student

8.

$$\begin{aligned}\int_a^b x^5\,dx &= \left[\frac{1}{6}x^6\right]_a^b \\ &= \frac{b^6 - a^6}{6}\end{aligned}$$

9.

$$\begin{aligned}\int_{-1}^{1} x^3 - x^4\,dx &= \left[\frac{1}{4}x^4 - \frac{1}{5}x^5\right]_{-1}^{1} \\ &= \frac{1}{4} - \frac{1}{5} - \frac{1}{4} - \frac{1}{5} \\ &= -\frac{2}{5}\end{aligned}$$

10.

$$\begin{aligned}\int_0^1 e^x + x\,dx &= \left[e^x + \frac{1}{2}x^2\right]_0^1 \\ &= e + \frac{1}{2} - 1 - 0 \\ &= e - \frac{1}{2}\end{aligned}$$

11.

$$\begin{aligned}\int_1^3 \frac{3}{x}\,dx &= [3\,\ln x]_1^3 \\ &= 3\,\ln 3 - 3\,\ln 1 \\ &= 3\,\ln 3\end{aligned}$$

12.

$$\begin{aligned}\int_0^{\pi/6} sec^2\theta\,d\theta &= [tan\ \theta]_0^{\pi/6} \\ &= \frac{1}{\sqrt{3}} - 0 \\ &= \frac{1}{\sqrt{3}}\end{aligned}$$

13. 0

14. Negative

15. Positive

16.

$$\begin{aligned}\int_0^3 9x - 3x^2\,dx &= \left[\frac{9}{2}x^2 - x^3\right]_0^3 \\ &= \frac{81}{2} - 27 - 0 \\ &= \frac{27}{2}\end{aligned}$$

17. Let $u = x^4$ then $du = 4x^3\,dx$

$$\begin{aligned}\int x^3 e^{x^4}\,dx &= \frac{1}{4}\int e^u\,du \\ &= \frac{1}{4}\,e^u + C \\ &= \frac{1}{4}\,e^{x^4} + C\end{aligned}$$

18. Let $u = 4t^6 + 3$ then $du = 24t^5\,dt$

$$\begin{aligned}\int \frac{24t^5}{4t^6+3}\,dt &= \int \frac{du}{u} \\ &= \ln u + C \\ &= \ln(4t^6 + 3) + C\end{aligned}$$

19. Let $u = \ln 4x$ then $du = \dfrac{dx}{x}$

$$\begin{aligned}\int \frac{\ln 4x}{2x}\,dx &= \int \frac{1}{2}\,u\,du\\ &= \frac{1}{4}u^2 + C\\ &= \frac{(\ln 4x)^2}{4} + C\end{aligned}$$

20. Let $u = e^{-3x}$ then $du = -3e^{-3x}\,dx$

$$\begin{aligned}\int 2e^{-3x}\,dx &= -\frac{2}{3}\int e^u\,du\\ &= -\frac{2}{3}\,e^u + C\\ &= -\frac{2}{3}\,e^{-3x} + C\end{aligned}$$

21. Let $u = x^3$ then $du = 3x^2\,dx$

$$\begin{aligned}\int csc^2(x^3)\,dx &= \frac{1}{3}\int csc^2 u\,du\\ &= \frac{1}{3}\,cot\ u + C\\ &= \frac{1}{3}\,cot\ x^3 + C\end{aligned}$$

22. Let $u = 3x$ and $dv = e^{3x}\,dx$
Then $du = 3dx$ and $v = \frac{1}{3}e^{3x}$

$$\begin{aligned}\int 3xe^{3x}\,dx &= xe^{3x} - \int e^{3x}\,dx\\ &= xe^{3x} - \frac{1}{3}\,e^{3x} + C\end{aligned}$$

23. Let $u = \ln x^7$ and $dv = dx$
Then $du = \dfrac{7}{x}\,dx$ and $v = x$

$$\begin{aligned}\int \ln x^7\,dx &= x\ \ln x^7 - \int 7\,dx\\ &= x\ \ln x^7 - 7x + C\end{aligned}$$

24. Let $u = \ln x$ and $dv = 3x^2\,dx$
Then $du = \dfrac{dx}{x}$ and $v = x^3$

$$\begin{aligned}\int 3x^2\ \ln x\,dx &= x^3\ \ln x - \int x^2\,dx\\ &= x^3\ \ln x - \frac{1}{3}\,x^3\end{aligned}$$

25. Let $u = 2x^2$ and $dv = sin\ x\,dx$
Then $du = 4xdx$ and $v = -cos\ x$

$$\int 2x^2\ sin\ x\,dx = -2x^2\ cos\ x + \int 4x\ cos\ x\,dx$$

Let $u = 4x$ and $dv = cos\ x\,dx$
Then $du = 4dx$ and $v = sin\ x$

$$\begin{aligned}\int 3xe^{3x}\,dx &= -2x^2\ cos\ xe^{3x} + 4x\ sin\ x - 4\int sin\ x\,dx\\ &= -2x^2\ cos\ xe^{3x} + 4x\ sin\ x - 4\ cos\ x + C\end{aligned}$$

26. $\displaystyle\int \frac{1}{49 - x^2}\,dx = \frac{1}{14}\ \ln(x+7) - \frac{1}{14}\ \ln(x-7) + C$

27. $\displaystyle\int x^2e^{5x}\,dx = \frac{1}{5}\ x^2e^{5x} - \frac{2}{25}\ xe^{5x} + \frac{2}{125}\ e^{5x} + C$

28. $\displaystyle\int \frac{x}{7x+1}\,dx = \frac{1}{7}\ x - \frac{1}{49}\ \ln(49x+7) + C$

29. $\displaystyle\int \frac{dx}{\sqrt{x^2-36}} = \ln\left(x + \sqrt{x^2-36}\right) + C$

30. $\displaystyle\int x^6\ \ln x\,dx = \frac{1}{7}\ x^7\ \ln x - \frac{1}{49}\ x^7 + C$

31. $\displaystyle\int xe^{8x}\,dx = \frac{1}{8}\ xe^{8x} - \frac{1}{64}\ e^{8x} + C$

32. $\displaystyle\int x^5\ cos\ x\,dx$

$$= x^5\ sin\ x + 5x^4\ cos\ x - 20x^3\ sin\ x - 60x^2\ cos\ x + 120\ cos\ x + 120x\ sin\ x + C$$

33. $\Delta x = \dfrac{4-1}{3} = 1$

$$\begin{aligned}\int_1^4 \frac{2}{x}\,dx &= f(2)\Delta x + f(3)\Delta x + f(4)\Delta x\\ &= 1\cdot 1 + \frac{2}{3}\cdot 1 + \frac{1}{2}\cdot 1\\ &= \frac{13}{6}\\ &\approx 2.16667\end{aligned}$$

34. The average is given by $\displaystyle\frac{1}{b-a}\int_a^b f(x)\,dx$

$$\begin{aligned}&= \frac{1}{2-0}\int_0^2 e^{-x} + 5\,dx\\ &= \frac{1}{2}\Big[-e^{-x} + 5x\Big]_0^2\\ &= \frac{1}{2}\Big[-\frac{1}{e^2} + 10 + 1 - 0\Big]\\ &= \frac{1}{2}\left(11 - \frac{1}{e^2}\right)\end{aligned}$$

35. The distance traveled is given by $\displaystyle\int_0^4 v(t)\,dt$

$$\begin{aligned}&= \int_0^4 3t^2 + 2t\,dt\\ &= \Big[t^3 + t^2\Big]_0^4\\ &= 64 + 16 - 0 - 0\\ &= 80\ km\end{aligned}$$

36.

$$\int_0^{30} 0.1e^{0.03t}\,dt$$

$$= \left[\frac{10}{3}\, e^{0.03t}\right]_0^{30}$$
$$= \frac{10}{3}\left(e^{0.9} - 1\right)$$
$$= 4.8653$$

37. Method used: tabular itegration

$10x^3e^{0.1x} - 300x^2e^{0.1x} + 6000xe^{0.1x} - 60000e^{0.1x} + C$

38. Method used: substitution

$\ln(4x^3 + 7) + C$

39. Method used: by parts

$\frac{2}{72}(4 + 5x)^{3/2} - \frac{8}{25}\sqrt{4 + 5x} + C$

40. Method used: substitution

$e^{x^5} + C$

41. Method used: substitution

$\ln(x + 9) + C$

42. Method used: substitution

$\frac{1}{96}(t^8 + 3)^{12} + C$

43. Method used: by parts

$x\ \ln 7x - x + C$

44. Method used: by parts

$\frac{1}{2}\, x^2\, \ln(8x) - \frac{1}{4}\, x^2 + C$

45. Method used: substitution

$\frac{1}{5}sec(t^5 + 1) + C$

46. Method used: substitution

$\frac{1}{3}\, \ln(cos(3x + 7) + C$

47.
$$\int_1^{\infty} \frac{1}{x^2}\, dx$$
$$= \lim_{b\to\infty} \int_1^b \frac{1}{x^2}\, dx$$
$$= \lim_{b\to\infty} \left[-\frac{1}{x}\right]_1^b$$
$$= \lim_{b\to\infty} \left[-\frac{1}{b} + 1\right]$$
$$= 1$$

The limit does exist. Thus the improper integral is convergent.

48.
$$\int_1^{\infty} e^{4x}\, dx$$
$$= \lim_{b\to\infty} \int_1^b e^{4x}\, dx$$
$$= \lim_{b\to\infty} \left[\frac{1}{4}e^{4x}\right]_1^b$$
$$= \lim_{b\to\infty} \left[\frac{1}{4}e^{4b} - \frac{1}{4}e^4\right]$$

The limit does not exist. Thus the improper integral is divergent.

49.
$$\int_0^{\infty} xe^{-2x}\, dx$$
$$= \lim_{b\to\infty} \int_0^b xe^{-2x}\, dx$$
$$= \lim_{b\to\infty} \left[-\frac{1}{2}xe^{-2x} - \frac{1}{4}e^{-2x}\right]_0^b$$
$$= \lim_{b\to\infty} \left[-\frac{1}{2}be^{-2b} - \frac{1}{4}e^{-2b} + 0 + \frac{1}{4}\right]$$
$$= \frac{1}{4}$$

The limit does exist. Thus the improper integral is convergent.

50.
$$\int_1^{\infty} \frac{3}{x^{1.01}}\, dx$$
$$= \lim_{b\to\infty} \int_1^b \frac{3}{x^{1.01}}\, dx$$
$$= \lim_{b\to\infty} \left[-\frac{300}{x^{0.01}}\right]_1^b$$
$$= \lim_{b\to\infty} \left[-\frac{300}{b^{0.01}} + 300\right]$$
$$= 300$$

The limit does exist. Thus the improper integral is convergent.

51.
$$V = \int_1^2 \pi\, (x^3)^2\, dx$$
$$= \frac{\pi}{7}\left[x^7\right]_1^2$$
$$= \frac{\pi}{7}(2^7 - 1)$$
$$= \frac{127\,\pi}{7}$$

52.
$$V = \int_0^1 \pi\left(\frac{1}{x+2}\right)^2 dx$$
$$= \int_0^1 \frac{\pi}{(x+2)^2}\, dx$$
$$= -\left[\frac{\pi}{x+2}\right]_0^1$$
$$= -\frac{\pi}{3} + \frac{\pi}{2}$$
$$= \frac{\pi}{6}$$

53. $40 = 50K \to K = \frac{4}{5}$

$$A(x) = \left(\frac{4}{5}x\right)^2 = \frac{16}{25}x^2$$

$$\begin{aligned} V &= \int_0^{50} \frac{16}{25}\, x^2\, dx \\ &= \left[\frac{16}{75}\, x^3\right]_0^{50} \\ &= \frac{16 \cdot 50^3}{75} - 0 \\ &= \frac{80000}{3} \\ &\approx 26666.7\ ft^3 \end{aligned}$$

54. $3 = 4K \to K = \dfrac{3}{4}$

$$A(x) = \frac{1}{2} \cdot \frac{3}{4}x \cdot \frac{\sqrt{3}}{2}x = \frac{3\sqrt{3}}{16}x^2$$

$$\begin{aligned} V &= \int_0^4 \frac{3\sqrt{3}}{16}\, x^2\, dx \\ &= \left[\frac{\sqrt{3}}{16}\, x^3\right]_0^4 \\ &= 4\sqrt{3} - 0 \\ &= 4\sqrt{3} \approx 6.9282 \end{aligned}$$

55. Let $u = \ln(t^5 + 3)$ then $du = \dfrac{5t^4}{t^5 + 3}\, dx$

$$\int \frac{t^4\ \ln(t^5+3)}{t^5+3}\, dt$$

$$\begin{aligned} &= \frac{1}{5}\int u\, du \\ &= \frac{1}{10}u^2 + C \\ &= \frac{1}{10}\left(\ln(t^5+3)\right)^2 + C \end{aligned}$$

56. Let $u = e^x$ then $\dfrac{du}{u} = dx$

$$\int \frac{1}{e^x + 2}\, dx$$

$$= \int \frac{1}{u+2} \cdot \frac{1}{u}\, du$$

Partial fraction decomposition

$$\begin{aligned} &= \int \frac{1}{-2(u+2)} + \frac{1}{2u}\, du \\ &= -\frac{1}{2}\ln(u+2) + \frac{1}{2}\ln u + C \\ &= -\frac{1}{2}\ln(e^x+2) + \frac{1}{2}\ln e^x + C \\ &= -\frac{1}{2}\ln(e^x+2) + \frac{1}{2}x + C \end{aligned}$$

57. Let $u = \ln\sqrt{x}$ then $du = \dfrac{dx}{2\sqrt{x} \cdot \sqrt{x}}$

$$\int \frac{\ln\sqrt{x}}{x}\, dx$$

$$\begin{aligned} &= 2u\, du \\ &= u^2 + C \\ &= \ln^2\sqrt{x} + C \end{aligned}$$

58. Let $u = \ln x$ and $dv = x^{91}\, dx$

Then $du = \dfrac{dx}{x}$ and $v = \dfrac{x^{92}}{92}$

$$\int x^{91} \ln x\, dx$$

$$\begin{aligned} &= \frac{1}{92}\, x^{92}\, \ln x - \int \frac{1}{92}x^{91}\, dx \\ &= \frac{1}{92}\, x^{92}\, \ln x - \frac{1}{8464}\, x^{92} + C \end{aligned}$$

59. Let $u = \ln\dfrac{x-3}{x-4}$ and $dv = dx$

Then $du = -\dfrac{dx}{(x-3)(x-4)}$ and $v = x$

$$\int \ln\left(\frac{x-3}{x-4}\right)\, dx$$

$$= x\ \ln\left(\frac{x-3}{x-4}\right) + \int \frac{x}{(x-3)(x-4)}\, dx$$

Partial fraction decomposition

$$\begin{aligned} &= x\ \ln\left(\frac{x-3}{x-4}\right) + \int \frac{-3}{(x-3)} - \frac{4}{(x-4)}\, dx \\ &= x\ \ln\left(\frac{x-3}{x-4}\right) - 3\ln(x-3) - 4\ln(x-4) + C \end{aligned}$$

60. Let $u = \ln x$ then $du = \dfrac{dx}{x}$

$$\int \frac{dx}{x(\ln x)^4}$$

$$\begin{aligned} &= \int u^{-4}\, du \\ &= \frac{-1}{3u^3} + C \\ &= \frac{-1}{3\ (\ln x)^3} + C \end{aligned}$$

61. $\dfrac{4}{3} \approx 1.3333334$

62. 102.045

63. 100.511

Chapter 5 Test

1. $\displaystyle\int dx = x + C$

2. $\displaystyle\int 1000x^4\, dx = 200x^5 + C$

3. $\displaystyle\int \left(e^x + \frac{1}{x} + x^{3/8}\right)\, dx = e^x + \ln x + \frac{8}{11}\, x^{11/8} + C$

4. The area is given by

$$\begin{aligned} \int_0^1 x - x^2\, dx &= \left[\frac{x^2}{2} - \frac{x^3}{3}\right]_0^1 \\ &= \frac{1}{2} - \frac{1}{3} - 0 + 0 \\ &= \frac{1}{6} \end{aligned}$$

5. The area is given by

$$\int_1^3 \frac{4}{x}\,dx = \Big[4\ \ln x\Big]_1^3 = 4\ \ln 3 - 0 = 4\ \ln 3$$

6. Left to the student

7.

$$\int_{-1}^{2}(2x+3x^2)\,dx = \Big[x^2+x^3\Big]_{-1}^{2} = 4+8-1+1 = 12$$

8.

$$\int_0^1 e^{-2x}\,dx = \left[-\frac{1}{2}\,e^{-2x}\right]_0^1 = -\frac{1}{2\,e^2}+\frac{1}{2} = \frac{1}{2}-\frac{1}{2\,e^2}$$

9.

$$\int_a^b \frac{dx}{x} = \Big[\ln x\Big]_a^b = \ln b - \ln a = \ln\left|\frac{b}{a}\right|$$

10. Positive

11. Let $u = x+8$ then $du = dx$

$$\int \frac{dx}{x+8} = \frac{du}{u} = \ln u + C = \ln|\,x+8\,| + C$$

12. Let $u = -0.5x$ then $du = -0.5\,dx$

$$\int e^{-0.5x}\,dx = -2\,e^u\,du = -2\,e^u + C = -2\,e^{-0.5x}+C$$

13. Let $u = e^x$ then $du = e^x\,dx$

$$\int e^x\,\sin e^x\,dx = \sin u\,du = -\cos u + C = -\cos e^x + C$$

14. Let $u = x$ and $dv = \sin 5x\,dx$
Then $du = dx$ and $v = -\frac{1}{5}\cos 5x$

$$\int x\,\sin 5x\,dx = -\frac{x}{5}\cos 5x - \int \frac{-1}{5}\cos 5x\,dx = -\frac{x}{5}\cos 5x + \frac{1}{25}\sin 5x + C = \frac{1}{25}\sin 5x - \frac{x}{5}\cos 5x + C$$

15. Let $u = \ln x^4$ and $dv = x^3\,dx$
Then $du = \frac{4\,dx}{x}$ and $v = \frac{1}{4}\,x^4$

$$\int x^3\,\ln x^4\,dx = \frac{x^4}{4}\,\ln x^4 - \int x^3\,dx = \frac{x^4}{4}\,\ln x^4 - \frac{1}{4}\,x^4 + C = \frac{1}{4}\,x^4\left[\ln x^4 - 1\right]+C$$

16. $\int 2^x\,dx = \frac{2^x}{\ln 2} + C$

17. $\int \frac{dx}{x(7-x)} = \frac{1}{7}\,\ln\left|\frac{x}{7-x}\right| + C$

18. The average is given by

$$\frac{1}{2-(-1)}\int_{-1}^{2} 4t^3+2t\,dt = \frac{1}{3}\Big[t^4+t^2\Big]_{-1}^{2} = \frac{1}{3}\,[16+4-1-1] = \frac{20}{3}$$

19. The area is given by

$$\int_0^1 x - x^5\,dx = \left[\frac{x^2}{2}-\frac{x^6}{6}\right]_0^1 = \frac{1}{2}-\frac{1}{6}-0+0 = \frac{1}{3}$$

20. $\Delta x = \frac{5-0}{5} = 1$

$$\int_0^5 25-x^2\,dx = \sum_1^5 f(x_i)\Delta x = f(1)\cdot 1 + f(2)\cdot 1 + f(3)\cdot 1 + f(4)\cdot 1 + f(5)\cdot 1 = 24+21+16+9+0 = 70$$

21.

$$\int_0^{100} 0.35+0.001t\,dt = \Big[0.35t+0.0005t^2\Big]_0^{100} = 35+5-0+0 = 40\ in$$

22.

$$\int_1^2 -6t^2+12t+90\,dt = \Big[-2t^3+6t^2+90t\Big]_1^2 = -16+24+180+2-6-90 = 94\ words$$

23. $\frac{1}{10}\ln\left|\frac{x}{10-x}\right|+C$

24. $x^5e^x-5x^4e^x+20x^3e^x-60x^2e^x+120xe^x-120e^x+C$

25. $\frac{1}{6}e^{x^6}+C$

26. $\frac{2}{3}x^{3/2}\ \ln x-\frac{4}{9}x^{3/2}+C$

27. $\frac{1}{5}x^2(x^2+4)^{3/2}-\frac{8}{15}(x^2+4)^{3/2}+C$

28. $\frac{1}{16}\ln\left|\frac{x+8}{x-8}\right|+C$

29. $10x^4e^{0.1x}-400x^3e^{0.1x}+12000x^2e^{0.1x}$
$-240000xe^{0.1x}+2400000e^{0.1x}+C$

30. $x^2\ sin\ x-2\ sin\ x+2x\ cos\ x+C$

31. The volume is given by

$$\begin{aligned}V &= \int_1^5 \pi\cdot\frac{1}{x}\,dx\\ &= \pi\Big[\ln x\Big]_1^5\\ &= \pi\ \ln 5-0\\ &= \pi\ \ln 5\end{aligned}$$

32. The volume is given by

$$\begin{aligned}V &= \int_0^1 \pi\ (2+x)\,dx\\ &= \pi\left[2x+\frac{x^2}{2}\right]_0^1\\ &= \pi\ \cdot\frac{5}{2}-0\\ &= \frac{5\pi}{2}\end{aligned}$$

33. $4=7K\rightarrow K=\frac{4}{7},\ A(x)=\frac{16x^2}{49}$

$$\begin{aligned}&\int_0^7\frac{16}{49}\,x^2\,dx\\ &= \frac{16}{49}\left[\frac{x^3}{3}\right]_0^7\\ &= \frac{16}{49}\left[\frac{7^3}{3}-0\right]\\ &= \frac{112}{3}\\ &\approx 37.33333\end{aligned}$$

34.

$$\begin{aligned}&\int_1^\infty\frac{1}{x^3}\,dx\\ &= \lim_{b\to\infty}\int_1^b\frac{1}{x^3}\,dx\\ &= \lim_{b\to\infty}\left[-\frac{1}{2x^2}\right]_1^b\\ &= \lim_{b\to\infty}\left[-\frac{1}{2b^2}+\frac{1}{2}\right]\\ &= \frac{1}{2}\end{aligned}$$

The limit does exist. Thus the improper integral is convergent.

35.

$$\begin{aligned}&\int_0^\infty\frac{3}{1+x}\,dx\\ &= \lim_{b\to\infty}\int_0^b\frac{3}{1+x}\,dx\\ &= \lim_{b\to\infty}\Big[3\ \ln(1+x)\Big]_0^b\\ &= \lim_{b\to\infty}\ [3\ \ln(1+b)-0]\end{aligned}$$

The limit does not exist. Thus the improper integral is divergent.

36. $\frac{1}{4}\ (\ln x)^4-\frac{4}{3}\ (\ln x)^3+5\ \ln x+C$

37. $3\ \ln|\,x+3\,|-5\ \ln|\,x+5\,|+x\ \ln\left(\frac{x+3}{x+5}\right)+C$

38. $\frac{3(5x-4)^{2/3}}{34375}\ \ (5000x^3+4500x^2+4320x+39559)+C$

39. 8

40. **a)** 0.88142
b) 0.88203

Technology Connection

- **Page 353**

1. 35750000
2. 35937500
3. 181.44
4. 161.1225

- **Page 358**

1. 0
2. 13.75
3. 0.53471
4. 27.97178
5. -260

- **Page 384**

  1. 62.0303

- **Page 391**

  1. 1.940727672

- **Page 403**

  1. 64

  2. 65

  3. Very accurate even with a small number of intervals

- **Page 414**

  1. $\lim_{x\to\infty}\left(1-\frac{1}{x}\right)=1$

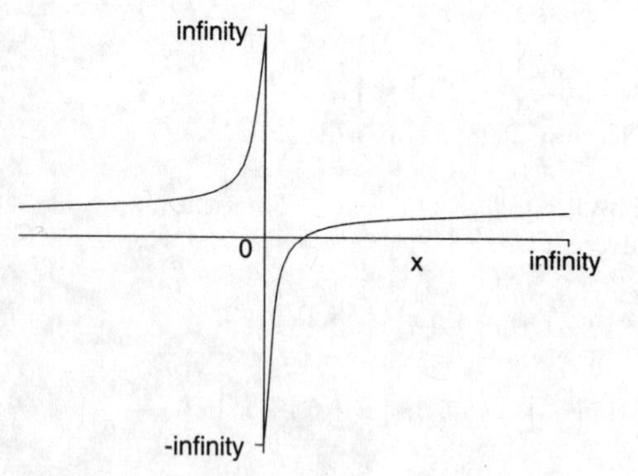

  2. $\lim_{x\to\infty}\ln x$ does not exist

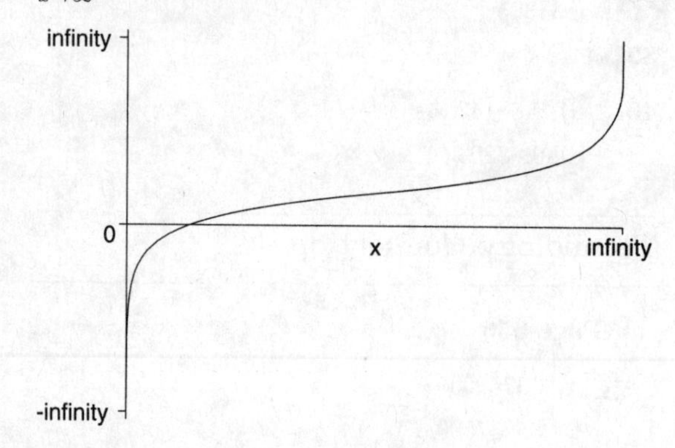

Extended Life Science Connection

1. **a)** $n=11$ and $\Delta x=0.25$

b) $0, 0.039, 0.092, 0.145, 0.1255, 0.0505$
$0.05, 0.0235, 0.0125, 0.0055, 0$

c) 2.174

d) 2.152

2. **a)** $n=23$ and $\Delta x=1$

b) $0, 0.154, 0.205, 0.235, 0.278, 0.3205$
$0.3505, 0.5555, 1.0895, 1.2395, 1.0045, 0.748$
$0.513, 0.393, 0.3205, 0.278, 0.2135, 0.1495$
$0.0855, 0.064, 0.0425, 0.032, 0$

c) 8.2715

d) 8.198

3.

$$
\begin{aligned}
1-ax^2 &= -1+ax^2\\
ax^2 &= 1\\
x^2 &= \frac{1}{a}\\
x &= \pm\frac{1}{\sqrt{a}}\\
y &= 1-a\left(\frac{1}{\sqrt{a}}\right)^2\\
y &= 1-1\\
y &= 0
\end{aligned}
$$

4. **a)** $\frac{400}{169}$

b) $\frac{100}{81}$

5. **a)** 1.451991667

b) 2.3652

6. **a)** 0.46218

b) 0.75287

7. Left to the student

8. **a)** 0.99461 m^3

b) 6.17203 m^3

9. **a)** 994.6 kg

b) 617.2 kg

10. Left to the student

Chapter 6

Matrices

Exercise Set 6.1

1.

$$A + B = \begin{bmatrix} 4+3 & -1+9 \\ 7+2 & -9+(-2) \end{bmatrix} = \begin{bmatrix} 7 & 8 \\ 9 & -11 \end{bmatrix}$$

$$B + A = \begin{bmatrix} 3+4 & 9+(-1) \\ 2+7 & -2+(-9) \end{bmatrix} = \begin{bmatrix} 7 & 8 \\ 9 & -11 \end{bmatrix}$$

2.

$$A + C = \begin{bmatrix} 4+8 & -1+3 \\ 7+0 & -9+(-3) \end{bmatrix} = \begin{bmatrix} 12 & 2 \\ 7 & -12 \end{bmatrix}$$

3.

$$B + C = \begin{bmatrix} 3+8 & 9+3 \\ 2+0 & -2+(-3) \end{bmatrix} = \begin{bmatrix} 11 & 12 \\ 2 & -5 \end{bmatrix}$$

4. Not possible, dimensions not the same

5. Not possible, dimensions not the same

6.

$$3A = \begin{bmatrix} 3\cdot 4 & 3\cdot -1 \\ 3\cdot 7 & 3\cdot -9 \end{bmatrix} = \begin{bmatrix} 12 & -3 \\ 21 & -27 \end{bmatrix}$$

7.

$$B2 = \begin{bmatrix} 3\cdot 2 & 9\cdot 2 \\ 2\cdot 2 & -2\cdot 2 \end{bmatrix} = \begin{bmatrix} 6 & 19 \\ 4 & -4 \end{bmatrix}$$

8. $B(-1) = \begin{bmatrix} -3 & -9 \\ -2 & 2 \end{bmatrix} = (-1)B$

9.

$$A - B = \begin{bmatrix} 4-3 & -1-9 \\ 7-2 & -9-(-2) \end{bmatrix} = \begin{bmatrix} 1 & -10 \\ 5 & -7 \end{bmatrix}$$

$$\begin{aligned} A + (-1)B &= \begin{bmatrix} 4+(-1)(3) & -1+(-1)(9) \\ 7+(-1)(2) & -9+(-1)(-2) \end{bmatrix} \\ &= \begin{bmatrix} 1 & -10 \\ 5 & -7 \end{bmatrix} \end{aligned}$$

The answers are identical.

10. $B - C = \begin{bmatrix} -5 & 6 \\ 2 & 1 \end{bmatrix} = B + (-1)C$

11.

$$\begin{aligned} AD &= \begin{bmatrix} 20-10 & -24-3 & 4+1 \\ 35+90 & -42-27 & 7+9 \end{bmatrix} \\ &= \begin{bmatrix} 10 & -27 & 5 \\ -55 & -69 & 16 \end{bmatrix} \end{aligned}$$

12. $A^2v = \begin{bmatrix} 3 \\ -292 \end{bmatrix}$

13.

$$\begin{aligned} A^3v &= \begin{bmatrix} 71 & -54 \\ 378 & -631 \end{bmatrix} \begin{bmatrix} 2 \\ -3 \end{bmatrix} \\ &= \begin{bmatrix} (71)(2)+(-54)(-3) \\ (378)(2)+(-631)(-3) \end{bmatrix} = \begin{bmatrix} 304 \\ 2649 \end{bmatrix} \end{aligned}$$

14. $B^2v = \begin{bmatrix} 27 \\ -62 \end{bmatrix}$

15.

$$\begin{aligned} B^3v &= \begin{bmatrix} 99 & 225 \\ 50 & -26 \end{bmatrix} \begin{bmatrix} 2 \\ -3 \end{bmatrix} \\ &= \begin{bmatrix} (99)(2)+(225)(-3) \\ (50)(2)+(-26)(-3) \end{bmatrix} = \begin{bmatrix} -477 \\ 178 \end{bmatrix} \end{aligned}$$

16. Not possible, dimensions do not match

17.

$$DE = \begin{bmatrix} -35+18+2 & 20-12-1 \\ -70-9-2 & 40+6+1 \end{bmatrix} = \begin{bmatrix} -15 & 7 \\ -81 & 47 \end{bmatrix}$$

$$\begin{aligned} A + DE &= \begin{bmatrix} 4 & -1 \\ 7 & -9 \end{bmatrix} + \begin{bmatrix} -15 & 7 \\ -81 & 47 \end{bmatrix} \\ &= \begin{bmatrix} -11 & 6 \\ -74 & 38 \end{bmatrix} \end{aligned}$$

18. $(DE)A = \begin{bmatrix} -11 & -48 \\ 5 & -342 \end{bmatrix} = D(EA)$

19.

$$BD = \begin{bmatrix} 15+90 & -18+27 & 3-9 \\ 10-20 & -12-6 & 2+2 \end{bmatrix} = \begin{bmatrix} 105 & 9 & -6 \\ -10 & -18 & 4 \end{bmatrix}$$

$$\begin{aligned} (BD)F &= \begin{bmatrix} 105 & 9 & -6 \\ -10 & -18 & 4 \end{bmatrix} \begin{bmatrix} -4 & 2 & 3 \\ 0 & -1 & 2 \\ -7 & -2 & 5 \end{bmatrix} \\ &= \begin{bmatrix} -420+0+42 & 210-9+12 & 315+18-30 \\ 40+0-28 & -20+18-8 & -30-36+20 \end{bmatrix} \\ &= \begin{bmatrix} -378 & 213 & 303 \\ 12 & -10 & -46 \end{bmatrix} \end{aligned}$$

$$\begin{aligned} DF &= \begin{bmatrix} -20+0-7 & 10+6-2 & 15-12+5 \\ -40+0+7 & 20-3+2 & 30+6-5 \end{bmatrix} \\ &= \begin{bmatrix} -27 & 14 & 8 \\ -33 & 19 & 31 \end{bmatrix} \end{aligned}$$

$$B(DF) = \begin{bmatrix} 3 & 9 \\ 2 & -2 \end{bmatrix}\begin{bmatrix} -27 & 14 & 8 \\ -33 & 19 & 31 \end{bmatrix} = \begin{bmatrix} -81-297 & 42+171 & 24+279 \\ -54+66 & 28-38 & 16-62 \end{bmatrix} = \begin{bmatrix} -378 & 213 & 303 \\ 12 & -10 & -46 \end{bmatrix}$$

20. $AI = \begin{bmatrix} 4 & -1 \\ 7 & -9 \end{bmatrix} = IA$

21. $(B+C) = \begin{bmatrix} 11 & 12 \\ 2 & -5 \end{bmatrix}$

$$A(B+C) = \begin{bmatrix} 4 & -1 \\ 7 & -9 \end{bmatrix}\begin{bmatrix} 11 & 12 \\ 2 & -5 \end{bmatrix} = \begin{bmatrix} 44-2 & 48+5 \\ 77-18 & 84+45 \end{bmatrix} = \begin{bmatrix} 42 & 53 \\ 59 & 129 \end{bmatrix}$$

$$AB = \begin{bmatrix} 12-2 & 36+2 \\ 21-18 & 63+18 \end{bmatrix} = \begin{bmatrix} 10 & 38 \\ 3 & 81 \end{bmatrix}$$

$$AC = \begin{bmatrix} 32+0 & 12+3 \\ 56+0 & 21+27 \end{bmatrix} = \begin{bmatrix} 32 & 15 \\ 56 & 48 \end{bmatrix}$$

$$AB + AC = \begin{bmatrix} 10+32 & 38+15 \\ 3+56 & 81+48 \end{bmatrix} = \begin{bmatrix} 42 & 53 \\ 59 & 129 \end{bmatrix}$$

22. $(B+C)D = \begin{bmatrix} 175 & -30 & -1 \\ -40 & -27 & 7 \end{bmatrix} = BD + CD$

23.

$$AB = \begin{bmatrix} 0+30-12 & 4+0-4 & -2-18-2 \\ 0-10-6 & 16+0-2 & -8+6-1 \\ 0+15-30 & 0+0-10 & 0-9-5 \end{bmatrix} = \begin{bmatrix} 18 & 0 & -22 \\ -16 & 14 & -3 \\ -15 & -10 & -14 \end{bmatrix}$$

$$BA = \begin{bmatrix} 0+16+0 & 0-8-6 & 0-4+10 \\ 5+0+0 & 30+0-9 & -10+0+15 \\ 6+8+0 & 36-4+3 & -12-2-5 \end{bmatrix} = \begin{bmatrix} 16 & -14 & 6 \\ 5 & 21 & 5 \\ 14 & 35 & -19 \end{bmatrix}$$

Clearly, $AB \neq BA$

24. $A(B+C) = \begin{bmatrix} 8 & 28 & 14 \\ 3 & 27 & -25 \\ -26 & 14 & -20 \end{bmatrix} = AB + AC$

25.

$$A(B-C) = \begin{bmatrix} 1 & 6 & -2 \\ 4 & -2 & -1 \\ 0 & 3 & -5 \end{bmatrix}\begin{bmatrix} -4 & 0 & -2 \\ 7 & -3 & -11 \\ 5 & 5 & -5 \end{bmatrix} = \begin{bmatrix} -4+42-10 & 0-18-10 & -2-66+10 \\ 16-14-5 & 0+6-5 & -8+22+5 \\ 0+21-25 & 0-9-25 & 0-33+25 \end{bmatrix} = \begin{bmatrix} 28 & -28 & -58 \\ -35 & 1 & 19 \\ -4 & -34 & -8 \end{bmatrix}$$

$$AB = \begin{bmatrix} 18 & 0 & -22 \\ -16 & 14 & -3 \\ -15 & -10 & -14 \end{bmatrix}$$

$$AC = \begin{bmatrix} 4-12-2 & 4+18+6 & 0+48-12 \\ 16+4-1 & 16-6+3 & 0-16-6 \\ 0-6-5 & 0+9+15 & 0+24-30 \end{bmatrix} = \begin{bmatrix} -10 & 28 & 36 \\ 19 & 13 & -22 \\ -11 & 24 & -6 \end{bmatrix}$$

$$AB - BC = \begin{bmatrix} 18 & 0 & -22 \\ 16 & 14 & -3 \\ -15 & -10 & -14 \end{bmatrix} - \begin{bmatrix} -10 & 28 & 36 \\ 19 & 13 & -22 \\ -11 & 24 & -6 \end{bmatrix} = \begin{bmatrix} 28 & -28 & -58 \\ -35 & 1 & 19 \\ -4 & -34 & -8 \end{bmatrix}$$

26. a) $A\begin{bmatrix} 0 \\ 5 \\ 6 \end{bmatrix} = \begin{bmatrix} 18 \\ -16 \\ -15 \end{bmatrix}$

b) $A\begin{bmatrix} 4 \\ 0 \\ 2 \end{bmatrix} = \begin{bmatrix} 0 \\ 14 \\ -10 \end{bmatrix}$

c) $A\begin{bmatrix} -2 \\ -3 \\ 1 \end{bmatrix} = \begin{bmatrix} -22 \\ -3 \\ -14 \end{bmatrix}$

d) The matrix multiplication is only defined in the direction stated in parts a), b) and c).

27. a)

$$\begin{bmatrix} 0 & 4 & -2 \\ 5 & 0 & -3 \\ 6 & 2 & 1 \end{bmatrix}\begin{bmatrix} 1 \\ 4 \\ 0 \end{bmatrix} = \begin{bmatrix} 0+16+0 \\ 5+0+0 \\ 6+8+0 \end{bmatrix} = \begin{bmatrix} 16 \\ 5 \\ 14 \end{bmatrix}$$

b)

$$\begin{bmatrix} 0 & 4 & -2 \\ 5 & 0 & -3 \\ 6 & 2 & 1 \end{bmatrix}\begin{bmatrix} 6 \\ -2 \\ 3 \end{bmatrix} = \begin{bmatrix} 0-8-6 \\ 30+0-9 \\ 36-4+3 \end{bmatrix} = \begin{bmatrix} -14 \\ 21 \\ 35 \end{bmatrix}$$

c)

$$\begin{bmatrix} 0 & 4 & -2 \\ 5 & 0 & -3 \\ 6 & 2 & 1 \end{bmatrix}\begin{bmatrix} -2 \\ -1 \\ -5 \end{bmatrix} = \begin{bmatrix} 0-4+10 \\ -10+0+15 \\ -12-2-5 \end{bmatrix} = \begin{bmatrix} 6 \\ 5 \\ -19 \end{bmatrix}$$

d) The multiplication is only defined in the direction stated in parts a), b) and c)

28. $A - 3I = \begin{bmatrix} -2 & 6 & -2 \\ 4 & -5 & -1 \\ 0 & 3 & -8 \end{bmatrix}$

29.

$$\begin{aligned} B - 4I &= \begin{bmatrix} 0 & 4 & -2 \\ 5 & 0 & -3 \\ 6 & 2 & 1 \end{bmatrix} - \begin{bmatrix} 4 & 0 & 0 \\ 0 & 4 & 0 \\ 0 & 0 & 4 \end{bmatrix} \\ &= \begin{bmatrix} 0-4 & 4-0 & -2-0 \\ 5-0 & 0-4 & -3-0 \\ 6-0 & 2-0 & 1-4 \end{bmatrix} \\ &= \begin{bmatrix} -4 & 4 & -2 \\ 5 & -4 & -3 \\ 6 & 2 & -3 \end{bmatrix} \end{aligned}$$

30. **a)** Left to the student

b) $\begin{bmatrix} 1.2 & 2.5 \\ 0.8 & 0.8 \end{bmatrix} \begin{bmatrix} 240 \\ 124 \end{bmatrix} = \begin{bmatrix} 598 \\ 291 \end{bmatrix}$

c) $\begin{bmatrix} 1.2 & 2.5 \\ 0.8 & 0.8 \end{bmatrix} \begin{bmatrix} 598 \\ 291 \end{bmatrix} = \begin{bmatrix} 1446 \\ 711 \end{bmatrix}$

31. **a)** Left to the student

b) $\begin{bmatrix} 1.4 & 1.8 \\ 0.5 & 0.4 \end{bmatrix} \begin{bmatrix} 56 \\ 20 \end{bmatrix} = \begin{bmatrix} 114 \\ 36 \end{bmatrix}$

c) $\begin{bmatrix} 1.4 & 1.8 \\ 0.5 & 0.4 \end{bmatrix} \begin{bmatrix} 114 \\ 20 \end{bmatrix} = \begin{bmatrix} 225 \\ 72 \end{bmatrix}$

32. **a)** $\begin{bmatrix} 0.5 & 1.25 \\ 0.75 & 0.25 \end{bmatrix}$

b) $\begin{bmatrix} 0.5 & 1.25 \\ 0.75 & 0.25 \end{bmatrix} \begin{bmatrix} 100 \\ 48 \end{bmatrix} = \begin{bmatrix} 110 \\ 87 \end{bmatrix}$

c) $\begin{bmatrix} 0.5 & 1.25 \\ 0.75 & 0.25 \end{bmatrix} \begin{bmatrix} 110 \\ 87 \end{bmatrix} = \begin{bmatrix} 164 \\ 102 \end{bmatrix}$

33. **a)** $\begin{bmatrix} 0.8 & 2 \\ 0.5 & 0.4 \end{bmatrix}$

b) $\begin{bmatrix} 0.8 & 2 \\ 0.5 & 0.4 \end{bmatrix} \begin{bmatrix} 1200 \\ 1520 \end{bmatrix} = \begin{bmatrix} 4000 \\ 1208 \end{bmatrix}$

c) $\begin{bmatrix} 0.8 & 2 \\ 0.5 & 0.4 \end{bmatrix} \begin{bmatrix} 4000 \\ 1208 \end{bmatrix} = \begin{bmatrix} 5616 \\ 2483 \end{bmatrix}$

34. **a)** $\begin{bmatrix} 0 & 0.76 & 0.95 \\ 0.2 & 0 & 0 \\ 0 & 0.91 & 0.91 \end{bmatrix}$

b) $\begin{bmatrix} 0 & 0.76 & 0.95 \\ 0.2 & 0 & 0 \\ 0 & 0.91 & 0.91 \end{bmatrix} \begin{bmatrix} 579 \\ 80 \\ 609 \end{bmatrix} = \begin{bmatrix} 639 \\ 116 \\ 627 \end{bmatrix}$

c) $\begin{bmatrix} 0.8 & 2 \\ 0.5 & 0.4 \end{bmatrix} \begin{bmatrix} 639 \\ 116 \\ 627 \end{bmatrix} = \begin{bmatrix} 684 \\ 128 \\ 676 \end{bmatrix}$

35. **a)** Left to the student

b) $\begin{bmatrix} 15 & 7.5 \\ 0.8 & 0 \end{bmatrix}$

c) $\begin{bmatrix} 0.8 & 2 \\ 0.5 & 0.4 \end{bmatrix} \begin{bmatrix} 20 \\ 0 \end{bmatrix} = \begin{bmatrix} 300 \\ 16 \end{bmatrix}$

d) $\begin{bmatrix} 0.8 & 2 \\ 0.5 & 0.4 \end{bmatrix} \begin{bmatrix} 300 \\ 16 \end{bmatrix} = \begin{bmatrix} 4620 \\ 240 \end{bmatrix}$

36. **a)** $\begin{bmatrix} 3.6 & 0.98 & 0.65 \\ 0.5 & 0 & 0 \\ 0 & 0.5 & 0.49 \end{bmatrix}$

b) $\begin{bmatrix} 3.6 & 0.98 & 0.65 \\ 0.5 & 0 & 0 \\ 0 & 0.5 & 0.49 \end{bmatrix} \begin{bmatrix} 20 \\ 0 \\ 0 \end{bmatrix} = \begin{bmatrix} 72 \\ 10 \\ 0 \end{bmatrix}$

c) $\begin{bmatrix} 3.6 & 0.98 & 0.65 \\ 0.5 & 0 & 0 \\ 0 & 0.5 & 0.49 \end{bmatrix} \begin{bmatrix} 72 \\ 10 \\ 0 \end{bmatrix} = \begin{bmatrix} 269 \\ 36 \\ 5 \end{bmatrix}$

d) $\begin{bmatrix} 3.6 & 0.98 & 0.65 \\ 0.5 & 0 & 0 \\ 0 & 0.5 & 0.49 \end{bmatrix} \begin{bmatrix} 269 \\ 36 \\ 5 \end{bmatrix} = \begin{bmatrix} 1007 \\ 135 \\ 20 \end{bmatrix}$

37. **a)**

$$\begin{aligned} A^2 &= \begin{bmatrix} 3 & -2 \\ 4 & 5 \end{bmatrix} \begin{bmatrix} 3 & -2 \\ 4 & 5 \end{bmatrix} \\ &= \begin{bmatrix} 9-8 & -6-10 \\ 12+20 & -8+25 \end{bmatrix} = \begin{bmatrix} 1 & -16 \\ 32 & 17 \end{bmatrix} \end{aligned}$$

b)

$$\begin{aligned} B^2 &= \begin{bmatrix} -2 & 7 \\ 1 & -3 \end{bmatrix} \begin{bmatrix} -2 & 7 \\ 1 & -3 \end{bmatrix} \\ &= \begin{bmatrix} 4+7 & -14-21 \\ -2-3 & 7+9 \end{bmatrix} = \begin{bmatrix} 11 & -35 \\ -5 & 16 \end{bmatrix} \end{aligned}$$

c)

$$\begin{aligned} (A+B)^2 &= \begin{bmatrix} 1 & 5 \\ 5 & 2 \end{bmatrix} \begin{bmatrix} 1 & 5 \\ 5 & 2 \end{bmatrix} \\ &= \begin{bmatrix} 1+25 & 5+10 \\ 5+10 & 25+4 \end{bmatrix} = \begin{bmatrix} 26 & 15 \\ 15 & 29 \end{bmatrix} \end{aligned}$$

d)

$$\begin{aligned} A^2 + 2AB + B^2 &= \begin{bmatrix} 1 & -16 \\ 32 & 17 \end{bmatrix} + 2\begin{bmatrix} -8 & 27 \\ -3 & 13 \end{bmatrix} + \\ &\quad \begin{bmatrix} 11 & -35 \\ -5 & 16 \end{bmatrix} \\ &= \begin{bmatrix} 1 & -16 \\ 32 & 17 \end{bmatrix} + \begin{bmatrix} -16 & 54 \\ -6 & 26 \end{bmatrix} + \\ &\quad \begin{bmatrix} 11 & -35 \\ -5 & 16 \end{bmatrix} \\ &= \begin{bmatrix} 1-16+11 & -16+54-35 \\ 32-6-5 & 17+26+16 \end{bmatrix} \\ &= \begin{bmatrix} -4 & 3 \\ 21 & 59 \end{bmatrix} \end{aligned}$$

e - f) $(A+B)^2 = A^2 + AB + BA + B^2$ the middle two terms are not equal since matrix multiplication is not commutative in all cases. Therefore $(A+B)^2 \neq A^2 + 2AB + B^2$

38. **a)** $(A-B)^2 = \begin{bmatrix} -2 & -117 \\ 39 & 37 \end{bmatrix}$

b) $A^2 - 2AB + B^2 = \begin{bmatrix} 28 & -105 \\ 33 & 7 \end{bmatrix}$

c - d) $(A-B)^2 = A^2 - AB - BA + B^2$ the middle two terms are not equal since matrix multiplication is not commutative in all cases. Therefore $(A-B)^2 \neq A^2 - 2AB + B^2$

39. **a)** A^T is $1 \times n$ and B^T is $n \times 1$

b) $A \cdot B$ will have dimensions of 1×1

$$\begin{aligned} A \cdot B &= \begin{bmatrix} a_{11} & a_{12} & a_{13} & \cdots & a_{1n} \end{bmatrix} \cdot \begin{bmatrix} b_{11} \\ b_{21} \\ b_{31} \\ \vdots \\ b_{n1} \end{bmatrix} \\ &= a_{11}b_{11} + a_{12}b_{21} + a_{13}b_{31} + \cdots + a_{1n}b_{n1} \end{aligned}$$

$B^T \cdot A^T$ will have dimensions of 1×1

$$\begin{aligned} B^T \cdot A^T &= \begin{bmatrix} b_{11} & b_{12} & b_{13} & \cdots & b_{1n} \end{bmatrix} \cdot \begin{bmatrix} a_{11} \\ a_{21} \\ a_{31} \\ \vdots \\ a_{n1} \end{bmatrix} \\ &= b_{11}a_{11} + b_{12}a_{21} + b_{13}a_{31} + \cdots + b_{1n}a_{n1} \end{aligned}$$

Thus, $A \cdot B = B^T \cdot A^T$

40. Let A be a $m \times n$ matrix and B a $n \times k$ matrix Then

a) $A^T B^T$ is only defined if the number of rows in A is the same as the number of columns in B. That is if $m = k$

b) $B^T A^T$ is always defined. B^T has dimension of $k \times n$ and A^T has dimensions of $n \times m$ which allows for matrix multiplication

41. This is true due to the associative property of addition for real numbers

42. This is true due to the commutative property of addition for real numbers

43. Since each entry in A is added to its additive inverse each entry in $A + (-A)$ will be zero

44. Since the matrices A and B are square both A^T and B^T will have the same dimensioins as A and B. The entries on the diagonal are clearly the same. Using the commutative property of matrix addition the off diagonal entries can be shown to be equal.

45. - 49.] Left to the student

50. $\begin{bmatrix} 1748798 \\ 859210 \end{bmatrix}$

51. $\begin{bmatrix} 200502 \\ 63757 \end{bmatrix}$

52. $\begin{bmatrix} 969 \\ 659 \end{bmatrix}$

53. $\begin{bmatrix} 3.414 \times 10^7 \\ 1.399 \times 10^7 \end{bmatrix}$

54. $\begin{bmatrix} 1270 \\ 235 \\ 1256 \end{bmatrix}$

55. $\begin{bmatrix} 7.644 \times 10^9 \\ 1.021 \times 10^9 \\ 1.568 \times 10^8 \end{bmatrix}$

Exercise Set 6.2

1. From the second equation we have $y = 2x$. Then

$$\begin{aligned} x + 2y &= 5 \\ x + 2(2x) &= 5 \\ x + 4x &= 5 \\ 5x &= 5 \\ x &= 1 \end{aligned}$$

Which means $y = 2(1) = 2$

2. From the second equation we have $y = 4x - 35$. Then

$$\begin{aligned} 3x - 2y &= 20 \\ 3x - 2(4x - 35) &= 20 \\ 3x - 8x + 70 &= 20 \\ -5x &= -50 \\ x &= 10 \end{aligned}$$

Which means $y = 4(10) - 35 = 5$

3. From the first equation we have $z = 5w - 14$. Then

$$\begin{aligned} 2w + 3z &= 26 \\ 2w + 3(5w - 14) &= 26 \\ 2w + 15w - 42 &= 26 \\ 17w - 42 &= 26 \\ 17w &= 26 + 42 = 68 \\ w &= 4 \end{aligned}$$

Which means $z = 5(4) - 14 = 6$

4. From the second equation we have $s = \frac{1}{2}(1 - 3r)$. Then

$$\begin{aligned} -2r - \frac{5}{2}(1 - 3r) &= -8 \\ -4r - 5(1 - 3r) &= -16 \\ -4r - 5 + 15r &= -16 \\ 11r &= -11 \\ r &= -1 \end{aligned}$$

Which means $s = \frac{1}{2}(1 - 3) = -1$

5. From the first equation we have $s = t + 7$. Then

$$\begin{aligned} -2s + 2t &= -5 \\ -2(t+7) + 2t &= -5 \\ -2t - 14 + 2t &= -5 \\ -14 &= -5 \end{aligned}$$

This is a contradiction, which means there is no solution.

6. From the first equation we have $b = 4a - 2$. Then

$$\begin{aligned} 12a - 3b &= 6 \\ 12a - 3(4a - 2) &= 6 \\ 12a - 12a + 6 &= 6 \\ 6 &= 6 \end{aligned}$$

The is an identity, which means there are many solutions

7. From the first equation we have $x = y + 7$. Then

$$\begin{aligned} -2x + 2y &= -14 \\ -2(y+7) + 2y &= -14 \\ -2y - 14 + 2y &= -14 \\ -14 &= -14 \end{aligned}$$

This is an identity, which means there are many solutions

8. From the first equation we have $y = 4x - 2$. Then

$$\begin{aligned} 12x - 3y &= 5 \\ 12x - 3(4x - 2) &= 5 \\ 12x - 12x + 6 &= 5 \\ 6 &= 5 \end{aligned}$$

This is a contradiction, which means there is no solution

9. - 16.] Left to the student

17. Write the augmented matrix

$$\left[\begin{array}{ccc|c} 0 & 1 & 3 & -1 \\ 1 & 0 & 6 & 37 \\ 0 & 2 & 1 & -2 \end{array}\right]$$

Interchange $R1$ and $R2$

$$\left[\begin{array}{ccc|c} 1 & 0 & 6 & 37 \\ 0 & 1 & 3 & -1 \\ 0 & 2 & 1 & -2 \end{array}\right]$$

$-2R2 + R3$

$$\left[\begin{array}{ccc|c} 1 & 0 & 6 & 37 \\ 0 & 1 & 3 & -1 \\ 0 & 0 & -5 & 0 \end{array}\right]$$

This means the $z = 0$
$y + 3(0) = -1 \to y = -1$
$x + 6(0) = 37 \to x = 37$

18. $x = -4/19$, $y = 25/38$, and $z = 10/19$

19. Write the augmented matrix

$$\left[\begin{array}{ccc|c} 7 & -1 & -9 & 1 \\ 2 & 0 & -4 & -4 \\ -4 & 0 & 6 & -3 \end{array}\right]$$

Interchange $R2$ and $R1$

$$\left[\begin{array}{ccc|c} 2 & 0 & -4 & -4 \\ 7 & -1 & -9 & 1 \\ -4 & 0 & 6 & -3 \end{array}\right]$$

$R1/2$

$$\left[\begin{array}{ccc|c} 1 & 0 & -2 & -2 \\ 7 & -1 & -9 & 1 \\ -4 & 0 & 6 & -3 \end{array}\right]$$

$-7R1 + R2$

$$\left[\begin{array}{ccc|c} 1 & 0 & -2 & -2 \\ 0 & -1 & 5 & 15 \\ -4 & 0 & 6 & -3 \end{array}\right]$$

$4R1 + R3$

$$\left[\begin{array}{ccc|c} 1 & 0 & -2 & -2 \\ 0 & -1 & 5 & 15 \\ 0 & 0 & -2 & -11 \end{array}\right]$$

$-R2$

$$\left[\begin{array}{ccc|c} 1 & 0 & -2 & -2 \\ 0 & 1 & -5 & -15 \\ 0 & 0 & -2 & -11 \end{array}\right]$$

This means $z = 11/2 = 5.5$
$y - 5(5.5) = -15 \to y = 12.5$
$x - 2(5.5) = -2 \to x = 9$

20. $x = 16$, $y = 20$, and $z = -6$

21. Write the augmented matrix

$$\left[\begin{array}{ccc|c} 2 & -2 & 3 & 3 \\ 4 & -3 & 3 & 2 \\ -1 & 1 & -1 & 4 \end{array}\right]$$

Interchange $R1$ and $-R3$

$$\left[\begin{array}{ccc|c} 1 & -1 & 1 & -4 \\ 4 & -3 & 3 & 2 \\ 2 & -2 & 3 & 3 \end{array}\right]$$

$-4R1 + R2$

$$\left[\begin{array}{ccc|c} 1 & -1 & 1 & -4 \\ 0 & 1 & -1 & 18 \\ 2 & -2 & 3 & 3 \end{array}\right]$$

$-2R1 + R3$

$$\left[\begin{array}{ccc|c} 1 & -1 & 1 & -4 \\ 0 & 1 & -1 & 18 \\ 0 & 0 & 1 & 11 \end{array}\right]$$

This means $z = 11$
$y - 11 = 18 \to y = 29$
$x - 29 + 11 = -4 \to x = 14$

22. $x = 1$, $y = 5$, $z = 5$

23. - 28.] Left to the student

29. Write the augmented matrix

$$\left[\begin{array}{ccc|c} 1 & 1 & 2 & 5 \\ 1 & 1 & 1 & -10 \\ 2 & 3 & 4 & 2 \end{array}\right]$$

$-R1 + R2$

$$\left[\begin{array}{ccc|c} 1 & 1 & 2 & 5 \\ 0 & 0 & -1 & -15 \\ 2 & 3 & 4 & 2 \end{array}\right]$$

Interchange $-R2$ and $R3$

$$\left[\begin{array}{ccc|c} 1 & 1 & 2 & 5 \\ 2 & 3 & 4 & 2 \\ 0 & 0 & 1 & 15 \end{array}\right]$$

$-2R1 + R2$

$$\left[\begin{array}{ccc|c} 1 & 1 & 2 & 5 \\ 0 & 1 & 0 & -5 \\ 0 & 0 & 1 & 15 \end{array}\right]$$

This means $z = 15$
$y = -5$
$x + (-5) + 2(15) = 5 \rightarrow x = -17$

30. $x = -122/23$, $y = 58/23$, and $z = -39/23$

31. Write the augmented matrix

$$\left[\begin{array}{ccc|c} 1 & 1 & -2 & 4 \\ 4 & 7 & 3 & 3 \\ 14 & 23 & 5 & 17 \end{array}\right]$$

$-4R1 + R2$

$$\left[\begin{array}{ccc|c} 1 & 1 & -2 & 4 \\ 0 & 3 & 11 & -13 \\ 14 & 23 & 5 & 17 \end{array}\right]$$

$-14R1 + R3$

$$\left[\begin{array}{ccc|c} 1 & 1 & -2 & 4 \\ 0 & 3 & 11 & -13 \\ 0 & 9 & 33 & -39 \end{array}\right]$$

$-3R2 + R3$

$$\left[\begin{array}{ccc|c} 1 & 1 & -2 & 4 \\ 0 & 3 & 11 & -13 \\ 0 & 0 & 0 & 0 \end{array}\right]$$

This means the system has many solutions.
Let $z = z$, then $y = -\frac{11}{3}z - \frac{13}{3}$
$x = \frac{11}{3}z + \frac{13}{3} + 2z + 4 = \frac{17}{3}z + \frac{25}{3}$

32. The system has no solution

33. Write the augmented matrix

$$\left[\begin{array}{ccc|c} 1 & -1 & 3 & 2 \\ 2 & 3 & -1 & 5 \\ -1 & -9 & 11 & 1 \end{array}\right]$$

$-2R1 + R2$

$$\left[\begin{array}{ccc|c} 1 & -1 & 3 & 2 \\ 0 & 5 & -7 & 1 \\ -1 & -9 & 11 & 1 \end{array}\right]$$

$R1 + R3$

$$\left[\begin{array}{ccc|c} 1 & -1 & 3 & 2 \\ 0 & 5 & -7 & 1 \\ 0 & -10 & 14 & 3 \end{array}\right]$$

$2R2 + R3$

$$\left[\begin{array}{ccc|c} 1 & -1 & 3 & 2 \\ 0 & 5 & -7 & 1 \\ 0 & 0 & 0 & 5 \end{array}\right]$$

This means the system has no solution

34. The system has many solutions.
Let $z = z$, then $y = \frac{7}{5}z + \frac{1}{5}$
$x = \frac{8}{5}z + \frac{11}{5}$

35. Write the augmented matrix

$$\left[\begin{array}{ccc|c} 1 & -2 & -5 & 0 \\ 2 & 3 & 15 & 0 \\ -2 & -1 & -8 & 1 \end{array}\right]$$

$-2R1 + R2$

$$\left[\begin{array}{ccc|c} 1 & -2 & -5 & 0 \\ 0 & 7 & 25 & 0 \\ -2 & -1 & -8 & 1 \end{array}\right]$$

$2R1 + R3$

$$\left[\begin{array}{ccc|c} 1 & -2 & -5 & 0 \\ 0 & 7 & 25 & 0 \\ 0 & -5 & -18 & 1 \end{array}\right]$$

$\frac{1}{7}R2$

$$\left[\begin{array}{ccc|c} 1 & -2 & -5 & 0 \\ 0 & 1 & \frac{25}{7} & 0 \\ 0 & -5 & -18 & 1 \end{array}\right]$$

$5R2 + R3$

$$\left[\begin{array}{ccc|c} 1 & -2 & -5 & 0 \\ 0 & 1 & \frac{25}{7} & 0 \\ 0 & 0 & -\frac{1}{7} & 1 \end{array}\right]$$

This means $z = -7$
$y = -\frac{25}{7} \cdot -7 = 25$
$x = 2(25) + 5(-7) = 15$

36. $x = 40$, $y = -30$, $z = 0$

37. Write the augmented matrix

$$\left[\begin{array}{cccc|c} 1 & 1 & 1 & 1 & 5 \\ 1 & 0 & 1 & 1 & 6 \\ 0 & 1 & 1 & 1 & 4 \\ 1 & 0 & 1 & 0 & 3 \end{array}\right]$$

$-R1 + R2$

$$\left[\begin{array}{cccc|c} 1 & 1 & 1 & 1 & 5 \\ 0 & -1 & 0 & 0 & 1 \\ 0 & 1 & 1 & 1 & 4 \\ 1 & 0 & 1 & 0 & 3 \end{array}\right]$$

$-R1 + R4$ and $-R2$

$$\left[\begin{array}{cccc|c} 1 & 1 & 1 & 1 & 5 \\ 0 & 1 & 0 & & -1 \\ 0 & 1 & 1 & 1 & 4 \\ 0 & -1 & 0 & -1 & -2 \end{array}\right]$$

$-R2 + R3$

$$\left[\begin{array}{cccc|c} 1 & 1 & 1 & 1 & 5 \\ 0 & 1 & 0 & & -1 \\ 0 & 0 & 1 & 1 & 5 \\ 0 & -1 & 0 & -1 & -2 \end{array}\right]$$

$R2 + R4$

$$\left[\begin{array}{cccc|c} 1 & 1 & 1 & 1 & 5 \\ 0 & 1 & 0 & & -1 \\ 0 & 0 & 1 & 1 & 5 \\ 0 & 0 & 0 & -1 & -3 \end{array}\right]$$

This means $w = 3$
$z = -3 + 5 = 2$
$y = -1$
$x = 1 - 2 - 3 + 5 = 1$

38. $x = -3$, $y = 2$, $z = -1$, $w = 4$

39. Write the augmented matrix

$$\left[\begin{array}{cccc|c} -2 & -1 & 6 & -1 & 2 \\ -3 & -5 & 6 & 1 & -3 \\ 1 & 1 & -2 & 0 & 4 \\ 0 & 1 & -1 & 0 & 1 \end{array}\right]$$

Interchange $R1$ and $R3$

$$\left[\begin{array}{cccc|c} 1 & 1 & -2 & 0 & 4 \\ -3 & -5 & 6 & 1 & -3 \\ -2 & -1 & 6 & -1 & 2 \\ 0 & 1 & -1 & 0 & 1 \end{array}\right]$$

$3R1 + R2$

$$\left[\begin{array}{cccc|c} 1 & 1 & -2 & 0 & 4 \\ 0 & -2 & 0 & 1 & 9 \\ -2 & -1 & 6 & -1 & 2 \\ 0 & 1 & -1 & 0 & 1 \end{array}\right]$$

$2R1 + R3$

$$\left[\begin{array}{cccc|c} 1 & 1 & -2 & 0 & 4 \\ 0 & -2 & 0 & 1 & 9 \\ 0 & 1 & 2 & -1 & 10 \\ 0 & 1 & -1 & 0 & 1 \end{array}\right]$$

Interchange $R2$ and $R4$

$$\left[\begin{array}{cccc|c} 1 & 1 & -2 & 0 & 4 \\ 0 & 1 & -1 & 0 & 1 \\ 0 & 1 & 2 & -1 & 10 \\ 0 & -2 & 0 & 1 & 9 \end{array}\right]$$

$-R2 + R3$ and $2R2 + R4$

$$\left[\begin{array}{cccc|c} 1 & 1 & -2 & 0 & 4 \\ 0 & 1 & -1 & 0 & 1 \\ 0 & 0 & 3 & -1 & 9 \\ 0 & 0 & -2 & 1 & 11 \end{array}\right]$$

$R3 + R4$

$$\left[\begin{array}{cccc|c} 1 & 1 & -2 & 0 & 4 \\ 0 & 1 & -1 & 0 & 1 \\ 0 & 0 & 3 & -1 & 9 \\ 0 & 0 & 1 & 0 & 20 \end{array}\right]$$

This means that $z = 20$
$w = 3(20) - 9 = 51$
$y = 20 + 1 = 21$
$x = -21 + 2(20) + 4 = 23$

40. $x = 15$, $y = 31$, $z = -8$, $w = -3$

41. We are looking for the values of B and F for which $B' = 0$ and $F' = 0$. Thus,

$$0 = -0.01B + 0.1 \to B = 10$$

and

$$0 = 0.01(10) - 0.02F \to F = 5$$

42. $\begin{bmatrix} 1.2 & 2.5 \\ 0.8 & 0.8 \end{bmatrix} \begin{bmatrix} 240 \\ 124 \end{bmatrix} = \begin{bmatrix} 598 \\ 291 \end{bmatrix}$

43. Write the augmented matrix

$$\left[\begin{array}{ccc|c} 0 & 0.76 & 0.95 & 579 \\ 0.2 & 0 & 0 & 80 \\ 0 & 0.91 & 0.91 & 609 \end{array}\right]$$

Interchange the $R1$ and $R2$

$$\left[\begin{array}{ccc|c} 0.2 & 0 & 0 & 80 \\ 0 & 0.76 & 0.95 & 579 \\ 0 & 0.91 & 0.91 & 609 \end{array}\right]$$

$R1/0.2$

$$\left[\begin{array}{ccc|c} 1 & 0 & 0 & 400 \\ 0 & 0.76 & 0.95 & 579 \\ 0 & 0.91 & 0.91 & 609 \end{array}\right]$$

$R2/0.76$

$$\left[\begin{array}{ccc|c} 1 & 0 & 0 & 400 \\ 0 & 1 & 1.25 & 761.842 \\ 0 & 0.91 & 0.91 & 609 \end{array}\right]$$

$-0.91R2 + R3$

$$\left[\begin{array}{ccc|c} 1 & 0 & 0 & 400 \\ 0 & 1 & 1.25 & 761.842 \\ 0 & 0 & -0.2275 & -84.276 \end{array}\right]$$

$R3/-0.2275$

$$\left[\begin{array}{ccc|c} 1 & 0 & 0 & 400 \\ 0 & 1 & 1.25 & 761.842 \\ 0 & 0 & 1 & 370.45 \end{array}\right]$$

This means that $A = 370$, $S = 761.842 - 1.25(370) = 299$, and $H = 400$

44. $\left[\begin{array}{ccc|c} 3.6 & 0.98 & 0.65 & 1000 \\ 0.5 & 0 & 0 & 2317 \\ 0 & 0.5 & 0.49 & 1493 \end{array}\right]$

This leads to

$$\left[\begin{array}{ccc|c} 1 & 49/180 & 13/72 & 2500/9 \\ 0 & 1 & 49/50 & 2986 \\ 0 & 0 & 1 & 59950.64 \end{array}\right]$$

Which means $3DP = 59951$, $2DP = -55766$, $DP = 4634$. Note, we cannot have negative numbers of shrimp, so we can conclude that $2DP = 0$.

45. Left to the student. [Answers may vary]

46. **a)** $AB = \begin{bmatrix} 4 & 5 & 6 \\ 1 & 2 & 3 \\ 7 & 8 & 9 \end{bmatrix}$

b) $AC = \begin{bmatrix} 1 & 1 & -3 \\ 4 & -2 & 7 \\ -4 & 3 & 6 \end{bmatrix}$

c) Matrix A is the identity matrix with first and second rows swapped. Therefore, the effect of multiplying A on the left yields a swap of the first and second rows of the matrix being multiplied by A.

47. **a)** $AB = \begin{bmatrix} 1 & 2 & 3 \\ 16 & 20 & 24 \\ 7 & 8 & 9 \end{bmatrix}$

b) $AC = \begin{bmatrix} 4 & -2 & 7 \\ 4 & 4 & -12 \\ -4 & 3 & 6 \end{bmatrix}$

c) Matrix A is the identity matrix with second row multiplied by 4. Therefore, the effect of multiplying A on the left with another 3×3 matrix will give a matrix with the second row multiplied by 4.

48. **a)** $AB = \begin{bmatrix} 1 & 2 & 3 \\ 7 & 11 & 15 \\ 7 & 8 & 9 \end{bmatrix}$

b) $AC = \begin{bmatrix} 4 & -2 & 7 \\ 13 & -5 & 18 \\ -4 & 3 & 6 \end{bmatrix}$

c) Matrix A is the identity matrix with the second row replaced by $3R1 + R2$. Therefore, the effect of multiplying A on the left with another 3×3 matrix will give a matrix with the second row replaced by three times the first row plus the second row.

49. **a)** Since (x_0, y_0, z_0) and (x_1, y_1, z_1) are solutions then

$$\begin{aligned} ax_0 + by_0 + cz_0 &= d \\ ax_1 + by_1 + cz_1 &= d \end{aligned}$$

Now consider $(tx_0 + (1-t)x_1, ty_0 + (1-t)y_1, tz_0 + (1-t)z_1)$

$a[tx_0+(1-t)x_1]+b[ty_0+(1-t)y_1]+c[tz_0+(1-t)z_1] = atx_0+ax_1-bty_1+aty_0+by_1-bty_1+ctz_0+cz_1-ctz_1$
Rearranging the terms we get

$$\begin{aligned} &= (atx_0 + bty_0 + ctz_0) + (ax_1 + by_1 + cz_1) \\ &\quad -(atx_1 + bty_1 + ctz_1) \\ &= t(ax_0 + by_0 + z_0) + (ax_1 + by_1 + cz_1) \\ &\quad -t(ax_1 + by_1 + cz_1) \\ &= t(d) + d - t(d) \\ &= d \end{aligned}$$

Therefore, $(tx_0+(1-t)x_1, ty_0+(1-t)y_1, tz_0+(1-t)z_1)$ is also a solution to $ax + by + cz = d$ for any value of t.

b) As seen in the previous part, since any value of t could be used to find a solution of the form $(tx_0 + (1-t)x_1, ty_0 + (1-t)y_1, tz_0 + (1-t)z_1)$, then if a system has two solutions then it has infinitely many solutions.

c) The answers in part b) may be generalized since the technique used in part a) can be easily extended to any number of variables.

50. - 55. Left to the student

Exercise Set 6.3

1.

$$\left[\begin{array}{cc|cc} 1 & 1 & 1 & 0 \\ -1 & 0 & 0 & 1 \end{array}\right] = \left[\begin{array}{cc|cc} 1 & 1 & 1 & 0 \\ 0 & 1 & 1 & 1 \end{array}\right] = \left[\begin{array}{cc|cc} 1 & 0 & 0 & -1 \\ 0 & 1 & 1 & 1 \end{array}\right]$$

Thus, the inverse is $\begin{bmatrix} 0 & -1 \\ 1 & 1 \end{bmatrix}$

2. $\begin{bmatrix} 5 & -3 \\ -3 & 2 \end{bmatrix}$

3.

$$\left[\begin{array}{cc|cc} 0 & 1 & 1 & 0 \\ 1 & 0 & 0 & 1 \end{array}\right] = \left[\begin{array}{cc|cc} 1 & 0 & 0 & 1 \\ 0 & 1 & 1 & 0 \end{array}\right]$$

Thus, the inverse is $\begin{bmatrix} 0 & 1 \\ 1 & 0 \end{bmatrix}$

4. $\begin{bmatrix} -1 & 0 \\ 0 & 1/2 \end{bmatrix}$

5. $det = (3 \cdot 7) - (5 \cdot 4) = 1$

Thus, the inverse is $\frac{1}{1} \cdot \begin{bmatrix} 7 & -4 \\ -5 & 3 \end{bmatrix} = \begin{bmatrix} 7 & -4 \\ -5 & 3 \end{bmatrix}$

6. $\begin{bmatrix} 7/18 & -1/18 \\ 2/9 & 1/9 \end{bmatrix}$

7. $det = (3 \cdot -2) - (8 \cdot 7) = 62$

Thus, the inverse is $\frac{1}{-62} \cdot \begin{bmatrix} -2 & -7 \\ -8 & -3 \end{bmatrix} = \begin{bmatrix} 1/31 & 7/62 \\ 4/31 & -3/62 \end{bmatrix}$

8. $\begin{bmatrix} 5/12 & 2/3 \\ 1/12 & 1/3 \end{bmatrix}$

9.

$$\left[\begin{array}{ccc|ccc} -2 & 2 & 1 & 1 & 0 & 0 \\ 1 & 2 & 0 & 0 & 1 & 0 \\ 0 & -1 & 0 & 0 & 0 & 1 \end{array}\right] = \left[\begin{array}{ccc|ccc} 1 & 2 & 0 & 0 & 1 & 0 \\ 0 & -1 & 0 & 0 & 0 & 1 \\ -2 & 2 & 1 & 1 & 0 & 0 \end{array}\right]$$

$$= \left[\begin{array}{ccc|ccc} 1 & 2 & 0 & 0 & 1 & 0 \\ 0 & -1 & 0 & 0 & 0 & 1 \\ 0 & 6 & 1 & 1 & 2 & 0 \end{array}\right]$$

$$= \left[\begin{array}{ccc|ccc} 1 & 2 & 0 & 0 & 1 & 0 \\ 0 & -1 & 0 & 0 & 0 & 1 \\ 0 & 0 & 1 & 1 & 2 & 6 \end{array}\right]$$

$$= \left[\begin{array}{ccc|ccc} 1 & 0 & 0 & 0 & 1 & 2 \\ 0 & -1 & 0 & 0 & 0 & 1 \\ 0 & 0 & 1 & 1 & 2 & 6 \end{array}\right]$$

$$= \left[\begin{array}{ccc|ccc} 1 & 0 & 0 & 0 & 1 & 2 \\ 0 & 1 & 0 & 0 & 0 & -1 \\ 0 & 0 & 1 & 1 & 2 & 6 \end{array}\right]$$

Thus, the inverse is $\begin{bmatrix} 0 & 1 & 2 \\ 0 & 0 & -1 \\ 1 & 2 & 6 \end{bmatrix}$

10. $\begin{bmatrix} -2 & 2 & 1 \\ 1 & 2 & 0 \\ 0 & -1 & 0 \end{bmatrix}$

11. Technology was used. Inverse is $\begin{bmatrix} 2 & 1 & -2 \\ 2 & 0 & 1 \\ -1 & 0 & 0 \end{bmatrix}$

12. $\begin{bmatrix} 2 & 0 & 1 \\ -1 & 0 & 0 \\ 6 & 1 & 2 \end{bmatrix}$

13. Technology was used. Inverse is $\begin{bmatrix} 8 & 2 & -1 \\ 8 & 3 & -3 \\ -3 & -1 & 1 \end{bmatrix}$

14. $\begin{bmatrix} -61 & -75 & -81 \\ 21 & 26 & 28 \\ 31 & 38 & 41 \end{bmatrix}$

15. Technology was used. Inverse is $\begin{bmatrix} -20 & -27 & -14 \\ 7 & 10 & 5 \\ 10 & 13 & 7 \end{bmatrix}$

16. $\begin{bmatrix} 15/2 & 17/2 & 8 \\ -1 & -2 & -2 \\ -9/2 & -9/2 & -4 \end{bmatrix}$

17. Technology was used. Inverse is $\begin{bmatrix} 0 & -3/2 & -1 \\ -1 & -3/2 & -5/2 \\ 0 & -1 & -1/2 \end{bmatrix}$

18. $\begin{bmatrix} 2 & 1 & -3/2 \\ -1/2 & 1/2 & -1/2 \\ 3/2 & 1/2 & -1 \end{bmatrix}$

19. Technology was used. Inverse is $\begin{bmatrix} -7 & -5 & -6 \\ 1 & 1 & 2 \\ 13/2 & 9/2 & 9/2 \end{bmatrix}$

20. $\begin{bmatrix} 9 & 10 & 11 \\ -5 & -7 & -5 \\ -1 & 0 & -2 \end{bmatrix}$

21. Technology was used. Inverse is $\begin{bmatrix} 8/3 & 4/3 & -3/2 \\ 7/3 & 5/3 & -3/2 \\ 19/3 & 8/3 & -7/2 \end{bmatrix}$

22. $\begin{bmatrix} 31/15 & 4/3 & -6/5 \\ 26/15 & 5/3 & -6/5 \\ 68/15 & 8/3 & -13/5 \end{bmatrix}$

23. Technology was used. Inverse is $\begin{bmatrix} 7 & 7 & 37 & 8 \\ 10 & 10 & 53 & 10 \\ -3 & -3 & -16 & -3 \\ 2 & 3 & 13 & 4 \end{bmatrix}$

24. $\begin{bmatrix} -2 & 5 & 5 & 8 \\ -3 & 7 & 7 & 10 \\ 1 & -2 & -2 & -3 \\ 0 & 2 & 3 & 4 \end{bmatrix}$

25. $det = (2 \cdot 6) - (3 \cdot 1) = 12 - 3 = 9$ The matrix is invertible

26. $det = (4 \cdot 3) - (-4 \cdot -2) = 12 - 8 = 4$ The matrix is invertible

27. $det = (9 \cdot 3) - (-5 \cdot 2) = 27 + 10 = 37$ The matrix is invertible

28. $det = (7 \cdot -9) - (8 \cdot 2) = -63 - 16 = -79$ The matrix is invertible

29. $det = (2 \cdot 9) - (6 \cdot 3) = 18 - 18 = 0$ The matrix is not invertible

30. $det = (-12 \cdot 2) - (3 \cdot -8) = 24 - (-24) = 0$ The matrix is not invertible

31.

$$\begin{aligned} det &= 1[(5)(2) - (-4)(-3)] - 8[(-2)(2) - (7)(-3)] + \\ &\quad 3[(-2)(-4) - (7)(5)] \\ &= -2 - 136 - 81 \\ &= -219 \end{aligned}$$

The matrix is invertible

32. $det = -157$ The matrix is invertible

33.

$$\begin{aligned} det &= 1[(2)(1) - (1)(3)] - 1[(2)(1) - (3)(1)] + \\ &\quad 1[(2)(3) - (2)(3)] \\ &= -1 + 1 - 0 \\ &= 0 \end{aligned}$$

The matrix is not invertible

34. $det = -77$ The matrix is invertible

35.

$$\begin{aligned} det &= 0[(6)(6) - (14)(2)] - 4[(-1)(6) - (-1)(2)] + \\ &\quad 2[(-1)(14) - (-1)(6)] \\ &= 0 + 16 - 16 \\ &= 0 \end{aligned}$$

The matrix is not invertible

36. $det = 504$ The matrix is invertible

37.

$$\begin{aligned} det &= 2[(-5)(5) - (-1)(7)] + 4[(-3)(5) - (-4)(7)] + \\ &\quad 8[(-3)(-1) - (-5)(-4)] \\ &= -36 + 52 - 136 \\ &= -120 \end{aligned}$$

The matrix is invertible

38. $det = 0$ The matrix is not invertible

39. $det = -96$ (technology used) The matrix is invertible

40. $det = 218$ The matrix is invertible

41. $det = 12$ The matrix is invertible

42. $det = 50$ The matrix is invertible

43.

$$P_1 = \begin{bmatrix} 0.5 & 1.25 \\ 0.75 & 0.25 \end{bmatrix}^{-1} \begin{bmatrix} 156 \\ 48 \end{bmatrix} = \begin{bmatrix} -4/13 & 20/13 \\ 12/13 & -8/13 \end{bmatrix} \begin{bmatrix} 156 \\ 48 \end{bmatrix} = \begin{bmatrix} 26 \\ 114 \end{bmatrix}$$

44. **a)** $G^{-1} = \begin{bmatrix} 0 & 10/3 & 0 \\ -20/3 & 0 & 20/3 \\ 120/19 & 0 & -100/19 \end{bmatrix}$

b)

$$G_2 = \begin{bmatrix} 0 & 0.75 & 0.95 \\ 0.3 & 0 & 0 \\ 0 & 0.9 & 0.95 \end{bmatrix} \begin{bmatrix} 206 \\ 55 \\ 211 \end{bmatrix} = \begin{bmatrix} 242 \\ 1 \\ 250 \end{bmatrix}$$

c)

$$G_1 = \begin{bmatrix} 0 & 10/3 & 0 \\ -20/3 & 0 & 20/3 \\ 120/19 & 0 & -100/19 \end{bmatrix} \begin{bmatrix} 206 \\ 55 \\ 211 \end{bmatrix} = \begin{bmatrix} 184 \\ 1389 \\ 0 \end{bmatrix}$$

45. **a)** $G^{-1} = \begin{bmatrix} 0 & 5/4 \\ 2/15 & -5/2 \end{bmatrix}$

b)

$$G_2 = \begin{bmatrix} 0 & 5/4 \\ 2/15 & -5/ \end{bmatrix} \begin{bmatrix} 58815 \\ 3060 \end{bmatrix} = \begin{bmatrix} 3825 \\ 192 \end{bmatrix}$$

c)

$$G_1 = \begin{bmatrix} 0 & 5/4 \\ 2/15 & -5/2 \end{bmatrix} \begin{bmatrix} 3825 \\ 192 \end{bmatrix} = \begin{bmatrix} 240 \\ 30 \end{bmatrix}$$

46. **a)** $G^{-1} = \begin{bmatrix} 0 & 2 & 0 \\ 3.157 & -22.732 & -4.188 \\ -3.222 & 23.196 & 6.314 \end{bmatrix}$

b)

$$G_2 = \begin{bmatrix} 0 & 2 & 0 \\ 3.157 & -22.732 & -4.188 \\ -3.222 & 23.196 & 6.314 \end{bmatrix} \begin{bmatrix} 8365 \\ 1095 \\ 310 \end{bmatrix} = \begin{bmatrix} 2190 \\ 220 \\ 408 \end{bmatrix}$$

c)

$$G_1 = \begin{bmatrix} 0 & 2 & 0 \\ 3.157 & -22.732 & -4.188 \\ -3.222 & 23.196 & 6.314 \end{bmatrix} \begin{bmatrix} 2190 \\ 220 \\ 408 \end{bmatrix} = \begin{bmatrix} 441 \\ 198 \\ 630 \end{bmatrix}$$

47. Left to the student

48.

$$\begin{aligned} ABB^{-1}A^{-1} &= A(BB^{-1})A^{-1} \\ &= A(I)A^{-1} \\ &= AA^{-1} \\ &= I \end{aligned}$$

Therefore, $B^{-1}A^{-1}$ is the inverse of (AB)

49. **a)** $AB = \begin{bmatrix} -8 & 25 \\ -10 & 41 \end{bmatrix}$ Thus

$$\begin{aligned} det(AB) &= -328 + 250 = -78 \\ &= -6 \cdot 13 \\ &= det(A)\ det(B) \end{aligned}$$

b) $AB = \begin{bmatrix} 8 & -26 & 20 \\ -2 & -7 & 0 \\ 10 & -34 & 16 \end{bmatrix}$ Thus

$$\begin{aligned} det(AB) &= 1032 \\ &= 12 \cdot 86 \\ &= det(A)\ det(B) \end{aligned}$$

c) Left to the student (answers vary)

50. When we use Theorem 6, since all a_{ij}, $i > j$ are zero therefore for $i > j$, $a_{ij}C_{ij} = 0$ which means

$$det(A) = \sum_i a_{ii}C_{ii}$$

which simplify to

$$det(A) = a_{11}a_{22} \cdots a_{nn}$$

50. - 54. Left to the student

55. - 58. Left to the student

Exercise Set 6.4

1.

$$\begin{bmatrix} 2 & 0 \\ 0 & 3 \end{bmatrix} \begin{bmatrix} 1 \\ 0 \end{bmatrix} = \begin{bmatrix} 2 \\ 0 \end{bmatrix} = 2 \cdot \begin{bmatrix} 1 \\ 0 \end{bmatrix}$$

Thus, the vector is an eigenvector with eigenvalue of 2

2.

$$\begin{bmatrix} 2 & 0 \\ 0 & 3 \end{bmatrix}\begin{bmatrix} 0 \\ 1 \end{bmatrix} = \begin{bmatrix} 0 \\ 3 \end{bmatrix} = 3 \cdot \begin{bmatrix} 0 \\ 1 \end{bmatrix}$$

Thus, the vector is an eigenvector with eigenvalue of 3

3.

$$\begin{bmatrix} 3 & 2 \\ 0 & 4 \end{bmatrix}\begin{bmatrix} 0 \\ 1 \end{bmatrix} = \begin{bmatrix} 2 \\ 4 \end{bmatrix}$$

The vector is not an eigenvector

4.

$$\begin{bmatrix} 5 & 0 \\ 3 & -2 \end{bmatrix}\begin{bmatrix} 1 \\ 0 \end{bmatrix} = \begin{bmatrix} 5 \\ 3 \end{bmatrix}$$

The vector is not an eigenvector

5.

$$\begin{bmatrix} 5 & 0 \\ 3 & -2 \end{bmatrix}\begin{bmatrix} 0 \\ 1 \end{bmatrix} = \begin{bmatrix} 0 \\ -2 \end{bmatrix} = -2 \cdot \begin{bmatrix} 0 \\ 1 \end{bmatrix}$$

Thus, the vector is an eigenvector with eigenvalue -2

6.

$$\begin{bmatrix} 10 & 0 \\ 42 & -4 \end{bmatrix}\begin{bmatrix} 1 \\ 3 \end{bmatrix} = \begin{bmatrix} 10 \\ 30 \end{bmatrix} = 10 \cdot \begin{bmatrix} 1 \\ 3 \end{bmatrix}$$

Thus, the vector is an eigenvector with eigenvalue 10

7.

$$\begin{bmatrix} -8.5 & -4.5 \\ 21 & 11 \end{bmatrix}\begin{bmatrix} 2 \\ -7 \end{bmatrix} = \begin{bmatrix} 15.5 \\ -35 \end{bmatrix}$$

The vector is not an eigenvector

8.

$$\begin{bmatrix} 5 & 0 & 0 \\ 0 & 3 & 0 \\ 0 & 0 & -2 \end{bmatrix}\begin{bmatrix} 1 \\ 0 \\ 0 \end{bmatrix} = \begin{bmatrix} 5 \\ 0 \\ 0 \end{bmatrix} = 5 \cdot \begin{bmatrix} 1 \\ 0 \\ 0 \end{bmatrix}$$

Thus, the vector is an eigenvector with eigenvalue 5

9.

$$\begin{bmatrix} 5 & 0 & 0 \\ 0 & 3 & 0 \\ 0 & 0 & -2 \end{bmatrix}\begin{bmatrix} 0 \\ 1 \\ 0 \end{bmatrix} = \begin{bmatrix} 0 \\ 3 \\ 0 \end{bmatrix} = 3 \cdot \begin{bmatrix} 1 \\ 0 \\ 0 \end{bmatrix}$$

Thus, the vector is an eigenvector with eigenvalue 5

10.

$$\begin{bmatrix} 5 & 0 & 0 \\ 0 & 3 & 0 \\ 0 & 0 & -2 \end{bmatrix}\begin{bmatrix} 0 \\ 0 \\ 1 \end{bmatrix} = \begin{bmatrix} 0 \\ 0 \\ -2 \end{bmatrix} = -2 \cdot \begin{bmatrix} 1 \\ 0 \\ 0 \end{bmatrix}$$

Thus, the vector is an eigenvector with eigenvalue -2

11.

$$\begin{bmatrix} -25 & 40 & 39 \\ -32 & 47 & 39 \\ 16 & -20 & -1 \end{bmatrix}\begin{bmatrix} -13 \\ -13 \\ 4 \end{bmatrix} = \begin{bmatrix} -39 \\ -39 \\ 48 \end{bmatrix}$$

The vector is not an eigenvector

12.

$$\begin{bmatrix} -25 & 40 & 39 \\ -32 & 47 & 39 \\ 16 & -20 & -1 \end{bmatrix}\begin{bmatrix} 3 \\ 7 \\ -2 \end{bmatrix} = \begin{bmatrix} 127 \\ 155 \\ -90 \end{bmatrix}$$

The vector is not an eigenvector

13. $v = 3w + 4u$

14. $v = -3w + \frac{4}{3}u$

15. $v = 2w + u$

16. $v = 5w - 2u$

17. $v = -w + 3u$

18. $v = -2w + 4u$

19.

$$\begin{aligned} \det(A - rI) &= 0 \\ \begin{vmatrix} 1-r & 0 \\ -1 & 2-r \end{vmatrix} &= 0 \\ (1-r)(2-r) - 0 &= 0 \\ r &= 1 \\ r &= 2 \end{aligned}$$

For $r = 1$

$$\begin{aligned} \begin{bmatrix} 0 & 0 \\ -1 & 1 \end{bmatrix}\begin{bmatrix} x \\ y \end{bmatrix} &= \begin{bmatrix} 0 \\ 0 \end{bmatrix} \\ \begin{bmatrix} 0 \\ -x+y \end{bmatrix} &= \begin{bmatrix} 0 \\ 0 \end{bmatrix} \\ x &= y \\ x &= t \\ y &= t \end{aligned}$$

Thus for the eigenvalue of $r = 1$ the eigenvector is $\begin{bmatrix} t \\ t \end{bmatrix}, (t \neq 0)$

For $r = 2$

$$\begin{aligned} \begin{bmatrix} -1 & 0 \\ -1 & 0 \end{bmatrix} \begin{bmatrix} x \\ y \end{bmatrix} &= \begin{bmatrix} 0 \\ 0 \end{bmatrix} \\ \begin{bmatrix} -x \\ -y \end{bmatrix} &= \begin{bmatrix} 0 \\ 0 \end{bmatrix} \\ x &= 0 \\ y &= t \end{aligned}$$

Thus, for the eigenvalue of $r = 2$ the eigenvector is $\begin{bmatrix} 0 \\ t \end{bmatrix}, (t \neq 0)$

20. Eigenvalue $r = 2$, eigenvector $\begin{bmatrix} -t \\ t \end{bmatrix}, (t \neq 0)$

Eigenvalue $r = 3$, eigenvector $\begin{bmatrix} 0 \\ t \end{bmatrix}, (t \neq 0)$.

21.

$$\begin{aligned} \det(A - rI) &= 0 \\ \begin{vmatrix} 5-r & 2 \\ -24 & -9-r \end{vmatrix} &= 0 \\ (5-r)(-9-r) - (-48) &= 0 \\ r^2 + 4r + 3 &= 0 \\ r &= -1 \\ r &= -3 \end{aligned}$$

For $r = -1$

$$\begin{aligned} \begin{bmatrix} 6 & 2 \\ -24 & -8 \end{bmatrix} \begin{bmatrix} x \\ y \end{bmatrix} &= \begin{bmatrix} 0 \\ 0 \end{bmatrix} \\ \begin{bmatrix} 6x+2y \\ -24x-8y \end{bmatrix} &= \begin{bmatrix} 0 \\ 0 \end{bmatrix} \\ 3x &= y \end{aligned}$$

Thus for the eigenvalue of $r = -1$ the eigenvector is $\begin{bmatrix} t \\ 3t \end{bmatrix}, (t \neq 0)$

For $r = -3$

$$\begin{aligned} \begin{bmatrix} 8 & 2 \\ -24 & -6 \end{bmatrix} \begin{bmatrix} x \\ y \end{bmatrix} &= \begin{bmatrix} 0 \\ 0 \end{bmatrix} \\ \begin{bmatrix} 8x+2y \\ -24x-6y \end{bmatrix} &= \begin{bmatrix} 0 \\ 0 \end{bmatrix} \\ 4x &= y \end{aligned}$$

Thus, for the eigenvalue of $r = -3$ the eigenvector is $\begin{bmatrix} t \\ -4t \end{bmatrix}, (t \neq 0)$

22. Eigenvalue $r = \frac{3}{2} + \frac{5\sqrt{97}}{2}$,

eigenvector $\begin{bmatrix} \left(\frac{-7}{6} + \frac{\sqrt{97}}{6}\right) t \\ t \end{bmatrix}, (t \neq 0)$

Eigenvalue $r = \frac{3}{2} - \frac{5\sqrt{97}}{2}$,

eigenvector $\begin{bmatrix} \left(\frac{-7}{6} - \frac{\sqrt{97}}{6}\right) t \\ t \end{bmatrix}, (t \neq 0)$

23.

$$\begin{aligned} \det(A - rI) &= 0 \\ \begin{vmatrix} -7.5-r & -15.75 \\ 6 & 12-r \end{vmatrix} &= 0 \\ (-7.5-r)(12-r) - (-94.5) &= 0 \\ r^2 - 4.5r + 4.5 &= 0 \\ r &= 1.5 \\ r &= 3 \end{aligned}$$

For $r = 1.5$

$$\begin{aligned} \begin{bmatrix} -9 & -15.75 \\ 6 & 10.5 \end{bmatrix} \begin{bmatrix} x \\ y \end{bmatrix} &= \begin{bmatrix} 0 \\ 0 \end{bmatrix} \\ \begin{bmatrix} -9x-15.75y \\ 6x+10.5y \end{bmatrix} &= \begin{bmatrix} 0 \\ 0 \end{bmatrix} \\ 4x &= -7y \end{aligned}$$

Thus for the eigenvalue of $r = 1.5$ the eigenvector is $\begin{bmatrix} 7t \\ -4t \end{bmatrix}, (t \neq 0)$

For $r = 3$

$$\begin{aligned} \begin{bmatrix} -10.5 & -15.75 \\ 6 & 9 \end{bmatrix} \begin{bmatrix} x \\ y \end{bmatrix} &= \begin{bmatrix} 0 \\ 0 \end{bmatrix} \\ \begin{bmatrix} -10.5x-15.75y \\ 6x+9y \end{bmatrix} &= \begin{bmatrix} 0 \\ 0 \end{bmatrix} \\ 2x &= -3y \end{aligned}$$

Thus, for the eigenvalue of $r = 3$ the eigenvector is $\begin{bmatrix} 3t \\ -2t \end{bmatrix}, (t \neq 0)$

24. Eigenvalue $r = 3$,

eigenvector $\begin{bmatrix} \frac{5}{3}t \\ t \end{bmatrix}, (t \neq 0)$

Eigenvalue $r = 7$,

eigenvector $\begin{bmatrix} t \\ -2t \end{bmatrix}, (t \neq 0)$

25.

$$\begin{aligned} \det(A - rI) &= 0 \\ \begin{vmatrix} 9.5-r & -4.5 \\ 15 & -7-r \end{vmatrix} &= 0 \\ (9.5-r)(-7-r) - (-67.5) &= 0 \\ r^2 - 2.5r + 1 &= 0 \\ r &= 0.5 \\ r &= 2 \end{aligned}$$

For $r = 0.5$

$$\begin{aligned} \begin{bmatrix} 9 & -4.5 \\ 15 & -7.5 \end{bmatrix} \begin{bmatrix} x \\ y \end{bmatrix} &= \begin{bmatrix} 0 \\ 0 \end{bmatrix} \\ \begin{bmatrix} 9x-4.5y \\ 15x-7.5y \end{bmatrix} &= \begin{bmatrix} 0 \\ 0 \end{bmatrix} \\ 2x &= y \end{aligned}$$

Thus for the eigenvalue of $r = 0.5$ the eigenvector is $\begin{bmatrix} t \\ 2t \end{bmatrix}, (t \neq 0)$

For $r = 2$

$$\begin{bmatrix} 7.5 & -4.5 \\ 15 & -9 \end{bmatrix}\begin{bmatrix} x \\ y \end{bmatrix} = \begin{bmatrix} 0 \\ 0 \end{bmatrix}$$
$$\begin{bmatrix} 7.5x-4.5y \\ 15x-9y \end{bmatrix} = \begin{bmatrix} 0 \\ 0 \end{bmatrix}$$
$$5x = 3y$$

Thus, for the eigenvalue of $r = 2$ the eigenvector is $\begin{bmatrix} 3t \\ 5t \end{bmatrix}, (t \neq 0)$

26. Eigenvalue $r = -0.5$,
eigenvector $\begin{bmatrix} -3t \\ t \end{bmatrix}, (t \neq 0)$

Eigenvalue $r = 3$,
eigenvector $\begin{bmatrix} 5t \\ 2t \end{bmatrix}, (t \neq 0)$

27. $\begin{bmatrix} 10-r & -4 & 15 \\ 8 & -2-r & 15 \\ -4 & 2 & -5-r \end{bmatrix}$

The characteristic equation is $r^3 - 3r^2 + 2r = 0$
The eigenvalues are $r = 0$, $r = 1$, and $r = 2$
For $r = 0$

$$\begin{bmatrix} 10 & -4 & 15 \\ 8 & -2 & 15 \\ -4 & 2 & -5 \end{bmatrix}\begin{bmatrix} x \\ y \\ z \end{bmatrix} = \begin{bmatrix} 0 \\ 0 \\ 0 \end{bmatrix}$$
$$10x - 4y + 15z = 0$$
$$8x - 2y - 15z = 0$$
$$-4x + 2y - 5z = 0$$

Solving the matrix above using any method discussed earlier this chapter yields that $x = 5t$, $y = 5t$, and $z = 2t$, which are the eigenvectors for $r = 0$.

For $r = 1$

$$\begin{bmatrix} 9 & -4 & 15 \\ 8 & -3 & 15 \\ -4 & 2 & -6 \end{bmatrix}\begin{bmatrix} x \\ y \\ z \end{bmatrix} = \begin{bmatrix} 0 \\ 0 \\ 0 \end{bmatrix}$$
$$9x - 4y + 15z = 0$$
$$8x - 3y - 15z = 0$$
$$-4x + 2y - 6z = 0$$

Solving the matrix above using any method discussed earlier this chapter yields that $x = 3t$, $y = 3t$, and $z = t$, which are the eigenvectors for $r = 1$.

For $r = 2$

$$\begin{bmatrix} 8 & -4 & 15 \\ 8 & -4 & 15 \\ -4 & 2 & -7 \end{bmatrix}\begin{bmatrix} x \\ y \\ z \end{bmatrix} = \begin{bmatrix} 0 \\ 0 \\ 0 \end{bmatrix}$$
$$8x - 4y + 15z = 0$$
$$8x - 4y - 15z = 0$$
$$-4x + 2y - 7z = 0$$

Solving the matrix above using any method discussed earlier this chapter yields that $x = t$, $y = 2t$, and $z = 0$, which are the eigenvectors for $r = 2$.

28. The characteristic equation is $(r + 3)(r^2 - r) = 0$
The eigenvalues are $r = 0$, $r = 1$, and $r = -3$
The eigenvectors are $[0, 5t, t]$, $[0, -6t, t]$, and $[-t, t, 0]$ respectively.

29. $\begin{bmatrix} -5-r & 8 & 2 \\ -15 & 18-r & 4 \\ 45 & -48 & -10-r \end{bmatrix}$

The characteristic equation is $r^3 - 3r^2 + 2r = 0$
The eigenvalues are $r = 0$, $r = 1$, and $r = 2$
For $r = 0$

$$\begin{bmatrix} -5 & 8 & 2 \\ -15 & 18 & 4 \\ 45 & -48 & -10 \end{bmatrix}\begin{bmatrix} x \\ y \\ z \end{bmatrix} = \begin{bmatrix} 0 \\ 0 \\ 0 \end{bmatrix}$$
$$-5x + 8y + 2z = 0$$
$$-15x - 18y + 4z = 0$$
$$45x - 48y - 10z = 0$$

Solving the matrix above using any method discussed earlier this chapter yields that $x = 15t$, $y = -5t$, and $z = -2t$, which are the eigenvectors for $r = 0$.

For $r = 1$

$$\begin{bmatrix} -6 & 8 & 2 \\ -15 & 17 & 4 \\ 45 & -48 & -11 \end{bmatrix}\begin{bmatrix} x \\ y \\ z \end{bmatrix} = \begin{bmatrix} 0 \\ 0 \\ 0 \end{bmatrix}$$
$$-6x + 8y + 2z = 0$$
$$-15x + 17y + 4z = 0$$
$$45x - 48y - 11z = 0$$

Solving the matrix above using any method discussed earlier this chapter yields that $x = -t$, $y = -3t$, and $z = 9t$, which are the eigenvectors for $r = 1$.

For $r = 2$

$$\begin{bmatrix} -7 & 8 & 2 \\ -15 & 16 & 4 \\ 45 & -48 & -12 \end{bmatrix}\begin{bmatrix} x \\ y \\ z \end{bmatrix} = \begin{bmatrix} 0 \\ 0 \\ 0 \end{bmatrix}$$
$$-7x + 8y + 2z = 0$$
$$-15x + 16y + 4z = 0$$
$$45x - 48y - 12z = 0$$

Solving the matrix above using any method discussed earlier this chapter yields that $x = 0$, $y = -t$, and $z = 4t$, which are the eigenvectors for $r = 2$.

30. The characteristic equation is $r^3 + 6r^2 + 11r + 6 = 0$
The eigenvalues are $r = -1$, $r = -2$, and $r = -3$
The
eigenvectors are $[0, 0, t]$, $[t, -3t, -9t]$, and $[2t, -5t, -15t]$ respectively

31. The characteristic equation is $r^3 + 2r^2 - r - 2 = 0$
The eigenvalues are $r = -2$, $r = -1$, and $r = 1$
The eigenvectors are $[-t, t, 2t]$, $[t, -2t, -4t]$, and $[0, t, t]$ respectively

32. The characteristic equation is $r^3 - 3r^2 - 4r + 12 = 0$
The eigenvalues are $r = -2$, $r = 2$, and $r = 3$
The eigenvectors are $[0, t, 0]$, $[-t, -t, t]$, and $[t, 2t, -2t]$ respectively

33. The characteristic equation is $r^3 - 2r^2 - 16r + 32 = 0$
The eigenvalues are $r = -4$, $r = 2$, and $r = 4$
The eigenvectors are $[t, t, t]$, $[0, 0, t]$, and $[2t, t, 0]$ respectively

34. The characteristic equation is $r^3 - 6r^2 + 8r = 0$
The eigenvalues are $r = 0$, $r = 2$, and $r = 4$
The eigenvectors are $[0, t, 0]$, $[-2t, 2t, t]$, and $[-3t, 3t, t]$ respectively

35. The characteristic equation is $r^3 - 2r^2 - r + 2 = 0$
The eigenvalues are $r = -1$, $r = 1$, and $r = 2$
The eigenvectors are $[t, t, -1]$, $[2t, t, 0]$, and $[-2t, -2t, t]$ respectively

36. The characteristic equation is $r^3 + r^2 - 4r - 4 = 0$
The eigenvalues are $r = -2$, $r = -1$, and $r = 2$
The eigenvectors are $[0, t, -2t]$, $[t, 2t, -2t]$, and $[t, t, -t]$ respectively

37.

$$\begin{aligned} A^n w &= 2(2)^{10}\begin{bmatrix} 1 \\ 0 \end{bmatrix} + 3(1)^{10}\begin{bmatrix} 0 \\ 1 \end{bmatrix} \\ &= \begin{bmatrix} 2048 \\ 0 \end{bmatrix} + \begin{bmatrix} 0 \\ 3 \end{bmatrix} \\ &= \begin{bmatrix} 2048 \\ 3 \end{bmatrix} \end{aligned}$$

38.

$$\begin{aligned} A^n w &= 2(1)^{100}\begin{bmatrix} 1 \\ 1 \end{bmatrix} + 3(-1)^{100}\begin{bmatrix} 0 \\ 1 \end{bmatrix} \\ &= \begin{bmatrix} 2 \\ 2 \end{bmatrix} + \begin{bmatrix} 0 \\ 3 \end{bmatrix} \\ &= \begin{bmatrix} 2 \\ 5 \end{bmatrix} \end{aligned}$$

39.

$$\begin{aligned} A^n w &= 2(-2)^{10}\begin{bmatrix} 1 \\ 1 \end{bmatrix} + 3(0)^{10}\begin{bmatrix} 0 \\ 1 \end{bmatrix} \\ &= \begin{bmatrix} 2048 \\ 2048 \end{bmatrix} + \begin{bmatrix} 0 \\ 0 \end{bmatrix} \\ &= \begin{bmatrix} 2048 \\ 2048 \end{bmatrix} \end{aligned}$$

40.

$$\begin{aligned} A^n w &= 2(3)^4\begin{bmatrix} 1 \\ -1 \end{bmatrix} + 4(0)^4\begin{bmatrix} 1 \\ 1 \end{bmatrix} \\ &= \begin{bmatrix} 162 \\ -162 \end{bmatrix} + \begin{bmatrix} 0 \\ 0 \end{bmatrix} \\ &= \begin{bmatrix} 162 \\ -162 \end{bmatrix} \end{aligned}$$

41. Long-term growth rate: 1.5
Long-term growth rate percentage: 50% (See page 483 for a reference; $Percentage = 100 * Long - termrate - 100\%$

42. Long-term growth rate: 1.01
Long-term growth rate percentage: 1% (See page 483 for a reference; $Percentage = 100 * Long - termrate - 100\%$

43. Long-term growth rate: 2.1
Long-term growth rate percentage: 110% (See page 483 for a reference; $Percentage = 100 * Long - termrate - 100\%$

44. Long-term growth rate: 4
Long-term growth rate percentage: 300% (See page 483 for a reference; $Percentage = 100 * Long - term\,rate - 100\%$

45. The characteristic equation for the Leslie matrix is

$$\begin{aligned} (0.5 - r)^2 - 1 &= 0 \\ r^2 - r + 0.25 - 1 &= 0 \\ r^2 - r - 0.75 &= 0 \\ r &= 1.5 \\ r &= -0.5 \end{aligned}$$

Thus, the long-term growth rate is 1.5

46. The characteristic equation for the Leslie matrix is

$$\begin{aligned} r^3 - 0.95r^2 - 0.225r + 0.01425 &= 0 \\ r &= -.2392697121 \\ r &= 1.136884236 \\ r &= .05238547577 \end{aligned}$$

Thus, the long-term growth rate is ≈ 1.14

47. The characteristic equation for the Leslie matrix is

$$\begin{aligned} (15 - r)(0 - r) - 6 &= 0 \\ r^2 - 15r - 6 &= 0 \\ r &= 15.38986692 \\ r &= -0.3898669190 \end{aligned}$$

Thus, the long-term growth rate is ≈ 15.39

48. The characteristic equation for the Leslie matrix is

$$\begin{aligned} r^3 - 4.09r^2 + 1.274r + 0.07760 &= 0 \\ r &= -0.05208901953 \\ r &= 0.3978834280 \\ r &= 3.744205592 \end{aligned}$$

Thus, the long-term growth rate is ≈ 3.74

49. **a)**

$$\begin{aligned} p &= \begin{bmatrix} 3 \\ 4 \end{bmatrix} \\ G^n p &= 3(1.1)^n\begin{bmatrix} 1 \\ 1 \end{bmatrix} + 4(0.8)^n\begin{bmatrix} 2 \\ 3 \end{bmatrix} \\ \lim_{n\to\infty}\left(\frac{1}{1.1}\right)^n G^n p &= \lim_{n\to\infty}\left(\frac{1}{1.1}\right)^n 3(1.1)^n\begin{bmatrix} 1 \\ 1 \end{bmatrix} + \\ &\quad \lim_{n\to\infty}\left(\frac{1}{1.1}\right)^n 4(0.8)^n\begin{bmatrix} 2 \\ 3 \end{bmatrix} \\ &= 3\begin{bmatrix} 1 \\ 1 \end{bmatrix} \end{aligned}$$

b) If the long term rate is not 1.1 then the $\lim_{n\to\infty}\left(\frac{1}{1.1}\right)^n G^n p$ may not converge

50. **a)**

$$\begin{aligned} p &= \begin{bmatrix} 4 \\ 7 \end{bmatrix} \\ G^n p &= 4(1.4)^n \begin{bmatrix} 3 \\ 5 \end{bmatrix} + 4(0.1)^n \begin{bmatrix} 1 \\ 2 \end{bmatrix} \\ \lim_{n\to\infty}\left(\frac{1}{1.4}\right)^n G^n p &= \lim_{n\to\infty}\left(\frac{1}{1.4}\right)^n 4(1.4)^n \begin{bmatrix} 3 \\ 5 \end{bmatrix} + \\ & \quad \lim_{n\to\infty}\left(\frac{1}{1.4}\right)^n 7(0.1)^n \begin{bmatrix} 1 \\ 2 \end{bmatrix} \\ &= 4\begin{bmatrix} 3 \\ 5 \end{bmatrix} \end{aligned}$$

b) If the long term rate is not 1.4 then the $\lim_{n\to\infty}\left(\frac{1}{1.4}\right)^n G^n p$ may not converge

51. $B^{-1} = \begin{bmatrix} 3 & -5 \\ -1 & 2 \end{bmatrix}$

a) $BA = \begin{bmatrix} 2 & 5 \\ 0 & 1 \end{bmatrix}$

$$\begin{aligned} B^{-1}BA &= \begin{bmatrix} 3 & -5 \\ -1 & 2 \end{bmatrix}\begin{bmatrix} 2 & 5 \\ 0 & 1 \end{bmatrix} \\ &= \begin{bmatrix} 6 & 10 \\ -2 & -3 \end{bmatrix} \end{aligned}$$

b) For matrix $A: (1-r)(2-r) - 0 = r^2 - 3r + 2$
For matrix $B^{-1}AB: (6-r)(-3-r)+20 = r^2-3r+2$
The answers are identical

c) Since the two matrices have the same characteristic equation, they will have the same eigenvalues

52. **a)**

$$B^{-1}BA = \begin{bmatrix} \frac{1}{3} & -1 \\ \frac{55}{9} & \frac{17}{3} \end{bmatrix}$$

b) For matrix $A: r^2 - 6r + 8$
For matrix $B^{-1}AB: r^2 - 6r + 8$
The answers are identical

c) Since the two matrices have the same characteristic equation, they will have the same eigenvalues

53. Left to the student

54. Yes, see Exercise 51 for an explanation

55. **a)** $(2-r)(-5-r) - 3 = r^2 + 3r - 13$

b)

$$\begin{aligned} A^2 + 3A - 13I &= \begin{bmatrix} 7 & -9 \\ -3 & 28 \end{bmatrix} + \begin{bmatrix} 6 & 9 \\ 3 & -15 \end{bmatrix} - \\ & \quad \begin{bmatrix} 13 & 0 \\ 0 & 13 \end{bmatrix} \\ &= \begin{bmatrix} 0 & 0 \\ 0 & 0 \end{bmatrix} \end{aligned}$$

c) Left to student

Exercise Set 6.5

1. Not linear

2. Linear, but not homogeneous

3. Linear and homogeneous

4. Linear, but not homogeneous

5. Linear and homogeneous

6. Not linear

7. $a = 2$ and $b = 3$
$r^2 - 2r - 3 = 0$
$r = -1$ and $r = 3$
$x_n = c_1(-1)^n + c_2(3)^n$

8. $a = 8$ and $b = -15$
$r^2 - 8r + 15 = 0$
$r = 3$ and $r = 5$
$x_n = c_1(3)^n + c_2(5)^n$

9. $a = -1$ and $b = 6$
$r^2 + r - 6 = 0$
$r = -3$ and $r = 2$
$x_n = c_1(-3)^n + c_2(2)^n$

10. $a = 5$ and $b = 6$
$r^2 - 5r - 6 = 0$
$r = 2$ and $r = 3$
$x_n = c_1(2)^n + c_2(3)^n$

11. $a = 1$ and $b = -2$
$r^2 - r + 2 = 0$
$r = -1$ and $r = 2$
$x_n = c_1(-1)^n + c_2(2)^n$

12. $a = 3$ and $b = 4$
$r^2 - 3r - 4 = 0$
$r = -1$ and $r = 4$
$x_n = c_1(-1)^n + c_2(4)^n$

13. $r^2 - r - 2 = 0$
$r = -1$ and $r = 2$
$x_n = c_1(-1)^n + c_2(2)^n$
For $n = 0$, $c_1 + c_2 = 2$
For $n = 1$, $-c_1 + 2c_2 = 1$
Solving the system gives $c_1 = 1$ and $c_2 = 1$
Therefore, $x_n = (-1)^n + (2)^n$

14. $r^2 - 2r - 8 = 0$
$r = -2$ and $r = 4$
$x_n = c_1(-2)^n + c_2(4)^n$
For $n = 0$, $c_1 + c_2 = 1$
For $n = 1$, $-2c_1 + 4c_2 = 4$
Solving the system gives $c_1 = 0$ and $c_2 = 1$
Therefore, $x_n = (4)^n$

15. $r^2 - \frac{5}{2}r + 1 = 0$
$r = \frac{1}{2} = 2^{-1}$ and $r = 2$
$x_n = c_1(2)^{-n} + c_2(2)^n$

For $n = 0$, $c_1 + c_2 = 0$
For $n = 1$, $\frac{1}{2}c_1 + 2c_2 = -\frac{3}{2}$
Solving the system gives $c_1 = 1$ and $c_2 = -1$
Therefore, $x_n = (2)^{-n} - (2)^n$

16. $r^2 - 2r - 3 = 0$
$r = -1$ and $r = 3$
$x_n = c_1(-1)^n + c_2(3)^n$
For $n = 0$, $c_1 + c_2 = 7$
For $n = 1$, $-c_1 + 2c_2 = 10$
Solving the system gives $c_1 = \frac{11}{4}$ and $c_2 = \frac{17}{4}$
Therefore, $x_n = \frac{11}{4}(-1)^n + \frac{17}{4}(2)^n$

17. $r^2 - 2r - 8 = 0$
$r = -2$ and $r = 4$
$x_n = c_1(-2)^n + c_2(4)^n$
For $n = 0$, $c_1 + c_2 = 1$
For $n = 1$, $-2c_1 + 4c_2 = -2$
Solving the system gives $c_1 = 1$ and $c_2 = 0$
Therefore, $x_n = (-2)^n$

18. $r^2 - 4r - 5 = 0$
$r = -1$ and $r = 5$
$x_n = c_1(-1)^n + c_2(5)^n$
For $n = 0$, $c_1 + c_2 = 7$
For $n = 1$, $-c_1 + 5c_2 = 17$
Solving the system gives $c_1 = 3$ and $c_2 = 4$
Therefore, $x_n = 3(-1)^n + 4(5)^n$

19. $r^2 + \frac{3}{2}r - 1 = 0$
$r = -2$ and $r = \frac{1}{2} = 2^{-1}$
$x_n = c_1(-2)^n + c_2(2)^{-n}$
For $n = 0$, $c_1 + c_2 = 5$
For $n = 1$, $\frac{1}{2}c_1 - 2c_2 = -\frac{5}{2}$
Solving the system gives $c_1 = 2$ and $c_2 = 3$
Therefore, $x_n = 2(-2)^n + 3(2)^{-n}$

20. $r^2 - 5r - 6 = 0$
$r = 2$ and $r = 3$
$x_n = c_1(2)^n + c_2(3)^n$
For $n = 0$, $c_1 + c_2 = 17$
For $n = 1$, $2c_1 + 3c_2 = 32$
Solving the system gives $c_1 = 19$ and $c_2 = -2$
Therefore, $x_n = 19(2)^n - 2(3)^n$

21. $c = 5c - 6c + 8 \rightarrow c = 4$
Homogeneous case:
$r^2 - 5r + 6 = 0$
$r = 2$ and $r = 3$
Therefore, $x_n = c_1(2)^n + c_2(3)^n + 4$

22. $c = c + 2c - 12 \rightarrow c = 1$
Homogeneous case:
$r^2 - r - 2 = 0$
$r = -1$ and $r = 2$
Therefore, $x_n = c_1(-1)^n + c_2(2)^n + 4$

23. $c = 9c - 20c + 12 \rightarrow c = 1$
Homogeneous case:
$r^2 - 9r + 20 = 0$
$r = 4$ and $r = 5$
Therefore, $x_n = c_1(4)^n + c_2(5)^n + 1$

24. $c = -5c - 6c + 108 \rightarrow c = 9$
Homogeneous case:
$r^2 + 5r + 6 = 0$
$r = 2$ and $r = 3$
Therefore, $x_n = c_1(2)^n + c_2(3)^n + 9$

25. $c = 7c - 12c + 12 \rightarrow c = 2$
Homogeneous case:
$r^2 - 7r + 12 = 0$
$r = 3$ and $r = 4$
Therefore, $x_n = c_1(3)^n + c_2(4)^n + 2$

26. $c = c + 6c - 36 \rightarrow c = 6$
Homogeneous case:
$r^2 - r - 6 = 0$
$r = -2$ and $r = 3$
$x_n = c_1(-2)^n + c_2(3)^n + 6$
For $n = 0$, $c_1 + c_2 + 6 = 8$
For $n = 1$, $-2c_1 + 3c_2 + 6 = 7$
Solving the system gives $c_1 = 1$ and $c_2 = 1$
Therefore, $x_n = (-2)^n + (3)^n + 6$

27. $c = -2c + 8c - 45 \rightarrow c = 9$
Homogeneous case:
$r^2 + 2r - 8 = 0$
$r = -4$ and $r = 2$
$x_n = c_1(-4)^n + c_2(2)^n + 9$
For $n = 0$, $c_1 + c_2 + 9 = 11$
For $n = 1$, $-2c_1 + 3c_2 + 9 = -5$
Solving the system gives $c_1 = 3$ and $c_2 = -1$
Therefore, $x_n = 3(-4)^n - (2)^n + 9$

28. $c = 3c + 28c - 150 \rightarrow c = 5$
Homogeneous case:
$r^2 - 3r + 28 = 0$
$r = -4$ and $r = 7$
$x_n = c_1(-4)^n + c_2(7)^n + 5$
For $n = 0$, $c_1 + c_2 + 5 = 13$
For $n = 1$, $-4c_1 + 7c_2 + 5 = -49$
Solving the system gives $c_1 = 10$ and $c_2 = -2$
Therefore, $x_n = 10(-4)^n - 2(7)^n + 5$

29. $c = 9c - 18c + 20 \rightarrow c = 2$
Homogeneous case:
$r^2 - 9r + 18 = 0$
$r = 3$ and $r = 6$
$x_n = c_1(3)^n + c_2(6)^n + 2$
For $n = 0$, $c_1 + c_2 + 2 = 7$
For $n = 1$, $3c_1 + 6c_2 + 2 = 26$
Solving the system gives $c_1 = 2$ and $c_2 = 3$
Therefore, $x_n = 2(3)^n + 3(6)^n + 2$

30. $c = c + 2c - 2 \rightarrow c = 1$
Homogeneous case:
$r^2 - r - 2 = 0$
$r = -1$ and $r = 2$
$x_n = c_1(-1)^n + c_2(2)^n + 1$
For $n = 0$, $c_1 + c_2 + 1 = 1$
For $n = 1$, $-c_1 + 2c_2 + 1 = -2$

Solving the system gives $c_1 = 1$ and $c_2 = -1$
Therefore, $x_n = (-1)^n - (2)^n + 1$

31. $a = 0.35$ and $b = 0.45$
Since $a + b < 1$
The population decreases exponentially to 0

32. $a = 0.1$ and $b = 0.95$
Since $a + b > 1$
The population grows exponentially to ∞

33. $a = 2.3$ and $b = 1.5$
Since $a + b > 1$
The population grows exponentially to ∞

34. $a = 0.92$ and $b = 0.2$
Since $a + b > 1$
The population grows exponentially to ∞

35. $a = 0.01$ and $b = 1.5$
Since $a + b > 1$
The population grows exponentially to ∞

36. $a = 0.2$ and $b = 0.5$
Since $a + b < 1$
The population decreases exponentially to 0

37. **a)** $r^2 - 0.92r - 0.15 = 0$
$r = -0.14133$ and $r = 1.06133$
$x_n = c_1(-0.14133)^n + c_2(1.06133)^n$
For $n = 0$, $c_1 + c_2 = 0$
For $n = 1$, $-0.14133c_1 + 1.06133c_2 = 50$
Solving the system gives
$c_1 = -41.57451$ and $c_2 = 41.57451$
Therefore,
$x_n = -41.57451(-0.14133)^n + 41.57451(1.06133)^n$

b) Since $0.92 + 0.15 > 1$, the population will grow

38. **a)** $r^2 - 0.81r - 0.13 = 0$
$r = -0.13724$ and $r = 0.94724$
$x_n = c_1(-0.13724)^n + c_2(0.94724)^n$
For $n = 0$, $c_1 + c_2 = 0$
For $n = 1$, $-0.13724c_1 + 0.94724c_2 = 50$
Solving the system gives
$c_1 = -41.57451$ and $c_2 = 41.57451$
Therefore,
$x_n = -46.10498(-0.13724)^n + 46.10498(0.94724)^n$

b) Since $0.81 + 0.13 < 1$, the population will die

39.

$$\begin{aligned} x_{n+1} &= ax_n + az_n \\ &= ax_n + ab(M - x_{n-1}) \\ &= ax_n - abx_{n-1} + abM \end{aligned}$$

40. $x_{n+1} = 1.15x_n - 0.0115x_{n-1} + 11.5$

a) $c = 1.15c - 0.0115c + 11.5 \to c = -83.03249$
Homogeneous case:
$r^2 - 1.15r + 0.0115 = 0$
$r = 0.01009$ and $r = 1.13991$
$x_n = c_1(0.01009)^n + c_2(1.13991)^n - 83.03249$

b) For $n = 0$, $c_1 + c_2 - 83.03249 = 100$
For $n = 1$, $0.01009c_1 + 1.13991c_2 - 83.03249 = 125$
Solving the system gives
$c_1 = .53845$ and $c_2 = 182.49404$
Therefore,
$x_n = .53845(0.01009)^n + 182.49404(1.13991)^n - 83.03249$

c) Since $1.15 + 0.0115 > 1$ the limit grows to ∞

41. $x_{n+1} = 0.95x_n - 0.095x_{n-1} + 95$

a) $c = 0.95c - 0.095c + 95 \to c = 655.17241$
Homogeneous case:
$r^2 - 0.95r + 0.095 = 0$
$r = 0.11358$ and $r = 0.83642$
$x_n = c_1(0.11358)^n + c_2(0.83642)^n + 655.17241$

b) For $n = 0$, $c_1 + c_2 + 655.17241 = 50$
For $n = 1$, $0.11358c_1 + 0.83642c_2 + 655.17241 = 130$
Solving the system gives
$c_1 = 26.27632$ and $c_2 = -631.44873$
Therefore,
$x_n = 26.27632(0.11385)^n - 631.44873(0.83642)^n + 655.17241$

c) Since $0.95 + 0.095 > 1$ the limit grows to ∞

42. **a)** $x_{n+1} = 1 - x_n$

b) $a = -1$ and $b = 0$
$c = 1 - c \to c = \dfrac{1}{2}$
$r^2 + r = 0$
$x_n = c_1(-1)^n + c_2(0)^n + \dfrac{1}{2}$
$x_n = c_1(-1)^n + \dfrac{1}{2}$
$x_0 = 0 \to c_1 = -\dfrac{1}{2}$
Thus,
$x_n = \dfrac{1}{2} - \dfrac{1}{2}(-1)^n$

43. **a)** $x_{n+1} = x_{n-1}$

b) $a = 0$ and $b = 1$
$r^2 - 1 = 0$
$x_n = c_1(-1)^n + c_2(1)^n$
$x_0 = 2 \to c_1 + c_2 = 2$
$x_1 = 4 \to -c_1 + c_2 = 4$
Solving the system gives
$c_1 = -1$ and $c_2 = 3$
Thus,
$x_n = 3(1)^n - (-1)^n$

44. **a)** $a = 1$ and $b = 1$
$r^2 - r - 1 = 0$
$r = \dfrac{1 \pm \sqrt{5}}{2}$
$$x_n = c_1\left(\frac{1-\sqrt{5}}{2}\right)^n + c_2\left(\frac{1+\sqrt{5}}{2}\right)^n$$

b) $x_0 = 0 \to c_1 + c_2 = 0$
$$x_1 = 1 \to \left(\frac{1-\sqrt{5}}{2}\right)c_1 + \left(\frac{1+\sqrt{5}}{2}\right)c_2 = 1$$
Solving the system gives

$c_1 = -\frac{\sqrt{5}}{5}$ and $c_2 = \frac{\sqrt{5}}{5}$

Thus,

$$x_n = -\frac{\sqrt{5}}{5}\left(\frac{1-\sqrt{5}}{2}\right)^n + \frac{\sqrt{5}}{5}\left(\frac{1+\sqrt{5}}{2}\right)^n$$

c) $x_3 = 2$, $x_{20} = 6765$

d) The number of digits of the nth term in the Fibonacci sequence can be given by $\frac{\left(\frac{1+\sqrt{5}}{2}\right)^n - \left[-\left(\frac{1+\sqrt{5}}{2}\right)\right]^{-n}}{\sqrt{5}}$ which could be approximated to $\frac{n}{5}$.

45. $a = 4$ and $b = -4$
$r^2 - 4r + 4 = 0$
$r = 2$ Repeated
$x_n = c_1(2)^n + c_2 n(2)^n$

46. $a = 2$ and $b = -1$
$r^2 - 2r + 1 = 0$
$r = 1$ Repeated
$x_n = c_1(1)^n + c_2 n(1)^n$
$x_n = c_1 + c_2 n$

47. $c = -2c - c + 12 \to c = 3$ $a = -2$ and $b = -1$
$r^2 + 2r + 1 = 0$
$r = -1$ Repeated
$x_n = c_1(-1)^n + c_2 n(-1)^n + 3$

48. $a = 6$ and $b = -9$
$r^2 - 6r + 9 = 0$
$r = 3$ Repeated
$x_n = c_1(3)^n + c_2 n(3)^n$
For $n = 0$, $c_1 = 7$
For $n = 1$, $3(7) + 3c_2 = 21 \to c_2 = 0$
$x_n = 7(3)^n$

49. - .54 Left to the student

55. a) $n = 1$, $M_1 = 1.2987$
$n = 2$, $M_2 = 1.6861$
$n = 3$, $M_3 = 2.1883$
$n = 4$, $M_4 = 2.8385$
$n = 5$, $M_5 = 3.6797$

b) The results from part (a) are very close since the ratio $\frac{(1+\rho)M}{1+\theta M} \approx 1.3$

56. a)

$$\begin{aligned} M &= \frac{(1+\rho)M}{1+\theta M} \\ M + \theta M^2 &= 1 + \rho M \\ \theta M^2 + (1-\rho)M - 1 &= 0 \\ 0.001M^2 + 0.7M - 1 &= 0 \\ M &= -701.4256678 \text{ not acceptable} \\ M &= 1.425667816 \end{aligned}$$

b) $n = 1$, $M_1 = 296.14$
$n = 2$, $M_2 = 297.02$
$n = 3$, $M_3 = 297.7$
$n = 4$, $M_4 = 298.23$
$n = 5$, $M_5 = 298.64$

c) The answers in part (b) are expected since they vary by a value close to the equilibrium value found in part (a)

57. a)

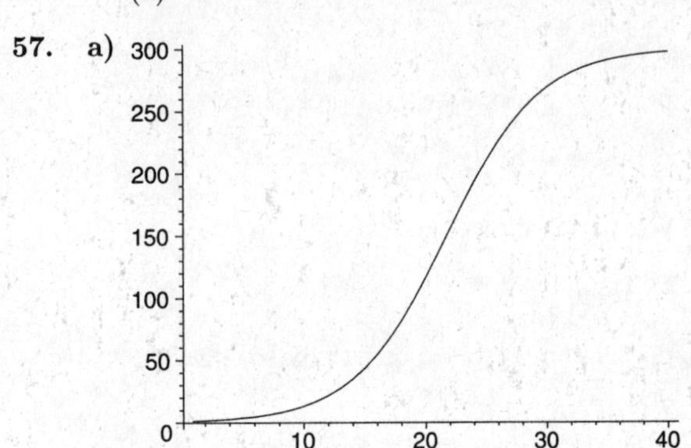

b) The graph looks like the logistic model discussed in section 4.3

Chapter Review Exercises

1. $3B = \begin{bmatrix} 3 & -9 & 6 \\ -12 & 21 & 15 \end{bmatrix}$

2. Not possible, dimensions do not match

3. BC

$$= \begin{bmatrix} (1)(2)+(-3)(-2)+(2)(7) & (1)(8)+(-3)(1)+(2)(8) \\ (-4)(2)+(7)(-2)+(5)(7) & (-4)(8)+(7)(1)+(5)(4) \end{bmatrix}$$

$$= \begin{bmatrix} 22 & 13 \\ 13 & -5 \end{bmatrix}$$

4. Not possible, dimensions do not match

5. Not possible, dimensions do not match

6. $CB = \begin{bmatrix} -30 & 50 & 44 \\ -6 & 13 & 1 \\ -9 & 7 & 34 \end{bmatrix}$

7. $2A + 3BC$

$$= \begin{bmatrix} 4 & -8 \\ 14 & 6 \end{bmatrix} + \begin{bmatrix} 66 & 39 \\ 39 & -15 \end{bmatrix}$$

$$= \begin{bmatrix} 70 & 31 \\ 53 & -9 \end{bmatrix}$$

8. $BD = \begin{bmatrix} 34 & -9 & 8 \\ -78 & -18 & 33 \end{bmatrix}$

9. ABC

$$= \begin{bmatrix} 2 & -4 \\ 7 & 3 \end{bmatrix}\begin{bmatrix} 22 & 13 \\ 13 & -5 \end{bmatrix}$$

$$= \begin{bmatrix} (2)(22)+(-4)(13) & (2)(13)+(-4)(-5) \\ (7)(22)+(3)(13) & (7)(13)+(3(-5) \end{bmatrix}$$

$$= \begin{bmatrix} -8 & 46 \\ 193 & 76 \end{bmatrix}$$

10. $Av = \begin{bmatrix} -14 \\ 36 \end{bmatrix}$

11. A^3v

$$= \begin{bmatrix} -188 & 36 \\ -63 & -197 \end{bmatrix} \begin{bmatrix} 3 \\ 5 \end{bmatrix}$$
$$= \begin{bmatrix} (-188)(3)+(36)(5) \\ (-63)(3)+(-197)(5) \end{bmatrix}$$
$$= \begin{bmatrix} -384 \\ -1174 \end{bmatrix}$$

12. D^2Cv

$$= \begin{bmatrix} -31 & 54 & -20 \\ -72 & -45 & 18 \\ 37 & -18 & 23 \end{bmatrix} \begin{bmatrix} 2 & 8 \\ -2 & 1 \\ 7 & 4 \end{bmatrix} \begin{bmatrix} 3 \\ 5 \end{bmatrix}$$
$$= \begin{bmatrix} -31 & 54 & -20 \\ -72 & -45 & 18 \\ 37 & -18 & 23 \end{bmatrix} \begin{bmatrix} 46 \\ -1 \\ 41 \end{bmatrix}$$
$$= \begin{bmatrix} -2300 \\ -2529 \\ 2663 \end{bmatrix}$$

13. Write the augmented matrix
$$\left[\begin{array}{cc|c} 2 & 3 & 11 \\ 5 & -2 & -39 \end{array}\right]$$
$\frac{1}{2}R1 \rightarrow$
$$\left[\begin{array}{cc|c} 1 & 3/2 & 11/2 \\ 5 & -2 & -39 \end{array}\right]$$
$-5R1 + R2 \rightarrow$
$$\left[\begin{array}{cc|c} 1 & 3/2 & 11/2 \\ 0 & -19/2 & -133/2 \end{array}\right]$$
$-\frac{2}{19}R2 \rightarrow$
$$\left[\begin{array}{cc|c} 1 & 3/2 & 11/2 \\ 0 & 1 & 7 \end{array}\right]$$
Thus, $y = 7$ and $x = \dfrac{11-21}{2} = -5$

14. Write the augmented matrix

$$\left[\begin{array}{ccc|c} 1 & 1 & 1 & 4 \\ 2 & -1 & -3 & -5 \\ 3 & -3 & -3 & -6 \end{array}\right]$$
$$\left[\begin{array}{ccc|c} 1 & -1 & -1 & -2 \\ 0 & 1 & 1 & 3 \\ 0 & 0 & 1 & 2 \end{array}\right]$$
Thus, $z = 2$, $y = 3 - 2 = 1$, and $x = -2 + 2 + 1 = 1$

15. Write the augmented matrix

$$\left[\begin{array}{ccc|c} 4 & 2 & 3 & 9 \\ 2 & -1 & -3 & -1 \\ 3 & -3 & -1 & 15 \end{array}\right]$$

$\frac{1}{4}R1 \rightarrow$

$$\left[\begin{array}{ccc|c} 1 & 1/2 & 3/4 & 9/4 \\ 2 & -1 & -3 & -1 \\ 3 & -3 & -1 & 15 \end{array}\right]$$

$-2R1 + R2$ and $-3R1 + R3 \rightarrow$

$$\left[\begin{array}{ccc|c} 1 & 1/2 & 3/4 & 9/4 \\ 0 & -2 & -9/2 & -11/2 \\ 0 & -9/2 & -13/4 & 33/4 \end{array}\right]$$
$-\frac{1}{2}R2 \rightarrow$
$$\left[\begin{array}{ccc|c} 1 & 1/2 & 3/4 & 9/4 \\ 0 & 1 & 9/4 & 11/4 \\ 0 & -9/2 & -13/4 & 33/4 \end{array}\right]$$
$\frac{9}{2}R2 + R3 \rightarrow$
$$\left[\begin{array}{ccc|c} 1 & 1/2 & 3/4 & 9/4 \\ 0 & 1 & 9/4 & 11/4 \\ 0 & 0 & 55/8 & 165/8 \end{array}\right]$$
$\frac{8}{55}R3 \rightarrow$
$$\left[\begin{array}{ccc|c} 1 & 1/2 & 3/4 & 9/4 \\ 0 & 1 & 9/4 & 11/4 \\ 0 & 0 & 1 & 3 \end{array}\right]$$
Thus, $z = 3$, $y = 11/4 - 27/4 = -4$, and $x = 9/4 - 9/4 + 2 = 2$

16. $\det[-3] = -3$, matrix is invertible

17. $\det[0] = 0$, matrix is not invertible

18. $\det = -3(2) - 6(-5) = 24$, matrix is invertible

19. $\det = 0.5(2) - 5(0.2) = 0$, matrix is not invertible

20. $\det = 1[(3)(1) - (0)(-4)] - 0[(-2)(1) - (0)(-1)] + 2[(-2)(-4) - (3)(-1)] = 25$, matrix is invertible

21. $\det = 7[(6)(11) - (0)(1)] - 2[(-3)(11) - (0)(5)] + 16[(-3)(1) - (6)(5)] = 0$, matrix is not invertible

22. $\det = -3$ thus $A^{-1} = \begin{bmatrix} 1/3 & 0 \\ 1/3 & -1 \end{bmatrix}$

23. $\det = 5$ thus $A^{-1} = \begin{bmatrix} -1/5 & -3/5 \\ 2/5 & 1/5 \end{bmatrix}$

24. $\left[\begin{array}{ccc|ccc} -5 & 1 & 7 & 1 & 0 & 0 \\ -4 & 2 & 6 & 0 & 1 & 0 \\ -1 & 1 & 1 & 0 & 0 & 1 \end{array}\right]$

Swap R1 and -R3 $\rightarrow$

$$\left[\begin{array}{ccc|ccc} 1 & -1 & -1 & 0 & 0 & -1 \\ -4 & 2 & 6 & 0 & 1 & 0 \\ -5 & 1 & 7 & 1 & 0 & 0 \end{array}\right]$$

$4R1 + R2$ and $5R1 + R3 \rightarrow$

$$\left[\begin{array}{ccc|ccc} 1 & -1 & -1 & 0 & 0 & -1 \\ 0 & -2 & 2 & 0 & 1 & -4 \\ 0 & -4 & -2 & 1 & 0 & -5 \end{array}\right]$$

$-2R2 + R3 \rightarrow$

$$\left[\begin{array}{ccc|ccc} 1 & -1 & -1 & 0 & 0 & -1 \\ 0 & -2 & 2 & 0 & 1 & -4 \\ 0 & 0 & -2 & 1 & -2 & 3 \end{array}\right]$$

$R3 + R2 \rightarrow$

$$\left[\begin{array}{ccc|ccc} 1 & -1 & -1 & 0 & 0 & -1 \\ 0 & -2 & 0 & 1 & -1 & -1 \\ 0 & 0 & -2 & 1 & -2 & 3 \end{array}\right]$$

$-\frac{1}{2}R2 + R1$ and $\frac{1}{2}R3 + R1 \rightarrow$

$$\left[\begin{array}{ccc|ccc} 1 & 0 & 0 & -1 & 3/2 & -2 \\ 0 & -2 & 0 & 1 & -1 & -1 \\ 0 & 0 & -2 & 1 & -2 & 3 \end{array}\right]$$

$\frac{R2}{-2}$ and $\frac{R3}{-2} \rightarrow$

$$\left[\begin{array}{ccc|ccc} 1 & 0 & 0 & -1 & 3/2 & -2 \\ 0 & 1 & 0 & -1/2 & 1/2 & 1/2 \\ 0 & 0 & 1 & -1/2 & 1 & -3/2 \end{array}\right]$$

Thus $A^{-1} = \left[\begin{array}{ccc} -1 & 3/2 & -2 \\ -1/2 & 1/2 & 1/2 \\ -1/2 & 1 & -3/2 \end{array}\right]$

25. $\left[\begin{array}{ccc|ccc} -20 & 7 & 10 & 1 & 0 & 0 \\ -27 & 10 & 13 & 0 & 1 & 0 \\ -14 & 5 & 7 & 0 & 0 & 1 \end{array}\right]$

Swap R3 and R2 $\rightarrow$

$$\left[\begin{array}{ccc|ccc} -20 & 7 & 10 & 1 & 0 & 0 \\ -14 & 5 & 7 & 0 & 0 & 1 \\ -27 & 10 & 13 & 0 & 1 & 0 \end{array}\right]$$

$-2R2 + R3 \rightarrow$

$$\left[\begin{array}{ccc|ccc} -20 & 7 & 10 & 1 & 0 & 0 \\ -14 & 5 & 7 & 0 & 0 & 1 \\ 1 & 0 & -1 & 0 & 1 & -2 \end{array}\right]$$

Swap R3 and R1 $\rightarrow$

$$\left[\begin{array}{ccc|ccc} 1 & 0 & -1 & 0 & 1 & -2 \\ -14 & 5 & 7 & 0 & 0 & 1 \\ -20 & 7 & 10 & 1 & 0 & 0 \end{array}\right]$$

$14R1 + R2$ and $20R1 + R3 \rightarrow$

$$\left[\begin{array}{ccc|ccc} 1 & 0 & -1 & 0 & 1 & -2 \\ 0 & 5 & -7 & 0 & 14 & -27 \\ 0 & 7 & 10 & 1 & 20 & -40 \end{array}\right]$$

$-7R2 + 5R3 \rightarrow$

$$\left[\begin{array}{ccc|ccc} 1 & 0 & -1 & 0 & 1 & -2 \\ 0 & 5 & -7 & 0 & 14 & -27 \\ 0 & 0 & -1 & 5 & 2 & -11 \end{array}\right]$$

$-R3 \rightarrow$

$$\left[\begin{array}{ccc|ccc} 1 & 0 & -1 & 0 & 1 & -2 \\ 0 & 5 & -7 & 0 & 14 & -27 \\ 0 & 0 & 1 & -5 & -2 & 11 \end{array}\right]$$

$7R3 + R2$ and $R3 + R1 \rightarrow$

$$\left[\begin{array}{ccc|ccc} 1 & 0 & 0 & -5 & -1 & 9 \\ 0 & 5 & 0 & -35 & 0 & 50 \\ 0 & 0 & 1 & -5 & -2 & 11 \end{array}\right]$$

$\frac{R2}{5} \rightarrow$

$$\left[\begin{array}{ccc|ccc} 1 & 0 & 0 & -5 & -1 & 9 \\ 0 & 1 & 0 & -7 & 0 & 10 \\ 0 & 0 & 1 & -5 & -2 & 11 \end{array}\right]$$

Thus $A^{-1} = \left[\begin{array}{ccc} -5 & -1 & 9 \\ -7 & 0 & 10 \\ -5 & -2 & 11 \end{array}\right]$

26. $\left[\begin{array}{cc} 3-r & 0 \\ 2 & 4-r \end{array}\right]$

$r^2 - 7r + 12 = 0$
Eigenvalues are $r = 3$ and $r = 4$
Eigenvectors are $[t, -2t]$ and $[0, t]$ respectively

27. $\left[\begin{array}{cc} 4-r & 2 \\ 0 & -1-r \end{array}\right]$

$r^2 - 3r - 4 = 0$
$r = -1$ and $r = 4$
For $r = -1$

$$\left[\begin{array}{cc} 5 & 2 \\ 0 & 0 \end{array}\right]\left[\begin{array}{c} x \\ y \end{array}\right] = \left[\begin{array}{c} 0 \\ 0 \end{array}\right]$$

$$\left[\begin{array}{c} x \\ y \end{array}\right] = \left[\begin{array}{c} 2t \\ -5t \end{array}\right]$$

For $r = 4$

$$\left[\begin{array}{cc} 0 & 2 \\ 0 & -5 \end{array}\right]\left[\begin{array}{c} x \\ y \end{array}\right] = \left[\begin{array}{c} 0 \\ 0 \end{array}\right]$$

$$\left[\begin{array}{c} x \\ y \end{array}\right] = \left[\begin{array}{c} t \\ 0 \end{array}\right]$$

28. $\left[\begin{array}{cc} 15-r & 28 \\ -6 & -11-r \end{array}\right]$

$r^2 - 4r + 3 = 0$
Eigenvalues are $r = 1$ and $r = 3$
Eigenvectors are $[-7t, 3t]$ and $[-2t, t]$ respectively

29. $\left[\begin{array}{cc} 15-r & -42 \\ 4 & -11-r \end{array}\right]$

$r^2 - 4r + 3 = 0$
$r = 1$ and $r = 3$
For $r = 1$

$$\begin{bmatrix} 14 & -42 \\ 4 & -12 \end{bmatrix} \begin{bmatrix} x \\ y \end{bmatrix} = \begin{bmatrix} 0 \\ 0 \end{bmatrix}$$
$$\begin{bmatrix} x \\ y \end{bmatrix} = \begin{bmatrix} 3t \\ t \end{bmatrix}$$

For $r = 3$

$$\begin{bmatrix} 12 & -42 \\ 4 & -14 \end{bmatrix} \begin{bmatrix} x \\ y \end{bmatrix} = \begin{bmatrix} 0 \\ 0 \end{bmatrix}$$
$$\begin{bmatrix} x \\ y \end{bmatrix} = \begin{bmatrix} 7t \\ 2t \end{bmatrix}$$

30. $\begin{bmatrix} 5-r & 0 & 0 \\ 0 & -2-r & 4 \\ 0 & 0 & -3-r \end{bmatrix}$

$r^3 - 6r^2 - r + 30 = 0$
Eigenvalues are $r = -2$, $r = 3$, and $r = 5$
Eigenvectors are $[0, t, 0]$, $[0, 4t, 5t]$ and $[t, 0, 0]$ respectively

31. $\begin{bmatrix} -7-r & 0 & 0 \\ 3 & 2-r & 0 \\ 4 & 6 & -1-r \end{bmatrix}$

$r^3 + 6r^2 - 9r - 14 = 0$
Eigenvalues are $r = -7$, $r = -1$, and $r = 2$

For $r = -7$

$$\begin{bmatrix} 0 & 0 & 0 \\ 3 & 9 & 0 \\ 4 & 6 & 6 \end{bmatrix} \begin{bmatrix} x \\ y \\ z \end{bmatrix} = \begin{bmatrix} 0 \\ 0 \\ 0 \end{bmatrix}$$
$$\begin{bmatrix} x \\ y \\ z \end{bmatrix} = \begin{bmatrix} -3t \\ t \\ t \end{bmatrix}$$

For $r = -1$

$$\begin{bmatrix} -6 & 0 & 0 \\ 3 & 3 & 0 \\ 4 & 6 & 0 \end{bmatrix} \begin{bmatrix} x \\ y \\ z \end{bmatrix} = \begin{bmatrix} 0 \\ 0 \\ 0 \end{bmatrix}$$
$$\begin{bmatrix} x \\ y \\ z \end{bmatrix} = \begin{bmatrix} 0 \\ 0 \\ t \end{bmatrix}$$

For $r = 2$

$$\begin{bmatrix} -9 & 0 & 0 \\ 3 & 0 & 0 \\ 4 & 6 & -3 \end{bmatrix} \begin{bmatrix} x \\ y \\ z \end{bmatrix} = \begin{bmatrix} 0 \\ 0 \\ 0 \end{bmatrix}$$
$$\begin{bmatrix} x \\ y \\ z \end{bmatrix} = \begin{bmatrix} 0 \\ t \\ 2t \end{bmatrix}$$

32. $\begin{bmatrix} 4-r & 2 & 0 \\ -1 & 1-r & 0 \\ -5 & -10 & -3-r \end{bmatrix}$

$r^3 - 2r^2 - 9r + 18 = 0$
Eigenvalues are $r = -3$, $r = 2$, and $r = 3$
Eigenvectors are $[0, 0, t]$, $[t, -t, t]$ and $[-2t, t, 0]$ respectively

33. $\begin{bmatrix} 3-r & 0 & 2 \\ -4 & -r & -3 \\ -4 & 0 & -3-r \end{bmatrix}$

$r^3 - r = 0$
Eigenvalues are $r = -1$, $r = 0$, and $r = 1$

For $r = -1$

$$\begin{bmatrix} 4 & 0 & 2 \\ -4 & 1 & -3 \\ -4 & 0 & -2 \end{bmatrix} \begin{bmatrix} x \\ y \\ z \end{bmatrix} = \begin{bmatrix} 0 \\ 0 \\ 0 \end{bmatrix}$$
$$\begin{bmatrix} x \\ y \\ z \end{bmatrix} = \begin{bmatrix} t \\ -2t \\ -2t \end{bmatrix}$$

For $r = 0$

$$\begin{bmatrix} 3 & 0 & 2 \\ -4 & 0 & -3 \\ -4 & 0 & -3 \end{bmatrix} \begin{bmatrix} x \\ y \\ z \end{bmatrix} = \begin{bmatrix} 0 \\ 0 \\ 0 \end{bmatrix}$$
$$\begin{bmatrix} x \\ y \\ z \end{bmatrix} = \begin{bmatrix} 0 \\ t \\ 0 \end{bmatrix}$$

For $r = 1$

$$\begin{bmatrix} 2 & 0 & 2 \\ -4 & 0 & -3 \\ -4 & 0 & -4 \end{bmatrix} \begin{bmatrix} x \\ y \\ z \end{bmatrix} = \begin{bmatrix} 0 \\ 0 \\ 0 \end{bmatrix}$$
$$\begin{bmatrix} x \\ y \\ z \end{bmatrix} = \begin{bmatrix} -t \\ t \\ t \end{bmatrix}$$

34. $\begin{bmatrix} 4-r & 1 & -1 \\ 6 & -9-r & 9 \\ 12 & -18 & 18-r \end{bmatrix}$

$r^3 - 13r^2 + 42r = 0$
Eigenvalues are $r = 0$, $r = 6$, and $r = 7$
Eigenvectors are $[0, 1t, 1t]$, $[t, -2t, -4t]$ and $[t, -3t, -6t]$ respectively

35. $r^2 + 2r - 8 = 0$
$r = -4$ and $r = 2$
$x_n = c_1(-4)^n + c_2(2)^n$
For $n = 0$, $c_1 + c_2 = 3$
For $n = 1$, $-4c_1 + 2c_2 = 0$
Solving the system gives
$c_1 = 1$ and $c_2 = 2$
Thus,
$x_n = (-4)^n + 2(2)^n$

36. $c = \frac{-3}{2}c + c + \frac{3}{2} \to c = 1$

$r^2 + \frac{3}{2}r - 1 = 0$

$r = \frac{1}{2}$ and $r = -2$
$x_n = c_1(2)^{-n} + c_2(-2)^n + 1$
For $n = 0$, $c_1 + c_2 = 0$
For $n = 1$, $\frac{1}{2}c_1 - 2c_2 = 2$
Solving the system gives
$c_1 = -\frac{2}{5}$ and $c_2 = -\frac{3}{5}$
Thus,
$x_n = -\frac{2}{5}(2)^{-n} - \frac{3}{5}(-2)^n + 1$

37. Solve the system $2 = B + L$ and $0 = -0.06B + 0.03L$
From the first equation $B = 2 - L$

$$\begin{aligned} 0 &= -0.06(2-L) + 0.03L \\ 0 &= -0.12_0.06L_0.03L \\ 0.09L &= 0.12 \\ L &= \frac{4}{3} \\ B &= 2 - \frac{4}{3} \\ &= \frac{2}{3} \end{aligned}$$

38. **a)** $\begin{bmatrix} 0.7 & 0.4 \\ 0.6 & 1 \end{bmatrix}$

b) $\begin{bmatrix} 0.7 & 0.4 \\ 0.6 & 1 \end{bmatrix}\begin{bmatrix} 60 \\ 40 \end{bmatrix} = \begin{bmatrix} 58 \\ 76 \end{bmatrix}$

c) $\begin{bmatrix} 0.7 & 0.4 \\ 0.6 & 1 \end{bmatrix}\begin{bmatrix} 58 \\ 76 \end{bmatrix} = \begin{bmatrix} 71 \\ 111 \end{bmatrix}$

d)

$$\begin{aligned} (0.7-r)(1-r) - (0.6)(0.4) &= 0 \\ r^2 - 1.7r + .46 &= 0 \\ r &= 0.33765 \\ r &= 1.36235 \end{aligned}$$

Thus the long term rate is 1.36235

39. **a)** $\begin{bmatrix} 0.8 & 2 \\ 0.5 & 0.4 \end{bmatrix}\begin{bmatrix} 100 \\ 60 \end{bmatrix} = \begin{bmatrix} 200 \\ 74 \end{bmatrix}$

b) $\begin{bmatrix} 0.8 & 2 \\ 0.5 & 0.4 \end{bmatrix}\begin{bmatrix} 200 \\ 74 \end{bmatrix} = \begin{bmatrix} 308 \\ 130 \end{bmatrix}$

c)

$$\begin{aligned} (0.8-r)(0.4-r) - (2)(0.5) &= 0 \\ r^2 - 1.2r - .68 &= 0 \\ r &= -0.41980 \\ r &= 1.61980 \end{aligned}$$

Thus the long term rate is 1.61980

40. Left to the student

41. **a)** $r = \frac{1}{2}$ and $r = 1$
b) $[0, t]$ and $[t, 6t]$
c) $\begin{bmatrix} 1 & 0 \\ 6 & 7.88861 \times 10^-31 \end{bmatrix}$

42. **a)** $r = 0.9$ and $r = 1$
b) $[0, t]$ and $[t, 40t]$
c) $\begin{bmatrix} 1 & 0 \\ 39.99894 & 2.65614 \times 10^-31 \end{bmatrix}$

Chapter 6 Test

1.

$$\begin{bmatrix} 7 & -2 \\ 5 & 3 \end{bmatrix}\begin{bmatrix} 1 & 6 & -3 \\ 5 & -1 & 4 \end{bmatrix} = \begin{bmatrix} 7-10 & 42+2 & -21-8 \\ 5+15 & 30-3 & -15+12 \end{bmatrix} = \begin{bmatrix} -3 & 44 & -29 \\ 20 & 27 & -3 \end{bmatrix}$$

2.

$$\begin{bmatrix} 10 & 4 \\ 2 & -5 \\ 7 & 2 \end{bmatrix}\begin{bmatrix} -5 & 6 & 0 \\ -2 & 1 & 8 \end{bmatrix} = \begin{bmatrix} -50+8 & 60-4 & 0-32 \\ -10+10 & 12-5 & 0-40 \\ -35-4 & 42+2 & 0+16 \end{bmatrix} = \begin{bmatrix} -42 & 56 & -32 \\ 0 & 13 & -40 \\ -39 & 44 & 16 \end{bmatrix}$$

3.

$$\begin{bmatrix} 1 & 6 & -3 \\ 5 & -1 & 4 \end{bmatrix}\begin{bmatrix} 10 & -4 \\ 2 & -5 \\ 7 & 2 \end{bmatrix} - \begin{bmatrix} 7 & -2 \\ 5 & 3 \end{bmatrix} = \begin{bmatrix} 1 & -40 \\ 76 & -7 \end{bmatrix} - \begin{bmatrix} 7 & -2 \\ 5 & 3 \end{bmatrix} = \begin{bmatrix} -6 & -38 \\ 71 & -10 \end{bmatrix}$$

4. Write the augmented matrix

$$\left[\begin{array}{cc|c} 2 & 3 & -1 \\ 3 & 2 & 6 \end{array}\right]$$

$-3R1 + 2R2 \to$

$$\left[\begin{array}{cc|c} 2 & 3 & -1 \\ 0 & -5 & 15 \end{array}\right]$$

$\frac{1}{5}R3 \to$

$$\left[\begin{array}{cc|c} 2 & 3 & -1 \\ 0 & 1 & -3 \end{array}\right]$$

Thus, $y = -3$ and $x = \frac{-1+9}{2} = 4$

5. Write the augmented matrix

$$\left[\begin{array}{ccc|c} -1 & 3 & 2 & -3 \\ 3 & -4 & 2 & 26 \\ 4 & 2 & -3 & -10 \end{array}\right]$$

$3R1 + R2$ and $4R1 + R3 \rightarrow$

$$\left[\begin{array}{ccc|c} -1 & 3 & 2 & -3 \\ 0 & 5 & 8 & 17 \\ 0 & 14 & 5 & -22 \end{array}\right]$$

$-14R2 + 5R3 \rightarrow$

$$\left[\begin{array}{ccc|c} -1 & 3 & 2 & -3 \\ 0 & 5 & 8 & 17 \\ 0 & 0 & -87 & -384 \end{array}\right]$$

$-\frac{1}{87}R3 \rightarrow$

$$\left[\begin{array}{ccc|c} -1 & 3 & 2 & -3 \\ 0 & 5 & 8 & 17 \\ 0 & 0 & 1 & 4 \end{array}\right]$$

Thus, $z = 4$, $y = \frac{17-32}{5} = -3$, and $x = \frac{-3-8+9}{-1} = 2$

6. det $= 2(4) - 1(8) = 0$, the matrix is not invertible

7. det $= -3(2) - 6(-1) = 0$, the matrix is not invertible

8. det $= 0.5[0.5(7.3) - 0.2(3.4)] - 3[2.4(7.3) - 1.3(0.2)] + 1.2[2.4(3.4) - 1.3(0.5)] = -41.283$, the matrix is invertible

9. det $= 2[4(1)-8(-1)]-(-5)[3(4)-8(1)]+0[3(-1)-1(1)] = 44$, the matrix is invertible

10. det $= 3(1) - 4(1) = -1$

Thus, the inverse is given by $\begin{bmatrix} -1 & 4 \\ 1 & -3 \end{bmatrix}$

11. $\left[\begin{array}{ccc|ccc} 0 & -1 & -2 & 1 & 0 & 0 \\ 1 & 1 & 2 & 0 & 1 & 0 \\ 0 & 1 & 1 & 0 & 0 & 1 \end{array}\right]$

Swap R1 and R2 $\rightarrow$

$$\left[\begin{array}{ccc|ccc} 1 & 1 & 2 & 0 & 1 & 0 \\ 0 & -1 & -2 & 1 & 0 & 0 \\ 0 & 1 & 1 & 0 & 0 & 1 \end{array}\right]$$

Swap R2 and R3 $\rightarrow$

$$\left[\begin{array}{ccc|ccc} 1 & 1 & 2 & 0 & 1 & 0 \\ 0 & 1 & 1 & 0 & 0 & 1 \\ 0 & -1 & -2 & 1 & 0 & 0 \end{array}\right]$$

$-R2 + R1$ and $R2 + R3 \rightarrow$

$$\left[\begin{array}{ccc|ccc} 1 & 0 & 1 & 0 & 1 & -1 \\ 0 & 1 & 1 & 0 & 0 & 1 \\ 0 & 0 & -1 & 1 & 0 & 1 \end{array}\right]$$

$R3 + R1$ and $R3 + R2 \rightarrow$

$$\left[\begin{array}{ccc|ccc} 1 & 0 & 0 & 1 & 1 & 0 \\ 0 & 1 & 0 & 1 & 0 & 2 \\ 0 & 0 & -1 & 1 & 0 & 1 \end{array}\right]$$

$-R3 \rightarrow$

$$\left[\begin{array}{ccc|ccc} 1 & 0 & 0 & 1 & 1 & 0 \\ 0 & 1 & 0 & 1 & 0 & 2 \\ 0 & 0 & 1 & -1 & 0 & -1 \end{array}\right]$$

Thus, the inverse is given by

$$\begin{bmatrix} 1 & 1 & 0 \\ 1 & 0 & 2 \\ -1 & 0 & -1 \end{bmatrix}$$

12. $\begin{bmatrix} 11\text{-r} & 20 \\ -6 & -11\text{-r} \end{bmatrix}$

$r^2 - 1 = 0$

The eigenvalues are $r = -1$ and $r = 1$

The eigenvectors are $[-5t, 3t]$ and $[-2t, t]$ respectively

13. $\begin{bmatrix} 15\text{-r} & 4 \\ -42 & -11\text{-r} \end{bmatrix}$

$r^2 - 4r + 3 = 0$

The eigenvalues are $r = 1$ and $r = 3$

For $r = 1$, $\begin{bmatrix} 14 & 4 \\ -42 & -12 \end{bmatrix}\begin{bmatrix} x \\ y \end{bmatrix} = \begin{bmatrix} 0 \\ 0 \end{bmatrix}$

$\begin{bmatrix} x \\ y \end{bmatrix} = \begin{bmatrix} -2t \\ 7t \end{bmatrix}$

For $r = 3$, $\begin{bmatrix} 12 & 4 \\ -42 & -14 \end{bmatrix}\begin{bmatrix} x \\ y \end{bmatrix} = \begin{bmatrix} 0 \\ 0 \end{bmatrix}$

$\begin{bmatrix} x \\ y \end{bmatrix} = \begin{bmatrix} -t \\ 3t \end{bmatrix}$

14. $\begin{bmatrix} -5\text{-r} & -4 & -8 \\ -2 & 5\text{-r} & 4 \\ 2 & -2 & -1\text{-R} \end{bmatrix}$

$r^3 + r^2 - 9r - 9 = 0$

The eigenvalues are $r = -3$, $r = -1$ and $r = 3$

The eigenvectors are $[2t, t, -t]$, $[t, t, -t]$ and $[0, -2t, t]$ respectively

15. $\begin{bmatrix} -1\text{-r} & 0 & -8 \\ -6 & -7\text{-r} & 8 \\ 4 & 0 & 11\text{-r} \end{bmatrix}$

$r^3 - 3r^2 - 49r + 147 = 0$

The eigenvalues are $r = -7$, $r = 3$ and $r = 7$

For $r = -7$, $\begin{bmatrix} 6 & 0 & -8 \\ -6 & 0 & 8 \\ 4 & 0 & 18 \end{bmatrix}\begin{bmatrix} x \\ y \\ z \end{bmatrix} = \begin{bmatrix} 0 \\ 0 \\ 0 \end{bmatrix}$

$\begin{bmatrix} x \\ y \\ z \end{bmatrix} = \begin{bmatrix} 0 \\ t \\ 0 \end{bmatrix}$

For $r = 3$, $\begin{bmatrix} -4 & 0 & -8 \\ -6 & -10 & 8 \\ 4 & 0 & 8 \end{bmatrix}\begin{bmatrix} x \\ y \\ z \end{bmatrix} = \begin{bmatrix} 0 \\ 0 \\ 0 \end{bmatrix}$

$\begin{bmatrix} x \\ y \\ z \end{bmatrix} = \begin{bmatrix} -2t \\ 2t \\ t \end{bmatrix}$

For $r = 7$, $\begin{bmatrix} -8 & 0 & -8 \\ -6 & -14 & 8 \\ 4 & 0 & 4 \end{bmatrix}\begin{bmatrix} x \\ y \\ z \end{bmatrix} = \begin{bmatrix} 0 \\ 0 \\ 0 \end{bmatrix}$

$\begin{bmatrix} x \\ y \\ z \end{bmatrix} = \begin{bmatrix} -t \\ t \\ t \end{bmatrix}$

16. $\begin{bmatrix} -6-r & -11 & -22 \\ 0 & 7-r & 2 \\ 0 & -1 & 4-r \end{bmatrix}$

$r^3 - 5r^2 - 36r + 180 = 0$
The eigenvalues are $r = -6$, $r = 5$ and $r = 6$
The eigenvectors are $[t, 0, 0]$, $[0, -2t, t]$, and $[-t, -t, t]$ respectively

17. $r^2 - 3r - 10 = 0$
$r = -2$ and $r = 5$
$x_n = c_1(-2)^n + c_2(5)^n$
For $n = 0$, $c_1 + c_2 = -1$
For $n = 1$, $-2c_1 + 5c_2 = 16$ Solving the system gives
$c_1 = -3$ and $c_2 = 2$
Thus,
$x_n = -3(-2)^n + 2(5)^n$

18. $c = 2c + 8c - 27 \rightarrow c = 3$
$r^2 - 2r - 8 = 0$
$r = -2$ and $r = 4$
$x_n = c_1(-2)^n + c_2(4)^n + 3$
For $n = 0$, $c_1 + c_2 + 3 = 1$
For $n = 1$, $-2c_1 + 5c_2 + 3 = 2$ Solving the system gives
$c_1 = -1$ and $c_2 = -1$
Thus,
$x_n = -(-2)^n - (5)^n + 3$

19. Solve the system
$0 = -0.015B + 0.09$ and
$0 = 0.01B - 0.025F$
From the first equation we get $B = \dfrac{0.09}{0.015} = 6$
Thus, $F = \dfrac{0.01(6)}{0.025} = \dfrac{12}{5}$

20. **a)** $\begin{bmatrix} 0.6 & 3 \\ 0.4 & 0.2 \end{bmatrix}\begin{bmatrix} 150 \\ 140 \end{bmatrix} = \begin{bmatrix} 660 \\ 116 \end{bmatrix}$

b) $\begin{bmatrix} 0.6 & 3 \\ 0.4 & 0.2 \end{bmatrix}\begin{bmatrix} 660 \\ 116 \end{bmatrix} = \begin{bmatrix} 744 \\ 287 \end{bmatrix}$

21. The long term rate is $r = 1.1$

22. **a)** det $= -29$ thus, $A^{-1} = \begin{bmatrix} -1/29 & 6/29 \\ 5/29 & -1/29 \end{bmatrix}$

b) det $= -11$ thus, $B^{-1} = \begin{bmatrix} 1/11 & 3/11 \\ 4/11 & -1/11 \end{bmatrix}$

c) $\begin{bmatrix} -1/29 & 6/29 \\ 5/29 & -1/29 \end{bmatrix}\begin{bmatrix} 1/11 & 3/11 \\ 4/11 & -1/11 \end{bmatrix} = \begin{bmatrix} 25/319 & -9/319 \\ -9/319 & 16/319 \end{bmatrix}$

d) $AB = \begin{bmatrix} 25 & 9 \\ 9 & 16 \end{bmatrix}$, det$= 25(16) - 81 = 319$

$(AB)^{-1} = \begin{bmatrix} 16/319 & -9/319 \\ -9/319 & 25/319 \end{bmatrix}$

e) $BA = \begin{bmatrix} 16 & 9 \\ 9 & 25 \end{bmatrix}$, det$= 25(16) - 81 = 319$

$(BA)^{-1} = \begin{bmatrix} 25/319 & -9/319 \\ -9/319 & 16/319 \end{bmatrix}$

f) Left to the student

23. **a)** $\begin{bmatrix} 439 \\ 132 \end{bmatrix}$

b) $\begin{bmatrix} 563 \\ 303 \end{bmatrix}$

c) $\begin{bmatrix} 3174909 \\ 1407942 \end{bmatrix}$

d) $r = 1.652996$

Technology Connection

- **Page 433**

1. $\begin{bmatrix} 53 & -76 \\ 84 & -120 \end{bmatrix}$

2. $\begin{bmatrix} 7 & -87 & 3 \\ 49 & -18 & -104 \end{bmatrix}$

3. [15]

- **Page 436**

1. $\begin{bmatrix} -3 & 3 \\ 6 & 7 \end{bmatrix}$

2. $\begin{bmatrix} 7 & 9 \\ 2 & 5 \end{bmatrix}$

3. $\begin{bmatrix} -14 & 6 \\ -4 & 36 \end{bmatrix}$

4. $\begin{bmatrix} 6 & -9 \\ 12 & 18 \end{bmatrix}$

5. $\begin{bmatrix} -2944 & -192 \\ 256 & -2688 \end{bmatrix}$

- **Page 453**

1. $x = 3$, $y = 4$, and $z = 3$

2. $x = \dfrac{24}{47} + \dfrac{13}{47}$, $y = -\dfrac{28}{47} + \dfrac{24t}{47}$, and $z = t$

3. $x = \dfrac{2675}{744}$, $y = \dfrac{617}{372}$, and $z = -\dfrac{413}{744}$

- **Page 459**

1. $\begin{bmatrix} -7/150 & 2/15 & -23/150 \\ -17/300 & 7/30 & -13/300 \\ 11/75 & -2/15 & 4/75 \end{bmatrix}$

2. $\begin{bmatrix} 35/129 & 59/258 & -15/43 \\ 40/129 & 48/258 & -11/43 \\ 28/129 & 73/258 & -12/43 \end{bmatrix}$

3.
$$\begin{bmatrix} 235/1478 & -136/2217 & -203/4434 & -269/4434 \\ 567/1478 & 51/739 & -201/1478 & -361/4434 \\ -34/739 & -77/2217 & 130/2217 & -68/2217 \\ -157/739 & -73/2217 & 296/2217 & 425/2217 \end{bmatrix}$$

- **Page 462**

1. $x = 3$, $y = 4$, and $z = 3$

2. $x = -\frac{83}{46}$, $y = -\frac{112}{23}$, and $z = -\frac{385}{46}$

3. $x = \frac{2675}{744}$, $y = \frac{617}{372}$, and $z = -\frac{413}{744}$

- **Page 468**

1. -92

2. -2147

- **Page 478** NOTE: neither 1 nor 3 is an eigenvalue of the given matrix.

1. $[t, t, t]$

2. $[t, t, t]$

Extended Life Science Connection

1. Since S is a proportion of organisms that survive, you cannot have less than 0 organisms surviving and you cannot have more than the entire population or organismsm survive. $S = 0$ means no organisms survived while $S = 1$ means all the organisms survived.

2. An organism cannot have negative productive or creative powers

3. $\begin{bmatrix} F_1 & F_2 \\ \text{S} & 0 \end{bmatrix}$

4.

$$\begin{aligned} (0.5 - r)(-r) - 2(0.25) &= 0 \\ r^2 - 0.5r - 0.5 &= 0 \\ r &= -0.5 \\ r &= 1 \end{aligned}$$

Thus, the long term rate is $r = 1$

5.

$$\begin{aligned} r^2 - F_1 r - 0.5 &= 0 \\ r &= \frac{F_1 + \sqrt{F_1^2 + 2F_1}}{2} \end{aligned}$$

Based on the formula above the long term rate will be less than 1 for $F_1 < 0.5$ and greater than 1 for $F_1 > 0.5$

6. - 7.

$$\begin{aligned} (F_1 - r)(-r) - F_2 S &= 0 \\ r^2 - F_1 r - F_2 S &= 0 \\ r &= \frac{F_1 \pm \sqrt{F_1^2 + 4F_2 S}}{2} \\ r_1 &= \frac{F_1}{2} + \frac{\sqrt{F_1^2 + 4F_2 S}}{2} \\ r_2 &= \frac{F_1}{2} - \frac{\sqrt{F_1^2 + 4F_2 S}}{2} \end{aligned}$$

8. r_1 is positive since it is the sum of positive terms. Since r_2 involves the same terms as r_1 but with a subtraction instead of addition then $r_1 > | r_2 |$

9.

$$\begin{aligned} r_1 &= \frac{F_1 + \sqrt{F_1^2 + 4F_2 S}}{2} \\ &= \frac{1 - F_2 S + \sqrt{(1 - F_2 s)^2 + 4F_2 S}}{2} \\ &= \frac{1 - F_2 S + \sqrt{1 - 2F_2 S + F_2^2 S^2 + 4F_2 S}}{2} \\ &= \frac{1 - F_2 S + 1 + F_2 S}{2} \\ &= 1 \end{aligned}$$

10. - 11. Recall that $r_1 = \frac{F_1 + \sqrt{F_1^2 + 4F_2 S}}{2}$
If $F_1 > 1 - F_2 S$ then $F_1 + F_2 S > 1$
Plug the last result into the radical and simplify

$$\begin{aligned} r_1 &= \frac{F_1 + \sqrt{F_1^2 + 4F_2 S}}{2} \\ &= \frac{F_1 + \sqrt{(1 - F_2 S)^2 + 4F_2 S}}{2} \\ &= \frac{F_1 + \sqrt{F_2^2 S^2 + 2F_2 S + 1}}{2} \\ &= \frac{F_1 + F_2 S + 1}{2} \end{aligned}$$

Since $F_1 + F_2 S > 1$ it follows that $r_1 > 1$
Similarly, if $F_1 < 1 - F_2 S$ then $F_1 + F_2 S < 1$ which would mean that $r_1 < 1$

12. Since $F_1 > 1 - F_2 S$ gives the larger eigenvalue then $F_1 + F_2 S$ determines the largest eigenvalue which is the long term growth rate

Chapter 7

Functions of Several Variables

Exercise Set 7.1

1. $f(x,y) = x^2 - 2xy$

$f(0,-2) = 0^2 - 2(0)(-2) = 0 - 0 = 0$

$f(2,3) = 2^2 - 2(2)(3) = 4 - 12 = -8$

$f(10,-5) = 10^2 - 2(10)(-5) = 100 + 100 = 200$

2. $f(x,y) = (y^2 + 3xy)^3$

$f(-2,0) = (0^2 + 3(-2)(0))^3 = 0$

$f(3,2) = (2^2 + 3(3)(2))^3 = (4 + 18)^3 = 10648$

$f(-5,10) = (10^2 + 3(-5)(10))^3 = (100 - 150)^3 = 125000$

3. $f(x,y) = 3^x + 7xy$

$f(0,-2) = 3^0 + 7(0)(-2) = 1 + 0 = 1$

$f(-2,1) = 3^{-2} + 7(-2)(1) = \frac{1}{9} - 14 = -\frac{125}{9}$

$f(2,1) = 3^2 + 7(2)(1) = 9 + 14 = 23$

4. $f(x,y) = log_{10}(x+y)$

$f(3,7) = log_{10}(3+7) = log_{10}(10) = 1$

$f(1,99) = log_{10}(1+99) = log_{10}(100) = 2$

$f(2,-1) = log_{10}(2-1) = log_{10}(1) = 0$

5. $f(x,y) = sin\ x\ tan\ y$

$f(\pi/2,0) = sin\ \pi/2\ tan\ 0 = (1)(0) = 0$

$f(3\pi/4, 2\pi/3) = sin\ 3\pi/4\ tan\ 2\pi/3 = (-1)(\frac{\sqrt{3}}{2}) = -\frac{\sqrt{3}}{2}$

$f(\pi/6,\pi/4) = sin\ \pi/6\ tan\ \pi/4 = (\frac{1}{2})(1) = \frac{1}{2}$

6. $f(x,y) = sin(x\ cos\ y)$

$f(\pi,\pi/3) = sin(\pi\ cos\ \pi/3) = sin(\pi/2) = 1$

$f(\pi/3,0) = sin(\pi/3\ cos\ 0) = sin(\pi/3) = \frac{\sqrt{3}}{2}$

$f(7\pi/6,\pi) = sin(7\pi/6\ cos\ \pi) = sin(-7\pi/6) = \frac{1}{2}$

7. $f(x,y,z) = x^2 - y^2 + z^2$

$f(-1,2,3) = (-1)^2 - (2)^2 + (3)^2 = 1 - 4 + 9 = 6$

$f(2,-1,3) = (2)^2 - (-1)^2 + (3)^2 = 4 - 1 + 9 = 12$

8. $f(x,y,z) = 2^x + 5zy - x$

$f(0,1,-3) = 2^0 + 5(-3)(1) - 0 = 1 - 15 = -14$

$f(1,0,-3) = 2^1 + 5(-3)(0) - 1 = 2 - 1 = 1$

9. $S(h,w) = \frac{\sqrt{hw}}{60}$

$$\begin{aligned} S(165,80) &= \frac{\sqrt{165 \cdot 80}}{60} \\ &= \frac{\sqrt{13200}}{60} \\ &\approx 1.915\ m^2 \end{aligned}$$

10. $S(h,w) = 0.024265h^{0.3964}w^{0.5378}$

$$\begin{aligned} S(165,80) &= 0.024265(165)^{0.3964}(80)^{0.5378} \\ &= 1.939\ m^2 \end{aligned}$$

11. $w(x,s,h,m,r,p) = x(9.38 + 0.264s + 0.000233hm + 4.62r[p+1])$

a) $w(275,1,160,71,0.068,0)$

$$\begin{aligned} w(275,1,160,71,0.068,0) &= 275(9.38 + 0.264(1) \\ &\quad +0.000233(160)(71) + \\ &\quad 4.62(0.068)[0+1]) \\ &= 3466.386 \end{aligned}$$

b) $w(282,-1,171,76,0.085,3)$

$$\begin{aligned} w(282,-1,171,76,0.085,3) &= 282(9.38 + 0.264(-1) \\ &\quad +0.000233(171)(76) + \\ &\quad 4.62(0.085)[3+1]) \\ &= 3771.589 \end{aligned}$$

12. $P(m,T) = \dfrac{1}{1 + e^{3.222 - 31.669m + 0.083T}}$

a) $P(0.1, 10)$

$$P(0.1, 10) = \frac{1}{1 + e^{3.222 - 31.669(0.1) + 0.083(10)}} = 0.292$$

b) $P(0.2, 25)$

$$P(0.2, 25) = \frac{1}{1 + e^{3.222 - 31.669(0.2) + 0.083(25)}} = 0.738$$

13. $S(d, V, a) = \frac{aV}{0.51d^2}$

$$S(100, 1600000, 0.78) = \frac{0.78 \cdot 1600000}{0.51(100)^2} = \frac{1248000}{5100} = 244.706$$

14. $Q(m, c) = 100 \cdot \frac{m}{c}$

a) $Q(21, 20)$

$$Q(21, 20) = 100 \cdot \frac{21}{20} = 105$$

b) $Q(19, 20)$

$$Q(19, 20) = 100 \cdot \frac{19}{20} = 95$$

15. $V(L, p, R, r, v) = \frac{p}{4Lv}(R^2 - r^2)$

$$V(1, 100, 0.0075, 0.0025, 0.05) = \frac{100}{4(1)(0.05)} \times [(0.0075)^2 - (0.0025)^2] = 500(0.00005) = 0.025$$

16. A function of one variable has only one independent variable, while a function of two variables has two independent variables.

17. Left to the student (answer may vary)

18. $W(20, 20) = -10$

19. $W, 40, 20) = -22$

20. $W(30, -10) = -64$

21.

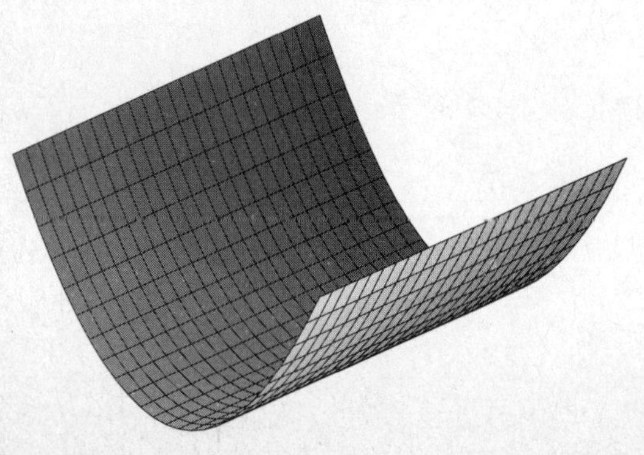

22.

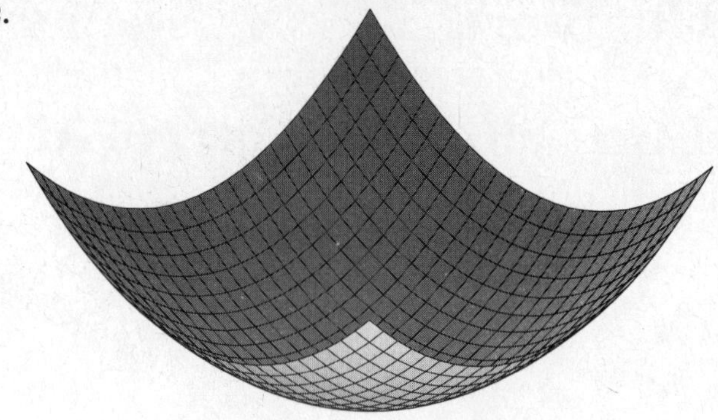

23.

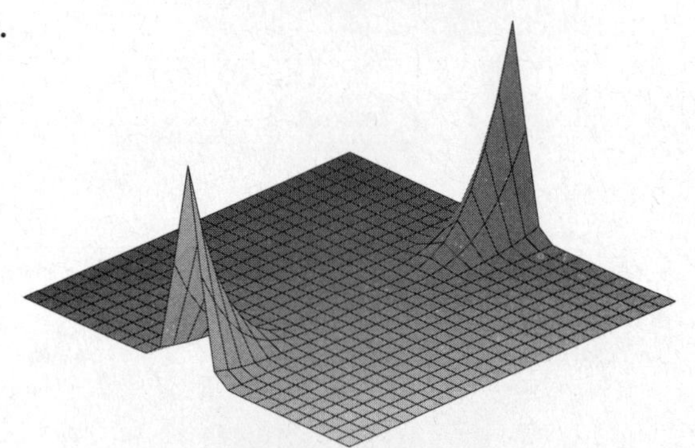

24.

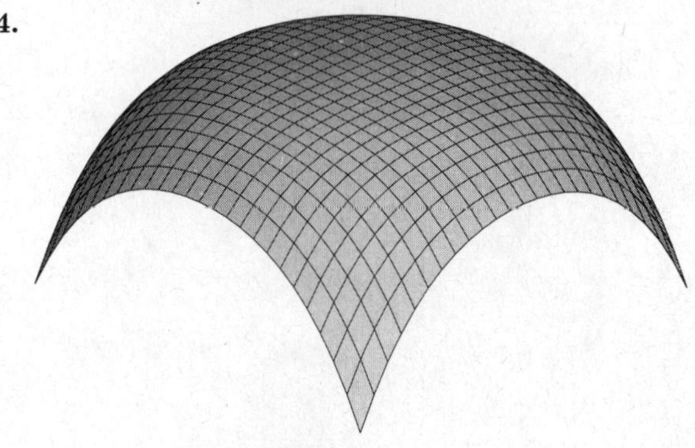

25.

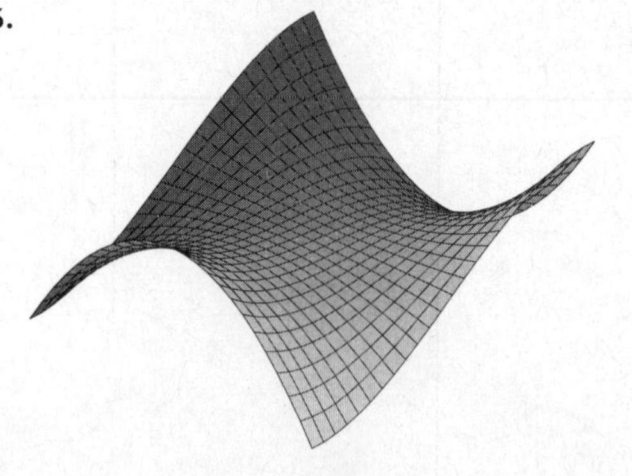

26.

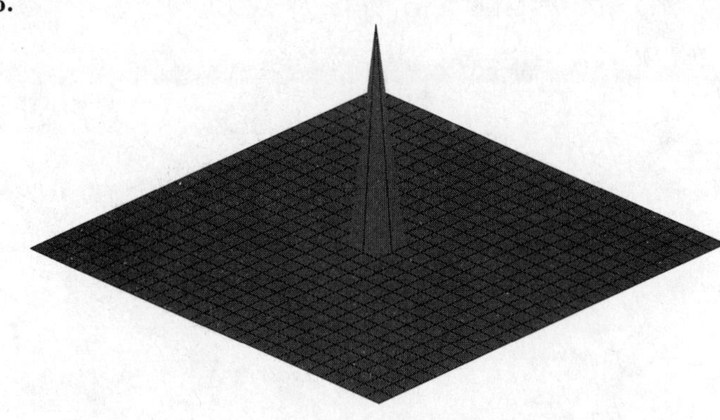

Exercise Set 7.2

1. $z = 2x - 3xy$

$\frac{dz}{dx} = 2 - 3y$

$\frac{dz}{dy} = -3x$

$\frac{dz}{dx}|_{(-2,-3)} = 2 - 3(-3) = -11$

$\frac{dz}{dy}|_{(0,-5)} = -3(0) = 0$

2. $z = (x - y)^3$

$\frac{dz}{dx} = 3(x-y)^2(1) = 3(x-y)^2$

$\frac{dz}{dx} = 3(x-y)^2(-1) = -3(x-y)^2$

$\frac{dz}{dx}|_{(-2,-3)} = 3(-2+3)^2 = 3(1)^2 = 3$

$\frac{dz}{dy}|_{(0,-5)} = -3(0+5)^2 = -75$

3. $z = 3x^2 - 2xy + y$

$\frac{dz}{dx} = 6x - 2y$

$\frac{dz}{dx} = -2x + 1$

$\frac{dz}{dx}|_{(-2,-3)} = 6(-2) - 2(-3) = -12 + 6 = -6$

$\frac{dz}{dy}|_{(0,-5)} = -2(0) + 1 = 0 + 1 = 1$

4. $z = 2x^3 + 3xy - x$

$\frac{dz}{dx} = 6x^2 + 3y - 1$

$\frac{dz}{dx} = 3x$

$\frac{dz}{dx}|_{(-2,-3)} = 6(-2)^2 + 3(-3) - 1 = 24 - 9 - 1 = 14$

$\frac{dz}{dy}|_{(0,-5)} = 3(0) = 0$

5. $f(x,y) = 2x - 3y$

$f_x = 2$ (a constant independent of x or y)

$f_y = -3$ (a constant independent of x or y)

$f_x(-2,1) = 2$

$f_y(-3,-2) = -3$

6. $f(x,y) = 5x + 7y$

$f_x = 5$ (a constant independent of x or y)

$f_y = 7$ (a constant independent of x or y)

$f_x(-2,1) = 5$

$f_y(-3,-2) = 7$

7. $f(x,y) = (x^2 + y^2)^{1/2}$

$f_x = \frac{1}{2}(x^2+y^2)^{-1/2}(2x) = \frac{x}{\sqrt{x^2+y^2}}$

$f_x = \frac{1}{2}(x^2+y^2)^{-1/2}(2y) = \frac{y}{\sqrt{x^2+y^2}}$

$f_x(-2,1) = \frac{-2}{\sqrt{4+1}} = \frac{-2}{\sqrt{5}}$

$f_y(-3,-2) = \frac{-2}{\sqrt{9+4}} = \frac{-2}{\sqrt{13}}$

8. $f(x,y) = (x^2 - y^2)^{1/2}$

$f_x = \frac{1}{2}(x^2+y^2)^{-1/2}(2x) = \frac{x}{\sqrt{x^2+y^2}}$

$f_x = \frac{1}{2}(x^2+y^2)^{-1/2}(-2y) = \frac{-y}{\sqrt{x^2+y^2}}$

$f_x(-2,1) = \frac{-2}{\sqrt{4-1}} = \frac{-2}{\sqrt{3}}$

$f_y(-3,-2) = \frac{-(-2)}{\sqrt{9-4}} = \frac{2}{\sqrt{5}}$

9. $f(x,y) = 2x - 3y$

$f_x = 2$

$f_y = -3$

10. $f(x,y) = e^{2x-y}$

$f_x = e^{2x-y}(2) = 2e^{2x-y}$

$f_y = e^{2x-y}(-1) = -e^{2x-y}$

11. $f(x,y) = \sqrt{x} + sin(xy)$

$f_x = \frac{1}{2\sqrt{x}} + cos(xy)(y) = \frac{1}{2\sqrt{x}} + y\ cos(xy)$

$f_y = cos(xy)(x) = x\ cos(xy)$

12. $f(x,y) = e^{2xy}$

$f_x = e^{2xy}(2y) = 2y\ e^{2xy}$

$f_y = e^{2xy}(2x) = 2x\ e^{2xy}$

13. $f(x,y) = x\ ln\ y$

$f_x = ln\ y$

$f_y = x \cdot \frac{1}{y} = \frac{x}{y}$

14. $f(x,y) = \frac{x}{y} - \frac{y}{x}$

$f_x = \frac{1}{y} - y[-1x^{-2}] = \frac{1}{y} + \frac{y}{x^2}$

$f_y = x[-1y^{-2}] - \frac{1}{x} = \frac{-x}{y^2} - \frac{1}{x}$

15. $f(x,y) = x^3 - 4xy + y^2$

$f_x = 3x^2 - 4y$

$f_y = -4x + 2y$

16. $f(x,y) = x^5 - 4x^2y^2 + 5xy^3 - 2y$

$f_x = 5x^4 - 8xy^2 + 5y^3$

$f_y = -8x^2y + 15xy^2 - 2$

17. $f(x,y) = (x^2 + 2y + 2)^4$

$f_x = 4(x^2 + 2y + 2)^3(2x) = 8x(x^2 + 2y + 2)^3$

$f_y = 4(x^2 + 2y + 2)^3(2) = 8(x^2 + 2y + 2)$

18. $f(x,y) = y\ ln\ (x^2 + y)$

$f_x = y \cdot \frac{2x}{(x^2+y)} = \frac{2xy}{x^2+y}$

$f_y = y \cdot \frac{1}{x^2+y} + ln\ (x^2 + y)(1) = \frac{y}{x^2+y} + ln\ (x^2 + y)$

19. $f(x,y) = sin(e^{x+y})$

$f_x = cos(e^{x+y})(e^{x+y}(1)) = e^{x+y}\ cos(e^{x+y})$

$f_x = cos(e^{x+y})(e^{x+y}(1)) = e^{x+y}\ cos(e^{x+y})$

20. $f(x,y) = x\sqrt{2x + y^3}$

$f_x = x \cot \frac{2}{2\sqrt{2x+y^3}} + \sqrt{2x + y^3} = \frac{x}{\sqrt{2x+y^3}} + \sqrt{2x + y^3}$

$f_y = x \cdot \frac{3y^2}{2\sqrt{2x+y^3}} = \frac{3xy^2}{2\sqrt{2x+y^3}}$

21. $f(x,y) = \frac{e^x}{y^2+1}$

$f_x = \frac{1}{y^2+1} \cdot e^x = \frac{e^x}{y^2+1}$

$f_y = e^x[-1(y^2 + 1)^{-2}(2y)] = \frac{-2ye^x}{(y^2+1)^2}$

22. $f(x,y) = \frac{cos(xy)}{x-y}$

$f_x = \frac{(x-y)(-sin(xy)(y) - cos(xy)(1))}{(x-y)^2} = \frac{(y^2-xy)\ sin(xy) - cos(xy)}{(x-y)^2}$

$f_y = \frac{(x-y)(-sin(xy)(x) - cos(xy)(-1))}{(x-y)^2} = \frac{(y^2-xy)\ sin(xy) + cos(xy)}{(x-y)^2}$

23. $f(x,y) = [x^5 + tan(y^2)]^4$

$f_x = 4[x^5 + tan(y^2)]^3(5x^4) = 20x^4[x^5 + tan(y^2)]^3$

$f_y = 4[x^5 + tan(y^2)]^3(sec^2(y^2)(2y))$
$= 8y\ sec^2(y^2)\ [x^5 + tan(y^2)]^3$

24. $f(x,y) = (sin(2x + 3y))^{1/2}$

$f_x = \frac{1}{2}(sin(2x + 3y))^{-1/2}(cos(2x + 3y)(2)) = \frac{cos(2x+3y)}{\sqrt{sin(2x+3y)}}$

$f_y = \frac{1}{2}(sin(2x + 3y))^{-1/2}(cos(2x + 3y)(3)) = \frac{3\ cos(2x+3y)}{2\sqrt{sin(2x+3y)}}$

25. $f(x,y) = \frac{y\ ln\ x}{y^3-1}$

$f_x = \frac{y}{y^3-1} \cdot \frac{1}{x} = \frac{y}{x(y^3-1)}$

$f_y = \frac{(y^3-1)(ln\ x) - y\ ln\ x(3y^2)}{(y^3-1)^2}$
$= \frac{-2y^3\ -1)\ ln\ x}{(y^3-1)^2}$

26. $f(x,y) = \frac{e^{x+2y}}{x+2y}$

$f_x = \frac{(x+2y)[e^{x+2y}(1)] - e^{x+2y}(1)}{(x+2y)^2} = \frac{e^{x+2y}[x+2y-1]}{(x+2y)^2}$

$f_y = \frac{(x+2y)[e^{x+2y}(2)] - e^{x+2y}(2)}{(x+2y)^2} = \frac{2e^{x+2y}[x+2y-1]}{(x+2y)^2}$

27. $f(b,m) = (m + b - 4)^2 + (2m + b - 5)^2 +$
$3m + b - 6)^2$

$$\begin{aligned} \frac{\partial f}{\partial b} &= 2(m + b - 4)(1) + 2(2m + b - 5)(1) + \\ &\quad 2(3m + b - 6)(1) \\ &= 2m + 2b - 8 + 4m + 2b - 10 + 6m + 2b - 12 \\ &= 12m + 6b - 30 \end{aligned}$$

$$\begin{aligned} \frac{\partial f}{\partial m} &= 2(m + b - 4)(1) + 2(2m + b - 5)(2) + \\ &\quad 2(3m + b - 6)(3) \\ &= 2m + 2b - 8 + 8m + 4b - 20 + 18m + 6b - 36 \\ &= 28m + 12b - 64 \end{aligned}$$

28. $f(b,m) = (m+b-6)^2 + (2m+b-8)^2 + (3m+b-9)^2$

$$\begin{aligned}\frac{\partial f}{\partial b} &= 2(m+b-6)(1) + 2(2m+b-8)(1) + \\ &\quad 2(3m+b-9)(1) \\ &= 2m+2b-12+4m+2b-12+6m+2b-18 \\ &= 12m+6b-42\end{aligned}$$

$$\begin{aligned}\frac{\partial f}{\partial m} &= 2(m+b-6)(1) + 2(2m+b-8)(2) + \\ &\quad 2(3m+b-9)(3) \\ &= 2m+2b-12+8m+4b-32+18m+6b-54 \\ &= 28m+12b-98\end{aligned}$$

29. $z = \frac{x^2+t^2}{x^2-t^2}$

$$\begin{aligned}z_x &= \frac{(x^2-t^2)(2x)-(x^2+t^2)(2x)}{(x^2-t^2)^2} \\ &= \frac{2x^3-2t^2x-2x^3-2t^2x}{(x^2-t^2)^2} \\ &= \frac{-4t^2x}{(x^2-t^2)^2}\end{aligned}$$

$$\begin{aligned}z_t &= \frac{(x^2-t^2)(2t)-(x^2+t^2)(-2t)}{(x^2-t^2)^2} \\ &= \frac{2tx^2-2t^3+2tx^2+2t^3}{(x^2-t^2)^2} \\ &= \frac{4tx^2}{(x^2-t^2)^2}\end{aligned}$$

30. $z = \frac{x^2-t}{x^3+t}$

$$\begin{aligned}z_x &= \frac{(x^3+t)(2x)-(x^2-t)(3x^2)}{(x^3+t)^2} \\ &= \frac{2x^4+2tx-3x^4-3tx^2}{(x^3+t)^2} \\ &= \frac{-x^4-3tx^2+2tx}{(x^3+t)^2}\end{aligned}$$

$$\begin{aligned}z_t &= \frac{(x^3+t)(-1)-(x^2-t)(1)}{(x^3+t)^2} \\ &= \frac{-x^3-t-x^2+t}{(x^3+t)^2} \\ &= \frac{-x^3-x^2}{(x^3+t)^2}\end{aligned}$$

31. $z = \frac{2\sqrt{x}-2\sqrt{t}}{1+2\sqrt{t}}$

$$\begin{aligned}z_x &= \frac{2}{1+\sqrt{t}} \cdot \frac{1}{2\sqrt{x}} \\ &= \frac{1}{(1+2\sqrt{t})\sqrt{x}}\end{aligned}$$

$$\begin{aligned}z_t &= \frac{(1+2\sqrt{t})(-2\cdot\frac{1}{2\sqrt{t}})-(2\sqrt{x}-2\sqrt{t})(-2\cdot\frac{1}{2\sqrt{t}})}{(1+2\sqrt{t})^2} \\ &= \frac{2\sqrt{x}+2\sqrt{t}-1-2\sqrt{t}}{\sqrt{t}\,(1+2\sqrt{t})^2} \\ &= \frac{2\sqrt{x}-1}{\sqrt{t}\,(1+2\sqrt{t})^2}\end{aligned}$$

32. $z = \left(\frac{x^2+t^2}{x^2-t^2}\right)^5$.

We will use the result obtained from Exercise 29.

$$\begin{aligned}z_x &= 5\left(\frac{x^2+t^2}{x^2-t^2}\right)^4 \cdot \frac{-4t^2x}{(x^2-t^2)^2} \\ &= \frac{-20t^2x}{(x^2-t^2)^2}\left(\frac{x^2+t^2}{x^2-t^2}\right)^4\end{aligned}$$

$$\begin{aligned}z_t &= 5\left(\frac{x^2+t^2}{x^2-t^2}\right)^4 \cdot \frac{4tx^2}{(x^2-t^2)^2} \\ &= \frac{20tx^2}{(x^2-t^2)^2}\left(\frac{x^2+t^2}{x^2-t^2}\right)^4\end{aligned}$$

33. $z = (x^3t^5)^{1/4}$

$$\begin{aligned}z_x &= \frac{1}{4}(x^3t^5)^{-3/4}(3x^2t^5) \\ &= \frac{3x^2t^5}{4\sqrt[4]{(x^3t^5)^3}}\end{aligned}$$

$$\begin{aligned}z_t &= \frac{1}{4}(x^3t^5)^{-3/4}(5x^3t^4) \\ &= \frac{5x^3t^4}{4\sqrt[4]{(x^3t^5)^3}}\end{aligned}$$

34. $z = 6x^{2/3} - 8x^{1/4}t^{1/2} - 12x^{-1/2}t^{3/2}$

$$z_x = 4x^{-1/3} - 2x^{-3/4}t^{1/2} + 6x^{-3/2}t^{3/2}$$

$$z_t = -4x^{1/4}t^{-1/2} - 18x^{-1/2}t^{1/2}$$

35. $f(x,y) = x+3y$, $g(x,y) = x-2y$

$$\begin{aligned}J &= \begin{bmatrix} \partial f/\partial x & \partial f/\partial y \\ \partial g/\partial x & \partial g/\partial y \end{bmatrix} \\ &= \begin{bmatrix} 1 & 3 \\ 1 & \text{-}2 \end{bmatrix}\end{aligned}$$

36. $f(x,y) = xy^2$, $g(x,y) = x^2y^{-1}$

$$J = \begin{bmatrix} \partial f/\partial x & \partial f/\partial y \\ \partial g/\partial x & \partial g/\partial y \end{bmatrix} = \begin{bmatrix} y^2 & 2xy \\ \frac{1}{y} & -\frac{x^2}{y^2} \end{bmatrix}$$

37. $f(x,y) = \sqrt{x+3y}$, $g(x,y) = e^{-x-y}$

$$J = \begin{bmatrix} \partial f/\partial x & \partial f/\partial y \\ \partial g/\partial x & \partial g/\partial y \end{bmatrix} = \begin{bmatrix} \frac{1}{2}(x+3y)^{-1/2}(1) & \frac{1}{2}(x+3y)^{-1/2}(3) \\ e^{-x-y}(-1) & e^{-x-y}(-1) \end{bmatrix} = \begin{bmatrix} \frac{1}{2\sqrt{x+3y}} & \frac{3}{2\sqrt{x+3y}} \\ -e^{-x-y} & -e^{-x-y} \end{bmatrix}$$

38. $f(x,y) = ln\ (2x^2y^3)$, $g(x,y) = (x + cos\ y)^2$

$$J = \begin{bmatrix} \partial f/\partial x & \partial f/\partial y \\ \partial g/\partial x & \partial g/\partial y \end{bmatrix} = \begin{bmatrix} \frac{1}{2x^2y^3} \cdot 4x & \frac{1}{2x^2y^3} \cdot 6y^2 \\ 2(x + cos\ x)(1) & 2(x + cos\ y)(-sin\ y) \end{bmatrix} = \begin{bmatrix} \frac{2}{xy^3} & \frac{3}{x^2y} \\ 2(x + cos\ y) & -2\ sin\ y\ (x + cos\ y) \end{bmatrix}$$

39. We will use the results from Exercise 9.

$f_x = 2 \to f_{xx} = 0$ and $f_{xy} = 0$

$f_y = -3 \to f_{yx} = 0$ and $f_{yy} = 0$

40. $f_x = 2e^{2x-y} \to f_{xx} = 2 \cdot 2e^{2x-y} = 4e^{2x-y}$

$f_{xy} = 2 \cdot e^{2x-y}(-1) = -2e^{2x-y}$

$f_y = -e^{2x-y} \to f_{yx} = -e^{2x-y}(2) = -2e^{2x-y}$

$f_{yy} = -e^{2x-y}(-1) = e^{2x-y}$

41. We will use the results from Exercise 11.

$f_x = \frac{1}{2}x^{-1/2} + y\ cos(xy) \to$

$f_{xx} = \frac{1}{4}x^{-3/2} - y(sin(xy)(y)) = \frac{1}{4\sqrt{x^3}} - y^2 sin(xy)$

$f_{xy} = y[-sin(xy)(x)] + cos(xy)(1) = -xy\ sin(xy) + cos(xy)$

$f_y = x\ cos(xy) \to$

$f_{yx} = x[-sin(xy)(y)] + cos(xy)(1) = -xy\ sin(xy) + cos(xy)$

$f_{yy} = x[-sin(xy)(x)] = -x^2\ sin(xy)$

42. $f_x = 2y\ e^{2xy} \to f_{xx} = 4y^2\ e^{2xy}$

$f_{xy} = 2y[e^{2xy}(2x)] + 2e^{2xy} = 2\ e^{2xy}(2xy + 1)$

$f_y = 2x\ e^{2xy} \to$

$f_{yx} = 2x[e^{2xy}(2y)] + 2\ e^{2xy} = 2\ e^{2xy}(2xy + 1)$

$f_{yy} = 4x^2\ e^{2xy}$

43. We will use the results from Exercise 13.

$f_x = ln\ y \to f_{xx} = 0$

$f_{xy} = \frac{1}{y}$

$f_y = \frac{x}{y} \to f_{yx} = \frac{1}{y}$

$f_{yy} = x \cdot -1y^{-2} = \frac{-x}{y^2}$

44. $f_x = \frac{1}{y} + \frac{y}{x^2} \to f_{xx} = \frac{-2y}{x^3}$

$f_{xy} = -\frac{1}{y^2} + \frac{1}{x^2}$

$f_y = \frac{-x}{y^2} - \frac{1}{x} \to f_{yx} = -\frac{1}{y^2} + \frac{1}{x^2}$

$f_{yy} = \frac{2x}{y^3}$

45. We will use the results from Exercise 15.

$f_x = 3x^2 - 4y \to f_{xx} = 6x$ and $f_{xy} = -4$

$f_y = -4x + 2y \to f_{yx} = -4$ and $f_{yy} = 2$

46. $f_x = 5x^4 - 8xy^2 + 5y^3 \to$

$f_{xx} = 20x^3 - 8y$ and $f_{xy} = -16xy + 15y^2$

$f_y = -8x^2y + 15xy^2 - 2 \to$

$f_{yx} = -16xy + 15y^2$ and $f_{yy} = -8x^2 + 30xy$

47. $f(x, y, z) = x^2y^3z^4$

$f_x = 2xy^3z^4$

$f_y = 3x^2y^2z^4$

$f_z = 4x^2y^3z^3$

48. $f(x, y, z) = xy^3z^{-4}$

$f_x = y^3z^{-4} = \frac{y^3}{z^4}$

$f_y = 3xy^2z^{-4} = \frac{3xy^2}{z^4}$

$f_z = -4xy^3z^{-5} = \frac{-4xy^3}{z^5}$

49. $f(x,y,z) = e^{x+y^2+z^3}$

$f_x = e^{x+y^2+z^3}(1) = e^{x+y^2+z^3}$

$f_y = e^{x+y^2+z^3}(2y) = 2y\ e^{x+y^2+z^3}$

$f_z = e^{x+y^2+z^3}(3z^2) = 3z^2\ e^{x+y^2+z^3}$

50. $f(x,y,z) = ln\ (xy^3z^5)$

$f_x = \frac{1}{xy^3z^5}(y^3z^5) = \frac{1}{x}$

$f_y = \frac{1}{xy^3z^5}(3xy^2z^5) = \frac{3}{y}$

$f_z = \frac{1}{xy^3z^5}(5x^3z^4) = \frac{5}{z}$

51. $z = f(x,y) = xy^2$

$f(2,3) = (2)(3)^2 = 4 \cdot 9 = 36$
$f_x = y^2$, so, $f_x(2,3) = 3^2 = 9$
$f_y = 2xy$, so, $f_y(2,3) = 2(2)(3) = 12$

$$\begin{aligned} z(x,y) &= f(a,b) + f_x(a,b)(x-a) + f_y(a,b)(y-b) \\ z(2.01,3.02) &= f(2,3) + f_x(2,3)(2.01-2) + \\ &\quad f_y(2,3)(3.02-3) \\ &= 36 + 9(0.01) + 12(0.02) \\ &= 18.33 \end{aligned}$$

52. $z = f(x,y) = x^2/y$

$f(4,2) = 4^2/2 = 8$
$f_x = 2x/y$, so $f_x(4,2) = 2(4)/2 = 4$
$f_y = -x^2/y^2$, so $f_y(4,2) = -(4^2)/(2^2) = -4$

$$\begin{aligned} z(4.02,1.97) &= 8 + 4(4.02-4) - 4(1.97-2) \\ &= 8 + 0.08 + 0.12 \\ &= 8.2 \end{aligned}$$

53. $z = f(x,y) = x\ sin(xy)$

$f(1,0) = (1)\ sin(0) = 0$
$f_x = x[ycos(xy)] + sin(xy)$, so $f_x(1,0) = 1[(0)cos(0)] + sin(0) = 0$
$f_y = x[cos(xy)(x)] = x^2\ cos(xy)$, so $f_(1,0) = 1^2\ cos(0) = 1$

$$\begin{aligned} z(x,y) &= f(a,b) + f_x(a,b)(x-a) + f_y(a,b)(y-b) \\ z(0.99,0.02) &= f(1,0) + f_x(1,0)(0.99-1) + \\ &\quad f_y(1,0)(0.02-0) \\ &= 0 + 0(-0.01) + 1(0.02) \\ &= 0.02 \end{aligned}$$

54. $z = f(x,y) = y^2\ e^x$

$f(0,4) = 4^2\ e^0 = 16$
$f_x = y^2\ e^x$, so $f_x(0,4) = 16$
$f_y = 2y\ e^x$, so $f_y(0,4) = 8$

$$\begin{aligned} z(-0.02,3.99) &= 16 + 16(-0.02-0) + 8(3.99-4) \\ &= 16 - 0.32 - 0.08 \\ &= 15.6 \end{aligned}$$

55. $S(h,w) = \sqrt{hw}/60$

a) $S(100,28)$

$$\begin{aligned} S(100,28) &= \sqrt{(100)(28)}/60 \\ &= \sqrt{2800}/60 \\ &= 0.881917 \end{aligned}$$

b) Use linearization to estimate $S(102,30)$

$S(100,28) = 0.881917$
$S_h = \frac{w}{120\sqrt{hw}}$, so
$S_h(100,28) = \frac{28}{120\ \sqrt{2800}} = 0.00441$
$S_w = \frac{h}{120\sqrt{hw}}$, so
$S_w(100,28) = \frac{100}{120\ \sqrt{2800}} = 0.01575$

$$\begin{aligned} S(102,30) &= S(100,28) + S_h(100,28)(102-100) + \\ &\quad S_w(100,28)(30-28) \\ &= 0.88917 + (0.00441)(2) + (0.01575)(2) \\ &= 0.9294 \end{aligned}$$

56. $S = \frac{aV}{0.51d^2}$

a)

$$\begin{aligned} S &= \frac{0.78 \cdot 1000000}{0.51(300)^2} \\ &= 16.99346 \end{aligned}$$

b) $S_V = \frac{a}{0.51d^2}$
$S_V|_{1000000,300} = 0.00001699$
$S_d = \frac{-2aV}{0.51d^3}$
$S_d|_{1000000,300} = -0.1132898$

$$\begin{aligned} S(1050000,305) &= 16.99346 + 0.00001699(50000) \\ &\quad -0.1132898(5) \\ &= 17.27651 \end{aligned}$$

57. $w(x,s,h,m,r,p) = x(9.38 + 0.264s + 0.000233hm + 4.62r[p+1])$

a) $w(280,1,150,65,0.08,0)$

$$\begin{aligned} w &= 280(9.38 + 0.264(1) + 0.000233(150)(65) + \\ &\quad 4.62(0.08)[0+1]) \\ &= 280(9.38 + 0.264 + 2.27175 + 0.3696 \\ &= 3439.9 \end{aligned}$$

b) $w_x = (9.38 + 0.264s + 0.000233hm + 4.62r[p+1])$
$w_r = 4.62x[p+1]$

$$\begin{aligned} w_x(280,1,150,65,0.08,0) &= 9.38 + 0.264(1) + \\ &\quad 0.0000233(150)(65) + 4.62(0.08)[0+1] \\ &= 9.38 + 0.264 + 0.227175 + 0.3696 \\ &= 12.28535 \end{aligned}$$

$$\begin{aligned} w_r(280,1,150,65,0.08,0) &= 4.62(280)[0+1] \\ &= 1293.6 \end{aligned}$$

$$\begin{aligned} w(276,1,150,65,0.081,0) &= 3439.9 + 12.28535(276-280) \\ &\quad +1293.6(0.081-0.08) \\ &= 3439.9 - 49.1414 + 1.2936 \\ &= 3392.05 \end{aligned}$$

58. **a)**

$$\begin{aligned} W(20,25) &= 9.14 - \frac{1}{110}[(10.45 + 6.68\sqrt{20} - 0.447(20)) \\ &\quad \times(457 - 5(25))] \\ &= 91.4 - \frac{1}{110}[31.38387 \cdot 332] \\ &= 91.4 - \frac{1}{110}[10419.44484] \\ &= -3.32223 \end{aligned}$$

b) $W_v = (457 - 5T)\left(\frac{0.447}{110} - \frac{3.34}{110\sqrt{v}}\right)$
$W_T = \frac{5(10.45+6.68\sqrt{v}-0.447v)}{110}$

$$\begin{aligned} W_v(20,25) &= (457 - 5(25))\left(\frac{0.447}{110} - \frac{3.34}{110\sqrt{20}}\right) \\ &= -0.904992 \end{aligned}$$

$$\begin{aligned} W_T(20,25) &= \frac{5(10.45 = 6.68\sqrt{20} - 0.447(20)}{110} \\ &= 1.426539 \end{aligned}$$

$$\begin{aligned} W(21,24) &= -3.32223 - 0.904992(21-20) + \\ &\quad 1.426539(24-25) \\ &= -5.653761 \end{aligned}$$

59. $P(m,T) = \frac{1}{1+e^{3.222-31.669m+0.083T}}$

a) $P(0.15, 20)$

$$\begin{aligned} P(0.15,20) &= \frac{1}{1+e^{3.222-31.669(0.15)+0.083(20)}} \\ &= \frac{1}{1+e^{0.13165}} \\ &= 0.467135 \end{aligned}$$

b) $P_m = -(1 + e^{3.222-31.669m+0.083T})^{-2}(31.669) = \frac{-31.669}{(1+e^{3.222-31.669m+0.083T})^2}$
$P_T = -(1 + e^{3.222-31.669m+0.083T})^{-2}(0.083) = \frac{-0.083}{(1+e^{3.222-31.669m+0.083T})^2}$

$$\begin{aligned} P_m(0.15,20) &= \frac{-31.669}{(1+e^{3.222-31.669(0.15)+0.083(20)})^2} \\ &= \frac{31.669}{(2.140709)^2} \\ &= 6.91065 \end{aligned}$$

$$\begin{aligned} P_T(0.15,20) &= \frac{-0.083}{(1+e^{3.222-31.669(0.15)+0.083(20)})^2} \\ &= \frac{-0.083}{(2.140709)^2} \\ &= -0.018112 \end{aligned}$$

$$\begin{aligned} P(0.155,19) &= 0.467135 + 6.91065(0.155 - 0.15) - \\ &\quad 0.018112(19.5 - 20) \\ &= 0.467135 + 0.03455325 + 0.009056 \\ &= 0.5107 \end{aligned}$$

60.

$$\begin{aligned} T_h(90,0.9) &= 1.98(90) - 1.09(1-.9)(90-58) - 56.9 \\ &= 117.8 \end{aligned}$$

61.

$$\begin{aligned} T_h(90,1) &= 1.98(90) - 1.09(1-1)(90-58) - 56.9 \\ &= 121.3 \end{aligned}$$

62.

$$\begin{aligned} T_h(78,1) &= 1.98(78) - 1.09(1-1)(78-58) - 56.9 \\ &= 97.5 \end{aligned}$$

63. $\frac{\partial T_h}{\partial H} = 1.09(T - 58)$
This means that for every 1 point change in humidity at a specific temperature T, the Temperature-Humidity Heat index changes by $1.09(T - 58)$

64. $\frac{\partial T_h}{\partial T} = 1.98 - -1.09(1 - H)$
This means that for every change in temperature of 1 degree at a specific humidity H, the Temperature-Humidity Heat index changes by $1.98 - 1.09(1 - H)$

65.

$$\begin{aligned} E(146,5) &= 206.835 - 0.846(146) - 1.015(5) \\ &= 206.835 - 123.516 - 5.075 \\ &= 78.244 \end{aligned}$$

66.

$$\begin{aligned} E(180,6) &= 206.835 - 0.846(180) - 1.015(6) \\ &= 48.465 \end{aligned}$$

67. $\frac{\partial E}{\partial w} = -0.846$. This means that for every increase of one syllable there is a decrease of 0.846 in the reading ease of a 100-word section.

68. $\frac{\partial E}{\partial s} = -1.015$. This means that for every increase of one word for the average words per sentence there is a decrease of 1.015 in the reading ease of a 100-word section.

69. $f(x,y) = ln\ (x^2+y^2)$

$f_x = \frac{2x}{(x^2+y^2)}$
$f_x x = \frac{(x^2+y^2)(2)-2x(2x)}{(x^2+y^2)^2} = \frac{2(y^2-x^2)}{(x^2+y^2)^2}$

$f_y = \frac{2y}{(x^2+y^2)}$
$f_x x = \frac{(x^2+y^2)(2)-2y(2y)}{(x^2+y^2)^2} = \frac{2(x^2-y^2)}{(x^2+y^2)^2}$

$$\begin{aligned}\frac{\partial^2 f}{\partial x^2} + \frac{\partial^2 f}{\partial x^2} &= \frac{2(y^2-x^2)}{(x^2+y^2)^2} + \frac{2(x^2-y^2)}{(x^2+y^2)^2} \\ &= \frac{2y^2-2x^2+2x^2-2y^2}{(x^2+y^2)^2} \\ &= 0\end{aligned}$$

70. $f(x,y) = x^3 - 5xy^2$

$f_y = -10xy$
$f_{xy} = f_{yx} = -10y$
$xf_{xy} - f_y = -10xy - (-10xy) = 0$

71. a)

$$\begin{aligned}\lim_{h\to 0}\frac{f(h,y)-f(0,y)}{h} &= \lim_{h\to 0}\frac{\frac{hy(h^2-y^2)}{(h^2+y^2)} - 0}{h} \\ &= \lim_{h\to 0}\frac{y(h^2-y^2)}{h^2+y^2} \\ &= \frac{-y^3}{y^2} \\ &= -y\end{aligned}$$

a)

$$\begin{aligned}\lim_{h\to 0}\frac{f(h,y)-f(0,y)}{h} &= \lim_{h\to 0}\frac{\frac{hy(h^2-y^2)}{(h^2+y^2)} - 0}{h} \\ &= \lim_{h\to 0}\frac{y(h^2-y^2)}{h^2+y^2} \\ &= \frac{-y^3}{y^2} \\ &= -y\end{aligned}$$

b)

$$\begin{aligned}\lim_{h\to 0}\frac{f(x,h)-f(x,0)}{h} &= \lim_{h\to 0}\frac{\frac{xh(x^2-h^2)}{(x^2+h^2}-0}{h} \\ &= \lim_{h\to 0}\frac{x(x^2-h^2)}{x^2+h^2} \\ &= \frac{x^3}{x^2} \\ &= x\end{aligned}$$

c) Using the results from the previous two parts we have $f_y(x,0) = x$, which means $f_{yx}(x,0) = 1$ and thus $f_{yx}(0,0) = 1$. Also, $f_y(0,y) = -y$, which means $f_{yx}(0,y) = -1$ and thus $f_{xy}(0,0) = -1$. We see that $f_{xy}(0,0) = -f_{yx}(0,0)$.

72. The partial derivatives of a function of two variables represent the slope of the tangent line for each variable. For example, $\partial f/\partial x$ at a point (x,y) represents the slope of tangent line at (x,y) in the x direction.

Exercise Set 7.3

1. $f(x,y) = x^2 + xy + y^2 - y$

- Find the partial derivatives:
  $f_x = 2x + y$, $f_{xx} = 2$

  $f_y = x + 2y - 1$, $f_{yy} = 2$

  $f_{xy} = 1$
- We solve $f_x = 0$ and $f_y = 0$. We use the substitution method, from $f_x = 0$ we can write the $y = -2x$. Thus

$$\begin{aligned}x+2y-1 &= 0 \\ x+2(-2x)-1 &= 0 \\ x-4x &= 1 \\ x &= -\frac{1}{3}\end{aligned}$$

and therefore

$$y = 2\cdot -\frac{1}{3} = -\frac{2}{3}$$

- Find D for $(-1/3,-2/3)$

$$\begin{aligned}D &= f_{xx}(-1/3,-2/3)\cdot f_{yy}(-1/3,-2/3) - \\ &\quad [f_{xy}(-1/3,-2/3)]^2 \\ &= 2\cdot 2 - [1]^2 \\ &= 4-1 \\ &= 3\end{aligned}$$

- Since $D > 0$ and $f_{xx}(-1/3,-2/3) > 0$, $f(x,y)$ has a relative minimum at $(-1/3,-2/3)$

2. $f(x,y) = x^2 + xy + y^2 - 5y$

- Find the partial derivatives:
  $f_x = 2x + y$, $f_{xx} = 2$

  $f_y = x + 2y - 5$, $f_{yy} = 2$

  $f_{xy} = 1$
- We solve $f_x = 0$ and $f_y = 0$. We use the substitution method, from $f_x = 0$ we can write the $y = -2x$. Thus

$$\begin{aligned}x+2y-5 &= 0 \\ x+2(-2x)-5 &= 0\end{aligned}$$

$$\begin{aligned} x - 4x &= 5 \\ x &= -\frac{1}{5} \end{aligned}$$

and therefore

$$y = 2 \cdot -\frac{1}{5} = -\frac{2}{5}$$

- Find D for $(-1/5, -2/5)$

$$\begin{aligned} D &= f_{xx}(-1/5,-2/5) \cdot f_{yy}(-1/5,-2/5) - \\ &\quad [f_{xy}(-1/5,-2/5)]^2 \\ &= 2 \cdot 2 - [1]^2 \\ &= 4 - 1 \\ &= 3 \end{aligned}$$

- Since $D > 0$ and $f_{xx}(-1/5,-2/5) > 0$, $f(x,y)$ has a relative minimum at $(-1/5,-2/5)$

3. $f(x,y) = 2xy - x^3 - y^2$

- Find the partial derivatives:
  $f_x = 2y - 3x^2$, $f_{xx} = -6x$

  $f_y = 2x - 2y$, $f_{yy} = -2$

  $f_{xy} = 2$
- We solve $f_x = 0$ and $f_y = 0$. We use the substitution method, from $f_y = 0$ we can write the $y = x$. Thus

$$\begin{aligned} 2y - 3x^2 &= 0 \\ 2(x) - 3x^2 &= 0 \\ x(2 - 3x) &= 1 \\ x &= 0 \end{aligned}$$

and therefore

$$y = 0$$

and

$$x = 2/3$$

and therefore

$$y = 2/3$$

- Find D for $(0,0)$ and $(2/3, 2/3)$

$$\begin{aligned} D &= f_{xx}(0,0) \cdot f_{yy}(0,0) - [f_{xy}(0,0)]^2 \\ &= 0 \cdot -2 - [2]^2 \\ &= -4 \\ D &= f_{xx}(2/3,2/3) \cdot f_{yy}(2/3,2/3) - \\ &\quad [f_{xy}(2/3,2/3)]^2 \\ &= -4 \cdot -2 - [2]^2 \\ &= 4 \end{aligned}$$

- Since $D < 0$ at $(0,0)$, $f(x,y)$ has a saddle at $(0,0)$ Since $D > 0$ and $f_{xx}(2/3,2/3) < 0$, $f(x,y)$ has a relative maximum at $(2/3,2/3)$

4. $f(x,y) = 4xy - x^3 - y^2$

- Find the partial derivatives:
  $f_x = 4y - 3x^2$, $f_{xx} = -6x$

  $f_y = 4x - 2y$, $f_{yy} = -2$

  $f_{xy} = 4$
- We solve $f_x = 0$ and $f_y = 0$. We use the substitution method, from $f_y = 0$ we can write the $y = 2x$. Thus

$$\begin{aligned} 4y - 3x^2 &= 0 \\ 4(2x) - 3x^2 &= 0 \\ x(8 - 3x) &= 1 \\ x &= 0 \end{aligned}$$

and therefore

$$y = 0$$

and

$$x = 8/3$$

and therefore

$$y = 16/3$$

- Find D for $(0,0)$ and $(8/3, 16/3)$

$$\begin{aligned} D &= f_{xx}(0,0) \cdot f_{yy}(0,0) - [f_{xy}(0,0)]^2 \\ &= 0 \cdot -2 - [4]^2 \\ &= -16 \\ D &= f_{xx}(8/3,16/3) \cdot f_{yy}(8/3,16/3) - \\ &\quad [f_{xy}(8/3,16/3)]^2 \\ &= -16 \cdot -2 - [4]^2 \\ &= 16 \end{aligned}$$

- Since $D < 0$ at $(0,0)$, $f(x,y)$ has a saddle at $(0,0)$ Since $D > 0$ and $f_{xx}(8/3,16/3) < 0$, $f(x,y)$ has a relative maximum at $(8/3,16/3)$

5. $f(x,y) = x^3 + y^3 - 3xy$

- Find the partial derivatives:
  $f_x = 3x^2 - 3y$, $f_{xx} = 6x$

  $f_y = 3y^2 - 3x$, $f_{yy} = 6y$

  $f_{xy} = -3$
- We solve $f_x = 0$ and $f_y = 0$. We use the substitution method, from $f_x = 0$ we can write the $y = x^2$. Thus

$$\begin{aligned} -3x + 3y^2 &= 0 \\ -3x + 3(x^2) &= 0 \\ 3x(x - 1) &= 0 \\ x &= 0 \end{aligned}$$

and therefore

$$y = 0$$

and

$$x = 1$$

and therefore

$$y = 1$$

- Find D for $(0,0)$ and $(1,1)$

$$\begin{aligned} D &= f_{xx}(0,0) \cdot f_{yy}(0,0) - [f_{xy}(0,0)]^2 \\ &= 0 \cdot 0 - [-3]^2 \\ &= -9 \\ D &= f_{xx}(1,1) \cdot f_{yy}(1,1) - \\ & \quad [f_{xy}(1,1)]^2 \\ &= 6 \cdot 6 - [-3]^2 \\ &= 27 \end{aligned}$$

- Since $D < 0$ at $(0,0)$, $f(x,y)$ has a saddle at $(0,0)$ Since $D > 0$ and $f_{xx}(1,1) > 0$, $f(x,y)$ has a relative minimum at $(1,1)$

6. $f(x,y) = x^3 + y^3 - 6xy$

- Find the partial derivatives:
  $f_x = 3x^2 - 6y$, $f_{xx} = 6x$

  $f_y = 3y^2 - 6x$, $f_{yy} = 6y$

  $f_{xy} = -6$
- We solve $f_x = 0$ and $f_y = 0$. We use the substitution method, from $f_x = 0$ we can write the $y = \frac{1}{2}x^2$. Thus

$$\begin{aligned} -3x + 3y^2 &= 0 \\ -3x + 3(\frac{1}{2}x^2) &= 0 \\ 3x(\frac{1}{2}x - 1) &= 0 \\ x &= 0 \end{aligned}$$

and therefore

$$y = 0$$

and

$$x = 2$$

and therefore

$$y = 2$$

- Find D for $(0,0)$ and $(2,2)$

$$\begin{aligned} D &= f_{xx}(0,0) \cdot f_{yy}(0,0) - [f_{xy}(0,0)]^2 \\ &= 0 \cdot 0 - [-3]^2 \\ &= -9 \\ D &= f_{xx}(2,2) \cdot f_{yy}(2,2) - \\ & \quad [f_{xy}(2,2)]^2 \\ &= 12 \cdot 12 - [-3]^2 \\ &= 135 \end{aligned}$$

- Since $D < 0$ at $(0,0)$, $f(x,y)$ has a saddle at $(0,0)$ Since $D > 0$ and $f_{xx}(1,1) > 0$, $f(x,y)$ has a relative minimum at $(2,2)$

7. $f(x,y) = x^2 + y^2 - 2x + 4y - 2$

- Find the partial derivatives:
  $f_x = 2x - 2$, $f_{xx} = 2$

  $f_y = 2y + 4$, $f_{yy} = 2$

  $f_{xy} = 0$
- We solve $f_x = 0$ and $f_y = 0$.

$$\begin{aligned} 2x - 2 &= 0 \\ 2x &= 2 \\ x &= 1 \end{aligned}$$

and

$$\begin{aligned} 2y + 4 &= 0 \\ 2y &= -4 \\ y &= -2 \end{aligned}$$

- Find D for $(1,-2)$

$$\begin{aligned} D &= f_{xx}(1,-2) \cdot f_{yy}(1,-2) - [f_{xy}(1,-2)]^2 \\ &= 2 \cdot 2 - [0]^2 \\ &= 4 \end{aligned}$$

- Since $D > 0$ and $f_{xx}(1,-2) > 0$, $f(x,y)$ has a relative minimum at $(1,-2)$

8. $f(x,y) = x^2 + 2xy + 2y^2 - 6y + 2$

- Find the partial derivatives:
  $f_x = 2x + 2y$, $f_{xx} = 2$

  $f_y = 2x + 4y - 6$, $f_{yy} = 4$

  $f_{xy} = 2$
- We solve $f_x = 0$ and $f_y = 0$. We use the substitution method, from $f_x = 0$ we can write the $y = -x$. Thus

$$\begin{aligned} 2x + 4y - 6 &= 0 \\ 2x + 4(-x) - 6 &= 0 \\ -2x - 6 &= 0 \\ x &= -3 \end{aligned}$$

and therefore

$$y = 3$$

- Find D for $(-3,3)$

$$\begin{aligned} D &= f_{xx}(-3,3) \cdot f_{yy}(-3,3) - [f_{xy}(-3,3)]^2 \\ &= 2 \cdot 4 - [2]^2 \\ &= 4 \end{aligned}$$

- Since $D > 0$ and $f_{xx}(-3,3) > 2$, $f(x,y)$ has a relative minimum at $(0,0)$

9. $f(x,y) = x^2 + y^2 + 2x - 4y$

- Find the partial derivatives:
  $f_x = 2x + 2$, $f_{xx} = 2$

  $f_y = 2y - 4$, $f_{yy} = 2$

  $f_{xy} = 0$

- We solve $f_x = 0$ and $f_y = 0$.

$$\begin{aligned} 2x + 2 &= 0 \\ 2x &= -2 \\ x &= -1 \\ 2y - 4 &= 0 \\ 2y &= 4 \\ y &= 2 \end{aligned}$$

- Find D for $(-1, 2)$

$$\begin{aligned} D &= f_{xx}(-1,2) \cdot f_{yy}(-1,2) - [f_{xy}(-1,2)]^2 \\ &= 2 \cdot 2 - [0]^2 \\ &= 4 \end{aligned}$$

- Since $D > 0$ and $f_{xx}(-1, 2) > 0$, $f(x, y)$ has a relative minimum at $(-1, 2)$

10. $f(x, y) = 4y + 6x - x^2 - y^2$

- Find the partial derivatives:
$f_x = 6 - 2x,\ f_{xx} = -2$

$f_y = 4 - 2y,\ f_{yy} = -2$

$f_{xy} = 0$
- We solve $f_x = 0$ and $f_y = 0$.

$$\begin{aligned} 6 - 2x &= 0 \\ -2x &= -6 \\ x &= 3 \\ 4 - 2y &= 0 \\ -2y &= -4 \\ y &= 2 \end{aligned}$$

- Find D for $(3, 2)$

$$\begin{aligned} D &= f_{xx}(3,2) \cdot f_{yy}(3,2) - [f_{xy}(3,2)]^2 \\ &= -2 \cdot -2 - [0]^2 \\ &= 4 \end{aligned}$$

- Since $D < 0$ and $f_{xx}(3, 2) < 0$, $f(x, y)$ has a relative maximum at $(0, 0)$

11. $f(x, y) = 4x^2 - y^2$

- Find the partial derivatives:
$f_x = 8x,\ f_{xx} = 8$

$f_y = -2y,\ f_{yy} = -2$

$f_{xy} = 0$
- We solve $f_x = 0$ and $f_y = 0$.

$$\begin{aligned} 8x &= 0 \\ x &= 0 \\ -2y &= 0 \\ y &= 0 \end{aligned}$$

- Find D for $(0, 0)$

$$\begin{aligned} D &= f_{xx}(0,0) \cdot f_{yy}(0,0) - [f_{xy}(0,0)]^2 \\ &= 8 \cdot -2 - [0]^2 \\ &= -16 \end{aligned}$$

- Since $D < 0$ at $(0, 0)$, $f(x, y)$ has a saddle at $(0, 0)$

12. $f(x, y) = x^2 - y^2$

- Find the partial derivatives:
$f_x = 2x,\ f_{xx} = 2$

$f_y = -2y,\ f_{yy} = -2$

$f_{xy} = 0$
- We solve $f_x = 0$ and $f_y = 0$.

$$\begin{aligned} 8x &= 0 \\ x &= 0 \\ -2y &= 0 \\ y &= 0 \end{aligned}$$

- Find D for $(0, 0)$

$$\begin{aligned} D &= f_{xx}(0,0) \cdot f_{yy}(0,0) - [f_{xy}(0,0)]^2 \\ &= 2 \cdot -2 - [0]^2 \\ &= -4 \end{aligned}$$

- Since $D < 0$ at $(0, 0)$, $f(x, y)$ has a saddle at $(0, 0)$

13. $f(x, y) = e^{x^2+y^2+1}$

- Find the partial derivatives:
$f_x = 2x\ e^{x^2+y^2+1},\ f_{xx} = 4x^2\ e^{x^2+y^2+1} + e^{x^2+y^2+1}$

$f_y = 2y\ e^{x^2+y^2+1},\ f_{yy} = 4y^2\ e^{x^2+y^2+1} + e^{x^2+y^2+1}$

$f_{xy} = 4xy\ e^{x^2+y^2+1}$
- We solve $f_x = 0$ and $f_y = 0$.

$$\begin{aligned} 2x\ e^{x^2+y^2+1} &= 0 \\ 2x &= 0 \\ x &= 0 \\ 2y\ e^{x^2+y^2+1} &= 0 \\ 2y &= 0 \\ y &= 0 \end{aligned}$$

- Find D for $(0, 0)$

$$\begin{aligned} D &= f_{xx}(0,0) \cdot f_{yy}(0,0) - [f_{xy}(0,0)]^2 \\ &= 1 \cdot 1 - [0]^2 \\ &= 1 \end{aligned}$$

- Since $D > 0$ and $f_{xx}(0, 0) > 0$, $f(x, y)$ has a relative minimum at $(0, 0)$

14. $f(x,y) = e^{x^2-2x+y^2-4y+2}$

- Find the partial derivatives:
  $f_x = (2x-2)\ e^{x^2-2x+y^2-4y+2}$
  $f_{xx} = (2x-2)^2\ e^{x^2-2x+y^2-4y+2} + e^{x^2-2x+y^2-4y+2}$

  $f_y = (2y-4)\ e^{x^2-2x+y^2-4y+2}$
  $f_{yy} = (2y-4)^2\ e^{x^2-2x+y^2-4y+2} + e^{x^2-2x+y62-4y+2}$

  $f_{xy} = (2x-2)(2y-4)\ e^{x^2-2x+y^2-4y+2}$
- We solve $f_x = 0$ and $f_y = 0$.

$$\begin{aligned} 2x-2 &= 0 \\ x &= 1 \\ 2y-4 &= 0 \\ y &= 2 \end{aligned}$$

- Find D for $(1,2)$

$$\begin{aligned} D &= f_{xx}(1,2)\cdot f_{yy}(1,2) - [f_{xy}(1,2)]^2 \\ &= 0.09957 \cdot 0.09957 - [0]^2 \\ &= 0.009914 \end{aligned}$$

- Since $D > 0$ and $f_{xx} > 0$, $f(x,y)$ has a relative minimum at $(1,2)$

15. We need to find the point at which $P_x = 0$ and $P_y = 0$

$P_x = 0.0345 - 0.000230x + 0.109y$

$P_y = 25.6 + 0.109x - 126.4y$

Solve $P_x = 0$ and $P_y = 0$. We can use the substitution method. From $P_x:\ y = 0.00211x - 0.31651$

$$\begin{aligned} 25.6 + 0.109x - 126.4y &= 0 \\ 25.6 + 0.109x - 126.2(0.00211x - 0.31651) &= 0 \\ -0.1577x + 65.60686 &= 0 \\ x &= 416.0127 \\ &\text{and therefore} \\ 0.00211(416.0127) - 0.31651 &= y \\ &= 0.56218 \end{aligned}$$

16. $P = 2ap + 80p - 15p^2 - \frac{1}{10}a^2p - 100$

$P_a = 2p - \frac{1}{5}ap$
$P_{aa} = -\frac{1}{5}p$

$P_p = 2a + 80 - 30p - \frac{1}{10}a$
$P_{pp} = -30$

$P_{ap} = 2 - \frac{1}{5}a$

Solving $P_a = 0$ and $P_p = 0$

$$\begin{aligned} 2p - \frac{1}{5}ap &= 0 \\ p(2 - \frac{1}{5}a) &= 0 \\ p &= 0 \\ &\text{therefore} \\ 2a + 80 - 30p - \frac{1}{10}a &= 0 \\ \frac{19}{20}a + 80 - 30(0) &= 0 \\ a &= -84.211 \\ &\text{and} \\ a &= 10 \\ &\text{therefore} \\ \frac{19}{20}(10) + 80 - 30(p) &= 0 \\ \frac{19}{20} + 80 &= 30p \\ 2.6983 &= p \end{aligned}$$

We have two points to check for a maximum $(-84.211, 0)$ and $(10, 2.6983)$. Note that $P_{aa} = 0$ at $(-84.211, 0)$, and therefore the point cannot be a relative maximum according to the D test
Find D for $(10, 2.6983)$

$$\begin{aligned} D &= [-\frac{1}{5}(2.6983)]\cdot[-30] - [2 - \frac{1}{5}(10)]^2 \\ &= 16.1898 - 0 \\ &= 16.1898 \end{aligned}$$

Since $D > 0$ and $P_{aa} < 0$ then $P(a,p)$ has a relative maximum at $(10, 2.6983)$. To find the value of the maximum we find $P(10, 2.6983)$

$$\begin{aligned} P(10, 2.6983) &= 2(10)(2.6983) + 80(2.6983) - \\ &\quad 15(2.6983)^2 - \frac{1}{10}(10)^2(2.6983) - 100 \\ &= 33.635 \end{aligned}$$

17. $P = -5a^2 - 3n^2 + 48a - 4n + 2an + 300$

$P_a = -10a + 48 + 2n$
$P_{aa} = -10$

$P_{an} = 2$

$P_n = -6n - 4 + 2a$
$P_{nn} = -6$

Solve $P_a = 0$ and $P_n = 0$. From $P_a = 0$, $n = 5a - 24$

$$\begin{aligned} -6n - 4 + 2a &= 0 \\ -6(5a - 24) - 4 + 2a &= 0 \\ -30a + 144 - 4 + 2a &= 0 \\ -28a &= -140 \\ a &= 5 \end{aligned}$$

therefore

$$\begin{aligned} n &= 5(5) - 24 \\ &= 1 \end{aligned}$$

Find D for $(5, 1)$

$$\begin{aligned} D &= -10 \cdot -6 - [2]^2 \\ &= 60 - 4 \\ &= 56 \end{aligned}$$

Since $D > 0$ and $P_{aa}(5, 1) < 0$ then $P(a, n)$ has a maximum at $(5, 1)$. To find the maximum value we have to find $P(5, 1)$

$$\begin{aligned} P(5,1) &= -5(5)^2 - 3(1)^2 + 48(5) + \\ & \quad -4(1) + 2(5)(1) + 300 \\ &= -125 - 3 + 240 - 4 + 10 + 300 \\ &= 418 \end{aligned}$$

18. The volume of the container is given by

$$320 = LWH$$

where L is the length, W is the width, and H is the height of the container. The cost function is given by

$$C = 5LW + 2LH + 2WH$$

From the volume equation we can write $H = 320/LW$ therfore the cost function becomes

$$C = 5LW + \frac{640}{W} + \frac{640}{L}$$

We need to find the minimum of C

$C_L = 5W - \frac{640}{L^2}$
$C_{LL} = \frac{1280}{L^3}$

$C_{LW} = 5$

$C_W = 5L - \frac{640}{W^2}$
$C_{WW} = \frac{1280}{W^3}$

$C_L = 0$ and $C_W = 0$ give $L = 4.9866$ and $W = 4.9866$

We find D for $(4.9866, 4.9866)$
$D = 10..323 \cdot 10.323 - [5]^2 = 81.56$

Since $D > 0$ and $C_{LL}(4.9866, 4.9866) > 0$ the C has a minimum at $(4.9866, 4.9866)$. To find the dimensions of the container we need to find H

$$\begin{aligned} H &= \frac{320}{LW} \\ &= \frac{320}{4.9866 \cdot 4.9866} \\ &= 12.8689 \end{aligned}$$

The dimensions of the container are

$$4.9866\ ft \times 4.9866\ ft \times 12.8689\ ft$$

19. $T(x, y) = x^2 + 2y^2 - 8x + 4y$

$T_x = 2x - 8$
$T_{xx} = 2$

$T_{xy} = 0$

$T_y = 4y + 4$
$T_{yy} = 4$

Solve $T_x = 0$ and $T_y = 0$

$$\begin{aligned} 2x - 8 &= 0 \\ 2x &= 8 \\ x &= 4 \\ 4y + 4 &= 0 \\ 4y &= -4 \\ y &= -1 \end{aligned}$$

Find D for $(4, -1)$

$$\begin{aligned} D &= 2 \cdot 4 - [0]^2 \\ &= 8 \end{aligned}$$

Since $D > 0$ and $T_{xx}(4, -1) > 0$ then $T(x, y)$ has a minimum at $(4, -1)$. To find the value of the minimum we need to find $T(4, -1)$

$$\begin{aligned} T(4,-1) &= (4)^2 + 2(-1)^2 - 8(4) + 4(-1) \\ &= 16 + 2 - 32 - 4 \\ &= -18 \end{aligned}$$

There is no maximum value.

20. $T_x = 4x - 2$
$T_{xx} = 4$

$T_{xy} = 0$

$T_y = 6y + 6$
$T_{yy} = 6$

Solve $T_x = 0$ and $T_y = 0$

$$\begin{aligned} 4x - 2 &= 0 \\ 4x &= 2 \\ x &= 1/2 \\ 6y + 6 &= 0 \\ 6y &= -6 \\ y &= -1 \end{aligned}$$

Find D for $(1/2, -1)$

$$\begin{aligned} D &= 4 \cdot 6 - [0]^2 \\ &= 24 \end{aligned}$$

Since $D > 0$ and $T_{xx}(1/2, -1) > 0$ then $T(x, y)$ has a minimum at $(1/2, -1)$. To find the value of the minimum we need to find $T(1/2, -1)$

$$\begin{aligned} T(1,-1) &= 2(1/2)^2 + 3(-1)^2 - 2(1/2) + 6(-1) \\ &= 1/2 + 3 - 1 - 6 \\ &= -7/2 \end{aligned}$$

21. $f(x, y) = e^x + e^y - e^{x+y}$

$f_x = e^x - e^{x+y}$
$f_{xx} = e^x - e^{x+y}$

$f_{xy} = -e^{x+y}$

$f_y = e^y - e^{x+y}$
$f_{yy} = e^y - e^{x+y}$

Solve $f_x = 0$ and $f_y = 0$

$$\begin{aligned} e^x - e^{x+y} &= 0 \\ e^x &= e^{x+y} \\ x &= x + y \\ 0 &= y \\ e^y - e^{x+y} &= 0 \\ e^y &= e^{x+y} \\ y &= x + y \\ 0 &= x \end{aligned}$$

Find D for $(0, 0)$

$$\begin{aligned} D &= f_{xx}(0,0) \cdot f_{yy}(0,0) - [f_{xy}(0,0)]^2 \\ &= 0 \cdot 0 - [-1]^2 \\ &= -1 \end{aligned}$$

Since $D < 0$ at $(0, 0)$ then $f(x, y)$ has a saddle point at $(0, 0)$

22. $f_x = y - \frac{2}{x^2}$
$f_{xx} = \frac{4}{x^3}$

$f_{xy} = 1$

$f_y = x - \frac{4}{y^2}$
$f_{yy} = \frac{8}{y^3}$

Solve $f_x = 0$ and $f_y = 0$ gives $x = 0$, which is not in the domain of the function, and $x = 1$, which gives $y = 2$. We find D for $(1, 2)$

$$\begin{aligned} D &= 4 \cdot 1 - [1]^2 \\ &= 3 \end{aligned}$$

Since $D > 0$ and $f_{xx}(1, 2) > 0$ then $f(x, y)$ has a relative minimum at $(1, 2)$

23. $f(x, y) = 2y^2 + x^2 - x^2y$

$f_x = 2x - 2xy$
$f_{xx} = 2 - 2y$

$f_{xy} = -2x$

$f_y = 4y - x^2$
$f_{yy} = 4$

Solve $f_x = 0$ and $f_y = 0$

$$\begin{aligned} 2x - 2xy &= 0 \\ 2x(1 - y) &= 0 \\ x &= 0 \\ &\text{and} \\ y &= 1 \end{aligned}$$

From $f_y = 0$ we get $y = x^2/4$ which means when $x = 0$, $y = 0$ and when $y = 1$, $x = \pm 2$. Find D for $(0, 0)$

$$\begin{aligned} D &= 2 \cdot 4 - [0]^2 \\ &= 8 \end{aligned}$$

Since $D > 0$ and $f_{xx}(0, 0) > 0$ then $f(x, y)$ has a relative minimum at $(0, 0)$ Find D for $(2, 1)$

$$\begin{aligned} D &= 0 \cdot 4 - [-4]^2 \\ &= -16 \end{aligned}$$

Since $D < 0$ at $(2, 1)$ then $f(x, y)$ has a saddle point at $(2, 1)$ Find D for $(-2, 1)$

$$\begin{aligned} D &= 0 \cdot 4 - [4]^2 \\ &= -16 \end{aligned}$$

Since $D < 0$ at $(-2, 1)$ then $f(x, y)$ has a saddle point at $(-2, 1)$

24. $S_b = 2(m + b - 72) + 2(2m + b - 73) + 2(3m + b - 75) = 12m + 6b - 440$
$S_{bb} = 2 + 2 + 2 = 6$

$S_{bm} = 2 + 4 + 6 = 12$

$S_m = 2(m + b - 72) + 4(2m + b - 73) + 6(3m + b - 75) = 28m + 12b - 886$
$S_{mm} = 2 + 8 + 18 = 28$

Solving the system $12m + 6b - 440 = 0$ and $28m + 12b - 886 = 0$ we get $m = 3/2$ and $b = 211/3$ Find D for $(3/2, 211/3)$

$$\begin{aligned} D &= 6 \cdot 28 - [12]^2 \\ &= 168 - 144 \\ &= 24 \end{aligned}$$

Since $D > 0$ and $S_{bb}(211/3, 3/2) > 0$ then $S(b, m)$ has a relative minimum at $(211/3, 3/2)$

25. The D-Test is a method similar to the second derivative test for functions of one variables. It computes the points were the first partial derivatives are zero and then computes the value of D to determine the nature of the zeros f the first partial derivatives

26. The relative minimum of a function of two variables is only the minimum in a local neighborhood around the minimum point while the absolute minimum is the smallest value the function has in its entire domain.

27. $R = e^{-1.236+1.35\ cos\ l\ cos\ s\ -1.707\ sin\ l\ sin\ s}$

a) $\partial R/\partial l$

$$\begin{aligned}\partial R/\partial l &= e^{-1.236+1.35\ cos\ l\ cos\ s\ -1.707\ sin\ l\ sin\ s} \times \\ &\quad (-1.35\ sin\ l\ cos\ s\ -1.707\ cos\ l\ sin\ s)\end{aligned}$$

b) The sign of $\partial R/\partial l$ is determined by the sign of the term in parenthesis in part a) since e^{anything} is positive for all values permissible. Since the trigonometric functions are positive in the first quadrant (the permissible values for l and s fall in the first quadrant) then the term in parenthesis will always be negative. Thus, $\partial R/\partial l$ is always negative for the permissible values of l and s

c - d) Since $\partial R/\partial l$ is negative for $0 \leq l \leq 90$ then the function is decreasing in the l -direction which means that the maximum occurs when $l = 0$ (The beginning of the interval)

28. **a)** $\partial R/\partial s$

$$\begin{aligned}\partial R/\partial s &= e^{-1.236+1.35\ cos\ l\ cos\ s\ -1.707\ sin\ l\ sin\ s} \times \\ &\quad (-1.35\ cos\ l\ sin\ s\ -1.707\ sin\ l\ cos\ s)\end{aligned}$$

b) The sign of $\partial R/\partial s$ is determined by the sign of the term in parenthesis in part a) since e^{anything} is positive for all values permissible. Since the trigonometric functions are positive in the first quadrant (the permissible values for l and s fall in the first quadrant) then the term in parenthesis will always be negative. Thus, $\partial R/\partial s$ is always negative for the permissible values of l and s

c - d) Since $\partial R/\partial s$ is negative for $0 \leq s \leq 60$ then the function is decreasing in the s -direction which means that the maximum occurs when $s = 0$ (The beginning of the interval)

29. Relative minimum of -5 at $(0,0)$

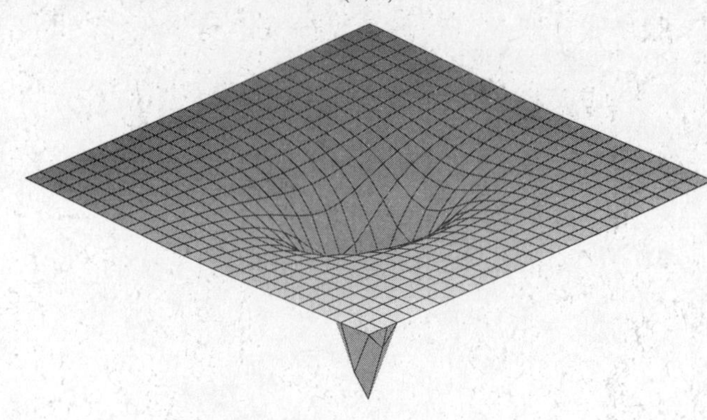

30. Relative minimum of $\dfrac{-55}{4}$ at $(4,-4)$

Relative maximum of $\dfrac{1}{2}$ at $(2,-2)$

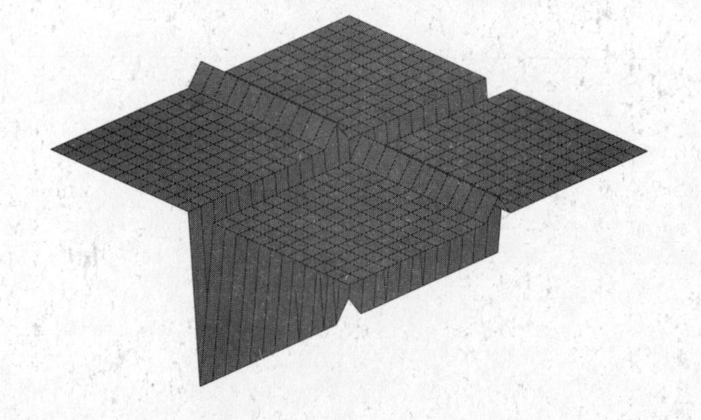

Exercise Set 7.4

1. $\overline{x} = \dfrac{0+1+\cdots+6}{7} = 3$

$$\overline{y} = \frac{33.49+34.72+\cdots+47.70}{7} = 40.75$$

$$m = \frac{(0-3)(33.49-40.75)+\cdots+(6-3)(47.70-40.75)}{(0-3)^2+(1-3)^2+\cdots+(6-3)^2}$$

$m = 2.618$

a) The regression line is

$$\begin{aligned} y - \overline{y} &= m(x-\overline{x}) \\ y - 40.75 &= 2.618(x-3) \\ y &= 2.618x - 7.854 + 40.75 \\ y &= 2.618x + 32.896 \end{aligned}$$

b) Find y when $x = 16$

$$\begin{aligned} y &= 2.618(16) + 32.896 \\ &= 74.78 \end{aligned}$$

Find y when $x = 21$

$$\begin{aligned} y &= 2.618(21) + 32.896 \\ &= 87.87 \end{aligned}$$

2. **a)** $y = 0.000466x + 0.535$

b) When $x = 4000$, $y = 2.399$

3. $\overline{x} = \dfrac{0 + 10 + \cdots + 50}{6} = 25$

$$\overline{y} = \frac{71.1 + 73.1 + \cdots + 79.5}{6} = 75.8$$

$$m = \frac{(0-25)(71.1-75.8) + \cdots + (50-25)(79.5-75.8)}{(10-25)^2 + (20-25)^2 + \cdots + (50-25)^2}$$
$$m = 0.177$$

a) The regression line is

$$\begin{aligned} y - \overline{y} &= m(x - \overline{x}) \\ y - 75.8 &= 0.177(x - 25) \\ y &= 0.177x - 4.425 + 75.8 \\ y &= 0.177x + 71.375 \end{aligned}$$

b) Find y when $x = 60$

$$\begin{aligned} y &= 0.177(60) + 71.375 \\ &= 81.995 \end{aligned}$$

Find y when $x = 65$

$$\begin{aligned} y &= 0.177(65) + 71.375 \\ &= 82.88 \end{aligned}$$

4. **a)** $0.174x + 64.843$

b) When $x = 60$, $y = 75.283$
When $x = 65$, $y = 76.153$

5. $\overline{x} = \dfrac{0 + 1 + \cdots + 4}{5} = 2$

$$\overline{y} = \frac{86 + 83.6 + \cdots + 79.5}{5} = 82.28$$

$$m = \frac{(0-2)(86-82.28) + \cdots + (4-2)(79.5-82.28)}{(0-2)^2 + (1-2)^2 + \cdots + (4-2)^2}$$
$$m = -1.63$$

a) The regression line is

$$\begin{aligned} y - \overline{y} &= m(x - \overline{x}) \\ y - 82.28 &= -1.63(x - 2) \\ y &= -1.63x + 3.26 + 82.28 \\ y &= -1.63x + 85.54 \end{aligned}$$

b) Find y when $x = 14$

$$\begin{aligned} y &= -1.63(14) + 85.54 \\ &= 62.72 \end{aligned}$$

6. **a)** $-1.6x + 33.7$

b) When $x = 14$, $y = 11.3$

7. $\overline{x} = \dfrac{70 + 60 + 85}{3} = 71.67$

$$\overline{y} = \frac{75 + 62 + 89}{3} = 75.34$$

$$\begin{aligned} m &= \frac{(70-71.67)(75-75.34) + (60-71.67)(62-75.34)}{(70-75.67)^2 + (60-71.67)^2 + (85-71.67)^2} \\ &\quad + \frac{(85-75.67)(89-71.34)}{(70-75.67)^2 + (60-71.67)^2 + (85-71.67)^2} \\ &= 1.07 \end{aligned}$$

a) The regression line is

$$\begin{aligned} y - \overline{y} &= m(x - \overline{x}) \\ y - 75.34 &= 1.07(x - 71.67) \\ y &= 1.07x - 76.69 + 75.34 \\ y &= 1.07x - 1.35 \end{aligned}$$

b) Find y when $x = 81$

$$\begin{aligned} y &= 1.07(81) - 1.35 \\ &= 85.32 \end{aligned}$$

8. **a)** $0.227x - 357.556$

b) When $x = 2010$, $y = 98.7$
When $x = 2050$, $y = 107.8$

c) The world record for $x = 2050$ is a little unrealistic since it is close to 9 feet. This unrealistic answer occurs because the regression line is calculated for a small range of years and therefore accurate for years relatively close to that range. When we consider the year 2050 the regression equation becomes less accurate in predicting the world record since 2050 is 61 years from the highest year used to calculate the regression line.

9. Linear regression is a method for finding an equation to model a data set obtained from an experiment.

10. Left to the student

11. **a)**

$X = \log\ x$	$Y = \log\ y$
1.4771	1.3979
2.301	1.4771
4.301	1.9031
7.3979	2.2304
9	2.3979

b) $\overline{x} = 4.89542$ and $\overline{y} = 1.88131$

$$\begin{aligned} m &= \frac{(1.4771-4.89542)(1.3979-1.88131)}{(1.4771-4.89542)^2 + \cdots (9-4.89542)^2} \\ &\quad + \cdots + \frac{(9-4.89542)(2.3979-1.88131)}{(1.4771-4.89542)^2 + \cdots (9-4.89542)^2} \\ &= 0.13568 \end{aligned}$$

The regression line is

$$\begin{aligned} Y - \overline{Y} &= m(X - \overline{X}) \\ Y - 1.88131 &= 0.13568(X - 4.89542) \\ Y &= 0.13568X - 0.66421 + 1.88131 \\ Y &= 0.13568X + 1.21710 \end{aligned}$$

c)

$$\begin{aligned}
Y &= 0.13568X + 1.21710 \\
\log(y) &= 0.13568\log(x) + 1.21710 \\
\log(y) &= \log(x^0.13568) + 1.21710 \\
\log(y) - \log(x^{0.13568}) &= 1.21710 \\
\log\frac{y}{x^{0.13568}} &= 1.2170 \\
\frac{y}{x^{0.13568}} &= 10^{1.21710} \\
\frac{y}{x^{0.13568}} &= 16.48542 \\
y &= 16.48542\ x^{0.13568}
\end{aligned}$$

d) Find y when $x = 1000000$

$$\begin{aligned}
y &= 16.48542(1000000)^{0.13568} \\
&= 107
\end{aligned}$$

12. a)

$X = \log\ x$	$Y = \log\ y$
0.30103	1.1761
0.30103	1.4771
0.95424	1.699
1.2304	2.0414
2.2553	2.243
3.3979	2.4314

b) $Y = 0.36427X + 1.33226$

c)

$$\begin{aligned}
Y &= 0.36427X + 1.33226 \\
\log(y) &= 0.36427\log(x) + 1.33226 \\
\log(y) &= \log(x^0.36427) + 1.33226 \\
\log(y) - \log(x^{0.36427}) &= 1.33226 \\
\log\frac{y}{x^{0.36427}} &= 1.33226 \\
\frac{y}{x^{0.36427}} &= 10^{1.33226} \\
\frac{y}{x^{0.36427}} &= 21.49117 \\
y &= 21.49117\ x^{0.36427}
\end{aligned}$$

d) Find y when $x = 5$

$$\begin{aligned}
y &= 21.49117(5)^{0.36427} \\
&= 38
\end{aligned}$$

13. a) $y = -0.005938x + 15.571914$

b) In 2010
$y = -0.005938(2010) + 15.571914 = 3.636534 = 3 : 38 : 19$
In 2015
$y = -0.005938(2015) + 15.571914 = 3.606844 = 3 : 36 : 41$

c) In 1999, the predicted value for the record is
$y = -0.005938(1999) + 15.571914 = 3.701852 = 3 : 42 : 11$

Exercise Set 7.5

1.

$$\begin{aligned}
\int_0^1\int_0^1 2y\ dxdy &= \int_0^1 2yx\ |_0^1\ dy \\
&= \int_0^1 2y(1-0)\ dy \\
&= \int_0^1 2y\ dy \\
&= y^2\ |_0^1 \\
&= 1-0 \\
&= 1
\end{aligned}$$

2.

$$\begin{aligned}
\int_0^1\int_0^1 2x\ dxdy &= \int_0^1 x^2\ |_0^1\ dy \\
&= \int_0^1 (1-0)\ dy \\
&= \int_0^1 dy \\
&= y|_0^1 \\
&= 1-0 \\
&= 1
\end{aligned}$$

3.

$$\begin{aligned}
\int_{-1}^1\int_x^1 xy\ dydx &= \int_{-1}^1 x\frac{y^2}{2}\ |_x^1\ dx \\
&= \int_{-1}^1 (\frac{x}{2} - \frac{x^3}{2})\ dx \\
&= \left(\frac{x^2}{4} - \frac{x^4}{8}\right|_{-1}^1 \\
&= (\frac{1}{4} - \frac{1}{8}) - (\frac{1}{4} - \frac{1}{8}) \\
&= 0
\end{aligned}$$

4.

$$\begin{aligned}
\int_{-1}^1\int_x^2 (x+y)\ dydx &= \int_{-1}^1 \left(xy + \frac{y^2}{2}\right)\ |_x^2\ dx \\
&= \int_{-1}^1 \left(2x + 2 - x^2 - \frac{x^2}{2}\right)\ dx \\
&= \int_{-1}^1 \left(2x + 2 - \frac{x^2}{2}\right)\ dx \\
&= \left(x^2 + 2x - \frac{x^3}{6}\right|_{-1}^1 \\
&= \left(1 + 2 - \frac{1}{6}\right) - \left(1 - 2 + \frac{1}{6}\right) \\
&= 3\ \frac{2}{3}
\end{aligned}$$

5.

$$\begin{aligned}\int_0^1\int_{-1}^3 (x+y)\,dydx &= \int_0^1 \left(xy+\frac{y^2}{2}\Big|_{-1}^3\right. dx\\ &= \int_0^1 \left(3x+\frac{9}{2}-x-\frac{1}{2}\right) dx\\ &= \int_0^1 (2x+4)\,dx\\ &= \left(x^2+4x\right|_0^1\\ &= 1+4-0\\ &= 5\end{aligned}$$

6.

$$\begin{aligned}\int_0^1\int_{-1}^1 (x+y)\,dydx &= \int_{-1}^1 \left(xy+\frac{y^2}{2}\Big|_{-1}^3\right. dx\\ &= \int_0^1 \left(x+\frac{1}{2}+x-\frac{1}{2}\right) dx\\ &= \int_0^1 2x\,dx\\ &= x^2|_0^1\\ &= 1-0\\ &= 1\end{aligned}$$

7.

$$\begin{aligned}\int_0^1\int_{x^2}^x (x+y)\,dydx &= \int_0^1 \left(xy+\frac{y^2}{2}\Big|_{x^2}^x\right. dx\\ &= \int_0^1 \left(x^2+\frac{x^2}{2}-x^3-\frac{x^4}{2}\right) dx\\ &= \int_0^1 \left(\frac{3x^2}{2}-x^3-\frac{x^4}{2}\right) dx\\ &= \left(\frac{x^3}{2}-\frac{x^4}{4}-\frac{x^5}{10}\Big|_0^1\right.\\ &= \frac{1}{2}-\frac{1}{4}-\frac{1}{10}\\ &= \frac{3}{20}\end{aligned}$$

8.

$$\begin{aligned}\int_0^2\int_0^x e^{x+y}dydx &= \int_0^2 \left(e^{x+y}\right|_0^x dx\\ &= \int_0^2 e^{2x}-e^x\,dx\\ &= \left(\frac{1}{2}e^{2x}-e^x\Big|_0^2\right.\\ &= \frac{1}{2}e^4-e^2-\frac{1}{2}+1\\ &= \frac{1}{2}e^4-e^2+\frac{1}{2}\end{aligned}$$

9.

$$\begin{aligned}\int_0^1\int_1^{e^x}\frac{1}{y}\,dydx &= \int_0^1 \ln\,y|_1^{e^x}\,dx\\ &= \int_0^1 (x-0)\,dx\\ &= \int_0^1 x\,dx\\ &= \left(\frac{x^2}{2}\Big|_0^1\right.\\ &= \frac{1}{2}-0\\ &= \frac{1}{2}\end{aligned}$$

10.

$$\begin{aligned}\int_0^1\int_{-1}^x (x^2+y^2)\,dydx &= \int_0^1 \left(x^2y+\frac{y^3}{3}\Big|_{-1}^x\right. dx\\ &= \int_0^1 \left(x^3+\frac{x^3}{3}+x^2+\frac{1}{3}\right) dx\\ &= \int_0^1 \left(\frac{4x^3}{3}+x^2+\frac{1}{3}\right) dx\\ &= \left(\frac{x^4}{3}+\frac{x^3}{3}+\frac{x}{3}\Big|_0^1\right.\\ &= \frac{1}{3}+\frac{1}{3}+\frac{1}{3}\\ &= 1\end{aligned}$$

11.

$$\begin{aligned}\int_0^2\int_0^x (x+y^2)\,dydx &= \int_0^2 \left(xy+\frac{y^3}{3}\Big|_0^x\right. dx\\ &= \int_0^2 \left(x^2+\frac{x^3}{3}\right) dx\\ &= \left(\frac{x^3}{3}+\frac{x^4}{12}\Big|_0^2\right.\\ &= \frac{8}{3}+\frac{4}{3}\\ &= 4\end{aligned}$$

12.

$$\begin{aligned}\int_1^3\int_0^x 2\,e^{x^2}\,dydx &= \int_1^3 \left(2\,e^{x^2}y\right|_0^x\,dx\\ &= \int_1^3 \left(2x\,e^{x^2}\right)\,dx\\ &= \left(e^{x^2}\right|_1^3\\ &= e^9-e\end{aligned}$$

13.

$$\begin{aligned}\int_0^1\int_0^{1-x^2} (1-y-x^2)\,dydx &= \int_0^1 \left(y-\frac{y^2}{2}-x^2y\Big|_0^{1-x^2}\right. dx\\ &= \int_0^1 (1-x^2)dx\\ &\quad -\int_0^1 \frac{(1-x^2)^2}{2}dx\\ &\quad -\int_0^1 (1-x^2)x^2dx\end{aligned}$$

$$= \int_0^1 \left(1 - x^2 - \frac{1}{2} + x^2\right) dx$$
$$+ \int_0^1 \left(-x^4 - x^2 + x^4\right) dx$$
$$= \int_0^1 \left(\frac{1}{2} - x^2\right) dx$$
$$= \left(\frac{x}{2} - \frac{x^3}{3}\right|_0^1$$
$$= \frac{1}{2} - \frac{1}{3}$$
$$= \frac{1}{6}$$

14.

$$\int_0^1 \int_0^{1-x} (x+y)\, dydx = \int_0^1 \left(xy + \frac{y^2}{2}\right|_0^{1-x} dx$$
$$= \int_0^1 \left(\frac{1}{2} - \frac{x^2}{2}\right) dx$$
$$= \left(\frac{x}{2} - \frac{x^3}{6}\right|_0^1$$
$$= \frac{1}{2} - \frac{1}{6}$$
$$= \frac{1}{3}$$

15.

$$\int_0^1 \int_x^{3x} y\, e^{x^3}\, dydx = \int_0^1 \left(\frac{y^2}{2}\, e^{x^3}\right|_x^{3x} dx$$
$$= \int_0^1 \left(\frac{9x^2\, e^{x^3}}{2} - \frac{x^2\, e^{x^3}}{2}\right) dx$$
$$= 4x^2\, e^{x^3}\, dx$$
$$= \left(\frac{4}{3}\, e^{x^3}\right|_0^1$$
$$= \frac{4}{3}e - \frac{4}{3}$$

16.

$$\int_0^2 \int_0^{x^2} x^{1/2} y^{1/2}\, dydx = \int_0^2 \left(\frac{2x^{1/2}}{3} y^{1/2}\right|_0^{x^2} dx$$
$$= \int_0^2 \left(\frac{2}{3} x^{7/2}\right) dx$$
$$= \left(\frac{4}{29} x^{9/2}\right|_0^2$$
$$= \frac{4}{27} \cdot 2^{9/2}$$
$$= \frac{64}{27}\sqrt{2}$$

17.

$$\int_0^3 \int_0^{3x-x^2} (7-2x)y\, dydx = \int_0^3 \left(\frac{(7-2x)y^2}{2}\right|_0^{3x-x^2} dx$$
$$= \int_0^3 \left(\frac{(7-2x)}{2}(3x-x^2)^2\right) dx$$
$$= \frac{1}{2}\int_0^3 4(3x-x^2)2\; dx +$$
$$\frac{1}{2}\int_0^3 (3-2x)(3x-x^2)^2\; dx$$
$$= \frac{1}{2}\int_0^3 (36x^2 - 24x^3 + 4x^4)\; dx$$
$$+ \frac{1}{2}\int_0^3 (3-2x)(3x-x^2)^2\; dx$$
$$= \frac{1}{2}\left[12x^3 - 6x^4 + \frac{4}{5}x^5\right|_0^3 +$$
$$\left[\frac{(3x-x^2)^3}{3}\right|_0^3$$
$$= 6(3)^3 - 3(3)^3 + \frac{2(3)^5}{5} +$$
$$\frac{\frac{1}{2}(9-9)^3}{3} - 0$$
$$= 16.2$$

18.

$$\int_1^3 \int_{1-x}^{x} \frac{\ln(x+y)}{x(x+y)}\, dydx = \int_1^3 \left(\frac{[\ln(x+y)]^2}{x}\right|_{1-x}^{x} dx$$
$$= \int_1^3 \frac{[\ln(2x)]^2}{x}\, dx$$
$$= \left(\frac{[\ln(2x)]^3}{3}\right|_1^3$$
$$= \frac{1}{3}\left(\ln(6)\right)^3 - \left(\ln(2)\right)^3$$
$$= 1.806$$

19.

$$\int_0^1 \int_0^{1-x} x^2 y\, dydx = \int_0^1 \left(\frac{x^2y^2}{2}\right|_0^{1-x} dx$$
$$= \frac{1}{2}\int_0^1 x^2(1-x)^2\; dx$$
$$= \frac{1}{2}\int_0^1 \left[x^2 - 2x^3 + x^4\right]$$
$$= \left(\frac{x^3}{6} - \frac{x^4}{6} + \frac{x^5}{10}\right|_0^1$$
$$= \frac{1}{6} - \frac{1}{4} + \frac{1}{10}$$
$$= \frac{1}{60}$$
$$= 0.01\overline{6}$$

20.

$$\int_0^1 \int_0^{x-x^2} (x+y)\, dydx = \int_0^1 \left(xy + \frac{y^2}{2}\right|_0^{x-x^2} dx$$
$$= \int_0^1 \frac{3x^2}{2} - 2x^3 + \frac{x^4}{2}\, dx$$
$$= \left(\frac{x^3}{2} - \frac{x^4}{2} + \frac{x^5}{10}\right|_0^1$$

$$\begin{aligned} &= \frac{1}{2} - \frac{1}{2} + \frac{1}{10} \\ &= \frac{1}{10} \\ &= 0.1 \end{aligned}$$

21.

$$\begin{aligned} \int_0^{\pi/2}\int_{-sin\ x}^{sin\ x} y^2 \cos x\ dydx &= \int_0^{\pi/2}\left(\frac{y^3}{3}\cos x\Big|_{-sin\ x}^{sin\ x}\right) dx \\ &= \int_0^{\pi/2} \frac{2}{3}\sin^3(x)\cos x\ dx \\ &= \left(\frac{\sin^4(x)}{6}\Big|_0^{\pi/2}\right) \\ &= \frac{1}{6} - 0 \\ &= \frac{1}{6} \end{aligned}$$

22.

$$\begin{aligned} \int_0^{16}\int_0^{9} (x+y)^{1/2}\ dydx &= \int_0^{16} \frac{2}{3}\left((x+y)^{3/2}\Big|_0^9\right) dx \\ &= \frac{2}{3}\int_0^{16}\left[(x+9)^{3/2} - x^{3/2}\right] dx \\ &= \frac{2}{3}\left(\frac{2}{5}\left((x+9)^{5/2} - x^{5/2}\right)\right)\Big|_0^{16} \\ &= \frac{4}{15}\left[5^5 - 2^5 - 3^5\right] \\ &= 760 \end{aligned}$$

23. a)

$$\begin{aligned} A &= \int_2^4 (x^3 - x)\ dx \\ &= \left(\frac{x^4}{4} - \frac{x^2}{2}\Big|_2^4\right) \\ &= (64-8) - (4-2) \\ &= 54 \end{aligned}$$

b)

$$\begin{aligned} V &= \int_2^4\int_x^{x^3} 1\ dydx \\ &= \int_2^4 y|_x^{x^3}\ dx \\ &= \int_2^4 (x^3 - x)\ dx \\ &= \left(\frac{x^4}{4} - \frac{x^2}{2}\Big|_2^4\right) \\ &= (64-8) - (4-2) \\ &= 54 \end{aligned}$$

c) The area found in part (a) equals the volume in part (b) because the "thickness" is 1. The volume equals the product of the area and the thickness.

24. The integral in Example 1 could not be solve if we switch the order of integration because then the volume the solid generated would depend on value of x and for values of x greater than 1 we would get a negative volume.

25. $\int_0^1\int_1^3\int_{-1}^2 (2x + 3y - z)\ dxdydz$

$$\begin{aligned} &= \int_0^1\int_1^3 x^2 + 3x - xz\Big|_{-1}^2\ dydz \\ &= \int_0^1\int_1^3 (3 + 3y - z)\ dydz \\ &= \int_0^1\left(3y + \frac{3}{2}y - yz\Big|_1^3\right) dz \\ &= \int_0^1 15 - 2z\ dz \\ &= 15z - z^2\Big|_0^1 \\ &= (15-1) - (0-0) \\ &= 14 \end{aligned}$$

26. $\int_0^2\int_1^4\int_{-1}^2 (8x - 2y + z)\ dxdydz$

$$\begin{aligned} &= \int_0^2\int_1^4 (12 - 6y + 3z)dydz \\ &= \int_0^2 (9z - 9)\ dz \\ &= 0 \end{aligned}$$

27.

$$\begin{aligned} \int_0^1\int_0^{1-x}\int_0^{2-x} xyz\ dzdydx &= \int_0^1\left(\frac{xyz^2}{2}\Big|_0^{2-x}\right) \\ &= \int_0^1\int_0^{1-x}\frac{x(2-x)^2y}{2}\ dydx \\ &= \int_0^1\left(\frac{x(2-x)^2y^2}{4}\Big|_0^{1-x}\right) dx \\ &= \int_0^1 \frac{x(2-x)^2(1-x)^2}{4}\ dx \\ &= \int_0^1\left(\frac{2x - 8x^2 + 11x^3}{4}\right) + \\ &\quad \int_0^1\left(\frac{-6x^4 + x^5}{4}\right) dx \\ &= \left(\frac{x^2 - \frac{8}{3}x^3 + \frac{11}{4}x^4}{4}\Big|_0^1\right. \\ &\quad +\left(\frac{-\frac{6}{5}x^5 + \frac{1}{6}x^6}{4}\Big|_0^1\right. \\ &= \frac{1}{80} \end{aligned}$$

28.

$$\int_0^2\int_{2-y}^{6-2y}\int_0^{\sqrt{4-y^2}} z\ dzdxdy = \int_{2-y}^{6-2y}\left(2 - \frac{y}{2}\right) dxdy$$

$$= \int_0^2 \left(8 - 2y - 2y^2 - \frac{1}{2}y^3\right) dy$$
$$= \left(8y - y^2 - \frac{2}{3}y^3 - \frac{1}{8}y^4\right|_0^2$$
$$= 16 - 4 - \frac{16}{3} - 2$$
$$= \frac{14}{3}$$

29. The geometric meaning of the multiple integral of a function of two variables is the volume of the solid generated from the function bounded by the limits of the multiple integrals.

30. Left to the student

31. 2.957335369

32. 11.61953492

33. 0.3353157821

34. −0.1202884987

Chapter Review Exercises

1. $f(2,0) = e^0 + 3(2)(0)^3 + 2(0) = 1$

2. $f_x = 3y^3$

3. $f_y = e^y + 9xy^2 + 2$

4. $f_{xy} = 9y^2$

5. $f_{yx} = 9y^2$

6. $f_{xx} = 0$

7. $f_{yy} = e^y + 18xy$

8.

$$\frac{dz}{dx} = \frac{1}{2x^3+y} \cdot 6x^2 + y^2 \cos(xy^2)$$
$$= \frac{6x^2}{2x^3+y} + y^2 \cos(xy^2)$$

9.

$$\frac{dz}{dy} = \frac{1}{2x^3+y} \cdot 1 + 2xy \cos(xy^2)$$
$$= \frac{1}{2x^3+y} + 2xy \cos(xy^2)$$

10.

$$\frac{\partial^2 z}{\partial x \partial y} = \frac{-6x^2}{(2x^3+y)^2} \cdot 1 + 2y \cos(xy^2) - xy^2 \sin(xy^2)$$
$$= \frac{-6x^2}{(2x^3+y)^2} + 2y \cos(xy^2) - xy^2 \sin(xy^2)$$

11.

$$\frac{\partial^2 z}{\partial x^2} = \frac{(2x^3+y)(12x) - 6x^2(6x^2)}{(2x^3+y)^2} - y^4 \sin(xy^2)$$
$$= \frac{12x(y-x^3)}{(2x^3+y)^2} - y^4 \sin(xy^2)$$

12.

$$\frac{\partial^2 z}{\partial y^2} = \frac{-1}{(2x^3+y)^2} + 2x \cos(xy^2) - 2x^2 y \sin(xy^2)$$

13.

$$J = \begin{bmatrix} \frac{\partial f}{\partial x} & \frac{\partial f}{\partial y} \\ \frac{\partial g}{\partial x} & \frac{\partial g}{\partial y} \end{bmatrix}$$
$$= \begin{bmatrix} 1 + 2xy^3 & 3x^2y^2 \\ e^{x-5y} & -5\, e^{x-5y} \end{bmatrix}$$

14. $f(1,0) = 1$
$f_x = e^y \to f_x(1,0) = 1$
$f_y = x\, e^y \to f_y(1,0) = 1$

$$f(1.01, 0.02) = 1 + 1(1.01 - 1) + 1(0.02 - 0)$$
$$= 1.03$$

15. $f(2,2) = \sqrt{2(8)} = 4$
$f_x = \frac{1}{2}(xy^3)^{-1/2}y^3 \to f_x(2,2) = \frac{1}{2} \cdot \frac{1}{\sqrt{16}} \cdot 8 = 1$
$f_y = \frac{1}{2}(xy^3)^{-1/2}(3xy^2) \to f_y(2,2) = \frac{1}{2} \cdot \frac{1}{\sqrt{16}} \cdot 24 = 3$

$$f(1.99, 2.02) = f(2,2) + f_x(2,2)(1.99 - 2) + f_y(2,2)(2.02 - 2)$$
$$= 4 + 1(-0.01) + 3(0.02)$$
$$= 4.05$$

16. $f_x = 3x^2 - 6y + 6,\ f_{xx} = 6x$
$f_y = -6x + 2y + 3,\ f_{yy} = 2$
$f_{xy} = -6$
Solving the system $3x^2 - 6y + 6 = 0$ and $-6x + 2y + 3 = 0$ gives two ordered pairs: $(1, 1.5)$ and $(5, 13.5)$

- For $(1, 1.5)$:

$$D = f_{xx}(1,1.5) f_{yy}(1,1.5) - [f_{xy}(1,1.5)]^2$$
$$= 6 \cdot 2 - (-6)^2$$
$$= -24$$

There is a saddle point at $(1, 1.5)$

- For $(5, 13.5)$:

$$D = f_{xx}(5,13.5) f_{yy}(5,13.5) - [f_{xy}(5,13.5)]^2$$
$$= 30 \cdot 2 - (-6)^2$$
$$= 24$$

Since $D > 0$ and $f_{xx}(5, 13.5) > 0$, the function has a relative minimum at $(5, 13.5)$. The value of the minimum is

$$f(5, 13.5) = 3(5)^2 - 6(13.5) + 6$$
$$= 0$$

17. $f_x = 2x - y - 2$, $f_{xx} = 2$
$f_y = -x + 2y + 4$, $f_{yy} = 2$
$f_{xy} = -1$
Solving the system $2x - y - 2 = 0$ and $-x + 2y + 4 = 0$ gives the ordered pair $(0, -2)$

$$\begin{aligned} D &= f_{xx}(0,-2)f_{yy}(0,-2) - [f_{xy}(0,-2)]^2 \\ &= 2 \cdot 2 - (-1)^2 \\ &= 3 \end{aligned}$$

Since $D > 0$ and $f_{xx}(0,-2) > 0$, the function has a relative minimum at $(0,-2)$. The value of the minimum is

$$\begin{aligned} f(0,-2) &= 0^2 - 0(-2) + (-2)^2 - 2(0) + 4(-2) \\ &= -4 \end{aligned}$$

18. $f_x = 3 - 2x$, $f_{xx} + -2$
$f_y = -6 - 2y$, $f_{yy} = -2$
$f_{xy} = 0$
Solving $f_x = 0$ and $f_y = 0$ gives the ordered pair $(1.5, -3)$

$$\begin{aligned} D &= -2 \cdot -2 - 0^2 \\ &= 4 \end{aligned}$$

Since $D > 0$ and $f_{xx}(1.5,-3) < 0$, the function has a relative maximum at $(1.5,-3)$. The value of the maximum is

$$\begin{aligned} f(1.5,-3) &= 3(1.5) - 6(-3) - (1.5)^2 - (-3)^2 \\ &= 11.25 \end{aligned}$$

19. $f_x = 4x^3 + 4$, $f_{xx} = 12x^2$
$f_y = 4y^3 - 32$, $f_{yy} = 12y^2$
$f_{xy} = 0$
Solving the system $f_x = 0$ and $f_y = 0$ gives the ordered pair $(-1, 2)$

$$\begin{aligned} D &= f_{xx}(-1,2)f_{yy}(-1,2) - [f_{xy}(-1,2)]^2 \\ &= 12 \cdot 48 - 0^2 \\ &= 576 \end{aligned}$$

Since $D > 0$ and $f_{xx}(-1,2) > 0$, then function has a relative minimum at $(-1,2)$. The value of the minimum is

$$\begin{aligned} f(-1,2) &= (-1)^4 + (2)^4 + 4(-1) - 32(2) + 80 \\ &= 29 \end{aligned}$$

20. **a)** $y = 0.670545x + 0.442$
b) When $x = 16$, $y = 1.5707$
When $x = 19$, $y = 1.7824$

21. **a)** $y = \frac{3}{5}x + \frac{20}{3}$
b) When $x = 4$, $y = 9.066$

22.

$$\begin{aligned} \int_0^1 \int_{x^2}^{3x} (x^3 + 2y)\, dydx &= \int_0^1 (x^3y + y^2|_{x^2}^{3x} \\ &= \int_0^1 (9x^2 + 2x^4 - x^5)dx \\ &= \left(3x^3 + \frac{2}{5}x^5 - \frac{x^6}{6}\right|_0^1 \\ &= 3 + \frac{2}{5} - \frac{1}{6} \\ &= \frac{97}{30} \end{aligned}$$

23.

$$\begin{aligned} \int_0^1 \int_{x^2}^{x} (x - y)\, dydx &= \int_0^1 \left(xy - \frac{y^2}{2}\right|_{x^2}^{x} \\ &= \int_0^1 (\frac{1}{2}x^2 - x^3 + \frac{1}{2}x^4)dx \\ &= \left(\frac{1}{6}x^3 - \frac{1}{4}x^4 \frac{1}{10}x^5\right|_0^1 \\ &= \frac{1}{6} - \frac{1}{4} + \frac{1}{10} \\ &= \frac{1}{60} \end{aligned}$$

24.

$$\begin{aligned} \int_0^2 \int_{1-2x}^{1-x} \int_0^{\sqrt{2-x^2}} zdzdydx &= \int_0^2 \int_{1-2x}^{1-x} \frac{1}{2}(2 - x^2)dydx \\ &= \int_0^2 \frac{1}{2}x(2 - x^2)dx \\ &= \int_0^2 x - \frac{1}{2}x^3 dx \\ &= \left(\frac{x^2}{2} - \frac{1}{8}x^4\right|_0^2 \\ &= 2 - 2 \\ & 0 \end{aligned}$$

25.

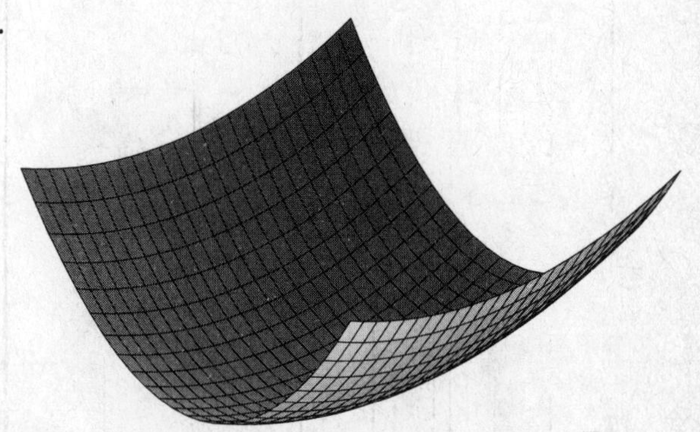

Chapter 7 Test

1. $f(-1,2) = cos((-1)^2e^4) = cos(e^4) = -0.370662$

2. $f_x = -2x\ e^{2y} sin(x^2\ e^{2y})$

3. $f_y = -2x^2\ e^{2y} sin(x^2 e^{2y})$

4.

$$\begin{aligned} f_{xx} &= -2x\ e^{2y} \cdot 2x\ e^{2y} cos(x^2 e^{2y}) - 2\ e^{2y} sin(x^2\ e^{2y}) \\ &= -4x^2\ e^{4y} cos(x^2\ e^{2y}) - 2\ e^{2y} sin(x^2\ e^{2y}) \end{aligned}$$

5.

$$\begin{aligned} f_{xy} &= -2x\ e^{2y} \cdot 2x^2\ e^{2y} cos(x^2\ e^{2y}) - 4x\ e^{2y} sin(x^2\ e^{2y}) \\ &= -4x^3\ e^{4y} cos(x^2\ e^{2y}) - 4xe^{2y} sin(x^2\ e^{2y}) \end{aligned}$$

6.

$$\begin{aligned} f_{yy} &= -2x\ e^{2y} \cdot 2x^2\ e^{2y} cos(x^2\ e^{2y}) - 4x\ e^{2y} sin(x^2\ e^{2y}) \\ &= -4x^4\ e^{4y} cos(x^2\ e^{2y}) - 4x^2\ e^{2y} sin(x^2\ e^{2y}) \end{aligned}$$

7. $f(2,4) = 2$
$f_x = \frac{3x^2}{y} \rightarrow f_x(2,4) = 3$
$f_y = \frac{-x^3}{y^2} \rightarrow f_y(2,4) = \frac{-1}{2}$

$$\begin{aligned} f(2.02, 4.01) &= f(2,4) + f_x(2,4)(2.02 - 2) + \\ &\quad f_y(2,4)(4.01 - 4) \\ &= 2 + 3(0.02) + (\frac{-1}{2})(0.01) \\ &= 2.055 \end{aligned}$$

8. $f(3,9) = 9$
$f_x = \sqrt{y} \rightarrow f_x(3,9) = 3$
$f_y = \frac{x}{2\sqrt{y}} \rightarrow f_y(3,9) = \frac{1}{2}$

$$\begin{aligned} f(2.99, 9.03) &= 9_3(2.99 - 3) + 0.5(9.03 - 9) \\ &= 8.985 \end{aligned}$$

9.

$$\begin{aligned} J &= \begin{bmatrix} \frac{\partial f}{\partial x} & \frac{\partial f}{\partial y} \\ \frac{\partial g}{\partial x} & \frac{\partial g}{\partial y} \end{bmatrix} \\ &= \begin{bmatrix} \frac{1}{y^2+1} & \frac{2xy}{(y^2+1)^2} \\ -y\ e^x sin(e^x y) & -e^x sin(e^x y) \end{bmatrix} \end{aligned}$$

10. $f_x = 2x - y - 1$, $f_{xx} + 2$
$f_y = -x + 3y^2$, $f_{yy} = 6y$
$f_{xy} = -1$

Solving the system $2x - y - 1 = 0$ and $-x + 3y^2 = 0$ gives the two ordered pairs $(1/3, -1/3)$ and $(3/4, 1/2)$

- For $(1/3, -1/3)$:

$$\begin{aligned} D &= 2 \cdot -2 - (-1)^2 \\ &= -5 \end{aligned}$$

There is a saddle point at $(1/3, -1/3)$

- For $(3/4, 1/2)$:

$$\begin{aligned} D &= 2 \cdot 3 - (-1)^2 \\ &= 5 \end{aligned}$$

Since $D > 0$ and $f_{xx}(3/4, 1/2) > 0$, the function has a relative minimum at $(3/4, 1/2)$. The value of the minimum is

$$\begin{aligned} f(3/4, 1/2) &= (\frac{3}{4})^2 - \frac{3}{4} \cdot \frac{1}{2} + (\frac{1}{3})^3 - \frac{3}{4} \\ &= \frac{-7}{16} \end{aligned}$$

11. $f_x = -2x$, $f_{xx} + -2$
$f_y = 8y$, $f_{yy} = 8$
$f_{xy} = 0$
Solving $f_x = 0$ and $f_y = 0$ gives the ordered pair $(0,0)$

$$\begin{aligned} D &= -2 \cdot 8 - 0^2 \\ &= -16 \end{aligned}$$

There is a saddle point at $(0,0)$

12. **a)** $y = -9.45x + 88.2667$
b) When $x = 13$, $y = -34.583$

13.

$$\begin{aligned} \int_0^2 \int_1^x (x^2 - y) dy dx &= \int_0^2 \left(x^2 y - \frac{y^2}{2} \right|_1^x dx \\ &= \int_0^2 x^3 - \frac{3}{2}x + \frac{1}{2} dx \\ &= \left(\frac{x^4}{4} - \frac{1}{2}x^3 + \frac{1}{2}x \right|_0^2 \\ &= 4 - 4 + 1 \\ &= 1 \end{aligned}$$

14.

$$\begin{aligned} \int_0^1 \int_0^x \int_0^{x+y} (x + y + z) dz dy dx &= \int_0^1 \int_0^x \frac{3}{2}(x + y)^2 dy dx \\ &= \int_0^1 \frac{7}{2} x^3 dx \\ &= \left(\frac{7}{8} x^4 \right|_0^1 \\ &= \frac{7}{8} \end{aligned}$$

15.

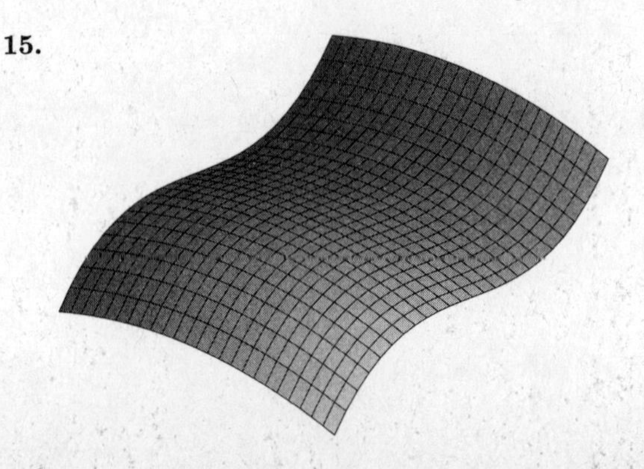

Extended Life Science Connection

1. $$\frac{\partial f}{\partial t} = \frac{N\,e^{-x^2/2\sigma^2 t}(-\sigma^2 t + x^2)}{4\pi\sigma^3 t^{5/2}}$$

$$\frac{\partial^2 f}{\partial x^2} = \frac{N\,e^{-x^2/2\sigma^2 t}(-\sigma^2 t + x^2)}{2\pi\sigma^5 t^{5/2}}$$

2. **a)** For $N = 1000$, $\sigma = 100$, $x = 100$ and $t = 3$
$f(x,t) = 7.7676$

b)

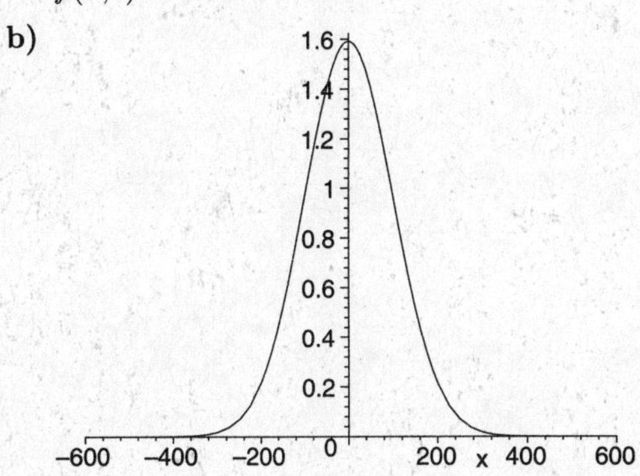

c)

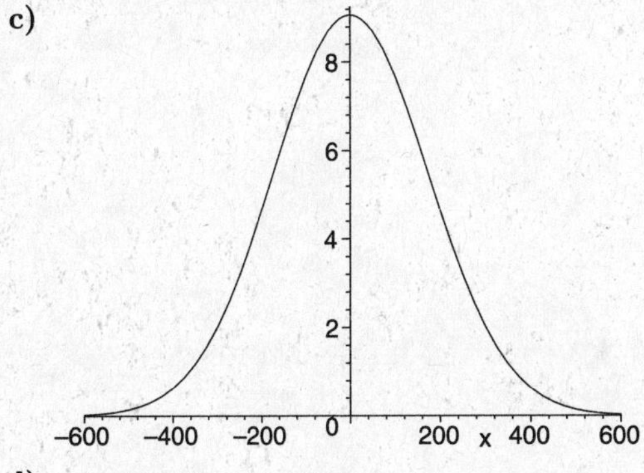

d)

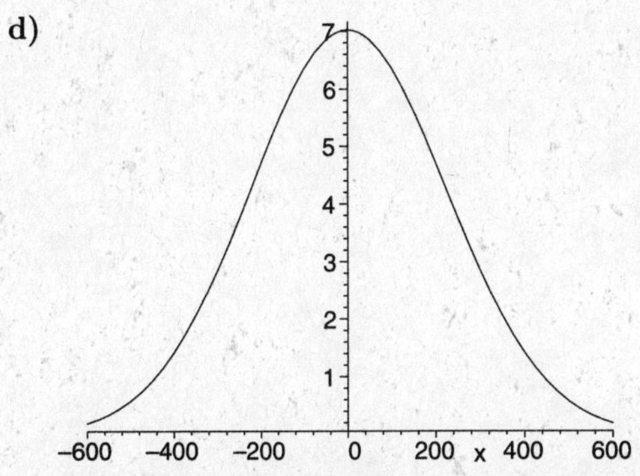

e) As t increases, the graph of the probabilty density function stretches horizontally. The density of the fish is decreasing as time increases.

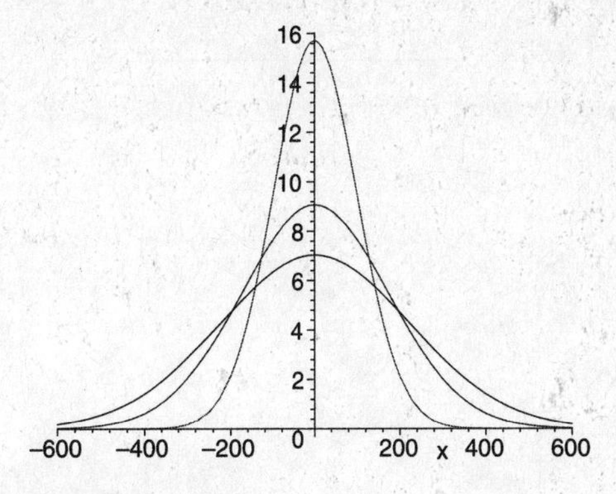

3. **a)** The fish seem to the largest at $t = 1$

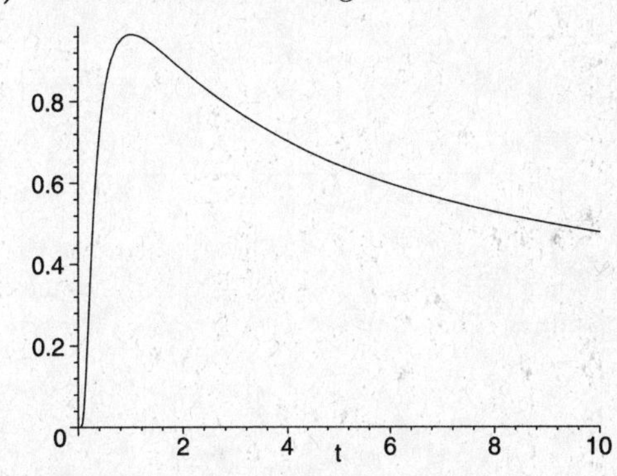

b) $\lim_{t \to 0^+} h(t) = 0$

c) The fish seem to the largest at $t = 4$

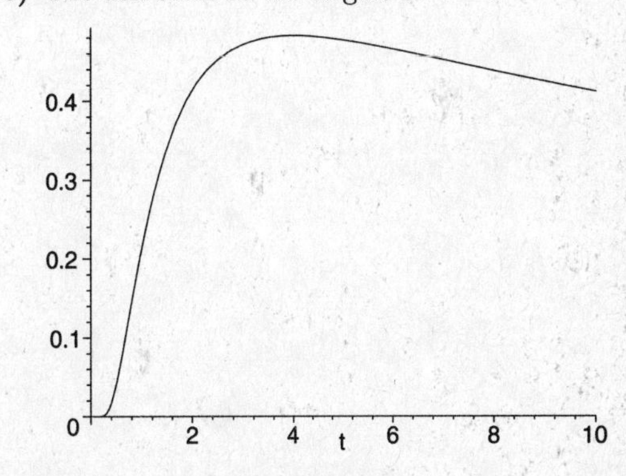

d) The fish seem to the largest at $t = 10$

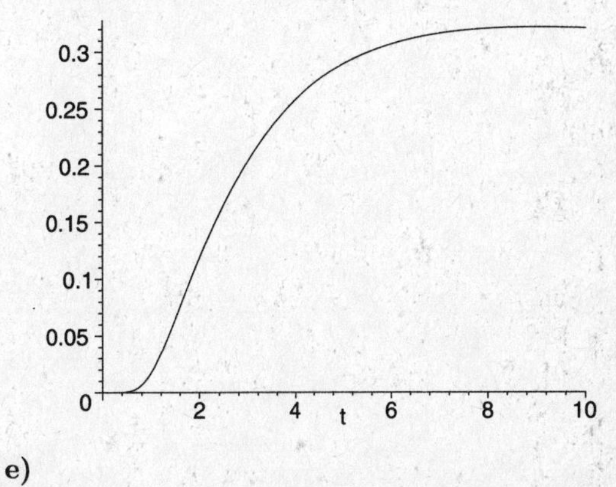

e)

$$\frac{\partial f}{\partial t} = 0$$

$$\begin{aligned}
\frac{Ne^{-x^2/2\sigma^2\sqrt{t}}(-\sigma^2 t + x^2)}{4\pi\sigma^3 t^{5/2}} &= 0 \\
-\sigma^2 t + x^2 &= 70 \\
t &= \frac{x^2}{\sigma^2}
\end{aligned}$$

f) The fish seem to the largest at $t = 1$

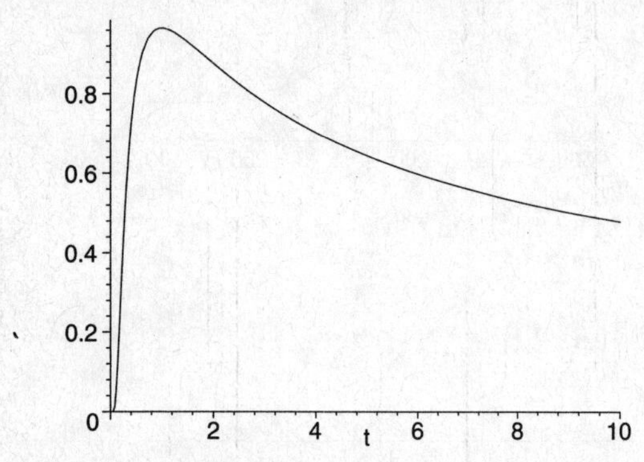

The answer for $x = -100$ is the same as the answer when $x = 100$ because the $f(x,t)$ is symmetric for x values. That is to say $f(x,t) = f(-x,t)$.

Chapter 8

First Order Differential Equations

Exercise Set 8.1

1. $y' = 4x^3$

$$\begin{aligned} y &= \int y'\,dx \\ &= \int 4x^3\,dx \\ &= x^4 + C \end{aligned}$$

2. $y' = \sqrt{2x-3}$

$$\begin{aligned} y &= \int y'\,dx \\ &= \int (2x-3)^{1/2}\,dx \\ &= \frac{1}{3}(2x-3)^{3/2} + C \end{aligned}$$

3. $y' = \frac{3}{x} - x^2 + x^5$

$$\begin{aligned} y &= \int y'\,dx \\ &= \int \frac{3}{x} - x^2 + x^5\,dx \\ &= 3\,\ln x - \frac{1}{3}x^3 + \frac{1}{6}x^6 + C \end{aligned}$$

4. $y' = 5\,sin\ x - 4$

$$\begin{aligned} y &= \int y'\,dx \\ &= \int 5\,sin\ x - 4\,dx \\ &= -5\,cos\ x - 4x + C \end{aligned}$$

5. $y' = 4e^{3x} + \sqrt{x}$

$$\begin{aligned} y &= \int y'\,dx \\ &= \int 4e^{3x} + x^{1/2}\,dx \\ &= \frac{4}{3}e^{3x} + \frac{2}{3}x^{3/2} + C \end{aligned}$$

6. $y' = \frac{2}{\sqrt{x}} + sec^2 x$

$$\begin{aligned} y &= \int y'\,dx \\ &= \int 2x^{-1/2} + sec^2 x\,dx \\ &= 4x^{1/2} + tan\ x + C \end{aligned}$$

7. $y' = x^2\sqrt{3x^3 - 5}$

$$\begin{aligned} y &= \int y'\,dx \\ &= \int x^2\sqrt{3x^3-5}\,dx \\ &= \frac{2}{27}(3x^3-5)^{3/2} + C \end{aligned}$$

8. $y' = \frac{\ln x}{x}$

$$\begin{aligned} y &= \int y'\,dx \\ &= \int \frac{\ln x}{x}\,dx \\ &= \frac{1}{2}(\ln x)^2 + C \end{aligned}$$

9. $y' = \frac{sin\ 2x}{(4 + cos\ 2x)^3}$

$$\begin{aligned} y &= \int y'\,dx \\ &= \int sin\ 2x(4 + cos\ 2x)^{-3}\,dx \\ &= \frac{1}{4}(4 + cos\ 2x)^{-2} + C \end{aligned}$$

10. $y' = x\ cos\ x$

$$\begin{aligned} y &= \int y'\,dx \\ &= \int x\ cos\ x\,dx \\ &= cos\ x + x\ sin\ x + C \end{aligned}$$

11. $y' = \frac{1}{1-x^2}$

$$\begin{aligned} y &= \int y'\,dx \\ &= \int \frac{1}{1-x^2}\,dx \\ &= \int \frac{1}{2(1+x)} + \frac{1}{2(1-x)}\,dx \\ &= \frac{\ln|1+x|}{2} - \frac{\ln|1-x|}{2} + C \\ &= \frac{1}{2}\ln\left|\frac{1+x}{1-x}\right| + C \end{aligned}$$

12. $y' = \frac{x}{(3x+2)^2}$

$$y = \int y'\,dx$$

$$= \int \frac{x}{3x+2)^2}\,dx$$
$$= \frac{1}{9}\left[\frac{2x}{(3x+2)} + \ln|3x+2|\right] + C$$

13. $y' = x^2 + 2x - 3$

$$\begin{aligned} y &= \int y'\,dx \\ &= \int x^2 + 2x - 3\,dx \\ &= \frac{1}{3}x^3 + x^2 - 3x + C \\ 4 &= 0 + 0 - 0 + C \\ 4 &= C \\ y &= \frac{1}{3}x^3 + x^2 - 3x + 4 \end{aligned}$$

14. $y' = 2x$

$$\begin{aligned} y &= \int y'\,dx \\ &= \int 2x\,dx \\ &= x^2 + C \\ 7 &= 1 + C \\ 6 &= C \\ y &= x^2 = 6 \end{aligned}$$

15. $y' = e^{3x} + 1$

$$\begin{aligned} y &= \int y'\,dx \\ &= \int e^{3x} + 1\,dx \\ &= \frac{1}{3}\,e^{3x} + x + C \\ 2 &= \frac{1}{3} + 0 + C \\ \frac{5}{3} &= C \\ y &= \frac{1}{3}\,e^{3x} + x + \frac{5}{3} \end{aligned}$$

16. $y' = sin\ x - x$

$$\begin{aligned} y &= \int y'\,dx \\ &= \int sin\ x - x\,dx \\ &= -cos\ x - \frac{1}{2}\,x^2 + C \\ 3 &= -1 + 0 + C \\ 4 &= C \\ y &= -cos\ x - \frac{1}{2}\,x^2 + 4 \end{aligned}$$

17. $f'(x)' = x^{2/3} - x$

$$\begin{aligned} f(x) &= \int f'(x)\,dx \\ &= \int x^{2/3} - x\,dx \\ &= \frac{3}{5}x^{5/3} - \frac{1}{2}\,x^2 + C \\ -6 &= \frac{3}{5} - \frac{1}{2} + C \\ -\frac{61}{10} &= C \\ f(x) &= \frac{3}{5}\,x^{5/3} - \frac{1}{2}\,x^2 - \frac{61}{10} \end{aligned}$$

18. $f'(x) = 1/x - 2x + x^{1/2}$

$$\begin{aligned} f(x) &= \int f'(x)\,dx \\ &= \int 1/x - 2x + x^{1/2}\,dx \\ &= \ln x - x^2 + \frac{2}{3}\,x^{3/2} + C \\ 2 &= \ln 4 - 16 + \frac{16}{3} + C \\ \frac{38}{3} - \ln 4 &= C \\ f(x) &= \ln x - x^2 + \frac{2}{3}\,x^{3/2} + \frac{38}{3} - \ln 4 \end{aligned}$$

19. $y' = x\sqrt{x^2+1}$

$$\begin{aligned} y &= \int y'\,dx \\ &= \int x\sqrt{x^2+1}\,dx \\ &= \frac{1}{3}\,(x^2+1)^{3/2} + C \\ 3 &= \frac{1}{3} + C \\ \frac{8}{3} &= C \\ y &= \frac{1}{3}\,(x^2+1)^{3/2} + \frac{8}{3} \end{aligned}$$

20. $y' = xe^{x^2}$

$$\begin{aligned} y &= \int y'\,dx \\ &= \int xe^{x^2}\,dx \\ &= \frac{1}{2}\,e^{x^2} + C \\ 1 &= \frac{1}{2} + C \\ \frac{1}{2} &= C \\ y &= \frac{1}{2}\,e^{x^2} + \frac{1}{2} \end{aligned}$$

21. $y' = x^3/(x^4+1)^2$

$$\begin{aligned} y &= \int y'\,dx \\ &= \int x^3(x^4+1)^{-2}\,dx \\ &= -\frac{1}{4}(x^4+1)^{-1} \\ -2 &= \frac{-1}{8} + C \\ -\frac{16}{8} &= C \\ y &= -\frac{1}{4(x^4+1)} - \frac{15}{8} \end{aligned}$$

22. $y' = \sin x/(2+\cos x)$

$$\begin{aligned} y &= \int y'\,dx \\ &= \int \sin x/(2+\cos x)\,dx \\ &= \ln|\,2+\cos x\,| + C \\ 3 &= \ln 1 + C \\ 3 &= C \\ y &= \ln|\,2+\cos x\,| + 3 \end{aligned}$$

23. $y' = xe^x$

$$\begin{aligned} y &= \int y'\,dx \\ &= \int xe^x\,dx \\ &= xe^x - e^x + C \\ 2 &= -1 + C \\ 3 &= C \\ y &= xe^x - e^x + 3 \end{aligned}$$

24. $y' = x\sin x$

$$\begin{aligned} y &= \int y'\,dx \\ &= \int x\sin x\,dx \\ &= \sin x - x\cos x + C \\ 3 &= 0 - 0 + C \\ 3 &= C \\ y &= \sin x - x\cos x + 3 \end{aligned}$$

25. $y' = \ln x$

$$\begin{aligned} y &= \int y'\,dx \\ &= \int \ln x\,dx \\ &= x\ln x - x + C \\ 2 &= 0 - 1 + C \\ 3 &= C \\ y &= x\ln x - x + 3 \end{aligned}$$

26. $y' = x\ln x$

$$\begin{aligned} y &= \int y'\,dx \\ &= \int x\ln x\,dx \\ &= \frac{1}{2}x^2\ln x - \frac{1}{4}x^2 + C \\ 0 &= 0 - \frac{1}{4} + C \\ \frac{1}{4} &= C \\ y &= \frac{1}{2}x^2\ln x - \frac{1}{4}x^2 + \frac{1}{4} \end{aligned}$$

27. $y' = x\sin(x^2)$

$$\begin{aligned} y &= \int y'\,dx \\ &= \int x\sin(x^2)\,dx \\ &= -\frac{1}{2}\cos(x^2) + C \\ 3 &= -\frac{1}{2} + C \\ -\frac{5}{2} &= C \\ y &= -\frac{1}{2}\cos(x^2) - \frac{5}{2} \end{aligned}$$

28. $y' = e^x\sqrt{e^x+3}$

$$\begin{aligned} y &= \int y'\,dx \\ &= \int e^x\sqrt{e^x+3}\,dx \\ &= \frac{2}{3}(e^x+3)^{3/2} + C \\ 1 &= \frac{16}{3} + C \\ -\frac{13}{3} &= C \\ y &= \frac{2}{3}(e^x+3)^{3/2} - \frac{13}{3} \end{aligned}$$

29. $f''(x) = 2$

$$\begin{aligned} f'(x) &= \int f''(x)\,dx \\ &= \int 2\,dx \\ &= 2x + C \\ 4 &= 0 + C \\ 4 &= C \\ f'(x) &= 2x + 4 \\ f(x) &= \int f'(x)\,dx \\ &= \int 2x + 4\,dx \\ &= x^2 + 4x + K \end{aligned}$$

$$\begin{aligned} 3 &= 0+0+K \\ 3 &= K \\ f(x) &= x^2+4x+3 \end{aligned}$$

30. $f''(x) = 12x$

$$\begin{aligned} f'(x) &= \int f''(x)\,dx \\ &= \int 12x\,dx \\ &= 6x^2+C \\ -2 &= 6+C \\ -8 &= C \\ f'(x) &= 6x^2-8 \\ f(x) &= \int f'(x)\,dx \\ &= \int 6x^2-8\,dx \\ &= 2x^3-8x+K \\ 2 &= 2-8+K \\ 8 &= K \\ f(x) &= 2x^3-8x+8 \end{aligned}$$

31. $f''(x) = x + 1/x^3$

$$\begin{aligned} f'(x) &= \int f''(x)\,dx \\ &= \int x+x^{-3}\,dx \\ &= \frac{1}{2}\,x^2-\frac{1}{2}\,x^{-2}+C \\ 0 &= 2-\frac{1}{8}+C \\ -\frac{15}{8} &= C \\ f'(x) &= \frac{1}{2}\,x^2-\frac{1}{2}\,x^{-2}-\frac{15}{8} \\ f(x) &= \int f'(x)\,dx \\ &= \int \frac{1}{2}\,x^2-\frac{1}{2}\,x^{-2}-\frac{15}{8}\,dx \\ &= \frac{1}{6}\,x^3+\frac{1}{2x}-\frac{15}{8}\,x+K \\ 1 &= \frac{8}{6}+\frac{1}{4}-\frac{15}{4}+K \\ \frac{19}{6} &= K \\ f(x) &= \frac{1}{6}\,x^3+\frac{1}{2x}-\frac{15}{8}\,x+\frac{19}{6} \end{aligned}$$

32. $f''(x) = e^{2x}$

$$\begin{aligned} f'(x) &= \int f''(x)\,dx \\ &= \int e^{2x}\,dx \\ &= \frac{1}{2}\,e^{2x}+C \\ 2 &= \frac{1}{2}+C \\ 4 &= C \\ f'(x) &= \frac{1}{2}\,e^{2x}+4 \\ f(x) &= \int f'(x)\,dx \\ &= \int \frac{1}{2}\,e^{2x}+4\,dx \\ &= \frac{1}{4}\,e^{2x}+4x+K \\ 4 &= \frac{1}{4}+K \\ \frac{15}{4} &= K \\ f(x) &= \frac{1}{4}\,e^{2x}+4x+\frac{15}{4} \end{aligned}$$

33. $f''(x) = sin\ 3x$

$$\begin{aligned} f'(x) &= \int f''(x)\,dx \\ &= \int sin\ 3x\,dx \\ &= -\frac{1}{3}\,cos\ 3x+C \\ -3 &= \frac{1}{3}+C \\ -\frac{10}{3} &= C \\ f'(x) &= -\frac{1}{3}\,cos\ 3x-\frac{10}{3} \\ f(x) &= \int f'(x)\,dx \\ &= \int -\frac{1}{3}\,cos\ 3x-\frac{10}{3}\,dx \\ &= -\frac{1}{9}\,sin\ 3x-\frac{10}{3}\,x+K \\ -2 &= 0-\frac{10\pi}{3}+K \\ -2+\frac{10\pi}{3} &= K \\ f(x) &= -\frac{1}{9}\,sin\ 3x-\frac{10}{3}\,x-2+\frac{10\pi}{3} \end{aligned}$$

34. $f'''(x) = x$

$$\begin{aligned} f''(x) &= \int f'''(x)\,dx \\ &= \int x\,dx \\ &= \frac{1}{2}\,x^2+C \\ \frac{1}{2} &= 0+C \\ \frac{1}{2} &= C \\ f''(x) &= \frac{1}{2}\,x^2+\frac{1}{2} \\ f'(x) &= \int f''(x)\,dx \end{aligned}$$

$$
\begin{aligned}
&= \int \frac{1}{2}\,x^2 + \frac{1}{2}\,dx \\
&= \frac{1}{6}\,x^3 + \frac{1}{2}\,x + K \\
-3 &= 0 + 0 + K \\
-3 &= K \\
f'(x) &= \frac{1}{6}\,x^3 + \frac{1}{2}\,x - 3 \\
f(x) &= \int f'(x)\,dx \\
&= \int \frac{1}{6}\,x^3 + \frac{1}{2}\,x - 3 \\
&= \frac{1}{24}\,x^4 + \frac{1}{4}\,x^2 - 3x + G \\
2 &= 0 + 0 - 0 + G \\
2 &= G \\
f(x) &= \frac{1}{24}\,x^4 + \frac{1}{4}\,x^2 - 3x + 2
\end{aligned}
$$

35. **a)** Slope at $(-2, 1)$ is $= (-2)^2 - 1 = 3$

b)

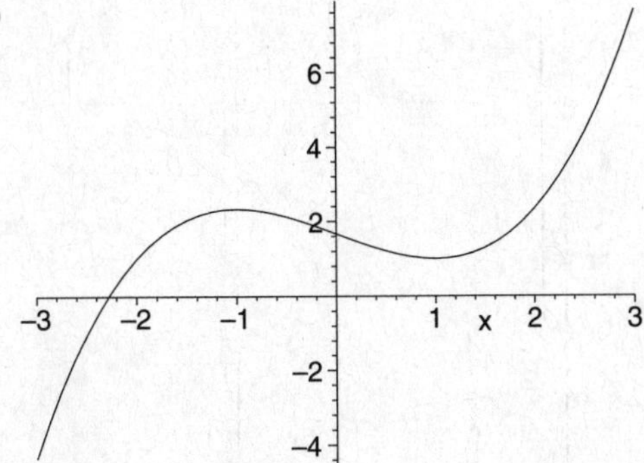

36. **a)** Slope at $(-2, 1)$ is $= \frac{1}{2}$

b)

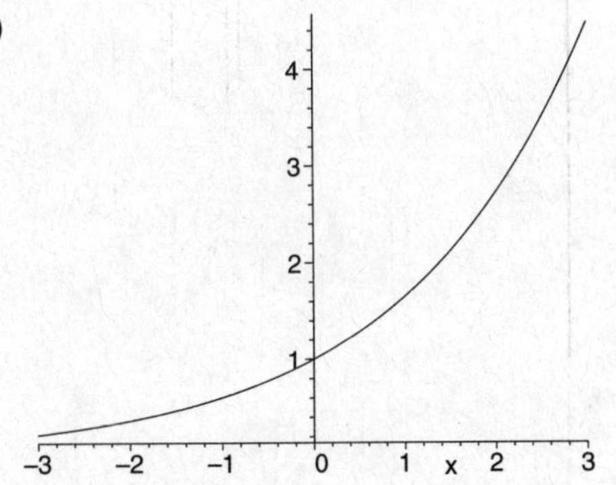

37. **a)** Slope at $(-2, 1)$ is $= \frac{-4}{3} + 1 = \frac{-1}{3}$

b)

38. **a)** Slope at $(-2, 1)$ is $= -2(1) = -2$

b)

39. $y = x\,\ln x + 3x - 2$
$y' = 1 + \ln x + 3 = \ln x + 4$
$y'' = \frac{1}{x}$
$y'' - 1/x = 1/x - 1/x = 0$
Thus, y is a solution to the differential equation

40. $y = x\,\ln x - 5x + 7$
$y' = 1 + \ln x - 5 = \ln x - 4$
$y'' = \frac{1}{x}$
$y'' - 1/x = 1/x - 1/x = 0$
Thus, y is a solution to the differential equation

41. $y = e^x + 3xe^x$
$y' = e^x + 3xe^x + 3e^x = 3xe^x + 4e^x$
$y'' = 3xe^x + 3e^x + 4e^x = 3xe^x + 7e^x$
$y'' - 2y' + y = 3xe^x + 7e^x - 6xe^x - 8e^x + 3xe^x + e^x = 0$
Thus, y is a solution to the differential equation

42. $y = -2e^x + xe^x$
$y' = -2e^x + xe^x + e^x = xe^x - e^x$
$y'' = xe^x + e^x - e^x = xe^x$
$y'' - 2y' + y = xe^x - 2xe^x + 2e^x - 2e^x + xe^x = 0$
Thus, y is a solution to the differential equation

43. $y = 2e^x - 7e^{3x}$
$y' = 2e^x - 21e^{3x}$
$y'' = 2e^x - 61e^{3x}$
$y'' - 4y' + 3y = 2e^x - 61e^{3x} - 8e^x + 84e^{3x} + 6e^x - 21e^{3x} = 0$
Thus, y is a solution to the differential equation

44. $y = xe^{2x}$
$y' = 2xe^{2x} + e^{2x}$

$y' - 2y = 2xe^{2x} + e^{2x} - 2xe^{2x} = e^{2x}$
Thus, y is a solution to the differential equation

45. $y = e^x \sin 2x$
$y' = 2e^x \cos 2x + e^x \sin 2x$
$y'' = -4e^x \sin 2x + 2e^x \cos 2x + 2e^x \cos 2x + e^x \sin 2x$
$y'' = 4e^x \cos 2x - 3e^x \sin 2x$
$y'' - 2y' + 5y$

$$\begin{aligned} &= 4e^x \cos 2x - 3e^x \sin 2x - 4e^x \cos 2x - \\ &\quad 2e^x \sin 2x + 5e^x \sin 2x \\ &= 0 \end{aligned}$$

Thus, y is a solution to the differential equation

46. $y = (1 - x) \cos 2x$
$y' = 2(x - 1) \sin 2x - \cos 2x$
$y'' = 4(x - 1) \cos 2x + 2 \sin 2x + 2 \sin 2x$
$y'' = 4x \cos 2x - 4 \cos 2x + 4 \sin 2x$
$y'' + 4y$

$$\begin{aligned} &= 4x \cos 2x - 4 \cos 2x + 4 \sin 2x + \\ &\quad 4(1 - x) \cos 2x \\ &= 4x \cos 2x - 4 \cos 2x + 4 \sin 2x + \\ &\quad 4 \cos 2x - 4x \cos 2x \\ &= 4 \sin 2x \end{aligned}$$

Thus, y is a solution to the differential equation

47.

$$\begin{aligned} \frac{dR}{dS} &= \frac{k}{S} \\ dR &= \frac{k}{S}\, dS \\ R &= \int \frac{k}{S}\, dS \\ &= k \ln \mid S \mid + C \\ 0 &= k \ln \mid S_0 \mid + C \\ -k \ln \mid S_0 \mid &= C \\ R(S) &= k \ln \mid S \mid - k \ln \mid S_0 \mid \end{aligned}$$

Thus, $R(S) = k \ln \left| \frac{S}{S_0} \right|$

48.

$$\begin{aligned} T''(x) &= -\frac{1}{D} \\ T'(x) &= \int T''(x)\, dx \\ &= -\frac{1}{D}\, x + C \\ 0 &= -\frac{L}{D} + C \\ \frac{L}{D} &= C \\ T'(x) &= -\frac{1}{D}\, x + \frac{L}{D} \\ T(x) &= \int T'(x)\, dx \\ &= -\frac{1}{2D}\, x^2 + \frac{L}{D}\, x + K \\ 0 &= 0 + 0 + K \\ 0 &= K \end{aligned}$$

Thus, $T(x) = -\frac{1}{2D}\, x^2 + \frac{L}{D}\, x$

49. **a)** $y = 0$, $y' = 0$, $0 = \sqrt{0}$
Thus, $y = 0$ is a solution of the initial value problem

b) $y = x^2/4$, $y' = x/2$, $x/2 = \sqrt{x^2/2}$
Thus, $y = x^2/4$ is a solution of the initial value problem

c)

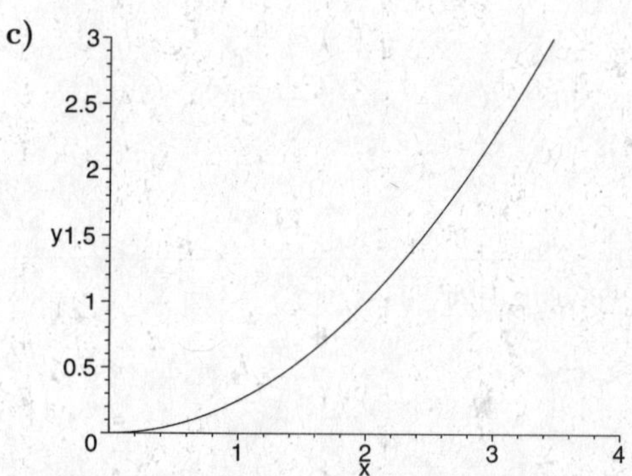

d) $x^2/4 = 0 \to x = 0 \to y = 0$
The initial value problem does not have a unique solution

e) $\partial f / \partial y = \frac{1}{2\sqrt{y}}$

f) The function does not satisfy the continuity criteria near the origin since the function is does not have "real" values to the left of the origin

50. The average time

$$\begin{aligned} &= \frac{1}{L} \int_0^L -\frac{1}{2D}\, x^2 + \frac{L}{D}\, x\, dx \\ &= \frac{1}{L} \left[-\frac{1}{6D}\, x^3 + \frac{L}{2D}\, x^2 \right]_0^L \\ &= \frac{1}{L} \left[-\frac{L^3}{6D} + \frac{L^3}{2D} \right] \\ &= \frac{L^2}{3D} \end{aligned}$$

51. **a)** $x^2 - 1 = 0 \to x = -1$ and$x = 1$

b) The slope of the tangent line at the point where $y' = 0$ is 0

c) Left to the student

52. **a)** $y/2 = 0 \to y = 0$

b) The slope of the tangent line at the point where $y' = 0$ is 0

c) Left to the student

53. **a)** $2x/3 + y = 0 \to y = -2x/3$

b) The slope of the tangent line at the point where $y' = 0$ is 0

c) Left to the student

54. a) $0 = xy \to x = 0$ or $y = 0$

b) The slope of the tangent line at the point where $y' = 0$ is 0

c) Left to the student

Exercise Set 8.2

1. Both $-x^2$ and x^3 are continuous for all real numbers. Therefore the unique solution will exist on all real numbers

2. Discontinuity at $x = 2$. Therefore the unique solution will exist on $(-\infty, 2)$

3. Discontinuities at $x = \pm\frac{\pi}{2}$. Therefore the unique solution will exist on $(-\pi/2, \pi/2)$

4. Discontinuities at $x = 2$ and $x = -1$. Therefore the unique solution will exist on $(-1, 2)$

5. Discontinuities at $x = 2$ and $x = -1$. Therefore the unique solution will exist on $(2, \infty)$

6. Discontinuities at $t = 0$, $t = -1$, and $t = 6$. Therefore the unique solution will exist on $(0, 6)$

7.

$$\begin{aligned}
\int -3\,dx &= -3x + C \\
G(x) &= e^{-3x} \\
e^{-3x}y &= \int 0\,dx \\
e^{-3x}y &= C \\
y &= C\,e^{3x}
\end{aligned}$$

8.

$$\begin{aligned}
\int 4\,dx &= 4x + C \\
G(x) &= e^{4x} \\
e^{4x}y &= \int 2e^{4x}\,dx \\
e^{-3x}y &= \frac{1}{2}\,e^{4x} + C \\
y &= \frac{1}{2}\,e^{7x} + C\,e^{3x}
\end{aligned}$$

9.

$$\begin{aligned}
\int \cos 2x\,dx &= \frac{1}{2}\sin 2x + C \\
G(x) &= e^{1/2\sin 2x} \\
e^{1/2\sin 2x}y &= \int \cos 2x\,e^{1/2\sin 2x}\,dx \\
e^{1/2\sin 2x}y &= e^{1/2\sin 2x} + C \\
y &= C\,e^{-1/2\sin 2x} + 1
\end{aligned}$$

10. $y' + 3y = 6x^2 - 4$

$$\begin{aligned}
\int 3\,dx &= 3x + C \\
G(x) &= e^{3x} \\
e^{3x}y &= \int e^{3x}(6x^2 - 4)\,dx \\
e^{3x}y &= 2x^2e^{3x} - \frac{4}{3}\,xe^{3x} - \frac{8}{9}\,e^{3x} + C \\
y &= 2x^2 - \frac{4}{3}\,x - \frac{8}{9} + Ce^{-3x}
\end{aligned}$$

11.

$$\begin{aligned}
\int -2t\,dt &= -t^2 + C \\
G(t) &= e^{-t^2} \\
e^{-t^2}y &= \int e^{-t^2}(2t)\,dt \\
e^{-t^2}y &= -e^{-t^2} + C \\
y &= Ce^{t^2} - 1
\end{aligned}$$

12. $y' + \dfrac{2t}{t^2+1} = \dfrac{t^3+1}{t^2+1}$

$$\begin{aligned}
\int \frac{2t}{t^2+1}\,dt &= \ln(t^2+1) + C \\
G(t) &= t^2 + 1 \\
(t^2+1)y &= \int (t^3+1)\,dt \\
(t^2+1)y &= \frac{1}{4}\,t^4 + t + C \\
y &= \frac{t^4+t}{4(t^2+1)} + \frac{C}{t^2+1}
\end{aligned}$$

13.

$$\begin{aligned}
\int -\,dt &= -t + C \\
G(t) &= e^{-t} \\
e^{-t}y &= \int e^{-t}e^t\,dt \\
e^{-t}y &= t + C \\
y &= t\,e^t + Ce^t
\end{aligned}$$

14.

$$\begin{aligned}
\int 4t^3\,dt &= t^4 + C \\
G(t) &= e^{t^4} \\
e^{t^4}y &= \int e^{t^4}4t^4e^{-4t}\,dt \\
e^{t^4}y &= \frac{4}{5}\,t^4 + C \\
y &= \frac{4}{5}\,t^5e^{-t^4} + Ce^{-t^4}
\end{aligned}$$

15. $y' - \frac{4}{x}\ y = x^5 e^{x^2} + 3x^3 - 6x^{-2}$

$$\begin{aligned} \int \frac{4}{x}\, dx &= -4\ \ln|\,x\,| + C = \ln x^{-4} + C \\ G(x) &= x^{-4} \\ x^{-4}y &= \int xe^{x^2} + \frac{3}{x} - 6x^{-6}\, dx \\ x^{-4}y &= \frac{1}{2}\ e^{x^2} + 3\ \ln|\,x\,| + \frac{6}{5}\ x^{-5} + C \\ y &= \frac{1}{2}\ x^4 e^{x^2} 3x^4\ \ln|\,x\,| + \frac{6}{5x} + cx^4 \end{aligned}$$

16. $y' + \frac{1}{x}\ y = sin\ x$

$$\begin{aligned} \int \frac{1}{x}\, dx &= \ln|\,x\,| + C \\ G(x) &= x \\ xy &= \int x\ sin\ x\, dx \\ xy &= sin\ x - x\ cos\ x + C \\ y &= \frac{sin\ x}{x} - cos\ x \end{aligned}$$

17. $y' + \frac{3}{x}\ y = \ln x$

$$\begin{aligned} \int \frac{3}{x}\, dx &= 3\ \ln|\,x\,| + C = \ln|\,x^3\,| + C \\ G(x) &= x^3 \\ x^3 y &= \int x^3\ \ln x\, dx \\ x^3 y &= \frac{1}{4}\ x^4\ \ln x - \frac{1}{16}\ x^4 + C \\ y &= \frac{1}{4}\ x \ln x - \frac{1}{16}\ x + Cx^{-3} \end{aligned}$$

18.

$$\begin{aligned} \int dx &= x + C \\ G(x) &= e^x \\ e^x y &= \int xe^x\, dx \\ e^x y &= xe^x - e^x + C \\ y &= x - 1 + Ce^{-x} \end{aligned}$$

19.

$$\begin{aligned} \int dt &= t + C \\ G(t) &= e^t \\ e^t y &= \int \ln t\, dt \\ e^t y &= t\ \ln t - t + C \\ y &= (t\ \ln t - t + C)e^{-t} \end{aligned}$$

20. $y' - \frac{sin\ 2t}{cos\ 2t} y = 3\ tan\ 2x + 1$

$$\begin{aligned} \int \frac{-sin\ 2t}{cos\ 2t}\, dt &= \ln cos\ 2t + C \\ G(t) &= cos\ 2t \\ cos\ 2t\ y &= \int 3\ sin\ 2t + cos\ 2t\ dt \\ cos\ 2t\ y &= -\frac{3}{2}\ cos\ 2t + \frac{1}{2}\ sin\ 2t + C \\ y &= \frac{1}{2}\ tan\ 2t - \frac{3}{2} + C\ sec\ 2t \end{aligned}$$

21.

$$\begin{aligned} \int 4\, dx &= 4x + C \\ G(x) &= e^{4x} \\ e^{4x} y &= \int 6e^{4x}\, dx \\ e^{4x} y &= \frac{3}{2}\ e^{4x} + C \\ y &= \frac{3}{2} + Ce^{-4x} \\ 2 &= \frac{3}{2} + C \\ \frac{1}{2} &= C \\ y &= \frac{3}{2} + \frac{1}{2}\ e^{-4x} \end{aligned}$$

The solution exists for all real numbers

22. $y' + \frac{1}{x}\ y = 0$

$$\begin{aligned} \int \frac{1}{x}\, dx &= \ln|\,x\,| + C \\ G(x) &= x \\ xy &= \int 0\, dx \\ xy &= C \\ y &= \frac{C}{x} \\ 3 &= C \\ y &= \frac{3}{x} \end{aligned}$$

The solutions exists on $(0, \infty)$

23. $y' + \frac{cos\ x}{1 + sin\ x}\ y = cos\ x$

$$\begin{aligned} \int \frac{cos\ x}{1 + sin\ x}\, dx &= \ln|\,1 + sin\ x\,| + C \\ G(x) &= 1 + sin\ x \\ (1 + sin\ x)y &= \int (1 + sin\ x) cos\ x\, dx \\ (1 + sin\ x)y &= \frac{1}{2}(1 + sin\ x)^2 + C \\ y &= \frac{1}{2}\ (1 + sin\ x) + \frac{C}{1 + sin\ x} \\ -1 &= 1 + \frac{C}{2} \end{aligned}$$

$$\begin{aligned} -4 &= C \\ y &= \frac{1}{2}(1+\sin x) - \frac{4}{1+\sin x} \\ &= \frac{2\sin^2 x + 2\sin x - 7}{2(1+\sin x)} \end{aligned}$$

The solution exists on $(-\pi/2, 3\pi/2)$

24.

$$\begin{aligned} \int \sin\, dx &= -\cos x + C \\ G(x) &= e^{-\cos x} \\ e^{-\cos x} y &= \int e^{-\cos x} \sin x\, dx \\ e^{-\cos x} y &= e^{-\cos x} + C \\ y &= 1 + C\, e^{-\cos x} \\ 4 &= 1 + C \\ 3 &= C \\ y &= 1 + C\, e^{-\cos x} \end{aligned}$$

The solution exists on all real numbers

25. $y' + \frac{1}{t}\, y = t^2$

$$\begin{aligned} \int \frac{1}{t}\, dt &= \ln|\,t\,| + C \\ G(t) &= t \\ ty &= \int t^3\, dt \\ ty &= \frac{1}{4}\, t^4 + C \\ y &= \frac{1}{4}\, t^3 + \frac{C}{t} \\ 5 &= 2 + \frac{C}{2} \\ 6 &= C \\ y &= \frac{1}{4}\, t^3 - \frac{6}{t} \end{aligned}$$

The solution exists on $(0, \infty)$

26. $y' + \frac{3t^2}{(t^3+1)}\, y = 1$

$$\begin{aligned} \int \frac{3t^2}{t^3+1}\, dt &= \ln|\,t^3+1\,| + C \\ G(t) &= t^3 + 1 \\ (t^3+1)y &= \int (t^3+1)\, dt \\ (t^3+1)y &= \frac{1}{4}\, t^4 + t + C \\ y &= \frac{t^4}{4(t^3+1)} + \frac{t}{(t^3+1)} + \frac{C}{t^3+1} \\ 2 &= C \\ y &= \frac{t^4}{4(t^3+1)} + \frac{t}{(t^3+1)} + \frac{2}{t^3+1} \end{aligned}$$

The solution exists on $(-1, \infty)$

27.

$$\begin{aligned} \int -5\, dx &= -5x + C \\ G(x) &= e^{-5x} \\ e^{-5x} y &= \int x e^{-5x} + e^{-5x}\, dt \\ e^{-5x} y &= \frac{-1}{5}\, x e^{-5x} - \frac{6}{25}\, e^{-5x} + C \\ y &= \frac{-1}{5}\, x - \frac{6}{25} + C\, e^{5x} \\ 1 &= \frac{6}{25} + C \\ \frac{31}{25} &= C \\ y &= \frac{-1}{5}\, x - \frac{6}{25} + \frac{31}{25}\, e^{5x} \end{aligned}$$

The solution exists for all real numbers

28.

$$\begin{aligned} \int e^{2x}\, dx &= \frac{1}{2}\, e^{2x} + C \\ G(x) &= e^{e^{2x}/2} \\ e^{e^{2x}/2} y &= \int e^{e^{2x}/2} e^{2x}\, dt \\ e^{e^{2x}/2} y &= e^{e^{2x}} + C \\ y &= 2 + C\, e^{-e^{2x}/2} \\ 0 &= 2 + C \\ -2 &= C \\ y &= 2 - 2\, e^{e^{2x}/2} \end{aligned}$$

The solution exists for all real numbers

29. $y' + \frac{e^x}{(e^x-2)}\, y = \frac{2e^{-2x} - e^{-3x}}{e^x - 2}$

$$\begin{aligned} \int \frac{e^x}{e^x-2}\, dx &= \ln|\,e^x - 2\,| + C \\ G(x) &= e^x - 2 \\ (e^x-2)y &= \int 2e^{-2x} - e^{-3x}\, dx \\ (e^x-2)y &= -e^{-2x} + \frac{1}{3}\, e^{-3x} + C \\ y &= \frac{e^{-3x}}{3(e^x-2)} - \frac{e^{-2x}}{e^x-2} + \frac{C}{e^x-2} \\ 3 &= \frac{-1}{3} + 1 - C \\ -\frac{7}{3} &= C \\ y &= \frac{e^{-3x}}{3(e^x-2)} - \frac{e^{-2x}}{e^x-2} - \frac{7}{3(e^x-2)} \\ &= \frac{e^{-3x} - 3e^{-2x} - 7}{3e^x - 6} \end{aligned}$$

The solution exists on $(-\infty, \ln 2)$

30.

$$\begin{aligned}
\int \cot x\, dx &= -\ln|\sin x| + C = \ln|\csc x| + C \\
G(x) &= -\csc x \\
\csc x\, y &= \int \cot x\, dx \\
\csc x\, y &= -\ln|\sin x| + C \\
y &= \frac{\ln|\csc x|}{\csc x} + C \sin x \\
-1 &= C \\
y &= \frac{\ln|\csc x|}{\csc x} - \sin x
\end{aligned}$$

The solution exists on $[0, \pi]$

31.

$$\begin{aligned}
\int 3t^2\, dt &= t^3 + C \\
G(t) &= e^{t^3} \\
e^{t^3} y &= \int t^2 e^{2t^3}\, dt \\
e^{t^3} y &= \frac{1}{6} e^{2t^3} + C \\
y &= \frac{1}{6} e^{2t^3} + C\, e^{-t^3} \\
3 &= \frac{1}{6} + C \\
\frac{17}{6} &= C \\
y &= \frac{1}{6} e^{2t^3} + \frac{17}{6} e^{-t^3}
\end{aligned}$$

The solution exists for all real numbers

32. $y' + \frac{1}{t^2}\, y = \frac{1}{t^2}$

$$\begin{aligned}
\int t^{-2}\, dt &= -t^{-1} + C \\
G(t) &= e^{-t^{-1}} \\
e^{-t^{-1}} y &= \int \frac{e^{-t^{-1}}}{t^2}\, dx \\
e^{-t^{-1}} y &= e^{t^{-1}} + C \\
y &= 1 + C\, e^{t^{-1}} \\
4 &= 1 + Ce \\
\frac{3}{e} &= C \\
y &= 1 + \frac{3}{e}\, e^{t^{-1}} \\
&= 1 + 3e^{t^{-1}-1}
\end{aligned}$$

The solution exists for all real numbers

33.

$$\begin{aligned}
\int 4t\, dt &= 2t^2 + C \\
G(t) &= e^{2t^2} \\
e^{2t^2} y &= \int t\, e^{2t^2}\, dx \\
e^{2t^2} y &= \frac{1}{4} e^{2t^2} + C \\
y &= \frac{1}{4} + C\, e^{-2t^2} \\
3 &= \frac{1}{4} + C \\
\frac{11}{4} &= C \\
y &= \frac{1}{4} + \frac{11}{4} e^{-2t^2}
\end{aligned}$$

The solution exists for all real numbers

34.

$$\begin{aligned}
\int \tan t\, dt &= -\ln|\cos t| + C = \ln|\sec t| + C \\
G(t) &= \sec t \\
\sec t\, y &= \int dt \\
\sec t\, y &= t + C \\
y &= t \cos t + C \cos t \\
\pi &= \frac{\pi}{2} + \frac{C}{2} \\
\pi &= C \\
y &= t \cos t + \pi \cos t
\end{aligned}$$

The solution exists on $[0, \pi/2])$

35. $P' = -0.2P + 3 \to P' + 0.2P = 3$

$$\begin{aligned}
\int 0.2\, dt &= 0.2t + C \\
G(t) &= e^{0.2t} \\
e^{0.2t} P &= \int 3e^{0.2t}\, dt \\
e^{0.2t} P &= 15e^{0.2t} + C \\
P &= 15 + C\, e^{-0.2t}
\end{aligned}$$

36. $T' = -0.5T + 2 \to T' + 0.5T = 2$

$$\begin{aligned}
\int 0.5\, dt &= 0.5t + C \\
G(t) &= e^{0.5t} \\
e^{0.5t} T &= \int 2e^{0.5t}\, dt \\
e^{0.5t} T &= 4e^{0.5t} + C \\
T &= 4 + C\, e^{-0.5t}
\end{aligned}$$

37. $Q' = -0.1Q - 5 \to Q' + 0.1Q = -5$

$$\begin{aligned}
\int 0.1\, dt &= 0.1t + C \\
G(t) &= e^{0.1t} \\
e^{0.1t} Q &= \int -5e^{0.1t}\, dt \\
e^{0.1t} Q &= -50e^{0.1t} + C \\
Q &= -50 + C\, e^{-0.1t}
\end{aligned}$$

38. $y' = (0.1 - 0.3)y + 3 \to y' + 0.2y = 3$

$$\begin{aligned} \int 0.2\,dt &= 0.2t + C \\ G(t) &= e^{0.2t} \\ e^{0.2t}y &= \int 3e^{0.2t}\,dt \\ e^{0.2t}y &= 15e^{0.2t} + C \\ y &= 15 + C\,e^{-0.2t} \end{aligned}$$

39. $P' = kP \to P' - kP = 0$

$$\begin{aligned} \int -k\,dt &= -kt + C \\ G(t) &= e^{-kt} \\ e^{-kt}P &= \int 0\,dt \\ e^{-kt}P &= C \\ P &= C\,e^{kt} \\ P_0 &= C \\ P &= P_0e^{kt} \end{aligned}$$

40. $S' = -0.006\,S + 1.5 \to S' + 0.006S = 1.5$

$$\begin{aligned} \int 0.006\,dt &= 0.006t + C \\ G(t) &= e^{0.006t} \\ e^{0.006t}S &= \int 1.5e^{0.006t}\,dt \\ e^{0.006t}S &= 250e^{0.006t} + C \\ S &= 250 + C\,e^{-0.006t} \\ 180 &= 250 + C \\ -70 &= C \\ S &= 250 - 70\,e^{-0.006t} \end{aligned}$$

After one hour, $t = 60$

$$\begin{aligned} S(60) &= 250 - 70e^{-0.006(60)} \\ &= 201.16 \text{ lb} \end{aligned}$$

41. $S' = -0.005S + 4 \to S' + 0.005S = 4$

$$\begin{aligned} \int 0.005\,dt &= 0.005t + C \\ G(t) &= e^{0.005t} \\ e^{0.005t}S &= \int 4e^{0.005t}\,dt \\ e^{0.005t}S &= 800e^{0.005t} + C \\ S &= 800 + C\,e^{-0.005t} \\ 100 &= 800 + C \\ -700 &= C \\ S &= 800 - 700\,e^{-0.005t} \end{aligned}$$

After two hours, $t = 120$

$$\begin{aligned} S(120) &= 800 - 700e^{-0.005(120)} \\ &= 415.83 \text{ lbs} \end{aligned}$$

42. **a)** $Y' + kY = 60k$

b)

$$\begin{aligned} \int k\,dx &= kx + C \\ G(x) &= e^{kx} \\ e^{kx}Y &= \int 60k\,e^{kx}\,dx \\ e^{kx}Y &= 60e^{kx} + C \\ Y &= 60 + C\,e^{-kx} \\ 15 &= 60 + C \\ -45 &= C \\ Y &= 60 - 45e^{kx} \end{aligned}$$

c)

$$\begin{aligned} 21 &= 60 - 45e^{10k} \\ \frac{39}{45} &= e^{10k} \\ \ln\frac{39}{45} &= 10k \\ k &= \frac{1}{10}\ln\frac{39}{45} \\ &= -0.01431 \end{aligned}$$

d)

$$\begin{aligned} Y &= 60 - 45e^{-0.01431(5)} \\ &= 18.11 \end{aligned}$$

e)

$$\begin{aligned} 42 &= 60 - 45e^{-0.01431x} \\ \frac{18}{45} &= e^{-0.01431x} \\ \ln\frac{18}{45} &= -0.01431x \\ x &= \frac{1}{-0.01431}\ln\frac{18}{45} \\ &= 64.03 \text{ pounds per acre} \end{aligned}$$

43. $T' + kT = kC$

a)

$$\begin{aligned} \int k\,dt &= kt + C \\ G(t) &= e^{kt} \\ e^{kt}T &= \int kCe^{kt}\,dt \\ e^{kt}T &= C\,e^{kt} + K \\ T &= C + Ke^{-kt} \\ T_0 &= C + K \\ K &= T_0 - C \\ T &= C + (T_0 - C)e^{-kt} \end{aligned}$$

b)

$$\begin{aligned}
117 &= 70+(143-70)e^{-30k}\\
\frac{47}{73} &= e^{-30k}\\
\ln\frac{47}{73} &= -30k\\
k &= \frac{1}{-30}\ln\frac{47}{73}\\
&= 0.0146771\\
T &= C+(T_0-C)e^{-0.0146771t}\\
90 &= 70+(143-70)e^{-0.0146771t}\\
\ln\frac{20}{73} &= -0.0146771t\\
t &= \frac{1}{-0.0146771}\ln\frac{20}{73}\\
&= 88.2 \text{ min}
\end{aligned}$$

44.

$$\begin{aligned}
S' &= 0\\
-\frac{1}{20}S+10 &= 0\\
S &= 200
\end{aligned}$$

This value agrees with the result obtained in Exmaple 7

45. $y'+\frac{2}{x}y=5x^2$

a) The solution will exist on $(0,\infty)$

b)

$$\begin{aligned}
\int\frac{2}{x}\,dx &= 2\ln x+C=\ln x^2+C\\
G(x) &= x^2\\
x^2y &= \int 5x^4\,dx\\
x^2y &= x^5+C\\
y &= x^3+Cx^{-2}\\
1 &= 1+C\\
0 &= C\\
y &= x^3
\end{aligned}$$

c) The domain of x^3 is all real numbers where the domain of $x^3+\frac{1}{4x^2}$ is $(0,\infty)$

d) Left to the student

46. $y'+\frac{1}{x}y=x^2$

a)

$$\begin{aligned}
\int\frac{1}{x}\,dx &= \ln x+C\\
G(x) &= x\\
xy &= \int x^3\,dx\\
xy &= \frac{1}{4}x^4+C\\
y &= \frac{1}{4}x^3+Cx^{-3}
\end{aligned}$$

b) Let the wanted value be k

$$\begin{aligned}
k &= 2+\frac{C}{8}\\
8(k-2) &= C
\end{aligned}$$

In order for the solution to exits for all real numbers $8(k-2)=0$ which means $k=2$

47. There is a net flow into the tank of 3 gallons per minute and no salt is added. Therefore

$$S'=0-\frac{2S}{500+3t}=-\frac{2}{500+3t}S$$

is the differential equation associated with this problem

$$\begin{aligned}
S' &= -\frac{2}{500+3t}S\\
\int\frac{dS}{S} &= \int\frac{2}{500+3t}\,dt\\
\ln S &= -\frac{2}{3}\ln(500+3t)+C\\
\ln S &= \ln(500+3t)^{-2/3}+C\\
S &= \frac{C}{(500+3t)^{2/3}}\\
200 &= \frac{C}{(500)^{2/3}}\\
S &= \frac{200(500)^{2/3}}{(500+3t)^{2/3}}
\end{aligned}$$

It will take $2000=500+3t\to t=500$ minutes to fill the tank

$$\begin{aligned}
S(500) &= \frac{200(500)^{2/3}}{(2000)^{2/3}}\\
&= 79.37 \text{ pounds}
\end{aligned}$$

48.

$$\begin{aligned}
P' &= 1.2(1-P)\\
P'+1.2P &= 1.2\\
\int 1.2\,dt &= 1.2t+C\\
G(t) &= e^{1.2t}\\
e^{1.2t}P &= \int 1.2e^{1.2t}\,dt\\
e^{1.2t}P &= e^{1.2t}+C\\
P &= 1+Ce^{-1.2t}\\
0 &= 1+C\\
C &= -1\\
P &= 1-e^{-1.2t}
\end{aligned}$$

49.

$$\begin{aligned}
P' &= 1.2(1-e^{-1.2t}-P)\\
P'+1.2P &= 1.2-1.2e^{-1.2t}\\
\int 1.2\,dt &= 1.2t+C\\
G(t) &= e^{1.2t}
\end{aligned}$$

$$\begin{aligned}
e^{1.2t}P &= \int 1.2e^{1.2t} - 1.2\,dt \\
e^{1.2t}P &= e^{1.2t} - 1.2t + C \\
P &= 1 - 1.2te^{-1.2t} + Ce^{-1.2t} \\
0 &= 1 - 0 + C \\
C &= -1 \\
P &= 1 - 1.2te^{-1.2t} - e^{-1.2t}
\end{aligned}$$

50.

$$\begin{aligned}
P' &= 1.2(\sin t - P) \\
P' + 1.2P &= 1.2 \sin t \\
\int 1.2\,dt &= 1.2t + C \\
G(t) &= e^{1.2t} \\
e^{1.2t}P &= \int 1.2e^{1.2t} \sin t\,dt \\
e^{1.2t}P &= 0.59016\,e^{1.2t} \sin t - 0.4918\,e^{1.2t} \cos t + C \\
P &= 0.59016 \sin t - 0.4918 \cos t + Ce^{-1.2t} \\
0 &= 0 - 0.4918 + C \\
C &= 0.4918 \\
P &= 0.59016 \sin t - 0.4918 \cos t + 0.4918e^{-1.2t}
\end{aligned}$$

51. **a)** Left to the student

b) $Q' + bQ = a$

$$\begin{aligned}
\int b\,dt &= bt + C \\
G(t) &= e^{bt} \\
e^{bt}Q &= \int ae^{bt}\,dt \\
e^{bt}Q &= \frac{a}{b}\,e^{bt} + C \\
Q &= \frac{a}{b} + C\,e^{-bt} \\
0 &= \frac{a}{b} + C \\
C &= -\frac{a}{b} \\
Q &= \frac{a}{b} - \frac{a}{b}\,e^{-bt}
\end{aligned}$$

c)

$$\begin{aligned}
Q' + bQ &= a \\
0 + bQ &= a \\
Q &= \frac{a}{b}
\end{aligned}$$

52. **a)** Left to the student

b) $C' + (b+k)C = a$

$$\begin{aligned}
\int b + k\,dt &= (b+k)t + C \\
G(t) &= e^{(b+k)t} \\
e^{(b+k)t}C &= \int a\,e^{(b+k)t}\,dt \\
e^{(b+k)t}C &= \frac{a}{b+k}\,e^{(b+k)t} + K \\
C &= \frac{a}{b+k} + K\,e^{-(b+k)t} \\
0 &= \frac{a}{b+k} + K \\
K &= -\frac{a}{b+k} \\
Q &= \frac{a}{b+k} - \frac{a}{b+k}\,e^{-(b+k)t}
\end{aligned}$$

c)

$$\begin{aligned}
Q' + (b+k)Q &= a \\
0 + (b+k)Q &= a \\
Q &= \frac{a}{b+k}
\end{aligned}$$

Exercise Set 8.3

1. **a)** $2 - y = 0 \rightarrow y = 2$

b) $y'' = -1 < 0$ Therefore the equilibrium point is asymptotically stabe

c) No inflection points since y'' does not change signs

d)

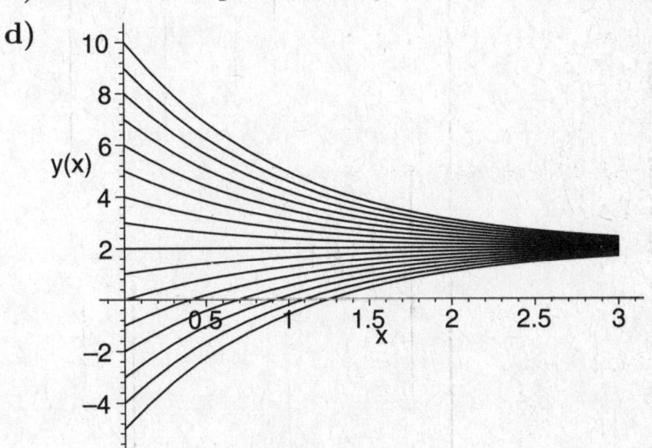

2. **a)** $5y + 4 = 0 \rightarrow y = \frac{4}{5}$

b) $y'' = 5 > 0$ Therefore the equilibrium point is unstabe

c) No inflection points since y'' does not change signs

d)

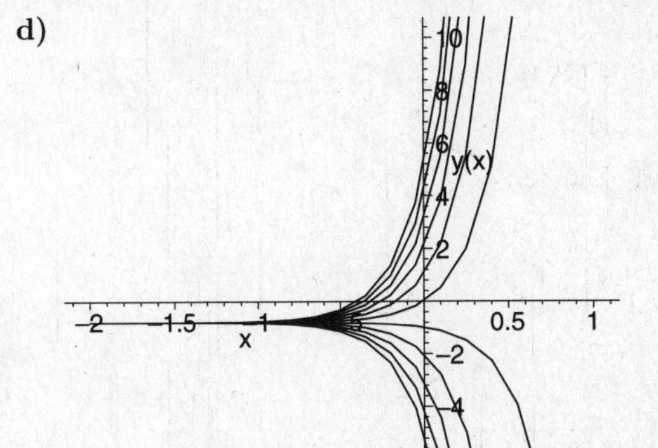

3. **a)**

$$\begin{aligned}
y^2 - 5y + 4 &= 0 \\
(y-1)(y-4) &= 0 \\
y &= 1 \\
y &= 4
\end{aligned}$$

b) $y'' = 2y - 5$
$y''(1) = -3$ Therefore $y = 1$ is an asymptotically stable equilibrium point
$y''(4) = 3$ Therefore $y = 4$ is an unstable equilibrium point

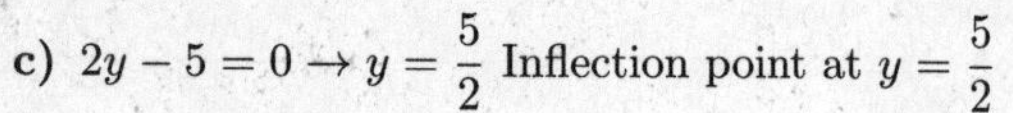

c) $2y - 5 = 0 \rightarrow y = \frac{5}{2}$ Inflection point at $y = \frac{5}{2}$

d)

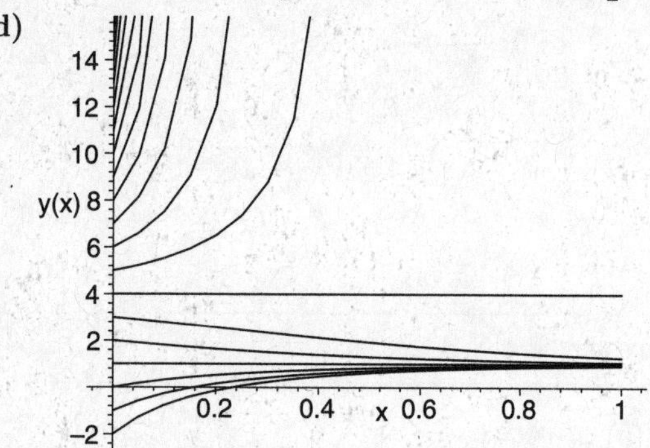

4. a)

$$\begin{aligned} (1-y)^2 &= 0 \\ y &= 1 \end{aligned}$$

b) $y'' = 2y - 2$
$y''(1) = 0$ Therefore $y = 1$ is a semistable equilibrium point

c) $2y - 2 = 0 \rightarrow y = 1$ No inflection points since $y = 1$ is an extremum point

d)

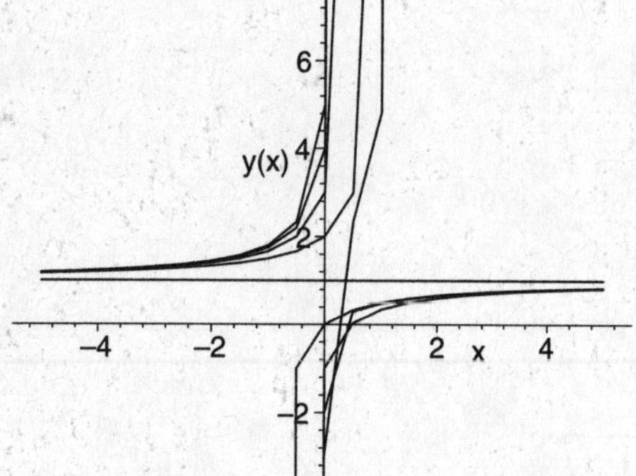

5. a)

$$\begin{aligned} y^3 - 2y^2 &= 0 \\ y^2(y-2) &= 0 \\ y &= 0 \\ y &= 2 \end{aligned}$$

b) $y'' = 3y^2 - 4y$
$y''(0) = 0$ Therefore $y = 0$ is a semistable equilibrium point
$y''(2) = 4$ Therefore $y = 2$ is an unstable equilibrium point

c) $y(3y - 4) = 0 \rightarrow y = 0,\ y = \frac{4}{3}$ Inflection point at $y = \frac{4}{3}$

d)

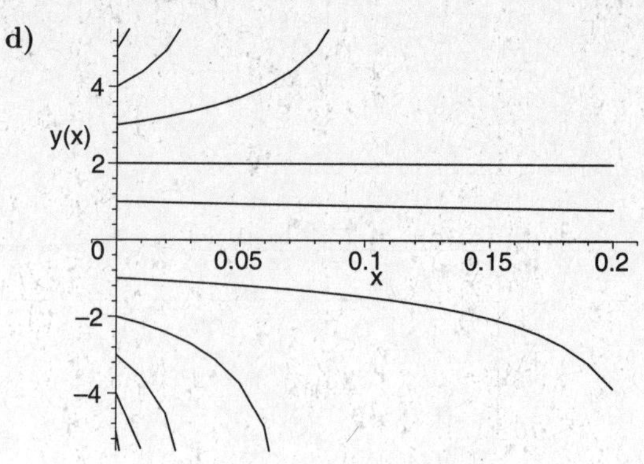

6. a)

$$\begin{aligned} -y^3 + 4y^2 - 4y &= 0 \\ -y(y-2)^2 &= 0 \\ y &= 0 \\ y &= 2 \end{aligned}$$

b) $y'' = -6y^2 + 8y - 4$
$y''(0) = -4$ Therefore $y = 0$ is an asymptotically stable equilibrium point
$y''(2) = -12$ Therefore $y = 2$ is an asymptotically stable equilibrium point

c) $-6y^2 + 8y - 4 = 0$ does not have real solutions therefore there are no inflection points

d)

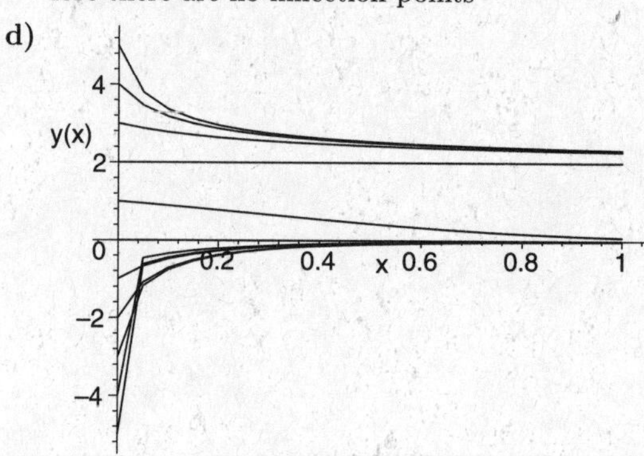

7. a)

$$\begin{aligned} y^3 + 8y^2 + 15y &= 0 \\ y(y+3)(y+5) &= 0 \\ y &= -5 \\ y &= -3 \\ y &= 0 \end{aligned}$$

b) $y'' = 3y^2 + 16y + 15$
$y''(-5) = 10$ Therefore $y = -5$ is an unstable equilibrium point
$y''(-3) = -6$ Therefore $y = -3$ is an asymptotically stable equilibrium point
$y''(0) = 15$ Therefore $y = 0$ is an unstable equilibrium point

c) $3y^2 + 16y + 15 = 0 \rightarrow y = \frac{-8 \pm \sqrt{19}}{3}$
Inflection points at $y = \frac{-8 \pm \sqrt{19}}{3}$

d)

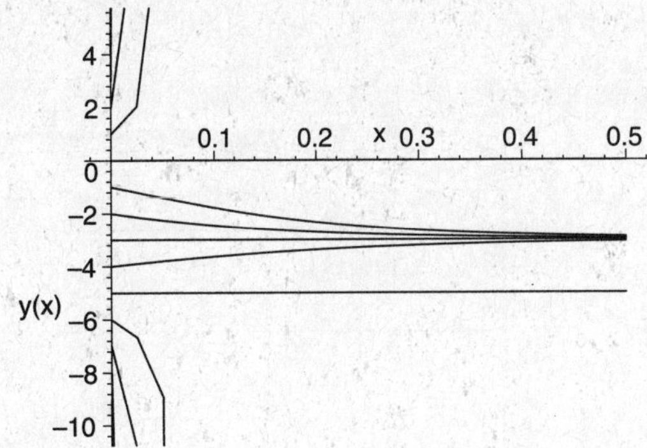

8. a)

$$\begin{aligned} 5 - e^y &= 0 \\ e^y &= 5 \\ y &= \ln 5 \end{aligned}$$

b) $y'' = -e^y$
$y''(-\ln 5) = -5$ Therefore $y = -3$ is an asymptotically stable equilibrium point

c) $-e^y = 0$ has no real solutions, therefore there are no inflection points

d)

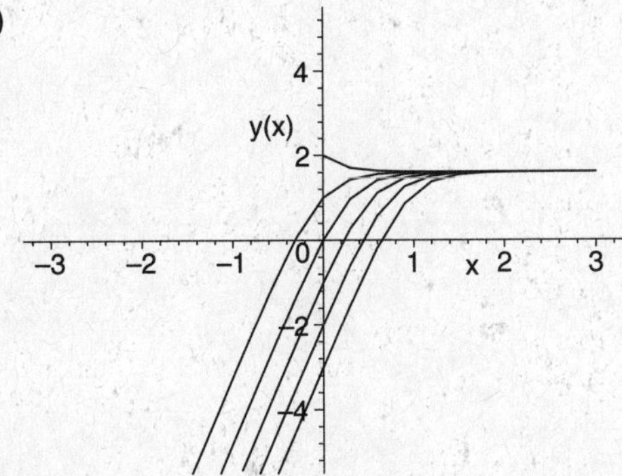

9. a)

$$\begin{aligned} e^{2y} - e^y &= 0 \\ e^y(e^y - 1) &= 0 \\ y &= 0 \end{aligned}$$

b) $y'' = 2e^{2y} - e^y$
$y''(0) = -1$ Therefore $y = -3$ is an unstable equilibrium point

c) $2e^{2y} - e^y = 0 \to y = -\ln 2$
Inflection point at $y = -\ln 2$

d)

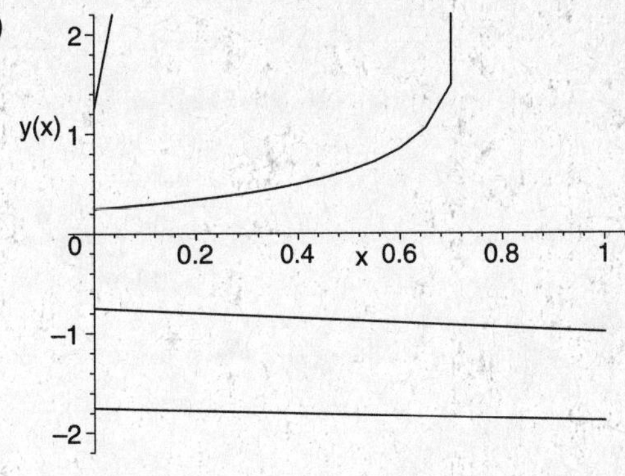

10. a)

$$\begin{aligned} e^{2y} - 5e^y + 6 &= 0 \\ (e^y - 3)(e^y - 2) &= 0 \\ y &= \ln 2 \\ y &= \ln 3 \end{aligned}$$

b) $y'' = 2e^{2y} - 5e^y$
$y''(\ln 2) = -2$ Therefore $y = -3$ is an asymptotically stable equilibrium point
$y''(\ln 3) = 3$ Therefore $y = -3$ is an unstable equilibrium point

c) $2e^{2y} - 5e^y = 0 \to y = \ln \dfrac{5}{2}$
Inflection point at $y = \ln \frac{5}{2}$

d)

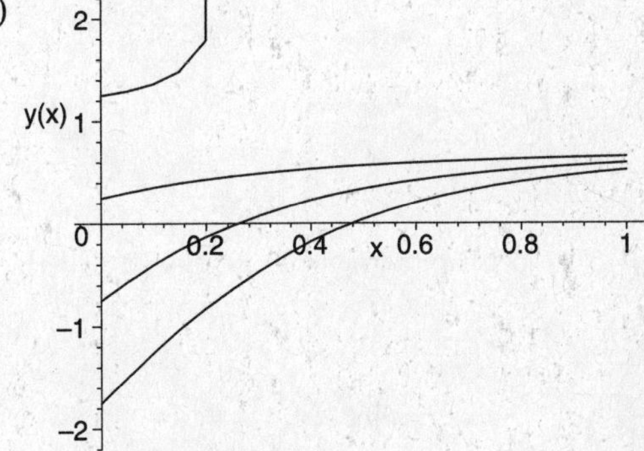

11. The population satisfies the logistic growth model with $k = 0.4$ and $L = 12500$.

a)

$$\begin{aligned} P(t) &= \frac{12500(9)}{9 + (12500 - 9)e^{-0.4t}} \\ &= \frac{112500}{9 + 12491e^{-0.4t}} \\ P(7) &= \frac{112500}{9 + 12491e^{-0.4(7)}} \\ &= 146.37 \text{ million} \end{aligned}$$

b)

$$\begin{aligned} 900 &= \frac{112500}{9 + 12491e^{-0.4t}} \\ \frac{112500}{900} &= 9 + 12491e^{-0.4t} \end{aligned}$$

$$\begin{aligned}
\frac{112500}{900} - 9 &= 12491e^{-0.4t} \\
116 &= 12491e^{-0.4t} \\
-0.4t &= \ln \frac{116}{12491} \\
t &= \frac{-1}{0.4} \ln \frac{116}{12491} \\
&= 11.70 \text{ hours}
\end{aligned}$$

12. The population satisfies the logistic growth model with $k = 0.14$ and $L = 20$.

a)

$$\begin{aligned}
P(t) &= \frac{28}{1.4 + 18.6e^{-0.14t}} \\
P(8) &= \frac{28}{1.4 + 18.6e^{-0.14(8)}} \\
&= 3.75 \text{ million}
\end{aligned}$$

b)

$$\begin{aligned}
8 &= \frac{28}{1.4 + 18.6e^{-0.14t}} \\
\frac{28}{8} &= 1.4 + 18.6e^{-0.14t} \\
2.1 &= 18.6e^{-0.14t} \\
-0.14t &= \ln \frac{2.1}{18.6} \\
t &= \frac{-1}{0.14} \ln \frac{2.1}{18.6} \\
&= 15.58 \text{ hours}
\end{aligned}$$

13. $y' = ky\left(1 - \left[\frac{y}{L}\right]\right)^{\theta}$

a) Left to the student

b) The per capita graowth rate is $k\left(1 - \left[\frac{y}{L}\right]^{\theta}\right)$

c)

$$\begin{aligned}
ky\left(1 - \left[\frac{y}{L}\right]^{\theta}\right) &= 0 \\
y &= 0 \\
\left(1 - \left[\frac{y}{L}\right]^{\theta}\right) &= 0 \\
y &= L
\end{aligned}$$

$y'' = k - k\left(\frac{y}{L}\right)^{\theta} - k\left(\frac{y}{L}\right)^{\theta}\theta$
$y''(0) = k > 0$
Therefore $y = 0$ is an unstable equilibrium point
$y''(L) = -k\theta < 0$
Therefore $y = L$ is an asyptotically stable equilibrium point

d)

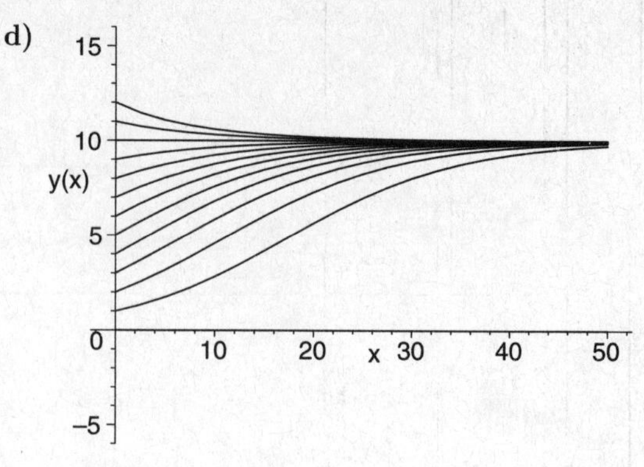

14. $y' = ky\ \ln(L/y)$

a) The per capita growth rate is $k\ \ln(L/y)$

b) $y' = 0 \rightarrow y = L$
$y'' = k\ \ln(L/y) - k$
$y''(L) = -k < 0$
Therefore $y = L$ is an asymptotically stabe equilibrium point

c)

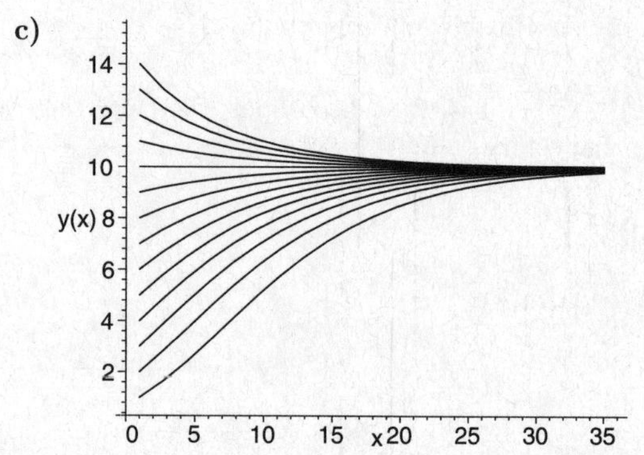

15. $y' = ky(P - y)$

a)

$$\begin{aligned}
ky(P - y) &= 0 \\
y &= 0 \\
y &= P
\end{aligned}$$

$y'' = kP - 2ky$
$y''(0) = kP > 0$
Therefore $y = 0$ is an unstable equilibrium point
$y''(P) = -kP < 0$
Therefore $y = P$ is an asymptotically stabe equilibrium point

b)

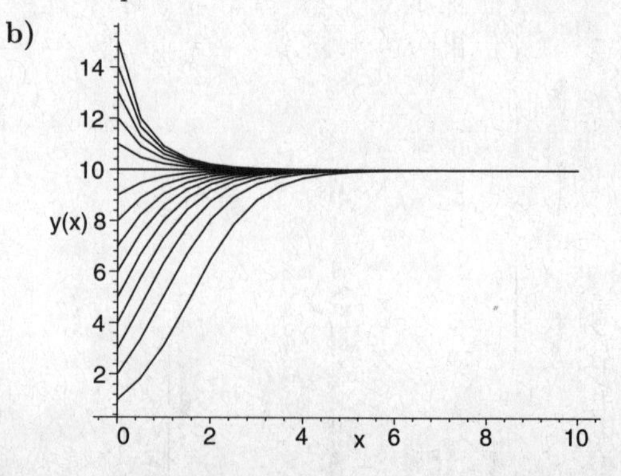

c) As the graph suggests, the entire population gets infected

16. **a - b)**

$$\begin{aligned} ky\left(1-\frac{y}{L}\right)-sy &= 0 \\ y[k-\frac{ky}{L}-s] &= 0 \\ y &= 0 \\ k-\frac{ky}{L}-s &= 0 \\ \frac{ky}{L} &= (k-s) \\ y &= \frac{(k-s)L}{k}=c \end{aligned}$$

c) If $s<k$ then $k-s>0 \to c>0$

d) The inequality states that the rate of harvesting is less than the growth rate

17. **a)** $y''=k-s-2\frac{k}{L}y$
$y''(0)=k-s>0$, thus $y=0$ is an unstable equilibrium point
$y''(c)=-2(k-s)<0$, thus $y=c$ is an asymptotically stable equilibrium point

b)

$$\begin{aligned} H &= sc \\ &= s\cdot\frac{(k-s)L}{k} \\ &= Ls-\frac{Ls^2}{k} \end{aligned}$$

c) $H'=L-\frac{2Ls}{k}=0\to s=\frac{k}{2}$

18. **a)** The equation follows the logistic growth model with a threshold with $k=2$, $L=20$, $T=5$ and a harvesting rate of $\frac{5}{25}$.

b)

$$\begin{aligned} 2y\left(1-\frac{y}{20}\right)\left(\frac{y}{5}-1\right)-\frac{7y}{25} &= 0 \\ y &= 0 \\ y &= 9 \\ y &= 16 \end{aligned}$$

$y=0$ is an asymptotically stable equilibrium point
$y=9$ is an unstable equilibrium point
$y=16$ is an asymptotically stable equilibrium point

c) It is less than 20 to ensure the survival of the population

d) It is greater than 5 to ensure the survival of the population

19. $y'=y^2-6y$

$$\begin{aligned} y^2-6y &= 0 \\ y(y-6) &= 0 \\ y &= 0 \\ y &= 6 \\ y'' &= 2y-6 \\ y''(0) &= -6<0 \text{ stable} \\ y''(6) &= 6>0 \text{ unstable} \end{aligned}$$

$$\lim_{t\to\infty} y(t)=0$$

20. $y'=y^2+y-20$

$$\begin{aligned} y^2+y-20 &= 0 \\ (y-4)(y+5) &= 0 \\ y &= -5 \\ y &= 4 \\ y'' &= 2y+1 \\ y''(-5) &= -9<0 \text{ stable} \\ y''(4) &= 9>0 \text{ unstable} \end{aligned}$$

$$\lim_{t\to\infty} y(t)=-9$$

21. **a)** By Theorem 4 it follows that c is semistable

b) The result follows from Part (a) and the definition on page 564

22. **a)** By Theorem 4 it follows that c is semistable

b) The result follows from Part (a) and the definition on page 564

23. $y'=\dfrac{ky^2}{T}-ky-\dfrac{ky^3}{LT}+ky^2/L$

$$y''=\frac{2ky}{T}-k-\frac{3ky^2}{LT}+\frac{2ky}{L}$$

$y''(0)=-k<0$ asymptotically stabe

$$y''(L)=\frac{2kL}{T}-k-\frac{3kL}{T}+2k=$$

$$k-\frac{kL}{T}<0 \text{ asymptotically stable}$$

24. $y''=\dfrac{2ky}{T}-k-\dfrac{3ky^2}{LT}+\dfrac{2ky}{L}$

$$y''(T)=2k-k-\frac{3kT}{L}+\frac{2kT}{L}=$$

$$k-\frac{kT}{L}>0 \text{ unstable}$$

25.

$$\begin{aligned} \frac{2ky}{T}-k-\frac{3ky^2}{LT}+\frac{2ky}{L} &= 0 \\ \frac{3k}{LT}y^2-\left(\frac{2k}{T}+\frac{2k}{L}\right)y+k &= 0 \\ \frac{\frac{2k}{T}+\frac{2k}{L}\pm\sqrt{\left(\frac{2k}{T}+\frac{2k}{L}\right)^2-\frac{12k^2}{LT}}}{\frac{6k}{LT}} &= y \\ \frac{L+T\pm\sqrt{L^2-LT+T^2}}{3} &= \end{aligned}$$

26. $y' = ky - \frac{k}{L}\, y^2 - s$

a)

$$\begin{aligned} \frac{k}{L}y^2 - ky + s &= 0 \\ \frac{k \pm \sqrt{k^2 - \frac{4ks}{L}}}{\frac{2k}{L}} &= y \\ \frac{kL \pm \sqrt{k^2L^2 - 4ksL}}{2k} &= \end{aligned}$$

b)

$$\begin{aligned} k - \frac{2ky}{L} &= y'' \\ k - \frac{2k}{L}\left(\frac{kL - \sqrt{k^2L^2 - 4ksL}}{2k}\right) &= y'' \\ \frac{\sqrt{k^2L^2 - 4ksL}}{2kL} &= \end{aligned}$$

Which is positive, thus the smaller equilibrium point is unstable

c)

$$\begin{aligned} k - \frac{2ky}{L} &= y'' \\ k - \frac{2k}{L}\left(\frac{kL + \sqrt{k^2L^2 - 4ksL}}{2k}\right) &= y'' \\ -\frac{\sqrt{k^2L^2 - 4ksL}}{2kL} &= \end{aligned}$$

Which is negative, thus the larger equilibrium point is asymptotically stable

d) The population grows indefinitely

27. a)

$$\begin{aligned} \frac{k}{L}y^2 - ky + s &= 0 \\ \frac{k \pm \sqrt{k^2 - \frac{4ks}{L}}}{\frac{2k}{L}} &= y \end{aligned}$$

The radicand is negative and therefore there are no real equilibrium points

b) $s > kL/4 \to s - kL/4 > 0$ but from Part (a) we know that there are no equilibrium points therefore the most the right hand side can get is $-(s - kl/4)$

c) Because 0 is the only physical value the population can approach.

28. $s = kL/4$

$$\begin{aligned} \frac{k}{L}y^2 - ky + s &= 0 \\ \frac{k \pm \sqrt{k^2 - \frac{4ks}{L}}}{\frac{2k}{L}} &= y \\ \frac{k \pm \sqrt{k^2 - k^2}}{\frac{2k}{L}} &= \\ \frac{L}{2} &= y \\ k - \frac{2k}{L}\, y &= y'' \\ 0 &= y''\left(\frac{L}{2}\right) \end{aligned}$$

Therefore, $y = \dfrac{L}{2}$ is a semistable equilibrium point

29.

$$\begin{aligned} y'' &= nky^{n-1} - k \\ y''(0) &= -k < 0 \end{aligned}$$

$y = 0$ is an asymptotically stable equilibrium point

30.

$$\begin{aligned} y'' &= nky^{n-1} - k \\ y''(0) &= nk - k > 0 \end{aligned}$$

$y = 0$ is an unstable equilibrium point

31.

$$\begin{aligned} nky^{n-1} - k &= 0 \\ y^{n-1} &= \frac{1}{n} \\ y &= \left(\frac{1}{n}\right)^{1/(n-1)} \end{aligned}$$

32. Left to the student

33. a)

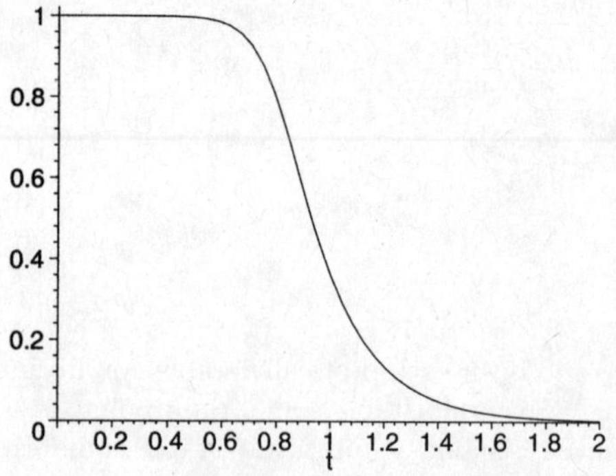

b)

$$\begin{aligned} y(1) &= \left(1 + 0.00000624e^{15}\right)^{-1/3} \\ &= 0.3602 \end{aligned}$$

c)

$$\begin{aligned} 0.5 &= \left(1 + 0.00000624e^{15t}\right)^{-1/3} \\ 0.5^{-3} - 1 &= 0.00000624e^{15t} \\ \frac{0.5^{-3} - 1}{0.00000624} &= e^{15t} \end{aligned}$$

$$\begin{aligned}
\ln\left(\frac{0.5^{-3}-1}{0.00000624}\right) &= 15t \\
\frac{1}{15}\ln\left(\frac{0.5^{-3}-1}{0.00000624}\right) &= t \\
0.93 &= t
\end{aligned}$$

34. **a)**

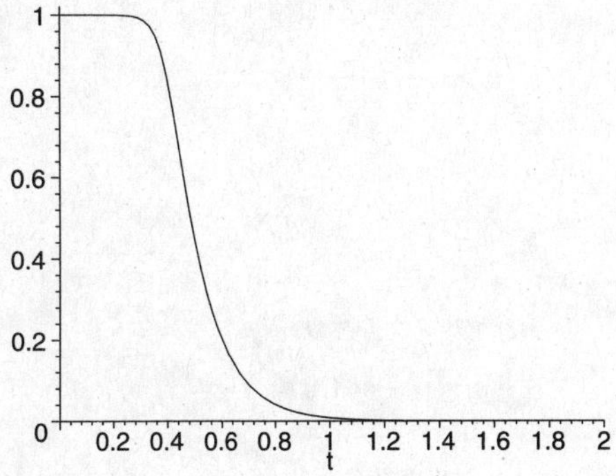

b)

$$\begin{aligned}
y(1) &= \left(1+0.000002849e^{32}\right)^{-1/4} \\
&= 0.0082
\end{aligned}$$

c)

$$\begin{aligned}
0.5 &= \left(1+0.000002849e^{32t}\right)^{-1/4} \\
0.5^{-4}-1 &= 0.000002849e^{32t} \\
\frac{0.5^{-4}-1}{0.000002849} &= e^{32t} \\
\ln\left(\frac{0.5^{-4}-1}{0.000002849}\right) &= 32t \\
\frac{1}{32}\ln\left(\frac{0.5^{-4}-1}{0.000002849}\right) &= t \\
0.48 &= t
\end{aligned}$$

Exercise Set 8.4

1.

$$\begin{aligned}
\frac{dy}{dx} &= 4x^3y \\
\frac{dy}{y} &= 4x^3\,dx \\
\int\frac{dy}{y} &= \int 4x^3\,dx \\
\ln y &= x^4+C \\
y &= Ce^{x^4}
\end{aligned}$$

2.

$$\begin{aligned}
3y^2\,\frac{dy}{dx} &= 5x \\
3y^2\,dy &= 5x\,dx \\
\int 3y^2\,dy &= \int 5x\,dx \\
y^3 &= \frac{5}{2}\,x^2+C \\
y &= \sqrt[3]{\frac{5}{2}\,x^2+C}
\end{aligned}$$

3.

$$\begin{aligned}
\frac{dy}{dx} &= \frac{x}{2y} \\
2y\,dy &= x\,dx \\
\int 2y\,dy &= \int x\,dx \\
y^2 &= \frac{1}{2}\,x^2+C \\
y &= \pm\sqrt{\frac{1}{2}\,x^2+C}
\end{aligned}$$

4.

$$\begin{aligned}
\frac{dy}{dx} &= x^2y^3 \\
\int y^{-3}\,dy &= \int x^2\,dx \\
\frac{-1}{2y^2} &= \frac{1}{3}\,x^3+C \\
y &= \sqrt{\frac{-3}{2(x^3+C)}}
\end{aligned}$$

5.

$$\begin{aligned}
\frac{dy}{dx} &= \frac{x\sqrt{y^2+1}}{y} \\
\frac{y}{\sqrt{y^2+1}}\,dy &= x\,dx \\
\sqrt{y^2+1} &= \frac{1}{2}\,x^2+C \\
y^2 &= \left(\frac{1}{2}\,x^2+C\right)^2-1 \\
y &= \pm\sqrt{\left(\frac{1}{2}\,x^2+C\right)^2-1}
\end{aligned}$$

6.

$$\begin{aligned}
\frac{dy}{dx} &= \frac{e^y+1}{e^y} \\
\int\frac{e^y}{e^y+1}\,dy &= \int dx \\
\ln|\,e^y+1\,| &= x+C \\
e^y+1 &= Ce^x \\
e^y &= Ce^x-1 \\
y &= \ln|\,Ce^x-1\,|
\end{aligned}$$

7.

$$\begin{aligned}
\frac{dy}{dx} &= \frac{x^2\sec y}{(x^3+1)^3} \\
\cos y\,dy &= \frac{x^2}{(x^3+1)^3}\,dx
\end{aligned}$$

$$\begin{aligned}
\int \cos y \, dy &= \int \frac{x^2}{(x^3+1)^3}\, dx \\
\sin y &= -\frac{3}{2}\,(x^3+1)^{-2} + C \\
\sin y + \frac{3}{2(x^3+1)^2} &= C
\end{aligned}$$

8.

$$\begin{aligned}
\frac{dy}{dx} &= e^x y\sqrt{e^x+4} \\
\int \frac{dy}{y} &= \int e^x\sqrt{e^x+4} \\
\ln y &= \frac{2}{3}\,(e^x+4)^{3/2} + C \\
\ln y - \frac{2}{3}\,(e^x+4)^{3/2} &= C
\end{aligned}$$

9.

$$\begin{aligned}
\frac{dy}{dt} &= \frac{y^3+1}{y^2} \\
\frac{y^2}{y^3+1}\, dy &= dt \\
\int \frac{y^2}{y^3+1}\, dy &= \int dt \\
\frac{1}{3}\,\ln|\,y^3+1\,| &= t + C \\
\ln|\,(y^3+1)^{1/3}\,| &= t + C \\
(y^3+1)^{1/3} &= Ce^t \\
y^3+1 &= Ce^{3t} \\
y &= \sqrt[3]{Ce^{3t}-1}
\end{aligned}$$

10.

$$\begin{aligned}
\frac{dy}{dt} &= t^2 e^{t^3} y^3 \\
\int y^{-3}\, dy &= \int t^2 e^{t^3}\, dt \\
-\frac{1}{2y^2} &= \frac{1}{3}\, e^{t^3} + C
\end{aligned}$$

$$\frac{1}{2y^2} - \frac{1}{3}\, e^{t^3} = C$$

11.

$$\begin{aligned}
\frac{dy}{dx} &= x\cos^2 y \\
\sec^2 y\, dy &= x\, dx \\
\int \sec^2 y\, dy &= \int x\, dx \\
\tan y &= \frac{1}{2}\, x^2 + C \\
\tan y - \frac{1}{2}\, x^2 &= C
\end{aligned}$$

NOTE: $y = \dfrac{(2n+1)\pi}{2}$ are constant solutions as well

12.

$$\begin{aligned}
\frac{dy}{dx} &= \frac{x}{y+y^5} \\
\int y + y^5\, dy &= \int x\, dx \\
\frac{1}{2}\, y^2 + \frac{1}{6}\, y^6 &= \frac{1}{2}\, x^2 + C \\
\frac{1}{2}\, y^2 + \frac{1}{6}\, y^6 - \frac{1}{2}\, x^2 &= C
\end{aligned}$$

13.

$$\begin{aligned}
\frac{dy}{dx} &= \frac{\sqrt{x}}{\sin y + \cos y} \\
\sin y + \cos y\, dy &= \sqrt{x}\, dx \\
\int \sin y + \cos y\, dy &= \int \sqrt{x}\, dx \\
-\cos y + \sin y &= \frac{2}{3}\, x^{3/2} + C \\
\sin y - \cos y - \frac{2}{3}\, x^{3/2} &= C
\end{aligned}$$

14.

$$\begin{aligned}
\frac{dy}{dx} &= \frac{e^x e^{2y}}{e^y+1} \\
\int e^{-y} + e^{-2y}\, dy &= \int e^x\, dx \\
-e^{-y} - \frac{1}{2}\, e^{-2y} &= e^x + C \\
e^{-y} + \frac{1}{2}\, e^{-2y} + e^x &= C
\end{aligned}$$

15.

$$\begin{aligned}
\frac{dy}{dt} &= \frac{t}{(t^2+1)(y^4+1)} \\
y^4 + 1\, dy &= \frac{t}{t^2+1}\, dt \\
\frac{1}{5}\, y^5 + y &= \frac{1}{2}\,\ln(t^2+1) + C \\
\frac{1}{5}\, y^5 + y - \frac{1}{2}\,\ln(t^2+1) &= C
\end{aligned}$$

16.

$$\begin{aligned}
\frac{dy}{dt} &= \frac{\sin^2 y}{t} \\
\int \csc^2 y\, dy &= \int \frac{dt}{t} \\
-\cot y &= \ln t + C \\
\cot y + \ln t &= C
\end{aligned}$$

17.

$$\begin{aligned}
\frac{dy}{dx} &= 3x^2(y-2)^2 \\
(y-2)^{-2}\, dy &= 3x^2\, dx \\
\int (y-2)^{-2}\, dy &= \int 3x^2\, dx
\end{aligned}$$

$$\begin{aligned} \frac{-1}{(y-2)} &= x^3 + C \\ y - 2 &= \frac{-1}{x^3+C} \\ y &= 2 - \frac{1}{x^3+C} \end{aligned}$$

NOTE $y = 2$ is a constant solution as well

18.

$$\begin{aligned} \frac{dy}{dx} &= y^2 - 1 \\ \int \frac{1}{y^2-1}\,dy &= \int dx \\ \frac{1}{2}\ln\left|\frac{y-1}{y+1}\right| &= x + C \\ \frac{y-1}{y+1} &= Ce^{2x} \end{aligned}$$

19.

$$\begin{aligned} \int 3y^2\,dy &= \int 2x\,dx \\ y^3 &= x^2 + C \\ 125 &= 4 + C \\ C &= 121 \\ y^3 &= x^2 + 121 \\ y &= \sqrt[3]{x^2+121} \end{aligned}$$

20.

$$\begin{aligned} \int \frac{dy}{2+y} &= \int x\,dx \\ \ln|\,y+2\,| &= \frac{1}{2}\,x^2 + C \\ \ln 2 &= C \\ \ln|\,y+2\,| &= \frac{1}{2}\,x^2 + \ln 2 \\ y + 2 &= 2e^{x^2/2} \\ y &= 2e^{x^2/2} - 2 \end{aligned}$$

21.

$$\begin{aligned} \int csc^2 y\,dy &= \int e^{2t}\,dt \\ -cot\ y &= \frac{1}{2}\,e^{2t} + C \\ -1 &= \frac{1}{2} + C \\ -\frac{3}{2} &= C \\ cot\ y + \frac{1}{2}\,e^{2t} &= \frac{3}{2} \end{aligned}$$

22.

$$\begin{aligned} \int \frac{y}{y^2+4} &= \int dt \\ \frac{1}{2}\ln(y^2+4) &= t + C \\ \frac{1}{2}\ln 13 &= C \\ \frac{1}{2}\ln(y^2+4) &= t + \ln\sqrt{13} \\ \ln(y^2+4) &= 2t + \ln 13 \\ y^2 + 4 &= 13e^{2t} \\ y &= \sqrt{13e^{2t} - 4} \end{aligned}$$

23.

$$\begin{aligned} \int y^{-2}\,dy &= \int \frac{\ln t}{t}\,dt \\ -\frac{1}{y} &= \frac{1}{2}\,\ln^2 t + C \\ \frac{1}{4} &= C \\ \frac{1}{y} &= -\frac{1}{2}\,\ln^2 t - \frac{1}{4} \\ y &= \frac{-4}{2\ln^2 t + 1} \end{aligned}$$

24.

$$\begin{aligned} \int y + \frac{1}{y}\,dy &= \int cos\ x\,dx \\ \frac{1}{2}\,y^2 + \ln|\,y\,| &= sin\ x + C \\ \frac{1}{2} &= 1 + C \\ -\frac{1}{2} &= C \\ \frac{1}{2}\,y^2 + \ln|\,y\,| &= sin\ x - \frac{1}{2} \end{aligned}$$

25.

$$\begin{aligned} \int 3y^2(y^3+2)^{-2}\,dy &= \int x\,dx \\ -(y^3+2)^{-1} &= \frac{1}{2}\,x^2 + C \\ -\frac{1}{10} &= \frac{1}{2} + C \\ -\frac{3}{5} &= C \\ \frac{1}{y^3+2} &= -\frac{1}{2}\,x^2 + \frac{3}{5} \\ \frac{1}{y^3+2} &= \frac{-5x^2+6}{10} \\ y^3 + 2 &= \frac{-10}{6-5x^2} \\ y^3 &= \frac{-10+10x^2-12}{6-5x^2} \\ y &= \sqrt[3]{\frac{10x^2-2}{6-5x^2}} \end{aligned}$$

26.

$$\begin{aligned} \int e^{-2y}\,dy &= \int x\,dx \\ -\frac{1}{2}\,e^{-2y} &= \frac{1}{2}\,x^2 + C \\ -\frac{1}{2}e^{-8} &= C \\ e^{-2y} &= e^{-8} - x^2 \end{aligned}$$

27.

$$\begin{aligned} \int \frac{\cos y}{(2+\sin y)^2}\,dy &= \int t^2\,dt \\ -\frac{1}{2+\sin y} &= \frac{1}{3}\,t^3 + C \\ -\frac{1}{5/2} &= -\frac{1}{3} + C \\ -\frac{1}{15} &= C \\ \frac{1}{2+\sin y} &= -\frac{1}{3}\,t^3 + \frac{1}{15} \\ \frac{1}{2+\sin y} &= \frac{-5t^3+1}{15} \\ \sin y + 2 &= \frac{15}{1-5t^3} \end{aligned}$$

28.

$$\begin{aligned} \int e^{-y}\,dy &= \int e^{2t}\,dt \\ -e^{-y} &= \frac{1}{2}\,e^{2t} + C \\ -e^2 &= \frac{1}{2}\,e^2 + C \\ \frac{3}{2}\,e^2 &= C \\ e^{-y} &= \frac{1}{2}\left[e^{2t} + 3e^2\right] \\ y &= -\ln\left|\frac{1}{2}\left[e^{2t} + 3e^2\right]\right| \end{aligned}$$

29.

$$\begin{aligned} \int e^{4y} + e^{5y}\,dy &= \int \sqrt{t} \\ \frac{1}{4}\,e^{4y} + \frac{1}{5}\,e^{5t} &= \frac{2}{3}\,t^{3/2} + C \\ \frac{1}{4} + \frac{1}{5} &= \frac{2}{3} + C \\ \frac{-13}{60} &= C \\ \frac{1}{4}\,e^{4y} + \frac{1}{5}\,e^{5y} &= \frac{2}{3}\,t^{3/2} - \frac{13}{60} \end{aligned}$$

30.

$$\begin{aligned} \int \cot y\,dy &= \int dt \\ \ln|\sin y| &= t + C \\ \ln\frac{1}{2} &= C \\ -\ln 2 &= C \\ \ln|\sin y| &= t - \ln 2 \\ \sin y &= e^{t-\ln 2} \\ \sin y &= \frac{1}{2}\,e^t \end{aligned}$$

31.

$$\begin{aligned} \int y\,dy &= \int x\,dx \\ \frac{1}{2}\,y^2 &= \frac{1}{2}\,x^2 + C \\ \frac{21}{2} - \frac{25}{2} &= C \\ -2 &= C \\ \frac{1}{2}\,y^2 &= \frac{1}{2}\,x^2 - 2 \\ y &= \sqrt{x^2-4} \end{aligned}$$

Which has a domain of $(2, \infty)$

32.

$$\begin{aligned} \int y\,dy &= \int x\,dx \\ \frac{1}{2}\,y^2 &= \frac{1}{2}\,x^2 + C \\ \frac{29}{2} - \frac{25}{2} &= C \\ 2 &= C \\ \frac{1}{2}\,y^2 &= \frac{1}{2}\,x^2 + 2 \\ y &= \sqrt{x^2+4} \end{aligned}$$

Which has a domain of $(-\infty, \infty)$

33. **a)**

$$\begin{aligned} \int \frac{dy}{y} &= \int k\,dt \\ \ln|y| &= kt + C \\ y &= Ce^{kt} \\ y_0 &= C \\ y &= y_0 e^{kt} \end{aligned}$$

b) Left to the student

34. **a)** Left to the student

b) $P' = k - \left(\frac{r}{V}\right)\,P$

c) P constant means $P' = 0$

$$\begin{aligned} k - \left(\frac{r}{V}\right)\,P &= 0 \\ k &= \left(\frac{r}{V}\right)\,P \end{aligned}$$

35.

$$\begin{aligned} y &= y_0e^{kt} \\ 4.404 &= 5e^{11k} \\ \frac{4.404}{5} &= e^{11k} \\ \frac{\ln 4.404}{5} &= 11k \\ \frac{1}{11}\ln\frac{4.404}{5} &= k \\ -0.01154 &= k \end{aligned}$$

The decay rate is 1.154% per day

36. **a)**

$$\begin{aligned} \int \frac{dT}{T-C} &= \int -k\,dt \\ \ln|T-C| &= -kt+K \\ T-C &= Ke^{-kt} \\ T_0-C &= K \\ T-C &= (T_0-C)e^{-kt} \\ T &= (T_0-C)e^{-kt}+C \end{aligned}$$

b) Left to the student

37. - 39. Left to the student

40. **a)**

$$\begin{aligned} \int y^{-2}\,dy &= \int dt \\ -\frac{1}{y} &= t+C \\ -1 &= C \\ -\frac{1}{y} &= t-1 \\ \frac{1}{y} &= 1-t \\ y &= \frac{1}{1-t} \end{aligned}$$

Doamin of solution is $t \neq 1$

b)

$$\begin{aligned} \int y^{-2}\,dy &= \int dt \\ -\frac{1}{y} &= t+C \\ -\frac{1}{2} &= C \\ -\frac{1}{y} &= t-\frac{1}{2} \\ \frac{1}{y} &= \frac{1-2t}{2} \\ y &= \frac{2}{1-2t} \end{aligned}$$

Doamin of solution is $t \neq 1/2$

c)

$$\begin{aligned} \int y^{-2}\,dy &= \int dt \\ -\frac{1}{y} &= t+C \\ -\frac{1}{k} &= C \\ -\frac{1}{y} &= t-\frac{1}{k} \\ \frac{1}{y} &= \frac{1-kt}{k} \\ y &= \frac{k}{1-kt} \end{aligned}$$

Doamin of solution is $t \neq 1/k$

d) There is no solution for the initial condition $y(0) = 0$ since we get a division by zero when we try to find the constant of integration

41. Left to the student

42. Left to the student

43. **a)**

$$\begin{aligned} \int \frac{L}{R(L-R)}\,dR &= \int k\,dt \\ \frac{1}{L}\ln\left(\frac{R}{L-R}\right) &= kt+C \\ \frac{1}{L}\ln\left(\frac{R_0}{L-R_0}\right) &= C \\ kt+\frac{1}{L}\ln\left(\frac{R_0}{L-R_0}\right) &= \frac{1}{L}\ln\left(\frac{R}{L-R}\right) \\ \frac{1}{L}\left[\ln\left(\frac{R}{L-R}\right)-\ln\left(\frac{R_0}{L-R_0}\right)\right] &= kt \\ \frac{1}{L}\left[\ln\frac{\left(\frac{R}{L-R}\right)}{\left(\frac{R_0}{L-R_0}\right)}\right] &= kt \\ \left[\ln\frac{\left(\frac{R}{L-R}\right)}{\left(\frac{R_0}{L-R_0}\right)}\right] &= Lkt \\ \frac{\left(\frac{R}{L-R}\right)}{\left(\frac{R_0}{L-R_0}\right)} &= e^{Lkt} \\ R(L-R_0) &= LR_0e^{Lkt} \\ R(L-R_0)+RR_0e^{Lkt} &= LR_0e^{Lkt} \\ R\left[(L-R_0)+R_0e^{Lkt}\right] &= LR_0e^{Lkt} \\ \frac{LR_0e^{Lkt}}{(L-R_0)+R_0e^{Lkt}} &= R \\ \frac{LR_0}{(L-R_0)e^{-Lkt}+R_0} &= \end{aligned}$$

b) Left to the student

44. **a)**

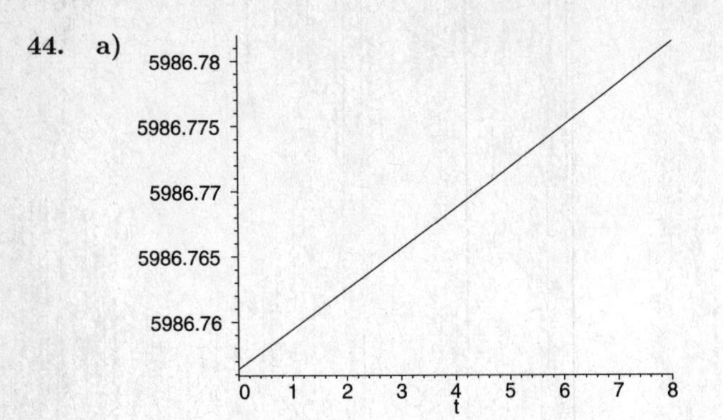

b) Left to the student

Exercise Set 8.5

1. **a)** $y(1) \approx -1.2$

b) $y(1) \approx -1$

c)

$$\begin{aligned} \int dy &= \int 2x\,dx \\ y &= x^2 + C \\ -2 &= C \\ y &= x^2 - 2 \end{aligned}$$

d) The exact value is $y(1) = -1$

2. **a)** $y(2) \approx 0.429$

b) $y(2) \approx 0.541$

c)

$$\begin{aligned} \int \frac{dy}{y} &= \int -1\,dx \\ \ln y &= -x + C \\ y &= Ce^{-x} \\ 4 &= C \\ y &= 4e^{-x} \end{aligned}$$

d) The exact value is $y(2) = 0.541$

3. **a)** $y(2) \approx 18.235$

b) $y(2) \approx 40.067$

c)

$$\begin{aligned} \int \frac{dy}{y} &= \int 2x\,dx \\ \ln y &= x^2 + C \\ y &= Ce^{x^2} \\ 2 &= Ce \\ C &= 2/e \\ y &= 2e^{x^2-1} \end{aligned}$$

d) The exact value is $y(2) = 40.171$

4. **a)** $y(3) \approx 7.850$

b) $y(3) \approx 8.9999$

c)

$$\begin{aligned} \int \frac{dy}{\sqrt{y}} &= \int x\,dx \\ 2\sqrt{y} &= \frac{x^2}{2} + C \\ 2 &= \frac{1}{2} + C \\ \frac{3}{2} &= C \\ \sqrt{y} &= \frac{x^2}{4} + \frac{3}{4} \\ y &= \left(\frac{x^2+3}{4}\right)^2 \end{aligned}$$

d) The exact value is $y(3) = 9$

5. **a)** $y(2) \approx 28.8988$

b) $y(2) \approx 34.1710$

c) $y' - y = 2x$

$$\begin{aligned} \int -1\,dx &= -x + C \\ G(x) &= e^{-x} \\ e^{-x}y &= \int 2xe^{-x}\,dx \\ e^{-x}y &= -2xe^{-x} - 2e^{-x} + C \\ y &= -2x - 2 + Ce^{x} \\ 2 &= 2 - 2 + Ce{-}1 \\ 2e &= C \\ y &= 2e^{x+1} - 2xe^{-x} - 2e^{-x} \end{aligned}$$

d) The exact value is $y(2) = 34.1711$

6. **a)** $y(1) \approx 3.004$

b) $y(1) \approx 3.354$

c)

$$\begin{aligned} \int \frac{y}{\sqrt{y^2+1}}\,dy &= \int x\,dx \\ \sqrt{y^2+1} &= \frac{x^2}{2} + C \\ 3 &= C \\ y^2+1 &= \frac{(x^2+6)^2}{4} \\ y &= \sqrt{\frac{(x^2+6)}{4}} - 1 \end{aligned}$$

d) The exact value is $y(1) = 2.5$

7. **a)** $y(2) \approx 9.304$

b) $y(2) \approx 14.390$

c)

$$\int \frac{dy}{y} = \int x^2\,dx$$

$$\begin{aligned} \ln y &= \frac{x^3}{3} + C \\ 0 &= C \\ \ln y &= \frac{x^3}{3} \\ y &= e^{x^3/3} \end{aligned}$$

d) The exact value is $y(2) = e^{8/3} \approx 14.392$

8. **a)** $y(3) \approx 16.909$

b) $y(3) \approx 17.056$

c)

$$\begin{aligned} \int \frac{dy}{y} &= 2\int \frac{dx}{x} \\ \ln y &= \ln x^2 + C \\ \ln 2 &= C \\ \ln y &= \ln x^2 + \ln 2 = \ln 2x^2 \\ y &= 2x^2 \end{aligned}$$

d) The exact value is $y(3) = 18$

9. **a)** $P(5) \approx 101.782$
$P(8) \approx 404.5580472$

b) $P(5) \approx 108.307$
$P(8) \approx 436.449$

c) $P(5) = 108.307$
$P(8) = 436.449$

10. **a)** $Q(30) \approx 0.6596$
$Q(90) \approx 0.9993$

b) $Q(30) \approx 0.6574$
$Q(90) \approx 0.9990$

c) $Q(30) = 0.657$
$Q(90) = 0.999$

11. **a)** $P(2) \approx 0.6975$
$P(4) \approx 0.9589$

b) $P(2) \approx 0.6916$
$P(4) \approx 0.9523$

c) $P(2) = 0.6916$
$P(4) = 0.9523$

12. **a)** $h(1) \approx 0.8757$
$h(2) \approx 0.00509$

b) $h(1) \approx 0.4079$
$h(2) \approx 0.00289$

c) $h(1) = 0.3647$
$h(2) = 0.0025$

13. **a)** $Q(6) \approx 3.0746$

b) $Q(6) \approx 3.1553$

14. **a)** $Q(20) \approx 34.6821$
$Q(40) \approx 158.5996$
$Q(60) \approx 206.5281$
$Q(80) \approx 215.0765$
$Q(100) \approx 216.4213$

b)

$$\begin{aligned} \frac{75Q}{Q+200} - 0.18Q &= 0 \\ 75Q - 0.18Q(200+Q) &= 0 \\ -0.18Q^2 + 39Q &= 0 \\ Q(-0.18Q + 39) &= 0 \\ Q &= 0 \\ Q &= \frac{39}{0.18} = 216.6667 \end{aligned}$$

Now we find the stability of the equilibrium point above

$$\begin{aligned} Q'' &= \frac{75(Q+200) - 75Q}{(Q+200)^2} - 0.18 \\ &= \frac{200}{(Q+200)^2} - 0.18 \\ Q''(0) &= -0.175 \text{ asymptotically stable} \\ Q''(216.6667) &= -0.1788 \text{ asymptotically stable} \end{aligned}$$

c) About 60 days

15. **a)** No

b) No

c) Yes

d) $\Delta x = \frac{1}{n}$ where n is an integer

e) $\Delta x = \frac{a}{n}$ where n is an integer

16. **a)** $y(2) = 2.4522$

b) $y(2) = 2.4290$

c) $y(2) = 2.1475$

d) $|\, 2.1475 - 2.4522 \,| = 0.3047$
$|\, 2.1475 - 2.4290 \,| = 0.2815$

e) 1.0824

17. **a)** $y(2) = 2.47523873$

b) $y(2) = 2.47522970$

c) $y(2) = 2.47522913$

d) $|\, 2.47522913 - 2.47523873 \,| = 0.00000917$
$|\, 2.47522913 - 2.47522970 \,| = 0.000000577$

e) 16

Chapter Review Exercises

1.

$$\begin{aligned} y &= \int 6x^2\,dx \\ &= 2x^3 + C \end{aligned}$$

2.

$$\begin{aligned} y &= \int \frac{\ln x}{x}\,dx \\ &= \frac{\ln^2 x}{2} + C \end{aligned}$$

3.

$$\begin{aligned} y &= \int xe^x\,dx \\ &= xe^x - e^x + C \end{aligned}$$

4. $y' + 6x^2y = 5x^2$

$$\begin{aligned} \int 6x^2\,dx &= 2x^3 + C \\ G(x) &= e^{2x^3} \\ e^{2x^3}y &= \int 5x^2e^{2x^3} \\ e^{2x^3}y &= \frac{5}{6}\,e^{2x^3} + C \\ y &= \frac{5}{6} + Ce^{-2x^3} \end{aligned}$$

5. $y' + \frac{2}{x}\,y = x^2 + 1$

$$\begin{aligned} \int \frac{2dx}{x} &= \ln x^2 + C \\ G(x) &= x^2 \\ x^2y &= \int x^4 + x^2\,dx \\ x^2y &= \frac{x^5}{5} + \frac{x^3}{3} + C \\ y &= \frac{x^3}{5} + \frac{x}{3} + \frac{C}{x^2} \end{aligned}$$

6.

$$\begin{aligned} \int x^2\,dx &= \frac{x^3}{3} + C \\ G(x) &= e^{x^3/3} \\ e^{x^3/3}y &= \int x^2e^{4x^3/3}\,dx \\ e^{x^3/3}y &= \frac{e^{4x^3/3}}{4} + C \\ y &= \frac{e^{x^3/3}}{4} + Ce^{-x^3/3} \end{aligned}$$

7.

$$\begin{aligned} \int y\,dy &= -\int x\,dx \\ \frac{y^2}{2} &= -\frac{x^2}{2} + C \\ y^2 &= C - x^2 \\ y &= \pm\sqrt{C - x^2} \end{aligned}$$

8.

$$\begin{aligned} \int sin\ y\,dy &= \int x\,dx \\ -cos\ y &= \frac{x^2}{2} + C \\ x^2 + 2\ cos\ y &= C \end{aligned}$$

9.

$$\begin{aligned} \int y + \sqrt{y}\,dy &= \int x^3 - 3x + 1\,dx \\ \frac{y^2}{2} + \frac{2y^{3/2}}{3} &= \frac{x^4}{4} - \frac{3x^2}{2} + x + C \\ 6y^2 + 8y^{3/2} &= 3x^4 - 18x^2 + 12x + C \end{aligned}$$

10.

$$\begin{aligned} y &= \int 4x - 5\,dx \\ &= 2x^2 - 5x + C \\ 3 &= 2 + 5 + C \\ -4 &= C \\ y &= 2x^2 - 5x - 4 \end{aligned}$$

11.

$$\begin{aligned} y &= \int cos\ x\sqrt{4 - 3\ sin\ x}\,dx \\ &= -\frac{2}{9}(4 - 3\ sin\ x)^{3/2} + C \\ -1 &= -\frac{2}{9} + C \\ -\frac{7}{9} &= C \\ y &= -\frac{2}{9}\,(4 - 3\ sin\ x)^{3/2} - \frac{7}{9} \end{aligned}$$

12. $y' + 5y = 5x$

$$\begin{aligned} \int 5\,dx &= 5x + C \\ G(x) &= e^{5x} \\ e^{5x}y &= \int 5xe^{5x}\,dx \\ e^{5x}y &= xe^{5x} - \frac{1}{5}\,e^{5x} + C \\ y &= x - \frac{1}{5} + Ce^{-5x} \\ 2 &= -\frac{1}{5} + C \\ \frac{11}{5} &= C \\ y &= x - \frac{1}{5} + \frac{11}{5}\,e^{-5x} \end{aligned}$$

13. $y' + \frac{4}{x}\,y = \frac{8}{x}$

$$\begin{aligned}
\int \frac{4\,dx}{x} &= 4\ln x + C \\
&= \ln x^4 + C \\
G(x) &= x^4 \\
x^4 y &= \int 8x^3\,dx \\
x^4 y &= 2x^4 + C \\
y &= 2 + \frac{C}{x^4} \\
4 &= 2 + C \\
2 &= C \\
y &= 2 + \frac{2}{x^4}
\end{aligned}$$

14.

$$\begin{aligned}
\int 1 + 4y^7\,dy &= \int x^3\,dx \\
y + \frac{y^8}{2} &= \frac{x^4}{4} + C \\
-1 + \frac{1}{2} &= 4 + C \\
-\frac{9}{2} &= C \\
y + \frac{y^8}{2} &= \frac{x^4}{4} - \frac{9}{2} \\
4y + 2y^8 &= x^4 - 18
\end{aligned}$$

15.

$$\begin{aligned}
\int y\,dy &= \int 3x^2\,dx \\
\frac{y^2}{2} &= x^3 + C \\
2 &= 1 + C \\
1 &= C \\
\frac{y^2}{2} &= x^3 + 1 \\
y^2 &= 2x^3 + 2 \\
y &= -\sqrt{2x^3 + 2}
\end{aligned}$$

16. $y' = y^2 - 2y - 24$
$y'' = 2y - 2$

a - b)

$$\begin{aligned}
y^2 - 2y - 24 &= 0 \\
y &= 6 \\
y &= -4 \\
y''(6) &= 10 \text{ unstable} \\
y''(-4) &= -10 \text{ asymptotically stable}
\end{aligned}$$

c)

$$\begin{aligned}
2y - 2 &= 0 \\
y &= 1 y''(0) = -2 < 0
\end{aligned}$$

$y''(3) = 4 > 0$

Inflection point at $y = 1$

d)

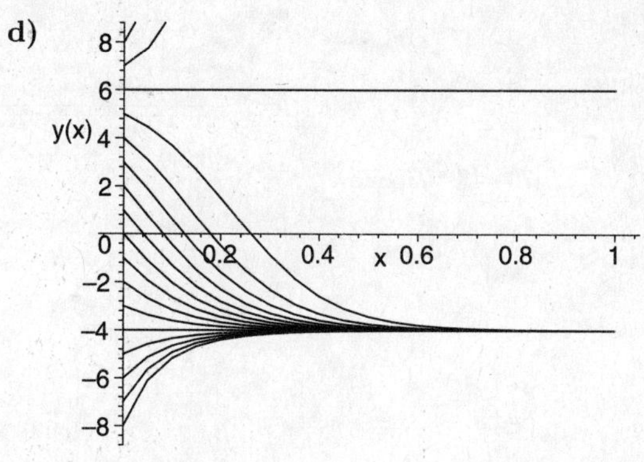

17. $y' = y^3 - 6y^2 + 9y$
$y'' = 3y^2 - 12y + 9$

a - b)

$$\begin{aligned}
y^3 - 6y^2 + 9y &= 0 \\
y(y-3)^2 &= 0 \\
y &= 0 \\
y &= 3 \\
y''(0) &= 9 \text{ unstable} \\
y''(3) &= 0 \text{ semistable}
\end{aligned}$$

c)

$$\begin{aligned}
3y^2 - 12y + 9 &= 0 \\
3(y-1)(y-3) &= 0 \\
y &= 1 \\
y &= 3 \\
y''(0) &= 9 > 0 \\
y''(2) &= -3 < 0 \\
y''(4) &= 9 > 0
\end{aligned}$$

Inflection points at $y = 1$ and $y = 3$

d)

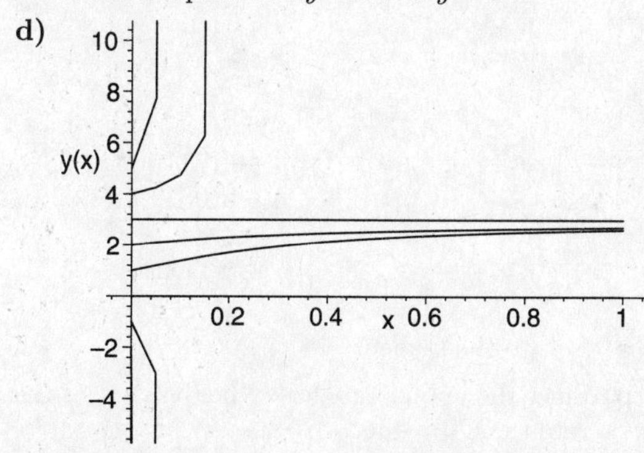

18. This follows the logistic growth model with $k = 0.62$ and $L = 4800$
The solution is given by (see page 570)

$$y(t) = \frac{33600}{7 + 4799.38e^{-0.62t}}$$

a)

$$\begin{aligned}
y(4) &= \frac{33600}{7 + 4799.38e^{-0.62(4)}} \\
&= 82.2 \text{ million}
\end{aligned}$$

b)

$$\begin{aligned}
500 &= \frac{33600}{7+4799.38e^{-0.62t}}\\
7+4799.38e^{-0.62t} &= \frac{33600}{500}\\
4799.38e^{-0.62t} &= \frac{33600}{500}-7\\
-0.62t &= \ln\left(\frac{\frac{33600}{500}-7}{4799.38}\right)\\
t &= \frac{-1}{0.62}\ln\left(\frac{\frac{33600}{500}-7}{4799.38}\right)\\
&= 7.06 \text{ hours}
\end{aligned}$$

19. a)

$$\begin{aligned}
\int \frac{dP}{(1-P)} &= \int k\,dt\\
-\ln(1-P) &= kt+C\\
\ln(1-P) &= -kt+C\\
1-P &= Ce^{-kt}\\
P &= Ce^{-kt}+1\\
0 &= C+1\\
C &= -1\\
P &= 1-e^{-kt}
\end{aligned}$$

b)

$$\begin{aligned}
0.5 &= 1-e^{-4k}\\
0.5 &= e^{-4k}\\
\ln 0.5 &= -4k\\
k &= \frac{-\ln 0.5}{4}\\
&= 0.1733
\end{aligned}$$

c)

$$\begin{aligned}
0.9 &= 1-e^{-0.1733t}\\
0.1 &= e^{-0.1733t}\\
\ln 0.1 &= -0.1733t\\
t &= \frac{-\ln 0.1}{0.1733}\\
&= 13.3 \text{ months}
\end{aligned}$$

20. a) Left to the student

b) Since the uptake rate has to be the same as the decay rate the $C' = -aC + b$

c)

$$\begin{aligned}
C'+aC &= b\\
\int a\,dt &= at+C\\
G(t) &= e^{at}\\
e^{at}C &= \int be^{at}\,dt\\
e^{at}C &= \frac{b}{a}e^{at}+K\\
K &= \frac{b}{a}+Ce^{-at}\\
C_0 &= \frac{b}{a}+K\\
C_0-\frac{b}{a} &= K\\
C &= \frac{b}{a}-\left(C_0-\frac{b}{a}\right)e^{-at}\\
&= \left(\frac{b}{a}-C_0\right)e^{-at}+\frac{b}{a}
\end{aligned}$$

d)

$$\begin{aligned}
\lim_{t\to\infty} C &= \lim_{t\to\infty}\left(\frac{b}{a}-C_0\right)e^{-at}+\frac{b}{a}\\
&= 0+\frac{b}{a}\\
&= \frac{b}{a}
\end{aligned}$$

21. a)

$$\begin{aligned}
r\frac{dx}{dt}+\frac{1}{c}x &= pd+\frac{x_0}{c}\\
x'+\frac{1}{rc}x &= \frac{cdp+x_0}{cr}\\
\int \frac{1}{cr}\,dt &= \frac{t}{cr}+C\\
G(x) &= e^{t/cr}\\
e^{t/cr}x &= \int e^{t/cr}\frac{cdp+x_0}{cr}\,dt\\
e^{t/cr}x &= (cdp+x_0)e^{t/cr}+C\\
x &= cdp+x_0+Ce^{-t/cr}
\end{aligned}$$

b)

$$\begin{aligned}
x &= cdp+x_0+Ce^{-t/cr}\\
0 &= cdp+x_0+C\\
C &= -cdp-x_0\\
x &= (cdp+x_0)(e^{-t/cr}-1)
\end{aligned}$$

22. a)

$$\begin{aligned}
r\frac{dx}{dt}+\frac{1}{c}x &= \frac{x_0}{c}\\
x'+\frac{1}{rc}x &= \frac{x_0}{cr}\\
\int \frac{1}{cr}\,dt &= \frac{t}{cr}+C\\
G(x) &= e^{t/cr}\\
e^{t/cr}x &= \int e^{t/cr}\frac{x_0}{cr}\,dt\\
e^{t/cr}x &= x_0e^{t/cr}+C\\
x &= x_0+Ce^{-t/cr}
\end{aligned}$$

b)

$$x = x_0+Ce^{-t/cr}$$

$$\begin{aligned} a &= x_0 + Ce^{-T/cr} \\ C &= (a - x_0)\, e^{T/cr} \\ x &= x_0 + e^{-t/cr} + (a - x_0)\, e^{T/cr} \\ &= x_0 + (a - x_0) e^{\frac{T-t}{cr}} \end{aligned}$$

23. **a)** $y(2) = 1.4285$
b) $y(2) = 1.5414$

24. **a)** $y(2) = 1.3649$
b) $y(2) = 1.3605$

Chapter 8 Test

1.

$$\begin{aligned} y &= \int 2x^2 - 4\,dx \\ &= \frac{2}{3}\,x^3 - 4x + C \end{aligned}$$

2.

$$\begin{aligned} y &= \int e^{2x}(e^{2x} - 2)^4\,dx \\ &= \frac{1}{5}(e^{2x} - 2)^5 + C \end{aligned}$$

3.

$$\begin{aligned} y &= \int x \; sin\; x\,dx \\ &= sin\; x - x\; cos\; x + C \end{aligned}$$

4.

$$\begin{aligned} \int 4\,dx &= 4x + C \\ G(x) &= e^{4x} \\ e^{4x} y &= \int 6e^{4x}\,dx \\ e^{4x} y &= \frac{3}{2}\,e^{4x} + C \\ y &= \frac{3}{2} + Ce^{-4x} \end{aligned}$$

5. $y' + \frac{2}{x}\,y = \frac{1}{x^3}$

$$\begin{aligned} \int \frac{2dx}{x} &= 2\ln x + C = \ln x^2 + C \\ G(x) &= x^2 \\ x^2 y &= \int \frac{dx}{x} \\ x^2 y &= \ln x + C \\ y &= \frac{\ln x + C}{x^2} \end{aligned}$$

6.

$$\begin{aligned} \int x\,dx &= \frac{x^2}{2} + C \\ G(x) &= e^{x^2/2} \\ e^{x^2/2} y &= \int 3xe^{x^2/2}\,dx \\ x^2 y &= 3e^{x^2/2} + C \\ y &= 3 + Ce^{-x^2/2} \end{aligned}$$

7. $y' = e^{3x} e^{-2y}$

$$\begin{aligned} \int e^{2y}\,dy &= \int e^{3x}\,dx \\ \frac{1}{2}\,e^{2y} &= \frac{1}{3}\,e^{3x} + C \\ e^{2y} &= \frac{2}{3}\,e^{3x} + C \\ 2y &= \ln\left|\frac{2}{3}e^{3x} + C\right| \\ y &= \frac{1}{2}\ln\left|\frac{2}{3}e^{3x} + C\right| \end{aligned}$$

8. Note $y = 0$ is a constant solution

$$\begin{aligned} \int y^{-3}\,dy &= \int x\,dx \\ \frac{-1}{2y^2} &= \frac{x^2}{2} + C \\ y^2 &= \frac{-1}{x^2 + C} \\ y &= \pm\frac{1}{\sqrt{C - x^2}} \end{aligned}$$

9.

$$\begin{aligned} \int y^3 - 2y\,dy &= \int x^2\,dx \\ \frac{y^4}{4} - y^2 &= \frac{x^3}{3} + C \\ 3y^4 - 12y^2 &= 4x^3 + C \end{aligned}$$

10.

$$\begin{aligned} y &= \int x^2 + 3x + 5\,dx \\ &= \frac{x^3}{3} + \frac{3x^2}{2} + 5x + C \\ 7 &= \frac{1}{3} + \frac{3}{2} + 5 + C \\ \frac{1}{6} &= C \\ y &= \frac{x^3}{3} + \frac{3x^2}{2} + 5x + \frac{1}{6} \end{aligned}$$

11.

$$y = \int x^3\; sin(x^4 + \pi)\,dx$$

$$\begin{aligned} &= \frac{-1}{4}\cos(x^4+\pi)+C \\ -1 &= \frac{-1}{4}(-1)+C \\ \frac{-5}{4} &= C \\ y &= -\frac{1}{4}\cos(x^4+\pi)-\frac{5}{4} \end{aligned}$$

12.

$$\begin{aligned} \int \frac{dy}{y} &= -3\int \frac{dx}{x} \\ \ln y &= -3\ \ln x + C \\ \ln y + \ln x^3 &= C \\ \ln x^3 y &= C \\ \ln 64 &= C \\ \ln x^3 y &= \ln 64 \\ x^3 y &= 64 \\ y &= \frac{64}{x^3} \end{aligned}$$

13.

$$\begin{aligned} \int -\tan x\, dx &= \ln|\cos x| + C \\ G(x) &= \cos x \\ \cos xy &= \int \cos x\, dx \\ \cos xy &= \sin x + C \\ y &= \tan x + C \sec x \\ 2 &= 1 + \frac{\sqrt{2}}{2}C \\ C &= 1 - \frac{\sqrt{2}}{2} \\ y &= \tan x + \sec x - \frac{\sqrt{2}}{2}\sec x \end{aligned}$$

14.

$$\begin{aligned} \int y\, dy &= \int x+1\, dx \\ \frac{y^2}{2} &= \frac{x^2}{2}+x+C \\ 8 &= 2+2+C \\ 4 &= C \\ \frac{y^2}{2} &= \frac{x^2}{2}+x+4 \\ y^2 &= x^2+2x+8 \\ y &= \sqrt{x^2+2x+8} \end{aligned}$$

15.

$$\begin{aligned} \int e^y+1\, dy &= \int dx \\ e^y+y &= x+C \\ e^{-1}-1 &= C \\ e^y+y &= x+e^{-1}-1 \\ e^y+y &= x-1+\frac{1}{e} \end{aligned}$$

16. $y' = 2y^2+6y$
$y'' = 4y+6$

a - b)

$$\begin{aligned} 2y^2+6y &= 0 \\ 2y(y+3) &= 0 \\ y &= 0 \\ y &= -3 \\ y''(0) &= 6 \text{ unstable} \\ y''(-3) &= -6 \text{ asymptotically stable} \end{aligned}$$

c)

$$\begin{aligned} 4y+6 &= 0 \\ y &= -\frac{3}{2} \\ y''(-2) &= -2<0 \\ y''(0) &= 6>0 \end{aligned}$$

Inflection point at $y = \frac{-3}{2}$

d)

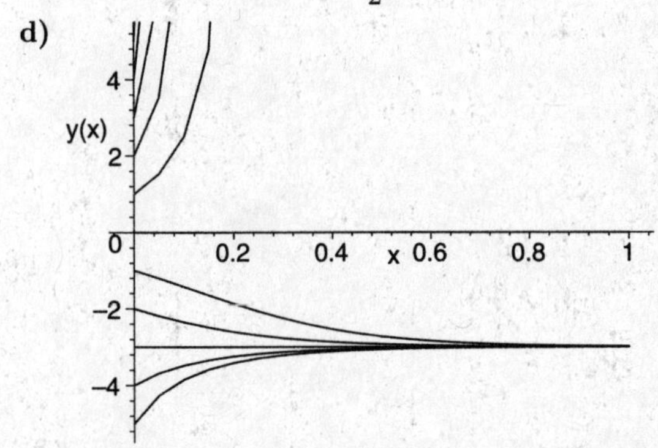

17. $y' = y^2-9$
$y'' = 2y$

a - b)

$$\begin{aligned} y^2-9 &= 0 \\ (y-3)(y+3) &= 0 \\ y &= -3 \\ y &= 3 \\ y''(-3) &= -6 \text{ asymptotically stable} \\ y''(3) &= 6 \text{ unstable} \end{aligned}$$

c)

$$\begin{aligned} 2y &= 0 \\ y &= 0 \\ y''(-2) &= -4<0 \\ y''(2) &= 4>0 \end{aligned}$$

Inflection point at $y = 0$

d)

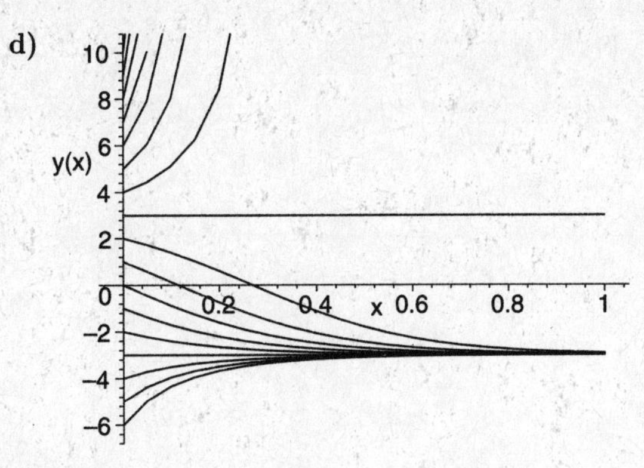

18. a)

$$\begin{aligned} \frac{dP}{dt} &= k(1-P) \\ P' &= k - kP \\ P' + kP &= k \end{aligned}$$

b)

$$\begin{aligned} \int k\,dt &= kt + C \\ G(t) &= e^{kt} \\ e^{kt}P &= \int ke^{kt}\,dt \\ e^{kt}P &= e^{kt} + C \\ P &= 1 + Ce^{-kt} \\ 0 &= 1 + C \\ C &= -1 \\ P &= 1 - e^{-kt} \end{aligned}$$

c)

$$\begin{aligned} 0.6 &= 1 - e^{-5k} \\ 0.4 &= e^{-5k} \\ \ln 0.4 &= -5k \\ k &= \frac{\ln 0.4}{5} \\ &= 0.1833 \end{aligned}$$

d)

$$\begin{aligned} P &= 1 - e^{-0.1833t} \\ P(10) &= 1 - e^{1.833} \\ &= 0.84 \end{aligned}$$

e)

$$\begin{aligned} 0.9 &= 1 - e^{-0.1833t} \\ 0.1 &= e^{-0.1833t} \\ \ln 0.1 &= -0.1833t \\ t &= \frac{\ln 0.1}{-0.1833} \\ &= 12.6 \approx 13 \end{aligned}$$

19. $S' = -\frac{4}{400}S + 8 \to S' + 0.01S = 8$

$$\begin{aligned} \int 0.01\,dt &= 0.01t + C \\ G(t) &= e^{0.01t} \\ e^{0.01t}S &= \int 8e^{0.01t}\,dt \\ e^{0.01t}S &= 800e^{0.01t} + C \\ S &= 800 + Ce^{-0.01t} \\ 100 &= 800 + C \\ C &= -700 \\ S &= 800 - 700e^{-0.01t} \\ S(60) &= 800 - 700e^{-0.6} \\ &= 415.83\text{pounds} \end{aligned}$$

20. a) Left to the student

b) There are two downtake rates and no uptake rate. $D' = -0.226D - (3 \times 10^{-8})$

c)

$$\begin{aligned} D' + 0.226D &= 3 \times 10^{-8} \\ \int 0.226\,dt &= 0.226t + C \\ G(t) &= e^{0.226t} \\ e^{0.226t}D &= \int (3 \times 10^{-8})e^{0.226t}\,dt \\ e^{0.226t}D &= (1.33 \times 10^{-7})e^{0.226t} + C \\ D &= (1.33 \times 10^{-7}) + Ce^{-0.226t} \\ D_0 &= (1.33 \times 10^{-7}) + C \\ C &= D_0 - (1.33 \times 10^{-7}) \\ D &= (1.33 \times 10^{-7}) + \\ & \left(D_0 - (1.33 \times 10^{-7})\right)e^{-0.226t} \end{aligned}$$

21. This population follows the logistic growth model with $k = 0.13$ and $L = 16$. The population is given by

$$\frac{27.2}{1.7 + 14.3e^{-0.13t}}$$

a)

$$\begin{aligned} P(6) &= \frac{27.2}{1.7 + 14.3e^{-0.13(6)}} \\ &= 3.29 \text{ million} \end{aligned}$$

b)

$$\begin{aligned} 10 &= \frac{27.2}{1.7 + 14.3e^{-0.13t}} \\ 1.7 + 14.3e^{-0.13t} &= 2.72 \\ 14.3e^{-0.13t} &= 1.02 \\ e^{-0.13t} &= \frac{1.02}{14.3} \\ -0.13t &= \ln\left(\frac{1.02}{14.3}\right) \\ t &= \frac{-1}{0.13}\ln\left(\frac{1.02}{14.3}\right) \\ &= 20.3\text{hours} \end{aligned}$$

22. Using the logistic growth model

$$\begin{aligned} P &= \frac{7000}{10+690e^{-kt}} \\ 300 &= \frac{7000}{10+690e^{-8k}} \\ 10+690e^{-8k} &= \frac{70}{3} \\ 690e^{-8k} &= \frac{70}{3}-10 \\ e^{-8k} &= \frac{1}{690}\left(\frac{40}{3}\right) \\ -8k &= \ln\left(\frac{40}{2070}\right) \\ k &= \frac{-1}{8}\ln\left(\frac{40}{2070}\right) \\ &= 0.4933 \end{aligned}$$

23. **a)** $y(2) = 28.9587$

b) $y(2) = 34.9444$

23. **a)** $y(2) = 1.8421$

b) $y(2) = 1.8422$

Technology Connection

- **Page 586** Left to the student
- **Page 589** Left to the student

Extended Life Science Connection

1. **a)** Left to the student

b) There is no growth rates to consider only decay due to loss of moisture

c)

$$\begin{aligned} \int \frac{dc}{c} &= \int -b\,dt \\ \ln c &= -bt+K \\ c(t) &= Ke^{-bt} \\ c_0 &= K \\ c(t) &= c_0e^{-bt} \end{aligned}$$

d) $c(T) = c_0e^{-bT}$

e)

$$\begin{aligned} c &= c_0e^{-bT} \\ \frac{c}{c_0} &= e^{-bT} = p \\ \ln p &= -bT \\ b &= -\frac{\ln p}{T} \end{aligned}$$

2. **a)** Left to the student

b) There are no growth rates only decay rates due to loss of moisture and mass eaten by the larva

c)

$$\begin{aligned} f'+bf &= -aN \\ \int b\,dt &= bt+C \\ G(t) &= e^{bt} \\ e^{bt}f &= \int -aNe^{bt}\,dt \\ e^{bt}f &= -\frac{aN}{b}e^{bt}+C \\ f &= -\frac{aN}{b}+C\,e^{-bt} \\ F_0 &= -\frac{aN}{b}+C \\ C &= F_0+\frac{aN}{b} \\ f(t) &= -\frac{aN}{b}+\frac{aN}{b}e^{-bt}+F_0e^{-bt} \\ &= -\frac{aN}{b}\left(1-e^{-bt}\right)+F_oe^{-bt} \end{aligned}$$

d) $F_1 = -\frac{aN}{b}\left(1-e^{-bT}\right)+F_oe^{-bT}$

e)

$$\begin{aligned} F_1 &= -\frac{aN}{b}\left(1-e^{-bT}\right)+F_oe^{-bT} \\ &= -\frac{aN}{b}\left(1-e^{-T(-\ln p/t)}\right)+F_0e^{-T(\ln p/T)} \\ &= -\frac{aN}{b}(1-p)+F_0p \end{aligned}$$

3. **a)** The mass of the leafs consumed has to equal the the number of larva times the rate at which the consume the leafs times the times they spend consuming the leaves.

b)

$$\begin{aligned} F_1 &= -\frac{aN}{b}\left(1-e^{-bT}\right)+F_oe^{-bT} \\ F_1-F_0e^{-bT} &= -\frac{aN}{b}\left(1-e^{-bT}\right) \\ F_1-F_0p &= -\frac{aN}{b}(1-p) \\ a &= \frac{-(F_1-F_0p)\,b}{(1-p)N} \end{aligned}$$

c)

$$\begin{aligned} M &= aNT \\ &= \frac{-(F_1-F_0p)\,bNT}{(1-p)N} \\ &= \frac{(F_1-F_0p)\ln p}{1-p} \\ &= \frac{-(pF_0-F_1)\ln p}{1-p} \end{aligned}$$

4. **a)**

$$\begin{aligned} M &= \frac{-(pF_0-F_1)\,\ln p}{1-p} \\ &= \frac{-\left[\left(\frac{0.324}{0.352}\right)0.346-0.307\right]\,\ln\left(\frac{0.324}{0.352}\right)}{1-\left(\frac{0.324}{0.352}\right)} \\ &= 0.01196 \text{ grams} \end{aligned}$$

b)

$$\begin{aligned} \%M &= \frac{0.01196}{0.346} \\ &= 0.034565 \approx 3.5\% \end{aligned}$$

c)

$$\begin{aligned} \overline{M} &= \frac{0.01196}{2 \cdot 20} \\ &= 0.0002989876 \text{ grams per day per larva} \end{aligned}$$

5. a)

$$\begin{aligned} M &= \frac{-(pF_0 - F_1)\ \ln p}{1-p} \\ &= \frac{-\left[\left(\frac{1.085}{1.282}\right) 1.165 - 0.365\right]\ \ln\left(\frac{1.085}{1.282}\right)}{1 - \left(\frac{1.085}{1.282}\right)} \\ &= 0.67422 \text{ grams} \end{aligned}$$

b)

$$\begin{aligned} \%M &= \frac{0.67422}{1.165} \\ &= 0.57873 \approx 57.9\% \end{aligned}$$

c)

$$\begin{aligned} \overline{M} &= \frac{.67422}{2 \cdot 30} \\ &= 0.011237 \text{ grams per day per larva} \end{aligned}$$

Chapter 9

Higher Order and Systems of Differential Equations

Exercise Set 9.1

1.

$$\begin{aligned} r^2 - 6r + 5 &= 0 \\ (r-1)(r-5) &= 0 \\ r &= 1 \\ r &= 5 \\ y &= C_1 e^{x} + C_2 e^{5x} \end{aligned}$$

2.

$$\begin{aligned} r^2 - 8r + 15 &= 0 \\ (r-3)(r-5) &= 0 \\ r &= 3 \\ r &= 5 \\ y &= C_1 e^{3x} + C_2 e^{5x} \end{aligned}$$

3.

$$\begin{aligned} r^2 - r - 2 &= 0 \\ (r-2)(r+1) &= 0 \\ r &= -1 \\ r &= 2 \\ y &= C_1 e^{-x} + C_2 e^{2x} \end{aligned}$$

4.

$$\begin{aligned} r^2 + 2r - 3 &= 0 \\ (r+3)(r-1) &= 0 \\ r &= -3 \\ r &= 1 \\ y &= C_1 e^{-3x} + C_2 e^{x} \end{aligned}$$

5.

$$\begin{aligned} r^2 + 3r + 2 &= 0 \\ (r+1)(r+2) &= 0 \\ r &= -1 \\ r &= -2 \\ y &= C_1 e^{-2x} + C_2 e^{-1x} \end{aligned}$$

6.

$$\begin{aligned} r^2 + 5r + 6 &= 0 \\ (r+2)(r+3) &= 0 \\ r &= -2 \\ r &= -3 \\ y &= C_1 e^{-3x} + C_2 e^{-2x} \end{aligned}$$

7.

$$\begin{aligned} 2r^2 - 5r + 2 &= 0 \\ (2r-1)(r-2) &= 0 \\ r &= \frac{1}{2} \\ r &= 2 \\ y &= C_1 e^{x/2} + C_2 e^{2x} \end{aligned}$$

8.

$$\begin{aligned} 2r^2 - r + 1 &= 0 \\ (2r+1)(r-1) &= 0 \\ r &= \frac{1}{2} \\ r &= 1 \\ y &= C_1 e^{x/2} + C_2 e^{x} \end{aligned}$$

9.

$$\begin{aligned} r^2 - 9 &= 0 \\ (r-3)(r+3) &= 0 \\ r &= -3 \\ r &= 3 \\ y &= C_1 e^{-3x} + C_2 e^{3x} \end{aligned}$$

10.

$$\begin{aligned} r^2 - 3 &= 0 \\ (r-\sqrt{3})(r+\sqrt{3}) &= 0 \\ r &= -\sqrt{3} \\ r &= \sqrt{3} \\ y &= C_1 e^{-\sqrt{x}} + C_2 e^{\sqrt{x}} \end{aligned}$$

11.

$$\begin{aligned} r^2 + 10r + 25 &= 0 \\ (r+5)^2 &= 0 \\ r &= -5 \text{ repeated} \\ y &= C_1 e^{-5x} + C_2 x e^{-5x} \end{aligned}$$

12.

$$\begin{aligned} r^2 - 8r + 16 &= 0 \\ (r-4)^2 &= 0 \\ r &= 4 \text{ repeated} \\ y &= C_1 e^{4x} + C_2 x e^{4x} \end{aligned}$$

13.

$$\begin{aligned} 4r^2+12r+9 &= 0\\ (2r+3)^2 &= 0\\ r &= -\frac{3}{2} \text{ repeated}\\ y &= C_1e^{-3x/2}+C_2xe^{-3x/2} \end{aligned}$$

14.

$$\begin{aligned} r^3+8r^2+16 &= 0\\ r(r+4)^2 &= 0\\ r &= 0\\ r &= -4 \text{ repeated}\\ y &= C_1+C_2e^{-4x}+C_3xe^{-4x} \end{aligned}$$

15.

$$\begin{aligned} r^3+r^2+4r+4 &= 0\\ r^2(r+1)+4(r+1) &= 0\\ (r+1)(r^2+4) &= 0\\ r &= -1\\ r &= \pm 2i\\ y &= C_1e^{-x}+C_2 \sin 2x+C_3 \cos 2x \end{aligned}$$

16.

$$\begin{aligned} r^3+2 &= 0\\ (r+2)(r^2-2r+4) &= 0\\ r &= -2\\ r &= 1\pm i\sqrt{3}\\ y &= C_1e^{-2x}+C_2e^{x} \sin \sqrt{3}x+\\ &\quad C_3e^{x} \cos \sqrt{3}x \end{aligned}$$

17.

$$\begin{aligned} r^3+6r^2+12r+8 &= 0\\ (r+2)^3 &= 0\\ r &= -2 \text{ repeated twice}\\ y &= C_1e^{-2x}+C_2xe^{-2x}+C_3x^2e^{-2x} \end{aligned}$$

18.

$$\begin{aligned} r^4+3r^3+3r^2+r &= 0\\ r(r+1)^3 &= 0\\ r &= 0\\ r &= -1 \text{ repeated twice}\\ y &= C_1+C_2e^{-x}+C_3xe^{-x}+C_4x^2e^{-x} \end{aligned}$$

19.

$$\begin{aligned} r^3-6r^2+3r-18 &= 0\\ (r-6)(r^2+3) &= 0\\ r &= 6\\ r &= \pm i\sqrt{3}\\ y &= C_1e^{6x}+C_2 \sin \sqrt{3}x+C_3 \cos \sqrt{3}x \end{aligned}$$

20.

$$\begin{aligned} r^4-4r^3 &= 0\\ r^3(r-4) &= 0\\ r &= 0 \text{ repeated twice}\\ r &= 4\\ y &= C_1+C_2x+C_3x^2+C_4e^{4x} \end{aligned}$$

21.

$$\begin{aligned} r^4-5r^3+4r^2 &= 0\\ r^2(r-4)(r-1) &= 0\\ r &= 0 \text{ repeated}\\ r &= 4\\ r &= 1\\ y &= C_1+C_2x+C_3e^{x}+C_4e^{4x} \end{aligned}$$

22.

$$\begin{aligned} r^4+3r^3-r^2-3r &= 0\\ (r+1)(r^3-1) &= 0\\ (r+1)(r-1)(r^2+r+1) &= 0\\ r &= -1\\ r &= 1\\ r &= -\frac{1}{2}\pm i\frac{\sqrt{3}}{2}\\ y &= C_1e^{-x}+C_2e^{x}+\\ &\quad C_3e^{-x/2} \sin \frac{\sqrt{3}x}{2}+\\ &\quad C_4e^{-x/2} \cos \frac{\sqrt{3}x}{2} \end{aligned}$$

23.

$$\begin{aligned} r^2+36 &= 0\\ r &= \pm 6i\\ y &= C_1 \sin 6x+C_2 \cos 6x \end{aligned}$$

24.

$$\begin{aligned} r^2+4r+13 &= 0\\ r &= -2\pm 3i\\ y &= e^{2x}\left(C_1 \sin 3x+C_2 \cos 3x\right) \end{aligned}$$

25.

$$\begin{aligned} r^2+8r+41 &= 0\\ r &= 4\pm 5i\\ y &= e^{4x}\left(C_1 \sin 5x+C_2 \cos 5x\right) \end{aligned}$$

26.

$$\begin{aligned} r^2+r+1 &= 0\\ r &= -\frac{1}{2}\pm\frac{\sqrt{3}}{2}\\ y &= e^{-x/2}\left(C_1 \sin \frac{\sqrt{3}x}{2}+C_2 \cos \frac{\sqrt{3}x}{2}\right) \end{aligned}$$

27.

$$\begin{aligned} r^3 + 2r^2 + 5r &= 0 \\ r(r^2 + 2r + 5) &= 0 \\ r &= 0 \\ r &= -1 \pm 2i \\ y &= C_1 + e^{-x}(C_1 \sin 2x + C_2 \cos 2x) \end{aligned}$$

28.

$$\begin{aligned} r^3 - 2r^2 + 4r - 8 &= 0 \\ (r-2)(r^2+2) &= 0 \\ r &= 2 \\ r &= \pm i\sqrt{2} \\ y &= C_1 e^{2x} + C_2 \sin \sqrt{2}x + C_3 \cos \sqrt{2}x \end{aligned}$$

29.

$$\begin{aligned} r^3 - 1 &= 0 \\ (r-1)(r^2+r+1) &= 0 \\ r &= 1 \\ r &= -\frac{1}{2} \pm \frac{i\sqrt{3}}{2} \\ y &= C_1 e^x + C_2 e^{-x/2} \sin \frac{\sqrt{3}x}{2} + \\ & C_3 e^{-x/2} \cos \frac{\sqrt{3}x}{2} \end{aligned}$$

30.

$$\begin{aligned} r^4 + 8r^2 + 16 &= 0 \\ (r^2+4)^2 &= 0 \\ r &= \pm 2i \text{ repeated} \\ y &= C_1 \sin 2x + C_2 \cos 2x + \\ & C_3 x \sin 2x + C_4 x \cos 2x \end{aligned}$$

31.

$$\begin{aligned} 2r^2 + 2r - 5 &= 0 \\ r &= -\frac{1 \pm \sqrt{11}}{2} \\ y &= C_1 e^{(-1-\sqrt{11})x/2} + C_2 e^{(-1+\sqrt{11})x/2} \end{aligned}$$

32.

$$\begin{aligned} r^2 - 5r + 1 &= 0 \\ r &= \frac{5 \pm \sqrt{17}}{4} \\ y &= C_1 e^{(5-\sqrt{17})x/4} + C_2 e^{(5+\sqrt{17})x/2} \end{aligned}$$

33.

$$\begin{aligned} 3r^2 - 2r + 10 &= 0 \\ r &= \frac{2 \pm i\sqrt{116}}{6} \\ y &= e^{x/3}\left(C_1 \sin \frac{\sqrt{116}x}{6} + C_2 \cos \frac{\sqrt{116}x}{6}\right) \\ &= e^{x/3}\left(C_1 \sin \frac{\sqrt{29}x}{3} + C_2 \cos \frac{\sqrt{29}x}{3}\right) \end{aligned}$$

34.

$$\begin{aligned} 2r^2 + 6r + 5 &= 0 \\ r &= \frac{-3 \pm 2i}{2} \\ y &= e^{-3x/2}(C_1 \sin x + C_2 \cos x) \end{aligned}$$

35.

$$\begin{aligned} r(r-1) &= 0 \\ r &= 0 \\ r &= 1 \\ y &= C_1 + C_2 e^x \\ y' &= C_2 e^x \\ y(0) &= 0 \to C_1 = -C_2 \\ y'(0) &= -1 \to C_2 = -1 \\ \to & \\ C_1 &= 1 \\ y &= 1 - e^x \end{aligned}$$

36.

$$\begin{aligned} r^2 + 1 &= 0 \\ r &= \pm i \\ y &= C_1 \sin x + C_2 \cos x \\ y' &= C_1 \cos x - C_2 \sin x \\ y(\pi) &= 4 \to C_1 = -4 \\ y'(\pi) &= 3 \to C_2 = -3 \\ y &= -4 \sin x - 3 \cos x \end{aligned}$$

37.

$$\begin{aligned} r^2 - 1 &= 0 \\ r &= \pm 1 \\ y &= C_1 e^{-x} + C_2 e^x \\ y' &= -C_1 e^{-x} + C_2 e^x \\ y(0) &= 1 \to C_1 + C_2 = 1 \\ y'(0) &= 2 \to -C_1 + C_2 = 3 \end{aligned}$$

Solving the system

$$\begin{aligned} C_1 &= -1 \\ C_2 &= 2 \\ y &= 2e^x - e^{-x} \end{aligned}$$

38.

$$\begin{aligned} (r-1)^2 &= 0 \\ r &= 1 \text{ repeated} \\ y &= C_1 e^x + C_2 x e^x \\ y' &= C_1 e^x + C_2 x e^x + C_2 e^x \\ y(0) &= 1 \to C_1 = 1 \\ y'(0) &= 2 \to C_2 = 1 \\ y &= e^x + x e^x \end{aligned}$$

39.

$$\begin{aligned}(r+2)^2 &= 0\\ r &= -2 \text{ repeated}\\ y &= C_1e^{-2x} + C_2xe^{-2x}\\ y' &= -2C_1e^{-x} - 2C_2xe^{-x} + C_2e^{-x}\\ y(0) &= 2 \to C_1 = 2\\ y'(0) &= 3 \to C_2 = 7\\ y &= e^x + xe^x\end{aligned}$$

40.

$$\begin{aligned}r^2 - 2r + 2 &= 0\\ r &= 1 \pm 2i\\ y &= e^x(C_1 \sin 2x + C_2 \cos 2x)\\ y' &= e^x(2C_1 \cos 2x - 2C_2 \sin 2x) +\\ &\quad e^x(C_1 \sin 2x + C_2 \cos 2x)\\ y(0) &= 1 \to C_1 = 1\\ y'(0) &= -1 \to C_2 = -1\\ y &= e^x(\sin 2x - \cos 2x)\end{aligned}$$

41.

$$\begin{aligned}r(r^2+1) &= 0\\ r &= 0\\ r &= \pm i\\ y &= C_1 + C_2 \sin x + C_3 \cos x\\ y' &= C_2 \cos x - C_3 \sin x\\ y'' &= -C_2 \sin x - C_3 \cos x\\ y(\pi) &= 1 \to C_1 - C_2 = 1\\ y'(\pi) &= 8 \to C_2 = 4\\ y''(\pi) &= 4 \to C_3 = -8\\ &\to\\ C_1 = 5\\ y &= 5 - 8 \sin x + 4 \cos x\end{aligned}$$

42.

$$\begin{aligned}r(r+3)^2 &= 0\\ r &= 0\\ r &= -3 \text{ repeated}\\ y &= C_1 + C_2e^{-3x} + C_3xe^{-3x}\\ y' &= -3C_2e^{-3x} - 3C_3xe^{-3x} + C_3e^{-3x}\\ y'' &= 9C_2e^{-3x} + 9C_3xe^{-3x} - 6C_3e^{-3x}\\ y(0) &= 2 \to C_1 + C_2 = 0\\ y'(0) &= -2 \to -3C_2 + C_3 = -2\\ y''(0) &= 3 \to 9C_2 - 6C_3 = 3\end{aligned}$$

Solving the system

$$\begin{aligned}C_3 &= -2\\ C_2 &= -1\\ C_1 &= 3\\ y &= 3 - e^{-3x} - 2xe^{-3x}\end{aligned}$$

43.

$$\begin{aligned}r^2 - 3r + 2 &= 0\\ r &= 1\\ r &= 2\\ N &= C_1e^t + C_2e^{2t}\\ N' &= C_1e^t + 2C_2e^{2t}\\ N(0) &= 5 \to C_1 + C_2 = 5\\ N'(0) &= 15 \to C_1 + 2C_2 = 15\end{aligned}$$

Solving the system

$$\begin{aligned}C_1 &= -5\\ C_2 &= 10\\ N(t) &= 10e^{2t} - 5e^t\\ N(4) &= 10e^8 - 5e^4\\ &= 29537\end{aligned}$$

44.

$$\begin{aligned}r^2 + (c+m+2a)r + a(c+m+a) &= 0\\ \frac{-(c+m+2a) \pm \sqrt{(c+m+2a)^2 - 4a(c+m+a)}}{2} &= r\end{aligned}$$

Since the radicand is always smaller than $(c+m+2a)$ the two roots will be negative.

45. $2.7 + 3.94 + 2(0.0576) = 6.7552$
$0.0576(2.7 + 3.94 + 0.0576) = 0.38578176$

$$\begin{aligned}r^2 + 6.7552r + 0.38578176 &= 0\\ r &= -0.0576\\ r &= -6.6976\\ N(t) &= C_1\, e^{-0.0576t} + C_2\, e^{-6.6976t}\end{aligned}$$

46. $1.64 + 2.2 + 2(0.0576) = 3.9552$
$0.0576(1.64 + 2.2 + 0.0576) = 0.22450176$

$$\begin{aligned}r^2 + 3.9552r + 0.22450176 &= 0\\ r &= -0.0576\\ r &= -3.8976\\ N(t) &= C_1\, e^{-0.0576t} + C_2\, e^{-3.8976t}\end{aligned}$$

47. $0.31 + 1.64 + 2(0.51) = 2.97$
$0.51(0.31 + 1.64 + 0.51) = 1.2546$

$$\begin{aligned}r^2 + 2.97r + 1.2546 &= 0\\ r &= -0.51\\ r &= -2.46\\ N(t) &= C_1\, e^{-0.51t} + C_2\, e^{-2.46t}\end{aligned}$$

48.

$$\begin{aligned}y(x) &= C_1e^{2x} + C_2xe^{2x} + C_3e^{3x} +\\ &\quad C_4xe^{3x} + C_5x^2e^{3x} + C_6x^3e^{3x}\end{aligned}$$

49.

$$\begin{aligned}y(x) &= C_1e^{7x} + C_2e^{9x} + C_3e^{2x} \sin 4x +\\ &\quad C_4e^{2x} \cos 4x + C_5xe^{2x} \sin 4x + C_6xe^{2x} \cos 4x\end{aligned}$$

50. **a)**

$$\begin{aligned} r^2+\left(\frac{2\pi k}{c}\right)^2 &= 0 \\ r &= \pm i\left(\frac{2\pi k}{c}\right) \\ p(x) &= C_1 \sin\left(\frac{2\pi kx}{c}\right)+C_2 \cos\left(\frac{2\pi kx}{c}\right) \end{aligned}$$

b)

$$\begin{aligned} p(0) &= 0 \\ C_1 \sin 0 + C_2 &= 0 \\ C_1 \text{ free variable} & \\ p(x) &= C \sin\left(\frac{2\pi kx}{c}\right) \end{aligned}$$

c)

$$\begin{aligned} p'(x) &= \left(\frac{2\pi k}{c}\right) C_1 \cos\left(\frac{2\pi kx}{c}\right) - \\ & \left(\frac{2\pi k}{c}\right) C_2 \cos\left(\frac{2\pi kx}{c}\right) \\ 0 &= \left(\frac{2\pi k}{c}\right) C_2 \cos\left(\frac{-2\pi kL}{c}\right) \\ 0 &= \left(\frac{2\pi k}{c}\right) C_2 \cos\left(\frac{2\pi kL}{c}\right) \\ \to & \\ \frac{2\pi kL}{c} &= \frac{(2n-1)\pi}{2} \\ \to & \\ k &= \frac{(2n-1)c}{4L} \end{aligned}$$

51. $y = C_1y_1 + C_2y_2$

$y' = C_1y_1' + C_2y_2'$

$y'' = C_1y_1'' + C_2y_2''$

$$\begin{aligned} ay''+by'+cy &= a(C_1y_1''+C_2y_2'')+ \\ & b(C_1y_1'+C_2y_2')+ \\ & c(C_1y_1''+C_2y_2'') \\ &= aC_1y_1''+bC_1y_1'+cC_1y+ \\ & aC_2y_2''+bC_2y_2'+cC_2y_2 \\ &= 0+0 \\ &= 0 \end{aligned}$$

52. Left to the student

53. **a)** If the quadratic equation has only one root then the discriminant must equal 0. Thus

$$\begin{aligned} x &= \frac{-b\pm\sqrt{0}}{2a} \\ &= \frac{-b}{2a} \end{aligned}$$

b) Since $x = r$ is a solution to the equation then substitution r for x in the equation ax^2+bx+c yields the desired result.

c)

$$\begin{aligned} r &= \frac{-b}{2a} \\ 2ar &= -b \\ 2ar+b &= 0 \end{aligned}$$

d)

$$\begin{aligned} y &= xe^{rx} \\ y' &= rxe^{rx}+e^{rx} \\ y'' &= r^2xe^{rx}+re^{rx}+re^{rx} \\ ay''+by'+cy &= \\ &= a(r^2xe^{rx}+re^{rx}+re^{rx})+ \\ & b(rxe^{rx}+e^{rx})+cxe^{rx} \\ &= (ar^2+br+c)xe^{rx}+(2ar+b)e^{rx} \\ &= 0+0 \\ &= 0 \end{aligned}$$

54. **a)** Let $r = p+iq$ then

$$\begin{aligned} a(p+iq)^2+b(p+iq)+c &= 0 \\ a(p^2+2ipq-q^2)+bp+ibq+c &= 0 \\ (ap^2-aq^2+bp+c)+i(2apq+bq) &= 0 \\ \to & \\ a(p^2-q^2)+bp+c &= 0 \\ \text{and} & \\ 2apq+bq &= 0 \end{aligned}$$

b)

$$\begin{aligned} y &= e^{px} \sin qx \\ y' &= qe^{px} \cos qx + pe^{px} \sin qx \\ y'' &= -q^2e^{px} \sin qx + 2qpe^{px} \cos qx + \\ & p^2e^{px} \sin qx \\ ay''+by'+cy &= \\ &= a(-q^2e^{px} \sin qx + 2qpe^{px} \cos qx + \\ & p^2e^{px} \sin qx) + b(qe^{px} \cos qx + \\ & pe^{px} \sin qx) + c(e^{px} \sin qx) \\ &= e^{px} \sin qx(ap^2-aq^2+bp+c)+ \\ & e^{px} \cos qx(2apq+bq) \\ &= 0+0 \\ &= 0 \end{aligned}$$

c) Follow similar steps to part (c)

Exercise Set 9.2

1.

$$\begin{aligned} r^2+1 &= 0 \\ r &= \pm\, i \\ y_h &= C_1\ sin\ x + C_2\ cos\ x \\ y_p &= A \\ y_p' &= 0 \\ y_p'' &= 0 \\ y_p''+y_p &= 7 \\ 0+A &= 7 \\ y_p &= 7 \\ y &= C_1\ sin\ x + C_2\ cos\ x + 7 \end{aligned}$$

2.

$$\begin{aligned} r^2-3r+2 &= 0 \\ r &= 1 \\ r &= 2 \\ y_h &= C_1\ e^x + C_2\ e^{2x} \\ y_p &= A \\ y_p' &= 0 \\ y_p'' &= 0 \\ y_p'' + -3y_p' + 2y_p &= 4 \\ 0++0+2A &= 4 \\ A &= 2 \\ y &= C_1\ e^x + C_2\ e^{2x} + 4 \end{aligned}$$

3.

$$\begin{aligned} r^2-2r+1 &= 0 \\ r &= 1 \text{ repeated} \\ y_h &= C_1\ e^x + C_2 x\ e^x \\ y_p &= A \\ y_p' &= 0 \\ y_p'' &= 0 \\ y_p''-2y_p'+y_p &= 3 \\ 0-0+A &= 3 \\ y_p &= 3 \\ y &= C_1\ e^x + C_2 x\ e^x + 3 \end{aligned}$$

4.

$$\begin{aligned} r^2-7r+10 &= 0 \\ r &= 2 \\ r &= 5 \\ y_h &= C_1\ e^{2x} + C_2\ e^{5x} \\ y_p &= Ax+B \\ y_p' &= A \\ y_p'' &= 0 \\ 0-7(A)+10(Ax+B) &= 10x-27 \\ 10Ax+(10B-7A) &= 10x-27 \\ 10A &= 10 \to A=1 \\ 10B-7 &= -27 \to B=-2 \\ y_p &= x-2 \\ y &= C_1\ e^{2x} + C_2\ e^{5x} + x - 2 \end{aligned}$$

5.

$$\begin{aligned} r^2+4r+4 &= 0 \\ r &= -2 \text{ repeated} \\ y_h &= C_1\ e^{-2x} + C_2 x\ e^{-2x} \\ y_p &= Ax+B \\ y_p' &= A \\ y_p'' &= 0 \\ 0+4(A)+4(Ax+B) &= 8-12x \\ 4Ax+(4A+4B) &= 8-12x \\ 4A &= -12 \to A=-3 \\ -12+4B &= 8 \to B=5 \\ y_p &= 5-3x \\ y &= C_1\ e^{-2x} + C_2 x\ e^{-2x} - 3x + 5 \end{aligned}$$

6.

$$\begin{aligned} r^2-4r+5 &= 0 \\ r &= 2\pm\ i \\ y_h &= e^{2x}\,(C_1\ sin\ x + C_2\ cos\ x) \\ y_p &= Ax+B \\ y_p' &= A \\ y_p'' &= 0 \\ 0-4(A)+5(Ax+B) &= 5x+1 \\ 5Ax+(-4A+5B) &= 5x+1 \\ 5A &= 5 \to A=1 \\ -4+5B &= 1 \to B=1 \\ y_p &= x+1 \\ y &= e^{2x}\,(C_1\ sin\ x + C_2\ cos\ x) + x + 1 \end{aligned}$$

7.

$$\begin{aligned} r^2-4r+3 &= 0 \\ r &= -3 \\ r &= -1 \\ C_1\ e^{-3x} + C_2\ e^{-x} &= y_h \\ y_p &= Ax^2+Bx+C \\ y_p' &= 2Ax+B \\ y_p'' &= 2A \\ 2A-4(2Ax+B)+3(Ax^2+Bx+C) &= 6x^2-4 \\ 3Ax^2+(-8A+3B)x+(2A-4B+3C) &= 6x^2-4 \\ 3A &= 6 \to A=2 \\ -8A+3B &= 0 \to B=-\frac{16}{3} \\ 2A-4B+3C &= -4 \to C=\frac{40}{9} \end{aligned}$$

$$\begin{aligned} 2x^2 - \frac{16}{3}x + \frac{40}{9} &= y_p \\ C_1\, e^{-3x} + C_2\, e^{-x} + 2x^2 - \frac{16}{3}x + \frac{40}{9} &= y \end{aligned}$$

8.

$$\begin{aligned} r^2 + 4 &= 0 \\ r &= \pm\, 2i \\ C_1 \sin 2x + C_2 \cos 2x &= y_h \\ y_p &= Ax^2 + Bx + C \\ y_p' &= 2Ax + B \\ y_p'' &= 2A \\ 2A + 4(Ax^2 + Bx + C) &= 8x^2 - 12x \\ 4Ax^2 + 4Bx + (2A + 4C) &= 8x^2 - 12x \\ 4A &= 8 \to A = 2 \\ -4B &= -12 \to B = 3 \\ 2A + 4C &= 0 \to C = 1 \\ 2x^2 + 3x + 1 &= y_p \\ C_1 \sin 2x + C_2 \cos 2x + 2x^2 + 3x + 1 \end{aligned}$$

9.

$$\begin{aligned} r^2 - r - 2 &= 0 \\ r &= 2 \\ r &= -1 \\ C_1\, e^{-x} + C_2\, e^{2x} &= y_h \\ Ax^3 + Bx^2 + Cx + D &= y_p \\ 3Ax^2 + 2Bx + C &= y_p' \\ 6Ax + 2B &= y_p'' \\ x^3 - 1 &= 6Ax + 2B - 3Ax^2 - 2Bx - \\ &\quad C + 2(Ax^3 + Bx^2 + Cx + D) \\ x^3 - 1 &= 2Ax^3 + (2B - 3A)x^2 + \\ &\quad (2C - 2B + 6A)x + \\ &\quad (2D + 2B - C) \\ A &= \frac{1}{2} \\ 2B - 3A &= 0 \to B = \frac{3}{4} \\ 2C - 2B + 6A &= 0 \to C = \frac{-9}{4} \\ 2D - C + 2B &= -1 \to D = \frac{19}{8} \\ y_p &= \frac{x^3}{2} + \frac{3x^2}{4} - \frac{9x}{4} + \frac{19}{8} \\ y &= C_1\, e^{-x} + C_2\, e^{2x} + \\ &\quad \frac{x^3}{2} + \frac{3x^2}{4} - \frac{9x}{4} + \frac{19}{8} \end{aligned}$$

10.

$$\begin{aligned} r^2 + 2r + 1 &= 0 \\ r &= -1 \text{ repeated} \\ y_h &= C_1\, e^{-x} + C_2 x\, e^{-x} \\ Ax^3 + Bx^2 + Cx + D &= y_p \\ 3Ax^2 + 2Bx + C &= y_p' \\ 6Ax + 2B &= y_p'' \\ x^3 + 2x - 1 &= 6Ax + 2B + 2(3Ax^2 + 2Bx + C) \\ &\quad + Ax^3 + Bx^2 + Cx + D \\ x^3 + 2x - 1 &= Ax^3 + (B + 6A)x^2 + \\ &\quad (C + 4B + 6A)x + (D + 2B + 2C) \\ A &= 1 \\ B + 6A &= 0 \to B = -6 \\ C + 4B + 6A &= 2 \to C = 20 \\ D + 2B + 2C &= -1 \to D = -29 \\ y_p &= x^3 - 6x^2 + 20x - 29 \\ y &= C_1\, e^{-x} + C_2 x\, e^{-x} + \\ &\quad x^3 - 6x^2 + 20x - 29 \end{aligned}$$

11.

$$\begin{aligned} r^2 - 3r &= 0 \\ r &= 0 \\ r &= 3 \\ y_h &= C_1 + C_2\, e^{3x} \\ y_p &= Ax + B \\ y_p' &= A \\ y_p'' &= 0 \\ 0 - 3A &= 4 \\ y_p &= -\frac{4}{3} \\ y &= C_1 + C_2\, e^{3x} - \frac{4x}{3} \end{aligned}$$

Note: The particular solution had to be $Ax + B$ since the homogeneous solution already contained the constant solution.

12.

$$\begin{aligned} r^3 + r &= 0 \\ r &= 0 \\ r &= \pm\, i \\ y_h &= C_1 + C_2 \sin x + C_3 \cos x \\ y_p &= Ax + B \\ y_p' &= A \\ y_p'' &= 0 \\ y_p''' &= 0 \\ 0 + A &= -2 \\ y_p &= -2x \\ y &= C_1 + C_2 \sin x + C_3 \cos x - 2x \end{aligned}$$

Note: The particular solution had to be $Ax + B$ since the homogeneous solution already contained the constant solution.

13.

$$\begin{aligned} r^3 + r^2 &= 0 \\ r &= 0 \text{ repeated} \end{aligned}$$

$$
\begin{aligned}
r &= -1 \\
y_h &= C_1 + C_2 x + C_3\, e^{-x} \\
y_p &= Ax^2 + Bx + C \\
y_p' &= 2Ax + B \\
y_p'' &= 2A \\
y_p''' &= 0 \\
0 + 2A &= -2 \\
A &= -1 \\
y_p &= -x^2 \\
y &= C_1 + C_2 x + C_3\, e^{-x} - x^2
\end{aligned}
$$

Note: The particular solution had to be $Ax^2 + Bx + C$ since the homogeneous solution already contained the linear solution.

14.

$$
\begin{aligned}
r^2 - 2r &= 0 \\
r &= 0 \\
r &= 2 \\
y_h &= C_1 + C_2\, e^{2x} \\
y_p &= Ax^2 + Bx + C \\
y_p' &= 2Ax + B \\
y_p'' &= 2A \\
2A - 4Ax - 2B &= -4x \\
-4Ax + (2A - 2B) &= -4x \\
A &= 1 \\
2A - 2B &= 0 \to B = 1 \\
y_p &= x^2 + x \\
y &= C_1 + C_2\, e^{2x} + x^2 + x
\end{aligned}
$$

Note: Note: The particular solution had to be $Ax^2 + Bx + C$ since the associated auxiliary equation has 0 as a root.

15.

$$
\begin{aligned}
r^3 + 4r^2 + 20r &= 0 \\
r &= 0 \\
r &= -2 \pm 4i \\
y_h &= C_1 + e^{-2x}\,(C_1\, sin\ 4x + C_2\, cos\ 4x) \\
y_p &= Ax^2 + Bx + C \\
y_p' &= 2Ax + B \\
y_p'' &= 2A \\
y_p''' &= 0 \\
0 + 8A + 40Ax + 20B &= 40x - 12 \\
40Ax + (8A + 20B) &= 40x - 12 \\
40A &= 40 \to A = 1 \\
8A + 20B &= -12 \to B = -1 \\
y_p &= x^2 - x \\
y &= C_1 + e^{-2x}\,(C_2\, sin\ 4x + C_3\, cos\ 4x) \\
&\quad + x^2 - x
\end{aligned}
$$

Note: The particular solution had to be $Ax^2 + Bx + C$ since the associated auxiliary equation has 0 as a root.

16.

$$
\begin{aligned}
r^2 - r &= 0 \\
r &= 0 \\
r &= 1 \\
y_h &= C_1 + C_2\, e^{x} \\
y_p &= Ax^3 + Bx^2 + Cx + D \\
y_p' &= 3Ax^2 + 2Bx + C \\
y_p'' &= 6Ax + 2B \\
y_p''' &= 6A \\
6Ax + 2B - 3Ax^2 + 2Bx + C &= 3x^2 - 8x + 5 \\
-3Ax^2 + (6A + 2B)x + (2B + C) &= 3x^2 - 8x + 5 \\
A &= -1 \\
6A + 2B &= -8 \to B = -1 \\
2B + C &= 5 \to 7 \\
y_p &= -x^3 - x^2 + 7x \\
y &= C_1 + C_2\, e^{x} -\ x^3 \\
&\quad -x^2 + 7x
\end{aligned}
$$

Note: The particular solution had to be $Ax^2 + Bx + C$ since the associated auxiliary equation has 0 as a root.

17.

$$
\begin{aligned}
r^2 - r - 2 &= 0 \\
r &= 2 \\
r &= -1 \\
y_h &= C_1\, e^{-x} + C_2\, e^{2x} \\
y_p &= Ax + B \\
y_p' &= A \\
y_p'' &= 0 \\
0 - A - 2Ax - 2B &= 2x - 1 \\
-2Ax + (-A - 2B) &= 2x - 1 \\
A &= -1 \\
B &= 1 \\
y_p &= -x + 1 \\
y &= C_1\, e^{-x} + C_2\, e^{2x} - x \\
y' &= -C_1\, e^{-x} + 2C_2\, e^{2x} - 1 \\
y(0) &= 6 \to C_1 + C_2 = 6 \\
y'(0) &= 0 \to -C_1 + 2C_2 = 0
\end{aligned}
$$

Solving the system

$$
\begin{aligned}
C_1 &= 4 \\
C_2 &= 2 \\
y &= 4\, e^{-x} + 2\, e^{2x} - x
\end{aligned}
$$

18.

$$
\begin{aligned}
r^2 + 9 &= 0 \\
r &= \pm\, 3i \\
y_h &= C_1\, sin\ 3x + C_2\, cos\ 3x \\
y_p &= Ax + B \\
y_p' &= A
\end{aligned}
$$

$$
\begin{aligned}
y_p'' &= 0\\
0-9Ax-9B &= -9x+9\\
A &= 1\\
B &= -1\\
y_p &= x-1\\
y &= C_1\ sin\ 3x + C_2\ cos\ 3x + x - 1\\
y' &= 3C_1\ cos\ 3x - 3C_2\ sin\ 3x + 1\\
y(0) &= 3 \to C_2 - 1 = 3\\
C_2 &= 4\\
y'(0) &= 2 \to 3C_1 + 1 = 2\\
C_1 &= \frac{1}{3}\\
y &= \frac{1}{3}\ sin\ 3x + 2\ cos\ 3x + x - 1
\end{aligned}
$$

19.

$$
\begin{aligned}
r^2+2r+1 &= 0\\
r &= -1 \text{ repeated}\\
y_h &= C_1\ e^{-x} + C_2 x\ e^{-x}\\
y_p &= Ax^2+Bx+C\\
y_p' &= 2Ax+B\\
y_p'' &= 2A\\
2A+4Ax+2B+Ax^2+Bx+C &= x^2\\
Ax^2+(4A+B)x+(2A+2B+C) &= x^2\\
A &= 1\\
4A+B &= 0 \to B=-4\\
2A+2B+C &= 0 \to C = 6\\
y_p &= x^2-4x+6\\
C_1\ e^{-x} + C_2 x\ e^{-x} + x^2 - 4x + 6 &= y\\
-C_1\ e^{x} - C_2 x\ e^{x} + C_2\ e^{x} + 2x - 4 &= y'\\
C_1 = -5 \leftarrow 1 &= y(0)\\
4 = -C_1 + C_2 \leftarrow 0 &= y'(0)\\
C_2 &= -1\\
-5\ e^{x} - x\ e^{x} + x^2 - 4x + 6 &= y
\end{aligned}
$$

20.

$$
\begin{aligned}
r^2-5r+6 &= 0\\
r &= 2\\
r &= 3\\
y_h &= C_1\ e^{2x} + C_2\ e^{3x}\\
y_p &= Ax^3+Bx^2+Cx+D\\
y_p' &= 3Ax^2+2Bx+C\\
y_p'' &= 6Ax+2B\\
18x^2+12x+1 &= 6Ax+2B-5(3Ax^2+2Bx+C)+\\
&\quad 6(Ax^3+Bx^2+Cx+D)\\
18x^2+12x+1 &= 6Ax^3+(6B-15A)x^2+\\
&\quad (6C-10B+6A)x+(6D-5C+2B)\\
A &= 0\\
6B-15A &= 18 \to B = 3\\
6C-10B+6A &= 12 \to C = 7\\
6D-5C+2B &= 1 \to D = 5\\
y &= C_1\ e^{2x} + C_2\ e^{3x} + 3x^2 + 7x + 5\\
y' &= 2C_1\ e^{2x} + 3C_2\ e^{3x} + 6x + 7\\
y(0) &= 3 \to C_1 + C_2 = -2\\
y'(0) &= 2 \to 2C_1 + 3C_2 = -5
\end{aligned}
$$

Solving the system

$$
\begin{aligned}
C_1 &= -1\\
C_2 &= -1\\
y &= -\ e^{2x} -\ e^{3x} + 3x^2 + 7x + 5
\end{aligned}
$$

21.

$$
\begin{aligned}
r^2+4r &= 0\\
r &= 0\\
r &= -4\\
y_h &= C_1 + C_2\ e^{-4x}\\
y_p &= Ax^2+Bx+C\\
y_p' &= 2Ax+B\\
y_p'' &= 2A\\
2A+8Ax+4B &= 16x\\
8Ax+(2A+4B) &= 16x\\
8A &= 16 \to A = 2\\
2A+4B &= 0 \to B = -1\\
y_p &= 2x^2-x\\
y &= C_1 + C_2\ e^{-4x} + 2x^2 - x\\
y' &= -4C_2\ e^{-4x} + 4x - 1\\
y(0) &= 2 \to C_1 + C_2 = 2\\
y'(0) &= -3 \to -4C_2 = -2\\
C_2 &= \frac{1}{2}\\
C_1 &= \frac{3}{2}\\
y &= \frac{3}{2} + \frac{1}{2}\ e^{-4x} + 2x - 1
\end{aligned}
$$

Note: $r = 0$ is a root to the auxiliary equation and the effected our selection of y_p

22.

$$
\begin{aligned}
r^3+3r^2 &= 0\\
r &= 0 \text{ repeated}\\
r &= -3\\
y_h &= C_1 + C_2 x + C_3\ e^{-3x}\\
y_p &= Ax^3+Bx^2+Cx+D\\
y_p' &= 3Ax^2+2Bx+C\\
y_p'' &= 6Ax+2B\\
y_p''' &= 6A\\
6A+18Ax+6B &= 18x-18\\
18A &= 18 \to A = 1\\
6A+6B &= -18 \to B = -2
\end{aligned}
$$

$$\begin{aligned}
y_p &= x^3 - 2x^2 \\
y &= C_1 + C_2 x + C_3\, e^{-3x} + x^3 - 2x^2 \\
y' &= C_2 - 3C_3\, e^{-3x} + 3x^2 - 4x \\
y'' &= 9C_3\, e^{-3x} + 6x - 4 \\
y(0) &= 3 \to C_1 + C_3 = 3 \\
y'(0) &= -1 \to C_2 - 3C_3 = -1 \\
y''(0) &= 10 \to 9C_3 - 4 = 10 \\
C_3 &= \frac{14}{9} \\
C_2 &= \frac{11}{3} \\
C_1 &= \frac{13}{9} \\
y &= \frac{13}{9} + \frac{11}{3}\, x + \frac{14}{9}\, e^{-3x} + x^3 - 2x^2
\end{aligned}$$

Note: $r = 0$ is a repeated root to the auxiliary equation and the effected our selection of y_p

23.

$$\begin{aligned}
r^2 + 2r + 2 &= 0 \\
r &= -1 \pm\, i \\
y_h &= e^{-x}\,(C_1\, sin\; x + C_2\, cos\; x) \\
y_p &= Ax + B \\
y_p' &= A \\
y_p'' &= 0 \\
0 + 2A + 2Ax + 2B &= 2 \\
2Ax + (2A + 2B) &= 2 \\
A &= 0 \\
2A + 2B &= 2 \to B = 1 \\
y_p &= 1 \\
y &= e^{-x}\,(C_1\, sin\; x + C_2\, cos\; x) + 1 \\
y' &= e^{-x}\,(C_1\, cos\; x - C_2\, sin\; x) \\
&\quad -e^{-x}\,(C_1\, sin\; x + C_2\, cos\; x) \\
y(0) &= 2 \to C_2 + 1 = 2 \to C_2 = 1 \\
y'(0) &= 1 \to C_1 - C_2 = 1 \to C_1 = 2 \\
y &= e^{-x}\,(2\; sin\; x + cos\; x) + 1
\end{aligned}$$

24.

$$\begin{aligned}
r^2 + 4r + 13 &= 0 \\
r &= -2 \pm\, 3i \\
y_h &= e^{-2x}\,(C_1\, sin\; 3x + C_2\, cos\; 3x) \\
y_p &= Ax + B \\
y_p' &= A \\
y_p'' &= 0 \\
0 + 4A + 13Ax + 13B &= 26 \\
A &= 0 \\
4A + 13B &= 26 \to B = 2 \\
y_p &= 2 \\
y &= e^{-2x}\,(C_1\, sin\; 3x + C_2\, cos\; 3x) + 2 \\
y' &= e^{-2x}\,(3C_1\, cos\; 3x - 3C_2\, sin\; 3x) \\
&\quad -2\, e^{-2x}\,(C_1\, sin\; 3x + C_2\, cos\; 3x) \\
y(0) &= 1 \to C_2 + 2 = 1 \to C_2 = -1 \\
y'(0) &= 0 \to 3C_1 - 2C_2 = 0 \to C_1 - 2 \\
y &= e^{-2x}\,(-2\; sin\; 3x -\; cos\; 3x) + 2
\end{aligned}$$

25.

$$\begin{aligned}
r^3 + 4r^2 + 5r &= 0 \\
r &= 0 \\
r &= -2 \pm\, i \\
y_h &= C_1 + e^{-2x}\,(C_2\, sin\; x + C_3\, cos\; x) \\
y_p &= Ax^2 + Bx + C \\
y_p' &= 2Ax + B \\
y_p'' &= 2A \\
y_p''' &= 0 \\
0 + 8A + 10Ax + 5B &= 25x - 5 \\
10Ax + (8A + 5B) &= 25x - 5 \\
10A &= 25 \to A = \frac{5}{2} \\
8A + 5B &= -5 \to B = 5 \\
y_p &= \frac{5}{2}\, x^2 + 5x \\
y &= C_1 + e^{-2x}\,(C_2\, sin\; x + C_3\, cos\; x) \\
&\quad + \frac{5}{2} x^2 + 5x \\
y' &= e^{-2x}\,(C_2\, cos\; x - C_3\, sin\; x) \\
&\quad -2\, e^{-2x}\,(C_2\, sin\; x + C_3\, cos\; x) \\
&\quad +5x \\
y'' &= 4\, e^{-2x}\,(C_2\, sin\; x + C_3\, cos\; x) - \\
&\quad 4\, e^{-2x}\,(C_2\, cos\; x - C_3\, sin\; x) \\
&\quad + e^{-2x}\,(-C_2\, sin\; x - C_3\, cos\; x) \\
&\quad +5 \\
y(0) &= 0 \to C_1 + C_3 = 0 \\
y'(0) &= 0 \to C_2 - 2C_3 = 0 \\
y''(0) &= 1 \to 4C_3 - 4C_2 - C_3 = 1
\end{aligned}$$

Solving the system

$$\begin{aligned}
C_1 &= \frac{21}{5} \\
C_2 &= -\frac{7}{5} \\
C_3 &= -\frac{16}{5} \\
y &= \frac{21}{5} + e^{-2x}\left(-\frac{7}{5}\; sin\; x - \frac{16}{5}\; cos\; x\right) \\
&\quad + \frac{5}{2}\, x^2 + 5x
\end{aligned}$$

26.

$$\begin{aligned}
r^3 - 6r^2 + 9r &= 0 \\
r &= 0 \\
r &= 3 \text{ repeated} \\
C_1 + C_2\, e^{3x} + C_3 x\, e^{3x} &= y_h \\
Ax^3 + Bx^2 + Cx + D &= y_p \\
3Ax^2 + 2Bx + C &= y_p'
\end{aligned}$$

$$\begin{aligned}
y_p'' &= 6Ax + 2B \\
y_p''' &= 6A \\
6A - 36Ax - 12B + 27Ax^2 + 18Bx + 9C &= 27x^2 \\
A &= 1 \\
18B - 36A = 0 \to B &= 2 \\
9C - 12B + 6A = 0 \to C &= 2 \\
x^3 + 2x^2 + 2x &= y_p \\
C_1 + C_2\, e^{3x} + C_3 x\, e^{3x} + 2x + 2x^2 + x^3 &= y \\
3C_2\, e^{3x} + 3C_3 x\, e^{3x} + C_3 e^{3x} + 2 + 4x + 3x^2 &= y' \\
9C_2\, e^{3x} + 9C_3 x\, e^{3x} + 6C_3\, e^{3x} + 4 + 6x &= y'' \\
27C_2\, e^{3x} + 27C_3 x\, e^{3x} + 27C_3\, e^{3x} + 6 &= y''' \\
2 = C_1 + C_2 \leftarrow 2 &= y(0) \\
-3 = 2 + 3C_2 + C_3 \leftarrow -3 &= y'(0) \\
1 = 4 + 9C_2 + 6C_3 \leftarrow 1 &= y''(0) \\
\text{Solving the system} & \\
C_1 &= 5 \\
C_2 &= -3 \\
C_3 &= 4 \\
2x + 2x^2 + x^3 + 5 - 3\, e^{3x} + 4x\, e^{3x} &= y
\end{aligned}$$

27. $x'' + 16x = 1$

$$\begin{aligned}
r^2 + 16 &= 0 \\
r &= \pm\, 4i \\
x_h &= C_1 \sin 4t + C_2 \cos 4t \\
x_p &= A \\
x_p' &= 0 \\
x_p'' &= 0 \\
0 + 16A &= 1 \to A = \frac{1}{16} \\
x_p &= \frac{1}{16} \\
x &= C_1 \sin 4t + C_2 \cos 4t + \frac{1}{16} \\
x' &= 4C_1 \cos 4t - 4C_2 \sin 4t \\
x(0) &= 0 \to C_2 + \frac{1}{16} \to C_2 = -\frac{1}{16} \\
x'(0) &= 0 \to 4C_1 = 0 \to C_1 = 0 \\
x &= \frac{1}{16} - \frac{1}{16} \cos 4t \\
&= \frac{1 - \cos 4t}{16}
\end{aligned}$$

28. $x'' + 4x = 2$

$$\begin{aligned}
r^2 + 4 &= 0 \\
r &= \pm\, 2i \\
x_h &= C_1 \sin 2t + C_2 \cos 2t \\
x_p &= A \\
x_p' &= 0 \\
x_p'' &= 0 \\
0 + 4A &= 2 \to A = \frac{1}{2} \\
x_p &= \frac{1}{2} \\
x &= C_1 \sin 2t + C_2 \cos 2t + \frac{1}{2} \\
x' &= 2C_1 \cos 2t - 2C_2 \sin 2t \\
x(0) &= 0 \to C_2 + \frac{1}{2} \to C_2 = -\frac{1}{2} \\
x'(0) &= 0 \to 2C_1 = 0 \to C_1 = 0 \\
x &= \frac{1}{2} - \frac{1}{2} \cos 2t \\
&= \frac{1 - \cos 2t}{2}
\end{aligned}$$

29. $x'' + 2x' + 5x = 10$

$$\begin{aligned}
r^2 + 2r + 5 &= 0 \\
r &= -1 \pm 2i \\
e^{-t}\,(C_1 \sin 2t + C_2 \cos 2t) &= x_h \\
x_p &= A \\
x_p' &= 0 \\
x_p'' &= 0 \\
0 + 0 + 5A = 10 \to A &= 2 \\
x_p &= 2 \\
e^{-t}\,(C_1 \sin 2t + C_2 \cos 2t) + 2 &= x \\
e^{-t}\,(2C_1 \cos 2t - 2C_2 \sin 2t) - & \\
e^{-t}\,(C_1 \sin 2t + C_2 \cos 2t) &= x' \\
0 = C_2 + 2 \leftarrow 0 &= x(0) \\
C_2 &= -2 \\
0 = 2C_1 - C_2 \leftarrow 0 &= x'(0) \\
C_1 &= -1 \\
e^{-t}\,(-\sin 2t - 2 \cos 2t) + 2 &= x
\end{aligned}$$

30. $2x'' + 20x' + 50x = 20$

$$\begin{aligned}
2r^2 + 20r + 50 &= 0 \\
r &= -5 \text{ repeated} \\
C_1\, e^{-5t} + C_2 t\, e^{-5t} &= x_h \\
x_p &= At + B \\
x_p' &= A \\
x_p'' &= 0 \\
0 + 20A + 50At + 50B = 25t \to A &= \frac{1}{2} \\
20A + 50B = 0 \to \frac{1}{5} & \\
x_p &= \frac{1}{2}\, t + \frac{1}{5} \\
C_1\, e^{-5t} + C_2 t\, e^{-5t} + \frac{1}{2}\, t + \frac{1}{5} &= x \\
-5C_1\, e^{-5t} - 5C_2 t\, e^{-5t} + C_2\, e^{-5t} + \frac{1}{2} &= x' \\
0 = \frac{1}{5} + C_2 + C_1 \leftarrow 0 &= x(0) \\
0 = \frac{1}{2} - 5C_1 + C_2 \leftarrow 0 &= x'(0) \\
\text{Solving the system} &= \\
C_1 &= -\frac{11}{30}
\end{aligned}$$

$$C_2 = -\frac{1}{6}$$

$$\frac{1}{5}+\frac{1}{2}t-\frac{11}{30}e^{-5t}-\frac{1}{6}t\,e^{-5t} = x$$

31. $cm\,x''+x = cpd+x_0$

a)

$$\begin{aligned}
cm\,r^2+1 &= 0\\
r &= \pm\frac{1}{\sqrt{cm}}\,i\\
x_h &= C_1 \sin\frac{t}{\sqrt{cm}}+C_2\cos\frac{t}{\sqrt{cm}}\\
x_p &= A\\
x_p' &= 0\\
x_p'' &= 0\\
A &= cpd+x_0\\
x_p &= cpd+x_0\\
x &= C_1 \sin\frac{t}{\sqrt{cm}}+C_2\cos\frac{t}{\sqrt{cm}}\\
&\quad +cpd+x_0
\end{aligned}$$

b)

$$\begin{aligned}
\frac{C_1}{\sqrt{cm}}\cos\frac{t}{\sqrt{cm}}-\frac{C_2}{\sqrt{cm}}\sin\frac{t}{\sqrt{cm}} &= x'\\
C_2+cpd+x_0=0\leftarrow 0 &= x(0)\\
-cpd-x_0 &= C_2\\
0=\frac{C_1}{\sqrt{cm}}\leftarrow 0 &= x'(0)\\
C_1 &= 0\\
cpd+x_0-(cpd+x_0)\cos\frac{t}{\sqrt{cm}} &= x\\
(cpd+x_0)\left(1-\cos\frac{t}{\sqrt{cm}}\right) &=
\end{aligned}$$

32. $F''+3.5F'+1.5F = 1$

$$\begin{aligned}
r^2+3.5r+1.5 &= 0\\
r &= -\frac{3}{4}\\
r &= -1\\
F_h &= C_1\,e^{-t}+C_2\,e^{-3t/4}\\
F_p &= A\\
F_p' &= 0\\
F_p'' &= 0\\
1.5A &= 1\rightarrow A=\frac{2}{3}\\
F &= C_1\,e^{-t}+C_2\,e^{-3t/4}+\frac{2}{3}\\
F' &= -C_1\,e^{-t}-\frac{3C_2}{4}\,e^{-3t/4}\\
F(0) &= 0\rightarrow C_1+C_2=-\frac{2}{3}\\
F'(0) &= 1\rightarrow -C_1-\frac{3C_2}{4}=1
\end{aligned}$$

Solving the system

$$\begin{aligned}
C_1 &= -2\\
C_2 &= \frac{4}{3}\\
F &= \frac{4}{3}\,e^{-3t/4}-2\,e^{-t}+\frac{2}{3}
\end{aligned}$$

33. $F''+1.05F'+0.05F = 0.05$

$$\begin{aligned}
r^2+1.05r+0.05 &= 0\\
r &= -\frac{1}{20}\\
r &= -1\\
F_h &= C_1\,e^{-t}+C_2\,e^{-t/20}\\
F_p &= A\\
F_p' &= 0\\
F_p'' &= 0\\
0.05A &= 0.05\rightarrow A=1\\
F &= C_1\,e^{-t}+C_2\,e^{-t/20}+1\\
F' &= -C_1\,e^{-t}-\frac{C_2}{20}\,e^{-t/20}\\
F(0) &= 0\rightarrow C_1+C_2=-1\\
F'(0) &= 0.05\rightarrow -C_1-\frac{C_2}{20}=0.05
\end{aligned}$$

Solving the system

$$\begin{aligned}
C_1 &= 0\\
C_2 &= -1\\
F &= 1-e^{-t/20}
\end{aligned}$$

34. $F''+4.1F'+2.1F = 8.7$

$$\begin{aligned}
r^2+4.1r+2.1 &= 0\\
r &= -\frac{7}{2}\\
r &= -\frac{3}{5}\\
F_h &= C_1\,e^{-7t/2}+C_2\,e^{-3t/5}\\
F_p &= A\\
F_p' &= 0\\
F_p'' &= 0\\
2.1A &= 8.7\rightarrow A=\frac{29}{7}\\
F &= C_1\,e^{-7t/2}+C_2\,e^{-3t/5}+\frac{29}{7}\\
F' &= -\frac{7C_1}{2}\,e^{-7t/2}-\frac{3C_2}{5}\,e^{-3t/5}\\
F(0) &= 0\rightarrow C_1+C_2=-\frac{29}{7}\\
F'(0) &= 8.7\rightarrow -\frac{7C_1}{2}-\frac{3C_2}{5}=8.7
\end{aligned}$$

Solving the system

$$\begin{aligned}
C_1 &= -\frac{15}{7}\\
C_2 &= -2\\
F &= \frac{29}{7}-\frac{15}{7}\,e^{-7t/2}-2\,e^{-3t/5}
\end{aligned}$$

35. $y' + 2y = x^2$

$$\begin{aligned} r+2 &= 0 \\ r &= -2 \\ y_h &= C\,e^{-2x} \\ y_p &= Ax^2 + Bx + C \\ y_p' &= 2Ax + B \\ y_p'' &= 2A \\ x^2 &= 2Ax + B + 2Ax^2 + 2Bx + 2C \\ A &= \frac{1}{2} \\ 2A + 2B &= 0 \to B = -\frac{1}{2} \\ B + 2C &= 0 \to C = \frac{1}{4} \\ y &= C\,e^{-2x} + \frac{1}{2}\,x^2 - \frac{1}{2}\,x + \frac{1}{4} \end{aligned}$$

36. Left to the student

Exercise Set 9.3

1.

$$\begin{aligned} x'' &= -2x + 3x' \\ x'' - 3x' + 2x &= 0 \\ r^2 - 3r + 2 &= 0 \\ r &= 2 \\ r &= 1 \\ x &= C_1\,e^t + C_2\,e^{2t} \\ y &= x' = C_1\,e^t + 2C_2\,e^{2t} \end{aligned}$$

2.

$$\begin{aligned} y &= \frac{x' - x}{2} \\ 2y' &= x'' - x' \\ \frac{x'' - x}{2} &= x' \\ x'' - 2x' - 1 &= 0 \\ r^2 - 2r - 1 &= 0 \\ r &= 1 \pm i\sqrt{2} \\ x &= e^t \left(C_1\,sin\,\sqrt{2}\,t + C_2\,cos\,\sqrt{2}\,t\right) \\ x' &= e^t \left(\sqrt{2}\,C_1\,cos\,\sqrt{2}\,t - \sqrt{2}\,C_2\,sin\,\sqrt{2}\,t\right) + \\ &\quad e^t \left(C_1\,sin\,\sqrt{2}\,t + C_2\,cos\,\sqrt{2}\,t\right) \\ y &= \frac{x' - x}{2} \\ &= \frac{e^t\,(\sqrt{2}\,C_1\,cos\,\sqrt{2}\,t - \sqrt{2}\,C_2\,sin\,\sqrt{2}\,t)}{2} \end{aligned}$$

3. $y = x' - 2x$ and $x'' = 2x + y'$

$$\begin{aligned} x'' &= 2x' + 3x + 4(x' - 2x) \\ x'' - 6x' + 5x &= 0 \\ r^2 - 6r + 5 &= 0 \\ r &= 5 \\ r &= 1 \\ x &= C_1\,e^t + C_2\,e^{5t} \\ x' &= C_1\,e^t + 5\,C_2\,e^{5t} \\ y &= x' - 2x \\ &= 3\,C_2\,e^{5t} - C_1\,e^t \end{aligned}$$

4. $y = \frac{1}{4}(3x - x')$ and $x'' = 3x' - 4y'$

$$\begin{aligned} x'' &= 3x' - 4\left[2x - \frac{3(3x - x')}{4}\right] \\ x'' - 7x &= 0 \\ r &= \pm\sqrt{7} \\ x &= C_1\,e^{-\sqrt{7}\,t} + C_2\,e^{\sqrt{7}\,t} \\ x' &= -\sqrt{7}\,c_1\,e^{\sqrt{7}\,t} + \sqrt{7}\,C_2\,e^{\sqrt{7}\,t} \\ y &= \frac{3x - x'}{4} \\ &= \frac{(3 + \sqrt{7})\,C_1\,e^{-\sqrt{7}\,t} + (3 - \sqrt{7})\,C_2\,e^{\sqrt{7}\,t}}{4} \end{aligned}$$

5. $x' = -0.5x + y \to y = x' + 0.5x$ and $y' = 0.5x$

$$\begin{aligned} x'' &= -0.5x' + y' \\ x'' &= -0.5x' + 0.5x \\ x'' + 0.5x' - 0.5x &= 0 \\ r^2 + 0.5r - 0.5 &= 0 \\ r &= 1/2 \\ r &= -1 \\ x &= C_1\,e^{t/2} + C_2\,e^{-t} \\ x &= \frac{C_1}{2}\,e^{t/2} - C_2\,e^{-t} \\ y &= x' + 0.5x \\ &= C_1\,e^{t/2} - C_2\,e^{-t} \end{aligned}$$

6. $y = -\frac{x' + x}{3}$

$$\begin{aligned} -\frac{x' + x}{3} &= \frac{x' + x}{3} \\ 6x' + 6x &= 0 \\ r &= -1 \\ x &= C_1\,e^{-t} \\ x' &= -C_1\,e^{-t} \\ y &= -\frac{x' + x}{3} \\ &= 0 \end{aligned}$$

7. $y = x' - 4x$

$$\begin{aligned} x'' &= 4x' + y' \\ &= 4x' - x + 2y \\ x'' &= 4x' - x + 2x' - 8x \\ x'' - 6x' + 9x &= 0 \end{aligned}$$

$$\begin{aligned} r^2 - 6r + 9 &= 0 \\ r &= 3 \text{ repeated} \\ x &= C_1\, e^{-3t} + C_2 t\, e^{-3t} \\ x' &= -3C_1\, e^{-3t} - 3C_2 t\, e^{-3t} + C_2\, e^{-3t} \\ y &= x' - 4x \\ &= (C_2 - C_1)\, e^{-3t} - 3C_2 t\, e^{-3t} \end{aligned}$$

8. $y = 2x - x'$

$$\begin{aligned} x'' &= 2x' - y' \\ x'' &= 2x' - 4x + 2(2x - x') = 0 \\ x' &= t + C_1 \\ x &= \frac{t^2}{2} + C_1\, t + C_2 \\ y &= 2x - x' \\ &= t^2 + 2C_1\, t + 2C_2 - t - C_1 \\ &= t^2 + 2(C_1 - 1)\, t + (2C_2 - C_1) \end{aligned}$$

9. $y = \frac{x'}{2}$

$$\begin{aligned} x'' &= 2y' = 2(-18x) \\ x'' + 36x &= 0 \\ r^2 + 36 &= 0 \\ r &= \pm\, 6i \\ x &= C_1\, sin\ 6x + C_2\, cos\ 6x \\ x' &= 6C_1\, cos\ 6x - C_2\, sin\ 6x \\ y &= \frac{x}{2} \\ &= 3C_1\, cos\ 6x - 3C_2\, sin\ 6x \end{aligned}$$

10. $y = x - x'$

$$\begin{aligned} x'' &= x' - y' \\ x'' &= x' - 5x - x' + x \\ x'' + 4x &= 0 \\ r &= \pm\, 2i \\ x &= C_1\, sin\ 2t + C_2\, cos\ 2t \\ x' &= 2C_1\, cos\ 2t - 2C_2\, sin\ 2t \\ y &= x - x' \\ &= (C_1 + 2C_2)\, sin\ 2t + (C_2 - 2C_1)\, cos\ 2t \end{aligned}$$

11. $5y = 3x - x'$

$$\begin{aligned} x'' &= 3x' - 5(x - y) \\ &= 3x' - 5x + 3x - x' \\ x'' - 2x' + 2 &= 0 \\ r^2 - 2r + 2 &= 0 \\ r &= 1 \pm\, i \\ x &= e^t\,(C_1\, sin\ t + C_2\, cos\ t) \\ x' &= e^t\,(C_1\, cos\ t - C_2\, sin\ t) \\ &\quad + e^t\,(C_1\, sin\ t + C_2\, cos\ t) \\ y &= \frac{e^t\,[(2C_1 - C_2)\, sin\ t + (2C_2 - C_1)\, cos\ t]}{5} \end{aligned}$$

12. $5y = 3x + x'$

$$\begin{aligned} x'' &= 3x' + 5(-x + y) \\ &= 3x' - 5x + 3x + x' \\ x'' + 2x' - 2 &= 0 \\ r^2 - 2r + 2 &= 0 \\ r &= -1 \pm\, i\sqrt{3} \\ x &= e^{-t}\left(C_1\, sin\ \sqrt{3}\, t + C_2\, cos\ \sqrt{3}\, t\right) \\ x' &= e^{-t}\left(\sqrt{3}\, C_1\, cos\ t - \sqrt{3}\, C_2\, sin\ t\right) \\ &\quad - e^{-t}\,(C_1\, sin\ t + C_2\, cos\ t) \\ y &= \frac{e^{-t}\left[(2C_1 - \sqrt{3}C_2)\, sin\ t\right]}{5} + \\ &\quad \frac{e^{-t}\left[(2C_2 + \sqrt{3}C_1)\, cos\ t\right]}{5} \end{aligned}$$

13. $y = 5x - x'$

$$\begin{aligned} x'' &= 5x' - (2x + (10x - 2x') + 4) \\ x'' - 7x' + 12 &= -4 \\ r^2 - 7r + 12 &= 0 \\ r &= 3 \\ r &= 4 \\ x_h &= C_1\, e^{3t} + C_2\, e^{4t} \\ x_p &= A \\ x_p' &= 0 \\ x_p'' &= 0 \\ 0 - 0 + 12A &= -4 \\ A = -\frac{1}{3} & \\ x_p &= -\frac{1}{3} \\ x &= C_1\, e^{3t} + C_2\, e^{4t} - \frac{1}{3} \\ x' &= 3C_1\, e^{3t} + 4C_2\, e^{4t} \\ y &= 5x - x' \\ &= 2C_1\, e^{3t} + C_2\, e^{4t} - \frac{5}{3} \end{aligned}$$

14. $y = x' - 2x - 1$

$$\begin{aligned} x'' &= 2x' - 2(x' - 2x - 1) - 22 \\ x'' - 4x &= -20 \\ r^2 - 4 &= 0 \\ r &= 2 \text{ repeated} \\ x_h &= C_1\, e^{2t} + C_2 t\, e^{2t} \\ x_p &= A \\ x_p' &= 0 \\ x_p'' &= 0 \\ 0 - 4A &= -20 \\ A &= 5 \\ x_p &= 5 \\ x &= C_1\, e^{2t} + C_2 t\, e^{2t} + 5 \\ x' &= 2C_1\, e^{2t} + 2C_2 t\, e^{2t} + C_2\, e^{2t} \end{aligned}$$

$$\begin{aligned} y &= x' - 2x - 1 \\ &= C_2\, e^{2t} - 11 \end{aligned}$$

15.

$$\begin{aligned} x'' &= y' + 2 \\ &= -x + 4t - 2 + 2 \\ x'' + x &= 4t \\ r^2 + 1 &= 0 \\ r &= \pm\, i \\ x_h &= C_1\, sin\ t + C_2\, cos\ t \\ x_p &= At + B \\ x_p' &= A \\ x_P'' &= 0 \\ 0 + At + B &= 4t \\ A &= 4 \\ B &= 0 \\ x_p &= 4t \\ x &= C_1\, sin\ t + C_2\, cos\ t + 4t \\ x' &= C_1\, cos\ t - C_2\, sin\ t + 4 \\ y &= x' - 2t - 3 \\ &= C_1\, cos\ t - C_2\, sin\ t - 2t + 1 \end{aligned}$$

16.

$$\begin{aligned} x'' &= -x' + (2x + 2t - 1) + 2 \\ x'' + x' - 2x &= 2t + 1 \\ r^2 + r - 2 &= 0 \\ r &= -2 \\ r &= 1 \\ x_h &= C_1\, e^{-2t} + C_2\, e^t \\ x_p &= At + B \\ x_p' &= A \\ x_p'' &= 0 \\ (A - 2B) - 2At &= 2t + 1 \\ A &= -1 \\ B &= 1 \\ x_p &= 1 - t \\ x &= C_1\, e^{-2t} + C_2\, e^t - t + 1 \\ x' &= -2C_1\, e^{-2t} + C_2\, e^t - 1 \\ y &= x' + x - 2t - 2 \\ &= 2C_2\, e^t - C_1\, e^{-2t} - 3t - 2 \end{aligned}$$

17.

$$\begin{aligned} x'' &= 2x' - (2x + 10x - 5x') \\ x'' - 7x' + 12x &= 0 \\ r^2 - 7r + 12 &= 0 \\ r &= 3 \\ r &= 4 \\ x &= C_1\, e^{3t} + C_2\, e^{4t} \\ x' &= 3C_1\, e^{3t} + 4C_2\, e^{4t} \end{aligned}$$

$$\begin{aligned} y &= 2x - x' \\ &= -C_1\, e^{3t} - 2C_2\, e^{4t} \\ x(0) &= 3 \to C_1 + C_2 = 3 \\ y(0) &= -5 \to -C_1 - 2C_2 = -5 \end{aligned}$$

Solving the system

$$\begin{aligned} C_1 &= 1 \\ C_2 &= 2 \\ x &= e^{3t} + 2\, e^{4t} \\ y &= -e^{3t} - 4\, e^{4t} \end{aligned}$$

18.

$$\begin{aligned} x'' &= x' + x + x' - x \\ x'' - 2x' &= 0 \\ r^2 - 2r &= 0 \\ r &= 0 \\ r &= 2 \\ x &= C_1 + C_2\, e^{2t} \\ x' &= 2C_2\, e^{2t} \\ y &= x' - x \\ &= C_2\, e^{2t} - C_1 \\ x(0) &= 2 \to C_1 + C_2 = 2 \\ y(0) &= 0 \to -C_1 + C_2 = 0 \end{aligned}$$

Solving the system

$$\begin{aligned} C_1 &= 1 \\ C_2 &= 1 \\ x &= 1 + e^{2t} \\ y &= e^{2t} - 1 \end{aligned}$$

19.

$$\begin{aligned} x'' &= 2x' + 3(-3x + 8y) \\ x'' &= 2x' - 9x + 8x' - 16x \\ x'' - 10x' + 25x &= 0 \\ r^2 - 10r + 25 &= 0 \\ r &= 5 \text{ repeated} \\ x &= C_1\, e^{5t} + C_2 t\, e^{5t} \\ x' &= 5C_1\, e^{5t} + 5C_2 t\, e^{5t} + C_2\, e^{5t} \\ y &= \frac{x' - 2x}{3} \\ &= \frac{(3C_1 + C_2)\, e^{5t}}{3} + C_2 t\, e^{5t} \\ x(0) &= 1 \to C_1 + C_2 = 1 \\ y(0) &= 1 \to C_1 + \frac{1}{3}\, C_2 = 1 \end{aligned}$$

Solving the system

$$\begin{aligned} C_1 &= 1 \\ C_2 &= 0 \\ x &= e^{5t} \\ y &= e^{5t} \end{aligned}$$

20.

$$\begin{aligned} x'' &= 2x' + 3(-3x + 8y) \\ x'' &= 2x' - 9x + 8x' - 16x \\ x'' - 10x' + 25x &= 0 \\ r^2 - 10r + 25 &= 0 \\ r &= 5 \text{ repeated} \\ x &= C_1\, e^{5t} + C_2 t\, e^{5t} \\ x' &= 5C_1\, e^{5t} + 5C_2 t\, e^{5t} + C_2\, e^{5t} \\ y &= \frac{x' - 2x}{3} \\ &= \frac{(3C_1 + C_2)\, e^{5t}}{3} + C_2 t\, e^{5t} \\ x(0) &= 1 \rightarrow C_1 + C_2 = 1 \\ y(0) &= 2 \rightarrow C_1 + \frac{1}{3}\, C_2 = 2 \end{aligned}$$

Solving the system

$$\begin{aligned} C_1 &= \frac{5}{2} \\ C_2 &= -\frac{3}{2} \\ x &= \frac{5}{2}\, e^{5t} - \frac{3}{2}\, t\, e^{5t} \\ y &= 6\, e^{5t} - 3t\, e^{5t} \end{aligned}$$

21.

$$\begin{aligned} x'' &= -4x \\ x'' + 4x &= 0 \\ r^2 + 4 &= 0 \\ r &= \pm\, 2i \\ x &= C_1 \sin 2x + C_2 \cos 2x \\ y &= x' = 2C_1 \cos 2x - 2C_2 \sin 2x \\ x(0) &= 1 \rightarrow C_2 = 1 \\ y(0) &= 2 \rightarrow C_1 = 1 \\ x &= \sin 2x + \cos 2x \\ y &= 2 \cos 2x - 2 \sin 2x \end{aligned}$$

22.

$$\begin{aligned} x'' &= x' - 5x - 15y \\ x'' &= x' - 5x - 3x' + 3x \\ x'' + 2x' + 2x &= 0 \\ r^2 + 2r + 2 &= 0 \\ r &= -1 \pm\, i \\ x &= e^{-t}\,(C_1 \sin t + C_2 \cos t) \\ x' &= e^{-t}\,(C_1 \cos t - C_2 \sin t) - \\ &\quad e^{-t}\,(C_1 \sin t + C_2 \cos t) \\ y &= \frac{x' - x}{5} \\ &= \frac{(-2C_1 - C_2)}{5}\, e^{-t} \sin t + \\ &\quad \frac{(C_1 - 2C_2)}{5}\, e^{-t} \cos t \\ x(0) &= 10 \rightarrow C_2 = 10 \\ y(0) &= -3 \rightarrow C_1 = 5 \\ x &= e^{-t}\,(5 \sin t + 10 \cos t) \\ y &= e^{-t}\,(4 \sin t - 3 \cos t) \end{aligned}$$

23.

$$\begin{aligned} x'' &= 2x' + (5x - 2y + 12) \\ &= 2x' + 5x - 2x' + 4x + 6 + 12 \\ x'' - 9x &= 18 \\ r^2 - 9 &= 0 \\ r &= -3 \\ r &= 3 \\ x_h &= C_1\, e^{-3t} + C_2\, e^{3t} \\ x_p &= A \\ x_p' &= 0 \\ x_p'' &= 0 \\ 0 - 9A &= 18 \\ A &= -2 \\ x &= C_1\, e^{-3t} + C_2\, e^{3t} - 2 \\ x' &= -3C_1\, e^{-3t} + 3\, e^{3t} \\ y &= x' - 2x - 3 \\ &= -5C_1\, e^{-3t} + C_2\, e^{3t} + 1 \\ x(0) &= 6 \rightarrow C_1 + C_2 = 8 \\ y(0) &= -3 \rightarrow -5C_1 + C_2 = -4 \end{aligned}$$

Solving the system

$$\begin{aligned} C_1 &= 2 \\ C_2 &= 6 \\ x &= 2\, e^{-3t} + 6\, e^{3t} - 2 \\ y &= -10\, e^{3-t} + 6\, e^{3t} + 1 \end{aligned}$$

24.

$$\begin{aligned} x'' &= -x + 2(x' - 5) + 10 \\ x'' - 2x' + x &= 0 \\ r^2 - 2r + 1 &= 0 \\ r &= 1 \text{ repeated} \\ x &= C_1\, e^t + C_2 t\, e^t \\ x' &= C_1\, e^t + C_2 t\, e^t + C_2\, e^t \\ y &= x' - 5 \\ &= C_1\, e^t + C_2 t\, e^t + C_2\, e^t - 5 \\ x(0) &= 1 \rightarrow C_1 = 1 \\ y(0) &= -2 \rightarrow C_1 + C_2 = 3 \\ C_2 &= 2 \\ x &= e^t + 2t\, e^t \\ y &= 3e^t + 2t\, e^t - 5 \end{aligned}$$

25. $P' = -3P + 2Q$ and $Q' = 3P - 2Q$

$$\begin{aligned} P'' &= -3P' + 2(3P - 2Q) \\ P'' &= -3P' + 6P - 2(P' + 3P) \\ P'' + 5P' &= 0 \\ r^2 + 5r &= 0 \end{aligned}$$

$$\begin{aligned} r &= 0 \\ r &= -5 \\ P &= C_1 + C_2\, e^{-5t} \\ P' &= -5C_2\, e^{-5t} \\ Q &= \frac{P' + 3P}{2} \\ &= \frac{3}{2}\, C_1 + C_2\, e^{-5t} \end{aligned}$$

26. $x' = -x + 4y$ and $y' = x - 4y$

$$\begin{aligned} x'' &= -x' + 4x - 4(x' + x) \\ x'' + 5x' &= 0 \\ r^2 + 5r &= 0 \\ x &= C_1 + C_2\, e^{-5t} \\ x' &= -5C_2\, e^{-5t} \\ y &= \frac{x' + x}{4} \\ &= \frac{1}{4}\, C_1 - C_2\, e^{-5t} \end{aligned}$$

27. $x' = -2x + 2x - 10$ and $y' = 2x - 2y$

$$\begin{aligned} x'' &= -2x' + 2y' \\ &= -2x' + 4x - 2(x' + 2x + 10) \\ x'' + 4x' &= -20 \\ r^2 + 4r &= 0 \\ r &= 0 \\ r &= -4 \\ x_h &= C_1 + C_2\, e^{-4t} \\ x_p &= At + B \\ x_p' &= A \\ x_p'' &= = 0 \\ 0 + 4A &= -20 \\ A &= -5 \\ x_p &= -5t \\ x &= C_1 + C_2\, e^{-4t} - 5t \\ x' &= -4C_2\, e^{-4t} - 5 \\ y &= \frac{x' + 2x + 10}{2} \\ &= C_1 - C_2\, e^{-4t} - 5t + \frac{5}{2} \end{aligned}$$

28. $x' = -3x + y - 2t - 6$ and $y' = 3x - y - 2t + 1$

$$\begin{aligned} x'' &= -3x' + 3x - y - 2t + 1 \\ &= -3x' + 3x - 3x - x' - 2t - 6 - 2t + 1 \\ x'' + 4x' &= -4t - 5 \\ r^2 + 4r &= 0 \\ r &= 0 \\ r &= -4 \\ x_h &= C_1 + C_2\, e^{-4t} \\ x_p &= At^2 + Bt + C \\ x_p' &= 2At + B \\ x_p'' &= 2A \\ 2A + 8At + 4B &= -4t - 5 \\ 8A &= -4 \to A = -\frac{1}{2} \\ 2A + 4B &= -5 \to B = -1 \\ x_p &= -\frac{t^2}{2} - t \\ x &= C_1 + C_2\, e^{-4t} - \frac{t^2}{2} - t \\ x' &= -4C_2\, e^{-4t} - t - 1 \\ y &= x' + 3x + 2t + 6 \\ &= 3C_1 - C_2\, e^{-4t} - \frac{3}{2}\, t^2 - 4t - 1 \end{aligned}$$

29. **a)** $A(t)/2$ per hour and $B(t)/2$ per hour

b) Left to the student

c) It follows from the statement of the problem

d)

$$\begin{aligned} A'' &= -0.5A' + 0.5B' \\ &= -0.5A' + 0.5(0.5A - 0.5B) \\ &= -0.5A' + 0.25A - \\ & \quad 0.25\left(\frac{A' + 0.5A'}{0.5}\right) \\ &= -0.5A' + 0.25A - 0.5A' - 0.25A \\ A'' + A' &= 0 \\ r^2 + r &= 0 \\ r &= 0 \\ r &= -1 \\ A &= C_1 + C_2\, e^{-t} \\ A' &= -C_2\, e^{-t} \\ B &= 2A' + A \\ &= C_1 - C_2\, e^{-t} \\ A(0) &= 2000 \to C_1 + C_2 = 2000 \\ B(0) &= 1000 \to C_1 - C_2 = 1000 \end{aligned}$$

Solving the system

$$\begin{aligned} C_1 &= 1500 \\ C_2 &= 500 \\ A(t) &= 1500 + 500\, e^{-t} \\ B(t) &= 1500 - 500\, e^{-t} \end{aligned}$$

e)

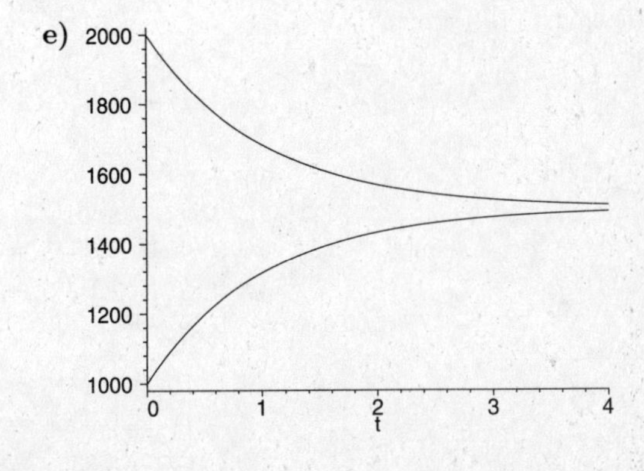

f) As the limit approaches infinity both functions approach 1500, which is the equillibrium point.

30. **a)** Left to the student

b) It follows from the statement of the problem

c)

$$\begin{aligned}
-0.1A' + 0.2B' &= A'' \\
-0.1A' + 0.2(0.1A - 0.27B) &= \\
-0.1A' + 0.02A - 0.054\left(\frac{A' + 0.1A'}{0.2}\right) &= \\
-0.1A' + 0.02A - 0.27A' - 0.027A &= \\
A'' + 0.37A' + 0.007 &= 0 \\
r^2 + 0.37r + 0.007 &= 0 \\
-0.02 &= r \\
-0.35 &= r \\
C_1\, e^{-0.02t} + C_2\, e^{-0.35t} &= A \\
-0.02C_1\, e^{-0.02t} - 0.35C_2\, e^{-0.35t} &= A' \\
\frac{A' + 0.1A}{0.2} &= B \\
0.4\, C_1\, e^{-0.02t} - 1.25\, C_2\, e^{-0.35t} &= B \\
C_1 + C_2 = 0 \leftarrow 0 &= A(0) \\
0.4C_1 - 1.25C_2 = 33 \leftarrow 33 &= B(0) \\
\text{Solving the system} & \\
38.8235 &= C_1 \\
-38.8235 &= C_2 \\
38.8235\, e^{-0.02t} - 38.8235\, e^{-0.35t} &= A(t) \\
15.5294\, e^{-0.02t} + 48.5294\, e^{-0.35t} &= B(t)
\end{aligned}$$

d)

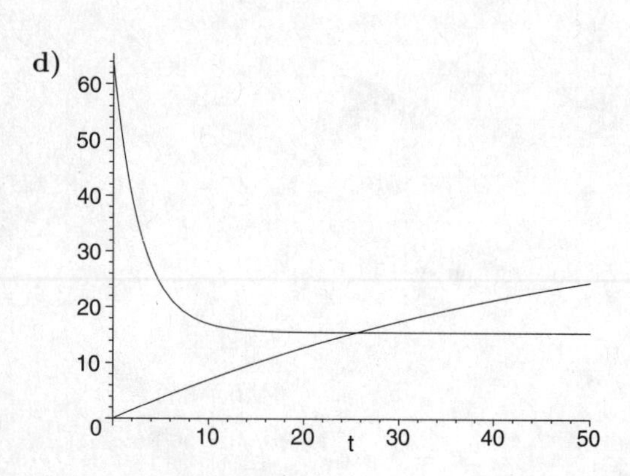

31. **a)** Left to the student

b) $P' = -3P + 0.6L - 2.4P$
$\rightarrow P' = -5.4P + 0.6L$ and $L' = 3P - 0.6L$

c)

$$\begin{aligned}
-5.4P' + 0.6(3P - 0.6L) &= P'' \\
-5.4P' + 1.8P - 0.6(-5.4P - P') &= \\
-4.8P' + 5.04P &= \\
P'' + 4.8P' - 5.04P &= 0 \\
r^2 + 4.8r - 5.04 &= 0 \\
0.886 &= r \\
-5.686 &= r \\
C_1\, e^{-5.686t} + C_2\, e^{-0.886t} &= P \\
-5.686C_1\, e^{-5.686t} - 0.886C_2\, e^{-0.886t} &= P' \\
\frac{P' + 5.4P}{0.6} &= L \\
-0.47667C_1\, e^{-5.686t} + 7.52333C_2\, e^{-0.886t} &= \\
C_1 + C_2 = 0 \leftarrow 0 &= P(0) \\
-0.47667C_1 + 7.52333C_2 = 241/15 \leftarrow 241/15 &= L(0) \\
\text{Solving the system} & \\
2.28004 &= C_1 \\
-2.28004 &= C_2 \\
2.28004\, e^{-5.686t} - 2.28004\, e^{-0.886t} &= P \\
1.086827\, e^{-5.686t} + 17.15349\, e^{-0.886t} &= L
\end{aligned}$$

32. **a)** Left to the student

b) It follows from the statement of the problem

c) $B' = -0.01B + 0.1$ and $F' = 0.01B - 0.02F$

$$\begin{aligned}
0.01B' - 0.02F' &= F'' \\
0.01(-0.01B + 0.1) - 0.02F' &= \\
0.01(-F' - 0.02F + 0.1) - 0.02F' &= \\
F'' + 0.03F' + 0.0002F &= 0.0001 \\
r^2 + 0.03r + 0.0002 &= 0 \\
r &= -0.01 \\
r &= -0.02 \\
C_1\, e^{-0.02t} + C_2\, e^{-0.01t} &= F \\
F_p &= A \\
F_p' &= 0 \\
F_p'' &= 0 \\
0 + 0 + 0.0002A &= 0.0001 \\
A &= 0.5 \\
F_p &= 0.5 \\
C_1\, e^{-0.02t} + C_2\, e^{-0.01t} + 0.5 &= F \\
-0.02C_1\, e^{-0.02t} - 0.01C_2\, e^{-0.01t} &= F' \\
100F' + 2F &= B \\
C_2\, e^{-0.01t} + 1 &= \\
C_1 + C_2 = -0.5 \leftarrow 0 &= F(0) \\
C_2 = -1 \leftarrow 0 &= B(0) \\
C_1 &= 0.5 \\
0.5\, e^{-0.02t} - 0.5\, e^{-0.01t} + 0.5 &= F \\
1 - 0.5\, e^{-0.01t} &= B
\end{aligned}$$

d)

$$\begin{aligned}
B' &= 0 \\
B &= \frac{0.1}{0.01} = 10 \\
F' &= 0 \\
F &= \frac{0.01 \cdot 10}{0.02} = 5
\end{aligned}$$

33. **a)** Left to the student

b) Left to the student

c)

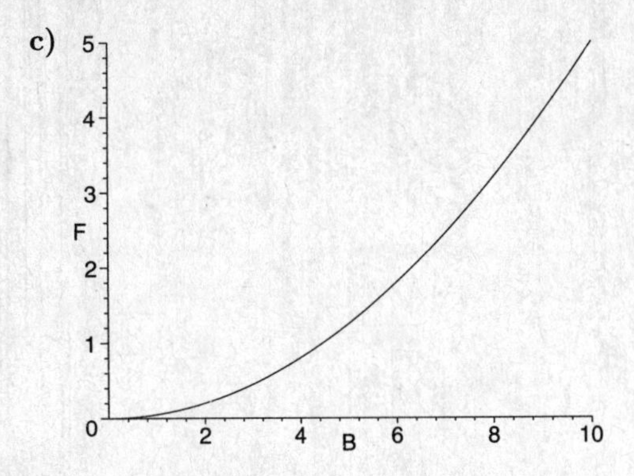

34. Left to the student

35. $L' = -aL - mL + cM + p$ and $M' = mL - aM - cM$
$L' = -(a+m)L + cM + p$ and $M' = mL - (a+c)M$

36.

$$\begin{aligned} -(a+m)L' + cM' &= L'' \\ L'' = -(a+m)L' + cmL - (a+c)L' - & \\ (a+c)(a+m)L - (a+c)p & \\ L'' = -(a+m+a+c)L' + cmL - a^2L & \\ -amL - caL - cmL - (a+c)p & \\ -(2a+c+m)L' - aL(a+m+c) - (a+c)p &= \\ L'' + (2a+c+m)L' + a(a+c+m)L &= (a+c)p \end{aligned}$$

37.

$$\begin{aligned} L_p &= A \\ L_p' &= 0 \\ L_p'' &= 0 \\ 0 + 0 + a(a+c+m)A &= (a+c)p \\ A &= \frac{(a+c)p}{a(a+c+m)} \end{aligned}$$

38. At equilibrium both $L'' = 0$ and $L' = 0$. Therefore the equation given for L'' in Exercise 36 becomes

$$\begin{aligned} 0 + 0 + a(c+m+a)L &= (c+a)p \\ L &= \frac{(c+a)p}{a(c+m+a)} \end{aligned}$$

39. $N'' + (c+m+2a)N' + a(c+m+a)N = 0$

$$\begin{aligned} L(t) - \frac{(c+a)p}{a(c+m+a)} &= N(t) \\ N'(t) &= L'(t) \\ N'' &= L''(t) \end{aligned}$$

Thus, $L'' + (c+m+2a)L' + a(c+m+a)\left(L - \frac{(c+a)p}{a(c+m+a)}\right) \to$

$$L'' + (c+m+2a)L' + a(c+m+a)L = (c+a)p$$

Which has L_p as a solution

40. a)

$$\begin{aligned} B' &= -0.06B + 0.03L \\ L' &= 0.06B - 0.03L \\ B' &= -0.06B + (0.06B - L') \\ B' &= -L' \\ B(t) &= -L(t) + C \\ B(t) + L(t) &= C \end{aligned}$$

b) The amount of serum available is 3 mg therefore $B(t) + L(t) = 3$

c)

$$\begin{aligned} B' &= -0.06B + 0.03L \\ B' &= -0.06B + 0.03(3 - B) \\ B' &= -0.09B + 0.09 \\ B' + 0.09B &= 0.09 \\ B_h &= C\,e^{-0.09t} \\ B_p &= A \\ B_p' &= 0 \\ 0 + 0.09A &= 0.09 \\ A &= 1 \\ B_p = 1 & \\ B &= 1 + C\,e^{-0.09t} \end{aligned}$$

d)

$$\begin{aligned} B &= 1 + C\,e^{-0.09t} \\ 3 &= 1 + C \\ C &= 2 \\ B &= 1 + 2\,e^{-0.09t} \end{aligned}$$

41. Left to the student

42. $B' = -pB + qL$
$L' = pB - qL$

$$\begin{aligned} B'' &= -pB' + qpB + q(-B' - pB) \\ B'' + (p-q)B' &= 0 \\ r^2 + (p-q)r &= 0 \\ r &= 0 \\ r &= -(p-q) = (q-p) \\ B &= C_1 + C_2\,e^{(q-p)t} \\ B' &= (q-p)C_2\,e^{(q-p)t} \\ L &= \frac{B' + pB}{q} \\ &= pC_1 + qC_2\,e^{(q-p)t} \end{aligned}$$

43.

$$\begin{aligned} b(t) &= \frac{C_1 + C_2\,e^{(q-p)t}}{V_B} \\ &= \frac{C_1}{V_B} + \frac{C_2}{V_B}\,e^{(q-p)t} \\ &= C_1 + C_2\,e^{-rt} \end{aligned}$$

44. $r = -(q-p) = p - q$

$$\begin{aligned} C_1 + C_2 &= \frac{D}{V_B} \\ \frac{pC_1}{V_B} + \frac{qC_2}{V_B} &= 0 \end{aligned}$$

Solving the system

$$\begin{aligned} C_1 &= -\frac{qD}{(p-q)V_B} = -\frac{qD}{rV_B} \\ C_2 &= \frac{pD}{(p-q)V_B} = \frac{pD}{rV_B} \end{aligned}$$

45.

$$\begin{aligned} C_1 + C_2 &= -\frac{qD}{rV_B} + \frac{pD}{rV_B} \\ C_1 + C_2 &= \frac{D(p-q)}{rV_B} \\ C_1 + C_2 &= \frac{D}{V_B} \\ V_B &= \frac{D}{C_1 + C_2} \end{aligned}$$

$$\begin{aligned} q &= -\frac{C_1 r V_B}{D} \\ &= -\frac{C_1 r}{C_1 + C_2} \\ p &= \frac{C_2 r V_B}{D} \\ &= \frac{C_2 r}{C_1 + C_2} \end{aligned}$$

46.

$$\begin{aligned} \frac{V_B}{V_L} &= \frac{q}{p} \\ V_L &= \frac{p}{q} V_B \\ &= \left(\frac{C_2 r}{C_1 + C_2} \cdot \frac{C_1 + C_2}{-C_1 r} \right) \frac{D}{C_1 + C_2} \\ &= -\frac{C_2 D}{C_1(C_1 + C_2)} \end{aligned}$$

Exercise Set 9.4

1. $\begin{bmatrix} x \\ y \end{bmatrix}' = \begin{bmatrix} 1 & -1 \\ 3 & -2 \end{bmatrix} \begin{bmatrix} x \\ y \end{bmatrix}$

2. $\begin{bmatrix} x \\ y \end{bmatrix}' = \begin{bmatrix} 0 & 1 \\ 2 & 0 \end{bmatrix} \begin{bmatrix} x \\ y \end{bmatrix}$

3. $\begin{bmatrix} x \\ y \end{bmatrix}' = \begin{bmatrix} 4 & 2 \\ 0 & 1 \end{bmatrix} \begin{bmatrix} x \\ y \end{bmatrix}$

4. $\begin{bmatrix} x \\ y \end{bmatrix}' = \begin{bmatrix} -1 & 3 \\ 3 & 1 \end{bmatrix} \begin{bmatrix} x \\ y \end{bmatrix}$

5. $x' = x + 3y$ and $y' = 5x + 7y$

6. $x' = 2x - 3y$ and $y' = -x + 5y$

7. $x' = 3y$ and $y' = x - 2y$

8. $x' = 4x$ and $y' = 2x - 3y$

9. $\begin{bmatrix} x \\ y \end{bmatrix}' = \begin{bmatrix} (2e^{3t} - e^{2t})' \\ (-2e^{3t} + 2e^{2t})' \end{bmatrix} = \begin{bmatrix} 6e^{3t} - 2e^{2t} \\ -6e^{3t} + 4e^{2t} \end{bmatrix}$

$\begin{bmatrix} 4 & 1 \\ -2 & 1 \end{bmatrix} \begin{bmatrix} 2e^{3t} - e^{2t} \\ -2e^{3t} + 2e^{2t} \end{bmatrix} = \begin{bmatrix} 6e^{3t} - 2e^{2t} \\ -6e^{3t} + 4e^{2t} \end{bmatrix}$

10. $\begin{bmatrix} x \\ y \end{bmatrix}' = \begin{bmatrix} (-e^{2t} + 2e^{-t})' \\ (-2e^{2t} - 2e^{-t})' \end{bmatrix} = \begin{bmatrix} -2e^{2t} - 2e^{-t} \\ -4e^{2t} + 2e^{-t} \end{bmatrix}$

$\begin{bmatrix} -4 & -3 \\ 6 & 5 \end{bmatrix} \begin{bmatrix} -e^{2t} + 2e^{-t} \\ -2e^{2t} - 2e^{-t} \end{bmatrix} = \begin{bmatrix} -2e^{2t} - 2e^{-t} \\ -4e^{2t} + 2e^{-t} \end{bmatrix}$

11. $\begin{bmatrix} x \\ y \end{bmatrix}' = \begin{bmatrix} (-2e^t \sin 2t)' \\ (3e^t \sin 2t + e^t \cos 2t)' \end{bmatrix} =$

$\begin{bmatrix} -4e^t \cos 2t - 2e^t \sin 2t \\ 6e^t \cos 2t + 3e^t \sin 2t - e^t \sin t + e^t \cos t \end{bmatrix}$

$\begin{bmatrix} -5 & -4 \\ 10 & 7 \end{bmatrix} \begin{bmatrix} -2e^t \sin 2t \\ 3e^t \sin 2t + e^t \cos 2t \end{bmatrix} =$

$\begin{bmatrix} -4e^t \cos 2t - 2e^t \sin 2t \\ 6e^t \cos 2t + 3e^t \sin 2t - e^t \sin t + e^t \cos t \end{bmatrix}$

12. $\begin{bmatrix} x \\ y \end{bmatrix}' = \begin{bmatrix} (-3 \sin t + \cos t)' \\ (2 \sin t + \cos t)' \end{bmatrix} =$

$\begin{bmatrix} -3 \cos t - \sin t \\ 2 \cos t - \sin t \end{bmatrix}$

$\begin{bmatrix} -1 & -2 \\ 1 & 1 \end{bmatrix} \begin{bmatrix} -3 \sin t + \cos t \\ 2 \sin t + \cos t \end{bmatrix} =$

$\begin{bmatrix} -\sin t - 3 \cos t \\ -\sin t + 2 \cos t \end{bmatrix}$

13. $A = \begin{bmatrix} 0 & -2 \\ 1 & 3 \end{bmatrix}$ $\det(A) = 2$ and $\mathrm{trace}(A) = 3$

$$\begin{aligned} r^2 - 3r + 2 &= 0 \\ (r-1)(r-2) &= 0 \\ r &= 1 \\ r &= 2 \end{aligned}$$

Then (by Theorem 9 of Chapter 6) the eigenvectors

For $r = 1$ are $\begin{bmatrix} -2 \\ 1 \end{bmatrix}$

For $r = 2$ are $\begin{bmatrix} -1 \\ 1 \end{bmatrix}$

Therefore, the general solution is given by

$$\begin{aligned} \begin{bmatrix} x \\ y \end{bmatrix} &= C_1 e^t \begin{bmatrix} -2 \\ 1 \end{bmatrix} + C_2 e^{2t} \begin{bmatrix} -2 \\ 1 \end{bmatrix} \\ &= \begin{bmatrix} -2C_1 e^t - C_2 e^{2t} \\ C_1 e^t + C_2 e^{2t} \end{bmatrix} \end{aligned}$$

14. $A = \begin{bmatrix} 4 & 1 \\ -2 & 1 \end{bmatrix}$ $\det(A) = 6$ and $\text{trace}(A) = 5$

$$\begin{aligned} r^2 - 5r + 6 &= 0 \\ (r-3)(r-2) &= 0 \\ r &= 2 \\ r &= 3 \end{aligned}$$

Then (by Theorem 9 of Chapter 6) the eigenvectors

For $r = 2$ are $\begin{bmatrix} 1 \\ 1 \end{bmatrix}$

For $r = 3$ are $\begin{bmatrix} 2 \\ -2 \end{bmatrix}$

Therefore, the general solution is given by

$$\begin{aligned} \begin{bmatrix} x \\ y \end{bmatrix} &= C_1\, e^{2t} \begin{bmatrix} 1 \\ 1 \end{bmatrix} + C_2\, e^{3t} \begin{bmatrix} 2 \\ -2 \end{bmatrix} \\ &= \begin{bmatrix} C_1\, e^{2t} + 2C_2\, e^{3t} \\ C_1\, e^{2t} - 2C_2\, e^{3t} \end{bmatrix} \end{aligned}$$

15. $A = \begin{bmatrix} 2 & 4 \\ 3 & -2 \end{bmatrix}$ $\det(A) = -16$ and $\text{trace}(A) = 0$

$$\begin{aligned} r^2 - 16 &= 0 \\ (r-4)(r+4) &= 0 \\ r &= -4 \\ r &= 4 \end{aligned}$$

Then (by Theorem 9 of Chapter 6) the eigenvectors

For $r = -4$ are $\begin{bmatrix} -4 \\ 6 \end{bmatrix} = \begin{bmatrix} -2 \\ 3 \end{bmatrix}$

For $r = 4$ are $\begin{bmatrix} 6 \\ 3 \end{bmatrix} = \begin{bmatrix} 2 \\ 1 \end{bmatrix}$

Therefore, the general solution is given by

$$\begin{aligned} \begin{bmatrix} x \\ y \end{bmatrix} &= C_1\, e^{-4t} \begin{bmatrix} -2 \\ 3 \end{bmatrix} + C_2\, e^{4t} \begin{bmatrix} 2 \\ 1 \end{bmatrix} \\ &= \begin{bmatrix} -2C_1\, e^{-4t} + 2C_2\, e^{4t} \\ 3C_1\, e^{-4t} + C_2\, e^{4t} \end{bmatrix} \end{aligned}$$

16. $A = \begin{bmatrix} 1 & -2 \\ -4 & -1 \end{bmatrix}$ $\det(A) = -9$ and $\text{trace}(A) = 0$

$$\begin{aligned} r^2 - 9 &= 0 \\ r &= -3 \\ r &= 3 \end{aligned}$$

Then (by Theorem 9 of Chapter 6) the eigenvectors

For $r = -3$ are $\begin{bmatrix} -2 \\ -4 \end{bmatrix} = \begin{bmatrix} 1 \\ 2 \end{bmatrix}$

For $r = 3$ are $\begin{bmatrix} -2 \\ 2 \end{bmatrix} = \begin{bmatrix} -1 \\ 1 \end{bmatrix}$

Therefore, the general solution is given by

$$\begin{aligned} \begin{bmatrix} x \\ y \end{bmatrix} &= C_1\, e^{-3t} \begin{bmatrix} 1 \\ 2 \end{bmatrix} + C_2\, e^{3t} \begin{bmatrix} -1 \\ 1 \end{bmatrix} \\ &= \begin{bmatrix} C_1\, e^{-3t} + 2C_2\, e^{3t} \\ -C_1\, e^{-3t} + C_2\, e^{3t} \end{bmatrix} \end{aligned}$$

17. $A = \begin{bmatrix} -2 & 4 \\ -1 & -7 \end{bmatrix}$ $\det(A) = 18$ and $\text{trace}(A) = -9$

$$\begin{aligned} r^2 - 9r + 18 &= 0 \\ r &= 3 \\ r &= 6 \end{aligned}$$

Then (by Theorem 9 of Chapter 6) the eigenvectors

For $r = 3$ are $\begin{bmatrix} -4 \\ -1 \end{bmatrix}$

For $r = 6$ are $\begin{bmatrix} -1 \\ 1 \end{bmatrix}$

Therefore, the general solution is given by

$$\begin{aligned} \begin{bmatrix} x \\ y \end{bmatrix} &= C_1\, e^{3t} \begin{bmatrix} -4 \\ -1 \end{bmatrix} + C_2\, e^{6t} \begin{bmatrix} -1 \\ 1 \end{bmatrix} \\ &= \begin{bmatrix} -4C_1\, e^{3t} - C_2\, e^{6t} \\ -C_1\, e^{3t} + C_2\, e^{6t} \end{bmatrix} \end{aligned}$$

18. $A = \begin{bmatrix} -10 & 4 \\ -3 & -2 \end{bmatrix}$ $\det(A) = 32$ and $\text{trace}(A) = -12$

$$\begin{aligned} r^2 + 12r + 32 &= 0 \\ r &= -4 \\ r &= -8 \end{aligned}$$

Then (by Theorem 9 of Chapter 6) the eigenvectors

For $r = -4$ are $\begin{bmatrix} 2 \\ 3 \end{bmatrix}$

For $r = -8$ are $\begin{bmatrix} 2 \\ 1 \end{bmatrix}$

Therefore, the general solution is given by

$$\begin{aligned} \begin{bmatrix} x \\ y \end{bmatrix} &= C_1\, e^{-4t} \begin{bmatrix} 2 \\ 3 \end{bmatrix} + C_2\, e^{-8t} \begin{bmatrix} 2 \\ 1 \end{bmatrix} \\ &= \begin{bmatrix} 2C_1\, e^{-4t} + 2C_2\, e^{-8t} \\ 2C_1\, e^{-4t} + C_2\, e^{-8t} \end{bmatrix} \end{aligned}$$

19. $A = \begin{bmatrix} -5 & 10 \\ -4 & 7 \end{bmatrix}$ $\det(A) = 5$ and $\text{trace}(A) = 2$

$$\begin{aligned} r^2 - 2r + 5 &= 0 \\ r &= 1 \pm 2i \\ x &= e^t\,(C_1 \sin 2t + C_2 \cos 2t) \\ x' &= e^t\,(2C_1 \cos 2t - 2C_2 \sin 2t) \\ &\quad + e^t\,(C_1 \sin 2t + C_2 \cos 2t) \\ y &= \frac{x' + 5x}{10} \\ &= \frac{(3C_1 - C_2)}{5}\, e^t \sin 2t + \frac{(C_1 + 3C_2)}{5}\, e^t \cos 2t \end{aligned}$$

20. $A = \begin{bmatrix} -2 & -2 \\ 1 & 0 \end{bmatrix}$ $\det(A) = 2$ and $\text{trace}(A) = -2$

$$\begin{aligned} r^2 + 2r + 2 &= 0 \\ r &= -1 \pm i \\ x &= e^{-t}\,(C_1 \; sin\; t + C_2 \; cos\; t) \\ x' &= e^{-t}\,(C_1 \; cos\; 2t - C_2 \; sin\; t) \\ &\quad -e^{-t}\,(C_1 \; sin\; t + C_2 \; cos\; t) \\ y &= \frac{-x' - 2x}{2} \\ &= \frac{(-C_1 + 2C_2)}{2}\; e^{-t}\; sin\; t + \\ &\quad \frac{(C_1 - C_2)}{2}\; e^{-t}\; cos\; t \end{aligned}$$

21. $A = \begin{bmatrix} 0 & -1 \\ 1 & 2 \end{bmatrix}$ $\det(A) = 1$ and $\text{trace}(A) = 2$

$$\begin{aligned} r^2 - 2r + 1 &= 0 \\ r &= 1 \text{ repeated} \\ x &= C_1\; e^t + C_2 t\; e^t \\ x' &= C_1\; e^t + C_2 t\; e^t + C_2\; e^t \\ y &= -x' = -(C_1 + C_2)\; e^t - C_2 t\; e^t \end{aligned}$$

22. $A = \begin{bmatrix} 1 & 1 \\ -9 & -5 \end{bmatrix}$ $\det(A) = 4$ and $\text{trace}(A) = -4$

$$\begin{aligned} r^2 + 4r + 4 &= 0 \\ r &= -2 \text{ repeated} \\ x &= C_1\; e^{-2t} + C_2 t\; e^{-2t} \\ x' &= -2C_1\; e^t - 2C_2 t\; e^t + C_2\; e^t \\ y &= x' - x = (-3C_1 + C_2)\; e^t - 2C_2 t\; e^t \end{aligned}$$

23. Since the eigenvalues have different signs, then the origin is an unstable saddle point

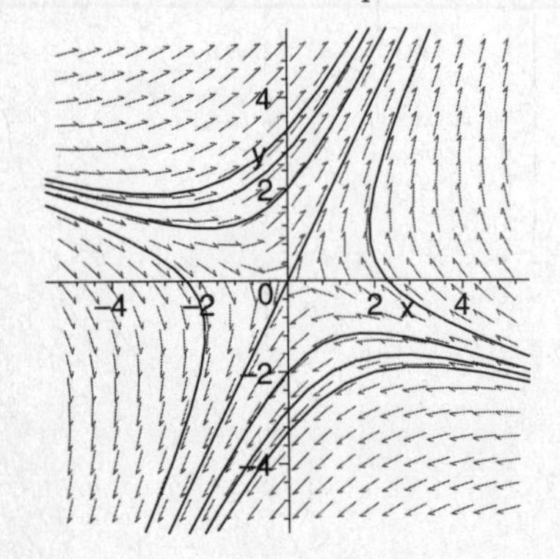

24. Since the eigenvalues have different signs, then the origin is an unstable saddle point

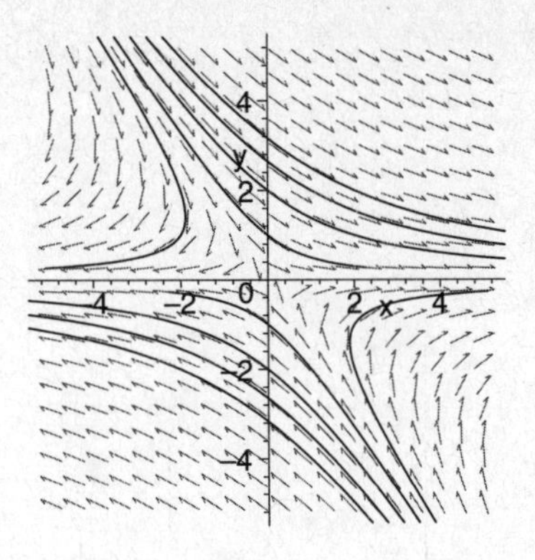

25. Since the eigenvalues are both positive, then the origin is an unstable node

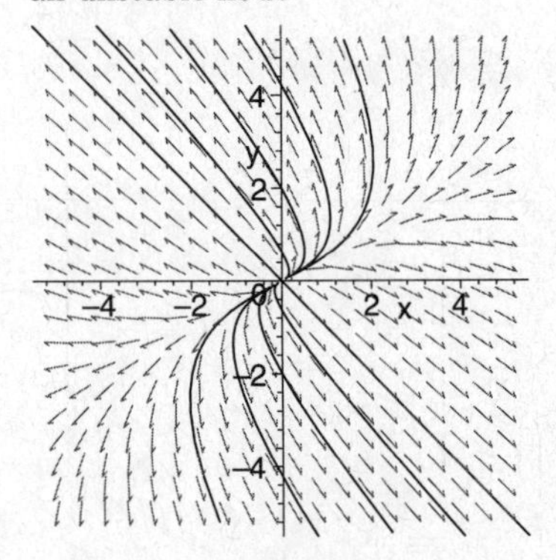

26. Since the eigenvalues are both positive, then the origin is an unstable node

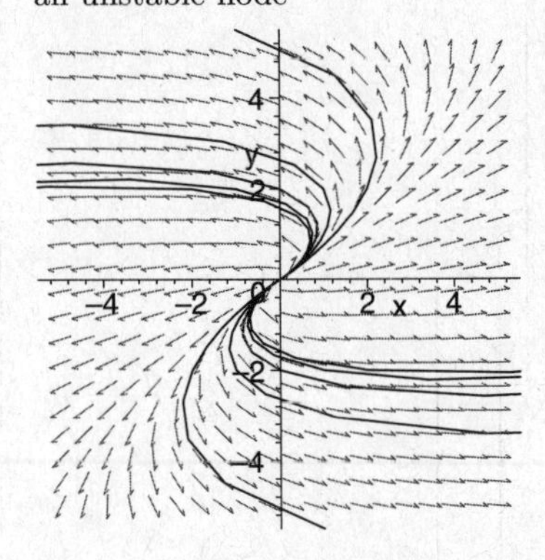

27. Since the eigenvalues are both negative, then the origin is an asymptotically stable node

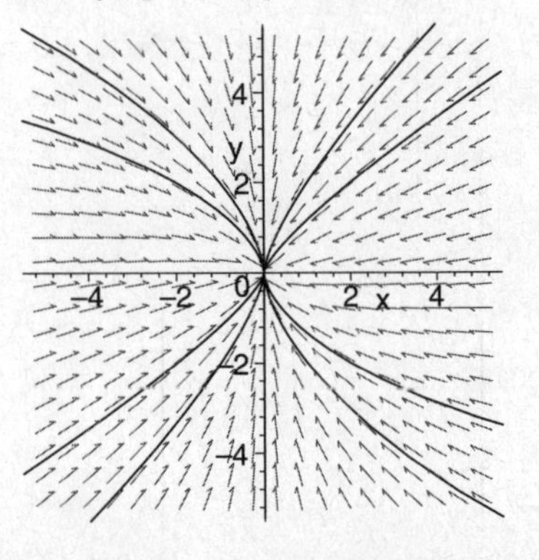

28. Since the eigenvalues are both negative, then the origin is an asymptotically stable node

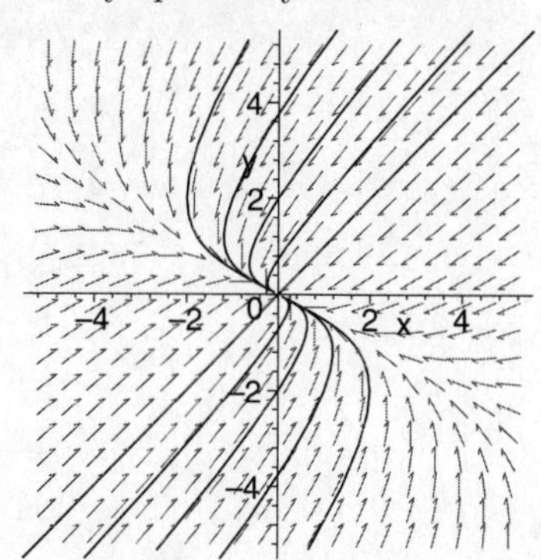

29. Since the eigenvalues are both positive, then the origin is an unstable node

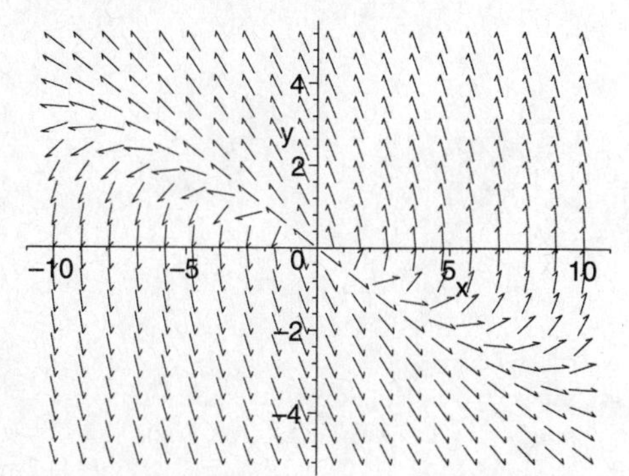

30. Since the eigenvalues are both positive, then the origin is an unstable node

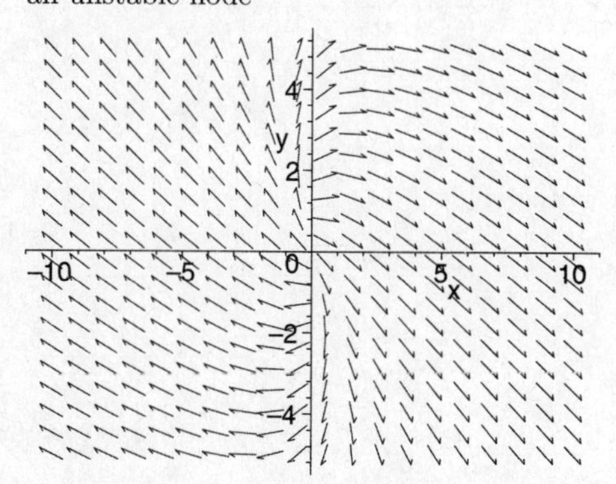

31. Since the eigenvalues have different signs, then the origin is an unstable saddle point

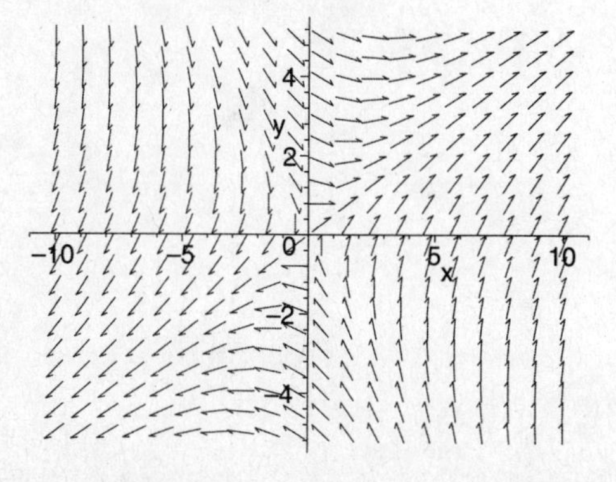

32. Since the eigenvalues have different signs, then the origin is an unstable saddle point

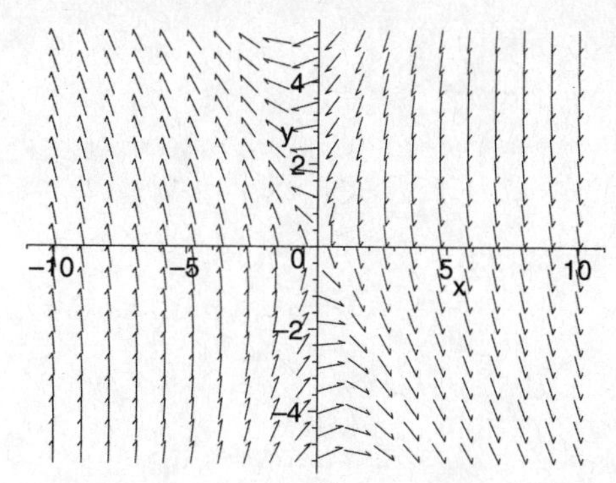

33. Since the eigenvalues are both positive, then the origin is an asymptotically stable node

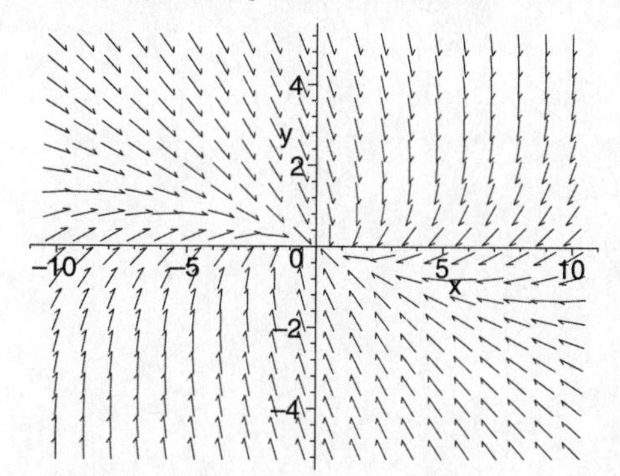

34. Since the eigenvalues are both negative, then the origin is an asymptotically stable node

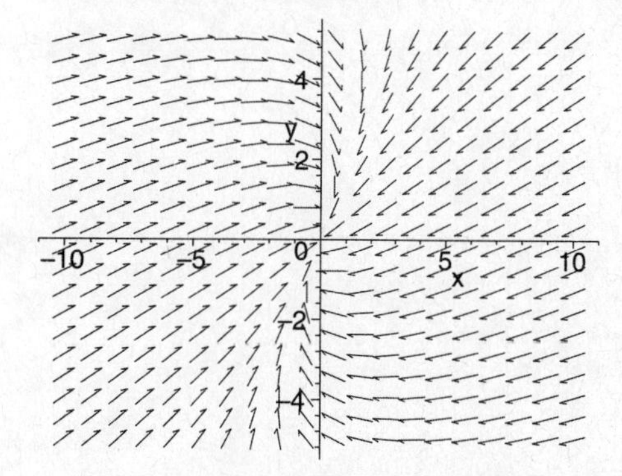

35. Since the eigenvalues are complex with a positive real part, then the prigin is an unstable spiral point

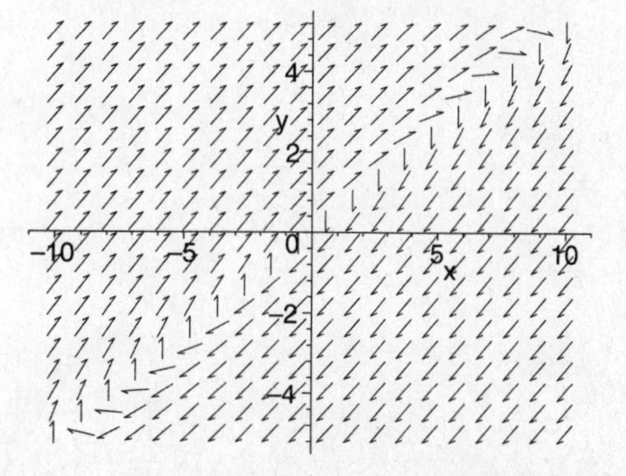

36. Since the eigenvalues are complex with a negative real part, then the prigin is an asymptotically stable spiral point

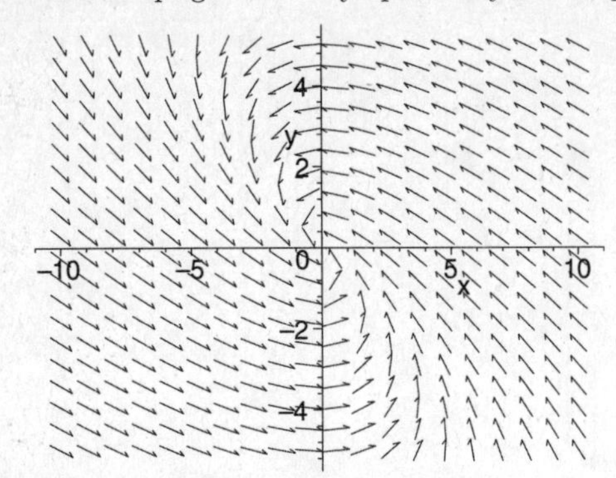

37. Since the eigenvalues are positive and equal, then the origin is an unstable improper node

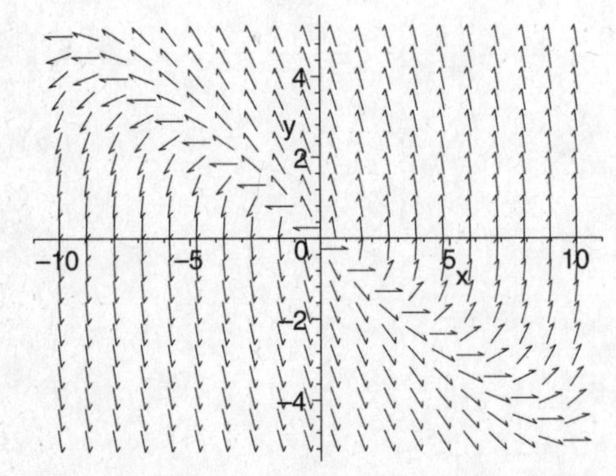

38. Since the eigenvalues are negative and equal, then the origin is an asymptotically stable improper node

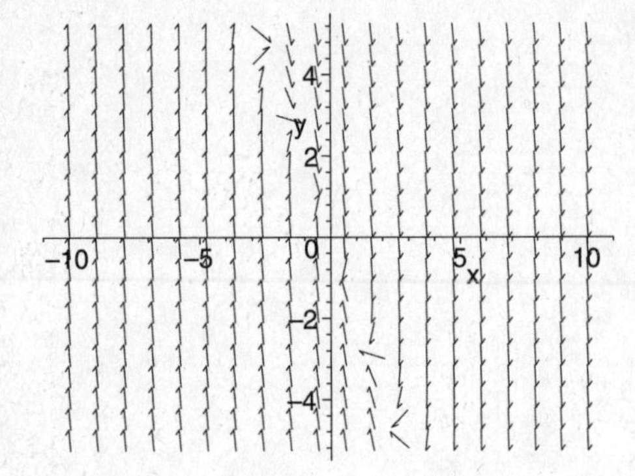

39. $A = \begin{bmatrix} 0 & 1 \\ -1 & 2 \end{bmatrix}$

a)

$$\begin{aligned} x'' &= -x + 2y \\ &= -x + 2x' \\ x'' - 2x' + x &= 0 \\ r^2 - 2r + 1 &= 0 \\ (r-1)^2 &= 0 \\ r &= 1 \text{ repeated} \end{aligned}$$

$$\begin{bmatrix} -1 & 1 \\ -1 & 1 \end{bmatrix} \begin{bmatrix} x \\ y \end{bmatrix} = \begin{bmatrix} 0 \\ 0 \end{bmatrix}$$

$$\begin{bmatrix} x \\ y \end{bmatrix} = \begin{bmatrix} 1 \\ 1 \end{bmatrix}$$

b) Left to the student

40. $A = \begin{bmatrix} r & 0 \\ 0 & r \end{bmatrix}$

a)

$$\begin{aligned} (r-\lambda)(r-\lambda) - 0 &= 0 \\ \lambda &= r \text{ repeated} \end{aligned}$$

$$\begin{bmatrix} 0 & 0 \\ 0 & 0 \end{bmatrix} \begin{bmatrix} x \\ y \end{bmatrix} = \begin{bmatrix} 0 \\ 0 \end{bmatrix}$$

The above is true for any vector

b)

$$\begin{aligned} \frac{x'}{x} &= r \\ \frac{dx}{x} &= r\,dt \\ \ln x &= rt + C \\ x &= C_1\, e^{rt} \end{aligned}$$

Similarly,

$$y = C_2\, e^{rt}$$

c) Note that $x = 0$ is the constant solution

$$e^{rt} = \frac{x}{C_1} = \frac{y}{C_2}$$

Thus,

$$\frac{x}{C_1} = \frac{y}{C_2}$$

$$\begin{aligned} y &= \frac{C_2}{C_1}\, x \\ &= mx \end{aligned}$$

d) Left to the student

e)

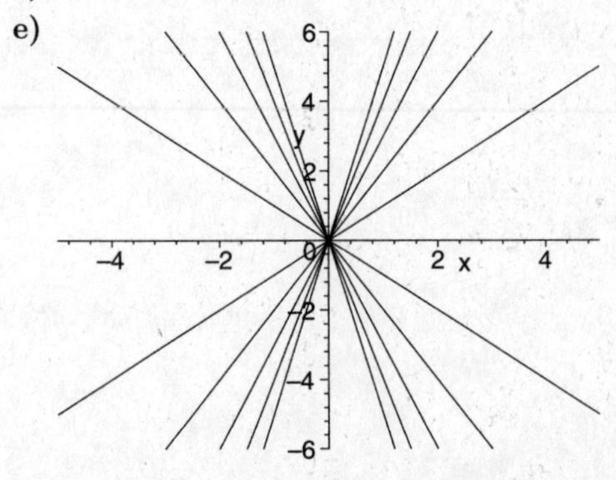

41.

$$\begin{aligned} x'' &- ax' + by' \\ &= ax' + b(cx + dy) \\ &= ax' + bcx + d(x' - ax) \\ &= ax' + bcx + dx' - adx \\ &= (a+d)x' - (ad - bc)x \\ x'' - (a+d)x' + (ad - bc)x &= 0 \\ r^2 - (a+d)r + (ad - bc) &= 0 \end{aligned}$$

42. Let $r = a$ and $r = b$, $a, b \neq 0$ be the two eigenvalues, then

$$\begin{aligned}(r-a)(r-b) &= 0\\ r^2-(a+b)r+ab &= 0\\ x''-(a+b)x'+abx &= 0\end{aligned}$$

We have a matrix with a trace of $a+b$ and a determinant of ab which means we can wrtie the matrix as follows

$$A = \begin{bmatrix} \text{a} & 0 \\ 0 & \text{b} \end{bmatrix}$$

Then, $x' = ax$ and $y' = by$ which means the origin is the only equilibrium point

43. **a)**

$$\begin{aligned}x'' &= -2x'+6(2x-6y)\\ &= -2x'+12x+6(-x'-2x)\\ &= -2x'+12x-6x'-12x\\ x''+8x' &= 0\\ r^2+8r &= 0\\ r &= -8\\ r &= 0\end{aligned}$$

b) The eigenvectors associated with $r=-8$ are $\begin{bmatrix} -1 \\ 1 \end{bmatrix}$ and the eigenvectors associated with $r=0$ are $\begin{bmatrix} 3 \\ 1 \end{bmatrix}$. Thus

$$\begin{aligned}\begin{bmatrix} \text{x} \\ \text{y} \end{bmatrix} &= \begin{bmatrix} 3 \\ 1 \end{bmatrix} C_1 + \begin{bmatrix} -1 \\ 1 \end{bmatrix} C_2\, e^{-8}\\ &= \begin{bmatrix} 3C_1 - C_2\, e^{-8t} \\ C_1 + C_2\, e^{-8t} \end{bmatrix}\end{aligned}$$

c)

$$\begin{aligned}y+x &= 3C_1 - C_2\, e^{-8t} + C_1 + C_2\, e^{-8t}\\ y+x &= 4C_1\\ y &= 4C_1 - x\\ y &= b-x\end{aligned}$$

d)

$$\begin{aligned}x' &= 0\\ -2x+6y &= 0\\ y &= \frac{x}{3}\end{aligned}$$

Similarly,

$$\begin{aligned}y' &= 0\\ 2x-6y &= 0\\ y &= \frac{x}{3}\end{aligned}$$

e)

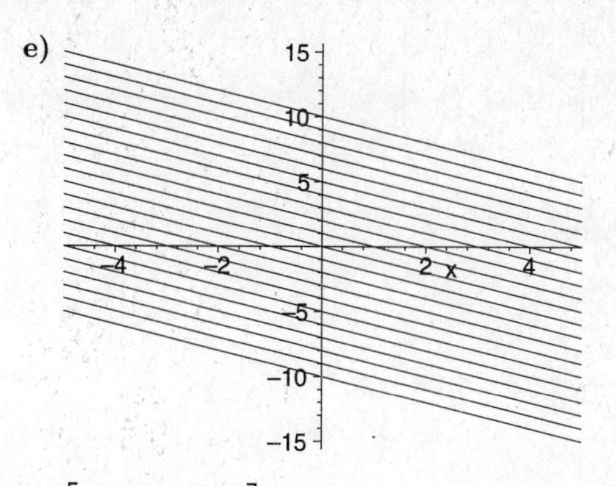

44. $A = \begin{bmatrix} 0 & -1 & -3 \\ 2 & 3 & 3 \\ -2 & 1 & 1 \end{bmatrix}$
The eigenvalues of matrix A are $r=-2$, $r=2$ and $r=4$ which have the following eigenvectors $\begin{bmatrix} 1 \\ -1 \\ 1 \end{bmatrix}$, $\begin{bmatrix} -1 \\ -1 \\ 1 \end{bmatrix}$, and $\begin{bmatrix} -1 \\ 1 \\ 1 \end{bmatrix}$ respectively. Therefore, the general solution is given by

$$\begin{bmatrix} C_1\, e^{-2t} - C_2\, e^{2t} - C_3\, e^{4t} \\ -C_1\, e^{-2t} - C_2\, e^{2t} + C_3\, e^{4t} \\ C_1\, e^{-2t} + C_2\, e^{2t} + C_3\, e^{4t} \end{bmatrix}$$

45. $A = \begin{bmatrix} 1 & -3 & 0 \\ -3 & -3 & -4 \\ 3 & 5 & 6 \end{bmatrix}$
The eigenvalues of matrix A are $r=-2$, $r=2$ and $r=4$ which have the following eigenvectors $\begin{bmatrix} -1 \\ -1 \\ 1 \end{bmatrix}$, $\begin{bmatrix} -3 \\ 1 \\ 1 \end{bmatrix}$, and $\begin{bmatrix} 1 \\ -1 \\ 1 \end{bmatrix}$ respectively. Therefore, the general solution is given by

$$\begin{bmatrix} -C_1\, e^{-2t} - 3C_2\, e^{2t} + C_3\, e^{4t} \\ -C_1\, e^{-2t} + C_2\, e^{2t} - C_3\, e^{4t} \\ C_1\, e^{-2t} + C_2\, e^{2t} + C_3\, e^{4t} \end{bmatrix}$$

46. $A = \begin{bmatrix} -3 & 12 & 6 \\ -2 & -9 & -6 \\ 4 & 0 & 3 \end{bmatrix}$
The eigenvalues of matrix A are $r=-9$, $r=-3$ and $r=3$ which have the following eigenvectors $\begin{bmatrix} -3 \\ 1 \\ 1 \end{bmatrix}$, $\begin{bmatrix} 3 \\ 1 \\ -2 \end{bmatrix}$, and $\begin{bmatrix} 0 \\ 1 \\ -2 \end{bmatrix}$ respectively. Therefore, the general solution is given by

$$\begin{bmatrix} -3C_1\, e^{-9t} + 3C_2\, e^{-3t} \\ C_1\, e^{-9t} + C_2\, e^{-3t} + C_3\, e^{3t} \\ C_1\, e^{-9t} - 2C_2\, e^{2t} - 2C_3\, e^{4t} \end{bmatrix}$$

47. $A = \begin{bmatrix} 3 & 0 & 2 \\ -1 & 2 & -2 \\ -1 & 0 & 0 \end{bmatrix}$
The eigenvalues of matrix A are $r = 1$, $r = 2$ and $r = 2$ which have the following eigenvectors $\begin{bmatrix} -1 \\ 1 \\ 1 \end{bmatrix}$, $\begin{bmatrix} 0 \\ 1 \\ 0 \end{bmatrix}$, and $\begin{bmatrix} -2 \\ 0 \\ 1 \end{bmatrix}$ respectively. Therefore, the general solution is given by

$$\begin{bmatrix} C_1\, e^{t} - 2C_3\, e^{2t} \\ C_1\, e^{t} + C_2\, e^{2t} \\ C_1\, e^{t} + C_3\, e^{2t} \end{bmatrix}$$

48. $A = \begin{bmatrix} 2 & 7 & 7 \\ 1 & 2 & 1 \\ -1 & -7 & -6 \end{bmatrix}$
The eigenvalues of matrix A are $r = -5$, $r = 1$ and $r = 2$ which have the following eigenvectors $\begin{bmatrix} -1 \\ 0 \\ 1 \end{bmatrix}$, $\begin{bmatrix} 0 \\ 1 \\ -1 \end{bmatrix}$, and $\begin{bmatrix} -1 \\ -1 \\ 1 \end{bmatrix}$ respectively. Therefore, the general solution is given by

$$\begin{bmatrix} -C_1\, e^{-5t} - C_3\, e^{2t} \\ C_2\, e^{t} + C_3\, e^{2t} \\ C_1\, e^{-5t} - C_2\, e^{t} + C_3\, e^{2t} \end{bmatrix}$$

49. $A = \begin{bmatrix} 2 & 0 & 6 \\ 1 & 1 & 0 \\ -1 & 1 & -4 \end{bmatrix}$
The eigenvalues of matrix A are $r = -2$, $r = -1$ and $r = 2$ which have the following eigenvectors $\begin{bmatrix} -3 \\ 1 \\ 2 \end{bmatrix}$, $\begin{bmatrix} -2 \\ 1 \\ 1 \end{bmatrix}$, and $\begin{bmatrix} 1 \\ 1 \\ 0 \end{bmatrix}$ respectively. Therefore, the general solution is given by

$$\begin{bmatrix} -3C_1\, e^{-2t} - 2C_2\, e^{-t} + C_3\, e^{2t} \\ C_1\, e^{-2t} + C_2\, e^{-t} + C_3\, e^{2t} \\ 2C_1\, e^{-2t} + C_2\, e^{-t} \end{bmatrix}$$

50. Left to the student

51. $S' = -aS$
$B' = aS - (b + c)B + dH$
$H' = cB - dH$

52. $A = \begin{bmatrix} -5 & 0 & 0 \\ 5 & -15/2 & 1 \\ 0 & 7/2 & -1 \end{bmatrix}$ which has eigenvalues $r = -8$, $r = -5$, and $r = \dfrac{-1}{2}$

53. Since all the eigenvalues are negative then the origin will be an asymptotically stable node

54. As time goes on the amount of carbon-14 will continue to decrease until is vanishes eventually

55. The eigenvalues -8, -5, and $\dfrac{-1}{2}$ have the eigenvectors $\begin{bmatrix} 0 \\ -2 \\ 1 \end{bmatrix}$, $\begin{bmatrix} 1 \\ 40/27 \\ -35/27 \end{bmatrix}$, and $\begin{bmatrix} 0 \\ 1 \\ 7 \end{bmatrix}$ respectively. Therefore, the general solution is given by

$$\begin{bmatrix} C_2\, e^{-5t} \\ -2C_1\, e^{-8t} + 40/27C_2\, e^{-5t} + C_3\, e^{-t/2} \\ C_1\, e^{-8t} - 35/27C_2\, e^{-5t} + 7C_3\, e^{t/2} \end{bmatrix}$$

Applying the initial conditions

$$\begin{aligned} 1 &= C_2 \\ -2C_1 + \frac{40}{27}C_2 + C_3 &= 0 \\ C_1 - \frac{35}{27}C_2 + 7C_3 &= 0 \end{aligned}$$

Solving the system

$$\begin{aligned} C_1 &= \frac{7}{9} \\ C_2 &= 1 \\ C_3 &= \frac{2}{27} \end{aligned}$$

Thus the solution is

$$S(t) = e^{-5t}$$

$$B(t) = -\frac{14}{9}\, e^{-8t} + \frac{40}{27}\, e^{-5t} + \frac{2}{7}\, e^{-t/2}$$

$$H(t) = \frac{7}{9}\, e^{-8t} - \frac{35}{27}\, e^{-5t} + \frac{14}{27}\, e^{-t/2}$$

Exercise Set 9.5

1. Jacobian $J = \begin{bmatrix} 2x + y + 4 & \text{x} \\ \text{y} & x - 4y + 1 \end{bmatrix}$

$$\begin{aligned} x(x + y + 4) &= 0 \\ x &= 0 \\ y(-2y + 1) &= 0 \\ y &= 0 \\ y &= \frac{1}{2} \\ x + y + 4 &= 0 \\ y &= -x - 4 \\ (-x - 4)(x + 2x + 8 + 1) &= 0 \\ x &= -4 \\ y &= 0 \\ x &= -3 \\ y &= -1 \end{aligned}$$

The equilibrium points are:
$(0, 0) \rightarrow$ Jacobian has two positive eigenvalues which means the equilibrium point is an unstable node
$(-4, 0) \rightarrow$ Jacobian has two negative eigenvalues which means the equilibrium point is an asymptotically stable node

$(-3,-1) \to$ Jacobian has eigenvalues with opposite signs which means the equilibrium point is an unstable saddle point
$\left(0, \frac{1}{2}\right) \to$ Jacobian has eigenvalues with opposite signs which means the equilibrium point is an unstable saddle point

2. Jacobian = $\begin{bmatrix} \text{y-2} & \text{x} \\ \text{y} & \text{x-3} \end{bmatrix}$
The equilibrium points are:
$(0,0) \to$ Jacobian has two negative eigenvalues which means the equilibrium point is an asymptotically stable node
$(3,2) \to$ Jacobian has eigenvalues with opposite signs which means the equilibrium point is an unstable saddle point

3. Jacobian = $\begin{bmatrix} \text{2x} & \text{2y} \\ 1 & 1 \end{bmatrix}$

$$\begin{aligned} y &= 7-x \\ x^2+(7-x)^2-25 &= 0 \\ x^2+49-14x+x^2-25 &= 0 \\ 2x^2-14x+24 &= 0 \\ x &= 3 \to y=4 \\ x &= 4 \to y=3 \end{aligned}$$

The equilibrium points are:
$(3,4) \to$ Jacobian has eigenvalues with opposite signs which means the equilibrium point is an unstable saddle point
$(4,3) \to$ Jacobian has eigenvalues with opposite signs which means the equilibrium point is an unstable saddle point

4. Jacobian = $\begin{bmatrix} \text{2x} & \text{2y} \\ \text{2x-4} & \text{-1} \end{bmatrix}$
The equilibrium points are:
$(1,2) \to$ Jacobian has complex eigenvalues with positive real part which means the equilibrium point is an unstable spiral point
$(2,1) \to$ Jacobian has eigenvalues with opposite signs which means the equilibrium point is an unstable saddle point

5. Jacobian = $\begin{bmatrix} -\sqrt{y} & (2-x)/\sqrt{y} \\ \text{y} & \text{x} \end{bmatrix}$

$$\begin{aligned} (2-x)\sqrt{y}7 &= 0 \\ x &= 0 \to y=4 \\ y &= 0 \to \text{ no solution} \end{aligned}$$

The equilibrium point is $(2,4) \to$ Jacobian has eigenvalues with opposite signs which means the equilibrium point is an unstable saddle point

6. Jacobian = $\begin{bmatrix} 1 & 1 \\ \text{2x} & \text{6y} \end{bmatrix}$
The equilibrium points are:
$(2,1) \to$ Jacobian has complex eigenvalues with positive real part which means the equilibrium point is an unstable spiral point
$\left(\frac{5}{2}, \frac{1}{2}\right) \to$ Jacobian has complex eigenvalues with positive real part which means the equilibrium point is an unstable spiral point

7. Jacobian = $\begin{bmatrix} \text{-1} & 2y \\ 1 & 4y^3 \end{bmatrix}$

$$\begin{aligned} -x+y^2 &= 0 \\ x+y^4-2 &= 0 \\ y^4+y^2-2 &= 0 \\ (y^2-1)(y^2+2) &= 0 \\ y &= -1 \to x=1 \\ y &= 1 \to x=1 \end{aligned}$$

The equilibrium point are:
$(1,1) \to$ Jacobian has eigenvalues with opposite signs which means the equilibrium point is an unstable saddle point
$(1,-1) \to$ Jacobian has two negative eigenvalues which means the equilibrium point is an asymptotically stable node

8. Jacobian = $\begin{bmatrix} -4x & \text{-1} \\ 4x^3 & 1 \end{bmatrix}$
The equilibrium points are:
$(-2,-8) \to$ Jacobian has eigenvalues with opposite signs which means the equilibrium point is an unstable saddle point
$(2,-8) \to$ Jacobian has complex eigenvalues with negative real part which means the equilibrium point is an asymptotically stable spiral point

9. Jacobian = $\begin{bmatrix} 1 & -e^y \\ 1 & 2e^{2y} \end{bmatrix}$

$$\begin{aligned} x-e^y &= 0 \\ x+e^{2y}-2 &= 0 \\ e^{2y}+e^y-2 &= 0 \\ (e^y-1)(e^y+2) &= 0 \\ e^y &= -2 \text{ no solution} \\ e^y-1 &= 0 \\ y &= 0 \to x=1 \end{aligned}$$

The equilibrium point is $(1,0) \to$ Jacobian has complex eigenvalues with positive real part which means the equilibrium point is an unstable spiral point

10. Jacobian = $\begin{bmatrix} 1 & -e^y \\ 2/x & 1 \end{bmatrix}$
The equilibrium points are:
$(-3,e^{-3}) \to$ Jacobian has complex eigenvalues with positive real part which means the equilibrium point is an unstable spiral point
$(2,e^2) \to$ Jacobian has complex eigenvalues with positive real part which means the equilibrium point is an unstable spiral point

11. $J = \begin{bmatrix} \text{0.1-0.02x -0.005y} & \text{-0.005x} \\ \text{-0.001y} & \text{0.05-0.001x-0.004y} \end{bmatrix}$

a)

$$\begin{aligned} x(0.1-0.01x-0.005y) &= 0 \\ y(0.05-0.001x-0.002y) &= 0 \\ x &= 0 \\ y(0.05-0.002y) &= 0 \\ y &= 0 \\ y &= 25 \\ y &= 0 \\ x(0.1-0.01x) &= 0 \\ x &= 0 \\ x &= 10 \\ 0.1-0.01x-0.005y &= 0 \\ 20-2x &= y \\ (20-2x)(0.05-0.001x-0.002(20-2x)) &= 0 \\ y=0 \leftarrow x &= 10 \\ \text{Only non-negative values accepted } \leftarrow x &= \frac{-4}{3} \end{aligned}$$

The non-negative equilibrium points are:
$(0,0) \rightarrow$ Jacobian has two positive eigenvalues which means the equilibrium point is an unstable node
$(0,25) \rightarrow$ Jacobian has two negative eigenvalues which means the equilibrium point is an asym[totically stable node
$(10,0) \rightarrow$ Jacobian has eigenvalues with opposite signs which means the equilibrium point is an unstable saddle point

b) Only the second species survives

12. $J = \begin{bmatrix} \text{0.04-0.0016x -0.0024y} & \text{-0.005x} \\ \text{-0.0012y} & \text{0.02-0.0012x-0.0008y} \end{bmatrix}$

a) The non-negative equilibrium points are:
$(0,0) \rightarrow$ Jacobian has two positive eigenvalues which means the equilibrium point is an unstable node
$(0,50) \rightarrow$ Jacobian has two negative eigenvalues which means the equilibrium point is an asymptotically stable node
$(50,0) \rightarrow$ Jacobian has two negative eigenvalues which means the equilibrium point is an asymptotically stable node
$(12.5,12.5)$ Jacobian has eigenvalues with opposite signs which means the equilibrium point is an unstable saddle point

b) The stable equilibrium point $(0,50)$ indicates that only the second species survives, while the stable equilibrium point $(50,0)$ indicates that only the first species survives

13. $J = \begin{bmatrix} \text{0.1-0.01x -0.002y} & \text{-0.002x} \\ \text{-0.001y} & \text{0.05-0.001x-0.004y} \end{bmatrix}$

a)

$$\begin{aligned} x(0.1-0.005x-0.002y) &= 0 \\ y(0.05y-0.001x-0.002y) &= 0 \\ x &= 0 \\ y(0.05-0.002y) &= 0 \\ y &= 0 \\ y &= 25 \\ y &= 0 \\ x(0.1-0.01x) &= 0 \\ x &= 0 \\ x &= 10 \\ 0.1-0.005x-0.002y &= 0 \\ 50-2.5x &= y \\ (50-2.5x)(0.05-0.001x-0.002(50-2.5x)) &= 0 \\ (50-2.5x)(-0.5+0.04x) &= 0 \\ y=0 \leftarrow x &= 25 \\ y=18.75 \leftarrow x &= 12.25 \end{aligned}$$

The non-negative equilibrium points are:
$(0,0) \rightarrow$ Jacobian has two positive eigenvalues which means the equilibrium point is an unstable node
$(0,25) \rightarrow$ Jacobian has two negative eigenvalues which means the equilibrium point is an asymptotically stable node
$(10,0) \rightarrow$ Jacobian has eigenvalues with opposite signs which means the equilibrium point is an unstable saddle point
$(12.5,18.75) \rightarrow$ Jacobian has two negative eigenvalues which means the equilibrium point is an asymptotically stable node

b) The stable equilibrium point indicates there is coexistence

14. $J = \begin{bmatrix} \text{0.01-0.0002x} & \text{-0.0007x} \\ \text{-0.0006y} & \text{0.02-0.0006x-0.0004y} \end{bmatrix}$

a) The non-negative equilibrium points are:
$(0,0) \rightarrow$ Jacobian has two positive eigenvalues which means the equilibrium point is an unstable node
$(0,100) \rightarrow$ Jacobian has two negative eigenvalues which means the equilibrium point is an asymptotically stable node
$(30,10) \rightarrow$ Jacobian has eigenvalues with opposite signs which means the equilibrium point is an unstable saddle point
$(100,0) \rightarrow$ Jacobian has two negative eigenvalues which means the equilibrium point is an asymptotically stable node

b) For the equilibrium point $(0,100)$ only the second species survives while for the equilibrium point $(100,0)$ only the first species survives

15. $J = \begin{bmatrix} \text{0.1-0.000x-0.0008y} & \text{-0.0008x} \\ \text{-0.002y} & \text{0.1-0.002x-0.01y} \end{bmatrix}$

a) The non-negative equilibrium points are:
$(0,0) \rightarrow$ Jacobian has two positive eigenvalues which

means the equilibrium point is an unstable node
$(0, 20)$ $\rightarrow$ Jacobian has eigenvalues with opposite signs which means the equilibrium point is an unstable saddle point
$(100, 0)$ $\rightarrow$ Jacobian has two negative eigenvalues which means the equilibrium point is an asymptotically stable node
$(300, -100)$ $\rightarrow$ Jacobian has eigenvalues with opposite signs which means the equilibrium point is an unstable saddle point

b) The stable equilibrium point indicates only the first species survives

16. $J = \begin{bmatrix} \text{0.1-0.012x-0.0008y} & \text{-0.0008x} \\ \text{-0.001y} & \text{0.2-0.001x-0.012y} \end{bmatrix}$

a) The non-negative equilibrium points are:
$(0, 0) \rightarrow$ Jacobian has two positive eigenvalues which means the equilibrium point is an unstable node
$\left(0, \frac{100}{3}\right)$ $\rightarrow$ Jacobian has eigenvalues with opposite signs which means the equilibrium point is an unstable saddle point
$\left(\frac{50}{3}, 0\right)$ $\rightarrow$ Jacobian has eigenvalues with opposite signs which means the equilibrium point is an unstable saddle point
$(12.5, 31.25) \rightarrow$ Jacobian has two negative eigenvalues which means the equilibrium point is an asymptotically stable node

b) The stable equilibrium point indicates there is coexistence

17. $J = \begin{bmatrix} \text{0.1 -0.02x -0.005y} & \text{-0.005x} \\ \text{-0.015x} & \text{0.2-0.015x-0.04y} \end{bmatrix}$

$$\begin{aligned}
x(0.1 - 0.01x - 0.005y) &= 0 \\
y(0.2 - 0.015x - 0.02y) &= 0 \\
x &= 0 \rightarrow y(0.2 - 0.02y) = 0 \\
y &= 0 \\
y &= 10 \\
y &= 0 \rightarrow x(0.1 - 0.01x) = 0 \\
x &= 0 \\
x &= 10 \\
0 &= (0.1 - 0.01x - 0.005y) \\
y &= 20 - 2x \\
0 &= (20 - 2x)(0.2 - 0.015x - \\
&\quad 0.02(20 - 2x)) \\
x &= 10 \\
x &= 8 \rightarrow y = 4
\end{aligned}$$

$(0, 0)$ gives an unstable node
$(0, 10)$ gives an unstable node
$(10, 0)$ gives an unstable saddle point $(8, 4)$ gives an asymptotically stable node, and it indicates coexistence

18. **a)** $x(0.1 - 0.01x - 0.005y) - 0.01x = 0$
$y(0.2 - 0.015x - 0.02y) = 0$

$$\begin{aligned}
0.2 - 0.015x - 0.02y &= 0 \\
10 - 0.75x &= y \\
x(0.1 - 0.01x - 0.02(10 - 0.75x) - 0.01x &= 0 \\
x\,[0.1 - 0.01x - 0.2 + 0.015x - 0.01] &= 0 \\
x &= 0 \\
x &= 6.4 \\
y &= 5.2
\end{aligned}$$

b) The new equilibrium point is $(6.4, 5.2) \rightarrow$ Jacobian has eigenvalues with different sign. This means the new equilibrium point is an unstable node. It is different from $(8, 4)$ due to the addition of the new predation rate $0.01x$ which is the increase of fishing for the first species

19. $J = \begin{bmatrix} \text{0.1-0.2x -0.005y} & \text{-0.005x} \\ \text{-0.015y} & \text{0.2-0.015x -0.4y} \end{bmatrix}$

a) $x(0.1 - 0.01x - 0.005y) - 0.08x = 0$
$y(0.2 - 0.015x - 0.02y) = 0$

$$\begin{aligned}
0.2 - 0.015x - 0.02y &= 0 \\
10 - 0.75x &= y \\
x(0.1 - 0.01x - 0.02(10 - 0.75x) - 0.08x &= 0 \\
x\,[0.1 - 0.01x - 0.2 + 0.015x - 0.08] &= 0 \\
x &= -4.8
\end{aligned}$$

not acceptable

$$\begin{aligned}
y &= 0 \\
x &= 0 \\
x &= 2
\end{aligned}$$

There are no coexistence equilibrium points

b) $(0, 0)$ gives an unstable node
$(0, 10)$ gives an asymptotically stable node
$(2, 0)$ gives an unstable saddle point

c) The excessive fishing (of the first species) helps eliminate the first species

20. $J = \begin{bmatrix} \text{0.5-0.4y} & \text{-0.4x} \\ \text{0.2y} & \text{-0.4+0.2x} \end{bmatrix}$

$$\begin{aligned}
x(0.5 - 0.4y) &= 0 \\
-y(0.4 - 0.2x) &= 0 \\
x &= 0 \rightarrow y = 0 \\
y &= 1.25 \rightarrow x = 2
\end{aligned}$$

$(0, 0)$ gives an unstable saddle point
$(2, 1.25)$ gives a center

21. $J = \begin{bmatrix} \text{0.6-0.3y} & \text{-0.3x} \\ \text{0.2y} & \text{-1+0.2x} \end{bmatrix}$

$$\begin{aligned}
x(0.6 - 0.3y) &= 0 \\
-y(1 - 0.2x) &= 0 \\
x &= 0 \rightarrow y = 0 \\
y &= 2 \rightarrow x = 5
\end{aligned}$$

$(0,0)$ gives an unstable saddle point
$(5,2)$ gives a center

22. $J = \begin{bmatrix} \text{0.8-0.2y} & \text{-0.2x} \\ \text{0.1y} & \text{-0.6+0.1x} \end{bmatrix}$

$$\begin{aligned} x(0.8-0.2y) &= 0 \\ -y(0.6-0.1x) &= 0 \\ x &= 0 \rightarrow y = 0 \\ y &= 4 \rightarrow x = 6 \end{aligned}$$

$(0,0)$ gives an unstable saddle point
$(6,4)$ gives a center

23. $J = \begin{bmatrix} \text{0.5-0.2y} & \text{-0.2x} \\ \text{0.1y} & \text{-0.4+0.1x} \end{bmatrix}$

$$\begin{aligned} x(0.5-0.2y) &= 0 \\ -y(0.4-0.1x) &= 0 \\ x &= 0 \rightarrow y = 0 \\ y &= 2.5 \rightarrow x = 4 \end{aligned}$$

$(0,0)$ gives an unstable saddle point
$(4,2.5)$ gives a center

24. $J = \begin{bmatrix} \text{0.5-0.1x -0.4y} & \text{-0.4x} \\ \text{0.2y} & \text{-0.4+0.2x} \end{bmatrix}$

$$\begin{aligned} x\,[0.5-0.05x-0.4y] &= 0 \\ -y(0.4-0.2x) &= 0 \\ x &= 0 \rightarrow y = 0 \\ x &= 2 \rightarrow y = 1 \\ x &= 10 \rightarrow y = 0 \end{aligned}$$

$(0,0)$ gives an unstable saddle point
$(2,1)$ gives an asymptotically stable spiral point $(10,0)$ gives an unstable saddle point

25. $J = \begin{bmatrix} \text{0.6-0.06x -0.3y} & \text{-0.3x} \\ \text{0.2y} & \text{-1+0.2x} \end{bmatrix}$

$$\begin{aligned} x\,[0.6-0.06x-0.3y] &= 0 \\ -y(1-0.2x) &= 0 \\ x &= 0 \rightarrow y = 0 \\ x &= 5 \rightarrow y = 1.5 \\ x &= 20 \rightarrow y = 0 \end{aligned}$$

$(0,0)$ gives an unstable saddle point
$(5,1.5)$ gives a center $(20,0)$ gives an unstable node

26. $J = \begin{bmatrix} \text{0.8-0.067x -0.2y} & \text{-0.2x} \\ \text{0.1y} & \text{-0.6+0.1x} \end{bmatrix}$

$$\begin{aligned} x\,[0.8-0.067x-0.2y] &= 0 \\ -y(0.6-0.1x) &= 0 \\ x &= 0 \rightarrow y = 0 \\ x &= 6 \rightarrow y = 3 \\ x &= 24 \rightarrow y = 0 \end{aligned}$$

$(0,0)$ gives an unstable saddle point
$(6,3)$ gives an asymptotically stable spiral point $(24,0)$ gives an unstable saddle point

27. $J = \begin{bmatrix} \text{0.5-0.1x -0.2y} & \text{-0.2x} \\ \text{0.1y} & \text{-0.4+0.1x} \end{bmatrix}$

$$\begin{aligned} x\,[0.5-0.05x-0.2y] &= 0 \\ -y(0.4-0.1x) &= 0 \\ x &= 0 \rightarrow y = 0 \\ x &= 4 \rightarrow y = 1.5 \\ x &= 10 \rightarrow y = 0 \end{aligned}$$

$(0,0)$ gives an unstable saddle point
$(4,1.5)$ gives an asymptotically stable spiral point $(10,0)$ gives an unstable saddle point

28. The coexistence equilibrium points are close for the two sets of exercises, but the most of the coexistence equilibrium points in the inhibited prey growth model are stable while in the uninhibited model they were centers.

29. **a)** When $x = 0$ and $y = 0$, then $ax - by = 0$ and $-dy + \dfrac{cx}{\sqrt{c^2x^2+1}} = 0$

b) $x = \dfrac{\sqrt{65}}{12}$, $y = \dfrac{\sqrt{65}}{9}$

c) Left to the student

30. **a)** When $x = 0$ and $y = 0$, then $ax - by = 0$ and $-dy + \dfrac{cx}{\sqrt{c^2x^2+1}} = 0$

b) $x = \dfrac{\sqrt{65}}{12}$, $y = \dfrac{\sqrt{65}}{9}$

c) Left to the student

31. $J = \begin{bmatrix} 0.2-0.1x-0.02y & -0.02x \\ -0.01y & 0.1-0.01x-0.04y \end{bmatrix}$

a) The eigenvalues for $(0,5)$ are -0.1 and 0.1 which have the corresponding eigenvectors $[0,1]$ and $[4,-1]$ respectively. The trajectories seems to approach parallel to $[0,1]$ and parallel to $[4,-1]$ as they leave

b) The eigenvalues for $(4,0)$ are -0.2 and 0.06 which have the corresponding eigenvectors $[1,0]$ and $[-4,13]$ respectively. The trajectories seems to approach parallel to $[1,0]$ and parallel to $[-4,13]$ as they leave

c) The eigenvalues for $(2.5,3.75)$ are -0.05 and -0.15 which have the corresponding eigenvectors $[-2,3]$ and $[2,1]$ respectively. The trajectories seems to approach parallel to $[-2,3]$

32. **a)** Exercises 13 and 16

b) Exercises 12,14, and 15

33. **a)** In Exercises 11, 13, 15 and 16, a_1a_2 is larger and in Exercises 12 and 14 b_1b_2 is larger

b) Left to the student

34. The term represents the amount for the prey that is over the "normal" or expected value for that species

35. If $a = 0$ then

$$\begin{aligned} x' &= px\left(1-\frac{x}{L}\right) - \frac{qxy}{1+0} \\ &= px\left(1-\frac{x}{L}\right) - qxy \\ y' &= -ry + \frac{sxy}{1+0} \\ &= -ry + sxy \end{aligned}$$

36. $x' = 0.2x\left(1-\frac{x}{10}\right) - \frac{0.04xy}{1+0.1x}$

$y' = -0.05y + \frac{0.03xy}{1+0.1x}$

The coexistence equilibrium point is $(2, 4.8)$ which is an unstable saddle point

37. $x' = 0.2x\left(1-\frac{x}{10}\right) - \frac{0.04xy}{1+0.1x}$

$y' = -0.05y + \frac{0.03xy}{1+0.1x}$

The coexistence equilibrium point is $(2, 5.7)$ which is an unstable spiral point

38. **a)**

$$\begin{aligned} x(p-qy) &= 0 \\ y(-r+sx) &= 0 \\ x &= 0 \to y = 0 \\ x &= \frac{r}{s} \to y = \frac{p}{q} \end{aligned}$$

b) $(0,0)$ gives an unstable saddle point

$\left(\frac{r}{s}, \frac{p}{q}\right)$ gives a center

39. **a)** Left to the student

b) When $x = \frac{r}{s}$ is plugged into the formula the numerator becomes zero

c) When $y = \frac{p}{q}$ is plugged into the formula the denominator becomes zero

d) One population is maximized when the other population agrees with the coexistence equilibrium point value

40.

$$\begin{aligned} \frac{dy}{dx} &= \frac{y(-r+sx)}{x(p-qy)} \\ \frac{p-qy}{y}\,dy &= \frac{-r+sx}{x}\,dx \\ \int \frac{p}{y} - q\,dy &= \int -\frac{r}{x} + s\,dx \\ p\ \ln y - qy &= -r\ \ln x + sx + C \\ p\ \ln y - qy + r\ \ln x - sx &= C \end{aligned}$$

41. $x' = 2x - xy$ and $y' = -y + 0.4xy$
This means $p = 2$, $q = 1$, $r = 1$, and $s = 0.4 \to$

$$\begin{aligned} 2\ \ln y - y + \ln x - 0.4x &= C \\ 2\ \ln 1 - 1 + \ln 5 - 0.4(5) &= C \\ -1 + \ln 5 - 2 &= C \\ \ln 5 - 3 &= C \\ 2\ \ln y - y + \ln x - 0.4x &= \ln 5 - 3 \end{aligned}$$

42. $x = 0.57990238$ and $x = 6.6958675$

43. $x = 1.0159393$ and $x = 5$

44. $x = 0.74988795$ and $x = 5.9118737$

45. $x = 1.6312622$ and $x = 3.6330767$

46. Left to the student

Exercise Set 9.6

1. $x(1) \approx 2.42368$
$y(1) \approx 1.048$

2. $x(1) \approx 12.41172$
$y(1) \approx 12.27048$

3. $x(2) \approx 38.21927$
$y(2) \approx 83.47852$

4. $x(2) \approx 14.337$
$y(2) \approx 14.20849$

5. $x(3) \approx 75.15387$
$y(3) \approx 212.32202$

6. $x(3) \approx -128.29883$
$y(3) \approx -239.83185$

7. **a)**

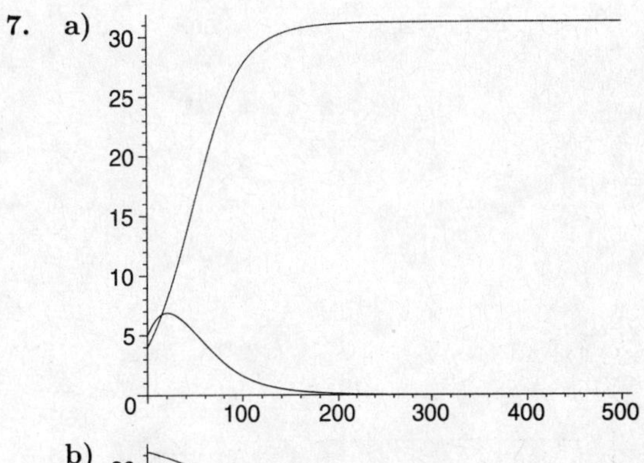

b)

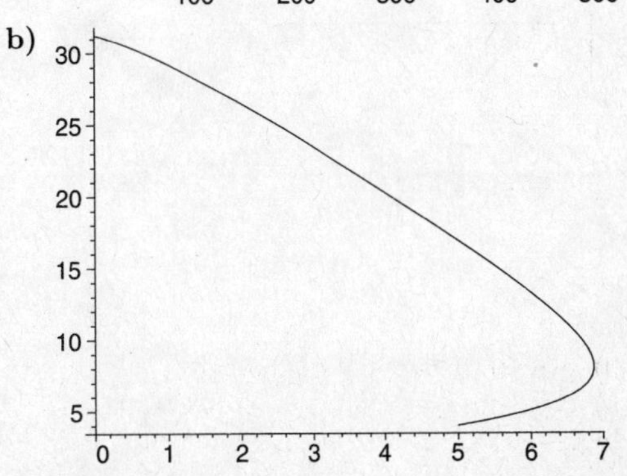

8. a)

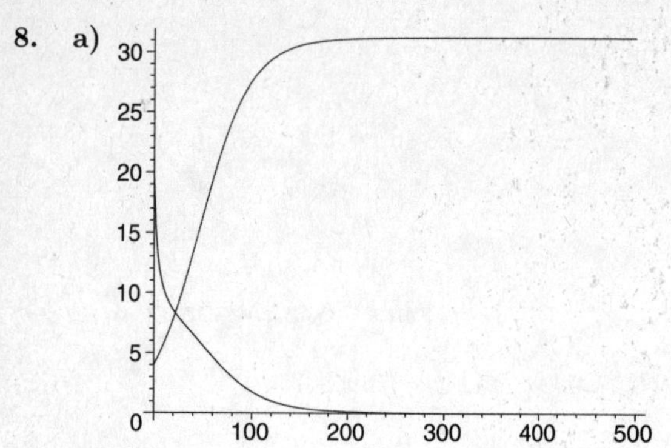

b)

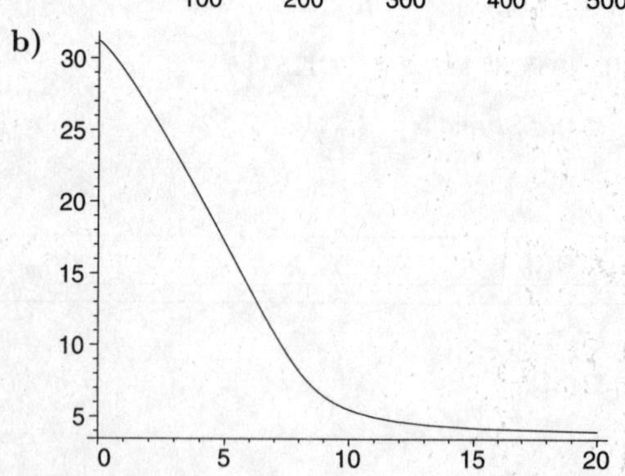

9. a)

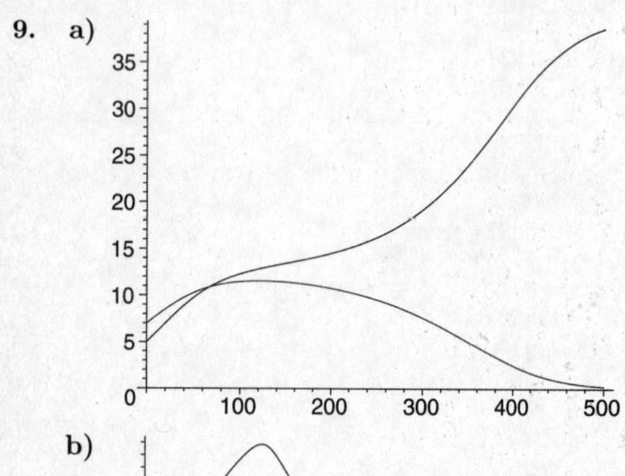

b)

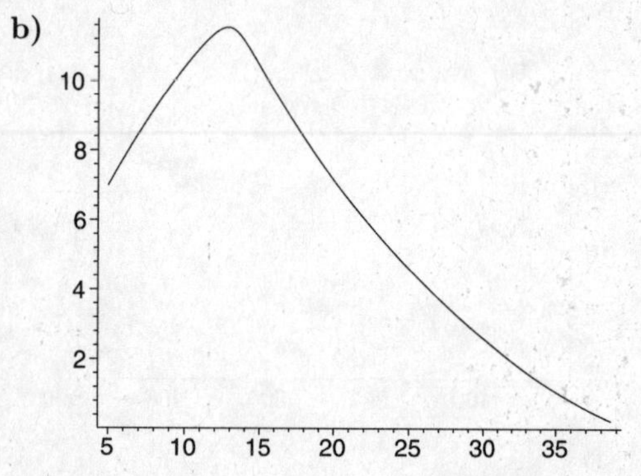

10. a)

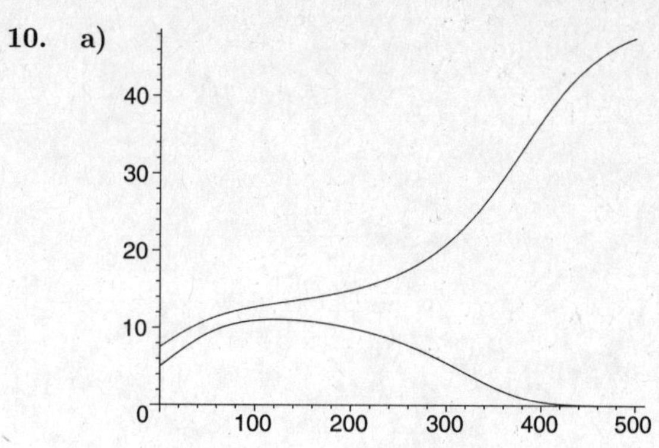

b)

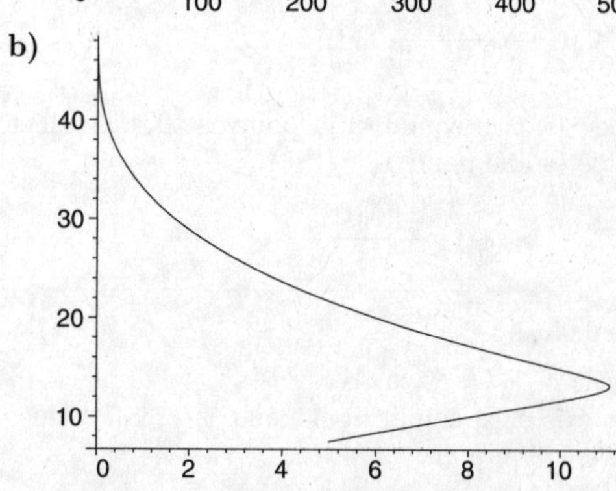

11. a)

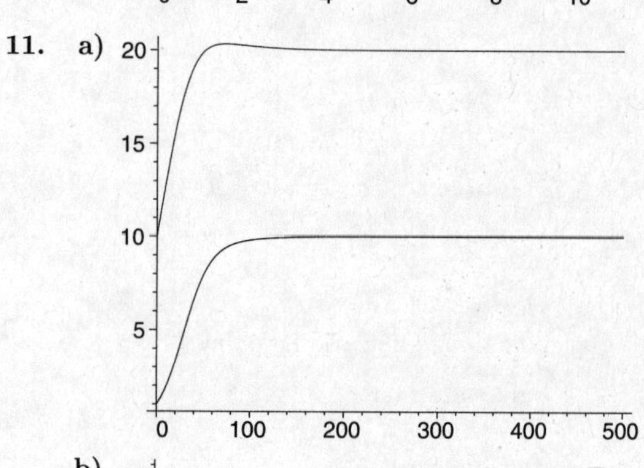

b)

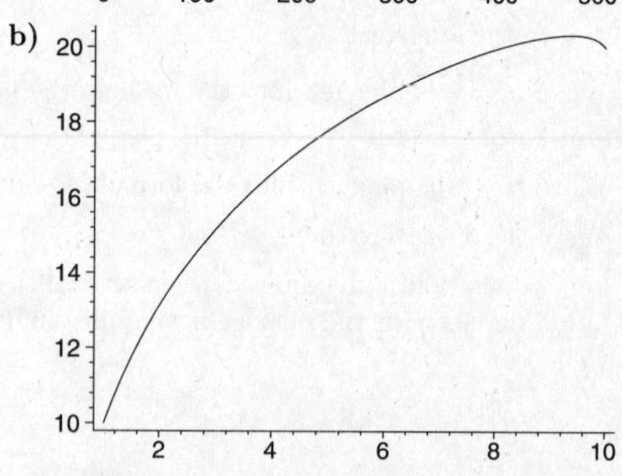

12. a)

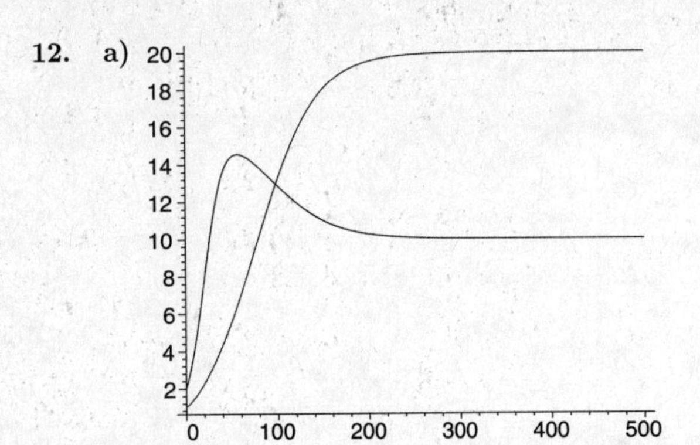

b)

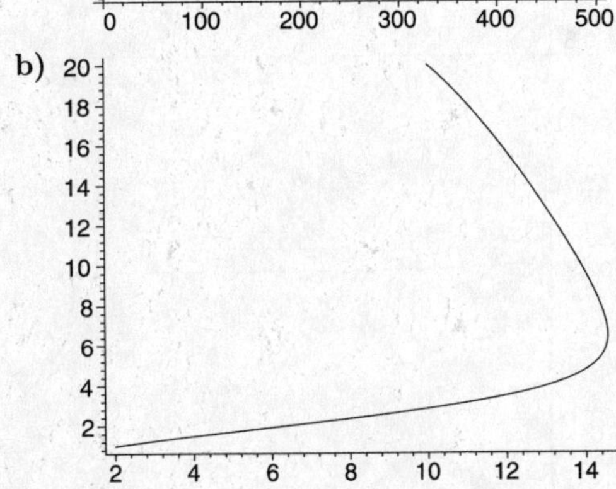

13. a)

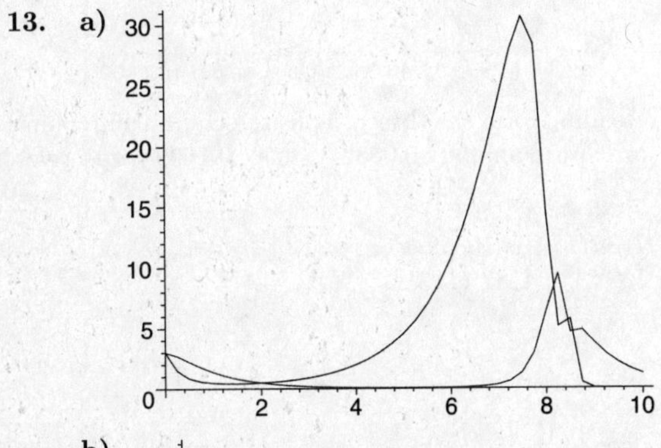

b)

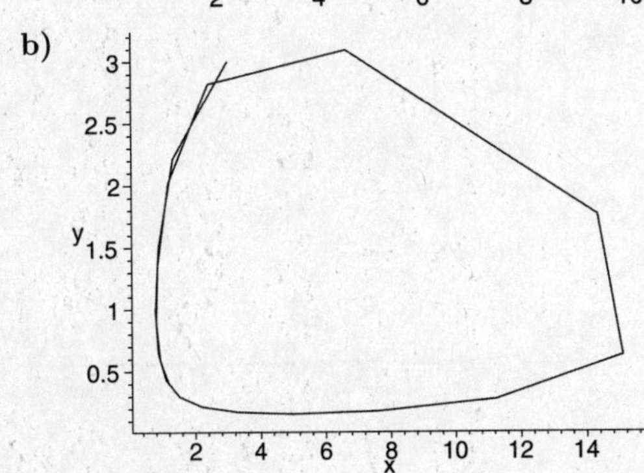

c) About 7.4 years

14. a)

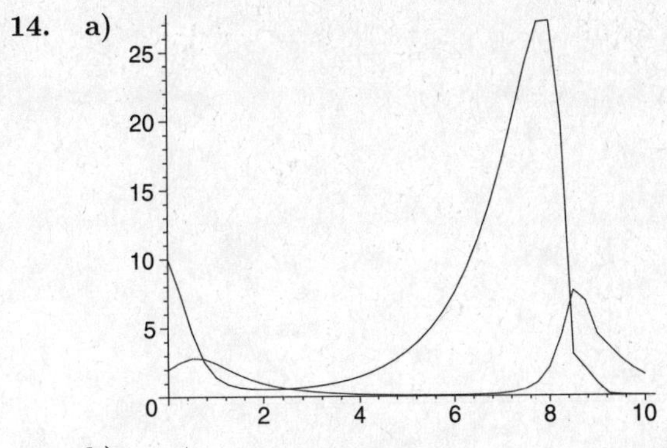

b)

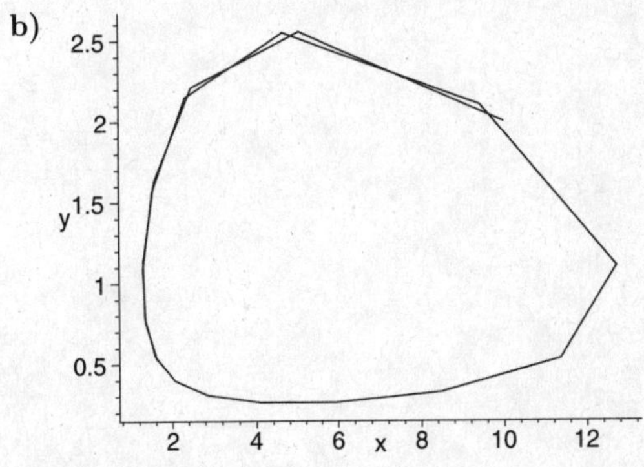

c) About 6.9 years

15. a)

b)

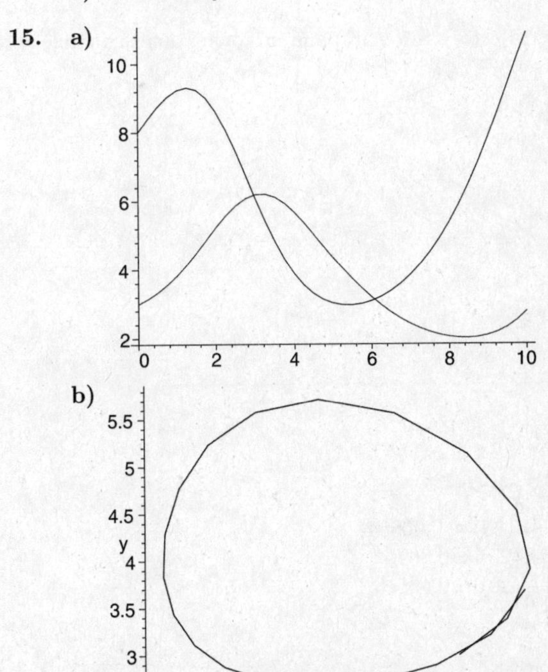

c) About 9.2 years

16. **a)**

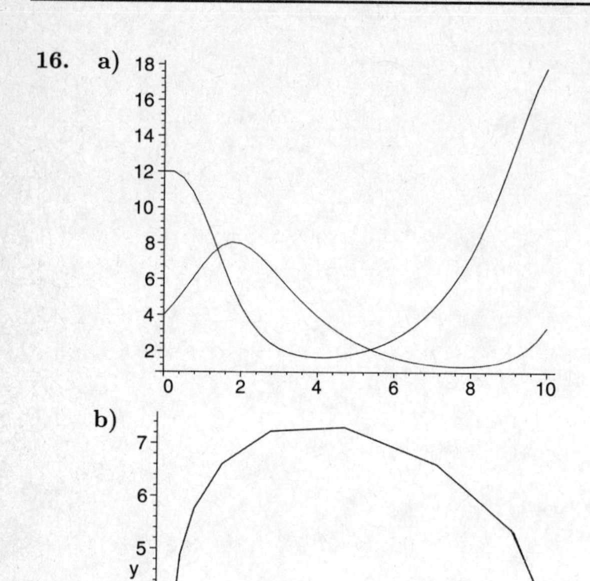

b)

c) About 9.5 years

17. The trajectory is a spiral and no longer a closed circle

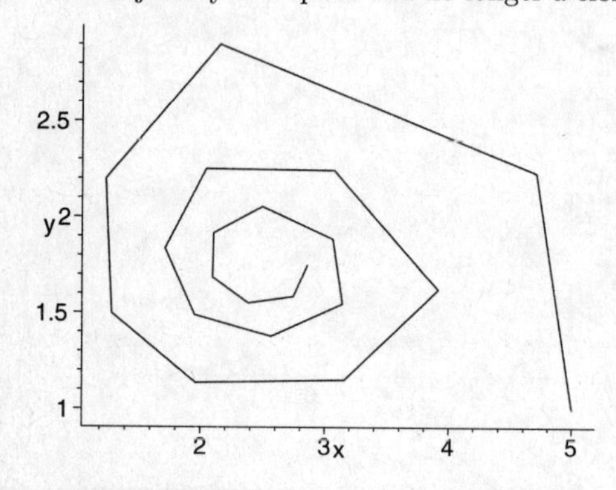

18. $S(10) \approx 2.13429$
$X(10) \approx 0.04225$

19. **a)** $W(20) \approx 0.00664$
$E(20) \approx 0.8620$
$U(20) \approx 0.19016$

b) $W(t)$

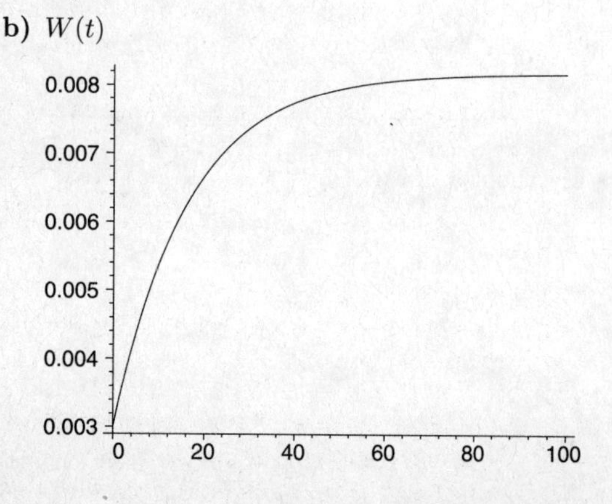

$E(t)$

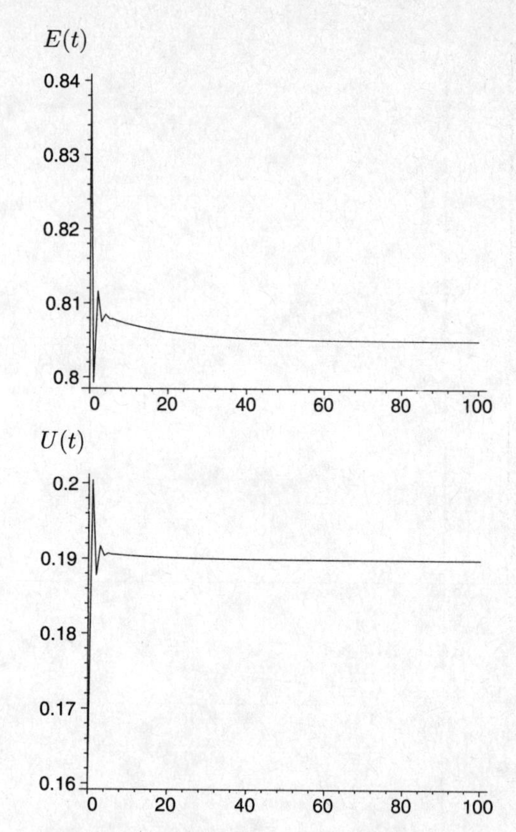

$U(t)$

c) Reading form the three graphs the equilibrium points are approximately 0.0082, 0.8050, 0.1899 respectively

20. **a)** $T(35) \approx 822.87647$
$I(35) \approx 100.10859$
$V(35) \approx 4187.11131$

b) $T(t)$

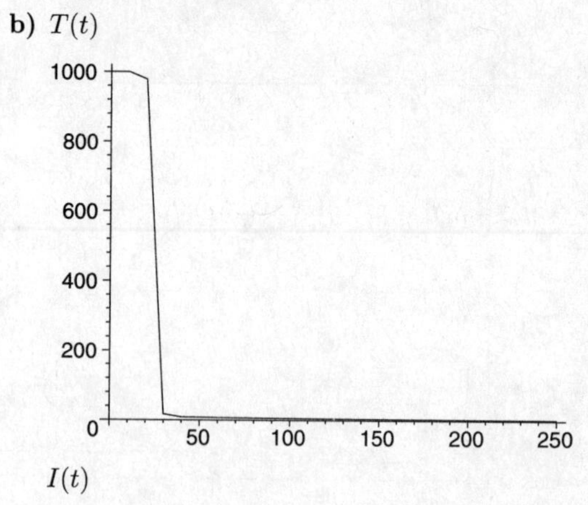

$I(t)$

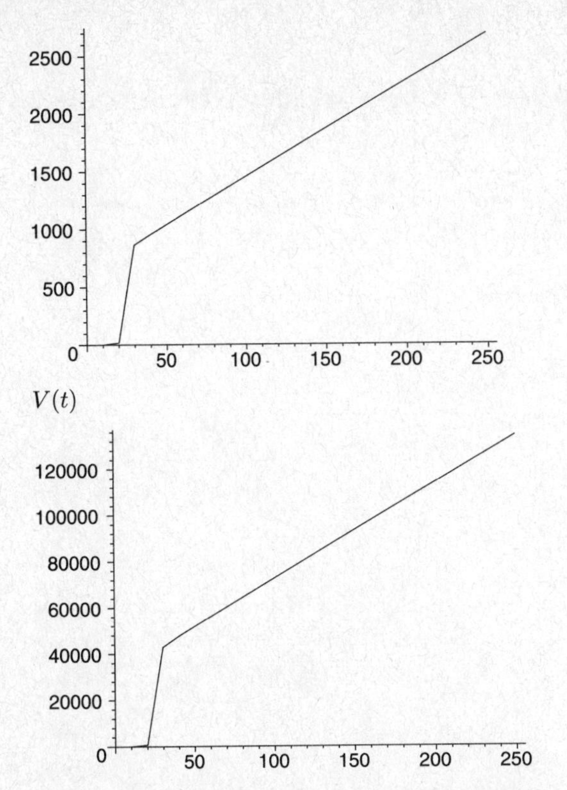

c) The equilibrium points are approximately 260.6778, 35.4579, 1768.2844 respectively

21. $J = \begin{bmatrix} 0.1 - 0.02x - 0.004y & -0.004x \\ -0.001y & 0.5 - 0.001x - 0.0032y \end{bmatrix}$

$$\begin{aligned} x(0.1 - 0.01x - 0.004y) &= 0 \\ x &= 0 \\ y &= 0 \\ (0.05 - 0.0016y) &= 0 \\ 31.25 &= y \\ 25 - 2.5x &= y \\ (25 - 2.5x)(0.05 - 0.001x - 0.0016(25 - 2.5x)) &= 0 \\ x &= 10 \\ y &= 0 \\ x &= -\frac{10}{3} \end{aligned}$$

Equilibrium points are:
$(0, 0)$ $\rightarrow$ Jacobian has two positive eigenvalues which means the equilibrium point is an unstable node
$(0, 31.25)$ $\rightarrow$ Jacobian has two negative eigenvalues which means the equilibrium point is an asymptotically stable node
$(10, 0)$ $\rightarrow$ Jacobian has eigenvalues with opposite signs which means the equilibrium point is an unstable saddle point

22. $J =$

$$\begin{bmatrix} 0.04 - 0.002x - 0.0022y & -0.0022x \\ -0.0012y & 0.02 - 0.0012x - 0.0008y \end{bmatrix}$$

The equilibrium points are:
$(0, 0)$ $\rightarrow$ Jacobian has two positive eigenvalues which means the equilibrium point is an unstable node
$(0, 50)$ $\rightarrow$ Jacobian has two negative eigenvalues which means the equilibrium point is an asymptotically stable node
$(12.5, 12.5)$ $\rightarrow$ Jacobian has eigenvalues with opposite signs which means the equilibrium point is an unstable saddle point
$(40, 0)$ $\rightarrow$ Jacobian has two negative eigenvalues which means the equilibrium point is an asymptotically stable node

23. $J = \begin{bmatrix} 0.1 - 0.012x - 0.002y & -0.002x \\ -0.001y & 0.05 - 0.001x - 0.004y \end{bmatrix}$

$$\begin{aligned} x(0.1 - 0.006x - 0.0002y) &= 0 \\ x &= 0 \\ y &= 0 \\ (0.05 - 0.002y) &= 0 \\ y &= 25 \\ y &= 50 - 3x \\ (50 - 3x)(0.05 - 0.001x - 0.002(50 - 0.5x)) &= 0 \\ x &= \frac{50}{3} \\ y &= 0 \\ x &= 10 \\ y &= 20 \end{aligned}$$

Equilibrium points are:
$(0, 0)$ $\rightarrow$ Jacobian has two positive eigenvalues which means the equilibrium point is an unstable node
$(0, 25)$ $\rightarrow$ Jacobian has two positive eigenvalues which means the equilibrium point is an unstable node
$(10, 20)$ $\rightarrow$ Jacobian has two negative eigenvalues which means the equilibrium point is an asymptotically stable node
$(\frac{50}{3}, 0)$ $\rightarrow$ Jacobian has eigenvalues with opposite signs which means the equilibrium point is an unstable saddle point

24. $J = \begin{bmatrix} 1 - y & -x \\ 0.2y & -1 + 0.2x \end{bmatrix}$

The equilibrium points are:
$(0, 0)$ $\rightarrow$ Jacobian has eigenvalues with opposite signs which means the equilibrium point is an unstable saddle point
$(5, 1)$ $\rightarrow$ Jacobian has two imaginary eigenvalues which means the equilibrium point is a center

25. $J = \begin{bmatrix} 0.8 - 0.2y & -0.2x \\ 0.1y & -0.6 + 0.1x \end{bmatrix}$

The equilibrium points are:
$(0, 0)$ $\rightarrow$ Jacobian has eigenvalues with opposite signs which means the equilibrium point is an unstable saddle point
$(6, 4)$ $\rightarrow$ Jacobian has two imaginary eigenvalues which means the equilibrium point is a center

26. $J = \begin{bmatrix} 2 - x/5 - y & -x \\ 0.4y & -1 + 0.4x \end{bmatrix}$

The equilibrium points are:
$(0, 0)$ → Jacobian has eigenvalues with opposite signs which means the equilibrium point is an unstable saddle point
$(2.5, 1.75)$ → Jacobian has complex eigenvalues with negative real part which means the equilibrium point is a stable spiral point

$(20, 0)$ → Jacobian has eigenvalues with opposite signs which means the equilibrium point is an unstable saddle point

27. Left to the student.

Chapter Review Exercises

1. $y'' - 7y' + 10y$

$$\begin{aligned} r^2 - 7r + 10 &= 0 \\ (r-2)(r-5) &= 0 \\ r &= 2 \\ r &= 5 \\ y &= C_1\, e^{2x} + C_2\, e^{5x} \end{aligned}$$

2. $y'' - 6y' + 9y$

$$\begin{aligned} r^2 - 6r + 9 &= 0 \\ (r-3)(r-3) &= 0 \\ r &= 3 \text{ repeated} \\ y &= C_1\, e^{3x} + C_2 x\, e^{3x} \end{aligned}$$

3. $y''' + 4y' = 0$

$$\begin{aligned} r^3 + 4r &= 0 \\ r(r^2 + 4) &= 0 \\ r &= 0 \\ r &= \pm\, 2i \\ y &= C_1 + C_2\, sin\ 2x + C_3\, cos\ 2x \end{aligned}$$

4. $y'' + 2y' + 26y = 100$

$$\begin{aligned} r^2 + 2r + 26 &= 0 \\ r &= -1 \pm\, 5i \\ y_h &= e^{-x}\,(C_1\, sin\ 5x + C_2\, cos\ 5x) \\ y_p &= A \\ y_p' &= 0 \\ y_p'' &= 0 \\ 0 + 0 + 26A &= 100 \\ A &= \frac{50}{13} \\ y_p &= \frac{50}{13} \\ y &= e^{-x}\,(C_1\, sin\ 5x + C_2\, cos\ 5x) + \frac{50}{13} \end{aligned}$$

5. $y'' - y' - 20y = 40x^2 - 16x + 15$

$$\begin{aligned} r^2 - r - 20 &= 0 \\ (r-5)(r+4) &= 0 \\ r &= -4 \\ r &= 5 \\ y_h &= C_1\, e^{-4x} + C_2\, e^{5x} \\ y_p &= Ax^2 + Bx + C \\ y_p' &= 2Ax + B \\ y_p'' &= 2A \\ 40x^2 - 16x + 15 &= 2A - 2Ax - B \\ &\quad -20Ax^2 - 20Bx - 20C \\ &= -20Ax^2 - (2A + 20B)x + \\ &\quad (2A - B - 20C) \\ 40 &= -20A \to A = -2 \\ -16 &= -2A - 20B \to B = 1 \\ 15 &= 2A - B - 20C \to C = -1 \\ y_p &= -2x^2 + x - 1 \\ y &= C_1\, e^{-4x} + C_2\, e^{5x} - 2x^2 + x - 1 \end{aligned}$$

6. $y'' + 5y' + 6 = 0$

$$\begin{aligned} r^2 + 5r + 6 &= 0 \\ r &= -2 \\ r &= -3 \\ y &= C_1\, e^{-3x} + C_2\, e^{-2x} \\ y' &= -3C_1\, e^{-3x} - 2C_2\, e^{-2x} \\ 2 &= C_1 + C_2 \\ 3 &= -3C_1 - 2C_2 \end{aligned}$$

Solving the system

$$\begin{aligned} C_1 &= -1 \\ C_2 &= 3 \\ y &= 12\, e^{-2x} - 10\, e^{-3x} \end{aligned}$$

7. $y'' + 4y' + 4 = 0$

$$\begin{aligned} r^2 + 4r + 4 &= 0 \\ r &= -2 \text{ repeated} \\ y &= C_1\, e^{-2x} + C_2 x\, e^{-2x} \\ y' &= -2C_1\, e^{-2x} + C_2\, e^{-2x} - 2C_2 x\, e^{-2x} \\ 3 &= C_1 \\ 6 &= -2C_1 + C_2 \end{aligned}$$

Solving the system

$$\begin{aligned} C_1 &= 3 \\ C_2 &= 12 \\ y &= 3\, e^{-2x} + 12x\, e^{-2x} \end{aligned}$$

8. $y'' + 8y' + 20y = 0$

$$\begin{aligned} r^2 + 8r + 20 &= 0 \\ r &= -4 \pm\, 2i \\ y &= e^{-4x}\,(C_1\, sin\ 2x + C_2\, cos\ 2x) \end{aligned}$$

$$\begin{aligned}
y' &= e^{-4x}(2C_1 \cos 2x - 2C_2 \sin 2x) \\
&\quad -4e^{-4x}(C_1 \sin 2x + C_2 \cos 2x) \\
-4 &= C_2 \\
1 &= 2C_1 - 4C_2 \\
C_1 &= -\frac{15}{2} \\
y &= e^{-4x}\left(-\frac{15}{2} \sin 2x - 4 \cos 2x\right)
\end{aligned}$$

9. $y'' - 8y' + 15y = 30t + 60$

$$\begin{aligned}
r^2 - 8r + 15 &= 0 \\
r &= 3 \\
r &= 5 \\
y_h &= C_1 e^{3t} + C_2 e^{5t} \\
y_p &= At + B \\
y_p' &= A \\
y_p'' &= 0 \\
0 - 8A + 15At + 15B &= 30t + 60 \\
15A &= 30 \rightarrow A = 2 \\
-8A + 15B &= 60 \\
B &= \frac{76}{15} \\
y_p &= 2t + \frac{76}{15} \\
y &= C_1 e^{3t} + C_2 e^{5t} + 2t + \frac{76}{15} \\
y' &= 3C_1 e^{3t} + 5C_2 e^{5t} + 2 \\
2 &= C_1 + C_2 + \frac{76}{15} \\
1 &= 3C_1 + 5C_2 + 2
\end{aligned}$$

Solving the system

$$\begin{aligned}
C_1 &= -\frac{43}{6} \\
C_2 &= \frac{41}{10} \\
y &= \frac{41}{10} e^{5t} - \frac{43}{6} e^{3t} + 2t + \frac{76}{15}
\end{aligned}$$

10. $y'' - 2y' + y = x^2$

$$\begin{aligned}
r^2 - 2r + 1 &= 0 \\
r &= 1 \text{ repeated} \\
y_h &= C_1 e^x + C_2 x e^x \\
y_p &= Ax^2 + Bx + C \\
y_p' &= 2Ax + B \\
y_p'' &= 2A \\
2A - 4Ax - 2B + Ax^2 + Bx + C &= x^2 \\
A &= 1 \\
-4A + B &= 0 \\
B &= 4 \\
2A - 2B + C &= 0 \\
C &= 6 \\
C_1 e^x + C_2 x e^x + x^2 + 4x + 6 &= y \\
(C_1 + C_2) e^x + C_2 x e^x + 2x + 4 &= y' \\
3 &= C_1 + 6 \\
C_1 &= -3 \\
-10 &= C_1 + C_2 + 4 \\
C_2 &= -11 \\
-3e^x - 11x e^x + x^2 + 4x + 6 &= y
\end{aligned}$$

11. $x' = 2x + 6y$ and $y' = 4x$

$$\begin{aligned}
x'' &= 2x' + 6(4x) \\
x'' - 2x' - 24x &= 0 \\
r^2 - 2r - 24 &= 0 \\
r &= 6 \\
r &= -4 \\
x &= C_1 e^{-4t} + C_2 e^{6t} \\
x' &= -4C_1 e^{-4t} + 6C_2 e^{6t} \\
y &= \frac{x' + 2x}{6} \\
&= \frac{-1}{3} C_1 e^{-4t} + \frac{4}{3} C_2 e^{6t}
\end{aligned}$$

12. $x' = 2x - y$ and $y' = x + 4y$

$$\begin{aligned}
x'' &= 2x' - x - 4y \\
&= 2x' - x + 4x' - 8x \\
x'' - 6x' + 9x &= 0 \\
r^2 - 6r + 9 &= 0 \\
r &= 3 \text{ repeated} \\
x &= C_1 e^{3t} + C_2 t e^{3t} \\
x' &= 3C_1 e^{3t} + C_2 e^{3t} + 3C_2 t e^{3t} \\
y &= 2x - x' \\
&= -C_1 e^{3t} - C_2 e^{3t} - C_2 t e^{3t}
\end{aligned}$$

13. $x' = x - 4y$ and $y' = 10x + 5y$

$$\begin{aligned}
x'' &= x' - 40x - 20y \\
&= x' - 40x + 5x' - 5x \\
x'' - 6x' + 45x &= 0 \\
r^2 - 6r + 45 &= 0 \\
r &= 3 \pm 6i \\
x &= e^{3t}(C_1 \sin 6x + C_2 \cos 6x) \\
x' &= e^{3t}(6C_1 \cos 6x - 6C_2 \sin 6x) \\
&\quad +3e^{3t}(C_1 \sin 6x + C_2 \cos 6x) \\
y &= \frac{x - x'}{4} \\
&= \frac{e^{3t}}{4}(6C_2 - 2C_1) \sin 6t - \\
&\quad \frac{e^{3t}}{4}(6C_1 + 2C_2) \cos 6t
\end{aligned}$$

14. $x' = 7x + 3y$ and $y' = -9x - 5y$

$$x'' = 7x' - 27x - 15y$$

$$\begin{aligned}
&= 7x' - 27x - 5x' + 35x \\
x'' - 2x' - 8x &= 0 \\
r^2 - 2r - 8 &= 0 \\
r &= 4 \\
r &= -2 \\
x &= C_1\, e^{-2t} + C_2\, e^{4t} \\
x' &= -2C_1\, e^{-2t} + 4C_2\, e^{4t} \\
y &= \frac{x' - 7x}{3} \\
&= 3C_1\, e^{-2t} - C_2\, e^{4t}
\end{aligned}$$

15. Since the eigenvalues have opposite signs, the origin is an unstable saddle point

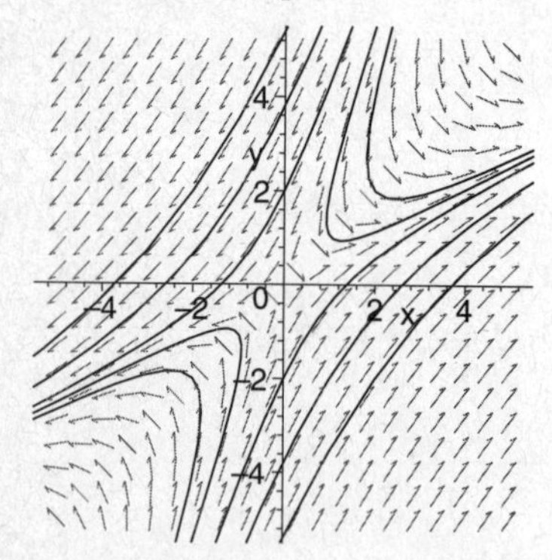

16. Since the eigenvalues are both positive the origin is an unstable node

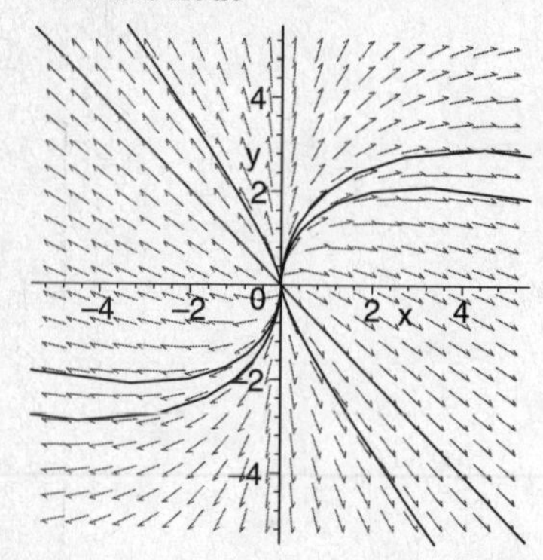

17. Since the eigenvalues are complex with a negative real part, the origin is a stable spiral point

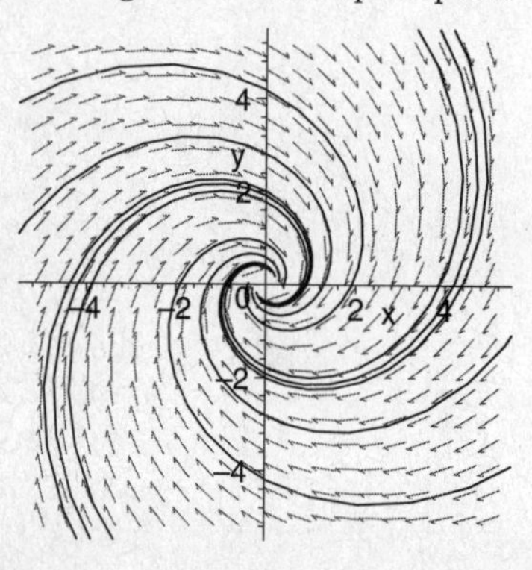

18. Since the eigenvalues are purely imaginary, the origin is a center

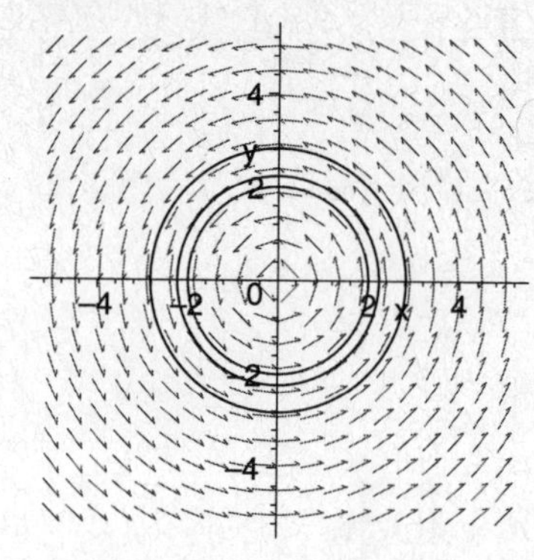

19. $x' = 3x + y$ and $y' = x + 3y$

$$\begin{aligned}
x'' &= 3x' + x + 3y \\
&= 3x' + x + 3x' - 9x \\
x'' - 6x' + 8x &= 0 \\
r^2 - 6r + 8 &= 0 \\
r &= 2 \\
r &= 4 \\
x &= C_1\, e^{2t} + C_2\, e^{4t} \\
x' &= 2C_1\, e^{2t} + 4C_2\, e^{4t} \\
y &= x' - 3x \\
&= C_2\, e^{4t} - C_1\, e^{2t} \\
2 &= C_1 + C_2 \\
0 &= C_2 - C_1 \\
C_1 &= 1 \\
C_2 &= 1 \\
x &= e^{2t} + e^{4t} \\
y &= e^{4t} - e^{2t}
\end{aligned}$$

20. $x' = 4x + y$ and $y' = -x + 2y$

$$\begin{aligned}
x'' &= 4x' - x + 2y \\
&= 4x' - x + 2x' - 8x \\
x'' - 6x' + 9x &= 0 \\
r^2 - 6r + 9 &= 0 \\
r &= 3 \text{ repeated} \\
x &= C_1\, e^{3t} + C_2 t\, e^{3t} \\
x' &= 3C_1\, e^{3t} + C_2\, e^{3t} + 3C_2 t\, e^{3t} \\
y &= x' - 4x \\
&= (C_2 - C_1)\, e^{3t} - C_2 t\, e^{3t} \\
3 &= C_1 \\
1 &= C_2 - C_1 \\
C_2 &= 4 \\
x &= 3\, e^{3t} + 4t\, e^{3t} \\
y &= e^{3t} - 4t\, e^{3t}
\end{aligned}$$

21. $x' = x(y+2)$ and $y' = x + y + 5$

$$J = \begin{bmatrix} y+2 & x \\ 1 & 1 \end{bmatrix}$$

$$\begin{aligned} x(y+2) &= 0 \\ x+y+5 &= 0 \\ x &= 0 \\ y &= -5 \\ y+2 &= 0 \\ y &= -2 \\ x &= -3 \end{aligned}$$

The equilibrium points are:
$(0,-5) \rightarrow$ Jacobian has eigenvalues with opposite signs which means that equilibrium point is an unstable saddle
$(-3,-2) \rightarrow$ Jacobian has complex eigenvales with positive real parts which means the equilibrium point is an unstable spiral point

22. $x' = -x + y - 1.5x = -2.5x + y$
and $y' = x - y$

$$\begin{aligned} x'' &= -2.5x' + x - x' - 2.5x \\ x'' + 3.5x' + 1.5x &= 0 \\ r^2 + 3.5r + 1.57 &= 0 \\ r &= -\frac{1}{2} \\ r &= -3 \\ x &= C_1\, e^{-3t} + C_2\, e^{-t/2} \\ x' &= -3C_1\, e^{-3t} - \frac{C_2}{2}\, e^{-t/2} \\ y &= x' + 2.5x \\ &= \frac{C_1}{2}\, e^{-3t} + 2C_2\, e^{-t/2} \end{aligned}$$

23. $A' = -0.2A + 0.2B$ and $B' = 0.2A - 0.2B$

a) Left to the student

b)

$$\begin{aligned} -0.2A' + 0.2(0.2A - 0.2B) &= A'' \\ -0.2A' + 0.04A + 0.2(-A' - 0.2A) &= A'' \\ A'' + 0.4A' &= 0 \\ r^2 + 0.4r &= 0 \\ r &= 0 \\ -0.4 = -\frac{2}{5} &= r \\ C_1 + C_2\, e^{-2t/5} &= A \\ -\frac{2C_2}{5}\, e^{-2t/5} &= A' \\ 5A' + A &= B \\ C_1 - C_2\, e^{-2t/5} &= \\ C_1 + C_2 &= 400 \\ C_1 - 2C_2 &= 500 \end{aligned}$$

Solving the system

$$\begin{aligned} C_1 &= 450 \\ C_2 &= -50 \\ 450 - 50\, e^{-2t/5} &= A \\ 450 + 50\, e^{-2t/5} &= B \end{aligned}$$

c) As time goes on the values of A and B both approach 450 pounds

24. $x' = x(0.1 - 0.0025x - 0.003y)$
$y' = y(0.05 - 0.002x - 0.0032y)$

$$J = \begin{bmatrix} 0.1 - 0.005x - 0.003y & -0.003x \\ -0.002y & 0.05 - 0.002x - 0.0064y \end{bmatrix}$$

$$\begin{aligned} x(0.1 - 0.0025x - 0.003y) &= 0 \\ x &= 0 \\ y &= 0 \\ 0.05 - 0.0032y &= 0 \\ y &= 15.625 \\ y(0.05 - 0.002x - 0.0032y) &= 0 \\ y &= 0 \\ x &= 0 \\ 0.05 - 0.002x &= 0 \\ x &= 40 \end{aligned}$$

The equilibrium points are:
$(0,0) \rightarrow$ Jacobian has two positive eigenvalues which means that equilibrium point is an unstable node
$(0, 15.625) \rightarrow$ Jacobian has eigenvalues with opposite signs which means the equilibrium point is an unstable saddle point
$(40,0) \rightarrow$ Joacobian has two negative eigenvalues which means the equilibrium is an asymptotically stable node

25. $x' = x - xy$ and $y' = -0.8y + 0.2xy$

$$J = \begin{bmatrix} 1-y & -x \\ 0.2y & -0.8+0.2x \end{bmatrix}$$

$$\begin{aligned} x - xy &= 0 \\ x &= 0 \\ y &= 0 \\ -0.8y + 0.2xy &= 0 \\ y &= 0 \rightarrow x = 0 \\ x &= 4 \\ y &= 1 \end{aligned}$$

The equilibrium points are:
$(0,0) \rightarrow$ Jacobian has eigenvalues with opposite signs which means the equilibrium point is an unstable saddle point
$(4,1) \rightarrow$ Jacobian has purely imaginary eigenvalues which means the equilibrium point is a center

26.

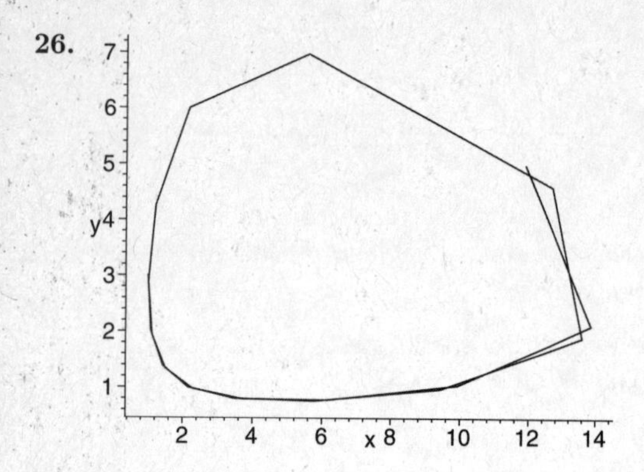

$$\begin{aligned} A - 12B &= -3 \\ B &= \frac{2}{9} \\ y_p &= -\frac{1}{3}\,x + \frac{2}{9} \\ y &= C_1\,e^{3x} + C_2\,e^{-4x} - \frac{x}{3} + \frac{2}{9} \end{aligned}$$

Chapter 9 Test

1. $y'' - 8y' + 15y = 0$

$$\begin{aligned} r^2 - 8r + 15 &= 0 \\ r &= 3 \\ r &= 5 \\ y &= C_1\,e^{3x} + C_2\,e^{5x} \end{aligned}$$

2. $y'' - 12y' + 36y = 0$

$$\begin{aligned} r^2 - 12r + 36 &= 0 \\ r &= 6 \text{ repeated} \\ y &= C_1\,e^{6x} + C_2x\,e^{6x} \end{aligned}$$

3. $y'' + 36y = 72$

$$\begin{aligned} r^2 + 36 &= 0 \\ r &= \pm\,6i \\ y_h &= C_1\,sin\ 6x + C_2\,cos\ 6x \\ y_p &= A \\ y_p' &= 0 \\ y_p'' &= 0 \\ 0 + 36A &= 72 \\ A &= 2 \\ y_p &= 2 \\ y &= C_1\,sin\ 6x + C_2\,cos\ 6x + 2 \end{aligned}$$

4. $y'' + y' - 12y = 4x - 3$

$$\begin{aligned} r^2 + r - 12 &= 0 \\ r &= 3 \\ r &= -4 \\ y_h &= C_1\,e^{3x} + C_2\,e^{-4x} \\ y_p &= Ax + B \\ y_p' &= A \\ y_p'' &= 0 \\ 0 + A - 12Ax - 12B &= 4x - 3 \\ -12A &= 4x \\ A &= -\frac{1}{3} \end{aligned}$$

5. $y'' - 6y' + 5y = 0$

$$\begin{aligned} r^2 - 6r + 5 &= 0 \\ r &= 1 \\ r &= 5 \\ y &= C_1\,e^{x} + C_2\,e^{5x} \\ y' &= C_1\,e^{x} + 5C_2\,e^{5x} \\ 1 &= C_1 + C_2 \\ 0 &= C_1 + 5C_2 \end{aligned}$$

Solving the system

$$\begin{aligned} C_1 &= \frac{5}{4} \\ C_2 &= -\frac{1}{4} \\ y &= \frac{5}{4}\,e^{x} - \frac{1}{4}\,e^{5x} \end{aligned}$$

6. $y'' + 4y' + 5y = 0$

$$\begin{aligned} r^2 + 4r + 5 &= 0 \\ r &= -2 \pm\ i \\ y &= e^{-2x}\,(C_1\,sin\ x + C_2\,cos\ x) \\ y' &= e^{-2x}\,(C_1\,cos\ x - C_2\,sin\ x)) \\ & \quad -2\,e^{-2x}\,(C_1\,sin\ x + C_2\,cos\ x) \\ 2 &= C_2 \\ -2 &= C_1 - 2C_2 \\ C_1 &= 2 \\ y &= e^{-2x}\,(2\,sin\ x + 2\,cos\ x) \end{aligned}$$

7. $y'' + 6y' + 9y = 12$

$$\begin{aligned} r^2 + 6r + 9 &= 0 \\ r &= -3 \text{ repeated} \\ y_h &= C_1\,e^{-3x} + C_2x\,e^{-3x} \\ y_p &= A \\ y_p' &= 0 \\ y'' &= 0 \\ 0 + 0 + 9A &= 12 \\ A &= \frac{4}{3} \\ y_p &= \frac{4}{3} \\ y &= C_1\,e^{-3x} + C_2x\,e^{-3x} + \frac{4}{3} \\ y' &= -3C_1\,e^{-3c} + C_2\,e^{-3x} - 3C_2x\,e^{-3x} \\ 0 &= C_1 + \frac{4}{3} \\ C_1 &= -\frac{4}{3} \end{aligned}$$

$$\begin{aligned} 3 &= -3C_1 + C_2 \\ C_2 &= -1 \\ y &= \frac{4}{3} - \frac{4}{3}\, e^{-3x} - x\, e^{-3x} \end{aligned}$$

8. $y'' + 25y = 15x - 5$

$$\begin{aligned} r^2 + 25 &= 0 \\ r &= \pm\, 5i \\ y_h &= C_1\, sin\ 5x + C_2\, cos\ 5x \\ y_p &= Ax + B \\ y_p' &= A \\ y_p'' &= 0 \\ 0 + 25Ax + 25B &= 15x - 5 \\ 25A &= 15 \\ A &= \frac{3}{5} \\ 25B &= -5 \\ B &= -\frac{1}{5} \\ y_p &= \frac{3x}{5} - \frac{1}{5} \\ y &= C_1\, sin\ 5x + C_2\, cos\ 5x + \frac{3x}{5} - \frac{1}{5} \\ y' &= 5C_1\, cos\ 5x - 5C_2\, sin\ 5x + \frac{3}{5} \\ 3 &= C_2 - \frac{1}{5} \\ C_2 &= \frac{16}{5} \\ -1 &= 5C_1 + \frac{3}{5} \\ C_1 &= -\frac{8}{25} \\ y &= \frac{16}{5}\, sin\ 5x - \frac{8}{25}\, cos\ 5x + \frac{3x}{5} - \frac{1}{5} \end{aligned}$$

9. $x' = 2x + 5y$ and $y' = 3x$

$$\begin{aligned} x'' &= 2x' + 15x \\ x'' - 2x' - 15x &= 0 \\ r^2 - 2r - 15 &= 0 \\ r &= -3 \\ r &= 5 \\ x &= C_1\, e^{-3t} + C_2\, e^{5t} \\ x' &= -3C_1\, e^{-3t} + 5C_2\, e^{5t} \\ y &= \frac{x' - 2x}{5} \\ &= 3C_2\, e^{5t} - C_1\, e^{-3t} \end{aligned}$$

10. $x' = -x - 3y$ and $y' = 2x + 4y$

$$\begin{aligned} x'' &= -x' - 6x + 4x' + 4x \\ x'' - 3x' + 2x &= 0 \\ r^2 - 3r + 2 &= 0 \\ r &= 1 \\ r &= 2 \\ x &= C_1\, e^{t} + C_2\, e^{2t} \\ x' &= C_1\, e^{t} + 2C_2\, e^{2t} \\ y &= \frac{-x' - x}{3} \\ &= \frac{-2}{3}\, C_1\, e^{t} - C_2\, e^{2t} \end{aligned}$$

11. $x' = 3x - y$ and $y' = -2x + 4y$

$$\begin{aligned} x'' &= 3x' + 2x + 4x' - 12x \\ x'' - 7x' + 10x &= 0 \\ r^2 - 7r + 10 &= 0 \\ r &= 2 \\ r &= 5 \\ x &= C_1\, e^{2t} + C_2\, e^{5t} \\ x' &= 2C_1\, e^{2t} + 5C_2\, e^{5t} \\ y &= 3x - x' \\ &= C_1\, e^{2t} - 2C_2\, e^{5t} \end{aligned}$$

12. $x' = -5x - 20y$ and $y' = 5x + 7y$

$$\begin{aligned} x'' &= -5x' - 100x + 35x + 7x' \\ x'' - 2x' + 65 &= 0 \\ r^2 - 2r + 65 &= 0 \\ r &= 1 \pm\, 8i \\ x &= e^{t}\,(C_1\, sin\ 8t + C_2\, cos\ 8t) \\ x' &= e^{t}\,(8C_1\, cos\ 8t - 8C_2\, sin\ 8t) \\ &\quad + e^{t}\,(C_1\, sin\ 8t + C_2\, cos\ 8t) \\ y &= \frac{-x' - 5x}{20} \\ &= e^{t}\left(\frac{(4C_2 - 3C_1)}{10}\, sin\ 8t\right) - \\ &\quad e^{t}\left(\frac{(4C_1 + 3C_2)}{10}\, cos\ 8t\right) \end{aligned}$$

13. Since the eigenvalues are complex with negative real part the origin is a stable spiral point

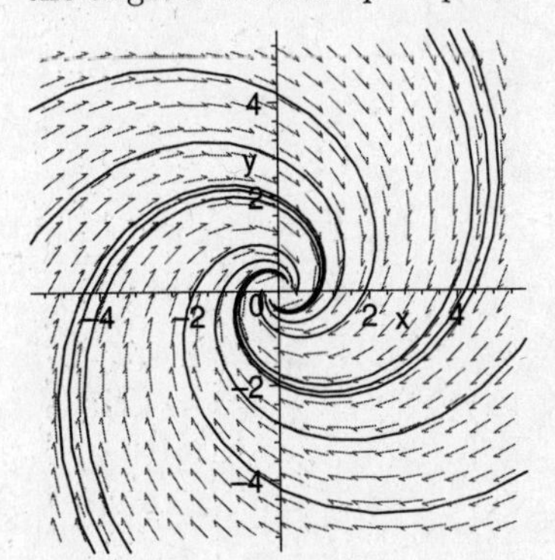

14. Since the eigen values have opposite signs the origin is an unstable saddle point

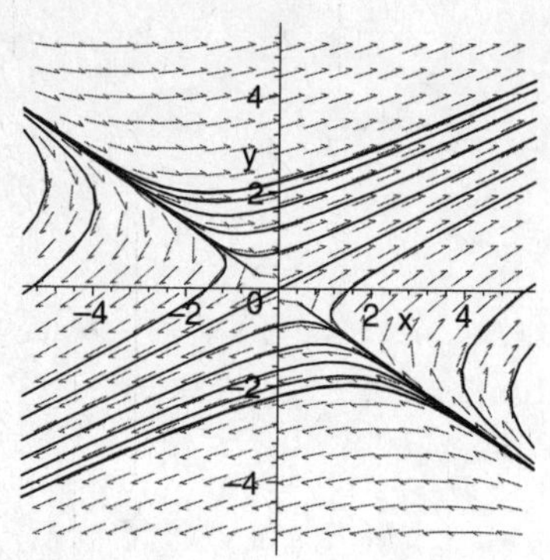

15. Since both eigenvalues are negative the origin is an asymptotically stable node

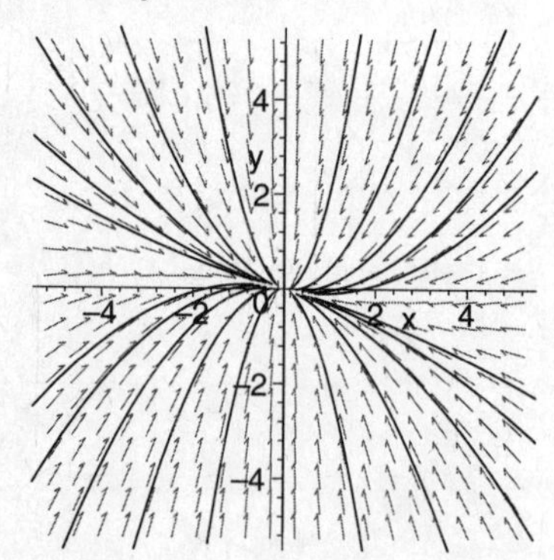

16. Since the eigenvalues are purely imaginary the origin is a center

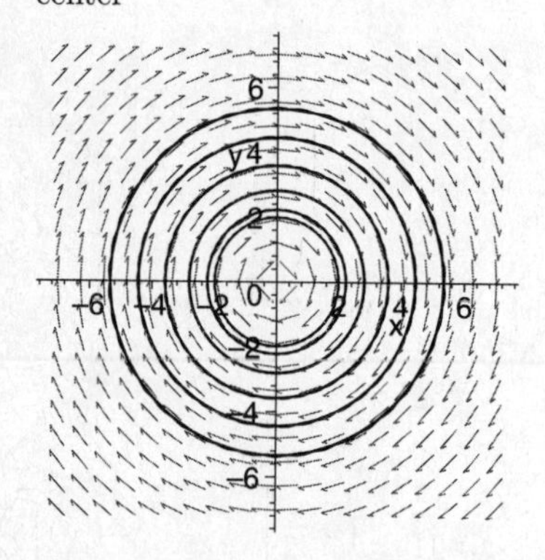

17. $x' = x^2 + y^2 - 13$ and
$y' = x - y + 5$

$$J = \begin{bmatrix} 2x & 2x \\ 1 & -1 \end{bmatrix}$$

$$\begin{aligned} x^2 + y^2 - 13 &= 0 \\ y &= x + 5 \\ x^2 + x^2 + 10x + 25 - 13 &= 0 \\ 2x^2 + 10x + 12 &= 0 \\ x^2 + 5x + 6 &= 0 \\ x &= -2 \rightarrow y = 3 \\ x &= 3 \rightarrow y = -2 \end{aligned}$$

The equilibrium points are:
$(-2, 3)$ $\rightarrow$ Jacobian has eigenvalues with opposite signs which means the equilibrium point is an unstable saddle point
$(3, -2)$ $\rightarrow$ Jacobian has two negative eigenvalues which means the equilibrium point is an asymptotically stable node

18. $x' = -0.5x + y$ and $y' = 0.5x - y$

$$\begin{aligned} x'' &= -0.5x' + 0.5x - 0.5x - x' \\ x'' + 1.5x' &= 0 \\ r^2 + 1.5r &= 0 \\ r &= 0 \\ r &= -\frac{3}{2} \\ x &= C_1 + C_2\, e^{-3t/2} \\ x' &= -\frac{3C_2}{2}\, e^{3t/2} \\ y &= x' + 0.5x \\ &= \frac{C_1}{2} - C_2\, e^{3t/2} \end{aligned}$$

19. $x' = -0.1x + 0.1y - 0.15x = -0.25x + 0.1y$ and
$y' = 0.1x - 0.1y$

a) Left to the student

b)

$$\begin{aligned} x'' &= -0.25x' + 0.01x + \\ &\quad 0.1(x' + 0.1(-0.25x)) \\ x'' + 0.35x' + 0.015 &= 0 \\ r^2 + \frac{35}{100}\, r + \frac{15}{1000} &= 0 \\ r &= -\frac{3}{10} \\ r &= \frac{1}{20} \\ x &= C_1\, e^{-3t/10} + C_2\, e^{-t/20} \\ x' &= -\frac{3C_1}{10}\, e^{-3t/10} - \frac{C_2}{20}\, e^{\ t/20} \\ y &= 10x' + \frac{5}{2}\, x \\ &= 2C_2\, e^{-t/20} - \frac{C_1}{2}\, e^{-3t/10} \\ 100 &= C_1 + C_2 \\ 150 &= 2C_2 - \frac{C_1}{2} \end{aligned}$$

Solving the system

$$\begin{aligned} C_1 &= 20 \\ C_2 &= 80 \\ x &= 20\, e^{-3t/10} + 80\, e^{-t/20} \\ y &= 160\, e^{-t/20} - 10\, e^{-3t/10} \end{aligned}$$

c) As time goes on both $x(t)$ and $y(t)$ approach 0

20. $x' = x(0.1 - 0.004x - 0.002y)$ and
$y' = y(0.06 - 0.002x - 0.003y)$

$$J = \begin{bmatrix} \text{0.1-0.008x -0.002y} & \text{-0.002x} \\ \text{-0.002y} & \text{0.06-0.002x -0.006y} \end{bmatrix}$$

$$\begin{aligned}
x(0.1 - 0.004x - 0.002y) &= 0 \\
y(0.06 - 0.002x - 0.003y) &= 0 \\
x &= 0 \\
y &= 0 \\
0.06 - 0.003y &= 0 \\
y &= 20 \\
y &= 0 \\
x &= 0 \\
0.1 - 0.004x &= 0 \\
x &= 25 \\
0.1 - 0.004x - 0.002y &= 0 \\
y &= 50 - 2x \\
(50 - 2x)(0.06 - 0.002x - 0.003(50 - 2x)) &= 0 \\
-0.09 - 0.004x &= 0 \\
x &= 22.5 \\
y &= 50 - 2(22.5) \\
&= 5
\end{aligned}$$

The equilibrium points are:
$(0,0)$ $\to$ Jacobian has two positive eigenvalues which means the equilibrium point is an unstable node
$(0,20)$ $\to$ Jaconian has eigenvalues with opposite signs which means the equilibrium point is an unstable saddle point
$(22.5,5)$ $\to$ Jacobian has two negative eigenvalues which means the equilibrium point is an asymptotically stable node
$(25,0)$ $\to$ Jacobian has eigenvalues with opposite signs which means the equilibrium point is an unstable saddle point

21. $x' = x - 0.5xy$ and $y' = -y + 0.1xy$

$$J = \begin{bmatrix} \text{1-0.5y} & \text{-0.5x} \\ \text{0.1y} & \text{-1+0.1x} \end{bmatrix}$$

$$\begin{aligned}
x - 0.5xy &= 0 \\
-y + 0.1xy &= 0 \\
x &= 0 \to y = 0 \\
-1 + 0.1x &= 0 \\
x &= 10 \\
10 - 5y &= 0 \\
y &= 2
\end{aligned}$$

The equilibrium points are:
$(0,0)$ $\to$ Jacobian has eigenvalues with opposite signs which means the equilibrium point is an unstable saddle point
$(10,2)$ $\to$ Jacobian has purely imaginary eigenvalues which means that equilibium point is a center

22.

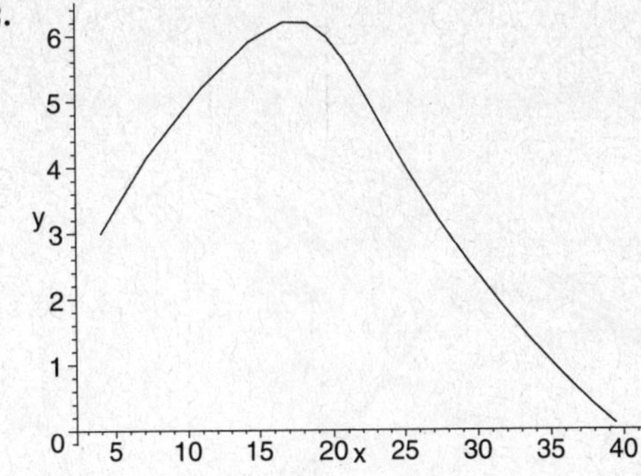

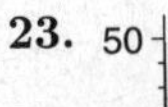

23.

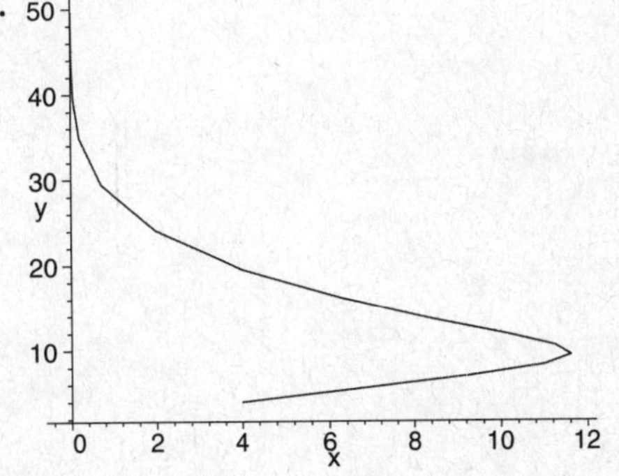

Technology Connection

- **Page 626**

1.

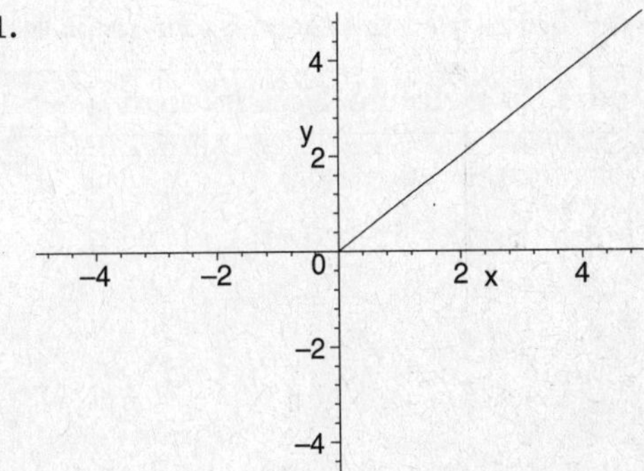

2.

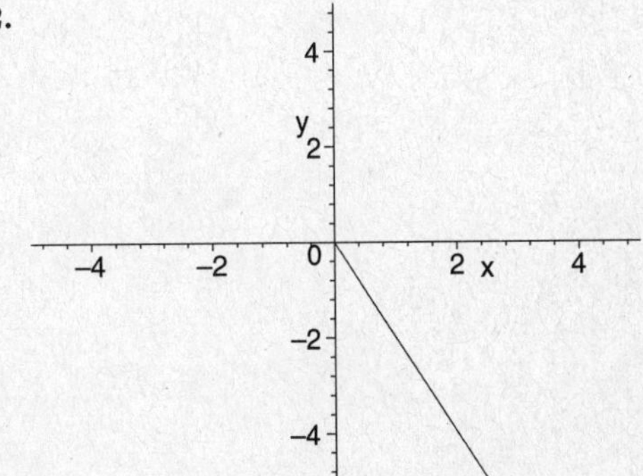

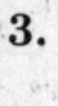
3.

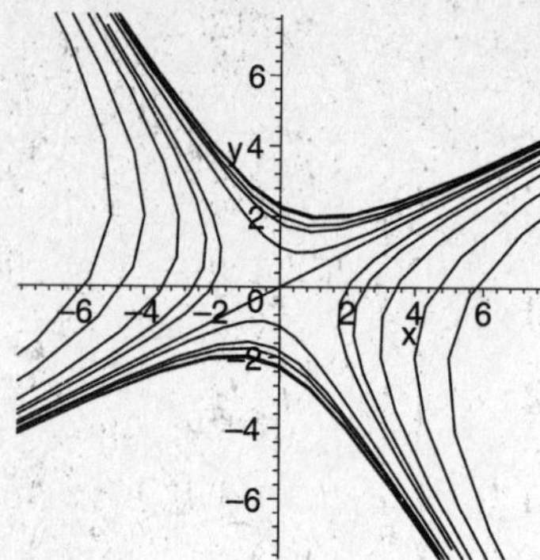

- **Page 646**

1. 101 rows are needed

2. $x(1) \approx 0.95472$
$y(1) \approx 0.91818$

3.

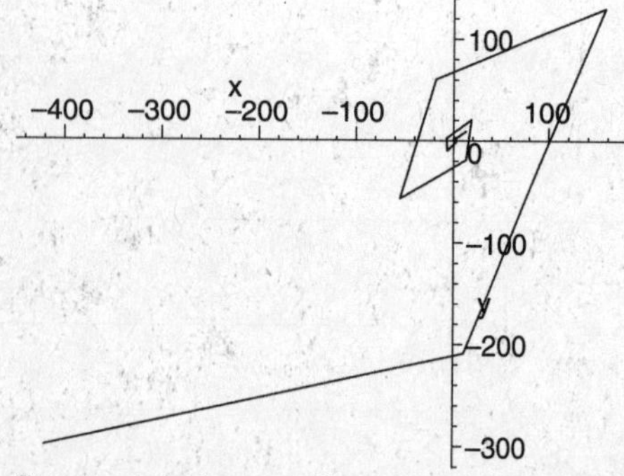

4. Since we took a smaller step size, we would expect the answer from this exercise to be more accurate.

Extended Life Science Connection

1. It follows from the statement of the problem

2. If P is constant then its derivative should be 0

$$\begin{aligned} P &= S + I + R \\ P' &= S' + I' + R' \\ &= -aSI + aSI - bl + bl \\ &= 0 \end{aligned}$$

3. **a)** $S(100) \approx 1415$
$I(100) \approx 0$
$R(100) \approx 585$

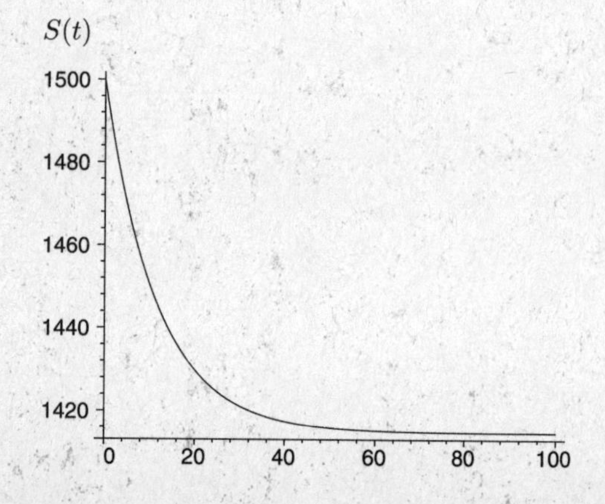

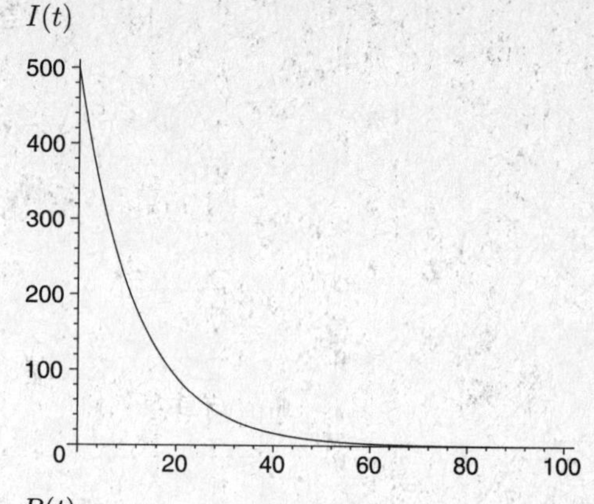

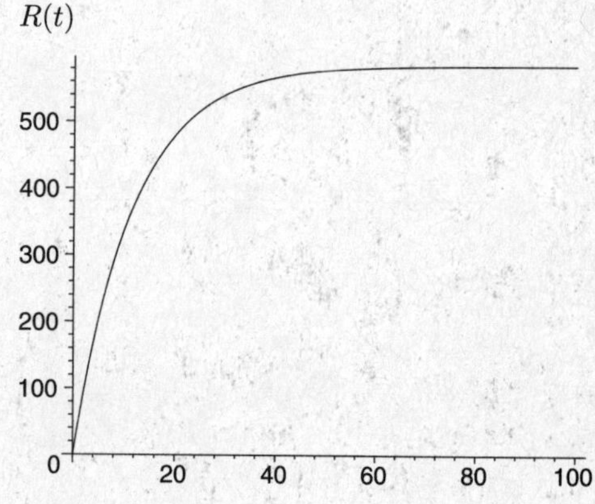

b) $S(100) \approx 4123$
$I(100) \approx 1$
$R(100) \approx 876$

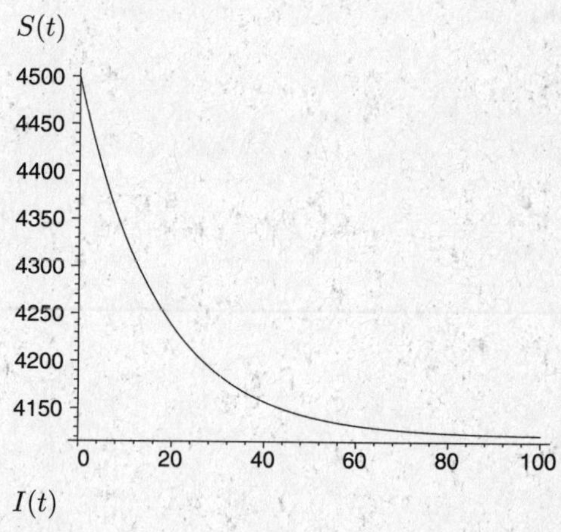

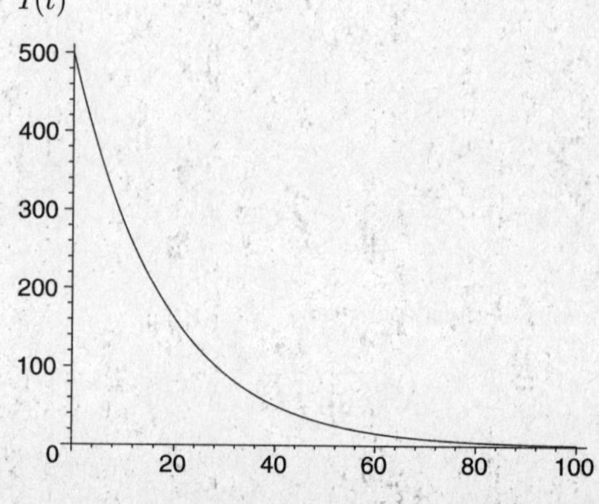

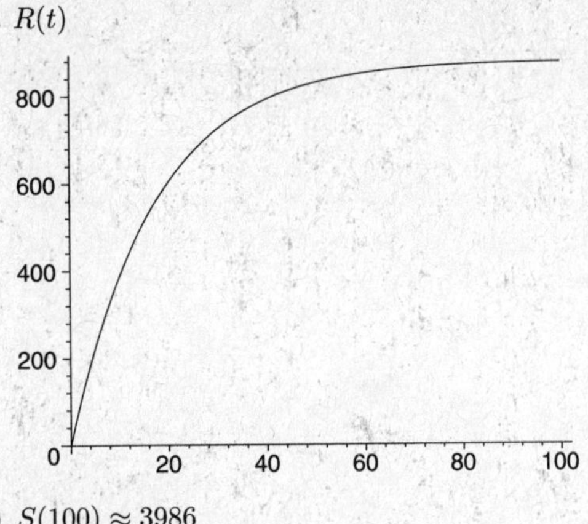

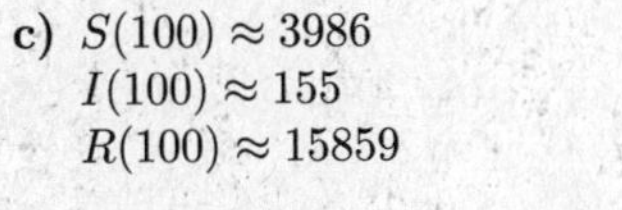

c) $S(100) \approx 3986$
$I(100) \approx 155$
$R(100) \approx 15859$

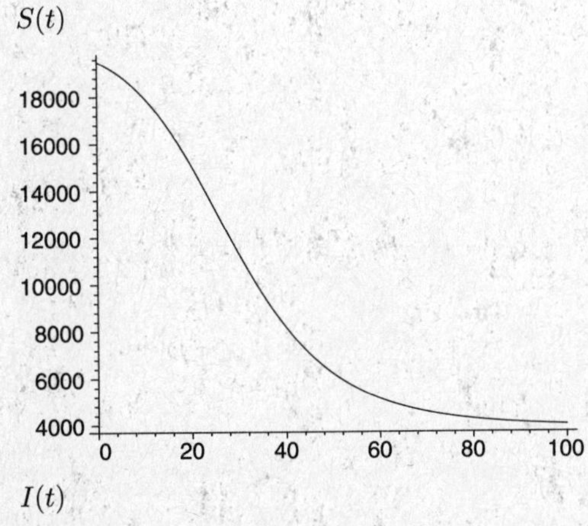

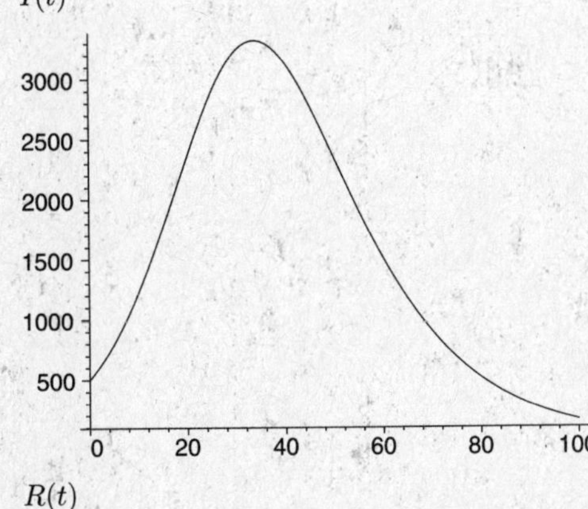

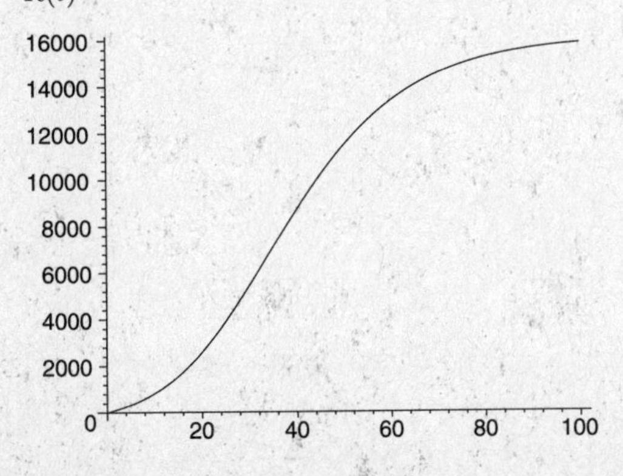

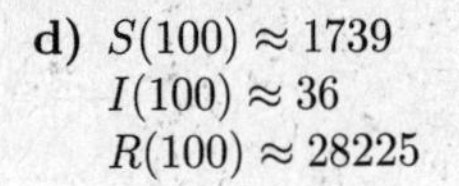

d) $S(100) \approx 1739$
$I(100) \approx 36$
$R(100) \approx 28225$

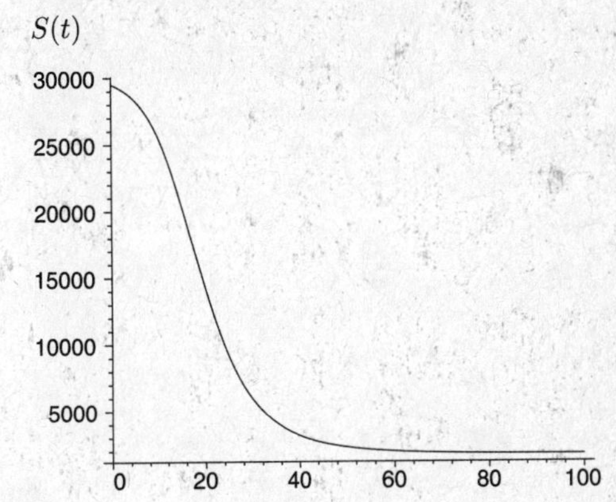

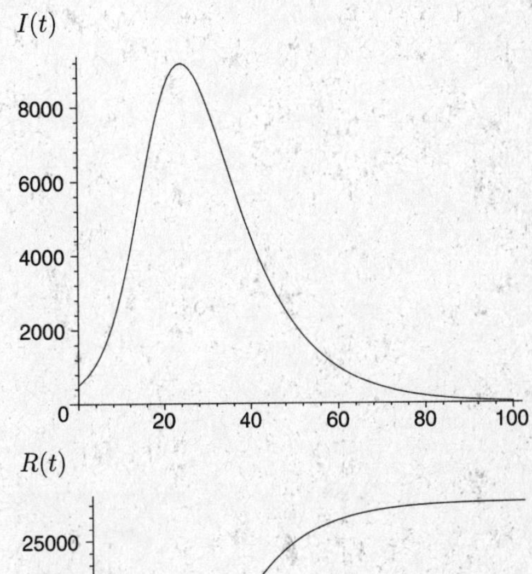

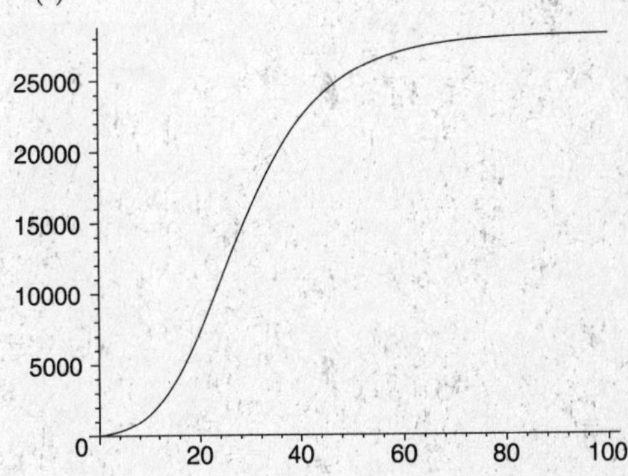

4. Populations in parts c) and d) will experience epidemics

5.

$$\begin{aligned} I' &= aSI - bI \\ &= I(aS - b) \end{aligned}$$

Since the sign of I is positive when an epidemic occurs it follows that the sign of I' agrees with the sign of $aS - b$

6. Since the value of a is much smaller than the value of b it will take a large population to overcome the difference and cause I' to become positive, which is the condition for an epidemic occuring

7. In part $c)$ $\frac{b}{a} = 10000 < 20000 = P$

In part $d)$ $\frac{b}{a} = 10000 < 30000 = P$

8. From the chain rule

$$\begin{aligned} \frac{dI}{dS} &= \frac{dI}{dt} \cdot \frac{1}{\frac{dS}{dt}} \\ &= (aSI - bI) \cdot \frac{1}{-aSI} \\ &= \frac{b}{aS} - 1 \end{aligned}$$

9.

$$\begin{aligned} \frac{dI}{dS} &= \frac{b}{aS} - 1 \\ \int dI &= \int \frac{b}{aS} - 1\, dS \\ I &= \frac{b}{a} \ln S - S + C \\ 500 &= \frac{b}{a} \ln[P - 500] - (P - 500) + C \\ C &= -\frac{b}{a} \ln[P - 500] + P \\ I &= \frac{b}{a} \ln S - \frac{b}{a} \ln[P - 500] - S + p \\ &= \frac{b}{a} (\ln S - \ln[P - 500]) - S + P \end{aligned}$$

10. After the pidemic has run its course the value of I is zero since there are no people infected thus

$$0 = \frac{b}{a} (\ln x - \ln[P - 500]) - x + P$$

11. $x = 3896.409 \approx 3896$

12. It is in agreement with the graph when smaller step sizes are considered

13. $x = 1749.521 \approx 1750$

Chapter 10

Probability

Exercise Set 10.1

1. $\frac{3}{50} = 0.06$

2. $\frac{10}{50} = 0.20$

3. $\frac{14}{50} = 0.28$

4. $\frac{0}{50} = 0.0$

5. $\frac{23}{50} = 0.46$

6. $\frac{10+3}{50} = \frac{13}{50} = 0.26$

7. $\frac{10+14}{50} = \frac{24}{50} = 0.48$

8. $\frac{50-3}{50} = \frac{47}{50} = 0.94$

9. $\frac{50-14}{50} = \frac{36}{50} = 0.72$

10. $\frac{50-23-3}{50} = \frac{24}{50} = 0.48$

11. The possible outcomes are **HH,HT,TH,TT**

a.

$$\begin{aligned} P(A) \cdot P(B) &= \frac{1}{2} \cdot \frac{1}{2} \\ &= \frac{1}{4} \\ P(A \,\&\, B) &= \frac{1}{4} \end{aligned}$$

Therefore, events A and B are independent

b. Events A and B are not disjoint. The outcome HT is common between them

12. **a.** Events A and B are not independent

b. Events A and B are disjoint

13. The possible outcomes are $\{1, 2, 3, 4, 5, 6\}$

a.

$$\begin{aligned} P(A) \cdot P(B) &= \frac{1}{6} \cdot \frac{1}{6} \\ &= \frac{1}{36} \\ P(A \,\&\, B) &= 0 \end{aligned}$$

Therefore, events A and B are independent

b. Events A and B are disjoint

14. **a.** Events A and B are independent

b. Events A and B are not disjoint

15. **a.**

$$\begin{aligned} P(A) \cdot P(B) &= \frac{4}{52} \cdot \frac{3}{51} \\ &= \frac{1}{221} \\ P(A \,\&\, B) &= \frac{12}{1326} = \frac{2}{221} \end{aligned}$$

Therefore, events A and B are not independent

b. Events A and B are not disjoint

16. **a.** Events A and B are independent

b. Events A and B are not disjoint

17. **a.**

$$\begin{aligned} P(A) \cdot P(B) &= \frac{4}{52} \cdot \frac{4}{52} \\ &= \frac{1}{169} \\ P(A \,\&\, B) &= \frac{1}{52} \end{aligned}$$

Therefore, events A and B are not independent

b. Events A and B are disjoint

18. **a.** Events A and B are not independent

b. Events A and B are not disjoint

19. $\frac{1}{5} = 0.2$

20. $\frac{1}{5} = 0.2$

21. $\frac{4}{5} = 0.8$

22. $\frac{4}{5} = 0.8$

23. $\frac{2}{6} = 0.4$

24. $\frac{3}{5} = 0.6$

25. $\frac{1}{5} \cdot \frac{1}{4} = 0.05$

26. $\frac{1}{5} \cdot \frac{1}{4} = 0.05$

27. $\frac{1}{5} \cdot \frac{3}{4} = 0.15$

28. $\frac{1}{5} \cdot \frac{3}{4} = 0.15$

29. $\frac{1}{5} \cdot \frac{1}{5} = 0.04$

30. $\frac{1}{5} \cdot \frac{1}{5} = 0.04$

31. $\frac{1}{5} \cdot \frac{4}{5} = 0.16$

32. $\frac{1}{5} \cdot \frac{4}{5} = 0.16$

33. $\frac{4}{10} = 0.4$

34. $\frac{3}{10} = 0.3$

35. $\frac{10-1}{10} = \frac{9}{10} = 0.9$

36. $\frac{8}{10} = 0.8$

37. $\frac{3+4}{10} = \frac{7}{10} = 0.7$

38. $\frac{7}{10} = 0.7$

39. $\frac{2}{10} \cdot \frac{1}{9} = \frac{1}{45}$

40. $\frac{4}{10} \cdot \frac{3}{9} = \frac{2}{15}$

41. $\frac{1}{10} \cdot \frac{6}{9} = \frac{1}{18}$

42. $\frac{3}{10} \cdot \frac{7}{9} = \frac{7}{30}$

43. $\frac{2}{10} \cdot \frac{2}{10} = 0.4$

44. $\frac{4}{10} \cdot \frac{3}{10} = 0.12$

45. $\frac{1}{10} \cdot \frac{6}{10} = 0.06$

46. $\frac{3}{10} \cdot \frac{7}{10} = 0.21$

47. $\frac{1}{6} \cdot \frac{1}{6} \cdot \frac{1}{6} = \frac{1}{216}$

48. $\frac{5}{6} \cdot \frac{5}{6} \cdot \frac{5}{6} = \frac{125}{216}$

49. $1 - \frac{1}{216} = \frac{215}{216}$

50. Exercise 48 asked for the probability of no sixes rolled, while Exercise 49 asked for the probabilty of at most 2 sixes are rolled

51. $\left(\frac{1}{2}\right)^5 = \frac{1}{32}$

52. $\left(\frac{1}{2}\right)^5 = \frac{1}{32}$

53. $1 - \frac{1}{32} = \frac{31}{32}$

54. Exercise 52 asked for the probability of no flips being heads, while Exercise 53 asked for the probability of at most 4 flips being heads

55. $\frac{1}{6} \cdot \frac{1}{6} \cdot \frac{1}{6} \cdot \frac{1}{6} = \frac{1}{1296}$

56. $6 \cdot \frac{1}{1296} = \frac{1}{216}$

57. $(6 \cdot 5 \cdot 4 \cdot 3) \cdot \frac{1}{1296} = \frac{5}{18}$

58.

59.

$$\begin{aligned} P(6 \text{ or } 4) &= P(6) + P(4) - P((6,4)) \\ &= \frac{6}{36} + \frac{6}{36} - \frac{1}{36} \\ &= \frac{11}{36} \end{aligned}$$

60.

$$\begin{aligned} P(H \text{ or } T) &= P(H) + P(T) - P((H,T)) \\ &= \frac{2}{4} + \frac{2}{4} - \frac{1}{2} \\ &= \frac{3}{4} \end{aligned}$$

61.

$$\begin{aligned} P(Ace \text{ or } King) &= P(Ace) + P(King) - P((Ace, King)) \\ &= \frac{4}{52} + \frac{4}{52} - \frac{4}{52} \cdot \frac{4}{51} \\ &= 0.14781 \end{aligned}$$

62.

$$\begin{aligned} P(not\ 6 \text{ or } not\ 4) &= P(not\ 6) + P(not\ 4) - P(not\ (6,4)) \\ &= \frac{30}{36} + \frac{30}{36} - \frac{1}{36} \\ &= \frac{29}{36} \end{aligned}$$

63. Answers Vary.

Exercise Set 10.2

1. The possible outcomes are (1, 2) or (3, 2) Thus the probability is
$\frac{1}{3} \cdot \frac{1}{2} + \frac{1}{3} \cdot \frac{1}{2} = \frac{1}{3}$

2. $\frac{1}{3} \cdot \frac{1}{2} + \frac{1}{3} \cdot \frac{1}{2} = \frac{1}{3}$

3. The possible outcomes are (1, 3) or (3, 1) Thus the probability is
$\frac{1}{3} \cdot \frac{1}{2} + \frac{1}{3} \cdot \frac{1}{2} = \frac{1}{3}$

4. 0. Not possible with drawing only two tickets.

5. The possible outcomes are (1, 2), (2, 2), or (3, 2) Thus the probability is
$\frac{1}{3} \cdot \frac{1}{3} + \frac{1}{3} \cdot \frac{1}{3} + \frac{1}{3} \cdot \frac{1}{3} = \frac{1}{3}$

6. $\frac{1}{3} \cdot \frac{1}{3} + \frac{1}{3} \cdot \frac{1}{3} = \frac{2}{9}$

7. The possible outcomes are (1, 3), (2, 2), or (3, 1) Thus the probability is
$\frac{1}{3} \cdot \frac{1}{3} + \frac{1}{3} \cdot \frac{1}{3} + \frac{1}{3} \cdot \frac{1}{3} = \frac{1}{3}$

8. $\frac{1}{3} \cdot \frac{1}{3} = \frac{1}{9}$

9. $\frac{1}{2} \cdot \frac{1}{2} \cdot \frac{1}{2} \cdot \frac{1}{2} = \frac{1}{16}$

10. $2 \cdot \frac{1}{16} = \frac{1}{8}$

11. $2 \cdot \frac{1}{16} = \frac{1}{8}$

12. $\frac{6}{16} = \frac{3}{8}$

13. $4 \cdot \frac{1}{2} \cdot \frac{1}{2} \cdot \frac{1}{2} \cdot \frac{1}{2} = \frac{1}{4}$

14. $\frac{4}{52} \cdot \frac{3}{51} = \frac{1}{221}$

15. $\frac{13}{52} \cdot \frac{39}{51} + \frac{39}{52} \cdot \frac{13}{51} = \frac{13}{34} = 0.38235$

16. $\frac{4}{52} \cdot \frac{48}{51} + \frac{48}{52} \cdot \frac{4}{51} = \frac{32}{221} = 0.144796$

17. **a.)** Possible genotype outcomes are $\{FF, Ff, ff\}$. Since FF cannot occur when we cross Ff and ff its probability is 0. and the other two outcomes have a probability of 1/2

b.) The genotypes in Example 2 did not allow for the ff type while in this Exercise the ff genotype is allowed. If a cross test is performed with the sperm allele transmitted from the ff plant and the ovum allele transmitted from the FF plant then we will be able to distinguish between the FF and the Ff genotypes.

18. **a.)** $\frac{3}{4} \cdot \frac{3}{4} = \frac{9}{16}$
b.) $\frac{1}{4} \cdot \frac{2}{4} = \frac{1}{8}$
c.) $\frac{1}{4} \cdot \frac{1}{4} = \frac{1}{16}$

19. **a.)** $\frac{3}{4} \cdot \frac{3}{4} = \frac{9}{16}$
b.) $\frac{1}{4} \cdot \frac{3}{4} = \frac{3}{16}$
c.) $\frac{3}{4} \cdot \frac{1}{4} = \frac{3}{16}$
d.) $\frac{1}{4} \cdot \frac{1}{4} = \frac{1}{16}$

20. **a.)** $P(T+|D+) = \frac{14}{17} = 0.824$
b.) $P(T-|D-) = \frac{45}{55} = 0.818$
c.) $P(D+ \text{ and } T+) = (0.4)(0.824) = 0.3296$
$P(T+) = (0.4)(0.824) + (0.6)(0.182) = 0.4388$
$P(D+|T+) = \frac{P(D+|T+)}{P(T+)} = 0.7511$

21. **a.)** $P(T+|D+) = \frac{37}{42} = 0.881$
b.) $P(T-|D-) = \frac{106}{110} = 0.964$
c.) $P(D+ \text{ and } T-) = (0.2)(0.119) = 0.0238$
$P(T-) = (0.2)(0.119) + (0.8)(0.964) = 0.795$
$P(D+|T-) = \frac{P(D+andT+)}{P(T+)} = \frac{0.0238}{0.795} = 0.0299$

22. **a.)** $P(T+|D+) = \frac{12}{29} = 0.414$
b.) $P(T-|D-) = \frac{102}{102} = 1.0$
c.) $P(D+ \text{ and } T+) = (0.012)(0.414) = 0.00497$
$P(T+) = (0.012)(0.414) + (0.988)(0.0) = 0.00497$
$P(D+|T+) = \frac{P(D+ \text{ and } T+)}{P(T+)} = \frac{0.00497}{0.00497} = 1$

23. **a.)** $P(D+ \text{ and } T+) = (0.1)(0.67) = 0.067$
$P(T+) = (0.1)(0.67) + (0.9)(0.24) = 0.283$
$P(D+|T+) = \frac{0.067}{0.283} = 0.237$
b.) $P(D+ \text{ and } T+) = (0.5)(0.67) = 0.335$
$P(T+) = (0.5)(0.67) + (0.5)(0.24) = 0.455$
$P(D+|T+) = \frac{0.335}{0.455} = 0.736$

24. **- 27.** Left to the student

28. $P(DD) = 1/4 = 0.25$, $P(Dd) = 2/4 = 0.5$, and $P(dd) = 1/4 = 0.25$

29.

$$\begin{aligned} P(DD|S) &= \frac{P(DD \text{ and } S)}{P(S)} \\ &= \frac{(0.25)(0.96)}{(0.25)(0.96) + (0.5)(0.96) + (0.25)(0.2)} \\ &= 0.312 \end{aligned}$$

$$\begin{aligned} P(Dd|S) &= \frac{P(Dd \text{ and } S)}{P(S)} \\ &= \frac{(0.5)(0.96)}{(0.25)(0.96) + (0.5)(0.96) + (0.25)(0.2)} \\ &= 0.623 \end{aligned}$$

$$\begin{aligned} P(dd|S) &= \frac{P(dd \text{ and } S)}{P(S)} \\ &= \frac{(0.25)(0.2)}{(0.25)(0.96) + (0.5)(0.96) + (0.25)(0.2)} \\ &= 0.065 \end{aligned}$$

Exercise Set 10.3

1. $\binom{4}{2} = 6$

2. $\binom{5}{3} = 10$

3. $\binom{3}{3} = 1$

4. $\binom{2}{2} = 1$

5. $\binom{8}{0} = 1$

6. $\binom{4}{5}$ is not possible

7. $\binom{5}{6}$ is not possible

8. $\binom{6}{0} = 1$

9. $\binom{9}{-4}$ is not possible

10. $\binom{12}{5} = 792$

11. $\binom{16}{10} = 8008$

12. $\binom{3}{-1}$ is not possible

13.

$$\begin{aligned} P(X=2) &= \binom{3}{2} \cdot \frac{1}{3}^2 \cdot \frac{2}{3}^1 \\ &= 3 \cdot \frac{1}{9} \cdot \frac{2}{3} \\ &= \frac{2}{9} \end{aligned}$$

14. $P(X = 4) = \frac{1215}{4096} = 0.29663$

15.

$$\begin{aligned} P(X+3) &= \binom{4}{3}(0.8)^3(0.2)^1 \\ &= 4 \cdot 0.512 \cdot 0.2 \\ &= 0.4096 \end{aligned}$$

16. $P(X = 1) = 0.33554$

17.

$$\begin{aligned} P(X=5) &= \binom{6}{5}(0.9)^5(0.1)^1 \\ &= 6 \cdot 0.59049 \cdot 0.1 \\ &= 0.35429 \end{aligned}$$

18. $P(X = 4) = 0.09842$

19.

$$\begin{aligned} P(X=10) &= \binom{20}{10}(0.6)^{10}(0.4)^{10} \\ &= 184756 \cdot 0.0060466176 \cdot 0.0001048576 \\ &= 0.11714 \end{aligned}$$

20. $P(X = 8) = 0.08113$

21. $P(X = 5) = 0$ since we cannot choose 5 out of 4 items

22. $P(X = 6) = 0$ since we cannot choose 6 out of 3 items

23.

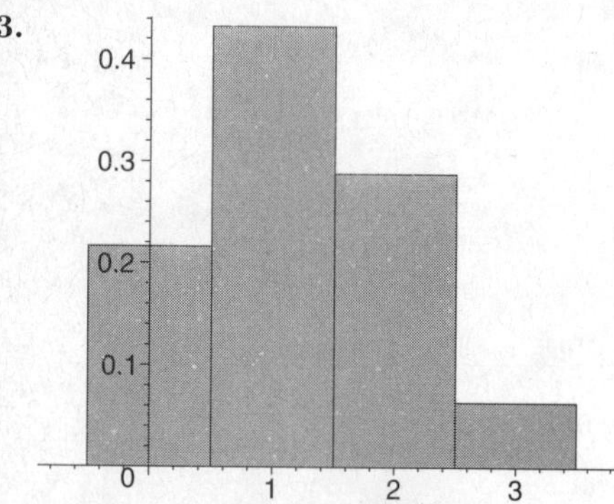

24.

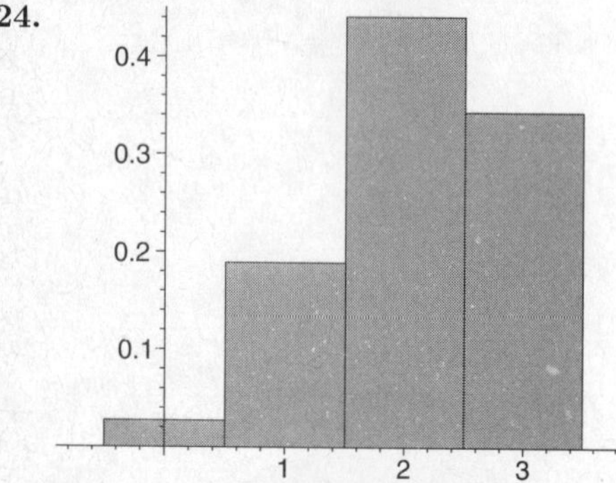

25.

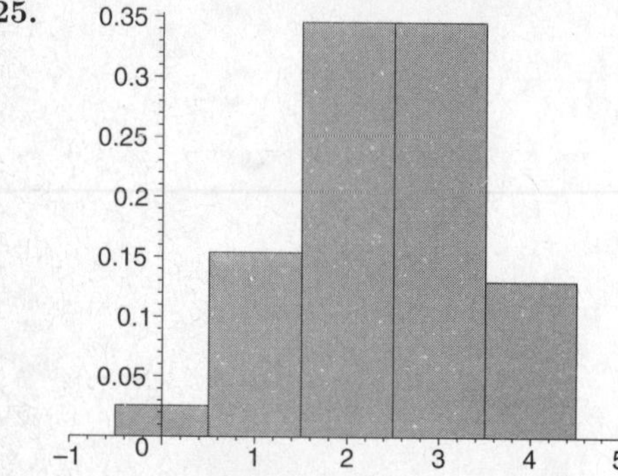

26.

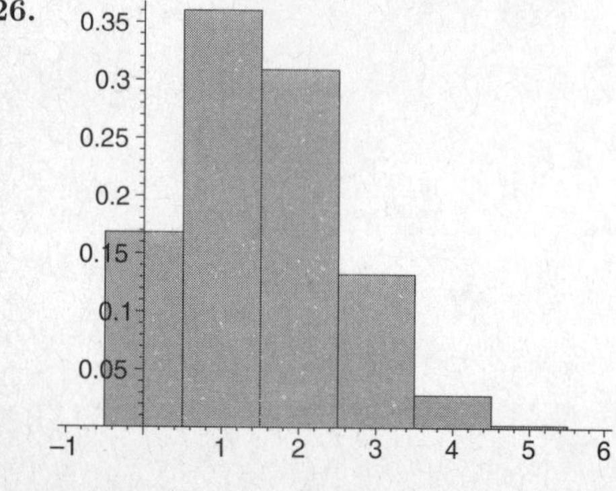

27.

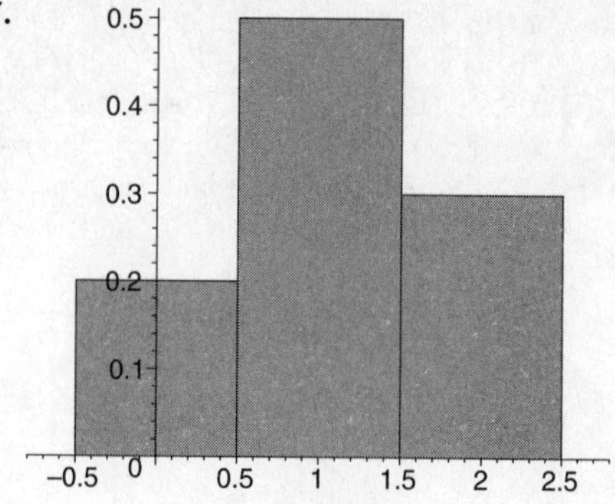

28.

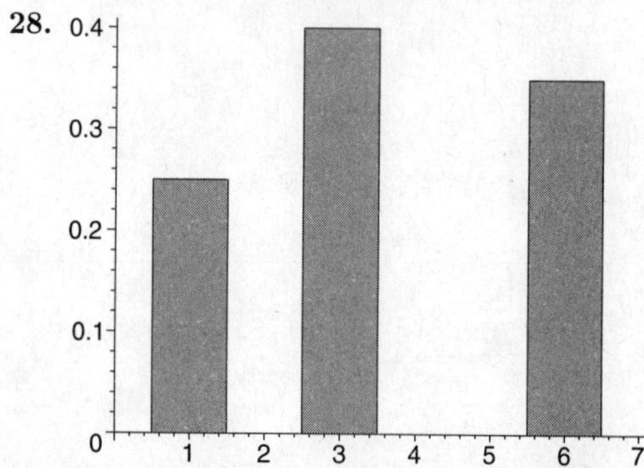

29.

$$\begin{aligned} P(X=5) &= \binom{8}{5}(0.5)^5(0.5)^3 \\ &= 56 \cdot 0.03125 \cdot 0.125 \\ &= 0.21875 \end{aligned}$$

30. $P(X \geq 1) = 1 - P(X < 1) = 1 - P(X = 0) = 1 - 0.125 = 0.875$

31.

$$\begin{aligned} P(X<2) &= P(X=1)+P(X=0) \\ &= \binom{6}{1}(\tfrac{1}{2})^1(\tfrac{1}{2})^5 + \binom{6}{0}(\tfrac{1}{2})^0(\tfrac{1}{2})^6 \\ &= \frac{6}{64}+\frac{1}{64} \\ &= \frac{7}{64} \end{aligned}$$

32.

$$\begin{aligned} P(X \geq 3) &= P(X=3)+P(X=4)+P(X=5) \\ &= 0.3125+0.15625+0.03125 \\ &= 0.5 \end{aligned}$$

33.

$$\begin{aligned} P(X=2) &= \binom{5}{2}\cdot(\tfrac{1}{6})^2\cdot(\tfrac{5}{6})^3 \\ &= 10\cdot\frac{1}{36}\cdot\frac{125}{216} \\ &= 0.16075 \end{aligned}$$

34.

$$\begin{aligned} P(X \geq 1) &= 1 - P(X = 0) \\ &= 1 - 0.25 \\ &= 0.75 \end{aligned}$$

35.

$$\begin{aligned} P(X \geq 1) &= 1 - P(X = 0) \\ &= 1 - \binom{10}{0} \cdot (\frac{1}{6})^0 \cdot (\frac{5}{6})^{10} \\ &= 1 - 0.16151 \\ &= 0.0.838494 \end{aligned}$$

36.

$$\begin{aligned} P(X \leq 1) &= P(X = 0) + P(X = 1) \\ &= 0.3348979767 + 0.401877572 \\ &= 0.736776 \end{aligned}$$

37.

$$\begin{aligned} P(X > 6) &= P(X = 7) + P(X = 8) + P(X = 9) \\ &= \binom{9}{7} \cdot (\frac{1}{6})^7 \cdot (\frac{5}{6})^2 + \\ &\quad \binom{9}{8} \cdot (\frac{1}{6})^8 \cdot (\frac{5}{6})^1 + \binom{9}{9} \cdot (\frac{1}{6})^9 \cdot (\frac{5}{6})^0 \\ &= 0.000089306 + 0.0000044653 + \\ &\quad 0.000000099229 \\ &= 0.000093871 \end{aligned}$$

38. $P(X = 2) = \binom{7}{2}(0.25)^2(0.75)^5 = 0.31146$

39.

$$\begin{aligned} P(X = 30) &= \binom{40}{30} \cdot (\frac{319159}{534694})^{30} \cdot (\frac{215535}{534694})^{10} \\ &= 0.018172 \end{aligned}$$

40.

$$\begin{aligned} P(X = 20) &= \binom{30}{20} \cdot (\frac{238351}{362470})^{20} \cdot (\frac{124119}{362470})^{10} \\ &= 0.15217 \end{aligned}$$

41.

$$\begin{aligned} P(X = 2) &= \binom{3}{2} \cdot (0.80)^2 \cdot (0.20)^1 \\ &= 0.384 \end{aligned}$$

42.

$$\begin{aligned} P(X = 3) &= \binom{3}{3} \cdot (0.60)^3 \cdot (0.40)^0 \\ &= 0.216 \end{aligned}$$

43. a.)

$$\begin{aligned} P(X \leq 2) &= P(X = 2) + P(X = 1) + P(X = 0) \\ &= \binom{12}{2} \cdot (0.346)^2 \cdot (0.654)^{10} + \\ &\quad \binom{12}{1} \cdot (0.346)^1 \cdot (0.654)^{11} + \\ &\quad \binom{12}{0} \cdot (0.346)^0 \cdot (0.654)^{12} \\ &= 0.11310 + 0.03887 + 0.0061226 \\ &= 0.15809 \end{aligned}$$

b.)

$$\begin{aligned} P(X \geq 9) &= P(X = 9) + P(X = 10) + \\ &\quad P(X = 11) + P(X = 12) \\ &= \binom{12}{9} \cdot (0.346)^9 \cdot (0.654)^3 + \\ &\quad \binom{12}{10} \cdot (0.346)^{10} \cdot (0.654)^2 + \\ &\quad \binom{12}{11} \cdot (0.346)^{11} \cdot (0.654)^1 + \\ &\quad \binom{12}{12} \cdot (0.346)^{12} \cdot (0.654)^0 \\ &= 0.004374 + 0.00069416 + \\ &\quad 0.000066772 + 0.00000294383 \\ &= 0.005138 \end{aligned}$$

44. a.)

$$\begin{aligned} P(X \leq 2) &= P(X = 2) + P(X = 1) + P(X = 0) \\ &= \binom{12}{2} \cdot (0.638)^2 \cdot (0.362)^{10} + \\ &\quad \binom{12}{1} \cdot (0.638)^1 \cdot (0.362)^{11} + \\ &\quad \binom{12}{0} \cdot (0.638)^0 \cdot (0.362)^{12} \\ &= 0.001038 + 0.00010710 + \\ &\quad 0.0000050641 \\ &= 0.00115 \end{aligned}$$

b.)

$$\begin{aligned} P(X \geq 9) &= P(X = 9) + P(X = 10) + \\ &\quad P(X = 11) + P(X = 12) \\ &= \binom{12}{9} \cdot (0.638)^9 \cdot (0.362)^3 + \\ &\quad \binom{12}{10} \cdot (0.638)^{10} \cdot (0.362)^2 + \\ &\quad \binom{12}{11} \cdot (0.638)^{11} \cdot (0.362)^1 + \\ &\quad \binom{12}{12} \cdot (0.638)^{12} \cdot (0.362)^0 \\ &= 0.182782 + 0.096642 + \\ &\quad 0.030968 + 0.0045483 \\ &= 0.31494 \end{aligned}$$

45.

$$\begin{aligned} P(X = CC) &= \binom{2}{2} \cdot (0.08)^2 \cdot (0.92)^0 \\ &= 0.0064 \\ P(X = CG) &= \binom{2}{1} \cdot (0.08)^1 \cdot (0.92)^1 \\ &= 0.1472 \\ P(X = GG) &= \binom{2}{2} \cdot (0.08)^0 \cdot (0.92)^2 \\ &= 0.8464 \end{aligned}$$

46.

$$\begin{aligned} P(X = CC) &= \binom{2}{2} \cdot (0.41)^2 \cdot (0.59)^0 \\ &= 0.1681 \\ P(X = CG) &= \binom{2}{1} \cdot (0.41)^1 \cdot (0.59)^1 \\ &= 0.4838 \\ P(X = GG) &= \binom{2}{2} \cdot (0.41)^0 \cdot (0.59)^2 \\ &= 0.3481 \end{aligned}$$

47. We seek the value of $2pq$
$p^2 = \frac{1}{2500} \rightarrow p = \frac{1}{50}$
$q = 1 - \frac{1}{50} = \frac{49}{50}$
Therefore, $2pq = 2(\frac{1}{50})(\frac{49}{50}) = 0.0392$

48. We seek the value of $2pq$
$p^2 = \frac{1}{3600} \rightarrow p = \frac{1}{60}$
$q = 1 - \frac{1}{60} = \frac{59}{60}$
Therefore, $2pq = 2(\frac{1}{60})(\frac{59}{60}) = 0.0328$

49. We seek the value of $2pq$
$p^2 = \frac{1}{360000} \rightarrow p = \frac{1}{600}$
$q = 1 - \frac{1}{600} = \frac{599}{600}$
Therefore, $2pq = 2(\frac{1}{600})(\frac{599}{600}) = 0.033278$

50. We seek the value of $2pq$
$p^2 = \frac{1}{400} \rightarrow p = \frac{1}{20}$
$q = 1 - \frac{1}{20} = \frac{19}{20}$
Therefore, $2pq = 2(\frac{1}{20})(\frac{19}{20}) = 0.095$

51.

$$\begin{aligned} P(X \geq 3) &= P(X = 3) + P(X = 4) + P(X = 5) \\ &= \binom{5}{3} \cdot (0.40)^3 (0.6)^2 + \binom{5}{4} \cdot (0.40)^4 (0.6)^1 \\ &\quad + \binom{5}{5} \cdot (0.40)^5 (0.6)^0 \\ &= 0.2304 + 0.0768 + 0.01024 \\ &= 0.31744 \end{aligned}$$

52. **a.)** $X = \{1, 2, 3, 4, 5, 6\}$
b.) $P(X = i) = \frac{1}{6}$, for $i = 1, 2, 3, 4, 5, 6$
c.)

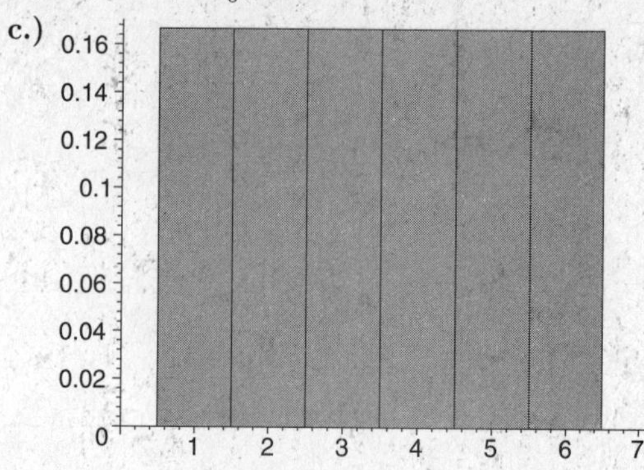

53. **a.)** $X = \{0, 1, 2\}$
b.) $P(X = 0) = \frac{39}{52} \cdot \frac{38}{51} = 0.558824$
$P(X = 1) = \frac{13}{52} \cdot \frac{39}{51} + \frac{39}{52} \cdot \frac{13}{51} = 0.382353$
$P(X = 2) = \frac{13}{52} \cdot \frac{12}{51} = 0.0588235$
c.)

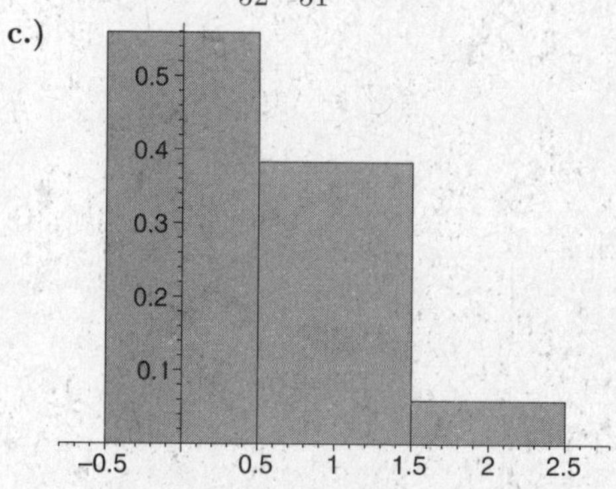

54. **a.)** $X = \{0, 1\}$
b.) $P(X = 0) = \frac{1}{3}$, $P(X = 1) = \frac{2}{3}$
c.)

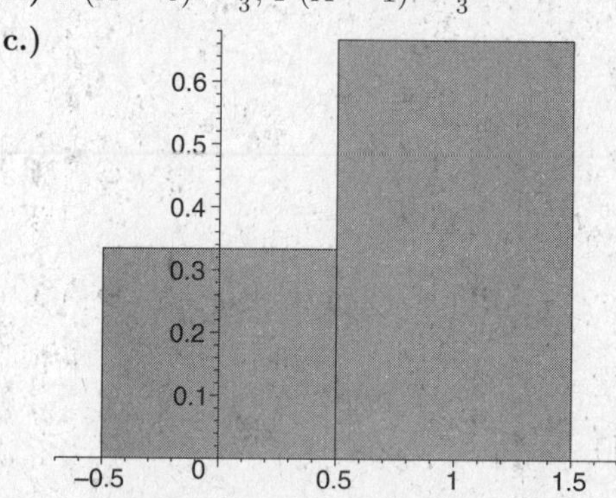

55. **a.)** $X = \{1, 2\}$
b.) $P(X = 1) = \frac{2}{3} \cdot \frac{1}{2} + \frac{1}{3} \cdot \frac{2}{2} = \frac{2}{3}$
$P(X = 2) = \frac{2}{3} \cdot \frac{1}{2} = \frac{1}{3}$

c.)

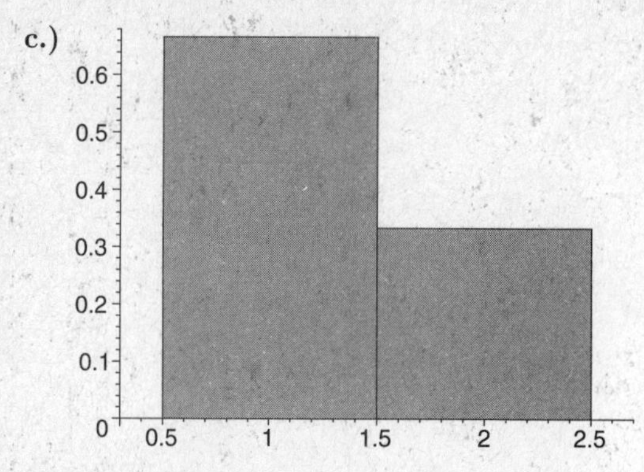

56. Left to the student

57. Left to the student

58. Left to the student

59. Left to the student

60. $\binom{7}{3} = \frac{7!}{4! \cdot 3!} = 35$ The same value as that in the seventh row third column

61. a.) $\binom{8}{5} = 56$ Eighth row fifth column

b.) $\binom{8}{5} = \frac{8!}{3! \cdot 5!} = 56$

62.

$$\begin{aligned} \binom{n}{k} &= \frac{n!}{(n-k)! \cdot k!} \\ \binom{n}{0} &= \frac{n!}{(n-0)! \cdot 0!} \\ &= \frac{n!}{n! \cdot 1} \\ &= 1 \end{aligned}$$

63.

$$\begin{aligned} \binom{n}{k} &= \frac{n!}{(n-k)! \cdot k!} \\ \binom{n}{n} &= \frac{n!}{(n-n)! \cdot n!} \\ &= \frac{n!}{0! \cdot n!} \\ &= \frac{n!}{1 \cdot n!} \\ &= 1 \end{aligned}$$

64.

$$\begin{aligned} \binom{n-1}{k-1} + \binom{n-1}{k} &= \frac{(n-1)!}{(n-k)! \cdot (k-1)!} + \\ &\quad \frac{(n-1)!}{(n-k-1)! \cdot k!} \\ &= \frac{(n-1)!}{(n-k)(n-k-1)! \cdot (k-1)!} \\ &\quad + \frac{(n-1)!}{(n-k-1)! \cdot k(k-1)!} \\ &= \frac{k(n-1)! + (n-k)(n-1)!}{(n-k)(n-k-1)! \cdot k(k-1)!} \\ &= \frac{[k+n-k](n-1)!}{(n-k)! \cdot k!} \\ &= \frac{n(n-1)!}{(n-k)! \cdot k!} \\ &= \frac{n!}{(n-k)! \cdot k!} \\ &= \binom{n}{k} \end{aligned}$$

65. Left to the student

66. a.) $\{SS, SF, FS, FF\}$

b.) 2

c.) $\binom{2}{1} = 2$

67. a.) $\{SSSS, SSSF, SSFS, SSFF, SFSS, SFSF, SFFS, SFFF, FSSS, FSSF, FSFS, FSFF, FFSS, FFSF, FFFS, FFFF\}$

b.) 6

c.) 4

d.) $\binom{4}{2} = 6$

e.) $\binom{4}{3} = 4$

68. $q = 1 - p$

$$\begin{aligned} p^2 + 2pq + q^2 &= p^2 + 2p(1-p) + (1-p)^2 \\ &= p^2 + 2p - 2p^2 + 1 - 2p + p^2 \\ &= 1 \end{aligned}$$

69.

$$\begin{aligned} P(i,j,k) &= \frac{n!}{i!j!k!}(P^2)^i(2pq)^j(q^2)^k \\ P(2,5,3) &= \frac{10!}{2!5!3!}(0.41^2)^2(2 \cdot 0.41 \cdot 0.59)^5(0.59^2)^3 \\ &= 2520 \cdot 0.02825761 \cdot 0.02650509 \cdot 0.04218053 \\ &= 0.0796118 \end{aligned}$$

70.

$$\begin{aligned} P(4,3,5) &= \frac{12!}{4!3!5!}(0.41^2)^4(2 \cdot 0.41 \cdot 0.59)^3(0.59^2)^5 \\ &= 0.01281096 \end{aligned}$$

71. a.) $P(Aa) = 2pq$, $P(aa) = q^2$

$P(A|Aa) = 1/2$, $P(a|Aa) = 1/2$

$P(A|aa) = 0$, $P(a|aa) = 1$

b.)

$$\begin{aligned} P(A) &= p^2 \cdot 1 + 2pq \cdot \frac{1}{2} + q^2 \cdot 0 \\ &= P^2 + p(1-p) + 0 \\ &= p^2 + p - p^2 \\ &= p \end{aligned}$$

c.) $q = 1 - p$

$$\begin{aligned} P(a) &= p^2 \cdot 0 + 2pq \cdot \frac{1}{2} + q^2 \cdot 1 \\ &= 0 + (1-q)q + q^2 \\ &= q - q^2 + q^2 \\ &= q \end{aligned}$$

d.) Left to the student

Exercise Set 10.4

1.

$$\begin{aligned} E(X) &= 0 \cdot P(X=0) + 1 \cdot P(X=1) + 2 \cdot P(X=2) \\ &= 0(0.2) + 1(0.3) + 2(0.5) \\ &= 0 + 0.3 + 1 \\ &= 1.3 \end{aligned}$$

2.

$$\begin{aligned} E(X) &= 1(0.4) + 3(0.4) + 5(0.2) \\ &= 2.6 \end{aligned}$$

3. From Exercise 1, $\mu = E(X) = 1.3$

$$\begin{aligned} Var(X) &= \sum (k-\mu)^2 P(X=k) \\ &= (0-1.3)^2(0.2) + (1-1.3)^2(0.3) + \\ &\quad (2-1.3)^2(0.5) \\ &= 0.338 + 0.027 + 0.245 \\ &= 0.61 \\ SD(X) &= \sqrt{Var(X)} \\ &= \sqrt{0.61} \\ &= 0.781025 \end{aligned}$$

4. From exercise 2, $E(X) = 2.6$

$$\begin{aligned} Var(X) &= \sum (k-\mu)^2 P(X=k) \\ &= (1-2.6)^2(0.4) + (3-2.6)^2(0.4) + \\ &\quad (5-2.6)^2(0.2) \\ &= 2.24 \\ SD(X) &= \sqrt{Var(X)} \\ &= \sqrt{2.24} \\ &= 1.496663 \end{aligned}$$

5. **a)** $E = 1 \cdot 0.25 + 1 \cdot 0.25 = 0.5$

b) $E = np = 2 \cdot 0.25 = 0.5$

6. **a)** $E = 1 \cdot 1/3 + 1 \cdot 1/3 = 2/3$

b) $E = np = 2 \cdot 1/3 = 2/3$

7. **a)** $E = 1 \cdot 0.5 + 1 \cdot 0.5 + 1 \cdot 0.5 = 1.5$

b) $E = np = 3 \cdot 0.5 = 1.5$

8. **a)** $E = 1 \cdot 0.25 + 1 \cdot 0.25 + 1 \cdot 0.25 = 0.75$

b) $E = np = 3 \cdot 0.25 = 0.75$

9. $SD = \sqrt{npq} = \sqrt{2(0.25)(0.75)} = 0.61237$

10. $SD = \sqrt{npq} = \sqrt{2(0.25)(0.75)} = 0.61237$

11. $SD = \sqrt{npq} = \sqrt{2(0.25)(0.75)} = 0.$

12. $SD = \sqrt{npq} = \sqrt{3(0.25)(0.75)} = 0.75$

13. $E(X) = np = 6 \cdot 0.2 = 1.2$
$SD(X) = \sqrt{npq} = \sqrt{6(0.2)(0.8)} = 0.979796$

14. $E(X) = np = 10 \cdot 0.5 = 5$
$SD(X) = \sqrt{npq} = \sqrt{10(0.5)(0.5)} = 1.58114$

15. $E(X) = np = 20 \cdot 0.1 = 2$
$SD(X) = \sqrt{npq} = \sqrt{20(0.1)(0.9)} = 1.34164$

16. $E(X) = np = 25 \cdot 0.36 = 9$
$SD(X) = \sqrt{npq} = \sqrt{25(0.36)(0.64)} = 2.4$

17. $E(X) = np = 50 \cdot 0.4 = 20$
$SD(X) = \sqrt{npq} = \sqrt{50(0.4)(0.6)} = 3.46410$

18. $E(X) = np = 100 \cdot 0.7 = 70$
$SD(X) = \sqrt{npq} = \sqrt{100(0.7)(0.3)} = 4.582576$

19.

$$\begin{aligned} z &= \frac{x - E(X)}{SD(X)} \\ &= \frac{20-16}{2} \\ &= 2 \end{aligned}$$

20.

$$\begin{aligned} z &= \frac{x - E(X)}{SD(X)} \\ &= \frac{10-12}{5} \\ &= -0.4 \end{aligned}$$

21.

$$\begin{aligned} z &= \frac{x - E(X)}{SD(X)} \\ &= \frac{13.1-13.5}{0.24} \\ &= \frac{-0.4}{0.24} \\ &= \frac{-5}{3} \end{aligned}$$

22.

$$\begin{aligned} z &= \frac{x - E(X)}{SD(X)} \\ &= \frac{-1.2 - (-1.2)}{0.9} \\ &= 0 \end{aligned}$$

23.

$$\begin{aligned} z &= \frac{x - E(X)}{SD(X)} \\ &= \frac{29.3 - 20.3}{4.5} \\ &= 2 \end{aligned}$$

24.

$$\begin{aligned} z &= \frac{x - E(X)}{SD(X)} \\ &= \frac{0.14 - 0.25}{0.04} \\ &= -2.75 \end{aligned}$$

25. **a)** $E(X) = np = 12(0.73) = 8.76$
$SD(X) = \sqrt{npq} = \sqrt{12(0.73)(0.27)} = 1.53792$

b)

$$\begin{aligned} P(X = k) &= \binom{n}{k} p^k q^{n-k} \\ P(X = 8) &= \binom{12}{8} (0.73)^8 (0.27)^4 \\ &= 495 \cdot 0.0806460092 \cdot 0.00531441 \\ &= 0.21215 \end{aligned}$$

26. $E(X) = 12 \cdot 0.346 = 4.152$
$SD(X) = \sqrt{12(0.346)(0.654)} = 1.64785$

27. $E(X) = 12 \cdot 0.638 = 7.656$
$SD(X) = \sqrt{12(0.638)(0.362)} = 1.66477$

28.

$$\begin{aligned} E(aX) &= \sum akP(X = k) \\ &= a \sum kP(X = k) \\ &= aE(X) \end{aligned}$$

29.

$$\begin{aligned} Var(aX) &= E(a^2X^2) - [E(ax)^2] \\ &= a^2E(X^2) - E(aX)E(aX) \\ &= a^2E(X^2) - aE(X)aE(X) \\ &= a^2E(X^2) - a^2E^2(X) \\ &= a^2\left(E(X^2) - E^2(X)\right) \\ &= a^2Var(X) \end{aligned}$$

30.

$$\begin{aligned} SD(aX) &= \sqrt{Var(aX)} \\ &= \sqrt{a^2Var(X)} \\ &= |\, a \,| \sqrt{Var(X)} \\ &= |\, a \,| \, SD(X) \end{aligned}$$

Exercise Set 10.5

1. $f(x) = 2x \geq 0$ on $[0, 1]$

$$\begin{aligned} \int_0^1 2x dx &= \left(x^2\right|_0^1 \\ &= 1^2 - 0^2 = 1 \end{aligned}$$

2. $f(x) = \frac{1}{3} \geq 0$ on $[4, 7]$

$$\begin{aligned} \int_4^7 \frac{1}{3} dx &= \left(\frac{1}{3}x\right|_4^7 \\ &= \frac{7}{3} - \frac{4}{3} = 1 \end{aligned}$$

3. $f(x) = \frac{3}{26}x^2 \geq 0$ on $[1, 3]$

$$\begin{aligned} \int_1^3 \frac{3}{26}x^2 dx &= \left(\frac{1}{26}x^3\right|_1^3 \\ &= \frac{27}{26} - \frac{1}{26} = 1 \end{aligned}$$

4. $f(x) = \frac{1}{1+x} \geq 0$ on $[0, e - 1]$

$$\begin{aligned} \int_0^1 \frac{1}{1+x} dx &= \left(\ln(x+1)\right|_0^{e-1} \\ &= 1 - 0 = 1 \end{aligned}$$

5.

$$\begin{aligned} P(1/4 \leq X \leq 3/4) &= \int_{1/4}^{3/4} 4x^3 dx \\ &= \left(x^4\right|_{1/4}^{3/4} \\ &= \left(\frac{3}{4}\right)^4 - \left(\frac{1}{4}\right)^4 \\ &= \frac{81}{256} - \frac{1}{256} \\ &= \frac{80}{256} = \frac{5}{16} \end{aligned}$$

6.

$$\begin{aligned} P(X \leq 1/2) &= \int_0^{1/2} 12x^2 - 12x^3 dx \\ &= \left(4x^3 - 3x^4\right|_0^{1/2} \\ &= [4 \cdot \frac{1}{8} - 3 \cdot \frac{1}{16}] - 0 \\ &= 0.3125 \end{aligned}$$

7.

$$\begin{aligned} P(1/4 \leq X) &= \int_{1/4}^1 20x(1-x)^3 dx \\ &= \int_{1/4}^1 20x(1 - 3x + 3x^2 - x^3) dx \end{aligned}$$

$$\begin{aligned} &= \int_{1/4}^{1} 20x - 60x^2 + 60x^3 - 20x^4 dx \\ &= (10x^2 - 20x^3 + 15x^4 - 4x^5|_{1/4}^{1} \\ &= [10 - 20 + 15 - 4] - [10(0.25)^2 - 20(0.25)^3 \\ &\quad +15(0.25)^4 - 4(0.25)^5] \\ &= 0.63281 \end{aligned}$$

8.

$$\begin{aligned} P(0 \le X \le \pi/4) &= \int_0^{\pi/4} \frac{1}{2} sin\ x\ dx \\ &= \left(-\frac{1}{2} cos\ x\right|_0^{\pi/4} \\ &= -\frac{1}{2} cos(\pi/4) + \frac{1}{2} cos(0) \\ &= 0.14645 \end{aligned}$$

9.

$$\begin{aligned} P(X \ge \pi/6) &= \int_{\pi/6}^{\pi/4} sec^2 x dx \\ &= (tan(x)|_{\pi/6}^{\pi/4} \\ &= tan(\pi/4) - tan(\pi/6) \\ &= 0.42265 \end{aligned}$$

10.

$$\begin{aligned} P(3 \le X \le 4) &= \int_3^4 \frac{1}{\sqrt{2x+3}} dx \\ &= (\sqrt{2x+3}|_3^4 \\ &= \sqrt{11} - \sqrt{9} \\ &= 0.31662 \end{aligned}$$

11.

$$\begin{aligned} P(0 \le X \le 1/2) &= \int_0^{1/2} \frac{e^x}{e-1} dx \\ &= \left(\frac{e^x}{e-1}\right|_0^{1/2} \\ &= -\frac{1}{2} cos(\pi/4) + \frac{1}{2} cos(0) \\ &= 0.14645 \end{aligned}$$

12.

$$\begin{aligned} P(0 \le X \le 1) &= \int_0^1 2xe^{-x^2} dx \\ &= \left(-e^{-x^2}\right|_0^1 \\ &= -e^{-1} + 1 \\ &= 0.63212 \end{aligned}$$

13.

$$\begin{aligned} P(X \ge 16) &= \int_{16}^{\infty} \frac{dx}{x^2} \\ &= \lim_{b \to \infty} \int_{16}^{b} \frac{dx}{x^2} \\ &= \lim_{b \to \infty} \left[\frac{-1}{x}\right]_{16}^{b} \\ &= \lim_{b \to \infty} \left[-\frac{1}{b} - \frac{-1}{16}\right] \\ &= \frac{1}{16} \end{aligned}$$

14.

$$\begin{aligned} P(X \ge 20) &= \int_{20}^{\infty} \frac{2dx}{x^3} \\ &= \lim_{b \to \infty} \int_{20}^{b} \frac{2dx}{x^3} \\ &= \lim_{b \to \infty} \left[\frac{-1}{x^2}\right]_{20}^{b} \\ &= \lim_{b \to \infty} \left[-\frac{1}{b^2} - \frac{-1}{(20)^2}\right] \\ &= \frac{1}{400} \end{aligned}$$

15.

$$\begin{aligned} P(X \ge 3.4) &= \int_{3.4}^{\infty} e^{-x} dx \\ &= \lim_{b \to \infty} \left[-e^{-x}\right]_{3.4}^{b} \\ &= \lim_{b \to \infty} \left[-e^{-b} - (-e^{-3.4})\right] \\ &= 0.03337 \end{aligned}$$

16.

$$\begin{aligned} P(X \ge 5) &= \int_{5}^{\infty} xe^{-x} dx \\ &= \lim_{b \to \infty} \left[-xe^{-x} - e^{-x}\right]_{5}^{b} \\ &= \lim_{b \to \infty} \left[5e^{-5} - e^{-5}\right] \\ &= 0.02695 \end{aligned}$$

17.

$$\begin{aligned} P(X \ge 4) &= \int_{4}^{\infty} \frac{2x}{(x^2+1)^2} dx \\ &= \lim_{b \to \infty} \left[\frac{-1}{(x^2+1)}\right]_{4}^{b} \\ &= \lim_{b \to \infty} \left[\frac{-1}{(b^2+1)} + \frac{1}{17}\right] \\ &= 0.058824 \end{aligned}$$

18.

$$\begin{aligned} P(X \geq 2) &= \int_2^\infty 3x^2 e^{-x^3} dx \\ &= \lim_{b\to\infty} \left[-e^{-x^3}\right]_2^b \\ &= \lim_{b\to\infty} \left[-e^{-b^3} - (-e^2)\right] \\ &= 0.13534 \end{aligned}$$

19.

$$\begin{aligned} P(X \geq c) &= \int_c^{\infty \frac{dx}{x^2}} \\ &= \lim_{b\to\infty} \int_c^b \frac{dx}{x^2} \\ &= \lim_{b\to\infty} \left[\frac{-1}{x}\right]_c^b \\ &= \lim_{b\to\infty} \left[\frac{-1}{b} + \frac{1}{c}\right] \\ &= \frac{1}{c} \end{aligned}$$

Thus,

$$\begin{aligned} \frac{1}{c} &= 0.05 \\ c &= \frac{1}{0.05} \\ &= 20 \end{aligned}$$

20.

$$\begin{aligned} P(X \geq c) &= \int_c^\infty \frac{3dx}{x^4} \\ &= \lim_{b\to\infty} \int_c^b \frac{3dx}{x^4} \\ &= \lim_{b\to\infty} \left[\frac{-1}{x^3}\right]_c^b \\ &= \lim_{b\to\infty} \left[\frac{-1}{b^3} + \frac{1}{c^3}\right] \\ &= \frac{1}{c^3} \end{aligned}$$

Thus,

$$\begin{aligned} \frac{1}{c^3} &= 0.01 \\ c &= \sqrt[3]{\frac{1}{0.01}} \\ &= 4.64159 \end{aligned}$$

21.

$$\begin{aligned} P(X \geq c) &= \int_c^\infty e^{-x}\, dx \\ &= \lim_{b\to\infty} \int_c^b e^{-x}\, dx \\ &= \lim_{b\to\infty} \left[-e^{-x}\right]_c^b \\ &= \lim_{b\to\infty} \left[-e^{-b} + e^{-c}\right] \\ &= e^{-c} \end{aligned}$$

Thus,

$$\begin{aligned} e^{-c} &= 0.05 \\ c &= -\ln(0.05) \\ &= 2.99573 \end{aligned}$$

22.

$$\begin{aligned} P(X \geq c) &= \int_c^\infty 2xe^{-x^2}\, dx \\ &= \lim_{b\to\infty} \int_c^b 2xe^{-x^2}\, dx \\ &= \lim_{b\to\infty} \left[-e^{-x^2}\right]_c^b \\ &= \lim_{b\to\infty} \left[-e^{-b^2} + e^{-c^2}\right] \\ &= e^{-c^2} \end{aligned}$$

Thus,

$$\begin{aligned} e^{-c^2} &= 0.01 \\ c &= \sqrt{-\ln(0.01)} \\ &= 2.14597 \end{aligned}$$

23.

$$\begin{aligned} P(X \geq c) &= \int_c^\infty (3.8/x^3 + 33.6/x^5)\, dx \\ &= \lim_{b\to\infty} \int_c^b (3.8/x^3 + 33.6/x^5)\, dx \\ &= \lim_{b\to\infty} \left[-1.9/x^2 - 8.4/x^4\right]_c^b \\ &= \lim_{b\to\infty} \left[\frac{-1.9}{b^2} - \frac{8.4}{b^4} + \frac{1.9}{c^2} + \frac{8.4}{c^4}\right] \\ &= \frac{1.9}{c^2} + \frac{8.4}{c^4} \end{aligned}$$

Thus,

$$\begin{aligned} \frac{1.9}{c^2} + \frac{8.4}{c^4} &= 0.05 \\ 0.05c^4 - 1.9c^2 - 8.4 &= 0 \\ \to & \\ c^2 &= -6.4807 \text{ not acceptable} \\ c^2 &= 6.4807 \to c = 2.5457 \end{aligned}$$

24.

$$\begin{aligned} P(X \geq c) &= \int_c^\infty (0.32e^{-x} + 1.36e^{-2x})\, dx \\ &= \lim_{b\to\infty} (0.32e^{-x} + 1.36e^{-2x})\, dx \\ &= \lim_{b\to\infty} \left[-0.32e^{-x} - 0.68e^{-2x}\right]_c^b \\ &= \lim_{b\to\infty} (-0.32e^{-b} - 0.68e^{-2b} + 0.32e^{-c} \\ &\quad +0.68e^{-2c}) \\ &= 0.32e^{-c} + 0.68e^{-2c} \end{aligned}$$

Thus,

$$\begin{aligned} 0.32e^{-c} + 0.68e^{-2c} &= 0.01 \\ \to c &= 3.52636 \end{aligned}$$

25. $f(x)$ is positive for all x in $[1,3]$

$$\begin{aligned} 1 &= \int_1^3 kx\, dx \\ &= \left[\frac{k}{2}x^2\right]_1^3 \\ &= \frac{9k}{2} - \frac{k}{2} \\ 1 &= 4k \end{aligned}$$

Thus,

$$k = \frac{1}{4}$$

26. Since $f(x)$ is negative at $x = -1$, we cannot make a probability density function of $f(x)$ on $[-1, 4]$

27. $f(x)$ is positive for all x in $[-1, 1]$

$$\begin{aligned} 1 &= \int_{-1}^1 kx^2\, dx \\ &= \left[\frac{k}{3}x^3\right]_{-1}^1 \\ &= \frac{k}{3} - \frac{-k}{3} \\ 1 &= \frac{2k}{3} \end{aligned}$$

Thus,

$$k = \frac{3}{2}$$

28. Since $f(x)$ is negative at $x = -2$ and $x = -1$, we cannot make a probability density function of $f(x)$ on $[-2, 2]$

29. Since $f(x)$ is negative on $[0, 1]$, we cannot make a probability density function of $f(x)$ on $[0, 2]$

30. $f(x)$ is positive for all x in $[0, 2]$

$$\begin{aligned} 1 &= \int_{-2}^2 kx^4 - 4kx^3 + 4kx^2)\, dx \\ &= \left[\frac{k}{5}x^5 - kx^4 + \frac{4}{3}x^3\right]_{-2}^2 \\ &= \left[\frac{32}{5}k - 16k + \frac{32}{3}k\right] \\ &\quad - \left[\frac{-32}{5}k - 16k - \frac{32}{3}k\right] \\ 1 &= \frac{512k}{15} \end{aligned}$$

Thus,

$$k = \frac{15}{512}$$

31. **a)** $f(x)$ is positive for all x in $[0, 2]$

$$\begin{aligned} 1 &= \int_0^2 k(2-x)^3\, dx \\ &= k\int_0^2 (8 - 12x + 6x^2 - x^3)\, dx \\ &= k\left[8x - 6x^2 + 2x^3 - \frac{1}{4}x^4\right]_0^2 \\ 1 &= k\,[16 - 24 + 16 - 4] \\ 1 &= 4k \end{aligned}$$

Thus,

$$\begin{aligned} k &= \frac{1}{4} \\ P(X = x) &= \frac{(2-x)^3}{4} \end{aligned}$$

b)

$$\begin{aligned} P(X \le 1) &= \int_0^1 \frac{1}{4}(2-x)^3\, dx \\ &= \frac{1}{4}\int_0^1 (8 - 12x + 6x^2 - x^3)\, dx \\ &= \frac{1}{4}\left[8x - 6x^2 + 2x^3 - \frac{1}{4}x^4\right]_0^1 \\ &= \frac{1}{4}\left[8 - 6 + 2 - \frac{1}{4}\right] \\ &= \frac{15}{16} \end{aligned}$$

32. **a)** $f(x)$ is positive for all x in $[1, 4]$

$$\begin{aligned} 1 &= \int_1^4 \frac{kdx}{x} \\ &= k\,[\ln(x)]_1^4 \\ &= k\ln(4) \end{aligned}$$

Thus,

$$\begin{aligned} k &= \frac{1}{\ln(4)} \\ P(X = x) &= \frac{1}{\ln(4)x} \end{aligned}$$

b)

$$\begin{aligned} P(X \le 3) &= \int_1^3 \frac{1}{\ln(4)x}\, dx \\ &= 0.79248 \end{aligned}$$

33. **a)** $f(x)$ is positive for all x in $[\pi/6, \pi/2]$

$$\begin{aligned} 1 &= \int_{\pi/6}^{\pi/2} k\ sin(x)\, dx \\ &= -k\,[cos(x)]_{\pi/6}^{\pi/2} \\ 1 &= -k\left[0 - \frac{\sqrt{3}}{2}\right] \\ 1 &= \frac{\sqrt{3}}{2}k \end{aligned}$$

Thus,

$$\begin{aligned} k &= \frac{2\sqrt{3}}{3} \\ P(X = x) &= \frac{2\sqrt{3}}{3}\ sin(x) \end{aligned}$$

b)

$$\begin{aligned}P(X \le 1) &= \int_{\pi/3}^{\pi/2} \frac{2\sqrt{3}}{3} \; sin(x)\; dx \\ &= \frac{2\sqrt{3}}{3} \int_{\pi/3}^{pi/2} sin(x)\; dx \\ &= \frac{2\sqrt{3}}{3} [-cos(x)]_{\pi/3}^{\pi/2} \\ &= \frac{2\sqrt{3}}{3} [0 - (-0.5)] \\ &= 0.57735\end{aligned}$$

34. **a)** $f(x)$ is positive for all x in $[0, \pi]$

$$\begin{aligned}1 &= \int_0^{\pi} k\; cos^2(x)\; dx \\ &= k \int_0^{\pi} \frac{1}{2}(1 + cos(2x))\; dx \\ &= \frac{k}{2}\left[x + \frac{1}{2} sin(2x)\right]_{\pi}^{0} \\ 1 &= \frac{k}{2}[\pi] \\ 1 &= \frac{\pi}{2} k\end{aligned}$$

Thus,

$$\begin{aligned}k &= \frac{2}{\pi} \\ P(X = x) &= \frac{2}{\pi} cos^2(x)\end{aligned}$$

b)

$$\begin{aligned}P(X \ge \frac{\pi}{4}) &= \frac{2}{4} \int_{\pi/4}^{\pi} cos^2(x)\; dx \\ &= \frac{2\sqrt{3}}{3} \int_{\pi/3}^{pi/2} sin(x)\; dx \\ &= 0.590845\end{aligned}$$

35. **a)** $f(x)$ is positive for all x in $[0, 5]$

$$\begin{aligned}1 &= \int_0^5 k \frac{2x + 3}{x2 + 3x + 4}\; dx \\ &= k\left[\ln(x^2 + 3x + 4)\right]_0^5 \\ &= k[\ln(44) - \ln(4)] = \ln(11)\end{aligned}$$

Thus,

$$\begin{aligned}k &= \frac{1}{\ln(11)} \\ P(X = x) &= \frac{2x + 3}{\ln(11)(x^2 + 3x + 4)}\end{aligned}$$

b)

$$\begin{aligned}P(X \le 3) &= \int_2^4 \frac{2x + 3}{\ln(11)(x^2 + 3x + 4)}\; dx \\ &= \frac{1}{\ln(11)} \left[\ln(x^2 + 3x + 4)\right]_2^4 \\ &= \frac{1}{\ln(11)} [\ln(32) - \ln(14)] \\ &= 0.34475\end{aligned}$$

36. **a)** $f(x)$ is positive for all x in $[0, \infty)$

$$\begin{aligned}1 &= k \int_0^{\infty} xe^{-x} - sin(x)e^{-x} \\ &= k| - xe^{-x} - e^{-x} + \frac{1}{2} e^{-x} cos(x) + \\ &\quad \frac{1}{2} e^{-x} sin(x)|_0^{\infty} \\ &= k\left[0 - \frac{1}{2}\right] \\ 1 &= \frac{-1}{2} k\end{aligned}$$

Thus,

$$\begin{aligned}k &= -2 \\ P(X = x) &= -2(x - sin(x))e^{-x}\end{aligned}$$

b)

$$\begin{aligned}P(X \le 3) &= -2 \int_3^{\infty} (xe{-}x - sin(x)e^{-x})\; dx \\ &= 0.22028\end{aligned}$$

37. **a)** $f(x)$ is positive for all x in $[-\pi, \pi)$

$$\begin{aligned}1 &= \int_{-\pi}^{\pi} kx\; sin(x) \\ &= k\left[sin(x) - x\; cos(x)\right]_{-\pi}^{\pi} \\ &= k[0 - \pi(-1) - (0 - (-\pi)(-1))] \\ 1 &= 2\pi k\end{aligned}$$

Thus,

$$\begin{aligned}k &= \frac{1}{2\pi} \\ P(X = x) &= \frac{x\; sin(x)}{2\pi}\end{aligned}$$

b)

$$\begin{aligned}P(X \le 3) &= \frac{1}{2pi} \int_{-pi}^{\pi/2} (x\; sin(x))\; dx \\ &= \frac{1}{2\pi} [sin(x) - x\; cos(x)]_{-\pi}^{\pi/2} \\ &= 0.65915\end{aligned}$$

38. $f(v) = \frac{1}{7-0} = \frac{1}{7}$

$$\begin{aligned}P(0 \le X \le 3) &= \int_0^3 \frac{1}{7}\; dx \\ &= \left[\frac{1}{7} x\right]_0^3 \\ &= \frac{3}{7}\end{aligned}$$

39. $f(v) = \frac{1}{4-2} = \frac{1}{2}$

$$\begin{aligned}P(3 \le X \le 4) &= \int_3^4 \frac{1}{2}\; dx \\ &= \left[\frac{1}{2} x\right]_3^4 \\ &= 2 - \frac{3}{2} \\ &= \frac{1}{2}\end{aligned}$$

40. $f(v) = \frac{1}{70-(-10)} = \frac{1}{80}$

$$\begin{aligned} P(10 \le X \le 30) &= \int_{10}^{30} \frac{1}{80}\, dx \\ &= \left[\frac{1}{80}x\right]_{10}^{30} \\ &= \frac{1}{4} \end{aligned}$$

41. $f(v) = \frac{1}{500-100} = \frac{1}{400}$

$$\begin{aligned} P(200 \le X \le 350) &= \int_{200}^{350} \frac{1}{400}\, dx \\ &= \left[\frac{1}{400}x\right]_{200}^{350} \\ &= \frac{350}{400} - \frac{200}{400} \\ &= \frac{3}{8} \end{aligned}$$

42.

$$\begin{aligned} E(X) &= \int_2^5 \frac{xdx}{3} \\ &= \left[\frac{x^2}{6}\right]_2^5 \\ &= \frac{1}{6}(25-4) \\ &= 3.5 \end{aligned}$$

$$\begin{aligned} Var(X) &= \int_2^5 (x-1)^2 \frac{1}{3}\, dx \\ &= \frac{1}{3}\int_2^5 (x^2-2x+1)\, dx \\ &= \frac{1}{3}\left[\frac{x^3}{3} - x^2 + x\right]_2^5 \\ &= \frac{1}{3}(21) \\ &= 7 \\ SD(X) &= \sqrt{7} \\ &= 2.64575 \end{aligned}$$

43.

$$\begin{aligned} E(X) &= \int_0^3 \frac{2x^2}{9}\, dx \\ &= \left[\frac{2x^3}{27}\right]_0^3 \\ &= \frac{2}{27}(27-0) \\ &= 2 \end{aligned}$$

$$\begin{aligned} Var(X) &= \int_0^3 (x-2)^2 \frac{2x}{9}\, dx \\ &= \frac{2}{9}\int_0^3 (x^3 - 4x^2 + 4x)\, dx \\ &= \frac{2}{9}\left[\frac{x^4}{4} - \frac{4}{3}x^3 + 2x^2\right]_0^3 \\ &= \frac{2}{9}(2.25-0) \\ &= 0.5 \\ SD(X) &= \sqrt{0.5} \\ &= 0.70711 \end{aligned}$$

44.

$$\begin{aligned} E(X) &= \int_{-2}^1 \frac{x^3}{3}\, dx \\ &= \left[\frac{x^4}{12}\right]_{-2}^1 \\ &= \frac{1}{12}(1-16) \\ &= -1.25 \end{aligned}$$

$$\begin{aligned} Var(X) &= \int_{-2}^1 (x+1.25)^2 \frac{x^2}{3}\, dx \\ &= \frac{1}{3}\int_{-2}^1 (x^4 + 2.5x^3 + 1.5625x^2)\, dx \\ &= \frac{1}{3}\left[\frac{x^5}{5} + 0.625x^4 + 0.5208333333333x^3\right]_{-2}^1 \\ &= \frac{1}{3}(1.9125) \\ &= 0.6375 \\ SD(X) &= \sqrt{0.6375} \\ &= 0.79844 \end{aligned}$$

45.

$$\begin{aligned} E(X) &= \int_1^2 \frac{dx}{\ln(2)} \\ &= \left[\frac{1}{\ln(2)}x\right]_1^2 \\ &= \frac{1}{\ln(2)}(2-1) \\ &= \frac{1}{\ln(2)} \\ &= 1.4427 \end{aligned}$$

$$\begin{aligned} Var(X) &= \int_1^2 (x - \frac{1}{\ln(2)})^2 \frac{1}{x\ \ln(2)}\, dx \\ &= \frac{1}{\ln(2)}\int_1^2 (x - \frac{2}{\ln(2)} + \frac{1}{x\ \ln^2(2)})\, dx \\ &= \frac{1}{\ln(2)}\left[\frac{x^2}{2} - \frac{2x}{\ln(2)} + \frac{1}{\ln^2(2)}\ \ln(x)\right]_1^2 \\ &= \frac{1}{\ln(2)}\left[2 - \frac{4}{\ln(2)} + \frac{1}{\ln^2(2)}\ln(2) - \frac{1}{2}\right] \\ &\quad + \frac{1}{ln(2)}\left[\frac{2}{\ln(2)} - \frac{1}{\ln^2(2)}\ln(1)\right] \\ &= 0.082674 \\ SD(X) &= \sqrt{0.082674} \\ &= 0.28753 \end{aligned}$$

46.

$$\begin{aligned}
E(X) &= \frac{1}{96}\int_0^6 x + x^2 + x^3\ dx \\
&= \frac{1}{96}\left[\frac{x^2}{2} + \frac{x^3}{3} + \frac{x^4}{4}\right]_0^6 \\
&= \frac{1}{96}(414) \\
&= 4.3125
\end{aligned}$$

$$\begin{aligned}
Var(X) &= \frac{1}{96}\int_0^6 (x - 4.3125)^2(1 + x + x^2)\ dx \\
&= \frac{1}{96}\int_0^6 (10.97265625x^2 - 7.6250x^3 + x^4 + \\
&\quad 9.97265625x + 18.59765625)\ dx \\
&= \frac{1}{96}\left[3.657552x^3 - 1.90625x^4 + \frac{x^5}{5}\right]_0^6 \\
&\quad + \frac{1}{96}\left[4.98638125 + 18.59765625x\right]_0^6 \\
&= \frac{1}{96}(165.825) \\
&= 1.727343750 \\
SD(X) &= \sqrt{1.727343750} \\
&= 1.314284501
\end{aligned}$$

47.

$$\begin{aligned}
E(X) &= \frac{4}{21}\int_1^2 (x^2 + x^4)\ dx \\
&= \frac{4}{21}\left[\frac{x^3}{3} + \frac{x^5}{5}\right]_1^2 \\
&= \frac{4}{21}\left(\frac{8}{3} + \frac{32}{5} - \frac{1}{3} - \frac{1}{5}\right) \\
&= 1.625396825
\end{aligned}$$

$$\begin{aligned}
Var(X) &= \frac{4}{21}\int_1^2 (x - 1.625396825)^2(x + x^3)\ dx \\
&= \frac{4}{21}\int_1^2 (3.641914839x^3 + x^5 - 3.25079365x^2 \\
&\quad -3.25079365x^4 + 2.641914839x)\ dx \\
&= \frac{4}{21}\left[0.9104787098x^4 + \frac{1}{6}x^6 - 1.083597883x^3\right]_1^2 \\
&\quad -\frac{4}{21}\left[0.65015873x^5 + 1.32095742x\right]_1^2 \\
&= \frac{4}{21}(0.3799470914) \\
&= 0.07237087455 \\
SD(X) &= \sqrt{0.07237087455} \\
&= 0.26990183536
\end{aligned}$$

48.

$$\begin{aligned}
E(X) &= \frac{12}{17}\int_0^1 (x^{3/2} + x^{4/3})\ dx \\
&= \frac{12}{17}\left[\frac{2x^{5/2}}{5} + \frac{3x^{7/3}}{7}\right]_0^1 \\
&= \frac{12}{17}\left(\frac{2}{5} + \frac{3}{7}\right) \\
&= 0.5848739496
\end{aligned}$$

$$\begin{aligned}
Var(X) &= \frac{12}{17}\int_0^1 (x - 0.5848739496)^2(x^{1/2} + x^{1/3})\ dx \\
&= 0.1011044418 \\
SD(X) &= \sqrt{0.1011044418} \\
&= 0.3179692466
\end{aligned}$$

49.

$$\begin{aligned}
E(X) &= \frac{1}{2}\int_0^\pi x\sin\ x\ dx \\
&= \frac{1}{2}\left[\sin\ x - x\cos\ x\right]_0^\pi \\
&= \frac{1}{2}(0 - (-\pi) - 0 + 0) \\
&= \frac{\pi}{2}
\end{aligned}$$

$$\begin{aligned}
Var(X) &= \frac{1}{2}\int_0^\pi (x - \frac{\pi}{2})^2 \sin\ x\ dx \\
&= \frac{1}{2}\int_0^\pi (x^2\sin\ x - \pi x\sin\ x + \frac{\pi^2}{4}\sin\ x)\ dx \\
&= \frac{1}{2}\left[-x^2\cos\ x + 2\cos\ x + 2x\sin\ x\right]_0^\pi - \\
&\quad \frac{1}{2}\left[\pi\sin\ x + \pi x\cos\ x - \frac{\pi^2}{4}\cos\ x\right]_0^\pi \\
&= \frac{1}{2}\left[(\pi^2 - 2 + 0 - 0 - \pi^2 - \frac{\pi^2}{4})\right] \\
&\quad -\frac{1}{2}\left[(0 + 2 + 0 - 0 + 0 - \frac{\pi^2}{4})\right] \\
&= \frac{1}{2}\left[\frac{\pi^2}{2} - 4\right] \\
&= 0.467401 \\
SD(X) &= \sqrt{0.467401} \\
&= 0.683667
\end{aligned}$$

50.

$$\begin{aligned}
E(X) &= \int_0^{\pi/2} x\cos\ x\ dx \\
&= \left[\cos\ x + x\sin\ x\right]_0^{\pi/2} \\
&= \frac{\pi}{2} - 1
\end{aligned}$$

$$\begin{aligned}
Var(X) &= \int_0^{\pi/2} (x - \frac{\pi}{2} + 1)^2 \cos\ x\ dx \\
&= \pi - 3 \\
&= 0.141592654 \\
SD(X) &= \sqrt{0.141592654} \\
&= 0.3762879934
\end{aligned}$$

51.

$$\begin{aligned}E(X) &= \frac{3}{14}\int_0^3 x(x+1)^{1/2})\,dx\\ &= \frac{3}{14}\left[\frac{2}{5}(x+1)^{5/2}-\frac{2}{3}(x+1)^{3/2}\right]_0^3\\ &= \frac{3}{14}\left(\frac{2}{5}(32)+\frac{2}{3}(8)\right)\\ &= 1.657142857\end{aligned}$$

$$\begin{aligned}Var(X) &= \frac{3}{14}\int_0^3 (x-1.657142857)^2(x+1)^{1/2})\,dx\\ &= \frac{3}{14}\left[\frac{2}{7}(x+1)^{7/2}-\frac{524}{75}(x+1)^{5/2}\right]_0^3\\ &\quad+\frac{3}{14}\left[\frac{34322}{675}(x+1)^{3/2}\right]_0^3\\ &= \frac{3}{14}(3.337142857)\\ &= 0.7151020408\\ SD(X) &= \sqrt{0.7151020408}\\ &= 0.84564\end{aligned}$$

52.

$$\begin{aligned}E(X) &= \int_1^e x\ \ln x\,dx\\ &= \left[\frac{x^2}{2}\ \ln x-\frac{x^2}{4}\right]_1^e\\ &- \frac{e^2}{4}+\frac{1}{4}\\ &= 2.09726\end{aligned}$$

$$\begin{aligned}Var(X) &= \int_1^e (x-2.09726)2\ \ln x\,dx\\ &= \left[\frac{1}{3}\ x^3\ \ln x-\frac{1}{9}\ x^3-x^2\ \ln x\right]_1^e\\ &\quad+\left[\frac{1}{2}\ x^2+x\ \ln x-x\right]_1^e\\ &= \left(\frac{x^3}{3}\ \ln x-\frac{x^3}{11}-2.09726x^2\ \ln x+\right.\\ &\quad 1.04863x^2+4.39850x\ \ln x-4.39850x\ \Big|_1^e\\ &\approx .17605\\ SD(X) &= \sqrt{.17605}\\ &= 0.41958\end{aligned}$$

53.

$$\begin{aligned}E(X) &= \int_0^\infty 3x\ e^{-3x}\,dx\\ &= \lim_{b\to\infty}\int_0^b 3x\ e^{-3x}\,dx\\ &= \lim_{b\to\infty}\left[-x\ e^{-3x}-\frac{1}{3}\ e^{-3x}\right]_0^b\\ &= \frac{1}{3}\end{aligned}$$

$$\begin{aligned}Var(X) &= \int_0^\infty 3\ e^{-3x}\ \left(x-\frac{1}{3}\right)^2\,dx\\ &= \lim_{b\to\infty}\int_0^b 3\ e^{-3x}\ \left(x-\frac{1}{3}\right)^2\,dx\\ &= \lim_{b\to\infty}\left[-x^3\ e^{-3x}-\frac{x^2\ e^{-3x}}{3}-\frac{x\ e^{-3x}}{3}-\frac{e^{-3x}}{9}\right]_0^b\\ &= \frac{1}{9}\\ SD(X) &= \sqrt{\frac{1}{9}}\\ &= \frac{1}{3}\end{aligned}$$

54.

$$\begin{aligned}E(X) &= \int_1^\infty 3x\ e^{-3(x-1)}\,dx\\ &= \lim_{b\to\infty}\int_1^b 3x\ e^{-3(x-1)}\,dx\\ &= \lim_{b\to\infty}\left[(1-x)\ e^{-3(x-1)}-\frac{4\ e^{-3(x-1)}}{3}\right]_1^b\\ &= \frac{4}{3}\end{aligned}$$

$$\begin{aligned}Var(X) &= \int_1^\infty 3\ e^{-3x}\ \left(x-\frac{4}{3}\right)^2\,dx\\ &= \lim_{b\to\infty}\int_1^b 3\ e^{-3x}\ \left(x-\frac{4}{3}\right)^2\,dx\\ &= \lim_{b\to\infty}\left[\frac{4(1-x)\ e^{-3(x-1)}}{3}-\frac{(3-3x)^2\ e^{-3(x-1)}}{9}\right]_1^b\\ &\quad-\lim_{b\to\infty}\left[\frac{e^{-3(x-1)}}{9}\right]_1^b\\ &= \frac{2}{9}\\ SD(X) &= \sqrt{\frac{1}{9}}\\ &= \frac{\sqrt{2}}{3}\end{aligned}$$

55.

$$\begin{aligned}E(X) &= \frac{3+9}{2}\\ &= 6\\ Var(X) &= \frac{(9-3)^2}{12}\\ &= 3\\ SD(X) &= \sqrt{3}\\ &= 1.73205\end{aligned}$$

56.

$$\begin{aligned}E(X) &= \frac{-2+5}{2}\\ &= 1.5\\ Var(X) &= \frac{(5-(-2))^2}{12}\end{aligned}$$

$$\begin{aligned} &= \frac{49}{12} \\ &\approx 4.083333333333 \\ SD(X) &= \sqrt{4.083333333333} \\ &= 2.02073 \end{aligned}$$

57.

$$\begin{aligned} E(X) &= \frac{10+20}{2} \\ &= 15 \\ Var(X) &= \frac{(20-10)^2}{12} \\ &= \frac{100}{12} = \frac{25}{3} \\ &\approx 8.3333333333 \\ SD(X) &= \sqrt{8.3333333333} \\ &= 2.88675 \end{aligned}$$

58.

$$\begin{aligned} E(X) &= \frac{0.001+0.002}{2} \\ &= 0.0015 \\ Var(X) &= \frac{(0.002-0.001)^2}{12} \\ &= \frac{0.000001}{12} \\ &\approx 0.000000083333333333 \\ SD(X) &= \sqrt{0.000000083333333333} \\ &= 0.000288675 \end{aligned}$$

59.

$$\begin{aligned} P &= \int_0^3 \frac{1}{5}\,dx \\ &= \left[\frac{x}{5}\right]_0^3 \\ &= \frac{3}{5} \end{aligned}$$

60.

$$\begin{aligned} P &= \int_1^4 \frac{1}{4}\,dx \\ &= \left[\frac{x}{4}\right]_1^4 \\ &= 1 - \frac{1}{4} \\ &= \frac{3}{4} \end{aligned}$$

61. $70 - 60 = 10$

a)

$$\begin{aligned} P &= \int_{60}^{65} \frac{1}{10}\,dx \\ &= \left[\frac{x}{10}\right]_{60}^{65} \\ &= \frac{1}{2} \end{aligned}$$

b)

$$\begin{aligned} P &= \int_{68}^{70} \frac{1}{10}\,dx \\ &= \left[\frac{x}{10}\right]_{68}^{70} \\ &= \frac{1}{5} \end{aligned}$$

62. $73 - 55 = 18$

a)

$$\begin{aligned} P &= \int_{55}^{61} \frac{1}{18}\,dx \\ &= \left[\frac{x}{18}\right]_{55}^{61} \\ &= \frac{1}{3} \end{aligned}$$

b)

$$\begin{aligned} P &= \int_{67}^{73} \frac{1}{18}\,dx \\ &= \left[\frac{x}{18}\right]_{67}^{73} \\ &= \frac{1}{3} \end{aligned}$$

63.

$$\begin{aligned} \int_0^b x^3\,dx &= 1 \\ \left.\frac{x^4}{4}\right|_0^b &= 1 \\ \frac{b^4}{4} &= 1 \\ b &= \sqrt[4]{4} = \sqrt{2} \end{aligned}$$

64.

$$\begin{aligned} \int_{-a}^{a} 12x^2\,dx &= 1 \\ \left.4x^3\right|_{-a}^{a} &= 1 \\ 4a^3 - (-4a^3) &= 1 \\ 8a^3 &= 1 \\ a &= \sqrt[3]{\frac{1}{8}} \\ a &= \frac{1}{2} \end{aligned}$$

65. a)

$$\begin{aligned} E(X) &= \int_a^b x\,f(x)\,dx \\ &= \int_a^b \frac{1}{b-a}\,x\,dx \\ &= \frac{1}{2(b-a)} \cdot \left.x^2\right|_a^b \end{aligned}$$

$$\begin{aligned} &= \frac{1}{2(b-a)} \cdot (b^2 - a^2) \\ &= \frac{1}{2(b-a)} \cdot (b-a)(b+a) \\ &= \frac{b+a}{2} \end{aligned}$$

b)

$$\begin{aligned} E(X^2) &= \int_a^b x^2 \, f(x)\, dx \\ &= \int_a^b \frac{x^2}{b-a}\, dx \\ &= \left.\frac{x^3}{3(b-a)}\right|_a^b \\ &= \frac{b^3 - a^3}{3(b-a)} \\ &= \frac{(b-a)(b^2+ab+b^2)}{3(b-a)} \\ &= \frac{1}{3}\,(b^2+ab+a^2) \end{aligned}$$

c)

$$\begin{aligned} Var(X) &= \int_a^b (x-\mu)^2\, f(x)\, dx \\ &= \int_a^b \left(x - \frac{(b+a)}{2}\right)^2 \frac{1}{b-a}\, dx \\ &= \frac{1}{b-a} \int_a^b x^2 - (b+a)x + \frac{(b+a)^2}{4}\, dx \\ &= \frac{1}{b-a}\left[\frac{x^3}{3} - \frac{(b+a)x^2}{2} + \frac{(b+a)^2}{4}\right]_a^b \\ &= \frac{1}{b-a}\left[\frac{(b^3-a^3}{3} - \frac{(b+a)(b^2-a^2)}{2}\right] \\ &\quad + \frac{1}{b-a}\left[\frac{(b+a)(b-a)}{4}\right] \\ &= \frac{b^2+ab+a^2}{3} - \frac{(b+a)^2}{2} + \frac{(b+a)^2}{4} \\ &= \frac{b^2+ab+a^2}{3} - \frac{(b+a)^2}{4} \\ &= \frac{4b^2+4ab+4a^2-3b^2-6ab-3a^2}{12} \\ &= \frac{b^2-2ab+a^2}{12} \\ &= \frac{(b-a)^2}{12} \end{aligned}$$

66.

$$\begin{aligned} Var(X) &= \int_a^b (x-\mu)^2\, f(x)\, dx \\ &= \int_a^b (x^2\, f(x) - 2\mu x\, f(x) + \mu^2\, f(x))\, dx \\ &= \int_a^b x^2\, f(x)\, dx - 2\mu \int_a^b x\, f(x)\, dx + \\ &\quad \mu^2 \int_a^b f(x)\, dx \\ &= E(X^2) - 2\mu(\mu) + \mu^2(1) \\ &= E(X^2) - \mu^2 \end{aligned}$$

67. a)

$$\begin{aligned} \int_1^{85} 1152.9k\; e^{0.051476x}\, dx &= 1 \\ 1152.9k \int_1^{85} e^{0.051476x}\, dx &= 1 \\ 1152.9k \cdot \left.\frac{e^{0.051476x}}{0.051476}\right|_1^{85} &= 1 \\ 1152.9k \cdot 1523.497676 &= 1 \\ k &= \frac{1}{1152.9 \cdot 1523.497676} \\ &= 5.69333 \times 10^{-7} \\ &= 0.000000569333 \end{aligned}$$

b) $k\; f(x) = 0.00065638\; e^{0.051476x}$

$$\begin{aligned} E(X) &= \int_1^{85} 0.00065638x\; e^{0.051476x}\, dx \\ &= (0.012751185x\; e0.051476x - \\ &\quad \left. 0.2477112639\; e^{0.051476x}\right|_1^{85} \\ &= 66.701 \end{aligned}$$

c)

$$\begin{aligned} Var(X) &= \int_1^{85} 0.00065638\;(x-66.701)^2\; e^{0.051476x}\, dx \\ &= 0.012751x^2\; e^{0.051476x} - 2.19645xe^{0.051476x} \\ &\quad \left. + 99.39922\; e^{0.051476x}\right|_1^{85} \\ &= 281.3908120 \\ SD(X) &= \sqrt{281.3908120} \\ &= 16.7747 \end{aligned}$$

68. a)

$$\begin{aligned} k\int_2^{125} (0.00394 - 0.00247x + \\ 0.00000190x^2 + 0.0365\;\ln x)\, dx &= 1 \\ k(4.606221576) &= 1 \\ k &= 0.217098 \end{aligned}$$

b)

$$\begin{aligned} P(13 \le X \le 30) &= 0.2170976762 \int_{13}^{30} 0.00394 - \\ &\quad 0.00247x + 0.00000190x^2 \\ &\quad + 0.0365\;\ln x)\, dx \\ &= 0.2171\;(0.00394x0.001235x^2 + \\ &\quad 0.000000633333x^3 + 0.0365x\;\ln x \\ &\quad \left. -0.0365x\right|_{13}^{30} \\ &= 0.2171(1.126173430) \\ &= 0.24449 \end{aligned}$$

69. **a)** Left to the student

b) $P(X \leq 2)$

$$\begin{aligned} &= \int_0^2 \frac{0.68}{0.89}\left(\frac{x}{0.89}\right)^{-0.32} e^{-(x/0.89)^{0.68}}\,dx \\ &= \frac{0.68}{0.89}\left[-1.308823528\, e^{-1.082467325x^{0.68}}\right]_0^2 \\ &= 0.82347 \end{aligned}$$

70. Answers vary

71. 0.028596

72. 0.069663

73. 0.018997

74. 0.013262

75. 0.004701

76. 0.013754

77. 0.265026

78. 0.171797

79. 9.488

80. 13.277

81. 12.592

82. 16.812

Exercise Set 10.6

1.

$$\begin{aligned} P(X=0) &= \frac{2.2^0}{0!}\, e^{-2.2} \\ &= 0.1108 \end{aligned}$$

2.

$$\begin{aligned} P(X=3) &= \frac{3.5^3}{3!}\, e^{-3.5} \\ &= 0.0.2158 \end{aligned}$$

3.

$$\begin{aligned} P(X=1) &= \frac{1.3^1}{1!}\, e^{-1.3} \\ &= 0.0.3543 \end{aligned}$$

4.

$$\begin{aligned} P(X=9) &= \frac{7^9}{9!}\, e^{-7} \\ &= 0.1014 \end{aligned}$$

5.

$$\begin{aligned} P(X=7) &= \frac{4^7}{7!}\, e^{-4} \\ &= 0.05954 \end{aligned}$$

6.

$$\begin{aligned} P(X=6) &= \frac{5.2^6}{6!}\, e^{-5.2} \\ &= 0.1515 \end{aligned}$$

7.

$$\begin{aligned} P(X \leq 2) &= P(X=0)+P(X=1)+P(X=2) \\ &= 0.2240+0.1494+0.04979 \\ &= 0.4232 \end{aligned}$$

8.

$$\begin{aligned} P(X \leq 3) &= P(X=0)+P(X=1)+P(X=2) \\ &\quad + P(X=3) \\ &= 0.06131+0.18394+0.36788+0.36788 \\ &= 0.9810 \end{aligned}$$

9.

$$\begin{aligned} P(X \geq 3) &= 1-P(X \leq 3) \\ &= 1-\Big[P(X=0)+P(X=1)+P(X=2) \\ &\quad +P(X=3)\Big] \\ &= 1-(0.22313+0.33470+0.25102+0.12551) \\ &= 0.0656 \end{aligned}$$

10.

$$\begin{aligned} P(X \geq 2) &= 1-P(X \leq 2) \\ &= 1-[P(X=0)+P(X=1)+P(X=2)] \\ &= 1-(0.01657+0.06795+0.13929) \\ &= 0.7762 \end{aligned}$$

11.

$$\begin{aligned} P(3 \leq X \leq 5) &= P(X=3)+P(X=4)+P(X=5) \\ &= 0.08674+0.02602+0.00625 \\ &= 0.1388 \end{aligned}$$

12.

$$\begin{aligned} P(2 \leq X \leq 5) &= P(X=2)+P(X=3)+P(X=4) \\ &\quad +P(X=5) \\ &= 0.27067+0.18045+0.09022+0.03610 \\ &= 0.57744 \end{aligned}$$

13. $E(X) = 2.3$

14. $E(X) = 4$

15. $SD(X) = \sqrt{4} = 2$

16. $SD(X) = \sqrt{5.7} = 2.3875$

17.

$$\begin{aligned} P(0 \le X \le 2) &= \int_0^2 e^{-x}\, dx \\ &= -e^{-x}\Big|_0^2 \\ &= 0.8647 \end{aligned}$$

18.

$$\begin{aligned} P(2 \le X \le 3) &= \int_2^3 e^{-x}\, dx \\ &= -e^{-x}\Big|_2^3 \\ &= 0.0856 \end{aligned}$$

19.

$$\begin{aligned} P(0 \le X \le 2) &= \int_0^2 3\, e^{-3x}\, dx \\ &= -e^{-3x}\Big|_0^2 \\ &= 0.9975 \end{aligned}$$

20.

$$\begin{aligned} P(2 \le X \le 3) &= \int_2^3 3e^{-3x}\, dx \\ &= -e^{-3x}\Big|_2^3 \\ &= 0.0024 \end{aligned}$$

21.

$$\begin{aligned} P(2 \le X) &= 1 - P(2 > X) \\ &= 1 - \int_0^2 2e^{-2x}\, dx \\ &= 1 - e^{-2x}\Big|_0^2 \\ &= 1 - 0.9817 \\ &= 0.0183 \end{aligned}$$

22.

$$\begin{aligned} P(3 \le X) &= 1 - P(3 > X) \\ &= 1 - \int_0^3 2e^{-2x}\, dx \\ &= 1 - e^{-2x}\Big|_0^3 \\ &= 1 - 0.9975 \\ &= 0.0025 \end{aligned}$$

23.

$$\begin{aligned} P(X \le 0.5) &= \int_0^{0.5} 2.5\, e^{-2.5x}\, dx \\ &= -e^{-2.5x}\Big|_0^{0.5} \\ &= 0.7135 \end{aligned}$$

24.

$$\begin{aligned} P(X \le 1) &= \int_0^1 2.5\, e^{-2.5x}\, dx \\ &= -e^{-2.5x}\Big|_0^1 \\ &= 0.9179 \end{aligned}$$

25. $E(X) = \frac{1}{4}$

26. $E(X) = \frac{1}{10}$

27. $SD(X) = \frac{1}{4}$

28. $E(X) = \frac{1}{10}$

29. - 40. Left to the student

41. - 48. Left to the student

49. $E(X) = 4$
$Var(X) = 4$
$SD(X) = \sqrt{4} = 2$

50. $E(X) = 3.5$
$Var(X) = 3.5$
$SD(X) = \sqrt{3.5} = 1.8708$

51. $E(X) = \frac{1}{3}$
$Var(X) = \frac{1}{9}$
$SD(X) = \sqrt{\frac{1}{9}} = \frac{1}{3}$

52. $E(X) = \frac{1}{5}$
$Var(X) = \frac{1}{25}$
$SD(X) = \sqrt{\frac{1}{25}} = \frac{1}{5}$

53. $E(X) = 22 \to \lambda = \frac{1}{22}$

$$\begin{aligned} P(X \le 20) &= \int_0^{20} \frac{1}{22}\, e^{-\frac{x}{22}}\, dx \\ &= -e^{-\frac{x}{22}}\Big|_{)}^{20} \\ &= 0.5971 \end{aligned}$$

54. $E(X) = 18 \to \lambda = \frac{1}{18}$

$$\begin{aligned} P(X \le 15) &= \int_0^{15} \frac{1}{18}\, e^{-\frac{x}{18}}\, dx \\ &= -e^{-\frac{x}{18}}\Big|_0^{15} \\ &= 0.5654 \end{aligned}$$

55. **a)** $\lambda = 3$, $\mu = 3(2) = 6$

$$\begin{aligned} P(Y=1) &= \frac{6^1}{1!}\, e^{-6} \\ &= 0.01487 \end{aligned}$$

b) $\mu = 3(3) = 9$

$$\begin{aligned} P(Y=5) &= \frac{9^5}{5!}\, e^{-9} \\ &= 0.0607 \end{aligned}$$

c) $\mu = 3$

$$\begin{aligned} P(Y=0) &= \frac{3^0}{0!}\, e^{-3} \\ &= 0.0498 \end{aligned}$$

56. **a)** $\lambda = 4$ and $\mu = 4$

$$\begin{aligned} P(Y=2) &= \frac{4^2}{2!}\, e^{-4} \\ &= 0.1465 \end{aligned}$$

b) $\mu = 4(2) = 8$

$$\begin{aligned} P(Y=10) &= \frac{8^{10}}{10!}\, e^{-8} \\ &= 0.0993 \end{aligned}$$

c) $\mu = 8$

$$\begin{aligned} P(Y<3) &= P(Y=0) + P(Y=1) + P(Y=2) \\ &= 0.0003355 + 0.002684 + 0.010735 \\ &= 0.01375 \end{aligned}$$

57. **a)** $\lambda = 3$, $\mu = \frac{4}{3}$

$$\begin{aligned} P(Y \geq 1) &= 1 - P(Y=0) \\ &= 1 - 0.2636 \\ &= 0.7364 \end{aligned}$$

b) $\mu = \frac{2}{3}$

$$\begin{aligned} P(Y=0) &= \frac{\left(\frac{2}{3}\right)^0}{0!}\, e^{-2/3} \\ &= 0.5134 \end{aligned}$$

c) $\mu = \frac{1}{3}$

$$\begin{aligned} P(Y=0) &= \frac{\left(\frac{1}{3}\right)^0}{0!}\, e^{-1/3} \\ &= 0.7165 \end{aligned}$$

d) $\mu = \frac{1}{3}$

$$\begin{aligned} P(Y=1) &= \frac{\left(\frac{1}{3}\right)^1}{1!}\, e^{-1/3} \\ &= 0.2388 \end{aligned}$$

e) $\mu = \frac{1}{3}$

$$\begin{aligned} P(Y \geq 2) &= 1 - P(Y<2) \\ &= 1 - (P(Y=0) + P(Y=1)) \\ &= 1 - (0.7165 - 0.2388) \\ &= 0.0447 \end{aligned}$$

58. **a)** $\lambda = \frac{1}{10}$, $\mu = \frac{5}{10} = \frac{1}{2}$

$$\begin{aligned} P(Y \geq 1) &= 1 - P(Y<1) \\ &= 1 - P(Y=0) \\ &= 1 - 0.6065 \\ &= 0.3935 \end{aligned}$$

b) $\mu = \frac{4}{5}$

$$\begin{aligned} P(Y=0) &= \frac{\left(\frac{4}{5}\right)^0}{0!}\, e^{-4/5} \\ &= 0.4493 \end{aligned}$$

c) $\mu = 2$

$$\begin{aligned} P(Y=0) &= \frac{2^0}{0!}\, e^{-2} \\ &= 0.1353 \end{aligned}$$

d) $\mu = 2.5$

$$\begin{aligned} P(Y=1) &= \frac{2.5^0}{0!}\, e^{-2.5} \\ &= 0.2052 \end{aligned}$$

e) $\mu = 1.5$

$$\begin{aligned} P(Y \geq 2) &= 1 - P(Y<2) \\ &= 1 - (P(Y=0) + P(Y=1)) \\ &= 1 - (0.2231 + 0.3347) \\ &= 0.4422 \end{aligned}$$

59. **a)** $E(X) = 6$
$SD(X) = \sqrt{6} = 2.4495$

b)

$$\begin{aligned} P(Y=0) &= \frac{6^0}{0!}\, e^{-6} \\ &= 0.00248 \end{aligned}$$

60. **a)** $E(X) = 13.1$
$SD(X) = \sqrt{13.1} = 3.6194$

b)

$$\begin{aligned} P(Y=10) &= \frac{13.1^{10}}{10!}\, e^{-13.1} \\ &= 0.08389 \end{aligned}$$

61. First, note that $f(x) > 0$ for all x values and positive $\lambda > 0$
Second,

$$\begin{aligned}\int_0^\infty f(x)\,dx &= \int_0^\infty \lambda\, e^{-\lambda\, x}\,dx \\ &= -\,e^{-\lambda\, x}\Big|_0^\infty \\ &= 0-(-1) \\ &= 1\end{aligned}$$

Therefore the function $f(x) = \lambda\, e^{-\lambda\, x}$ is a probability density function

62. a)

$$\begin{aligned}P(X > c) &= \int_c^\infty \lambda\, e^{-\lambda\, x}\,dx \\ &= -\,e^{-\lambda\, x}\Big|_c^\infty \\ &= 0-(-\,e^{\lambda\, c}) \\ &= e^{-\lambda\, c}\end{aligned}$$

b) $P(X > a + b|X > a)$

$$\begin{aligned}&= \frac{P(X > a+b \text{ and } X > a)}{P(X > a)} \\ &= \frac{e^{-\lambda(a+b)}}{e^{-\lambda\, a}} \\ &= e^{-\lambda\, b} \\ &= P(X > b)\end{aligned}$$

c) The events are independent

63.

$$\begin{aligned}\int_0^m \frac{1}{2}\,x\,dx &= \frac{1}{2} \\ \frac{x^2}{4}\Big|_0^m &= \frac{1}{2} \\ \frac{m^2}{4} &= \frac{1}{2} \\ m^2 &= 2 \\ m &= \sqrt{2}\end{aligned}$$

64.

$$\begin{aligned}\int_{-1}^m \frac{3}{2}\,x^2\,dx &= \frac{1}{2} \\ \frac{x^3}{2}\Big|_0^m &= \frac{1}{2} \\ \frac{m^3}{2} &= \frac{1}{2} \\ m^3 &= 1 \\ m &= 1\end{aligned}$$

65. $E(X) = \dfrac{3}{2}$

$$\begin{aligned}\int_0^m \frac{1}{3}\,dx &= \frac{1}{2} \\ \frac{x}{3}\Big|_0^m &= \frac{1}{2} \\ \frac{m}{3} &= \frac{1}{2} \\ m &= \frac{3}{2}\end{aligned}$$

The mean and median are equal

66. $E(X) = \dfrac{a+b}{2}$

$$\begin{aligned}\int_0^m \frac{1}{b-a}\,dx &= \frac{1}{2} \\ \frac{x}{b-a}\Big|_a^m &= \frac{1}{2} \\ \frac{m}{b-a} - \frac{a}{b-a} &= \frac{1}{2} \\ \frac{m-a}{b-a} &= \frac{1}{2} \\ m &= \frac{b-a}{2} + a \\ &= \frac{b+a}{2}\end{aligned}$$

The mean and median are equal

67. a)

$$\begin{aligned}\int_0^m e^{-x}\,dx &= \frac{1}{2} \\ -e^{-x}\Big|_0^m &= \frac{1}{2} \\ -e^{-m}+1 &= \frac{1}{2} \\ e^{-m} &= \frac{1}{2} \\ -m &= \ln\frac{1}{2} = -\ln 2 \\ m &= \ln 2\end{aligned}$$

b) The median is larger than the mean $E(X) = 1$. This makes sense since at $m = \ln 2$, for exponential given, the value of $f(x) = 0.5$

68. a)

$$\begin{aligned}\int_0^m \lambda\, e^{-\lambda\, x}\,dx &= \frac{1}{2} \\ -e^{-\lambda\, x}\Big|_0^m &= \frac{1}{2} \\ -e^{-m\,\lambda}+1 &= \frac{1}{2} \\ e^{-m\,\lambda} &= \frac{1}{2} \\ -m\,\lambda &= \ln\frac{1}{2} = -\ln 2 \\ m &= \frac{\ln 2}{\lambda}\end{aligned}$$

b) The median is larger than the mean $E(X) = \frac{1}{\lambda}$. This makes sense since at $m = \frac{\ln 2}{\lambda}$, for exponential given, the value of $f(x) = 0.5$

69. Left to the student

70. Binomial: $P(X = 3) = 0.180628$
Poisson: $P(X = 3) = 0.180447$

71. Binomial: $P(X = 4) = 0.168284$
Poisson: $P(X = 4) = 0.168031$

72. Binomial: $P(X = 5) = 0.175555$
Poisson: $P(X = 5) = 0.175467$

73. Binomial: $P(X = 2) = 0.256561$
Poisson: $P(X = 2) = 0.256516$

74. Binomial: $P(X = 3) = 0.140367$
Poisson: $P(X = 3) = 0.140374$

75. Binomial: $P(X = 2) = 0.183949$
Poisson: $P(X = 2) = 0.183940$

Exercise Set 10.7

1.

$$\begin{aligned} P(-2 \le Z \le 2) &= P(Z = 2) + P(Z = -2) \\ &= 0.47725 + 0.47725 \\ &= 0.9545 \end{aligned}$$

2.

$$\begin{aligned} P(-1.45 \le Z \le 1.45) &= P(Z = 1.45) + P(Z = -1.45) \\ &= 0.42647 + 0.42647 \\ &= 0.8529 \end{aligned}$$

3.

$$\begin{aligned} P(-0.26 \le Z \le 0.7) &= P(Z = 0.7) + P(Z = -0.26) \\ &= 0.25804 + 0.10257 \\ &= 0.3606 \end{aligned}$$

4.

$$\begin{aligned} P(-1.29 \le Z \le 2.04) &= P(Z = 2.04) + P(Z = -1.29) \\ &= 0.479325 + 0.401475 \\ &= 0.8808 \end{aligned}$$

5.

$$\begin{aligned} P(1.26 \le Z \le 1.43) &= P(Z = 1.43) - P(Z = 1.26) \\ &= 0.42364 - 0.396165 \\ &= 0.0275 \end{aligned}$$

6.

$$\begin{aligned} P(1.94 \le Z \le 2.93) &= P(Z = 2.93) - P(Z = 1.94) \\ &= 0.498305 - 0.473810 \\ &= 0.0245 \end{aligned}$$

7.

$$\begin{aligned} P(-2.47 \le Z \le -0.38) &= P(Z = -0.38) - P(Z = -2.47) \\ &= 0.493244 - 0.148027 \\ &= 0.3452 \end{aligned}$$

8.

$$\begin{aligned} P(-1.65 \le Z \le -1.08) &= P(Z = -1.08) - P(Z = -1.65) \\ &= 0.450529 - 0.359929 \\ &= 0.0906 \end{aligned}$$

9.

$$\begin{aligned} P(0 \le Z < \infty) &= P(Z = \infty) - P(Z = 0) \\ &= 0.5 - 0 \\ &= 0.5 \end{aligned}$$

10.

$$\begin{aligned} P(-\infty < Z \le 0) &= P(Z = 0) + P(Z = -\infty) \\ &= 0 + 0.5) \\ &= 0.5 \end{aligned}$$

11.

$$\begin{aligned} P(1.17 \le Z < \infty) &= P(Z = \infty) - P(Z = 1.17) \\ &= 0.5 - 0.378999 \\ &= .1210 \end{aligned}$$

12.

$$\begin{aligned} P(-\infty < Z \le 3.06) &= P(Z = 3.06) + P(Z = -\infty) \\ &= 0.49889 + 0.5 \\ &= 0.9989 \end{aligned}$$

13. **a)** $\mu = 2$

b) $\sigma = 5$

c) $Z = \frac{4-2}{5} = 0.4$

$$\begin{aligned} P(Z \ge 0.4) &= 1 - P(Z < 0.4) \\ &= 1 - 0.6554 \\ &= 0.3446 \end{aligned}$$

d) $Z = \frac{3-2}{5} = 0.2$

$$P(Z \le 0.3) = 0.5793$$

e) $Z = \frac{-8-2}{5} = -2$, $Z = \frac{7-2}{5} = 1$

$$\begin{aligned} P(-2 \le Z \le 1) &= P(Z = -2) + P(Z = 1) \\ &= 0.4772 + 0.3413 \\ &= 0.8185 \end{aligned}$$

14. **a)** $\mu = -3$

b) $\sigma = 3$

c) $Z = \frac{-5-(-3)}{3} = -0.67$

$$\begin{aligned} P(Z \geq -0.67) &= 1 - P(Z < -0.67) \\ &= 1 - 0.2514 \\ &= 0.7486 \end{aligned}$$

d) $Z = \frac{4-(-3)}{3} = 2.33$

$$P(Z \leq 2.33) = 0.9901$$

e) $Z = \frac{-4-(-3)}{3} = -0.33$

$Z = \frac{5-(-3)}{3} = 2.67$

$$\begin{aligned} &P(-0.33 \leq Z \leq 2.67) \\ &= P(Z = -0.33) + P(Z = 2.67) \\ &= 0.1293 + 0.4692 \\ &= 0.6255 \end{aligned}$$

15. a) $Z = \frac{-0.2-0}{0.1} = -2$

$$\begin{aligned} P(Z \geq -2) &= 1 - P(Z < -2) \\ &= 1 - 0.0228 \\ &= 0.9972 \end{aligned}$$

b) $Z = \frac{-0.05-0}{0.1} = -0.5$

$$P(Z \leq -0.5) = 0.3085$$

c) $Z = \frac{-0.08-0}{0.1} = -0.8$

$Z = \frac{0.09-0}{0.1} = 0.9$

$$\begin{aligned} &P(-0.8 \leq Z \leq 0.9) \\ &= P(Z = -0.8) + P(Z = 0.9) \\ &= 0.2881 + 0.3159 \\ &= 0.6041 \end{aligned}$$

16. a) $Z = \frac{2-0.1}{1} = 1.9$

$$\begin{aligned} P(Z \geq 1.9) &= 1 - P(Z < 1.9) \\ &= 1 - 0.9713 \\ &= 0.0287 \end{aligned}$$

b) $Z = \frac{1-0.1}{1} = 0.9$

$$P(Z \leq 0.9) = 0.8159$$

c) $Z = \frac{-1.1-0.1}{1} = -1.2$

$Z = \frac{-0.1-0.1}{1} = -0.2$

$$\begin{aligned} &P(-0.8 \leq Z \leq 0.9) \\ &= P(Z = -1.2) - P(Z = 0.2) \\ &= 0.3849 - 0.0793 \\ &= 0.3056 \end{aligned}$$

17. - 32. Left to the student (see answers to 1. - 16.)

33. $\mu = 100(0.2) = 20$

$\sigma = \sqrt{100(0.2)(0.8)} = 4$

$Z = \frac{18-20}{4} = -0.5$

$$\begin{aligned} P(Z \geq -0.5) &= 1 - P(Z < -0.5) \\ &= 1 - 0.3085 \\ &= 0.6915 \end{aligned}$$

34. $\mu = 400(0.1) = 40$

$\sigma = \sqrt{400(0.1)(0.9)} = 6$

$Z = \frac{35-40}{6} = -0.83$

$Z = \frac{48-40}{6} = 1.33$

$$\begin{aligned} &P(-0.83 \leq Z \leq 1.33) \\ &= P(Z = -0.83) + P(Z = 1.33) \\ &= 0.2967 + 0.4082 \\ &= 0.7049 \end{aligned}$$

35. $\mu = 64(0.5) = 32$

$\sigma = \sqrt{64(0.5)(0.5)} = 4$

$Z = \frac{30-32}{4} = -0.5$

$$P(Z \leq -0.5) = 0.3085$$

36. $\mu = 100(0.9) = 90$

$\sigma = \sqrt{100(0.9)(0.1)} = 3$

$Z = \frac{85-90}{3} = -1.67$

$$\begin{aligned} P(Z > -1.67) &= 1 - P(Z < -1.67) \\ &= 1 - 0.0475 \\ &= 0.9525 \end{aligned}$$

37. $\mu = 1000(0.3) = 300$

$\sigma = \sqrt{1000(0.3)(0.7)} = 14.4914$

$Z = \frac{280-300}{14.4914} = -1.38$

$Z = \frac{300-300}{14.4914} = 0$

$$\begin{aligned} P(-1.38 \leq Z \leq 0) &= P(Z = -1.38) + P(Z = 0) \\ &= 0.4162 + 0 \\ &= 0.4162 \end{aligned}$$

38. $\mu = 500(0.34) = 170$

$\sigma = \sqrt{500(0.34)(0.66)} = 10.5925$

$Z = \dfrac{158 - 170}{10.5925} = -1.13$

$P(Z \leq -1.13) \;=\; 0.1292$

39. **a)** $Z = \dfrac{5-4}{1} = 1$

$$\begin{aligned} P(Z \geq 1) &= 1 - P(Z < 1) \\ &= 1 - 0.8413 \\ &= 0.1587 \end{aligned}$$

b) $Z = \dfrac{2-4}{1} = -2$

$$\begin{aligned} P(Z \geq -2) &= 1 - P(Z < -2) \\ &= 1 - 0.0228 \\ &= 0.0.9772 \end{aligned}$$

c) $Z = \dfrac{1-4}{1} = -3$

$P(Z \leq -3) \;=\; 0.0013$

40. **a)** $Z = \dfrac{6-13}{3} = -2.33$

$$\begin{aligned} P(Z \geq -2.33) &= 1 - P(Z < -2.33) \\ &= 1 - 0.0099 \\ &= 0.9901 \end{aligned}$$

b) $Z = \dfrac{8-13}{3} = -1.67$

$P(Z \leq -1.67) \;=\; 0.0465$

c) $Z = \dfrac{5-13}{3} = -2.67$

$P(Z \leq -2.67) \;=\; 0.0038$

41. $\mu = (300)(0.25) = 75$

$\sigma = \sqrt{300(0.75)(0.25)} = 7.5$

$Z = \dfrac{80 - 75}{7.5} = 0.67$

$$\begin{aligned} P(Z \geq 0.67) &= 1 - P(Z < 0.67) \\ &= 1 - 0.7486 \\ &= 0.2514 \end{aligned}$$

42. $\mu = (190)(0.5) = 95$

$\sigma = \sqrt{190(0.5)(0.5)} = 6.8920$

$Z = \dfrac{80 - 95}{6.8920} = -2.18$

$Z = \dfrac{100 - 95}{6.8920} = 0.73$

$$\begin{aligned} P(-2.18 \leq Z \leq 0.73) &= P(Z = -2.18) + P(Z = 0.73) \\ &= 0.4854 + 0.2673 \\ &= 0.7527 \end{aligned}$$

43. $\mu = (70)(0.25) = 17.5$

$\sigma = \sqrt{70(0.25)(0.75)} = 3.6228$

$Z = \dfrac{20 - 17.5}{3.6228} = 0.69$

$P(Z < 0.69) \;=\; 0.7549$

44. $p = \dfrac{319159}{534694} = 0.5969$

$\mu = 400(0.5969) = 238.76$

$\sigma = \sqrt{400(0.5969)(0.4031)} = 9.81$

$Z = \dfrac{260 - 238.76}{9.81} = 2.17$

$$\begin{aligned} P(Z \geq 2.17) &= 1 - P(Z < 2.17) \\ &= 1 - 0.9850 \\ &= 0.0150 \end{aligned}$$

45. $p = \dfrac{238351}{362470} = 0.6576$

$\mu = 120(0.6576) = 78.91$

$\sigma = \sqrt{120(0.6576)(0.3424)} = 5.20$

$Z = \dfrac{80 - 78.91}{5.20} = 0.21$

$\dfrac{90 - 78.91}{5.20} = 2.13$

$$\begin{aligned} P(0.21 \leq Z \leq 2.13) &= P(Z < 2.13) - P(Z = 0.21) \\ &= 0.4834 - 0.0832 \\ &= 0.4002 \end{aligned}$$

46. $\mu = 100(9/19) = 47.37$

$\sigma = \sqrt{100(9/19)(10/19)} = 4.99$

$Z = \dfrac{51 - 47.37}{4.99} = 0.73$

$$\begin{aligned} P(Z \geq 0.73) &= 1 - P(Z < 0.73) \\ &= 1 - 0.7673 \\ &= 0.2327 \end{aligned}$$

47. **a)** $\mu = 1000(9/19) = 473.7$

$\sigma = \sqrt{1000(9/19)(10/19)} = 15.79$

$Z = \dfrac{501 - 473.7}{15.79} = 1.73$

$$\begin{aligned} P(Z \geq 1.73) &= 1 - P(Z < 1.73) \\ &= 1 - 0.9582 \\ &= 0.0418 \end{aligned}$$

b) The probability in part a)was larger. This makes sense since the probability of winning is less than the probability of winning.

48. **a)** $30\% \approx X = -0.5$

b) $50\% \approx X = 0$

c) $95\% \approx X = 1.67$

49. **a)** $35\% \approx X = -0.38$

$$\begin{aligned} -0.38 &= \frac{X' - 507}{111} \\ -0.38(111) + 507 &= X' \\ 464 &= \end{aligned}$$

b) $60\% \approx X = 0.25$

$$\begin{aligned} 0.25 &= \frac{X' - 507}{111} \\ 0.25(111) + 507 &= X' \\ 535 &= \end{aligned}$$

c) $92\% \approx X = 1.41$

$$\begin{aligned} 1.41 &= \frac{X' - 507}{111} \\ 1.41(111) + 507 &= X' \\ &= 664 \end{aligned}$$

50. Significance Level

$$\begin{aligned} &= 1 - P(X \geq 2.13) \\ &= 1 - 0.9834 \\ &= 0.0166 \end{aligned}$$

51. Significance Level

$$\begin{aligned} &= 1 - P(X \geq 0.93) \\ &= 1 - 0.8238 \\ &= 0.1762 \end{aligned}$$

52. Significance Level

$$= P(X \leq -1.96)$$

53. Significance Level

$$\begin{aligned} &= 1 - P(X \geq 2.33) \\ &= 1 - 0.9901 \\ &= 0.0099 \end{aligned}$$

54. Significance Level

$$\begin{aligned} &= 1 - P(X \geq 0.19) \\ &= 1 - 0.5753 \\ &= 0.4247 \end{aligned}$$

55. Significance Level

$$\begin{aligned} &= P(X \leq -1.24) \\ &= 0.1075 \end{aligned}$$

56. Critical Value is 1.64

57. Critical Value is 2.33

58. Critical Value is 2.58

59. Critical Value is -1.64

60. Critical Value is -2.33

61. Critical Value is -2.58

62. $Z_1 = \dfrac{3-3}{SD} = 0$

$Z_2 = \dfrac{5-3}{SD} = 0$

$$\begin{aligned} P(0 \leq Z \leq Z_2) &= 0.4332 \\ Z_2 &= 1.50 \\ 1.50 &= \frac{5-3}{SD} \\ SD &= \frac{2}{1.50} \\ &= 1.3333 \end{aligned}$$

63. $Z_1 = \dfrac{9-10}{SD} = \dfrac{-1}{SD}$

$Z_2 = \dfrac{10-10}{SD} = 0$

$$\begin{aligned} P(Z_1 \leq Z \leq 0) &= 0.4922 \\ Z_2 &= -2.42 \\ -2.42 &= \frac{-1}{SD} \\ SD &= \frac{-1}{-2.42} \\ &= .4132231405 \end{aligned}$$

64. $Z_1 = \dfrac{5-4}{SD} = \dfrac{1}{SD}$

$Z_2 = \dfrac{3-4}{SD} = -\dfrac{1}{SD}$

$$\begin{aligned} P(-\frac{1}{SD} \leq Z \leq \frac{1}{SD}) &= 0.7620 \\ P(0 \leq Z \leq \frac{1}{SD}) &= \frac{0.7620}{2} = 0.3810 \\ \frac{1}{SD} &= 1.18 \\ SD &= \frac{1}{1.18} \\ &= .8474576271 \end{aligned}$$

65. $Z_1 = \frac{-6-(-3)}{SD} = \frac{-3}{SD}, Z_2 = \frac{0-(-3)}{SD} = \frac{3}{SD}$

$$\begin{aligned} P(-\frac{3}{SD} \le Z \le \frac{3}{SD}) &= 0.3108 \\ P(0 \le Z \le \frac{3}{SD}) &= \frac{0.3108}{2} = 0.1554 \\ \frac{3}{SD} &= 0.40 \\ SD &= \frac{3}{0.40} \\ &= 7.5 \\ Var &= (7.5)^2 = 56.25 \end{aligned}$$

66.

$$\begin{aligned} P(Z \ge 2) &= 0.6293 \\ P(Z < 2) &= 1 - 0.6293 \\ &= 0.3707 \\ Z &= 1.13 \\ 1.13 &= \frac{2-E}{3} \\ E &= 2 - 3(1.13) \\ &= -1.39 \end{aligned}$$

67.

$$\begin{aligned} P(Z \ge -1) &= 0.409 \\ P(Z < -1) &= 1 - 0.409 \\ &= 0.591 \\ Z &= 0.23 \\ 0.23 &= \frac{-1-E}{2} \\ E &= -1 - 2(0.23) \\ &= -1.46 \end{aligned}$$

68. $Z_1 = \frac{4.5-E}{SD}, Z_2 = \frac{7-E}{SD}$

$$\begin{aligned} P(Z \ge Z_1) &= 1 - P(Z < Z_1) \\ &= 0.5 \\ Z_1 &= 0 \\ P(Z \ge Z_2) &= 1 - P(Z < Z_2) \\ &= 0.992 \\ Z_2 &= 2.41 \end{aligned}$$

Thus,

$$\frac{4.5-E}{SD} = 0$$

$$\begin{aligned} E &= 4.5 \\ \frac{7-E}{SD} &= 2.41 \\ 2.41SD + 4.5 &= 7 \\ SD &= 1.037344398 \end{aligned}$$

69. $Z_1 = \frac{2.8-E}{SD}, Z_2 = \frac{10.3-E}{SD}$

$$\begin{aligned} P(Z \ge Z_1) &= 1 - P(Z < Z_1) \\ &= 0.5 \\ Z_1 &= 0 \\ P(Z \le Z_2) &= 0.8944 \\ Z_2 &= 1.25 \end{aligned}$$

Thus,

$$\frac{2.8-E}{SD} = 0$$

$$\begin{aligned} E &= 2.8 \\ \frac{10.3-E}{SD} &= 1.25 \\ 1.25SD + 2.8 &= 10.3 \\ SD &= 6 \end{aligned}$$

70. $Z_1 = \frac{10-E}{SD}, Z_2 = \frac{6-E}{SD}$

$$\begin{aligned} P(Z \ge Z_1) &= 1 - P(Z < Z_1) \\ &= 0.6064 \\ Z_1 &= 0.27 \\ P(Z \le Z_2) &= 0.3936 \\ Z_2 &= -1.24 \end{aligned}$$

Thus,

$$\frac{10-E}{SD} = 0.27$$

$$\begin{aligned} 0.27SD + E &= 10 \\ \frac{6-E}{SD} &= 1.24 \\ 1.24SD + 2.8 &= 6 \end{aligned}$$

Solving the system

$$\begin{aligned} E &= 9.284768212 \\ SD &= 2.649006623 \end{aligned}$$

71. $Z_1 = \frac{-3.1-E}{SD}, Z_2 = \frac{1.4-E}{SD}$

$$\begin{aligned} P(Z \ge Z_1) &= 1 - P(Z < Z_1) \\ &= 0.1539 \\ Z_1 &= 0.4 \\ P(Z \le Z_2) &= 0.8461 \\ Z_2 &= 1 \end{aligned}$$

Thus,

$$\frac{-3.1-E}{SD} = 0.4$$

$$\begin{aligned} 0.4SD + E &= -3.1 \\ \frac{1.4-E}{SD} &= 1 \\ SD + E &= 1.4 \end{aligned}$$

Solving the system

$$\begin{aligned} E &= -0.85 \\ SD &= 2.25 \end{aligned}$$

72. The reason is the symmetric nature of the Normal$(0, 1)$ distribution about 0

73. The reason is the symmetric nature of the Normal(μ, σ) distribution about μ

74. **a)** $\mu = 20$, $\sigma = 4$
$P(14.5 \le X \le 20.5) = 0.46517$

b) $Z_1 = \dfrac{15-20}{4} = -1.25$
$Z_2 = \dfrac{20-20}{4} = 0$
$P(-1.25 \le Z \le 0) = 0.3944$

c) 0.07077

75. **a)** $\mu = 2000$, $\sigma = 40$
$P(1949.5 \le X \le 2000.5) = 0.40160$

b) $Z_1 = \dfrac{1950-2000}{40} = -1.25$
$Z_2 = \dfrac{2000-2000}{4} = 0$
$P(-1.25 \le Z \le 0) = 0.3944$

c) 0.00720

76. **a)** $\mu = 200000$, $\sigma = 400$
$P(199499.5 \le X \le 200000.5) = 0.39508$

b) $Z_1 = \dfrac{1950-2000}{40} = -1.25$
$Z_2 = \dfrac{2000-2000}{4} = 0$
$P(-1.25 \le Z \le 0) = 0.3944$

c) 0.000677

77. As n increases the difference between probability with the continuity correction and the probability without the continuity correction becomes smaller.

78. $n = 2 \to 0.842881$
$n = 4 \to 0.852459$
$n = 6 \to 0.854219$
$n = 8 \to 0.854834$

79. $n = 2 \to 0.341529$
$n = 4 \to 0.341355$
$n = 6 \to 0.341347$
$n = 8 \to 0.341345$

Review Exercises

1. $\dfrac{1}{4}$

2. $\dfrac{3}{4}$

3. $\dfrac{2}{4} = \dfrac{1}{2}$

4. $\dfrac{1}{4} \cdot \dfrac{1}{4} = \dfrac{1}{16}$

5.

$$\begin{aligned} P(X=2) &= \binom{3}{2}(0.4)^2(0.6) \\ &= 0.288 \end{aligned}$$

6.

$$\begin{aligned} P(X=4) &= \binom{6}{4}(0.3)^4(0.7)^2 \\ &= 0.059535 \end{aligned}$$

7.

$$\begin{aligned} P(1 \le X \le 5) &= \int_1^5 \frac{1}{10-0}\,dx \\ &= \frac{1}{10}\cdot[5-1] \\ &= \frac{2}{5} \end{aligned}$$

8.

$$\begin{aligned} P(X \ge 0) &= 1 - P(X < 0) \\ &= 1 - \int_0^{-4} \frac{1}{10-(-4)}\,dx \\ &= 1 - \frac{1}{14}\cdot 4 \\ &= \frac{5}{7} \end{aligned}$$

9. $Z_1 = \dfrac{2-4}{5} = -0.4$
$Z_2 = \dfrac{10-4}{5} = 1.2$

$$\begin{aligned} P(-0.4 \le Z \le 1.2) &= P(Z = -0.4) + P(Z = 1.2) \\ &= 0.5404 \end{aligned}$$

10. $Z = \dfrac{7-8.2}{1.3} = -0.92$

$$\begin{aligned} P(Z \ge -0.92) &= 0.5 + P(Z < -0.92) \\ &= 0.8212 \end{aligned}$$

11.

$$\begin{aligned} P(1.5 \le T \le 3) &= \int_{1.5}^3 0.5\, e^{-0.5x}\,dx \\ &= -\, e^{-0.5x}\Big|_0^{1.5} \\ &= 0.24924 \end{aligned}$$

12.

$$\begin{aligned} P(T \le 5) &= \int_0^5 0.1\, e^{-0.1x}\,dx \\ &= -\, e^{-0.1x}\Big|_0^{5} \\ &= 0.39347 \end{aligned}$$

13.

$$\begin{aligned} P(X=3) &= \frac{2.5^3}{3!}\, e^{-2.5} \\ &= 0.21376 \end{aligned}$$

14.

$$\begin{aligned} P(X \le 2) &= P(X=0) + P(X=1) + P(X=2) \\ &= \frac{4^0}{0!}\, e^4 + \frac{4^1}{1!}\, e^{-4} + \frac{4^2}{2!}\, e^{-4} \\ &= 0.23810 \end{aligned}$$

15. $E(X) = 6(0.12) = 0.72$

$Var(X) = 6(0.12)(0.88) = 0.6336$

$SD(X) = \sqrt{0.6336} = 0.79599$

16. $E(X) = 14(0.8) = 11.2$

$Var(X) = 14(0.8)(0.2) = 2.24$

$SD(X) = \sqrt{2.24} = 1.49666$

17. $E(X) = \frac{20+36}{2} = 28$

$Var(X) = \frac{(36-20)^2}{12} = \frac{64}{3}$

$SD(X) = \sqrt{64/3} = 4.6188$

18. $E(X) = \frac{5+(-3)}{2} = 1$

$Var(X) = \frac{(5-(-3))^2}{12} = \frac{16}{3}$

$SD(X) = \sqrt{16/3} = 2.3094$

19. $E(X) = 0$, $Var(X) = 1$, and $SD(X) = 1$

20. $E(X) = 5$, $Var(X) = 4$, and $SD(X) = 2$

21. $E(X) = \frac{1}{4} = 0.25$
$Var(X) = \frac{1}{16} = 0.0625$
and $SD(X) = \frac{1}{4} = 0.25$

22. $E(X) = \frac{1}{0.4} = 2.5$
$Var(X) = \frac{1}{0.16} = 6.25$
and $SD(X) = \frac{1}{0.4} = 2.5$

23. $E(X) = 4$, $Var(X) = 4$, and $SD(X) = 2$

24. $E(X) = 3$, $Var(X) = 3$, and $SD(X) = \sqrt{3}$

25.

$$\begin{aligned} P(X \geq 3) &= \int_3^4 \frac{x^3}{60}\,dx \\ &= \left.\frac{x^4}{240}\right|_3^4 \\ &= 0.729167 \end{aligned}$$

26.

$$\begin{aligned} P(1 \leq X \leq 2) &= \int_1^2 3x^2\, e^{-x^3}\,dx \\ &= \left.-\,e^{-x^3}\right|_1^2 \\ &= .36754 \end{aligned}$$

27.

$$\begin{aligned} P(X \geq 500) &= \int_{500}^{\infty} \frac{dx}{x^2} \\ &= \lim_{b\to\infty} \int_{500}^{b} \frac{dx}{x^2} \\ &= \lim_{b\to\infty} \left.\frac{-1}{x}\right|_{500}^{b} \\ &= \frac{1}{500} = 0.002 \end{aligned}$$

28.

$$\begin{aligned} P(X \geq c) &= \int_c^{\infty} 4\,e^{-4x}\,dx \\ &= \lim_{b\to\infty} \left.-e^{-4x}\right|_c^b \\ &= e^{-4c} \\ 0.05 &= e^{-4c} \\ \ln 0.05 &= -4c \\ c &= 0.74893 \end{aligned}$$

29.

$$\begin{aligned} k\int_0^4 3x^2 + 3\,dx &= 1 \\ k(x^3 + 3x|_0^4 &= 1 \\ 76k &= 1 \\ k &= \frac{1}{76} \end{aligned}$$

30.

$$\begin{aligned} k\int_1^{10} \frac{1}{x}\,dx &= 1 \\ k \ln x|_1^{10} &= 1 \\ \ln(10)k &= 1 \\ k &= \frac{1}{\ln(10)} \end{aligned}$$

31.

$$\begin{aligned} E(X) &= \frac{1}{60}\int_2^4 x^4\,dx \\ &= \left.\frac{x^5}{300}\right|_2^4 \\ &= 3.30667 \end{aligned}$$

$$\begin{aligned} Var(X) &= \frac{1}{60}\int_2^4 (x-3.30667)^2 x^3\,dx \\ &= \left..166667x^6 - 1.32267x^5 + 2.73352x^4\right|_2^4 \\ &= 0.26596 \\ SD(X) &= \sqrt{0.265958} \\ &= 0.51571 \end{aligned}$$

32. Since there are four possibilities for the offspring, the probability is $\frac{1}{4}$

33.

$$\begin{aligned} P(D+|T+) &= \frac{0.4\cdot 0.82}{0.4\cdot 0.82+0.6\cdot(1-0.9)} \\ &= \frac{0.328}{0.388} \\ &= 0.84536 \end{aligned}$$

34. This is a binomial random variable

$$\begin{aligned} P(X=2) &= \binom{5}{2}\left(\frac{1}{6}\right)^2\left(\frac{1}{6}\right)^3 \\ &= 0.16075 \end{aligned}$$

35. $\mu = 500\left(\frac{238351}{362470}\right) = 328.78721$

$\sigma = \sqrt{500\left(\frac{238351}{362470}\right)\left(1-\frac{238351}{362470}\right)} = 10.61062$

$$\begin{aligned} Z_1 &= \frac{300-328.78721}{10.61062} = -2.71 \\ Z_2 &= \frac{340-328.78721}{10.61062} = 1.06 \\ P(-2.71\le Z\le 1.06) &= 0.4966+0.3554 \\ &= 0.8623 \end{aligned}$$

36.

$$\begin{aligned} P(1.25\le X\le 3.75) &= \frac{1}{10}\int_{1.25}^{3.75} dx \\ &= \frac{1}{10}[3.75-1.25] \\ &= 0.25 \end{aligned}$$

37. **a)**

$$\begin{aligned} Z &= \frac{90-75}{8} = 1.88 \\ P(Z\ge 1.88) &= 1-P(Z<1.88) \\ &= 1-0.9699 \\ &= 0.0301 \end{aligned}$$

b) $Z_1 = \frac{80-75}{8} = 0.63$
$Z_2 = \frac{90-75}{8} = 1.88$

$$\begin{aligned} P(0.63\le X\le 1.88) &= 0.4699-0.2357 \\ &= 0.2342 \end{aligned}$$

c)

$$\begin{aligned} Z &= \frac{60-75}{8} = -1.88 \\ P(Z<-1.88) &= 0.0301 \end{aligned}$$

38. $Z_1 = \frac{-1-(-0.25)}{0.5} = -1.5$
$Z_2 = \frac{1-(-0.25)}{0.5} = 2.5$

$$\begin{aligned} P(-1.5\le X\le 2.5) &= 0.4332+0.4938 \\ &= 0.927 \end{aligned}$$

39. $E(X) = 50 \to \lambda = \frac{1}{50} = 0.02$

$$\begin{aligned} P(X\le 40) &= \int_0^{40} 0.02\, e^{-0.02x}\, dx \\ &= -\left. e^{-0.02x}\right|_0^{40} \\ &= .55067 \end{aligned}$$

40. $E(X) = 10 \to \lambda = \frac{1}{10} = 0.01$

$$\begin{aligned} P(X\le 7) &= \int_0^{7} 0.01\, e^{-0.01x}\, dx \\ &= -\left. e^{-0.01x}\right|_0^{7} \\ &= 0.50342 \end{aligned}$$

41. exactly 50 times

42. **a)** 0.49865
b) 0.001318
c) 0.0000314
d) 0.0000003

Chapter 10 Test

1. $\frac{13}{52} = \frac{1}{4}$

2. $\frac{48}{52} = \frac{12}{13}$

3. $\frac{13+13}{52} = \frac{1}{2}$

4. $\frac{13}{52}\cdot\frac{13}{52} = 0.063725$

5.

$$\begin{aligned} P(X=8) &= \binom{10}{8}(.7)^8(0.3)^3 \\ &= 0.23347 \end{aligned}$$

6.

$$\begin{aligned} P(N=4) &= \frac{1.5^4}{4!}e^{-1.5} \\ &= 0.04707 \end{aligned}$$

7.

$$\begin{aligned} P(2.3\le X\le 4.1) &= \frac{1}{10-1}\int_{2.3}^{4.1} dx \\ &= \frac{1}{9}[4.1-2.3] \\ &= 0.18 \end{aligned}$$

8.

$$\begin{aligned} P(2 \leq T \leq 5) &= \int_2^5 0.5\, e^{-0.5x}\, dx \\ &= -\, e^{-0.5x}\Big|_2^5 \\ &= 0.28579 \end{aligned}$$

9. $Z_1 = \dfrac{-1-3}{2} = -2$

$Z_2 = \dfrac{4-2}{2} = 1$

$$\begin{aligned} P(-2 \leq Z \leq 1) &= P(Z=-2) + P(Z=1) \\ &= 0.6687 \end{aligned}$$

10. $E(X) = 9(0.4) = 3.6$

$Var(X) = 9(0.4)(0.6) = 2.16$

$SD(X)\sqrt{2.16} = 1.46969$

11. $E(X) = 2.4$

$Var(X) = 2.4$

$SD(X)\sqrt{2.4} = 1.54919$

12. $E(X) = \dfrac{20+40}{2} = 30$

$Var(X) = \dfrac{(40-20)^2}{12} = \dfrac{100}{3}$

$SD(X) = \sqrt{\dfrac{100}{3}} = 5.7735$

13. $E(X) = \dfrac{1}{5} = 0.2$

$Var(X) = \dfrac{1}{25} = 0.04$

$SD(X) = \dfrac{1}{5} = 0.2$

14. $E(X) = 3.2$

$Var(X) = (0.5)^2 = 0.25$

$SD(X) = 0.5$

15.

$$\begin{aligned} P(X \leq 4) &= \int_2^4 \frac{24}{x^4}\, dx \\ &= \frac{-8}{x^3}\Big|_2^4 \\ &= -\frac{1}{8} + 1 \\ &= \frac{7}{8} = 0.875 \end{aligned}$$

16.

$$\begin{aligned} P(5 \leq X \leq 10) &= \int_5^{10} \frac{24}{x^4}\, dx \\ &= \frac{-8}{x^3}\Big|_5^{10} \\ &= -0.008 + 0.064 \\ &= 0.056 \end{aligned}$$

17.

$$\begin{aligned} E(X) &= \int_2^\infty \frac{24}{x^3}\, dx \\ &= \lim_{b\to\infty} \frac{-12}{x^2}\Big|_2^b \\ &= 3 - 0 \\ &= 3 \end{aligned}$$

18.

$$\begin{aligned} Var(X) &= \int_2^\infty (x-3)^2 \frac{24}{x^4}\, dx \\ &= 24 \int_2^\infty \frac{1}{x^2} - \frac{6}{x^3} + \frac{9}{x^4}\, dx \\ &= \lim_{b\to\infty} \frac{-24}{x} + \frac{72}{x^2} - \frac{72}{x^3}\Big|_2^b \\ &= 0 - (-12 + 18 - 9) \\ &= 3 \\ SD(X) &= \sqrt{3} \\ &= 1.73205 \end{aligned}$$

19.

$$\begin{aligned} P(X \geq 8) &= \int_8^\infty \frac{24}{x^4}\, dx \\ &= \lim_{b\to\infty} \int_8^b \frac{24}{x^4}\, dx \\ &= \lim_{b\to\infty} \frac{-8}{x^3}\Big|_8^b \\ &= 0 - \frac{-1}{64} \\ &= 0.0.015625 \end{aligned}$$

20.

$$\begin{aligned} P(X \geq c) &= \int_c^\infty \frac{24}{x^4}\, dx \\ &= \lim_{b\to\infty} \int_c^b \frac{24}{x^4}\, dx \\ &= \lim_{b\to\infty} \frac{-8}{x^3}\Big|_c^b \\ &= \frac{8}{c^3} \\ 0.05 &= \frac{8}{c^3} \\ c^3 &= \frac{8}{0.05} = 160 \\ c &= \sqrt[3]{160} \\ &= 5.42884 \end{aligned}$$

21.

$$\begin{aligned} P(D+|T+) &= \frac{0.3 \cdot 0.74}{0.3 \cdot 0.74 + 0.7 \cdot (1-0.85)} \\ &= \frac{0.222}{0.327} \\ &= 0.678899 \end{aligned}$$

22. This is a binomial random variable

$$\begin{aligned} P(X=4) &= \binom{5}{4}\left(\frac{1}{2}\right)^4\left(\frac{1}{2}\right) \\ &= 0.15625 \end{aligned}$$

23. $\mu = 200 \cdot \dfrac{319159}{534694} = 119.38$

$$\sigma = \sqrt{200 \cdot \frac{319159}{534694} \cdot \left(1 - \frac{319159}{534694}\right)} = 6.937$$

$$\begin{aligned} Z &= \frac{110-119.38}{6.739} = -1.35 \\ P(Z < -1.35) &= 0.08851 \end{aligned}$$

24.

$$\begin{aligned} P(45 \le X \le 48) &= \frac{1}{54-42}\int_{45}^{48} dx \\ &= \frac{1}{12}\,[48-45] \\ &= \frac{1}{4} = 0.25 \end{aligned}$$

25. $E(X) = 60 \to \lambda = \dfrac{1}{60}$

$$\begin{aligned} P(X < 45) &= \int_0^{45} \frac{1}{60}\, e^{-x/60}\, dx \\ &= -\, e^{-x/60}\Big|_0^{45} \\ &= 1 - e^{0.75} \\ &= 0.52763 \end{aligned}$$

26. $Z = \dfrac{75-70}{3} = 1.66$

$$\begin{aligned} P(Z \ge 1.66) &= 1 - P(Z < 1.66) \\ &= 1 - 0.9515 \\ &= 0.0485 \end{aligned}$$

27.

$$\begin{aligned} P(X > 0.5) &= 1 - P(X \le 0.5) \\ &= 1 - \int_0^{0.5} 25x\, e^{-5x}\, dx \\ &= 1 - \left[5x\, e^{-5x} - e^{-5x}\right|_0^{0.5} \\ &= 1 - 0.712703 \\ &= 0.287297 \end{aligned}$$

28. $E(X)$

$$\begin{aligned} &= \int_0^{\infty} 25x^2\, e^{-5x}\, dx \\ &= \lim_{b\to\infty} \int_0^b 25x\, e^{-5x}\, dx \\ &= \lim_{b\to\infty} -5x^2\, e^{-5x} - 2x\, e^{-5x} - \frac{2}{5}\, e^{-5x}\Big|_0^b \\ &= 0 - \left(-\frac{2}{5}\right) \\ &= \frac{2}{5} = 0.4 \end{aligned}$$

29.

$$\begin{aligned} Var(X) &= \int_0^{\infty} \left(x - \frac{2}{5}\right)^2 25x\, e^{-5x}\, dx \\ &= \lim_{b\to\infty} \int_0^b \left(x - \frac{2}{5}\right)^2 25x\, e^{-5x}\, dx \\ &= \lim_{b\to\infty} \left(\frac{-2}{25} - \frac{2}{5}\,x + x^2 - 5x^3\right) e^{-5x}\Big|_0^b \\ &= 0 - \left(-\frac{2}{25}\right) \\ &= 0.08 \\ SD(X) &= \sqrt{0.08} \\ &= 0.28284 \end{aligned}$$

30. **a)** 0.1504
b) 0.2406

Technology Connection

- **Page 674**
  1. 1
  2. 3003
  3. Not possible
  4. Not possible
- **Page 706**
  1. 0.2120211636
  2. 0.9996645374
  3. 0.3834004996
- **Page 709**
  1. 0.3678794412
  2. 0.1839.97206
  3. 0.0613132402
  4. 0.0067379470
  5. 0.0336897350
  6. 0.0842243375

- **Page 716**

  1. 0.6294894068
  2. 0.0227500620

- **Page 719**

  1. Left to the student
  2. Left to the student
  3. Left to the student
  4. Left to the student
  5. Left to the student
  6. Left to the student
  7. **a)** 0.4999683134
     b) 0.4999997124
     c) 0.4999999985
     d) Because the area under the curve (the value of the integral) is getting closer to $2(0.5) = 1$

Extended Life Science Connection

1. **a)** Because the number of algae and the number of bacteria in the sample depends on chance

b) Because the number of algae and the number of bacteria is affected by time

c) The axenic smaple would require some algae to be present with no contamination with bacteria or other matter, therefore $B = 0$ and $A \geq 1$

2. **a)** Since the number of algea depends only on the volume and b is a volume density then it follows that
$\lambda_A = v$ and
$\lambda_B = vb$

b)

$$\begin{aligned} P(A \geq 1) &= 1 - P(X < 1) \\ &= 1 - P(A = 0) \\ &= 1 - \frac{v^0}{0!} e^{-v} \\ &= 1 - e^{-v} \end{aligned}$$

c)

$$\begin{aligned} P(B = 0) &= \frac{(vb)^0}{0!} e^{-vb} \\ &= e^{-vb} \end{aligned}$$

d) Since A and B are independent then

$$\begin{aligned} p &= P(A \geq 1 \text{ and } B = 0) \\ &= P(A \geq 1) \cdot P(B = 0) \\ &= \left[1 - e^{-v}\right] e^{-vb} \\ &= e^{-vb} - e^{-v-vb} \\ &= e^{-vb} - e^{-v(1+b)} \end{aligned}$$

3. $x = e^{-v}$

$$\begin{aligned} p &= \left(e^{-v}\right)^b - \left(e^{-v}\right)^{(1+b)} \\ &= x^b - x^{b+1} \end{aligned}$$

4. **a)**

$$\begin{aligned} p' &= bx^{b-1} - (b+1)x^b \\ p'' &= b(b-1)x^{b-2} - b(b+1)x^{b-1} \\ p &= 0 \\ 0 &= x^{b-1}(b - (b+1)x) \\ x &= 0 \text{ trivial} \\ b - (b+1)x &= 0 \\ x &= \frac{b}{b+1} \end{aligned}$$

$$\begin{aligned} & p''\left(\tfrac{b}{b+1}\right) \\ &= b(b-1)\left(\frac{b}{b+1}\right)^{b-2} - b(b+1)\left(\frac{b}{b+1}\right)^{b-1} \\ &= b(b-1)\left(\frac{b}{b+1}\right)^{b-2} - b^2\left(\frac{b}{b+1}\right)^{b-2} \\ &= -b\left(\frac{b}{b+1}\right)^{b-2} < 0 \end{aligned}$$

Thus, $x = \frac{b}{b+1}$ maximizes p

b)

$$\begin{aligned} x &= e^{-v} \\ v &= -\ln x \\ &= \ln x^{-1} \\ &= \ln\left(\frac{b}{b+1}\right)^{-1} \\ &= \ln\left(\frac{b+1}{b}\right) \end{aligned}$$

$$\begin{aligned} p &= x^b - x^{b+1} \\ &= \left(\frac{b}{b+1}\right)^b - \left(\frac{b}{b+1}\right)^{b+1} \\ &= \left(\frac{b^b(b+1) - b^{b+1}}{(b+1)^{b+1}}\right) \\ &= \frac{b^b}{(b+1)^{b+1}} \end{aligned}$$

5. **a)** $v = 1.0986, \quad p = 0.3849$
b) $v = 0.40547, \quad p = 0.14815$
c) $v = 0.22314, \quad p = 0.08192$

6. It is reasonable since the there is no replacement after a selection is made

7. This could be represented by a binomial random variable

$$P(X = 0) = \binom{20}{0} p^0 (1-p)^{20}$$

$$\begin{aligned} &= 1 \cdot 1 \cdot (1-p)^{20} \\ &= (1-p)^{20} \\ &= \left(1 - \frac{b^b}{(b+1)^{b+1}}\right)^{20} \end{aligned}$$

8. **a)** 0.0000601055

b) 0.0404820554

c) 0.1809716439

9. **a)** $1 - 0.0000601055 = 0.9999398945$

b) $1 - 0.0404820554 = 0.9595179446$

c) $1 - 0.1809716439 = 0.8190283561$

d) The values of b is parts a) and b) are feasible

10. **a)** $b = 2.1603374$

b) Since we know that the value of b computed, call it b_m maximizes p it will minimize the value of q therefore maximize the probability of producing an axenic sample

c) Any value of at most b_m will produce an axenic sample since it those values of b will be less than the maximum value obtained from b_m